DEGREES AND RADIANS (1.1, 2.1)

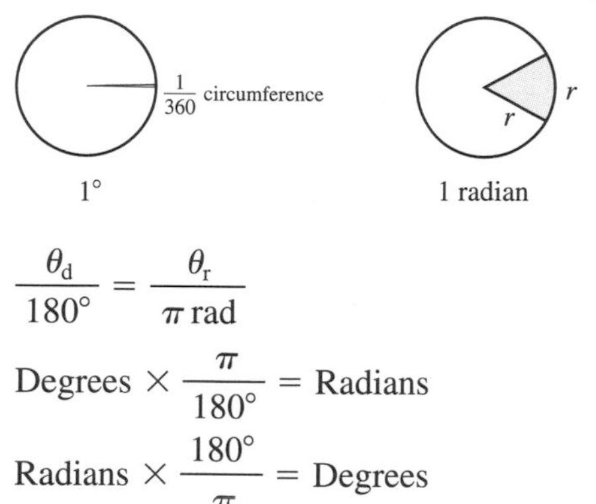

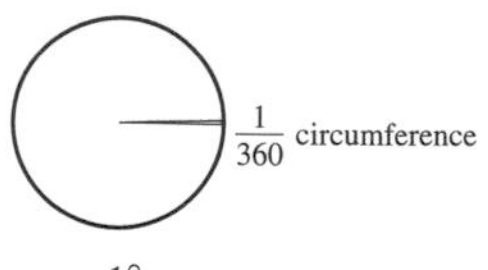

1°

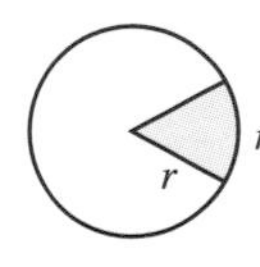

1 radian

$$\frac{\theta_d}{180^\circ} = \frac{\theta_r}{\pi \text{ rad}}$$

$$\text{Degrees} \times \frac{\pi}{180^\circ} = \text{Radians}$$

$$\text{Radians} \times \frac{180^\circ}{\pi} = \text{Degrees}$$

TRIGONOMETRIC FUNCTIONS (2.3, 2.6)

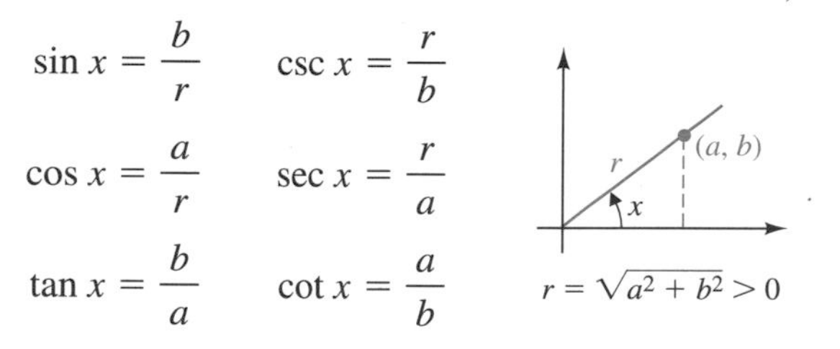

$$\sin x = \frac{b}{r} \qquad \csc x = \frac{r}{b}$$

$$\cos x = \frac{a}{r} \qquad \sec x = \frac{r}{a}$$

$$\tan x = \frac{b}{a} \qquad \cot x = \frac{a}{b}$$

(a, b), r, x

$$r = \sqrt{a^2 + b^2} > 0$$

(x in degrees or radians)

For x any real number and T any trigonometric function,

$$T(x) = T(x \text{ rad})$$

For a unit circle:

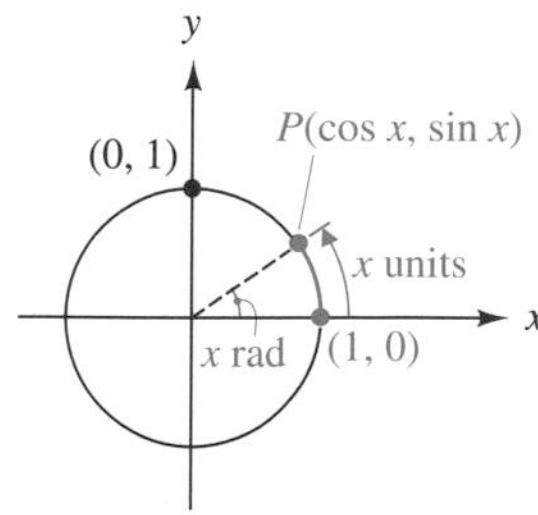

SPECIAL VALUES (2.5)

θ	$\sin\theta$	$\csc\theta$	$\cos\theta$	$\sec\theta$	$\tan\theta$	$\cot\theta$
0° or 0	0	N.D.	1	1	0	N.D.
30° or $\pi/6$	1/2	2	$\sqrt{3}/2$	$2/\sqrt{3}$	$1/\sqrt{3}$	$\sqrt{3}$
45° or $\pi/4$	$1/\sqrt{2}$	$\sqrt{2}$	$1/\sqrt{2}$	$\sqrt{2}$	1	1
60° or $\pi/3$	$\sqrt{3}/2$	$2/\sqrt{3}$	1/2	2	$\sqrt{3}$	$1/\sqrt{3}$
90° or $\pi/2$	1	1	0	N.D.	N.D.	0

N.D. = Not defined

PYTHAGOREAN THEOREM (1.3, C.2)

$$a^2 + b^2 = c^2$$

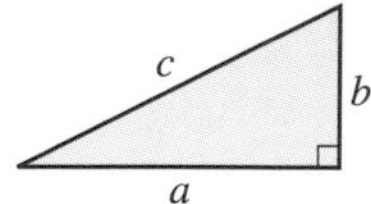

SPECIAL TRIANGLES (2.5)

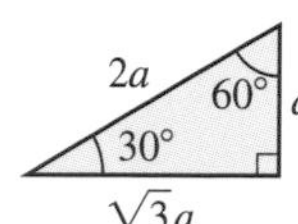

30°–60° triangle

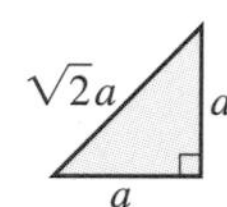

45° triangle

NUMBER OF SIGNIFICANT DIGITS (A.3)

Form of Number	Number of Significant Digits
x, with no decimal point	Count the digits of x from left to right, starting with the first digit and ending with the last nonzero digit.
x, with a decimal point	Count the digits of x from left to right, starting with the first nonzero digit and ending with the last digit (which may be zero).

ACCURACY OF CALCULATED VALUES (A.3)

The number of significant digits in a calculation involving multiplication, division, powers, and/or roots is the same as the number of significant digits in the number in the calculation with the smallest number of significant digits.

ACCURACY FOR TRIANGLES (1.3)

Angle to Nearest	Significant Digits for Side Measure
1°	2
10′ or 0.1°	3
1′ or 0.01°	4
10″ or 0.001°	5

(Continued inside back cover)

TRIGONOMETRIC IDENTITIES

Reciprocal Identities

$$\csc x = \frac{1}{\sin x} \qquad \sec x = \frac{1}{\cos x} \qquad \cot x = \frac{1}{\tan x}$$

Quotient Identities

$$\tan x = \frac{\sin x}{\cos x} \qquad \cot x = \frac{\cos x}{\sin x}$$

Identities for Negatives

$$\sin(-x) = -\sin x \qquad \cos(-x) = \cos x$$
$$\tan(-x) = -\tan x$$

Pythagorean Identities

$$\sin^2 x + \cos^2 x = 1 \qquad \tan^2 x + 1 = \sec^2 x$$
$$1 + \cot^2 x = \csc^2 x$$

Sum Identities

$$\sin(x + y) = \sin x \cos y + \cos x \sin y$$
$$\cos(x + y) = \cos x \cos y - \sin x \sin y$$
$$\tan(x + y) = \frac{\tan x + \tan y}{1 - \tan x \tan y}$$

Difference Identities

$$\sin(x - y) = \sin x \cos y - \cos x \sin y$$
$$\cos(x - y) = \cos x \cos y + \sin x \sin y$$
$$\tan(x - y) = \frac{\tan x - \tan y}{1 + \tan x \tan y}$$

Cofunction Identities

(Replace $\pi/2$ with 90° if x is in degree measure.)

$$\sin\left(\frac{\pi}{2} - x\right) = \cos x \qquad \cos\left(\frac{\pi}{2} - x\right) = \sin x$$
$$\tan\left(\frac{\pi}{2} - x\right) = \cot x \qquad \cot\left(\frac{\pi}{2} - x\right) = \tan x$$
$$\sec\left(\frac{\pi}{2} - x\right) = \csc x \qquad \csc\left(\frac{\pi}{2} - x\right) = \sec x$$

Product–Sum Identities

$$\sin x \cos y = \tfrac{1}{2}[\sin(x + y) + \sin(x - y)]$$
$$\cos x \sin y = \tfrac{1}{2}[\sin(x + y) - \sin(x - y)]$$
$$\sin x \sin y = \tfrac{1}{2}[\cos(x - y) - \cos(x + y)]$$
$$\cos x \cos y = \tfrac{1}{2}[\cos(x + y) + \cos(x - y)]$$

TRIGONOMETRIC IDENTITIES (cont'd) LAWS OF SINES AND COSINES

Sum–Product Identities

$$\sin x + \sin y = 2 \sin \frac{x + y}{2} \cos \frac{x - y}{2}$$
$$\sin x - \sin y = 2 \cos \frac{x + y}{2} \sin \frac{x - y}{2}$$
$$\cos x + \cos y = 2 \cos \frac{x + y}{2} \cos \frac{x - y}{2}$$
$$\cos x - \cos y = -2 \sin \frac{x + y}{2} \sin \frac{x - y}{2}$$

Double-Angle Identities

$$\sin 2x = 2 \sin x \cos x \qquad \cos 2x = \begin{cases} \cos^2 x - \sin^2 x \\ 1 - 2\sin^2 x \\ 2\cos^2 x - 1 \end{cases}$$
$$\tan 2x = \frac{2 \tan x}{1 - \tan^2 x} = \frac{2 \cot x}{\cot^2 x - 1} = \frac{2}{\cot x - \tan x}$$

Half-Angle Identities

$$\sin \frac{x}{2} = \pm\sqrt{\frac{1 - \cos x}{2}}$$
$$\cos \frac{x}{2} = \pm\sqrt{\frac{1 + \cos x}{2}}$$

Sign is determined by quadrant in which $x/2$ lies.

$$\tan \frac{x}{2} = \frac{1 - \cos x}{\sin x} = \frac{\sin x}{1 + \cos x} = \pm\sqrt{\frac{1 - \cos x}{1 + \cos x}}$$
$$\sin^2 x = \frac{1 - \cos 2x}{2} \qquad \cos^2 x = \frac{1 + \cos 2x}{2}$$
$$\tan^2 x = \frac{1 - \cos 2x}{1 + \cos 2x}$$

Law of Sines

$$\frac{\sin \alpha}{a} = \frac{\sin \beta}{b} = \frac{\sin \gamma}{c}$$

Law of Cosines

$$a^2 = b^2 + c^2 - 2bc \cos \alpha$$
$$b^2 = a^2 + c^2 - 2ac \cos \beta$$
$$c^2 = a^2 + b^2 - 2ab \cos \gamma$$

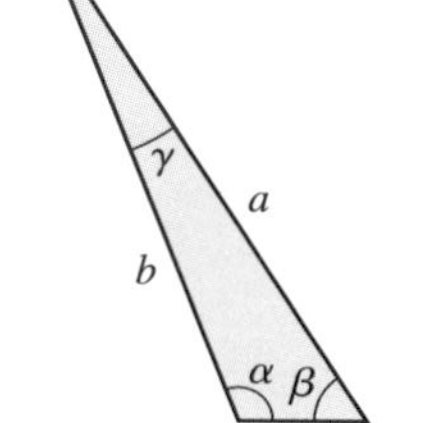

QUICK REFERENCE CARD

Raymond A. Barnett, Michael R. Ziegler, Karl E. Byleen, *Analytic Trigonometry with Applications,* Seventh Edition

Also available to help you succeed in this course:

Student Solutions Manual for Analytic Trigonometry with Applications, Seventh Edition
Contains solutions to all odd-numbered problems in the main text. Ask your bookstore manager for more details. (ISBN: 0-534-35839-X).

Boxer Trigonometry CD–ROM
This award-winning, multimedia Windows® CD–ROM provides a step-by-step instructional approach and visual simulations that will allow you to explore trigonometric concepts and comprehend them with ease. (Just call 1-800-736-2824)

For more information about this book or any Brooks/Cole Mathematics product, please visit our web site: **www.brookscole.com**

Removing this card may affect resale value.

Brooks/Cole Publishing Company
I(T)P® An International Thomson Publishing Company

ANGLES AND TRIANGLES

Degrees and Radians

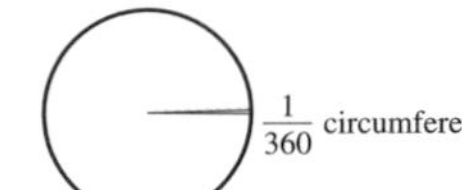

1°

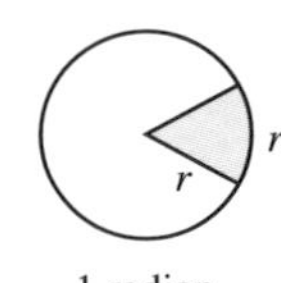

1 radian

$$\frac{\theta_d}{180°} = \frac{\theta_r}{\pi \text{ rad}}$$

$$\text{Degrees} \times \frac{\pi}{180°} = \text{Radians}$$

$$\text{Radians} \times \frac{180°}{\pi} = \text{Degrees}$$

Angles and Arcs

θ in Degrees

$$\frac{\theta}{360°} = \frac{s}{C}$$

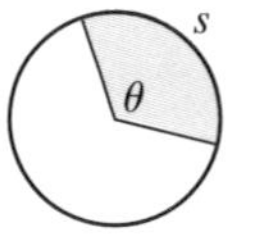

C (circumference)

θ in Radians

$$\theta = \frac{s}{r}$$

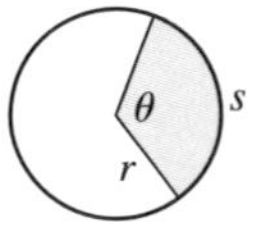

$$s = r\theta$$

Pythagorean Theorem

$$a^2 + b^2 = c^2$$

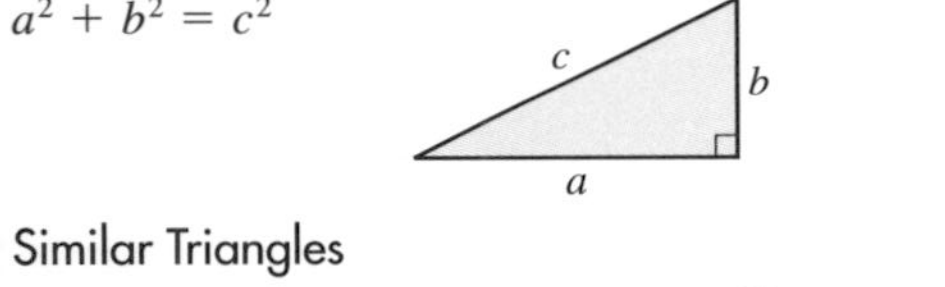

Similar Triangles

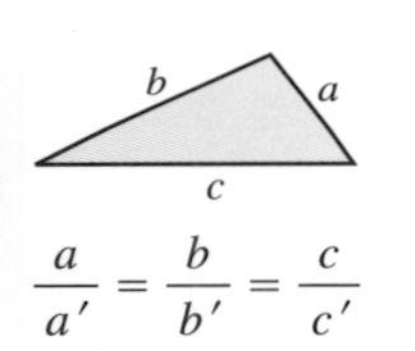

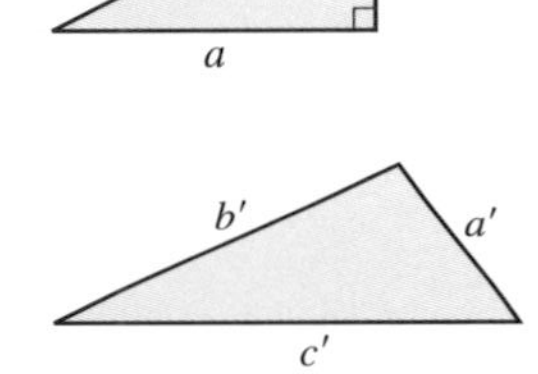

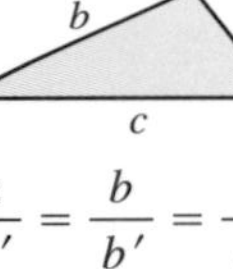

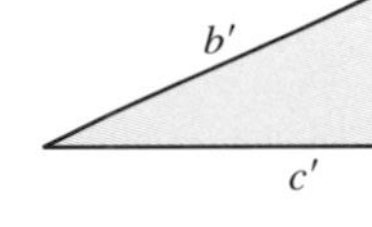

$$\frac{a}{a'} = \frac{b}{b'} = \frac{c}{c'}$$

TRIGONOMETRIC FUNCTIONS ACCURACY FOR TRIANGLES

Trigonometric Functions

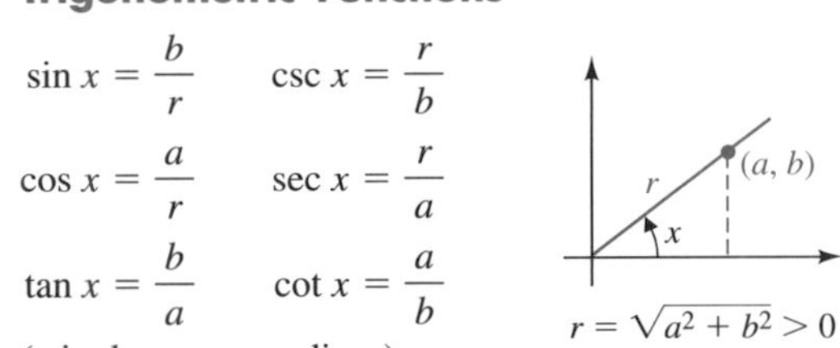

$$\sin x = \frac{b}{r} \qquad \csc x = \frac{r}{b}$$

$$\cos x = \frac{a}{r} \qquad \sec x = \frac{r}{a}$$

$$\tan x = \frac{b}{a} \qquad \cot x = \frac{a}{b}$$

$$r = \sqrt{a^2 + b^2} > 0$$

(x in degrees or radians)

For x any real number and T any trigonometric function,

$T(x) = T(x \text{ rad})$

For a unit circle:

$P(\cos x, \sin x)$, $(0, 1)$, $(1, 0)$, x units, x rad

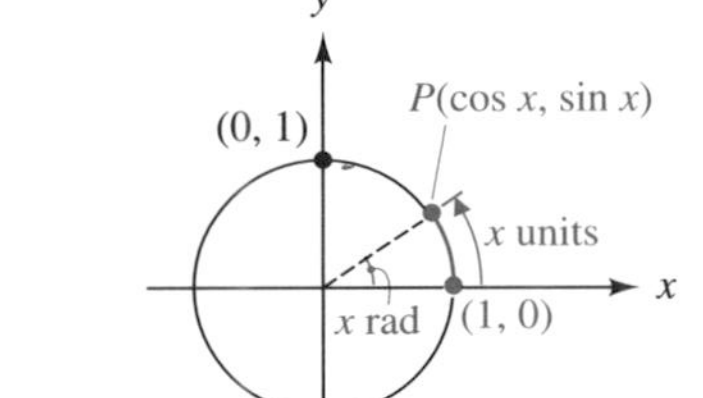

Special Triangles

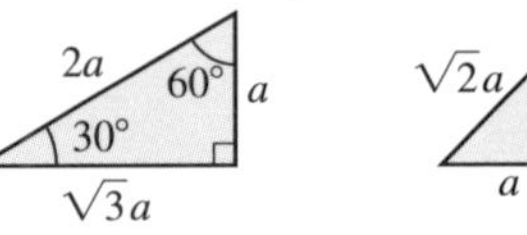

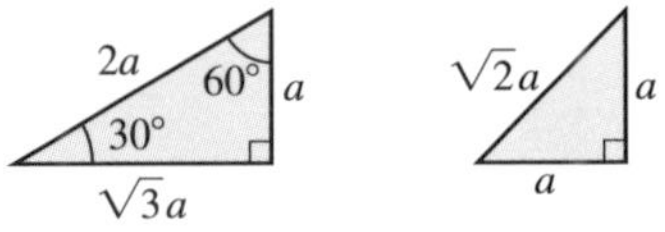

30°–60° triangle (sides $2a$, a, $\sqrt{3}a$)

45° triangle (sides $\sqrt{2}a$, a, a)

Special Values

θ	$\sin\theta$	$\csc\theta$	$\cos\theta$	$\sec\theta$	$\tan\theta$	$\cot\theta$
0° or 0	0	N.D.	1	1	0	N.D.
30° or $\pi/6$	1/2	2	$\sqrt{3}/2$	$2/\sqrt{3}$	$1/\sqrt{3}$	$\sqrt{3}$
45° or $\pi/4$	$1/\sqrt{2}$	$\sqrt{2}$	$1/\sqrt{2}$	$\sqrt{2}$	1	1
60° or $\pi/3$	$\sqrt{3}/2$	$2/\sqrt{3}$	1/2	2	$\sqrt{3}$	$1/\sqrt{3}$
90° or $\pi/2$	1	1	0	N.D.	N.D.	0

N.D. = Not defined

Accuracy for Triangles

Angle to Nearest	Significant Digits for Side Measure
1°	2
10′ or 0.1°	3
1′ or 0.01°	4
10″ or 0.001°	5

TRIGONOMETRIC FUNCTION GRAPHS INVERSE FUNCTIONS

Graphing Trigonometric Functions

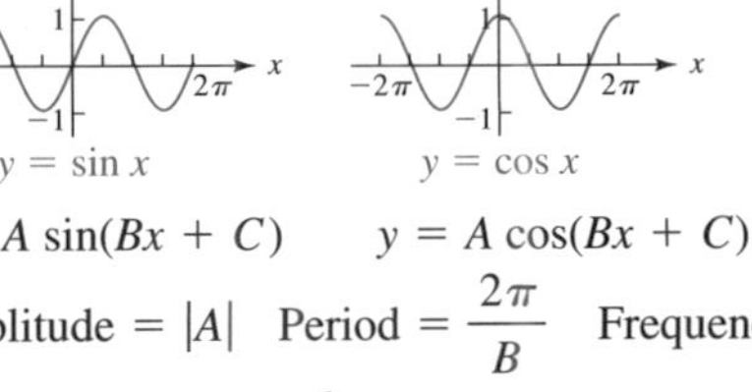

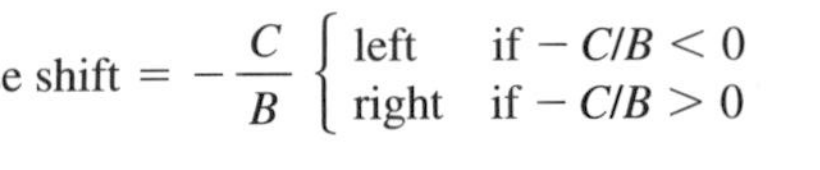

$y = A\sin(Bx + C) \qquad y = A\cos(Bx + C)$

$$\text{Amplitude} = |A| \quad \text{Period} = \frac{2\pi}{B} \quad \text{Frequency} = \frac{B}{2\pi}$$

$$\text{Phase shift} = -\frac{C}{B}\begin{cases}\text{left} & \text{if } -C/B < 0\\ \text{right} & \text{if } -C/B > 0\end{cases}$$

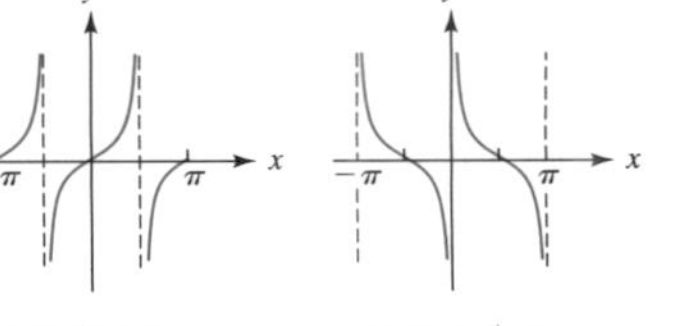

$y = A\tan(Bx + C) \qquad y = A\cot(Bx + C)$

$$\text{Period} = \frac{\pi}{B} \quad \text{Phase shift} = -\frac{C}{B}\begin{cases}\text{left} & \text{if } -C/B < 0\\ \text{right} & \text{if } -C/B > 0\end{cases}$$

Inverse Trigonometric Functions

$y = \sin^{-1} x$ means $x = \sin y$

where $-\frac{\pi}{2} \le y \le \frac{\pi}{2}$ and $-1 \le x \le 1$

$y = \cos^{-1} x$ means $x = \cos y$

where $0 \le y \le \pi$ and $-1 \le x \le 1$

$y = \tan^{-1} x$ means $x = \tan y$

where $-\frac{\pi}{2} < y < \frac{\pi}{2}$ and x is any real number

$y = \cot^{-1} x$ means $x = \cot y$

where $0 < y < \pi$ and x is any real number

$y = \sec^{-1} x$ means $x = \sec y$

where $0 \le y \le \pi$, $y \ne \frac{\pi}{2}$, and $x \le -1$ or $x \ge 1$

$y = \csc^{-1} x$ means $x = \csc y$

where $-\frac{\pi}{2} \le y \le \frac{\pi}{2}$, $y \ne 0$, and $x \le -1$ or $x \ge 1$

Applications Index

ANALYTIC TRIGONOMETRY
with Applications

ANALYTIC TRIGONOMETRY with Applications

SEVENTH EDITION

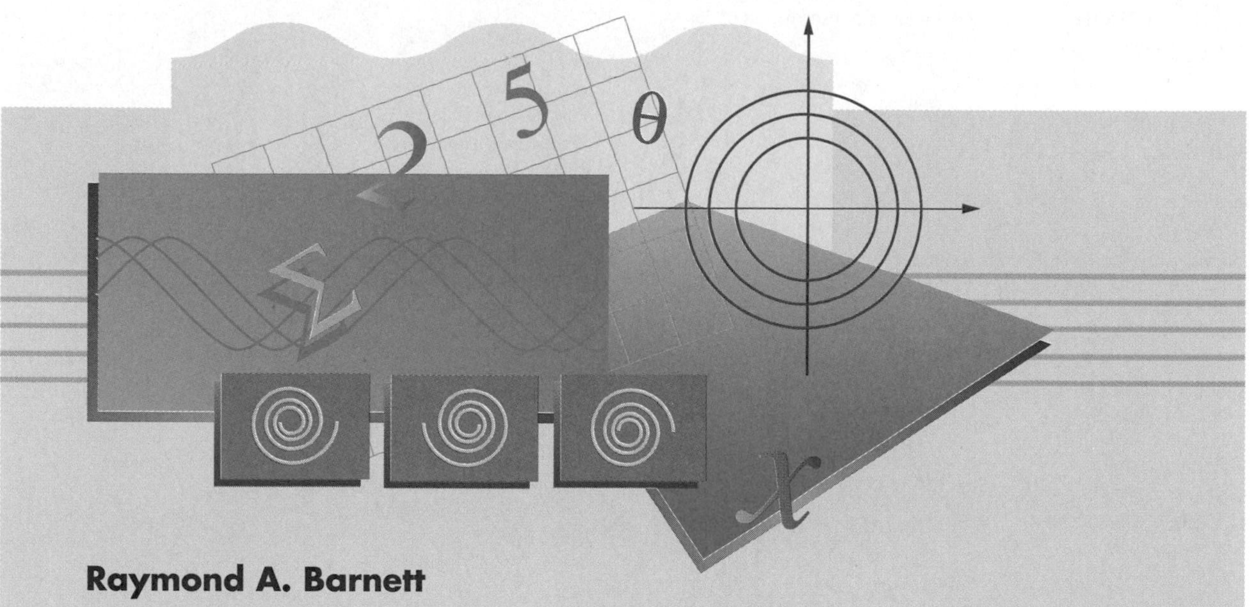

Raymond A. Barnett
Merritt College

Michael R. Ziegler
Marquette University

Karl E. Byleen
Marquette University

Brooks/Cole Publishing Company

I(T)P® *An International Thomson Publishing Company*

Pacific Grove • Albany • Belmont • Bonn • Boston • Cincinnati • Detroit • Johannesburg • London
Madrid • Melbourne • Mexico City • New York • Paris • Singapore • Tokyo • Toronto • Washington

Sponsoring Editor: *Margot Hanis*
Project Development Editor: *Elizabeth Rammel*
Marketing Team: *Caroline Croley, Donna Shore*
Marketing Assistant: *Debra Johnston*
Editorial Assistant: *Kimberly Raburn*
Production Service: *Phyllis Niklas*
Interior Design: *Julia Gecha*
Cover Art: *Istvan Bodoczky*
Cover Design: *Roger Knox*
Interior Illustration: *Scientific Illustrators*
Typesetting: *Progressive Information Technologies*
Cover Printing: *Phoenix Color Corporation*
Printing and Binding: *World Color*

For more information, contact:

BROOKS/COLE PUBLISHING COMPANY
511 Forest Lodge Road
Pacific Grove, CA 93950
USA

International Thomson Publishing Europe
Berkshire House 168–173
High Holborn
London WC1V 7AA
England

Thomas Nelson Australia
102 Dodds Street
South Melbourne, 3205
Victoria, Australia

Nelson Canada
1120 Birchmount Road
Scarborough, Ontario
Canada M1K 5G4

International Thomson Editores
Seneca 53
Col. Polanco
11560 México, D. F., México

International Thomson Publishing GmbH
Königswinterer Strasse 418
53227 Bonn
Germany

International Thomson Publishing Asia
60 Albert Street
#15-01 Albert Complex
Singapore 189969

International Thomson Publishing Japan
Hirakawacho Kyowa Building, 3F
2-2-1 Hirakawacho
Chiyoda-ku, Tokyo 102
Japan

Printed in the United States of America.

10 9 8 7 6 5 4

Library of Congress Cataloging-in-Publication Data

Barnett, Raymond A.
Analytic trigonometry with applications.—7th ed. / Raymond A. Barnett, Michael R. Ziegler, Karl E. Byleen.
p. cm.
Includes indexes.
ISBN 0-534-35838-1 (hc)
1. Trigonometry, Plane. I. Ziegler, Michael R. II. Byleen, Karl. III. Title.
QA533.B32 1999
516.3′4—dc21 98-27338
CIP

Dedicated to the memory of Don Dellen

Preface

The seventh edition of *Analytic Trigonometry with Applications* has benefited from thc generous response of the many users of the earlier editions. Prerequisites for the book are $1\frac{1}{2}$–2 years of high school algebra and 1 year of high school geometry or their equivalents. Great care has been taken to produce a book that students can actually read, understand, and enjoy.

Important Features Retained from the Sixth Edition

- The focus of this book is on student comprehension. An **informal style** is used for exposition, definitions, and theorems. Precision, however, is not compromised.
- To gain reader interest quickly, the text moves directly into trigonometric concepts and applications. **Review material** from prerequisite courses is either integrated in certain developments (particularly in Chapters 4 and 5) or can be found in the appendixes. This material can be reviewed as needed by the student or taught in class by an instructor.
- Concept development proceeds from the **concrete to the abstract. Trigonometric functions** are defined first in terms of angle domains using degree and radian measure side-by-side, and then in terms of real number domains. All of this is done early in the book and is reinforced throughout. By the end of the course, students should be comfortable with all three modes.
- Almost every concept is illustrated by an **example** followed by a **matching problem** (with answers given near the end of each section) to encourage an active rather than passive involvement in the learning process.
- There are enough relevant and interesting **applications** from diverse fields to convince even the most skeptical student that trigonometry is really useful. Many developments are motivated by interesting and relevant real-world applications. These applications are distributed uniformly throughout the book.
- The book includes **more than 2,000 carefully selected and graded problems.** The problems in most exercise sets are divided into A, B, and C groupings. The A problems are easy and routine, the B problems are more challenging but still emphasize mechanics, and the C problems are a mixture of difficult mechanics and theory. In short, the text is designed so that an average

or struggling student will be able to experience success and a very capable student will be challenged.

- **Answers** to most of the odd-numbered problems and all chapter and cumulative review exercises are included at the end of the book. Most of the even-numbered answers can be found in the *Instructor's Resource Manual.*
- **Cautions** alerting students to potential problem areas have been inserted where appropriate (see, for example, Sections 1.1, 2.3, and 4.2).
- **Dashed "think boxes"** are used to indicate steps that are usually performed mentally after a concept or procedure is understood (see, for example, Sections 1.1, 2.4, and 6.5).
- **Chapter reviews** are included at the end of each chapter, and **cumulative reviews** are included after Chapters 3, 5, and 7.
- **Formulas and symbols** (keyed to sections in which they are first introduced) and the **metric system** are summarized inside the front and back covers of the book for convenient reference.
- The content of the text **satisfies the requirements for many succeeding courses**, including calculus, analytic geometry, physics, and applied mathematics courses.
- **Pedagogical use of color:** Color is used not only to make the text more attractive, but more importantly it is used functionally to improve communications. For example, color is used in:

  1. Commentary that accompanies a solution process (see, for example, Sections 1.3, 4.2, and 5.3).
  2. Graphing to visually separate the various parts of a graph (especially in Chapters 3 and 7).
  3. Boxed highlighted material to distinguish among Assumptions/Definitions, Theorems, and Strategies/Processes (see, for example, Sections 1.3, 4.2, and 6.5).

ASSUMPTIONS/DEFINITIONS

THEOREMS

STRATEGIES/PROCESSES

New Features in the Seventh Edition

The impact of **mathematics reform** movements cannot be ignored. Curriculum and textbook changes resulting from mathematics reform include increased use of technology, cooperative learning, exploration and discovery, writing, and multiple interpretations of mathematical concepts and results (numeric, graphic, verbal, and symbolic). The following new text features were added to accommodate these reform changes.

- A **graphing calculator or utility** is not required to use this book; nevertheless, the use of a graphing calculator or utility will add significantly to the understanding of certain topics. For instructors who wish to require the use of a graphing utility with this book, there is more than enough material to justify this requirement. The expanded graphing utility material is interspersed in context throughout the book (see, for example, Section 3.1, Problems 21–26, and Section 5.4). All material requiring the use of a graphing calculator is identified by the calculator icon shown in the margin, and can be omitted without loss of continuity.

- Each section contains from one to four **Explore–Discuss** boxes that anticipate certain developments (for example, Explore–Discuss 1, Section 1.2), expand on developments (for example, Explore–Discuss 1, Section 1.1), or tie several developments together (for example, Explore–Discuss 2, Section 1.1). The Explore–Discuss material can be used for written student responses, in-class discussions, or as group activities. A comprehensive **Chapter Group Activity,** which ties together and extends several chapter concepts, has been added to the end of each chapter. Many problems and applications in the exercise sets can also be used for group activities (for example, Problems 27–30 in Exercise 1.4, Problems 45–50 in Exercise 3.3, and Problem 71 in Exercise 5.1). Any Explore–Discuss box or Chapter Group Activity may be omitted if time is of concern. Sketched solutions for the Explore–Discuss material and the Chapter Group Activities can be found in the *Instructor's Resource Manual.*
- There is now a little less emphasis on drill and **more emphasis on concept development, understanding, and communication.** Problems in exercise sets that require a **written response** (from a sentence to a paragraph) have the problem numeral printed in red for easy identification.
- More applications in exercise sets are **multistep discovery** problems that require **multiple interpretations** (for example, Problems 27 and 28 in Exercise 1.2, Problems 45–50 in Exercise 3.3, and Problem 71 in Exercise 5.1).
- More **historical remarks** have been added and others have been expanded (see, for example, Sections 1.1 and 2.4).

Ancillaries for Students

- A **student solutions manual** is available for student purchase. The manual includes detailed solutions to all odd-numbered problems and all chapter and cumulative review exercises.

- A **graphing calculator supplement** that coordinates the use of a graphing calculator with appropriate text topics is available for student purchase.
- A perforated **Quick Reference Card** is included in the text. This card can be removed, and places key equations and graphs at the student's fingertips.

Ancillaries for Instructors

- An ***Instructor's Resource Manual*** includes most of the answers that are not included in the text. It also includes outlines of solutions to the Explore–Discuss material and Chapter Group Activities.
- A **student solutions manual,** containing worked out solutions to all chapter review and cumulative review exercises and all odd-numbered problems in the book, is available without charge to any instructor adopting the book.
- **Thomson World Class Learning *Testing Tools.*** This integrated testing and tutorial software package features algorithmic test generation, on-line testing, class management capabilities, and tutorials.
- ***Boxer Trigonometry.*** This comprehensive tutorial on CD–ROM teaches the fundamentals of trigonometry in a creative and interactive learning environment.
- **Transparencies.** Reproductions of selected diagrams from the text are available for classroom presentation.
- **Test items.** A bound test bank is available free to adopters.

Error Check

Because of the careful independent checking and proofing by three competent college mathematics instructors, the authors and publisher believe this book to be substantially error-free. If any errors remain, the authors would be grateful if corrections were sent to: Reprints Coordinator, Brooks/Cole Publishing Company, 511 Forest Lodge Road, Pacific Grove, CA 93950-5098.

Acknowledgments

In addition to the authors, the publication of a book requires the effort and skills of many people. We would like to extend particular thanks to several very competent people: Brooks/Cole mathematics editors Elizabeth Rammel and Margot Hanis, for their skill, dependability, and support; production supervisor Phyllis Niklas, for her considerable expertise in preparing and guiding the book to publication; Fred Safier of City College of San Francisco, for his careful checking of the exercise sets and his masterly preparation of the solutions manual that accompanies this text; and Gholamhossein Hamedani, Robert Mullins, and Caroline Woods, for their careful checking and proofing of the entire manuscript, including examples and exercise sets.

We also wish to thank the many users and reviewers of the sixth edition and the post-revision reviewers for their helpful suggestions and comments. In particular, we wish to thank

George Alexander
University of Wisconsin—Madison

Charlene Beckman
Grand Valley State University

Steve Butcher
University of Central Arkansas

Ronald Bzoch
University of North Dakota

Robert A. Chaffer
Central Michigan University

Rodney Chase
Oakland Community College

Barbara Chudilowsky
De Anza College

Richard L. Francis
Southeast Missouri State University

Al Giambrone
Sinclair Community College

Frank Glaser
California State Polytechnic University—Pomona

Jack Goebel
Montana College of Mineral Science and Technology

Deborah Gunter
Langston University

James Hart
Middle Tennessee State University

Pamala Heard
Langston University

Steven Heath
Southern Utah State College

Robert Henkel
Pierce College

John G. Henson
Mesa College

Fred Hickling
University of Central Arkansas

Quin Higgins
Louisiana State University

JoAnnah Hill
Montana State University

Norma F. James
New Mexico State University—Las Cruces

Judith M. Jones
Valencia Community College

Stephanie H. Kurtz
Louisiana State University

Ted Laetsch
University of Arizona

Vernon H. Leach
Pierce College

Lauri Lindberg
Pierce College

Van Ma
Nicholls State University

LaVerne McFadden
Parkland College

John Martin
Santa Rosa Junior College

Susan Meriwether
University of Mississippi

Eldon Miller
University of Mississippi

Georgia Miller
Millsaps College

Margaret Ullmann Miller
Skyline College

Kathy Monaghan
American River College

Charles Mullins
University of Central Arkansas

Karen Murany
Oakland Community College

Harald M. Ness
University of Wisconsin

Nagendra N. Pandey
Montana College of Mineral Science and Technology

Howard Penn
El Centro College

Gladys Rockin
Oakland Community College

C. Robert Secrist
Kellogg Community College

Michael Seery
Scottsdale Community College

Robyn Elaine Serven
University of Central Arkansas

M. Vali Siadat
Richard J. Daley College

Thomas Spradley
American River College

Lowell Stultz
Kalamazoo Valley Community College

Richard Tebbs
Southern Utah State College

Roberta Tugenberg
Southwestern College

Steve Waters
Pacific Union College

Bobby Winters
Pittsburgh State University

Thomas Worthington
Grand Rapids Community College

Raymond A. Barnett
Michael R. Ziegler
Karl E. Byleen

CONTENTS

☆ Sections marked with a star may be omitted without loss of continuity.

[calculator icon] Sections marked with the calculator icon require the use of a graphing utility. Inclusion of this material will enrich the course, but its omission will not affect the continuity of the course.

6 ADDITIONAL TOPICS: TRIANGLES AND VECTORS 323

7 POLAR COORDINATES; COMPLEX NUMBERS 393

Appendix A COMMENTS ON NUMBERS 441

Appendix B FUNCTIONS AND INVERSE FUNCTIONS 456

Appendix C PLANE GEOMETRY: SOME USEFUL FACTS 472

A Note on Calculators

Use of calculators is emphasized throughout this book. Many brands and types of scientific calculators are available and can be found starting at about $10. Graphing calculators are more expensive; but, in addition to having most of the capabilities of a scientific calculator, they have very powerful graphing capabilities. Your instructor should help you decide on the type and model best suited to this course and the emphasis on calculator use he or she desires.

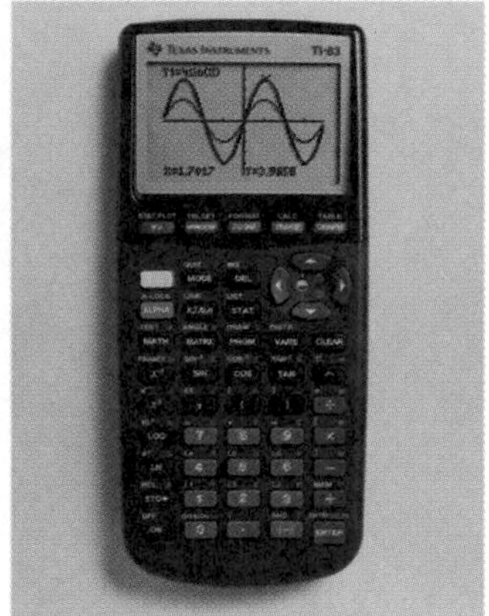

Scientific Calculator (left) and Graphing Calculator (right)
Courtesy Texas Instruments Incorporated

Whichever calculator you use, it is essential that you read the user's manual for your calculator. A large variety of calculators are on the market, and each is slightly different from the others. Therefore, take the time to read the manual. The first time through do not try to read and understand everything the calculator can do—that will tend to overwhelm and confuse you. Read only those sections pertaining to the operations you are or will be using; then return to the manual as necessary when you encounter new operations.

It is important to remember that a calculator is not a substitute for thinking. It can save you a great deal of time in certain types of problems, but you still must understand basic concepts so that you can interpret results obtained through the use of a calculator.

RIGHT TRIANGLE RATIOS

1

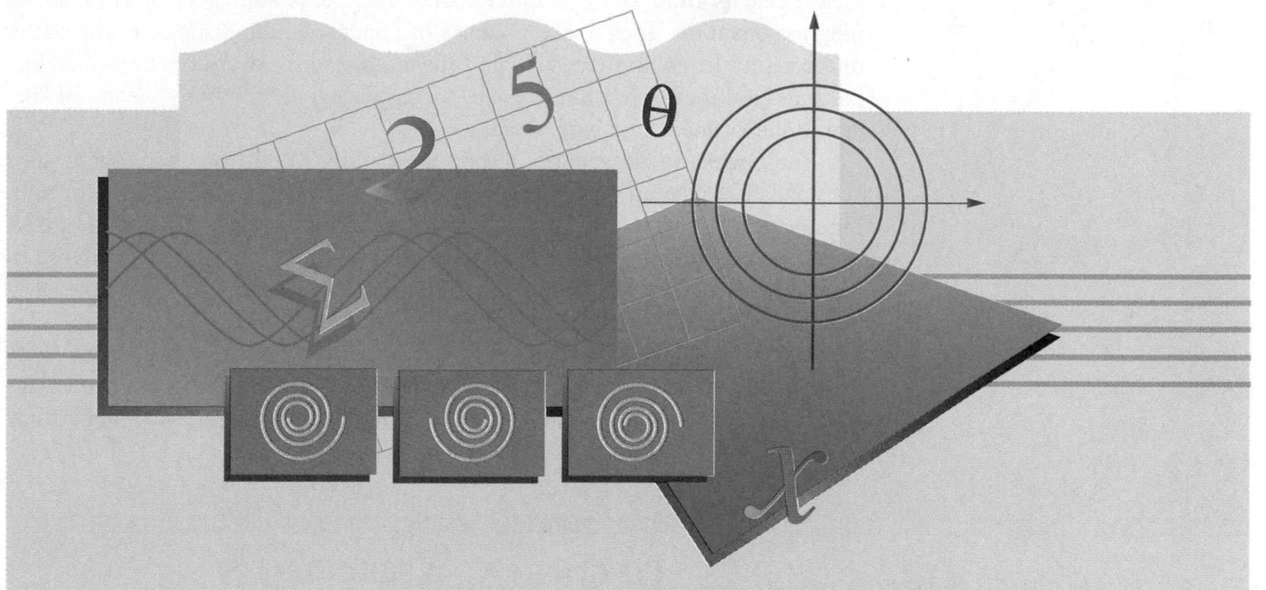

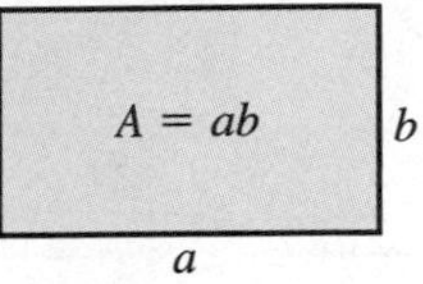

If you were asked to find your height, you would no doubt take a ruler or tape measure and measure it directly. But if you were asked to find the area of your bedroom floor in square feet, you would not be likely to measure the area directly by laying 1 ft squares over the entire floor and counting them. Instead, you would probably find the area **indirectly** by using the formula $A = ab$ from plane geometry, where A represents the area of the room and a and b are the lengths of its sides, as indicated in the figure in the margin.

In general, **indirect measurement** is a process of determining unknown measurements from known measurements by a reasoning process. How do we measure quantities such as the volumes of containers, the distance to the center of the earth, the area of the surface of the earth, and the distances to the sun and the stars? All these measurements are accomplished indirectly by the use of special formulas and deductive reasoning.

The Greeks in Alexandria, during the period 300 BC–AD 200, contributed substantially to the art of indirect measurement by developing formulas for finding areas, volumes, and lengths. Using these formulas, they were able to determine the circumference of the earth (with an error of only about 2%) and to estimate the distance to the moon. We will examine these measurements as well as others in the sections that follow.

It was during the early part of this Greek period that trigonometry, the study of triangles, was born. Hipparchus (160–127 BC), one of the greatest astronomers of the ancient world, is credited with making the first systematic study of the indirect measurement of triangles.

1.1 ANGLES, DEGREES, AND ARCS

- Angles
- Degree Measure of Angles
- Angles and Arcs
- Approximation of Earth's Circumference
- Approximation of the Diameters of the Sun and Moon; Total Solar Eclipse

Angles and the degree measure of an angle are the first concepts introduced in this section. Then, a very useful relationship between angles and arcs of circles is developed. This simple relationship between an angle and an arc will enable us to measure indirectly many useful quantities such as the circumference of the earth and the diameter of the sun. (If you are rusty on certain geometric relationships and facts, refer to Appendix C as needed.)

Angles

Central to the study of trigonometry is the concept of angle. An **angle** is formed by rotating a half-line, called a **ray**, around its end point. One ray k, called the **initial side** of the angle, remains fixed; a second ray l, called the **terminal side**

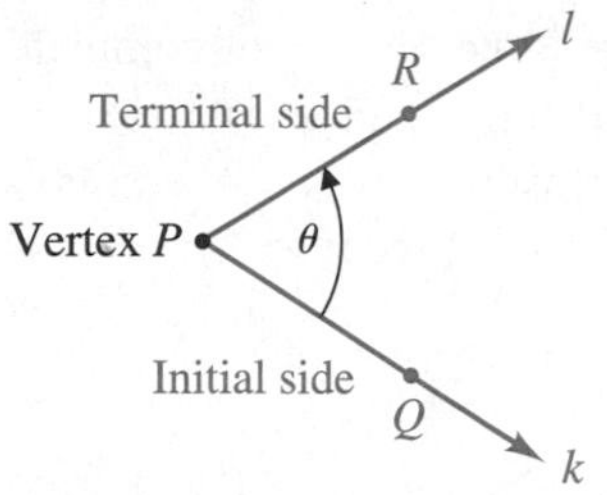

FIGURE 1
Angle θ: A ray rotated around its end point

of the angle, starts in the initial side position and rotates around the common end point P in a plane until it reaches its terminal position. The common end point P is called the **vertex.** (See Fig. 1.)

We may refer to the angle in Figure 1 in any of the following ways:

Angle θ	$\angle\theta$	Angle QPR	$\angle QPR$
Angle P	$\angle P$		

The symbol $\angle$ denotes *angle.* The Greek letters theta (θ), alpha (α), beta (β), and gamma (γ) are often used to name angles.

There is no restriction on the amount or direction of rotation in a given plane. When the terminal side is rotated counterclockwise, the angle formed is **positive** (see Figs. 1 and 2a); when it is rotated clockwise, the angle formed is **negative** (see Fig. 2b). Two different angles may have the same initial and terminal sides, as shown in Figure 2c. Such angles are said to be **coterminal.** In this chapter we will concentrate on positive angles; we will consider more general angles in detail in subsequent chapters.

FIGURE 2

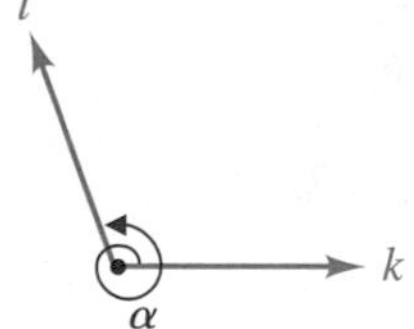

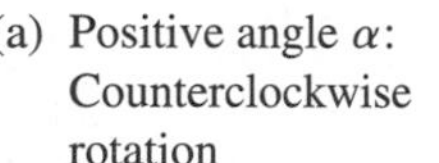

(a) Positive angle α: Counterclockwise rotation

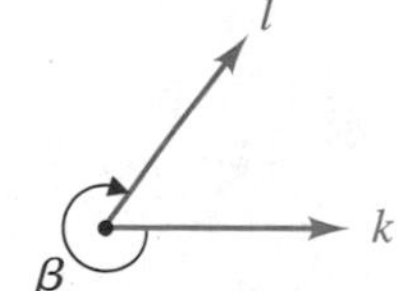

(b) Negative angle β: Clockwise rotation

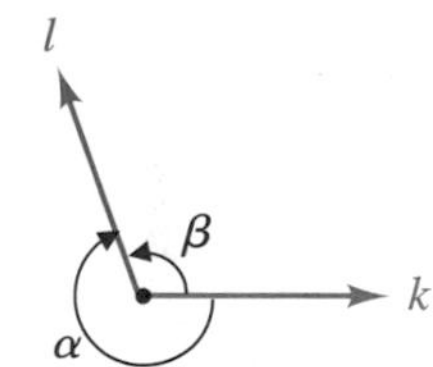

(c) α and β are coterminal

Degree Measure of Angles

To compare angles of different sizes a standard unit of measure is necessary. Just as a line segment can be measured in inches, meters, or miles, an angle is measured in *degrees* or *radians.* (We will postpone our discussion of radian measure of angles until Section 2.1.)

DEGREE MEASURE OF ANGLES

An angle formed by one complete revolution of the terminal side in a counterclockwise direction has **measure 360 degrees,** written **360°.** An angle of **1 degree measure,** written **1°**, is formed by $\frac{1}{360}$ of one complete revolution in a counterclockwise direction.

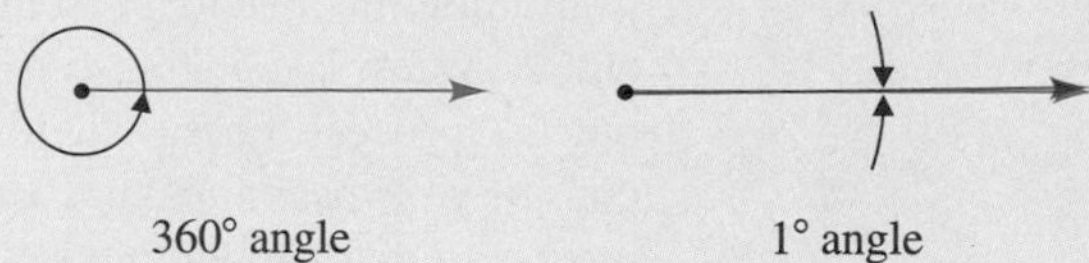

Angles of measure 90° and 180° represent $\frac{90}{360} = \frac{1}{4}$ and $\frac{180}{360} = \frac{1}{2}$ of complete revolutions, respectively. A 90° angle is called a **right angle** and a 180° angle is called a **straight angle.** An **acute angle** has angle measure between 0° and 90°. An **obtuse angle** has angle measure between 90° and 180°. (See Fig. 3.)

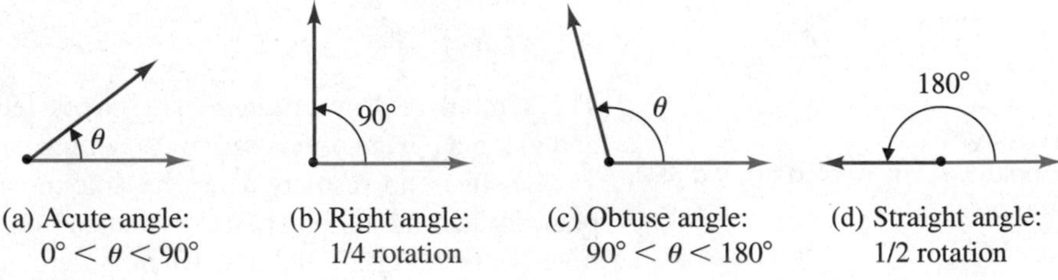

FIGURE 3
Special angles

Remark In Figure 3 we used θ in two different ways: to name an angle and to represent the measure of an angle. This usage is common; the context will dictate the interpretation. □

Two positive angles are **complementary** if the sum of their measures is 90°; they are **supplementary** if the sum of their measures is 180° (see Fig. 4).

FIGURE 4

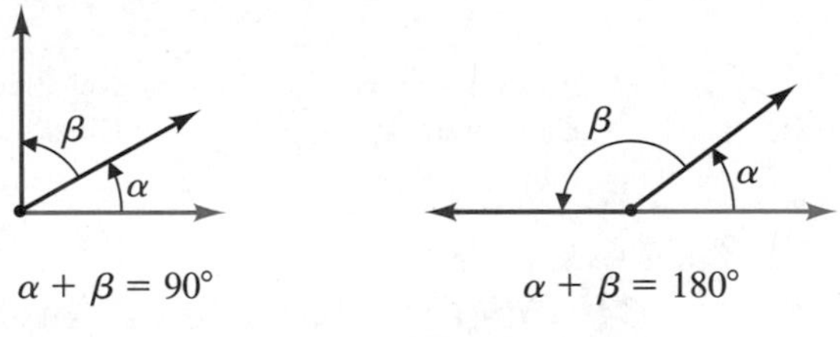

$\alpha + \beta = 90°$ (a) Complementary angles

$\alpha + \beta = 180°$ (b) Supplementary angles

A degree can be divided using decimal notation. For example, 36.25° represents an angle of degree measure 36 plus one-fourth of 1 degree. A degree can also be divided into minutes and seconds just as an hour is divided into minutes and seconds. Each degree is divided into 60 equal parts called minutes (′), and each minute is divided into 60 equal parts called seconds (″). Thus,

$$5°12'32''$$

is a concise way of writing 5 degrees, 12 minutes, and 32 seconds.

Degree measure in **decimal degree (DD)** form is useful in some instances, and degree measure in **degree-minute-second (DMS)** form is useful in others. You should be able to go from one form to the other as illustrated in Examples 1 and 2. Many calculators can perform the conversion either way automatically, but the process varies significantly among the various types of calculators—consult your user's manual. Examples 1 and 2 first illustrate a nonautomatic approach so that you will understand the process. This is followed by an automatic calculator approach that can be used for efficiency.

Conversion Accuracy If an angle is measured to the nearest second, the converted decimal form should not go beyond three decimal places, and vice versa. □

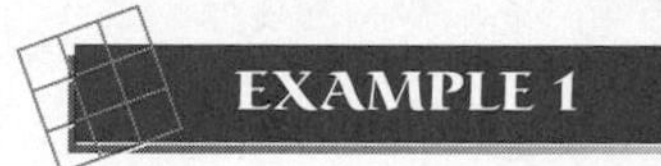

EXAMPLE 1 From DMS to DD

Convert 12°6′23″ to decimal degree form.

Solution **Method I** *Multistep conversion.* Since

$$6' = \left(\frac{6}{60}\right)^\circ \quad \text{and} \quad 23'' = \left(\frac{23}{3{,}600}\right)^\circ$$

then

$$12^\circ 6' 23'' = \boxed{\left(12 + \frac{6}{60} + \frac{23}{3{,}600}\right)^\circ}^{*}$$
$$= 12.106^\circ \qquad \text{To three decimal places}$$

Method II *Single-step calculator conversion.* Consult the user's manual for your particular calculator. The conversion shown in Figure 5 is from a graphing calculator.

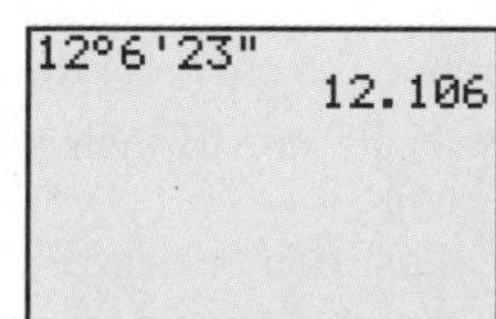

FIGURE 5
From DMS to DD

■

Matched Problem 1[†] Convert 128°42′8″ to decimal degree form. ■

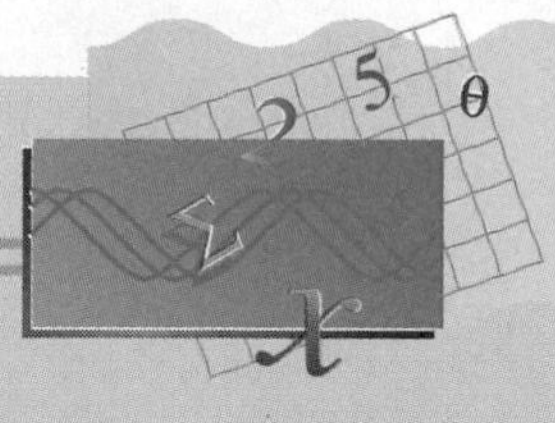

EXPLORE/DISCUSS 1

In 1988, Belayneh Densimo of Ethiopia ran the world's fastest recorded time for a marathon: 42.2 km in 2 hr, 6 min, 50 sec. Convert the hour-minute-second form to decimal hours. Explain the similarity of this conversion to the conversion of a degree-minute-second form to a decimal degree form.

---

* Dashed "think boxes" indicate steps that can be performed mentally once a concept or procedure is understood.

† Answers to matched problems are located at the end of each section just before the exercise set.

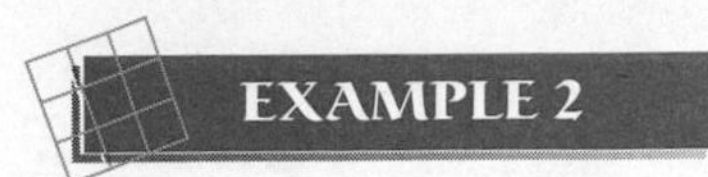

EXAMPLE 2 **From DD to DMS**

Convert 35.413° to a degree-minute-second form.

Solution **Method I** *Multistep conversion.*

$$\begin{aligned} 35.413° &= 35°\,(0.413 \cdot 60)' \\ &= 35°24.78' \\ &= 35°24'\,(0.78 \cdot 60)'' \\ &\approx 35°24'47'' \end{aligned}$$

```
35.413°▸DMS
      35°24'46.8"
```

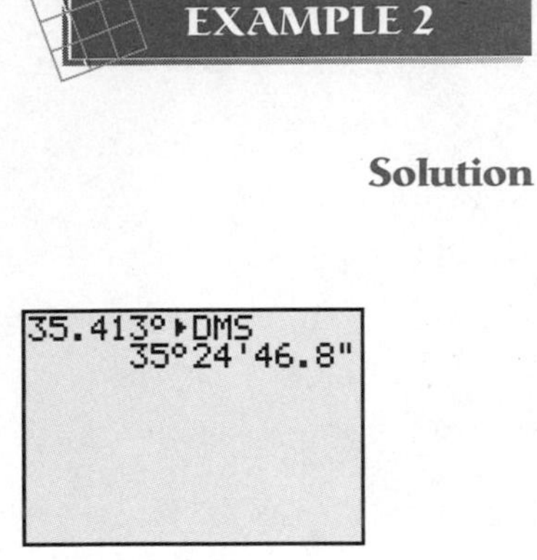

FIGURE 6
From DD to DMS

Method II *Single-step calculator conversion.* Consult the user's manual for your particular calculator. The conversion shown in Figure 6 is from a graphing calculator. Rounded to the nearest second, we get the same result as in method I: 35°24′47″.

Matched Problem 2 Convert 72.103° to degree-minute-second form.

Angles and Arcs

Given an arc RQ of a circle with center P, the angle RPQ is said to be the **central angle** that is **subtended** by the arc RQ. We also say that the arc RQ is subtended by the angle RPQ; see Figure 7.

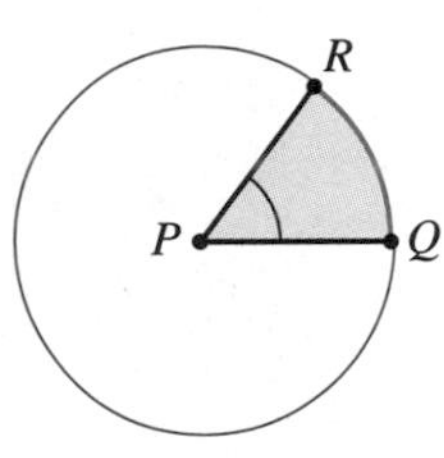

FIGURE 7
Central angle *RPQ* subtended by arc *RQ*

It follows from the definition of degree that a central angle subtended by an arc $\frac{1}{4}$ the circumference of a circle has degree measure 90; $\frac{1}{2}$ the circumference of a circle, degree measure 180, and the whole circumference of a circle, degree measure 360. In general, to determine the degree measure of an angle θ subtended by an arc of s units for a circle with circumference C units, use the following proportion:

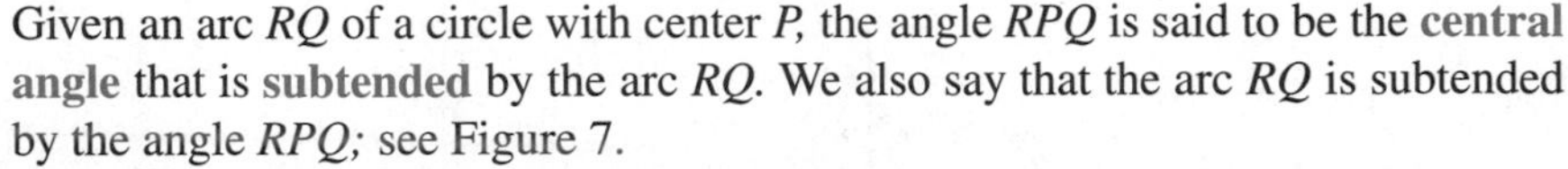

PROPORTION RELATING CENTRAL ANGLES AND ARCS

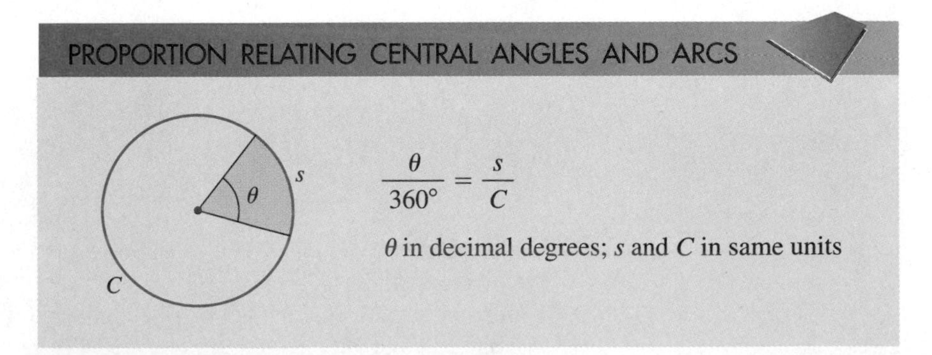

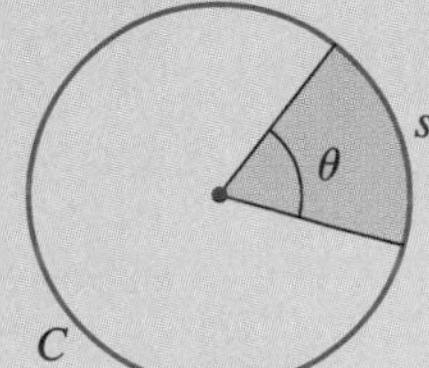

$$\frac{\theta}{360°} = \frac{s}{C}$$

θ in decimal degrees; s and C in same units

If we know any two of the three quantities, s, C, or θ, we can find the third by using simple algebra. For example, in a circle with circumference 72 in., the degree measure of a central angle θ subtended by an arc of length 12 in. is given by

$$\theta = \frac{12}{72} \cdot 360° = 60°$$

Remark Note that "an angle of 6°" means "an angle of degree measure 6," and "$\theta = 72°$" means "the degree measure of angle θ is 72." □

Approximation of Earth's Circumference

The early Greeks were aware of the proportion relating central angles and arcs, which Eratosthenes (240 BC) used in his famous calculation of the circumference of the earth. He reasoned as follows: It was well-known that at Syene (now Aswan), during the summer solstice, the noon sun was reflected on the water in a deep well (this meant the sun shone straight down the well and must be directly overhead). Eratosthenes reasoned that if the sun rays entering the well were continued down into the earth, they would pass through its center (see Fig. 8). On the same day at the same time, 5,000 stadia (approx. 500 mi) due north, in Alexandria, sun rays crossed a vertical pole at an angle of 7.5° as indicated in Figure 8. Since sun rays are very nearly parallel when they reach the earth, Eratosthenes concluded that $\angle ACS$ was also 7.5°. (Why?)*

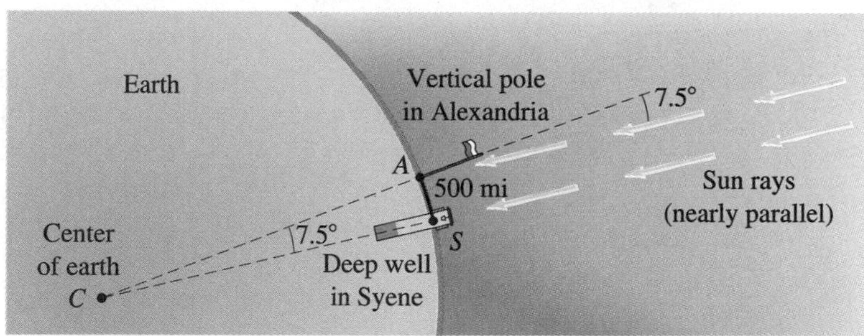

FIGURE 8
Estimating the earth's circumference

Even though Eratosthenes' reasoning was profound, his final calculation of the circumference of the earth requires only elementary algebra:

$$\frac{s}{C} = \frac{\theta}{360°}$$

$$\frac{500 \text{ mi}}{C} \approx \frac{7.5°}{360°}$$

$$C \approx \frac{360}{7.5}(500 \text{ mi}) = 24{,}000 \text{ mi}$$

The value calculated today is 24,875 mi.

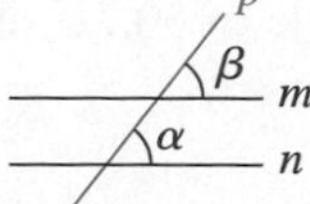

* If line p crosses parallel lines m and n, then angles α and β have the same measure.

The diameter of the earth (d) and the radius (r) can be found from this value using the formulas $C = 2\pi r$ and $d = 2r$ from plane geometry. [*Note:* $\pi = C/d$ for all circles.] The constant π has a long and interesting history; a few important dates are listed below:

1650 BC	Rhind Papyrus	$\pi \approx \frac{256}{81} = 3.16049...$
240 BC	Archimedes	$3\frac{10}{71} < \pi < 3\frac{1}{7}$ $(3.1408... < \pi < 3.1428...)$
AD 264	Liu Hui	$\pi \approx 3.14159$
AD 470	Tsu Ch'ung-chih	$\pi \approx \frac{355}{113} = 3.1415929...$
AD 1674	Leibniz	$\pi = 4(1 - \frac{1}{3} + \frac{1}{5} - \frac{1}{7} + \frac{1}{9} - \frac{1}{11} + \cdots)$ $\approx 3.1415926535897932384626$ (This and other series can be used to compute π to any decimal accuracy desired.)
AD 1761	Johann Lambert	Showed π to be irrational (π as a decimal is nonrepeating and nonterminating)

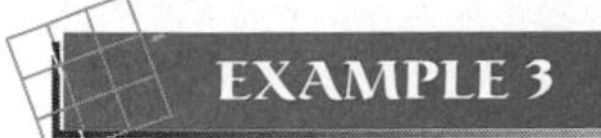

Arc Length

How large an arc is subtended by a central angle of 6.23° on a circle with radius 10 cm? (Compute the answer to two decimal places.)

Solution Since

$$\frac{s}{C} = \frac{\theta}{360°} \quad \text{and} \quad C = 2\pi r$$

then

$$\frac{s}{2\pi r} = \frac{\theta}{360°} \qquad \text{Replace } C \text{ with } 2\pi r.$$

$$\frac{s}{2(\pi)(10 \text{ cm})} = \frac{6.23°}{360°}$$

$$s = \frac{2(\pi)(10 \text{ cm})(6.23)}{360} = 1.09 \text{ cm}$$

■

Matched Problem 3 How large an arc is subtended by a central angle of 50.73° on a circle with radius 5 m? (Compute the answer to two decimal places.) ■

Approximation of the Diameters of the Sun and Moon; Total Solar Eclipse

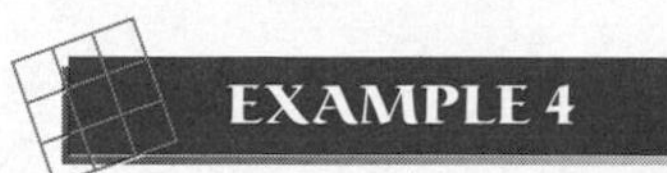

Sun's Diameter

If the distance from the earth to the sun is 93,000,000 mi, find the diameter of the sun (to the nearest thousand miles) if it subtends an angle of $0°31'55''$ on the surface of the earth.

Solution For small central angles in circles with very large radii, the **intercepted arc** (arc opposite the central angle) and its **chord** (the straight line joining the end points of the arc) are approximately the same length (see Fig. 9). We thus use the intercepted arc to approximate its chord in many practical problems, particularly when the length of the intercepted arc is easier to compute. We apply these ideas to finding the diameter of the sun as follows:

$$\theta = 0°31'55'' = 0.532°$$

$$\frac{s}{2\pi r} = \frac{\theta}{360°}$$

$$s = \frac{2\pi r\theta}{360} = \frac{2(\pi)(93{,}000{,}000 \text{ mi})(0.532)}{360} = 864{,}000 \text{ mi}$$

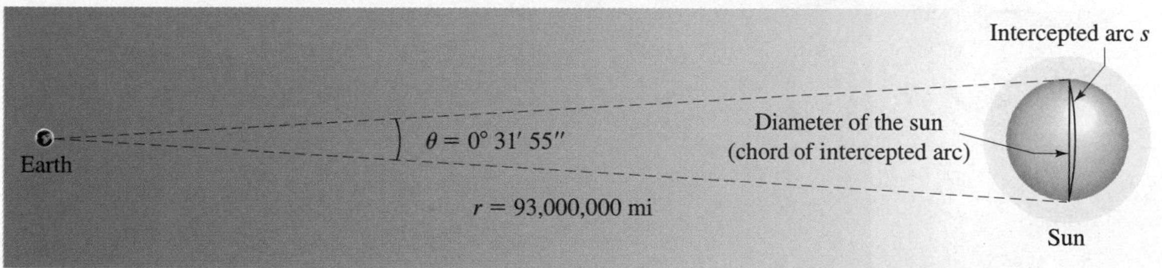

FIGURE 9

■

Matched Problem 4 If the moon subtends an angle of about $0°31'5''$ on the surface of the earth when it is 239,000 mi from the earth, estimate its diameter to the nearest 10 mi. ■

Answers to Matched Problems

1. 128.702°
2. $72°6'11''$
3. 4.42 m
4. 2,160 mi

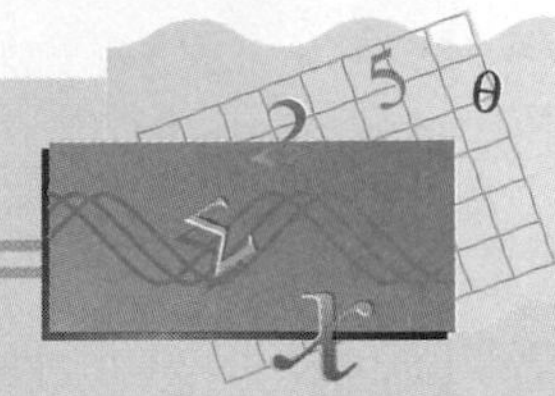

EXPLORE/DISCUSS 2

A total solar eclipse will occur when the moon passes between the earth and the sun and the angle subtended by the diameter of the moon is at least as large as the angle subtended by the diameter of the sun (see Fig. 10).

(A) Since the distances from the sun and moon to the earth vary with time, explain what happens to the angles subtended by the diameters of the sun and moon as their distances from the earth increase and decrease.

(B) For a total solar eclipse to occur when the moon passes between the sun and the earth, would it be better for the sun to be as far away as possible and the moon to be as close as possible, or vice versa? Explain.

(C) The diameters of the sun and moon are, respectively, 864,000 mi and 2,160 mi. Find the maximum distance that the moon can be from the earth for a total solar eclipse when the sun is at its maximum distance from the earth, 94,500,000 mi.

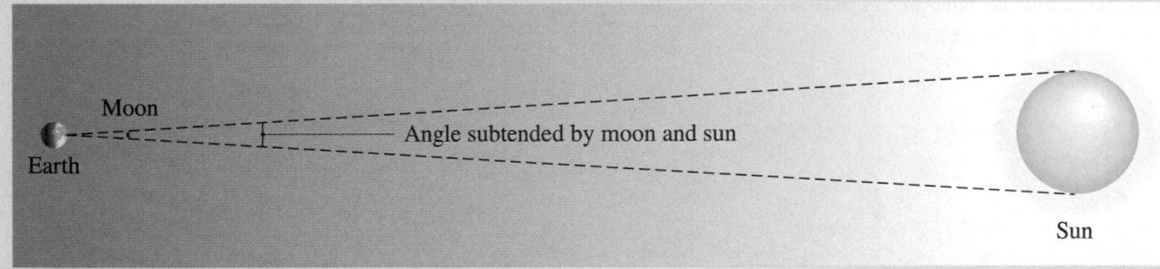

FIGURE 10
Total solar eclipse

EXERCISE 1.1

Throughout the text the problems in most exercise sets are divided into A, B, and C groupings: The A problems are easy and routine, the B problems are more challenging but still emphasize mechanics, and the C problems are a mixture of difficult mechanics and theory.

A *Indicate the number of degrees in the angle formed by the terminal side rotating counterclockwise through:*

1. $\frac{1}{2}$ revolution
2. $\frac{1}{4}$ revolution
3. $\frac{1}{8}$ revolution
4. $\frac{1}{3}$ revolution
5. $\frac{2}{3}$ revolution
6. $\frac{1}{12}$ revolution

Identify the following angles as acute, right, obtuse, or straight. If the angle is none of these, say so.

7. 123°
8. 18°
9. 180°

10. 90° **11.** 45° **12.** 91°

13. 270° **14.** 225°

* **15.** Describe the meaning of an angle of 1 degree.

16. Describe the meaning of minutes and seconds in angle measure.

B *Change to decimal degrees accurate to three decimal places:*

17. 43°21′4″ **18.** 61°52′11″ **19.** 2°12′47″

20. 23°5′21″ **21.** 103°17′41″ **22.** 228°40′51″

Change to degree-minute-second form:

23. 13.633° **24.** 22.767° **25.** 83.017°

26. 74.023° **27.** 187.204° **28.** 204.333°

29. Which of the angle measures, 47°33′41″ or 47.572°, is the larger? Explain how you obtained your answer.

30. Runner A ran a marathon in 3 hr, 42 min, 21 sec, and runner B in 3.712 hr. Which runner is faster? Explain how you obtained your answer.

Two angles, α and β, have degree measures as indicated. Write $\alpha < \beta$ or $\alpha > \beta$ as appropriate.

31. $\alpha = 27°9'17''$, $\beta = 27.163°$

32. $\alpha = 123°16'5''$, $\beta = 123.254°$

33. $\alpha = 12.807°$, $\beta = 12°47'13''$

34. $\alpha = 68.916°$, $\beta = 68°55'42''$

In Problems 35–38, perform the indicated operations directly on a calculator. Express the answers in DMS form. For example, a particular graphing calculator performs the following calculation (67°13′4″ − 45°27′32″) as follows:

```
67°13'4"-45°27'3
2"▸DMS
          21°45'32"
```

35. 47°37′49″ + 62°40′15″

36. 105°53′22″ + 26°38′55″

37. 90° − 67°37′29″

38. 180° − 121°51′22″

* A red problem number indicates a problem that requires a written interpretation or comment and not just a numerical answer.

In Problems 39–46, find C, θ, s, or r as indicated. Refer to the figure:

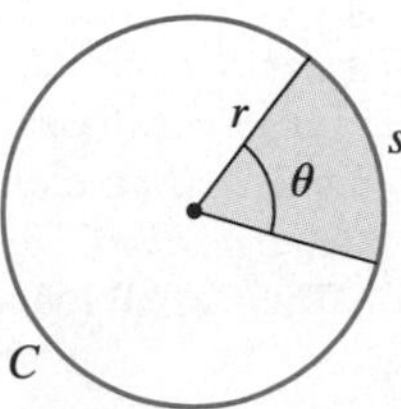

39. $C = 1{,}000$ cm, $\theta = 36°$, $s = ?$ (exact)

40. $s = 12$ m, $C = 108$ m, $\theta = ?$ (exact)

41. $s = 25$ km, $\theta = 20°$, $C = ?$ (exact)

42. $C = 740$ mi, $\theta = 72°$, $s = ?$ (exact)

C **43.** $r = 5{,}400{,}000$ mi, $\theta = 2.6°$, $s = ?$ (to the nearest 10,000 mi)

44. $s = 38{,}000$ cm, $\theta = 45.3°$, $r = ?$ (to the nearest 1,000 cm)

45. $\theta = 12°31'4''$, $s = 50.2$ cm, $C = ?$ (to the nearest 10 cm)

46. $\theta = 24°16'34''$, $s = 14.23$ m, $C = ?$ (to one decimal place)

In Problems 47–50, find A or θ as indicated. Refer to the figure: The ratio of the area of a sector of a circle (A) to the total area of the circle (πr^2) is the same as the ratio of the central angle of the sector (θ) to 360°:

$$\frac{A}{\pi r^2} = \frac{\theta}{360°}$$

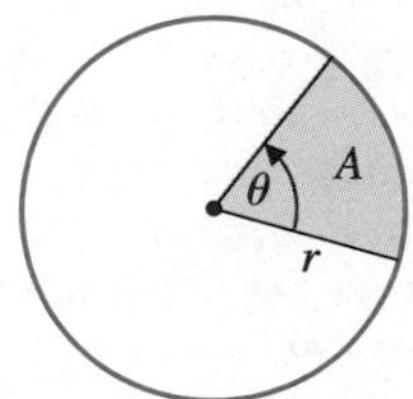

47. $r = 25.2$ cm, $\theta = 47.3°$, $A = ?$ (to the nearest unit)

48. $r = 7.38$ ft, $\theta = 24.6°$, $A = ?$ (to one decimal place)

49. $r = 12.6$ m, $A = 98.4$ m^2, $\theta = ?$ (to one decimal place)

50. $r = 32.4$ in., $A = 347$ in.^{2}, $\theta = ?$ (to one decimal place)

Applications

Biology: Eye *In Problems 51–54, find r, θ, or s as indicated. Refer to the following: The eye is roughly spherical with a spherical bulge in front called the cornea. (Based on the article, "The Surgical Correction of Astigmatism" by Sheldon Rothman and Helen Strassberg in* The UMAP Journal, *Vol. V, No. 2, 1984.)*

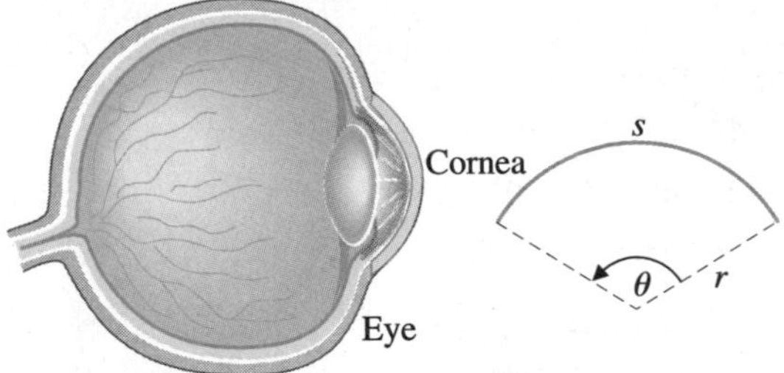

Figure for 51–54

51. $s = 11.5$ mm, $\theta = 118.2°$, $r = ?$ (to two decimal places)

52. $s = 12.1$ mm, $r = 5.26$ mm, $\theta = ?$ (to one decimal place)

53. $\theta = 119.7°$, $r = 5.49$ mm, $s = ?$ (to one decimal place)

54. $\theta = 117.9°$, $s = 11.8$ mm, $r = ?$ (to two decimal places)

Geography/Navigation *In Problems 55–58, find the distance on the surface of the earth between each pair of cities (to the nearest mile), given their respective latitudes. Latitudes are given to the nearest 10′. Note that each chosen pair of cities has approximately the same longitude (ie, lies on the same north–south line). Use $r = 3{,}960$ mi for the earth's radius. The figure shows the situation for San Francisco and Seattle. [Hint: $C = 2\pi r$]*

* **55.** San Francisco, CA, 37°50′N; Seattle, WA, 47°40′N

* **56.** Phoenix, AZ, 33°30′N; Salt Lake City, UT, 40°40′N

* **57.** Dallas, TX, 32°50′N; Lincoln, NE, 40°50′N

* **58.** Buffalo, NY, 42°50′N; Durham, NC, 36°0′N

* The most difficult problems are double-starred (**); moderately difficult problems are single-starred (*); easier problems are not marked.

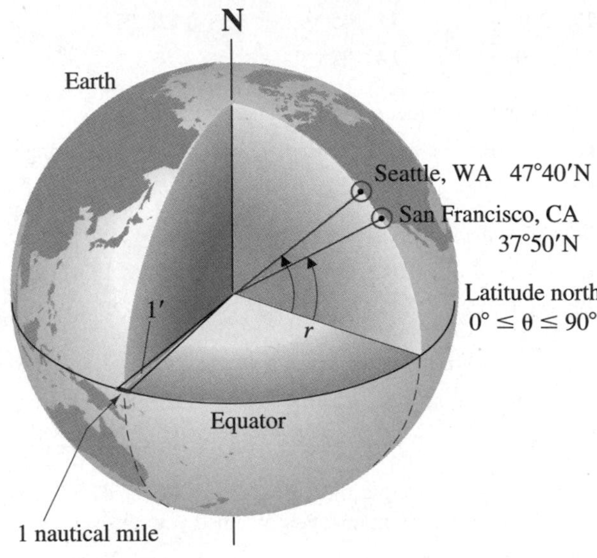

Figure for 55–62

Geography/Navigation *In Problems 59–62, find the distance on the earth's surface, to the nearest nautical mile, between each pair of cities. A* **nautical mile** *is the length of 1′ of arc on the equator or any other circle on the surface of the earth having the same center as the equator. See the figure. Since there are $60 \times 360 = 21{,}600''$ in 360°, the length of the equator is 21,600 nautical miles.*

59. San Francisco, CA, 37°50′N; Seattle, WA, 47°40′N

60. Phoenix, AZ, 33°30′N; Salt Lake City, UT, 40°40′N

61. Dallas, TX, 32°50′N; Lincoln, NE, 40°50′N

62. Buffalo, NY, 42°50′N; Durham, NC, 36°0′N

63. Photography
(A) The angle of view of a 300 mm lens is 8°. Approximate the width of the field of view to the nearest foot when the camera is at a distance of 500 ft.
(B) Explain the assumptions that are being made in the approximation calculation in part (A).

* **64. Satellite Telescopes** In the Orbiting Astronomical Observatory launched in 1968, ground personnel could direct telescopes in the vehicle with an accuracy of 1′ of arc. They claimed this corresponds to hitting a 25¢ coin at a distance of 100 yd (see the figure). Show that this claim is approximately correct; that is, show that the length of an arc subtended by 1′ at 100 yd is approximately the diameter of a quarter, 0.94 in.

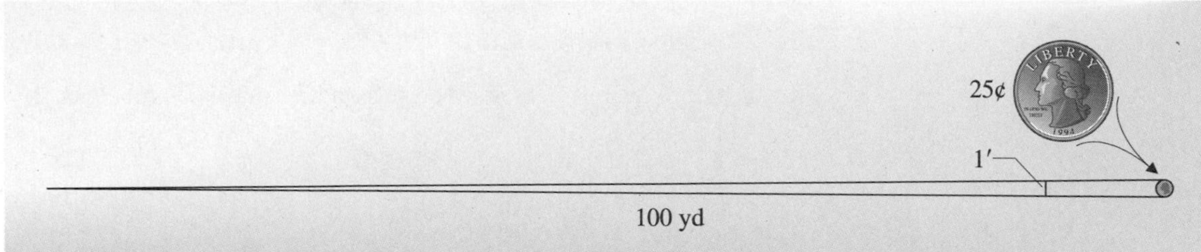

Figure for 64

1.2 SIMILAR TRIANGLES

- Euclid's Theorem and Similar Triangles
- Applications

Properties of similar triangles, stated in Euclid's theorem below, are central to this section and form a cornerstone for the development of trigonometry. Euclid (300 BC), a Greek mathematician who taught in Alexandria, was one of the most influential mathematicians of all time. He is most famous for writing the *Elements,* a collection of thirteen books (or chapters) on geometry, geometric algebra, and number theory. In the Western World, next to the Bible, the *Elements* is probably the most studied text of all time.

The ideas on indirect measurement presented here are included to help you understand basic concepts. More efficient methods will be developed in the next section.

Remark Since calculators routinely compute to eight or ten digits, one could easily believe that a computed result is far more accurate than warranted. **Generally, a final result cannot be any more accurate than the least accurate number used in the calculation. Regarding calculation accuracy, we will be guided by Appendix A.3 on significant digits throughout the text.** □

Euclid's Theorem and Similar Triangles

In Section 1.1, problems were included that required knowledge of the distances from the earth to the moon and the sun. How can inaccessible distances of this type be determined? Surprisingly, the ancient Greeks made fairly accurate calculations of these distances as well as many others. The basis for their methods is the following elementary theorem of Euclid:

EUCLID'S THEOREM

If two triangles are similar, their corresponding sides are proportional.

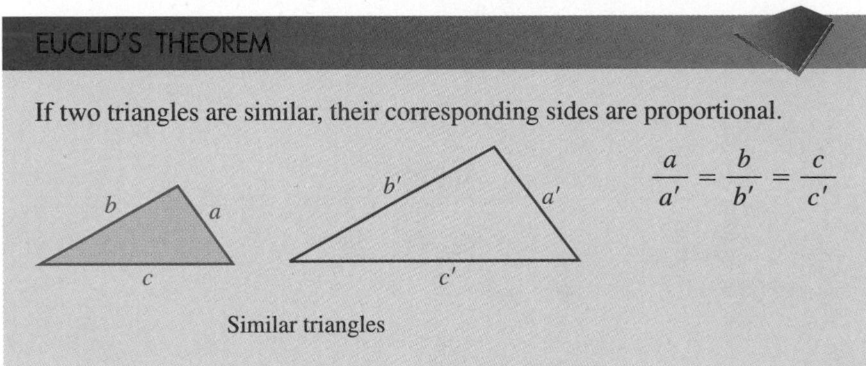

Similar triangles

Remark Recall from plane geometry (see Appendix C.2) that the sum of the measures of the three angles of any triangle is always 180° and that two triangles are similar if two angles of one triangle have the same measure as two angles of the other. If the two triangles happen to be right triangles, then they are similar if an acute angle in one has the same measure as an acute angle in the other. Take a moment to draw a few figures and to think about these statements. □

EXPLORE/DISCUSS 1

Identify pairs of triangles from the following that are similar and explain why:

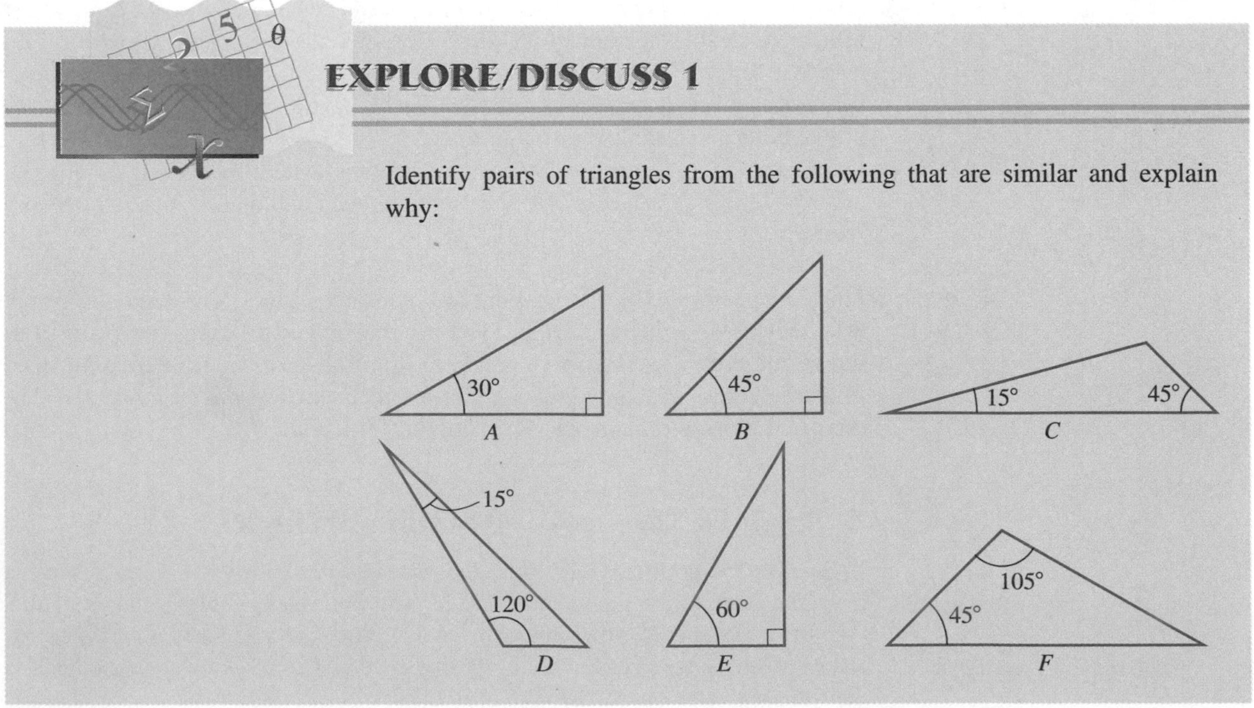

Applications

We now solve some elementary problems using Euclid's theorem.

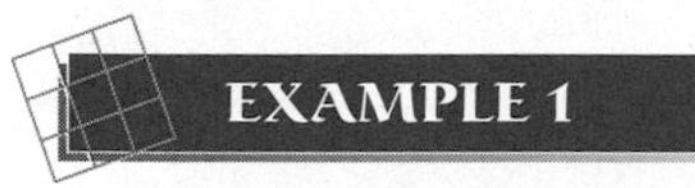

Height of a Tree

A tree casts a shadow of 32 ft at the same time a vertical yardstick (3.0 ft) casts a shadow of 2.2 ft (see Fig. 1). How tall is the tree?

FIGURE 1

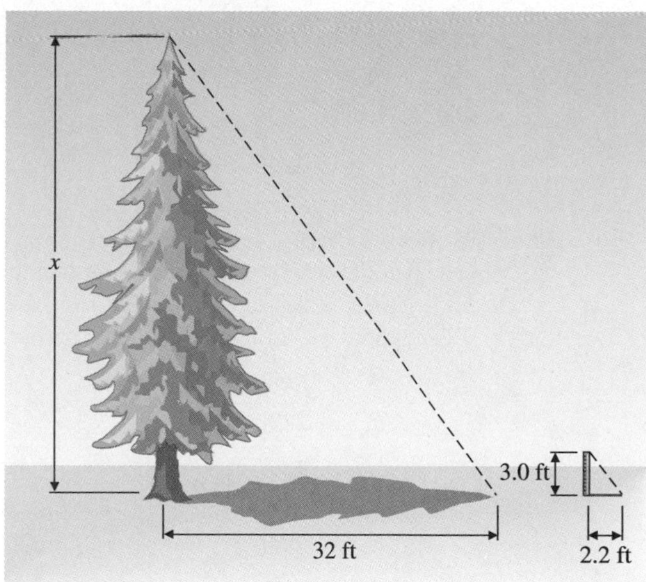

Solution The parallel sun rays make the same angle with the tree and the yardstick. Since both triangles are right triangles and have an acute angle of the same measure, the triangles are similar. The corresponding sides are proportional, and we can write

$$\frac{x}{3.0\text{ ft}} = \frac{32\text{ ft}}{2.2\text{ ft}}$$

$$x = \frac{3.0}{2.2}(32\text{ ft})$$

$$= 44\text{ ft} \qquad \text{To two significant digits}$$

■

Matched Problem 1 A tree casts a shadow of 31 ft at the same time a 5.0 ft vertical pole casts a shadow of 0.56 ft. How high is the tree? ■

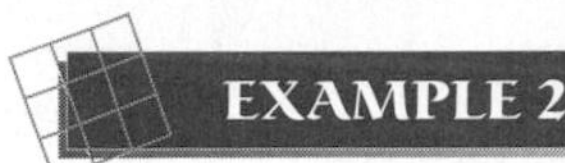

EXAMPLE 2

Length of an Air Vent

Find the length of the proposed air vent indicated in Figure 2.

FIGURE 2

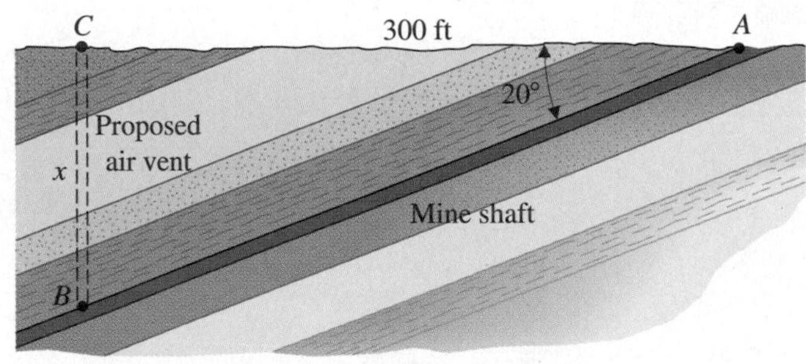

Solution We make a careful scale drawing of the main shaft relative to the proposed air vent as follows: Pick any convenient length, say 2.0 in., for $A'C'$; copy the 20° angle CAB and the 90° angle ACB using a protractor (see Fig. 3). Now measure $B'C'$ (approx. 0.7 in.), and set up an appropriate proportion:

$$\frac{x}{0.7 \text{ in.}} = \frac{300 \text{ ft}}{2.0 \text{ in.}}$$

$$x = \frac{0.7}{2.0}(300 \text{ ft})$$

$$= 100 \text{ ft} \qquad \text{To one significant digit}$$

FIGURE 3

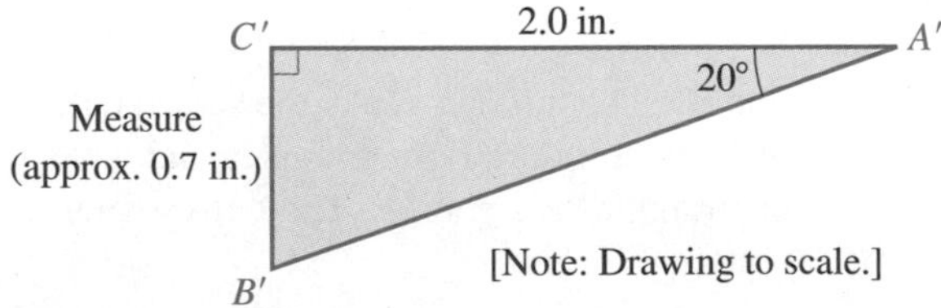

[*Note:* The use of scale drawings for finding indirect measurements is included here only to demonstrate basic ideas. A more efficient method will be developed in Section 1.3.] ■

Matched Problem 2 Suppose in Example 2 that $AC = 500$ ft and $\angle A = 30°$. If in a scale drawing $A'C'$ is chosen to be 3 in. and $B'C'$ is measured as 1.76 in., find BC, the length of the proposed mine shaft. ■

EXPLORE/DISCUSS 2

We want to measure the depth of a canyon from a point on its rim by using similar triangles and no scale drawings. The process illustrated in Figure 4 was used in medieval times.* A vertical pole of height a is moved back from the canyon rim so that the line of sight from the top of the pole passes the rim at D to a point G at the bottom of the canyon. The setback DC is then measured to be b. The same vertical pole is moved to the canyon rim at D and a horizontal pole is moved out from the rim through D until the line of sight from B to G includes the end of the pole at E. The pole overhang DE is measured to be c. We now have enough information to find y, the depth of the canyon.

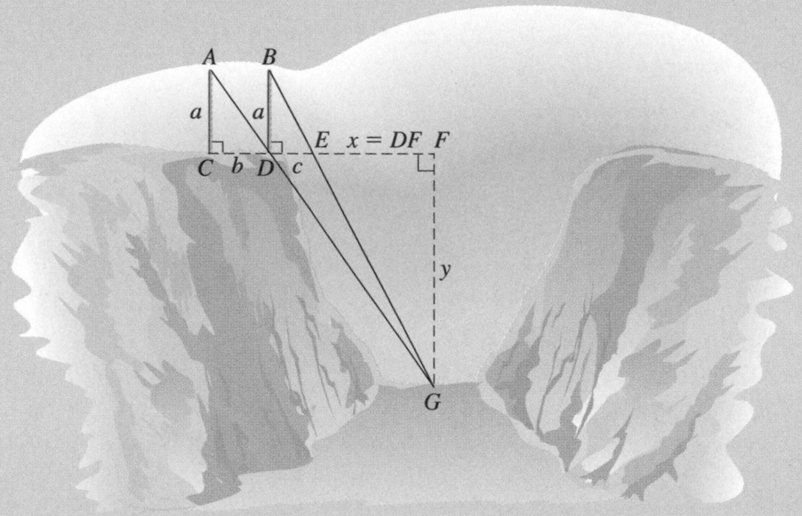

FIGURE 4
Depth of a canyon

(A) Explain why triangles ACD and GFD are similar.

(B) Explain why triangles BDE and GFE are similar.

(C) Set up appropriate proportions, and with the use of a little algebra, show that

$$x = \frac{bc}{b - c} \qquad y = \frac{ac}{b - c}$$

* This process is mentioned in an excellent article by Victor J. Katz (University of the District of Columbia), titled *The Curious History of Trigonometry,* in *The UMAP Journal,* Winter 1990.

Answers to Matched Problems

1. 280 ft (to two significant digits)
2. 300 ft (to one significant digit)

EXERCISE 1.2

Round each answer to an appropriate accuracy; use the guidelines from Appendix A.3.

A

1. If two angles of one triangle have the same measure as two angles of another triangle, what can you say about the measure of the third angle of each triangle? Why?

2. If an acute angle of one right triangle has the same measure as the acute angle of another right triangle, what can you say about the measure of the second acute angle of each triangle? Why?

Given two similar triangles, as in the figure, find the unknown length indicated.

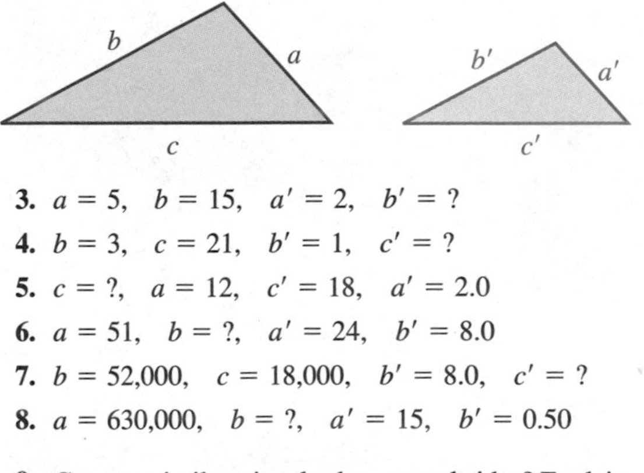

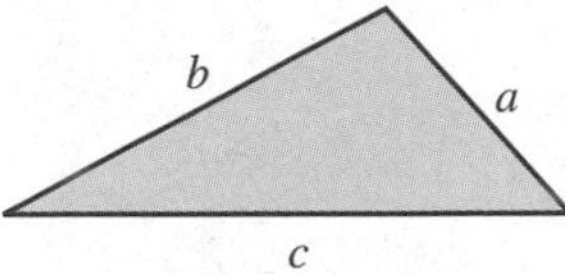

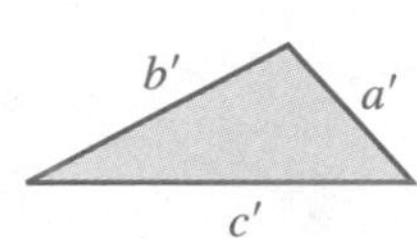

3. $a = 5, \quad b = 15, \quad a' = 2, \quad b' = ?$

4. $b = 3, \quad c = 21, \quad b' = 1, \quad c' = ?$

5. $c = ?, \quad a = 12, \quad c' = 18, \quad a' = 2.0$

6. $a = 51, \quad b = ?, \quad a' = 24, \quad b' = 8.0$

7. $b = 52{,}000, \quad c = 18{,}000, \quad b' = 8.0, \quad c' = ?$

8. $a = 630{,}000, \quad b = ?, \quad a' = 15, \quad b' = 0.50$

B

9. Can two similar triangles have equal sides? Explain.

10. If two triangles are similar and a side of one triangle is equal to the corresponding side of the other, are the remaining sides of the two triangles equal? Explain.

Given the similar triangles in the figure, find the indicated quantities to two significant digits.

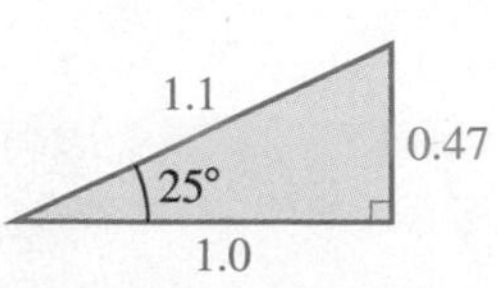

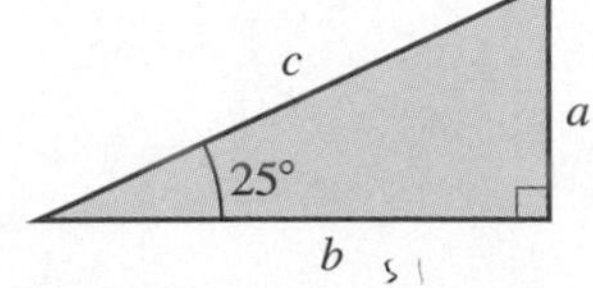

11. $b = 51$ in., $a = ?, \quad c = ?$

12. $b = 32$ cm, $a = ?, \quad c = ?$

13. $a = 23.4$ m, $b = ?, \quad c = ?$

14. $a = 63.19$ cm, $b = ?, \quad c = ?$

15. $b = 2.489 \times 10^{9}$ yd, $a = ?, \quad c = ?$

16. $b = 1.037 \times 10^{13}$ m, $a = ?, \quad c = ?$

17. $c = 8.39 \times 10^{-5}$ mm, $a = ?, \quad b = ?$

18. $c = 2.86 \times 10^{-8}$ cm, $a = ?, \quad b = ?$

C

Find the unknown quantities. (If you have a protractor, make a scale drawing and complete the problem using your own measurements and calculations. If you do not have a protractor, use the quantities given in the problem.)

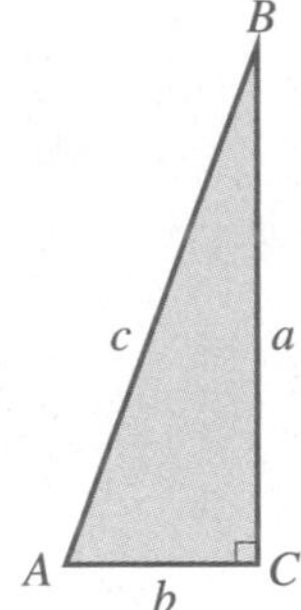

19. Suppose in the figure that $\angle A = 70°$, $\angle C = 90°$, and $a = 101$ ft. If a scale drawing is made of the triangle by choosing a' to be 2.00 in. and c' is then measured to be 2.13 in., estimate c in the original triangle.

20. Repeat Problem 19, choosing $a' = 5.00$ in. and c', a measured quantity, to be 5.28 in.

Applications

21. Tennis A ball is served from the center of the baseline into the deuce court. If the ball is hit 9 ft above the ground, travels in a straight line down the middle of the court, and the net is 3 ft high, how far from the base of the net will the ball land if it just clears the top of the net? (See the figure.) Assume all figures are exact and compute the answer to one decimal place.

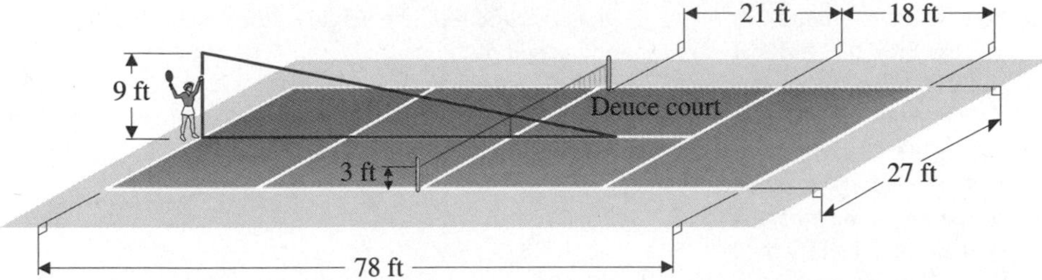

Figure for 21 and 22

22. **Tennis** If the ball in Problem 21 is hit 8.5 ft above the ground, how far away from the base of the net will the ball land? Assume all figures are exact and compute the answer to two decimal places. (Now you can see why tennis players try to spin the ball on a serve so that it curves downward.)

23. **Indirect Measurement** Find the height of the tree in the figure given that $AC = 24$ ft, $CD = 2.1$ ft, and $DE = 5.5$ ft.

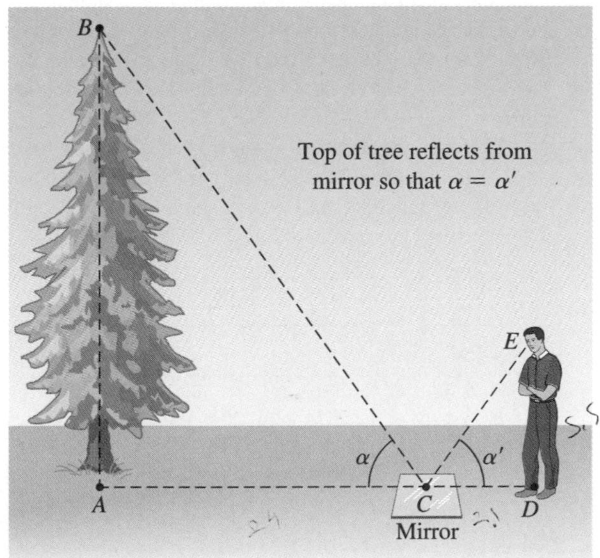

Figure for 23 and 24

24. **Indirect Measurement** Find the height of the tree in the figure given that $AC = 25$ ft, $CD = 2$ ft 3 in., and $DE = 5$ ft 9 in.

Problems 25 and 26 are optional for those who have protractors and can make scale drawings.

25. **Astronomy** The following figure illustrates a method that is used to determine depths of moon craters from observatories on earth. If sun rays strike the surface of the moon so that $\angle BAC = 15°$ and AC is measured to be 4.0 km, how high is the rim of the crater above its floor?

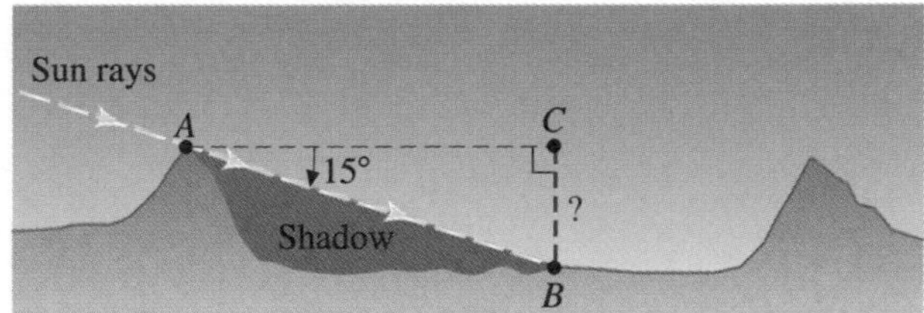

Depth of moon crater

Figure for 25

26. **Astronomy** The figure illustrates how the height of a mountain on the moon can be determined from earth. If sun rays strike the surface of the moon so that $\angle BAC = 28°$ and AC is measured to be 4.0×10^3 m, how high is the mountain?

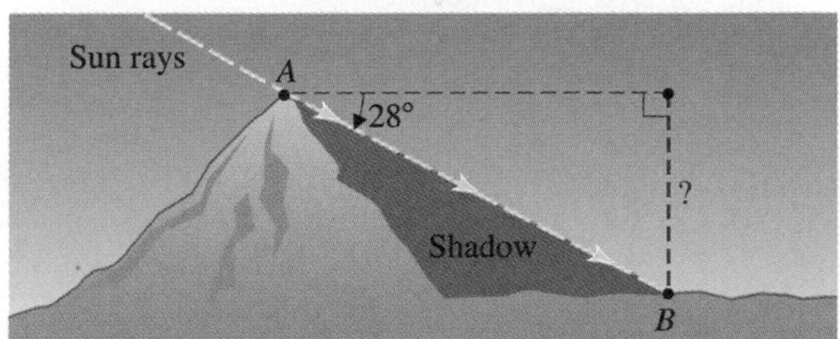

Height of moon mountain

Figure for 26

27. **Fundamental Lens Equation** Parallel light rays (light rays from infinity) approach a thin convex lens and are focused at a point on the other side of the lens. The distance f from the lens to the **focal point** F is called the

focal length of the lens [see part (a) of the figure]. A standard lens on many 35 mm cameras has a focal length of 50 mm (about 2 in.), a 28 mm lens is a wide-angle lens, and a 200 mm lens is telephoto. How does a lens focus an image of an object on the film in a camera? Part (b) of the figure shows the geometry involved for a thin convex lens. Point P at the top of the object is selected for illustration (any point on the object would do). Light rays from point P travel in all directions. Those that go through the lens are focused at P'. Since light rays PA and CP' are parallel, AP' and CP pass through focal points F' and F, respectively, which are equidistant from the lens; that is, $FB = BF' = f$. Also note that $AB = h$, the height of the object, and $BC = h'$, the height of the image.

(A) Explain why triangles PAC and FBC are similar, and why triangles ACP' and ABF' are similar.

(B) From the properties of similar triangles, show that

$$\frac{h + h'}{u} = \frac{h'}{f} \quad \text{and} \quad \frac{h + h'}{v} = \frac{h}{f}$$

(C) Combining the results in part (B), derive the important lens equation

$$\frac{1}{u} + \frac{1}{v} = \frac{1}{f}$$

where u is the distance between the object and the lens, v is the distance between the focused image on the film and the lens, and f is the focal length of the lens.

(D) How far (to three decimal places) must a 50 mm lens be from the film if the lens is focused on an object 3 m away?

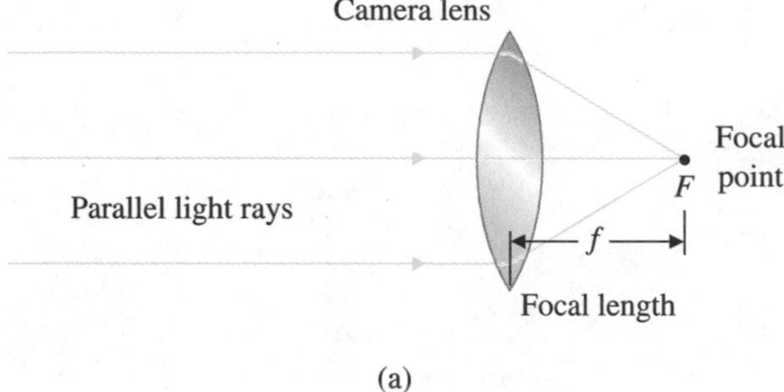

(a)

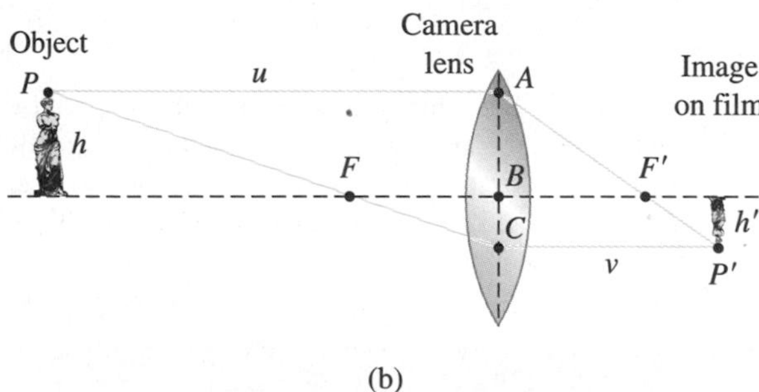

(b)

Figure for 27 and 28

28. Fundamental Lens Equation (Refer to Problem 27.) For a 50 mm lens it is said that if the object is more than 20 m from the lens, then the rays coming from the object are very close to being parallel, and v will be very close to f.

(A) Use the lens equation given in part (C) of Problem 27 to complete the following table (to two decimal places) for a 50 mm lens. (Convert meters to millimeters before using the lens equation.)

u (m)	10	20	30	40	50	60
v (m)						

(B) Referring to part (A), how does v compare with f as u increases beyond 20? (Note that 0.1 mm is less than the diameter of a period on this page.)

1.3 TRIGONOMETRIC RATIOS AND RIGHT TRIANGLES

- Trigonometric Ratios
- Complementary Angles and Cofunctions
- Calculator Evaluation
- Solving Right Triangles

In many applications of trigonometry, we are given two sides of a right triangle, or one side and an acute angle, and are asked to find the remaining sides and acute angles. This is called **solving a triangle.** In this section we show how this can be done without using scale drawings. The concepts introduced here will be generalized extensively as we progress through the book.

Trigonometric Ratios

We see in Figure 1 that there are six possible ratios of the sides of a right triangle that can be computed for each angle θ.

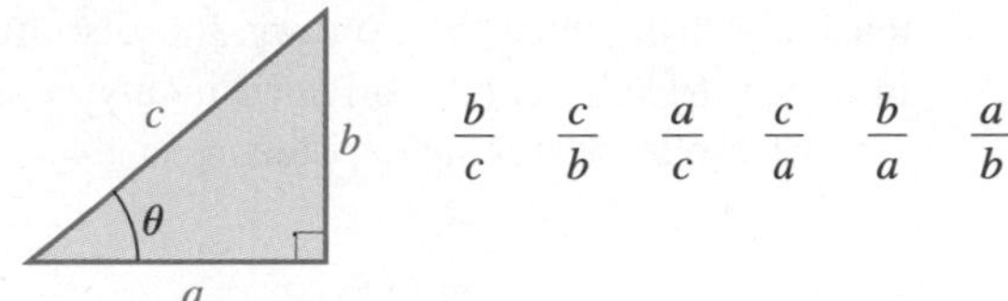

$$\frac{b}{c} \quad \frac{c}{b} \quad \frac{a}{c} \quad \frac{c}{a} \quad \frac{b}{a} \quad \frac{a}{b}$$

FIGURE 1
Six possible ratios of sides

These ratios are referred to as **trigonometric ratios,** and because of their importance, each is given a name: sine (sin), cosine (cos), tangent (tan), cosecant (csc), secant (sec), and cotangent (cot). And each is written in abbreviated form as follows:

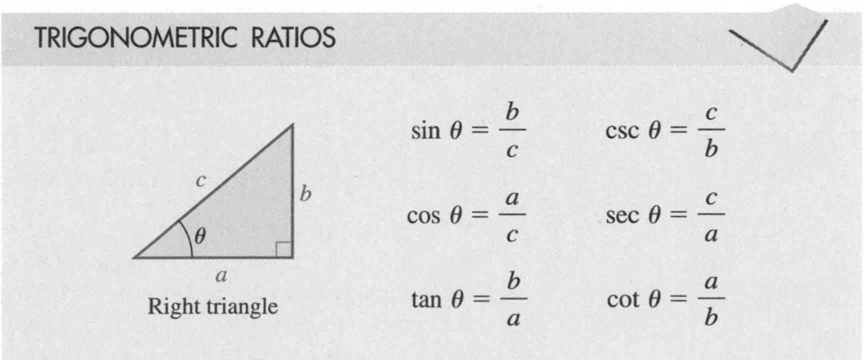

TRIGONOMETRIC RATIOS

Right triangle

$$\sin\theta = \frac{b}{c} \qquad \csc\theta = \frac{c}{b}$$

$$\cos\theta = \frac{a}{c} \qquad \sec\theta = \frac{c}{a}$$

$$\tan\theta = \frac{b}{a} \qquad \cot\theta = \frac{a}{b}$$

Side b is often referred to as the **side opposite** angle θ, a as the **side adjacent** to angle θ, and c as the **hypotenuse.** Using these designations, the ratios become:

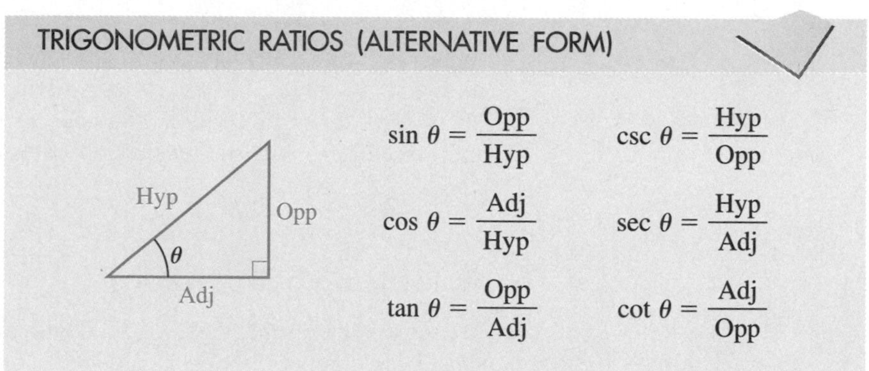

TRIGONOMETRIC RATIOS (ALTERNATIVE FORM)

$$\sin\theta = \frac{\text{Opp}}{\text{Hyp}} \qquad \csc\theta = \frac{\text{Hyp}}{\text{Opp}}$$

$$\cos\theta = \frac{\text{Adj}}{\text{Hyp}} \qquad \sec\theta = \frac{\text{Hyp}}{\text{Adj}}$$

$$\tan\theta = \frac{\text{Opp}}{\text{Adj}} \qquad \cot\theta = \frac{\text{Adj}}{\text{Opp}}$$

These trigonometric ratios should be learned, because they will be used extensively in the work that follows. It is also important to note that the right angle in a right triangle can be oriented in any position and that the names of the angles are arbitrary, but the hypotenuse is always opposite the right angle.

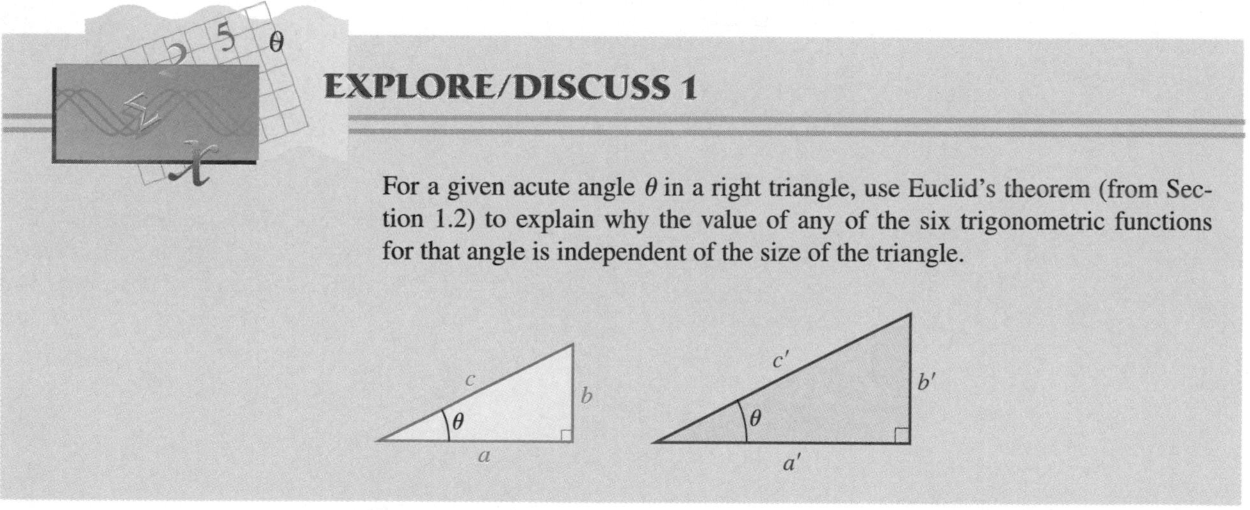

EXPLORE/DISCUSS 1

For a given acute angle θ in a right triangle, use Euclid's theorem (from Section 1.2) to explain why the value of any of the six trigonometric functions for that angle is independent of the size of the triangle.

Complementary Angles and Cofunctions

What is the meaning of the prefix *co-* in cosine, cosecant, and cotangent? The *co-* refers to a complementary angle relationship. Recall that two positive angles are complementary if their sum is 90°. The two acute angles in a right triangle are complementary. (Why?)* Referring to the definition of the six trigonometric ratios, we can see the complementary relationships given in the box at the top of the next page.

---

* Since the sum of the measures of all three angles in a triangle is 180°, and a right triangle has one 90° angle, the two remaining acute angles must have measures that sum to $180° - 90° = 90°$. Thus, the two acute angles in a right triangle are always complementary.

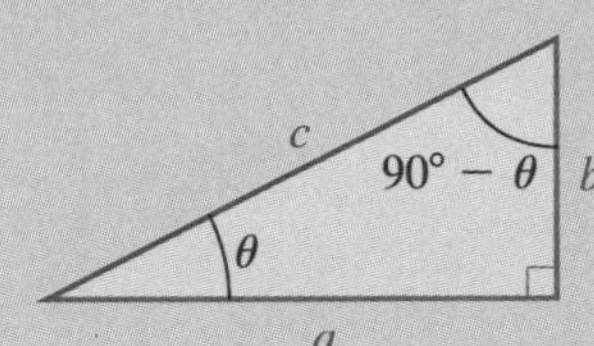

$$\sin\theta = \frac{b}{c} = \cos(90^\circ - \theta)$$

$$\tan\theta = \frac{b}{a} = \cot(90^\circ - \theta)$$

$$\sec\theta = \frac{c}{a} = \csc(90^\circ - \theta)$$

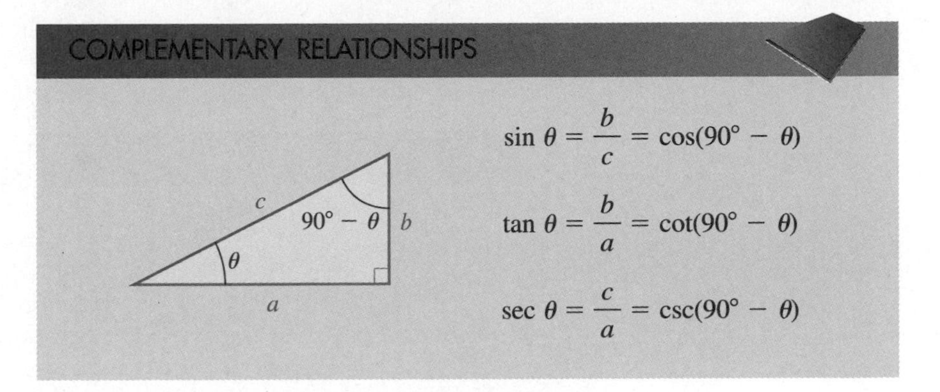

Thus, the sine of θ is the same as the cosine of the complement of θ (which is $90^\circ - \theta$ in the triangle shown), the tangent of θ is the cotangent of the complement of θ, and the secant of θ is the cosecant of the complement of θ. The trigonometric ratios cosine, cotangent, and cosecant are sometimes referred to as the **cofunctions** of sine, tangent, and secant, respectively.

Calculator Evaluation

For the trigonometric ratios to be useful in solving right triangle problems, we must be able to find each for any acute angle. Scientific and graphing calculators can approximate (almost instantly) these ratios to eight or ten significant digits. Scientific and graphing calculators generally use different sequences of steps. Consult the user's manual for your particular calculator.

The use of a scientific calculator is assumed throughout the book, and a graphing utility (graphing calculator or computer with graphing capability) is required for many optional problems. A graphing calculator is a scientific calculator with additional capabilities, including graphing.

Calculators have two trigonometric modes: degree and radian. Our interest now is in *degree mode*. Later we will discuss radian mode in detail.

Caution Refer to the user's manual accompanying your calculator to determine how it is to be set in degree mode, and set it that way. *This is an important step and should not be overlooked.* Many errors can be traced to calculators being set in the wrong mode. □

If you look at the function keys on your calculator, you will find three keys labeled

[sin] [cos] [tan]

These keys are used to find sine, cosine, and tangent ratios, respectively. The calculator also can be used to compute cosecant, secant, and cotangent ratios using the reciprocal* relationships, which follow directly from the definition of the six trigonometric ratios.

RECIPROCAL RELATIONSHIPS FOR $0° < \theta < 90°$

$$\csc\theta\sin\theta = \frac{c}{b}\cdot\frac{b}{c} = 1 \quad \text{thus} \quad \csc\theta = \frac{1}{\sin\theta}$$

$$\sec\theta\cos\theta = \frac{c}{a}\cdot\frac{a}{c} = 1 \quad \text{thus} \quad \sec\theta = \frac{1}{\cos\theta}$$

$$\cot\theta\tan\theta = \frac{a}{b}\cdot\frac{b}{a} = 1 \quad \text{thus} \quad \cot\theta = \frac{1}{\tan\theta}$$

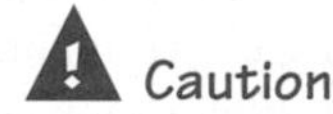

Caution When using reciprocal relationships, many students tend to associate cosecant with cosine and secant with sine: just the opposite is correct. □

EXAMPLE 1 **Calculator Evaluation**

Evaluate to four significant digits using a calculator:

(A) sin 23.72° (B) tan 54°37′

(C) sec 49.31° (D) cot 12.86°

Solution First, set the calculator in degree mode.

(A) sin 23.72° = 0.4023

(B) Some calculators require 54°37′ to be converted to decimal degrees first; others can do the calculation directly—check your user's manual.

$$\tan 54°37' = \tan 54.6166\ldots = 1.408$$

(C) $\sec 49.31° = \dfrac{1}{\cos 49.31°} = 1.534$

(D) $\cot 12.86° = \dfrac{1}{\tan 12.86°} = 4.380$ ■

Matched Problem 1 Evaluate to four significant digits using a calculator:

(A) cos 38.27° (B) sin 37°44′

(C) cot 49.82° (D) csc 77°53′ ■

---

* Recall that two numbers a and b are **reciprocals** of each other if $ab = 1$; then we may write $a = 1/b$ and $b = 1/a$.

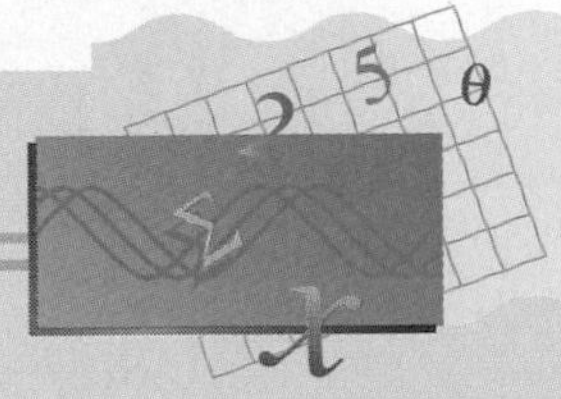

EXPLORE/DISCUSS 2

Experiment with your calculator to determine which of the following two window displays from a graphing calculator is the result of the calculator being set in degree mode and which is the result of the calculator being set in radian mode.

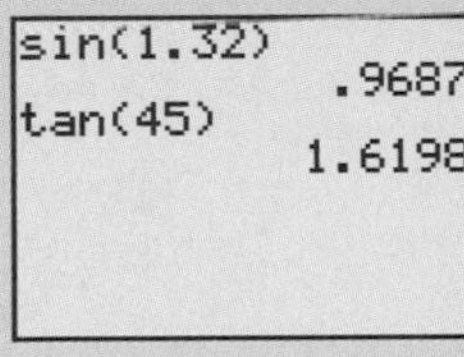

(a)

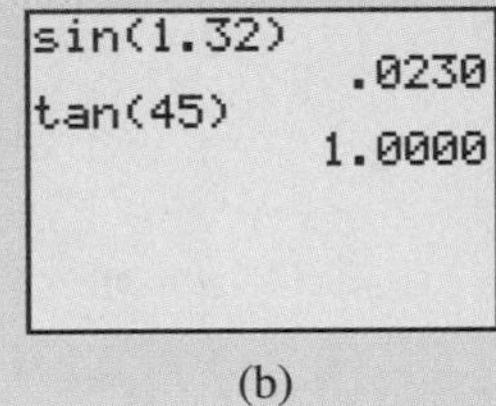

(b)

Now we reverse the process illustrated in Example 1. Suppose we are given

$$\sin\theta = 0.3174$$

How do we find θ? That is, how do we find the acute angle θ whose sine is 0.3174? The solution to this problem is written symbolically as either

$$\theta = \arcsin 0.3174$$ "arcsin" and "$\sin^{-1}$" both represent the same thing.

or

$$\theta = \sin^{-1} 0.3174$$

Both of these expressions are read "θ is the angle whose sine is 0.3174."

Caution It is important to note that $\sin^{-1} 0.3174$ does not mean $1/(\sin 0.3174)$; the -1 "exponent" is a superscript that is part of a *function symbol.* More will be said about this in Chapter 5, where a detailed discussion of these concepts is given. □

We can find θ directly using a calculator. The function key $\boxed{\sin^{-1}}$ or its equivalent takes us from a trigonometric sine ratio back to the corresponding acute angle in decimal degrees when the calculator is in degree mode. Thus, if $\sin\theta = 0.3174$, then we can write $\theta = \arcsin 0.3174$ or $\theta = \sin^{-1} 0.3174$. We choose the latter, and proceed as follows:

$$\theta = \sin^{-1} 0.3174$$
$$= 18.506^\circ$$ To three decimal places
$$\text{or}\quad 18^\circ 30' 21''$$ To the nearest second

✔ Check $\sin 18.506^\circ = 0.3174$

Finding Inverses

Find each acute angle θ to the accuracy indicated:

(A) $\cos\theta = 0.7335$ (to three decimal places)

(B) $\theta = \tan^{-1} 8.207$ (to the nearest minute)

(C) $\theta = \arcsin 0.0367$ (to the nearest 10′)

Solution First, set the calculator in degree mode.

(A) If $\cos\theta = 0.7335$, then

$$\theta = \cos^{-1} 0.7335$$
$$= 42.819° \quad \text{To three decimal places}$$

(B) $\theta = \tan^{-1} 8.207$

$$= 83.053$$
$$= 83°3' \quad \text{To the nearest minute}$$

(C) $\theta = \arcsin\ 0.0367$

$$= 2.103°$$
$$= 2°10' \quad \text{To the nearest } 10'$$

■

Matched Problem 2 Find each acute angle θ to the accuracy indicated:

(A) $\tan\theta = 1.739$ (to two decimal places)

(B) $\theta = \sin^{-1} 0.2571$ (to the nearest 10″)

(C) $\theta = \arccos 0.0367$ (to the nearest minute)

■

We postpone any further discussion of $\cot^{-1}$, $\sec^{-1}$, and $\csc^{-1}$ until Chapter 5. The preceding discussion will handle all our needs at this time.

Solving Right Triangles

To solve a right triangle is to find, given the measures of two sides or the measures of one side and an acute angle, the measures of the remaining sides and angles. Solving right triangles is best illustrated through examples. Note at the outset that accuracy of the computations is governed by Table 1 (which is also reproduced inside the front cover for easy reference).

TABLE 1

Angle to nearest	Significant digits for side measure
1°	2
10′ or 0.1°	3
1′ or 0.01°	4
10″ or 0.001°	5

Remark When we use the equal sign (=) in the following computations, it should be understood that equality holds only to the number of significant digits justified by Table 1. The approximation symbol (≈) is used only when we want to emphasize the approximation. □

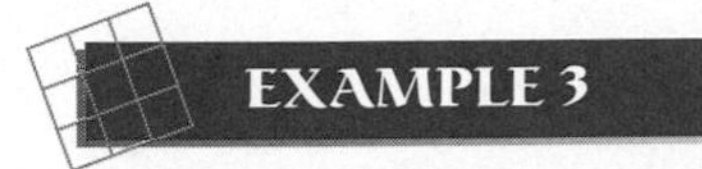

Solving a Right Triangle

Solve the right triangle in Figure 2.

Solution *Solve for the complementary angle.*

$$90° - \theta = 90° - 35.7° = 54.3° \qquad \text{Remember, } 90° \text{ is exact.}$$

124 m
b
35.7°
a

FIGURE 2

Solve for b. Since $\theta = 35.7°$ and $c = 124$ m, we look for a trigonometric ratio that involves θ and c (the known quantities) and b (an unknown quantity). Referring to the definition of the trigonometric ratios (page 21), we see that both sine and cosecant involve all three quantities. We choose sine, and proceed as follows:

$$\sin\theta = \frac{b}{c}$$

$$\begin{aligned} b &= c\sin\theta \\ &= (124\text{ m})(\sin 35.7°) \\ &= 72.4\text{ m} \end{aligned}$$

Solve for a. Now that we have b, we can use the tangent, cotangent, cosine, or secant to find a. We choose the cosine. Thus,

$$\cos\theta = \frac{a}{c}$$

$$\begin{aligned} a &= c\cos\theta \\ &= (124\text{ m})(\cos 35.7°) \\ &= 101\text{ m} \end{aligned}$$

■

Matched Problem 3 Solve the triangle in Example 3 with $\theta = 28.3°$ and $c = 62.4$ cm. ■

Solving a Right Triangle

Solve the right triangle in Figure 3 for θ and $90° - \theta$ to the nearest 10′ and for a to three significant digits.

Solution *Solve for θ.*

$$\sin\theta = \frac{b}{c} = \frac{42.7\text{ km}}{51.3\text{ km}}$$

$$\sin\theta = 0.832$$

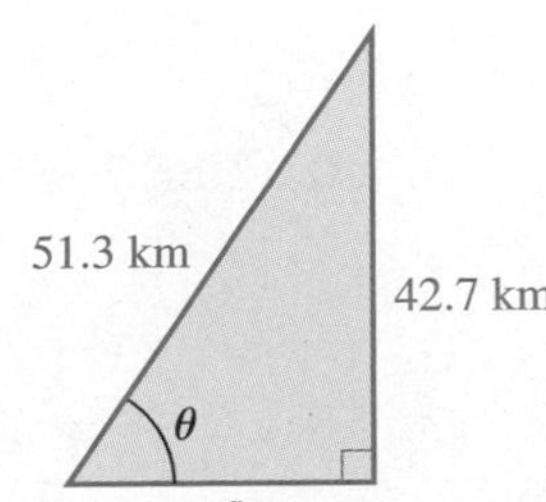

FIGURE 3

Given the sine of θ, how do we find θ? We can find θ directly using a calculator as discussed in Example 2.

The $\boxed{\sin^{-1}}$ key or its equivalent takes us from a trigonometric ratio back to the corresponding angle in decimal degrees (if the calculator is in degree mode).

$$\theta = \sin^{-1} 0.832$$

$$= 56.3° \qquad (0.3)(60) = 18' \approx 20'$$

$$= 56°20' \qquad \text{To the nearest } 10'$$

Solve for the complementary angle.

$$90° - \theta = 90° - 56°20'$$

$$= 33°40'$$

Solve for a. Use cosine, secant, cotangent, or tangent. We will use tangent:

$$\tan \theta = \frac{b}{a}$$

$$a = \frac{b}{\tan \theta}$$

$$= \frac{42.7 \text{ km}}{\tan 56°20'}$$

$$= 28.4 \text{ km}$$

✔ Check We check by using the Pythagorean theorem.* (See Fig. 4.)

$$28.4^2 + 42.7^2 \stackrel{?}{=} 51.3^2 \qquad \text{Compute both sides to three significant digits.}$$

$$2{,}630 \stackrel{\checkmark}{=} 2{,}630$$

FIGURE 4

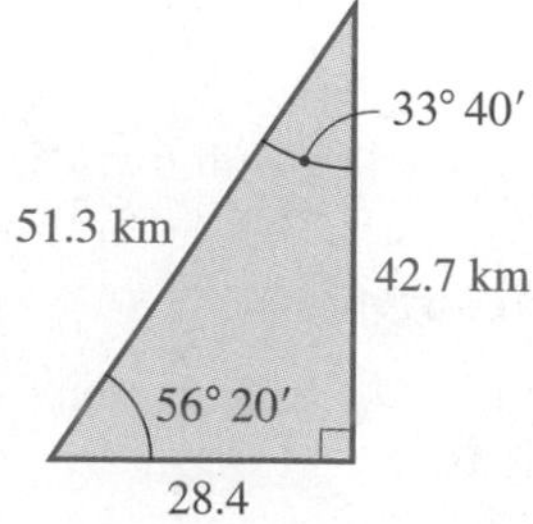

Matched Problem 4 Repeat Example 4 with $b = 23.2$ km and $c = 30.4$ km.

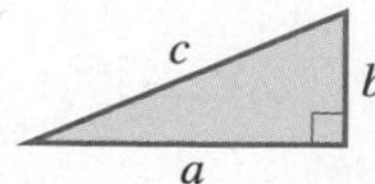

* **Pythagorean theorem:** A triangle is a right triangle if and only if the sum of the squares of the two shorter sides is equal to the square of the longest side:

$$a^2 + b^2 = c^2$$

In this section we have concentrated on technique. In the next section we will consider a large variety of applications involving the techniques discussed in this section.

Answers to Matched Problems

1. (A) 0.7851 (B) 0.6120 (C) 0.8445 (D) 1.023
2. (A) 60.10° (B) 14°53′50″ (C) 87°54′
3. $90° - \theta = 61.7°$, $b = 29.6$ cm, $a = 54.9$ cm
4. $\theta = 49°40'$, $90° - \theta = 40°20'$, $a = 19.7$ km

EXERCISE 1.3

A *In Problems 1–6, refer to the figure and identify each of the named trigonometric ratios with one of the following quotients: a/b, b/a, a/c, c/a, b/c, c/b. Do not look back in the text.*

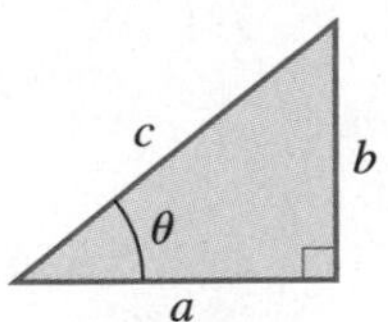

Figure for 1–12

1. $\cos\theta$ **2.** $\sin\theta$ **3.** $\tan\theta$

4. $\cot\theta$ **5.** $\sec\theta$ **6.** $\csc\theta$

In Problems 7–12, refer to the figure and identify each of the quotients with a named trigonometric ratio from the following list: sin θ, cos θ, sec θ, csc θ, tan θ, cot θ. Do not look back in the text.

7. b/c **8.** a/c **9.** b/a

10. a/b **11.** c/b **12.** c/a

Find each to three significant digits.

13. sin 25.6° **14.** cos 36.4° **15.** tan 35°20′

16. cot 12°40′ **17.** sec 44.8° **18.** csc 18.3°

19. cos 72.9° **20.** sin 63.1° **21.** cot 54.9°

22. tan 48.3° **23.** csc 67°30′ **24.** sec 51°40′

B *In Problems 25–30, find each acute angle θ to the accuracy indicated.*

25. $\sin\theta = 0.8032$ (to two decimal places)

26. $\tan\theta = 3.144$ (to two decimal places)

27. $\theta = \arccos 0.7153$ (to the nearest 10′)

28. $\theta = \arcsin 0.1152$ (to the nearest 10′)

29. $\theta = \tan^{-1} 1.948$ (to the nearest minute)

30. $\theta = \cos^{-1} 0.5509$ (to the nearest minute)

In Problems 31–34, if you are given the indicated measures in a right triangle, explain why you can or cannot solve the triangle.

31. The measures of two adjacent sides a and b

32. The measures of one side and one angle

33. The measures of two acute angles

34. The measure of the hypotenuse c

In Problems 35–44, solve the right triangle (labeled as in the figure at the beginning of the exercise) given the information in each problem.

35. $\theta = 58°40'$, $c = 15.0$ mm

36. $\theta = 62°10'$, $c = 33.0$ cm

37. $\theta = 83.7°$, $b = 3.21$ km

38. $\theta = 32.4°$, $a = 42.3$ m

39. $\theta = 71.5°$, $b = 12.8$ in.

40. $\theta = 44.5°$, $a = 2.30 \times 10^6$ m

41. $b = 63.8$ ft, $c = 134$ ft (angles to the nearest 10′)

42. $b = 22.0$ km, $a = 46.2$ km (angles to the nearest 10′)

43. $b = 132$ mi, $a = 108$ mi (angles to the nearest 0.1°)

44. $a = 134$ m, $c = 182$ m (angles to the nearest 0.1°)

45. The graphing calculator screen below shows the solution of the accompanying triangle for sides a and b. Clearly, something is wrong. Explain what is wrong, and find the correct measures for the two sides.

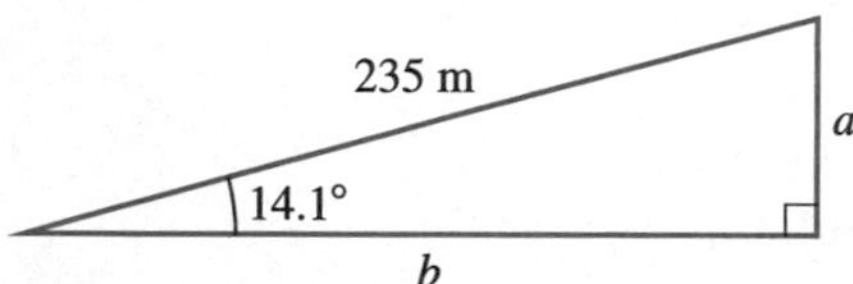

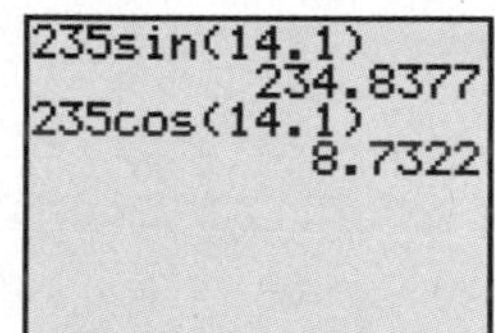

46. In the figure below, side a was found two ways: one using α and sine, and the other using β and cosine. The graphing calculator screen shows the two calculations, but there is an error. Explain the error and correct it.

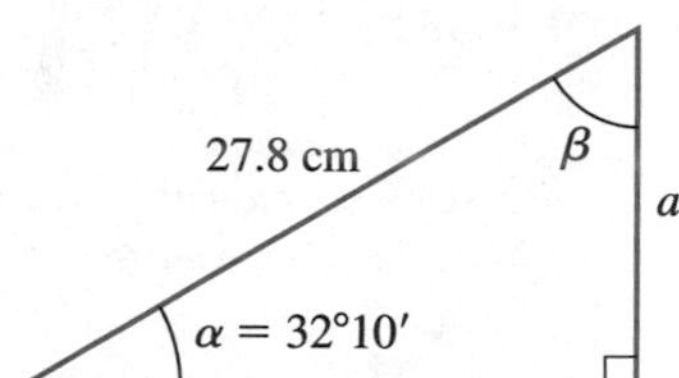

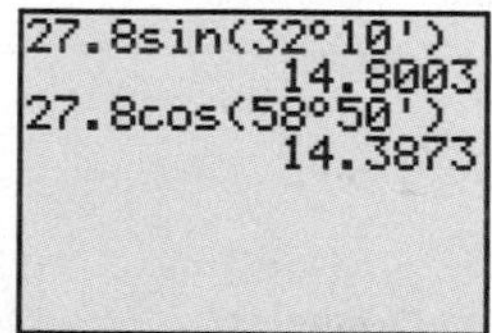

In Problems 47–50, verify each statement for the indicated values.

47. $(\sin\theta)^2 + (\cos\theta)^2 = 1$
(A) $\theta = 11°$ (B) $\theta = 6.09°$ (C) $\theta = 43°24'47''$

48. $(\sin\theta)^2 + (\cos\theta)^2 = 1$
(A) $\theta = 34°$ (B) $\theta = 37.281°$ (C) $\theta = 87°23'41''$

49. $\sin\theta - \cos(90° - \theta) = 0$
(A) $\theta = 19°$ (B) $\theta = 49.06°$ (C) $\theta = 72°51'12''$

50. $\tan\theta - \cot(90° - \theta) = 0$
(A) $\theta = 17°$ (B) $\theta = 27.143°$ (C) $\theta = 14°12'33''$

C *In Problems 51–56, solve the right triangles (labeled as in the figure at the beginning of the exercise).*

51. $a = 23.82$ mi, $\theta = 83°12'$

52. $a = 6.482$ m, $\theta = 35°44'$

53. $b = 42.39$ cm, $a = 56.04$ cm
(angles to the nearest 1′)

54. $a = 123.4$ ft, $c = 163.8$ ft
(angles to the nearest 1′)

55. $b = 35.06$ cm, $c = 50.37$ cm
(angles to the nearest 0.01°)

56. $b = 5.207$ mm, $a = 8.030$ mm
(angles to the nearest 0.01°)

57. Show that $(\sin\theta)^2 + (\cos\theta)^2 = 1$, using the definition of the trigonometric ratios (page 21) and the Pythagorean theorem.

58. Without looking back in the text, show that for each acute angle θ:

(A) $\csc\theta = \dfrac{1}{\sin\theta}$ (B) $\cos(90° - \theta) = \sin\theta$

59. Without looking back in the text, show that for each acute angle θ:

(A) $\cot\theta = \dfrac{1}{\tan\theta}$ (B) $\csc(90° - \theta) = \sec\theta$

60. Without looking back in the text, show that for each acute angle θ:

(A) $\sec\theta = \dfrac{1}{\cos\theta}$ (B) $\cot(90° - \theta) = \tan\theta$

Geometric Interpretation of Trigonometric Ratios *Problems 61–66 refer to the figure, where O is the center of a circle of radius 1, θ is the acute angle AOD, D is the intersection point of the terminal side of angle θ with the circle, and EC is tangent to the circle at D.*

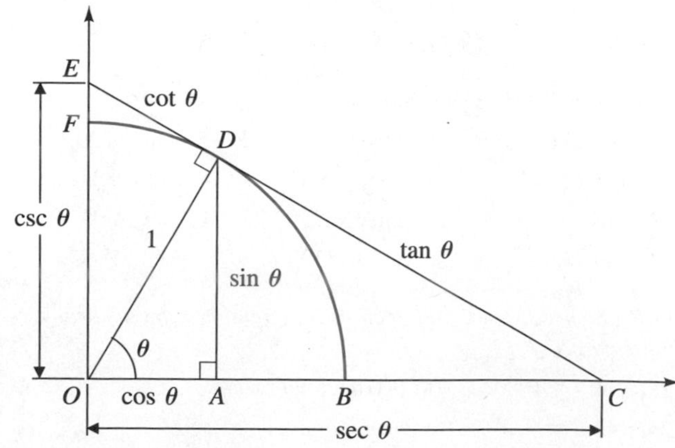

Figure for 61–66

61. Show that:
(A) $\sin \theta = AD$ (B) $\tan \theta = DC$ (C) $\csc \theta = OE$

62. Show that:
(A) $\cos \theta = OA$ (B) $\cot \theta = DE$ (C) $\sec \theta = OC$

63. Explain what happens to each of the following as the acute angle θ approaches 90°:
(A) $\sin \theta$ (B) $\tan \theta$ (C) $\csc \theta$

64. Explain what happens to each of the following as the acute angle θ approaches 90°:
(A) $\cos \theta$ (B) $\cot \theta$ (C) $\sec \theta$

65. Explain what happens to each of the following as the acute angle θ approaches 0°:
(A) $\cos \theta$ (B) $\cot \theta$ (C) $\sec \theta$

66. Explain what happens to each of the following as the acute angle θ approaches 0°:
(A) $\sin \theta$ (B) $\tan \theta$ (C) $\csc \theta$

1.4 RIGHT TRIANGLE APPLICATIONS

Now that you know how to solve right triangles, we can consider a variety of interesting and significant applications.

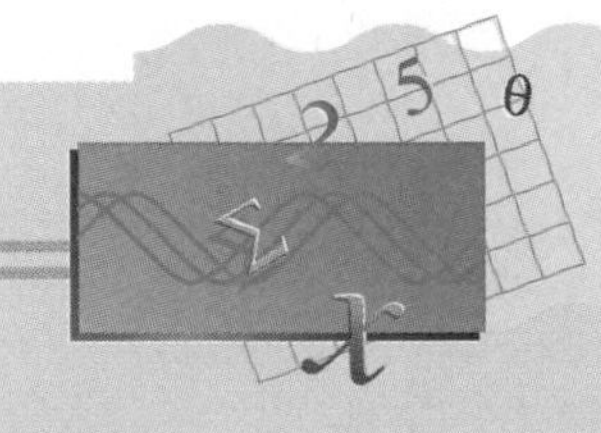

EXPLORE/DISCUSS 1

Discuss the minimum number of sides and/or angles that must be given in a right triangle in order for you to be able to solve for the remaining angles and sides.

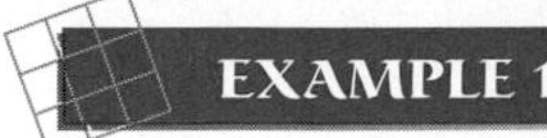

EXAMPLE 1 **Mine Shaft Application**

Solve the mine shaft problem in Example 2, Section 1.2, without using a scale drawing. See Figure 1.

FIGURE 1

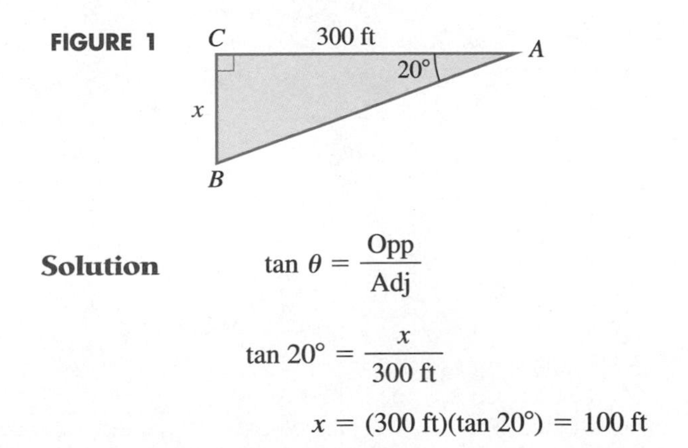

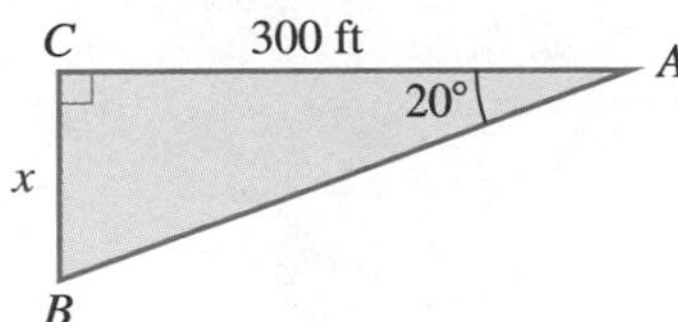

Solution

$$\tan \theta = \frac{\text{Opp}}{\text{Adj}}$$

$$\tan 20° = \frac{x}{300 \text{ ft}}$$

$$x = (300 \text{ ft})(\tan 20°) = 100 \text{ ft}$$ *To one significant digit* ■

Matched Problem 1 Solve the mine shaft problem in Example 1 if $AC = 500$ ft and $\angle A = 30°$. ■

Before proceeding further, we introduce two new terms: **angle of elevation** and **angle of depression.** An angle measured from the horizontal upward is called an angle of elevation; one measured from the horizontal downward is called an angle of depression (see Fig. 2).

FIGURE 2

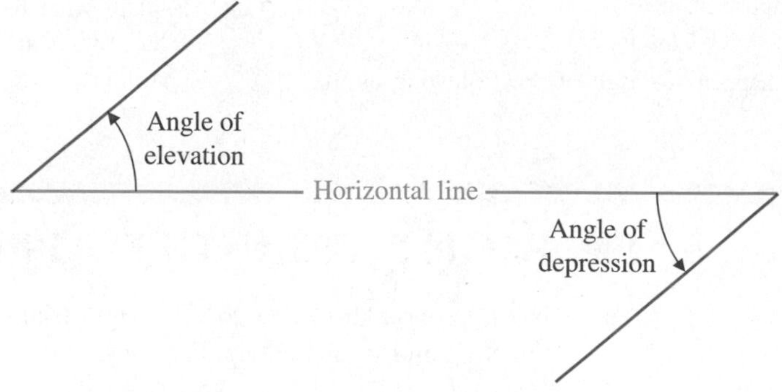

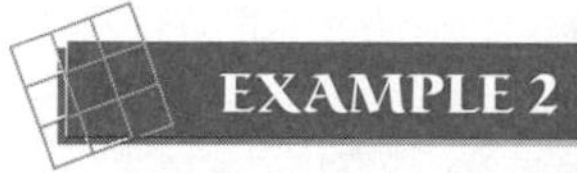

EXAMPLE 2 Length of Air-to-Air Fueling Hose

To save time or because of the lack of landing facilities for large jets in some parts of the world, the military and some civilian companies use air-to-air refueling for some planes (see Fig. 3). If the angle of elevation of the refueling plane's hose is $\theta = 32°$ and b in Figure 3 is 120 ft, how long is the hose?

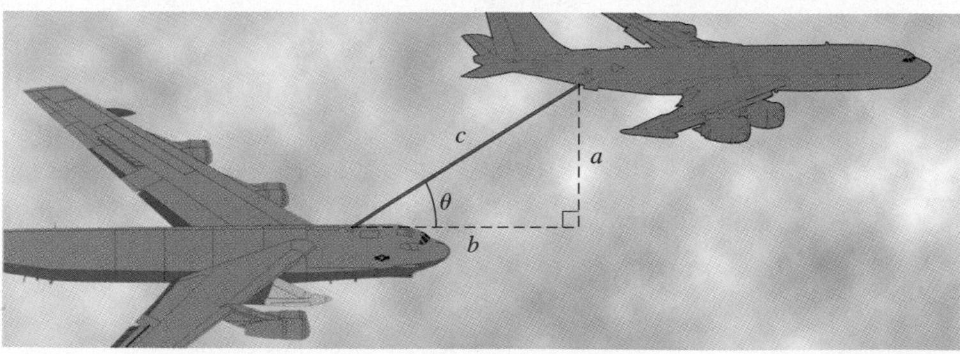

FIGURE 3
Air-to-air fueling

Solution

$$\sec\theta = \frac{c}{b}$$

$$\begin{aligned} c &= b\sec\theta \\ &= 120\sec 32° \\ &= 120(\cos 32°)^{-1} \\ &= 140\text{ ft} \end{aligned}$$

[To two significant digits] ■

Matched Problem 2 The horizontal shadow of a vertical tree is 23.4 m long when the angle of elevation of the sun is 56.3°. How tall is the tree? ■

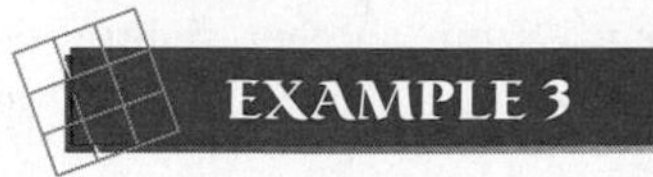
EXAMPLE 3

Astronomy

If we know that the distance from the earth to the sun is approximately 93,000,000 mi, and we find that the largest angle between the earth–sun line and the earth–Venus line is 46°, how far is Venus from the sun? (Assume that the earth and Venus have circular orbits around the sun—see Fig. 4.)

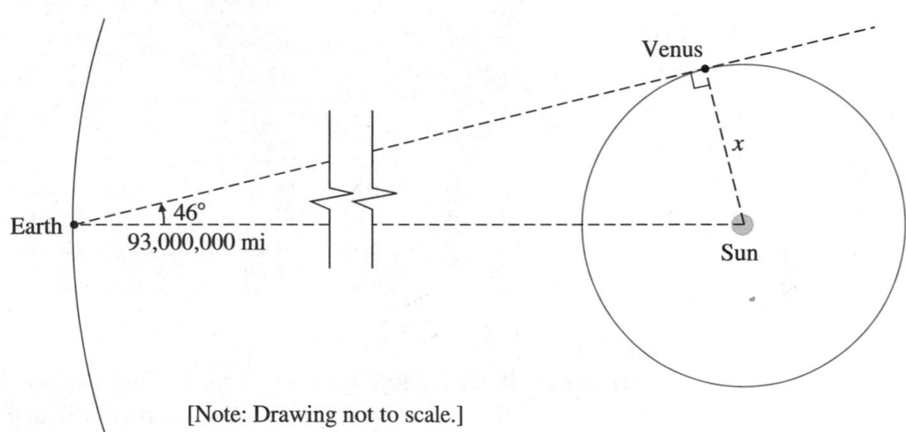

FIGURE 4

Solution The earth–Venus line at its largest angle to the earth–sun line must be tangent to Venus's orbit. Thus, from plane geometry, the Venus–sun line must be at right angles to the earth–Venus line at this time. The sine ratio involves two known quantities and the unknown distance from Venus to the sun. To find x we proceed as follows:

$$\sin 46° = \frac{x}{93{,}000{,}000}$$

$$x = 93{,}000{,}000 \sin 46°$$

$$= 67{,}000{,}000 \text{ mi}$$

■

Matched Problem 3 If the largest angle that the earth–Mercury line makes with the earth–sun line is 28°, how far is Mercury from the sun? (Assume circular orbits.) ■

Coastal Piloting

A boat is cruising along the coast on a straight course. A rocky point is sighted at an angle of 31° from the course. After continuing 4.8 mi, another sighting is taken and the point is found to be 55° from the course (see Fig. 5). How close will the boat come to the point?

FIGURE 5

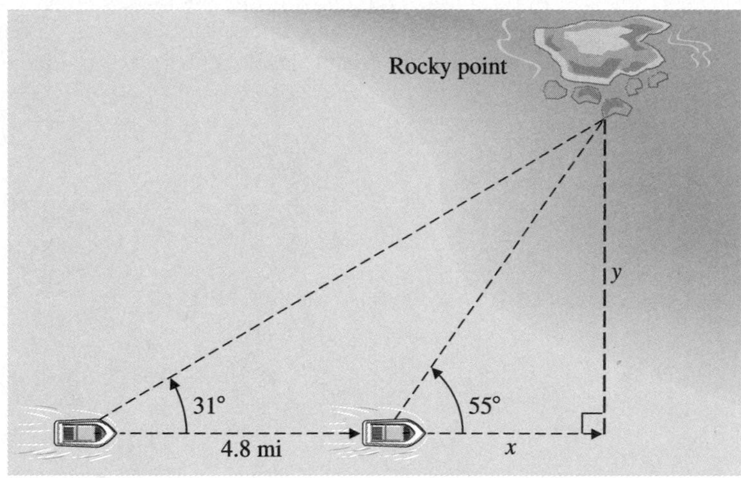

Solution Referring to Figure 5, y is the closest distance that the boat will be to the point. To find y we proceed as follows. From the small right triangle we note that

$$\cot 55^\circ = \frac{x}{y}$$

$$x = y \cot 55^\circ \qquad (1)$$

Now, from the large right triangle, we see that

$$\cot 31^\circ = \frac{4.8 + x}{y}$$

$$y \cot 31^\circ = 4.8 + x \qquad (2)$$

Substituting equation (1) into (2), we obtain

$$y \cot 31^\circ = 4.8 + y \cot 55^\circ$$

$$y \cot 31^\circ - y \cot 55^\circ = 4.8$$

$$y(\cot 31^\circ - \cot 55^\circ) = 4.8$$

$$y = \frac{4.8}{\cot 31^\circ - \cot 55^\circ}$$

$$y = 5.0 \text{ mi}$$

■

Matched Problem 4 Repeat Example 4 after replacing 31° with 28°, 55° with 49°, and 4.8 mi with 5.5 mi. ■

Answers to Matched Problems **1.** 300 ft (to one significant digit) **2.** 35.1 m **3.** 44,000,000 mi **4.** 5.4 mi

EXERCISE 1.4

Applications

A

1. Construction A ladder 8.0 m long is placed against a building as indicated in the figure. How high will the top of the ladder reach up the building?

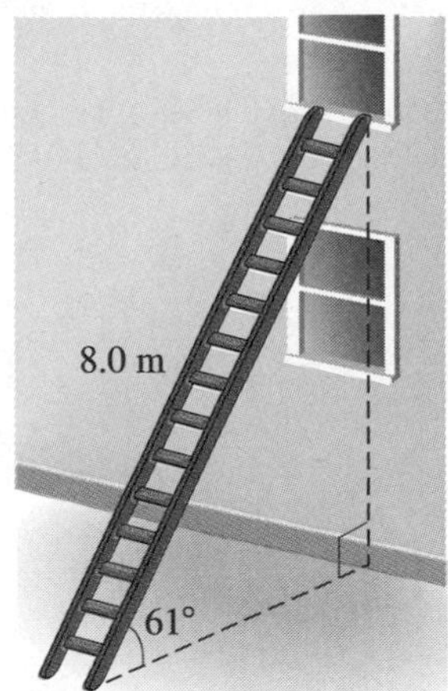

Figure for 1

2. Construction In Problem 1, how far is the foot of the ladder from the wall of the building?

3. Boat Safety Use the information in the figure to find the distance x from the boat to the base of the cliff.

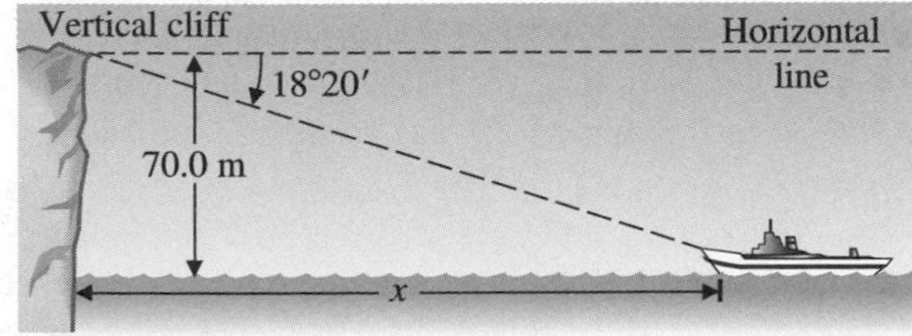

Figure for 3

4. Boat Safety In Problem 3, how far is the boat from the top of the cliff?

5. Geography on the Moon Find the depth of the moon crater in Problem 25, Exercise 1.2.

6. Geography on the Moon Find the height of the mountain on the moon in Problem 26, Exercise 1.2.

7. Flight Safety A glider is flying at an altitude of 8,240 m. The angle of depression from the glider to the control tower at an airport is 15°40′. What is the horizontal distance (in kilometers) from the glider to a point directly over the tower?

8. Flight Safety The height of a cloud or fog cover over an airport can be measured as indicated in the figure. Find h in meters if $b = 1.00$ km and $\alpha = 23.4°$.

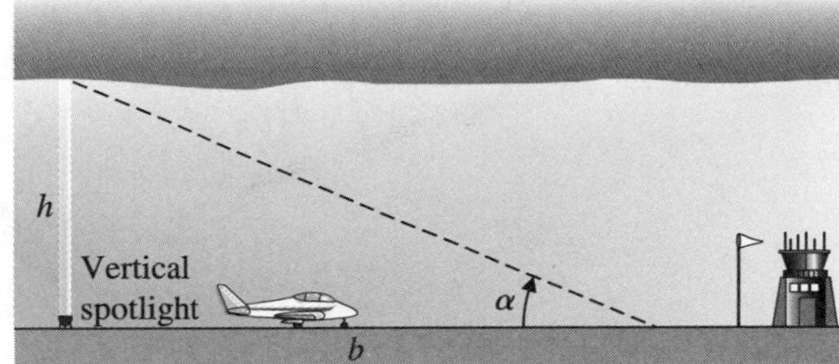

Figure for 8

9. Space Flight The figure at the top of the next page shows the reentry flight pattern of a space shuttle. If at the beginning of the final approach the shuttle is at an altitude of 3,300 ft and its ground distance is 8,200 ft

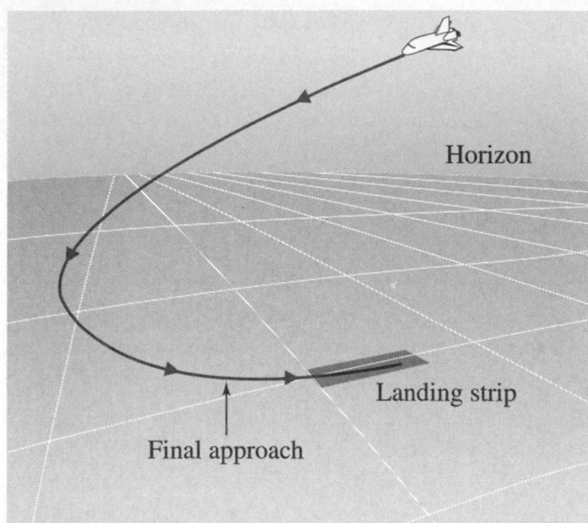

Figure for 9

from the beginning of the landing strip, what glide angle must be used for the shuttle to touch down at the beginning of the landing strip?

10. **Space Flight** If at the beginning of the final approach the shuttle in Problem 9 is at an altitude of 3,600 ft and its ground distance is 9,300 ft from the beginning of the landing strip, what glide angle must be used for the shuttle to touch down at the beginning of the landing strip?

B

11. **Architecture** An architect who is designing a two-story house in a city with a 40°N latitude wishes to control sun exposure on a south-facing wall. Consulting an architectural standards reference book, she finds that at this latitude the noon summer solstice sun has a sun angle of 75° and the noon winter solstice sun has a sun angle of 27° (see the figure).

    (A) How much roof overhang should she provide so that at noon on the day of the summer solstice the shadow of the overhang will reach the bottom of the south-facing wall?

    (B) How far down the wall will the shadow of the overhang reach at noon on the day of the winter solstice?

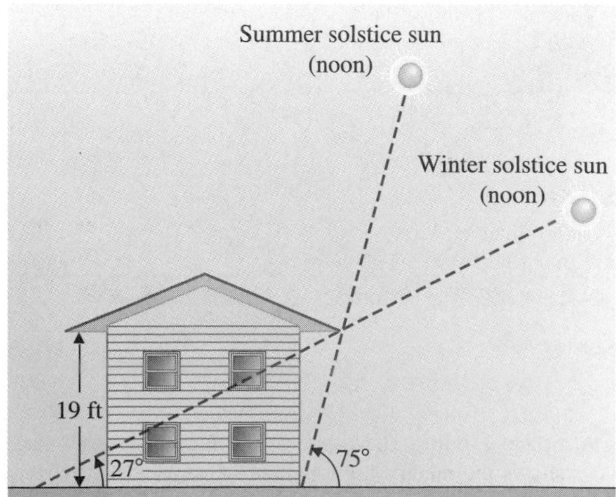

Figure for 11

12. **Architecture** Repeat Problem 11 for a house located at 32°N latitude, where the summer solstice sun angle is 82° and the winter solstice sun angle is 35°.

13. **Geometry** What is the altitude of an equilateral triangle with side 4.0 m? [An equilateral triangle has all sides (and all angles) equal.]

14. **Geometry** The altitude of an equilateral triangle is 5.0 cm. What is the length of a side?

15. **Geometry** Find the length of one side of a nine-sided regular polygon inscribed in a circle with radius 8.32 cm (see the figure).

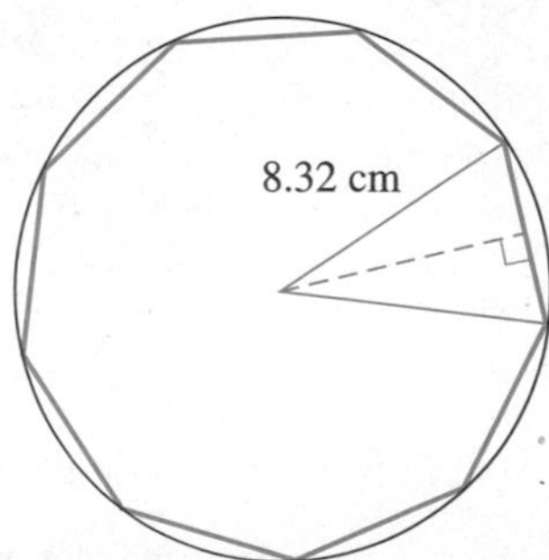

Figure for 15

16. **Geometry** What is the radius of a circle inscribed in the polygon in Problem 15? (The circle will be tangent to each side of the polygon and the radius will be perpendicular to the tangent line at the point of tangency.)

17. **Lightning Protection** A grounded lightning rod on the mast of a sailboat produces a cone of safety as indicated in the figure at the top of the next page. If the top of the rod is 67.0 ft above the water, what is the diameter of the circle of safety on the water?

Figure for 17

18. Lightning Protection In Problem 17, how high should the top of the lightning rod be above the water if the diameter of the circle on the water is to be 100 ft?

19. Diagonal Parking To accommodate cars of most sizes, a parking space needs to contain an 18 ft by 8.0 ft rectangle as shown in the figure. If a diagonal parking space makes an angle of 72° with the horizontal, how long are the sides of the parallelogram that contain the rectangle?

Figure for 19

20. Diagonal Parking Repeat Problem 19 using 68° instead of 72°.

21. Earth Radius A person in an orbiting spacecraft (see the figure) h mi above the earth sights the horizon on the earth at an angle of depression of α. (Recall from geometry that a line tangent to a circle is perpendicular to the radius at the point of tangency.) We wish to find an expression for the radius of the earth in terms of h and α.

(A) Express $\cos \alpha$ in terms of r and h.

(B) Solve the answer to part (A) for r in terms of h and α.

(C) Find the radius of the earth if $\alpha = 22°47'$ and $h = 335$ mi.

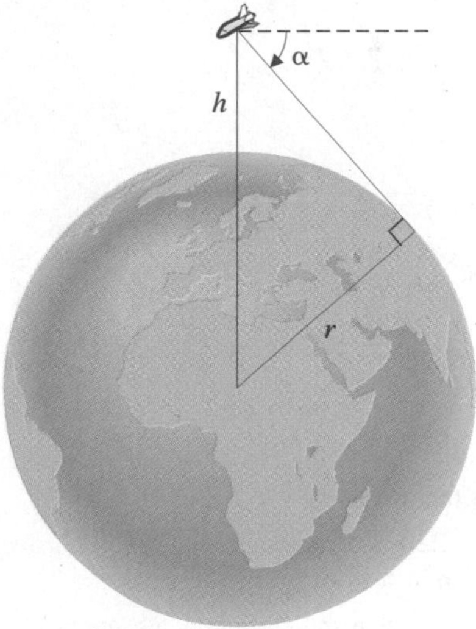

22. Orbiting Spacecraft Height A person in an orbiting spacecraft sights the horizon line on earth at an angle of depression α. (Refer to the figure in Problem 21.)

(A) Express $\cos \alpha$ in terms of r and h.

(B) Solve the answer to part (A) for h in terms of r and α.

(C) Find the height of the spacecraft h if the sighted angle of depression $\alpha = 24°14'$ and the known radius of the earth, $r = 3{,}960$ mi, is used.

23. Navigation Find the radius of the circle that passes through points P, A, and B in part (a) of the figure at the top of the next page. [*Hint:* The central angle in a circle subtended by an arc is twice any inscribed angle subtended by the same arc—see figure (b).] If A and B are known objects on a maritime navigation chart, then a person on a boat at point P can locate the position of P on a circle on the chart by sighting the angle APB and completing the calculations as suggested. By repeating the procedure with another pair of known points, the

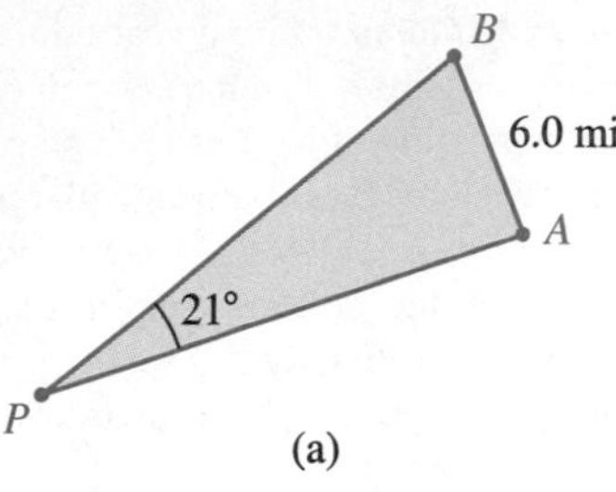

(a)

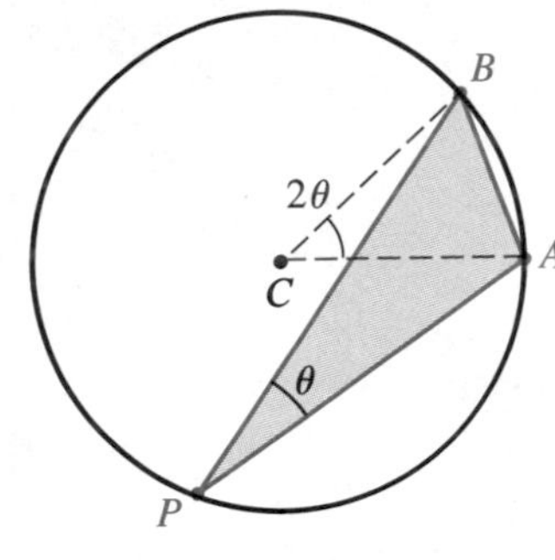

(b)

Figure for 23

position of the boat on the chart will be at an intersection point of the two circles.

24. Navigation Repeat Problem 23 using 33° instead of 21° and 7.5 km instead of 6.0 mi.

25. Geography Assume the earth is a sphere (it is nearly so) and that the circumference of the earth at the equator is 24,900 mi. A **parallel of latitude** is a circle around the earth at a given latitude that is parallel to the equator (see the figure). Approximate the length of a parallel of latitude passing through San Francisco, which is at a latitude of 38°N. See the figure, where θ is the latitude, R is the radius of the earth, and r is the radius of the parallel of latitude. In general, show that if E is the length of the equator and L is the length of a parallel of latitude at a latitude θ, then $L = E \cos \theta$.

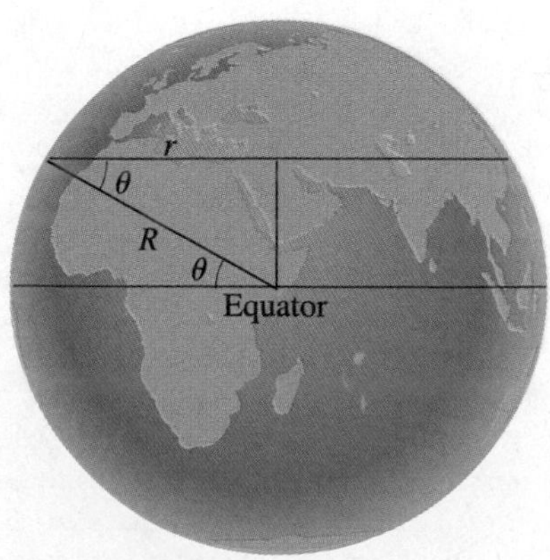

Figure for 25

26. Geography Using the information in Problem 25 and the fact that the circumference of the earth at the equator is 40,100 km, determine the length of the Arctic Circle (66°33′N) in kilometers.

27. Precalculus: Lifeguard Problem A lifeguard sitting in a tower spots a distressed swimmer, as indicated in the figure. To get to the swimmer, the lifeguard must run some distance along the beach at rate p, enter the water, and swim at rate q to the distressed swimmer.

Figure for 27 and 28

(A) To minimize the total time to the swimmer, should the lifeguard enter the water directly or run some distance along the shore and then enter the water? Explain your reasoning.

(B) Express the total time T it takes the lifeguard to reach the swimmer in terms of θ, d, c, p, and q.

(C) Find T (in seconds to two decimal places) if $\theta = 51°$, $d = 380$ m, $c = 76$ m, $p = 5.1$ m/sec, and $q = 1.7$ m/sec.

(D) The following table from a graphing calculator display shows the various times it takes to reach the swimmer for θ from 55° to 85° (X represents θ and Y1 represents T). Explain the behavior of T relative

X	Y1
55.000	118.65
60.000	117.53
65.000	116.89
70.000	116.66
75.000	116.80
80.000	117.28
85.000	118.08

X=55

to θ. For what value of θ in the table is the total time T minimum?

(E) How far (to the nearest meter) should the lifeguard run along the shore before swimming to achieve the minimal total time estimated in part (D)?

28. Precalculus: Lifeguard Problem Refer to Problem 27.

(A) Express the total distance D covered by the lifeguard from the tower to the distressed swimmer in terms of d, c, and θ.

(B) Find D (to the nearest meter) for the values of d, c, and θ in Problem 27C.

(C) Using the values for distances and rates in Problem 27C, what is the time (to two decimal places) it takes the lifeguard to get to the swimmer for the shortest distance from the lifeguard tower to the swimmer? Does going the shortest distance take the least time for the lifeguard to get to the swimmer? Explain. (See the graphing calculator table in Problem 27D.)

29. Precalculus: Pipeline An island is 4 mi offshore in a large bay. A water pipeline is to be run from a water tank on the shore to the island, as indicated in the figure. The pipeline costs $40,000 per mile in the ocean and $20,000 per mile on the land.

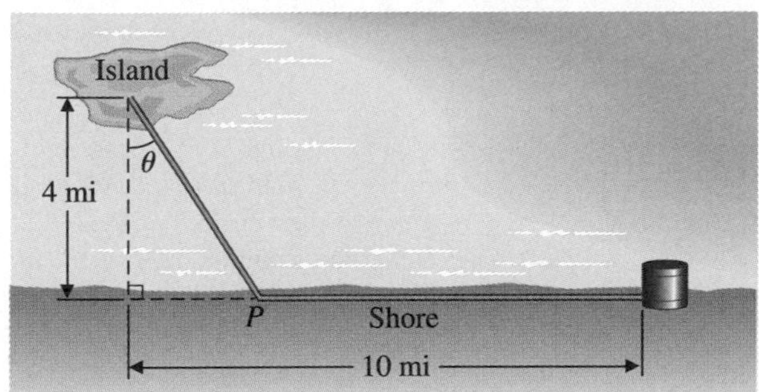

Figure for 29 and 30

(A) Do you think that the total cost is independent of the angle θ chosen, or does it depend on θ? Explain.

(B) Express the total cost C of the pipeline in terms of θ.

(C) Find C for $\theta = 15°$ (to the nearest hundred dollars).

(D) The following table from a graphing calculator display shows the various costs for θ from 15° to 45° (X represents θ and Y1 represents C). Explain the behavior of C relative to θ. For what value of θ in the table is the total cost C minimum? What is the minimum cost (to the nearest hundred dollars)?

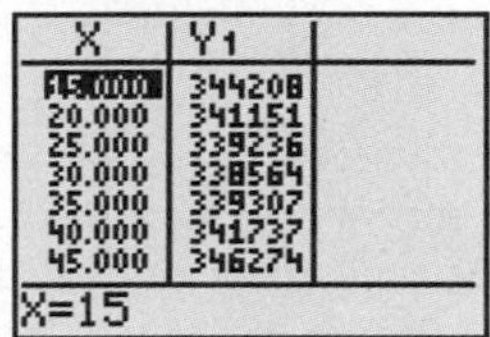

X	Y1
15.000	344208
20.000	341151
25.000	339236
30.000	338564
35.000	339307
40.000	341737
45.000	346274

X=15

(E) How many miles of pipe (to two decimal places) should be laid on land and how many miles placed in the water for the total cost to be minimum?

30. Precalculus: Pipeline Refer to Problem 29.

(A) Express the total length of the pipeline L in terms of θ.

(B) Find L (to two decimal places) for $\theta = 35°$.

(C) What is the cost (to the nearest hundred dollars) of the shortest pipeline from the island to the water tank? Is the shortest pipeline from the island to the water tank the least costly? Explain. (See the graphing calculator table in Problem 29D.)

31. Surveying Use the information in the figure to find the height y of the mountain.

$$\tan 42° = \frac{y}{x} \qquad \tan 25° = \frac{y}{1.0 + x}$$

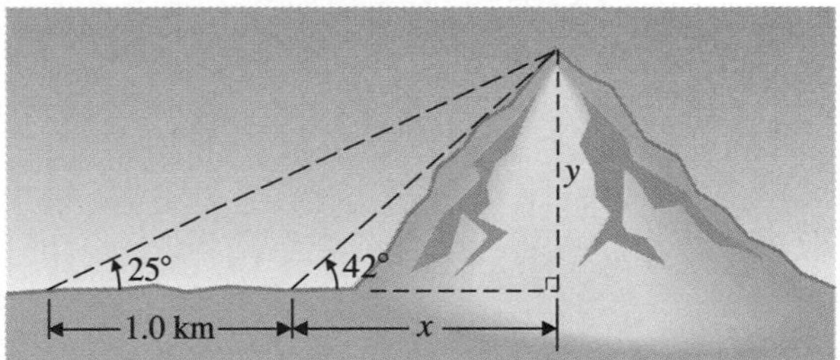

Figure for 31

32. Surveying

(A) Using the figure, show that: $h = \dfrac{d}{\cot \alpha - \cot \beta}$

Figure for 32

(B) Use the results in part (A) to find the height of the mountain in Problem 31.

33. Surveying From the sunroof of Janet's apartment building, the angle of depression to the base of an office building is 51.4° and the angle of elevation to the top of the office building is 43.2° (see the figure). If the office building is 847 ft high, how far apart are the two buildings and how high is the apartment building?

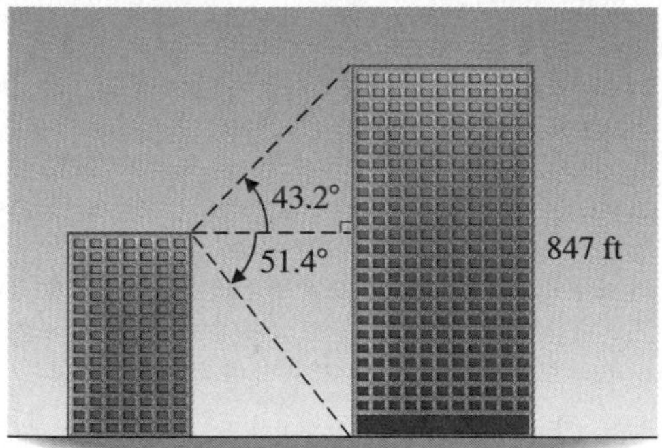

Figure for 33

34. Surveying

(A) Using the figure, show that

$$h = \frac{d}{\cot \alpha + \cot \beta}$$

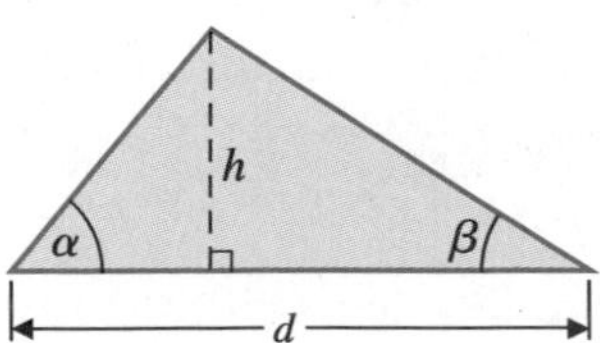

Figure for 34

(B) Use the results in part (A) to find the distance between the two buildings in Problem 33.

35. Precalculus: Physics In physics one can show that the velocity (v) of a ball rolling down an inclined plane (neglecting air resistance and friction) is given by

$$v = g(\sin \theta)t$$

where g is the gravitational constant and t is time (see the figure).

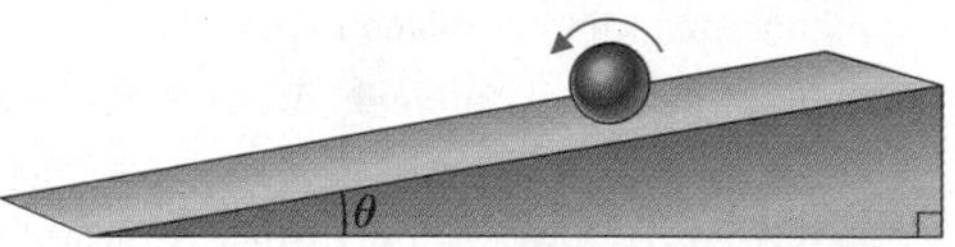

Figure for 35

Galileo (1564–1642) used this equation in the form

$$g = \frac{v}{(\sin \theta)t}$$

so he could determine g after measuring v experimentally. (There were no timing devices available then that were accurate enough to measure the velocity of a free-falling body. He had to use an inclined plane to slow the motion down, and then he was able to calculate an approximation for g.) Find g if at the end of 2.00 sec a ball is traveling at 11.1 ft/sec down a plane inclined at 10.0°.

36. Precalculus: Physics In Problem 35 find g if at the end of 1.50 sec a ball is traveling at 12.4 ft/sec down a plane inclined at 15.0°.

C **37. Geometry** In part (a) of the figure, M and N are midpoints to the sides of a square. Find the exact value of $\sin \theta$. [*Hint:* The solution utilizes the Pythagorean theorem, similar triangles, and the definition of sine. Some useful auxiliary lines are drawn in part (b) of the figure.]

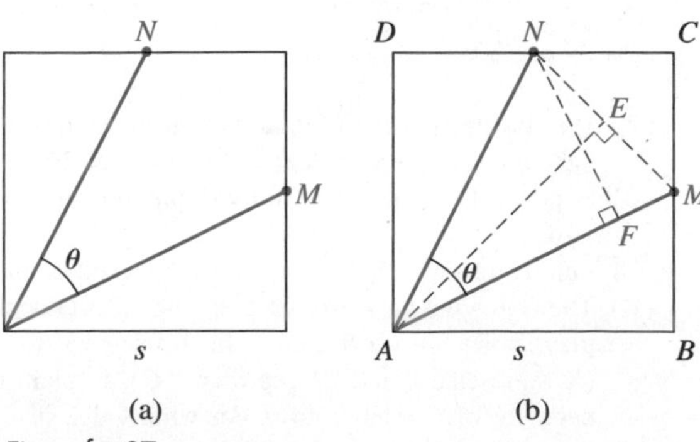

Figure for 37

38. Geometry Find r in the figure. The circle is tangent to all three sides of the isosceles triangle. (An isosceles triangle has two sides equal.) [*Hint:* The radius of a circle and a tangent line are perpendicular at the point of tangency. Also, the altitude of the isosceles triangle will pass through the center of the circle and will divide the original triangle into two congruent triangles.]

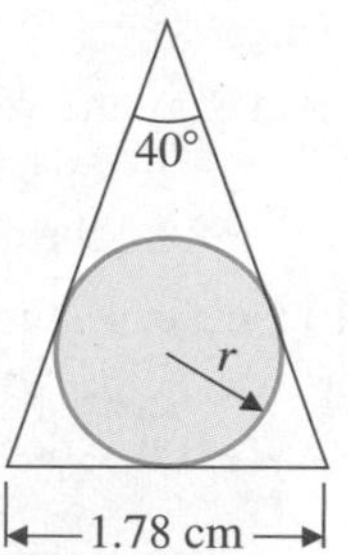

Figure for 38

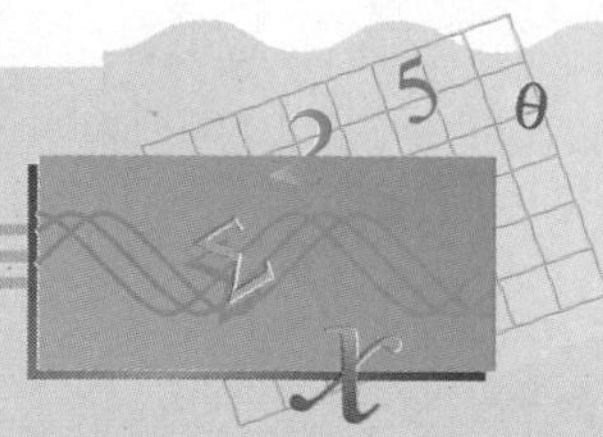

CHAPTER 1 GROUP ACTIVITY

A Logistics Problem

A log of length L floats down a canal that has a right angle turn, as indicated in Figure 1.

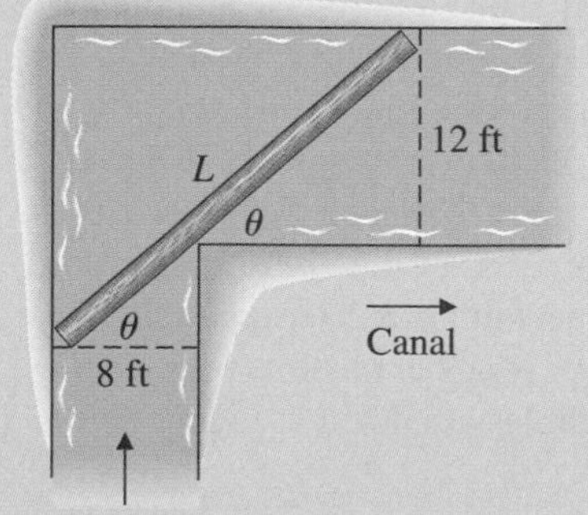

FIGURE 1

(A) Can logs of any length be floated around the corner? Discuss your first thoughts on this problem without any mathematical analysis.

(B) Express the length of the log L in terms of θ and the two widths of the canal (neglect the width of the log), assuming the log is touching the sides and corner of the canal as shown in Figure 1. (In calculus, this equation is used to find the longest log that will float around the corner. Here, we will try to approximate the maximum log length by using noncalculus tools.)

(C) An expression for the length of the log in part (B) is

$$L = 12 \csc \theta + 8 \sec \theta$$

What are the restrictions on θ? Describe what you think happens to the length L as θ increases from values close to 0° to values close to 90°.

(D) Complete Table 1, giving values of L to one decimal place.

TABLE 1

θ	30°	35°	40°	45°	50°	55°	60°
L	33.2						

(E) Using the values in Table 1, describe what happens to L as θ increases from 30° to 60°. Estimate the longest log (to one decimal place) that will go around the corner in the canal.

(F) We can refine the results in parts (D) and (E) further by using more values of θ close to the result found in part (E). Complete Table 2, giving values of L to two decimal places.

TABLE 2

θ	44°	46°	48°	50°	52°	54°	56°
L	28.40						

(G) Using the values in Table 2, estimate the longest log (to two decimal places) that will go around the corner in the canal.

(H) Discuss further refinements on this process. (A graphing utility with table-producing capabilities would be helpful in carrying out such a process.)

CHAPTER 1 REVIEW

1.1 ANGLES, DEGREES, AND ARCS

An **angle** is formed by rotating a half-line, called a **ray**, around its end point. See Figure 1: One ray k, called the **initial side** of the angle, remains fixed; a second ray l, called the **terminal side** of the angle, starts in the initial side position and is rotated around the common end point P in a plane until it reaches its terminal position. The common end point is called the **vertex.** An angle is **positive** if the terminal side is rotated counterclockwise and **negative** if the terminal side is rotated clockwise. Different angles with the same initial and terminal sides are called **coterminal.** An angle of **1 degree** is $\frac{1}{360}$ of a complete revolution in a counterclockwise direction. Names for special angles are noted in Figure 2.

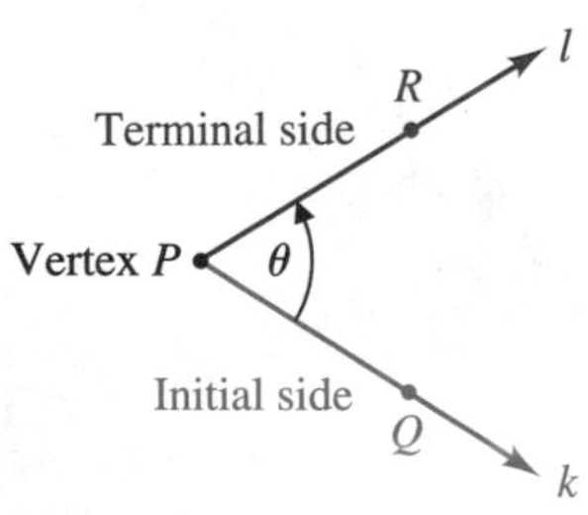

FIGURE 1

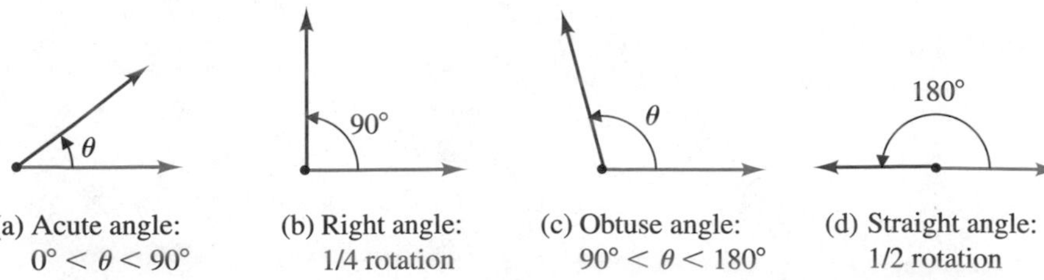

(a) Acute angle: $0° < \theta < 90°$ (b) Right angle: 1/4 rotation (c) Obtuse angle: $90° < \theta < 180°$ (d) Straight angle: 1/2 rotation

FIGURE 2

Two positive angles are **complementary** if the sum of their measures is 90°; they are **supplementary** if the sum of their measures is 180°.

Angles can be represented in terms of **decimal degrees,** or in terms of **minutes** ($\frac{1}{60}$ of a degree) and **seconds** ($\frac{1}{60}$ of a minute). Calculators can be used to convert from decimal degrees (DD) to degrees-minutes-seconds (DMS), and vice versa. The **arc length** s of an arc **subtended** by a **central angle** θ in a circle of radius r (Fig. 3) satisfies

FIGURE 3

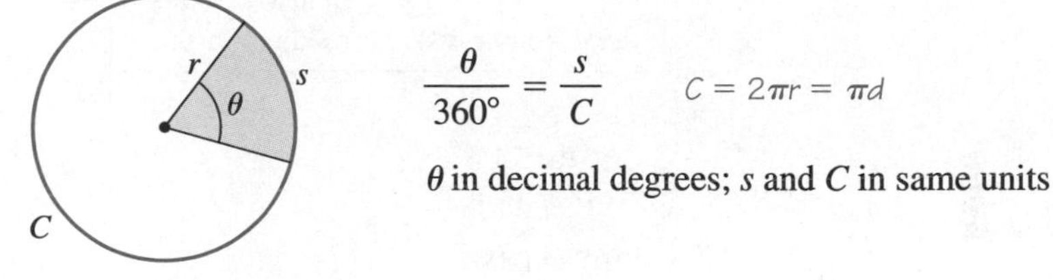

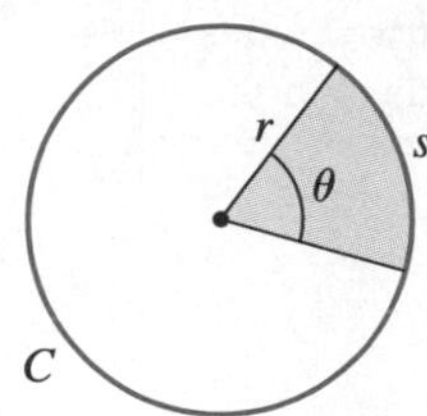

$$\frac{\theta}{360°} = \frac{s}{C} \qquad C = 2\pi r = \pi d$$

θ in decimal degrees; s and C in same units

1.2 SIMILAR TRIANGLES

The properties of similar triangles stated in **Euclid's theorem** are central to the development of trigonometry: If two triangles are similar, their corresponding sides are proportional. See Figure 4.

FIGURE 4

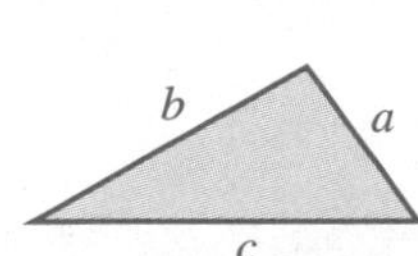

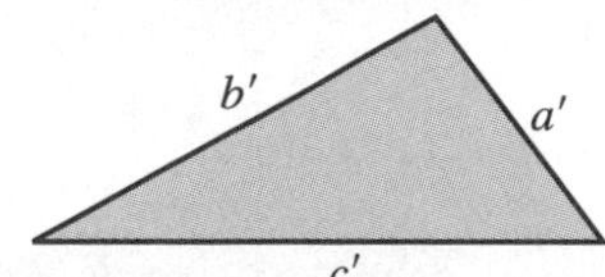

$$\frac{a}{a'} = \frac{b}{b'} = \frac{c}{c'}$$

Similar triangles

1.3 TRIGONOMETRIC RATIOS AND RIGHT TRIANGLES

The six **trigonometric ratios** for the angle θ in a right triangle (see Fig. 5) with **opposite side** b, **adjacent side** a, and **hypotenuse** c are:

$$\sin\theta = \frac{b}{c} = \frac{\text{Opp}}{\text{Hyp}} \qquad \csc\theta = \frac{c}{b} = \frac{\text{Hyp}}{\text{Opp}}$$

$$\cos\theta = \frac{a}{c} = \frac{\text{Adj}}{\text{Hyp}} \qquad \sec\theta = \frac{c}{a} = \frac{\text{Hyp}}{\text{Adj}}$$

$$\tan\theta = \frac{b}{a} = \frac{\text{Opp}}{\text{Adj}} \qquad \cot\theta = \frac{a}{b} = \frac{\text{Adj}}{\text{Opp}}$$

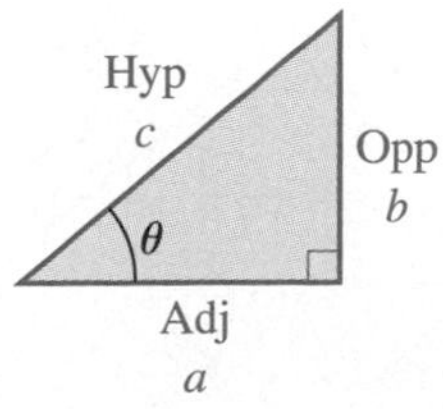

Right triangle

FIGURE 5

The complementary relationships shown below illustrate why cosine, cotangent, and cosecant are called the **cofunctions** of sine, tangent, and secant, respectively. See Figure 6.

Complementary Relationships	Reciprocal Relationships
$\sin\theta = \frac{b}{c} = \cos(90° - \theta)$	$\csc\theta = \frac{1}{\sin\theta}$
$\tan\theta = \frac{b}{a} = \cot(90° - \theta)$	$\sec\theta = \frac{1}{\cos\theta}$
$\sec\theta = \frac{c}{a} = \csc(90° - \theta)$	$\cot\theta = \frac{1}{\tan\theta}$

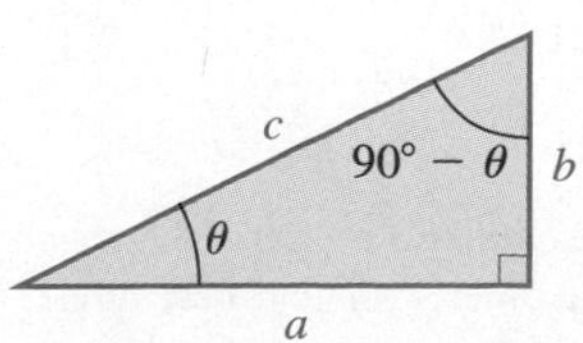

FIGURE 6

Solving a right triangle involves finding the measures of the remaining sides and acute angles when given the measure of two sides or the measure of one side and one acute angle. Accuracy of these computations is governed by the following table:

Angle to nearest	Significant digits for side measure
1°	2
10′ or 0.1°	3
1′ or 0.01°	4
10″ or 0.001°	5

1.4 RIGHT TRIANGLE APPLICATIONS

An angle measured upward from the horizontal is called an **angle of elevation** and one measured downward from the horizontal is called an **angle of depression.**

CHAPTER 1 REVIEW EXERCISE

Work through all the problems in this chapter review and check answers in the back of the book. Answers to all review problems are there, and following each answer is a number in italics indicating the section in which that type of problem is discussed. Where weaknesses show up, review appropriate sections in the text.

1. $2°1'20'' = ?''$
2. An arc of $\frac{1}{6}$ the circumference of a circle subtends a central angle of how many degrees?
3. Given two similar triangles, as shown in the figure, find a if $c = 20{,}000$, $a' = 2$, and $c' = 5$.

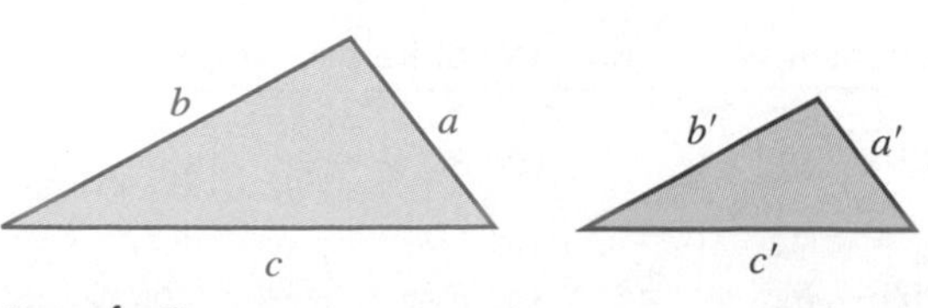

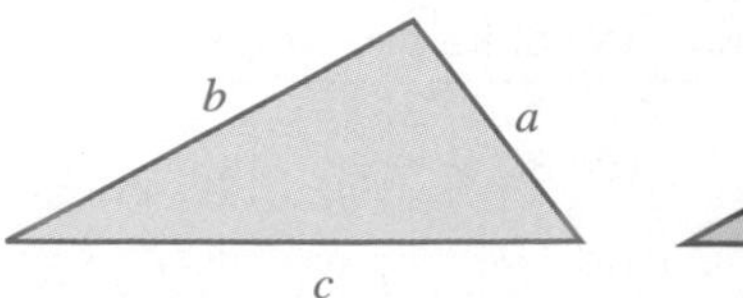

Figure for 3

4. Change 36°23′ to decimal degrees (to two decimal places).
5. Write a definition of an angle of degree measure 1.
6. Which of the following triangles are similar? Explain why.

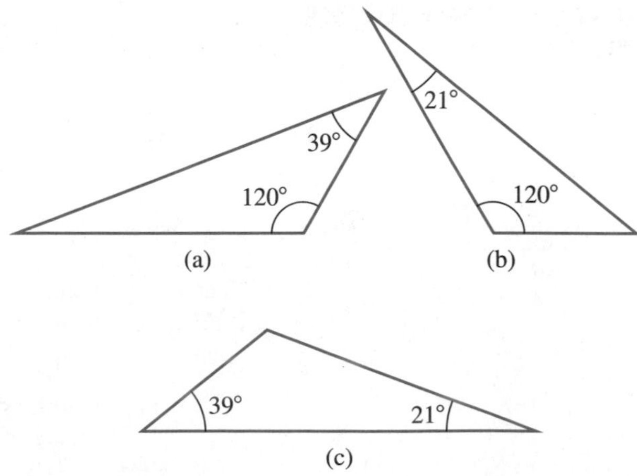

Figure for 6

7. Is it possible in two triangles that are similar to have one with an obtuse angle and the other with no obtuse angle? Explain.
8. Explain why a triangle cannot have more than one obtuse angle.

9. If an office building casts a shadow of 31 ft at the same time a vertical yardstick (36 in.) casts a shadow of 2.0 in., how tall is the building?

10. For the triangle shown here, identify each ratio:
(A) $\sin\theta$ (B) $\sec\theta$ (C) $\tan\theta$
(D) $\csc\theta$ (E) $\cos\theta$ (F) $\cot\theta$

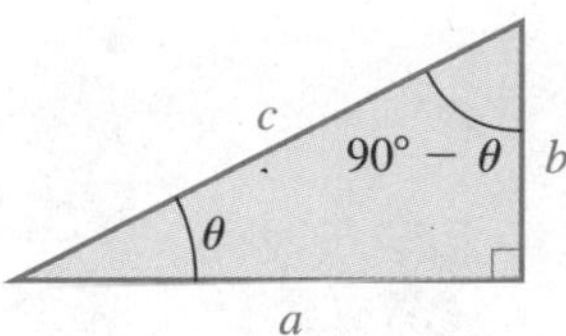

Figure for 10

11. Solve the right triangle in Problem 10, given $c = 20.2$ cm and $\theta = 35.2°$.

B

12. Find the degree measure of a central angle subtended by an arc of 8.00 cm in a circle with circumference 20.0 cm.

13. If the minute hand of a clock is 2.00 in. long, how far does the tip of the hand travel in exactly 20 min?

14. One angle has a measure of 27°14′ and another angle has a measure of 27.19°. Which is larger? Explain how you obtained your answer.

15. Use a calculator to:
(A) Convert 72°55′49″ to decimal degree form.
(B) Convert 113.837° to degree-minute-second form.

16. Perform the following calculations on a calculator and write the results in DMS form:
(A) $90° - 33°27'51''$
(B) $12(28°32'14'' - 13°40'22'')$

17. For the triangles in Problem 3, find b to two significant digits if $a = 4.1 \times 10^{-6}$ mm, $a' = 1.5 \times 10^{-4}$ mm, and $b' = 2.6 \times 10^{-4}$ mm.

18. For the triangle in Problem 10, identify by name each of the following ratios relative to angle θ:
(A) a/c (B) b/a (C) b/c
(D) c/a (E) c/b (F) a/b

19. For a given value θ, explain why $\sin\theta$ is independent of the size of a right triangle having θ as an acute angle.

20. Solve the right triangle in Problem 10, given $\theta = 62°20'$ and $a = 4.00 \times 10^{-8}$ m.

21. Find each θ to the accuracy indicated.
(A) $\tan\theta = 1.662$ (to two decimal places)
(B) $\theta = \arccos 0.5607$ (to the nearest 10′)
(C) $\theta = \sin^{-1} 0.0138$ (to the nearest second)

22. Which of the following window displays from a graphing calculator is the result of the calculator being set in degree mode and which is the result of the calculator being set in radian mode?

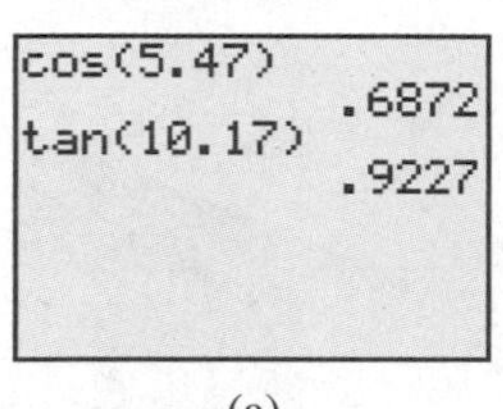

(a)

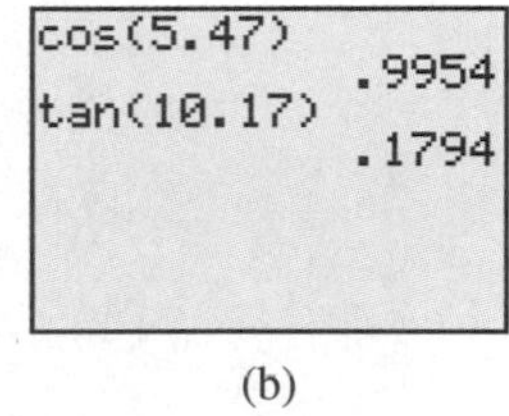

(b)

23. Solve the right triangle in Problem 10, given $b = 13.3$ mm and $a = 15.7$ mm. (Find angles to the nearest 0.1°.)

24. Find the angles in Problem 23 to the nearest 10′.

25. If an equilateral triangle has a side of 10 ft, what is its altitude (to two significant digits)?

C

26. A curve of a railroad track follows an arc of a circle of radius 1,500 ft. If the arc subtends a central angle of 36° how far will a train travel on this arc?

27. Find the area of a sector with central angle 36.5° in a circle with radius 18.3 ft. Compute your answer to the nearest unit.

28. Solve the triangle in Problem 10, given $90° - \theta = 23°43'$ and $c = 232.6$ km.

29. Solve the triangle in Problem 10, given $a = 2{,}421$ m and $c = 4{,}883$ m. (Find angles to the nearest 0.01°.)

30. Use a calculator to find csc 72.3142° to four decimal places.

Applications

31. Precalculus: Shadow Problem A person is standing 20 ft away from a lamppost. If the lamp is 18 ft above the ground and the person is 5 ft 6 in. tall, how long is the person's shadow?

32. **Construction** The front porch of a house is 4.25 ft high. The angle of elevation of a ramp from the ground to the porch is 10.0°. (See the figure.) How long is the ramp? How far is the end of the ramp from the porch?

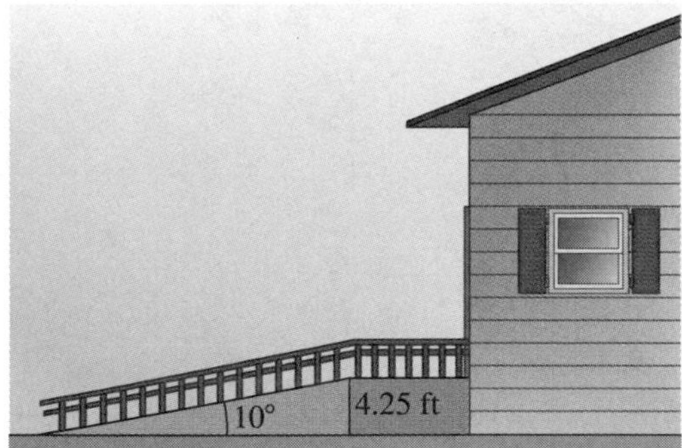

Figure for 32

33. **Medicine: Stress Test** Cardiologists give stress tests by having patients walk on a treadmill at various speeds and inclinations. The amount of inclination may be given as an angle or as a percentage (see the figure). Find the angle of inclination if the treadmill is set at a 4% incline. Find the percentage of inclination if the angle of inclination is 4°.

Angle of inclination: θ

Percentage of inclination: $\dfrac{a}{b}$

Figure for 33

34. **Geography/Navigation** Find the distance (to the nearest mile) between Green Bay, WI, with latitude 44°31′N and Mobile, AL, with latitude 30°42′N. (Both cities have approximately the same longitude.) Use $r = 3{,}960$ mi for the earth's radius.

***35.** **Precalculus: Balloon Flight** The angle of elevation from the ground to a hot air balloon at an altitude of 2,800 ft is 64°. What will be the new angle of elevation if the balloon descends straight down 1,400 ft?

***36.** **Surveying** Use the information in the figure to find the length x of the island.

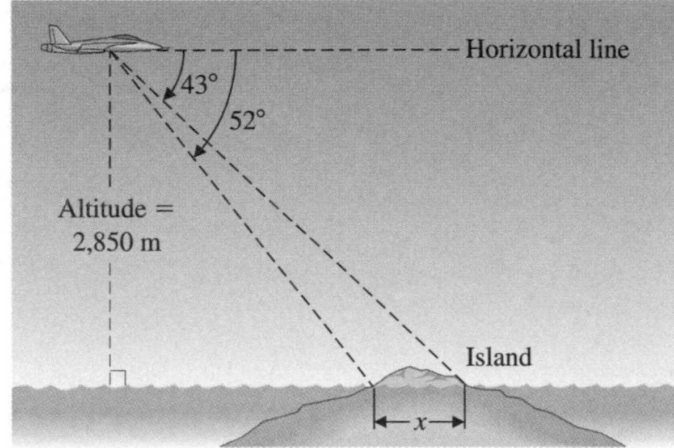

Figure for 36

***37.** **Precalculus: Balloon Flight** Two tracking stations 525 m apart measure angles of elevation of a weather balloon to be 73.5° and 54.2°, as indicated in the figure. How high is the balloon at the time of the measurements?

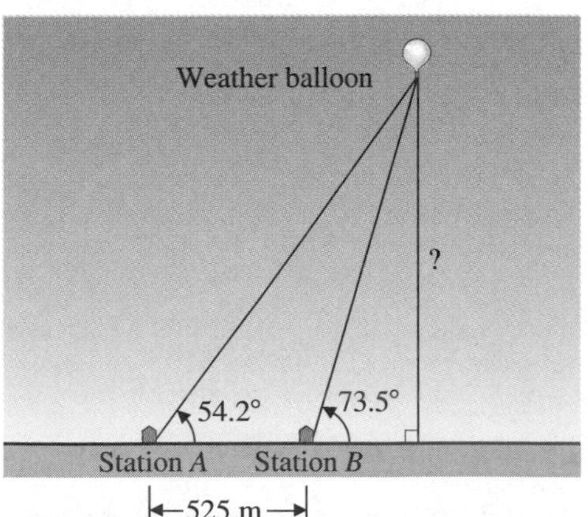

Figure for 37

***38.** **Navigation: Chasing the Sun** Your flight is westward from Buffalo, NY, and you notice the sun just above the horizon. How fast would the plane have to fly to keep the sun in the same position? (The latitude of

Buffalo is 42°50′N, the radius of the earth is 3,960 mi, and the earth makes a complete rotation about its axis in 24 hr.)

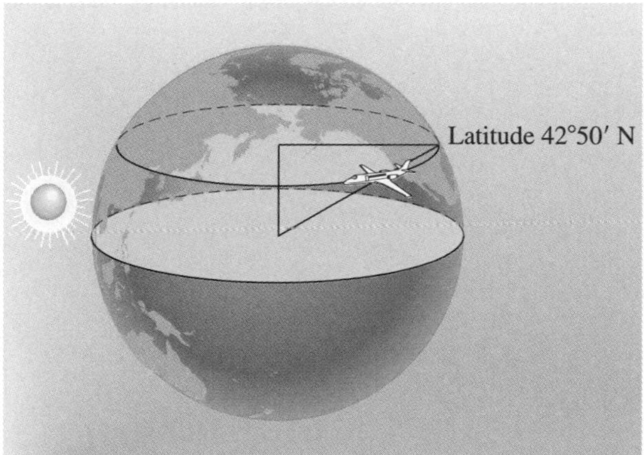

Figure for 38

39. Solar Energy A truncated conical solar collector is aimed directly at the sun as shown in part (a) of the figure. An analysis of the amount of solar energy absorbed by the collecting disk requires certain equations relating the quantities shown in part (b) of the figure. Find an equation that expresses

(A) β in terms of α

(B) r in terms of α and h

(C) $H - h$ in terms of r, R, and α

(Based on the article, "The Solar Concentrating Properties of a Conical Reflector," by Don Leake in *The UMAP Journal,* Vol. 8, No. 4, 1987.)

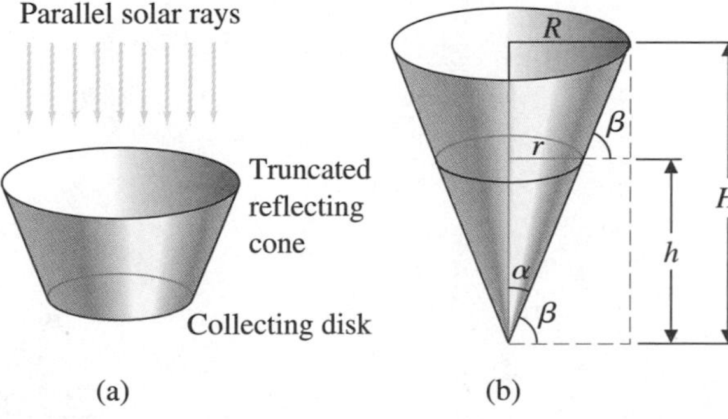

Figure for 39

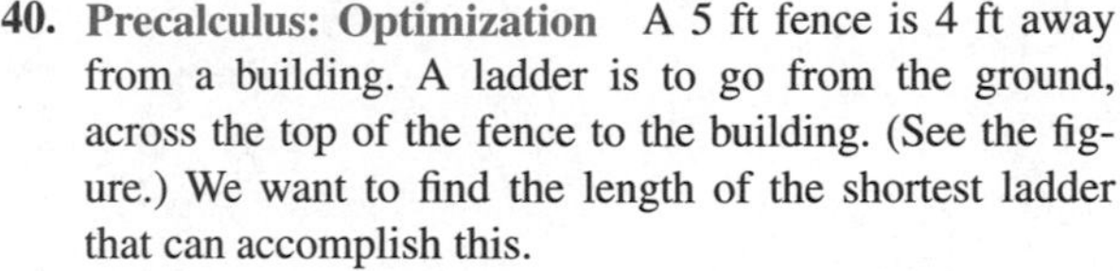

40. Precalculus: Optimization A 5 ft fence is 4 ft away from a building. A ladder is to go from the ground, across the top of the fence to the building. (See the figure.) We want to find the length of the shortest ladder that can accomplish this.

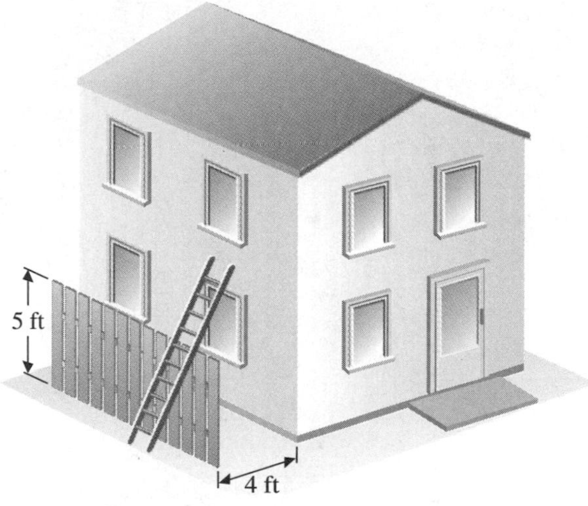

Figure for 40

(A) How is the required length of the ladder affected as the foot of the ladder moves away from the fence? How is the length affected as the foot moves closer to the fence?

(B) Express the length of the ladder in terms of the distance from the fence to the building, the height of the fence, and θ, the angle of elevation that the ladder makes with the level ground.

(C) Complete the table, giving values of L to two decimal places:

θ	25°	35°	45°	55°	65°	75°	85°
L	16.25						

(D) Explain what happens to L as θ moves from 25° to 85°. What value of θ produces the minimum value of L in the table? What is the minimum value?

(E) Describe how you can continue the process to get a better estimate of the minimum ladder length.

TRIGONOMETRIC FUNCTIONS

2

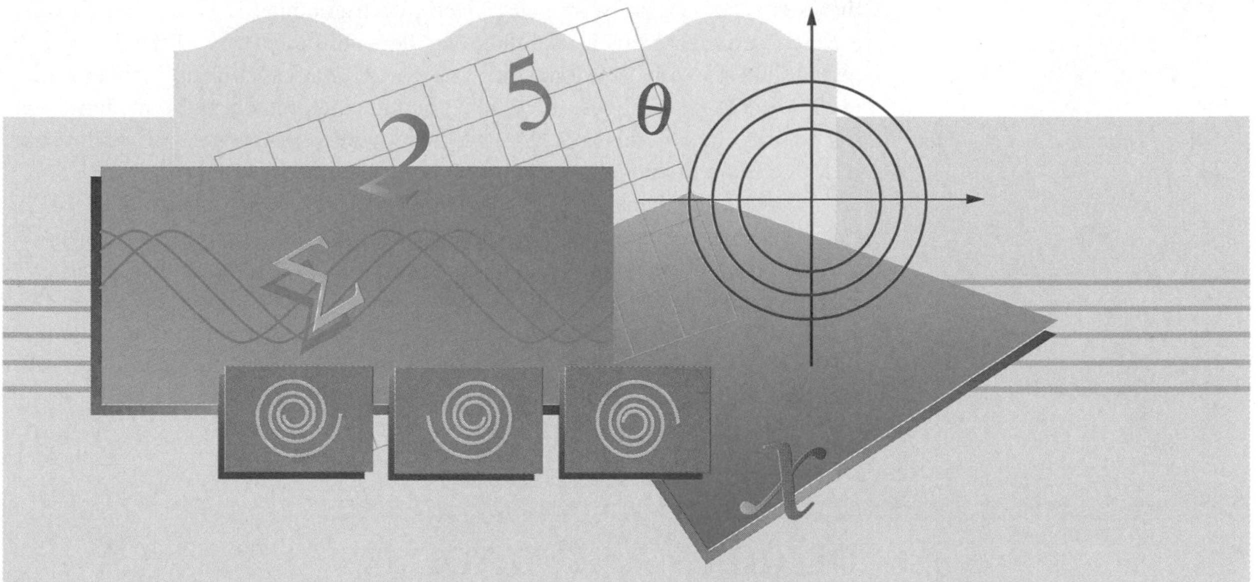

☆ Sections marked with a star may be omitted without loss of continuity.

The trigonometric ratios we studied in Chapter 1 provide a powerful tool for indirect measurement. It was for this purpose only that trigonometry was used for nearly 2,000 years. Astronomy, surveying, map-making, navigation, construction, and military uses catalyzed the extensive development of trigonometry as a tool for indirect measurement.

A turning point in trigonometry occurred after the development of the rectangular coordinate system (credited mainly to the French philosopher–mathematician René Descartes, 1596–1650). The trigonometric ratios, through the use of this system, were generalized into trigonometric functions. This generalization increased their usefulness far beyond the dreams of those originally responsible for this development. The Swiss mathematician Leonhard Euler (1707–1783), probably the greatest mathematician of his century, made substantial contributions in this area. (In fact, there were very few areas in mathematics in which Euler did not make significant contributions.)

Through the demands of modern science, the *periodic* nature of these new functions soon become apparent, and they were quickly put to use in the study of various types of periodic phenomena. The trigonometric functions began to be used on problems that had nothing whatsoever to do with angles and triangles.

In this chapter we generalize the concept of trigonometric ratios along the lines just suggested. Before we undertake this task, however, we will introduce another form of angle measure called the *radian.*

2.1 DEGREES AND RADIANS

- Degree and Radian Measure of Angles
- Angle in Standard Position
- Arc Length and Sector Area

Degree and Radian Measure of Angles

In Chapter 1 we defined an angle and its degree measure. Recall that a central angle in a circle has angle measure 1° if it subtends an arc $\frac{1}{360}$ of the circumference of the circle. Another angle measure that is of considerable use is *radian*

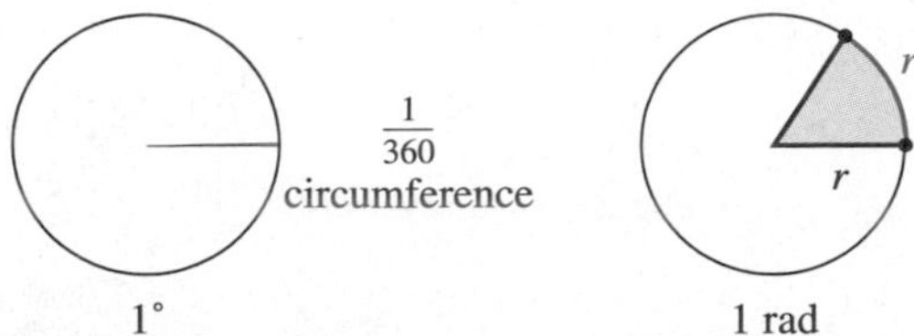

FIGURE 1
Degree and radian measure

measure. A central angle subtended by an arc of length equal to the radius of the circle is defined to be an angle of **radian measure 1** (see Fig. 1). Thus, when we write $\theta = 2°$, we are referring to an angle of degree measure 2. When we write $\theta = 2$ rad, we are referring to an angle of radian measure 2.

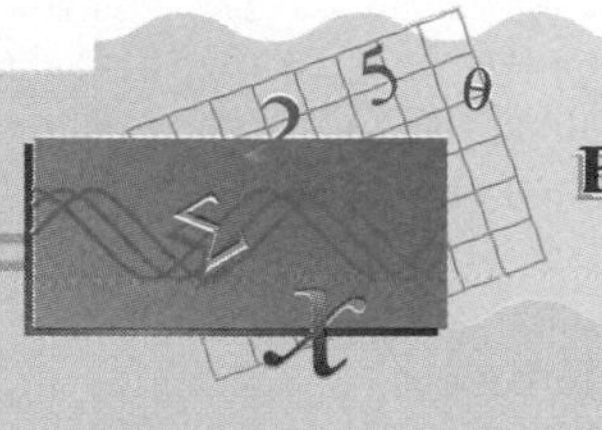

EXPLORE/DISCUSS 1

Discuss why the radian measure and degree measure of an angle are independent of the size of the circle having the angle as a central angle.

It follows from the definition that the radian measure of a central angle θ subtended by an arc of length s is found by determining how many times the length of the radius r, used as a unit length, is contained in the arc length s. In terms of a formula, we have the following:

RADIAN MEASURE OF CENTRAL ANGLES

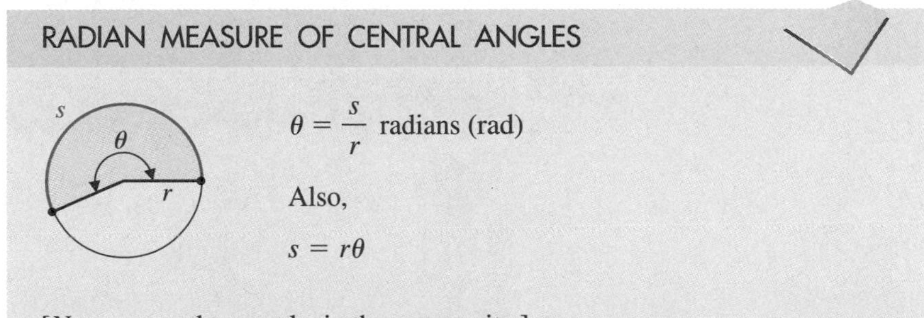

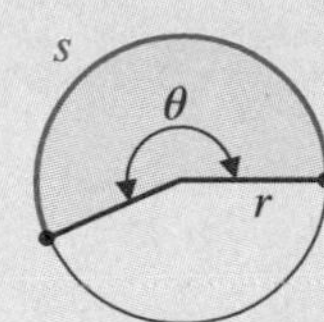

$$\theta = \frac{s}{r} \text{ radians (rad)}$$

Also,

$$s = r\theta$$

[*Note:* s and r must be in the same units.]

Because of their importance, the formulas in the above box should be understood before proceeding further.

What is the radian measure of a central angle subtended by an arc of 32 cm in a circle of radius 8 cm?

$$\theta = \frac{32 \text{ cm}}{8 \text{ cm}} = 4 \text{ rad}$$

Remark *Radian measure is a unitless number.* The units in which the arc length and radius are measured cancel; hence, we are left with a "unitless," or pure, number. For this reason, the word *radian* is often omitted when we are dealing with the radian measure of angles unless a special emphasis is desired. □

What is the radian measure of an angle of 180°? A central angle of 180° is subtended by an arc $\frac{1}{2}$ of the circumference of the circle. Thus, if C is the circumference of a circle, then $\frac{1}{2}$ of the circumference is given by

$$s = \frac{C}{2} = \frac{2\pi r}{2} = \pi r \qquad \text{and} \qquad \theta = \frac{s}{r} = \frac{\pi r}{r} = \pi \text{ rad}$$

Hence, 180° corresponds to π rad. This is important to remember, since the radian measures of many special angles can be obtained from this correspondence. For example, 90° is 180°/2; therefore, 90° corresponds to $\pi/2$ rad. Since 360° is twice 180°, 360° corresponds to 2π rad. Similarly, 60° corresponds to $\pi/3$ rad, 45° to $\pi/4$ rad, and 30° to $\pi/6$ rad. These special angles and their degree and radian measures will be referred to frequently throughout this book. Table 1 summarizes these special correspondences for ease of reference.

TABLE 1

Radians	$\pi/6$	$\pi/4$	$\pi/3$	$\pi/2$	π	2π
Degrees	30	45	60	90	180	360

In general, the following proportion can be used to convert degree measure to radian measure and vice versa.

RADIAN–DEGREE CONVERSION FORMULAS

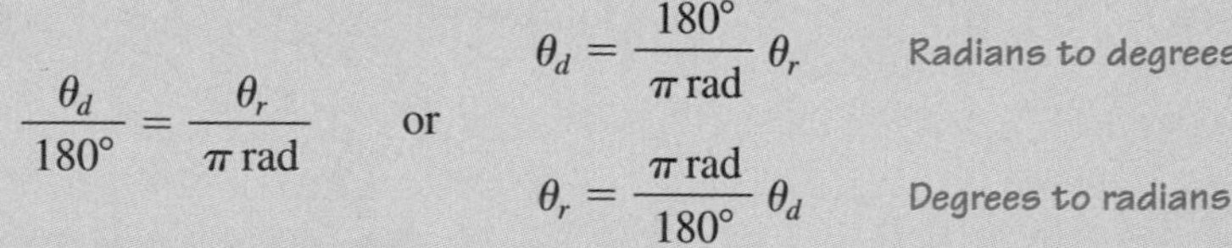

$$\frac{\theta_d}{180°} = \frac{\theta_r}{\pi \text{ rad}} \qquad \text{or} \qquad \begin{aligned} \theta_d &= \frac{180°}{\pi \text{ rad}}\,\theta_r \quad &\text{Radians to degrees} \\ \theta_r &= \frac{\pi \text{ rad}}{180°}\,\theta_d \quad &\text{Degrees to radians} \end{aligned}$$

[*Note:* The proportion is usually easier to remember. Also, we will omit units in calculations until the final answer.]

EXAMPLE 1 **Radian–Degree Conversion**

(A) Find the degree measure of -1.5 rad in exact form and in decimal form to four decimal places.

(B) Find the radian measure of 44° in exact form and in decimal form to four decimal places.

(C) Use a calculator with automatic conversion capability to perform the conversions in parts (A) and (B).

Solution (A) $\theta_d = \dfrac{180°}{\pi \text{ rad}}\theta_r$

$= \dfrac{180}{\pi}(-1.5)$

$= -\dfrac{270°}{\pi}$ Exact form

$= -85.9437°$ To four decimal places

(B) $\theta_r = \dfrac{\pi \text{ rad}}{180°}\theta_d$

$= \dfrac{\pi}{180}(44)$

$= \dfrac{11\pi}{45} \text{ rad}$ Exact form

$= 0.7679 \text{ rad}$ To four decimal places

(C) Check the user's manual for your calculator to find out how to convert parts (A) and (B) with an automatic routine. The following window display is from a graphing calculator with an automatic conversion routine:

```
(-1.5)ʳ
             -85.9437
44°
                .7679
```

■

Matched Problem 1 (A) Find the degree measure of 1 rad in exact form and in decimal form to four decimal places.

(B) Find the radian measure of $-120°$ in exact form and in decimal form to four decimal places.

(C) Use a calculator with automatic conversion capability to perform the conversions in parts (A) and (B). ■

■ Angle in Standard Position

To generalize the concept of trigonometric ratios, we first locate an angle in **standard position** in a rectangular coordinate system. To do this, we place the vertex at the origin and the initial side along the positive x axis. Recall that when the rotation is counterclockwise, the angle is positive and when the rotation is clockwise, the angle is negative. Figure 2 (at the top of the next page) illustrates several angles in standard position.

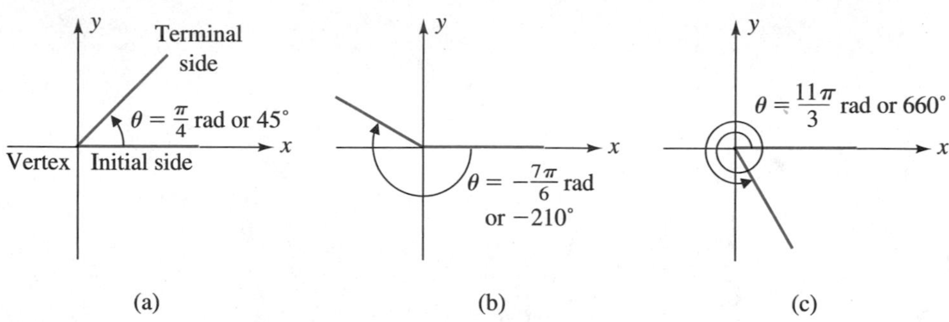

FIGURE 2
Angles in standard position

EXAMPLE 2 **Sketching Angles in Standard Position**

Sketch the following angles in their standard positions:

(A) −60° (B) $3\pi/2$ rad (C) -3π rad (D) 405°

Solution (A)

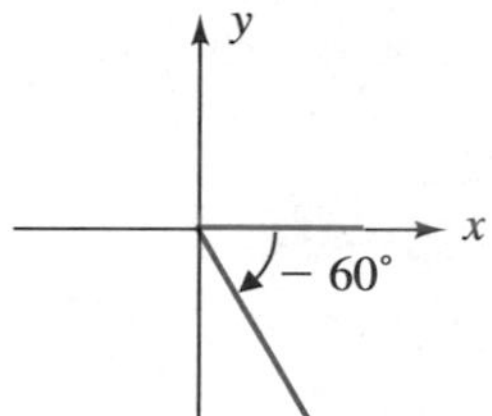

(B)

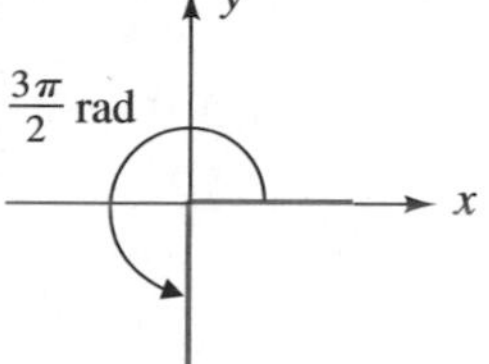

(C)

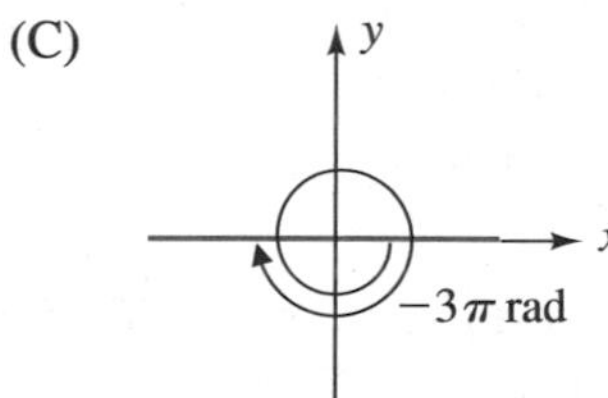

(D)

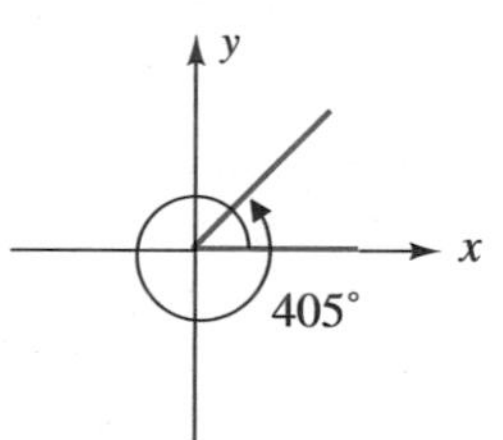

■

Matched Problem 2 Sketch the following angles in their standard positions:

(A) 120° (B) $-\pi/6$ rad (C) $7\pi/2$ rad (D) −495° ■

Two angles are said to be **coterminal** if their terminal sides coincide when both angles are placed in their standard positions in the same rectangular coordinate system. Figure 3 shows two pairs of coterminal angles.

Remarks

1. The degree measures of two coterminal angles differ by an integer* multiple of 360°.

---

* An integer is a positive or negative whole number or 0; that is, the set of integers is {. . . , −4, −3, −2, −1, 0, 1, 2, 3, 4, . . .}.

2. The radian measures of two coterminal angles differ by an integer multiple of 2π. □

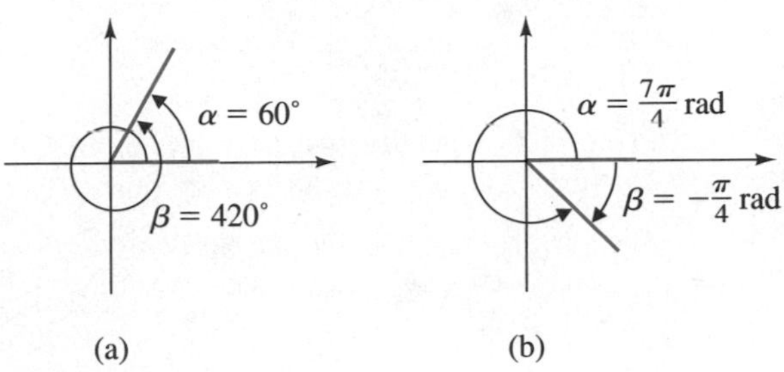

FIGURE 3
Coterminal angles

EXAMPLE 3

Recognizing Coterminal Angles

Which of the following pairs of angles are coterminal?

(A) $\alpha = -135°$, $\beta = 225°$

(B) $\alpha = 120°$, $\beta = -420°$

(C) $\alpha = -\pi/3$ rad, $\beta = 2\pi/3$ rad

(D) $\alpha = \pi/3$ rad, $\beta = 7\pi/3$ rad

Solution (A) The angles are coterminal if $\alpha - \beta$ is an integer multiple of 360°.

$$\alpha - \beta = (-135°) - 225° = -360° = -1(360°)$$

Thus, α and β are coterminal.

(B) $$\alpha - \beta = 120° - (-420°) = 540°$$

The angles are not coterminal, since 540° is not an integer multiple of 360°.

(C) The angles are coterminal if $\alpha - \beta$ is an integer multiple of 2π.

$$\alpha - \beta = \left(-\frac{\pi}{3}\right) - \frac{2\pi}{3} = -\frac{3\pi}{3} = -\pi$$

The angles are not coterminal, since $-\pi$ is not an integer multiple of 2π.

(D) $$\alpha - \beta = \frac{\pi}{3} - \frac{7\pi}{3} = -\frac{6\pi}{3} = -2\pi = (-1)(2\pi)$$

Thus, α and β are coterminal. ■

Matched Problem 3 Which of the following pairs of angles are coterminal?

(A) $\alpha = 90°$, $\beta = -90°$

(B) $\alpha = 750°$, $\beta = 30°$

(C) $\alpha = -\pi/6$ rad, $\beta = -25\pi/6$ rad

(D) $\alpha = 3\pi/4$ rad, $\beta = 7\pi/4$ rad ■

EXPLORE/DISCUSS 2

(A) List all angles that are coterminal with $\theta = \pi/3$ rad, $-5\pi \le \theta \le 5\pi$. Explain how you arrived at your answer.

(B) List all angles that are coterminal with $\theta = -45°$, $-900° \le \theta \le 900°$. Explain how you arrived at your answer.

Arc Length and Sector Area

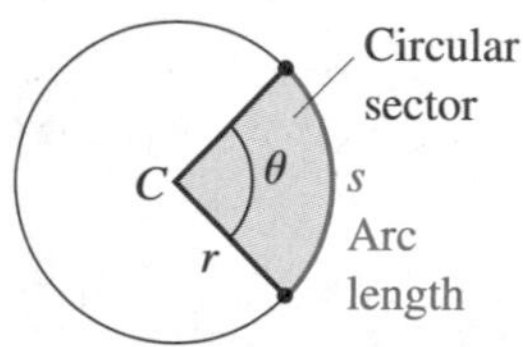

FIGURE 4
Circular sector

At first it may appear that radian measure of angles is more complicated and less useful than degree measure. However, just the opposite is true. The computation of arc length and the area of a circular sector, which we now discuss, should begin to convince you of some of the advantages of radian measure over degree measure. Refer to Figure 4 in the following discussion.

From the definition of radian measure of an angle.

$$\theta = \frac{s}{r} \text{ rad}$$

Solving for s, we obtain a **formula for arc length:**

$$s = r\theta \qquad \theta \text{ in radian measure} \tag{1}$$

If θ is in degree measure, we must multiply by $\pi/180$ first (to convert to radians); then formula (1) becomes

$$s = \frac{\pi}{180} r\theta \qquad \theta \text{ in degree measure} \tag{2}$$

We see that the formula for arc length is much simpler when θ is in radian measure.

EXAMPLE 4 **Arc Length**

In a circle of radius 4.00 cm, find the arc length subtended by a central angle of:

(A) 3.40 rad (B) 10.0°

Solution (A) $s = r\theta$
$= 4.00(3.40) = 13.6$ cm

(B) $s = \dfrac{\pi}{180} r\theta$
$= \dfrac{\pi}{180}(4.00)(10.0) = 0.698$ cm ■

Matched Problem 4 In a circle of radius 6.00 ft, find the arc length subtended by a central angle of:

(A) 1.70 rad (B) 40.0° ■

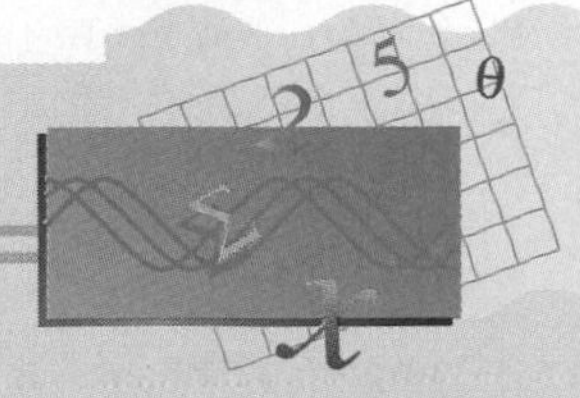

EXPLORE/DISCUSS 3

When you look at a full moon, the image appears on the retina of your eye as shown in Figure 5.

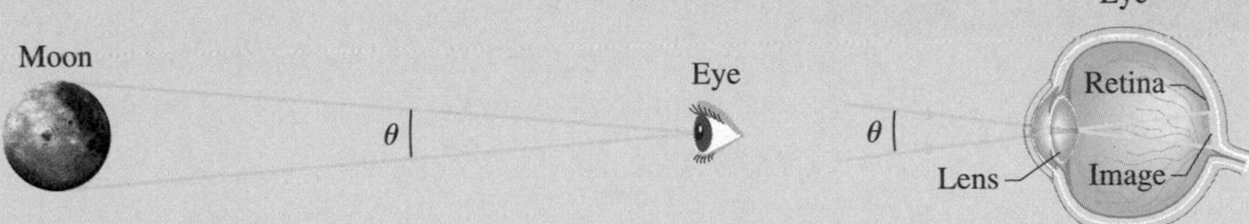

FIGURE 5
Vision

(A) Discuss how you would estimate the diameter of the moon's image d' on the retina, given the moon's diameter d, the distance u from the moon to the eye lens, and the distance v from the image to the eye lens. [Recall from Section 1.1: For small central angles in circles with very large radii, the intercepted arc and its chord are approximately the same length.]

(B) The moon's image on the retina is smaller than a period on this page! To verify this, find the diameter of the moon's image d' on the retina to two decimal places, given:

Moon diameter $= d = 2{,}160$ mi

Moon distance from the eye $= u = 239{,}000$ mi

Distance from eye lens to retina $= v = 17.4$ mm

The formula for the **area A of a circular sector** in a circle with radius r and central angle θ in radian measure (see Fig. 4) can be found by starting with the following proportion:

$$\frac{A}{\pi r^2} = \frac{\theta}{2\pi}$$

$$A = \frac{1}{2}r^2\theta \quad \theta \text{ in radian measure} \tag{3}$$

If θ is in degree measure, we must multiply by $\pi/180$ first (to convert to radians); then formula (3) becomes

$$A = \frac{\pi}{360} r^2\theta \qquad \theta \text{ in degree measure} \tag{4}$$

Again we see that the formula for sector area is much simpler when θ is in radian measure.

Area of a Sector

In a circle of radius 3 m, find the area (to three significant digits) of the circular sector with central angle:

(A) 0.4732 rad (B) 25°

Solution (A) $A = \frac{1}{2} r^2\theta$
$= \frac{1}{2}(3)^2(0.4732) = 2.13 \text{ m}^2$

(B) $A = \frac{\pi}{360} r^2\theta$
$= \frac{\pi}{360}(3)^2(25) = 1.96 \text{ m}^2$ ■

Matched Problem 5 In a circle of radius 7 in., find the area (to four significant digits) of the circular sector with central angle:

(A) 0.1332 rad (B) 110° ■

Angular velocity is another significant application of radian measure in engineering and physics problems. A detailed discussion of angular velocity is presented in the next section.

It is important to gain experience in the use of radian measure, since the concept will be used extensively in many developments that follow. By the time you finish this book you should feel as comfortable with radian measure as you now do with degree measure.

Answers to Matched Problems

1. (A) $180°/\pi$; 57.2958°
(B) $-2\pi/3$ rad; -2.0944 rad
(C)

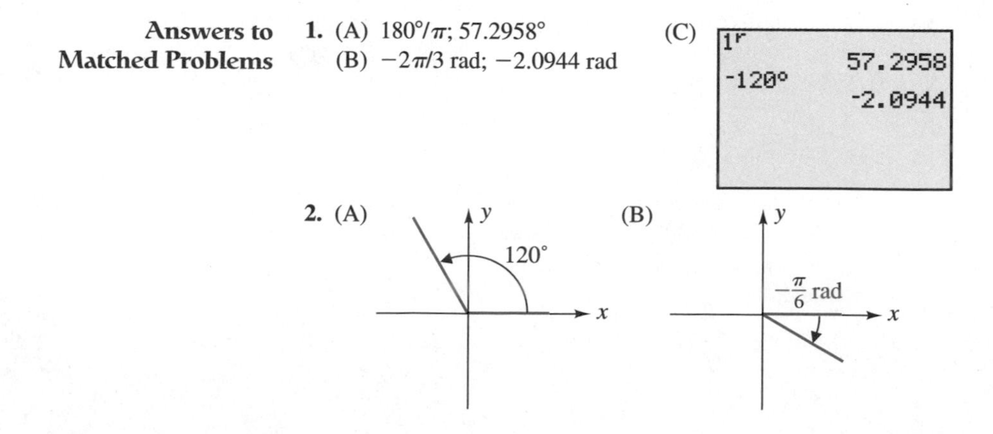

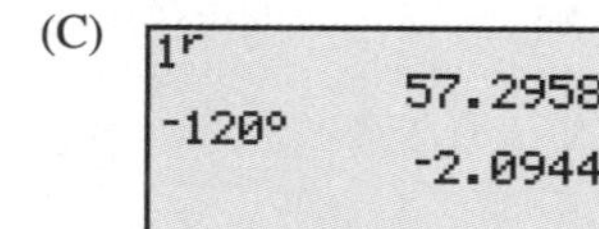

2. (A)

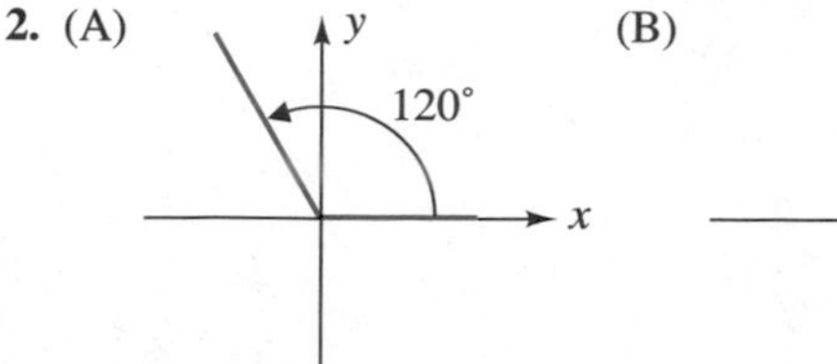

(B)

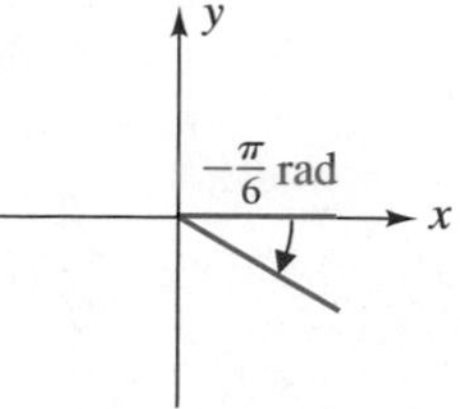

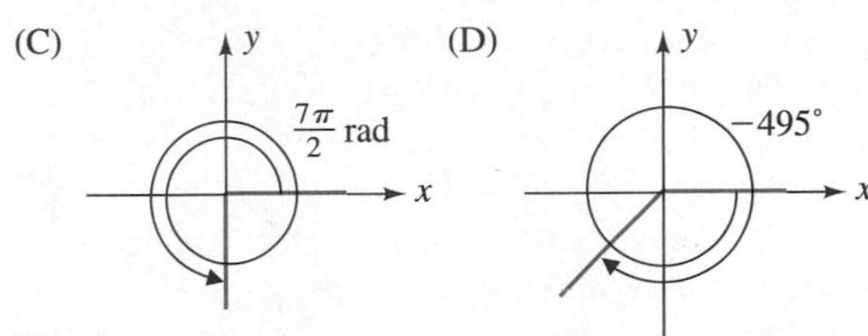

3. (A) Not coterminal (B) Coterminal (C) Coterminal (D) Not coterminal
4. (A) 10.2 ft (B) 4.19 ft 5. (A) 3.263 in.^{2} (B) 47.04 in.^{2}

EXERCISE 2.1

A

1. Explain the meaning of an angle of radian measure 1.

2. Explain the meaning of an angle of degree measure 1.

3. Mentally convert the following to exact radian measure by starting with 30° and taking multiples (remember that 180° corresponds to π radians):

30°, 60°, 90°, 120°, 150°, 180°, 210°, 240°, 270°, 300°, 330°, 360°

4. Mentally convert the following to exact radian measure by starting with 45° and taking multiples (remember that 180° corresponds to π radians):

45°, 90°, 135°, 180°, 225°, 270°, 315°, 360°

5. Which is larger: an angle of degree measure 50 or an angle of radian measure 1? Explain.

6. Which is smaller: an angle of degree measure 20 or an angle of radian measure $\frac{1}{2}$? Explain.

B

In Problems 7–10, sketch each angle in its standard position and find the degree measure of the two nearest angles (one negative and one positive) that are coterminal to the given angle.

7. 60° **8.** 45° **9.** − 30° **10.** − 45°

In Problems 11–14, find the radian measure for each angle. Express the answer in exact form and in approximate form to four significant digits.

11. 18° **12.** 9° **13.** 130° **14.** 140°

In Problems 15–18, find the degree measure for each angle. Express the answer in exact form and in approximate form with decimal degrees to four significant digits.

15. 1.6 rad **16.** 0.5 rad

17. $\pi/60$ rad **18.** $\pi/180$ rad

In Problems 19–24, sketch each angle in its standard position.

19. $-\pi/6$ rad **20.** $-\pi/3$ rad **21.** 300°

22. 390° **23.** $-7\pi/3$ rad **24.** $-11\pi/4$ rad

In Problems 25 and 26, use a calculator with an automatic radian–degree conversion routine.

25. (A) Find the degree measures of 8.30 rad and − 11.6 rad.
(B) Find the radian measures of 563° and − 1,230°.

26. (A) Find the degree measures of 12.04 rad and − 7.12 rad.
(B) Find the radian measures of 2,672° and − 431.8°.

27. If the radius of a circle is 4.0 m, find the radian measure of an angle subtended by an arc of length:
(A) 12 m (B) 18 m

28. If the radius of a circle is 4.0 cm, find the radian measure of an angle subtended by an arc of length:
(A) 6 cm (B) 4.5 cm

29. In a circle of radius 25.0 m, find the length of the arc subtended by a central angle of:
(A) 2.33 rad (B) 19.0°
(C) 0.821 rad (D) 108°

30. In a circle of radius 14.0 in., find the length of the arc subtended by a central angle of:
(A) 0.447 rad (B) 68.0°
(C) 2.68 rad (D) 212°

31. If the radian measure of an angle is doubled, is the degree measure of the same angle also doubled? Explain.

32. If the degree measure of an angle is cut in half, is the radian measure of the same angle also cut in half? Explain.

33. An arc length s on a circle is held constant while the radius of the circle r is doubled. Explain what happens to the central angle subtended by the arc.

34. The radius of a circle r is held constant while an arc length s on the circle is doubled. Explain what happens to the central angle subtended by the arc.

35. In a circle of radius 14.0 cm, find the area of the circular sector with central angle:
(A) 0.473 rad (B) 25.0°
(C) 1.02 rad (D) 112°

36. In a circle of radius 115 ft, find the area of the circular sector with central angle:
(A) 2.49 rad (B) 33.0°
(C) 0.382 rad (D) 204°

C 37. Explain why an angle in standard position and of radian measure m intercepts an arc of length m on a unit circle with center at the origin.

38. An angle in standard position intercepts an arc of length s on a unit circle with center at the origin. Explain why the radian measure of the angle is also s.

In which quadrant does the terminal side of each angle lie?*

39. 432° 40. 821° 41. $-14\pi/3$ rad
42. $-17\pi/4$ rad 43. 1,243° 44. $-942°$

Find each value to four decimal places. Convert radians to decimal degrees.

45. 57.3421° = ? rad 46. 103.2187° = ? rad
47. 0.3184 rad = ?° 48. 1.0394 rad = ?°
49. 26°23′14″ = ? rad 50. 179°3′43″ = ? rad

Applications

51. **Radian Measure** What is the radian measure of the smaller angle made by the hands of a clock at 2:30 (see the figure)? Express the answer in terms of π and as a decimal fraction to two decimal places.

Figure for 51 and 52

52. **Radian Measure** Repeat Problem 51 for 4:30.

53. **Pendulum** A clock has a pendulum 22 cm long. If it swings through an angle of 32°, how far does the bottom of the bob travel in one swing?

Figure for 53 and 54

54. **Pendulum** If the bob on the bottom of the 22 cm pendulum in Problem 53 traces a 9.5 cm arc on each swing, through what angle (in degrees) does the pendulum rotate on each swing?

55. **Engineering** Oil is pumped from some wells using a donkey pump as shown in the figure. Through how many

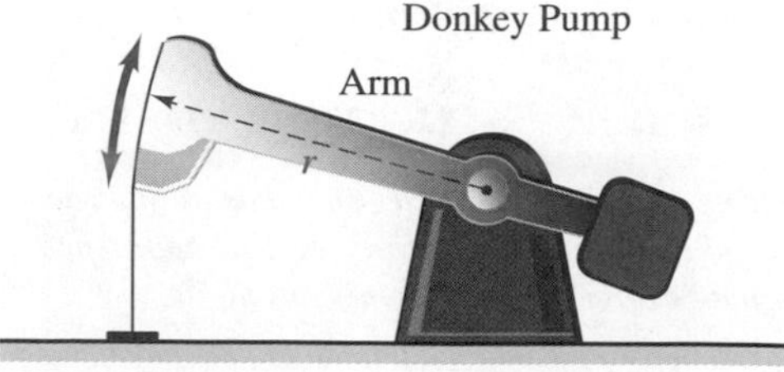

Figure for 55 and 56

* Recall that a rectangular coordinate system divides a plane into four parts called quadrants. These quadrants are numbered in a counterclockwise direction starting in the upper right-hand corner.

II	I
III	IV

degrees must an arm with a 72 in. radius rotate to produce a 24 in. vertical stroke at the pump down in the ground? Note that a point at the end of the arm must travel through a 24 in. arc to produce a 24 in. vertical stroke at the pump.

56. **Engineering** In Problem 55, find the arm length r that would produce an 18 in. vertical stroke while rotating through 21°.

57. **Bioengineering** A particular woman, when standing erect and facing forward, can swing an arm in a plane perpendicular to her shoulders through an angle of $3\pi/2$ rad (see the figure). Find the length of the arc (to the nearest centimeter) her fingertips trace out for one complete swing of her arm. The length of her arm from the pivot point in her shoulder to the end of her longest finger is 54.3 cm while it is kept straight and her fingers are extended with palm facing in.

Figure for 57

58. **Bioengineering** A particular man, when standing erect and facing forward, can swing a leg through an angle of $2\pi/3$ rad (see the figure). Find the length of the arc (to the nearest centimeter) his heel traces out for one complete swing of the leg. The length of his leg from the pivot point in his hip to the bottom of his heel is 102 cm while his leg is kept straight and his foot is kept at right angles to his leg.

Figure for 58

59. **Astronomy** The sun is about 1.5×10^8 km from the earth. If the angle subtended by the diameter of the sun on the surface of the earth is 9.3×10^{-3} rad, approximately what is the diameter of the sun? [*Hint:* Use the intercepted arc to approximate the diameter.]

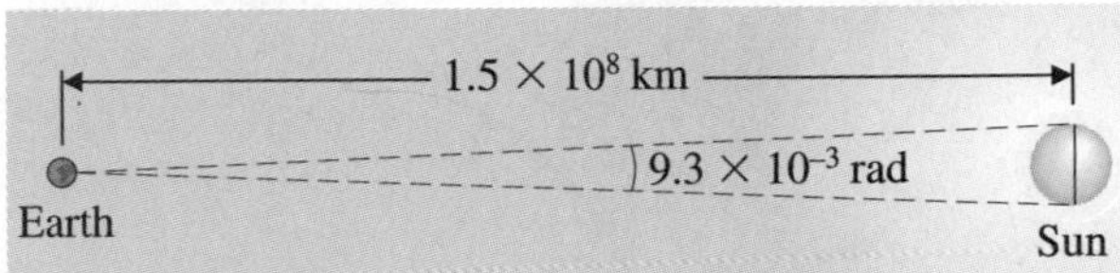

Figure for 59

* 60. **Surveying** If a natural gas tank 5.000 km away subtends an angle of 2.44°, approximate its height to the nearest meter. (See Problem 59.)

61. **Photography** The angle of view for a 300 mm telephoto lens is 8°. At 1,250 ft, what is the approximate width of the field of view? Use an arc length to approximate the chord length to the nearest foot.

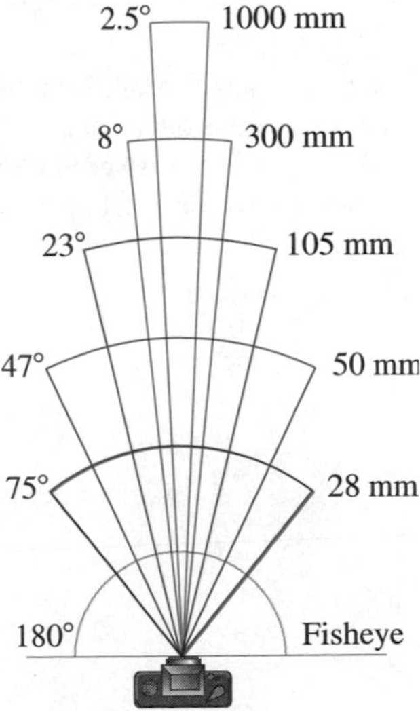

Figure for 61 and 62

62. **Photography** The angle of view for a 1,000 mm telephoto lens is 2.5°. At 865 ft, what is the approximate width of the field of view? Use an arc length to approximate the chord length to the nearest foot.

* 63. **Spy Satellites** Some spy satellites have cameras that can distinguish objects that subtend angles of as little as 5×10^{-7} rad. If such a satellite passes over a particular

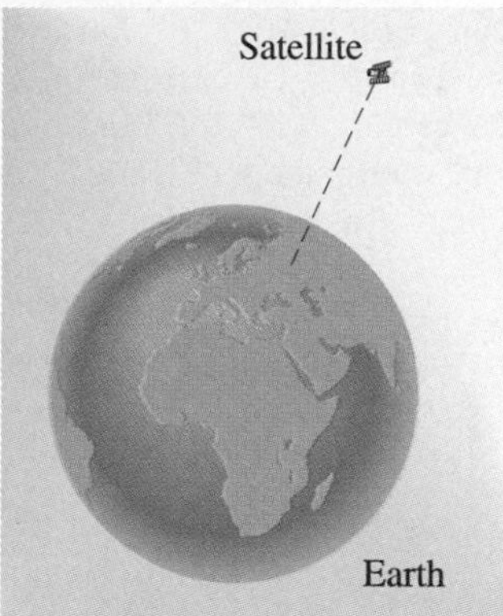

Figure for 63 and 64

country at an altitude of 250 mi, how small an object can the camera distinguish? Give the answer in meters to one decimal place and also in inches to one decimal place. (1 mi = 1,609 m; 1 m = 39.37 in.)

*64. **Spy Satellites** Repeat Problem 63 if the satellite passes over the country at an altitude of 155 mi. Give the answers to three decimal places.

65. **Astronomy** Assume that the earth's orbit is circular. A line from the earth to the sun sweeps out an angle of how many radians in 1 week? Express the answer in terms of π and as a decimal fraction to two decimal places. (Assume exactly 52 weeks in a year.)

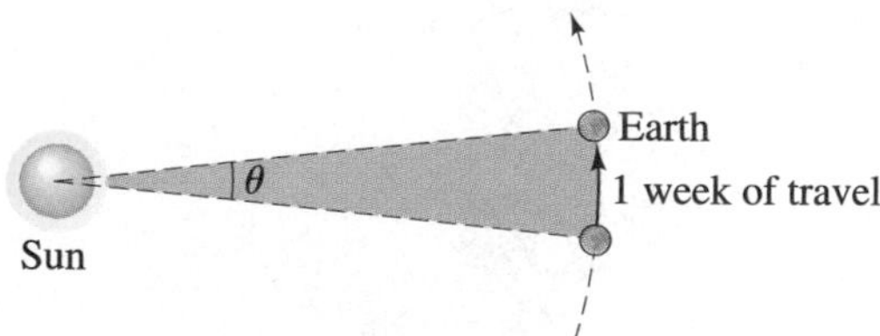

Figure for 65 and 66

66. **Astronomy** Repeat Problem 65 for 13 weeks.

*67. **Astronomy** When measuring time, an error of 1 sec per day may not seem like a lot. But suppose a clock is in error by at most 1 sec per day. Then in 1 year the accumulated error could be as much as 365 sec. If we assume the earth's orbit about the sun is circular, with a radius of 9.3×10^7 mi, what would be the maximum error (in miles) in computing the distance the earth travels in its orbit after 1 year?

*68. **Astronomy** Using the clock described in Problem 67, what would be the maximum error (in miles) in computing the distance that Venus travels in a "Venus year"? Assume Venus's orbit around the sun is circular, with a radius of 6.7×10^7 mi, and that Venus completes one orbit (a "Venus year") in 224 earth days.

*69. **Geometry** A circular sector has an area of 52.39 ft^2 and a radius of 10.5 ft. Calculate the perimeter of the sector to the nearest foot.

*70. **Geometry** A circular sector has an area of 145.7 cm^2 and a radius of 8.4 cm. Calculate the perimeter of the sector to the nearest centimeter.

*71. **Revolutions and Radians**
(A) Describe how you would find the number of radians generated by the spoke in a bicycle wheel that turns through n revolutions.
(B) Does your answer to part (A) depend on the size of the bicycle wheel? Explain.
(C) Find the number of radians generated for a wheel turning through 5 revolutions. Through 3.6 revolutions.

*72. **Revolutions and Radians**
(A) Describe how you would find the number of radians through which a 10 cm diameter pulley turns if u meters of rope are pulled through without slippage.
(B) Does your answer to part (A) depend on the diameter of the pulley? Explain.
(C) Through how many radians does the pulley in part (A) turn when 5.75 m of rope are pulled through without slippage?

*73. **Engineering** Rotation of a drive wheel causes a shaft to rotate (see the figure). If the drive wheel completes 3 revolutions, how many revolutions will the shaft complete? Through how many radians will the shaft turn? Compute answers to one decimal place.

*74. **Engineering** In Problem 73, find the radius (to the nearest millimeter) of the drive wheel required for the 12

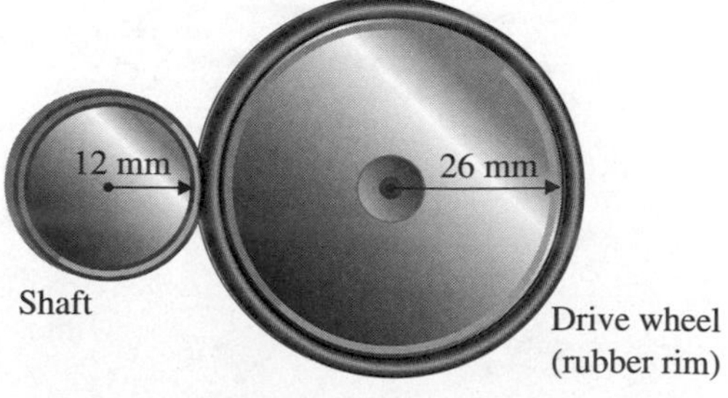

Figure for 73 and 74

mm shaft to make 7 revolutions when the drive wheel makes 3 revolutions.

75. **Radians and Arc Length** A bicycle wheel of diameter 32 in. travels a distance of 20 ft. Find the angle (to the nearest degree) swept out by one of the spokes.

76. **Radians and Arc Length** A bicycle has a front wheel with a diameter of 24 cm and a back wheel with a diameter of 60 cm. Through what angle (in radians) does the front wheel turn if the back wheel turns through 12 rad?

Figure for 75 and 76

☆2.2 LINEAR AND ANGULAR VELOCITY

We will put our notion of radian measure to use in defining two special kinds of velocity that involve rotating motion. These velocities, called **angular velocity** and **linear velocity on a circle,** are used extensively in engineering and physics.

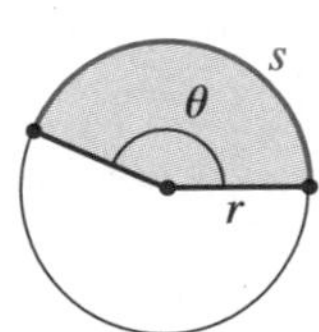

FIGURE 1

Starting with θ in radian measure (see Fig. 1 and recall Section 2.1),

$$\theta = \frac{s}{r}$$

we rewrite the formula in the equivalent form

$$s = r\theta \tag{1}$$

If we think of a point moving on the circumference of the circle with uniform speed, then a radial line from the center of the circle to this point will sweep out an angle θ at a uniform rate. Thus, if we divide both sides of equation (1) by time t, we obtain

$$\frac{s}{t} = r\frac{\theta}{t} \tag{2}$$

Now,

$$\begin{aligned}\frac{s}{t} &= \text{Change in arc length per unit change in time}\\ &= \text{Linear velocity of point on the circle}\\ &= V\end{aligned}$$

and

$$\begin{aligned}\frac{\theta}{t} &= \text{Change in angle in radian measure per unit change in time}\\ &= \text{Angular velocity}\\ &= \omega\end{aligned}$$

☆ Sections marked with a star may be omitted without loss of continuity.

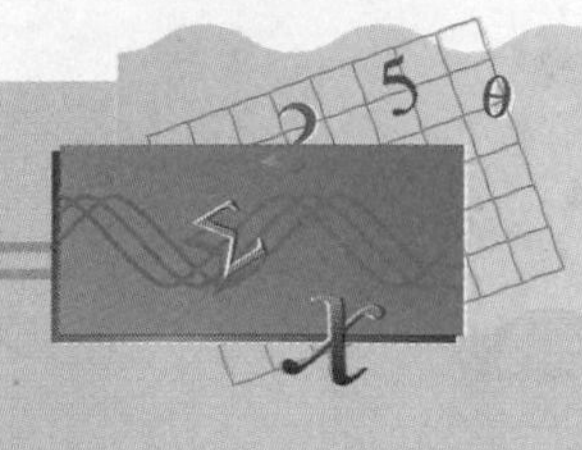

EXPLORE/DISCUSS 1

Substituting these last results into equation (2), show that

$$V = r\omega \qquad \text{and} \qquad \omega = \frac{V}{r}$$

Verbally explain what these formulas represent and how they might be used.

The above results are summarized in the following box for convenient reference:

LINEAR AND ANGULAR VELOCITY ON A CIRCLE

Angular velocity

$$\omega = \frac{\theta}{t} \qquad \omega = \frac{V}{r}$$ *Radians per unit time relative to a rotating circle*

Linear velocity

$$V = \frac{s}{t} \qquad V = r\omega$$ *Distance moved per unit time of a point on a rotating circle*

where r is the radius of the circle and s is arc length on the circle.

EXAMPLE 1 **Electrical Wind Generator**

An electrical wind generator (see Fig. 2) has propeller blades that are 5.00 m long. If the blades are rotating at 8π rad/sec, what is the linear velocity (to the nearest meter per second) of a point on the tip of one of the blades?

Solution

$$\begin{aligned} V &= r\omega \\ &= 5.00(8\pi) \\ &= 126 \text{ m/sec} \end{aligned}$$

■

Matched Problem 1 If a 3.0 ft diameter wheel turns at 12 rad/min, what is the velocity of a point on the wheel (in feet per minute)? ■

FIGURE 2
Electrical wind generator

EXAMPLE 2

Angular Velocity

A point on the rim of a 6.0 in. diameter wheel is traveling at 75 ft/sec. What is the angular velocity of the wheel (in radians per second)?

Solution

$$\omega = \frac{V}{r}$$

$$= \frac{75}{0.25} = 300 \text{ rad/sec} \qquad [\textit{Note:} \ 3.0 \text{ in.} = 0.25 \text{ ft}]$$

Matched Problem 2 A point on the rim of a 4.00 in. diameter wheel is traveling at 88.0 ft/sec. What is the angular velocity of the wheel (in radians per second)?

EXAMPLE 3

Linear Velocity

If a 6 cm shaft is rotating at 4,000 rpm (revolutions per minute), what is the speed of a particle on its surface (in centimeters per minute, to two significant digits)?

Solution Since 1 revolution is equivalent to 2π rad, we multiply 4,000 by 2π to obtain the angular velocity of the shaft in radians per minute:

$$\omega = 8{,}000\pi \text{ rad/min}$$

Now we use $V = r\omega$ to complete the solution:

$$V = 3(8{,}000\pi) = 75{,}000 \text{ cm/min}$$

Matched Problem 3 If an 8 cm diameter drive shaft in a boat is rotating at 350 rpm, what is the speed of a particle on its surface (in centimeters per second, to three significant digits)?

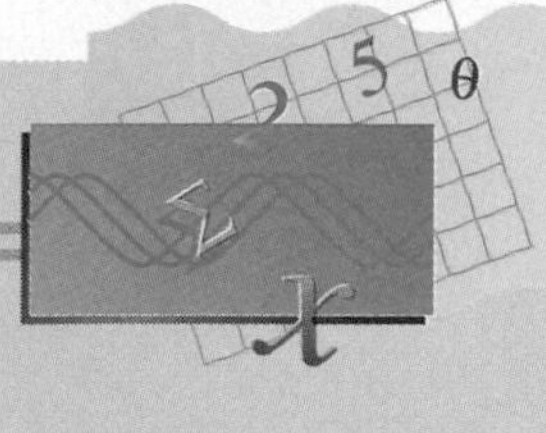

EXPLORE/DISCUSS 2

The earth follows an elliptical path around the sun while rotating on its axis. Relative to the sun, the earth completes one rotation every 24 hr, which is called the **mean solar day.** However, relative to certain fixed stars used in astronomy, the average time it takes the earth to rotate about its axis, called the **mean sidereal day,** is 23 hr 56 min 4.091 sec of mean solar time. The earth's radius at the equator is 3,963.205 mi. Use a mean sidereal day in the following discussions.

(A) Discuss how you would find the angular velocity ω of the earth's rotation in radians per hour; then find ω to three decimal places.

(B) Explain how you would find how fast a person standing on the equator is moving in miles per hour; then find this linear velocity.

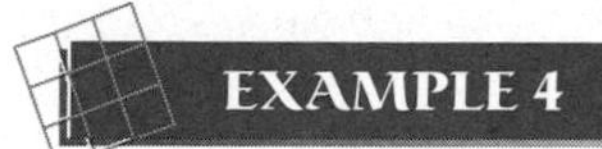

EXAMPLE 4 **Hubble Space Telescope**

The 25,000 lb Hubble space telescope (see Fig. 3) was launched April 1990 and placed in a 380 mi circular orbit above the earth's surface. It completes one orbit every 97 min, going from a dawn-to-dusk cycle nearly 15 times a day. If the radius of the earth is 3,964 mi, what is the linear velocity of the space telescope in miles per hour (mph)?

Solution The telescope completes 1 revolution (2π rad) in

$$\frac{97}{60}\text{ hr} = 1.6\text{ hr}$$

The angular velocity generated by the space telescope relative to the center of the earth is

$$\omega = \frac{\theta}{t} = \frac{2\pi}{1.6} = 3.9\text{ rad/hr}$$

The linear velocity of the telescope is

$$\begin{aligned} V &= r\omega \\ &= (3{,}964 + 380)(3.9) \\ &= 17{,}000\text{ mph} \end{aligned}$$

■

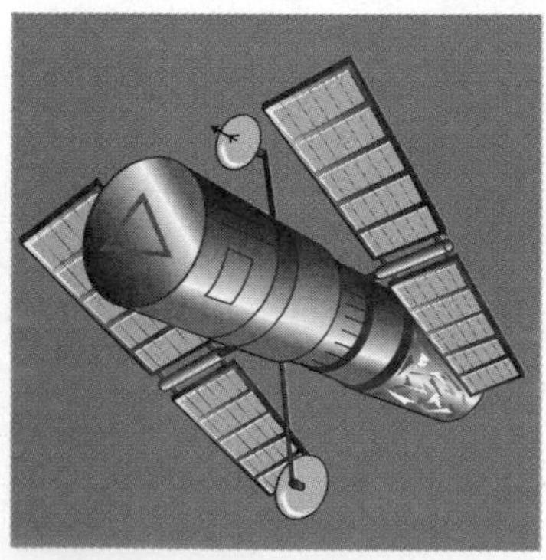

FIGURE 3
Hubble space telescope

Matched Problem 4 The space shuttle *Columbia* was placed in a circular orbit 250 mi above the earth's surface. One orbit is completed in 1.51 hr. If the radius of the earth is

3,964 mi, what is the linear velocity of the shuttle in miles per hour (to three significant digits)? ■

Answers to Matched Problems **1.** 18 ft/min **2.** 528 rad/sec **3.** 147 cm/sec **4.** 17,500 mph

EXERCISE 2.2

A *Find the velocity V of a point on the rim of a wheel, given the indicated information.*

1. $r = 6$ mm, $\omega = 0.5$ rad/sec

2. $r = 4{,}000$ cm, $\omega = 0.05$ rad/hr

Find the angular velocity ω given the following information.

3. $r = 6.0$ cm, $V = 102$ cm/sec

4. $r = 250$ km, $V = 500$ km/hr

B **5.** Describe the meaning of angular velocity.

6. Describe the meaning of linear velocity of a point on a rotating circle.

Find the angular velocity of a wheel turning through θ radians in time t.

7. $\theta = 2\pi$ rad, $t = 1.7$ hr

8. $\theta = 2\pi$ rad, $t = 3.04$ sec

9. $\theta = 8.07$ rad, $t = 13.6$ sec

10. $\theta = 13.67$ rad, $t = 21.03$ sec

Applications

11. Engineering A 16 mm diameter shaft rotates at 1,500 rps (revolutions per second). Find the speed of a particle on its surface (to the nearest meter per second).

12. Engineering A 6 cm diameter shaft rotates at 500 rps. Find the speed of a particle on its surface (to the nearest meter per second).

13. Space Science An earth satellite travels in a circular orbit at 20,000 mph. If the radius of the orbit is 4,300 mi, what angular velocity (in radians per hour, to three significant digits) is generated?

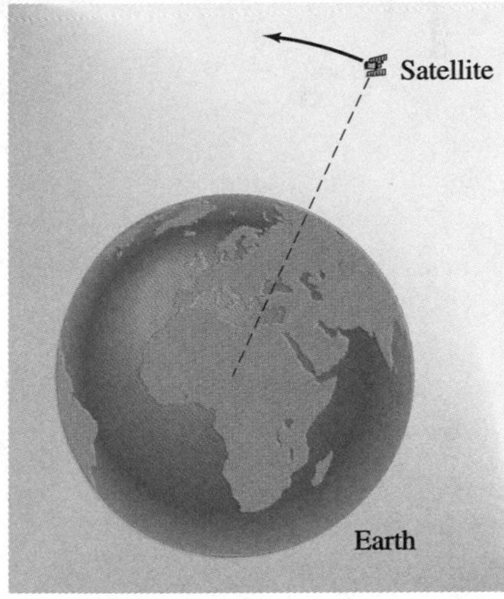

Figure for 13

14. Engineering A bicycle is ridden at a speed of 7.0 m/sec. If the wheel diameter is 64 cm, what is the angular velocity of the wheel in radians per second?

15. Physics The velocity of sound in air is approx. 335.3 m/sec. If an airplane has a 3.000 m diameter propeller, at what angular velocity will its tip pass through the sound barrier?

16. Physics If an electron in an atom travels around the nucleus in a circular orbit (see the figure) at 8.11×10^6

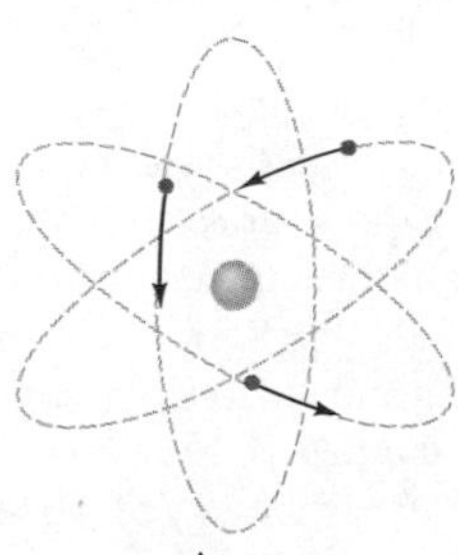

cm/sec, what angular velocity (in radians per second) does it generate, assuming the radius of the orbit is 5.00×10^{-9} cm?

17. **Astronomy** The earth revolves about the sun in an orbit that is approximately circular with a radius of 9.3×10^7 mi (see the figure). The radius of the orbit sweeps out an angle with what exact angular velocity (in radians per hour)? How fast (to the nearest hundred miles per hour) is the earth traveling along its orbit?

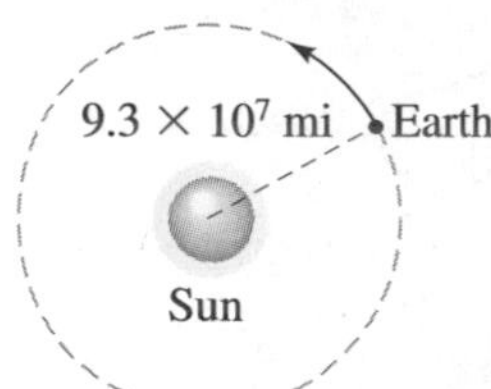

Velocity of the earth

Figure for 17

18. **Astronomy** Take into consideration only the daily rotation of the earth to find out how fast (in miles per hour) a person halfway between the equator and North Pole would be moving. The radius of the earth is approx. 3,964 mi, and a daily rotation takes 23.93 hr.

19. **Astronomy** Jupiter makes one full revolution about its axis every 9 hr 55 min. If Jupiter's equatorial diameter is 88,700 mi.:
   (A) What is its angular velocity relative to its axis of rotation (in radians per hour)?
   (B) What is the linear velocity of a point on Jupiter's equator?

20. **Astronomy** The sun makes one full revolution about its axis every 27.0 days. Assume 1 day = 24 hr. If its equatorial diameter is 865,400 mi:
   (A) What is the sun's angular velocity relative to its axis of rotation (in radians per hour)?
   (B) What is the linear velocity of a point on the sun's equator?

21. **Space Science** For an earth satellite to stay in orbit over a given stationary spot on earth, it must be placed in orbit 22,300 mi above the earth's surface (see the figure). It will then take the satellite the same time to complete one orbit as the earth, 23.93 hr. Such satellites are called **geostationary satellites** and are used for communications and tracking space shuttles. If the radius of the earth is 3,964 mi, what is the linear velocity of a geostationary satellite?

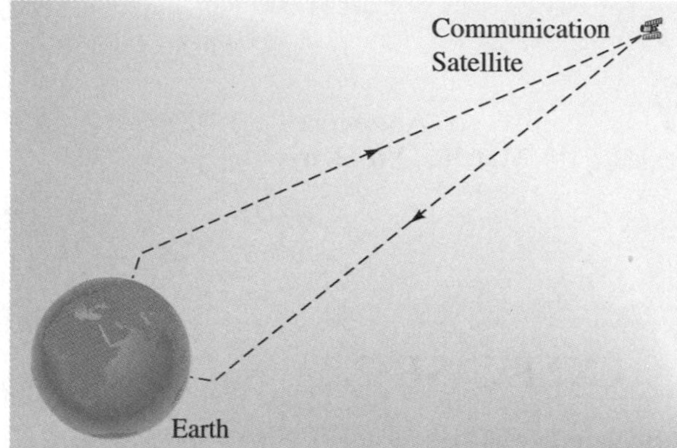

Figure for 21

22. **Astronomy** Until 1999 the planet Neptune is the planet farthest from the sun, 2.795×10^9 mi. (After 1999 and for the next 228 years Pluto will have this honor.) If Neptune takes 164 years to complete one orbit, what is its linear velocity in miles per hour?

*23. **Space Science** The earth rotates on its axis once every 23.93 hr, and a space shuttle revolves around the earth in the plane of the earth's equator once every 1.51 hr. Both are rotating in the same direction (see the figure). What is the length of time between consecutive passages of the shuttle over a particular point P on the equator? [*Hint:* The shuttle will make one complete revolution (2π rad) and a little more to be over the same point P again. Thus, the central angle generated by point P on earth must equal the central angle generated by the space shuttle minus 2π.]

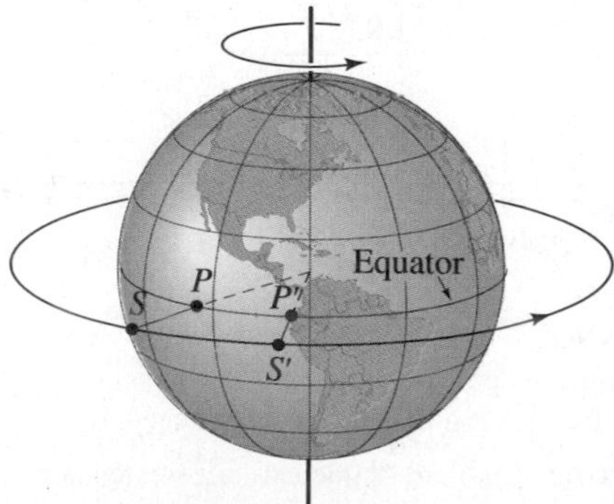

Figure for 23

*24. **Astronomy** One of the moons of Jupiter rotates around the planet in its equatorial plane once every 42 hr 30 min. Jupiter rotates around its axis once every 9 hr 55 min. Both are rotating in the same direction. What is the length of time between consecutive passages of the moon over the same point on Jupiter's equator? [See the hint for Problem 23.]

* **25. Precalculus: Rotating Beacon** A beacon light 15 ft from a wall rotates clockwise at the rate of exactly 1 rps (see the figure). To answer the following questions, start counting time (in seconds) when the light spot is at C.

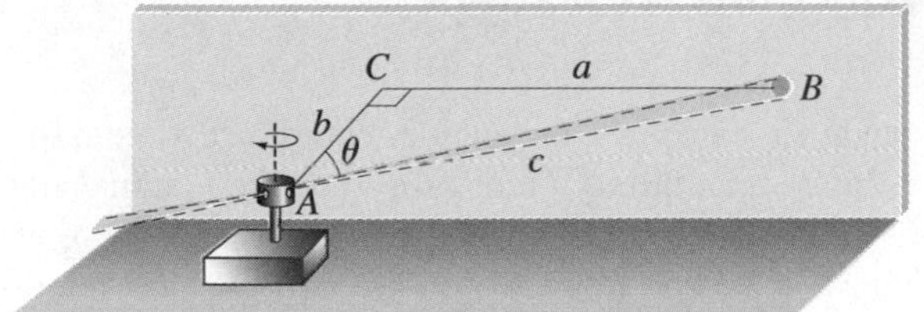

Figure for 25 and 26

(A) Describe how you can represent θ in terms of time t; then represent θ in terms of t.

(B) Describe how you can represent a, the distance the light travels along the wall, in terms of t; then represent a in terms of t.

(C) Complete Table 1, relating a and t, to two decimal places. (If your calculator has a table-producing capability, use it.) What does Table 1 seem to tell you about the motion of the light spot on the wall as t increases from 0.00 to 0.24? What happens to a when $t = 0.25$?

TABLE 1

t (sec)	0.00	0.04	0.08	0.12	0.16	0.20	0.24
a (ft)		3.85					

* **26. Precalculus: Rotating Beacon**

(A) Referring to the rotating beacon in Problem 25, describe how you can represent c, the length of the light beam, in terms of t; then represent c in terms of t.

(B) Complete Table 2, relating c and t, to two decimal places. (If your calculator has a table-producing capability, use it.) What does Table 2 seem to tell you about the rate of change of the length of the light beam as t increases from 0.00 to 0.24? What happens to c when $t = 0.25$?

TABLE 2

t (sec)	0.00	0.04	0.08	0.12	0.16	0.20	0.24
c (ft)		15.49					

2.3 TRIGONOMETRIC FUNCTIONS*

- Trigonometric Functions with Angle Domains
- Trigonometric Functions with Real Number Domains
- Calculator Evaluation
- Application
- Summary of Sign Properties

In Chapter 1 we introduced the concept of trigonometric ratios and tied this idea to right triangles. We were able to use these ratios to define six trigonometric functions with angle domains restricted to 0°–90°. In this section we introduce more general definitions that will apply to angle domains of arbitrary size, positive, negative, or zero, in degree or radian measure. We then move one (giant)

* A brief review of Appendix B.1 on functions might prove helpful before starting this section.

step further and define these functions for arbitrary real numbers. With these new functions we will be able to do everything we did with the trigonometric ratios in the first chapter, plus a great deal more. In Section 2.6 we will approach the subject from a more modern point of view, where angles are not a necessary part of the definition. Each approach has its advantages for certain applications and uses.

Trigonometric Functions with Angle Domains

We start with an arbitrary angle θ located in a rectangular coordinate system in a standard position. We then choose an arbitrary point $P(a, b)$ on the terminal side of θ, but away from the origin. If r is the distance of $P(a, b)$ from the origin, we can form six ratios involving r and the coordinates of P. We will use these six ratios to define six trigonometric functions, which are direct generalizations of the six trigonometric ratios given in Section 1.3. It is important to understand the definitions of the six trigonometric functions—a great deal depends on them.

Definition 1

TRIGONOMETRIC FUNCTIONS WITH ANGLE DOMAINS

For an arbitrary angle θ:

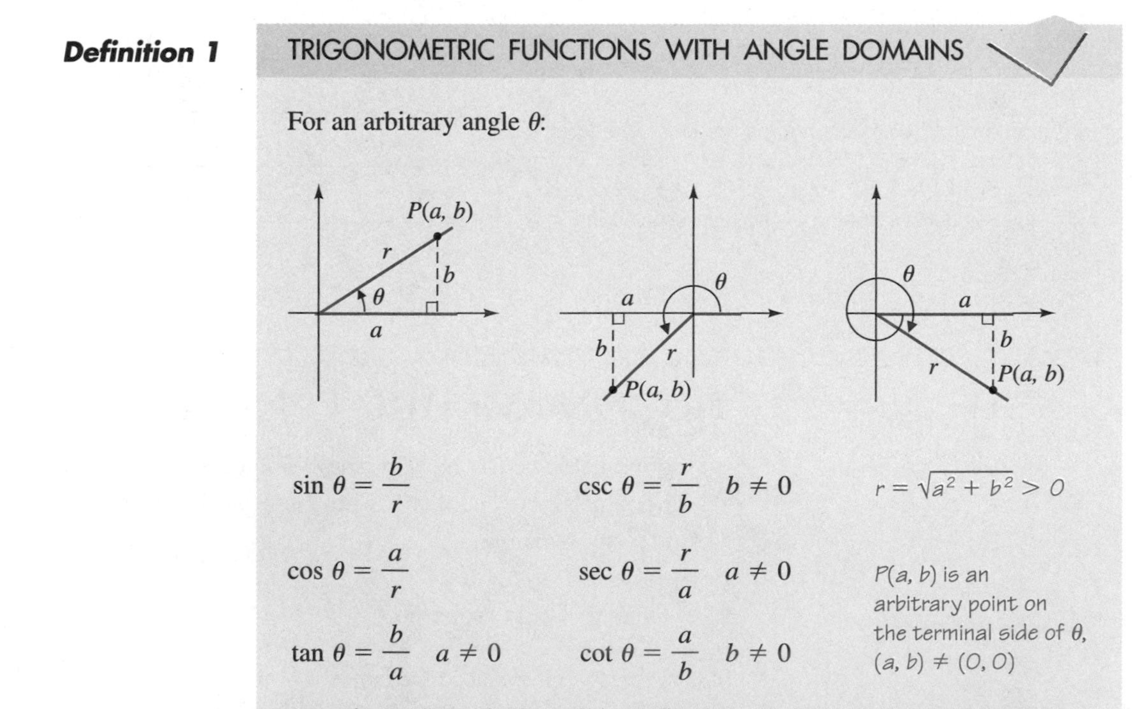

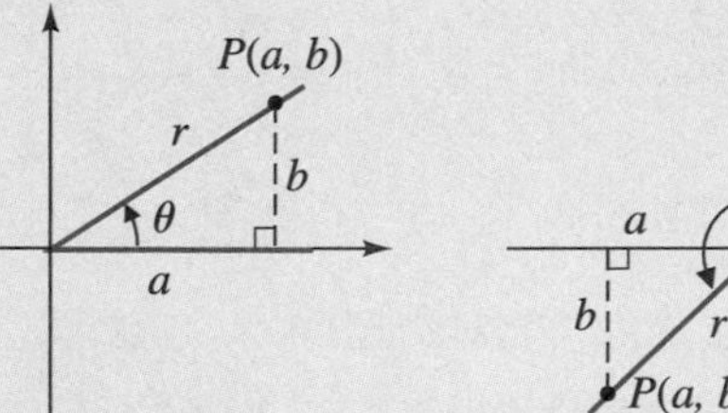

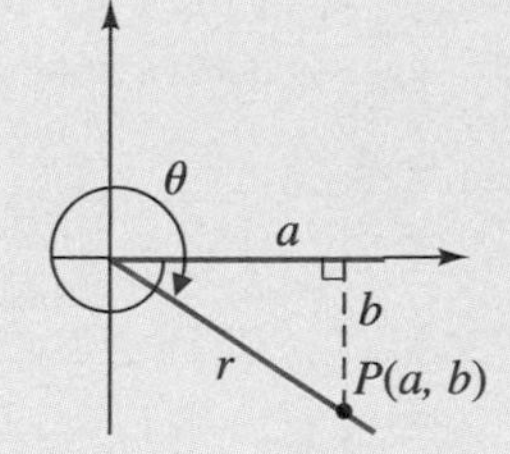

$$\sin\theta = \frac{b}{r} \qquad \csc\theta = \frac{r}{b} \quad b \neq 0 \qquad r = \sqrt{a^2 + b^2} > 0$$

$$\cos\theta = \frac{a}{r} \qquad \sec\theta = \frac{r}{a} \quad a \neq 0$$

$$\tan\theta = \frac{b}{a} \quad a \neq 0 \qquad \cot\theta = \frac{a}{b} \quad b \neq 0$$

$P(a, b)$ is an arbitrary point on the terminal side of θ, $(a, b) \neq (0, 0)$

Domains: Sets of all possible angles for which the ratios are defined
Ranges: Subsets of the set of real numbers

[*Note:* A trigonometric function has the same value at coterminal angles. Also, more precise statements about domains and ranges will be given in Chapter 3.]

Remarks

1. Definition 1 of the six trigonometric functions is stated in terms of an a,b coordinate system instead of an x,y coordinate system because when we generalize the definition of trigonometric functions still further (later in this section and in Section 2.6), we want to reserve x for an independent variable and y for a dependent variable.
2. In Definition 1 the right triangle formed by dropping a perpendicular from $P(a, b)$ to the horizontal axis is called the **reference triangle** associated with the angle θ. (A more complete discussion of reference triangles is given in Section 2.5.) We label the legs of the reference triangle with the coordinates of P, which means that one or both legs are labeled with negative numbers if P is in quadrant II, III, or IV. □

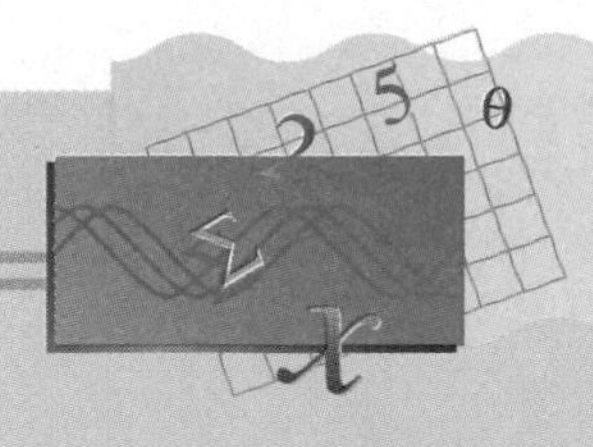

EXPLORE/DISCUSS 1

(A) Discuss why, for a given θ, the ratios in Definition 1 are independent of the choice of $P(a, b)$ on the terminal side of θ, as long as $(a, b) \neq (0, 0)$.

(B) Explain how Definition 1 includes, as a special case, the trigonometric functions of acute angles discussed in Section 1.3.

(C) For a given angle θ, discuss the relationship between the sign of a ratio in Definition 1 and the quadrant in which the terminal side of θ lies.

EXAMPLE 1 **Evaluating Trigonometric Functions, Given a Point on the Terminal Side of θ**

Find the exact value of each of the six trigonometric functions for the angle θ with terminal side containing $P(-4, -3)$. (See Fig. 1. Note that the legs of the reference triangle are labeled with negative numbers—see Remark 2 above.)

Solution

$(a, b) = (-4, -3)$

$r = \sqrt{a^2 + b^2} = \sqrt{(-4)^2 + (-3)^2} = \sqrt{25} = 5$

$$\sin\theta = \frac{b}{r} = \frac{-3}{5} = -\frac{3}{5} \qquad \csc\theta = \frac{r}{b} = \frac{5}{-3} = -\frac{5}{3}$$

$$\cos\theta = \frac{a}{r} = \frac{-4}{5} = -\frac{4}{5} \qquad \sec\theta = \frac{r}{a} = \frac{5}{-4} = -\frac{5}{4}$$

$$\tan\theta = \frac{b}{a} = \frac{-3}{-4} = \frac{3}{4} \qquad \cot\theta = \frac{a}{b} = \frac{-4}{-3} = \frac{4}{3}$$

■

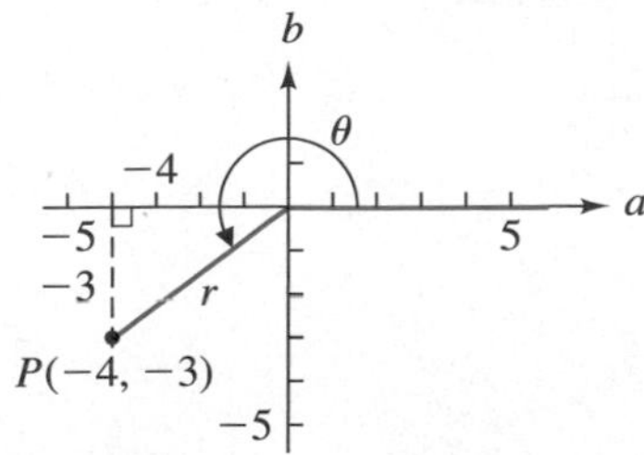

FIGURE 1

Matched Problem 1 Find the exact value of each of the six trigonometric functions if the terminal side of θ contains the point $(-8, -6)$. [*Note:* This point lies on the terminal side of the same angle as in Example 1.] ■

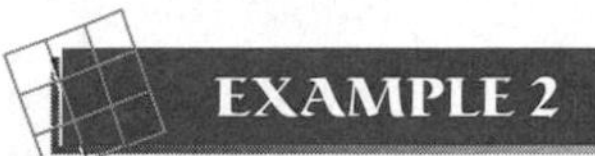

EXAMPLE 2

Using Given Information to Evaluate Trigonometric Functions

Find the exact value of each of the other five trigonometric functions for the given angle θ—without finding θ—given that the terminal side of θ is in quadrant III and

$$\cos\theta = \frac{-3}{5}$$

Solution The information given is sufficient for us to locate a reference triangle in quadrant III for θ even though we do not know θ (see Fig. 2). We sketch the reference triangle, label what we know, and then complete the problem as indicated.

Since

$$\cos\theta = \frac{a}{r} = \frac{-3}{5}$$

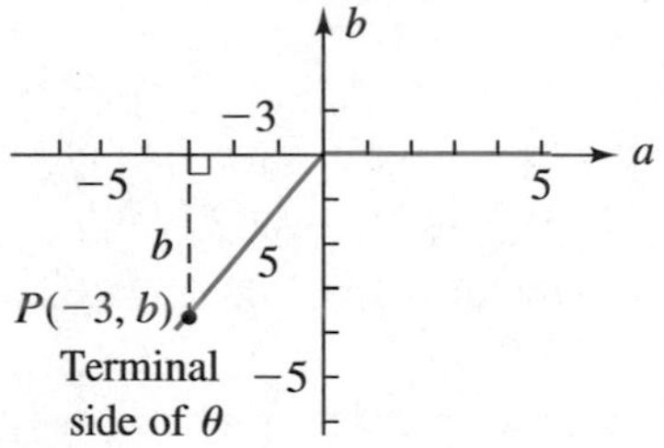

we know that $a = -3$ and $r = 5$ (r is never negative). If we can find b, we can determine the values of the other five functions using their definitions.

We use the Pythagorean theorem to find b:

$$(-3)^2 + b^2 = 5^2$$

$$b^2 = 25 - 9 = 16$$

b is negative since P(a, b) is in quadrant III.

$$b = -4$$

FIGURE 2

Thus,

$$(a, b) = (-3, -4) \qquad \text{and} \qquad r = 5$$

We can now find the other five functions using their definitions.

$$\sin\theta = \frac{b}{r} = \frac{-4}{5} = -\frac{4}{5} \qquad \csc\theta = \frac{r}{b} = \frac{5}{-4} = -\frac{5}{4}$$

$$\tan\theta = \frac{b}{a} = \frac{-4}{-3} = \frac{4}{3} \qquad \cot\theta = \frac{a}{b} = \frac{-3}{-4} = \frac{3}{4}$$

$$\sec\theta = \frac{r}{a} = \frac{5}{-3} = -\frac{5}{3}$$

■

Matched Problem 2 Repeat Example 2 for $\tan\theta = -\frac{3}{4}$ and the terminal side of θ in quadrant II. ■

Trigonometric Functions with Real Number Domains

We now turn to the problem of defining trigonometric functions for real number domains. First note that to each real number x there corresponds an angle of x radians, and to each angle of x radians there corresponds the real number x. We define trigonometric functions with real number domains in terms of trigonometric functions with angle domains.

Definition 2

TRIGONOMETRIC FUNCTIONS WITH REAL NUMBER DOMAINS

For x any real number:

$$\sin x = \sin(x \text{ rad}) \qquad \csc x = \csc(x \text{ rad})$$

$$\cos x = \cos(x \text{ rad}) \qquad \sec x = \sec(x \text{ rad})$$

$$\tan x = \tan(x \text{ rad}) \qquad \cot x = \cot(x \text{ rad})$$

Domains: Subsets of the set of real numbers
Ranges: Subsets of the set of real numbers

Thus, for example, $\sin 3 = \sin(3 \text{ rad})$, $\cos 1.23 = \cos(1.23 \text{ rad})$, $\tan(-9) = \tan(-9 \text{ rad})$, and so on.

Remark Because of Definition 2, we will often omit "rad" after x and interpret x as a real number or an angle with radian measure x, whichever fits the context in which x appears. □

At first glance, the definition of trigonometric functions with real number domains (Definition 2) appears artificial, but we will see that it frees the trigonometric functions from angles and opens them up to a large variety of significant applications not directly connected to angles.

Calculator Evaluation

We used a calculator in Section 1.3 to approximate trigonometric ratios for acute angles in degree measure. These same calculators are internally programmed to approximate (to eight or ten significant digits) trigonometric functions for *any* angle (however large or small, positive or negative) in degree or radian measure, or for *any* real number. (Remember, most graphing calculators use different sequences of steps than scientific calculators. Consult the owner's manual for your calculator.) In Section 2.5 we will show how to obtain exact values for certain special angles (integer multiples of 30° and 45° or integer multiples of $\pi/6$ and $\pi/4$) without the use of a calculator.

Caution

1. Set your calculator in **degree mode** when evaluating trigonometric functions of angles in degree measure.
2. Set your calculator in **radian mode** when evaluating trigonometric functions of angles in radian measure or trigonometric functions of real numbers. □

We generalize the reciprocal relationships stated in Section 1.3 to evaluate secant, cosecant, and cotangent.

RECIPROCAL RELATIONSHIPS

For x any real number or angle in degree or radian measure,

$$\csc x = \frac{1}{\sin x} \qquad \sin x \neq 0$$

$$\sec x = \frac{1}{\cos x} \qquad \cos x \neq 0$$

$$\cot x = \frac{1}{\tan x} \qquad \tan x \neq 0$$

EXAMPLE 3

Calculator Evaluation of Trigonometric Functions

With a calculator, evaluate each to four significant digits. [*Note:* When evaluating trigonometric functions with angle domains in degree measure, some calculators require decimal degrees (DD) and others can use either DD or DMS. Check your user's manual.]

(A) $\sin 286.38°$ (B) $\tan(3.472 \text{ rad})$ (C) $\cot 5.063$

(D) $\cos(-107°35')$ (E) $\sec(-4.799)$ (F) $\csc 192°47'22''$

Solution

(A) $\sin 286.38° = -0.9594$ — Degree mode

(B) $\tan(3.472 \text{ rad}) = 0.3430$ — Radian mode

(C) $\cot 5.063 = \dfrac{1}{\tan 5.063}$

$= -0.3657$ — Radian mode

(D) $\cos(-107°35') = \cos(-107.5833\ldots)$

$= -0.3021$ — Degree mode

(E) $\sec(-4.799) = \dfrac{1}{\cos(-4.799)}$

$= 11.56$ — Radian mode

(F) $\csc 192°47'22'' = \dfrac{1}{\sin 192.7894\ldots}$

$= -4.517$ — Degree mode ■

Matched Problem 3 With a calculator, evaluate to four significant digits:

(A) $\cos 303.73°$ (B) $\sec(-2.805)$ (C) $\tan(-83°29')$

(D) $\sin(12 \text{ rad})$ (E) $\csc 100°52'43''$ (F) $\cot 9$ ■

■ Application

An application of trigonometric functions that does not involve any angles will now be considered. More will be said about alternating current in subsequent sections.

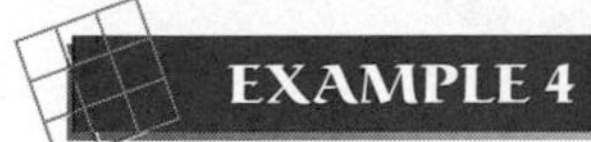

EXAMPLE 4 **Wind Generators**

The Department of Energy reports that wind-produced electricity will jump to close to 1% of the nation's total electrical output by the year 2010, about ten times that produced today. "Wind farms" are springing up in many parts of the United States (see Fig. 3). A particular wind generator can generate alternating current given by the equation

$$I = 50\cos(120\pi t + 45\pi)$$

where t is time in seconds and I is current in amperes. What is the current I (to two decimal places) when $t = 1.09$ sec? (More will be said about alternating current in subsequent sections.)

FIGURE 3
Wind generators

Solution Set the calculator in radian mode; then evaluate the equation for $t = 1.09$:

$$I = 50\cos[120\pi(1.09) + 45\pi] = 40.45 \text{ amperes}$$ ■

Matched Problem 4 Repeat Example 4 for $t = 2.17$ sec. ■

Summary of Sign Properties

We close this important section by having you summarize the sign properties of the six trigonometric functions in Table 1 in Explore–Discuss 2. Note that Table 1 does not need to be committed to memory, because particular cases are readily determined from the definitions of the functions involved (see Fig. 4).

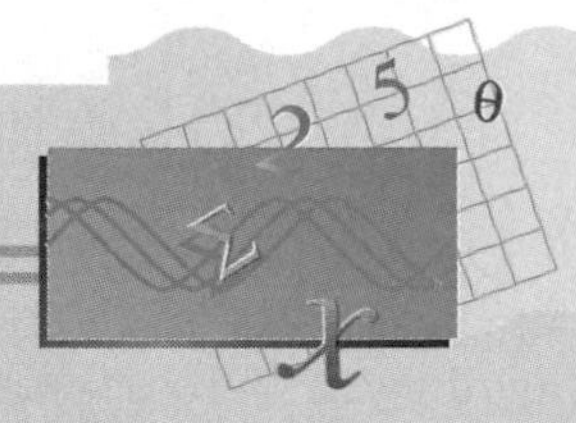

EXPLORE/DISCUSS 2

Using Figure 4 as an aid, complete Table 1 by determining the sign of each of the six trigonometric functions in each quadrant. In Table 1, x is associated with an angle that terminates in the respective quadrant, (a, b) is a point on the terminal side of the angle, and $r = \sqrt{a^2 + b^2} > 0$.

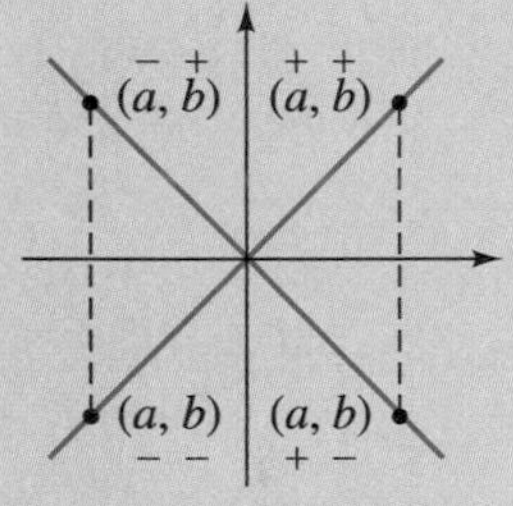

FIGURE 4

TABLE 1

	Quadrant I	Quadrant II	Quadrant III	Quadrant IV
	a b r + + +	a b r − + +	a b r − − +	a b r + − +
$\sin x = b/r$ $\csc x = r/b$				−
$\cos x = a/r$ $\sec x = r/a$				
$\tan x = b/a$ $\cot x = a/b$	+			

Answers to Matched Problems

1. $\sin \theta = -\frac{3}{5}$, $\cos \theta = -\frac{4}{5}$, $\tan \theta = \frac{3}{4}$, $\csc \theta = -\frac{5}{3}$, $\sec \theta = -\frac{5}{4}$, $\cot \theta = \frac{4}{3}$
[What do you think the values of the six trigonometric functions would be if we picked another point, say $(-12, -9)$, on the terminal side of the same angle θ as in Example 1? *Answer:* The same. (Why?)]
2. $\sin \theta = \frac{3}{5}$, $\csc \theta = \frac{5}{3}$, $\cot \theta = -\frac{4}{3}$, $\cos \theta = -\frac{4}{5}$, $\sec \theta = -\frac{5}{4}$
3. (A) 0.5553 (B) −1.059 (C) −8.754 (D) −0.5366
(E) 1.018 (F) −2.211
4. $I = -15.45$ amperes

EXERCISE 2.3

A *Find the exact value of each of the six trigonometric functions if the terminal side of θ contains the point $P(a, b)$. Do the same for $Q(a, b)$.*

1. $P(3, 4); Q(6, 8)$ **2.** $P(-3, -4); Q(-9, -12)$

3. $P(4, -3); Q(12, -9)$ **4.** $P(-3, 4); Q(-6, 8)$

In Problems 5–10, find the exact value of each of the other five trigonometric functions for the angle θ (without finding θ), given the indicated information. It would be helpful to sketch a reference triangle.

5. $\cos\theta = \frac{3}{5}$
θ is a quadrant I angle

6. $\sin\theta = \frac{3}{5}$
θ is a quadrant I angle

7. $\cos\theta = \frac{3}{5}$
θ is a quadrant IV angle

8. $\sin\theta = \frac{3}{5}$
θ is a quadrant II angle

9. $\csc\theta = -\frac{5}{4}$
θ is a quadrant III angle

10. $\tan\theta = -\frac{4}{3}$
θ is a quadrant II angle

11. Is it possible to find a real number x such that $\cos x$ is positive and $\sec x$ is negative? Explain.

12. Is it possible to find an angle θ such that $\tan\theta$ is negative and $\cot x$ is positive? Explain.

Use a calculator to find Problems 13–32 to four significant digits. Make sure the calculator is in the correct mode (degree or radian) for each problem.

13. $\tan 89°$ **14.** $\sin 37°$ **15.** $\cos(3 \text{ rad})$

16. $\sin(4 \text{ rad})$ **17.** $\csc 162°$ **18.** $\sec 283°$

19. $\cot 341°$ **20.** $\csc 269°$ **21.** $\sin 13$

22. $\cos 7$ **23.** $\cot 2$ **24.** $\tan 1$

25. $\sec 74$ **26.** $\cot 108$ **27.** $\sin 428°$

28. $\tan 269°$ **29.** $\cos(-12)$ **30.** $\csc(-72)$

31. $\cot(-167°)$ **32.** $\sec(-273°)$

B *Find the exact value of each of the six trigonometric functions for an angle θ that has a terminal side containing the indicated point.*

33. $(\sqrt{3}, 1)$ **34.** $(1, 1)$ **35.** $(1, -\sqrt{3})$

36. $(-1, \sqrt{3})$

In which quadrants must the terminal side of an angle θ lie in order for each of the following to be true?

37. $\cos\theta > 0$ **38.** $\sin\theta > 0$ **39.** $\tan\theta > 0$

40. $\cot\theta > 0$ **41.** $\sec\theta > 0$ **42.** $\csc\theta > 0$

43. $\sin\theta < 0$ **44.** $\cos\theta < 0$ **45.** $\cot\theta < 0$

46. $\tan\theta < 0$ **47.** $\csc\theta < 0$ **48.** $\sec\theta < 0$

In Problems 49–52, find the exact value of each of the other five trigonometric functions for an angle θ (without finding θ), given the indicated information. It would be helpful to sketch a reference triangle.

49. $\sin\theta = -\frac{2}{3}$; $\cot\theta > 0$ **50.** $\cos\theta = -\frac{3}{5}$; $\tan\theta < 0$

51. $\sec\theta = \sqrt{3}$; $\sin\theta < 0$ **52.** $\tan\theta = -2$; $\csc\theta > 0$

53. If angles α and β, $\alpha \neq \beta$, are in standard position and are coterminal, are $\cos\alpha$ and $\cos\beta$ equal? Explain.

54. If $\cos\alpha = \cos\beta$, $\alpha \neq \beta$, are α and β coterminal? Explain.

55. From the following display on a graphing calculator, explain how you would find $\cot x$ without finding x. Then find $\cot x$ to five decimal places.

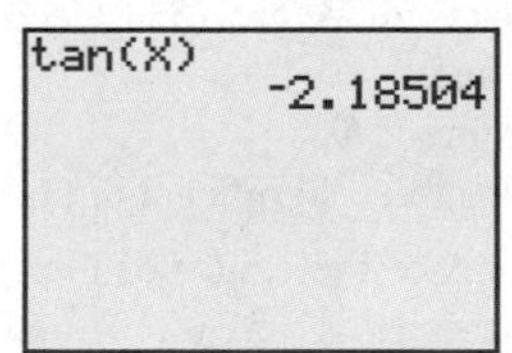

56. From the following display on a graphing calculator, explain how you would find $\sec x$ without finding x. Then find $\sec x$ to five decimal places.

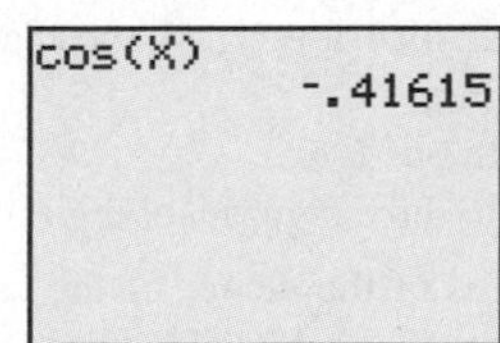

Use a calculator to find Problems 57–74 to four significant digits.

57. $\cos 308.25°$ **58.** $\sin 170.23°$

59. $\tan 1.371$ **60.** $\sin 3.519$

61. $\cot(-265.33°)$ **62.** $\csc(-45.27°)$

63. $\sec(-4.013)$ **64.** $\cot(-0.1578)$

65. $\cos 208°12'55''$ **66.** $\cos 192°45'13''$

67. csc 112°5′38″

68. cot 321°18′5″

69. sec(− 1,000)

70. cot(− 3,000)

71. sin 405.33°

72. tan 623.05°

73. cos(− 168°32′5″)

74. csc(− 263°6′12″)

C **75.** Which trigonometric functions are not defined when the terminal side of an angle lies along the positive or negative vertical axis? Explain.

76. Which trigonometric functions are not defined when the terminal side of an angle lies along the positive or negative horizontal axis? Explain.

For Problems 77–80, refer to the figure:

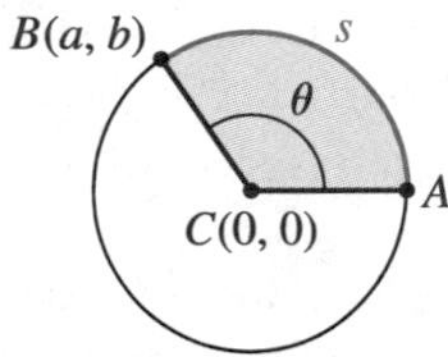

77. In the figure, the coordinates of the center of the circle are (0, 0). If the coordinates of A are (5, 0) and arc length s is exactly 6 units, find:
(A) The exact radian measure of θ
(B) The coordinates of B (to three significant digits)

78. In the figure, the coordinates of the center of the circle are (0, 0). If the coordinates of A are (4, 0) and arc length s is exactly 10 units, find:
(A) The exact radian measure of θ
(B) The coordinates of B (to three significant digits)

79. In the figure, the coordinates of the center of the circle are (0, 0). If the coordinates of A are (1, 0) and the arc length s is exactly 2 units, find:
(A) The exact radian measure of θ
(B) The coordinates of B (to three significant digits)

80. In the figure, the coordinates of the center of the circle are (0, 0). If the coordinates of A are (1, 0) and the arc length s is exactly 4 units, find:
(A) The exact radian measure of θ
(B) The coordinates of B (to three significant digits)

81. A circle with its center at the origin in a rectangular coordinate system passes through the point (4, 3). What is the length of the arc on the circle in the first quadrant between the positive horizontal axis and the point (4, 3)? Compute the answer to two decimal places.

82. Repeat Problem 81 with the circle passing through (3, 4).

Applications

83. Solar Energy Light intensity I on a solar cell changes with the angle of the sun and is given by the formula in the figure below. Find the intensity in terms of the constant k for $\theta = 0°$, $\theta = 20°$, $\theta = 40°$, $\theta = 60°$, and $\theta = 80°$. Compute each answer to two decimal places.

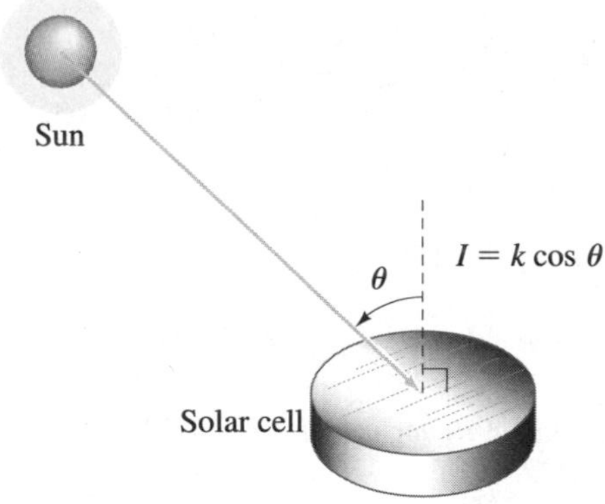

Figure for 83 and 84

84. Solar Energy In Problem 83, at what angle will the light intensity I be 50% of the vertical intensity?

Sun's Energy and Seasons The reason we have summers and winters is because the earth's axis of rotation tilts 23.5° away from the perpendicular, as indicated in the figure. The formula given in Problem 85 quantifies this phenomenon.

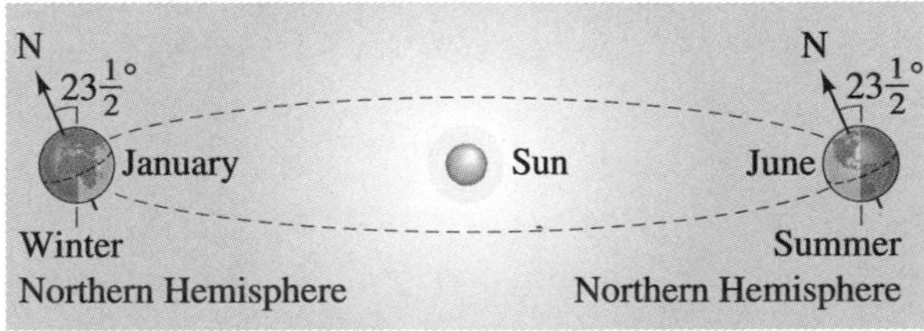

85. The amount of heat energy E from the sun received per square meter per unit of time in a given region on the surface of the earth is approximately proportional to the cosine of the angle θ that the sun makes with the vertical (see the figure on the next page). Thus,

$$E = k \cos \theta$$

where k is the constant of proportionality for a given region. For a region with a latitude of 40°N, compare the energy received at the summer solstice ($\theta = 15°$) with the energy received at the winter solstice ($\theta = 63°$). Express answers in terms of k to two significant digits.

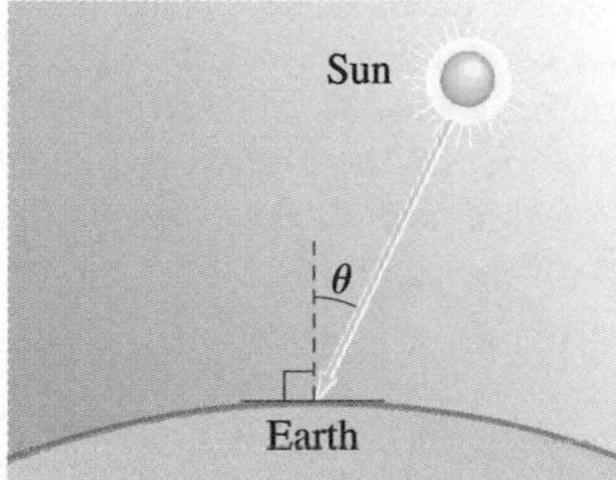

Figure for 85 and 86

86. For a region with a latitude of 32°N, compare the energy received at the summer solstice ($\theta = 8°$) with the energy received at the winter solstice ($\theta = 55°$). Refer to Problem 85, and express answers in terms of k to two significant digits.

For Problems 87 and 88, refer to the figure:

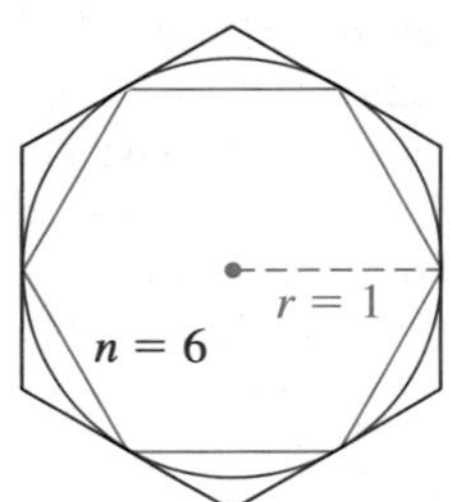

Figure for 87 and 88

87. Precalculus: Calculator Experiment It can be shown that the area of a polygon of n equal sides inscribed in a circle of radius 1 is given by

$$A_n = \frac{n}{2}\sin\left(\frac{360}{n}\right)^\circ$$

Refer to the figure.

(A) Complete the table, giving A_n to five decimal places:

n	6 10 100 1,000 10,000
A_n	

(B) As n gets larger and larger, what number does A_n seem to approach? [*Hint:* What is the area of a circle with radius 1?]

(C) Will an inscribed polygon ever be a circle for any n, however large? Explain.

88. Precalculus: Calculator Experiment It can be shown that the area of a polygon of n equal sides circumscribed around a circle of radius 1 is given by

$$A_n = n\tan\left(\frac{180}{n}\right)^\circ$$

(A) Complete the table, giving A_n to five decimal places:

n	6 10 100 1,000 10,000
A_n	

(B) As n gets larger and larger, what number does A_n seem to approach? [*Hint:* What is the area of a circle with radius 1?]

(C) Will a circumscribed polygon ever be a circle for any n, however large? Explain.

89. Engineering The figure shows a piston connected to a wheel that turns at 10 revolutions per second (rps). If P is at (1, 0) when $t = 0$, then $\theta = 20\pi t$, where t is time in seconds. Show that

$$x = a + \sqrt{5^2 - b^2} = \cos 20\pi t + \sqrt{25 - (\sin 20\pi t)^2}$$

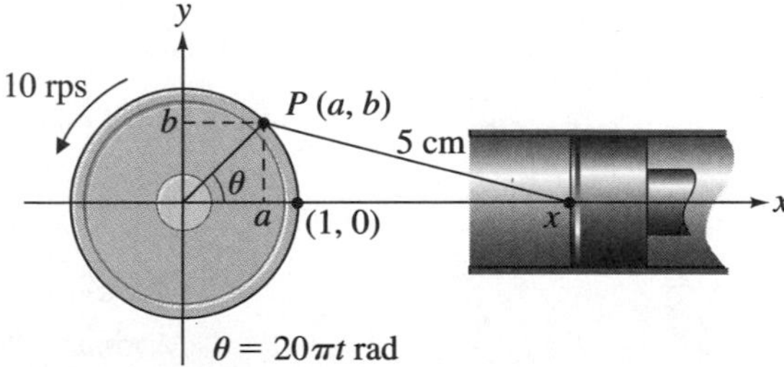

Figure for 89 and 90

***90. Engineering** In Problem 89, find the position (to two decimal places) of the piston (the value of x) for $t = 0$ and $t = 0.01$ sec.

91. Alternating Current An alternating current generator produces an electric current (measured in amperes) that is described by the equation

$$I = 35\sin(48\pi t - 12\pi)$$

where t is time in seconds. (See Example 4 and the figure at the top of the next page.) What is the current I when $t = 0.13$ sec?

Figure for 91 and 92

92. Alternating Current What is the current I in Problem 91 when $t = 0.310$ sec?

93. Precalculus: Angle of Inclination The **slope** of a nonvertical line passing through points $P_1(x_1, y_1)$ and $P_2(x_2, y_2)$ is given by the formula

$$\text{Slope} = m = \frac{y_2 - y_1}{x_2 - x_1}$$

The angle θ that the line L makes with the x axis, $0° \le \theta < 180°$, is called the **angle of inclination** of the line L (see the figure). Thus,

$$\text{Slope} = m = \tan\theta \qquad 0° \le \theta < 180°$$

(A) Compute the slopes (to two decimal places) of the lines with angles of inclination 63.5° and 172°.

(B) Find the equation of a line passing through $(-3, 6)$ with an angle of inclination 143°. [*Hint:* Recall, $y - y_1 = m(x - x_1)$.] Write the answer in the form $y = mx + b$, with m and b to two decimal places.

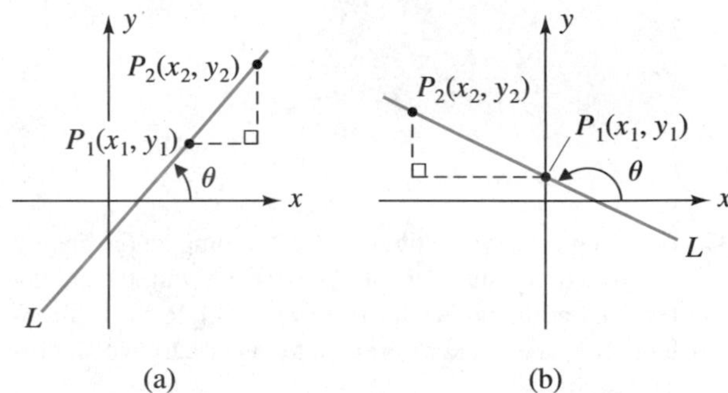

Figure for 93 and 94

94. Precalculus: Angle of Inclination Refer to Problem 93.

(A) Compute the slopes (to two decimal places) of the lines with angles of inclination 89.2° and 179°.

(B) Find the equation of a line passing through $(7, -4)$ with an angle of inclination 101°. Write the answer in the form $y = mx + b$, with m and b to two decimal places.

☆2.4 Additional Applications

- Modeling Light Waves and Refraction
- Modeling Bow Waves
- Modeling Sonic Booms
- High-Energy Physics: Modeling Particle Energy
- Psychology: Modeling Perception

If time permits, the material in this section will provide additional understanding of the use of trigonometry relative to several interesting applications. If the material must be omitted, at least look over the next few pages to gain a better appreciation of some additional applications of trigonometric functions.

☆ Sections marked with a star may be omitted without loss of continuity.

Modeling Light Waves and Refraction

Did you ever look at a pencil in a glass of water or a straight pole pushed into a clear pool of water? The objects appear to bend at the surface (see Fig. 1). This bending phenomenon is caused by *refracted light.* The principle behind refracted light is **Fermat's least-time principle.** In 1657, the French mathematician Pierre Fermat proposed the intriguing idea that light, in going from one point to another, travels along the path that takes the least time. In this single simple statement, Fermat captured a benchmark principle in optics. It can be shown, using more advanced mathematics, that the least-time path of light traveling from point A in medium M_1 to point B in medium M_2 is a bent line as shown in Figure 2. For the light reflected off the surface of medium M_2 back into medium M_1, the angle of incidence equals the angle of reflection (see Fig. 2). This principle, called the **law of reflection,** was known to Euclid in the third century BC.

FIGURE 1
One or two pencils? There is actually only one.

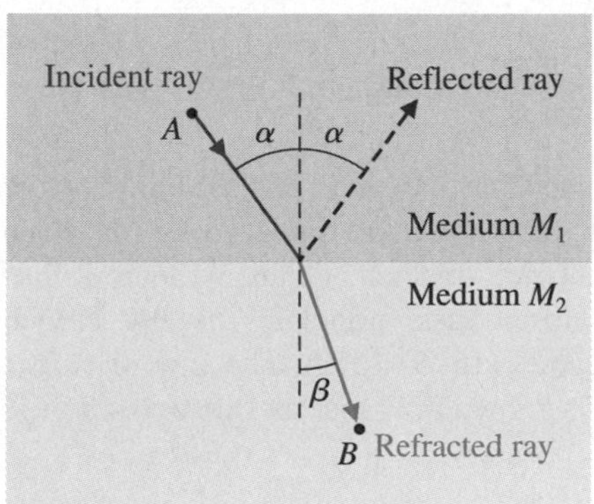

FIGURE 2
Refraction

In physics it is shown that

$$\frac{c_1}{c_2} = \frac{\sin \alpha}{\sin \beta} \tag{1}$$

where c_1 is the speed of light in medium M_1, c_2 is the speed of light in medium M_2, and α and β are as indicated in Figure 2.

A more convenient form of equation (1) uses the notion of the **index of refraction,** which is the ratio of the speed of light in a vacuum to the speed of light in a given substance:

$$\text{Index of refraction, } n = \frac{\text{Speed of light in a vacuum}}{\text{Speed of light in a substance}}$$

If we let c represent the speed of light in a vacuum, then

$$\frac{c_1}{c_2} = \frac{c_1/c}{c_2/c} = \frac{1/n_1}{1/n_2} = \frac{n_2}{n_1} \tag{2}$$

where n_1 is the index of refraction in medium M_1, and n_2 is the index of refraction in medium M_2. (The index of refraction of a substance, rather than the velocity of light in that substance, is the property that is generally tabulated.) Substituting (2) into (1), we obtain **Snell's law:**

$$\frac{n_2}{n_1} = \frac{\sin \alpha}{\sin \beta} \qquad \text{Snell's law} \tag{3}$$

[*Note:* n_2 is on the top and n_1 is on the bottom in (3), while c_1 is on the top and c_2 is on the bottom in (1).]

For any two given substances, the ratio n_2/n_1 is a constant. Thus, equation (3) is equivalent to

$$\frac{\sin \alpha}{\sin \beta} = \text{Constant}$$

The fact that the sines of the angles of incidence and refraction are in a constant ratio to each other was discovered experimentally by the Dutch astronomer–mathematician Willebrod Snell (1591–1626) and was deduced later from basic principles by the French mathematician–philosopher René Descartes (1596–1650). The **law of refraction** stated in equation (3), **Snell's law,** is known in France as **Descartes' law.**

Refracted Light

A spotlight shining on a pond strikes the water so that the angle of incidence α in Figure 3 is 23.5°. Find the refracted angle β. Use Snell's law and the fact that $n = 1.33$ for water and $n = 1.00$ for air.

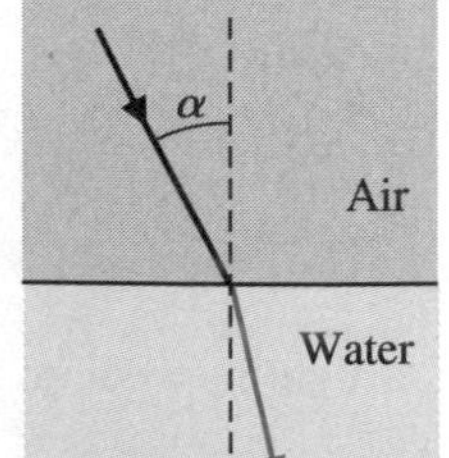

FIGURE 3
Refraction

Solution Use

$$\frac{n_2}{n_1} = \frac{\sin \alpha}{\sin \beta}$$

where $n_2 = 1.33$, $n_1 = 1.00$, and $\alpha = 23.5°$, and solve for β:

$$\frac{1.33}{1.00} = \frac{\sin 23.5°}{\sin \beta}$$

$$\sin \beta = \frac{\sin 23.5°}{1.33}$$

$$\beta = \sin^{-1}\left(\frac{\sin 23.5°}{1.33}\right) = 17.4°$$

■

Matched Problem 1 Repeat Example 1 with $\alpha = 18.4°$. ■

The fact that light bends when passing from one medium into another is what makes telescopes, microscopes, and cameras possible. It is the carefully controlled bending of light rays that produces the useful results in optical instruments. Figures 4–6 illustrate a variety of phenomena connected with refracted light.

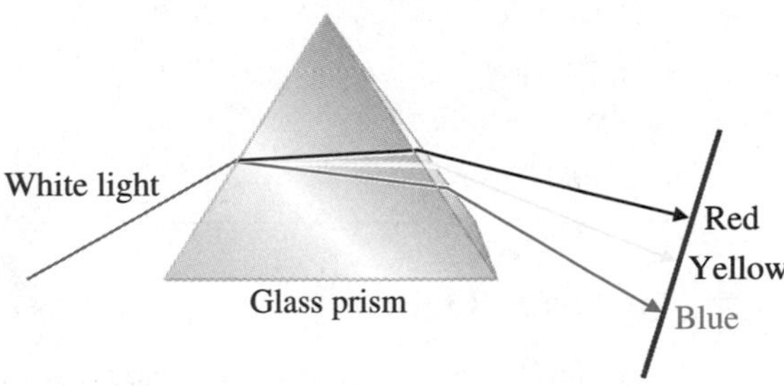

FIGURE 4
Light spectrum—different wavelengths, different indexes of refraction, different colors

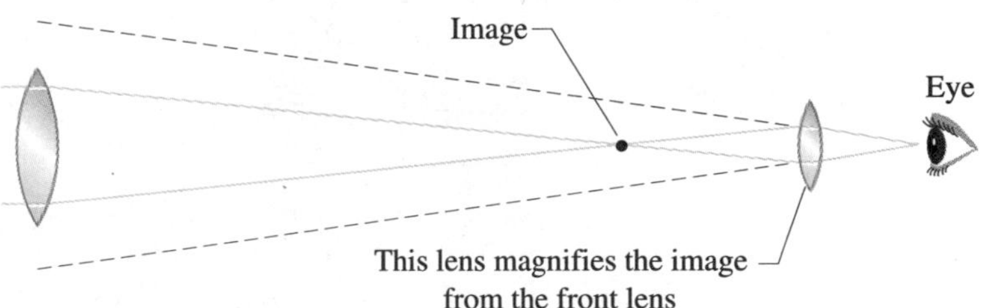

FIGURE 5
Telescope optics

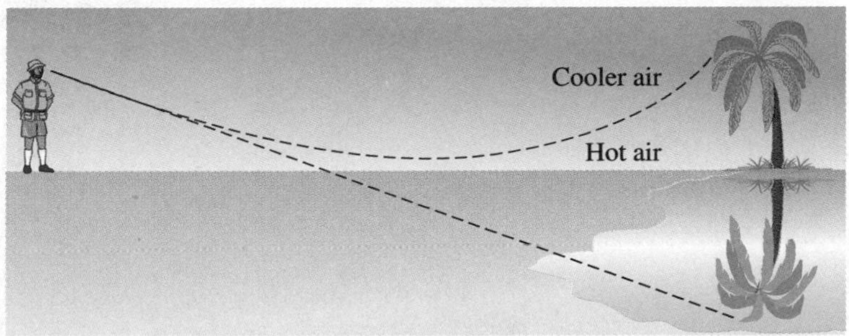

FIGURE 6
Mirage—light is refracted and bent to appear as if reflected off water

Table 1 lists refractive indexes for several common materials.

TABLE 1
Refractive Indexes

Material	Refractive index
Air	1.0003
Crown glass	1.52
Diamond	2.42
Flint glass	1.66
Ice	1.31
Water	1.33

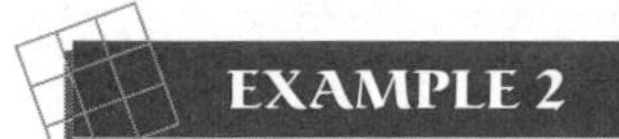

Reflected Light

If an underwater flashlight is directed toward the surface of a swimming pool, at what angle of incidence α will the light beam be totally reflected?

Solution The index of refraction for water is $n_1 = 1.33$ and that for air is $n_2 = 1.00$. Find the angle of incidence α in Figure 7 such that the angle of refraction β is 90°.

$$\frac{\sin \alpha}{\sin \beta} = \frac{n_2}{n_1}$$

$$\sin \alpha = \frac{1.00}{1.33} \sin 90^\circ$$

$$\boxed{\sin \alpha = \frac{1.00}{1.33}(1)}$$

$$\alpha = \sin^{-1} \frac{1.00}{1.33} = 48.8^\circ$$

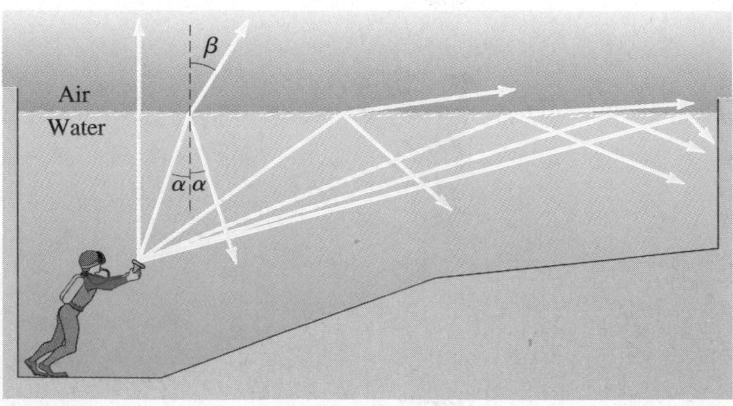

FIGURE 7
Underwater light refraction

Thus, the light will be totally reflected if $\alpha \geq 48.8°$ (no light will be transmitted through the surface). ■

Matched Problem 2 Show that the light beam passing through the flint glass prism shown in Figure 8 is totally reflected off the slanted surface.

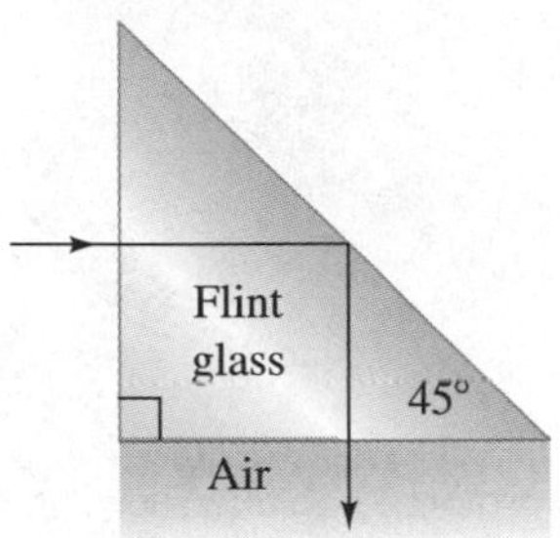

FIGURE 8
Flint glass prism

■

EXPLORE/DISCUSS 1

Figure 9 shows a person spear fishing from a rocky point over clear water. The real fish is the dark one and the apparent fish (the fish that is actually seen) is the light one. Discuss this phenomenon relative to refraction. Should the person aim high or low relative to the apparent image in order to spear the fish? Explain why.

FIGURE 9
Aim high or low?

In general, the angle of incidence α such that the angle of refraction β is 90° is called the **critical angle.** For any angle of incidence larger than the critical angle, a light ray will be totally reflected. This critical angle (for total reflection) is important in the design of many optical instruments (such as binoculars) and is

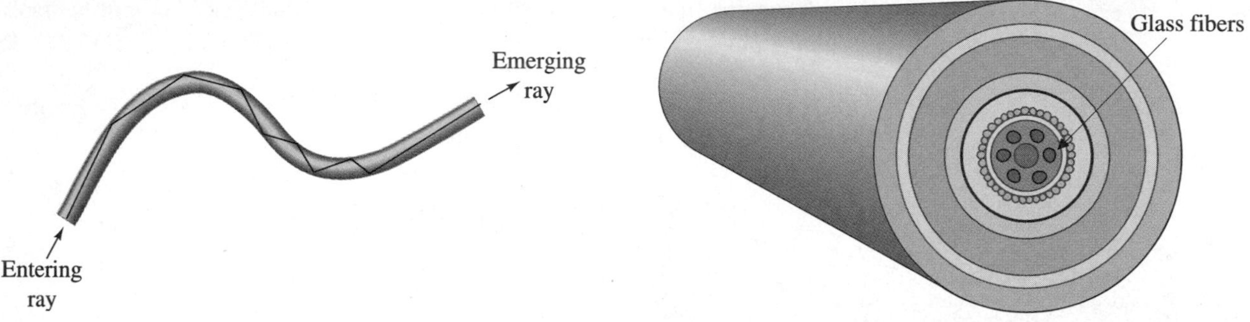

FIGURE 10
Fiber optics

FIGURE 11
Transoceanic fiber optic armored cable (about 1 in. diameter)

at the heart of the science of fiber optics. A small-diameter glass fiber bent in a curve (shown in Fig. 10) will "trap" a light ray entering one end, and the ray will be totally reflected and emerge out the other end.

Important uses of fiber optics are found in medicine and communications. Physicians use fiber optic instruments to see inside functioning organs. Surgeons use fiber optics to perform surgery involving only small incisions and outpatient facilities. A fiber optics communication cable carries information using high-speed pulses of laser light that is essentially distortion-free. Using underwater armored cable (Fig. 11), fiber optics communication networks are now in place worldwide. (The Atlantic cable was completed in 1988, and the Pacific cable was completed in 1989.) One fiber optic communication cable can carry 40,000 simultaneous telephone conversations, or transmit the entire contents of the Encyclopaedia Britannica in less than 1 minute.

Modeling Bow Waves

A boat moving at a constant rate, faster than the water waves it produces, generates a **bow wave** that extends back from the bow of the boat at a given angle (Fig. 12). If we know the speed of the boat and the speed of the waves produced by the boat, then we can determine the angle of the bow wave. Actually, if we

FIGURE 12
Bow waves of boats, sonic booms, and high-energy physics are related in a curious way; this section explains how

know any two of these quantities, we can always find the third. Surprisingly, the solution to this problem also can be applied to sonic booms and high-energy particle physics, as we will see later in this section.

Referring to Figure 13, we reason as follows: When a boat is at P_1, the water wave it produces will radiate out in a circle, and by the time the boat reaches P_2, the wave will have moved a distance of r_1, which is less than the distance between P_1 and P_2 since the boat is assumed to be traveling faster than the wave. By the time the boat reaches the apex position B in Figure 13, the wave motion at P_2 will have moved r_2 units, and the wave motion at P_1 will have continued on out to r_3. Because of the constant speed of the boat and the constant speed of the wave motion, these circles of wave radiation will all have a common tangent that passes through the boat. Of course, the motion of the boat is continuous, and what we have said about P_1 and P_2 applies to all points along the path of the boat. The result of this phenomenon is the clearly visible wave front produced by the bow of the boat. Refer again to Figure 13, and you will see that the boat travels from P_1 to B in the same time t that the bow wave travels from P_1 to A; hence, if S_b is the speed of the boat and S_w is the speed of the bow wave, then (using $d = rt$),

$$\text{Distance from } P_1 \text{ to } A = S_w t \qquad \text{Distance from } P_1 \text{ to } B = S_b t$$

and, since triangle P_1BA is a right triangle,

$$\sin\frac{\theta}{2} = \frac{S_w t}{S_b t} = \frac{S_w}{S_b}$$

Thus,

$$\sin\frac{\theta}{2} = \frac{S_w}{S_b}$$

where S_w is the speed of the bow wave, S_b is the speed of the boat, and $S_b > S_w$.

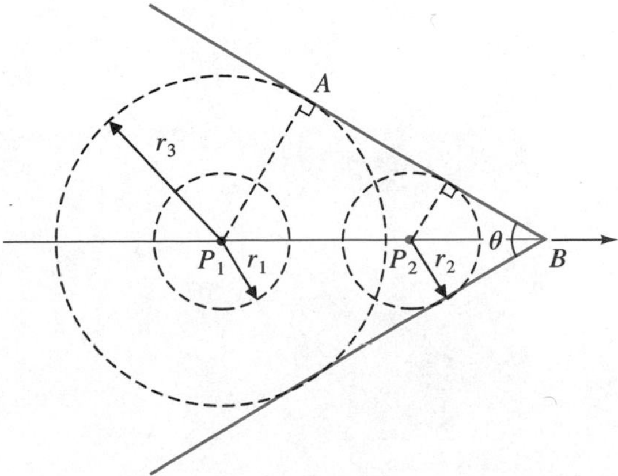

FIGURE 13

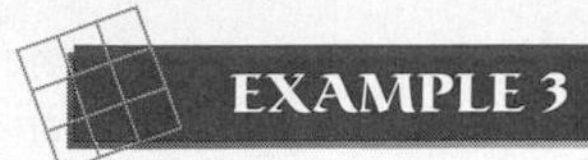

Bow Wave Speed

If a speedboat travels at 45 km/hr and the angle between the bow waves is 72°, how fast is the bow wave traveling?

Solution We use

$$\sin\frac{\theta}{2} = \frac{S_w}{S_b}$$

where $\theta = 72°$ and $S_b = 45$ km/hr. Then we solve for S_w:

$$S_w = (45 \text{ km/hr})(\sin 36°) \approx 26 \text{ km/hr}$$

■

Matched Problem 3 If a speedboat is traveling at 121 km/hr and the angle between the bow waves is 74.5°, how fast is the bow wave moving? ■

■ Modeling Sonic Booms

If we follow exactly the same line of reasoning as in the discussion of bow waves, we see that an aircraft flying faster than the speed of sound produces sound waves that pile up behind the aircraft in the form of a cone (see Fig. 14). The cone intersects the ground in the form of a hyperbola, and along this curve we experience a phenomenon called a **sonic boom.** As in the bow wave analysis, we have

$$\sin\frac{\theta}{2} = \frac{S_s}{S_a}$$

where S_s is the speed of the sound wave, S_a is the speed of the aircraft, and $S_a > S_s$.

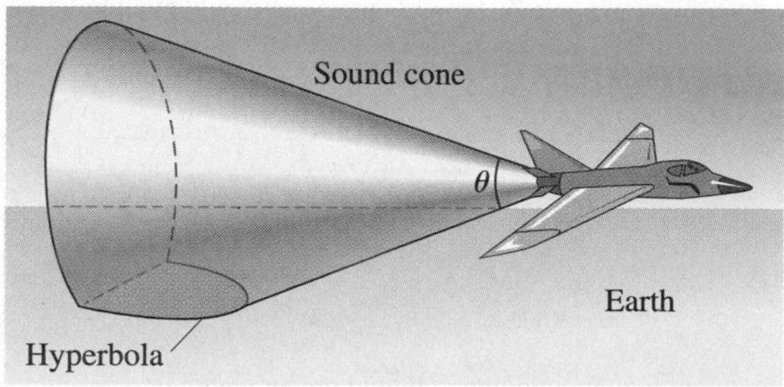

FIGURE 14
Sound cones and sonic booms

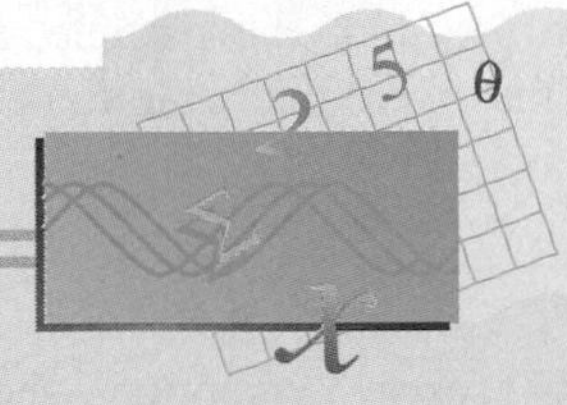

EXPLORE/DISCUSS 2

Chuck Yeager (1923–) was a 24-year-old U.S. Air Force test pilot when he broke the sound barrier in 1947 in a Bell X-1 rocket plane, ushering in the age of supersonic flight. The speed of supersonic aircraft is measured in terms of Mach numbers. [The prefix Mach is named for Ernst Mach (1838–1916), an Austrian physicist who made significant contributions to the study of sound.] A **Mach number** is the ratio of the speed of an object to the speed of sound in the surrounding medium. Mach 1 is the speed of an object flying at the speed of sound, approximately 750 mph at sea level.

(A) Write a formula for the Mach number M of a supersonic aircraft, if S_a is the speed of the aircraft and S_s is the speed of sound in the surrounding medium.

(B) Rewrite the formula for the angle θ of the sound cone in Figure 14 in terms of M, assuming $S_a > S_s$.

(C) Find the angle θ of the sound cone of the British–French commercial aircraft Concorde, flying at Mach 2.

High-Energy Physics: Modeling Particle Energy

Nuclear particles can be made to move faster than the velocity of light in certain materials, such as glass. In 1958, three physicists (Cerenkov, Frank, and Tamm) jointly received a Nobel prize for the work they did based on this fact. Interestingly, the bow wave analysis for boats applies equally well here. Instead of a sound cone, as in Figure 14, they obtained a light cone, as shown in Figure 15.

By measuring the cone angle θ, Cerenkov, Frank, and Tamm were able to determine the speed of the particle because the speed of light in glass is readily determined. They used the formula

$$\sin\frac{\theta}{2} = \frac{S_\ell}{S_p}$$

FIGURE 15
Particle energy

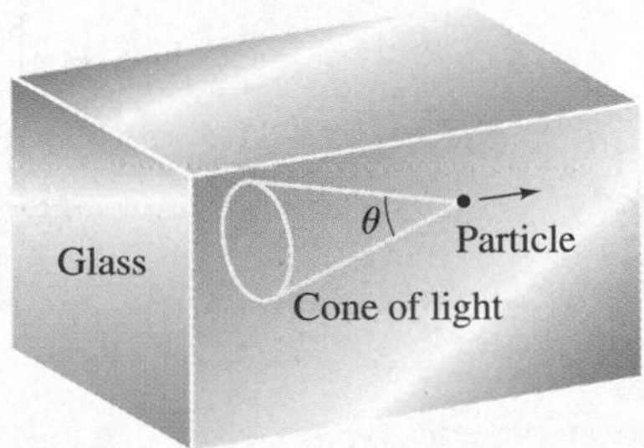

where S_ℓ is the speed of light in glass, S_p is the speed of the particle, and $S_p > S_\ell$. By determining the speed of the particle, they were then able to determine its energy by routine procedures.

Psychology: Modeling Perception

An important field of study in psychology concerns sensory perception—hearing, seeing, smelling, feeling, and tasting. It is well known that individuals see certain objects differently in different surroundings. Lines that appear to be parallel in one setting may appear to be curved in another. Lines of the same length may appear to have different lengths in two different settings. Is a square always a square? (See Fig. 16.)

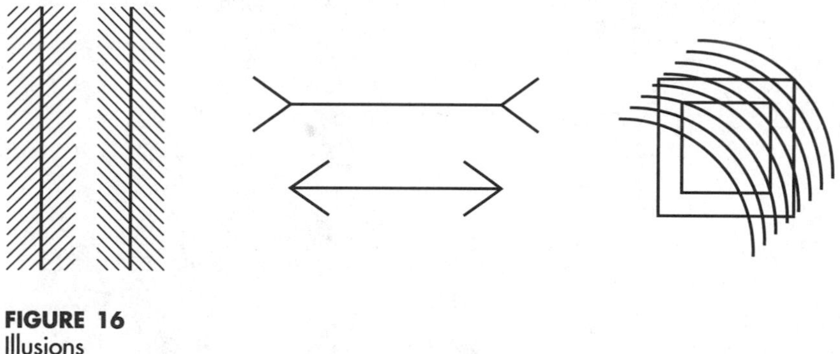

FIGURE 16
Illusions

Figure 17 illustrates a perspective illusion in which the people farthest away appear to be larger than those closest—they are actually all the same size.

FIGURE 17
Which person is tallest?

An interesting experiment in visual perception was conducted by psychologists Berliner and Berliner. A tilted field of parallel lines was presented to several subjects who were then asked to estimate the position of a horizontal line in the field. Berliner and Berliner (*American Journal of Psychology,* vol. 65, pp. 271–277, 1952) reported that most subjects were consistently off, and that the difference in degrees d between their estimates and the actual horizontal could be approximated by the equation

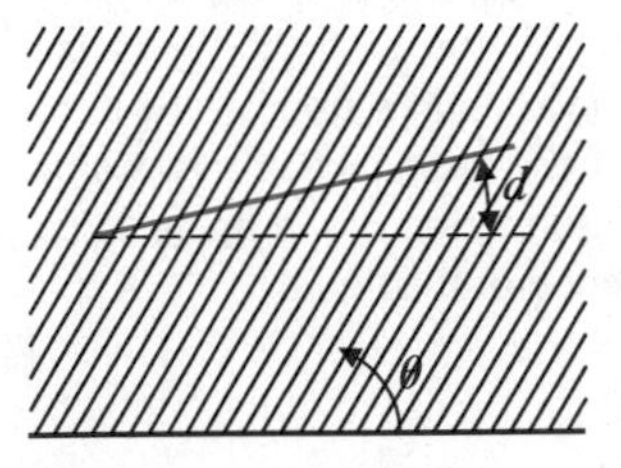

FIGURE 18
Visual perception

$$d = a + b \sin 4\theta$$

where a and b were constants associated with a particular individual and θ was the angle of tilt of the visual field in degrees (see Fig. 18).

Answers to Matched Problems

1. 13.7°
2. $\dfrac{\sin 45°}{\sin \beta} = \dfrac{1.00}{1.66}$

   $\sin \beta = 1.66 \sin 45°$

   $\sin \beta = 1.17$

   Since $\sin \beta$ cannot exceed 1 (see Definition 1 of trigonometric functions on page 70), the condition $\sin \beta = 1.17$ cannot physically happen! In other words, the light must be totally reflected, as indicated in Figure 8.
3. 73.2 km/hr

EXERCISE 2.4

Applications

Light Waves and Refraction *For Problems 1–8 refer to the following figure and table (repeated from the text for convenience):*

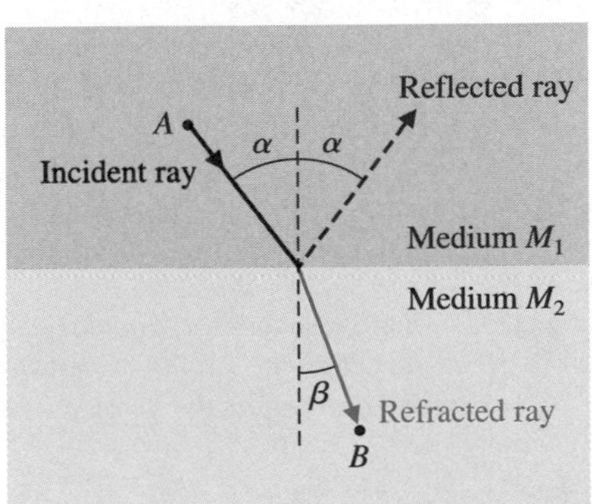

Refractive Indexes

Material	Refractive index
Air	1.0003
Crown glass	1.52
Diamond	2.42
Flint glass	1.66
Ice	1.31
Water	1.33

1. A light ray passing through air strikes the surface of a pool of water so that the angle of incidence is $\alpha = 40.6°$. Find the angle of refraction β.
2. Repeat Problem 1 with $\alpha = 34.2°$.
3. A light ray from an underwater spotlight passes through a porthole (of flint glass) of a sunken ocean liner. If the angle of incidence α is 32.0°, what is the angle of refraction β?
4. Repeat Problem 3 with $\alpha = 45.0°$.

5. If light inside a diamond strikes one of its facets (flat surfaces), what is the critical angle of incidence α for total reflection? (The diamond is surrounded by air.) Compute your answer in decimal degrees to three significant digits.

6. If light inside a triangular flint glass prism strikes one of the flat surfaces, what is the critical angle of incidence α for total reflection? (The prism is surrounded by air.)

7. A golfer hits a ball into a pond. From the side of the pond, the ball is spotted on the bottom. Does the ball appear to be above or below the real ball? Explain.

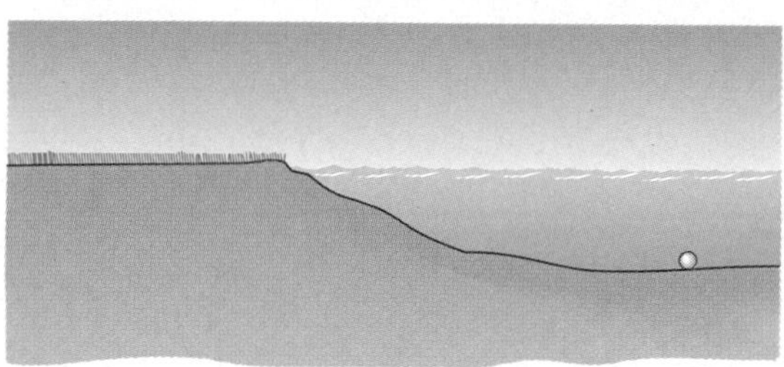

8. The figure shows a diver looking up on a very calm day. Explain why the diver sees a bright circular field surrounded by darkness. What does the diver see within this bright circular field? The bright circular field with the diver's eye at the apex forms a cone. What is the angle θ of the cone?

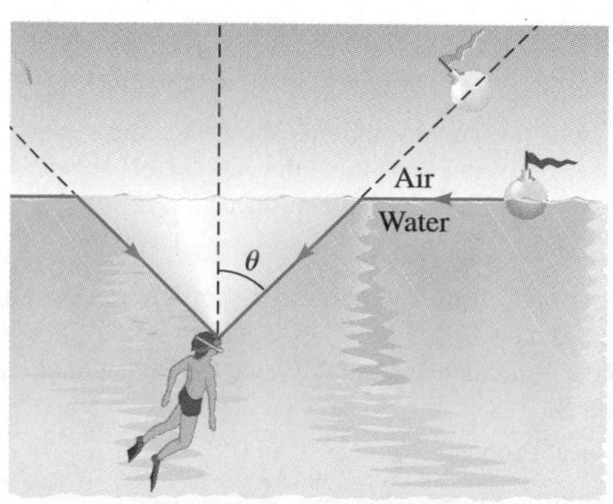

9. **Index of Refraction** An experiment is set up to find the index of refraction of an unknown liquid. A rectangular container is filled to the top with the liquid and the experimenter moves the container up and down until the far bottom edge of the container is just visible (see the figure). Use the information in the figure to determine the index of refraction of the liquid.

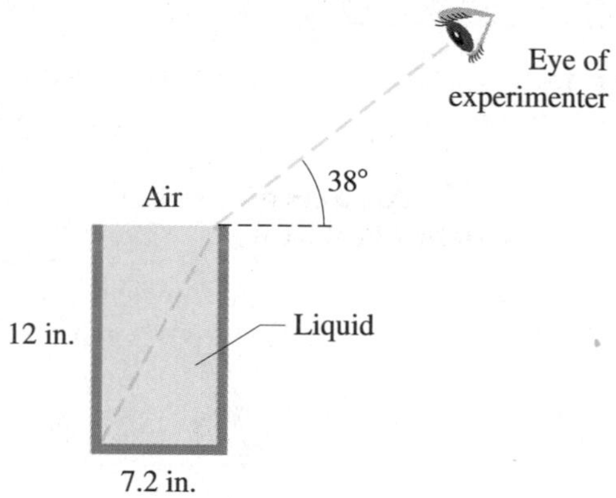

Figure for 9 and 10

10. **Index of Refraction** Repeat Problem 9 with 38° replaced by 43°.

11. **Bow Waves** If the bow waves of a boat travel at 20 km/hr and create an angle of 60°, how fast is the boat traveling?

12. **Bow Waves** A boat traveling at 55 km/hr produces bow waves that separate at an angle of 54°. How fast is a bow wave moving?

13. **Sound Cones** If a supersonic jet flies at Mach 1.5, what will be the cone angle (to the nearest degree)?

14. **Sound Cones** If a supersonic jet flies at Mach 3.2, what will be the cone angle (to the nearest degree)?

15. **Light Cones** In crown glass, light travels at approx. 2×10^{10} cm/sec. If a high-energy particle passing through this glass creates a light cone of 90°, how fast is it traveling?

16. **Light Cones** Repeat Problem 15 using a light cone angle of 60°.

Psychology: Perception *Using the empirical formula, $d = a + b \sin 4\theta$, from Berliner and Berliner's study of perception, determine d (to the nearest degree) for the given values of a, b, and θ.*

17. $a = -2.2,\quad b = -4.5,\quad \theta = 30°$

18. $a = -1.8,\quad b = -4.2,\quad \theta = 40°$

2.5 EXACT VALUE FOR SPECIAL ANGLES AND REAL NUMBERS

- Introduction
- Evaluation of Trigonometric Functions for Quadrantal Angles
- Reference Triangle and Angle
- Special 30°–60° and 45° Right Triangles
- Evaluation of Trigonometric Functions for Angles or Real Numbers with 30°–60° and 45° Reference Triangles

Introduction

Decimal approximations of many numbers are not exact, no matter how many decimal places are computed. For example, 0.5714 is a decimal approximation of $\frac{4}{7}$ and 1.414 is a decimal approximation of $\sqrt{2}$, but neither $\frac{4}{7}$ nor $\sqrt{2}$ can be represented exactly by a finite decimal no matter how many decimal places are computed. In certain formulas and computations it is better to use an exact form rather than a decimal approximation.

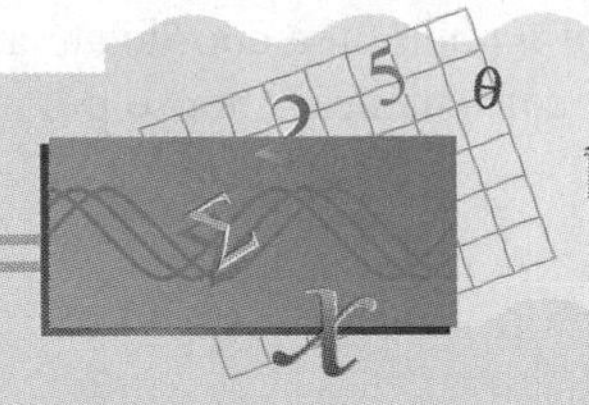

EXPLORE/DISCUSS 1

Which of the following graphing calculator displays represents cos 45° exactly? Experiment with your own calculator. Do you believe that cos 45° has an exact finite decimal representation? This section will provide an answer to this question.

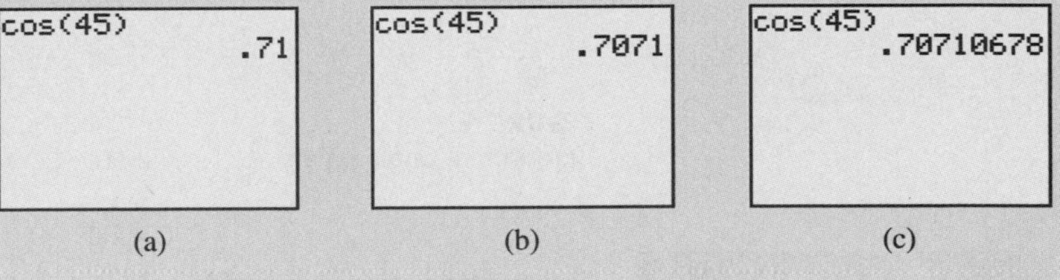

(a) (b) (c)

If an angle is an integer multiple of 30°, 45°, $\pi/6$ rad, or $\pi/4$ rad, and if a real number is an integer multiple of $\pi/6$ or $\pi/4$ (see Fig. 1), then, for those values for which each trigonometric function is defined, the function can be evalu-

FIGURE 1
Some multiples of 30°, 45°, $\pi/6$, and $\pi/4$

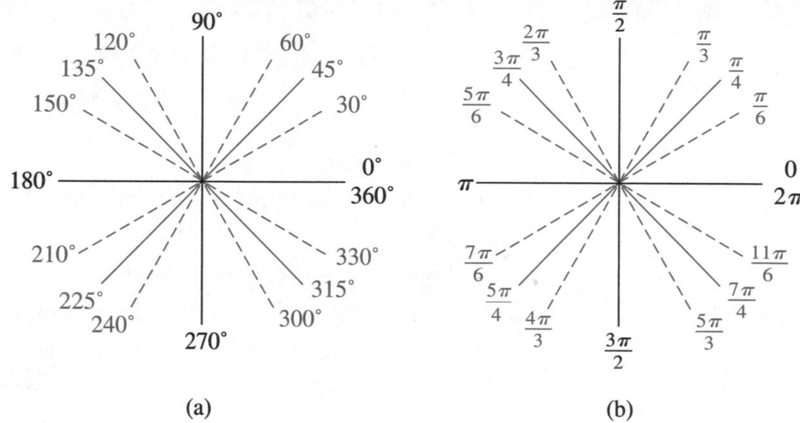

ated exactly without the use of any calculator (which is different from finding approximate values using a calculator). With a little practice, you will—mentally—be able to determine these exact values.

Evaluation of Trigonometric Functions for Quadrantal Angles

The easiest angles to deal with are **quadrantal angles**—that is, angles with their terminal side lying along a coordinate axis. These angles are integer multiples of 90° or $\pi/2$. It is easy to find coordinates of a point on a coordinate axis. Since any nonorigin point will do, we shall, for convenience, choose points 1 unit from the origin (Fig. 2).

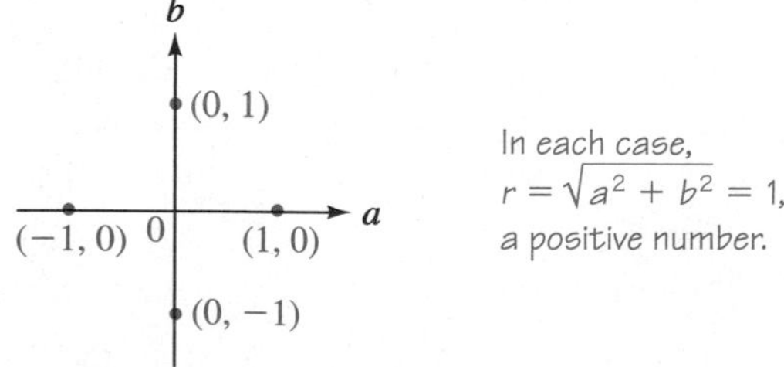

In each case, $r = \sqrt{a^2 + b^2} = 1$, a positive number.

FIGURE 2
Quadrantal angle points

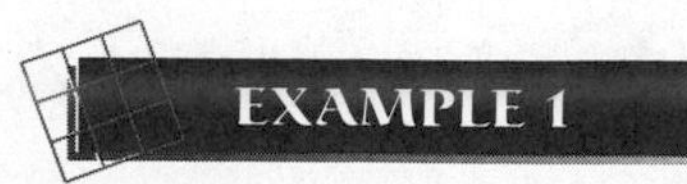

Evaluation Involving Quadrantal Angles

Find:

(A) $\sin 90°$ (B) $\cos \pi$ (C) $\tan(-2\pi)$ (D) $\cot(-180°)$

Solution For each, visualize the location of the terminal side of the angle relative to Figure 2. With a little practice, you should be able to do most of the following calculations mentally.

(A) $\sin 90° = \dfrac{b}{r} = \dfrac{1}{1} = 1$ $\quad (a, b) = (0, 1), \quad r = 1$

(B) $\cos \pi = \dfrac{a}{r} = \dfrac{-1}{1} = -1$ $\quad (a, b) = (-1, 0), \quad r = 1$

(C) $\tan(-2\pi) = \dfrac{b}{a} = \dfrac{0}{1} = 0$ $\quad (a, b) = (1, 0), \quad r = 1$

(D) $\cot(-180°) = \dfrac{a}{b} = \dfrac{-1}{0}$ $\quad (a, b) = (-1, 0), \quad r = 1$

Not defined ■

Matched Problem 1 Find:

(A) $\sin(3\pi/2)$ (B) $\sec(-\pi)$ (C) $\tan 90°$ (D) $\cot(-270°)$ ■

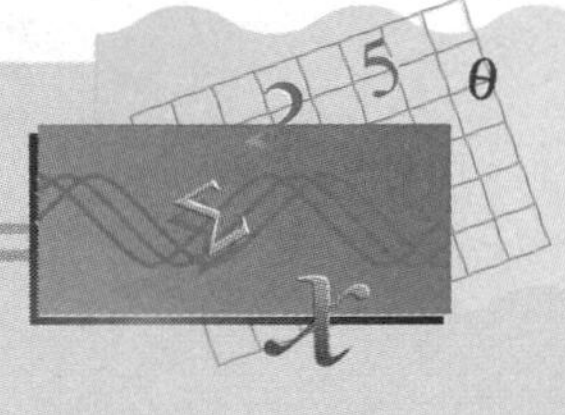

EXPLORE/DISCUSS 2

In Example 1D, notice that $\cot(-180°)$ is not defined. Discuss other angles in degree measure for which the cotangent is not defined. For what angles in degree measure is the cosecant function not defined?

Reference Triangle and Angle

Because the reference triangle is going to play a very important role in the work that follows, we now restate its definition as well as that of a reference angle.

REFERENCE TRIANGLE AND ANGLE

For a nonquadrantal angle θ:

1. To form a **reference triangle** for θ, drop a perpendicular from a point $P(a, b)$ on the terminal side of θ to the horizontal axis.
2. The **reference angle** α is the acute angle (always taken positive) between the terminal side of θ and the horizontal axis.

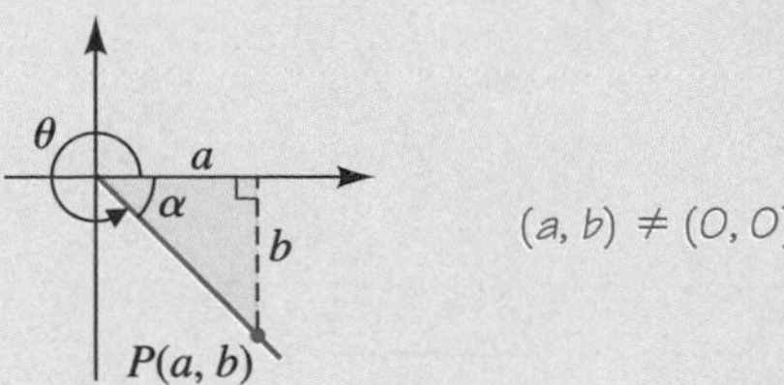

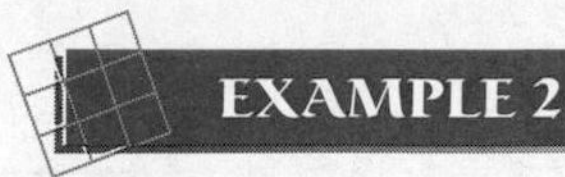

EXAMPLE 2 Reference Triangles and Angles

Sketch the reference triangle and find the reference angle α for each of the following angles:

(A) $\theta = 330°$ (B) $\theta = -315°$ (C) $\theta = -\pi/4$ (D) $\theta = 4\pi/3$

Solution (A)

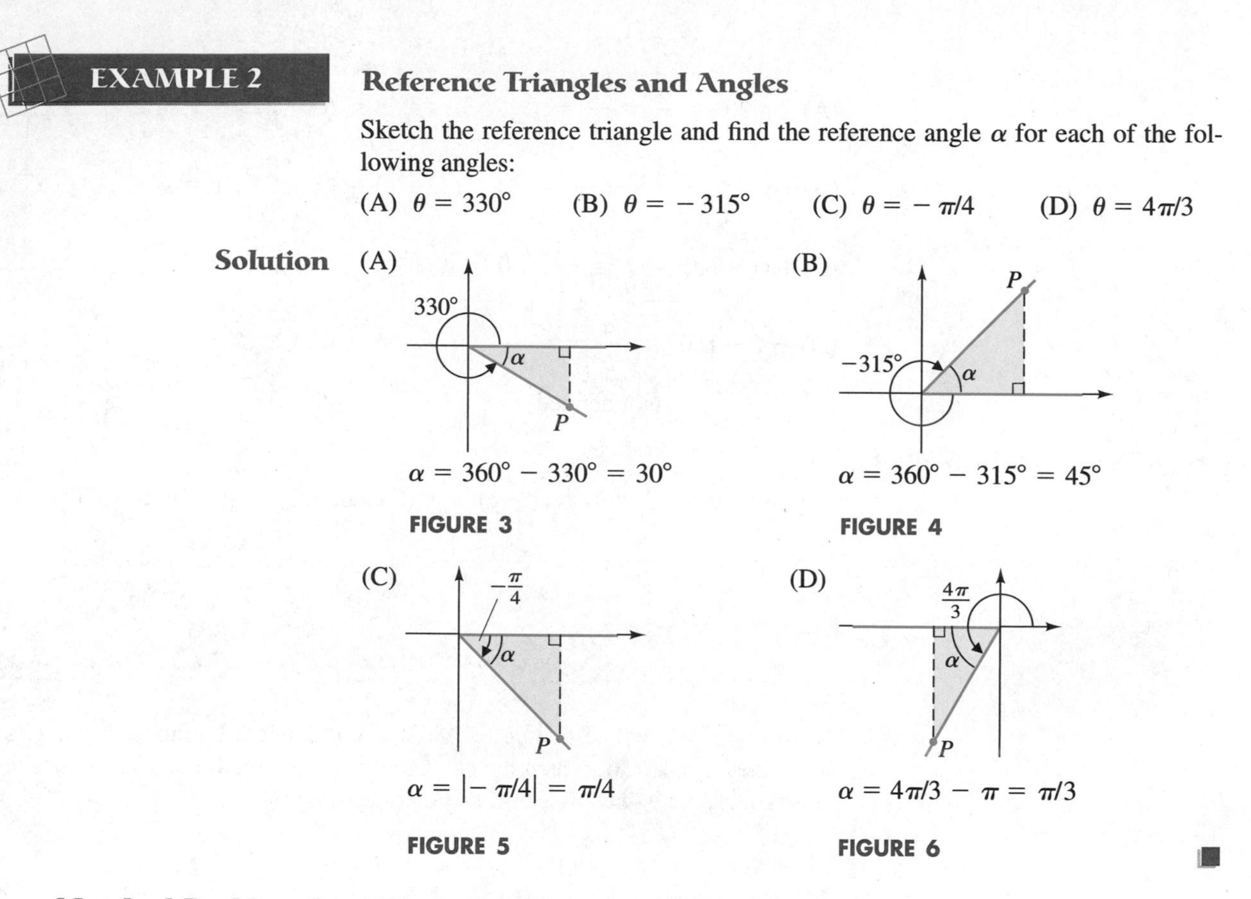

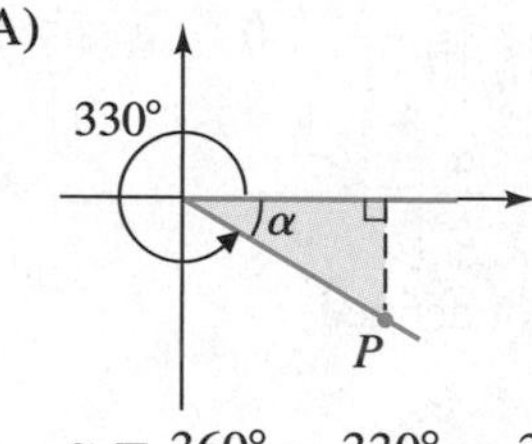

$\alpha = 360° - 330° = 30°$

FIGURE 3

(B)

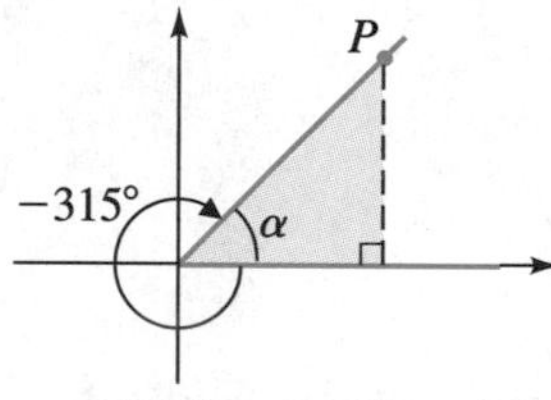

$\alpha = 360° - 315° = 45°$

FIGURE 4

(C)

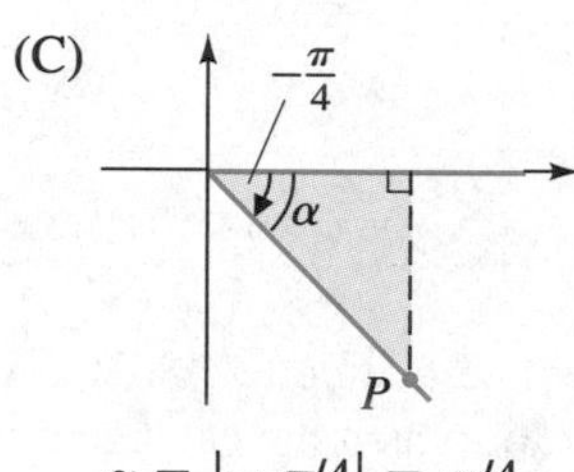

$\alpha = |-\pi/4| = \pi/4$

FIGURE 5

(D)

$\alpha = 4\pi/3 - \pi = \pi/3$

FIGURE 6

Matched Problem 2 Sketch the reference triangle and find the reference angle α for each of the following angles:

(A) $\theta = -225°$ (B) $\theta = 420°$ (C) $\theta = -5\pi/6$ (D) $\theta = 2\pi/3$

Special 30°–60° and 45° Right Triangles

If a reference triangle of a given angle is a 30°–60° right triangle or a 45° right triangle, then we will be able to find exact nonorigin coordinates on the terminal side of the given angle.

If we take a 30°–60° right triangle, we note that it is one-half of an equilateral triangle, as indicated in Figure 7. Since all sides are equal in an equilateral triangle, we can apply the Pythagorean theorem to obtain a useful relationship among the three sides of the original triangle:

FIGURE 7

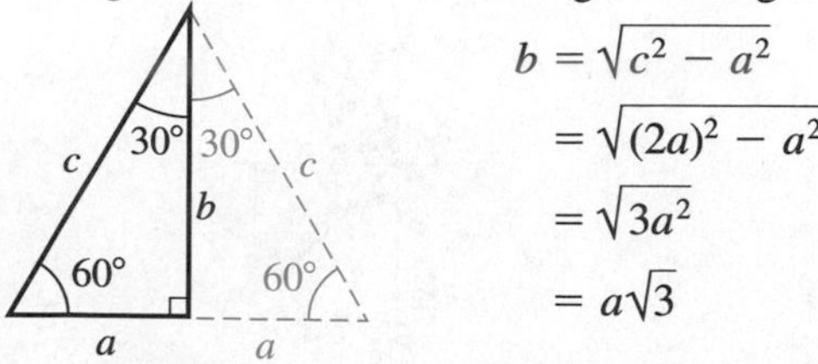

$$\begin{aligned} b &= \sqrt{c^2 - a^2} \\ &= \sqrt{(2a)^2 - a^2} \qquad \text{Since } c = 2a \\ &= \sqrt{3a^2} \\ &= a\sqrt{3} \end{aligned}$$

Similarly, using the Pythagorean theorem on a 45° right triangle, we obtain the following (see Fig. 8):

FIGURE 8

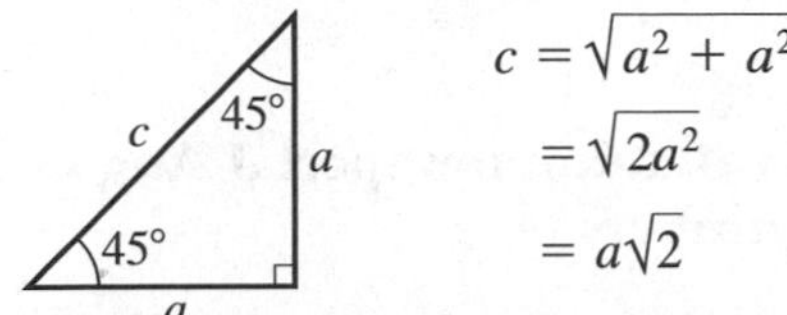

$$
\begin{aligned}
c &= \sqrt{a^2 + a^2} \\
&= \sqrt{2a^2} \\
&= a\sqrt{2}
\end{aligned}
$$

We summarize these results in Figure 9, along with some frequently used special cases. The ratios of these special triangles should be learned, since they will be used often in this and subsequent sections. The two triangles shown in blue are the easiest to remember. The others can be obtained from these by multiplying or dividing the length of each side by the same nonzero quantity.

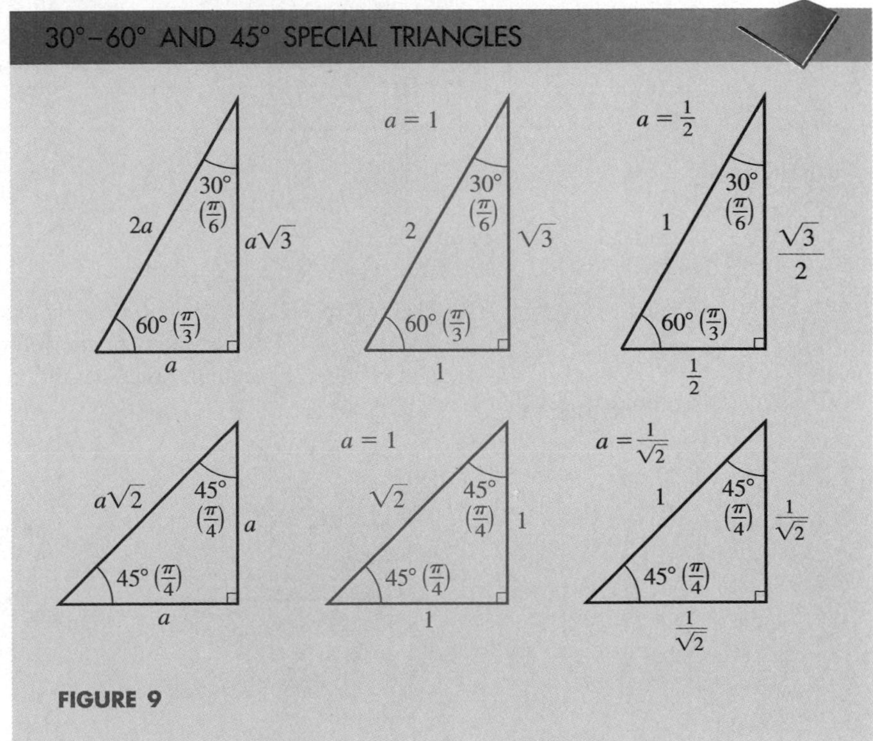

FIGURE 9

Evaluation of Trigonometric Functions for Angles or Real Numbers with 30°–60° and 45° Reference Triangles

If an angle θ has a 30°–60° or 45° reference triangle, then it is easy to find exact coordinates of a point P on the terminal side of θ and the exact distance of P from the origin. Then, using the definitions of the six trigonometric functions given in Section 2.3, we can find the exact value of any of the six functions for the given θ. Several examples will illustrate the process.

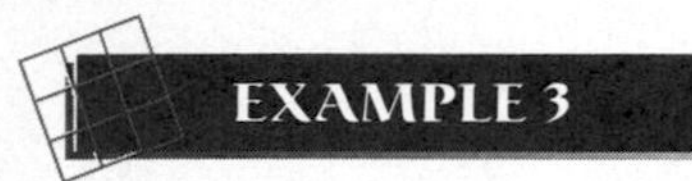

Exact Evaluation for Special Angles and Real Numbers

Evaluate exactly:

(A) $\sin 30°$, $\cos(\pi/6)$, $\cot(\pi/6)$ (B) $\cos 45°$, $\tan(\pi/4)$, $\csc(\pi/4)$

Solution (A) Use the special 30°–60° triangle (Fig. 10) as the reference triangle for $\theta = 30°$ and $\theta = \pi/6$. Use the sides of the reference triangle to determine $P(a, b)$ and r; then use Definition 1 from Section 2.3.

FIGURE 10

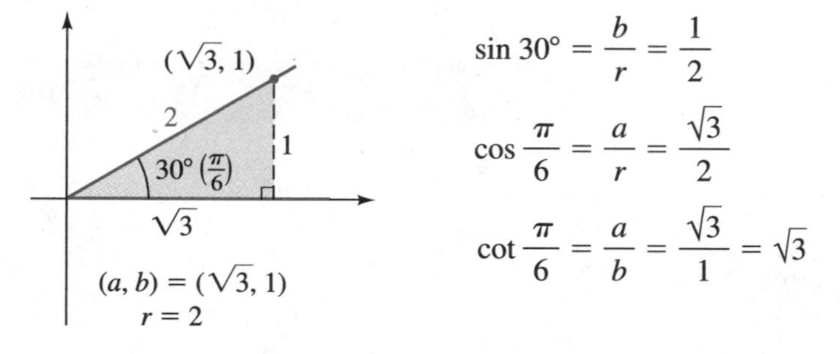

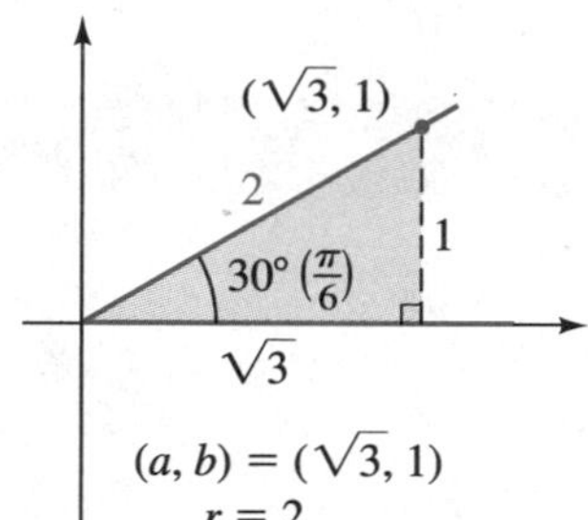

$$\sin 30° = \frac{b}{r} = \frac{1}{2}$$

$$\cos\frac{\pi}{6} = \frac{a}{r} = \frac{\sqrt{3}}{2}$$

$$\cot\frac{\pi}{6} = \frac{a}{b} = \frac{\sqrt{3}}{1} = \sqrt{3}$$

(B) Use the special 45° triangle (Fig. 11) as the reference triangle for $\theta = 45°$ and $\theta = \pi/4$. Use the sides of the reference triangle to determine $P(a, b)$ and r; then use the appropriate definition from Section 2.3.

FIGURE 11

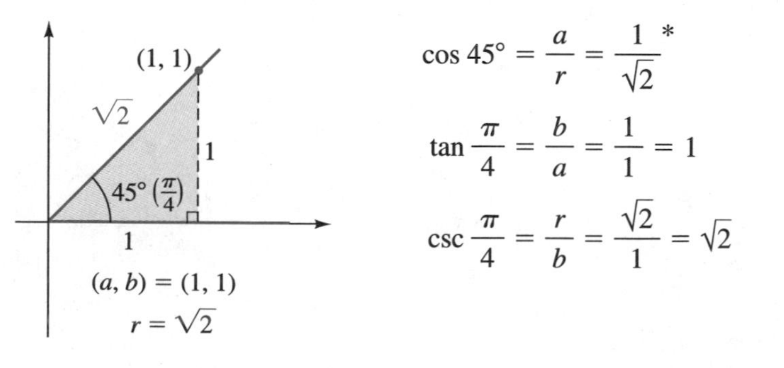

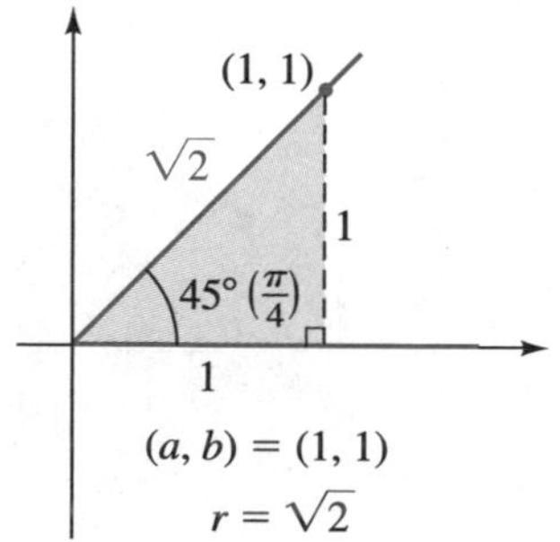

$$\cos 45° = \frac{a}{r} = \frac{1}{\sqrt{2}}\ *$$

$$\tan\frac{\pi}{4} = \frac{b}{a} = \frac{1}{1} = 1$$

$$\csc\frac{\pi}{4} = \frac{r}{b} = \frac{\sqrt{2}}{1} = \sqrt{2}$$

Matched Problem 3 Evaluate exactly:

(A) $\cos 60°$, $\sin(\pi/3)$, $\tan(\pi/3)$ (B) $\sin 45°$, $\cot(\pi/4)$, $\sec(\pi/4)$

---

* Whether we rationalize a denominator or not depends entirely on what we want to do with the answer. Leave answers in unrationalized form unless directed otherwise.

Before proceeding with examples of reference triangles in quadrants other than the first quadrant, it is useful to recall the multiples of $\pi/3$ (60°), $\pi/6$ (30°), and $\pi/4$ (45°). See Figure 12.

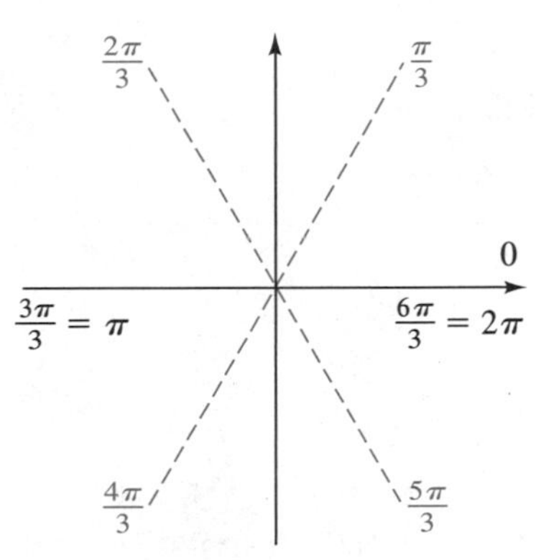

(a) Multiples of $\frac{\pi}{3}$ (60°)

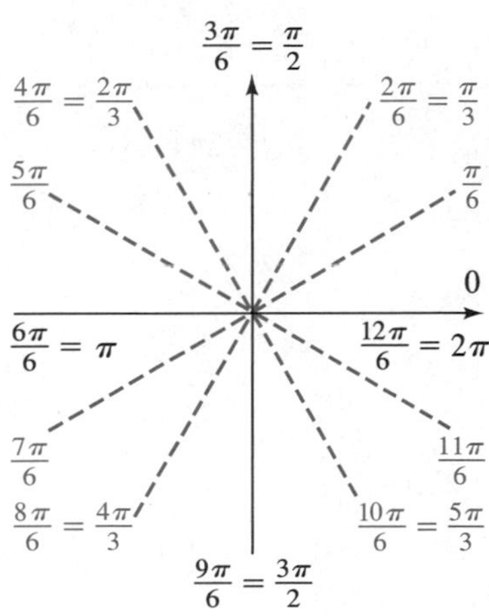

(b) Multiples of $\frac{\pi}{6}$ (30°)

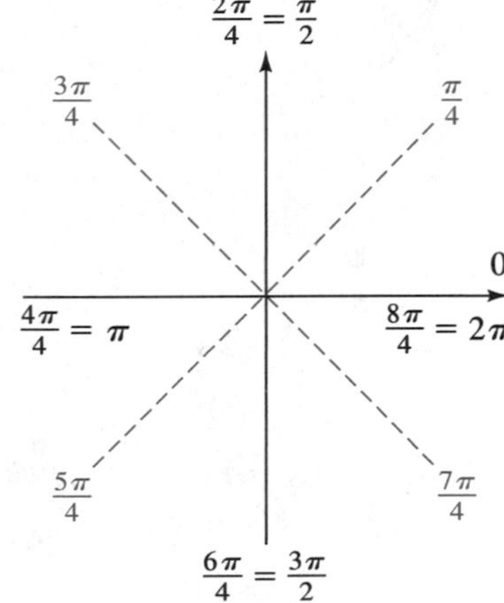

(c) Multiples of $\frac{\pi}{4}$ (45°)

FIGURE 12
Multiples of special angles

EXAMPLE 4 Exact Evaluation for Special Angles and Real Numbers

Evaluate exactly:

(A) $\sin 210°$ (B) $\cos(2\pi/3)$ (C) $\cot(-5\pi/6)$ (D) $\csc(-240°)$

Solution Each angle has a 30°–60° or 45° reference triangle. Locate it, determine $P(a, b)$ and r, and then evaluate. (Recall, we label the horizontal and vertical sides of the reference triangle as a and b, respectively; consequently, one or the other may be negative depending on the quadrant containing the terminal side of θ.)

(A) $\sin 210° = \dfrac{-1}{2} = -\dfrac{1}{2}$

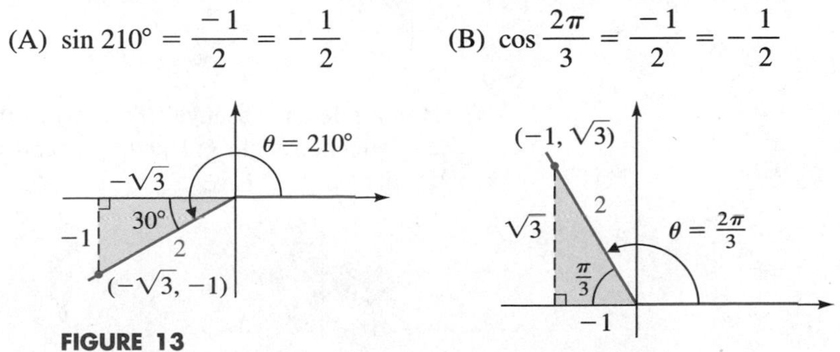

FIGURE 13

(B) $\cos \dfrac{2\pi}{3} = \dfrac{-1}{2} = -\dfrac{1}{2}$

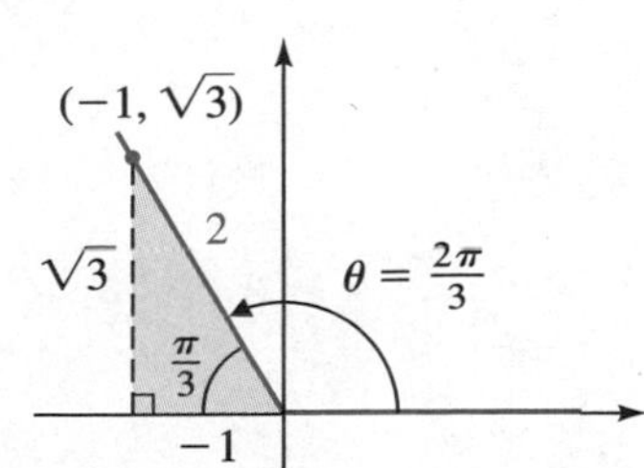

FIGURE 14

(C) $\cot\left(-\frac{5\pi}{6}\right) = \frac{-\sqrt{3}}{-1} = \sqrt{3}$ (D) $\csc(-240°) = \frac{2}{\sqrt{3}}$

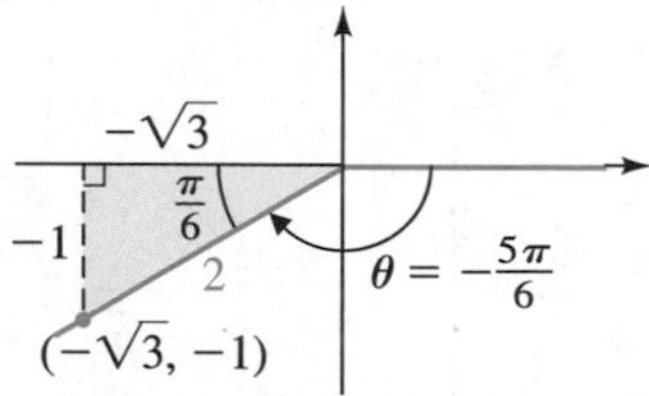

FIGURE 15

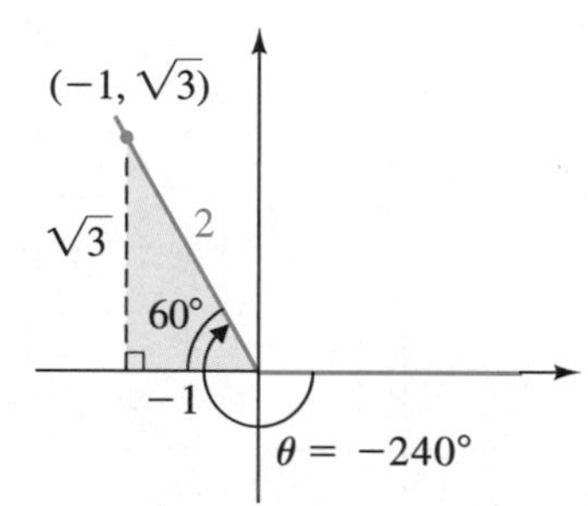

FIGURE 16

Matched Problem 4 Evaluate exactly:

(A) $\tan 210°$ (B) $\sin(2\pi/3)$ (C) $\csc(-5\pi/6)$ (D) $\sec(-240°)$

Now let us reverse the problem. That is, we are given the exact value of one of the six trigonometric functions that corresponds to one of the special reference triangles, and we must find θ.

Finding Special Angles θ

Find the least positive θ in degree and radian measure for which each is true.

(A) $\sin \theta = \sqrt{3}/2$ (B) $\cos \theta = -1/\sqrt{2}$

Solution (A) Draw a reference triangle (Fig. 17) in the first quadrant with side opposite reference angle $\sqrt{3}$ and hypotenuse 2. Observe that this is a special 30°–60° triangle:

FIGURE 17

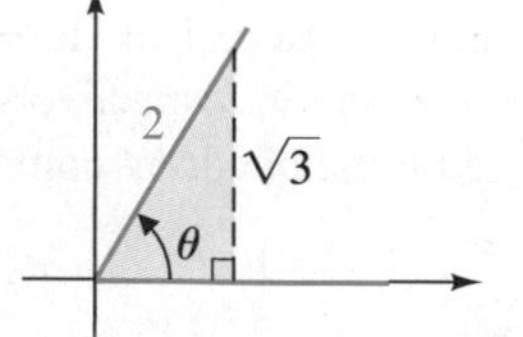

$\theta = 60°$ or $\frac{\pi}{3}$

(B) Draw a reference triangle (Fig. 18) in the second quadrant with side adjacent reference angle -1 and hypotenuse $\sqrt{2}$. Observe that this is a special 45° triangle:

FIGURE 18

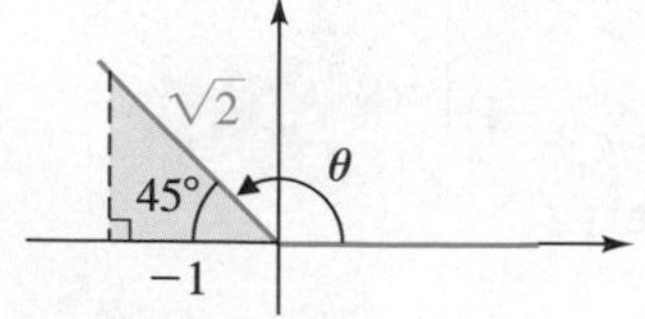

$\theta = 180° - 45° = 135°$ or $\frac{3\pi}{4}$

Matched Problem 5 Repeat Example 5 for:

(A) $\tan \theta = 1/\sqrt{3}$ (B) $\sec \theta = -\sqrt{2}$ ■

We conclude this section with a summary of special values in Table 1. Some people memorize this table; others use the definition in Section 2.3 and special triangles.

TABLE 1
Special Values

θ	$\sin \theta$	$\csc \theta$	$\cos \theta$	$\sec \theta$	$\tan \theta$	$\cot \theta$
0° or 0	0	N.D.	1	1	0	N.D.
30° or $\pi/6$	1/2	2	$\sqrt{3}/2$	$2/\sqrt{3}$	$1/\sqrt{3}$	$\sqrt{3}$
45° or $\pi/4$	$1/\sqrt{2}$	$\sqrt{2}$	$1/\sqrt{2}$	$\sqrt{2}$	1	1
60° or $\pi/3$	$\sqrt{3}/2$	$2/\sqrt{3}$	1/2	2	$\sqrt{3}$	$1/\sqrt{3}$
90° or $\pi/2$	1	1	0	N.D.	N.D.	0

N.D. = Not defined

The special angle values for sine and cosine are easily remembered if you observe the (unexpected) pattern after completing Table 2 in Explore–Discuss 3.

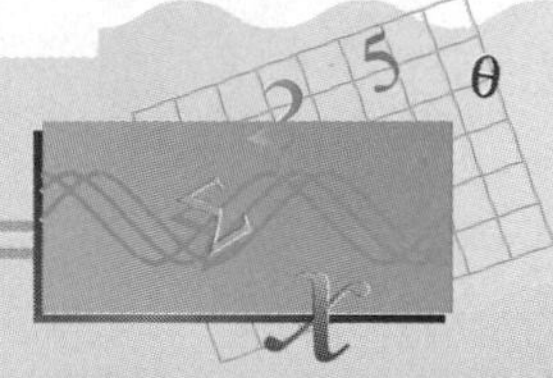

EXPLORE/DISCUSS 3

Fill in the sine column in Table 2 with a pattern of values similar to those in the cosine column. Discuss how the two columns are generated and how they are related.

TABLE 2
Memory Aid

θ	$\sin \theta$	$\cos \theta$
0° or 0		$\sqrt{4}/2 = 1$
30° or $\pi/6$		$\sqrt{3}/2$
45° or $\pi/4$		$\sqrt{2}/2 = 1\sqrt{2}$
60° or $\pi/3$		$\sqrt{1}/2 = 1/2$
90° or $\pi/2$		$\sqrt{0}/2 = 0$

Answers to Matched Problems

1. (A) -1 (B) -1 (C) Not defined (D) 0

2. (A)

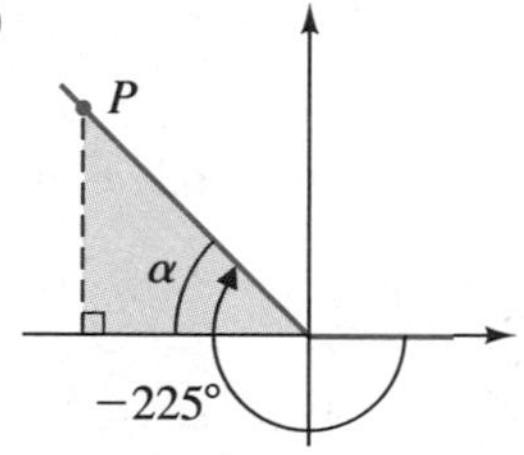

$\alpha = 225° - 180° = 45°$

(B)

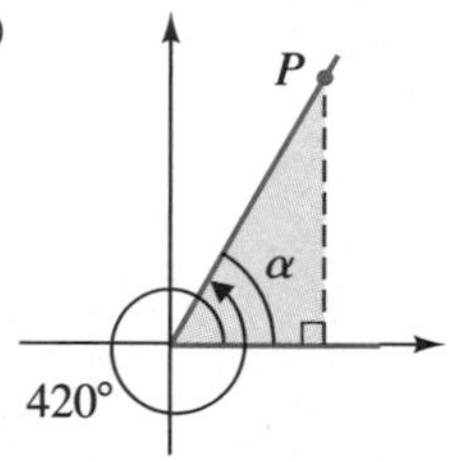

$\alpha = 420° - 360° = 60°$

(C)

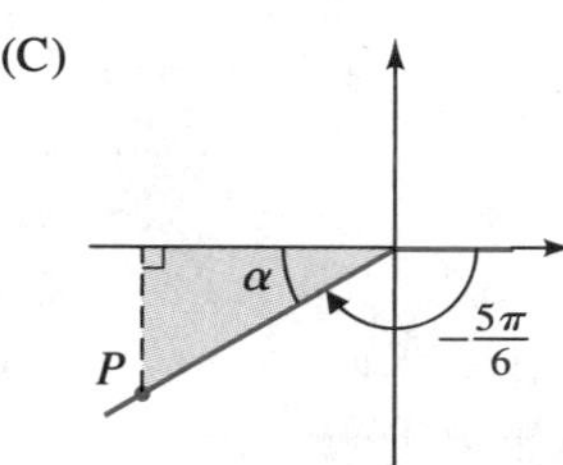

$\alpha = \pi - 5\pi/6 = \pi/6$

(D)

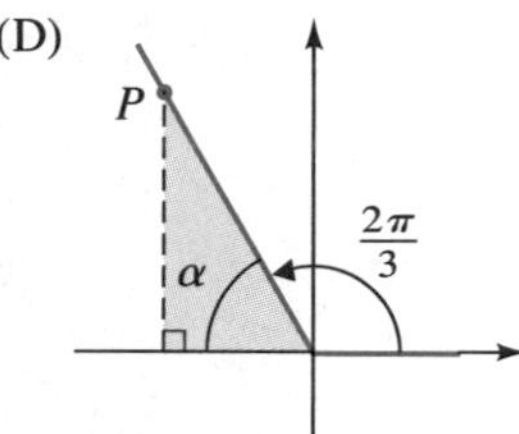

$\alpha = \pi - 2\pi/3 = \pi/3$

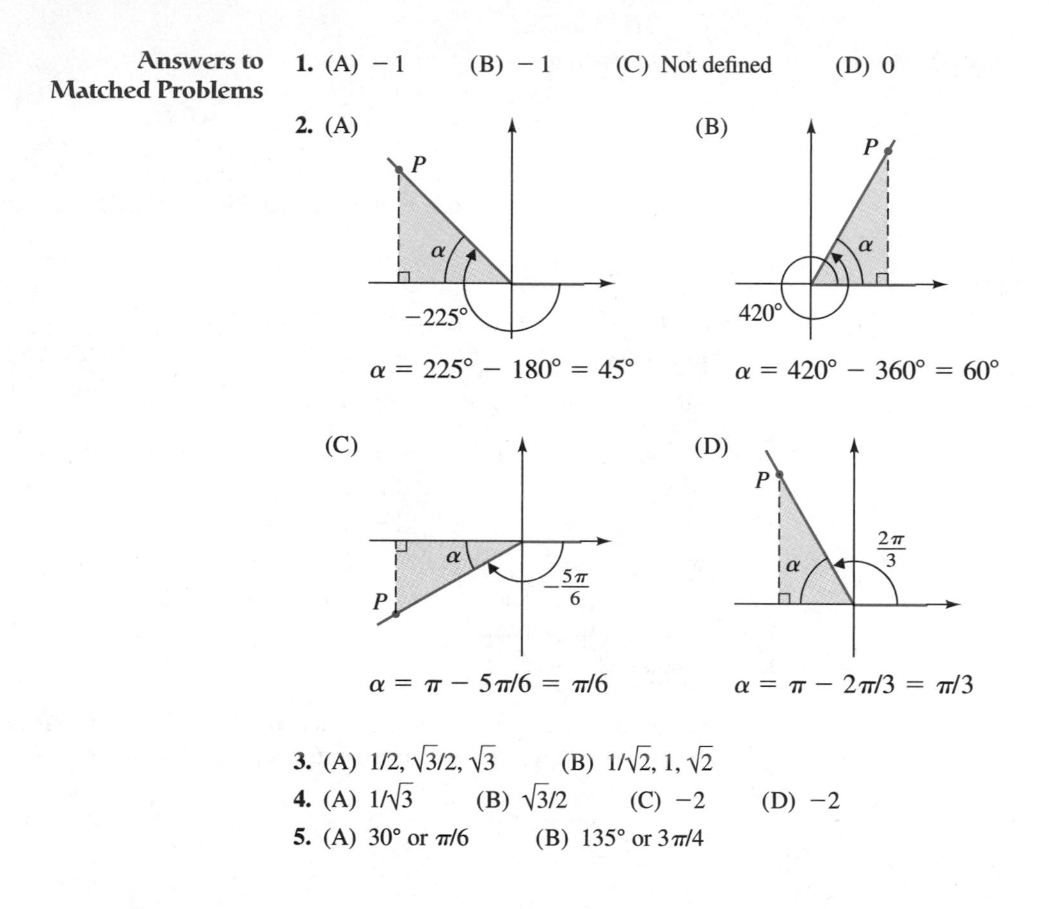

3. (A) $1/2, \sqrt{3}/2, \sqrt{3}$ (B) $1/\sqrt{2}, 1, \sqrt{2}$
4. (A) $1/\sqrt{3}$ (B) $\sqrt{3}/2$ (C) -2 (D) -2
5. (A) $30°$ or $\pi/6$ (B) $135°$ or $3\pi/4$

EXERCISE 2.5

Do not use a calculator for any of the problems in this exercise.

A *Sketch the reference triangle and find the reference angle α for each of the following angles.*

1. $\theta = 60°$ **2.** $\theta = 45°$ **3.** $\theta = -60°$

4. $\theta = -45°$ **5.** $\theta = \frac{-\pi}{3}$ **6.** $\theta = \frac{-\pi}{4}$

7. $\theta = \frac{3\pi}{4}$ **8.** $\theta = \frac{5\pi}{6}$ **9.** $\theta = -210°$

10. $\theta = -150°$ **11.** $\theta = \frac{-5\pi}{4}$ **12.** $\theta = \frac{-5\pi}{3}$

Find the exact value of each of the following.

13. $\cos 0°$ **14.** $\sin 0°$ **15.** $\sin 30°$

16. $\cos 45°$ **17.** $\sin \frac{\pi}{2}$ **18.** $\cot \frac{\pi}{4}$

19. $\tan 45°$ **20.** $\tan 0$ **21.** $\tan \frac{\pi}{6}$

22. $\cot 0$ **23.** $\sin(-30°)$ **24.** $\cos(-60°)$

25. $\cos \frac{-\pi}{2}$ **26.** $\tan \pi$ **27.** $\cos \frac{-\pi}{6}$

28. $\tan \frac{-\pi}{4}$ **29.** $\cot 150°$ **30.** $\cot(-60°)$

B *In Problems 31–42, find the exact value of each:*

31. $\sin \frac{7\pi}{6}$ **32.** $\sin \frac{3\pi}{4}$

33. $\sin \frac{3\pi}{2}$ **34.** $\cos \frac{11\pi}{6}$

35. $\sin 225°$ **36.** $\cos 300°$

37. $\cot \frac{-5\pi}{4}$ **38.** $\tan \frac{-4\pi}{3}$

39. $\cos \frac{-5\pi}{6}$ **40.** $\sin \frac{-5\pi}{3}$

41. $\csc 420°$ **42.** $\sec 390°$

In Problems 43–46, find all angles θ, $0 \le \theta \le 2\pi$, for which the following functions are not defined. Explain why.

43. tangent **44.** cotangent

45. cosecant **46.** secant

In each graphing calculator display in Problems 47–50, indicate which values are not exact and find the exact value.

47.
```
sin(π/6)
            .5000
sin(-45°)
           -.7071
cos(0)
           1.0000
```

48.
```
cos(90°)
           0.0000
sin(2π/3)
            .8660
sin(π/6)
            .5000
```

49.
```
tan(45°)
           1.0000
tan(180°)
           0.0000
tan(-π/3)
          -1.7321
```

50.
```
sin(-150°)
           -.5000
tan(-150°)
            .5774
tan(5π/4)
           1.0000
```

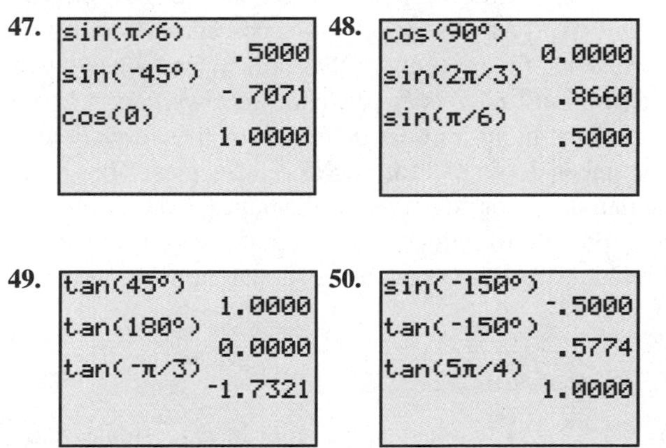

In Problems 51–56, find the least positive θ in
(A) Degree measure (B) Radian measure
for which each is true.

51. $\sin \theta = \frac{1}{2}$ **52.** $\cos \theta = \frac{1}{\sqrt{2}}$

53. $\cos \theta = \frac{-1}{2}$ **54.** $\sin \theta = \frac{-1}{2}$

55. $\tan \theta = -\sqrt{3}$ **56.** $\cot \theta = -1$

C **57.** Find the exact value of all the angles between 0° and 360° for which $\sin \theta = -\sqrt{3}/2$.

58. Find the exact value of all the angles between 0° and 360° for which $\tan \theta = -\sqrt{3}$.

59. Find the exact value of all the angles between 0 rad and 2π rad for which $\cot \theta = -\sqrt{3}$.

60. Find the exact value of all the angles between 0 rad and 2π rad for which $\cos \theta = -\sqrt{3}/2$.

For each graphing calculator display in Problems 61–64, find the least positive exact X in radian measure that produces the result shown.

61.
```
-√(2)/2
       -.70710678
cos(X)
       -.70710678
```

62.
```
√(2)/2
        .70710678
sin(X)
        .70710678
```

63.
```
√(3)/3
        .57735027
tan(X)
        .57735027
```

64.
```
-√(3)
      -1.73205081
tan(X)
      -1.73205081
```

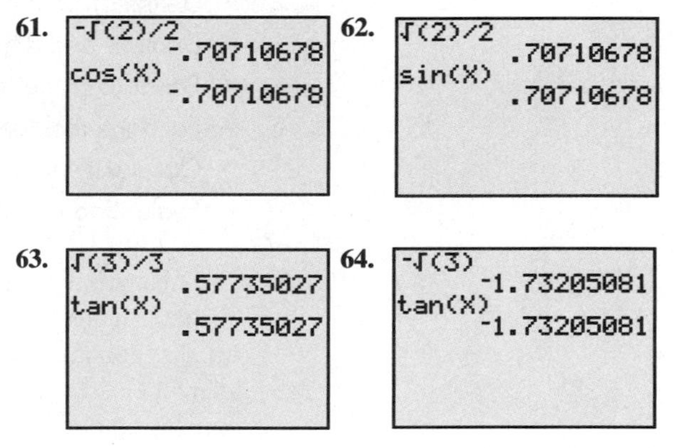

Find the exact values of x and y in Problems 65 and 66.

65. (A) (B) (C)

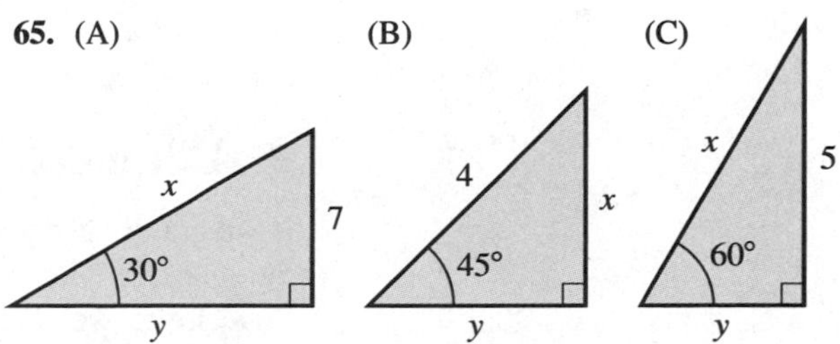

66. (A) (B) (C)

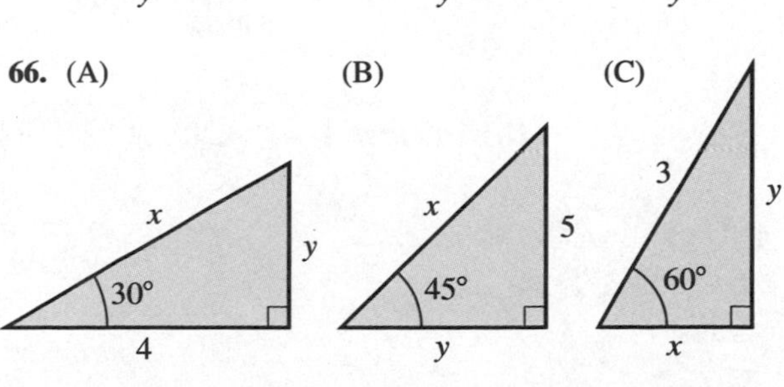

Later, we will show that the area of an n-sided regular polygon inscribed in a circle of radius r (such as in the figure) is given by

$$A = \frac{nr^2}{2} \sin \frac{2\pi}{n}$$

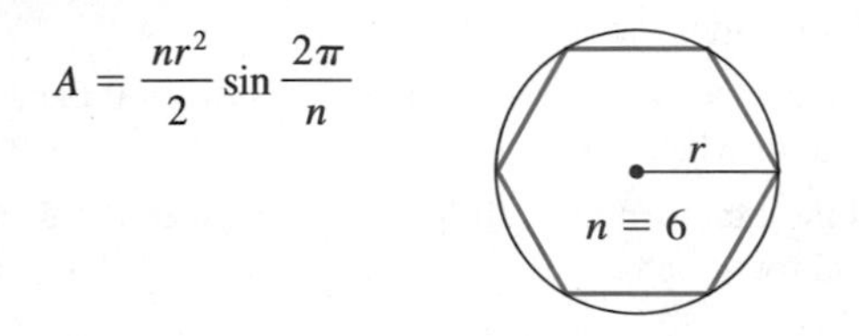

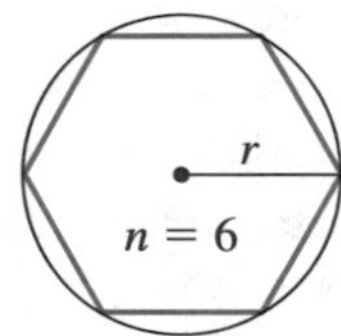

Use this formula to find the exact areas of the polygons defined in Problems 67–70. Assume each radius is exact.

67. $n = 3, \quad r = 2$ cm **68.** $n = 4, \quad r = 5$ ft

69. $n = 6, \quad r = 10$ in. **70.** $n = 8, \quad r = 4$ mm

2.6 CIRCULAR FUNCTIONS

- Definition of Circular Functions
- Domain and Range for Sine and Cosine Functions
- Periodic Properties
- Fundamental Identities
- Circular Functions and Trigonometric Functions
- Evaluating Circular Functions

Our treatment of trigonometric functions has progressed from the concrete to the abstract, which follows the historical development of the subject. We first defined trigonometric ratios relative to right triangles—a procedure used by the ancient Greeks. We then expanded the meaning of these functions by using generalized angles in standard positions in a rectangular coordinate system.

We now turn to the modern, more abstract definition of these functions, a definition involving real number domains, not angles or triangles. This is the preferred and most useful definition for advanced mathematics and the sciences, including calculus. We will be able to quickly observe some very useful trigonometric properties and relationships that were not as apparent using the angle approach.

Definition of Circular Functions

If we graph the equation $a^2 + b^2 = 1$ in a rectangular coordinate system, we obtain a circle with center at the origin and radius 1 called the **unit circle.** Using this circle, we define the **circular functions** with real number domains, as follows.

Definition 1 CIRCULAR FUNCTIONS

Let x be an arbitrary real number and let U be the unit circle with equation $a^2 + b^2 = 1$. Start at (1, 0) and proceed counterclockwise if x is positive and clockwise if x is negative around the unit circle until an arc length of $|x|$ has been covered. Let $P(a, b)$ be the point at the terminal end of the arc.

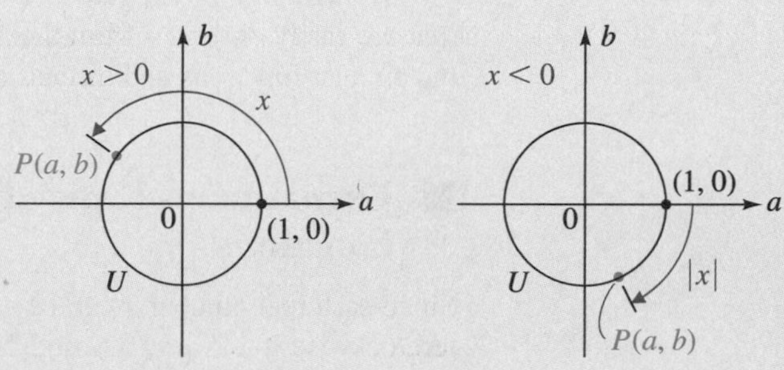

The following six **circular functions*** are defined in terms of the coordinates of P:

$$y = \sin x = b \qquad y = \cos x = a \qquad y = \tan x = \frac{b}{a}$$

$$y = \csc x = \frac{1}{b} \qquad y = \sec x = \frac{1}{a} \qquad y = \cot x = \frac{a}{b}$$

The independent variable is x and the dependent variable is y.

Remarks

1. The circular function definitions do not involve any angles.
2. The circular function definitions use standard function notation

$$y = f(x)$$

with f replaced by the name of a particular circular function; for example, $y = \sin x$ actually means $y = \sin(x)$. (See Appendix B.1 for a brief discussion of function and function notation.)

3. When we write $y = f(x) = x^2 - 1$, the expression on the right has a recipe for evaluating f (square x and subtract 1). When we write $y = f(x) = \sin x$, the expression on the right only identifies the function; for its evaluation we must refer to the circular function definition. □

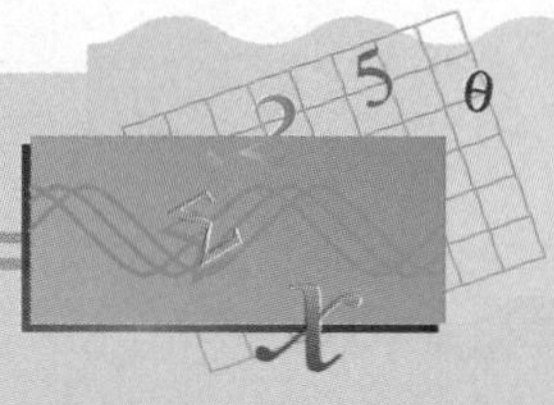

EXPLORE/DISCUSS 1

(A) For the function $y = f(x) = 3 - x^2$, describe the process in words for determining y for each real value x.

(B) For the function $y = h(x) = \sin x$, describe the process in words (in terms of the circle-based definition of the sine function) for determining y for each real value x.

---

* Following common usage, we also refer to circular functions as trigonometric functions.

We will now investigate a few important properties of the circular functions that are easily observed from their definitions. In the next chapter we will graph the circular functions and discuss many additional properties.

Domain and Range for Sine and Cosine Functions

Since each real number x can be associated with an arc $|x|$ units in length (counterclockwise if x is positive and clockwise if x is negative), the domain of each circular function is the set of real numbers for which the function is defined.

We now investigate the domains and ranges of the sine and cosine functions. (The domains and ranges of the other four circular functions will be covered in detail in the next chapter.) Referring to the definition of circular functions, observe that

$$a = \cos x \qquad \text{and} \qquad b = \sin x$$

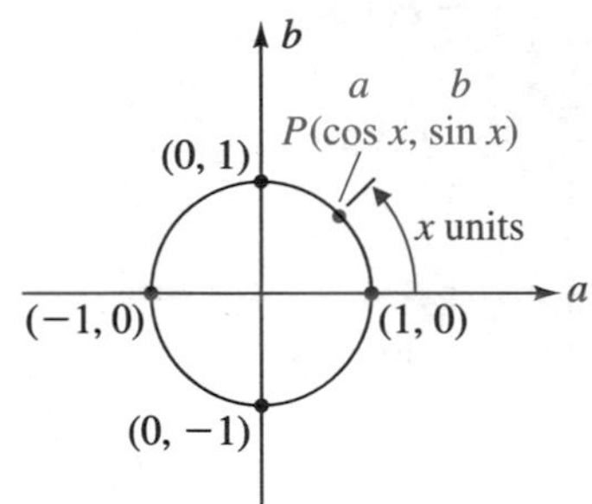

FIGURE 1

Thus, as in Figure 1, the coordinates of a point P at the end of an arc of length $|x|$, starting at (1, 0), are

$$P(\cos x, \sin x)$$

It is clear that $\cos x$ and $\sin x$ exist for each real number x, since $(\cos x, \sin x)$ are the coordinates of the point P on the unit circle corresponding to the arc length $|x|$. Thus, the domain of each function is the set of all real numbers.

What about the range of each function? As $P(\cos x, \sin x)$ moves around the unit circle, the abscissa of the point, $\cos x = a$, and the ordinate of the point, $\sin x = b$, both vary between -1 and 1. Thus, we conclude that the range of each function is the set of all real numbers y such that $-1 \le y \le 1$.

The above results are summarized in the box below:

DOMAIN AND RANGE FOR SINE AND FOR COSINE

Domain: All real numbers
Range: $-1 \le y \le 1$, y a real number

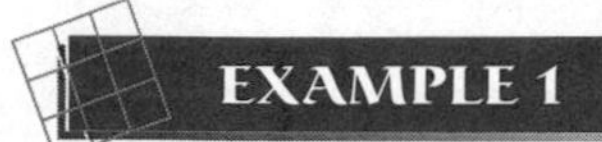

EXAMPLE 1

Finding Domain Values That Correspond to a Given Range Value

Find the domain values of the sine function, $-2\pi \le x \le 2\pi$, that have a range value -1. That is, find x, $-2\pi \le x \le 2\pi$, such that $\sin x = -1$.

Solution Refer to Figure 1 to see that $\sin x = -1$ at the point $(0, -1)$. Thus, for any domain value x associated with an arc starting at (1, 0) and terminating at $(0, -1)$, we have $\sin x = -1$. For the interval $-2\pi \le x \le 2\pi$, $\sin x = -1$ for $x = -\pi/2$ and $x = 3\pi/2$, as indicated in Figure 2.

FIGURE 2

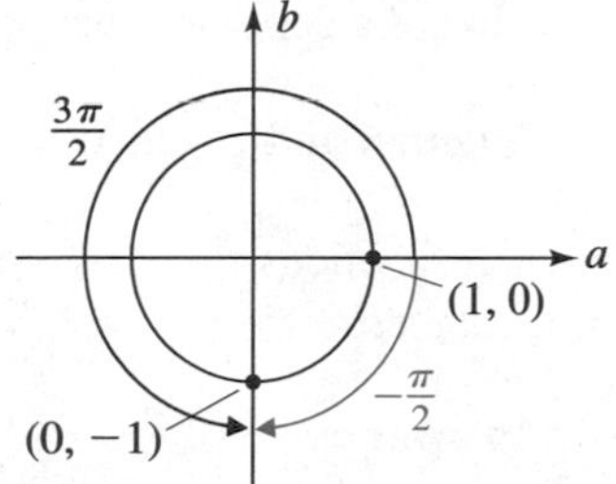

Matched Problem 1 Find the domain values of the cosine function, $-2\pi \le x \le 2\pi$, that have a range value -1. That is, find x, $-2\pi \le x \le 2\pi$, such that $\cos x = -1$.

Periodic Properties

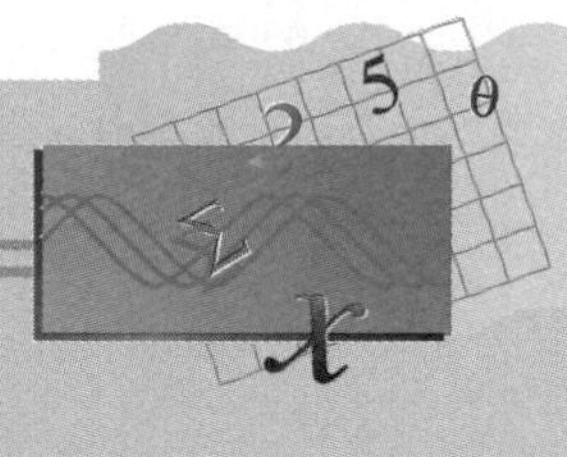

EXPLORE/DISCUSS 2

Using Figure 1, explore the behavior of $P(\cos x, \sin x)$ as x moves around the unit circle in either direction. For what values of x will the coordinates repeat? Conclusions?

The circumference of the unit circle is

$$C = 2\pi r = 2\pi(1) = 2\pi$$

Thus, for a given value x (see Fig. 1), if we add or subtract 2π, we will return to exactly the same point P with the same coordinates. And we conclude that

$$\sin(x + 2\pi) = \sin x \quad \text{and} \quad \cos(x + 2\pi) = \cos x$$
$$\sin(x - 2\pi) = \sin x \quad \text{and} \quad \cos(x - 2\pi) = \cos x$$

In general, if we add any integer multiple of 2π to x, we will return to exactly the same point P. Thus, for k any integer $(\ldots, -3, -2, -1, 0, 1, 2, 3, \ldots)$,

$$\sin(x + 2k\pi) = \sin x \quad \text{and} \quad \cos(x + 2k\pi) = \cos x$$

Functions with this kind of repetitive behavior are called **periodic functions.** In general:

PERIODIC FUNCTIONS

A function f is **periodic** if there is a positive real number p such that

$$f(x + p) = f(x)$$

for all x in the domain of f. The smallest such positive p, if it exists, is called **the period of f.**

From the definition of a periodic function, we conclude that:

Both the sine function and cosine function have a period of 2π.

The other four circular functions also have periodic properties, which we discuss in detail in the next chapter.

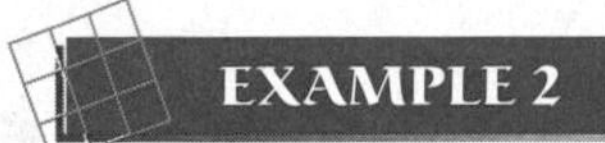

EXAMPLE 2 **Using Periodic Properties**

If $\cos x = -0.0315$, what is the value of each of the following?

(A) $\cos(x + 2\pi)$ (B) $\cos(x - 2\pi)$
(C) $\cos(x + 18\pi)$ (D) $\cos(x - 34\pi)$

Solution All are equal to -0.0315, because the cosine function is periodic with period 2π. That is, $\cos(x + 2k\pi) = \cos x$ for *all* integers k. In part (A), $k = 1$; in part (B), $k = -1$; in part (C), $k = 9$; and in part (D), $k = -17$. ■

Matched Problem 2 If $\sin x = 0.7714$, what is the value of each of the following?

(A) $\sin(x + 2\pi)$ (B) $\sin(x - 2\pi)$
(C) $\sin(x + 14\pi)$ (D) $\sin(x - 26\pi)$ ■

The periodic properties of the circular functions are of paramount importance in the development of further mathematics as well as the applications of mathematics. As we will see in the next chapter, the circular functions are made to order for the analysis of real-world periodic phenomena: light, sound, electrical, and water waves; motion in buildings during earthquakes; motion in suspension systems in automobiles, planetary motion; business cycles; and so on.

Fundamental Identities

Returning to the definition of the circular functions and noting that

$$\sin x = b \qquad \text{and} \qquad \cos x = a$$

we can obtain the following useful relationships among the six functions:

$$\csc x = \frac{1}{b} = \frac{1}{\sin x} \tag{1}$$

$$\sec x = \frac{1}{a} = \frac{1}{\cos x} \tag{2}$$

$$\cot x = \frac{a}{b} = \frac{1}{b/a} = \frac{1}{\tan x} \tag{3}$$

$$\tan x = \frac{b}{a} = \frac{\sin x}{\cos x} \tag{4}$$

$$\cot x = \frac{a}{b} = \frac{\cos x}{\sin x} \tag{5}$$

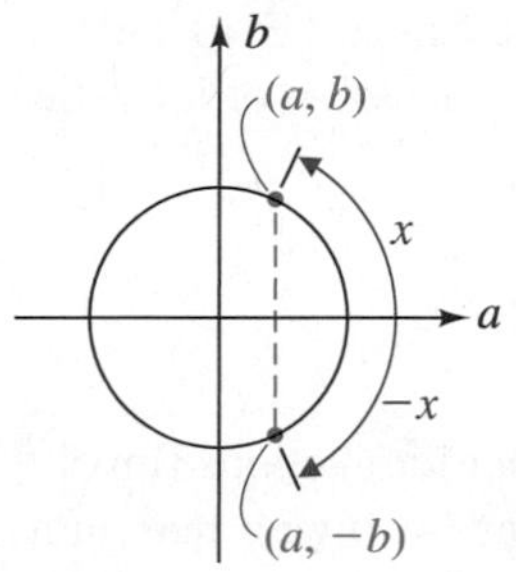

FIGURE 3

Because the terminal points of x and $-x$ are symmetric with respect to the horizontal axis (see Fig. 3), we have the following sign properties:

$$\sin(-x) = -b = -\sin x \tag{6}$$

$$\cos(-x) = a = \cos x \tag{7}$$

$$\tan(-x) = \frac{-b}{a} = -\frac{b}{a} = -\tan x \tag{8}$$

Finally, because $(a, b) = (\cos x, \sin x)$ is on the unit circle $a^2 + b^2 = 1$, it follows that

$$(\cos x)^2 + (\sin x)^2 = 1$$

which is usually written in the form

$$\sin^2 x + \cos^2 x = 1 \tag{9}$$

where $\sin^2 x$ and $\cos^2 x$ are concise ways of writing $(\sin x)^2$ and $(\cos x)^2$, respectively.

Caution Note that $(\cos x)^2 \neq \cos x^2$ and $(\sin x)^2 \neq \sin x^2$. □

Equations (1)–(9) are called **fundamental identities.** They hold true for all replacements of x by real numbers (or angles in degree or radian measure, as we will see before the conclusion of this section) for which both sides of an equation are defined.

EXAMPLE 3 **Use of Identities**

Simplify each expression using the fundamental identities.

(A) $\dfrac{\sin^2 x + \cos^2 x}{\tan x}$ (B) $\dfrac{\sin(-x)}{\cos(-x)}$

Solution (A) $\dfrac{\sin^2 x + \cos^2 x}{\tan x}$ Use identity (9).

$= \dfrac{1}{\tan x}$ Use identity (3).

$= \cot x$

(B) $\dfrac{\sin(-x)}{\cos(-x)}$ Use identity (4).

$= \tan(-x)$ Use identity (8).

$= -\tan x$ ■

Matched Problem 3 Simplify each expression using the fundamental identities.

(A) $\dfrac{1 - \cos^2 x}{\sin^3 x}$ (B) $\tan(-x)\cos(-x)$ ■

Some of the fundamental identities will be used in Chapter 3 as aids to graphing some of the trigonometric functions. A detailed discussion of identities is found in Chapter 4.

Circular Functions and Trigonometric Functions

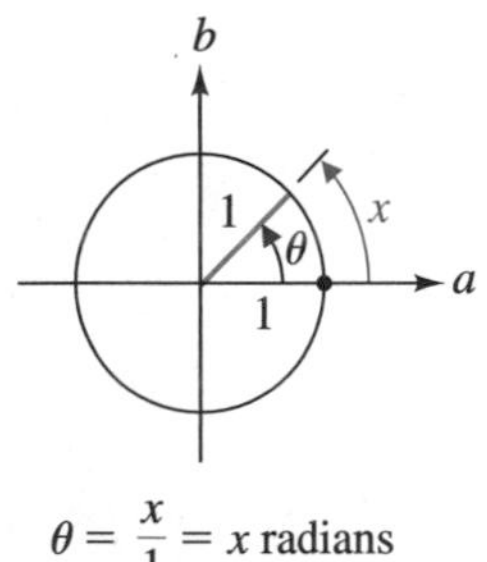

$\theta = \frac{x}{1} = x$ radians

FIGURE 4

We now show how the earlier definitions of the trigonometric functions (involving angle domains) can be related to the circular functions (involving real number domains and the unit circle). To this end, let us look at the radian measure of an angle θ subtended by an arc of x units on the unit circle (Fig. 4).

We see that for the unit circle, the angle subtended by an arc of x units has a radian measure of x. Thus, every real number can be associated with an arc of x units on the unit circle or a central angle of x radians on the same circle (if x is positive, we go counterclockwise; and if x is negative, we go clockwise). Notice that the point on the terminal end of the arc of x units is also on the terminal side of the angle of x radians. This provides the following very useful relationships between the previously defined trigonometric functions with angle domains and the circular functions with real number domains:

CIRCULAR FUNCTIONS AND TRIGONOMETRIC FUNCTIONS

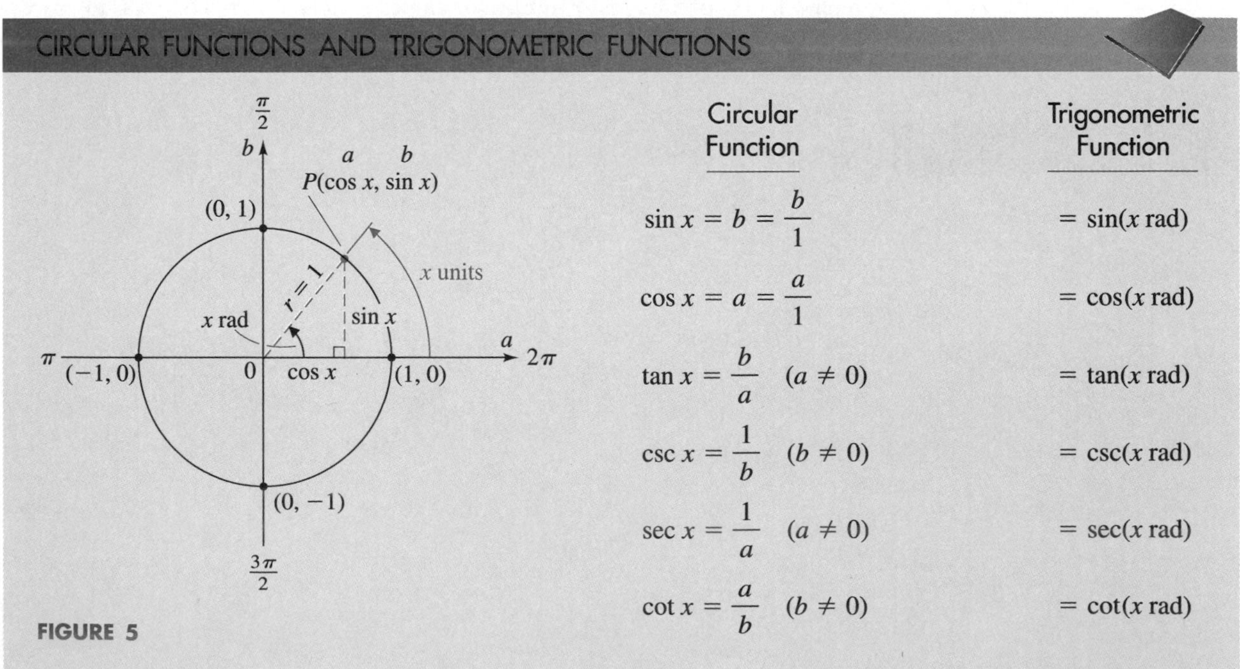

FIGURE 5

Circular Function	Trigonometric Function
$\sin x = b = \frac{b}{1}$	$= \sin(x \text{ rad})$
$\cos x = a = \frac{a}{1}$	$= \cos(x \text{ rad})$
$\tan x = \frac{b}{a} \quad (a \neq 0)$	$= \tan(x \text{ rad})$
$\csc x = \frac{1}{b} \quad (b \neq 0)$	$= \csc(x \text{ rad})$
$\sec x = \frac{1}{a} \quad (a \neq 0)$	$= \sec(x \text{ rad})$
$\cot x = \frac{a}{b} \quad (b \neq 0)$	$= \cot(x \text{ rad})$

The information in this box, which relates circular functions with trigonometric functions, is very useful and should be understood. We will use it to develop a number of properties that hold simultaneously for the circular functions and the trigonometric functions. (For example, the fundamental identities discussed above hold for all real numbers x or angles in degree or radian measure for which the functions are defined.)

Evaluating Circular Functions

Because circular functions are related to trigonometric functions, we can evaluate circular functions by the procedures we used to evaluate trigonometric functions.

If x is an integer multiple of $\pi/4$ or $\pi/6$, we can find exact values of the six circular functions when they are defined. Corresponding coordinates of points on the unit circle can be found by using appropriate reference triangles and angles. (Recall the special 45° and 30°–60° triangles on page 97 in Section 2.5.) Using these relationships, we can determine coordinates of points that correspond to integer multiples of $\pi/4$ or $\pi/6$. Figure 6 shows these coordinates for multiples ranging from 0 to 2π.

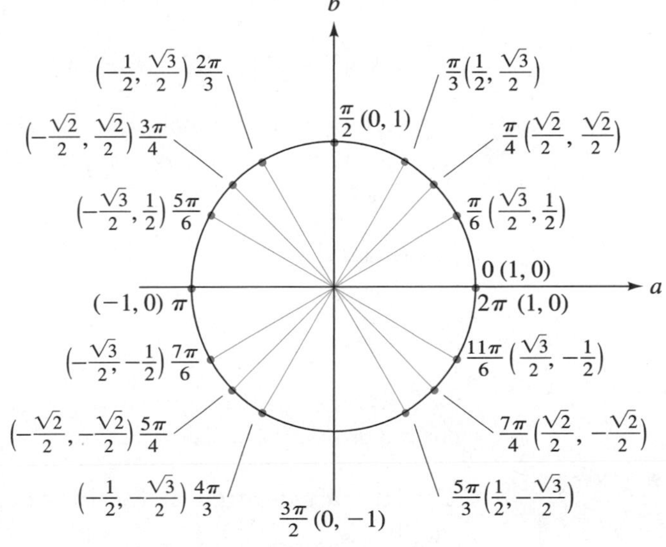

FIGURE 6
Multiples of $\pi/6$ and $\pi/4$ on a unit circle

Note that if you learn the coordinates for the points corresponding to $\pi/6$, $\pi/4$, and $\pi/3$ in the first quadrant, you can easily obtain the coordinates of the corresponding points in the other three quadrants by symmetry and reflection across the coordinate axis. To evaluate the six circular functions exactly for integer multiples of $\pi/6$ or $\pi/4$, refer directly to Figure 6 or proceed as in Section 2.5 to relate the real number x to an angle of x radians.

EXAMPLE 4 **Evaluating Circular Functions Exactly**

Find each exactly:

(A) $\sin \frac{8\pi}{3}$ (B) $\sec\left(-\frac{5\pi}{6}\right)$ (C) $\tan 7\pi$

Solution (A) Refer directly to Figure 6 or the fact that

$$\sin \frac{8\pi}{3} = \sin\left(\frac{8\pi}{3} \text{ rad}\right)$$

and proceed using reference triangles as in Section 2.5. In either case,

$$\sin \frac{8\pi}{3} = \frac{\sqrt{3}}{2}$$

(B) Refer directly to Figure 6 or the fact that

$$\sec\left(-\frac{5\pi}{6}\right) = \sec\left(-\frac{5\pi}{6} \text{ rad}\right)$$

and proceed using reference triangles as in Section 2.5. In either case,

$$\sec\left(-\frac{5\pi}{6}\right) = -\frac{2}{\sqrt{3}}$$

(C) 7π corresponds to the point $(-1, 0)$. Thus,

$$\tan 7\pi = \frac{0}{-1} = 0$$

■

Matched Problem 4 Find each exactly:

(A) $\cos \frac{13\pi}{6}$ (B) $\csc\left(-\frac{5\pi}{4}\right)$ (C) $\cot(-3\pi)$ ■

To evaluate circular functions using a calculator, set the calculator in radian mode and proceed as in Section 2.5.

EXAMPLE 5 **Calculator Evaluation of Circular Functions**

Evaluate to four significant digits:

(A) $\sin(-13.72)$ (B) $\sec 22.33$

Solution (A) $\sin(-13.72) = -0.9142$ *Set in radian mode.*

(B) $\sec 22.33 = \boxed{\frac{1}{\cos 22.33}} = -1.060$ *Set in radian mode.* ■

Matched Problem 5 Evaluate to four significant digits using a calculator:

(A) $\cos 505.3$ (B) $\cot(-0.003211)$ ■

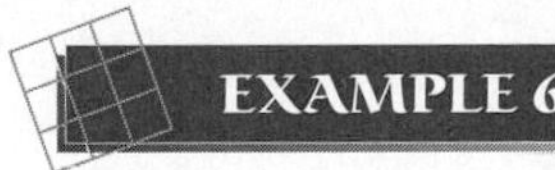

Periodic Properties

Evaluate $\cos x$ to two significant digits for:

(A) $x = 1.4$ (B) $x = 1.4 + 2\pi$ (C) $x = 1.4 - 2\pi$
(D) $x = 1.4 + 20\pi$ (E) $x = 1.4 - 8\pi$

Solution All have the same value because the cosine function is periodic with a period of 2π. That is, $\cos(x + 2k\pi) = \cos x$ for all integers k. Using a calculator (set in radian mode to evaluate real numbers), we obtain:

(A) $\cos 1.4 = 0.17$ (B) $\cos(1.4 + 2\pi) = 0.17$
(C) $\cos(1.4 - 2\pi) = 0.17$ (D) $\cos(1.4 + 20\pi) = 0.17$
(E) $\cos(1.4 - 8\pi) = 0.17$ ■

Matched Problem 6 Evaluate $\sin x$ to two significant digits for:

(A) $x = -3.2$ (B) $x = -3.2 + 2\pi$ (C) $x = -3.2 - 2\pi$
(D) $x = -3.2 + 22\pi$ (E) $x = -3.2 - 12\pi$ ■

Answers to Matched Problems
1. $-\pi, \pi$ **2.** All 0.7714 **3.** (A) $\csc x$ (B) $-\sin x$
4. (A) $\sqrt{3}/2$ (B) $\sqrt{2}$ (C) Not defined
5. (A) -0.8793 (B) -311.4 **6.** All 0.058

EXERCISE 2.6

Figure 5 is repeated here for convenient reference.

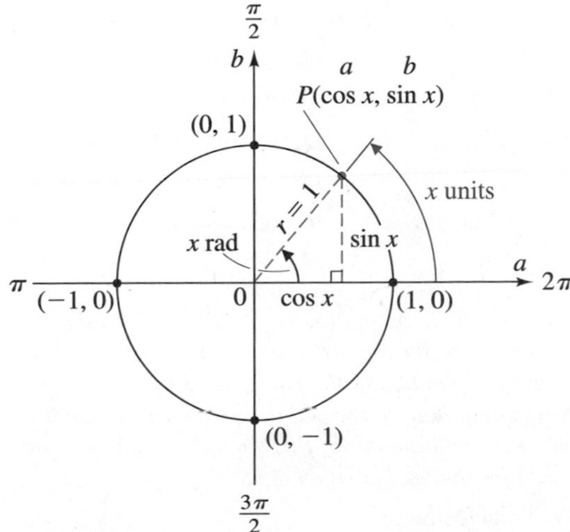

Figure for 1–24

A

1. Starting with the circumference of the unit circle, 2π, find:
(A) $\frac{1}{2}$ circumference (B) $\frac{3}{4}$ circumference

2. Starting with the circumference of the unit circle, 2π, find:
(A) $\frac{1}{4}$ circumference (B) $\frac{1}{8}$ circumference

3. Referring to the figure, state the coordinates of P for the indicated values of x:
(A) $x = 0$ (B) $x = \pi/2$ (C) $x = -3\pi/2$
(D) $x = \pi$ (E) $x = -2\pi$ (F) $x = 5\pi/2$

4. Referring to the figure, state the coordinates of P for the indicated values of x:
(A) $x = 2\pi$ (B) $x = -\pi/2$ (C) $x = -\pi$
(D) $x = 3\pi/2$ (E) $x = -5\pi/2$ (F) $x = -3\pi$

5. Given $y = \sin x$, how does y vary for the indicated variation in x?
(A) x varies from 0 to $\pi/2$
(B) x varies from $\pi/2$ to π

(C) x varies from π to $3\pi/2$
(D) x varies from $3\pi/2$ to 2π
(E) x varies from 2π to $5\pi/2$

6. Given $y = \cos x$, how does y vary for the indicated variation in x?
(A) x varies from 0 to $\pi/2$
(B) x varies from $\pi/2$ to π
(C) x varies from π to $3\pi/2$
(D) x varies from $3\pi/2$ to 2π
(E) x varies from 2π to $5\pi/2$

7. Given $y = \cos x$, how does y vary for the indicated variation in x?
(A) x varies from 0 to $-\pi/2$
(B) x varies from $-\pi/2$ to $-\pi$
(C) x varies from $-\pi$ to $-3\pi/2$
(D) x varies from $-3\pi/2$ to -2π
(E) x varies from -2π to $-5\pi/2$

8. Given $y = \sin x$, how does y vary for the indicated variation in x?
(A) x varies from 0 to $-\pi/2$
(B) x varies from $-\pi/2$ to $-\pi$
(C) x varies from $-\pi$ to $-3\pi/2$
(D) x varies from $-3\pi/2$ to -2π
(E) x varies from -2π to $-5\pi/2$

B *Find the exact values of x from the indicated interval that satisfy the indicated equation or condition. (Refer to the figure at the beginning of the exercise.)*

9. $\sin x = 1, \quad 0 \le x \le 4\pi$
10. $\cos x = 1, \quad 0 \le x \le 4\pi$
11. $\sin x = 0, \quad 0 \le x \le 4\pi$
12. $\cos x = 0, \quad 0 \le x \le 4\pi$
13. $\tan x = 0, \quad 0 \le x \le 4\pi$
14. $\cot x = 0, \quad 0 \le x \le 4\pi$
15. $\sin x = -1, \quad 0 \le x \le 4\pi$
16. $\cos x = -1, \quad 0 \le x \le 4\pi$
17. $\cos x = 1, \quad -2\pi \le x \le 2\pi$
18. $\sin x = 1, \quad -2\pi \le x \le 2\pi$
19. $\cos x = 0, \quad -2\pi \le x \le 2\pi$
20. $\sin x = 0, \quad -2\pi \le x \le 2\pi$
21. $\tan x$ not defined, $\quad 0 \le x \le 4\pi$
22. $\cot x$ not defined, $\quad 0 \le x \le 4\pi$
23. $\csc x$ not defined, $\quad 0 \le x \le 4\pi$
24. $\sec x$ not defined, $\quad 0 \le x \le 4\pi$

In Problems 25–36, determine each to one significant digit. Use only the figure below, the definition of circular functions, and a calculator if necessary for multiplication and division.

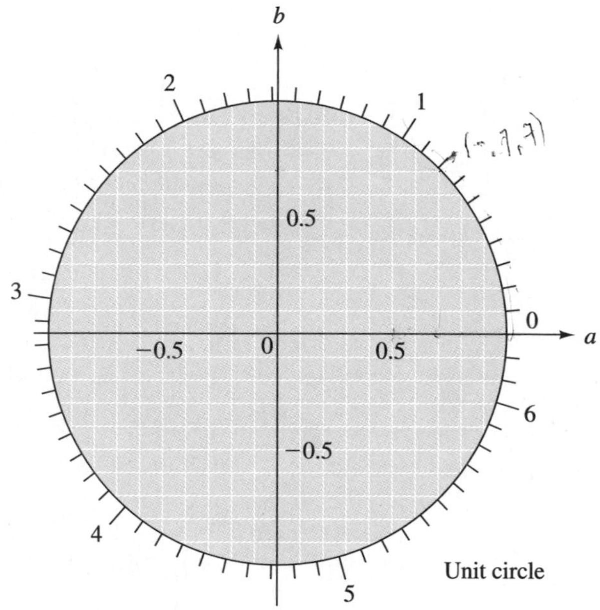

Figure for 25–36

25. sin 0.8	26. cos 0.8	27. cos 2.3
28. sin 5.5	29. sin(−0.9)	30. cos(−2)
31. sec 2.2	32. csc 3.8	33. tan 0.8
34. cot 2.8	35. cot(−0.4)	36. tan(−4)

Evaluate to four significant digits using a calculator.

37. sin(−0.2103)	38. cos 14.78
39. sec 1.432	40. cot(−7.809)
41. tan 4.704	42. csc(−3.109)
43. cos 105.2	44. sin 23.04
45. cot(−0.03333)	46. tan(−4.7)
47. csc 6.2	48. sec(−1.5)

In Problems 49–52, a point $P(a, b)$ starts at $(1, 0)$ and moves around a unit circle for the distance and direction indicated. Explain how you would find the coordinates of point P at its final position, and how you would determine which quadrant P is in. Find the coordinates of P to four decimal places and the quadrant for the final position of P.

49. 0.898 unit clockwise
50. 2.037 units clockwise

51. 26.77 units counterclockwise
52. 44.86 units counterclockwise

Evaluate Problems 53–58 exactly.

53. $\cos \frac{3\pi}{4}$ **54.** $\sin \frac{2\pi}{3}$ **55.** $\csc\left(-\frac{\pi}{4}\right)$

56. $\tan\left(-\frac{7\pi}{6}\right)$ **57.** $\tan\left(-\frac{5\pi}{2}\right)$ **58.** $\cot(-3\pi)$

59. If $\sin x = 0.9525$, what is the value of each of the following?
(A) $\sin(x + 2\pi)$ (B) $\sin(x - 2\pi)$
(C) $\sin(x + 10\pi)$ (D) $\sin(x - 6\pi)$

60. If $\cos x = -0.0379$, what is the value of each of the following?
(A) $\cos(x + 2\pi)$ (B) $\cos(x - 2\pi)$
(C) $\cos(x + 8\pi)$ (D) $\cos(x - 12\pi)$

61. Evaluate $\tan x$ and $(\sin x)/(\cos x)$ to two significant digits for:
(A) $x = 1$ (B) $x = 5.3$ (C) $x = -2.376$

62. Evaluate $\cot x$ and $(\cos x)/(\sin x)$ to two significant digits for:
(A) $x = -1$ (B) $x = 8.7$ (C) $x = -12.64$

63. Evaluate $\sin(-x)$ and $-\sin x$ to two significant digits for:
(A) $x = 3$ (B) $x = -12.8$ (C) $x = 407$

64. Evaluate $\cos(-x)$ and $\cos x$ to two significant digits for:
(A) $x = 5$ (B) $x = -13.4$ (C) $x = -1{,}003$

65. Evaluate $\sin^2 x + \cos^2 x$ to two significant digits for:
(A) $x = 1$ (B) $x = -8.6$ (C) $x = 263$

66. Evaluate $1 - \sin^2 x$ and $\cos^2 x$ to two significant digits for:
(A) $x = 14$ (B) $x = -16.3$ (C) $x = 766$

Simplify each expression using the fundamental identities.

67. $\sin x \csc x$ **68.** $\cos x \sec x$

69. $\cot x \sec x$ **70.** $\tan x \csc x$

71. $\frac{\sin x}{1 - \cos^2 x}$ **72.** $\frac{\cos x}{1 - \sin^2 x}$

73. $\cot(-x)\sin(-x)$ **74.** $\tan(-x)\cos(-x)$

C *Problems 75 and 76 show the coordinates of a point on a unit circle in a graphing calculator window. Let s be the length of the least positive arc from (1, 0) to the point. Find s to three decimal places.*

75.

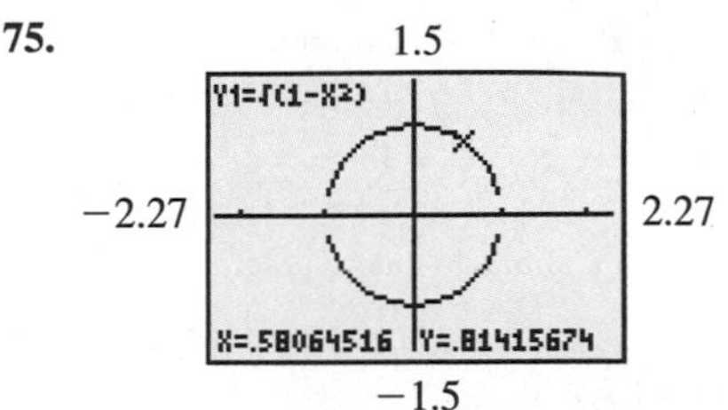

76.

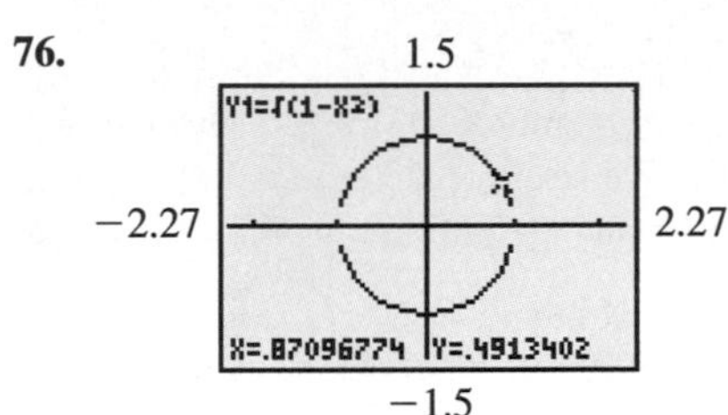

For Problems 77 and 78, fill the blanks in the Reason column with the appropriate identity, (1)–(9).

77.

Statement	Reason
$\tan^2 x + 1 = \left(\frac{\sin x}{\cos x}\right)^2 + 1$	(A) ______
$= \frac{\sin^2 x}{\cos^2 x} + 1$	Algebra
$= \frac{\sin^2 x + \cos^2 x}{\cos^2 x}$	Algebra
$= \frac{1}{\cos^2 x}$	(B) ______
$= \left(\frac{1}{\cos x}\right)^2$	Algebra
$= \sec^2 x$	(C) ______

78.

Statement	Reason
$\cot^2 x + 1 = \left(\frac{\cos x}{\sin x}\right)^2 + 1$	(A) ______
$= \frac{\cos^2 x}{\sin^2 x} + 1$	Algebra
$= \frac{\cos^2 x + \sin^2 x}{\sin^2 x}$	Algebra
$= \frac{1}{\sin^2 x}$	(B) ______
$= \left(\frac{1}{\sin x}\right)^2$	Algebra
$= \csc^2 x$	(C) ______

79. What is the period of the cosecant function?

80. What is the period of the secant function?

Applications

81. Precalculus: Pi Estimate With s_n as shown in the figure, a sequence of numbers is formed as indicated. Compute the first five terms of the sequence to six decimal places, and compare the fifth term with the value of $\pi/2$.

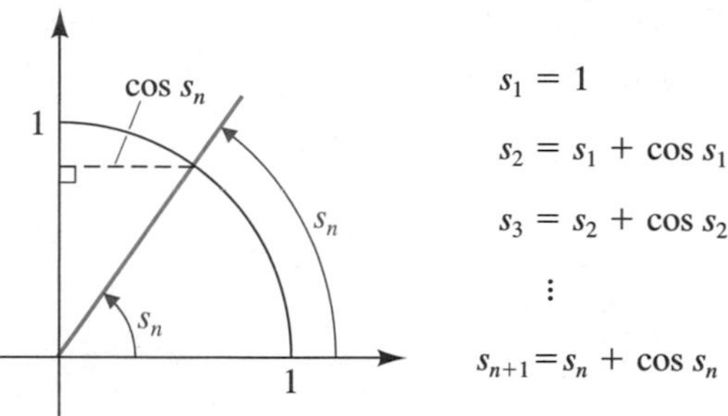

Figure for 81

82. Precalculus: Pi Estimate Repeat Problem 81 using $s_1 = 0.5$ as the first term of the sequence.

CHAPTER 2 GROUP ACTIVITY

Speed of Light in Water

Given the speed of light in air, how can you determine the speed of light in water by conducting a simple experiment? To start, read "Modeling Light Waves and Refraction" in Section 2.4. Snell's law for refracted light is stated here again for easy reference:

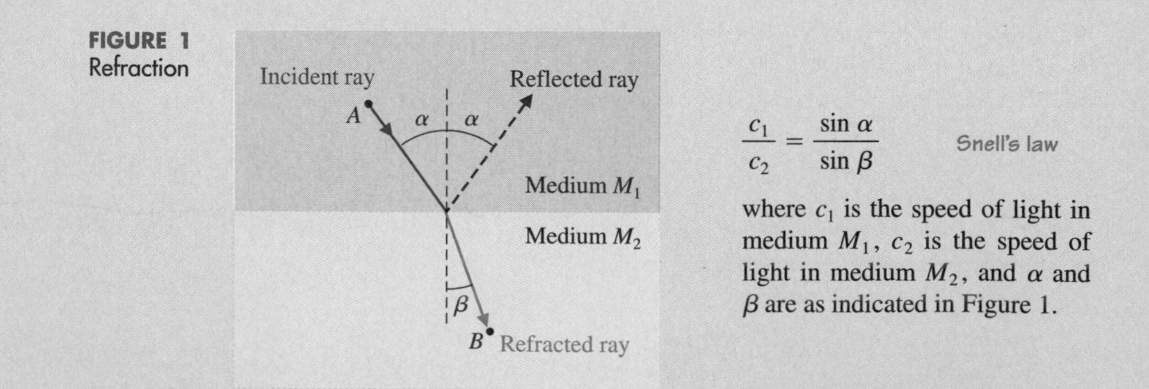

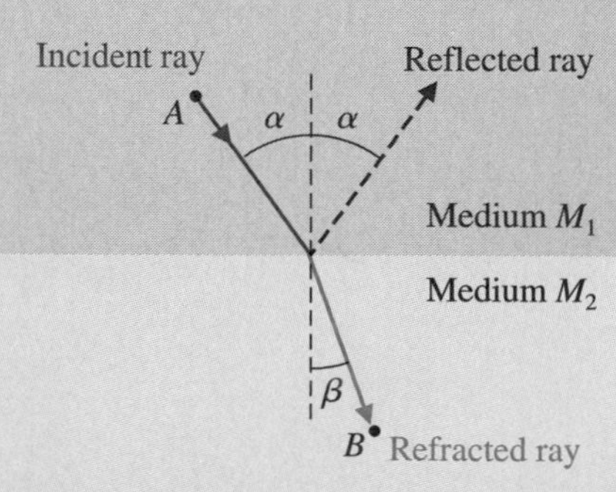

FIGURE 1 Refraction

$$\frac{c_1}{c_2} = \frac{\sin \alpha}{\sin \beta} \qquad \text{Snell's law}$$

where c_1 is the speed of light in medium M_1, c_2 is the speed of light in medium M_2, and α and β are as indicated in Figure 1.

The speed of light in clean air is $c_a = 186{,}225$ miles per second. The problem is to find the speed of light in clear water, c_w. Start by showing that

$$c_w = c_a \frac{\sin \beta}{\sin \alpha} \tag{1}$$

If you can find α and β for light going from a point in air to a point in water, you can calculate c_w using equation (1). The following simple experiment using readily available materials produces fairly good estimates for α and β, which in turn produce a surprisingly good estimate for the speed of light in water.

EXPERIMENT

Materials needed

- A large, straight-sided light-colored coffee mug—on the inside of the mug opposite the handle mark a small dark line along the edge where the bottom meets the side. Use a black or red marking pen.
- A centimeter scale, with each centimeter unit divided into tenths (millimeters).
- A piece of light-colored stiff cardboard approximately 8.5 in. by 11 in. (or 28 cm by 22 cm)—cut out of a box or use a manila file folder. Place the cardboard with the long side down and make a clear mark perpendicular to the bottom edge 15 cm from the lower right corner. Place a paper clip on the right side with the edge parallel to the bottom (see Fig. 2).

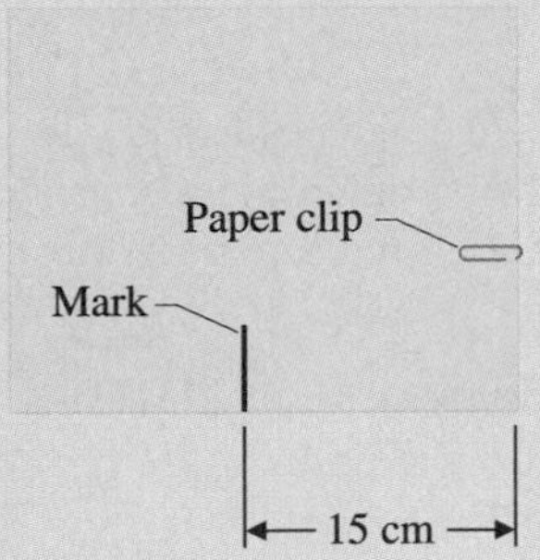

FIGURE 2

Procedure

Perform the experiment working in pairs or as individuals; then compare results with other members of your group and average your estimates for the speed of light in water. Discuss any problems and how you could improve your estimates.

Step 1 Place the mug on a flat surface where there is good overhead light and position the handle toward you. Fill the mug to the top with clear tap water.

Step 2 Place the cardboard vertically on top of the cup with the 15 cm mark at the inside edge of the cup. The cardboard should be lined up over the handle and the mark you made on the inside bottom edge of the cup opposite the handle (see Fig. 3).

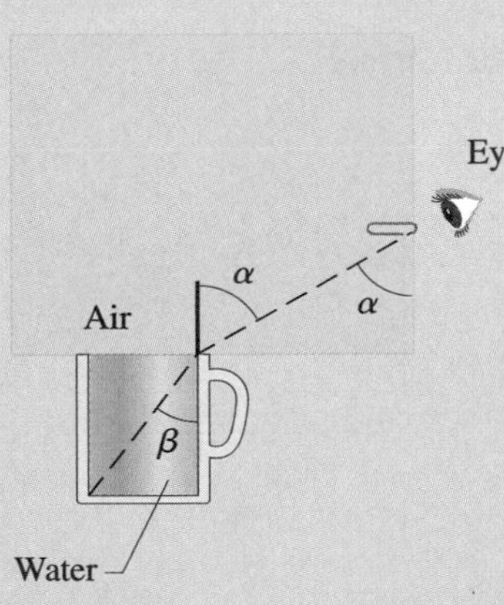

FIGURE 3

Step 3 Holding the vertical cardboard as instructed in step 2, and with your face 2 or 3 in. away from the right edge of the board, move your eye up and down until the near inside top edge of the cup lines up with the dark mark on the opposite bottom edge. Move the paper clip until the bottom right-hand corner is on the line of sight indicated above (see Fig. 3). Remove the cardboard and measure (to the nearest millimeter) the distance from the bottom right corner to the bottom edge of the paper clip.

Step 4 Discuss how you can find α, and find it.

Step 5 Discuss how you can find β, and find it.

Step 6 Use equation (1) to find your estimate of the speed of light in water, and compare your value with the values obtained by others in the group.

Step 7 The speed of light in clear water, determined through more precise measurements, is $c_w = 140{,}061$ miles per second. How does the value you found compare with this value?

Step 8 Discuss why you think the average of all the values found by your group should be more or less precise than your individual value. Calculate the average of all the individual values produced by the group. Conclusion?

CHAPTER 2 REVIEW

2.1 DEGREES AND RADIANS

An angle of **radian measure 1** is a central angle of a circle subtended by an arc having the same length as the radius. The **radian measure** of a central angle subtending an arc of length s in a circle of radius r is $\theta = s/r$ radians (rad). **Radian and degree measure** are related by

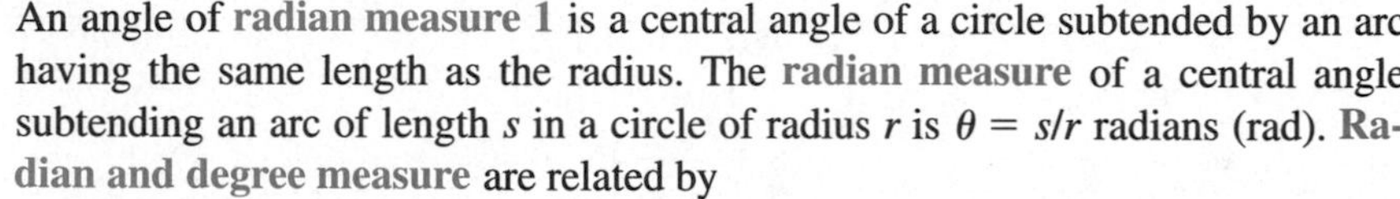

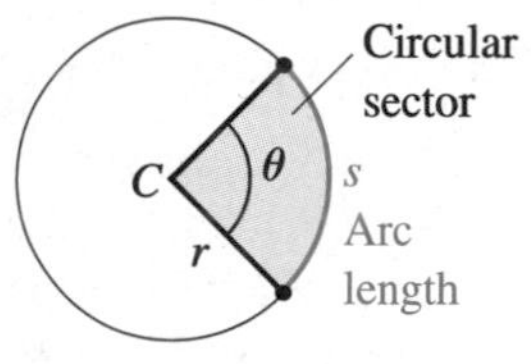

FIGURE 1

An angle with its vertex at the origin and initial side along the positive x axis is in **standard position.** Two angles are **coterminal** if their terminal sides coincide when both angles are placed in their standard position in the same coordinate system. The arc length and the area of a **circular sector** (see Fig. 1) are given by

	Radian Measure	Degree Measure
Arc length	$s = r\theta$	$s = \dfrac{\pi}{180} r\theta$
Area	$A = \dfrac{1}{2} r^2\theta$	$A = \dfrac{\pi}{360} r^2\theta$

☆2.2 LINEAR AND ANGULAR VELOCITY

The **linear velocity** V of a point moving on the circumference of a circle of radius r at a uniform rate, and the **angular velocity** ω (in radians per unit time) of the angle swept out by this point are related by $V = r\omega$ or, equivalently, $\omega = V/r$.

2.3 TRIGONOMETRIC FUNCTIONS

Trigonometric Functions with Angle Domains

For an arbitrary angle θ:

FIGURE 2

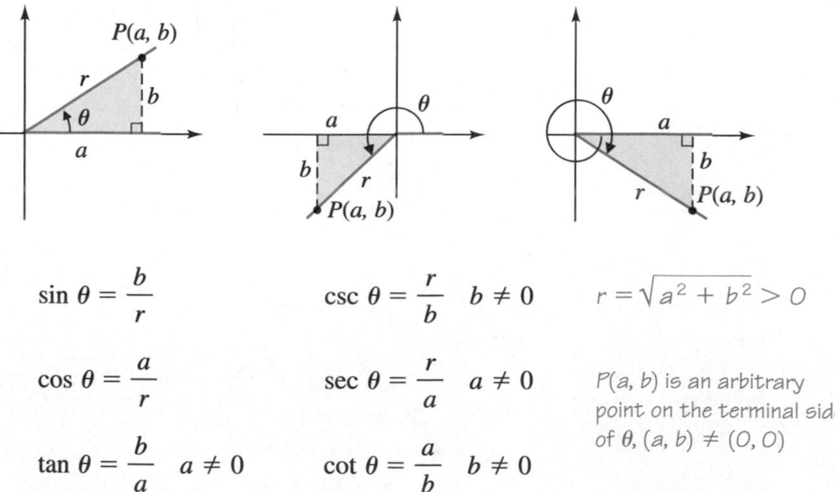

Domains: Sets of all possible angles for which the ratios are defined
Ranges: Subsets of the set of real numbers

The right triangle formed by dropping a perpendicular from $P(a, b)$ to the horizontal axis is called the **reference triangle** associated with the angle θ.

Trigonometric Functions with Real Number Domains

For x any real number, we define

$$\sin x = \sin(x \text{ rad}) \qquad \csc x = \csc(x \text{ rad})$$
$$\cos x = \cos(x \text{ rad}) \qquad \sec x = \sec(x \text{ rad})$$
$$\tan x = \tan(x \text{ rad}) \qquad \cot x = \cot(x \text{ rad})$$

Domains: Subsets of the set of real numbers
Ranges: Subsets of the set of real numbers

If x is any real number or any angle in degree or radian measure, then the following **reciprocal relationships** hold (division by 0 excluded).

$$\csc x = \frac{1}{\sin x} \qquad \sec x = \frac{1}{\cos x} \qquad \cot x = \frac{1}{\tan x}$$

☆2.4 ADDITIONAL APPLICATIONS

Refraction

If n_1 and n_2 represent the **index of refraction** for mediums M_1 and M_2, respectively, α is the angle of incidence, and β is the angle of refraction (see Fig. 3), then according to **Snell's law:**

FIGURE 3

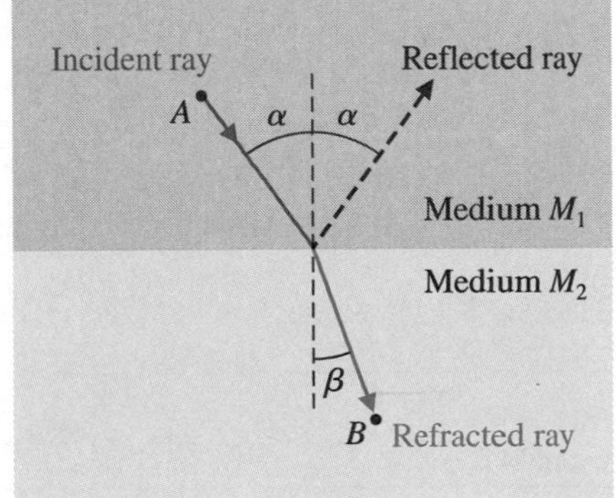

$$\frac{n_2}{n_1} = \frac{\sin \alpha}{\sin \beta}$$

The angle α for which $\beta = 90°$ is the **critical angle.**

Bow Waves

If S_b is the (uniform) speed of a boat, S_w is the speed of the **bow waves** generated by the boat, and θ is the angle between the bow waves, then

$$\sin \frac{\theta}{2} = \frac{S_w}{S_b} \qquad S_b > S_w$$

This relationship also applies to **sound waves** when planes travel faster than the speed of sound and to **nuclear particles** that travel faster than the speed of light.

Perception

When individuals try to estimate the position of a horizontal line in a field of parallel lines tilted at an angle θ (in degrees), the difference between their estimate and the actual horizontal in degrees d can be approximated by

$$d = a + b \sin 4\theta$$

where a and b are constants associated with a particular individual.

2.5 EXACT VALUE FOR SPECIAL ANGLES AND REAL NUMBERS

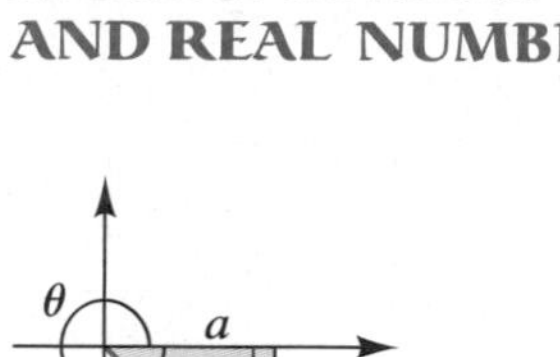

$(a, b) \neq (0, 0)$

FIGURE 4

Reference Triangle and Angle

For a nonquadrantal angle θ (see Fig. 4):

1. To form a **reference triangle** for θ, drop a perpendicular from a point $P(a, b)$ on the terminal side of θ to the horizontal axis.
2. The **reference angle** α is the acute angle (always taken positive) between the terminal side of θ and the horizontal axis.

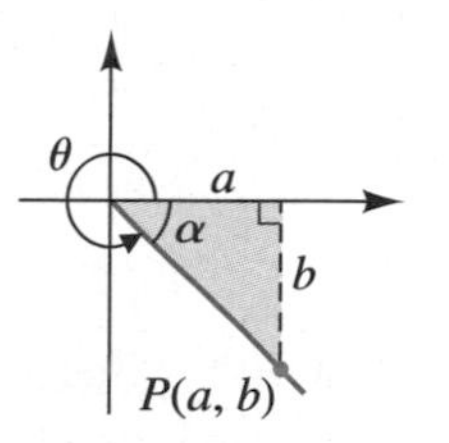

If the reference triangle for an angle or a real number θ is a 30°–60° right triangle or a 45° right triangle, then the relationships in Figure 5 can be used to find exact values of the trigonometric functions of θ.

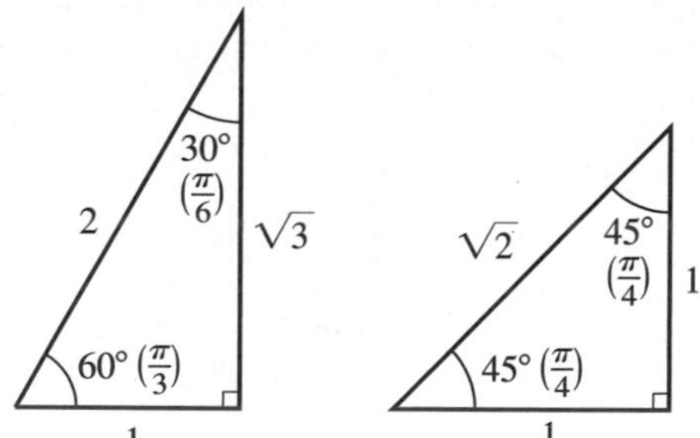

FIGURE 5

These exact values are summarized in Table 1.

TABLE 1
Special Values

θ	$\sin\theta$	$\csc\theta$	$\cos\theta$	$\sec\theta$	$\tan\theta$	$\cot\theta$
0° or 0	0	N.D.	1	1	0	N.D.
30° or $\pi/6$	1/2	2	$\sqrt{3}/2$	$2/\sqrt{3}$	$1/\sqrt{3}$	$\sqrt{3}$
45° or $\pi/4$	$1/\sqrt{2}$	$\sqrt{2}$	$1/\sqrt{2}$	$\sqrt{2}$	1	1
60° or $\pi/3$	$\sqrt{3}/2$	$2/\sqrt{3}$	1/2	2	$\sqrt{3}$	$1/\sqrt{3}$
90° or $\pi/2$	1	1	0	N.D.	N.D.	0

N.D. = Not defined

2.6 CIRCULAR FUNCTIONS

The **circular functions** for a real number x are defined in terms of the coordinates of a point on a **unit circle;** they are related to the trigonometric functions of an angle of x radians as follows (see Fig. 6):

Circular Functions and Trigonometric Functions

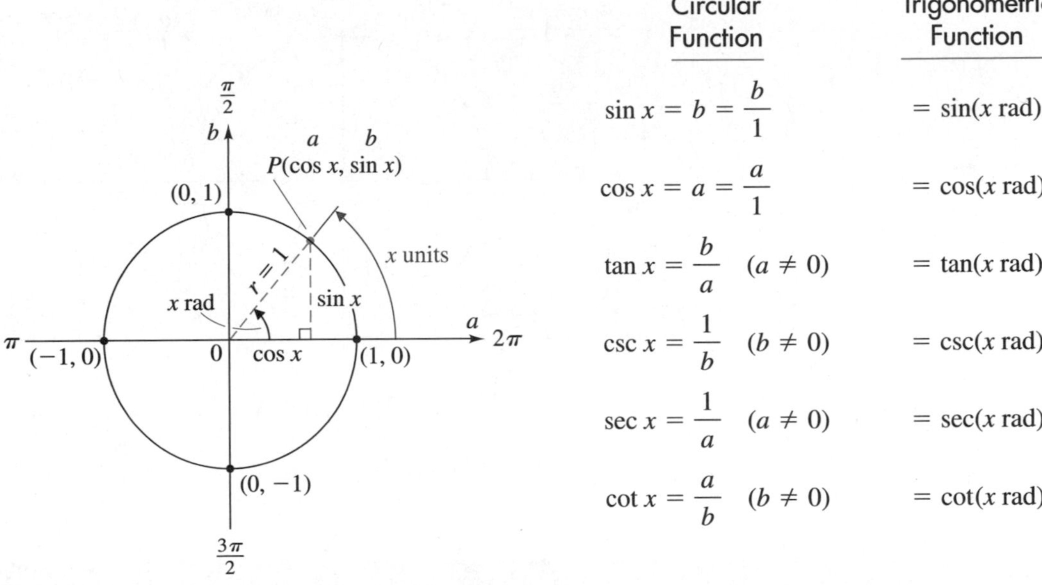

Circular Function		Trigonometric Function
$\sin x = b = \frac{b}{1}$		$= \sin(x \text{ rad})$
$\cos x = a = \frac{a}{1}$		$= \cos(x \text{ rad})$
$\tan x = \frac{b}{a}$	$(a \neq 0)$	$= \tan(x \text{ rad})$
$\csc x = \frac{1}{b}$	$(b \neq 0)$	$= \csc(x \text{ rad})$
$\sec x = \frac{1}{a}$	$(a \neq 0)$	$= \sec(x \text{ rad})$
$\cot x = \frac{a}{b}$	$(b \neq 0)$	$= \cot(x \text{ rad})$

FIGURE 6

A function f is **periodic** if there is a positive real number p such that $f(x + p) = f(x)$ for all x in the domain of f. The smallest such positive p, if it exists, is called **the period of f.** Both the sine function and the cosine function have a period of 2π.

The properties of circular functions can be used to establish the following **fundamental identities:**

1. $\csc x = \dfrac{1}{\sin x}$
2. $\sec x = \dfrac{1}{\cos x}$
3. $\cot x = \dfrac{1}{\tan x}$
4. $\tan x = \dfrac{\sin x}{\cos x}$
5. $\cot x = \dfrac{\cos x}{\sin x}$
6. $\sin(-x) = -\sin x$
7. $\cos(-x) = \cos x$
8. $\tan(-x) = -\tan x$
9. $\sin^2 x + \cos^2 x = 1$

Figure 7 is useful for finding exact values of the circular functions at multiples of $\pi/6$ and $\pi/4$.

FIGURE 7

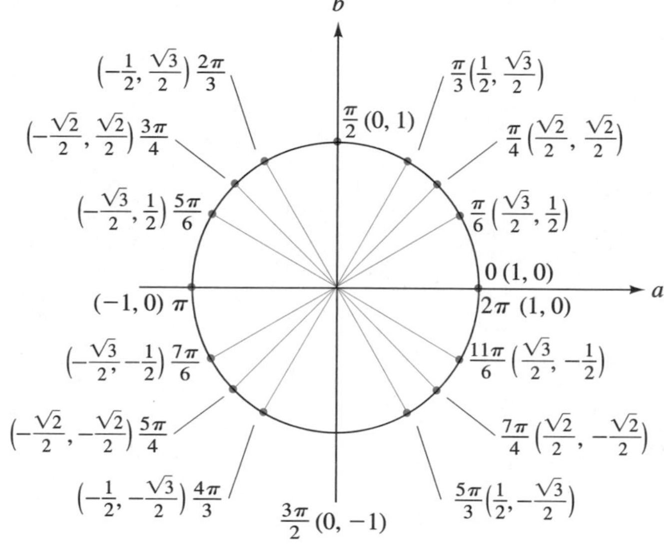

Unit circle $a^2 + b^2 = 1$

CHAPTER 2 REVIEW EXERCISE

Work through all the problems in this chapter review and check the answers. Answers to all review problems appear in the back of the book; following each answer is an italic number that indicates the section in which that type of problem is discussed. Where weaknesses show up, review the appropriate sections in the text. Review problems flagged with a star (☆) are from optional sections.

A

1. Convert to radian measure in terms of π.
(A) 60° (B) 45° (C) 90°

2. Convert to degree measure.
(A) $\pi/6$ (B) $\pi/2$ (C) $\pi/4$

3. Explain the meaning of a central angle of radian measure 2.

4. Which is larger: an angle of radian measure 1.5 or an angle of degree measure 1.5? Explain.

5. (A) Find the degree measure of 15.26 rad.
(B) Find the radian measure of −389.2°.

☆ **6.** Find the velocity V of a point on the rim of a wheel if $r = 25$ ft and $\omega = 7.4$ rad/min.

☆ **7.** Find the angular velocity ω of a point on the rim of a wheel if $r = 5.2$ m and $V = 415$ m/hr.

8. Find the value of $\sin\theta$ and $\tan\theta$ if the terminal side of θ contains $P(-4, 3)$.

9. Is it possible to find a real number x such that $\sin x$ is negative and $\csc x$ is positive? Explain.

Evaluate Problems 10–12 to four significant digits using a calculator.

10. (A) cot 53°40′ (B) csc 67°10′

11. (A) cos 23.5° (B) tan 42.3°

12. (A) cos 0.35 (B) tan 1.38

13. Sketch the reference triangle and find the reference angle α for:

(A) $\theta = 120°$ (B) $\theta = -\dfrac{7\pi}{4}$

14. Evaluate exactly without a calculator.
(A) sin 60° (B) $\cos(\pi/4)$ (C) tan 0°

The following figure from the text is repeated here for use in Problems 15 and 16.

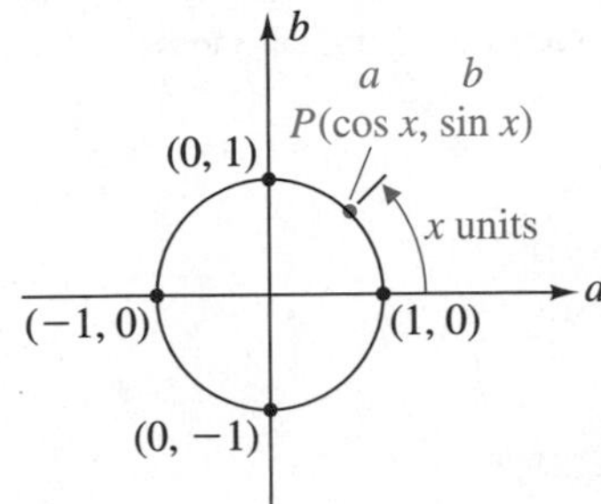

15. Refer to the figure and state the coordinates of P for the indicated values of x.
(A) $x = -2\pi$ (B) $x = \pi$
(C) $x = -3\pi/2$ (D) $x = \pi/2$
(E) $x = -5\pi$ (F) $x = 7\pi/2$

16. Refer to the figure: Given $y = \sin x$, how does y vary for the indicated variations in x?
(A) x varies from 0 to $\pi/2$
(B) x varies from $\pi/2$ to π
(C) x varies from π to $3\pi/2$
(D) x varies from $3\pi/2$ to 2π
(E) x varies from 2π to $5\pi/2$
(F) x varies from $5\pi/2$ to 3π

B **17.** List all angles that are coterminal with $\theta = \pi/6$ rad, $-3\pi \le \theta \le 3\pi$. Explain how you arrived at your answer.

18. What is the degree measure of a central angle subtended by an arc exactly $\frac{7}{60}$ of the circumference of a circle?

19. If the radius of a circle is 4 cm, find the length of an arc intercepted by an angle of 1.5 rad.

20. Convert 212° to radian measure in terms of π.

21. Convert $\pi/12$ rad to degree measure.

22. Use a calculator with an automatic radian–degree conversion routine to find:
(A) The radian measure (to two decimal places) of $-213.23°$
(B) The degree measure (to two decimal places) of 4.62 rad

23. If the radian measure of an angle is tripled, is the degree measure of the angle tripled? Explain.

24. If $\sin \alpha = \sin \beta$, $\alpha \ne \beta$, are angles α and β necessarily coterminal? Explain.

25. From the following display on a graphing calculator, explain how you would find csc x without finding x. Then find csc x to four decimal places.

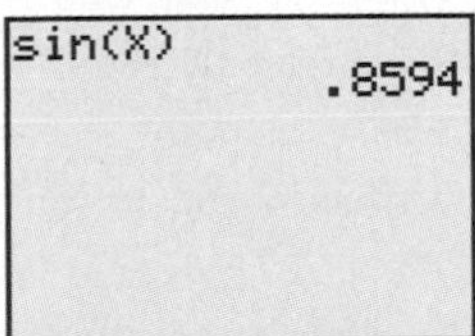

26. Find the tangent of 0, $\pi/2$, π, and $3\pi/2$.

27. In which quadrant does the terminal side of each angle lie?
(A) 732° (B) -7 rad

Evaluate Problems 28–33 to three significant digits using a calculator.

28. cos 187.4° **29.** sec 103°20′ **30.** cot(−37°40′)

31. sin 2.39 **32.** cos 5 **33.** cot(−4)

In Problems 34–42, find the exact value of each without using a calculator.

34. $\cos \frac{5\pi}{6}$ **35.** $\cot \frac{7\pi}{4}$ **36.** $\sin \frac{3\pi}{2}$

37. $\cos \frac{3\pi}{2}$ **38.** $\sin \frac{-4\pi}{3}$ **39.** $\sec \frac{-4\pi}{3}$

40. $\cos 3\pi$ **41.** $\cot 3\pi$ **42.** $\sin \frac{-11\pi}{6}$

43. In the following graphing calculator display, indicate which value(s) are not exact and find the exact value.

```
cos(π/3)
            .5000
sin(180°)
           0.0000
tan(-60°)
          -1.7321
```

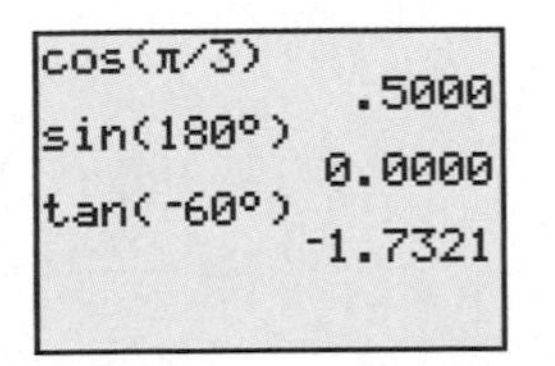

In Problems 44–47, use a calculator to evaluate each to five decimal places.

44. sin 384.0314° **45.** tan(− 198°43′6″)

46. cos 26 **47.** cot(− 68.005)

48. If $\sin \theta = -\frac{4}{5}$ and the terminal side of θ does not lie in the third quadrant, find the exact values of $\cos \theta$ and $\tan \theta$ without finding θ.

49. Find the least positive exact value of θ in radian measure such that $\sin \theta = -\frac{1}{2}$.

50. Find the exact value of each of the other five trigonometric functions if

$$\sin\theta = -\tfrac{2}{5} \qquad \text{and} \qquad \tan\theta < 0$$

51. Find all the angles exactly between 0° and 360° for which $\tan\theta = -1$.

52. Find all the angles exactly between 0 and 2π for which $\cos\theta = -\sqrt{3}/2$.

53. In a circle of radius 12.0 cm, find the length of an arc subtended by a central angle of:
(A) 1.69 rad (B) 22.5°

54. In a circle with diameter 80 ft, find the area (to three significant digits) of the circular sector with central angle:
(A) 0.773 rad (B) 135°

55. Find the distance between Charleston, West Virginia (38°21′N latitude) and Cleveland, Ohio (41°28′N latitude). Both cities have the same longitude and the radius of the earth is 3,964 mi.

☆ **56.** Find the angular velocity of a wheel turning through 6.43 rad in 15.24 sec.

☆ **57.** What is meant by a rotating object having an angular velocity of 12π rad/sec?

58. Evaluate $\cos x$ to three significant digits for:
(A) $x = 7$ (B) $x = 7 + 2\pi$
(C) $x = 7 - 30\pi$

59. Evaluate $\tan(-x)$ and $-\tan x$ to three significant digits for:
(A) $x = 7$ (B) $x = -17.9$ (C) $x = -2{,}135$

60. One of the following is not an identity. Indicate which one.

(A) $\csc x = \dfrac{1}{\sin x}$ (B) $\cot x = \dfrac{1}{\tan x}$

(C) $\tan x = \dfrac{\sin x}{\cos x}$ (D) $\sec x = \dfrac{1}{\sin x}$

(E) $\sin^2 x + \cos^2 x = 1$ (F) $\cot x = \dfrac{\cos x}{\sin x}$

61. Simplify: $(\csc x)(\cot x)(1 - \cos^2 x)$

62. Simplify: $\cot(-x)\sin(-x)$

63. A point $P(a, b)$ moves clockwise around a unit circle starting at (1, 0) for a distance of 29.37 units. Explain how you would find the coordinates of the point P at its final position, and how you would determine which quadrant P is in. Find the coordinates to four decimal places and the quadrant.

C **64.** An angle in standard position intercepts an arc of length 1.3 units on a unit circle with center at the origin. Explain why the radian measure of the angle is also 1.3.

65. Which circular functions are not defined for $x = k\pi$, k any integer? Explain.

66. In the following graphing calculator display, find the least positive exact value of x (in radian measure) that produces the indicated result.

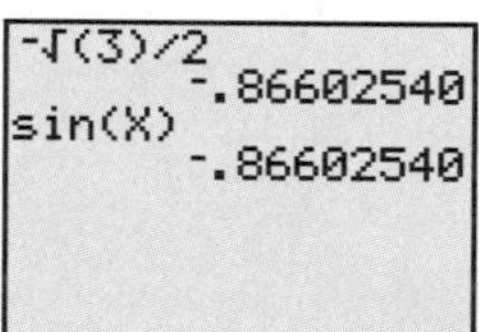

67. The following graphing calculator display shows the coordinates of a point on a unit circle. Let s be the length of the least positive arc from (1, 0) to the point. Find s to four decimal places.

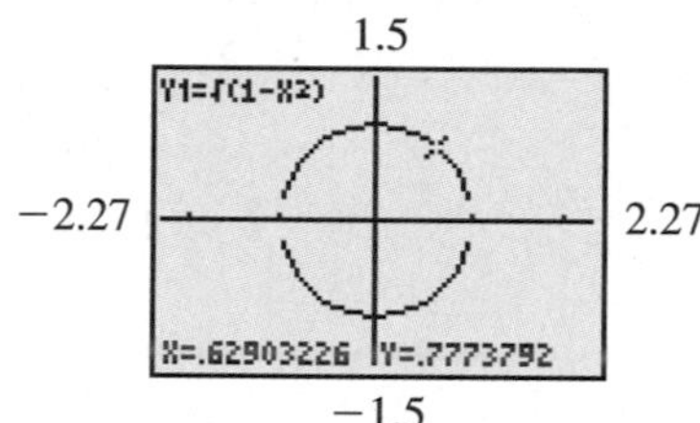

68. A circular sector has an area of 342.5 m² and a radius of 12 m. Calculate the arc length of the sector to the nearest meter.

69. A circle with its center at the origin in a rectangular coordinate system passes through the point (4, 5). What is the length of the arc on the circle in the first quadrant between the positive horizontal axis and the point (4, 5)? Compute the answer to two decimal places.

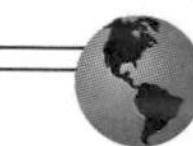

Applications

70. Engineering Through how many radians does a pulley with 10 cm diameter turn when 10 m of rope has been pulled through it without slippage? How many revolutions result? (Give answers to one decimal place.)

☆ **71. Engineering** If the large gear in the figure at the top of the next page completes 5 revolutions, how many revolutions will the middle gear complete? How many revolutions will the small gear complete?

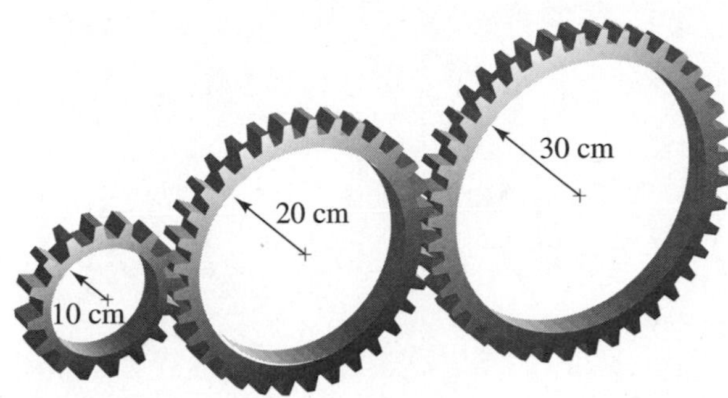

Figure for 71

☆ **72. Engineering** An automobile is traveling at 70 ft/sec. If the wheel diameter is 27 in., what is the angular velocity in rad/sec?

☆ **73. Space Science** A satellite is placed in a circular orbit 1,000 mi above the earth's surface. If the satellite completes one orbit every 114 min and the radius of the earth is 3,964 mi, what is the linear velocity (to three significant digits) of the satellite in miles per hour?

74. Electric Current An alternating current generator produces an electrical current (measured in amperes) that is described by the equation

$$I = 30 \sin(120\pi t - 60\pi)$$

where t is time in seconds. What is the current I when $t = 0.015$ sec? (Give answers to one decimal place.)

75. Precalculus A ladder of length L leaning against a building just touches a fence that is 10 ft high located 2 ft from the building (see the figure).

(A) Express the length of the ladder in terms of θ.

(B) Describe what you think happens to the length of the ladder L as θ varies between 0 and $\pi/2$ radians.

(C) Complete the table (to two decimal places) using a calculator. (If you have a table-generating calculator, use it.)

θ rad	0.70	0.80	0.90	1.00	1.10	1.20	1.30
L ft	18.14						

Figure for 75

(D) From the table, select the angle θ that produces the shortest ladder that satisfies the conditions in the problem. (Calculus techniques can be used on the equation from part (A) to find the angle θ that produces the shortest ladder.)

☆ **76. Light Waves** A light wave passing through air strikes the surface of a pool of water so that the angle of incidence is $\alpha = 31.7°$. Find the angle of refraction. (Water has a refractive index of 1.33 and air has a refractive index of 1.00.)

☆ **77. Light Waves** A triangular crown glass prism is surrounded by air. If light inside the prism strikes one of its facets, what is the critical angle of incidence α for total reflection? (The refractive index for crown glass is 1.52, and the refractive index for air is 1.00.)

☆ **78. Bow Waves** If a boat traveling at 25 mph produces bow waves that separate at an angle of 51°, how fast are the bow waves traveling?

GRAPHING TRIGONOMETRIC FUNCTIONS

3

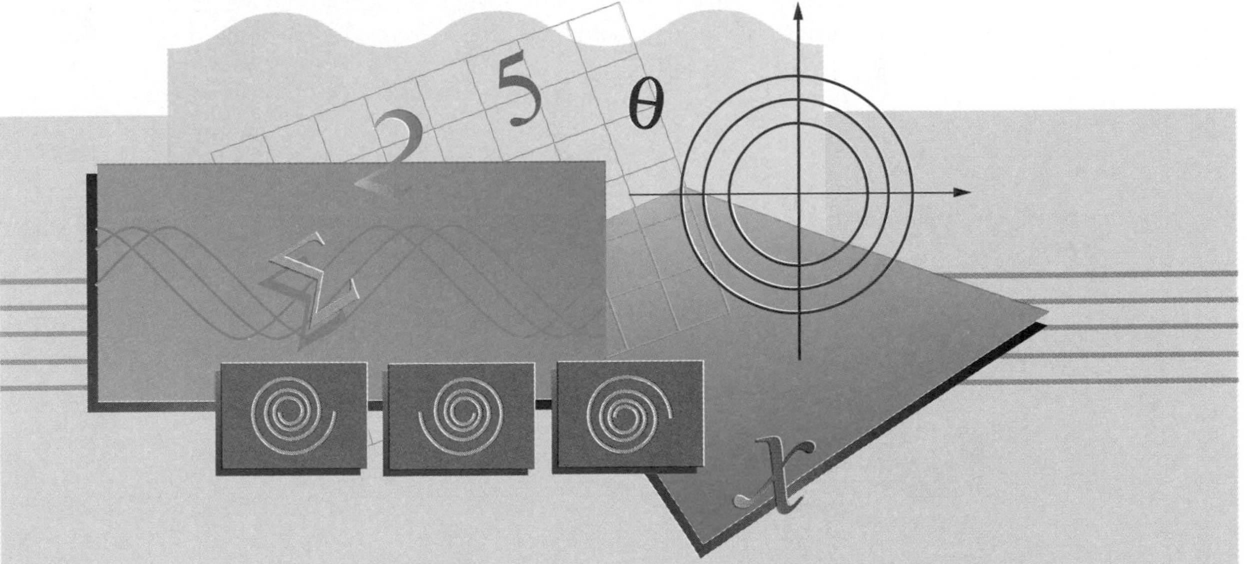

☆ Sections marked with a star may be omitted without loss of continuity.

With the trigonometric functions defined, we are now in a position to consider a substantially expanded list of applications and properties. As a brief preview, look at Figure 1.

What feature seems to be shared by the different illustrations? All appear to be repetitive—that is, periodic. The trigonometric functions, as we will see shortly, can be used to describe such phenomena with remarkable precision. Glance through Section 3.4, "Additional Applications," to preview an even greater variety of applications.

In this chapter you will learn how to sketch graphs of the trigonometric functions quickly and easily. You will also learn how to recognize certain fundamental and useful properties of these functions.

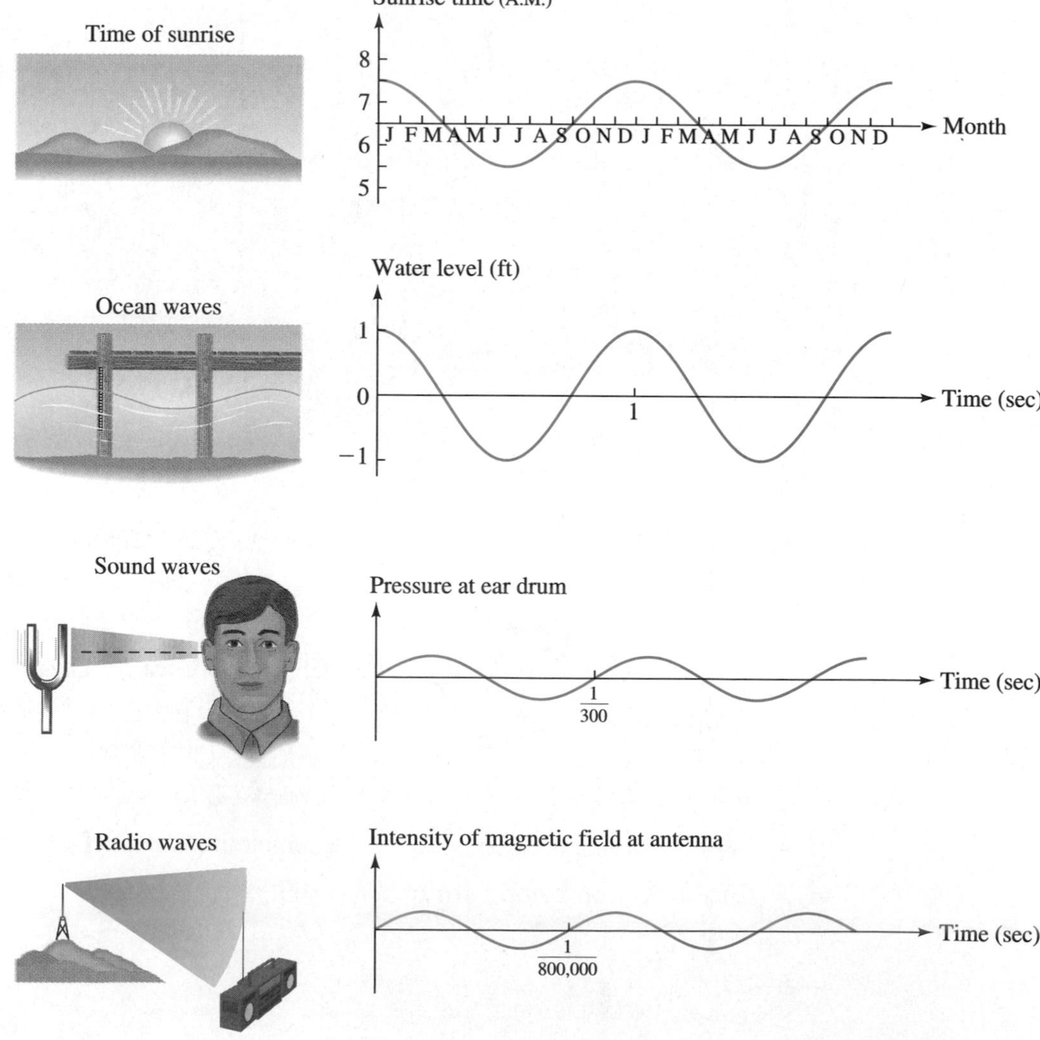

FIGURE 1

3.1 BASIC GRAPHS

- Graphs of $y = \sin x$ and $y = \cos x$
- Graphs of $y = \tan x$ and $y = \cot x$
- Graphs of $y = \csc x$ and $y = \sec x$
- Graphing with a Graphing Calculator

In this section we will discuss the graphs of the six trigonometric functions introduced in Chapter 2. We will also discuss the domains, ranges, and periodic properties of these functions. Section 2.6 on circular functions will prove particularly important in our work.

Although it appears that there is a lot to remember in this section, you mainly need to be familiar with the graphs and properties of the sine, cosine, and tangent functions. The reciprocal relationships we discussed in Section 2.6 will enable you to determine the graphs and properties of the other three trigonometric functions from the sine, cosine, and tangent functions.

Graphs of $y = \sin x$ and $y = \cos x$

First, we consider

$$y = \sin x \qquad x \text{ a real number} \tag{1}$$

The graph of the sine function is the graph of the set of all ordered pairs of real numbers (x, y) that satisfy equation (1). How do we find these pairs of numbers? To help make the process clear, we refer to a function machine with the unit circle definition inside (Fig. 1).

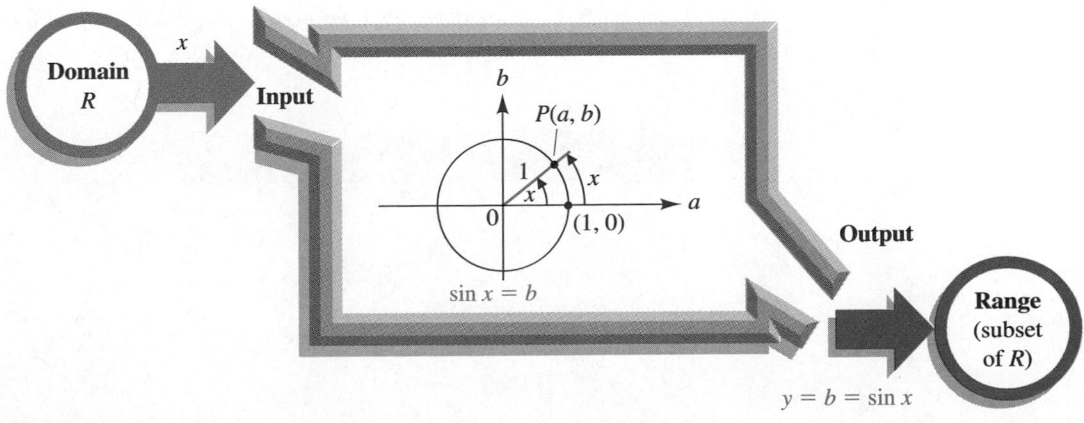

FIGURE 1
Sine function machine
(R = All real numbers)

We are interested in graphing, in an xy coordinate system, all ordered pairs of real numbers (x, y) produced by the function machine. We could resort to point-by-point plotting using a calculator, which becomes tedious and tends to obscure some important properties. Instead, we choose to speed up the process by using some of the properties discussed in Section 2.6 and by observing how $y = \sin x = b$ varies as $P(a, b)$ moves around the unit circle. We know that the domain of the sine function is the set of all real numbers R, its range is the set of all real numbers y such that $-1 \le y \le 1$, and its period is 2π.

Because the sine function is periodic with period 2π, we will concentrate on the graph over one period, from 0 to 2π. Once we have the graph for one period, we can complete as much of the rest of the graph as we wish by repeating the graph to the left and to the right.

Figure 2 illustrates how $y = \sin x = b$ varies as x increases from 0 to 2π and $P(a, b)$ moves around the unit circle.

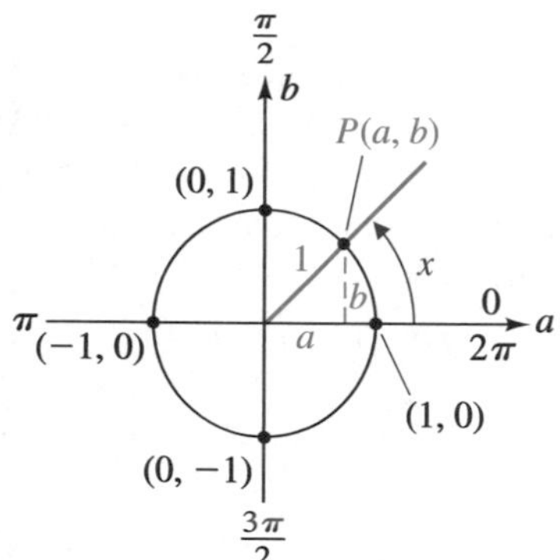

As x increases	$y = \sin x = b$
from 0 to $\pi/2$	increases from 0 to 1
from $\pi/2$ to π	decreases from 1 to 0
from π to $3\pi/2$	decreases from 0 to -1
from $3\pi/2$ to 2π	increases from -1 to 0

FIGURE 2
$y = \sin x = b$

The information in Figure 2 can be translated into a graph of $y = \sin x$, $0 \le x \le 2\pi$, as shown in Figure 3. (Where the graph is uncertain, fill in with calculator values.)

To complete the graph of $y = \sin x$ over any interval desired, we need only to repeat the final graph in Figure 3 to the left and to the right as far as we wish. The next box summarizes what we now know about the graph of $y = \sin x$ and its basic properties.

GRAPH OF $y = \sin x$

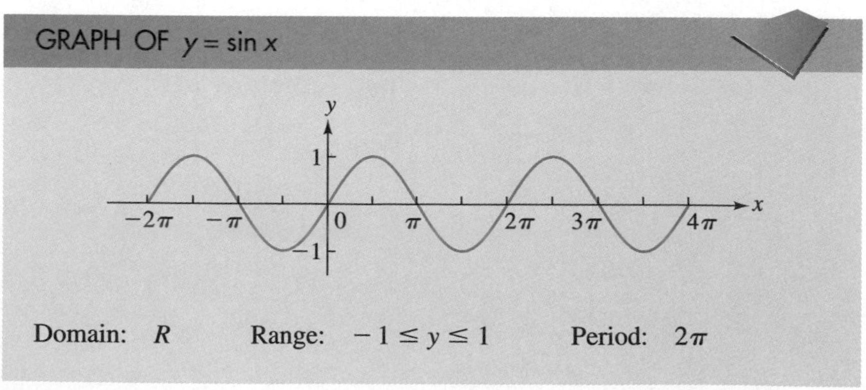

Domain: R Range: $-1 \le y \le 1$ Period: 2π

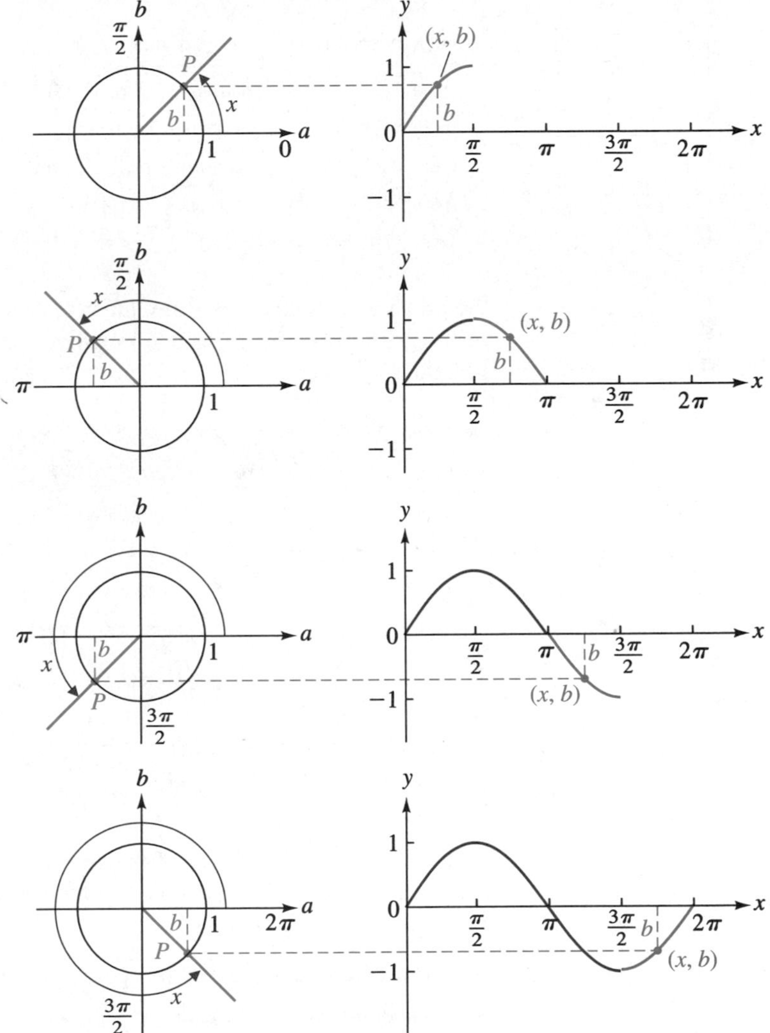

FIGURE 3
$y = \sin x,\ 0 \le x \le 2\pi$

Both the x and y axes are real number lines (see Appendix A.1). Because the domain of the sine function is all real numbers, the graph of $y = \sin x$ extends without limit in both horizontal directions. Also, because the range is $-1 \le y \le 1$, no point on the graph can have a y coordinate greater than 1 or less than -1.

Proceeding in the same way for the cosine function, we can obtain its graph. Figure 4 (on the next page) shows how $y = \cos x = a$ varies as x increases from 0 to 2π.

Using the results of Figure 4 and the fact that the cosine function is periodic with period 2π (and filling in with calculator values where necessary), we obtain

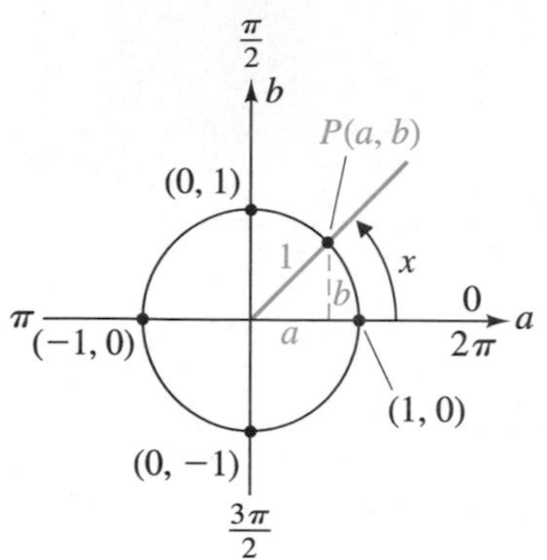

FIGURE 4
$y = \cos x = a$

As x increases	$y = \cos x = a$
from 0 to $\pi/2$	decreases from 1 to 0
from $\pi/2$ to π	decreases from 0 to -1
from π to $3\pi/2$	increases from -1 to 0
from $3\pi/2$ to 2π	increases from 0 to 1

the graph of $y = \cos x$ over any interval desired. The next box shows a portion of the graph of $y = \cos x$ and summarizes some of the properties of the cosine function.

GRAPH OF $y = \cos x$

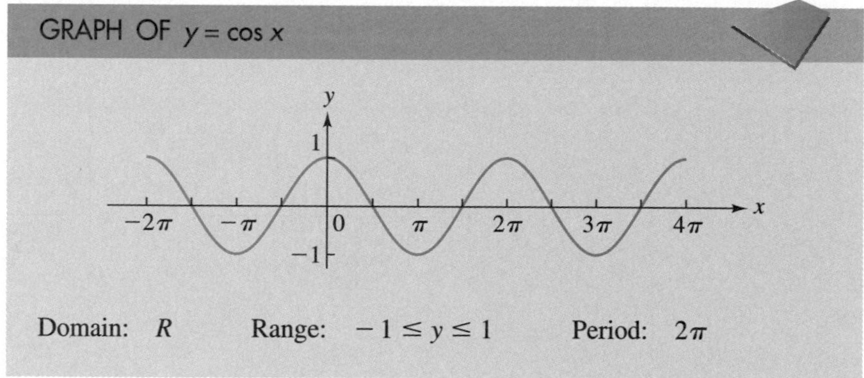

Domain: R Range: $-1 \le y \le 1$ Period: 2π

The basic characteristics of the sine and cosine graphs should be learned so that the curves can be sketched quickly. In particular, you should be able to answer the following questions:

(A) How often does the graph repeat (what is the period)?
(B) Where are the x intercepts?
(C) Where are the y intercepts?
(D) Where do the high and low points occur?
(E) What are the symmetry properties relative to the origin, y axis, x axis?

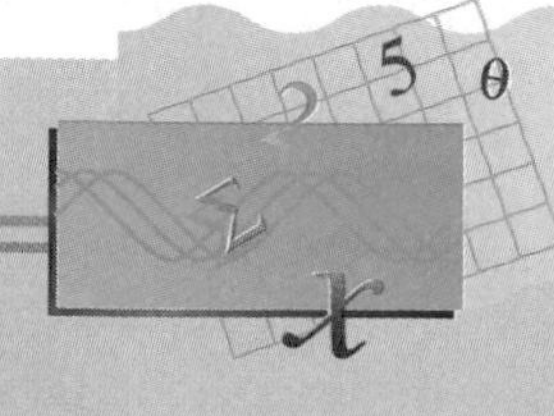

EXPLORE/DISCUSS 1

(A) Discuss how the graphs of the sine and cosine functions are related.
(B) Can one be obtained from the other by a horizontal shift? Explain how.

Graphs of $y = \tan x$ and $y = \cot x$

We first discuss the graph of $y = \tan x$. Later, because $\cot x = 1/(\tan x)$, we will be able to get the graph of $y = \cot x$ from the graph of $y = \tan x$ using reciprocals of ordinates.

From Figure 5 we can see that whenever $P(a, b)$ is on the horizontal axis of the unit circle (that is, whenever $x = k\pi$, k an integer), then $(a, b) = (\pm 1, 0)$ and $\tan x = b/a = 0/(\pm 1) = 0$. These values of x are the x intercepts of the graph of $y = \tan x$; that is, where the graph crosses the x axis.

x intercepts: $x = k\pi$ **k an integer**

Thus, as a first step in graphing $y = \tan x$, we locate the x intercepts with solid dots along the x axis, as indicated in Figure 6.

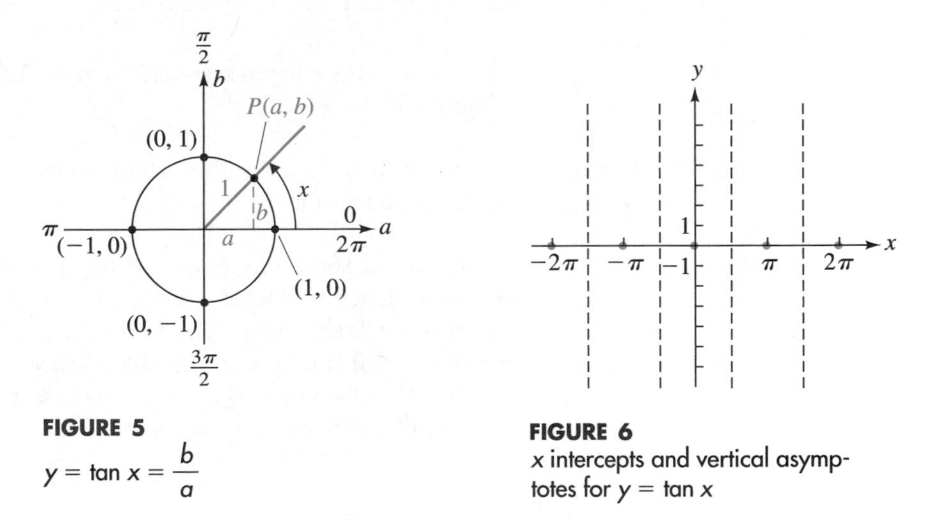

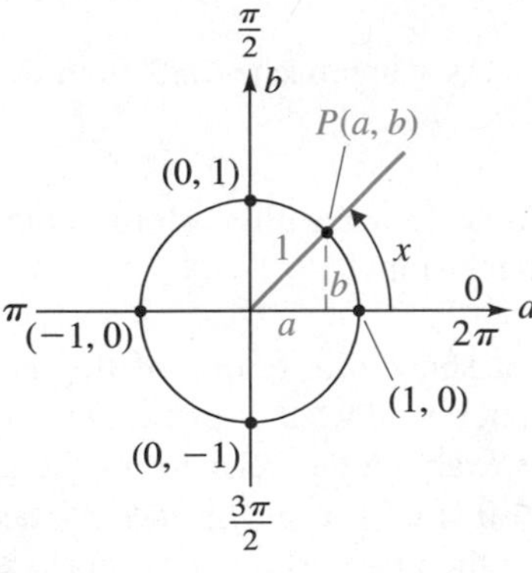

FIGURE 5

$y = \tan x = \dfrac{b}{a}$

FIGURE 6

x intercepts and vertical asymptotes for $y = \tan x$

Also, from Figure 5 we can see that whenever $P(a, b)$ is on the vertical axis of the unit circle (that is, whenever $x = \pi/2 + k\pi$, k an integer), then $(a, b) = (0, \pm 1)$ and $\tan x = b/a = \pm 1/0$, which is not defined. There can be no points plotted for these values of x. Thus, as a second step in graphing $y = \tan x$, we draw dashed vertical lines through each of these points on the x axis where $\tan x$ is not defined; the graph cannot touch these lines (see Fig. 6). These dashed lines will become guidelines, called *vertical asymptotes,* which are very helpful for sketching the graph of $y = \tan x$.

Vertical asymptotes: $x = \dfrac{\pi}{2} + k\pi$ **k an integer**

We next investigate the behavior of the graph of $y = \tan x$ over the interval $0 \le x < \pi/2$. Two points are easy to plot: $\tan 0 = 0$ and $\tan(\pi/4) = 1$. What happens to $\tan x$ as x approaches $\pi/2$ from the left? [Remember that $\tan(\pi/2)$ is not defined.] When x approaches $\pi/2$ from the left, $P(a, b)$ approaches $(0, 1)$ and stays in the first quadrant. Thus, a approaches 0 through positive values and b

approaches 1. What happens to $y = \tan x$ in this process? In Example 1, we perform a calculator experiment that suggests an answer.

EXAMPLE 1 **Calculator Experiment**

Form a table of values for $y = \tan x$ with x approaching $\pi/2 \approx 1.570\ 796\ 3$ from the left (through values less than $\pi/2$). Any conclusions?

Solution We create a table as follows:

x	0	0.5	1	1.5	1.57	1.5707	1.570 796 3
$\tan x$	0	0.5	1.6	14.1	1,256	10,381	37,320,535

Conclusion: As x approaches $\pi/2$ from the left, $y = \tan x$ appears to increase without bound. ■

Matched Problem 1 Repeat Example 1, but with x approaching $-\pi/2 \approx -1.570\ 796\ 3$ from the right. Any conclusions? ■

Figure 7a shows the results of the analysis in Example 1: $y = \tan x$ increases without bound when x approaches $\pi/2$ from the left.

Now we examine the behavior of the graph of $y = \tan x$ over the interval $-\pi/2 < x \le 0$. Because of the identity $\tan(-x) = -\tan x$ (see Section 2.6), we can reflect the graph of $y = \tan x$ for $0 \le x < \pi/2$ through the origin to obtain the full graph over the interval $-\pi/2 < x < \pi/2$ (Fig. 7b).

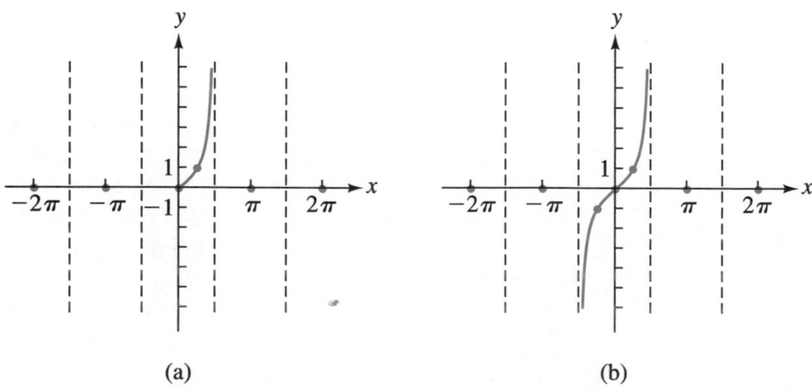

FIGURE 7
$y = \tan x$

Proceeding in the same way for the other intervals between the asymptotes (the dashed vertical lines), it appears that the tangent function is periodic with period π. We confirm this as follows: If (a, b) are the coordinates of P associated

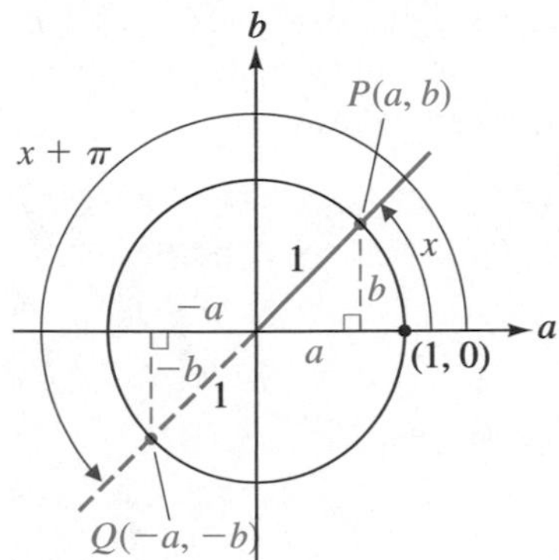

FIGURE 8
$\tan(x + \pi) = \tan x$

with x (see Fig. 8), then using unit circle symmetry and congruent reference triangles, $(-a, -b)$ are the coordinates of the point Q associated with $x + \pi$. Consequently,

$$\tan(x + \pi) = \frac{-b}{-a} = \frac{b}{a} = \tan x$$

We conclude that the tangent function is periodic with period π. In general,

$$\tan(x + k\pi) = \tan x \qquad k \text{ an integer}$$

for all values of x for which both sides of the equation are defined.

Now, to complete as much of the general graph of $y = \tan x$ as we wish, all we need to do is to repeat the graph in Figure 7b to the left and to the right over intervals of π units. The main characteristics of the graph of $y = \tan x$ should be learned so that the graph can be sketched quickly. The figure in the next box summarizes the above discussion.

GRAPH OF $y = \tan x$

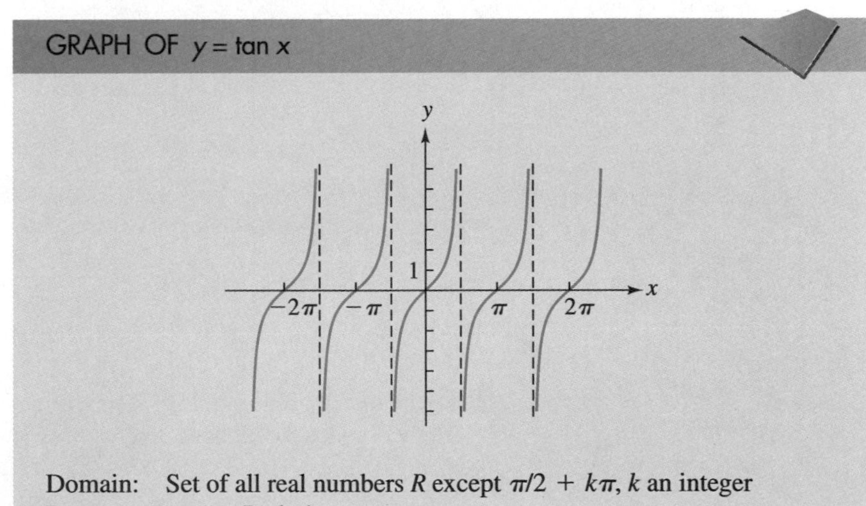

Domain: Set of all real numbers R except $\pi/2 + k\pi$, k an integer
Range: R Period: π

To graph $y = \cot x$, we recall that

$$\cot x = \frac{1}{\tan x}$$

and proceed by taking reciprocals of ordinate values in the graph of $y = \tan x$. Note that the x intercepts for the graph of $y = \tan x$ become vertical asymptotes for the graph of $y = \cot x$, and the vertical asymptotes for the graph of

$y = \tan x$ become x intercepts for the graph of $y = \cot x$. The graph of $y = \cot x$ is shown in the following box. Again, you should learn the main characteristics of this function so that its graph can be sketched readily.

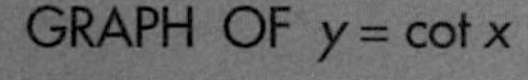

GRAPH OF $y = \cot x$

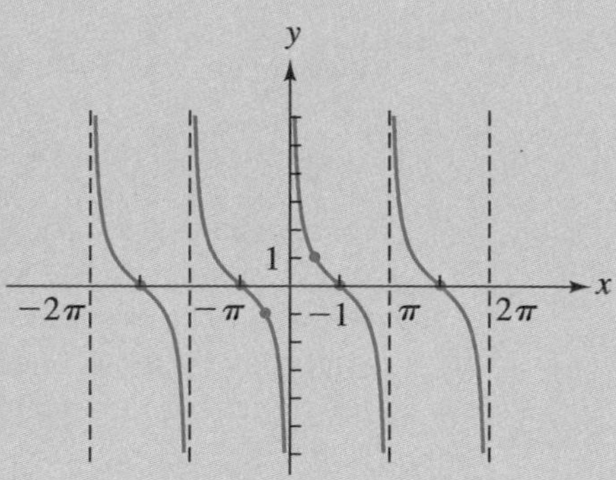

Domain: Set of all real numbers R except $k\pi$, k an integer
Range: R Period: π

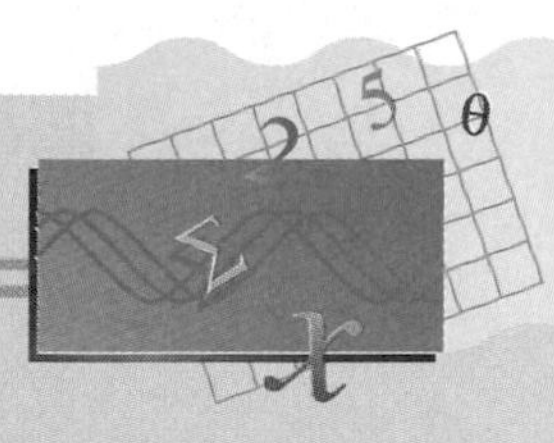

EXPLORE/DISCUSS 2

(A) Discuss how the graphs of the tangent and cotangent functions are related.

(B) Explain how the graph of one can be obtained from the graph of the other by horizontal shifts and/or reflections across an axis or through the origin.

Graphs of $y = \csc x$ and $y = \sec x$

Just as we obtained the graph of $y = \cot x$ by taking reciprocals of the ordinate values in the graph of $y = \tan x$, since

$$\csc x = \frac{1}{\sin x} \quad \text{and} \quad \sec x = \frac{1}{\cos x}$$

we can obtain the graphs of $y = \csc x$ and $y = \sec x$ by taking reciprocals of ordinate values in the respective graphs of $y = \sin x$ and $y = \cos x$.

The graphs of $y = \csc x$ and $y = \sec x$ are shown in the next two boxes. Note that, because of the reciprocal relationship, vertical asymptotes occur at the x intercepts of $\sin x$ and $\cos x$, respectively. It is helpful to first draw the graphs of $y = \sin x$ and $y = \cos x$, and then draw vertical asymptotes through the x intercepts.

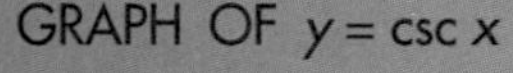

GRAPH OF $y = \csc x$

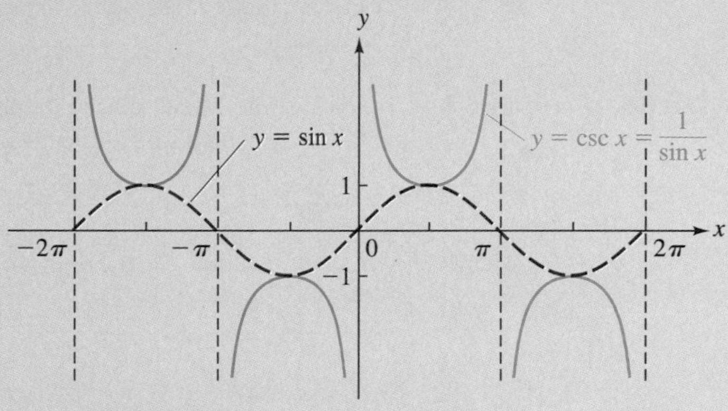

Domain: All real numbers x, except $x = k\pi$, k an integer
Range: All real numbers y such that $y \leq -1$ or $y \geq 1$
Period: 2π

GRAPH OF $y = \sec x$

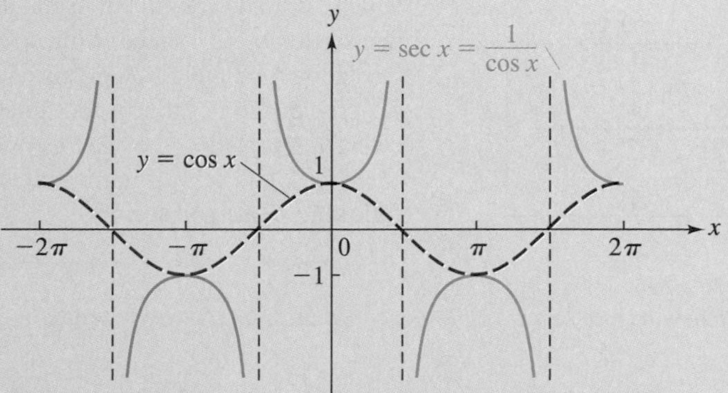

Domain: All real numbers x, except $x = \pi/2 + k\pi$, k an integer
Range: All real numbers y such that $y \leq -1$ or $y \geq 1$
Period: 2π

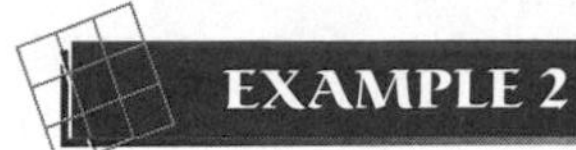

Calculator Experiment

To verify some points on the graph of $y = \csc x$, $0 < x < \pi$, we use a calculator to form the following table:

x	0.001	0.01	0.1	1.57	3.00	3.13	3.14
csc x	1,000	100	10	1	7.1	86.3	628

■

Matched Problem 2 To verify some points on the graph of $y = \sec x$, $-\pi/2 < x < \pi/2$, complete the following table using a calculator:

x	−1.57	−1.56	−1.4	0	1.4	1.56	1.57
sec x							

■

■ Graphing with a Graphing Calculator*

We determined the graphs of the six basic trigonometric functions by analyzing the behavior of the functions, exploiting relationships between the functions, and plotting only a few points. We refer to this process as curve sketching; one of the major objectives of this course is that you master this technique.

Graphing calculators also can be used to sketch graphs of functions; their accuracy depends on the screen resolution of the calculator. The smallest darkened rectangular area on the screen that the calculator can display is called a *pixel*. Most graphing calculators have a resolution of about 50 pixels per inch, which results in rough, but useful sketches. Note that the graphs shown earlier in this section were created using sophisticated computer software and printed at a resolution of about 1,000 pixels per inch.

The portion of the *xy* coordinate plane displayed on the screen of a graphing calculator is called the **viewing window** and is determined by the **range** and **scale** for *x* and for *y*. Figure 9 illustrates a **standard viewing window** using the following range and scale:

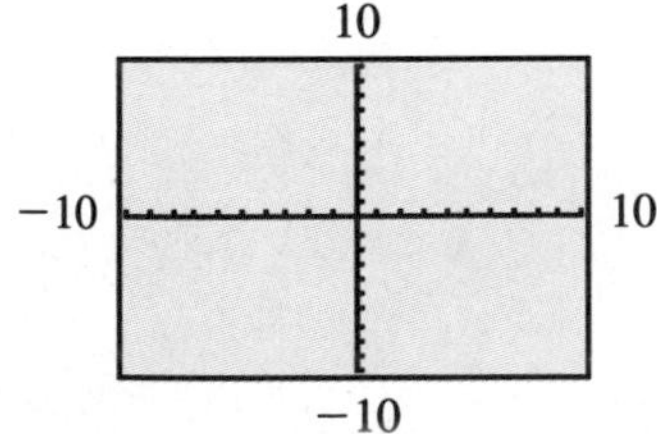

FIGURE 9
Standard viewing window

xmin = −10	xmax = 10	xscl = 1
ymin = −10	ymax = 10	yscl = 1

* Material that requires a graphing utility (that is, a graphing calculator or graphing software for a computer) is included in the text and exercise sets in this and subsequent chapters. This material is clearly identified with the icon shown in the margin. Any or all of the graphing utility material may be omitted without loss of continuity. Treatments are generic in nature. If you need help with your specific calculator, refer to your user's manual or to the graphing calculator supplement that accompanies this text.

Most graphing calculators do not display labels on the axes. We have added numeric labels on each side of the viewing window to make the graphs easier to read. Now we want to see what the graphs of the trigonometric functions will look like on a graphing calculator.

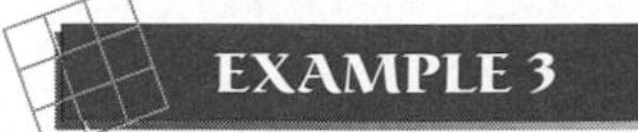

Trigonometric Graphs on a Graphing Calculator

Use a graphing calculator to graph the functions

$$y = \sin x \qquad y = \tan x \qquad y = \sec x$$

for $-2\pi \leq x \leq 2\pi$, $-5 \leq y \leq 5$. Display each graph in a separate viewing window.

Solution First, set the calculator to the radian mode. Most calculators remember this setting, so you should have to do this only once. Next, enter the following values:

$$\text{xmin} = -2\pi \qquad \text{xmax} = 2\pi \qquad \text{xscl} = 1$$

$$\text{ymin} = -5 \qquad \text{ymax} = 5 \qquad \text{yscl} = 1$$

This defines a viewing window ranging from -2π to 2π on the horizontal axis and from -5 to 5 on the vertical axis with tick marks one unit apart on each axis. Now enter the function $y = \sin x$ and draw the graph (see Fig. 10a). Repeat for $y = \tan x$ and $y = \sec x = 1/(\cos x)$ to obtain the graphs in Figures 10b and 10c.

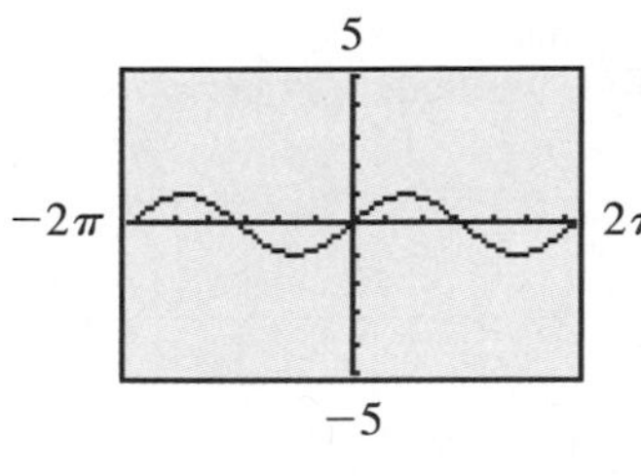

(a) $y = \sin x$

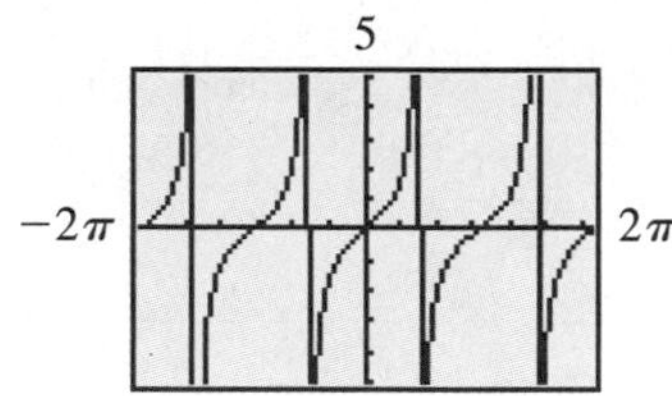

(b) $y = \tan x$

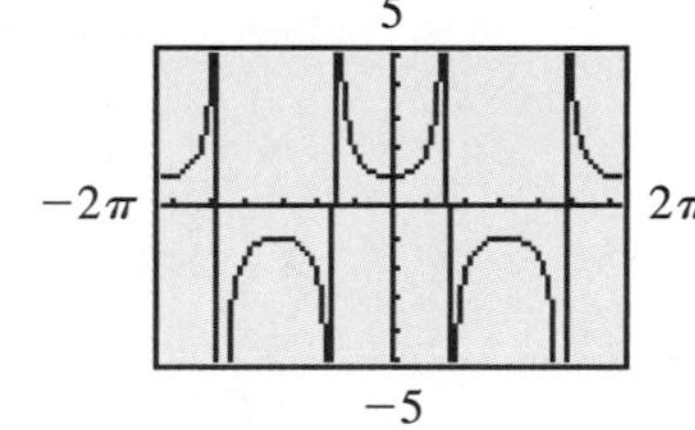

(c) $y = \sec x$

FIGURE 10
Graphing calculator graphs of trigonometric functions

■

Matched Problem 3 Repeat Example 3 for:

(A) $y = \cos x$ (B) $y = \cot x$ (C) $y = \csc x$ ■

In Figures 10b and 10c, it appears that the calculator has also drawn the vertical asymptotes for these functions, but this is not the case. Most graphing calculators calculate points on a graph and connect these points with line segments. The last point plotted to the left of the asymptote and the first plotted to the right of the asymptote will usually have very large y coordinates. If these y coordinates have opposite sign, then the calculator will connect the two points with a nearly vertical line segment, which gives the appearance of an asymptote. There is no harm in this as long as you understand that the calculator is not performing any analysis to identify asymptotes; it is simply connecting points with line segments. If you wish, you can set the calculator in dot mode to plot the points without the connecting line segments, as illustrated in Figure 11. Unless stated to the contrary, we will use connected mode.

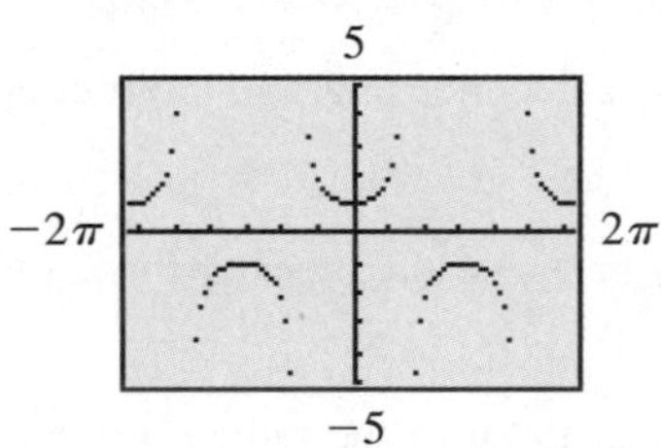

FIGURE 11
Graph of $y = \sec x$ in the plot-points-only mode

Answers to Matched Problems

1.

x	0	-0.5	-1	-1.5	-1.57	-1.5707	$-1.570\ 796\ 3$
$\tan x$	0	-0.5	-1.6	-14.1	$-1{,}256$	$-10{,}381$	$-37{,}320{,}535$

Conclusion: As x approaches $-\pi/2$ from the right, $y = \tan x$ appears to decrease without bound.

2.

x	-1.57	-1.56	-1.4	0	1.4	1.56	1.57
$\sec x$	1,256	92.6	5.9	1	5.9	92.6	1,256

3. (A) $y = \cos x$

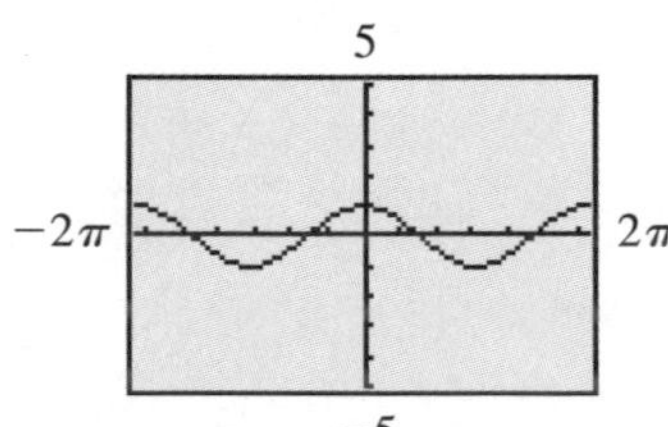

(B) $y = \cot x$

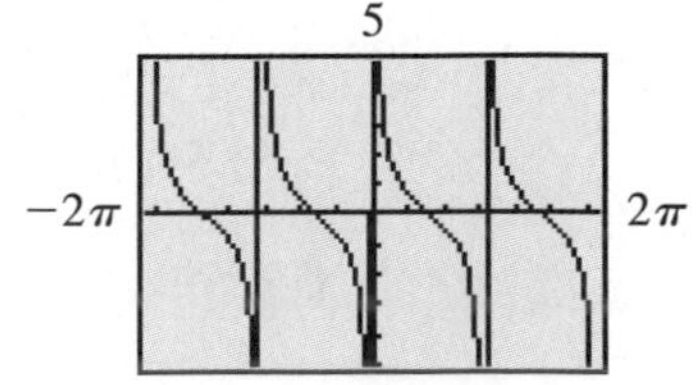

(C) $y = \csc x$

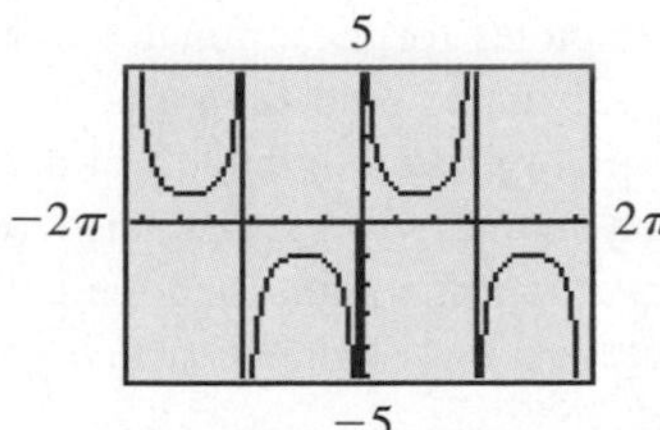

EXERCISE 3.1

A

1. What are the periods of the cosine, secant, and tangent functions?
2. What are the periods of the sine, cosecant, and cotangent functions?
3. How far does the graph of each of the following functions deviate from the x axis?
(A) $\sin x$ (B) $\cot x$ (C) $\sec x$
4. How far does the graph of each of the following functions deviate from the x axis?
(A) $\cos x$ (B) $\tan x$ (C) $\csc x$

In Problems 5–10, what are the x intercepts for the graph of each function over the interval $-2\pi \le x \le 2\pi$*?*

5. $\cos x$
6. $\sin x$
7. $\tan x$
8. $\cot x$
9. $\sec x$
10. $\csc x$

B

11. For what values of x, $-2\pi \le x \le 2\pi$, are the following not defined?
(A) $\sin x$ (B) $\cot x$ (C) $\sec x$
12. For what values of x, $-2\pi \le x \le 2\pi$, are the following not defined?
(A) $\cos x$ (B) $\tan x$ (C) $\csc x$
13. Use a calculator and point-by-point plotting to produce an accurate graph of $y = \cos x$, $0 \le x \le 1.6$, using domain values 0, 0.1, 0.2, . . . , 1.5, 1.6.
14. Use a calculator and point-by-point plotting to produce an accurate graph of $y = \sin x$, $0 \le x \le 1.6$, using domain values 0, 0.1, 0.2, . . . , 1.5, 1.6.

In Problems 15–20, make a sketch of each trigonometric function without looking at the text or using a calculator. Label each point where the graph crosses the x axis in terms of π*.*

15. $y = \sin x, \quad -2\pi \le x \le 2\pi$
16. $y = \cos x, \quad -2\pi \le x \le 2\pi$
17. $y = \tan x, \quad 0 \le x \le 2\pi$
18. $y = \cot x, \quad 0 < x < 2\pi$
19. $y = \csc x, \quad -\pi < x < \pi$
20. $y = \sec x, \quad -\pi \le x \le \pi$

Problems 21–26 require the use of a graphing utility. These problems provide a preliminary exploration into the relationships of the graphs of $y = A \sin x$, $y = A \cos x$, $y = \sin Bx$, $y = \cos Bx$, $y = \sin(x + C)$, *and* $y = \cos(x + C)$ *relative to the graphs of* $y = \sin x$ *and* $y = \cos x$*. The topic is discussed in detail in the next section.*

21. (A) Graph $y = A \sin x$ ($-2\pi \le x \le 2\pi$, $-3 \le y \le 3$) for $A = -2, 1, 3$, all in the same viewing window.
(B) Do the x intercepts change? If so, where?
(C) How far does each graph deviate from the x axis? Experiment with other values of A.
(D) Describe how the graph of $y = \sin x$ is changed by changing the values of A in $y = A \sin x$.
22. (A) Graph $y = A \cos x$ ($-2\pi \le x \le 2\pi$, $-3 \le y \le 3$) for $A = -3, 1, 2$, all in the same viewing window.
(B) Do the x intercepts change? If so, where?
(C) How far does each graph deviate from the x axis? Experiment with other values of A.
(D) Describe how the graph of $y = \cos x$ is changed by changing the values of A in $y = A \cos x$.
23. (A) Graph $y = \cos Bx$ ($-\pi \le x \le \pi$, $-2 \le y \le 2$) for $B = 1, 2, 3$, all in the same viewing window.
(B) How many periods of each graph appear in this viewing window? Experiment with additional positive values of B.
(C) Based on your experiments in part (B), how many periods of the graph of $y = \cos nx$, n a positive integer, would appear in this viewing window?
24. (A) Graph $y = \sin Bx$ ($-\pi \le x \le \pi$, $-2 \le y \le 2$) for $B = 1, 2, 3$, all in the same viewing window.
(B) How many periods of each graph appear in this viewing window? Experiment with additional positive values of B.
(C) Based on your experiments in part (B), how many periods of the graph of $y = \sin nx$, n a positive integer, would appear in this viewing window?
25. (A) Graph
$$y = \sin(x + C)$$
($-2\pi \le x \le 2\pi, -2 \le y \le 2$) for $C = -\pi/2, 0, \pi/2$, all in the same viewing window. Experiment with additional values of C.
(B) Describe how the graph of $y = \sin x$ is changed by changing the values of C in $y = \sin(x + C)$.
26. (A) Graph
$$y = \cos(x + C)$$
($-2\pi \le x \le 2\pi, -2 \le y \le 2$) for $C = -\pi/2, 0, \pi/2$, all in the same viewing window. Experiment with additional values of C.

(B) Describe how the graph of $y = \cos x$ is changed by changing the values of C in $y = \cos(x + C)$.

C **27.** Try to calculate each of the following on your calculator. Explain the problem.
(A) $\cot 0$ (B) $\tan(\pi/2)$ (C) $\csc \pi$

28. Try to calculate each of the following on your calculator. Explain the problem.
(A) $\tan(-\pi/2)$ (B) $\cot(-\pi)$ (C) $\sec(\pi/2)$

Problems 29–32 require the use of a graphing utility.

29. In applied mathematics certain formulas, derivations, and calculations are simplified by replacing $\tan x$ with x for small x. What justifies this procedure? To find out, graph $y_1 = \tan x$ and $y_2 = x$ in the same viewing window for $-1 \le x \le 1$ and $-1 \le y \le 1$.
(A) What do you observe about the two graphs when x is close to 0, say $-0.5 \le x \le 0.5$?
(B) Complete the table to three decimal places (use the table feature on your graphing utility if it has one):

x	−0.3	−0.2	−0.1	0.0	0.1	0.2	0.3
tan x							

(C) Is it valid to replace $\tan x$ with x for small x if x is in degrees? Graph $y_1 = \tan x$ and $y_2 = x$ in the same viewing window with the calculator set in degree mode for $-45° \le x \le 45°$ and $-5 \le y \le 5$, and explain the results after exploring the graphs using TRACE.

30. In applied mathematics certain formulas, derivations, and calculations are simplified by replacing $\sin x$ with x for small x. What justifies this procedure? To find out, graph $y_1 = \sin x$ and $y_2 = x$ in the same viewing window for $-1 \le x \le 1$ and $-1 \le y \le 1$.
(A) What do you observe about the two graphs when x is close to 0, say $-0.5 \le x \le 0.5$?
(B) Complete the table to three decimal places (use the table feature on your graphing utility if it has one):

x	−0.3	−0.2	−0.1	0.0	0.1	0.2	0.3
sin x							

(C) Is it valid to replace $\sin x$ with x for small x if x is in degrees? Graph $y_1 = \sin x$ and $y_2 = x$ in the same viewing window with the calculator set in degree mode for $-10° \le x \le 10°$ and $-1 \le y \le 1$, and explain the results after exploring the graphs using TRACE.

31. Set your graphing utility in radian and parametric (Par) modes. Make the entries as indicated in the figure to obtain the indicated graph (set Tmax and Xmax to 2π and set Xscl to $\pi/2$). The parameter T represents a central angle in the unit circle of T radians or an arc length of T on the unit circle starting at (1, 0). As T moves from 0 to 2π, the point $P(\cos T, \sin T)$ moves counterclockwise around the unit circle starting at (1, 0) and ending at (1, 0), and the point $P(T, \sin T)$ moves along the sine curve from (0, 0) to $(2\pi, 0)$. Use TRACE to observe this behavior on each curve.

```
Plot1 Plot2 Plot3
\X1T=cos(T)
 Y1T=sin(T)
\X2T=T
 Y2T=sin(T)
\X3T=
 Y3T=
\X4T=
```

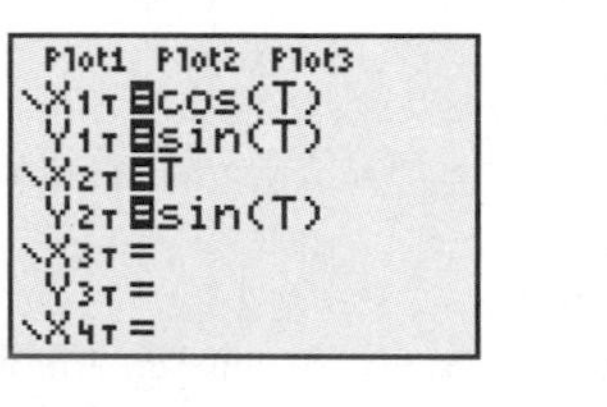

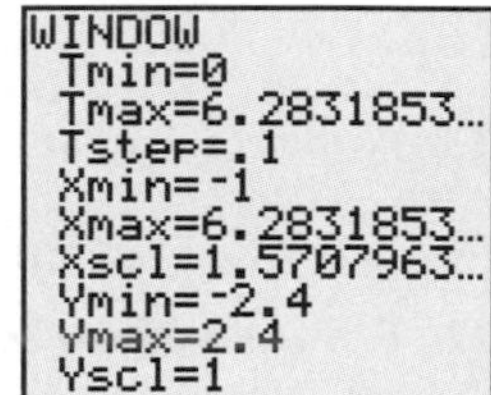

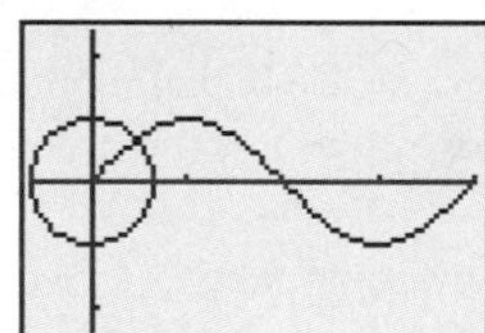

Figure for 31

Now use TRACE and move back and forth between the unit circle and the graph of the sine function for various values of T as T increases from 0 to 2π. Discuss what happens in each case.

32. Repeat Problem 31 with $y_{2T} = \cos T$.

3.2 GRAPHING $y = k + A \sin Bx$ AND $y = k + A \cos Bx$

- Graphs of $y = A \sin x$ and $y = A \cos x$
- Graphs of $y = \sin Bx$ and $y = \cos Bx$
- Graphs of $y = A \sin Bx$ and $y = A \cos Bx$
- Graphs of $y = k + A \sin Bx$ and $y = k + A \cos Bx$
- Application: Sound Frequency
- Application: Floating Objects

Graphing the equations $y = k + A \sin Bx$ and $y = k + A \cos Bx$ is not difficult if you have a clear understanding of the graphs of the basic equations $y = \sin x$ and $y = \cos x$ studied in Section 3.1. Being able to graph trigonometric forms involving the constants k, A, and B significantly increases the variety of applications that can be considered.

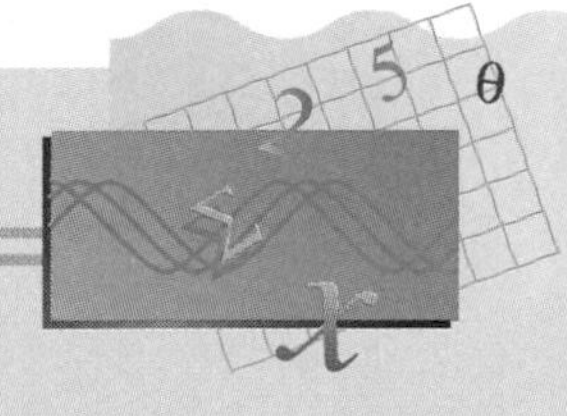

EXPLORE/DISCUSS 1

Describe how the graph of y_2 is related to the graph of y_1 in each of the following graphing calculator displays.

(A)

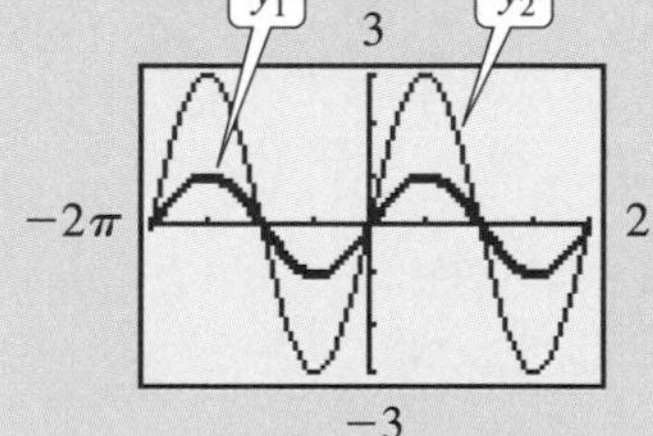

(B)

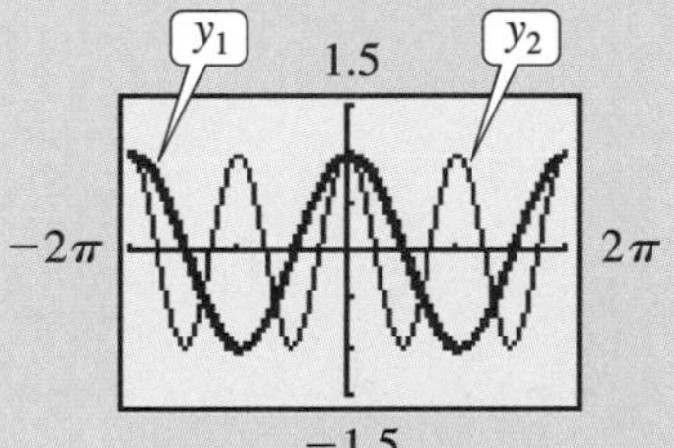

Graphs of $y = A \sin x$ and $y = A \cos x$

To start, we investigate the effect of the constant A by comparing

$$y = \sin x \qquad \text{and} \qquad y = A \sin x$$

We can obtain the graph of $y = A \sin x$ from the graph of $y = \sin x$ by multiplying each y value of $y = \sin x$ by the constant A. The graph of $y = A \sin x$ will cross the x axis everywhere the graph of $y = \sin x$ crosses the x axis, because $A \cdot 0 = 0$. Thus, the graphs of $y = A \sin x$ and $y = \sin x$ have the same x intercepts.

We know that the maximum deviation of the graph of $y = \sin x$ from the x axis is 1. Thus, the maximum deviation of the graph of $y = A \sin x$ from the x axis is $|A| \cdot 1 = |A|$. The constant $|A|$ is called the **amplitude** of the graph of $y = A \sin x$ and represents the maximum deviation of the graph from the x axis. Finally, the period of $y = A \sin x$ is also 2π, since $A \sin(x + 2\pi) = A \sin x$.

EXAMPLE 1

Comparing Amplitudes

Compare the graphs of $y = \frac{1}{3} \sin x$ and $y = -3 \sin x$ with the graph of $y = \sin x$ by graphing each on the same coordinate system for $0 \le x \le 2\pi$.

Solution We see that the graph of $y = \frac{1}{3} \sin x$ has an amplitude of $|\frac{1}{3}| = \frac{1}{3}$, the graph of $y = -3 \sin x$ has an amplitude of $|-3| = 3$, and the graph of $y = \sin x$ has an amplitude of $|1| = 1$. The negative sign in $y = -3 \sin x$ turns the graph of $y = 3 \sin x$ upside down. That is, the graph of $y = -3 \sin x$ is the same as the graph of $y = 3 \sin x$ reflected across the x axis. The graphs of all three equations, $y = \frac{1}{3} \sin x$, $y = -3 \sin x$, and $y = \sin x$, are shown in Figure 1.

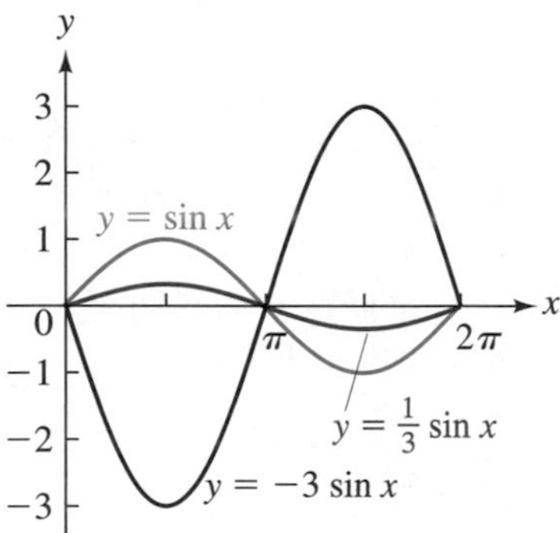

FIGURE 1
Comparing amplitudes

■

In summary, the results of Example 1 show that the effect of A in $y = A \sin x$ is to increase or decrease the y values of $y = \sin x$ without affecting the x values. A similar analysis applies to $y = A \cos x$; this function also has an amplitude of $|A|$ and a period of 2π.

Matched Problem 1 Compare the graphs of $y = \frac{1}{2} \cos x$ and $y = -2 \cos x$ with the graph of $y = \cos x$ by graphing each on the same coordinate system for $0 \le x \le 2\pi$. ■

■ Graphs of $y = \sin Bx$ and $y = \cos Bx$

We now examine the effect of B by comparing

$$y = \sin x \qquad \text{and} \qquad y = \sin Bx \qquad B > 0$$

Both have the same amplitude, 1, but how do their periods compare? Since $\sin x$

has a period of 2π, it follows that $\sin Bx$ completes one cycle as Bx varies from

$$Bx = 0 \quad \text{to} \quad Bx = 2\pi$$

or as x varies from

$$x = \frac{0}{B} = 0 \quad \text{to} \quad x = \frac{2\pi}{B}$$

Thus, the period of $\sin Bx$ is $2\pi/B$. We check this result as follows: If $f(x) = \sin Bx$, then

$$f\left(x + \frac{2\pi}{B}\right) = \sin\left[B\left(x + \frac{2\pi}{B}\right)\right] = \sin(Bx + 2\pi) = \sin Bx = f(x)$$

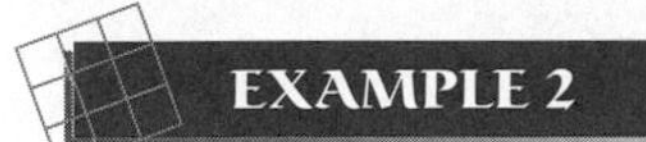

Comparing Periods

Compare the graphs of $y = \sin 2x$ and $y = \sin(x/2)$ with the graph of $y = \sin x$ by graphing each on the same coordinate system for one period starting at the origin.

Solution *Period for* $\sin 2x$: $\sin 2x$ completes one cycle as $2x$ varies from

$$2x = 0 \quad \text{to} \quad 2x = 2\pi$$

or as x varies from

$$x = 0 \quad \text{to} \quad x = \pi$$

Thus, the period for $\sin 2x$ is π.

Period for $\sin(x/2)$: $\sin(x/2)$ completes one cycle as $x/2$ varies from

$$\frac{x}{2} = 0 \quad \text{to} \quad \frac{x}{2} = 2\pi$$

or as x varies from

$$x = 0 \quad \text{to} \quad x = 4\pi$$

Thus, the period for $\sin(x/2)$ is 4π. The graphs of all three equations, $y = \sin 2x$, $y = \sin(x/2)$, and $y = \sin x$, are shown in Figure 2.

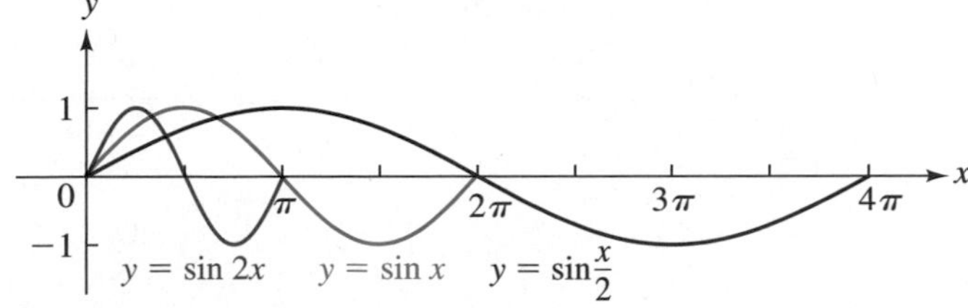

FIGURE 2
Comparing periods

■

We see from Example 2 that the effect of B is to compress or stretch the basic sine curve by changing the period of $\sin x$. A similar analysis applies to $y = \cos Bx$, where $B > 0$; its period is $2\pi/B$.

Matched Problem 2 Compare the graphs of $y = \cos 2x$ and $y = \cos(x/2)$ with the graph of $y = \cos x$ by graphing each on the same coordinate system for one period starting at the origin. ■

Graphs of $y = A \sin Bx$ and $y = A \cos Bx$

We summarize the results of the discussions of amplitude and period in the box below.

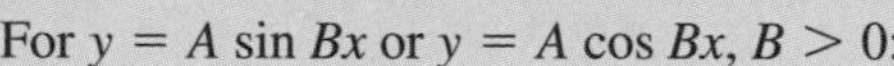

For $y = A \sin Bx$ or $y = A \cos Bx$, $B > 0$:

$$\text{Amplitude} = |A| \qquad \text{Period} = \frac{2\pi}{B}$$

If $B > 1$, the basic sine or cosine curve is horizontally compressed.
If $0 < B < 1$, the basic sine or cosine curve is horizontally stretched.

We now consider several examples where we show how graphs of $y = A \sin Bx$ and $y = A \cos Bx$ can be sketched rather quickly.

EXAMPLE 3

Graphing the Form $y = A \cos Bx$

State the amplitude and period for $y = 3 \cos 2x$, and graph the equation for $-\pi \leq x \leq 2\pi$.

Solution Amplitude $= |A| = |3| = 3$ Period $= \dfrac{2\pi}{B} = \dfrac{2\pi}{2} = \pi$

To sketch the graph, divide the interval of one period, from 0 to π, into four equal parts, locate x intercepts, and locate high and low points (see Fig. 3a). Sketch the graph for one period; then extend this graph to fill out the desired interval (Fig. 3b).

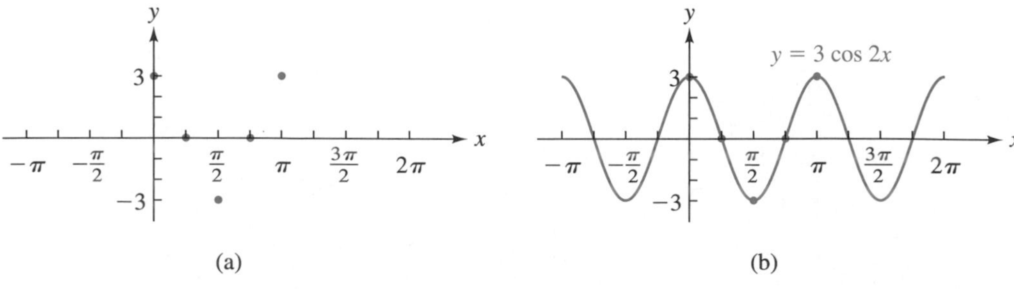

FIGURE 3

■

Notice that we scaled the x axis using the period divided by 4; that is, the basic unit on the x axis is $\pi/4$. Also, we adjusted the scale on the y axis to accommodate the amplitude 3.

The scales on both axes do not have to be the same.

Matched Problem 3 State the amplitude and period for $y = \frac{1}{3}\sin(x/2)$, and graph the equation for $-2\pi \le x \le 6\pi$. ■

EXAMPLE 4 **Graphing the Form $y = A \sin Bx$**

State the amplitude and period for $y = -\frac{1}{2}\sin(\pi x/2)$, and graph the equation for $-5 \le x \le 5$.

Solution Amplitude $= |A| = \left|-\frac{1}{2}\right| = \frac{1}{2}$ Period $= \frac{2\pi}{B} = \frac{2\pi}{\pi/2} = 4$

Because of the $-\frac{1}{2}$, the graph of $y = -\frac{1}{2}\sin(\pi x/2)$ is the graph of $y = \frac{1}{2}\sin(\pi x/2)$ reflected across the x axis (turned upside down). As before, we divide one period, from 0 to 4, into four equal parts, locate x intercepts, and locate high and low points (see Fig. 4a). Then we graph the equation for one period, and extend the graph over the desired interval (see Fig. 4b).

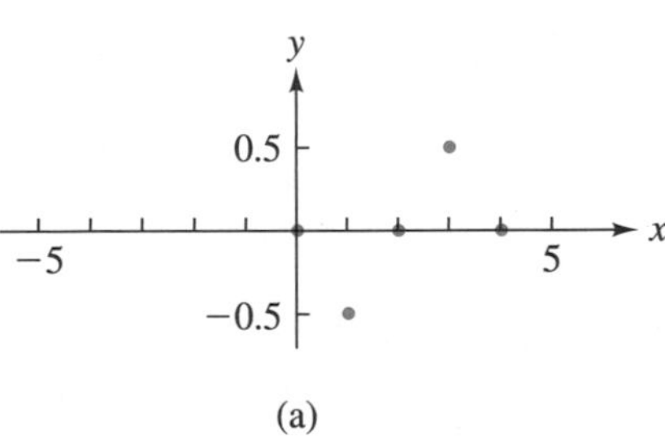

(a)

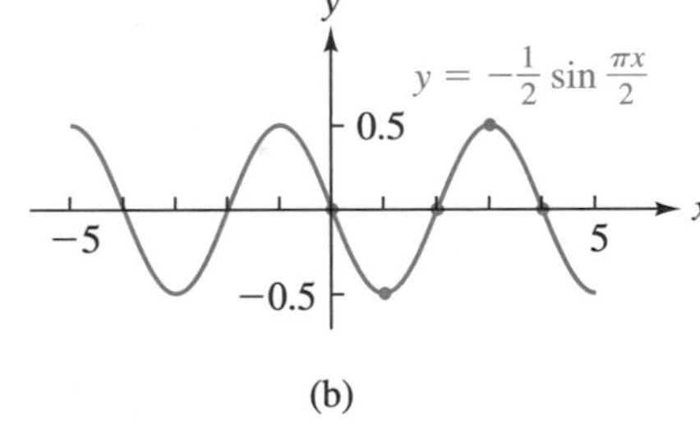

(b)

FIGURE 4 ■

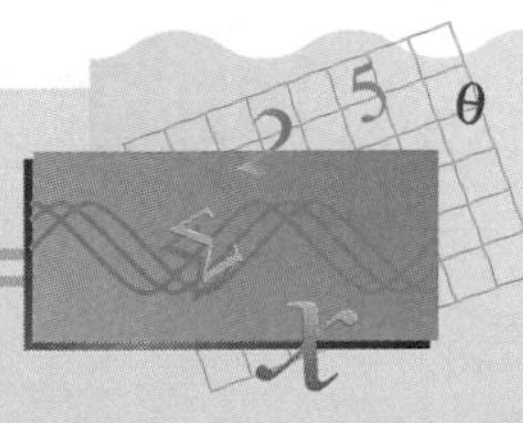

EXPLORE/DISCUSS 2

Find an equation of the form $y = A \sin Bx$ that produces the graph shown in the following graphing calculator display:

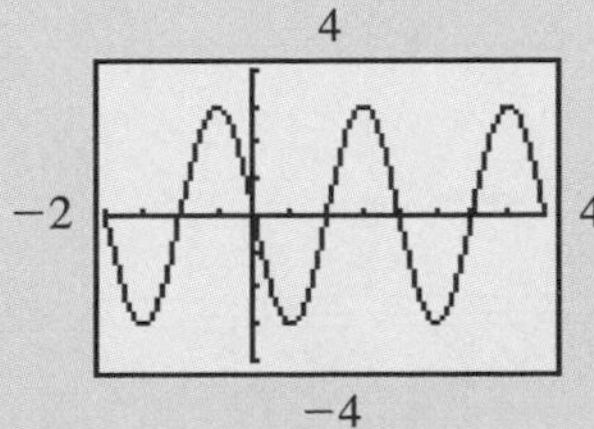

Is it possible for an equation of the form $y = A \cos Bx$ to produce the same graph? Explain.

Matched Problem 4 State the amplitude and period for $y = -2 \cos 2\pi x$, and graph the equation for $-2 \leq x \leq 2$. ■

Graphs of $y = k + A \sin Bx$ and $y = k + A \cos Bx$

By adding a constant k to either $A \sin Bx$ or $A \cos Bx$, we are simply adding k to the ordinates of the points on their graphs. That is, we are **vertically translating** the graphs of $y = A \sin Bx$ and $y = A \cos Bx$ up k units if $k > 0$ or down $|k|$ units if $k < 0$.

EXAMPLE 5 **Graphing the Form $y = k + A \cos Bx$**

Graph: $y = -2 + 3 \cos 2x, \quad -\pi \leq x \leq 2\pi$

Solution We first graph $y = 3 \cos 2x$ (as we did in Example 3), and then move the graph down $|k| = |-2| = 2$ units (since $k = -2 < 0$), as shown in Figure 5.

FIGURE 5

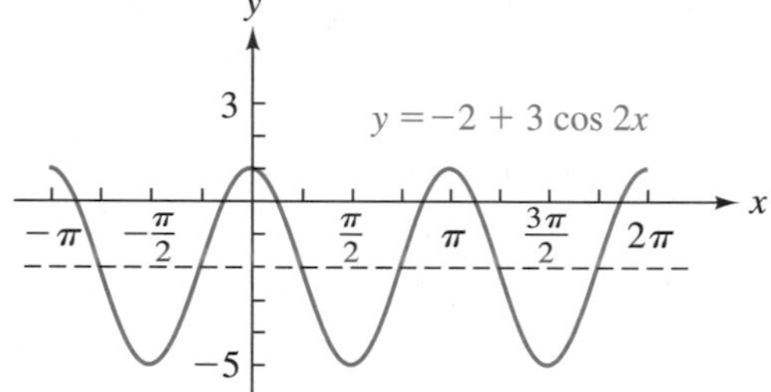

■

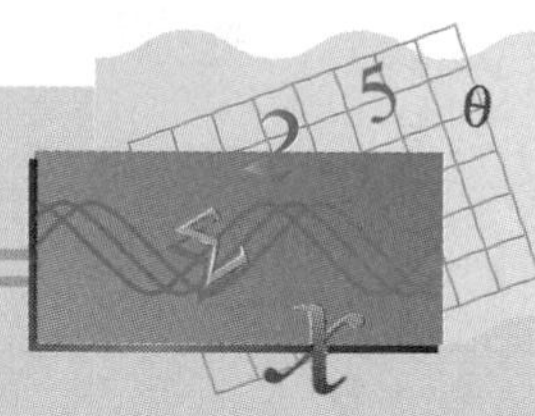

EXPLORE/DISCUSS 3

Find an equation of the form $y = k + A \cos Bx$ that produces the graph shown in the following graphing calculator display:

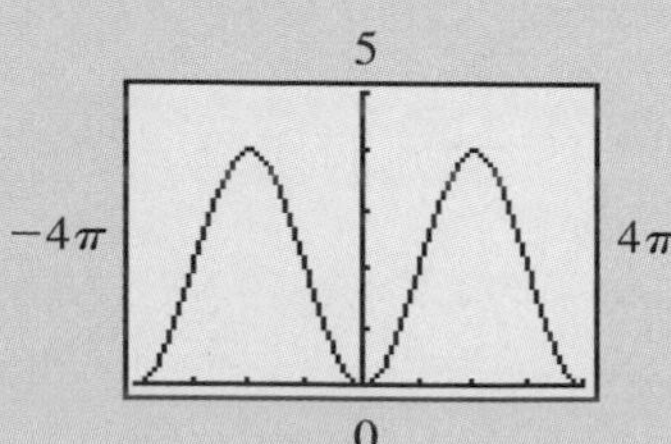

Is it possible for an equation of the form $y = k + A \sin Bx$ to produce the same graph? Explain.

You may find it helpful to first draw the horizontal dashed line shown in Figure 5. In this case, the line is 2 units below the x axis, which represents a vertical translation of -2. Then the graph of $y = 3 \cos 2x$ is drawn relative to the dashed line and the original y axis.

Matched Problem 5 Graph: $y = 3 - 2 \cos 2\pi x, \quad -2 \leq x \leq 2$ ■

Application: Sound Frequency

In many applications involving periodic phenomena and time (sound waves, pendulum motion, water waves, electromagnetic waves, and so on), we speak of the **period** as the length of time taken for one complete cycle of motion. For example, since each complete vibration of an A440 tuning fork (see Fig. 6) lasts $\frac{1}{440}$ sec, we say that its period is $\frac{1}{440}$ sec.

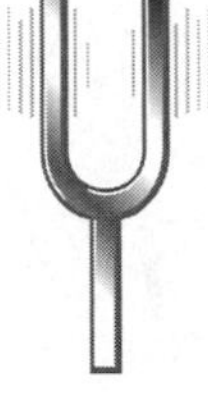

FIGURE 6
Tuning fork

Closely related to the period is the concept of **frequency,** which is the number of periods or cycles per second. The frequency of the A440 tuning fork is 440 cycles/sec, which is the reciprocal of the period. Instead of "cycles per second," we usually write "Hz," where Hz (read "hertz") is the standard unit for frequency (cycles per unit of time), and is named after the German physicist Heinrich Rudolph Hertz (1857–1894), who discovered and produced radio waves.

PERIOD AND FREQUENCY

For any periodic phenomenon, if P is the period and f is the frequency, then:

$$P = \frac{1}{f}$$ Period is the time for one complete cycle.

$$f = \frac{1}{P}$$ Frequency is the number of cycles per unit of time.

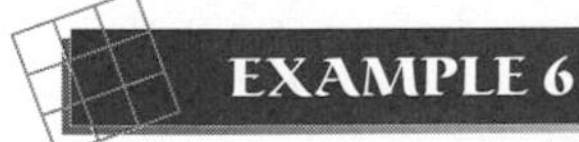

EXAMPLE 6 **An Equation of a Sound**

Referring to the tuning fork in Figure 6, we want to model the motion of the tip of one prong using a sine or cosine function. We choose deviation from the rest position to the right as positive and to the left as negative. If the prong motion has a frequency of 440 Hz (cycles/sec), an amplitude of 0.04 cm, and is 0.04 cm to the right when $t = 0$, find A and B so that $y = A \cos Bt$ is an approximate model for this motion.

Solution *Find A.* The amplitude $|A|$ is given to be 0.04. Since $y = 0.04$ when $t = 0$, $A = 0.04$ (and not -0.04).

Find B. We are given that the frequency, f, is 440 Hz; hence, the period is found using the reciprocal formula:

$$P = \frac{1}{f} = \frac{1}{440} \text{ sec}$$

But from the earlier discussion, $P = 2\pi/B$. Thus

$$B = \frac{2\pi}{P} = \frac{2\pi}{\frac{1}{440}} = 880\pi$$

Write the equation:

$$y = A \cos Bt = 0.04 \cos 880\,\pi t$$

✔ Check

$$\text{Amplitude} = |A| = |0.04| = 0.04$$

$$\text{Period} = \frac{2\pi}{B} = \frac{2\pi}{880\pi} = \frac{1}{440}\text{ sec}$$

$$\text{Frequency} = \frac{1}{\text{Period}} = \frac{1}{\frac{1}{440}} = 440\text{ Hz}$$

And when $t = 0$,

$$y = 0.04 \cos 880\pi(0)$$
$$= 0.04(1) = 0.04$$

■

Matched Problem 6 Repeat Example 6 if the prong motion has a frequency of 262 Hz (cycles/sec), an amplitude of 0.065 cm, and is 0.065 cm to the left of the rest position when $t = 0$. ■

■ Application: Floating Objects

Did you know that it is possible to determine the mass of a floating object (Fig. 7) simply by making it bob up and down in the water and timing its period of oscillation?

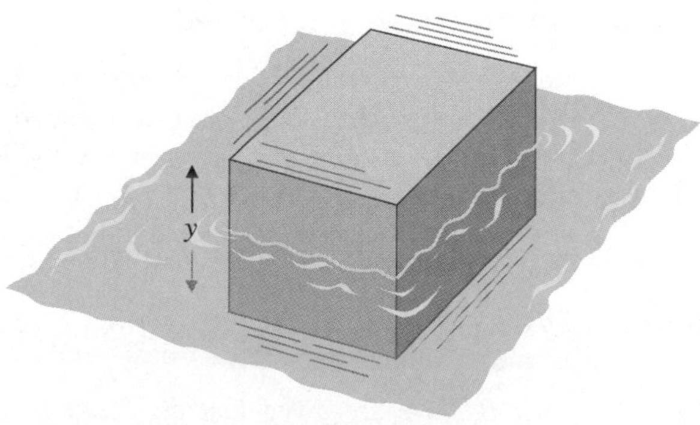

FIGURE 7
Floating object

If the rest position of the floating object is taken to be 0 and we start counting time as the object passes up through 0 when it is made to oscillate, then its equation of motion can be shown to be (neglecting water and air resistance)

$$y = D \sin \sqrt{\frac{1{,}000gA}{M}}\, t$$

where y and D are in meters, t is time in seconds, $g = 9.75 \text{ m/sec}^2$ (gravitational constant), A is the horizontal cross-sectional area in square meters, and M is mass in kilograms. The amplitude and period for the motion are given by

$$\text{Amplitude} = |D| \qquad \text{Period} = \frac{2\pi}{\sqrt{1{,}000gA/M}}$$

Mass of a Buoy

A cylindrical buoy with cross-sectional area 1.25 m^2 is observed (after being pushed) to bob up and down with a period of 0.50 sec. Approximately what is the mass of the buoy (in kilograms)?

Solution We use the formula

$$\text{Period} = \frac{2\pi}{\sqrt{1{,}000gA/M}}$$

with $g = 9.75 \text{ m/sec}^2$, $A = 1.25 \text{ m}^2$, and period $= 0.50$ sec, and solve for M. Thus,

$$0.50 = \frac{2\pi}{\sqrt{(1{,}000)(9.75)(1.25)/M}}$$

$$M \approx 77 \text{ kg}$$

■

Matched Problem 7 Write the equation of motion of the buoy in Example 7 in the form $y = D \sin Bt$, assuming its amplitude is 0.4 m. ■

Answers to Matched Problems

1.

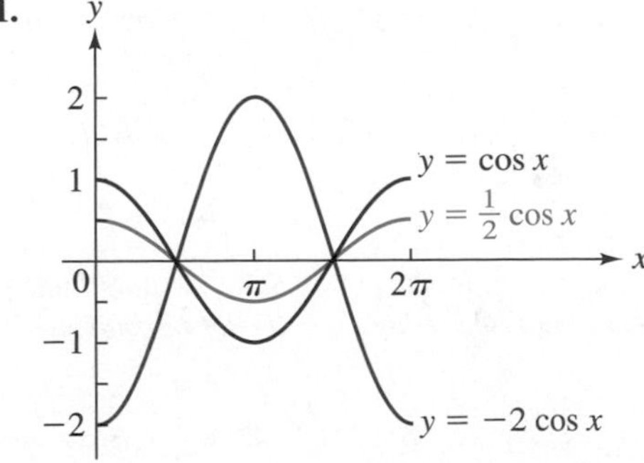

2.

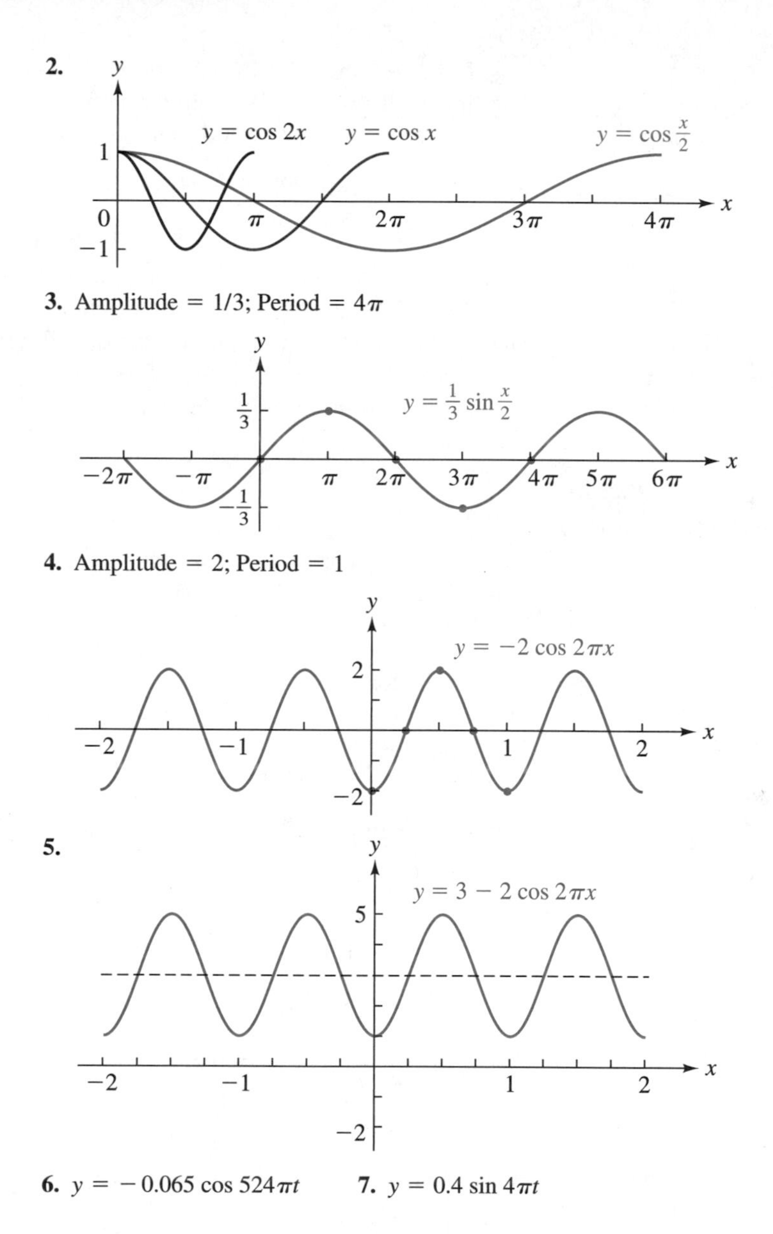

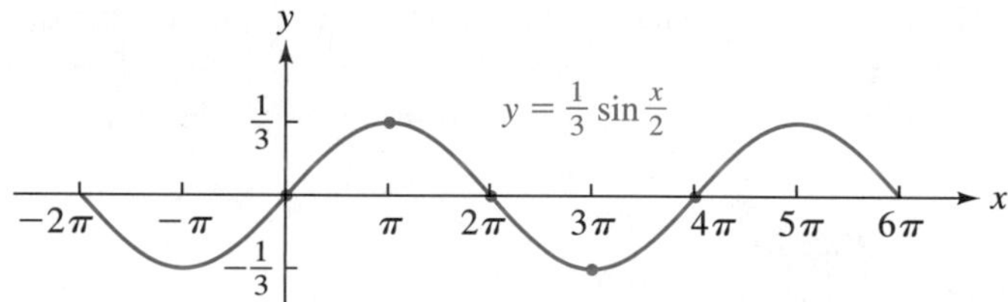

3. Amplitude = 1/3; Period = 4π

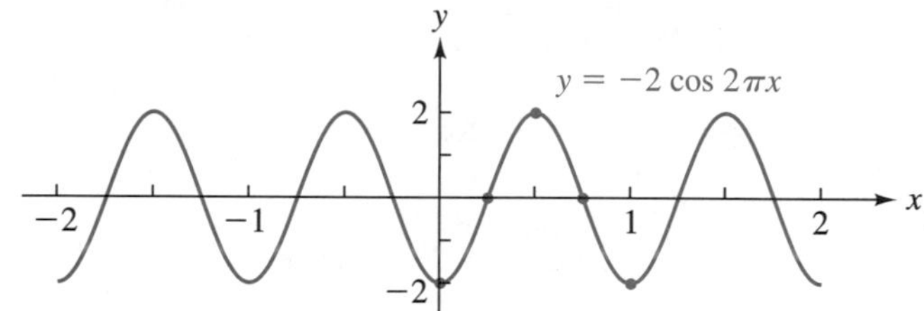

4. Amplitude = 2; Period = 1

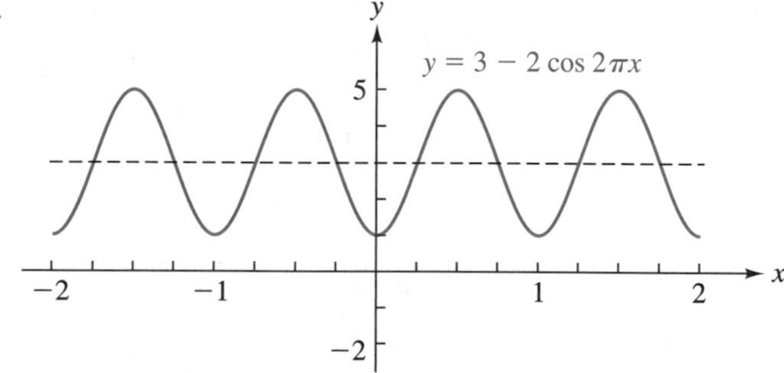

5.

6. $y = -0.065 \cos 524\pi t$ 7. $y = 0.4 \sin 4\pi t$

EXERCISE 3.2

A *Make a sketch of each trigonometric function without looking at the text or using a calculator. Label each point where the graph crosses the x axis.*

1. $y = \sin x, \quad -2\pi \leq x \leq 2\pi$
2. $y = \cos x, \quad -2\pi \leq x \leq 2\pi$

State the amplitude and period for each equation, and graph it over the indicated interval.

3. $y = -2 \sin x, \quad 0 \leq x \leq 4\pi$
4. $y = -3 \cos x, \quad 0 \leq x \leq 4\pi$
5. $y = \frac{1}{2} \sin x, \quad 0 \leq x \leq 2\pi$

6. $y = \frac{1}{3} \cos x, \quad 0 \le x \le 2\pi$
7. $y = \sin 2\pi x, \quad -2 \le x \le 2$
8. $y = \cos 4\pi x, \quad -1 \le x \le 1$
9. $y = \cos \frac{x}{4}, \quad 0 \le x \le 8\pi$
10. $y = \sin \frac{x}{2}, \quad 0 \le x \le 4\pi$

B *In Problems 11–16, state the amplitude and period for each equation, and graph it over the indicated interval.*

11. $y = 2 \sin 4x, \quad -\pi \le x \le \pi$
12. $y = 3 \cos 2x, \quad -\pi \le x \le \pi$
13. $y = \frac{1}{3} \cos 2\pi x, \quad -2 \le x \le 2$
14. $y = \frac{1}{2} \sin 2\pi x, \quad -2 \le x \le 2$
15. $y = -\frac{1}{4} \sin \frac{x}{2}, \quad -4\pi \le x \le 4\pi$
16. $y = -3 \cos \frac{x}{2}, \quad -4\pi \le x \le 4\pi$

Problems 17 and 18 involve oscillating objects. For each problem, find an equation of the form $y = A \sin Bt$ or $y = A \cos Bt$ that satisfies the given conditions (y is displacement from a central position at time t).

17. Displacement is 0 ft when t is 0, amplitude is 2 ft, and period is 2 sec.
18. Displacement from the t axis is 5 cm when t is 0, amplitude is 5 cm, and period is 0.1 sec.

19. Describe what happens to the size of the period of $y = A \sin Bx$ as B increases without bound.
20. Describe what happens to the size of the period of $y = A \cos Bx$ as B decreases through positive values toward 0.

Graph each equation over the indicated interval.

21. $y = -1 + \frac{1}{3} \cos 2\pi x, \quad -2 \le x \le 2$
22. $y = -\frac{1}{2} + \frac{1}{2} \sin 2\pi x, \quad -2 \le x \le 2$
23. $y = 2 - \frac{1}{4} \sin \frac{x}{2}, \quad -4\pi \le x \le 4\pi$
24. $y = 3 - 3 \cos \frac{x}{2}, \quad -4\pi \le x \le 4\pi$

In Problems 25–28, find the equation of the form $y = A \sin Bx$ that produces the indicated graph.

25.

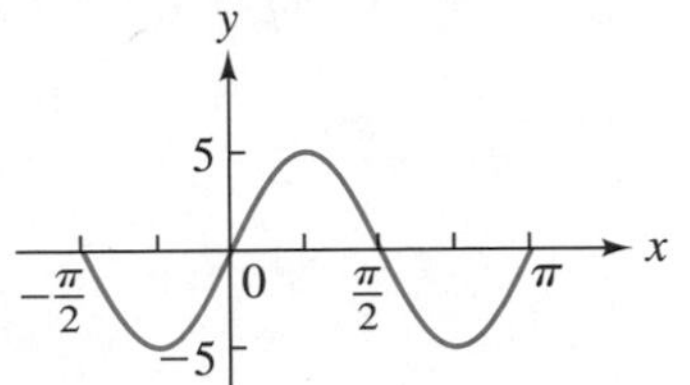

26.

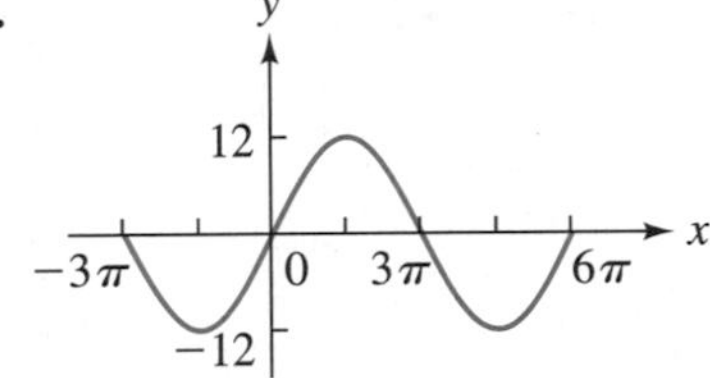

27.

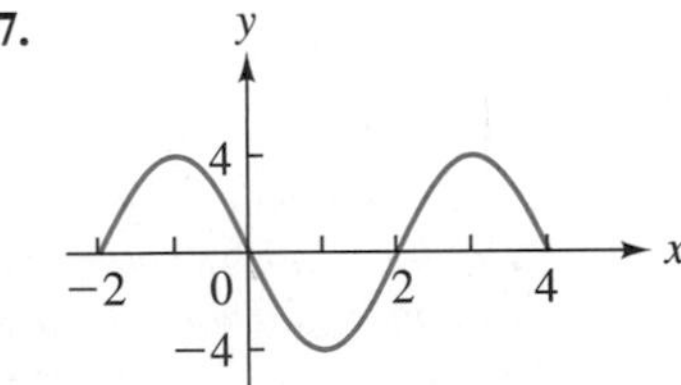

28.

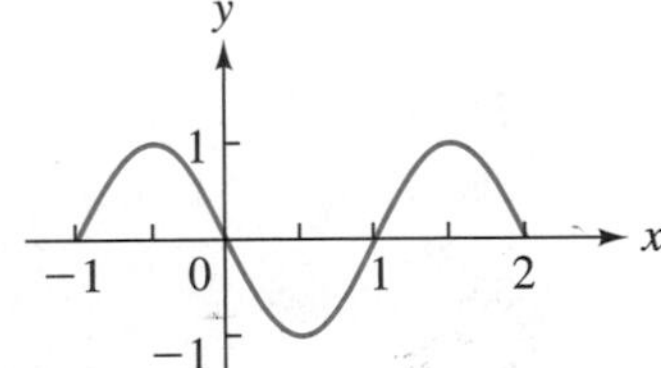

In Problems 29–32, find the equation of the form $y = A \cos Bx$ that produces the indicated graph.

29.

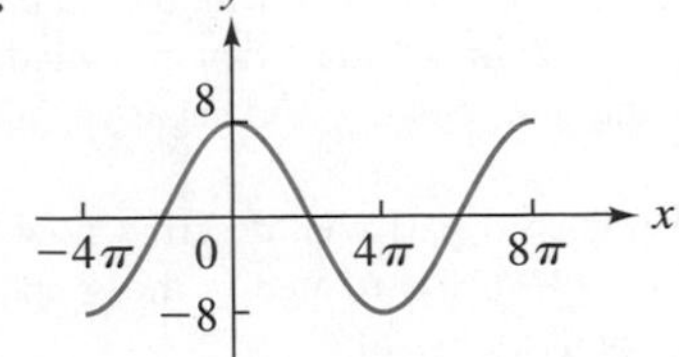

30.

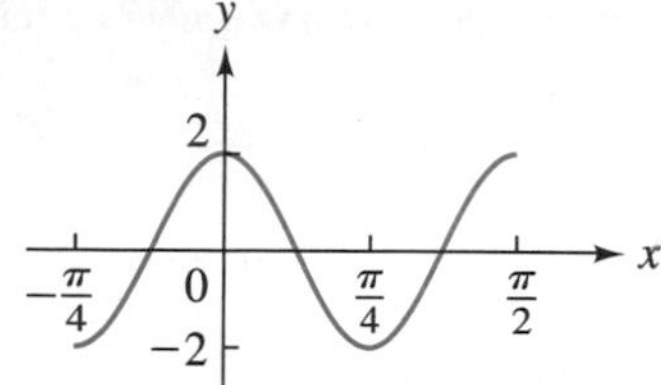

31.

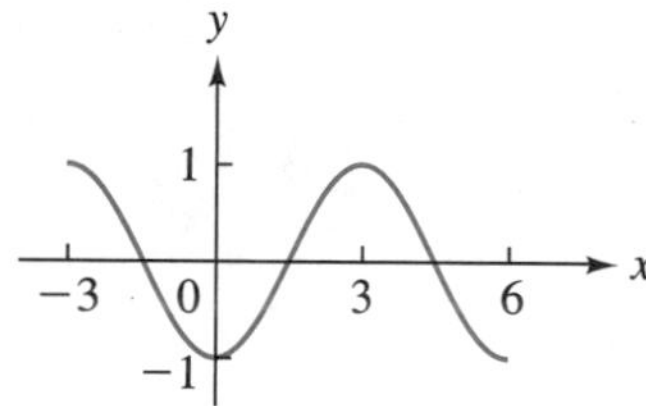

32.

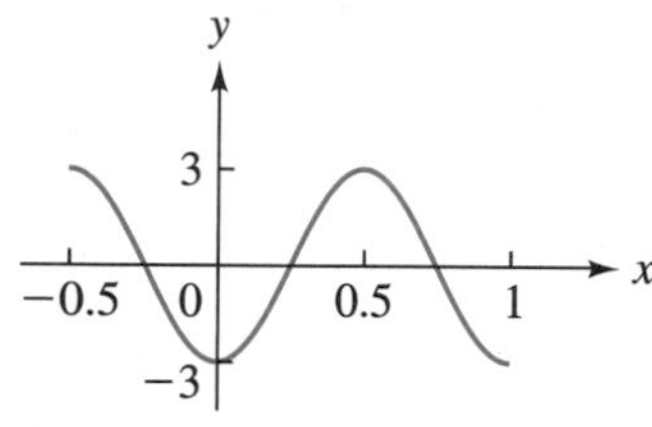

In Problems 33–38, graph the given equation on a graphing utility. (Adjust the ranges in the viewing windows so that you see at least two periods of a particular function.) Find an equation of the form $y = k + A \sin Bx$ or $y = k + A \cos Bx$ that has the same graph. These problems suggest the existence of further identities in addition to the basic identities discussed in Section 2.6. Further identities are discussed in detail in Chapter 4.

33. $y = \sin x \cos x$

34. $y = \cos^2 x - \sin^2 x$

35. $y = 2 \cos^2 x$

36. $y = 2 \sin^2 x$

37. $y = 2 - 4 \sin^2 2x$

38. $y = 6 \cos^2 \dfrac{x}{2} - 3$

Problems 39 and 40 graphically explore the relationships of the graphs of $y = \sin(x + C)$ and $y = \cos(x + C)$ relative to the corresponding graphs of $y = \sin x$ and $y = \cos x$. This topic is discussed in detail in the next section.

39. (A) Graph $y = \sin(x + C)$, $-2\pi \le x \le 2\pi$, for $C = 0$ and $C = -\pi/2$ in one viewing window, and for $C = 0$ and $C = \pi/2$ in another viewing window. (Experiment with other positive and negative values of C.)

(B) Based on the graphs in part (A), describe how the graph of $y = \sin(x + C)$ is related to the graph of $y = \sin x$ for various values of C.

40. (A) Graph $y = \cos(x + C)$, $-2\pi \le x \le 2\pi$, for $C = 0$ and $C = -\pi/2$ in one viewing window, and for $C = 0$ and $C = \pi/2$ in another viewing window. (Experiment with other positive and negative values of C.)

(B) Based on the graphs in part (A), describe how the graph of $y = \cos(x + C)$ is related to the graph of $y = \cos x$ for various values of C.

C

41. From the graph of $f(x) = \cos^2 x$, $-2\pi \le x \le 2\pi$, on a graphing utility, determine the period of f; that is, find the smallest positive number p such that $f(x + p) = f(x)$.

42. From the graph of $f(x) = \sin^2 x$, $-2\pi \le x \le 2\pi$, on a graphing utility, determine the period of f; that is, find the smallest positive number p such that $f(x + p) = f(x)$.

43. The table in the figure was produced using a table feature for a particular graphing calculator. Find a function of the form $y = A \sin Bx$ or $y = A \cos Bx$ that will produce the table.

X	Y1
0.0	2.0
.5	1.0
1.0	-1.0
1.5	-2.0
2.0	-1.0
2.5	1.0
3.0	2.0

X=0

44. The table in the figure was produced using a table feature for a particular graphing calculator. Find a function of the form $y = A \sin Bx$ or $y = A \cos Bx$ that will produce the table.

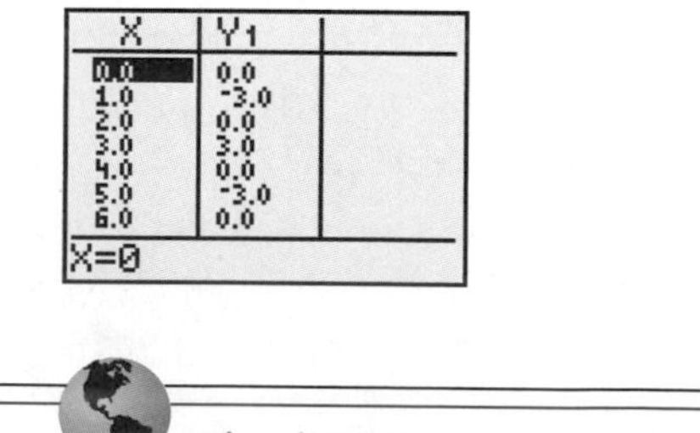

X	Y1
0.0	0.0
1.0	-3.0
2.0	0.0
3.0	3.0
4.0	0.0
5.0	-3.0
6.0	0.0

X=0

Applications

In these applications, assume all given values are exact unless indicated otherwise.

45. **Electrical Circuits** The voltage E in an electrical circuit is given by $E = 110 \sin 120\pi t$, where t is time in seconds. What are the amplitude and period of the function? What is the frequency of the function? Graph the function for $0 \le t \le \frac{3}{60}$.

46. Spring–Mass System The equation $y = -4 \cos 8t$, where t is time in seconds, represents the motion of a weight hanging on a spring after it has been pulled 4 cm below its equilibrium point and released (see the figure). What are the amplitude, period, and frequency of the function? [Air resistance and friction (damping forces) are neglected.] Graph the function for $0 \le t \le 3\pi/4$.

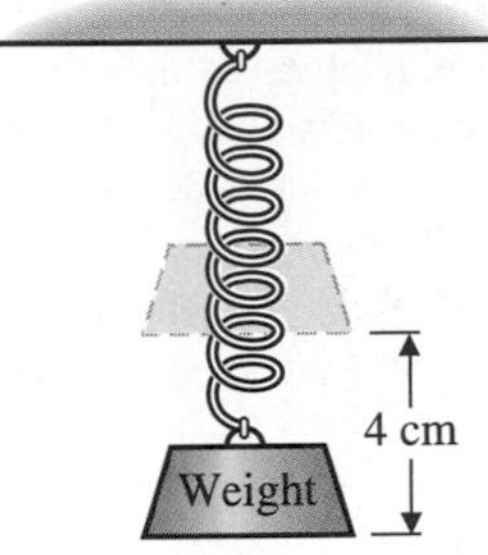

Figure for 46

***47. Electrical Circuits** If the voltage E in an electrical circuit has an amplitude of 12 V and a frequency of 40 Hz, and if $E = 12$ V when $t = 0$ sec, find an equation of the form $E = A \cos Bt$ that gives the voltage at any time t.

***48. Spring–Mass System** If the motion of the weight in Problem 46 has an amplitude of 6 in. and a frequency of 2 Hz, and if its position when $t = 0$ sec is 6 in. above its position at rest (above the rest position is positive and below is negative), find an equation of the form $y = A \cos Bt$ that describes the motion at any time t. (Neglect any damping forces—that is, air resistance and friction.)

***49. Floating Objects**

(A) A $3 \text{ m} \times 3 \text{ m} \times 1 \text{ m}$ float in the shape of a rectangular solid is observed to bob up and down with a period of 1 sec. What is the mass of the float (in kilograms, to three significant digits)?

(B) Write an equation of motion for the float in part (A) in the form $y = D \sin Bt$, assuming the amplitude of the motion is 0.2 m.

(C) Graph the equation found in part (B) for $0 \le t \le 2$.

***50. Floating Objects**

(A) A cylindrical buoy with diameter 0.6 m is observed (after being pushed) to bob up and down with a period of 0.4 sec. What is the mass of the buoy (to the nearest kilogram)?

(B) Write an equation of motion for the buoy in the form $y = D \sin Bt$, assuming the amplitude of the motion is 0.1 m.

(C) Graph the equation for $0 \le t \le 1.2$.

51. Physiology A normal seated adult breathes in and exhales about 0.80 liter of air every 4.00 sec. The volume of air $V(t)$ in the lungs (in liters) t seconds after exhaling is modeled approximately by

$$V(t) = 0.45 - 0.40 \cos \frac{\pi t}{2} \qquad 0 \le t \le 8$$

(A) What is the maximum amount of air in the lungs, and what is the minimum amount of air in the lungs? Explain how you arrived at these numbers.

(B) What is the period of breathing?

(C) How many breaths are taken per minute? Show your computation.

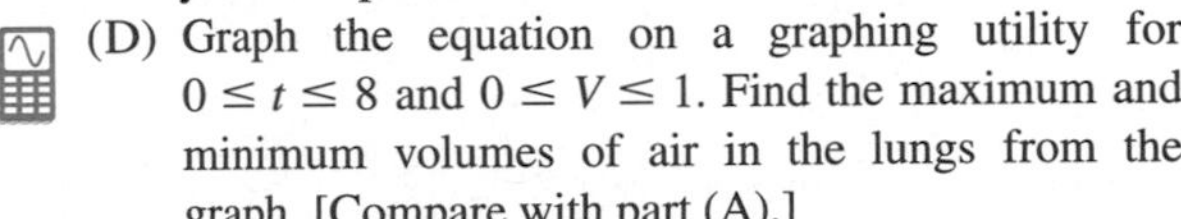

(D) Graph the equation on a graphing utility for $0 \le t \le 8$ and $0 \le V \le 1$. Find the maximum and minimum volumes of air in the lungs from the graph. [Compare with part (A).]

52. Physiology When you have your blood pressure measured during a physical examination, the nurse or doctor will record two numbers as a ratio—for example, 120/80. The first number (**systolic pressure**) represents the amount of pressure in the blood vessels when the heart contracts (beats) and pushes blood through the circulatory system. The second number (**diastolic pressure**) represents the pressure in the blood vessels at the lowest point between the heartbeats, when the heart is at rest. According to the National Institutes of Health, normal blood pressure is below 130/85, moderately high is from 160/100 to 179/109, and severe is higher than 180/110. The blood pressure of a particular person with moderately high blood pressure is modeled approximately by

$$P = 135 + 30 \cos 2.5\pi t \qquad t \ge 0$$

where P is pressure in millimeters of mercury t seconds after the heart contracts.

(A) Express the person's blood pressure as an appropriate ratio of two numbers. Explain how you arrived at these numbers.

(B) What is the period of the heartbeat?

(C) What is the pulse rate in beats per minute? Show your computation.

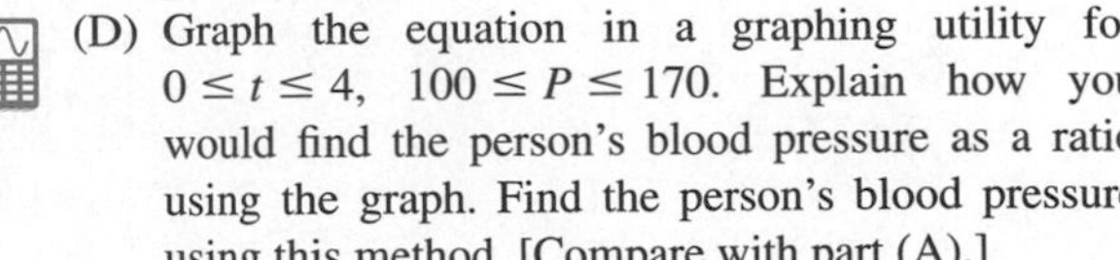

(D) Graph the equation in a graphing utility for $0 \le t \le 4$, $100 \le P \le 170$. Explain how you would find the person's blood pressure as a ratio using the graph. Find the person's blood pressure using this method. [Compare with part (A).]

TABLE 1
Leg Oscillation

t (sec)	0.0	0.5	1.0	1.5	2.0	2.5	3.0	3.5	4.0
θ (deg)	$-25°$	$0°$	$25°$	$0°$	$-25°$	$0°$	$25°$	$0°$	$-25°$

*53. **Rotary and Linear Motion** A Ferris wheel with a diameter of 40 m rotates counterclockwise at 4 rpm (revolutions per minute), as indicated in the figure. It starts at $\theta = 0$ when time t is 0. At the end of t minutes, $\theta = 8\pi t$. Why? Convince yourself that the position of the person's shadow (from the sun directly overhead) on the x axis is given by

$$x = 20 \sin 8\pi t$$

Graph this equation for $0 \le t \le 1$.

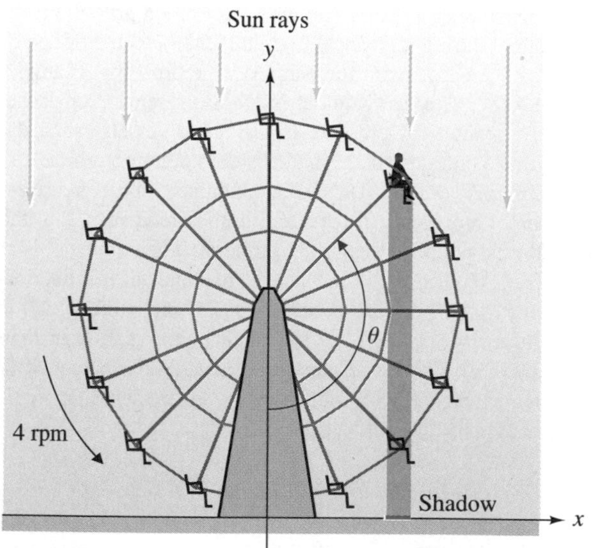

Figure for 53 and 54

*54. **Rotary and Linear Motion** Repeat Problem 53 for a Ferris wheel with a diameter of 30 m rotating at 6 rpm.

55. **Bioengineering** For a walking person (see the figure), a leg rotates back and forth relative to the hip socket through an angle θ. The angle θ is measured relative to the vertical, and is negative if the leg is behind the vertical and positive if the leg is in front of the vertical. Measurements of θ are calculated for a particular person taking one step per second, where t is time in seconds. The results for the first 4 sec are shown in Table 1 (above).

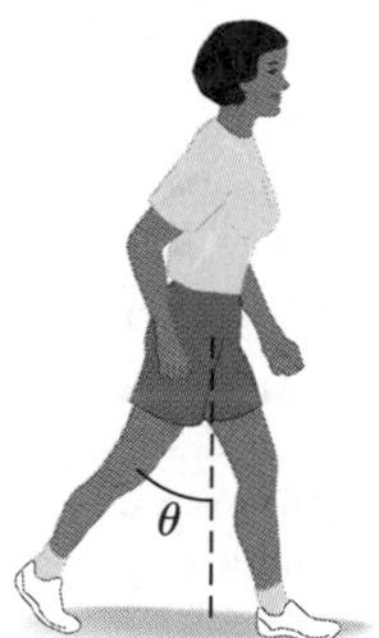

Figure for 55 and 56

(A) Verbally describe how you can find the period and amplitude of the data in Table 1, and find each.

(B) Which of the equations, $\theta = A \cos Bt$ or $\theta = A \sin Bt$, is the more suitable model for the data in Table 1? Why? Find the more suitable equation.

(C) Plot the points in Table 1 and the graph of the equation found in part (B), in the same coordinate system.

56. **Bioengineering** Repeat Problem 55 for the upper arm oscillation, using the data in Table 2.

TABLE 2
Upper Arm Oscillation

t (sec)	0.0	0.5	1.0	1.5	2.0	2.5	3.0	3.5	4.0
θ (deg)	$0°$	$18°$	$0°$	$-18°$	$0°$	$18°$	$0°$	$-18°$	$0°$

57. Engineering—Data Analysis A mill has a paddle wheel 30 ft in diameter that rotates counterclockwise at 5 rpm (see the figure). The bottom of the wheel is 2 ft below the surface of the water in the mill chase. We are interested in finding a function that will give the height h (in feet) of the point P above (or below) the water t seconds after P is at the top of the wheel.

Figure for 57

(A) Complete Table 3 using the information provided above:

TABLE 3

t (sec)	0	3	6	9	12	15	18	21	24
h (ft)			−2	13					

(B) Enter the data in Table 3 in a graphing utility and produce a scatter plot in the viewing window. (The scatter plot routine automatically selects the window dimensions.)

(C) The data are periodic with what period? Why does it appear that $h = k + A \cos Bt$ is a better model for the data than $h = k + A \sin Bt$?

(D) Find the equation to model the data. The constants k, A, and B are easily determined from the completed Table 3 as follows:

$$|A| = (\text{Max } h - \text{Min } h)/2$$

$$B = 2\pi/\text{Period}$$

$$k = |A| + \text{Min } h$$

Refer to the scatter plot to determine the sign of A.

(E) Plot both the scatter plot from part (B) and the equation from part (D) in the same viewing window.

58. Engineering—Data Analysis The top of a turning propeller of a large partially loaded ship protrudes 3 ft above the water as shown in the figure. The propeller is 12 ft in diameter and rotates clockwise at 15 rpm. We are interested in finding a function that will give the height h (in feet) of the point P above (or below) the water t seconds after P is at the bottom of the circle of rotation.

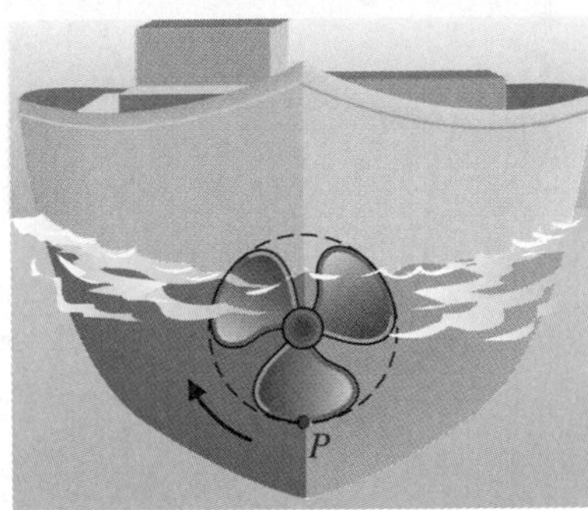

Figure for 58

(A) Complete Table 4 using the information provided above:

TABLE 4

t (sec)	0	1	2	3	4	5	6	7	8
h (ft)	−9		3						

(B) Enter the data in Table 4 in a graphing utility and produce a scatter plot in the viewing window. (The scatter plot routine automatically selects the window dimensions.)

(C) The data are periodic with what period? Why does it appear that $h = k + A \cos Bt$ is a better model for the data than $h = k + A \sin Bt$?

(D) Find the equation to model the data. The constants k, A, and B are easily determined from the completed Table 4 as described in Problem 57D.

(E) Plot both the scatter plot from part (B) and the equation from part (D) in the same viewing window.

3.3 GRAPHING $y = k + A \sin(Bx + C)$ AND $y = k + A \cos(Bx + C)$

- Graphing $y = A \sin(Bx + C)$ and $y = A \cos(Bx + C)$
- Graphing $y = k + A \sin(Bx + C)$ and $y = k + A \cos(Bx + C)$
- Finding the Equation of the Graph of a Simple Harmonic

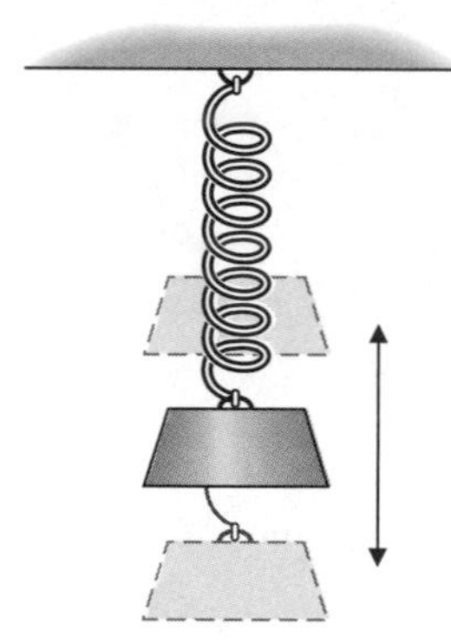

We are now ready to consider sine and cosine functions of a more general form. A function of the form $y = A \sin(Bx + C)$ or $y = A \cos(Bx + C)$ is said to be a **simple harmonic.** Imagine an object suspended from the ceiling by a spring. If it is pulled down and released, then, assuming no air resistance or friction, it would oscillate up and down forever with the same amplitude and frequency. Such idealized motion can be described by a simple harmonic and is called **simple harmonic motion.** These functions are used extensively to model real-world phenomena; for example, see Problems 41–50 in Exercise 3.3 and the applications discussed in Section 3.4.

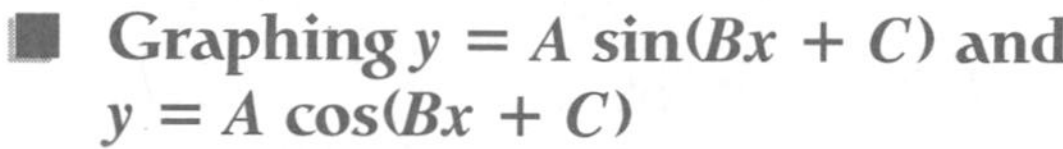

Graphing $y = A \sin(Bx + C)$ and $y = A \cos(Bx + C)$

We are interested in graphing equations of the form

$$y = A \sin(Bx + C) \qquad \text{and} \qquad y = A \cos(Bx + C)$$

We will find that the graphs of these equations are simply the graphs of

$$y = A \sin Bx \qquad \text{or} \qquad y = A \cos Bx$$

translated horizontally to the left or to the right, which can be seen as follows: Since $A \sin x$ has a period of 2π, it follows that $A \sin(Bx + C)$ completes one cycle as $Bx + C$ varies from

$$Bx + C = 0 \quad \text{to} \quad Bx + C = 2\pi$$

or, solving for x, as x varies from

Phase shift   Period

$$x = -\frac{C}{B} \quad \text{to} \quad x = -\frac{C}{B} + \frac{2\pi}{B}$$

Thus, $y = A \sin(Bx + C)$ has a period of $2\pi/B$, and its graph is the graph of $y = A \sin Bx$ translated horizontally $|-C/B|$ units to the right if $-C/B > 0$ and $|-C/B|$ units to the left if $-C/B < 0$. The horizontal translation, determined by the number $-C/B$, is often referred to as the **phase shift.**

What are the period and phase shift for $y = \sin(x + \pi/2)$? To answer this question, you can use the formulas above for period and phase shift or you can go through the process used to derive the formulas. It is probably easier for most people to remember and use the process. We set $x + \pi/2$ equal to the end points of one complete cycle of the sine function, 0 and 2π, and then solve for x.

$$x + \frac{\pi}{2} = 0 \qquad\qquad x + \frac{\pi}{2} = 2\pi$$

$$x = -\frac{\pi}{2} \qquad\qquad x = -\frac{\pi}{2} + 2\pi$$

Phase shift (both $-\frac{\pi}{2}$) — Period (2π)

Thus, $-\pi/2$ is the phase shift, and the graph of $y = \sin x$ is horizontally translated $|-\pi/2| = \pi/2$ units to the left (since $-\pi/2 < 0$). The period is 2π. Figure 1a shows the graphs of $y = \sin x$ and $y = \sin(x + \pi/2)$. Going through a similar process, Figure 1b shows the graphs of $y = \sin x$ and $y = \sin(x - \pi/2)$. Here, the phase shift is $\pi/2$, and the graph of $y = \sin x$ is horizontally translated $\pi/2$ units to the right (since $\pi/2 > 0$).

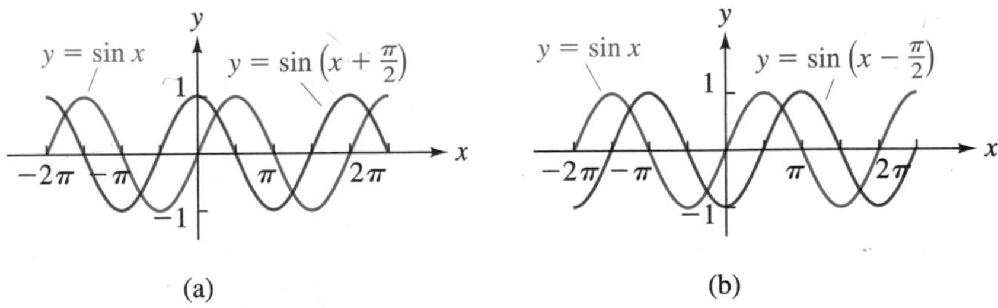

FIGURE 1
Phase shift

A similar analysis applies to $y = A \cos(Bx + C)$, and the results are summarized in the following box.

PROPERTIES OF $y = A \sin(Bx + C)$ AND $y = A \cos(Bx + C)$

For $B > 0$,

Amplitude $= |A|$ Period $= \dfrac{2\pi}{B}$ Phase shift $= -\dfrac{C}{B}$

As we have already indicated, it is not necessary to memorize the formulas for period and phase shift unless you wish to. The period and phase shift are easily found in the following steps for graphing:

STEPS FOR GRAPHING $y = A\sin(Bx + C)$ AND $y = A\cos(Bx + C)$

Step 1 Find the amplitude: $|A|$

Step 2 Solve $Bx + C = 0$ and $Bx + C = 2\pi$:

$$Bx + C = 0 \qquad \text{and} \qquad Bx + C = 2\pi$$

$$x = -\frac{C}{B} \qquad\qquad x = -\frac{C}{B} + \frac{2\pi}{B}$$

Phase shift: $-\frac{C}{B}$; Period: $\frac{2\pi}{B}$

The graph completes one full cycle as $Bx + C$ varies from 0 to 2π; that is, as x varies from $-C/B$ to $(-C/B) + (2\pi/B)$.

Step 3 Graph one cycle over the interval from $-C/B$ to $(-C/B) + (2\pi/B)$.

Step 4 Extend the graph in step 3 to the left or to the right as desired.

EXAMPLE 1 Graphing the Form $y = A\cos(Bx + C)$

Graph

$$y = 20\cos\left(\pi x - \frac{\pi}{2}\right) \qquad -1 \le x \le 3$$

Solution ***Step 1*** Find the amplitude:

$$\text{Amplitude} = |A| = |20| = 20$$

Step 2 Solve $Bx + C = 0$ and $Bx + C = 2\pi$:

$$\pi x - \frac{\pi}{2} = 0 \qquad \pi x - \frac{\pi}{2} = 2\pi$$

$$x = \frac{1}{2} \qquad x = \frac{1}{2} + 2$$

Phase shift: $\frac{1}{2}$; Period: 2

$$\text{Phase shift} = \frac{1}{2} \qquad \text{Period} = 2$$

Step 3 Graph one cycle over the interval from $\frac{1}{2}$ to $(\frac{1}{2} + 2) = \frac{5}{2}$ (Fig. 2).

FIGURE 2

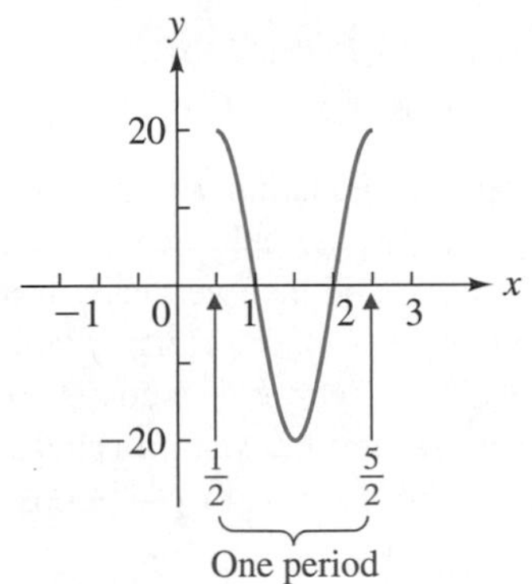

Step 4 Extend the graph from -1 to 3 (Fig. 3).

FIGURE 3

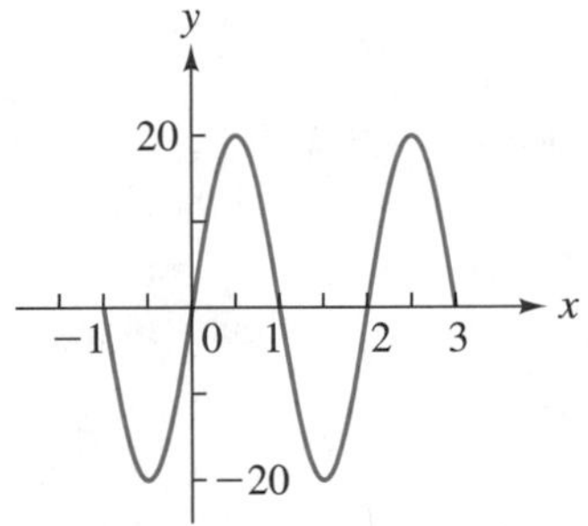

Matched Problem 1 State the amplitude, period, and phase shift for $y = -5 \sin(x/2 + \pi/2)$. Graph the equation for $-3\pi \leq x \leq 5\pi$.

EXPLORE/DISCUSS 1

Find an equation of the form $y = A \cos(Bx + C)$ that produces the graph in the following graphing calculator display (choose the smallest positive phase shift):

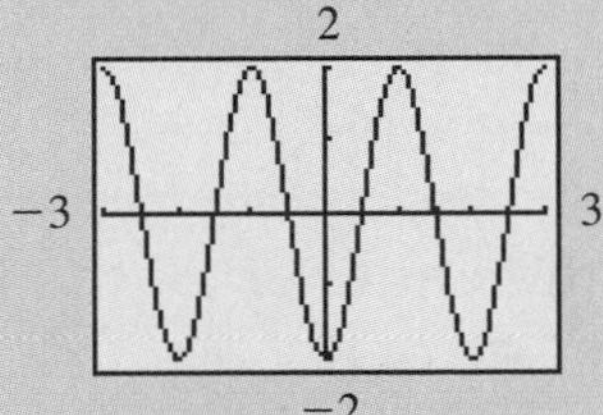

Is it possible for an equation of the form $y = A \sin(Bx + C)$ to produce the same graph? Explain. If it is possible, find the equation using the smallest positive phase shift.

Graphing $y = k + A\sin(Bx + C)$ and $y = k + A\cos(Bx + C)$

In order to graph an equation of the form $y = k + A\sin(Bx + C)$ or $y = k + A\cos(Bx + C)$, we graph $y = A\sin(Bx + C)$ or $y = A\cos(Bx + C)$, as outlined previously, and then vertically translate the graph up k units if $k > 0$ or down $|k|$ units if $k < 0$. Thus, graphing equations of the form $y = k + A\sin(Bx + C)$ or $y = k + A\cos(Bx + C)$, where $k \neq 0$ and $C \neq 0$, involves both a horizontal translation (phase shift) and a vertical translation of the basic equation $y = A\sin Bx$ or $y = A\cos Bx$.

EXAMPLE 2 **Graphing the Form $y = k + A\cos(Bx + C)$**

Graph

$$y = 10 + 20\cos\left(\pi x - \frac{\pi}{2}\right) \qquad -1 \leq x \leq 3$$

Solution We graph $y = 20\cos(\pi x - \pi/2)$, $-1 \leq x \leq 3$, as in Example 1, and then vertically translate the graph up 10 units (Fig. 4).

FIGURE 4

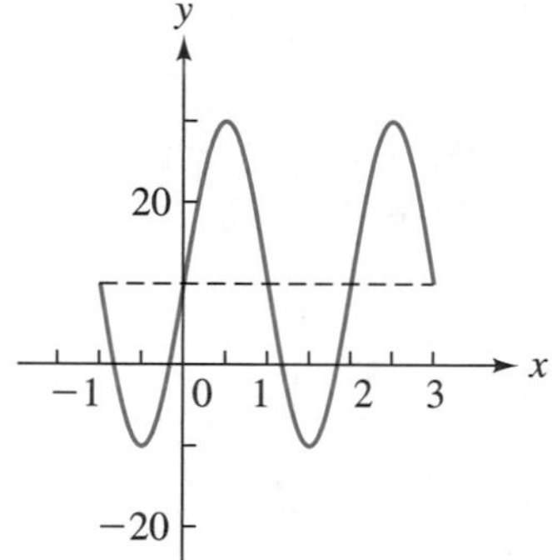

■

Matched Problem 2 Graph

$$y = -2 - 5\sin\left(\frac{x}{2} + \frac{\pi}{2}\right) \qquad -3\pi \leq x \leq 5\pi$$

■

Finding the Equation of the Graph of a Simple Harmonic

Given the graph of a simple harmonic, we wish to find an equation of the form

$$y = A\sin(Bx + C) \qquad \text{or} \qquad y = A\cos(Bx + C)$$

that produces the graph. An example illustrates the process with the aid of a graphing utility.

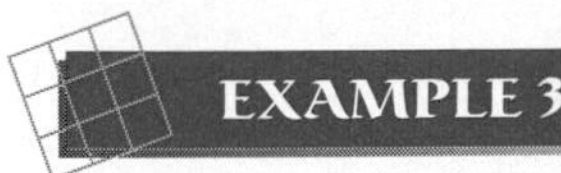

Finding the Equation of the Graph of a Simple Harmonic with the Aid of a Graphing Calculator

Graph $y_1 = 4\sin x - 3\cos x$ in a graphing utility, and find an equation of the form $y_2 = A\sin(Bx + C)$ that has the same graph. Find A and B exactly and C to three decimal places.

Solution The graph of y_1 is shown in Figure 5. This graph appears to be a sine curve, with amplitude 5 and period 2π, that has been shifted to the right. Thus, we conclude that $A = 5$ and $B = 2\pi/P = 2\pi/2\pi = 1$.

5

-2π 2π

-5

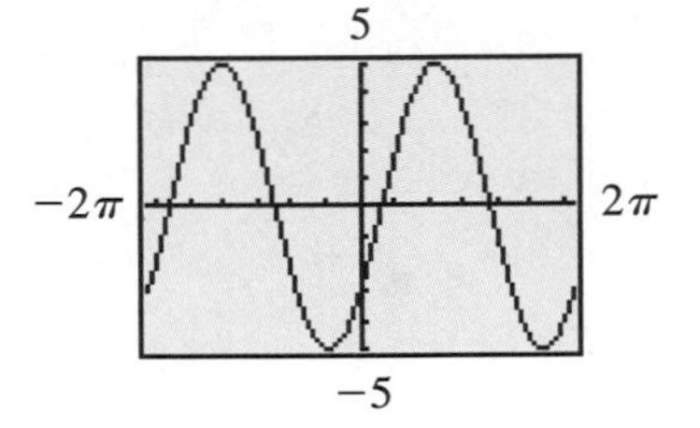

FIGURE 5

To determine C, first find the phase shift from the graph. We choose the smallest positive phase shift, which is the first point to the right of the origin where the graph crosses the x axis. This is the first positive zero of y_1. Most graphing utilities have a built-in routine for finding the zeros of a function. Figure 6 shows the result from a particular graphing calculator. Thus, the zero to three decimal places is $x = 0.644$. To find C, we substitute $B = 1$ and $x = 0.644$ in the phase shift equation $x = -C/B$ and solve for C:

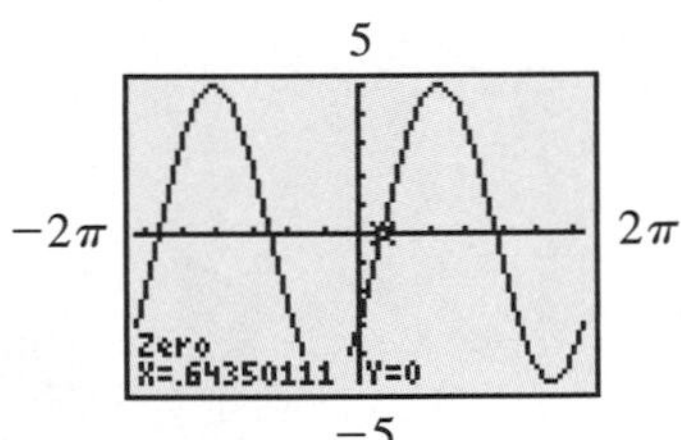

FIGURE 6

$$x = -\frac{C}{B}$$

$$0.644 = -\frac{C}{1}$$

$$C = -0.644$$

Thus,

$$y_2 = 5\sin(x - 0.644)$$

If greater accuracy is desired, choose the phase shift x from Figure 6 to more decimal places.

✔ *Check* Graph y_1 and y_2 in the same viewing window. If the graphs are the same, it appears that only one graph is drawn—the second graph is drawn over the first. To check further that the graphs are the same, use TRACE and switch back and forth between y_1 and y_2 at different values of x. Figure 7 shows a comparison at $x = 0$.

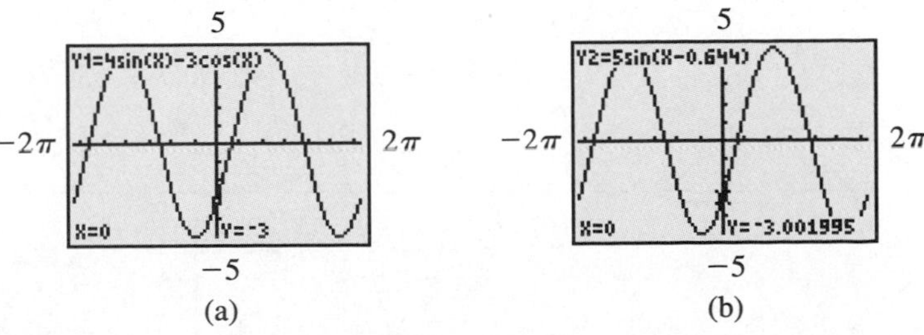

FIGURE 7

■

Matched Problem 3 Graph $y_1 = 3 \sin x + 4 \cos x$, find the x intercept closest to the origin (correct to three decimal places), and find an equation of the form $y_2 = A \sin(Bx + C)$ that has the same graph. ■

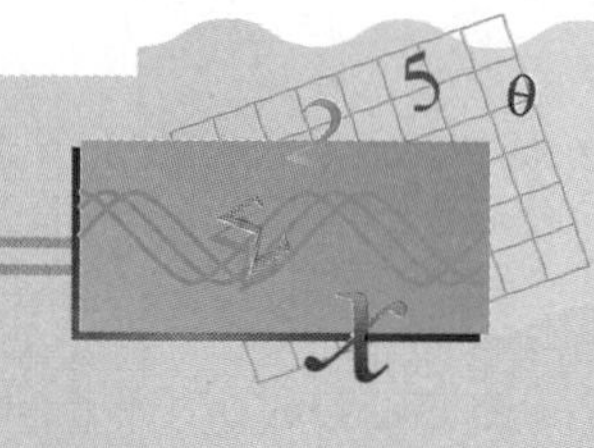

EXPLORE/DISCUSS 2

Explain why any function of the form $y = A \sin(Bx + C)$ can also be written in the form $y = A \cos(Bx + D)$ for an appropriate choice of D.

Answers to Matched Problems

1. Amplitude = 5; Period = 4π; Phase shift = $-\pi$

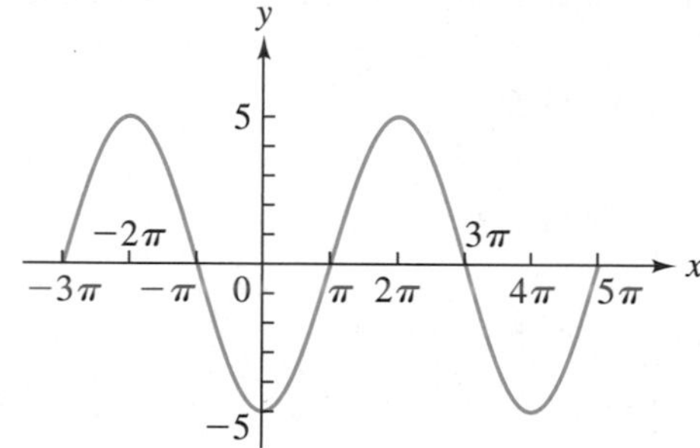

2.

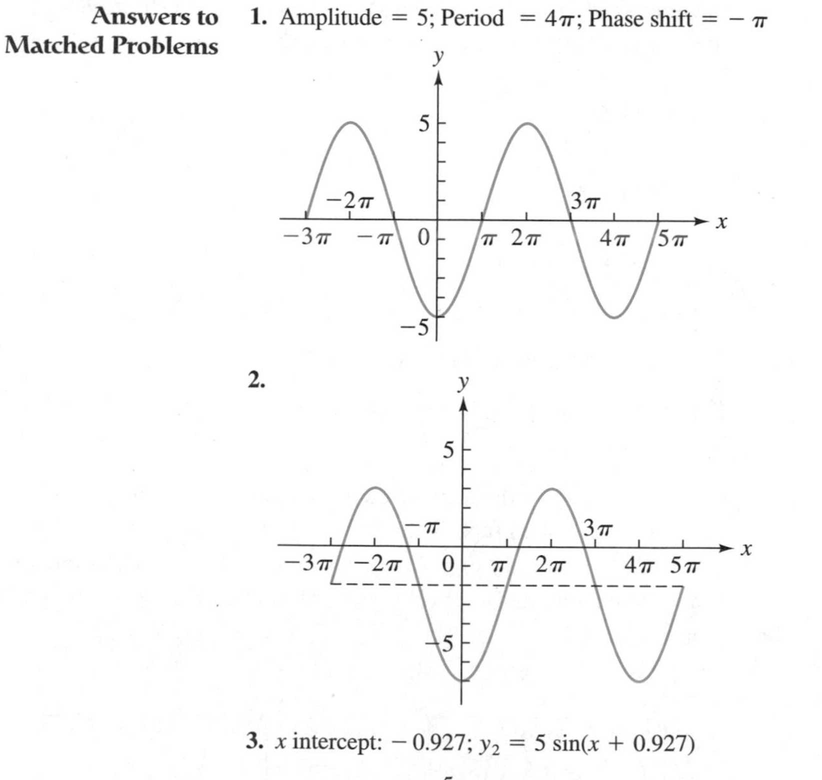

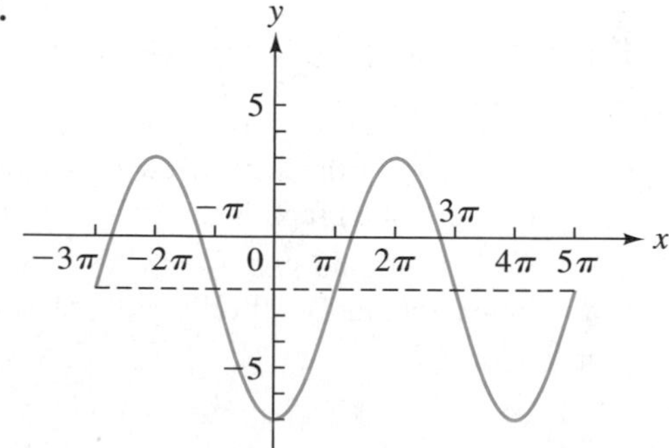

3. x intercept: -0.927; $y_2 = 5 \sin(x + 0.927)$

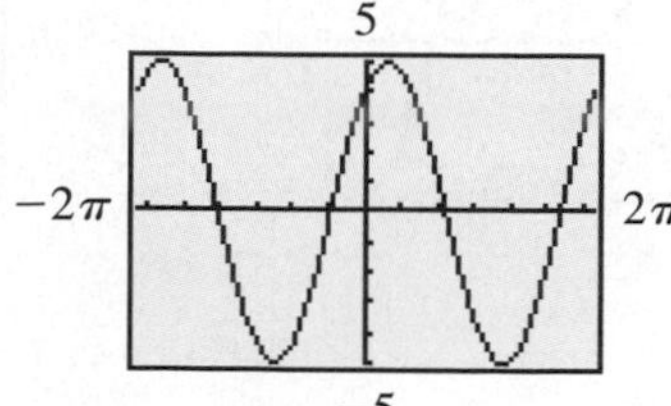

EXERCISE 3.3

A

1. Explain what is meant by the graph of a simple harmonic having a phase shift of -2.
2. Explain what is meant by the graph of a simple harmonic having a phase shift of 3.

In Problems 3–6, indicate the phase shift for each equation, and graph it over the stated region.

3. $y = \cos\left(x + \dfrac{\pi}{2}\right), \quad \dfrac{-\pi}{2} \le x \le \dfrac{3\pi}{2}$
4. $y = \cos\left(x - \dfrac{\pi}{2}\right), \quad \dfrac{\pi}{2} \le x \le \dfrac{5\pi}{2}$
5. $y = \sin\left(x - \dfrac{\pi}{4}\right), \quad -\pi \le x \le 2\pi$
6. $y = \cos\left(x + \dfrac{\pi}{4}\right), \quad -\pi \le x \le 2\pi$

B

In Problems 7–10, state the amplitude, period, and phase shift for each equation, and graph it over the indicated interval.

7. $y = 4\cos\left(\pi x + \dfrac{\pi}{4}\right), \quad -1 \le x \le 3$
8. $y = 2\sin\left(\pi x - \dfrac{\pi}{2}\right), \quad -2 \le x \le 2$
9. $y = -2\cos(2x + \pi), \quad -\pi \le x \le 3\pi$
10. $y = -3\sin(4x - \pi), \quad -\pi \le x \le \pi$
11. Graph

$$y = \cos\left(x - \frac{\pi}{2}\right) \quad \text{and} \quad y = \sin x$$

in the same coordinate system. Conclusion?

12. Graph

$$y = \sin\left(x + \frac{\pi}{2}\right) \quad \text{and} \quad y = \cos x$$

in the same coordinate system. Conclusion?

In Problems 13–16, graph each equation over the indicated interval.

13. $y = -2 + 4\cos\left(\pi x + \dfrac{\pi}{4}\right), \quad -1 \le x \le 3$
14. $y = -3 + 2\sin\left(\pi x - \dfrac{\pi}{2}\right), \quad -2 \le x \le 2$
15. $y = 3 - 2\cos(2x + \pi), \quad -\pi \le x \le 3\pi$
16. $y = 4 - 3\sin(4x - \pi), \quad -\pi \le x \le \pi$

In Problems 17–20, match each equation with one of the following graphing utility displays. Explain how you made the choice relative to period and phase shift.

(A)

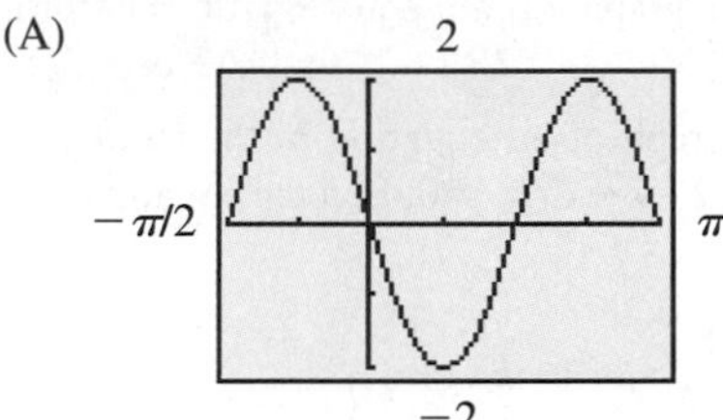

(B)

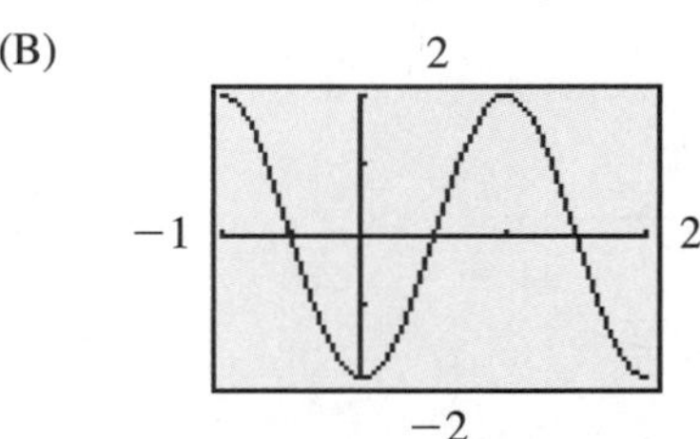

(C)

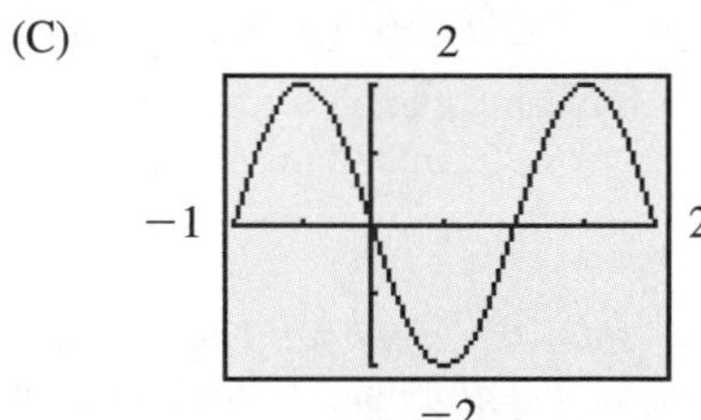

(D)

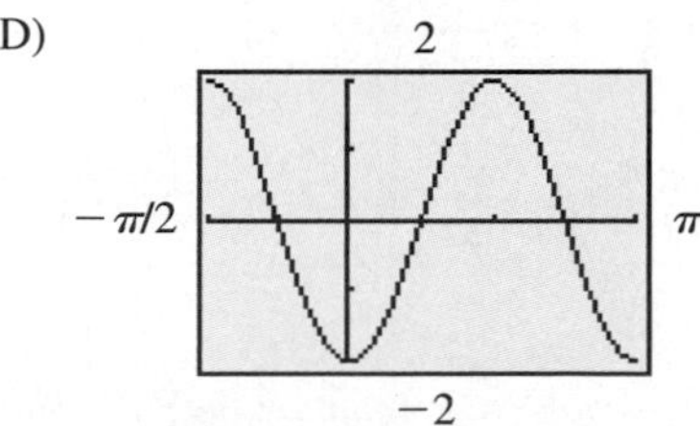

17. $y = 2\sin\left(\pi x - \dfrac{\pi}{2}\right)$
18. $y = 2\cos\left(\pi x + \dfrac{\pi}{2}\right)$
19. $y = 2\cos\left(2x + \dfrac{\pi}{2}\right)$
20. $y = 2\sin\left(2x - \dfrac{\pi}{2}\right)$

For Problems 21 and 22, refer to the graph:

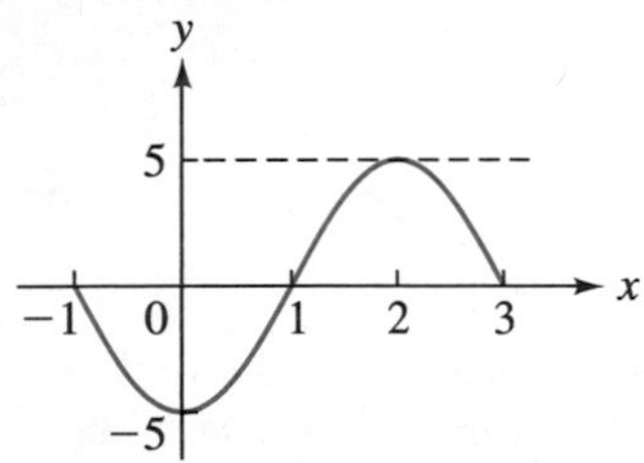

21. If the graph is a graph of an equation of the form $y = A \sin(Bx + C), 0 < -C/B < 2$, find the equation.

22. If the graph is a graph of an equation of the form $y = A \sin(Bx + C), -2 < -C/B < 0$, find the equation.

For Problems 23 and 24, refer to the graph:

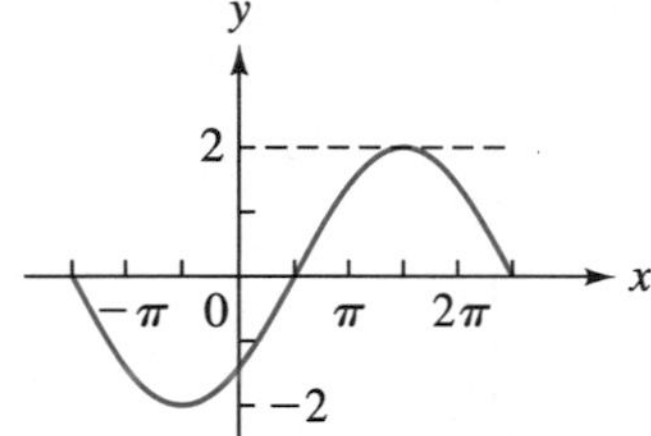

23. If the graph is a graph of an equation of the form $y = A \cos(Bx + C), -2\pi < -C/B < 0$, find the equation.

24. If the graph is a graph of an equation of the form $y = A \cos(Bx + C), 0 < -C/B < 2\pi$, find the equation.

C *In Problems 25 and 26, state the amplitude, period, and phase shift for each equation, and graph it over the indicated interval.*

25. $y = 2 \sin\left(3x - \frac{\pi}{2}\right), \quad \frac{-2\pi}{3} \le x \le \frac{5\pi}{3}$

26. $y = -4 \cos\left(4x + \frac{\pi}{2}\right), \quad \frac{-\pi}{2} \le x \le \pi$

In Problems 27 and 28, graph each equation over the indicated interval.

27. $y = 4 + 2 \sin\left(3x - \frac{\pi}{2}\right), \quad \frac{-2\pi}{3} \le x \le \frac{5\pi}{3}$

28. $y = 6 - 4 \cos\left(4x + \frac{\pi}{2}\right), \quad \frac{-\pi}{2} \le x \le \pi$

Problems 29–32 require the use of a graphing utility. First, state the amplitude, period, and phase shift of each function; then graph the function in a graphing utility.

29. $y = 2.3 \sin\left[\frac{\pi}{1.5}(x - 2)\right], \quad 0 \le x \le 6$

30. $y = -4.7 \sin\left[\frac{\pi}{2.2}(x + 3)\right], \quad 0 \le x \le 10$

31. $y = 18 \cos[4\pi(x + 0.137)], \quad 0 \le x \le 2$

32. $y = -48 \cos[2\pi(x - 0.205)], \quad 0 \le x \le 3$

Problems 33–40 require the use of a graphing utility. Graph the given equation and find the x intercept closest to the origin, correct to three decimal places. Use this intercept to find an equation of the form $y = A \sin(Bx + C)$ that has the same graph as the given equation.

33. $y = \sin x + \sqrt{3} \cos x$

34. $y = \sqrt{3} \sin x - \cos x$

35. $y = \sqrt{2} \sin x - \sqrt{2} \cos x$

36. $y = \sqrt{2} \sin x + \sqrt{2} \cos x$

37. $y = 1.4 \sin 2x + 4.8 \cos 2x$

38. $y = 4.8 \sin 2x - 1.4 \cos 2x$

39. $y = 2 \sin \frac{x}{2} - \sqrt{5} \cos \frac{x}{2}$

40. $y = \sqrt{5} \sin \frac{x}{2} + 2 \cos \frac{x}{2}$

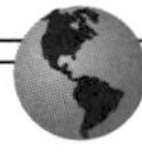

Applications

In these applications, assume all given values are exact unless indicated otherwise.

41. Water Waves (See Section 3.4) At a particular point in the ocean, the vertical change in the water due to wave action is given by

$$y = 5 \sin \frac{\pi}{6}(t + 3)$$

Figure for 41

where y is in meters and t is time in seconds. What are the amplitude, period, and phase shift? Graph the equation for $0 \le t \le 39$.

42. Water Waves Repeat Problem 41 if the wave equation is

$$y = 8\cos\frac{\pi}{12}(t - 6) \qquad 0 \le t \le 72$$

43. Electrical Circuit (See Section 3.4) The current I (in amperes) in an electrical circuit is given by $I = 30\sin(120\pi t - \pi)$, where t is time in seconds. State the amplitude, period, frequency (cycles per second), and phase shift. Graph the equation for the interval $0 \le t \le \frac{3}{60}$.

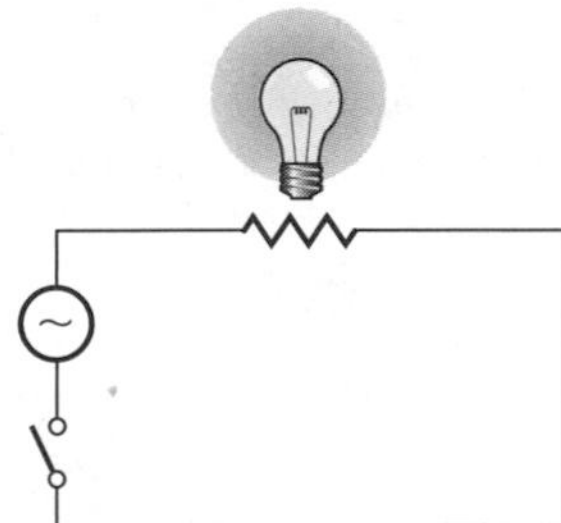

Figure for 43

44. Electrical Circuit Repeat Problem 43 for

$$I = 110\cos\left(120\pi t + \frac{\pi}{2}\right) \qquad 0 \le t \le \frac{2}{60}$$

*** 45. An Experiment with Phase Shift** The tip of the second hand on a wall clock is 6 in. from the center of the clock (see the figure). Let d be the (perpendicular) dis-

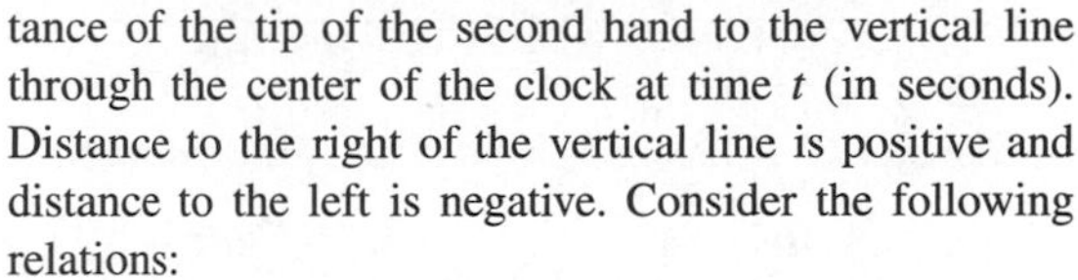

tance of the tip of the second hand to the vertical line through the center of the clock at time t (in seconds). Distance to the right of the vertical line is positive and distance to the left is negative. Consider the following relations:

(1) Distance d, t seconds after the second hand points to 12

(2) Distance d, t seconds after the second hand points to 9

(A) Complete Tables 1 and 2, giving values of d to one decimal place.

(B) Referring to Tables 1 and 2, and to the clock in the figure, discuss how much relation (2) is out of phase relative to relation (1).

(C) Relations (1) and (2) are simple harmonics. What are the amplitudes and periods of these relations?

(D) Both relations have equations of the form $y = A\sin(Bx + C)$. Find the equations, and check that the equations give the values in the tables. (Choose the phase shift so that $|C|$ is minimum.)

(E) Write an equation of the form $y = A\sin(Bx + C)$ that represents the distance, d, t seconds after the second hand points to 3. What is the phase shift? (Choose the phase shift so that $|C|$ is minimum.)

(F) Graph the three equations found in parts (D) and (E) in the same viewing window on a graphing utility for $0 \le t \le 120$.

Figure for 45 and 46

TABLE 1
Distance d, t Seconds After the Second Hand Points to 12

t (sec)	0	5	10	15	20	25	30	35	40	45	50	55	60
d (in.)	0.0			6.0				−3.0			−5.2		

TABLE 2
Distance d, t Seconds After the Second Hand Points to 9

t (sec)	0	5	10	15	20	25	30	35	40	45	50	55	60
d (in.)	−6.0			0.0				5.2			−3.0		

TABLE 3
Distance d, t Hours After the Hour Hand Points to 12

t (hr)	0	1	2	3	4	5	6	7	8	9	10	11	12
d (in.)	0.0			4.0				−2.0			−3.5		

TABLE 4
Distance d, t Hours After the Hour Hand Points to 3

t (hr)	0	1	2	3	4	5	6	7	8	9	10	11	12
d (in.)	4.0			0.0				−3.5			2.0		

* **46. An Experiment with Phase Shift** The tip of the hour hand on a wall clock is 4 in. from the center of the clock (see the figure on page 167). Let d be the distance of the tip of the hour hand to the vertical line through the center of the clock at time t (in hours). Distance to the right of the vertical line is positive and distance to the left is negative. Consider the following relations:

(1) Distance d, t hours after the hour hand points to 12

(2) Distance d, t hours after the second hand points to 3

(A) Complete Tables 3 and 4 above, giving values of d to one decimal place.
(B) Referring to Tables 3 and 4, and to the clock in the figure, discuss how much relation (2) is out of phase relative to relation (1).
(C) Relations (1) and (2) are simple harmonics. What are the amplitudes and periods of these relations?
(D) Both relations have equations of the form $y = A\sin(Bx + C)$. Find the equations, and check that the equations give the values in the tables. (Choose the phase shift so that $|C|$ is minimum.)
(E) Write an equation of the form $y = A\sin(Bx + C)$ that represents the distance, d, t hours after the hour hand points to 9. What is the phase shift? (Choose the phase shift so that $|C|$ is minimum.)
(F) Graph the three equations found in parts (D) and (E) in the same viewing window on a graphing utility for $0 \le t \le 24$.

* **47. Modeling Temperature Variation** The 30 yr average monthly temperatures (in °F) for each month of the year for San Antonio, TX, are given in Table 5.
(A) Enter the data for a 2 yr period ($1 \le x \le 24$) in your graphing calculator and produce a scatter plot in the viewing window.
(B) From the scatter plot in part (A), it appears that a sine curve of the form

$$y = k + A\sin(Bx + C)$$

will closely model the data. Discuss how the constants k, $|A|$, and B can be determined from Table 5, and find them. To estimate C, visually estimate (to one decimal place) the smallest positive phase shift from the scatter plot in part (A). Write the

TABLE 5

x (month)	1	2	3	4	5	6	7	8	9	10	11	12
y (temperature, °F)	50	54	62	70	76	82	85	84	79	70	60	53

Source: *World Almanac*

equation, and plot it in the same viewing window as the scatter plot. Adjust C as necessary to produce a better visual fit.

(C) Fit the data you have entered into your graphing utility from Table 5 with a **sinusoidal regression equation and curve.** This is a process many graphing utilities perform automatically (consult your user's manual for the process). The result is an equation of the form $y = k + A\sin(Bx + C)$ that gives the "best fit" to the data when graphed. Write the sinusoidal regression equation to one decimal place. Show the graphs of the scatter plot of the data in Table 5 and the regression equation that is already in your utility in the same viewing window. (Do not reenter the regression equation. Also, the window dimensions are automatically set.)

(D) What are the differences between the regression equation in part (C) and the equation obtained in part (B)? Which equation appears to give the best fit?

* **48. Modeling Sunrise Times** Sunrise times for the fifth of each month over a 1 yr period were taken from a tide booklet for San Francisco Bay to form Table 6 (daylight savings time was ignored). Repeat parts (A)–(D) in Problem 47 with Table 5 replaced by Table 6. (Before entering the data, convert sunrise times from hours and minutes to decimal hours rounded to two decimal places.)

* **49. Modeling Temperature Variation—Collecting Your Own Data** Replace Table 5 in Problem 47 with data for your own city or a city near you, using an appropriate reference. Then repeat parts (A)–(D).

* **50. Modeling Sunrise Times—Collecting Your Own Data** Replace Table 6 in Problem 48 with data for your own city or a city near you, using an appropriate reference. Then repeat parts (A)–(D).

TABLE 6

x (months)	1	2	3	4	5	6	7	8	9	10	11	12
y (sunrise time, A.M.)	7:26	7:11	6:36	5:50	5:10	4:48	4:53	5:16	5:43	6:09	6:39	7:10

☆3.4 ADDITIONAL APPLICATIONS

- Modeling Electric Current
- Modeling Light and Other Electromagnetic Waves
- Modeling Water Waves
- Simple and Damped Harmonic Motion; Resonance

Many types of applications of trigonometry have already been considered in this and the preceding chapters. This section provides a sampler of additional applications from several different fields. You do not need any prior knowledge of any particular topic to understand the discussion or to work any of the problems. Several of the applications considered use important properties of the sine and cosine functions, which are restated at the top of the next page for convenient reference.

☆ Sections marked with a star may be omitted without loss of continuity.

PROPERTIES OF SINE AND COSINE

For $y = A \sin(Bt + C)$ or $y = A \cos(Bt + C)$:

$$\text{Amplitude} = |A| \qquad \text{Period} = \frac{2\pi}{B} \qquad \text{Frequency} = \frac{1}{\text{Period}}$$

$$\text{Phase shift} = -\frac{C}{B}\begin{cases}\text{Right} & \text{if } -C/B > 0\\ \text{Left} & \text{if } -C/B < 0\end{cases}$$

As we indicated in the previous section, you do not need to memorize the formulas for period and phase shift. You can obtain the same results by solving the two equations

$$Bt + C = 0 \qquad \text{and} \qquad Bt + C = 2\pi$$

(We use the variable t, rather than x, because most of the applications here involve functions of time.)

Recall, functions of the form $y = A \sin(Bt + C)$ or $y = A \cos(Bt + C)$ are said to be **simple harmonics.**

Modeling Electric Current

In physics, a **field** is an area surrounding an object in which a gravitational or electromagnetic force is exerted on other objects. Fields were first introduced by the British physicist Sir Isaac Newton (1642–1727) to explain gravitational forces. Around 140 years later, the British physicist–chemist Michael Faraday (1791–1867) used the concept of fields to explain electromagnetic forces. In the 19th century, the British physicist James Maxwell (1831–1879) developed equations that described electromagnetic fields. And in the 20th century, the German–American physicist Albert Einstein (1879–1955) developed a set of equations for gravitational fields.

A particularly significant discovery by Faraday was the observation that a flow of electricity could be created by moving a wire in a magnetic field. This discovery was the start of the electronic revolution, which is continuing today with no slowdown in sight.

Suppose we bend a wire in the form of a rectangle, and we locate this wire between the south and north poles of magnets, as shown in Figure 1. We now rotate the wire at a constant counterclockwise speed, starting with BC in its lowest position. As BC turns toward the horizontal, an electrical current will flow from C to B (see Fig. 1a). The strength of the current (measured in amperes) will be 0 at the lowest position, and will increase to a maximum value at the horizontal position. As BC continues to turn from the horizontal to the top position, the cur-

FIGURE 1
Alternating current

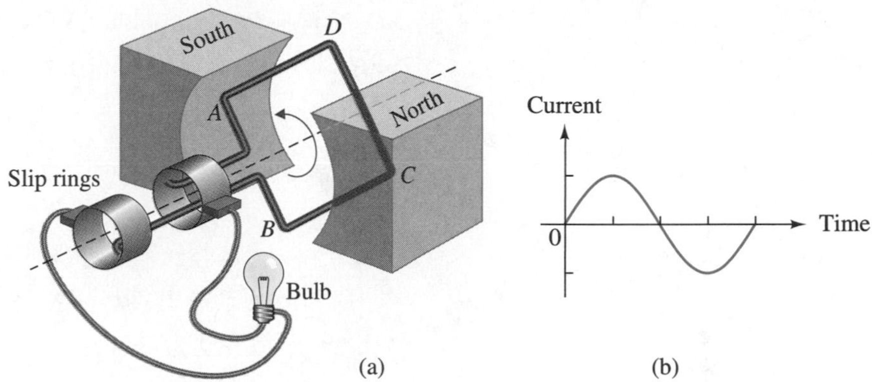

rent flow decreases to 0. As BC starts down from the top position in a counterclockwise direction, the current flow reverses, going from B to C, and again reaches a maximum at the left horizontal position. When BC moves from this horizontal position back to the original bottom position, the current flow again decreases to 0. This pattern repeats itself for each revolution, and hence is periodic.

Is it too much to expect that a trigonometric function can describe the relationship between current and time (as in Fig. 1b)? If we measure and graph the strength of the current I (in amperes) in the wire relative to time t (in seconds), we can indeed find an equation of the form

$$I = A \sin(Bt + C)$$

that will give us the relationship between current and time. For example,

$$I = 30 \sin 120\pi t$$

represents a 60 Hz alternating current flow with a maximum value of 30 amperes.

EXAMPLE 1

Alternating Current Generator

An alternating current generator produces an electrical current (measured in amperes) that is described by the equation

$$I = 35 \sin(40\pi t - 10\pi)$$

where t is time in seconds.

(A) What are the amplitude, period, frequency, and phase shift for the current?

(B) Graph the equation on a graphing utility for $0 \le t \le 0.2$.

Solution (A) $y = 35 \sin(40\pi t - 10\pi)$

$$\text{Amplitude} = |35| = 35 \text{ amperes}$$

To find the period and phase shift, solve $Bt + C = 0$ and $Bt + C = 2\pi$:

$$40\pi t - 10\pi = 0 \qquad\qquad 40\pi t - 10\pi = 2\pi$$

$$40\pi t = 10\pi \qquad\qquad 40\pi t = 10\pi + 2\pi$$

$$t = \frac{1}{4} \qquad\qquad t = \frac{1}{4} + \frac{1}{20}$$

Phase shift ↑ (under $\frac{1}{4}$); Period ↑ (under $\frac{1}{20}$)

Phase shift $= \frac{1}{4}$ sec (right) Period $= \frac{1}{20}$ sec

$$\text{Frequency} = \frac{1}{\text{Period}} = 20 \text{ Hz}$$

(B)

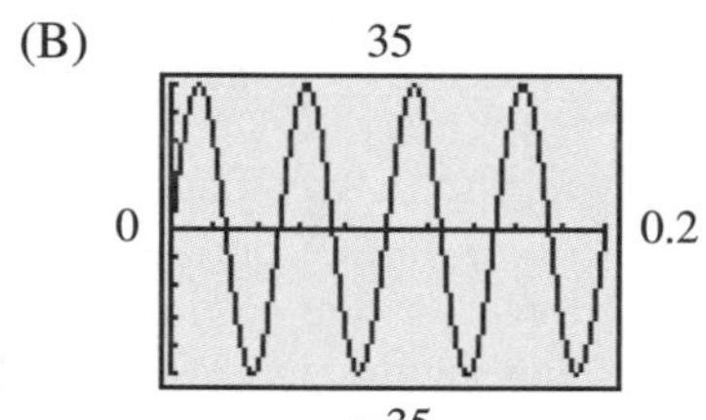

Matched Problem 1 If the alternating current generator in Example 1 produces an electrical current described by the equation $I = -25\sin(30\pi t + 5\pi)$:

(A) Find the amplitude, period, frequency, and phase shift for the current.

(B) Graph the equation on a graphing utility for $0 \le t \le 0.3$.

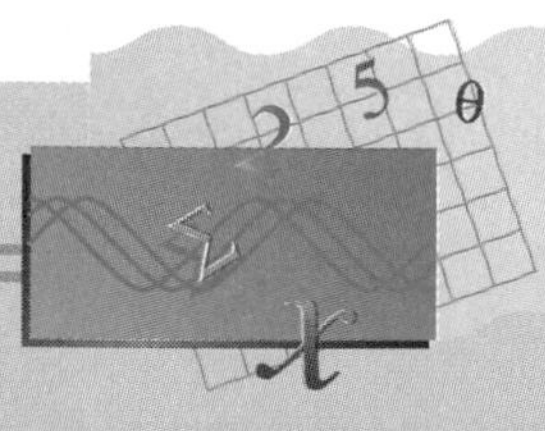

EXPLORE/DISCUSS 1

An electric garbage disposal unit in a house has a printed plate fastened to it that states: 120 V 60 Hz 7.6 A (that is, 120 volts, 60 Hz, 7.6 amperes). Discuss how you can arrive at an equation for the current I in the form $I = A\sin(Bt + C)$, where t is time in seconds, if the current is 5 amperes when $t = 0$ sec. Find the equation.

Modeling Light and Other Electromagnetic Waves

Visible light is a transverse wave form with a frequency range between 4×10^{14} Hz (red) and 7×10^{14} Hz (violet). The retina of the eye responds to these vibrations, and through a complicated chemical process, the vibrations are eventually perceived by the brain as light in various colors. Light is actually a small part of a continuous spectrum of electromagnetic wave forms—most of which are not visible (see Fig. 2). Included in the spectrum are radio waves (AM and FM), microwaves, x rays, and gamma rays. All these waves travel at the speed of light, approx. 3×10^{10} cm/sec (186,000 mi/sec), and many of them are either partially or totally adsorbed in the atmosphere of the earth, as indicated in Figure 2. Their distinguishing characteristics are wavelength and frequency, which are related

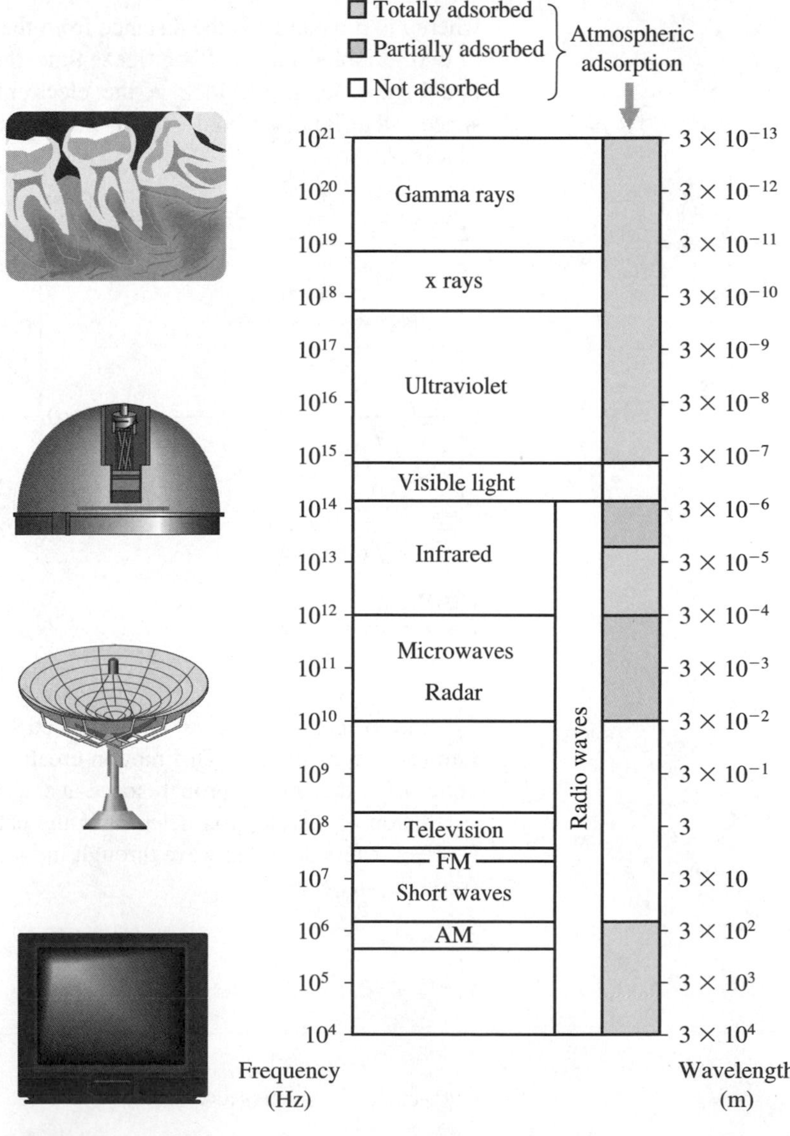

FIGURE 2
Electromagnetic wave spectrum

by the formula

$$\lambda\nu = c \tag{1}$$

where c is the speed of light, λ is the wavelength, and ν is the frequency.

Electromagnetic waves are traveling waves that can be described by an equation of the form

$$E = A \sin 2\pi\left(\nu t - \frac{r}{\lambda}\right) \quad \text{Traveling wave equation} \tag{2}$$

where t is time and r is the distance from the source. This equation is a function of two variables, t and r. If we freeze time, then the graph of (2) looks something like Figure 3a. If we look at the electromagnetic field at a single point in space—that is, if we hold r fixed—then the graph of (2) looks something like Figure 3b.

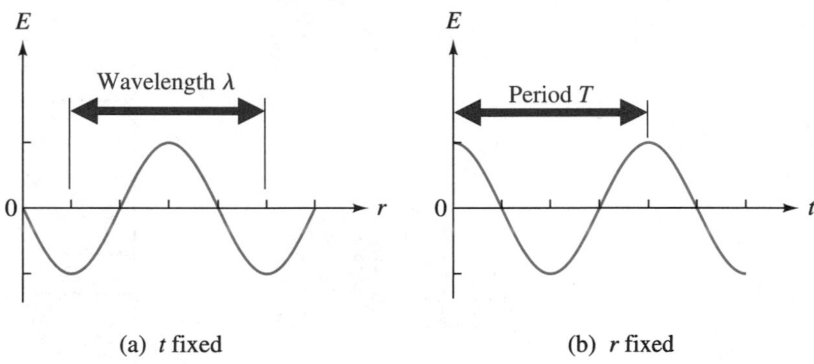

FIGURE 3
Electromagnetic wave

Electromagnetic waves are produced by electrons that have been excited into oscillatory motion. This motion creates a combination of electric and magnetic fields that move through space at the speed of light. The frequency of the oscillation of the electron determines the nature of the wave (see Fig. 2), and a receiver responds to the wave through induced oscillation of the same frequency (see Fig. 4).

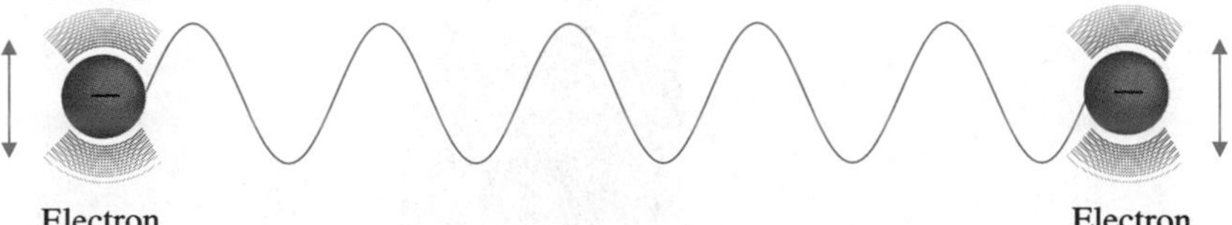

FIGURE 4
Electromagnetic field

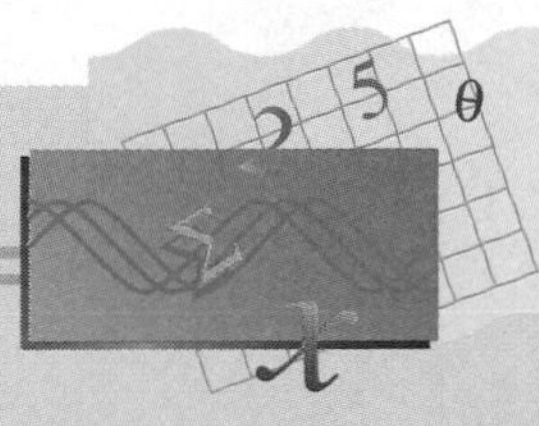

EXPLORE/DISCUSS 2

Commonly received radio waves in the home are FM waves and AM waves—both are electromagnetic waves. FM means **frequency modulated,** and AM means **amplitude modulated.** Figure 5 illustrates AM and FM waves at fixed points in space. Identify which is which, and explain your reasoning in your selection.

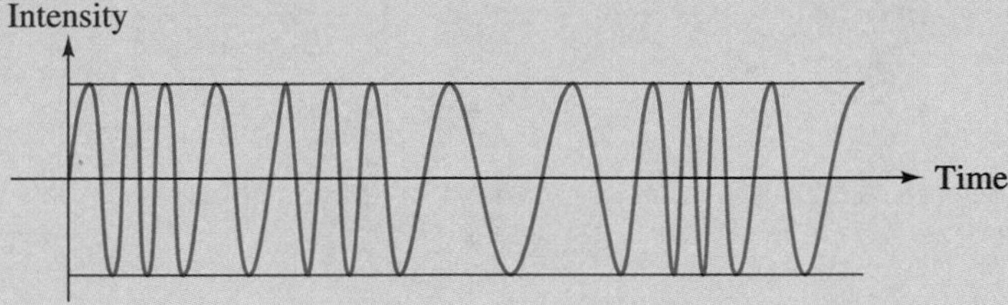

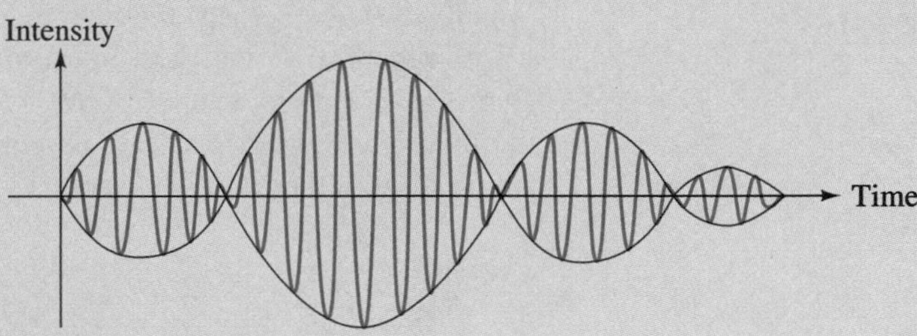

FIGURE 5
FM and AM electromagnetic waves

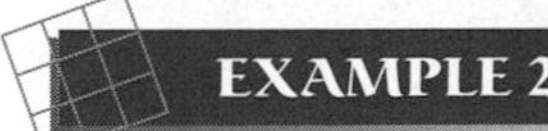

EXAMPLE 2 Electromagnetic Waves

If an electromagnetic wave has a frequency of $\nu = 10^{12}$ Hz, what is its period? What is its wavelength (in meters)?

Solution

$$\text{Period} = \frac{1}{\nu} = \frac{1}{10^{12}} = 10^{-12}\text{ sec}$$

To find the wavelength λ, we use the formula

$$\lambda\nu = c$$

with the speed of light $c \approx 3 \times 10^8$ m/sec:

$$\lambda = \frac{3 \times 10^8\text{ m/sec}}{10^{12}\text{ Hz}} = 3 \times 10^{-4}\text{ m}$$

■

Matched Problem 2 Repeat Example 2 for $\nu = 10^6$ Hz. ■

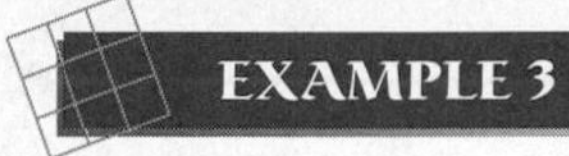

Ultraviolet Waves

An ultraviolet wave has an equation of the form

$$y = A \sin Bt$$

Find B if the wavelength is $\lambda = 3 \times 10^{-9}$ m.

Solution We first use $\lambda\nu = c$ ($c \approx 3 \times 10^8$ m/sec) to find the frequency, and then use $\nu = B/2\pi$ to find B:

$$\nu = \frac{c}{\lambda} = \frac{3 \times 10^8 \text{ m/sec}}{3 \times 10^{-9} \text{ m}} = 10^{17} \text{ Hz}$$

$$B = 2\pi\nu = 2\pi \times 10^{17}$$ ■

Matched Problem 3 A gamma ray has an equation of the form $y = A \sin Bt$. Find B if the wavelength is $\lambda = 3 \times 10^{-12}$ m. ■

Modeling Water Waves

Water waves are perhaps the most familiar wave form. You can actually see the wave in motion! Water waves are formed by particles of water rotating in circles (Fig. 6). A particle actually moves only a short distance as the wave passes through. These wave forms are moving waves, and it can be shown that in their simplest form, they are sine curves.

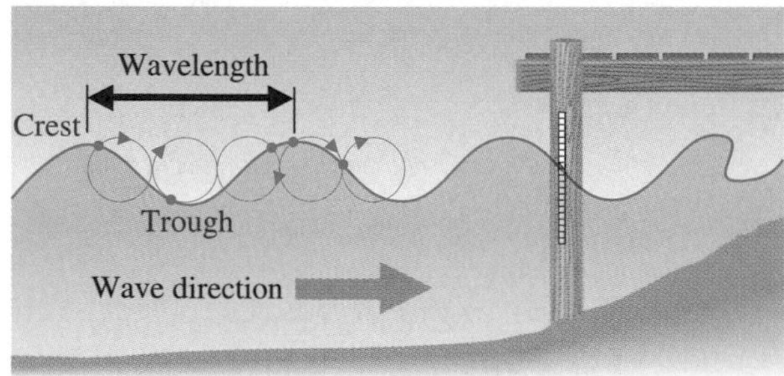

FIGURE 6
Water waves

The waves can be represented by an equation of the form

$$y = A \sin 2\pi\left(ft - \frac{r}{\lambda}\right) \qquad \text{Moving wave equation*} \qquad (3)$$

* Actually, this general wave equation can be used to describe any longitudinal (compressional) or transverse wave motion of one frequency in a nondispersive medium. A sound wave is an example of a longitudinal wave; water waves and electromagnetic waves are examples of transverse waves.

where f is frequency, t is time, r is distance from the source, and λ is wavelength. Thus, for a wave of a given frequency and wavelength, y is a function of the two variables, t and r.

Equation (3) is typical for moving waves. If a wave passes a pier piling with a vertical scale attached (Fig. 6), the graph of the water level on the scale relative to time would look something like Figure 7a, since r would be fixed. On the other hand, if we actually photograph a wave, freezing the motion in time, then the profile of the wave would look something like Figure 7b.

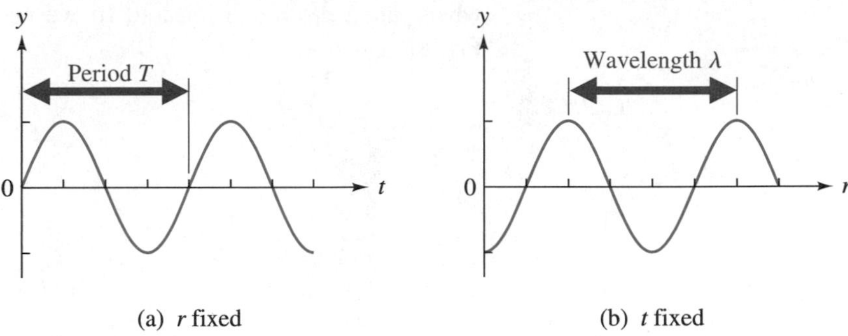

FIGURE 7
Water wave

Experiments have shown that the **wavelength λ** for water waves is given approximately by

$$\lambda = 5.12T^2 \qquad \text{In feet}$$

where T is the period of the wave in seconds (see Fig. 7a), and the **speed S of the wave** is given approximately by

$$S = \sqrt{\frac{g\lambda}{2\pi}} \qquad \text{In feet per second}$$

where $g = 32$ ft/sec^2 (gravitational constant). Thus, a wave with a period of exactly 8 sec and an amplitude of exactly 3 ft would have an equation of the form (holding r fixed)

$$y = 3 \sin 2\pi\left(\frac{1}{8}t - \frac{r}{\lambda}\right)$$
$$= 3 \sin\left(\frac{\pi}{4}t + c\right)$$

where c is a constant. Its wavelength would be

$$\lambda = 5.12(8^2) \approx 328 \text{ ft}$$

and it would move at a speed of approximately

$$S \approx \sqrt{\frac{32(328)}{2(3.14)}} \approx 41 \text{ ft/sec}$$

or, in miles per hour (mph),

$$(41 \text{ ft/sec}) \frac{3{,}600 \text{ sec/hr}}{5{,}280 \text{ ft/mi}} \approx 28 \text{ mph}$$

Simple and Damped Harmonic Motion; Resonance

An object of mass M hanging on a spring will produce simple harmonic motion when pulled down and released (if we neglect friction and air resistance; see Fig. 8).

FIGURE 8

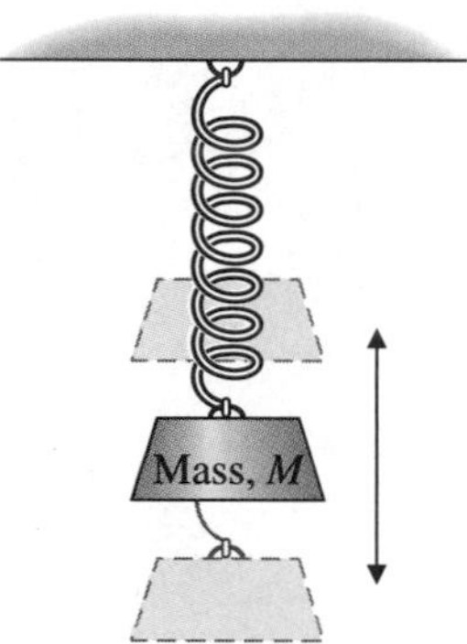

Figure 9 illustrates simple harmonic motion, damped harmonic motion, and resonance. **Damped harmonic motion** occurs when amplitude decreases to 0 as time increases. **Resonance** occurs when amplitude increases as time increases. Damped harmonic motion is essential in the design of suspension systems for cars, buses, trains, and motorcycles, as well as in the design of buildings, bridges, and aircraft. Resonance is useful in some electric circuits and in some mechanical systems, but it can be disastrous in bridges, buildings, and aircraft. Commercial jets have lost wings in flight, and large bridges have collapsed because of resonance. Marching soldiers must break step while crossing a bridge in order to avoid the creation of resonance—and the collapse of the bridge!

Damped Harmonic Motion

Graph

$$y = \frac{1}{t} \sin \frac{\pi}{2} t \qquad 1 \le t \le 8$$

and indicate the type of motion.

Solution The $1/t$ factor in front of $\sin(\pi/2)t$ affects the amplitude. Since the maximum and minimum that $\sin(\pi/2)t$ can assume are 1 and -1, respectively, if we graph $y = 1/t$ first $(1 \le t \le 8)$ and reflect the graph across the t axis, we will have upper and lower bounds for the graph of $y = (1/t)\sin(\pi/2)t$. We also note that $(1/t)\sin(\pi/2)t$ will still be 0 when $\sin(\pi/2)t$ is 0.

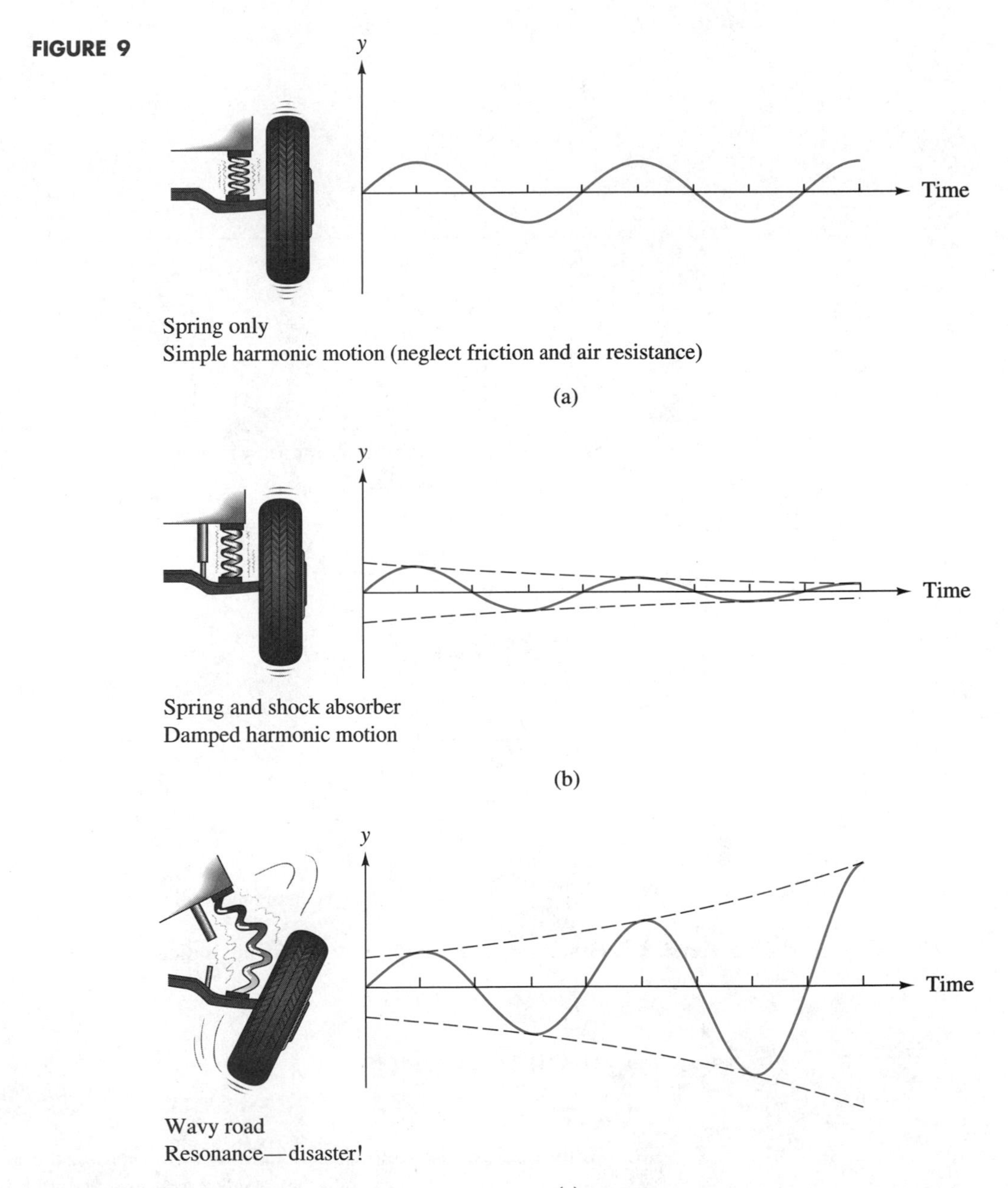

FIGURE 9

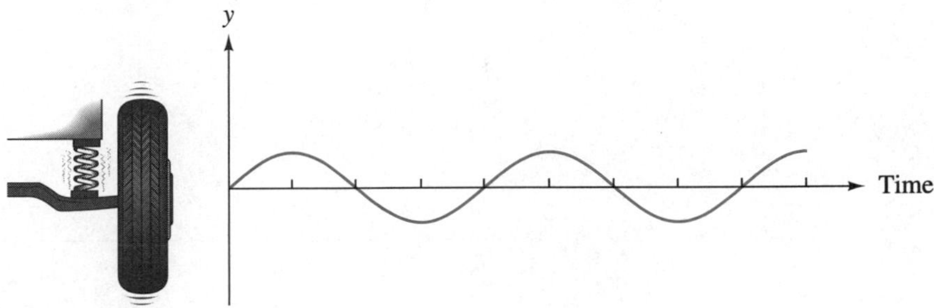

Spring only
Simple harmonic motion (neglect friction and air resistance)

(a)

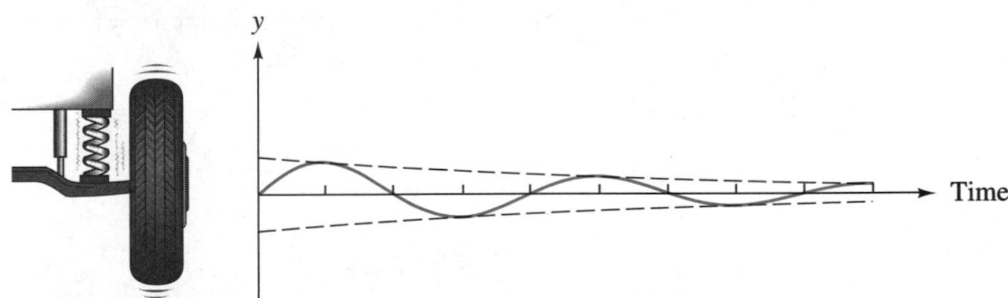

Spring and shock absorber
Damped harmonic motion

(b)

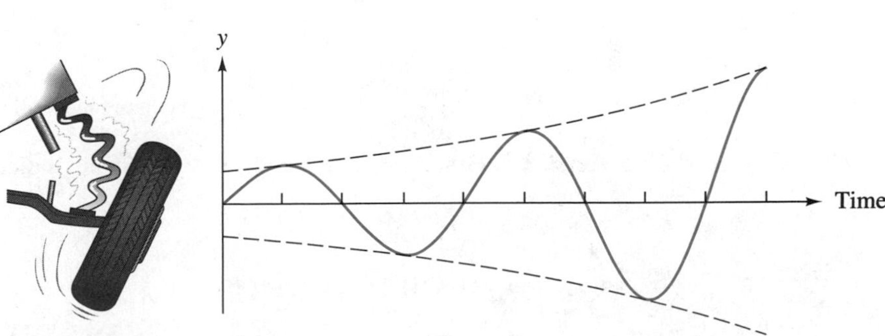

Wavy road
Resonance—disaster!

(c)

Step 1 Graph $y = 1/t$ and its reflection (Fig. 10)—called the **envelope** for the graph of $y = (1/t)\sin(\pi/2)t$.

FIGURE 10

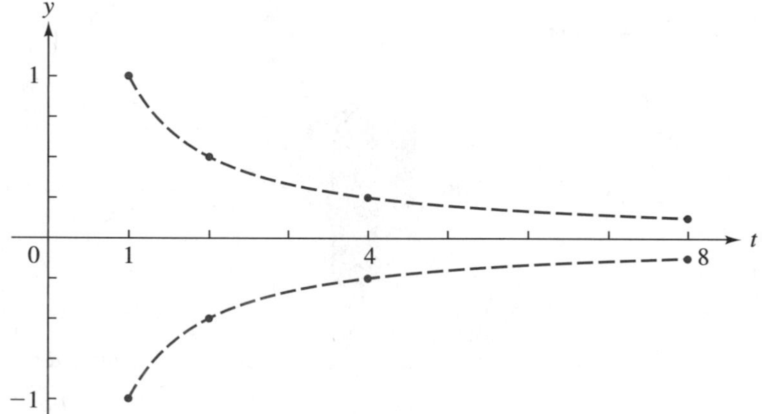

Step 2 Now sketch the graph of $y = \sin(\pi/2)t$, but keep high and low points within the envelope (Fig. 11).

FIGURE 11

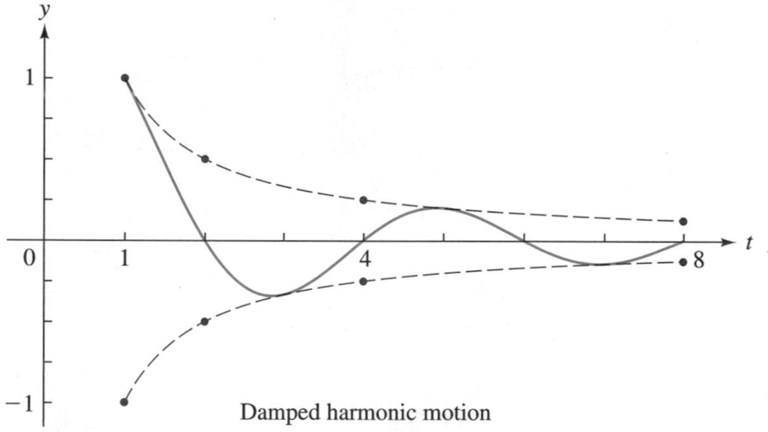

Matched Problem 4 Graph $y = t\sin(\pi t/2)$, $1 \le t \le 8$, and indicate the type of motion.

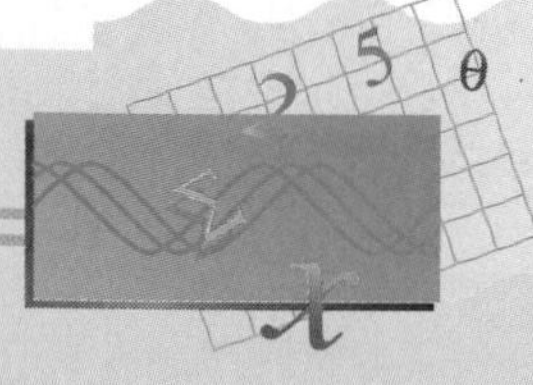

EXPLORE/DISCUSS 3

Graph each of the following equations in a graphing utility, along with their envelopes, and discuss whether each represents simple harmonic motion, damped harmonic motion, or resonance.

(A) $y = t^{0.7}\sin(\pi t/2)$, $\quad 1 \le t \le 10$

(B) $y = t^{-0.5}\sin(\pi t/2)$, $\quad 1 \le t \le 10$

Answers to Matched Problems

1. (A) Amplitude = 25 amperes; Period = $\frac{1}{15}$ sec; Frequency = 15 Hz; Phase shift = $-\frac{1}{6}$ sec

(B)

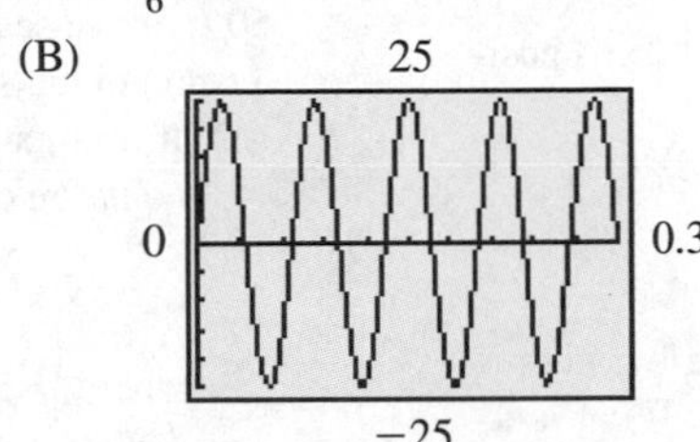

2. Period = 10^{-6} sec; λ = 300 m

3. $B = 2\pi \times 10^{20}$

4.

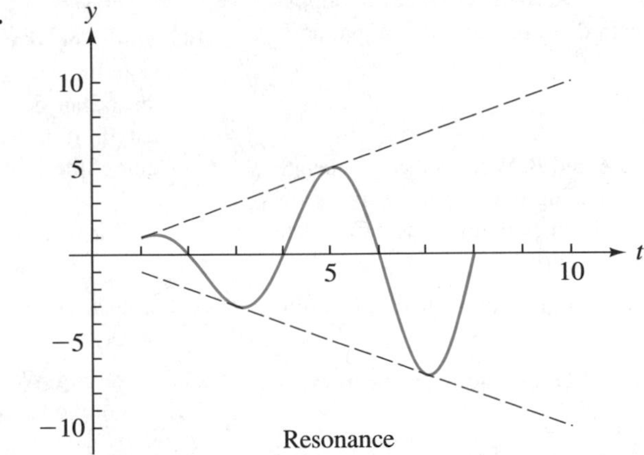

Resonance

EXERCISE 3.4

Applications

In these applications assume all given values are exact unless indicated otherwise.

1. **Modeling Electric Current** An alternating current generator produces a current given by

$$I = 10 \sin(120\pi t - \pi/2)$$

where t is time in seconds and I is in amperes.
(A) What are the amplitude, frequency, and phase shift for this current?
(B) Explain how you would find the maximum current in this circuit, and find it.

(C) Graph the equation in a graphing utility for $0 \le t \le 0.1$. How many periods are shown in the graph?

2. **Modeling Electric Current** An alternating current generator produces a current given by

$$I = -50 \cos(80\pi t + 2\pi/3)$$

where t is time in seconds and I is in amperes.
(A) What are the amplitude, frequency, and phase shift for this current?
(B) Explain how you would find the maximum current in this circuit, and find it.
(C) Graph the equation in a graphing utility for $0 \le t \le 0.1$. How many periods are shown in the graph?

3. **Modeling Electric Current** An alternating current generator produces a 30 Hz current flow with a maximum value of 20 amperes. Write an equation in the form $I = A \cos Bt$, $A > 0$, for this current.

4. **Modeling Electric Current** An alternating current generator produces a 60 Hz current flow with a maxi-

mum value of 10 amperes. Write an equation of the form $I = A \sin Bt, A > 0$, for this current.

5. Modeling Water Waves A water wave at a fixed position has an equation of the form

$$y = 15 \sin \frac{\pi}{8} t$$

where t is time in seconds and y is in feet. How high is the wave from trough to crest? (See Fig. 6). What is its wavelength in feet? How fast is it traveling in feet per second? Calculate your answers to the nearest foot.

6. Modeling Water Waves A water wave has an amplitude of 30 ft and a period of 14 sec. If its equation is given by

$$y = A \sin Bt$$

at a fixed position, find A and B. What is the wavelength in feet? How fast is it traveling in feet per second? How high would the wave be from trough to crest? Calculate your answers to the nearest foot.

7. Modeling Water Waves Graph the equation in Problem 5 for $0 \le t \le 32$.

8. Modeling Water Waves Graph the equation in Problem 6 for $0 \le t \le 28$.

9. Modeling Water Waves A **tsunami** is a sea wave caused by an earthquake. These are very long waves that can travel at nearly the rate of a jet airliner. (The speed of a tsunami, whose wavelength far exceeds the ocean's depth D, is approximately $\sqrt{gD}$, where g is the gravitational constant.) At sea, these waves have an amplitude of only 1 or 2 ft, so a ship would be unaware of the wave passing underneath. As a tsunami approaches a shore, however, the waves are slowed down and the water piles up to a virtual wall of water, sometimes over 100 ft high. When such a wave crashes into land, considerable destruction can result—a three story concrete lighthouse 50 ft above sea level was completely washed away when a tsunami hit Scotch Cap, Alaska, in 1946.

(A) If at a particular moment in time, a tsunami has an equation of the form

$$y = A \sin Br$$

where y is in feet and r is the distance from the source in miles, find the equation if the amplitude is 2 ft and the wavelength is 150 mi.

(B) Find the period (in seconds) of the tsunami in part (A).

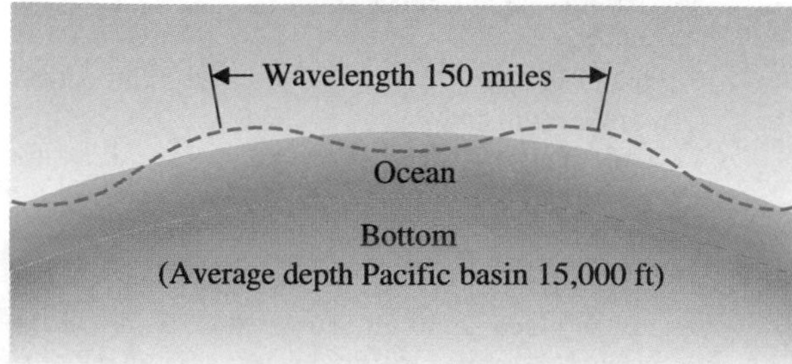

Tsunami (vertical exaggerated)

Figure for 9

10. Modeling Water Waves Because of increased friction from the ocean bottom, a wave will begin to break when the ocean depth becomes less than one-half the wavelength. If an ocean wave at a fixed position has an equation of the form

$$y = 8 \sin \frac{\pi}{5} t$$

at what depth (to the nearest foot) will the wave start to break?

***11. Modeling Water Waves** A water wave has an equation of the form

$$y = 25 \sin 2\pi\left(\frac{t}{10} + \frac{r}{512}\right)$$

where y and r are in feet and t is in seconds.

(A) Describe what the equation models if r (the distance from the source) is held constant at $r = 1{,}024$ ft and t is allowed to vary.

(B) Under the conditions in part (A), compute the period or wavelength, whichever is appropriate, and explain why you made your choice.

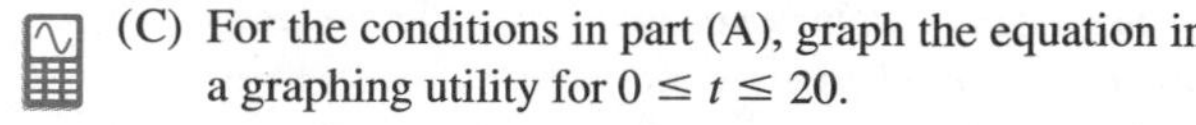
(C) For the conditions in part (A), graph the equation in a graphing utility for $0 \le t \le 20$.

***12. Modeling Water Waves** Referring to the equation in Problem 11:

(A) Describe what the equation models if t (time in seconds) is frozen at $t = 0$ and r is allowed to vary.

(B) Under the conditions in part (A), compute the period or wavelength, whichever is appropriate, and explain why you made your choice. What is the height of the wave from trough to crest?

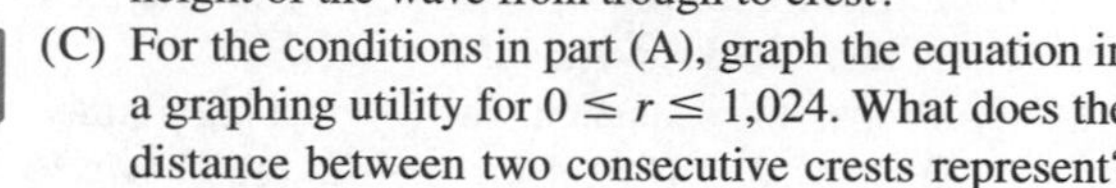
(C) For the conditions in part (A), graph the equation in a graphing utility for $0 \le r \le 1{,}024$. What does the distance between two consecutive crests represent?

13. **Modeling Electromagnetic Waves** Suppose an electron oscillates at 10^8 Hz ($\nu = 10^8$ Hz), creating an electromagnetic wave. What is its period? What is its wavelength (in meters)?

14. **Modeling Electromagnetic Waves** Repeat Problem 13 with $\nu = 10^{18}$ Hz.

15. **Modeling Electromagnetic Waves** An x ray has an equation of the form

$$y = A \sin Bt$$

Find B if the wavelength of the x ray, λ, is 3×10^{-10} m. [*Note:* $c \approx 3 \times 10^8$ m/sec]

16. **Modeling Electromagnetic Waves** A microwave has an equation of the form

$$y = A \sin Bt$$

Find B if the wavelength is $\lambda = 0.003$ m.

17. **Modeling Electromagnetic Waves** An AM radio wave for a given station has an equation of the form

$$y = A[1 + 0.02 \sin(2\pi \cdot 1{,}200t)] \sin(2\pi \cdot 10^6 t)$$

for a given 1,200 Hz tone. The expression in brackets modulates the amplitude A of the carrier wave

$$y = A \sin(2\pi \cdot 10^6 t)$$

(see Fig. 5). What are the period and frequency of the carrier wave for time t in seconds? Can a wave with this frequency pass through the atmosphere? (See Fig. 2.)

18. **Modeling Electromagnetic Waves** An FM radio wave for a given station has an equation of the form

$$y = A \sin[2\pi \cdot 10^9 t + 0.02 \sin(2\pi \cdot 1{,}200t)]$$

for a given tone of 1,200 Hz. The second term within the brackets modulates the frequency of the carrier wave $y = A \sin(2\pi \cdot 10^9 t)$ (see Fig. 5). What are the period and frequency of the carrier wave? Can a wave with this frequency pass through the atmosphere? (See Fig. 2.)

***19.** **Spring–Mass Systems** Graph each of the following equations representing the vertical motion y, relative to time t, of a mass attached to a spring. Indicate whether the motion is simple harmonic, damped harmonic, or illustrates resonance.

(A) $y = -5 \cos(4\pi t), \quad 0 \le t \le 2$

(B) $y = -5e^{-1.2t} \cos(4\pi t), \quad 0 \le t \le 2$

***20.** **Spring–Mass Systems** Graph each of the following equations representing the vertical motion y, relative to time t, of a mass attached to a spring. Indicate whether the motion is simple harmonic, damped harmonic, or illustrates resonance.

(A) $y = \sin(4\pi t), \quad 0 \le t \le 2$

(B) $y = e^{1.25t} \sin(4\pi t), \quad 0 \le t \le 2$

3.5 GRAPHING COMBINED FORMS

- Graphing by Hand Using Addition of Ordinates
- Graphing Combined Forms on a Graphing Utility
- Sound Waves
- Fourier Series: A Brief Look

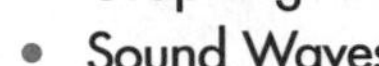

After having considered the trigonometric functions individually, we now consider them in combination with each other and with other functions. To help fix basic ideas, we start by hand graphing an easy combination using a technique called *addition of ordinates.* It soon becomes clear that for more complicated combinations, hand graphing is not practical, and we are happy to succumb to the use of a graphing utility.

Graphing by Hand Using Addition of Ordinates

The process of graphing by hand using **addition of ordinates** is best illustrated through an example. The process is used to introduce basic principles and is not practical for most problems of significance.

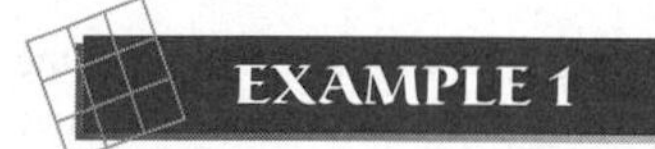

Addition of Ordinates

Graph

$$y = \frac{x}{2} + \sin x \qquad 0 \le x \le 2\pi$$

Solution A simple and fast method of hand graphing equations involving two or more terms is to graph each term separately on the same coordinate system, and then add ordinates. In this case, we form

$$y_1 = \frac{x}{2} \qquad \text{and} \qquad y_2 = \sin x$$

We sketch the graph of each equation in the same coordinate system, then use a compass, dividers, ruler, or eye to add the ordinates $y_1 + y_2$ (see Fig. 1). The final graph of $y = (x/2) + \sin x$ is shown in Figure 2. Use a calculator to determine specific points on the graph if greater accuracy is desired.

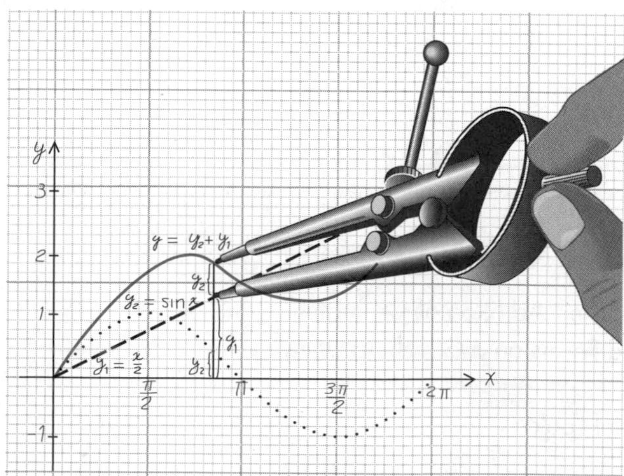

FIGURE 1
Addition of ordinates

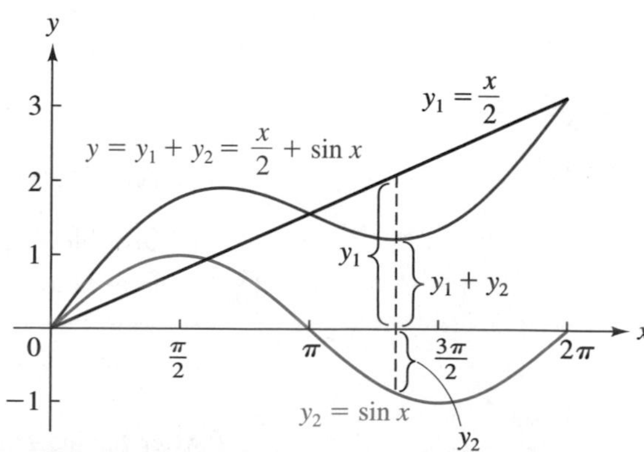

FIGURE 2
Addition of ordinates

Matched Problem 1 Graph $y = (x/2) + \cos x$, $0 \le x \le 2\pi$, using addition of ordinates.

Graphing Combined Forms on a Graphing Utility

We now repeat Example 1 on a graphing utility.

EXAMPLE 2 Addition of Ordinates on a Graphing Utility

Graph the following equations from Example 1 on a graphing utility. Graph all in the same viewing window for $0 \le x \le 2\pi$. Watch each graph as it is traced out. Plot y_3 with a darker line so that it stands out

$$y_1 = \frac{x}{2} \qquad y_2 = \sin x \qquad y_3 = \frac{x}{2} + \sin x$$

Solution

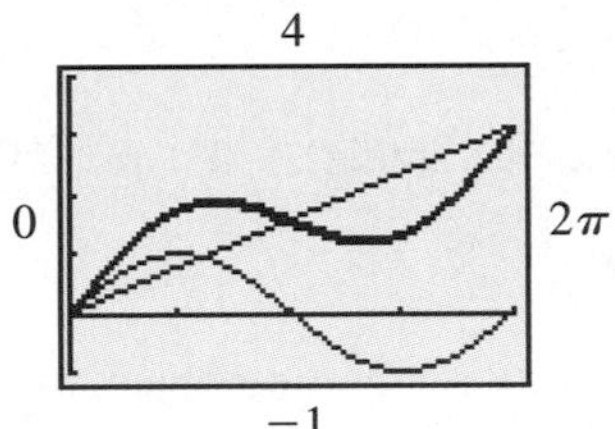

Matched Problem 2 Graph $y_1 = x/2$, $y_2 = \cos x$, and $y_3 = x/2 + \cos x$, $0 \le x \le 2\pi$, in the same viewing window. Plot y_3 with a darker line so that it stands out.

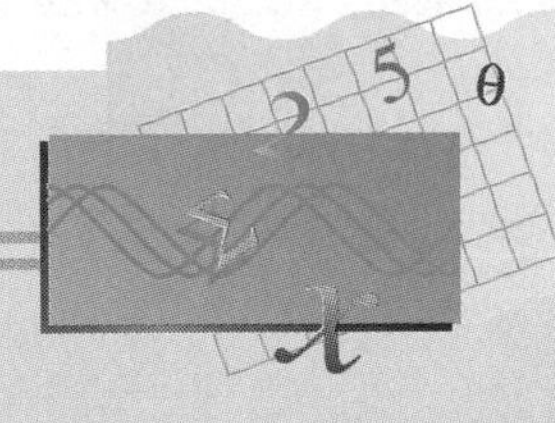

EXPLORE/DISCUSS 1

The graphing utility display in Figure 3 illustrates the graph of one of the following:

(1) $y_1 = y_2 + y_3$ (2) $y_2 = y_1 + y_3$ (3) $y_3 = y_1 + y_2$

Discuss which one and why.

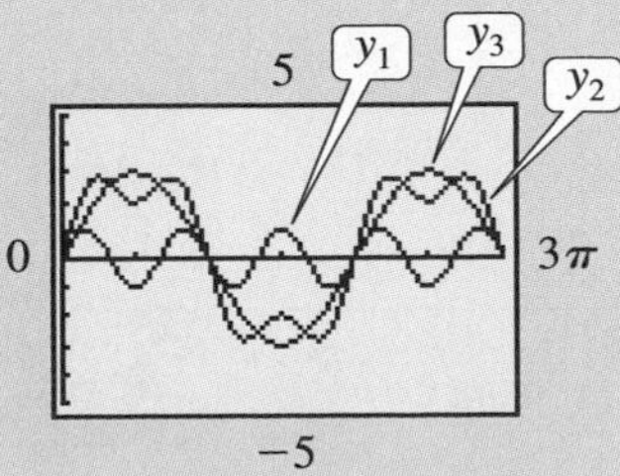

FIGURE 3

A graphing utility can plot complicated combined forms almost as easily and quickly as it plots simple forms.

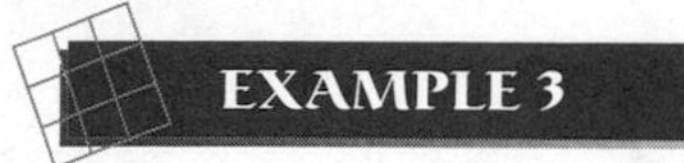

Graphing Combined Forms on a Graphing Utility

Graph $y = 3 \sin x + \cos(3x), \quad 0 \le x \le 3\pi.$

Solution

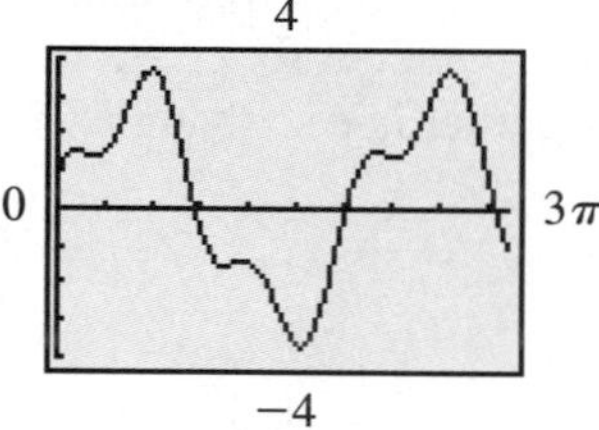

■

Matched Problem 3 Graph $y = 2 \cos x - 2 \sin(2x), \quad 0 \le x \le 4\pi.$ ■

Sound Waves

Sound is produced by a vibrating object that, in turn, excites air molecules into motion. The vibrating air molecules cause a periodic change in air pressure that travels through air at about 1,100 ft/sec. (Sound is not transmitted in a vacuum.) When this periodic change in air pressure reaches your eardrum, the drum vibrates at the same frequency as the source, and the vibration is transmitted to the brain as sound. The range of audible frequencies covers about ten octaves, extending from 20 Hz up to 20,000 Hz. Low frequencies are associated with low pitch and high frequencies with high pitch.

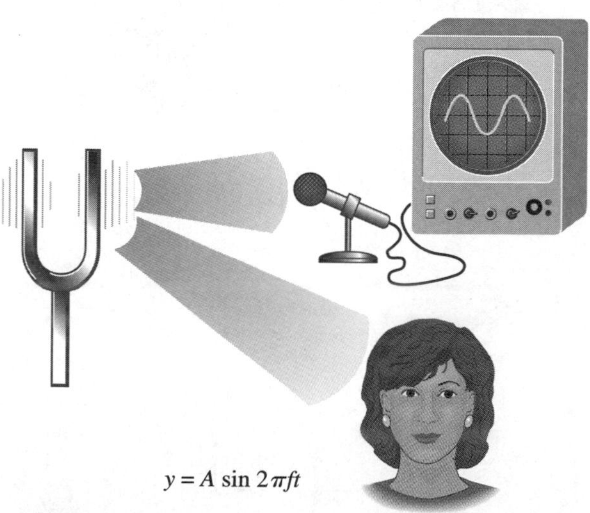

FIGURE 4
A simple sound wave

If in place of an eardrum we use a microphone, then the air disturbance can be changed into a pulsating electrical signal that can be visually displayed on an oscilloscope (Fig. 4). A pure tone from a tuning fork will look like a sine curve (also called a **sine wave**). The sound from a tuning fork can be accurately described by either the sine function or the cosine function. For example, a tuning fork vibrating at 264 Hz ($f = 264$) with an amplitude of 0.002 in., produces C on the musical scale, and the wave on an oscilloscope can be described by the simple harmonic

$$y = A \sin 2\pi f t$$
$$= 0.002 \sin 2\pi(264)t$$

Most sounds are more complex than that produced by a tuning fork. Figure 5 (page 188) illustrates a note produced by a guitar. You may be surprised to learn that even these more complex sound forms can be described in terms of simple harmonics by means of a *Fourier series* (which we discuss below). Theoretically, any sound can be reproduced by an appropriate combination of pure tones from tuning forks. The separate components of a complex sound are called **partial tones.** The partial tone with the smallest frequency is the **fundamental tone.** The other partial tones are called **overtones** (see Fig. 5).

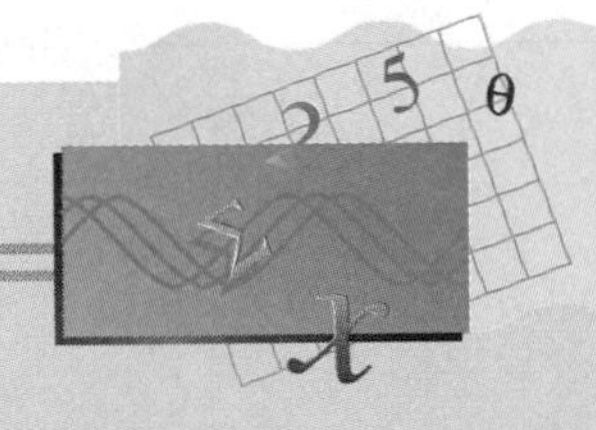

EXPLORE/DISCUSS 2

Use a graphing utility to graph

$$y = 0.12 \sin 400\pi t + 0.04 \sin 800\pi t + 0.02 \sin 1{,}200\pi t \qquad 0 \le t \le 0.01$$

which is a close approximation of the guitar note in Figure 5.

Fourier Series: A Brief Look

In the above discussion on sound waves, we mentioned a significant use of combinations of trigonometric functions called **Fourier series** (named after the French mathematician Joseph Fourier, 1768–1830). These combinations may be encountered in advanced applied mathematics in the study of sound, heat flow, electrical fields and circuits, and spring–mass systems. The following discussion is included only for illustrative purposes—the reader is not expected to become proficient in this area at this time.

The following are examples of Fourier series:

$$y = \sin x + \frac{\sin 3x}{3} + \frac{\sin 5x}{5} + \cdots \qquad (1)$$

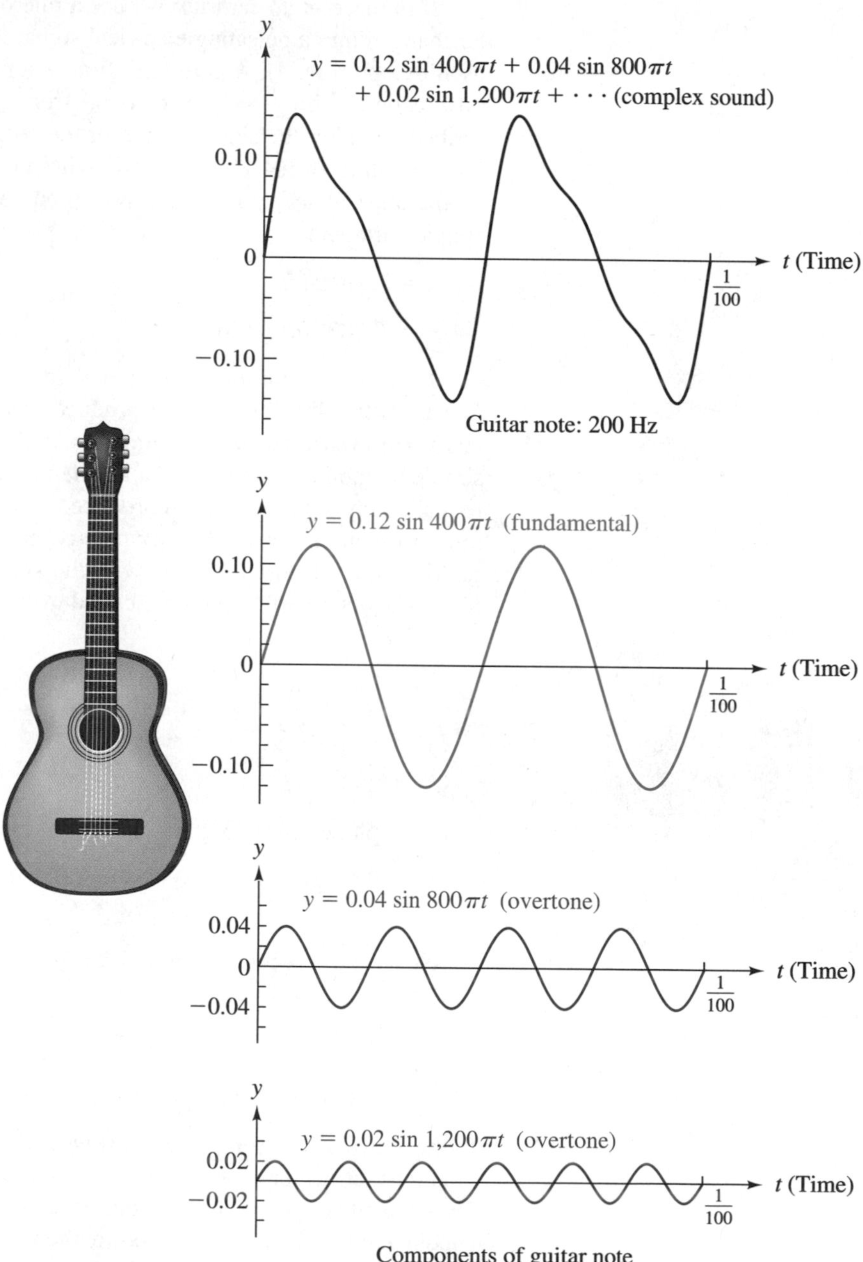

FIGURE 5
A complex sound wave—the guitar note shown at the top can be approximated very closely by adding the three simple harmonics (pure tones) below

$$y = \sin \pi x + \frac{\sin 2\pi x}{2} + \frac{\sin 3\pi x}{3} + \cdots \qquad (2)$$

The three dots at the end of each series indicate that the pattern established in the first three terms continues indefinitely. (Note that the terms of each series are simple harmonics.) If we graph the first term of each series, then the sum of the first two terms, and so on, we will obtain a sequence of graphs that will get closer and closer to a **square wave** for (1) and a **sawtooth wave** for (2). The greater the number of terms we take in the series, the more the graph will look like the indicated wave form. Figures 6 and 7 illustrate these phenomena.

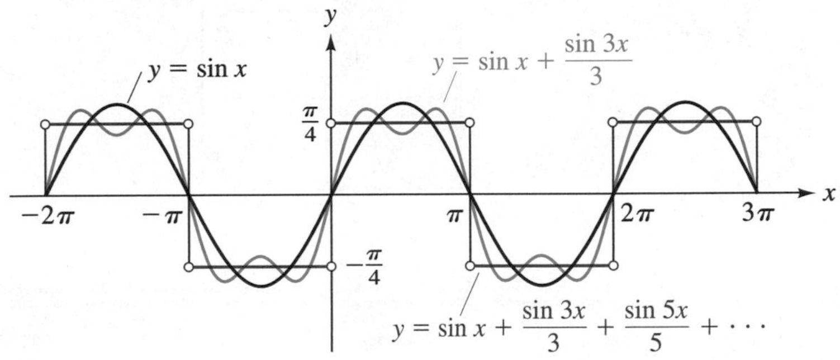

FIGURE 6
Square wave

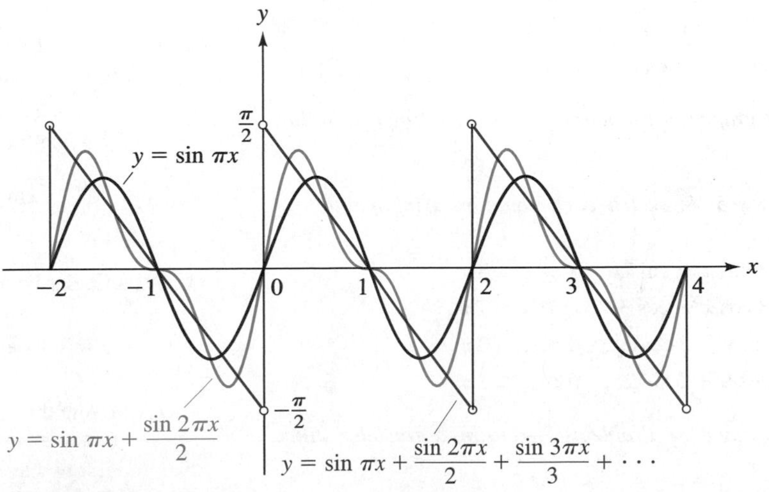

FIGURE 7
Sawtooth wave

Answers to Matched Problems

1.

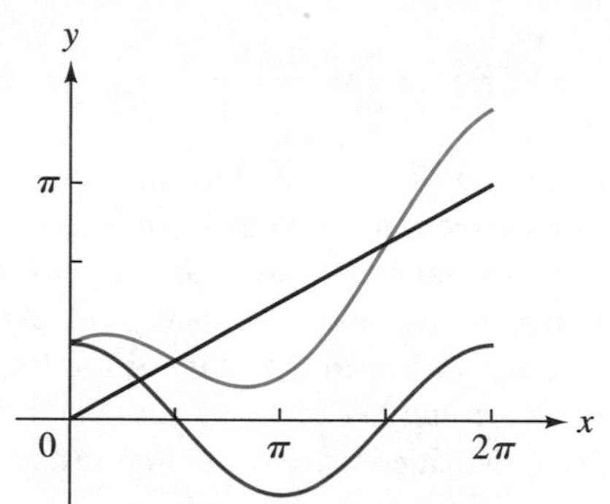

2.

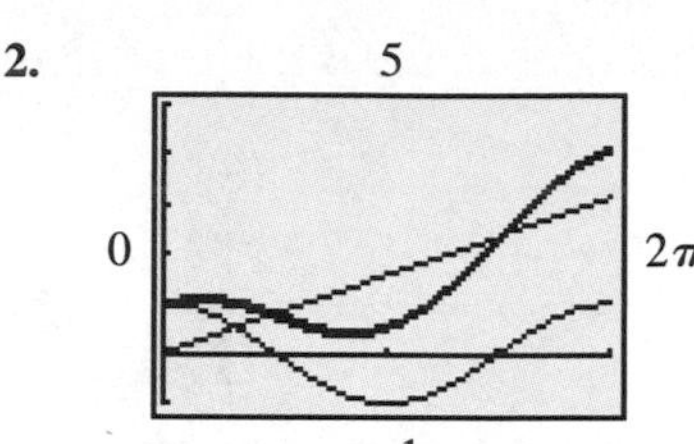

3.

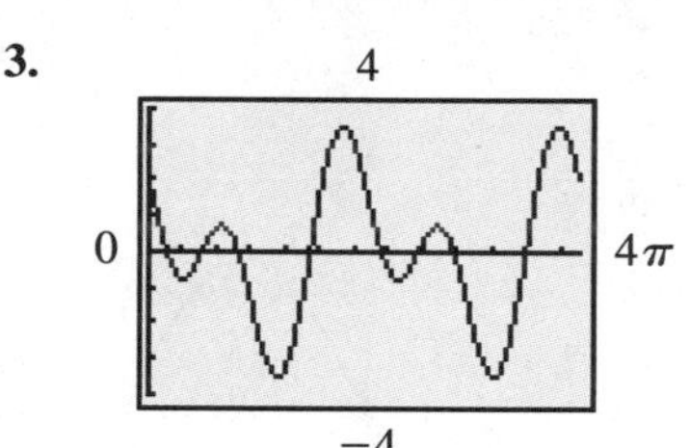

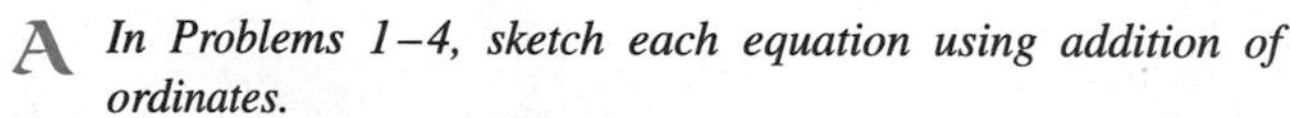

EXERCISE 3.5

A *In Problems 1–4, sketch each equation using addition of ordinates.*

1. $y = x + \cos x, \quad 0 \le x \le 5\pi/2$
2. $y = x + \sin x, \quad 0 \le x \le 2\pi$
3. $y = x/2 + \cos \pi x, \quad 0 \le x \le 3$
4. $y = x/2 - \sin 2\pi x, \quad 0 \le x \le 2$

Check the graphs in Problems 1–4 using a graphing utility.

B *In Problems 5–8, sketch each equation using addition of ordinates.*

5. $y = 3 \cos x + \sin 2x, \quad 0 \le x \le 3\pi$
6. $y = 3 \cos x + \cos 3x, \quad 0 \le x \le 2\pi$
7. $y = \sin x + 2 \cos 2x, \quad 0 \le x \le 3\pi$
8. $y = \cos x + 3 \sin 2x, \quad 0 \le x \le 3\pi$

Check the graphs in Problems 5–8 using a graphing utility.

C *Problems 9–14 require the use of a graphing utility.*

9. Graph the following equation for $-2\pi \le x \le 2\pi$ and $-1 \le y \le 1$ (compare to the square wave in Fig. 6).

$$y = \sin x + \frac{\sin 3x}{3} + \frac{\sin 5x}{5} + \frac{\sin 7x}{7} + \frac{\sin 9x}{9}$$

10. Graph the following equation for $-4 \le x \le 4$ and $-2 \le y \le 2$ (compare to the sawtooth wave in Fig. 7).

$$y = \sin \pi x + \frac{\sin 2\pi x}{2} + \frac{\sin 3\pi x}{3} + \frac{\sin 4\pi x}{4} + \frac{\sin 5\pi x}{5}$$

11. A particular Fourier series is given as follows:

$$y = 1.15 + \cos x + \frac{\cos 3x}{3^2} + \frac{\cos 5x}{5^2} + \cdots$$

(A) Using the window dimensions $-4\pi \le x \le 4\pi$ and $0 \le y \le 4$, graph in three separate windows: the first two terms of the series, the first three terms of the series, and the first four terms of the series.

(B) Describe the shape of the wave form the graphs tend to approach as more terms are added to the series.

(C) As additional terms are added to the series, the graphs appear to approach the wave form you described in part (B), and that form is comprised entirely of straight line segments. Hand-sketch a graph of this wave form.

12. A particular Fourier series is given as follows:

$$y = \cos\frac{x}{2} + \frac{1}{9}\cos\frac{3x}{2} + \frac{1}{25}\cos\frac{5x}{2} + \cdots$$

(A) Using the window dimensions $-4\pi \le x \le 4\pi$ and $-2 \le y \le 2$, graph in two separate windows: the first two terms of the series and the first three terms of the series.

(B) Describe the shape of the wave form the graphs tend to approach as more terms are added to the series.

(C) As additional terms are added to the series, the graphs appear to approach the wave form you described in part (B), and that form is comprised entirely of straight line segments. Hand-sketch a graph of this wave form.

13. Graph the sound wave given by

$$y = 0.08 \sin 400\pi t + 0.04 \sin 800\pi t + 0.02 \sin 1200\pi t \qquad 0 \le t \le 0.01$$

14. Graph the sound wave given by

$$y = 0.06 \sin 200\pi t + 0.05 \sin 400\pi t + 0.03 \sin 600\pi t \qquad 0 \le t \le 0.02$$

Applications

15. Physiology A normal seated adult breathes in and exhales about 0.80 liter of air every 4.00 sec. See the graph. The volume V of air in the lungs t seconds after exhaling can be approximated by an equation of the form $V = k + A\cos Bt$, $0 \le t \le 8$. Find the equation.

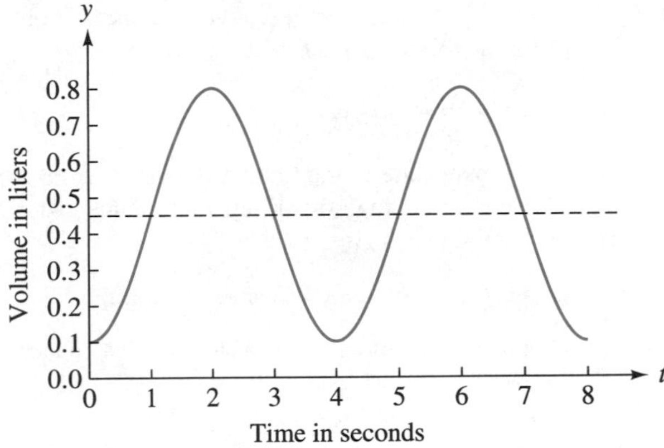

Figure for 15

16. Earth Science During the autumnal equinox, the surface temperature (in °C) of the water of a lake x hours after sunrise was recorded over a 24 hr period, and the results were recorded on the following graph. The surface temperature can be approximated by an equation of the form $T = k + A\cos Bx$. Find the equation.

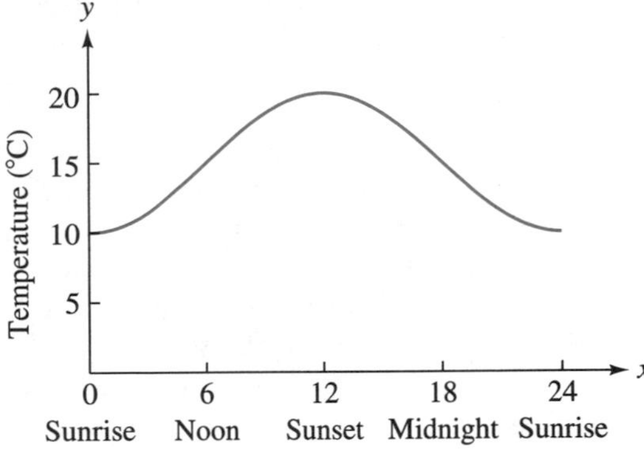

Figure for 16

17. Modeling a Seasonal Business Cycle A large soft drink company's sales vary seasonally, but, overall, the company is also experiencing a steady growth rate. The financial analysis department has developed the following mathematical model for sales:

$$S = 5 + \frac{t}{52} - 4\cos\frac{\pi t}{26}$$

where S represents sales in millions of dollars for a week of sales t weeks after January 1.

(A) Graph this function for a 3 yr period starting January 1.

(B) What are the sales for the 26th week in the 3rd year (to three significant digits)?

(C) What are the sales for the 52nd week in the 3rd year (to one significant digit)?

18. Modeling a Seasonal Business Cycle Repeat Problem 17 for the following mathematical model:

$$S = 4 + \frac{t}{52} - 2\cos\frac{\pi t}{26}$$

19. Sound A pure tone is transmitted through a speaker that is also emitting high-frequency, low-volume static. The resulting sound is given by

$$y = 0.1 \sin 500\pi t + 0.003 \sin 24{,}000\ \pi t$$

(A) Graph this equation, using a graphing calculator, for $0 \le t \le 0.004$ and $-0.3 \le y \le 0.3$. Notice that the result looks like a pure sine wave.

(B) Use the ZOOM feature (or its equivalent) to zoom in on the graph in the vicinity of (0.001, 0.1). You should now *see* the static element of the sound.

20. Sound Another pure tone is transmitted through the same speaker as in Problem 19. The resulting sound is given by

$$y = 0.08 \cos 500\pi t + 0.004 \sin 26{,}000\ \pi t$$

(A) Graph this equation, using a graphing calculator, for $0 \le t \le 0.004$ and $-0.3 \le y \le 0.3$. Notice that the sound appears to be a pure tone.

(B) Use the ZOOM feature (or its equivalent) to zoom in on the graph in the vicinity of (0.002, −0.08). Now you should be able to *see* the static.

21. Experiment on Wave Interference When wave forms are superposed, they may tend to reinforce or cancel each other. Such effects are called **interference.** If the resultant amplitude is increased, the interference is said to be **constructive;** if it is decreased, the interference is **destructive.** Interference applies to all wave forms, including sound, light, and water. In the following three equations, y_3 represents the superposition of the individual waves, y_1 and y_2:

$$y_1 = 2\sin t \qquad y_2 = 2\sin(t + C) \qquad y_3 = y_1 + y_2$$

For each given value of C below, graph these three equations in the same viewing window ($0 \le t \le 4\pi$), and indicate whether the interference shown by y_3 is constructive or destructive. (Graph y_3 with a darker line so it will stand out.)

(A) $C = 0$ (B) $C = -\pi/4$
(C) $C = -3\pi/4$ (D) $C = -\pi$

22. Experiment on Wave Interference Repeat Problem 21 with the following three equations:

$$y_1 = 2\cos t \qquad y_2 = 2\cos(t + C) \qquad y_3 = y_1 + y_2$$

23. Noise Control Jet engines, air-conditioning systems in commercial buildings, and automobile engines all have something in common—NOISE! Destructive interference (see Problem 21) provides an effective tool for noise control. Acoustical engineers have devised an Active Noise Control (ANC) system that is now widely used. The figure illustrates a system used to reduce noise in air-conditioning ducts. The input microphone senses the noise sound wave, processes it electronically, and produces the same sound wave in the loudspeaker, but out of phase with the original. The resulting destructive interference effectively reduces the noise in the system.

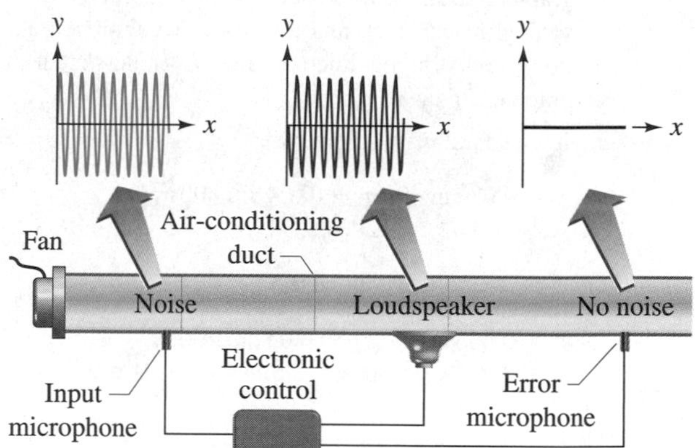

Figure for 23 and 24

(A) Suppose the input microphone indicates a sound wave given by $y_1 = 65 \sin 400\pi t$, where y_1 is sound intensity in decibels and t is time in seconds. Write an equation of the form $y_2 = A\sin(Bt + C)$, $C < 0$, that represents a sound wave that will have total destructive interference with y_1.

(B) Graph y_1, y_2, and $y_3 = y_1 + y_2$, $0 \le t \le 0.01$, in the same viewing window of a graphing utility, and explain the results.

24. Noise Control Repeat Problem 23 with the noise given by $y_1 = 30 \cos 600\pi t$.

3.6 TANGENT, COTANGENT, SECANT, AND COSECANT FUNCTIONS REVISITED

- Graphing $y = A\tan(Bx + C)$ and $y = A\cot(Bx + C)$
- Graphing $y = A\sec(Bx + C)$ and $y = A\csc(Bx + C)$

We now graph the more general forms of the tangent, cotangent, secant, and cosecant functions following essentially the same process we developed for graphing $y = A\sin(Bx + C)$ and $y = A\cos(Bx + C)$. The process is not difficult if you have a clear understanding of the basic graphs for these functions, including periodic properties.

Graphing $y = A\tan(Bx + C)$ and $y = A\cot(Bx + C)$

The basic graphs for $y = \tan x$ and $y = \cot x$ that were developed in Section 3.1 are repeated below for convenient reference.

GRAPH OF $y = \tan x$

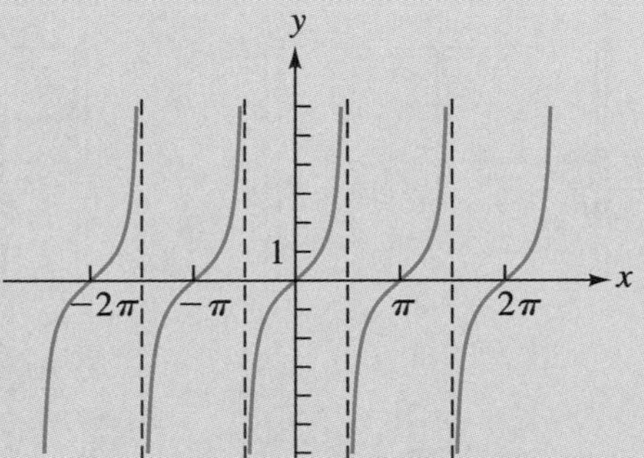

Domain: Set of all real numbers R except $\pi/2 + k\pi$, k an integer
Range: R Period: π

GRAPH OF $y = \cot x$

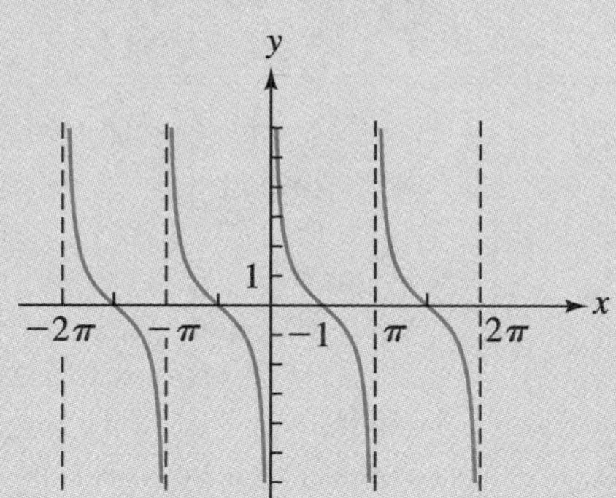

Domain: Set of all real numbers R except $k\pi$, k an integer
Range: R Period: π

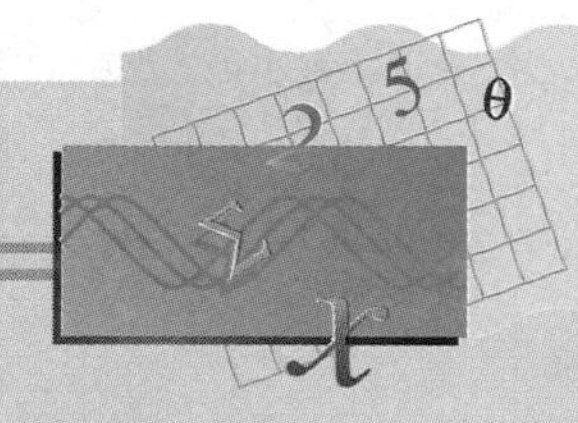

EXPLORE/DISCUSS 1

(A) Match each function to its graph in one of the graphing utility displays, and discuss how the graph compares to the graph of $y = \tan x$ or $y = \cot x$.

(1) $y = \cot 2x$ (2) $y = 3 \tan x$ (3) $y = \tan(x + \pi/2)$

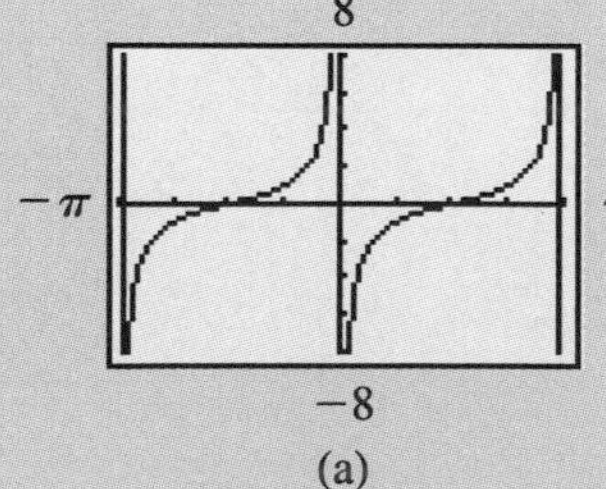

(a)

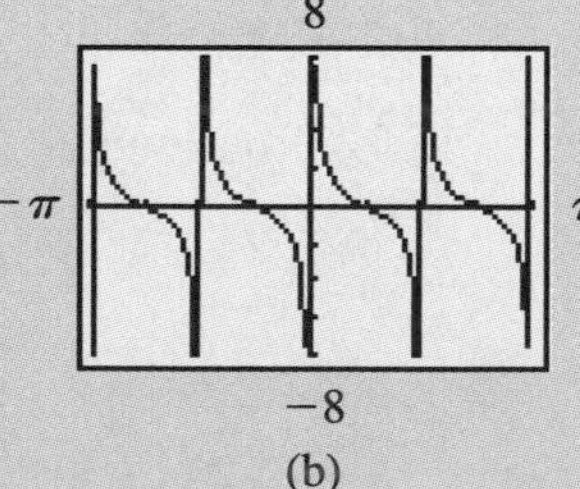

(b)

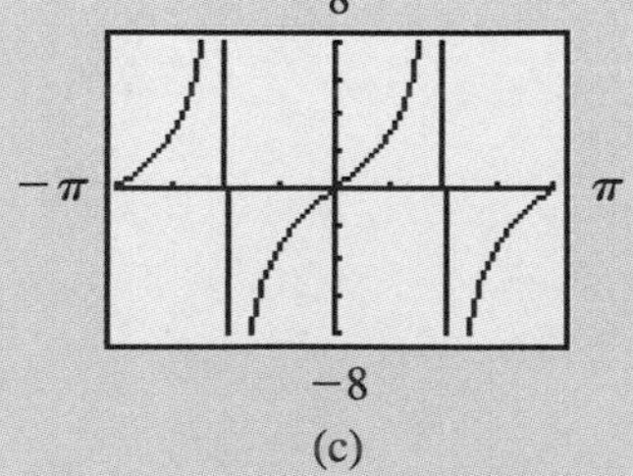

(c)

(B) Use a graphing utility to explore the nature of the changes in the graphs of the following functions when the values of A, B, and C are changed. Discuss what happens in each case.

(4) $y = A \tan x$ and $y = A \cot x$

(5) $y = \tan Bx$ and $y = \cot Bx$

(6) $y = \tan(x + C)$ and $y = \cot(x + C)$

To quickly sketch the graphs of equations of the form $y = A \tan(Bx + C)$ and $y = A \cot(Bx + C)$, you need to know how the constants A, B, and C affect the basic graphs $y = \tan x$ and $y = \cot x$, respectively.

First note that **amplitude is not defined for the tangent and cotangent functions,** since the deviation from the x axis for both functions is indefinitely far in both directions. The effect of A is to make the graph steeper if $|A| > 1$ or to make the curve less steep if $|A| < 1$. If A is negative, the graph is reflected across the x axis (turned upside down).

Just as with the sine and cosine functions, the constants B and C involve a period change and a phase shift. Since both $A \tan x$ and $A \cot x$ have a period of π, it follows that both $A \tan(Bx + C)$ and $A \cot(Bx + C)$ complete one cycle as $Bx + C$ varies from

$$Bx + C = 0 \quad \text{to} \quad Bx + C = \pi$$

or, solving for x, as x varies from

$$\underbrace{x = -\frac{C}{B}}_{\text{Phase shift}} \quad \text{to} \quad x = \underbrace{-\frac{C}{B}}_{\text{Phase shift}} + \underbrace{\frac{\pi}{B}}_{\text{Period}}$$

Thus, $y = A \tan(Bx + C)$ and $y = A \cot(Bx + C)$ each have a period of π/B, and their graphs are translated horizontally $-C/B$ units to the right if $-C/B > 0$ and $|-C/B|$ units to the left if $-C/B < 0$. As before, the horizontal translation determined by the number $-C/B$ is the phase shift.

Again, you do not need to memorize the formulas for period and phase shift. You need only remember the process used above.

EXAMPLE 1

Graphing $y = A \tan Bx$

Find the period and phase shift for $y = 3 \tan(\pi x/2)$. Then sketch its graph for $-3 < x < 3$.

Solution One cycle of $y = 3 \tan(\pi x/2)$ is completed as $\pi x/2$ varies from 0 to π. Solve for x:

$$\frac{\pi x}{2} = 0 \qquad \frac{\pi x}{2} = \pi$$

$$x = 0 \qquad x = 0 + 2$$

$$\text{Period} = 2 \qquad \text{Phase shift} = 0$$

[*Note:* In general, if $C = 0$, then there is no phase shift.] Graph the equation for one period, and extend this graph over the interval from -3 to 3 (Fig. 1). ■

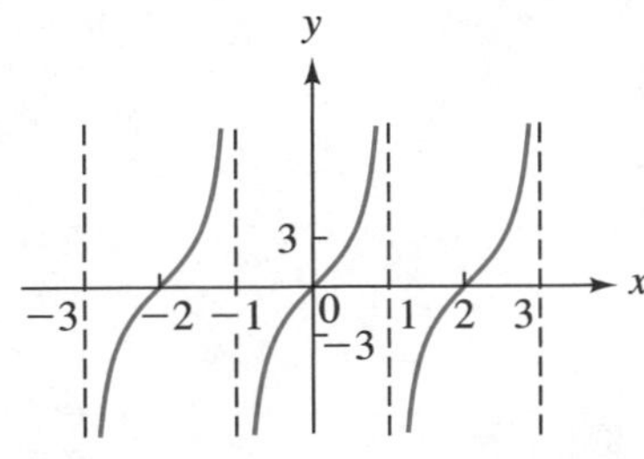

FIGURE 1

Matched Problem 1 Find the period and phase shift for $y = 2 \cot(x/2)$. Then sketch its graph for $-2\pi < x < 2\pi$. ■

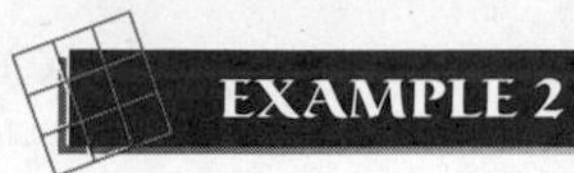

EXAMPLE 2 **Graphing $y = A \tan (Bx + C)$**

Find the period and phase shift for

$$y = \tan\left(\frac{\pi}{2}x + \frac{\pi}{4}\right)$$

Then sketch the graph for $-1.5 < x < 2.5$.

Solution ***Step 1*** Find the period and phase shift by solving $Bx + C = 0$ and $Bx + C = \pi$ for x:

$$\frac{\pi}{2}x + \frac{\pi}{4} = 0 \qquad\qquad \frac{\pi}{2}x + \frac{\pi}{4} = \pi$$

$$\frac{\pi}{2}x = -\frac{\pi}{4} \qquad\qquad \frac{\pi}{2}x = -\frac{\pi}{4} + \pi$$

$$x = -\frac{1}{2} \qquad\qquad x = -\frac{1}{2} + 2$$

$$\text{Period} = 2 \qquad \text{Phase shift} = -\frac{1}{2}$$

FIGURE 2

Step 2 Sketch one period of the graph starting at $x = -\frac{1}{2}$ (the phase shift) and ending at $x = (-\frac{1}{2}) + 2 = \frac{3}{2}$ (the phase shift plus one period), as shown in Figure 2. Note that there is a vertical asymptote at $x = \frac{1}{2}$.

Step 3 Extend the graph from -1.5 to 2.5 (Fig. 3).

FIGURE 3

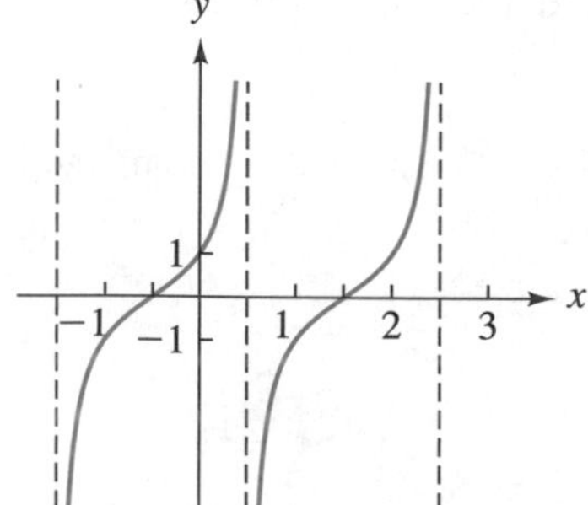

■

Matched Problem 2 Find the period and phase shift for

$$y = \cot\left(2x + \frac{\pi}{2}\right)$$

Then sketch the graph for $-\pi/2 \le x \le \pi$. ■

Graphing $y = A\sec(Bx + C)$ and $y = A\csc(Bx + C)$

For convenient reference, the basic graphs for $y = \sec x$ and $y = \csc x$ that were developed in Section 3.1 are repeated below.

GRAPH OF $y = \csc x$

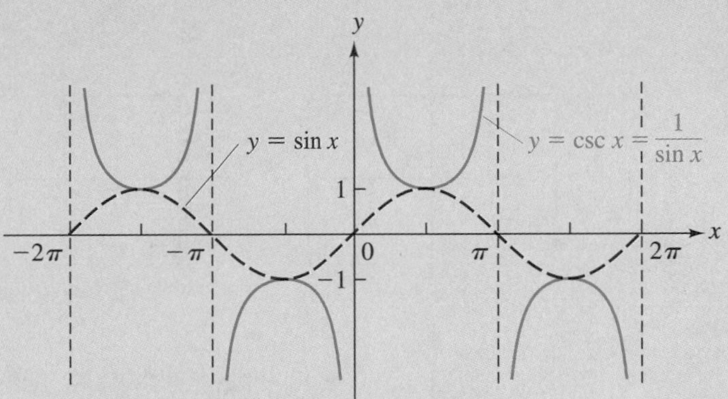

Domain: All real numbers x, except $x = k\pi$, k an integer
Range: All real numbers y such that $y \le -1$ or $y \ge 1$
Period: 2π

GRAPH OF $y = \sec x$

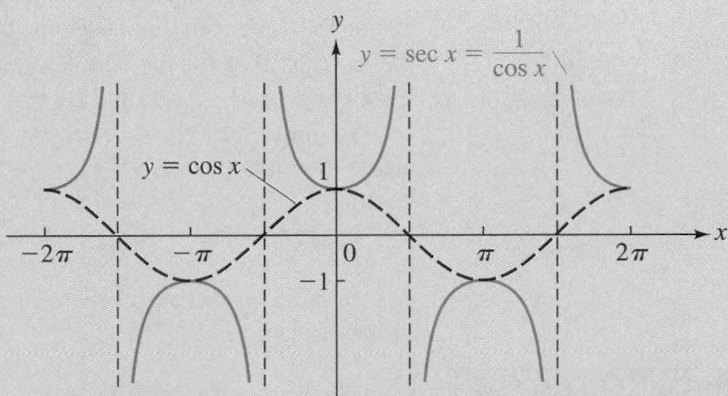

Domain: All real numbers x, except $x = \pi/2 + k\pi$, k an integer
Range: All real numbers y such that $y \le -1$ or $y \ge 1$
Period: 2π

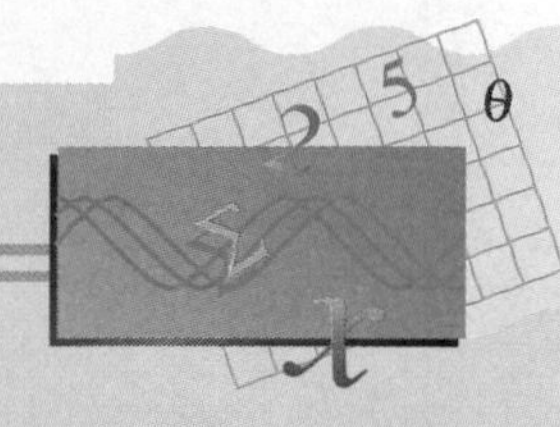

EXPLORE/DISCUSS 2

(A) Match each function to its graph in one of the graphing utility displays, and discuss how the graph compares to the graph of $y = \csc x$ or $y = \sec x$.

(1) $y = \sec 2x$ (2) $y = \frac{1}{2} \csc x$ (3) $y = \csc(x + \pi/2)$

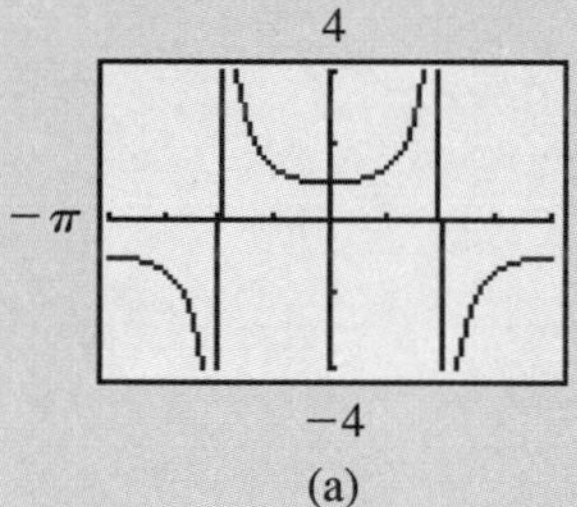

(a)

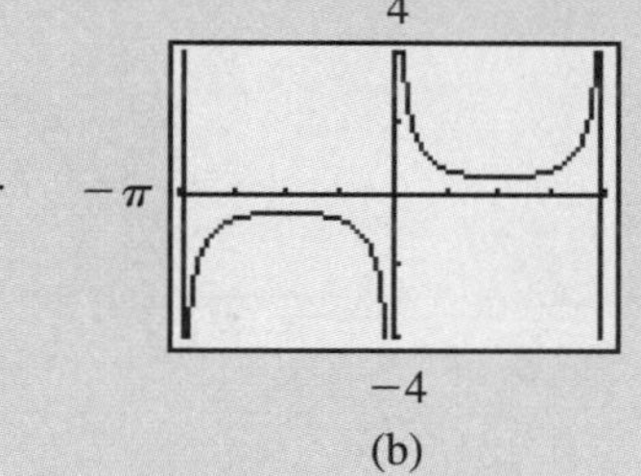

(b)

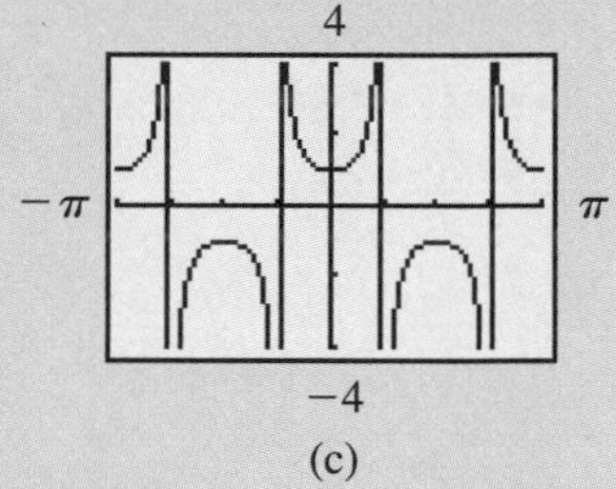

(c)

(B) Use a graphing utility to explore the nature of the changes in the graphs of the following functions when the values of A, B, and C are changed. Discuss what happens in each case.

(4) $y = A \csc x$ and $y = A \sec x$

(5) $y = \csc Bx$ and $y = \sec Bx$

(6) $y = \csc(x + C)$ and $y = \sec(x + C)$

As with the tangent and cotangent functions, **amplitude is not defined for either the secant or the cosecant** functions. Since the period of each function is the same as that for the sine and cosine, 2π, we find the period and phase shift by solving $Bx + C = 0$ and $Bx + C = 2\pi$.

To graph either $y = A \sec(Bx + C)$ or $y = A \csc(Bx + C)$, you may find it easier to graph

$$y = \frac{1}{A}\cos(Bx + C) \qquad \text{or} \qquad y = \frac{1}{A}\sin(Bx + C)$$

with a dashed curve and then take reciprocals. An example should make the process clear:

EXAMPLE 3

Graphing $y = A \csc (Bx + C)$

Find the period and phase shift for

$$y = 2\csc\left(\frac{\pi}{2}x - \pi\right)$$

Then sketch the graph for $-2 < x < 10$.

Solution *Step 1* Find the period and phase shift by solving $Bx + C = 0$ and $Bx + C = 2\pi$ for x:

$$\frac{\pi}{2}x - \pi = 0 \qquad \frac{\pi}{2}x - \pi = 2\pi$$

$$\frac{\pi}{2}x = \pi \qquad \frac{\pi}{2}x = \pi + 2\pi$$

$$x = 2 \qquad x = 2 + 4 = 6$$

Period = 4 Phase shift = 2

Step 2 Since

$$2\csc\left(\frac{\pi}{2}x - \pi\right) = \frac{1}{\frac{1}{2}\sin\left(\frac{\pi}{2}x - \pi\right)}$$

we graph

$$y = \frac{1}{2}\sin\left(\frac{\pi}{2}x - \pi\right)$$

for one cycle from 2 to 6 with a dashed curve, and then take reciprocals. From our work in Section 3.3, we know that this graph is the graph of

$$y = \frac{1}{2}\sin\frac{\pi}{2}x$$

shifted to the right 2 units, as shown in Figure 4. Notice that we also place vertical asymptotes through the x intercepts of the sine graph to guide us when we sketch the cosecant function.

FIGURE 4

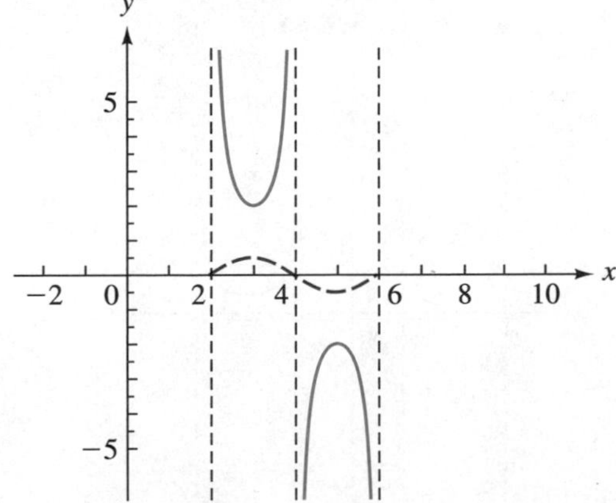

Step 3 Extend the one cycle from step 2 over the required interval from -2 to 10 (see Fig. 5).

FIGURE 5

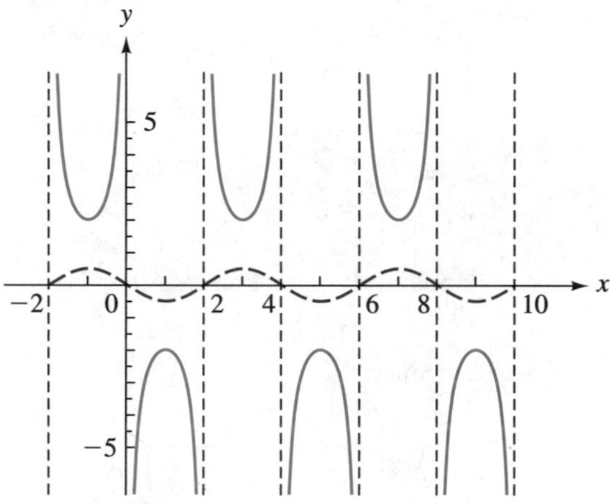

Matched Problem 3 Find the period and phase shift for $y = \frac{1}{2}\sec(2x + \pi)$. Then sketch the graph for $-3\pi/4 < x < 3\pi/4$.

Answers to Matched Problems

1. Period $= 2\pi$
   Phase shift $= 0$

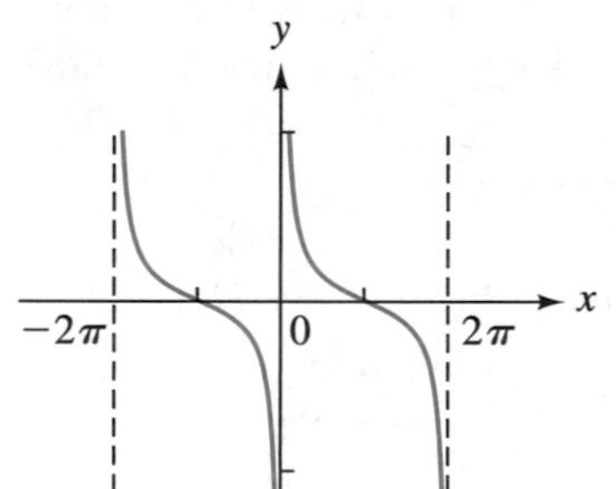

2. Period $= \pi/2$
   Phase shift $= -\pi/4$

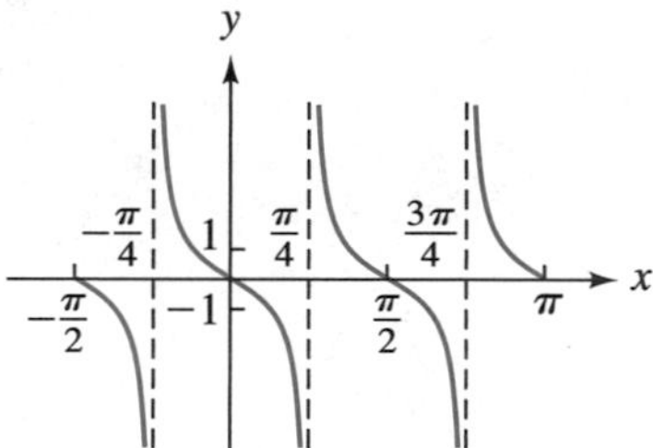

3. Period $= \pi$
   Phase shift $= -\pi/2$

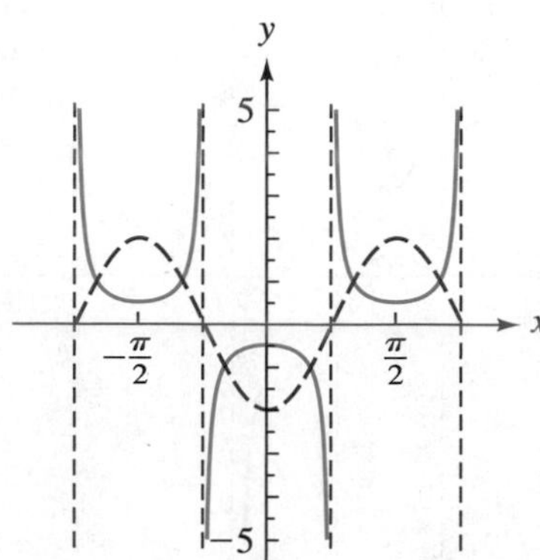

EXERCISE 3.6

A *Sketch a graph of each of the following without looking at the text or using a calculator.*

1. $y = \tan x, \quad 0 \le x \le 2\pi$

2. $y = \cot x, \quad 0 < x < 2\pi$

3. $y = \csc x, \quad -\pi < x < \pi$

4. $y = \sec x, \quad -\pi \le x \le \pi$

B *In Problems 5–10, indicate the period of each function, and sketch a graph of the function over the indicated interval.*

5. $y = 3 \tan 2x, \quad -\pi \le x \le \pi$

6. $y = 2 \cot 4x, \quad 0 < x < \pi/2$

7. $y = \frac{1}{2} \tan(x/2), \quad -\pi < x < 3\pi$

8. $y = \frac{1}{2} \cot(x/2), \quad 0 < x < 4\pi$

9. $y = 2 \csc(x/2), \quad 0 < x < 8\pi$

10. $y = 2 \sec \pi x, \quad -1 < x < 3$

Check the graphs in Problems 5–10 on a graphing utility.

In Problems 11–14, indicate the period and phase shift for each function, and sketch a graph of the function over the indicated interval.

11. $y = \cot(2x - \pi), \quad -\frac{\pi}{2} < x < \frac{\pi}{2}$

12. $y = \tan(2x + \pi), \quad -\frac{3\pi}{4} < x < \frac{3\pi}{4}$

13. $y = \csc\left(\pi x - \frac{\pi}{2}\right), \quad -\frac{1}{2} < x < \frac{5}{2}$

14. $y = \sec\left(\pi x + \frac{\pi}{2}\right), \quad -1 < x < 1$

Check the graphs in Problems 11–14 on a graphing utility.

Graph each equation in Problems 15–18 on a graphing utility; then find an equation of the form $y = A \tan Bx$, $y = A \cot Bx$, $y = A \csc Bx$, or $y = A \sec Bx$ that has the same graph. (These problems suggest additional identities beyond the fundamental ones that were discussed in Section 2.6—additional important identities will be discussed in detail in Chapter 4.)

15. $y = \csc x - \cot x$

16. $y = \csc x + \cot x$

17. $y = \cot x + \tan x$

18. $y = \cot x - \tan x$

C *In Problems 19–22, indicate the period and phase shift for each function, and sketch a graph of the function over the indicated interval.*

19. $y = -3 \cot(\pi x - \pi), \quad -2 < x < 2$

20. $y = -2 \tan\left(\frac{\pi}{4}x - \frac{\pi}{4}\right), \quad -1 < x < 7$

21. $y = 2 \sec\left(\pi x - \frac{\pi}{2}\right), \quad -1 < x < 3$

22. $y = 3 \csc\left(\frac{\pi}{2}x + \frac{\pi}{2}\right), \quad -1 < x < 3$

Check the graphs in Problems 19–22 on a graphing utility.

Graph each equation in Problems 23–26 on a graphing utility; then find an equation of the form $y = A \tan Bx$, $y = A \cot Bx$, $y = A \csc Bx$, or $y = A \sec Bx$ that has the same graph. (These problems suggest additional identities beyond the fundamental ones that were discussed in Section 2.6—additional important identities will be discussed in detail in Chapter 4.)

23. $y = \cos 2x + \sin 2x \tan 2x$

24. $y = \sin 3x + \cos 3x \cot 3x$

25. $y = \dfrac{\sin 6x}{1 - \cos 6x}$

26. $y = \dfrac{\sin 4x}{1 + \cos 4x}$

Applications

***27. Precalculus** A beacon light 15 ft from a wall rotates clockwise at the rate of exactly 1 revolution per second (rps); thus, $\theta = 2\pi t$ (see the figure).

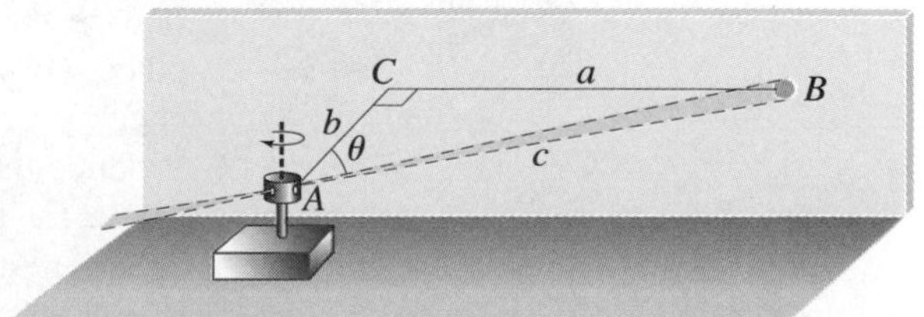

Figure for 27 and 28

(A) If we start counting time in seconds when the light spot is at C, write an equation for the distance a the light spot travels along the wall in terms of time t.
(B) Graph the equation found in part (A) for the time interval $0 \leq t < 0.25$.
(C) Explain what happens to a as t approaches 0.25 sec.

* **28. Precalculus** Refer to Problem 27.
(A) Write an equation for the length of the light beam c in terms of t.
(B) Graph the equation found in part (A) for the time interval $0 \leq t < 0.25$.
(C) Explain what happens to c as t approaches 0.25 sec.

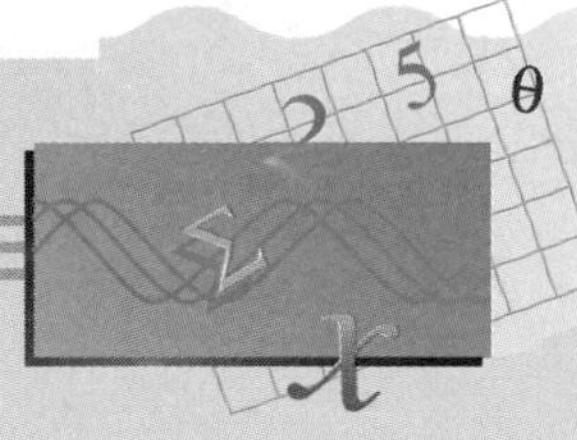

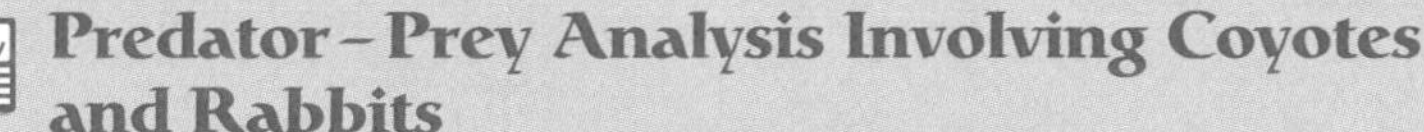

CHAPTER 3 GROUP ACTIVITY

Predator–Prey Analysis Involving Coyotes and Rabbits

In a national park in the western United States, natural scientists from a nearby university conducted a study on the interrelated populations of coyotes (sometimes called prairie wolves) and rabbits. They found that the population of each species goes up and down in cycles, but these cycles are out of phase with each other. Each year for 12 years the scientists estimated the number in each population with the results indicated in Table 1.

TABLE 1
Coyote/Rabbit Populations

Year	1	2	3	4	5	6	7	8	9	10	11	12
Coyotes (hundreds)	8.7	11.7	11.5	9.1	6.3	6.5	8.9	11.6	11.9	8.8	6.2	6.4
Rabbits (thousands)	54.5	53.7	38.0	25.3	26.8	41.2	55.4	53.1	38.6	26.2	28.0	40.9

(A) *Coyote population analysis*
(1) Enter the data for the coyote population in a graphing utility for the time interval $1 \leq t \leq 12$, and produce a scatter plot for the data.
(2) A function of the form $y = K + A \sin(Bx + C)$ can be used to model the data. Use the data for the coyote population to determine K, A, and B. Use the graph from part (1) to visually estimate C to one decimal place. (If your graphing utility has a sine regression feature, use it to check your work.)
(3) Produce the scatter plot from part (1) and graph the equation from part (2) in the same viewing window. Adjust C, if necessary, for a better equation fit of the data.
(4) Write a summary of the results, describing fluctuations and cycles of the coyote population.

(B) *Rabbit population analysis*

(1) Enter the data for the rabbit population in a graphing utility for the time interval $1 \leq t \leq 12$, and produce a scatter plot for the data.

(2) A function of the form $y = K + A \sin(Bx + C)$ can be used to model the data. Use the data for the rabbit population to determine K, A, and B. Use the graph from part (1) to visually estimate C to one decimal place. (If your graphing utility has a sine regression feature, use it to check your work.)

(3) Produce the scatter plot from part (1) and graph the equation from part (2) in the same viewing window. Adjust C, if necessary, for a better equation fit of the data.

(4) Write a summary of the results, describing fluctuations and cycles of the rabbit population.

(C) *Predator–prey interrelationship*

(1) Discuss the relationship of the maximum predator populations to the maximum prey populations relative to time.

(2) Discuss the relationship of the minimum predator populations to the minimum prey populations relative to time.

(3) Discuss the dynamics of the fluctuations of the two interdependent populations. What causes the two populations to rise and fall, and why are they out of phase from one another?

CHAPTER 3 REVIEW

3.1 BASIC GRAPHS

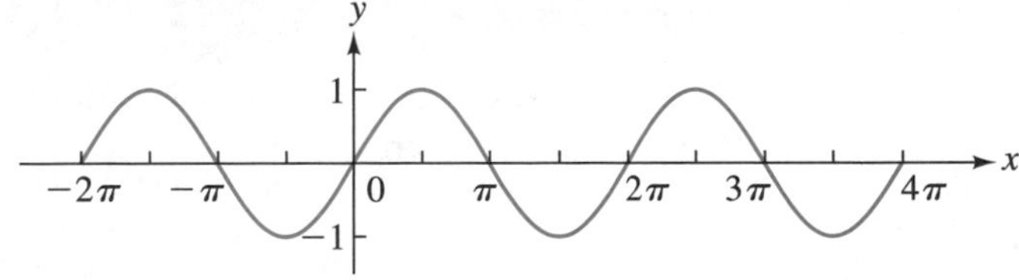

FIGURE 1
Graph of $y = \sin x$

Domain: R Range: $-1 \leq y \leq 1$ Period: 2π

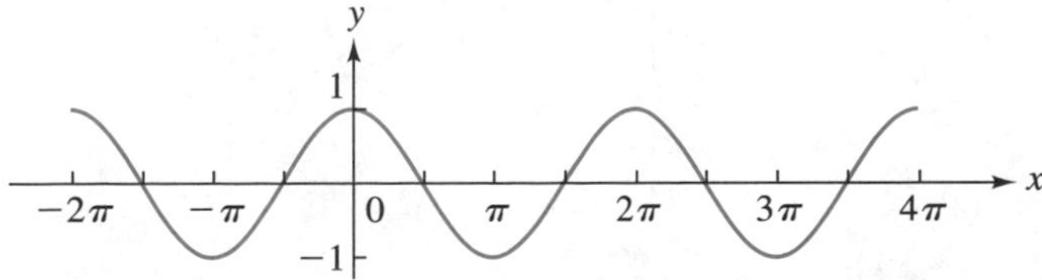

FIGURE 2
Graph of $y = \cos x$

Domain: R Range: $-1 \leq y \leq 1$ Period: 2π

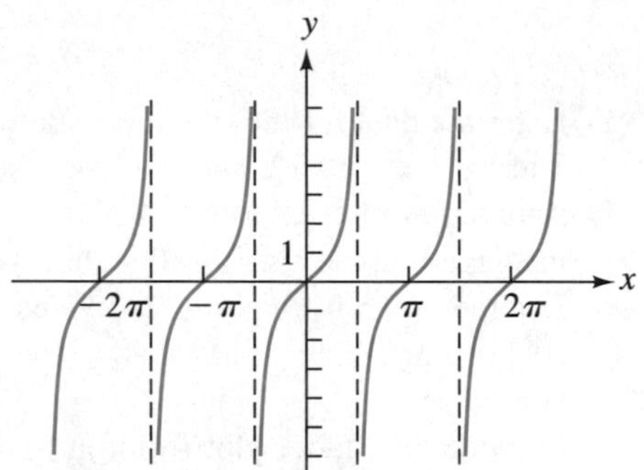

FIGURE 3
Graph of $y = \tan x$

Domain: Set of all real numbers R except $\pi/2 + k\pi$, k an integer
Range: R Period: π

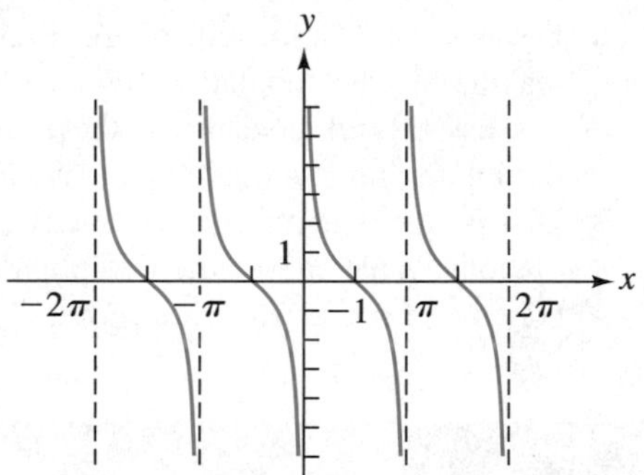

FIGURE 4
Graph of $y = \cot x$

Domain: Set of all real numbers R except $k\pi$, k an integer
Range: R Period: π

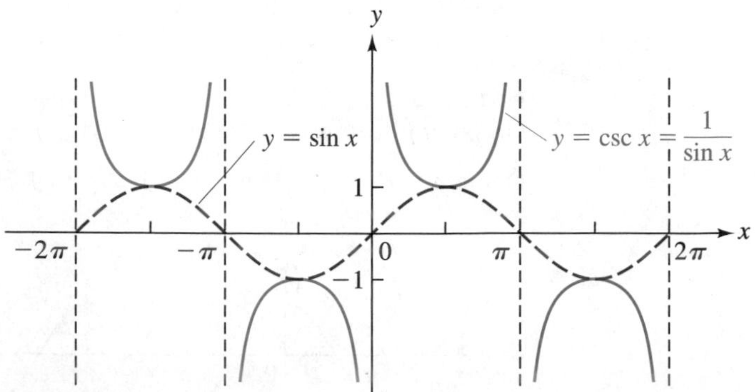

FIGURE 5
Graph of $y = \csc x$

Domain: All real numbers x, except $x = k\pi$, k an integer
Range: All real numbers y such that $y \leq -1$ or $y \geq 1$
Period: 2π

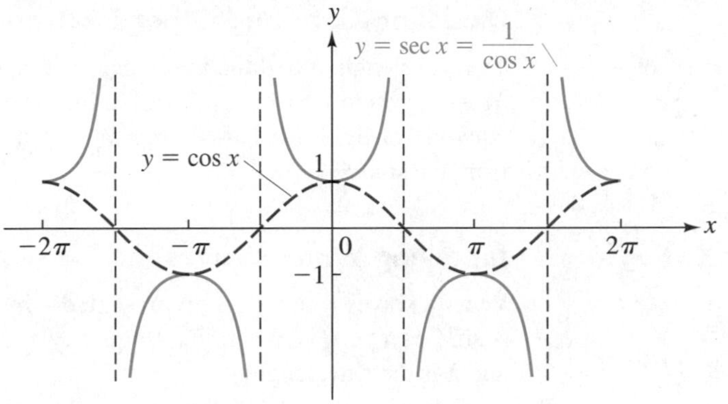

FIGURE 6
Graph of $y = \sec x$

Domain: All real numbers x, except $x = \pi/2 + k\pi$, k an integer
Range: All real numbers y such that $y \le -1$ or $y \ge 1$
Period: 2π

3.2 GRAPHING $y = k + A \sin Bx$ AND $y = k + A \cos Bx$

$$\textbf{Amplitude} = |A| \qquad \textbf{Period} = \frac{2\pi}{B}$$

The basic sine or cosine curve is compressed if $B > 1$ and stretched if $0 < B < 1$. The graph is translated up k units if $k > 0$ and down $|k|$ units if $k < 0$.

If P is the period and f is the **frequency** of a periodic phenomenon, then

$$P = \frac{1}{f}$$ Period is the time for one complete cycle.

$$f = \frac{1}{P}$$ Frequency is the number of cycles per unit of time.

The period for an object bobbing up and down in water is given by $2\pi/\sqrt{1{,}000gA/M}$ where $g = 9.75 \text{ m/sec}^2$, A is the horizontal cross-sectional area in square meters, and M is the mass in kilograms.

3.3 GRAPHING $y = k + A \sin(Bx + C)$ AND $y = k + A \cos(Bx + C)$

To find the **period** and the **phase shift,** solve $Bx + C = 0$ and $Bx + C = 2\pi$:

$$x = -\frac{C}{B} \qquad x = -\frac{C}{B} + \frac{2\pi}{B}$$

Phase shift ($-\frac{C}{B}$); Period ($\frac{2\pi}{B}$)

The graph completes one full cycle as x varies from $-C/B$ to $(-C/B) + (2\pi/B)$. A function of the form $y = A \sin(Bx + C)$ or $y = A \cos(Bx + C)$ is said to be a **simple harmonic.**

☆3.4 ADDITIONAL APPLICATIONS

Modeling Electric Current

Alternating current flows are represented by equations of the form $I = A \sin(Bt + C)$, where I is the current in amperes and t is time in seconds.

Modeling Light and Other Electromagnetic Waves

The wavelength λ and the frequency ν of an electromagnetic wave are related by $\lambda\nu = c$, where c is the speed of light. These waves can be represented by an equation of the form $E = A \sin 2\pi(\nu t - r/\lambda)$, where t is time and r is distance from the source.

Modeling Water Waves

Water waves can be represented by equations of the form $y = A \sin 2\pi(ft - r/\lambda)$, where f is frequency, t is time, r is distance from the source, and λ is the wavelength.

Simple and Damped Harmonic Motion; Resonance

An object hanging on a spring that is set in motion is said to exhibit **simple harmonic motion** if its amplitude remains constant, **damped harmonic motion** if its amplitude decreases to 0 as time increases, and **resonance** if its amplitude increases as time increases.

3.5 GRAPHING COMBINED FORMS

Functions combined using addition or subtraction, if simple enough, can be graphed by hand using **addition of ordinates;** but whether simple or not, such functions can always be graphed on a graphing utility. In the analysis of sound waves, it is also useful to graph the separate components of the wave, called **partial tones.** The partial tone with the smallest frequency is called the **fundamental tone,** and the other partial tones are called **overtones. Fourier series** use sums of trigonometric functions to approximate periodic wave forms such as a **square wave** or a **sawtooth wave.**

3.6 TANGENT, COTANGENT, SECANT, AND COSECANT FUNCTIONS REVISITED

Amplitude is not defined for the tangent, cotangent, secant, or cosecant functions.

To find the period and the phase shift for the graph of $y = A \tan(Bx + C)$ or $y = A \cot(Bx + C)$, solve $Bx + C = 0$ and $Bx + C = \pi$:

$$\underbrace{x = -\frac{C}{B} \qquad x = -\frac{C}{B}}_{\text{Phase shift}} + \underbrace{\frac{\pi}{B}}_{\text{Period}}$$

The graph completes one full cycle as x varies from $-C/B$ to $(-C/B) + (\pi/B)$.

Find the period and the phase shift for the graph of $y = A \sec(Bx + C)$ or $y = A \csc(Bx + C)$ by solving $Bx + C = 0$ and $Bx + C = 2\pi$. It is helpful to first graph

$$y = \frac{1}{A}\cos(Bx + C) \qquad \text{or} \qquad y = \frac{1}{A}\sin(Bx + C)$$

with a dashed curve and then take reciprocals.

CHAPTER 3 REVIEW EXERCISE

Work through all the problems in this chapter review and check the answers. Answers to all review problems appear in the back of the book; following each answer is an italic number that indicates the section in which that type of problem is discussed. Where weaknesses show up, review the appropriate sections in the text. Review problems flagged with a star (☆) are from optional sections.

A *Sketch a graph of each function for* $-2\pi \le x \le 2\pi$.

1. $y = \sin x$ **2.** $y = \cos x$ **3.** $y = \tan x$

4. $y = \cot x$ **5.** $y = \sec x$ **6.** $y = \csc x$

In Problems 7–9, sketch a graph of each function for the indicated interval.

7. $y = 3 \cos \frac{x}{2}, \quad -4\pi \le x \le 4\pi$

8. $y = \frac{1}{2} \sin 2x, \quad -\pi \le x \le \pi$

9. $y = 4 + \cos x, \quad 0 \le x \le 2\pi$

10. Describe any properties that all six trigonometric functions share.

B **11.** Explain how increasing or decreasing the size of B, $B > 0$, in $y = A \cos Bx$ affects the period.

12. Explain how changing the value of C in $y = A \sin(Bx + C)$ affects the original graph, assuming A and $B > 0$ are not changed.

13. Match one of the basic trigonometric functions with each description:
(A) Not defined at $x = n\pi$, n an integer; period 2π
(B) Not defined at $x = n\pi$, n an integer; period π
(C) Amplitude 1; graph passes through (0, 0)

In Problems 14–21, sketch a graph of each function for the indicated interval.

14. $y = -\frac{1}{3} \cos 2\pi x, \quad -2 \le x \le 2$

15. $y = -1 + \frac{1}{2} \sin 2x, \quad -\pi \le x \le \pi$

16. $y = 4 - 2 \sin(\pi x - \pi), \quad 0 \le x \le 2$

17. $y = \tan 2x, \quad -\pi \le x \le \pi$

18. $y = \cot \pi x, \quad -2 < x < 2$

19. $y = 3 \csc \pi x, \quad -1 < x < 2$

20. $y = 2 \sec \frac{x}{2}, \quad -\pi < x < 3\pi$

21. $y = \tan\left(x + \frac{\pi}{2}\right), \quad -\pi < x < \pi$

Check the graphs in Problems 14–21 on a graphing utility.

22. What are the period and amplitude of the function in Problem 14?

23. What are the periods of the functions in Problems 18 and 19?

24. Find the amplitude, period, and phase shift for

$$y = -3 \cos(\pi x + \pi)$$

25. Find the period and phase shift for

$$y = -2 \tan\left(\frac{\pi}{2}x + \frac{\pi}{2}\right)$$

26. Explain how the graph of y_2 is related to the graph of y_1 in each of the following graphing calculator displays:

(A)

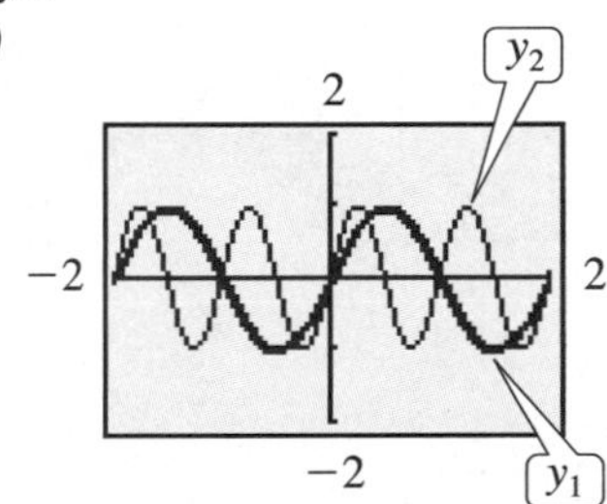

(B)

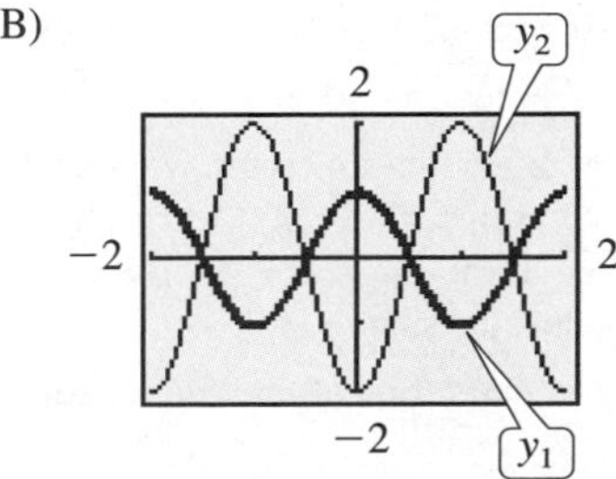

(C) Write the equations of the graphs of y_1 and y_2 in part (A) in the form $A \sin Bx$.

(D) Write the equations of the graphs of y_1 and y_2 in part (B) in the form $A \cos Bx$.

27. Find an equation of the form $y = k + A \sin Bx$ that produces the graph shown in the following graphing calculator display:

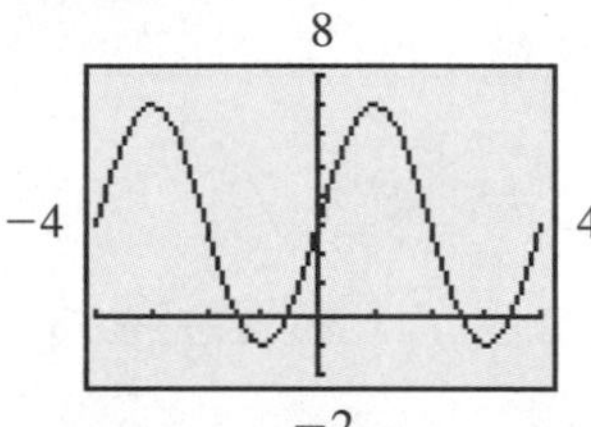

28. Find an equation of the form $y = A \sin Bt$ or $y = A \cos Bt$ for an oscillating system, if the displacement is 0 when $t = 0$, the amplitude is 65, and the period is 0.01.

29. The graphing utility display in the figure illustrates the graph of one of the following:

(1) $y_1 = y_2 + y_3$ (2) $y_2 = y_1 + y_3$
(3) $y_3 = y_1 + y_2$

Discuss which one and why.

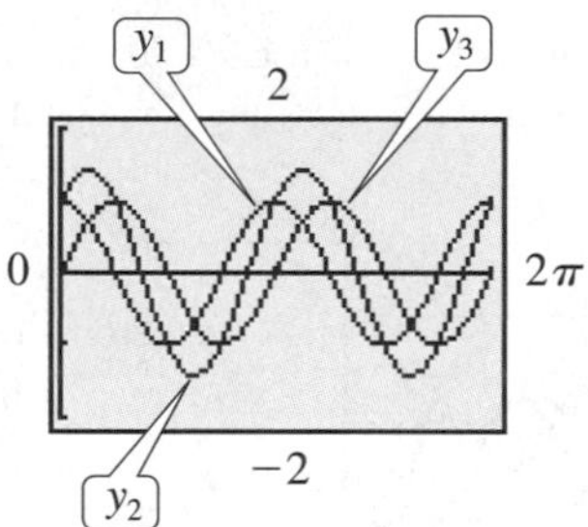

Figure for 29

Problems 30–34 require the use of a graphing utility.

30. Graph $y = x + \sin \pi x, \quad 0 \le x \le 2$.

31. Graph $y = 2 \sin x + \cos 2x, \quad 0 \le x \le 4\pi$.

32. Graph $y = 1/(1 + \tan^2 x)$ in a viewing window that displays at least two full periods of the graph. Find an equation of the form $y = k + A \sin Bx$ or $y = k + A \cos Bx$ that has the same graph.

33. The sum of the first five terms of the Fourier series for a wave form are

$$y = -\frac{\pi}{4} + \frac{4}{\pi}\cos\frac{x}{2} + \frac{2}{\pi}\cos x + \frac{4}{9\pi}\cos\frac{3x}{2} + \frac{4}{25\pi}\cos\frac{5x}{2}$$

Use a graphing utility to graph this equation for $-4\pi \le x \le 4\pi$ and $-3 \le y \le 3$. Then sketch by hand the corresponding wave form. (Assume that the graph of the wave form is comprised entirely of straight line segments.)

34. Graph each of the following equations and find an equation of the form $y = A \tan Bx$, $y = A \cot Bx$, $y = A \sec Bx$, or $y = A \csc Bx$ that has the same graph as the given equation.

(A) $y = \dfrac{2 \sin x}{\sin 2x}$ (B) $y = \dfrac{2 \cos x}{\sin 2x}$

(C) $y = \dfrac{2 \cos^2 x}{\sin 2x}$ (D) $y = \dfrac{2 \sin^2 x}{\sin 2x}$

C **35.** Describe the smallest horizontal shift and/or reflection that transforms the graph of $y = \cot x$ into the graph of $y = \tan x$.

In Problems 36 and 37, sketch a graph of each function for the indicated interval.

36. $y = 2 \tan\left(\pi x + \dfrac{\pi}{2}\right), \quad -1 < x < 1$

37. $y = 2 \sec(2x - \pi), \quad 0 \le x < \dfrac{5\pi}{4}$

38. The table in the figure was produced using a table feature for a particular graphing calculator. Find a function of the form $y = A \sin Bx$ or $y = A \cos Bx$ that will produce the table.

X	Y1
0.00	0.00
.25	2.00
.50	0.00
.75	-2.00
1.00	0.00
1.25	2.00
1.50	0.00

X=0

39. If the following graph is a graph of an equation of the form $y = A \sin(Bx + C)$, $0 < -C/B < 1$, find the equation.

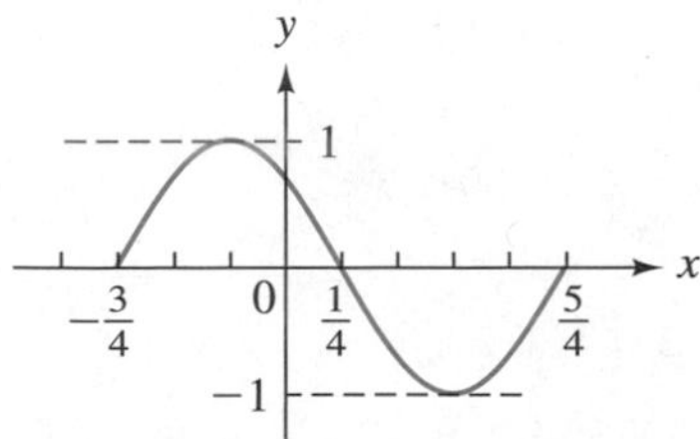

Problems 40–42 require the use of a graphing utility.

40. Graph $y = 1.2 \sin 2x + 1.6 \cos 2x$ and approximate the x intercept closest to the origin (correct to three decimal places). Use this intercept to find an equation of the form $y = A \sin(Bx + C)$ that has the same graph as the given equation.

41. In calculus it is shown that

$$\sin x \approx x - \frac{x^3}{3!} + \frac{x^5}{5!} - \frac{x^7}{7!}$$

(A) Graph $\sin x$ and the first term of the series in the same viewing window.

(B) Graph $\sin x$ and the first two terms of the series in the same viewing window.

(C) Graph $\sin x$ and the first three terms of the series in the same viewing window.

(D) Explain what is happening as more terms of the series are used to approximate $\sin x$.

42. From the graph of $f(x) = |\sin x|$ on a graphing utility, determine the period of f; that is, find the smallest positive number p such that $f(x + p) = f(x)$.

Applications

43. Spring–Mass System If the motion of a weight hung on a spring has an amplitude of 4 cm and a frequency of 8 Hz, and if its position when $t = 0$ sec is 4 cm below its position at rest (above the rest position is positive and below is negative), find an equation of the form $y = A \cos Bt$ that describes the motion at any time t (neglecting any damping forces such as air resistance and friction). Explain why an equation of the form $y = A \sin Bt$ cannot be used to model the motion.

44. Pollution In a large city the amount of sulfur dioxide pollutant released into the atmosphere due to the burning of coal and oil for heating purposes varies seasonally. If measurements over a 2 yr period produced the graph shown, find an equation of the form $P = k + A \cos Bn$, $0 \le n \le 104$, where P is the number of tons of pollutants released into the atmosphere during the nth week after January 31. Can an equation of the form $P = k + A \sin Bn$ model the situation? If yes, find it. If no, explain why.

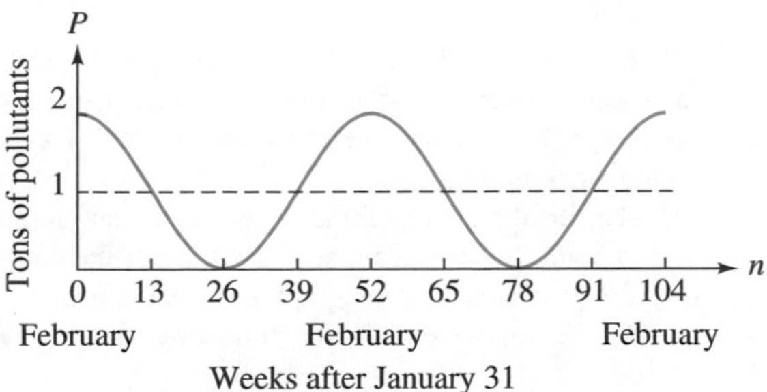

45. Floating Objects A cylindrical buoy with diameter 1.2 m is observed (after being pushed down) to bob up and down with a period of 0.8 sec and an amplitude of 0.6 m.

(A) Find the mass of the buoy to the nearest kilogram.

(B) Write an equation of motion for the buoy in the form $y = D \sin Bt$.

(C) Graph the equation in part (B) for $0 \le t \le 1.6$.

46. Sound Waves The motion of one tip of a tuning fork is given by an equation of the form $y = 0.05 \cos Bt$.

(A) If the frequency of the fork is 280 Hz, what is the period? What is the value of B?

(B) If the period of the motion of the fork is 0.0025 sec, what is the frequency? What is the value of B?

(C) If $B = 700\pi$, what is the period? What is the frequency?

47. Electrical Circuits If the voltage E in an electrical circuit has amplitude 18 and frequency 30 Hz, and $E = 18$ V when $t = 0$ sec, find an equation of the form $y = A \cos Bt$ that gives the voltage at any time t.

48. Water Waves At a particular point in the ocean, the vertical change in the water due to wave action is given by

$$y = 6 \cos \frac{\pi}{10}(t - 5)$$

where y is in meters and t is time in seconds. Find the amplitude, period, and phase shift. Graph the equation for $0 \le t \le 80$.

☆ **49. Water Waves** A water wave at a fixed position has an equation of the form

$$y = 12 \sin \frac{\pi}{3} t$$

where t is time in seconds and y is in feet. How high is the wave from trough to crest? What is its wavelength (in feet)? How fast is it traveling (in feet per second)? Compute answers to the nearest foot.

☆ **50. Electromagnetic Waves** An ultraviolet wave has a frequency of $\nu = 10^{15}$ Hz. What is its period? What is its wavelength (in meters)? (Speed of light $c \approx 3 \times 10^8$ m/sec.)

TABLE 1

t (sec)	0.00	0.25	0.50	0.75	1.00	1.25	1.50	1.75	2.00
θ (deg)	0°	36°	0°	– 36°	0°	36°	0°	– 36°	0°

☆ **51. Spring–Mass Systems** Graph each of the following equations for $0 \le t \le 2$, and identify each as an example of simple harmonic motion, damped harmonic motion, or resonance.

(A) $y = \sin 2\pi t$ (B) $y = (1 + t)\sin 2\pi t$

(C) $y = \dfrac{1}{1 + t}\sin 2\pi t$

52. Modeling a Seasonal Business Cycle The sales for a national chain of ice cream shops are growing steadily but are subject to seasonal variations. The following mathematical model has been developed to project the monthly sales for the next 24 months:

$$S = 5 + t + 5\sin\frac{\pi t}{6}$$

where S represents the sales in millions of dollars and t is time in months.

(A) Graph the equation for $0 \le t \le 24$.

(B) Describe what the graph shows regarding monthly sales over the 2 yr period.

53. Rocket Flight A camera recording the launch of a rocket is located 1,000 m from the launching pad (see the figure).

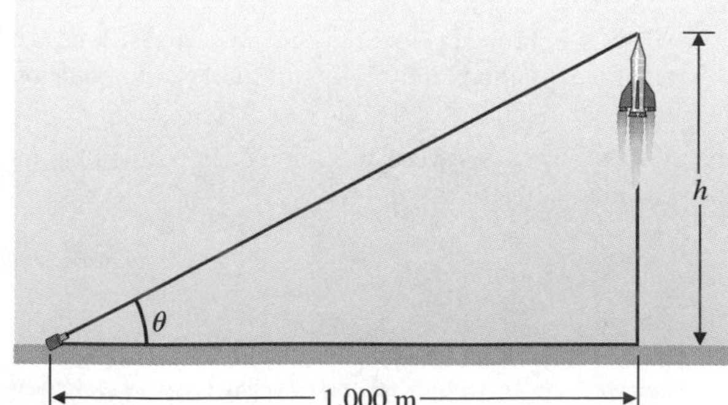

Figure for 53

(A) Write an equation for the altitude, h, of the rocket in terms of the angle of elevation, θ, of the camera.

(B) Graph this equation for $0 \le \theta < \pi/2$.

(C) Describe what happens to h as θ approaches $\pi/2$.

54. Bioengineering For a running person (see the figure), the upper leg rotates back and forth relative to the hip socket through an angle θ. The angle θ is measured relative to the vertical and is negative if the leg is behind the vertical and positive if the leg is in front of the vertical. Measurements of θ were calculated for a person taking two strides per second, where t is the time in seconds. The results for the first 2 sec are shown in Table 1 above.

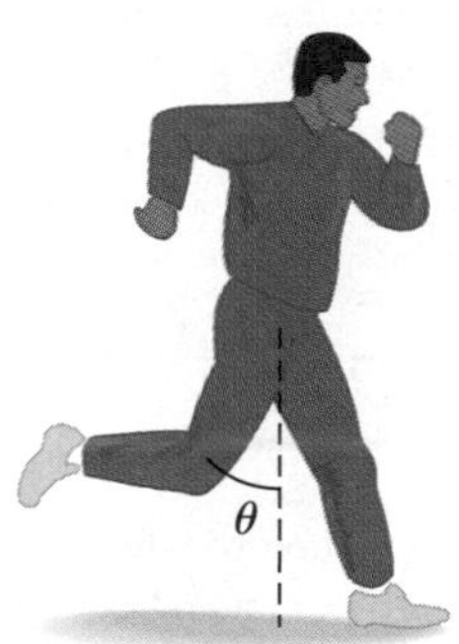

Figure for 54

(A) Verbally describe how you can find the period and amplitude of the data in Table 1, and find each.

(B) Which of the equations, $\theta = A\cos Bt$ or $\theta = A\sin Bt$, is the more suitable model for the data in Table 1? Why? Use the model you chose to find the equation.

(C) Plot the points in Table 1 and the graph of the equation found in part (B) in the same coordinate system.

* **55. Modeling Sunset Times** Sunset times for the fifth of each month over a 1 yr period were taken from a tide booklet for San Francisco Bay to form Table 2. Daylight savings time was ignored.

(A) Convert the data in Table 2 from hours and minutes to hours (to two decimal places). Enter the data for a 2 yr period in your graphing calculator and produce a scatter plot in the following viewing window: $1 \le x \le 24$, $16 \le y \le 20$.

TABLE 2

x (months)	1	2	3	4	5	6	7	8	9	10	11	12
y (sunset time, P.M.)	17:05	17:38	18:07	18:36	19:04	19:29	19:35	19:15	18:34	17:47	17:07	16:51

(B) A function of the form $y = k + A \sin(Bx + C)$ can be used to model the data. Use the converted data from Table 2 to determine k and A (to two decimal places), and B (exact value). Use the graph from part (A) to visually estimate C (to one decimal place).

(C) Plot the data from part (A) and the equation from part (B) in the same viewing window. If necessary, adjust your value of C to produce a better fit.

CUMULATIVE REVIEW EXERCISE CHAPTERS 1–3

Work through all the problems in this cumulative review and check the answers. Answers to all review problems are in the back of the book; following each answer is an italic number that indicates the section in which that type of problem is discussed. Where weaknesses show up, review the appropriate sections in the text. Review problems flagged with a star (☆) are from optional sections.

1. An arc of $\frac{1}{4}$ the circumference of a circle subtends a central angle of how many degrees? Of how many radians?
2. Change 21°47′ to decimal degrees to two decimal places.
3. Find the degree measure of 1.67 rad.
4. Find the radian measure of $-715.3°$.
5. The hypotenuse of a right triangle is 34 in. and one of the acute angles is 25°. Find the other acute angle and the other two sides.
6. Find the value of $\sin\theta$ and $\tan\theta$ if the terminal side of θ contains $P(8, -15)$.
7. Evaluate to four significant digits using a calculator:
   (A) $\sin 23°12'$ (B) $\sec 145.6°$ (C) $\cot 0.88$
8. Sketch the reference triangle and find the reference angle α for:
   (A) $\theta = \dfrac{11\pi}{6}$ (B) $\theta = -225°$
9. Sketch a graph of each function for $-2\pi \le x \le 2\pi$.
   (A) $y = \sin x$ (B) $y = \tan x$ (C) $y = \sec x$
10. Explain what is meant by an angle of radian measure 1.5.
11. Is it possible to find a real number x such that $\cos x$ is negative and $\csc x$ is positive? Explain.
12. Is it possible to construct a triangle with more than one obtuse angle? Explain.

B

13. Find θ to the nearest 10′ if $\tan\theta = 0.9465$ and $-90° \le \theta \le 90°$.
14. If the second hand of a clock is 5.00 cm long, how far does the tip of the hand travel in 40 sec?
15. ☆ How fast is the tip of the second hand in Problem 14 moving (in centimeters per minute)?
16. Find x in the following figure:

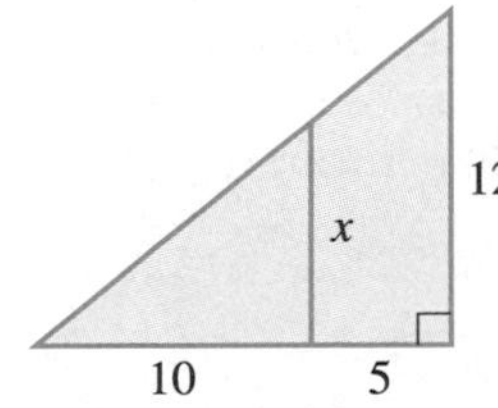

17. Convert 48° to exact radian measure in terms of π.
18. Which of the following graphing calculator displays is the result of the calculator being set in radian mode and which is the result of the calculator being set in degree mode?

(A)

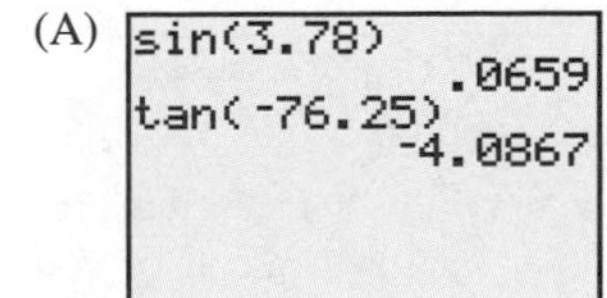

(B)

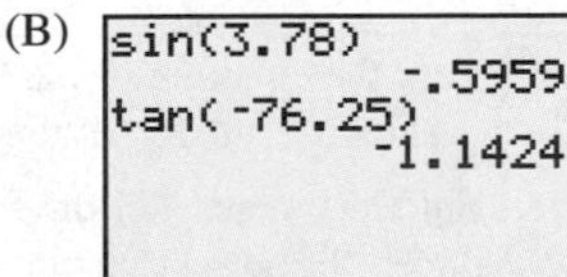

19. If the degree measure of an angle is doubled, is the radian measure of the angle doubled? Explain.

20. From the following graphing calculator display, explain how you would find cot x without finding x. Then find cot x to four decimal places.

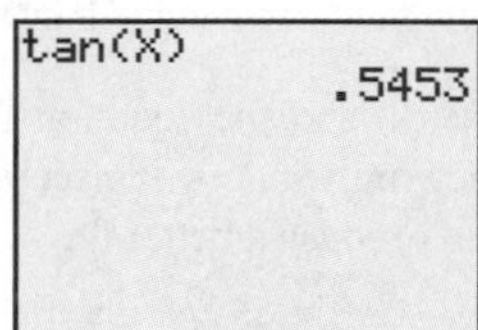

21. Find the exact value of each of the following without using a calculator.

(A) $\sin \frac{5\pi}{4}$ (B) $\cos \frac{7\pi}{6}$

(C) $\tan \frac{-5\pi}{3}$ (D) $\csc 3\pi$

22. Find the exact value of each of the other five trigonometric functions if $\cos \theta = -\frac{2}{3}$ and $\tan \theta < 0$.

23. In the following graphing calculator display, indicate which value(s) are not exact and find the exact form:

```
cos(0°)
          1.0000
sin(-π/3)
          -.8660
tan(135°)
         -1.0000
```

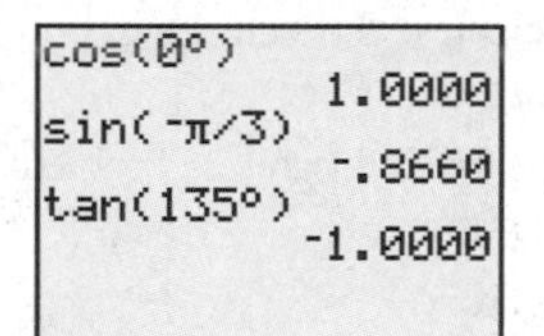

In Problems 24–29, sketch a graph of each function for the indicated interval. State the period and, if applicable, the amplitude and phase shift for each function.

24. $y = 1 - \frac{1}{2}\cos 2x, \quad -2\pi \le x \le 2\pi$

25. $y = 2\sin\left(x - \frac{\pi}{4}\right), \quad -\pi \le x \le 3\pi$

26. $y = 5\tan 4x, \quad 0 \le x \le \pi$

27. $y = \csc \frac{x}{2}, \quad -4\pi < x < 4\pi$

28. $y = 2\sec \pi x, \quad -2 \le x \le 2$

29. $y = \cot\left(\pi x + \frac{\pi}{2}\right), \quad -1 \le x \le 3$

30. Find the equation of the form $y = A \sin Bx$ whose graph is shown at the top of the next column.

31. Find an equation of the form $y = k + A\cos Bx$ that produces the graph shown in the graphing calculator display in the next column.

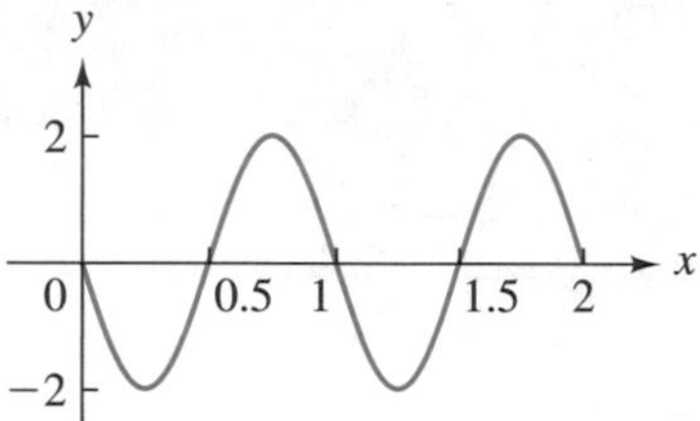

Figure for 30

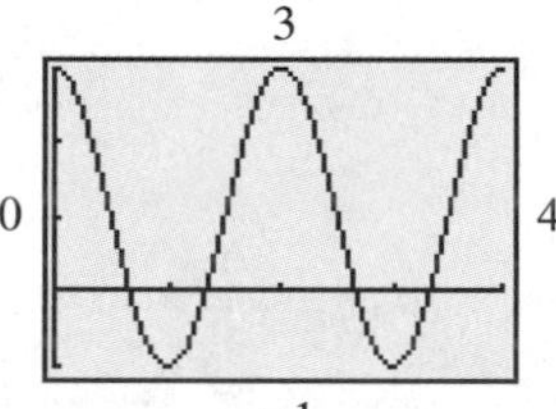

Figure for 31

32. Simplify: $(\tan x)(\sin x) + \cos x$

33. Find the exact value of all angles between 0° and 360° for which $\sin \theta = -\frac{1}{2}$.

34. If the sides of a right triangle are 23.5 in. and 37.3 in., find the hypotenuse and find the acute angles to the nearest 0.1°.

35. Graph $y = \sin x + \sin 2x$ for $0 \le x \le 2\pi$.

36. Graph $y = (\tan^2 x)/(1 + \tan^2 x)$ in a viewing window that displays at least two full periods of the graph. Find an equation of the form $y = k + A\sin Bx$ or $y = k + A\cos Bx$ that has the same graph as the given equation.

37. The sum of the first four terms of the Fourier series for a wave form are

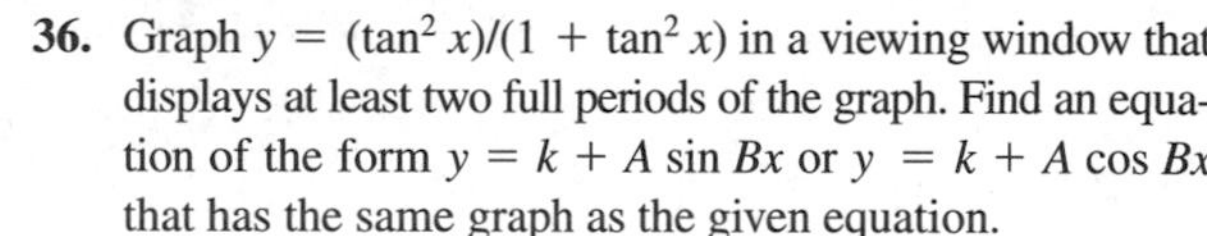

$$y = \frac{\pi}{2} - \frac{4}{\pi}\cos x - \frac{4}{9\pi}\cos 3x - \frac{4}{25\pi}\cos 5x$$

Use a graphing utility to graph this equation for $-2\pi \le x \le 2\pi$ and $-2 \le y \le 4$, and sketch by hand the corresponding wave form. (Assume that the graph of the wave form is comprised entirely of straight line segments.)

C

38. If θ is a first quadrant angle and $\tan \theta = a$, express the other five trigonometric functions of θ in terms of a.

39. A circle with its center at the origin in a rectangular coordinate system passes through the point (8, 15). What is the length of the arc on the circle in the first quadrant between the positive horizontal axis and (8, 15)? Compute the answer to two decimal places.

40. A point moves clockwise around a unit circle, starting at (1, 0), for a distance of 53.077 units. Explain how you would find the coordinates of the point P at its final position and how you would determine which quadrant P is in. Find the coordinates to four decimal places.

41. The following graphing calculator display shows the coordinates of a point on a unit circle. Let s be the length of the least positive arc from (1, 0) to the point. Find s to four decimal places.

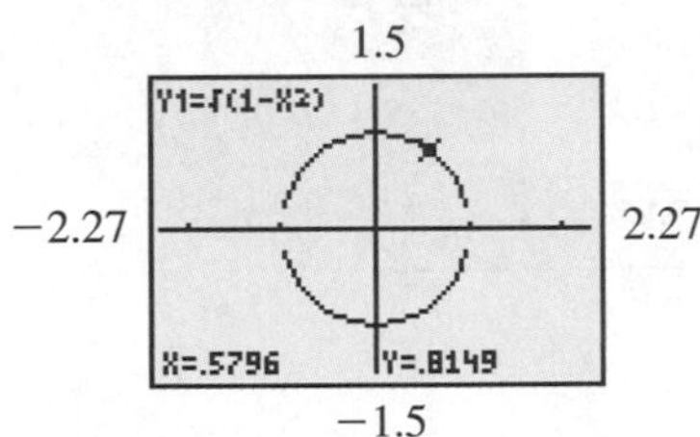

42. Find the equation of the form $y = A\sin(Bx + C)$, $0 < -C/B < 1$, whose graph is

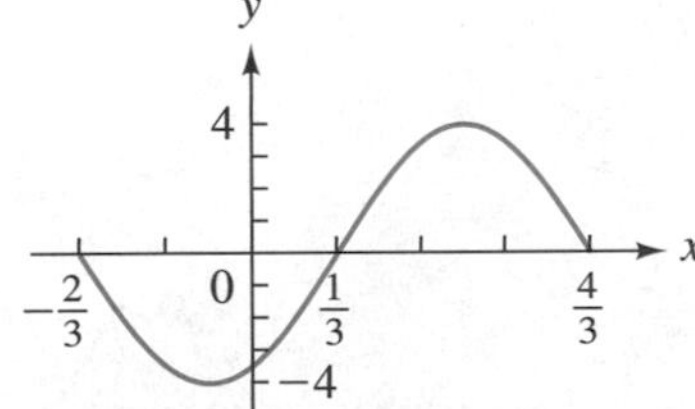

43. Graph $y = 2.4\sin(x/2) - 1.8\cos(x/2)$ and approximate the x intercept closest to the origin, correct to three decimal places. Use this intercept to find an equation of the form $y = A\sin(Bx + C)$ that has the same graph as the given equation.

44. Graph each of the following equations and find an equation of the form $y = A\tan Bx$, $y = A\cot Bx$, $y = A\sec Bx$, or $y = A\csc Bx$ that has the same graph as the given equation.

(A) $y = \dfrac{\sin 2x}{1 + \cos 2x}$ (B) $y = \dfrac{2\cos x}{1 + \cos 2x}$

(C) $y = \dfrac{2\sin x}{1 - \cos 2x}$ (D) $y = \dfrac{\sin 2x}{1 - \cos 2x}$

Applications

***45.** **Geography/Navigation** Find the distance (to the nearest mile) between Gary, IN, with latitude 41°36′N, and Pensacola, FL, with latitude 30°25′N. (Both cities have approximately the same longitude.) Use $r \approx 3{,}960$ mi for the radius of the earth.

46. **Volume of a Cone** A paper drinking cup has the shape of a right circular cone with altitude 9 cm and radius 4 cm (see the figure). Find the volume of the water in the cup (to the nearest cubic centimeter) when the water is 6 cm deep. [*Recall:* $V = \frac{1}{3}\pi r^2 h$.]

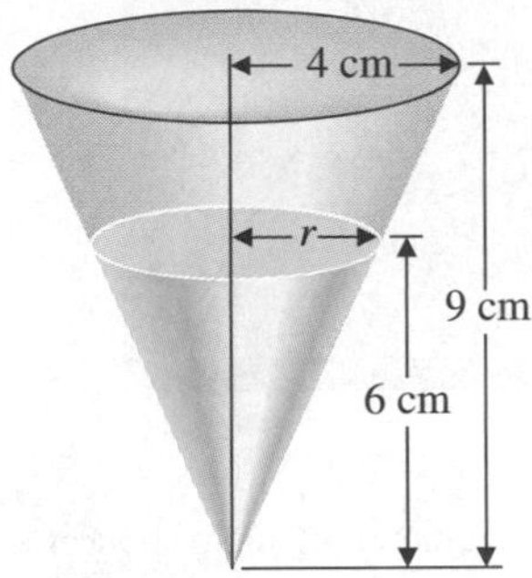

Figure for 46

47. **Construction** The base of a 40 ft arm on a crane is 10 ft above the ground (see the figure). The crane is picking up an object whose horizontal distance from the base of the arm is 30 ft. If the tip of the arm is directly above the object, find the angle of elevation of the arm and the altitude of the tip.

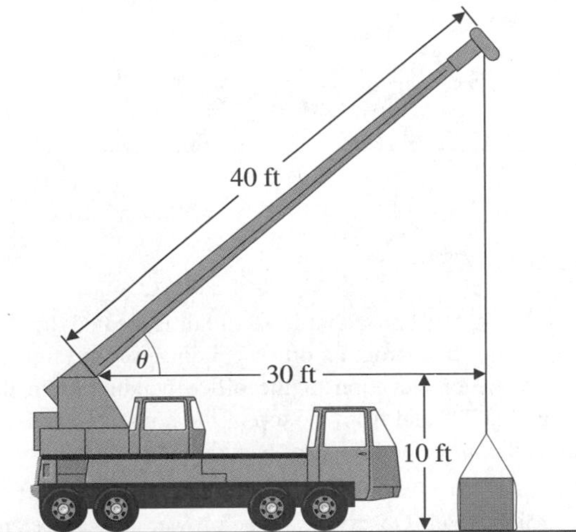

Figure for 47

48. **Billiards** Under ideal conditions, when a ball strikes a cushion on a pocket billiards table at an angle, it bounces off the cushion at the same angle (see the figure). Where should the ball in the center of the table strike the upper cushion so that it bounces off the cushion into the lower right corner pocket? Give the answer in terms of the distance from the center of the side pocket to the point of impact.

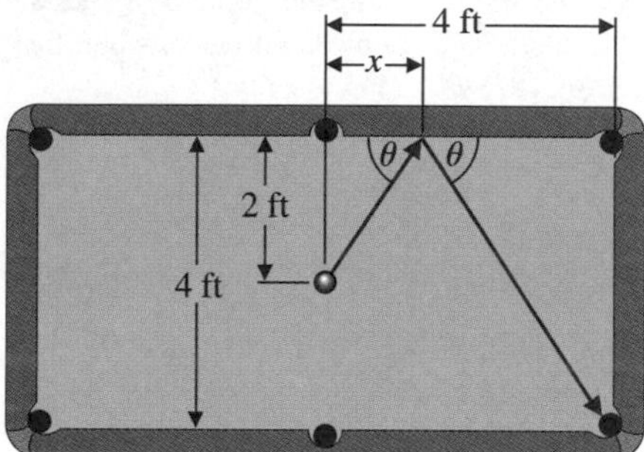

Figure for 48

49. Railroad Grades The amount of inclination of a railroad track is referred to as its *grade* and is usually given as a percentage (see the figure). Most major railroads limit grades to a maximum of 3%, although some logging and mining railroads have grades as high as 7%, requiring the use of specially geared locomotives. Find the angle of inclination (in degrees, to the nearest 0.1°) for a 3% grade. Find the grade (to the nearest 1%) if the angle of inclination is 3°.

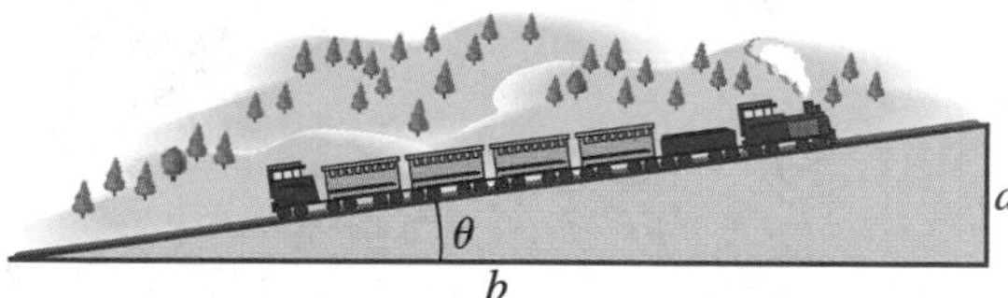

Angle of inclination: θ Grade: $\frac{a}{b}$

Figure for 49

***50.** Surveying A house that is 10 m tall is located directly across the street from an office building (see the figure). The angle of elevation of the office building from the ground is 72° and from the top of the house is 68°. How tall is the office building? How wide is the street?

***51.** Forest Fires Two fire towers are located 5 mi apart on a straight road. Observers at each tower spot a fire and report its location in terms of the angles in the figure.

(A) How far from tower B is the point on the road closest to the fire?

(B) How close is the fire to the road?

☆52. Engineering A large saw in a sawmill is driven by a chain connected to a motor (see the figure). The drive wheel on the motor is rotating at 300 rpm.

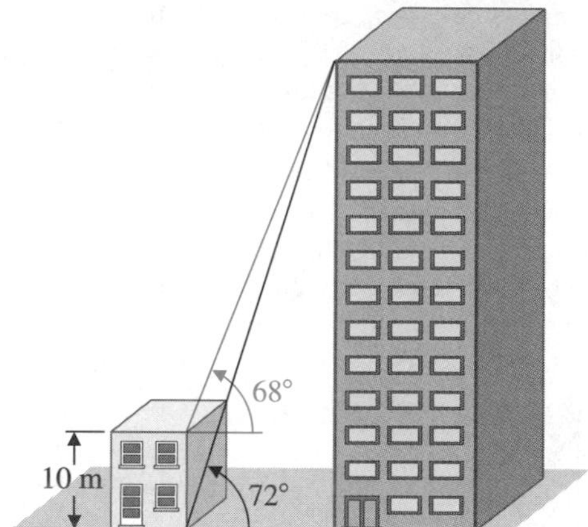

Figure for 50

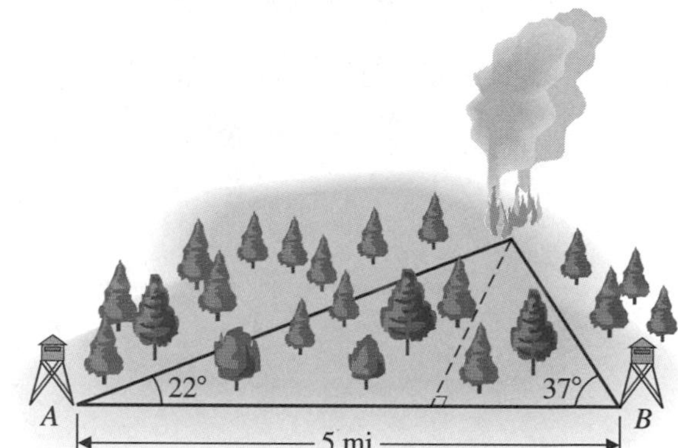

Figure for 51

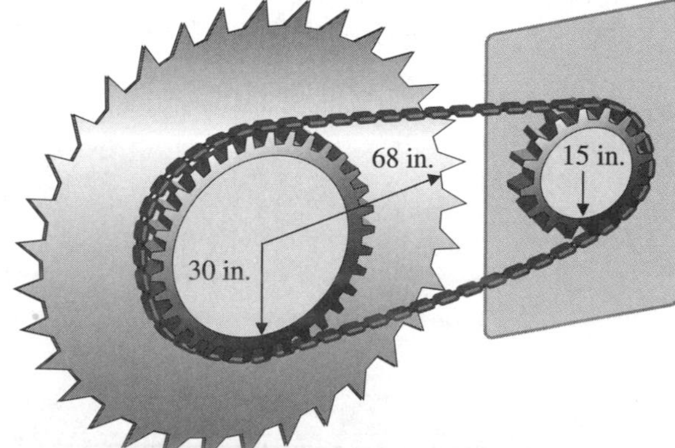

Figure for 52

(A) What is the angular velocity of the saw (in radians per minute)?

(B) What is the linear velocity of a point on the rim of the saw (in inches per minute)?

☆ **53. Light Waves** A light ray passing through air strikes the surface of a pool of water so that the angle of incidence is $\alpha = 38.4°$. Find the angle of refraction β. (Water has a refractive index of 1.33 and air has a refractive index of 1.00.)

☆ **54. Sonic Boom** Find the angle of the sound cone (to the nearest degree) for a plane traveling at 1.5 times the speed of sound.

* **55. Precalculus** In calculus, the ratio $(\sin x)/x$ comes up naturally, and the problem is to determine what the ratio approaches when x approaches 0. Can you guess? (Note that the ratio is not defined when x equals 0.) Here, our approach to the problem is geometric; in Problem 56, we will use a graphic approach.

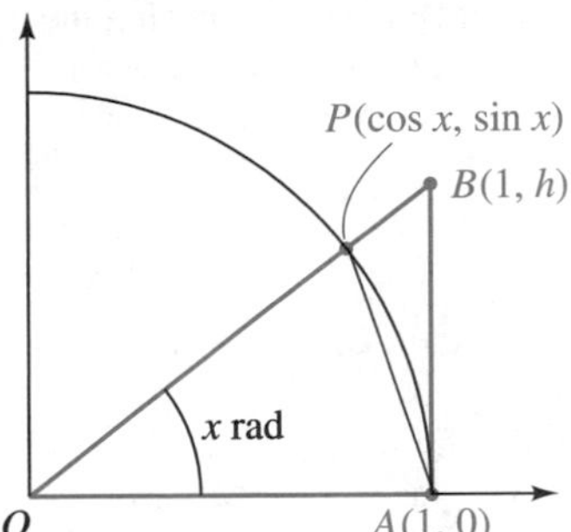

Figure for 55 and 56

(A) Referring to the figure, show the following:

$$A_1 = \text{Area of triangle } OAP = \frac{\sin x}{2}$$

$$A_2 = \text{Area of sector } OAP = \frac{x}{2}$$

$$A_3 = \text{Area of triangle } OAB = \frac{\tan x}{2}$$

(B) Using the fact that $A_1 < A_2 < A_3$, which we can see clearly in the figure, show that

$$\cos x < \frac{\sin x}{x} < 1 \qquad x > 0$$

(C) From the inequality in part (B), explain how you can conclude that for $x > 0$, $(\sin x)/x$ approaches 1 as x approaches 0.

56. Precalculus Refer to Problem 55. We now approach the problem using a graphing utility, and we allow x to be either positive or negative.

(A) Graph $y_1 = \cos x$, $y_2 = (\sin x)/x$, and $y_3 = 1$ in the same viewing window for $-1 \le x \le 1$ and $0 \le y \le 1.2$.

(B) From the graph, write an inequality of the form $a < b < c$ using y_1, y_2, and y_3.

(C) Use TRACE to investigate the values of y_1, y_2, and y_3 for x close to 0. Explain how you can conclude that $(\sin x)/x$ approaches 1 as x approaches 0 from either side of 0. (Remember, the ratio is not defined when $x = 0$.)

57. Spring–Mass System If the motion of a weight hung on a spring has an amplitude of 3.6 cm and a frequency of 6 Hz, and if its position when $t = 0$ sec is 3.6 cm above its rest position (above rest position is positive and below is negative), find an equation of the form $y = A \cos Bt$ that describes the motion at any time t (neglecting any damping forces such as air resistance and friction). Can an equation of the form $y = A \sin Bt$ be used to model the motion? Explain why or why not.

58. Electric Circuits The current (in amperes) in an electrical circuit is given by $I = 12 \sin(60\pi t - \pi)$, where t is time in seconds.

(A) State the amplitude, period, frequency, and phase shift.

(B) Graph the equation for $0 \le t \le 0.1$.

☆ **59. Electromagnetic Waves** An infrared wave has a wavelength of 6×10^{-5} m. What is its period? (The speed of light is $c \approx 3 \times 10^8$ m/sec.)

60. Precalculus: Surveying A university has just constructed a new building at the intersection of two perpendicular streets (see the figure). Now they want to connect the existing streets with a walkway that passes behind and just touches the building.

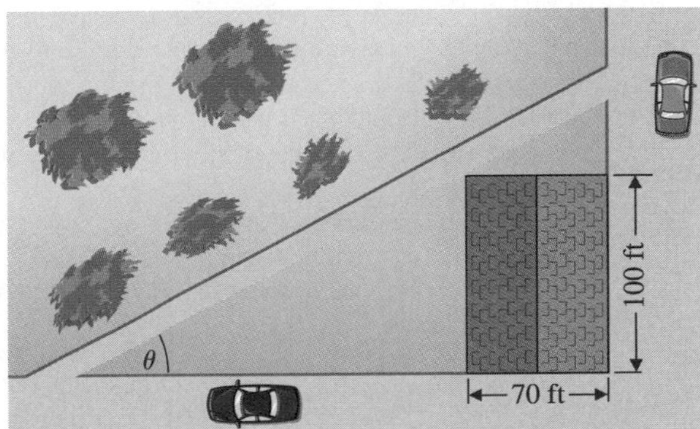

Figure for 60

(A) Express the length, L, of the walkway in terms of θ.

(B) From the figure, describe what you think happens to L as θ varies between 0° and 90°.

(C) Complete Table 1, giving values of L to one decimal

TABLE 1

θ (rad)	0.50	0.60	0.70	0.80	0.90	1.00	1.10
L (ft)	288.3						

place, using a calculator. (If you have a table-generating calculator, use it.) From the table, select the angle θ that produces the shortest walkway that satisfies the conditions of the problem. (In calculus, special techniques are used to find this angle.)

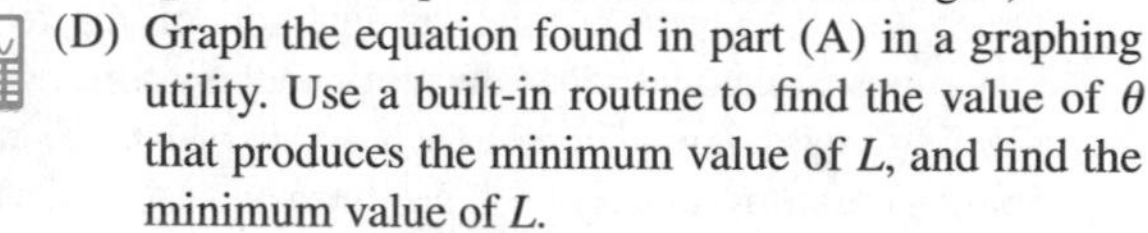

(D) Graph the equation found in part (A) in a graphing utility. Use a built-in routine to find the value of θ that produces the minimum value of L, and find the minimum value of L.

61. Emergency Vehicles An emergency vehicle is parked 50 ft from a building (see the figure). A warning light on top of the vehicle is rotating clockwise at exactly 20 rpm.

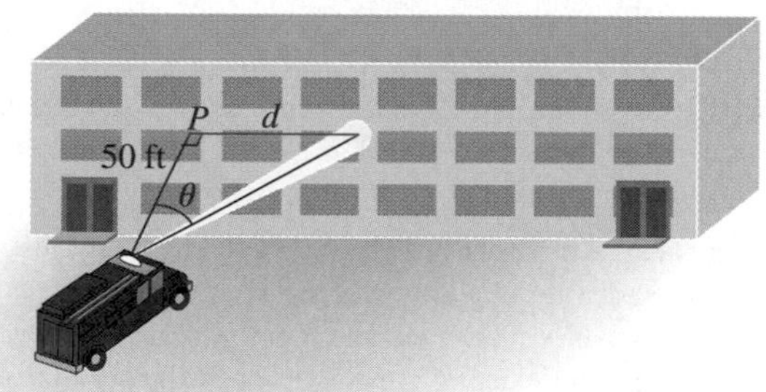

Figure for 61

(A) Write an equation for the distance d in terms of the angle θ.

(B) If $t = 0$ when the light is pointing at P, write an equation for θ in terms of t, where t is time in minutes.

(C) Write an equation for d in terms of t.

(D) Graph the equation from part (C) for $0 < t < \frac{1}{80}$, and describe what happens to d as t approaches $\frac{1}{80}$ th min.

☆ **62. Spring–Mass System** Graph each of the following equations for $0 \le t \le 4$, and identify each as an example of simple harmonic motion, damped harmonic motion, or resonance.

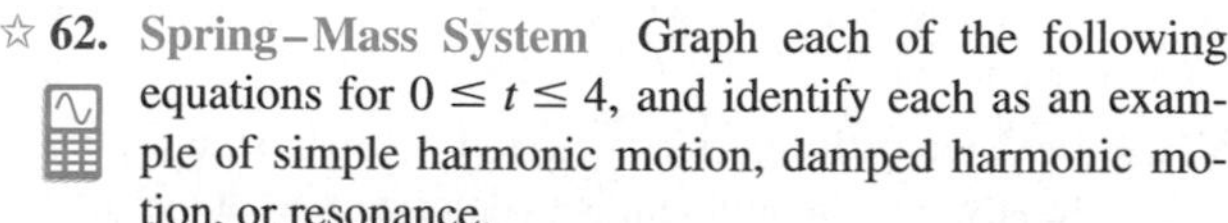

(A) $y = t^{-0.4} \cos \pi t$ (B) $y = 2^{-0.5} \cos \pi t$

(C) $y = t^{0.8} \cos \pi t$

63. Modeling a Seasonal Business Cycle The sales for a national soft drink company are growing steadily but are subject to seasonal variations. The following model has been developed to project the monthly sales for the next 36 months:

$$S = 30 + 0.5n + 8 \sin \frac{\pi n}{6}$$

where S represents the sales (in millions of dollars) for the nth month, $0 \le n \le 36$.

(A) Graph the equation for $0 \le n \le 36$.

(B) Describe what the graph shows regarding monthly sales over the 36 month period.

* **64. Modeling Temperature Variation** The 30 year average monthly temperature (in °F) for each month of the year for Milwaukee, WI, is given in Table 2.

(A) Enter the data in Table 2 for a 2 yr period in your graphing calculator and produce a scatter plot in the viewing window $1 \le x \le 24$, $15 \le y \le 75$.

(B) A function of the form $y = k + A \sin(Bx + C)$ can be used to model the data. Use the data in Table 2 to determine k, A, and B. Use the graph in part (A) to visually estimate C (to one decimal place).

(C) Plot the data from part (A) and the equation from part (B) in the same viewing window. If necessary, adjust the value of C to produce a better fit.

TABLE 2

x (months)	1	2	3	4	5	6	7	8	9	10	11	12
y (temperature, °F)	19	23	32	45	55	65	71	69	62	51	37	25

Source: World Almanac

IDENTITIES 4

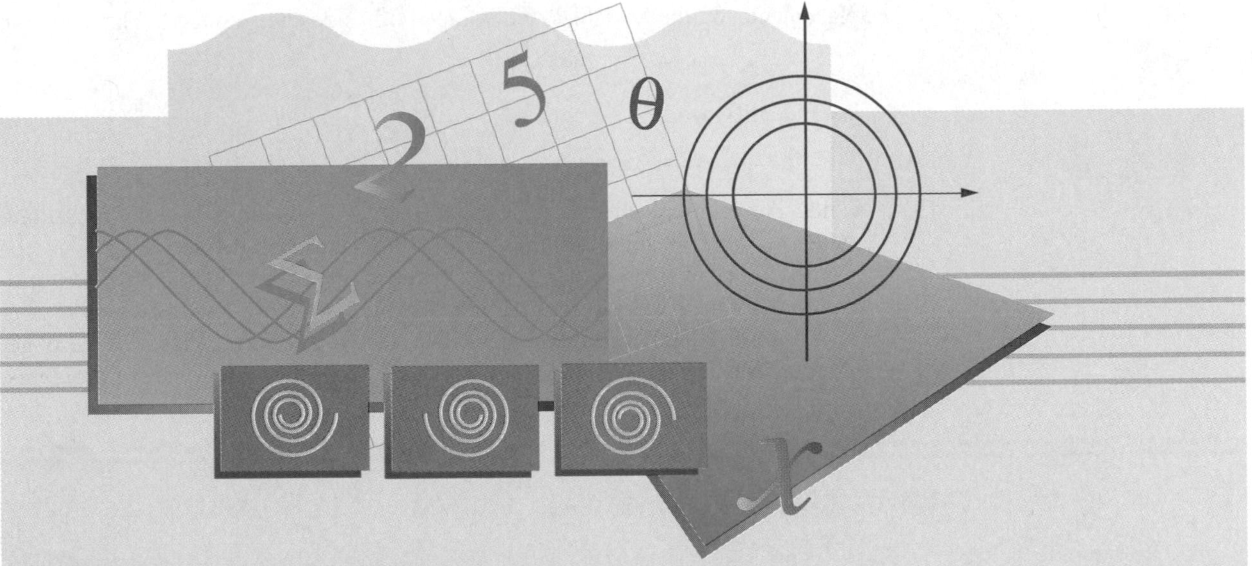

☆ Sections marked with a star may be omitted without loss of continuity.

Trigonometric functions have many uses. In addition to solving real-world problems, they are used in the development of mathematics—analytic geometry, calculus, and so on. Whatever their use, it is often of value to be able to change a trigonometric expression from one form to an equivalent form. This involves the use of identities. An equation in one or more variables is said to be an **identity** if the left side is equal to the right side for all replacements of the variables for which both sides are defined. The equation

$$x^2 - x - 6 = (x - 3)(x + 2)$$

is an identity, while

$$x^2 - x - 6 = 2x$$

is not. The latter is called a **conditional equation,** since it holds only for certain values of x and not for all values for which both sides are defined.

In this chapter we will develop a number of very useful classes of trigonometric identities, and you will get practice in using these identities to convert a variety of trigonometric expressions into equivalent forms. You will also get practice in using the identities directly to solve several other types of problems.

4.1 FUNDAMENTAL IDENTITIES AND THEIR USE

- Fundamental Identities
- Evaluating Trigonometric Functions
- Converting to Equivalent Forms

Fundamental Identities

Our first encounter with trigonometric identities was in Section 2.6, where we established several fundamental forms. We restate and name these identities in the box for convenient reference. These fundamental identities will be used very frequently in the work that follows.

FUNDAMENTAL TRIGONOMETRIC IDENTITIES

For x any real number or angle in degree or radian measure for which both sides are defined:

Reciprocal identities

$$\csc x = \frac{1}{\sin x} \qquad \sec x = \frac{1}{\cos x} \qquad \cot x = \frac{1}{\tan x}$$

Quotient identities

$$\tan x = \frac{\sin x}{\cos x} \qquad \cot x = \frac{\cos x}{\sin x}$$

Identities for negatives

$$\sin(-x) = -\sin x \qquad \cos(-x) = \cos x \qquad \tan(-x) = -\tan x$$

Pythagorean identities

$$\sin^2 x + \cos^2 x = 1 \qquad \tan^2 x + 1 = \sec^2 x \qquad 1 + \cot^2 x = \csc^2 x$$

The second and third Pythagorean identities were established in Probems 77 and 78 in Exercise 2.6. An easy way to remember them is suggested in Explore–Discuss 1.

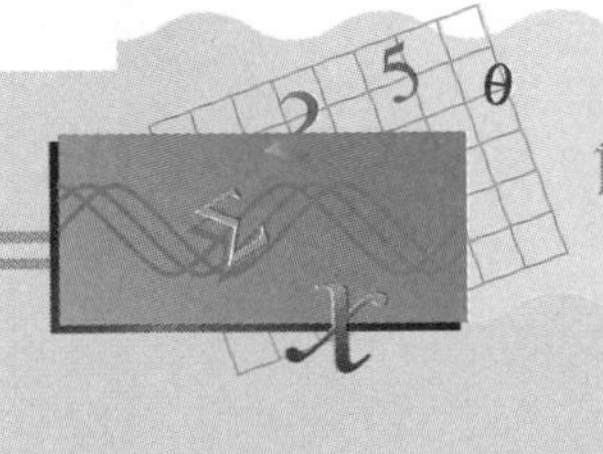

EXPLORE/DISCUSS 1

Discuss an easy way to remember the second and third Pythagorean identities based on the first. [*Hint:* Divide through the first Pythagorean identity by appropriate expressions.]

Evaulating Trigonometric Functions

Suppose we know that $\cos x = -\frac{4}{5}$ and $\tan x = \frac{3}{4}$. How can we find the exact values of the remaining trigonometric functions of x without finding x and without using reference triangles? We use fundamental identities.

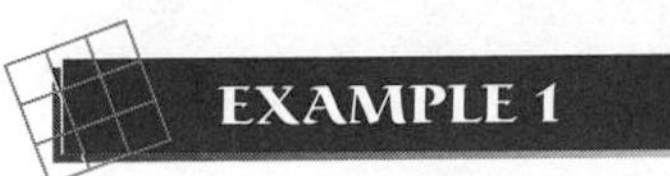

EXAMPLE 1 Using Fundamental Identities

If $\cos x = -\frac{4}{5}$ and $\tan x = \frac{3}{4}$, use the fundamental identities to find the exact values of the remaining four trigonometric functions at x.

Solution *Find sec x:* $\sec x = \dfrac{1}{\cos x} = \dfrac{1}{-\frac{4}{5}} = -\dfrac{5}{4}$

Find cot x: $\cot x = \dfrac{1}{\tan x} = \dfrac{1}{\frac{3}{4}} = \dfrac{4}{3}$

Find sin x: We can start with either a Pythagorean identity or a quotient identity. We choose the quotient identity $\tan x = (\sin x)/(\cos x)$ changed to the form

$$\sin x = (\cos x)(\tan x) = \left(-\frac{4}{5}\right)\left(\frac{3}{4}\right) = -\frac{3}{5}$$

Find csc x: $$\csc x = \frac{1}{\sin x} = \frac{1}{-\frac{3}{5}} = -\frac{5}{3}$$ ■

Matched Problem 1 If $\sin x = -\frac{4}{5}$ and $\cot x = -\frac{3}{4}$, use the fundamental identities to find the exact values of the remaining four trigonometric functions at x. ■

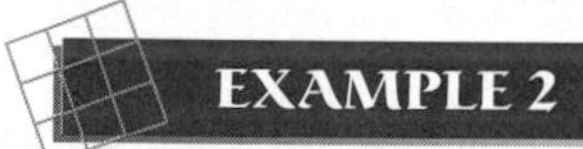

EXAMPLE 2 **Using Fundamental Identities**

Use the fundamental identities to find the exact values of the remaining trigonometric functions of x, given

$$\cos x = \frac{-4}{\sqrt{17}} \quad \text{and} \quad \tan x < 0$$

Solution *Find sin x:* We start with the Pythagorean identity

$$\sin^2 x + \cos^2 x = 1$$

and solve for $\sin x$:

$$\sin x = \pm\sqrt{1 - \cos^2 x}$$

Since both $\cos x$ and $\tan x$ are negative, x is associated with the second quadrant, where $\sin x$ is positive; hence,

$$\begin{aligned}\sin x &= \sqrt{1 - \cos^2 x}\\ &= \sqrt{1 - \left(\frac{-4}{\sqrt{17}}\right)^2}\\ &= \sqrt{\frac{1}{17}} = \frac{1}{\sqrt{17}}\,^*\end{aligned}$$

Find sec x: $$\sec x = \frac{1}{\cos x} = \frac{1}{-4/\sqrt{17}} = -\frac{\sqrt{17}}{4}$$

Find csc x: $$\csc x = \frac{1}{\sin x} = \frac{1}{1/\sqrt{17}} = \sqrt{17}$$

---

* An equivalent answer is $1/\sqrt{17} = \sqrt{17}/(\sqrt{17}\sqrt{17}) = \sqrt{17}/17$, a form in which we have rationalized (eliminated radicals in) the denominator. Whether we rationalize the denominator or not depends entirely on what we want to do with the answer—sometimes an unrationalized form is more useful than a rationalized form. For the remainder of this book, you should leave answers to matched problems and exercises unrationalized, unless directed otherwise.

Find tan x: $\tan x = \dfrac{\sin x}{\cos x} = \dfrac{1/\sqrt{17}}{-4/\sqrt{17}} = -\dfrac{1}{4}$

Find cot x: $\cot x = \dfrac{1}{\tan x} = \dfrac{1}{-\frac{1}{4}} = -4$ ■

Matched Problem 2 Use the fundamental identities to find the exact values of the remaining trigonometric functions of x, given:

$$\tan x = -\frac{\sqrt{21}}{2} \quad \text{and} \quad \cos x > 0$$ ■

■ Converting to Equivalent Forms

One of the most important and frequent uses of the fundamental identities is the conversion of trigonometric forms into equivalent simpler or more useful forms. A couple of examples will illustrate the process.

EXAMPLE 3 **Simplifying Trigonometric Expressions**

Use fundamental identities and appropriate algebraic operations to simplify the following expression:

$$\frac{1}{\cos^2 \alpha} - 1$$

Solution We start by forming a single fraction:

$$\frac{1}{\cos^2 \alpha} - 1 = \frac{1 - \cos^2 \alpha}{\cos^2 \alpha} \quad \text{Algebra}$$

$$= \frac{\sin^2 \alpha}{\cos^2 \alpha} \quad \text{Pythagorean identity}$$

$$= \left(\frac{\sin \alpha}{\cos \alpha}\right)^2 \quad \text{Algebra}$$

$$= \tan^2 \alpha \quad \text{Quotient identity}$$

Key Algebraic Steps:

$$\frac{1}{b^2} - 1 = \frac{1}{b^2} - \frac{b^2}{b^2} = \frac{1 - b^2}{b^2} \quad \text{and} \quad \frac{a^2}{b^2} = \left(\frac{a}{b}\right)^2$$ ■

Matched Problem 3 Use fundamental identities and appropriate algebraic operations to simplify the following expression:

$$\frac{\sin^2 \theta}{\cos^2 \theta} + 1$$ ■

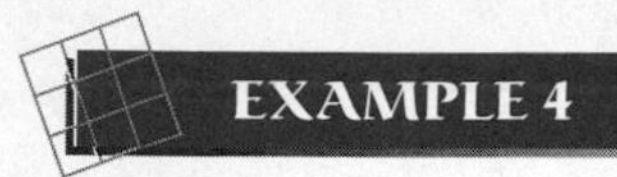

EXAMPLE 4

Converting a Trigonometric Expression to an Equivalent Form

Using fundamental identities, write the following expression in terms of sines and cosines, and then simplify:

$$\frac{\tan x - \cot x}{\tan x + \cot x}$$

Write the final answer in terms of the cosine function.

Solution

$$\frac{\tan x - \cot x}{\tan x + \cot x} = \frac{\dfrac{\sin x}{\cos x} - \dfrac{\cos x}{\sin x}}{\dfrac{\sin x}{\cos x} + \dfrac{\cos x}{\sin x}}$$ Change to sines and cosines.

$$= \frac{(\sin x \cos x)\left(\dfrac{\sin x}{\cos x} - \dfrac{\cos x}{\sin x}\right)}{(\sin x \cos x)\left(\dfrac{\sin x}{\cos x} + \dfrac{\cos x}{\sin x}\right)}$$ Multiply numerator and denominator by the least common denominator of all internal fractions.

$$= \frac{\sin^2 x - \cos^2 x}{\sin^2 x + \cos^2 x}$$ Algebra

$$= \frac{1 - \cos^2 x - \cos^2 x}{1}$$ Pythagorean identities

$$= 1 - 2\cos^2 x$$ Algebra

Key Algebraic Steps:

$$\frac{\dfrac{a}{b} - \dfrac{b}{a}}{\dfrac{a}{b} + \dfrac{b}{a}} = \frac{ab\left(\dfrac{a}{b} - \dfrac{b}{a}\right)}{ab\left(\dfrac{a}{b} + \dfrac{b}{a}\right)} = \frac{a^2 - b^2}{a^2 + b^2}$$

■

Matched Problem 4 Using fundamental identities, write the following expression in terms of sines and cosines, and then simplify:

$$1 + \frac{\tan z}{\cot z}$$

■

Answers to Matched Problems

1. $\csc x = -\frac{5}{4}$, $\tan x = -\frac{4}{3}$, $\cos x = \frac{3}{5}$, $\sec x = \frac{5}{3}$
2. $\cot x = -2/\sqrt{21}$, $\sec x = \frac{5}{2}$, $\cos x = \frac{2}{5}$, $\sin x = -\sqrt{21}/5$, $\csc x = -5/\sqrt{21}$
3. $\sec^2 \theta$
4. $\sec^2 z$

EXERCISE 4.1

A

1. List the reciprocal identities and identities for negatives without looking at the text.
2. List the quotient identities and Pythagorean identities without looking at the text.
3. One of the following equations is an identity and the other is a conditional equation. Identify each, and explain the difference between the two.

   (1) $3(2x - 3) = 3(3 - 2x)$

   (2) $3(2x - 3) = 6x - 9$

4. One of the following equations is an identity and the other is a conditional equation. Identify each, and explain the difference between the two.

   (1) $\sin x + \cos x = 1$

   (2) $\sin^2 x = 1 - \cos^2 x$

In Problems 5–8, use the fundamental identities to find the exact values of the remaining trigonometric functions of x, given the following:

5. $\sin x = -\frac{2}{3}$ and $\cos x = \sqrt{5}/3$
6. $\sin x = 2/\sqrt{5}$ and $\cos x = -1/\sqrt{5}$
7. $\tan x = 2$ and $\sin x = -2/\sqrt{5}$
8. $\cot x = -\frac{2}{3}$ and $\sec x = \sqrt{13}/2$

In Problems 9–20, simplify each expression using the fundamental identities.

9. $\tan u \cot u$
10. $\sec x \cos x$
11. $\tan x \csc x$
12. $\sec \theta \cot \theta$
13. $\dfrac{\sec^2 x - 1}{\tan x}$
14. $\dfrac{\csc^2 v - 1}{\cot v}$
15. $\dfrac{\sin^2 \theta}{\cos \theta} + \cos \theta$
16. $\dfrac{1}{\csc^2 x} + \dfrac{1}{\sec^2 x}$
17. $\dfrac{1}{\sin^2 \beta} - 1$
18. $\dfrac{1 - \sin^2 u}{\cos u}$
19. $\dfrac{(1 - \cos x)^2 + \sin^2 x}{1 - \cos x}$
20. $\dfrac{\cos^2 x + (\sin x + 1)^2}{\sin x + 1}$

B

21. If an equation has an infinite number of solutions, is it an identity? Explain.
22. Does an identity have an infinite number of solutions? Explain.

In Problems 23–28, use the fundamental identities to find the exact values of the remaining trigonometric functions of x, given the following:

23. $\sin x = \frac{1}{4}$ and $\tan x < 0$
24. $\cos x = \frac{2}{3}$ and $\csc x < 0$
25. $\tan x = -2$ and $\sin x < 0$
26. $\cot x = -3$ and $\cos x > 0$
27. $\csc x = \frac{3}{2}$ and $\tan x < 0$
28. $\sec x = -\frac{5}{3}$ and $\cot x > 0$

29. For the following graphing calculator displays, find the value of the final expression without finding x or using a calculator:

(A)

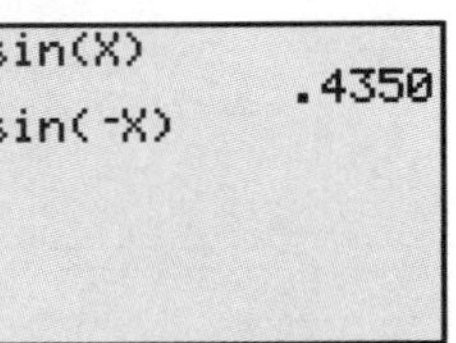

(B)

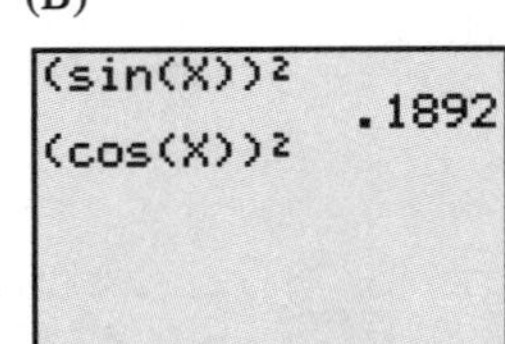

30. For the following graphing calculator displays, find the value of the final expression without finding x or using a calculator:

(A)

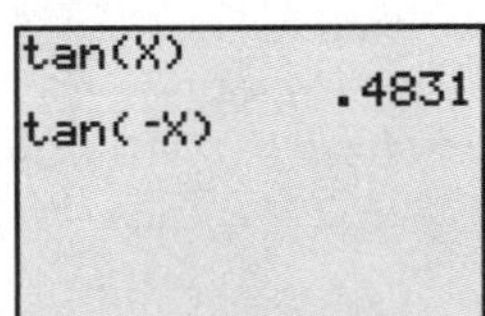

(B)

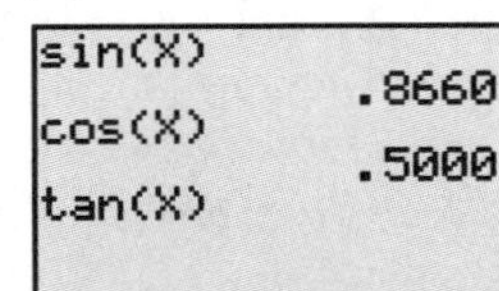

Using fundamental identities, write the expressions in Problems 31–40 in terms of sines and cosines, and then simplify.

31. $\csc(-y)\cos(-y)$
32. $\sin(-\alpha)\sec(-\alpha)$
33. $\cot x \cos x + \sin x$
34. $\cos u + \sin u \tan u$
35. $\dfrac{\cot(-\theta)}{\csc \theta} + \cos \theta$
36. $\sin y - \dfrac{\tan(-y)}{\sec y}$
37. $\dfrac{\cot x}{\tan x} + 1$
38. $\dfrac{1 + \cot^2 y}{\cot^2 y}$
39. $\sec w \csc w - \sec w \sin w$
40. $\csc \theta \sec \theta - \csc \theta \cos \theta$

C **41.** If $\sin x = \frac{2}{5}$, find:
(A) $\sin^2(x/2) + \cos^2(x/2)$ (B) $\csc^2(2x) - \cot^2(2x)$

42. If $\cos x = \frac{3}{7}$, find:
(A) $\sin^2(2x) + \cos^2(2x)$ (B) $\sec^2(x/2) - \tan^2(x/2)$

Each of the following is an identity in certain quadrants. Indicate which quadrants.

43. $\sqrt{1 - \cos^2 x} = \sin x$

44. $\sqrt{1 - \sin^2 x} = \cos x$

45. $\sqrt{1 - \sin^2 x} = -\cos x$

46. $\sqrt{1 - \cos^2 x} = -\sin x$

47. $\sqrt{1 - \sin^2 x} = |\cos x|$

48. $\sqrt{1 - \cos^2 x} = |\sin x|$

49. $\dfrac{\sin x}{\sqrt{1 - \sin^2 x}} = \tan x$

50. $\dfrac{\sin x}{\sqrt{1 - \sin^2 x}} = -\tan x$

Applications

Precalculus: Trigonometric Substitution *In calculus, problems are frequently encountered that involve radicals of the forms $\sqrt{a^2 - u^2}$ and $\sqrt{a^2 + u^2}$. It is very useful to be able to make trigonometric substitutions and use fundamental identities to transform these expressions into nonradical forms. Problems 51–54 involve such transformations. (Recall from algebra that $\sqrt{N^2} = N$ if $N \ge 0$ and $\sqrt{N^2} = -N$ if $N < 0$.)*

51. In the expression $\sqrt{a^2 - u^2}$, $a > 0$, let $u = a \sin x$, $-\pi/2 < x < \pi/2$. After using an appropriate fundamental identity, write the given expression in a final form free of radicals.

52. In the expression $\sqrt{a^2 - u^2}$, $a > 0$, let $u = a \cos x$, $0 < x < \pi$. After using an appropriate fundamental identity, write the given expression in a final form free of radicals.

53. In the expression $\sqrt{a^2 + u^2}$, $a > 0$, let $u = a \tan x$, $0 < x < \pi/2$. After using an appropriate fundamental identity, write the given expression in a final form free of radicals.

54. In the expression $\sqrt{a^2 + u^2}$, $a > 0$, let $u = a \cot x$, $0 < x < \pi/2$. After using an appropriate fundamental identity, write the given expression in a final form free of radicals.

Precalculus: Parametric Equations *Suppose we are given the parametric equations of a curve,*

$$\begin{cases} x = \cos t \\ y = \sin t \end{cases} \qquad 0 \le t \le 2\pi$$

[The parameter t is assigned values and the corresponding points (cos t, sin t) are plotted in a rectangular coordinate system.] These parametric equations can be transformed into a standard rectangular form free of the parameter t by use of the fundamental identities as follows:

$$x^2 + y^2 = \cos^2 t + \sin^2 t = 1$$

Thus,

$$x^2 + y^2 = 1$$

is the nonparametric equation for the curve. The latter is the equation of a circle with radius 1 and center at the origin. Refer to this discussion for Problems 55 and 56.

55. (A) Transform the parametric equations (by suitable use of a fundamental identity) into nonparametric form.

$$\begin{cases} x = 5 \cos t \\ y = 2 \sin t \end{cases} \qquad 0 \le t \le 2\pi$$

Hint: First write the parametric equations in the following form, then square and add:

$$\begin{cases} \dfrac{x}{5} = \cos t \\ \dfrac{y}{2} = \sin t \end{cases} \qquad 0 \le t \le 2\pi$$

(B) Graph the parametric equations from part (A) in a graphing utility. Observe that the graph is an ellipse that is wider than it is high.

56. (A) Transform the parametric equations (by suitable use of a fundamental identity) into nonparametric form.

$$\begin{cases} x = 3 \cos t \\ y = 4 \sin t \end{cases} \qquad 0 \le t \le 2\pi$$

(B) Graph the parametric equations from part (A) in a graphing utility. Observe that the graph is an ellipse that is higher than it is wide.

4.2 VERIFYING TRIGONOMETRIC IDENTITIES

- Verifying Identities
- Testing Identities Using a Graphing Utility

We now use the experience gained in the last section to verify identities. If we start with an equation that is not a known identity, how do we proceed? A graphing utility will be helpful in this case.

Verifying Identities

We will now verify (prove) some given trigonometric identities; this process will be helpful to you if you want to convert a trigonometric expression into a form that may be more useful. Verifying a trigonometric identity is different from solving an equation. When solving an equation you use properties of equality such as adding the same quantity to each side or multiplying both sides by a nonzero quantity. These operations are not valid in the process of verifying identities because, at the start, we do not know that the left and right expressions are equal.

VERIFYING AN IDENTITY

To verify an identity, start with the expression on one side and, through a sequence of valid steps involving the use of known identities or algebraic manipulation, convert that expression into the expression on the other side.

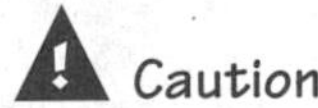

Caution When verifying an identity, *do not* add the same quantity to each side, multiply both sides by the same nonzero quantity, or square (or take the square root of) both sides. □

The following examples illustrate some of the techniques used to establish certain identities. To become proficient in the process, it is important that you work many problems on your own.

EXAMPLE 1 **Identity Verification**

Verify the identity: $\csc(-x) = -\csc x$

Verification

$$\begin{aligned}\csc(-x) &= \frac{1}{\sin(-x)} && \text{Reciprocal identity}\\ &= \frac{1}{-\sin x} && \text{Identity for negatives}\\ &= -\frac{1}{\sin x} && \text{Algebra}\\ &= -\csc x && \text{Reciprocal identity}\end{aligned}$$

■

Matched Problem 1 Verify the identity: $\sec(-x) = \sec x$ ■

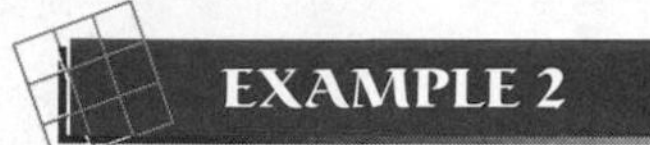

EXAMPLE 2

Identity Verification

Verify the identity: $\tan x \sin x + \cos x = \sec x$

Verification

$$\begin{aligned} \tan x \sin x + \cos x &= \frac{\sin x}{\cos x} \sin x + \cos x && \text{Quotient identity} \\ &= \frac{\sin^2 x + \cos^2 x}{\cos x} && \text{Algebra} \\ &= \frac{1}{\cos x} && \text{Pythagorean identity} \\ &= \sec x && \text{Reciprocal identity} \end{aligned}$$

Key Algebraic Steps:

$$\frac{a}{b} a + b = \frac{a^2}{b} + b = \frac{a^2 + b^2}{b}$$

■

Matched Problem 2 Verify the identity: $\cot x \cos x + \sin x = \csc x$ ■

To verify an identity, proceed from one side to the other, or both sides to the middle, making sure all steps are reversible. Even though there is no fixed method of verification that works for all identities, there are certain steps that help in many cases.

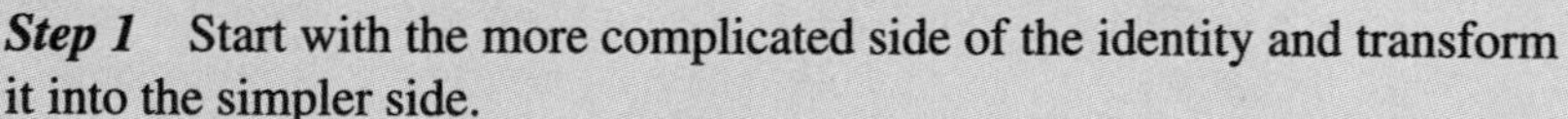

SOME SUGGESTIONS FOR VERIFYING IDENTITIES

Step 1 Start with the more complicated side of the identity and transform it into the simpler side.

Step 2 Try using basic or other known identities.

Step 3 Try algebraic operations such as multiplying, factoring, combining fractions, or splitting fractions.

Step 4 If other steps fail, try expressing each function in terms of sine and cosine functions; then perform appropriate algebraic operations.

Step 5 At each step, keep the other side of the identity in mind. This often reveals what you should do in order to get there.

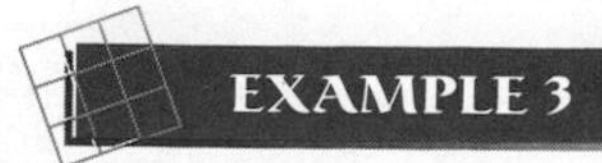

EXAMPLE 3

Identity Verification

Verify the identity: $\dfrac{\cot^2 x - 1}{1 + \cot^2 x} = 1 - 2 \sin^2 x$

Verification

$$\frac{\cot^2 x - 1}{1 + \cot^2 x} = \frac{\dfrac{\cos^2 x}{\sin^2 x} - 1}{1 + \dfrac{\cos^2 x}{\sin^2 x}}$$ Convert to sines and cosines.

$$= \frac{(\sin^2 x)\left(\dfrac{\cos^2 x}{\sin^2 x} - 1\right)}{(\sin^2 x)\left(1 + \dfrac{\cos^2 x}{\sin^2 x}\right)}$$ Multiply numerator and denominator by $\sin^2 x$, the LCD of all secondary fractions.

$$= \frac{\cos^2 x - \sin^2 x}{\sin^2 x + \cos^2 x}$$ Algebra

$$= \frac{1 - \sin^2 x - \sin^2 x}{1}$$ Pythagorean identities, twice

$$= 1 - 2\sin^2 x$$ Algebra

(See Explore–Discuss 1 for a shorter sequence of steps.)

Key Algebraic Steps:

$$\frac{\dfrac{b^2}{a^2} - 1}{1 + \dfrac{b^2}{a^2}} = \frac{a^2\left(\dfrac{b^2}{a^2} - 1\right)}{a^2\left(1 + \dfrac{b^2}{a^2}\right)} = \frac{b^2 - a^2}{a^2 + b^2}$$

■

Matched Problem 3 Verify the identity: $\dfrac{\tan^2 x - 1}{1 + \tan^2 x} = 1 - 2\cos^2 x$ ■

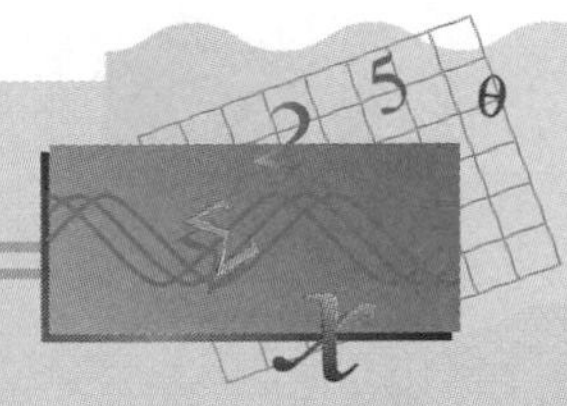

EXPLORE/DISCUSS 1

Can you verify the identity in Example 3,

$$\frac{\cot^2 x - 1}{1 + \cot^2 x} = 1 - 2\sin^2 x$$

using another sequence of steps? The following start, using Pythagorean identities, leads to a shorter verification:

$$\frac{\cot^2 x - 1}{1 + \cot^2 x} = \frac{\csc^2 x - 1 - 1}{\csc^2 x}$$

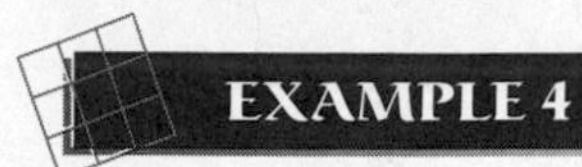

EXAMPLE 4 **Identity Verification**

Verify the identity: $\dfrac{1 + \cos x}{\sin x} + \dfrac{\sin x}{1 + \cos x} = 2 \csc x$

Verification

$$
\begin{aligned}
\frac{1 + \cos x}{\sin x} + \frac{\sin x}{1 + \cos x} &= \frac{(1 + \cos x)^2 + \sin^2 x}{(\sin x)(1 + \cos x)} && \text{Algebra} \\
&= \frac{1 + 2\cos x + \cos^2 x + \sin^2 x}{(\sin x)(1 + \cos x)} && \text{Algebra} \\
&= \frac{1 + 2\cos x + 1}{(\sin x)(1 + \cos x)} && \text{Pythagorean identity} \\
&= \frac{2 + 2\cos x}{(\sin x)(1 + \cos x)} && \text{Algebra} \\
&= \frac{2(1 + \cos x)}{(\sin x)(1 + \cos x)} && \text{Algebra} \\
&= \frac{2}{\sin x} && \text{Cancel common factor by division.} \\
&= 2 \csc x && \text{Reciprocal identity}
\end{aligned}
$$

Key Algebraic Steps:

$$\frac{1 + b}{a} + \frac{a}{1 + b} = \frac{(1 + b)^2 + a^2}{a(1 + b)} = \frac{1 + 2b + b^2 + a^2}{a(1 + b)}$$

and

$$\frac{2 + 2b}{a(1 + b)} = \frac{2(1 + b)}{a(1 + b)} = \frac{2}{a}$$

■

Matched Problem 4 Verify the identity: $\dfrac{1 + \sin x}{\cos x} + \dfrac{\cos x}{1 + \sin x} = 2 \sec x$ ■

EXAMPLE 5 **Identity Verification**

Verify the identity

$$\csc x + \cot x = \frac{\sin x}{1 - \cos x}$$

(A) Going from left to right (B) Going from right to left

Verification (A) Going from left to right:

$$\csc x + \cot x = \frac{1}{\sin x} + \frac{\cos x}{\sin x}$$ Convert to sines and cosines.

$$= \frac{1 + \cos x}{\sin x}$$ Algebra

$$= \frac{(\sin x)(1 + \cos x)}{\sin^2 x}$$ We need a sin x on top, so we multiply numerator and denominator by sin x.

$$= \frac{(\sin x)(1 + \cos x)}{1 - \cos^2 x}$$ Pythagorean identity

$$= \frac{(\sin x)(1 + \cos x)}{(1 - \cos x)(1 + \cos x)}$$ Factor denominator.

$$= \frac{\sin x}{1 - \cos x}$$ Cancel common factor by division.

Key Algebraic Steps:

$$\frac{1}{a} + \frac{b}{a} = \frac{1 + b}{a} = \frac{a(1 + b)}{a^2} \quad \text{and} \quad \frac{a(1 + b)}{1 - b^2} = \frac{a(1 + b)}{(1 - b)(1 + b)} = \frac{a}{1 - b}$$

(B) Going from right to left:

$$\frac{\sin x}{1 - \cos x} = \frac{(\sin x)(\mathbf{1 + \cos x})}{(1 - \cos x)(\mathbf{1 + \cos x})}$$ Multiply numerator and denominator by 1 + cos x so that we can take advantage of the Pythagorean identity.

$$= \frac{(\sin x)(1 + \cos x)}{1 - \cos^2 x}$$ Algebra

$$= \frac{(\sin x)(1 + \cos x)}{\sin^2 x}$$ Pythagorean identity

$$= \frac{1 + \cos x}{\sin x}$$ Cancel common factor.

$$= \frac{1}{\sin x} + \frac{\cos x}{\sin x}$$ Algebra

$$= \csc x + \cot x$$ Fundamental identities.

Key Algebraic Steps:

$$\frac{a}{1 - b} = \frac{a(1 + b)}{(1 - b)(1 + b)} = \frac{a(1 + b)}{1 - b^2} \quad \text{and} \quad \frac{a(1 + b)}{a^2} = \frac{1 + b}{a} = \frac{1}{a} + \frac{b}{a}$$

■

Matched Problem 5 Verify the identity

$$\sec m + \tan m = \frac{\cos m}{1 - \sin m}$$

(A) Going from left to right (B) Going from right to left ■

Testing Identities Using a Graphing Utility

Given an equation, it is not always easy to tell whether it is an identity or a conditional equation. A graphing utility puts us on the right track with little effort.

EXAMPLE 6

Testing Identities Using a Graphing Utility

Use a graphing utility to test whether each equation is an identity. If an equation appears to be an identity, verify it. If the equation does not appear to be an identity, find a value of x for which both sides are defined but are not equal.

(A) $\dfrac{\sin x}{1 - \cos^2 x} = \sec x$ (B) $\dfrac{\sin x}{1 - \cos^2 x} = \csc x$

Solution (A) Graph both sides of the equation in the same viewing window (Fig. 1). The graphs do not match; therefore, the equation is not an identity. The left side is not equal to the right side for $x = 1$, for example.

(B) Graph both sides of the equation in the same viewing window (Fig. 2). Use TRACE and check the values of each function for different values of x. The equation appears to be an identity, which we now verify:

$$\frac{\sin x}{1 - \cos^2 x} = \frac{\sin x}{\sin^2 x} = \frac{1}{\sin x} = \csc x$$

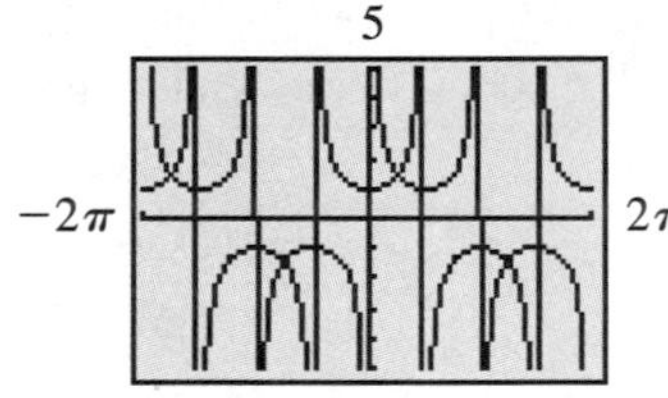

FIGURE 1

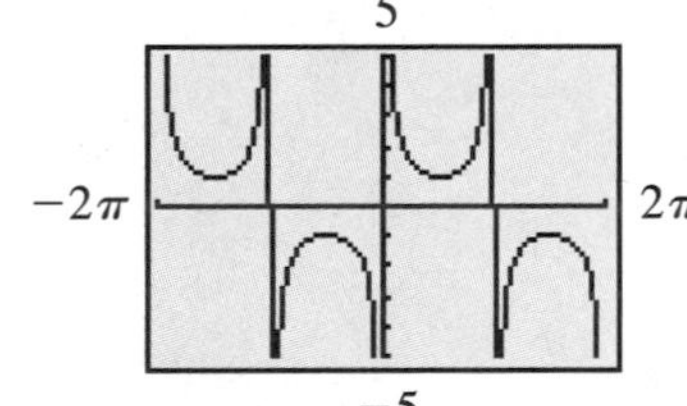

FIGURE 2

Matched Problem 6 Repeat Example 6 for the following two equations:

(A) $\tan x + 1 = (\sec x)(\sin x - \cos x)$

(B) $\tan x - 1 = (\sec x)(\sin x - \cos x)$

Answers to Matched Problems

1. $\sec(-x) = \dfrac{1}{\cos(-x)} = \dfrac{1}{\cos x} = \sec x$

2. $\cot x \cos x + \sin x = \dfrac{\cos^2 x}{\sin x} + \sin x = \dfrac{\cos^2 x + \sin^2 x}{\sin x}$

$$= \frac{1}{\sin x} = \csc x$$

3. $$\frac{\tan^2 x - 1}{1 + \tan^2 x} = \frac{\dfrac{\sin^2 x}{\cos^2 x} - 1}{1 + \dfrac{\sin^2 x}{\cos^2 x}} = \frac{(\cos^2 x)\left(\dfrac{\sin^2 x}{\cos^2 x} - 1\right)}{(\cos^2 x)\left(1 + \dfrac{\sin^2 x}{\cos^2 x}\right)}$$
$$= \frac{\sin^2 x - \cos^2 x}{\cos^2 x + \sin^2 x} = \frac{1 - \cos^2 x - \cos^2 x}{1}$$
$$= 1 - 2\cos^2 x$$

4. $$\frac{1 + \sin x}{\cos x} + \frac{\cos x}{1 + \sin x} = \frac{(1 + \sin x)^2 + \cos^2 x}{(\cos x)(1 + \sin x)}$$
$$= \frac{1 + 2\sin x + \sin^2 x + \cos^2 x}{(\cos x)(1 + \sin x)}$$
$$= \frac{2 + 2\sin x}{(\cos x)(1 + \sin x)} = \frac{2}{\cos x} = 2\sec x$$

5. (A) Going from left to right:

$$\sec m + \tan m = \frac{1}{\cos m} + \frac{\sin m}{\cos m} = \frac{1 + \sin m}{\cos m}$$
$$= \frac{(\cos m)(1 + \sin m)}{\cos^2 m} = \frac{(\cos m)(1 + \sin m)}{1 - \sin^2 m}$$
$$= \frac{(\cos m)(1 + \sin m)}{(1 - \sin m)(1 + \sin m)} = \frac{\cos m}{1 - \sin m}$$

(B) Going from right to left:

$$\frac{\cos m}{1 - \sin m} = \frac{(\cos m)(1 + \sin m)}{(1 - \sin m)(1 + \sin m)} = \frac{(\cos m)(1 + \sin m)}{1 - \sin^2 m}$$
$$= \frac{(\cos m)(1 + \sin m)}{\cos^2 m} = \frac{1 + \sin m}{\cos m}$$
$$= \frac{1}{\cos m} + \frac{\sin m}{\cos m} = \sec m + \tan m$$

6. (A) Not an identity; the left side is not equal to the right side for $x = 0$, for example:

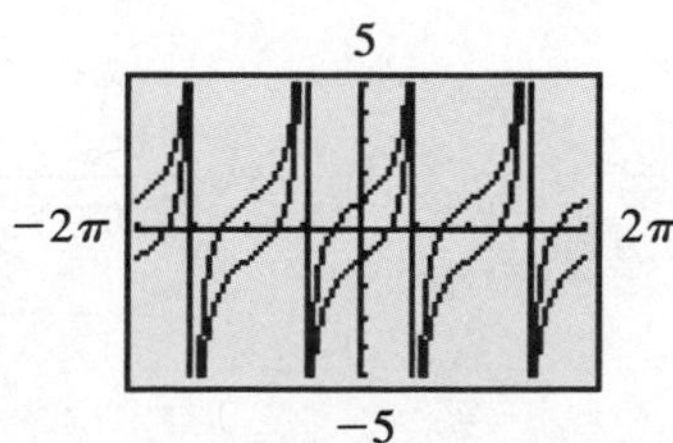

6. (B)

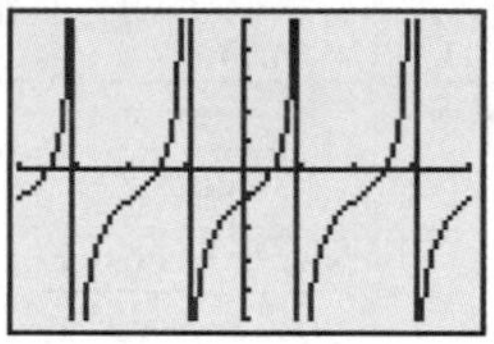

5

-2π 2π

-5

The equation appears to be an identity, which is verified as follows:

$$\begin{aligned}(\sec x)(\sin x - \cos x) &= \frac{1}{\cos x}(\sin x - \cos x) \\ &= \frac{\sin x}{\cos x} - \frac{\cos x}{\cos x} \\ &= \tan x - 1\end{aligned}$$

EXERCISE 4.2

A *In Problems 1–26, verify each identity.*

1. $\cos x \sec x = 1$

2. $\sin x \csc x = 1$

3. $\tan x \cos x = \sin x$

4. $\cot x \sin x = \cos x$

5. $\tan x = \sin x \sec x$

6. $\cot x = \cos x \csc x$

7. $\csc(-x) = -\csc x$

8. $\sec(-x) = \sec x$

9. $\dfrac{\sin \alpha}{\cos \alpha \tan \alpha} = 1$

10. $\dfrac{\cos \alpha}{\sin \alpha \cot \alpha} = 1$

11. $\dfrac{\cos \beta \sec \beta}{\tan \beta} = \cot \beta$

12. $\dfrac{\tan \beta \cot \beta}{\sin \beta} = \csc \beta$

13. $(\sec \theta)(\sin \theta + \cos \theta) = \tan \theta + 1$

14. $(\csc \theta)(\cos \theta + \sin \theta) = \cot \theta + 1$

15. $\dfrac{\cos^2 t - \sin^2 t}{\sin t \cos t} = \cot t - \tan t$

16. $\dfrac{\cos \alpha - \sin \alpha}{\sin \alpha \cos \alpha} = \csc \alpha - \sec \alpha$

17. $\dfrac{\cos \beta}{\cot \beta} + \dfrac{\sin \beta}{\tan \beta} = \sin \beta + \cos \beta$

18. $\dfrac{\tan u}{\sin u} - \dfrac{\cot u}{\cos u} = \sec u - \csc u$

19. $\sec^2 \theta - \tan^2 \theta = 1$

20. $\csc^2 \theta - \cot^2 \theta = 1$

21. $(\sin^2 x)(1 + \cot^2 x) = 1$

22. $(\cos^2 x)(\tan^2 x + 1) = 1$

23. $(\csc \alpha + 1)(\csc \alpha - 1) = \cot^2 \alpha$

24. $(\sec \beta - 1)(\sec \beta + 1) = \tan^2 \beta$

25. $\dfrac{\sin t}{\csc t} + \dfrac{\cos t}{\sec t} = 1$

26. $\dfrac{1}{\sec^2 m} + \dfrac{1}{\csc^2 m} = 1$

B **27.** How does solving a conditional equation differ from verifying an identity (both in one variable)?

28. Is $(1 - \cos^2 x)/(\sin x) = \sin x$ an identity for all real values of x? Explain.

In Problems 29–60, verify each identity.

29. $\dfrac{1 - (\cos \theta - \sin \theta)^2}{\cos \theta} = 2 \sin \theta$

30. $\dfrac{1 - (\sin \theta - \cos \theta)^2}{\sin \theta} = 2 \cos \theta$

31. $\dfrac{\tan w + 1}{\sec w} = \sin w + \cos w$

32. $\dfrac{\cot y + 1}{\csc y} = \cos y + \sin y$

33. $\dfrac{1}{1 - \cos^2 \theta} = 1 + \cot^2 \theta$

34. $\dfrac{1}{1 - \sin^2 \theta} = 1 + \tan^2 \theta$

35. $\dfrac{\sin^2 \beta}{1 - \cos \beta} = 1 + \cos \beta$

36. $\dfrac{\cos^2 \beta}{1 + \sin \beta} = 1 - \sin \beta$

37. $\dfrac{2 - \cos^2 \theta}{\sin \theta} = \csc \theta + \sin \theta$

38. $\dfrac{2 - \sin^2 \theta}{\cos \theta} = \sec \theta + \cos \theta$

39. $\tan x + \cot x = \sec x \csc x$

40. $\dfrac{\csc x}{\cot x + \tan x} = \cos x$

41. $\dfrac{1 - \csc x}{1 + \csc x} = \dfrac{\sin x - 1}{\sin x + 1}$

42. $\dfrac{1 - \cos x}{1 + \cos x} = \dfrac{\sec x - 1}{\sec x + 1}$

43. $\csc^2 \alpha - \cos^2 \alpha - \sin^2 \alpha = \cot^2 \alpha$

44. $\sec^2 \alpha - \sin^2 \alpha - \cos^2 \alpha = \tan^2 \alpha$

45. $(\sin x + \cos x)^2 - 1 = 2 \sin x \cos x$

46. $\sec x - 2 \sin x = \dfrac{(\sin x - \cos x)^2}{\cos x}$

47. $(\sin u - \cos u)^2 + (\sin u + \cos u)^2 = 2$

48. $(\tan x - 1)^2 + (\tan x + 1)^2 = 2 \sec^2 x$

49. $\sin^4 x - \cos^4 x = 1 - 2 \cos^2 x$

50. $\sin^4 x + 2 \sin^2 x \cos^2 x + \cos^4 x = 1$

51. $\dfrac{\sin \alpha}{1 - \cos \alpha} - \dfrac{1 + \cos \alpha}{\sin \alpha} = 0$

52. $\dfrac{1 + \cos \alpha}{\sin \alpha} + \dfrac{\sin \alpha}{1 + \cos \alpha} = 2 \csc \alpha$

53. $\dfrac{\cos^2 n - 3 \cos n + 2}{\sin^2 n} = \dfrac{2 - \cos n}{1 + \cos n}$

54. $\dfrac{\sin^2 n + 4 \sin n + 3}{\cos^2 n} = \dfrac{3 + \sin n}{1 - \sin n}$

55. $\dfrac{1 - \cot^2 x}{\tan^2 x - 1} = \cot^2 x$

56. $\dfrac{\tan^2 x - 1}{1 - \cot^2 x} = \tan^2 x$

57. $\sec^2 x + \csc^2 x = \sec^2 x \csc^2 x$

58. $\tan^2 x - \sin^2 x = \tan^2 x \sin^2 x$

59. $\dfrac{1 + \sin t}{\cos t} = \dfrac{\cos t}{1 - \sin t}$

60. $\dfrac{\sin t}{1 - \cos t} = \dfrac{1 + \cos t}{\sin t}$

61. (A) Graph both sides of the following equation in the same viewing window for $-\pi \le x \le \pi$. Is the equation an identity over the interval $-\pi \le x \le \pi$? Explain.

$$\sin x = x - \frac{x^3}{3!} + \frac{x^5}{5!} - \frac{x^7}{7!}$$

(B) Extend the interval in part (A) to $-2\pi \le x \le 2\pi$. Now, does the equation appear to be an identity? What do you observe?

62. (A) Graph both sides of the following equation in the same viewing window for $-\pi \le x \le \pi$. Is the equation an identity over the interval $[-\pi, \pi]$? Explain.

$$\cos x = 1 - \frac{x^2}{2!} + \frac{x^4}{4!} - \frac{x^6}{6!} + \frac{x^8}{8!}$$

(B) Extend the interval in part (A) to $-2\pi \le x \le 2\pi$. Now, does the equation appear to be an identity? What do you observe?

In Problems 63–70, use a graphing utility to test whether each equation is an identity. If an equation appears to be an identity, verify it. If an equation does not appear to be an identity, find a value of x for which both sides are defined but are not equal.

63. $\dfrac{\cos x}{\sin(-x)\cot(-x)} = 1$

64. $\dfrac{\sin x}{\cos x \tan(-x)} = -1$

65. $\dfrac{\cos(-x)}{\sin x \cot(-x)} = 1$

66. $\dfrac{\sin(-x)}{\cos(-x)\tan(-x)} = -1$

67. $\dfrac{\cos x}{\sin x + 1} - \dfrac{\cos x}{\sin x - 1} = 2 \csc x$

68. $\dfrac{\tan x}{\sin x + 2 \tan x} = \dfrac{1}{\cos x - 2}$

69. $\dfrac{\cos x}{1 - \sin x} + \dfrac{\cos x}{1 + \sin x} = 2 \sec x$

70. $\dfrac{\tan x}{\sin x - 2 \tan x} = \dfrac{1}{\cos x - 2}$

C *In Problems 71–76, verify each identity.*

71. $\dfrac{\sin x}{1 - \cos x} - \cot x = \csc x$

72. $\dfrac{\cos x}{1 - \sin x} - \tan x = \sec x$

73. $\dfrac{\cot \beta}{\csc \beta + 1} = \dfrac{\csc \beta - 1}{\cot \beta}$

74. $\dfrac{\tan \beta}{\sec \beta - 1} = \dfrac{\sec \beta + 1}{\tan \beta}$

75. $\dfrac{3 \cos^2 m + 5 \sin m - 5}{\cos^2 m} = \dfrac{3 \sin m - 2}{1 + \sin m}$

76. $\dfrac{2 \sin^2 z + 3 \cos z - 3}{\sin^2 z} = \dfrac{2 \cos z - 1}{1 + \cos z}$

In Problems 77 and 78, verify each identity. (The problems involve trigonometric functions with two variables. Be careful with the terms you combine and simplify.)

77. $\dfrac{\sin x \cos y + \cos x \sin y}{\cos x \cos y - \sin x \sin y} = \dfrac{\tan x + \tan y}{1 - \tan x \tan y}$

78. $\dfrac{\tan \alpha + \tan \beta}{1 - \tan \alpha \tan \beta} = \dfrac{\cot \alpha + \cot \beta}{\cot \alpha \cot \beta - 1}$

4.3 SUM, DIFFERENCE, AND COFUNCTION IDENTITIES

- Sum and Difference Identities for Cosine
- Cofunction Identities
- Sum and Difference Identities for Sine and Tangent
- Summary and Use

Sum and Difference Identities for Cosine

The fundamental identities we discussed in Section 4.1 involve only one variable. We now consider an important identity, called a **difference identity for cosine,** which involves two variables:

$$\cos(x - y) = \cos x \cos y + \sin x \sin y \tag{1}$$

Many other useful identities can be readily established from this particular one. We will sketch a proof of identity (1) in which we assume that x and y are restricted as follows: $0 < y < x < 2\pi$. Identity (1) holds, however, for all real numbers and angles in radian or degree measure. In the proof, we will make use of the **formula for the distance between two points,**

$$d(P_1, P_2) = \sqrt{(x_2 - x_1)^2 + (y_2 - y_1)^2}$$

for points $P_1(x_1, y_1)$ and $P_2(x_2, y_2)$ in a rectangular coordinate system.

We associate x and y with arcs and angles on a unit circle, as indicated in Figure 1a. Using the definitions of the circular functions in Section 2.6, the terminal points of x and y are labeled as indicated in Figure 1a.

Now, if we rotate the triangle AOB clockwise about the origin until the terminal point A coincides with $D(1, 0)$, then terminal point B will be at C (see Fig. 1b). Thus, since rotation preserves lengths, we have

FIGURE 1

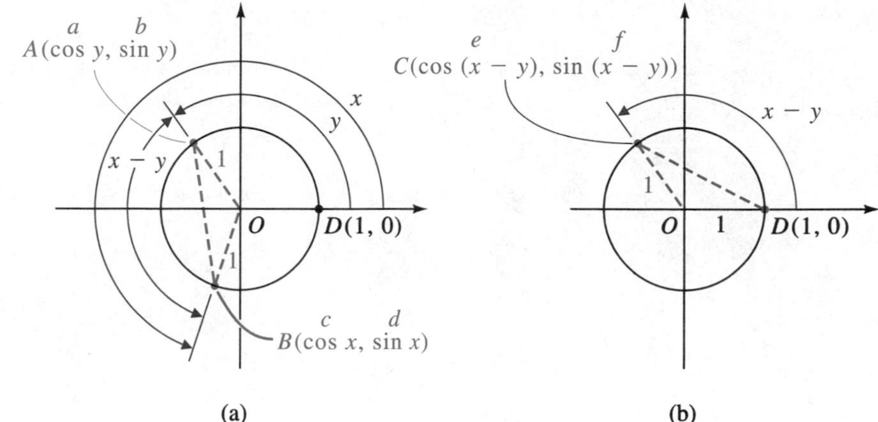

(a) (b)

$$d(A, B) = d(C, D)$$
$$\sqrt{(c-a)^2 + (d-b)^2} = \sqrt{(1-e)^2 + (0-f)^2}$$
$$(c-a)^2 + (d-b)^2 = (1-e)^2 + f^2$$
$$c^2 - 2ac + a^2 + d^2 - 2db + b^2 = 1 - 2e + e^2 + f^2$$
$$(c^2 + d^2) + (a^2 + b^2) - 2ac - 2db = 1 - 2e + (e^2 + f^2) \tag{2}$$

Since $c^2 + d^2 = 1$, $a^2 + b^2 = 1$, and $e^2 + f^2 = 1$ (Why?), equation (2) becomes

$$e = ac + bd \tag{3}$$

Replacing e, a, c, b, and d with $\cos(x - y)$, $\cos y$, $\cos x$, $\sin y$, and $\sin x$, respectively (see Fig. 1), we obtain

$$\begin{aligned}\cos(x - y) &= \cos y \cos x + \sin y \sin x \\ &= \cos x \cos y + \sin x \sin y\end{aligned} \tag{4}$$

If we replace y with $-y$ in (4) and use the identities for negatives, we obtain the **sum identity for cosine:**

$$\cos(x + y) = \cos x \cos y - \sin x \sin y \tag{5}$$

EXPLORE/DISCUSS 1

Find values of x and y such that:

(A) $\cos(x - y) \neq \cos x - \cos y$

(B) $\cos(x + y) \neq \cos x + \cos y$

Cofunction Identities

To obtain sum and difference identities for the sine and tangent functions, we first derive **cofunction identities** directly from identity (1), the difference identity for cosine:

$$\begin{aligned} \cos(x - y) &= \cos x \cos y + \sin x \sin y \qquad \text{Let } x = \pi/2. \\ \cos\left(\frac{\pi}{2} - y\right) &= \cos\frac{\pi}{2}\cos y + \sin\frac{\pi}{2}\sin y \\ &= 0 \cos y + 1 \sin y \\ &= \sin y \end{aligned}$$

Thus,

$$\cos\left(\frac{\pi}{2} - y\right) = \sin y \tag{6}$$

for y any real number or angle in radian measure. If y is in degree measure, replace $\pi/2$ with $90°$. Now, if in (6) we let $y = \pi/2 - x$, then we have

$$\begin{aligned} \cos\left[\frac{\pi}{2} - \left(\frac{\pi}{2} - x\right)\right] &= \sin\left(\frac{\pi}{2} - x\right) \\ \cos x &= \sin\left(\frac{\pi}{2} - x\right) \end{aligned}$$

or

$$\sin\left(\frac{\pi}{2} - x\right) = \cos x \tag{7}$$

where x is any real number or angle in radian measure. If x is in degree measure, replace $\pi/2$ with $90°$.

Finally, we state the cofunction identities for tangent and secant (and leave the derivations to Problems 9 and 11 in Exercise 4.3):

$$\tan\left(\frac{\pi}{2} - x\right) = \cot x \qquad \sec\left(\frac{\pi}{2} - x\right) = \csc x \tag{8}$$

for x any real number or angle in radian measure. If x is in degree measure, replace $\pi/2$ with $90°$.

Remark If $0 < x < 90°$, then x and $90° - x$ are complementary angles. Originally, *cosine, cotangent,* and *cosecant* meant, respectively, *complements sine, complements tangent,* and *complements secant.* Now we simply refer to cosine, cotangent, and cosecant as **cofunctions of** sine, tangent, and secant, respectively. □

Sum and Difference Identities for Sine and Tangent

To derive a **difference identity for sine,** we use (7), (1), and (6) as follows:

$$\begin{aligned}
\sin(x - y) &= \cos\left[\frac{\pi}{2} - (x - y)\right] && \text{Use (6).}\\
&= \cos\left[\left(\frac{\pi}{2} - x\right) - (-y)\right] && \text{Algebra}\\
&= \cos\left(\frac{\pi}{2} - x\right)\cos(-y) + \sin\left(\frac{\pi}{2} - x\right)\sin(-y) && \text{Use (1).}\\
&= \sin x \cos y - \cos x \sin y && \text{Use (6), (7), and identities for negatives.}
\end{aligned}$$

The same result is obtained by replacing $\pi/2$ with 90°. Thus,

$$\sin(x - y) = \sin x \cos y - \cos x \sin y \tag{9}$$

Now, if we replace y with $-y$ (a good exercise to do), we obtain

$$\sin(x + y) = \sin x \cos y + \cos x \sin y \tag{10}$$

It is not difficult to derive sum and difference identities for the tangent function. See if you can supply the reason for each step:

$$\begin{aligned}
\tan(x - y) &= \frac{\sin(x - y)}{\cos(x - y)}\\
&= \frac{\sin x \cos y - \cos x \sin y}{\cos x \cos y + \sin x \sin y}\\
&= \frac{\dfrac{\sin x \cos y}{\cos x \cos y} - \dfrac{\cos x \sin y}{\cos x \cos y}}{\dfrac{\cos x \cos y}{\cos x \cos y} + \dfrac{\sin x \sin y}{\cos x \cos y}}\\
&= \frac{\tan x - \tan y}{1 + \tan x \tan y}
\end{aligned}$$

Thus, for all angles or real numbers x and y,

$$\tan(x - y) = \frac{\tan x - \tan y}{1 + \tan x \tan y} \tag{11}$$

And if we replace y in (11) with $-y$ (another good exercise to do), we obtain

$$\tan(x + y) = \frac{\tan x + \tan y}{1 - \tan x \tan y} \tag{12}$$

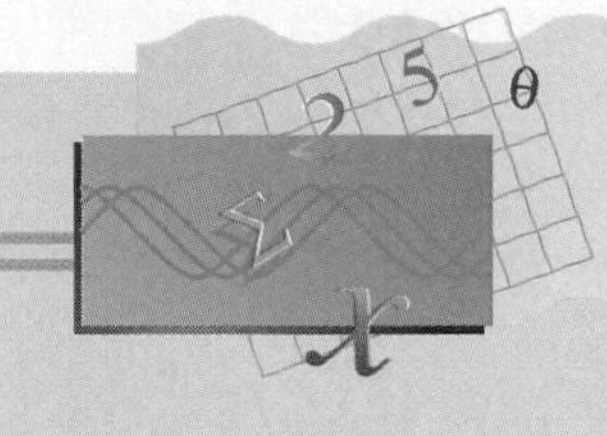

EXPLORE/DISCUSS 2

Find values of x and y such that:

(A) $\tan(x - y) \neq \tan x - \tan y$

(B) $\tan(x + y) \neq \tan x + \tan y$

Summary and Use

Before proceeding with examples that illustrate the use of these new identities, we list them and the other cofunction identities for convenient reference.

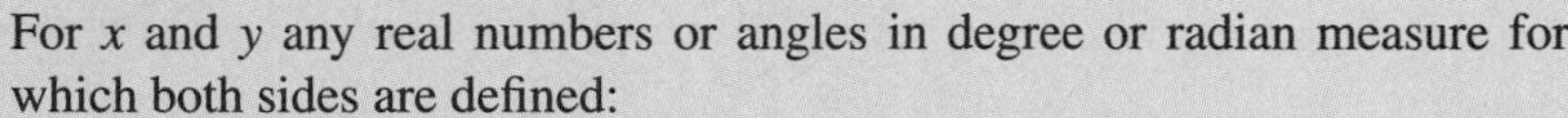

SUMMARY OF IDENTITIES

For x and y any real numbers or angles in degree or radian measure for which both sides are defined:

Sum identities

$$\sin(x + y) = \sin x \cos y + \cos x \sin y$$

$$\cos(x + y) = \cos x \cos y - \sin x \sin y$$

$$\tan(x + y) = \frac{\tan x + \tan y}{1 - \tan x \tan y}$$

Difference identities

$$\sin(x - y) = \sin x \cos y - \cos x \sin y$$

$$\cos(x - y) = \cos x \cos y + \sin x \sin y$$

$$\tan(x - y) = \frac{\tan x - \tan y}{1 + \tan x \tan y}$$

Cofunction identities

(Replace $\pi/2$ with 90° if x is in degree measure.)

$$\sin\left(\frac{\pi}{2} - x\right) = \cos x \qquad \cos\left(\frac{\pi}{2} - x\right) = \sin x$$

$$\tan\left(\frac{\pi}{2} - x\right) = \cot x \qquad \cot\left(\frac{\pi}{2} - x\right) = \tan x$$

$$\sec\left(\frac{\pi}{2} - x\right) = \csc x \qquad \csc\left(\frac{\pi}{2} - x\right) = \sec x$$

EXAMPLE 1 Using a Difference Identity

Simplify $\sin(x - \pi)$ using a difference identity.

Solution Use the difference identity for sine, replacing y with π:

$$\begin{aligned} \sin(x - y) &= \sin x \cos y - \cos x \sin y \\ \sin(x - \pi) &= \sin x \cos \pi - \cos x \sin \pi \\ &= (\sin x)(-1) - (\cos x)(0) \\ &= -\sin x \end{aligned}$$

■

Matched Problem 1 Simplify $\cos(x + 3\pi/2)$ using a sum identity. ■

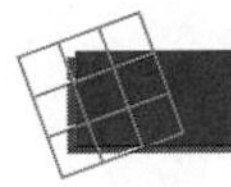

EXAMPLE 2 Checking the Use of an Identity on a Graphing Utility

Simplify $\cos(x + \pi)$ using an appropriate identity. Then check the result using a graphing utility.

Solution

$$\begin{aligned} \cos(x + \pi) &= \cos x \cos \pi - \sin x \sin \pi \\ &= (\cos x)(-1) - (\sin x)(0) \\ &= -\cos x \end{aligned}$$

To check with a graphing utility, graph $y_1 = \cos(x + \pi)$ and $y_2 = -\cos x$ in the same viewing window to see if they produce the same graph (Fig. 2). Use TRACE and move back and forth between y_1 and y_2 for different values of x to see that the corresponding y values are the same, or nearly the same.

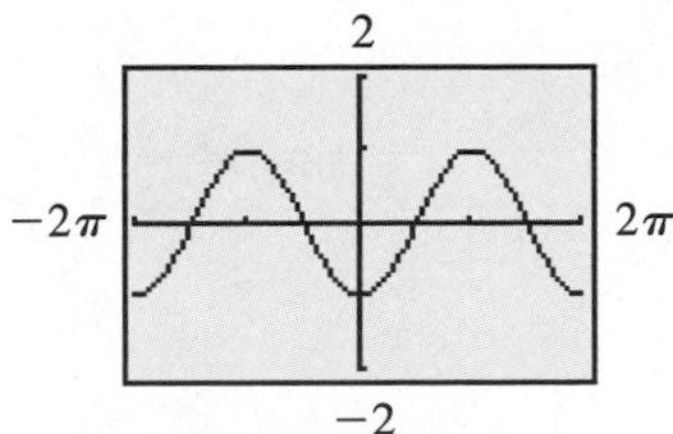

FIGURE 2
$y_1 = \cos(x + \pi)$, $y_2 = -\cos x$

■

Matched Problem 2 Simplify $\sin(x - \pi)$ using an appropriate identity. Then check the result using a graphing utility. ■

EXAMPLE 3 Finding Exact Values

Find the exact value of $\cos 15°$ in radical form.

Solution Note that $15° = 45° - 30°$, the difference of two special angles. Thus,

$$\begin{aligned}\cos 15° &= \cos(45° - 30°)\\ &= \cos 45° \cos 30° + \sin 45° \sin 30°\\ &= \frac{1}{\sqrt{2}} \cdot \frac{\sqrt{3}}{2} + \frac{1}{\sqrt{2}} \cdot \frac{1}{2} = \frac{\sqrt{3} + 1}{2\sqrt{2}}\end{aligned}$$

■

Matched Problem 3 Find the exact value of $\tan 75°$ in radical form. ■

EXAMPLE 4 Finding Exact Values

Find the exact value of $\cos(x + y)$, given $\sin x = \frac{3}{5}$, $\cos y = \frac{4}{5}$, x in quadrant II, and y in quadrant I. Do not use a calculator.

Solution We start with the sum identity for cosine:

$$\cos(x + y) = \cos x \cos y - \sin x \sin y$$

We know $\sin x$ and $\cos y$, but not $\sin y$ and $\cos x$. We find the latter two values by using reference triangles and the Pythagorean theorem. From Figure 3,

FIGURE 3

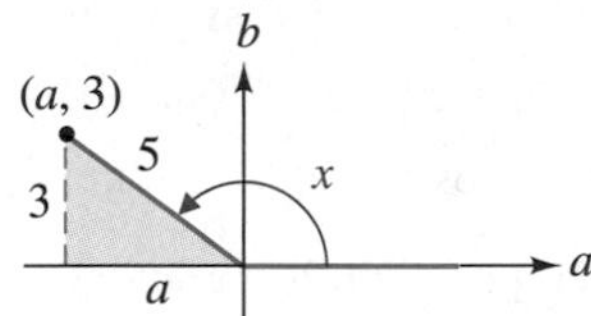

$$a = -\sqrt{5^2 - 3^2} = -4$$

$$\cos x = -\frac{4}{5}$$

From Figure 4,

FIGURE 4

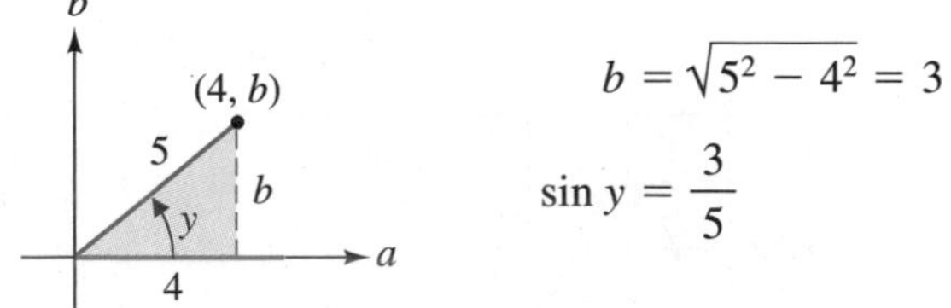

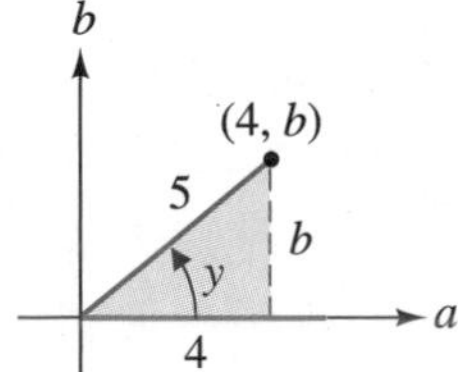

$$b = \sqrt{5^2 - 4^2} = 3$$

$$\sin y = \frac{3}{5}$$

Thus,

$$\begin{aligned}\cos(x + y) &= \cos x \cos y - \sin x \sin y\\ &= \left(-\frac{4}{5}\right)\left(\frac{4}{5}\right) - \left(\frac{3}{5}\right)\left(\frac{3}{5}\right) = \frac{-25}{25} = -1\end{aligned}$$

■

Matched Problem 4 Find the exact value of $\sin(x - y)$, given $\sin x = -\frac{2}{3}$, $\cos y = \sqrt{5}/3$, x in quadrant III, and y in quadrant IV. ■

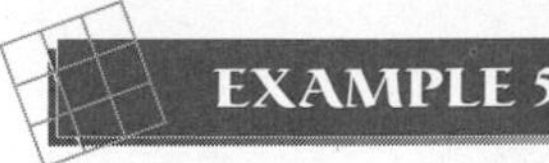

Verifying an Identity

Verify the identity: $\cot y - \cot x = \dfrac{\sin(x - y)}{\sin x \sin y}$

Verification

$$\begin{aligned}\frac{\sin(x-y)}{\sin x \sin y} &= \frac{\sin x \cos y - \cos x \sin y}{\sin x \sin y} && \text{Difference Identity}\\ &= \frac{\sin x \cos y}{\sin x \sin y} - \frac{\cos x \sin y}{\sin x \sin y} && \text{Algebra}\\ &= \cot y - \cot x && \text{Quotient identity}\end{aligned}$$

■

Matched Problem 5 Verify the identity: $\tan x + \cot y = \dfrac{\cos(x - y)}{\cos x \sin y}$ ■

Answers to Matched Problems

1. $\sin x$
2. $y_1 = \sin(x - \pi)$; $y_2 = -\sin x$

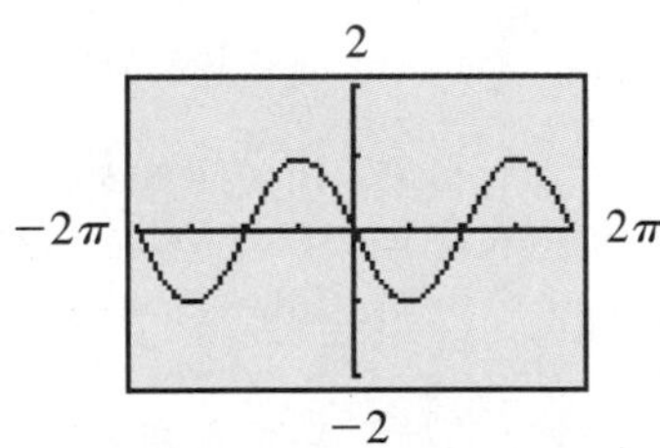

3. $2 + \sqrt{3}$
4. $\dfrac{-4\sqrt{5}}{9}$
5. $$\begin{aligned}\frac{\cos(x-y)}{\cos x \sin y} &= \frac{\cos x \cos y + \sin x \sin y}{\cos x \sin y}\\ &= \frac{\cos x \cos y}{\cos x \sin y} + \frac{\sin x \sin y}{\cos x \sin y}\\ &= \cot y + \tan x = \tan x + \cot y\end{aligned}$$

EXERCISE 4.3

A *We can use sum identities to verify periodic properties for the trigonometric functions. Verify the following identities using sum identities.*

1. $\cos(x + 2\pi) = \cos x$
2. $\sin(x + 2\pi) = \sin x$
3. $\cot(x + \pi) = \cot x$
4. $\tan(x + \pi) = \tan x$
5. $\sin(x + 2k\pi) = \sin x$, k an integer
6. $\cos(x + 2k\pi) = \cos x$, k an integer
7. $\tan(x + k\pi) = \tan x$, k an integer
8. $\cot(x + k\pi) = \cot x$, k an integer

Verify each identity using cofunction identities for sine and cosine and the fundamental identities discussed in Section 4.1.

9. $\tan\left(\dfrac{\pi}{2} - x\right) = \cot x$
10. $\cot\left(\dfrac{\pi}{2} - x\right) = \tan x$

11. $\sec\left(\frac{\pi}{2} - x\right) = \csc x$ **12.** $\csc\left(\frac{\pi}{2} - x\right) = \sec x$

Convert to forms involving sin x, cos x, and/or tan x using sum or difference identities.

13. $\sin(x - 45°)$ **14.** $\sin(30° - x)$

15. $\cos(x + 180°)$ **16.** $\sin(180° - x)$

17. $\tan\left(\frac{\pi}{4} - x\right)$ **18.** $\tan\left(x + \frac{\pi}{3}\right)$

B *Use appropriate identities to find exact values for each of the following. Do not use a calculator.*

19. $\sin 75°$ **20.** $\sec 75°$

21. $\cos \frac{\pi}{12}$ $\left[\textit{Hint: } \frac{\pi}{12} = \frac{\pi}{4} - \frac{\pi}{6}\right]$

22. $\sin \frac{7\pi}{12}$ $\left[\textit{Hint: } \frac{7\pi}{12} = \frac{\pi}{3} + \frac{\pi}{4}\right]$

23. $\sin 22° \cos 38° + \cos 22° \sin 38°$

24. $\cos 74° \cos 44° + \sin 74° \sin 44°$

25. $\dfrac{\tan 110° - \tan 50°}{1 + \tan 110° \tan 50°}$ **26.** $\dfrac{\tan 27° + \tan 18°}{1 - \tan 27° \tan 18°}$

Find sin(x − y) and tan(x + y) exactly without a calculator using the information given and appropriate identities.

27. $\sin x = \frac{2}{3}$, $\cos y = -\frac{1}{4}$, x in quadrant II, and y in quadrant III

28. $\sin x = -\frac{3}{5}$, $\sin y = \sqrt{8}/3$, x in quadrant IV, and y in quadrant I

29. $\cos x = -\frac{1}{3}$, $\tan y = \frac{1}{2}$, x in quadrant II, and y in quadrant III

30. $\tan x = \frac{3}{4}$, $\tan y = -\frac{1}{2}$, x in quadrant III, and y in quadrant IV

Verify each identity.

31. $\sin 2x = 2 \sin x \cos x$

32. $\cos 2x = \cos^2 x - \sin^2 x$

33. $\cot(x - y) = \dfrac{\cot x \cot y + 1}{\cot y - \cot x}$

34. $\cot(x + y) = \dfrac{\cot x \cot y - 1}{\cot x + \cot y}$

35. $\cot 2x = \dfrac{\cot^2 x - 1}{2 \cot x}$

36. $\tan 2x = \dfrac{2 \tan x}{1 - \tan^2 x}$

37. $\dfrac{\tan \alpha + \tan \beta}{\tan \alpha - \tan \beta} = \dfrac{\sin(\alpha + \beta)}{\sin(\alpha - \beta)}$

38. $\dfrac{\cot \alpha + \cot \beta}{\cot \alpha - \cot \beta} = \dfrac{\sin(\beta + \alpha)}{\sin(\beta - \alpha)}$

39. $\tan x - \tan y = \dfrac{\sin(x - y)}{\cos x \cos y}$

40. $\cot x - \tan y = \dfrac{\cos(x + y)}{\sin x \cos y}$

41. $\tan(x + y) = \dfrac{\cot x + \cot y}{\cot x \cot y - 1}$

42. $\tan(x - y) = \dfrac{\cot y - \cot x}{\cot x \cot y + 1}$

43. $\dfrac{\sin(x + h) - \sin x}{h} = (\sin x)\left(\dfrac{\cos h - 1}{h}\right) + (\cos x)\left(\dfrac{\sin h}{h}\right)$

44. $\dfrac{\cos(x + h) - \cos x}{h} = (\cos x)\left(\dfrac{\cos h - 1}{h}\right) - (\sin x)\left(\dfrac{\sin h}{h}\right)$

45. How would you show that $\csc(x - y) = \csc x - \csc y$ is not an identity?

46. How would you show that $\sec(x + y) = \sec x + \sec y$ is not an identity?

47. Use a graphing utility to show that $\sin(x - 2) = \sin x - \sin 2$ is not an identity. Then explain what you did.

48. Use a graphing utility to show that $\cos(x + 1) = \cos x + \cos 1$ is not an identity. Then explain what you did.

In Problems 49–52, use sum or difference identities to convert each equation to a form involving sin x, cos x, and/or tan x. To check your result, enter the original equation in a graphing utility as y_1 and the converted form as y_2. Then graph y_1 and y_2 in the same viewing window. Use TRACE *to compare the two graphs.*

49. $y = \cos(x + 5\pi/6)$ **50.** $y = \sin(x - \pi/3)$

51. $y = \tan(x - \pi/4)$ **52.** $y = \tan(x + 2\pi/3)$

In Problems 53–56, write each equation in terms of a single trigonometric function. Check the result by entering the original equation in a graphing utility as y_1 and the converted form as y_2. Then graph y_1 and y_2 in the same viewing window. Use TRACE *to compare the two graphs.*

53. $y = \sin 3x \cos x - \cos 3x \sin x$

54. $y = \cos 3x \cos x - \sin 3x \sin x$

55. $y = \sin \dfrac{\pi x}{4} \cos \dfrac{3\pi x}{4} + \cos \dfrac{\pi x}{4} \sin \dfrac{3\pi x}{4}$

56. $y = \cos(\pi x)\cos\dfrac{\pi x}{2} + \sin(\pi x)\sin\dfrac{\pi x}{2}$

C *Verify the identities in Problems 57 and 58.*
[*Hint:* $\sin(x + y + z) = \sin[(x + y) + z]$.]

57. $\sin(x + y + z) = \sin x \cos y \cos z + \cos x \sin y \cos z + \cos x \cos y \sin z - \sin x \sin y \sin z$

58. $\cos(x + y + z) = \cos x \cos y \cos z - \sin x \sin y \cos z - \sin x \cos y \sin z - \cos x \sin y \sin z$

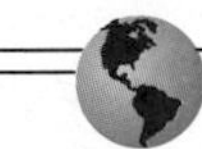

Applications

59. Precalculus: Angle of Intersection of Two Lines Use the information in the figure to show that

$$\tan(\theta_2 - \theta_1) = \frac{m_2 - m_1}{1 + m_1 m_2}$$

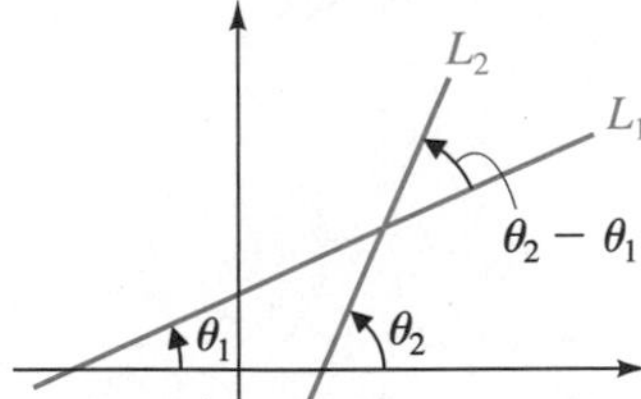

Figure for 59
$\tan\theta_1 = \text{Slope of } L_1 = m_1$
$\tan\theta_2 = \text{Slope of } L_2 = m_2$

60. Precalculus: Angle of Intersection of Two Lines Find the acute angle of intersection between the two lines $y = 3x + 1$ and $y = \frac{1}{2}x - 1$. (Use the results of Problem 59.)

*** 61. Analytic Geometry** Find the radian measure of the angle θ in the figure (to three decimal places), if A has coordinates (2, 4) and B has coordinates (3, 3). [*Hint:* Label the angle between OB and the x axis as α; then use an appropriate sum identity.

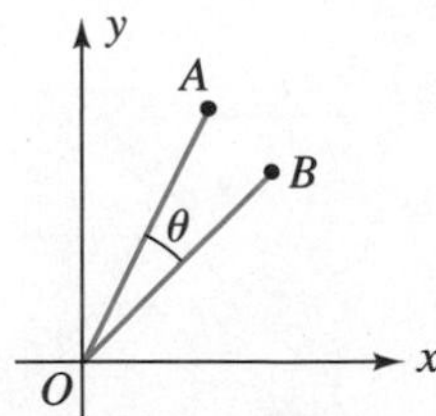

Figure for 61 and 62

*** 62. Analytic Geometry** Find the radian measure of the angle θ in the figure (to three decimal places), if A has coordinates (3, 9) and B has coordinates (6, 3).

*** 63. Surveying** A prominent geological feature of Yosemite National Park is the large monolithic granite peak called Half Dome. The dome rises straight up from the valley floor, where Mirror Lake provides an early morning mirror image of the dome. How can the height H of Half Dome be determined by using only a sextant h feet high to measure the angle of elevation, β, to the top of the dome, and the angle of depression, α, to the reflected dome top in the lake? (See the figure, which is not to scale.) [*Note:* AB and BC are not measured as in earlier problems of this type.]*

(A) Using right triangle relationships, show that

$$H = h\left(\frac{1 + \tan\beta\cot\alpha}{1 - \tan\beta\cot\alpha}\right)$$

(B) Using sum identities, show that the result in part (A) can be written in the form

$$H = h\left(\frac{\sin(\alpha + \beta)}{\sin(\alpha - \beta)}\right)$$

(C) If a sextant of height 5.50 ft measures α to be 45.00° and β to be 44.92°, compute the height H of Half Dome above Mirror Lake to three significant digits.

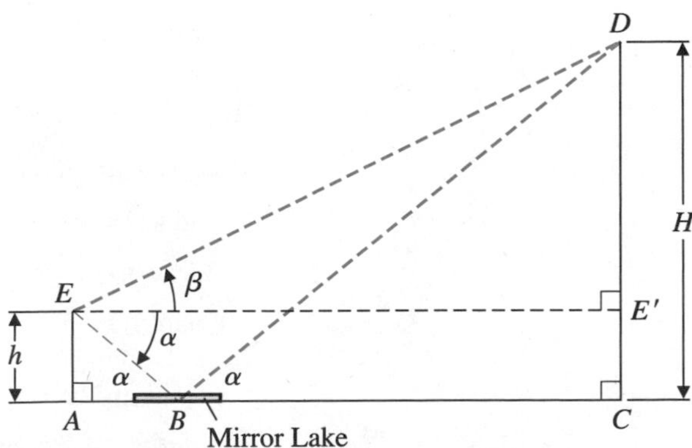

Figure for 63

* The solution outlined in parts (A) and (B) is due to Richard J. Palmaccio of Fort Lauderdale, Florida.

64. Light Refraction Light rays passing through a plate glass window are refracted when they enter the glass and again when they leave to continue on a path parallel to the entering rays (see the figure).

(A) If the plate glass is A inches thick, the parallel displacement of the light rays is B inches, the angle of incidence is α, and the angle of refraction is β, show that

$$\tan \beta = \tan \alpha - \frac{B}{A} \sec \alpha$$

[*Hint:* First use geometric relationships to obtain

$$\frac{A}{\sin(90° - \beta)} = \frac{B}{\sin(\alpha - \beta)}$$

Then use sum identities and fundamental identities to complete the task.]

(B) Using the results in part (A), find β to the nearest degree if $\alpha = 45°$, $A = 0.5$ in., and $B = 0.2$ in.

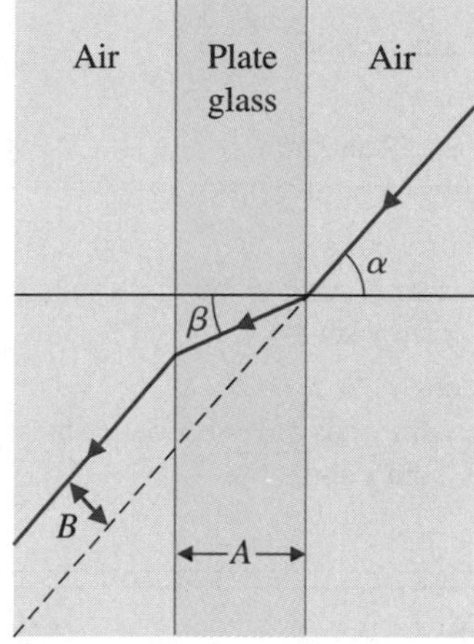

Figure for 64

4.4 DOUBLE-ANGLE AND HALF-ANGLE IDENTITIES

- Double-Angle Identities
- Half-Angle Identities

We now develop another important set of identities called **double-angle** and **half-angle identities.** We can obtain these identities directly from the sum and difference identities that were found in Section 4.3. In spite of names involving the word *angle,* the new identities hold for real numbers as well.

Double-Angle Identities

If we start with the sum identity for sine,

$$\sin(x + y) = \sin x \cos y + \cos x \sin y$$

and let $y = x$, we obtain

$$\sin(x + x) = \sin x \cos x + \cos x \sin x$$

or

$$\mathbf{\sin 2x = 2 \sin x \cos x} \qquad (1)$$

Similarly, if we start with the sum identity for cosine,

$$\cos(x + y) = \cos x \cos y - \sin x \sin y$$

and let $y = x$, we obtain

$$\cos(x + x) = \cos x \cos x - \sin x \sin x$$

or

$$\mathbf{\cos 2x = \cos^2 x - \sin^2 x} \qquad (2)$$

Now, using the Pythagorean identities in the two forms

$$\cos^2 x = 1 - \sin^2 x \tag{3}$$

$$\sin^2 x = 1 - \cos^2 x \tag{4}$$

and substituting (3) into (2), we obtain

$$\cos 2x = 1 - \sin^2 x - \sin^2 x$$

$$\mathbf{\cos 2x = 1 - 2\sin^2 x} \tag{5}$$

Substituting (4) into (2), we obtain

$$\cos 2x = \cos^2 x - (1 - \cos^2 x)$$

$$\mathbf{\cos 2x = 2\cos^2 x - 1} \tag{6}$$

A double-angle identity can be developed for the tangent function in the same way by starting with the sum identity for tangent. This is left as an exercise for you to do. We list these double-angle identities for convenient reference.

DOUBLE-ANGLE IDENTITIES

For x any real number or angle in degree or radian measure for which both sides are defined:

$$\sin 2x = 2\sin x\cos x \qquad \cos 2x = \cos^2 x - \sin^2 x$$

$$= 1 - 2\sin^2 x$$

$$\tan 2x = \frac{2\tan x}{1 - \tan^2 x} \qquad = 2\cos^2 x - 1$$

The double-angle formulas for cosine, written in the following forms, are used in calculus to transform power forms to nonpower forms:

$$\sin^2 x = \frac{1 - \cos 2x}{2} \qquad \cos^2 x = \frac{1 + \cos 2x}{2}$$

EXPLORE/DISCUSS 1

(A) Show that the following equations are **not** identities:

$$\sin 2x = 2\sin x \qquad \cos 2x = 2\cos x \qquad \tan 2x = 2\tan x$$

(B) Graph $y_1 = \sin 2x$ and $y_2 = 2\sin x$ in the same viewing window. What can you conclude? Repeat the process for the other two equations in part (A).

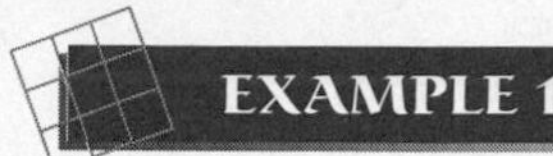

EXAMPLE 1 Verifying an Identity

Verify the identity: $\sin 2x = \dfrac{2\tan x}{1 + \tan^2 x}$

Verification We start with the right side:

$$\frac{2\tan x}{1+\tan^2 x} = \frac{2\left(\dfrac{\sin x}{\cos x}\right)}{1 + \dfrac{\sin^2 x}{\cos^2 x}} \qquad \text{Quotient identity}$$

$$= \frac{2\sin x\cos x}{\cos^2 x + \sin^2 x} \qquad \text{Multiply numerator and denominator by } \cos^2 x.$$

$$= \frac{\sin 2x}{1} \qquad \text{Double-angle and Pythagorean identities}$$

$$= \sin 2x$$

■

Matched Problem 1 Verify the identity: $\cos 2x = \dfrac{1 - \tan^2 x}{1 + \tan^2 x}$ ■

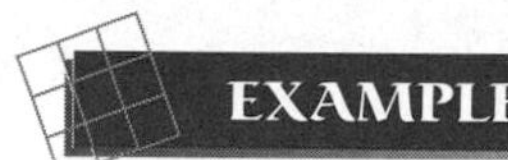

EXAMPLE 2 Using Double-Angle Identities

Find the exact value of $\cos 2x$ and $\tan 2x$ if $\sin x = \frac{4}{5}$, $\pi/2 < x < \pi$.

Solution First draw a reference triangle in the second quadrant, and find $\cos x$ and $\tan x$ (see Fig. 1).

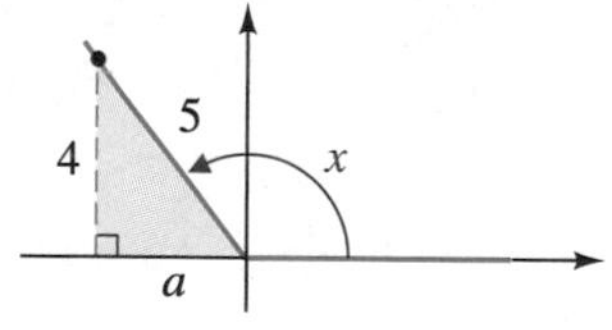

FIGURE 1

$$a = -\sqrt{5^2 - 4^2} = -3$$

$$\sin x = \tfrac{4}{5}$$

$$\cos x = -\tfrac{3}{5}$$

$$\tan x = -\tfrac{4}{3}$$

$$\cos 2x = 1 - 2\sin^2 x \qquad \text{Use double-angle identity and the results above.}$$

$$= 1 - 2\left(\tfrac{4}{5}\right)^2$$

$$= -\tfrac{7}{25}$$

$$\tan 2x = \frac{2\tan x}{1 - \tan^2 x} \qquad \text{Use double-angle identity and the preceding results.}$$

$$= \frac{2\left(-\frac{4}{3}\right)}{1 - \left(-\frac{4}{3}\right)^2}$$

$$= \frac{24}{7}$$

■

Matched Problem 2 Find the exact value of $\sin 2x$ and $\cos 2x$ if $\tan x = -\frac{3}{4}$, $-\pi/2 < x < 0$. ■

Half-Angle Identities

Half-angle identities are simply double-angle identities in an alternative form. We start with the double-angle identity for cosine in the form

$$\cos 2u = 1 - 2\sin^2 u$$

and let $u = x/2$. Then

$$\cos x = 1 - 2\sin^2 \frac{x}{2}$$

Now solve for $\sin(x/2)$ to obtain a half-angle formula for the sine function:

$$2\sin^2 \frac{x}{2} = 1 - \cos x$$

$$\sin^2 \frac{x}{2} = \frac{1 - \cos x}{2}$$

$$\sin \frac{x}{2} = \pm\sqrt{\frac{1 - \cos x}{2}} \qquad (7)$$

In identity (7), the choice of the sign is determined by the quadrant in which $x/2$ lies.

Now we start with the double-angle identity for cosine in the form

$$\cos 2u = 2\cos^2 u - 1$$

and let $u = x/2$. We then obtain a half-angle formula for the cosine function:

$$\cos x = 2\cos^2 \frac{x}{2} - 1$$

$$2\cos^2 \frac{x}{2} = 1 + \cos x$$

$$\cos^2 \frac{x}{2} = \frac{1 + \cos x}{2}$$

$$\cos \frac{x}{2} = \pm\sqrt{\frac{1 + \cos x}{2}} \qquad (8)$$

In identity (8), the choice of the sign is again determined by the quadrant in which $x/2$ lies.

To obtain a half-angle identity for the tangent function, we can use the quotient identity and the half-angle formulas for sine and cosine:

$$\tan \frac{x}{2} = \frac{\sin \frac{x}{2}}{\cos \frac{x}{2}} = \frac{\pm\sqrt{\frac{1 - \cos x}{2}}}{\pm\sqrt{\frac{1 + \cos x}{2}}} = \pm\sqrt{\frac{1 - \cos x}{1 + \cos x}} \qquad (9)$$

where the sign is determined by the quadrant in which $x/2$ lies.

We now list all the half-angle identities for convenient reference. Two of the half-angle identities for tangent are left as Problems 27 and 28 in Exercise 4.4.

HALF-ANGLE IDENTITIES

For x any real number or angle in degree or radian measure for which both sides are defined:

$$\sin \frac{x}{2} = \pm\sqrt{\frac{1 - \cos x}{2}} \qquad \cos \frac{x}{2} = \pm\sqrt{\frac{1 + \cos x}{2}}$$

$$\tan \frac{x}{2} = \pm\sqrt{\frac{1 - \cos x}{1 + \cos x}} = \frac{\sin x}{1 + \cos x} = \frac{1 - \cos x}{\sin x}$$

where the sign is determined by the quadrant containing $x/2$.

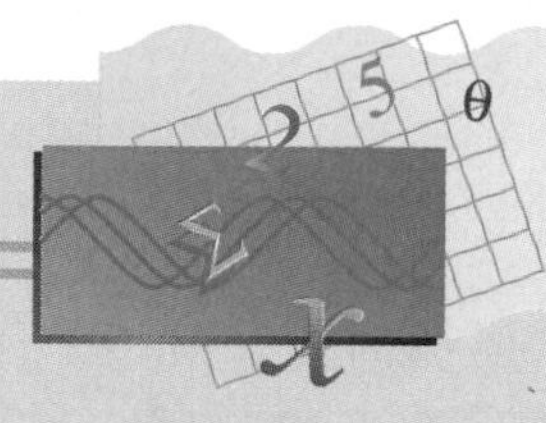

EXPLORE/DISCUSS 2

(A) Show that the following equations are **not** identities:

$$\sin \frac{x}{2} = \frac{1}{2} \sin x \qquad \cos \frac{x}{2} = \frac{1}{2} \cos x \qquad \tan \frac{x}{2} = \frac{1}{2} \tan x$$

(B) Graph $y_1 = \sin (x/2)$ and $y_2 = \frac{1}{2}\sin x$ in the same viewing window. What can you conclude? Repeat the process for the other two equations in part (A).

EXAMPLE 3 **Using a Half-Angle Identity**

Find cos 165° exactly by means of a half-angle identity.

Solution

$$\cos 165° = \cos \frac{330°}{2} = -\sqrt{\frac{1 + \cos 330°}{2}}$$

The negative square root is used since 165° is in the second quadrant and cosine is negative there. We complete the evaluation by noting that the reference triangle for 330° is a 30°–60° triangle in the fourth quadrant (see Fig. 2).

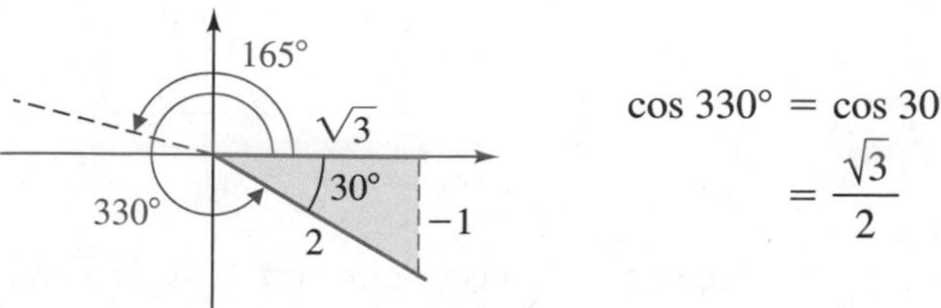

FIGURE 2

Thus,

$$\cos 165° = -\sqrt{\frac{1 + \sqrt{3}/2}{2}} = -\frac{\sqrt{2 + \sqrt{3}}}{2}$$ ■

Matched Problem 3 Find the exact value of sin 165° using a half-angle identity. ■

EXAMPLE 4 Using Half-Angle Identities

Find the exact value of $\sin(x/2)$, $\cos(x/2)$, and $\tan(x/2)$ if $\sin x = -\frac{3}{5}$, $\pi < x < 3\pi/2$.

Solution Draw a reference triangle in the third quadrant and find $\cos x$ (see Fig. 3).

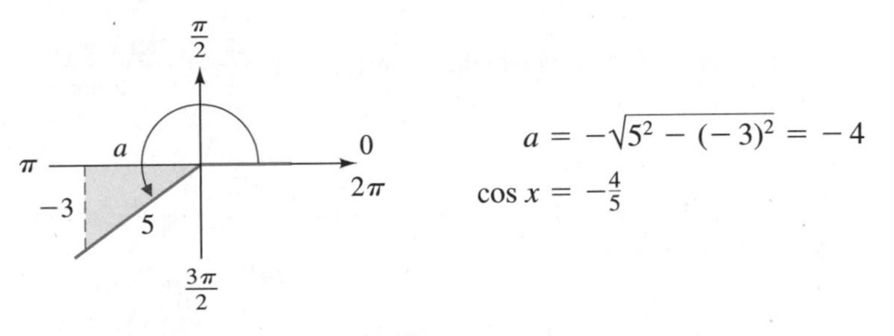

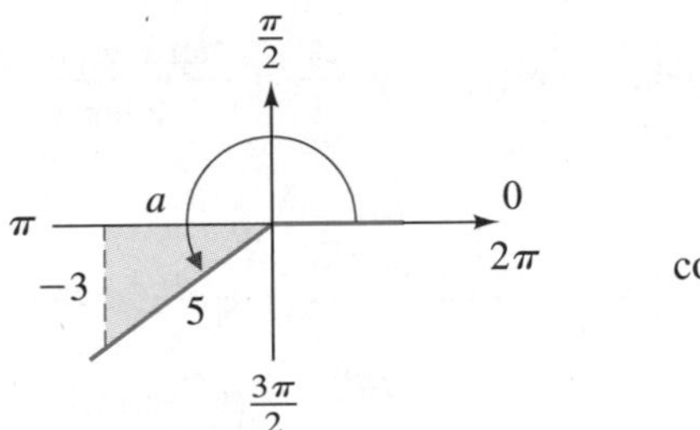

$$a = -\sqrt{5^2 - (-3)^2} = -4$$
$$\cos x = -\tfrac{4}{5}$$

FIGURE 3

If $\pi < x < 3\pi/2$, then

$\pi/2 < x/2 < 3\pi/4$ Divide each member of $\pi < x < 3\pi/2$ by 2.

Thus, $x/2$ is in the second quadrant, where sine is positive and cosine and tangent are negative. Using half-angle identities, we obtain

$$\sin\frac{x}{2} = \sqrt{\frac{1 - \cos x}{2}} = \sqrt{\frac{1 - (-\frac{4}{5})}{2}} = \sqrt{\frac{9}{10}} \text{ or } \frac{3\sqrt{10}}{10}$$

$$\cos\frac{x}{2} = -\sqrt{\frac{1 + \cos x}{2}} = -\sqrt{\frac{1 + (-\frac{4}{5})}{2}} = -\sqrt{\frac{1}{10}} \text{ or } \frac{-\sqrt{10}}{10}$$

$$\tan\frac{x}{2} = \frac{\sin(x/2)}{\cos(x/2)} = \frac{3\sqrt{10}/10}{-\sqrt{10}/10} = -3$$ ■

Matched Problem 4 Find the exact value for $\sin(x/2)$, $\cos(x/2)$, and $\tan(x/2)$ if $\cot x = -\frac{4}{3}$, $\pi/2 < x < \pi$. ■

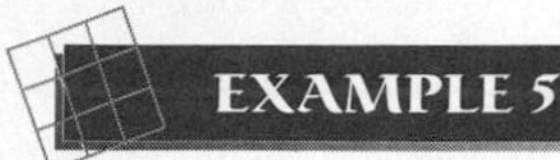

EXAMPLE 5 Verifying an Identity

Verify the identity: $\cos^2 \frac{x}{2} = \frac{\tan x + \sin x}{2 \tan x}$

Verification

$$\cos^2 \frac{x}{2} = \frac{1 + \cos x}{2}$$ Square both sides of the half-angle identity for cosine.

$$= \frac{\tan x}{\tan x} \cdot \frac{1 + \cos x}{2}$$ Algebra

$$= \frac{\tan x + \tan x \cos x}{2 \tan x}$$ Algebra

$$= \frac{\tan x + \sin x}{2 \tan x}$$ Quotient identity and algebra ■

Matched Problem 5 Verify the identity: $\sin^2 \frac{x}{2} = \frac{\tan x - \sin x}{2 \tan x}$ ■

Answers to Matched Problems

1. $\dfrac{1 - \tan^2 x}{1 + \tan^2 x} = \dfrac{1 - \dfrac{\sin^2 x}{\cos^2 x}}{1 + \dfrac{\sin^2 x}{\cos^2 x}} = \cos^2 x - \sin^2 x = \cos 2x$
2. $\sin 2x = -\frac{24}{25}$; $\cos 2x = \frac{7}{25}$
3. $\dfrac{\sqrt{2 - \sqrt{3}}}{2}$
4. $\sin(x/2) = 3\sqrt{10}/10$, $\cos(x/2) = \sqrt{10}/10$, $\tan(x/2) = 3$
5. $\sin^2 \frac{x}{2} = \frac{1 - \cos x}{2} = \frac{\tan x}{\tan x} \cdot \frac{1 - \cos x}{2} = \frac{\tan x - \sin x}{2 \tan x}$

EXERCISE 4.4

A *Evaluate each side of the indicated identity for $x = 60°$ (thus verifying it for one particular case).*

1. $\sin 2x = 2 \sin x \cos x$
2. $\cos 2x = \cos^2 x - \sin^2 x$
3. $\tan 2x = \dfrac{2 \tan x}{1 - \tan^2 x}$
4. $\sin \dfrac{x}{2} = \pm\sqrt{\dfrac{1 - \cos x}{2}}$

Use half-angle identities to find the exact value of Problems 5–8. Do not use a calculator.

5. $\sin 105°$
6. $\cos 105°$
7. $\tan 15°$
8. $\tan 75°$

In Problems 9–12, graph y_1 and y_2 in the same viewing window. Then use TRACE *to compare the two graphs.*

9. $y_1 = 2 \sin x \cos x$, $y_2 = \sin 2x$, $-2\pi < x < 2\pi$
10. $y_1 = \cos^2 x - \sin^2 x$, $y_2 = \cos 2x$, $-2\pi < x < 2\pi$
11. $y_1 = \dfrac{2 \tan x}{1 - \tan^2 x}$, $y_2 = \tan 2x$, $-\pi < x < \pi$
12. $y_1 = \dfrac{\sin x}{1 + \cos x}$, $y_2 = \tan \dfrac{x}{2}$, $-2\pi < x < 2\pi$

B *In Problems 13–30, verify each identity.*

13. $\sin 2x = (\tan x)(1 + \cos 2x)$
14. $(\sin x + \cos x)^2 = 1 + \sin 2x$

15. $2\sin^2\dfrac{x}{2} = \dfrac{\sin^2 x}{1+\cos x}$

16. $2\cos^2\dfrac{x}{2} = \dfrac{\sin^2 x}{1-\cos x}$

17. $(\sin\theta - \cos\theta)^2 = 1 - \sin 2\theta$

18. $\sin 2\theta = (\sin\theta + \cos\theta)^2 - 1$

19. $\cos^2\dfrac{w}{2} = \dfrac{1+\cos w}{2}$

20. $\sin^2\dfrac{w}{2} = \dfrac{1-\cos w}{2}$

21. $\cot\dfrac{\alpha}{2} = \dfrac{1+\cos\alpha}{\sin\alpha}$

22. $\cot\dfrac{\alpha}{2} = \dfrac{\sin\alpha}{1-\cos\alpha}$

23. $\dfrac{\cos 2t}{1-\sin 2t} = \dfrac{1+\tan t}{1-\tan t}$

24. $\cos 2t = \dfrac{1-\tan^2 t}{1+\tan^2 t}$

25. $\tan 2x = \dfrac{2\tan x}{1-\tan^2 x}$

26. $\sin 2x = \dfrac{2\tan x}{1+\tan^2 x}$

27. $\tan\dfrac{x}{2} = \dfrac{\sin x}{1+\cos x}$

28. $\tan\dfrac{x}{2} = \dfrac{1-\cos x}{\sin x}$

29. $\sec^2 x = (\sec 2x)(2 - \sec^2 x)$

30. $2\csc 2x = \dfrac{1+\tan^2 x}{\tan x}$

Find the exact value of $\sin 2x$, $\cos 2x$, and $\tan 2x$ for the information given in Problems 31–34. Do not use a calculator.

31. $\cos x = -\frac{4}{5}$, $\pi/2 < x < \pi$

32. $\sin x = \frac{3}{5}$, $\pi/2 < x < \pi$

33. $\cot x = -\frac{5}{12}$, $-\pi/2 < x < 0$

34. $\tan x = -\frac{5}{12}$, $-\pi/2 < x < 0$

Find the exact value of $\sin(x/2)$ and $\cos(x/2)$ for the information given in Problems 35–38. Do not use a calculator.

35. $\cos x = \frac{1}{3}$, $0° < x < 90°$

36. $\sin x = \frac{4}{5}$, $0° < x < 90°$

37. $\sin x = -\frac{1}{3}$, $\pi < x < 3\pi/2$

38. $\cos x = -\frac{1}{4}$, $\pi < x < 3\pi/2$

Your friend is having trouble finding exact values of $\sin\theta$ and $\cos\theta$ from the information given in Problems 39 and 40, and comes to you for help. Instead of just working the problems, you guide your friend through the solution process using the following questions (A)–(E). What is the correct response to each question for each problem?

(A) The angle 2θ is in which quadrant? How do you know?

(B) How can you find $\sin 2\theta$ and $\cos 2\theta$? Find each.

(C) Which identities relate $\sin\theta$ and $\cos\theta$ with either $\sin 2\theta$ or $\cos 2\theta$?

(D) How would you use the identities in part (C) to find $\sin\theta$ and $\cos\theta$ exactly, including the correct sign?

(E) What are the exact values for $\sin\theta$ and $\cos\theta$?

39. Find the exact values of $\sin\theta$ and $\cos\theta$, given $\sec 2\theta = -\frac{5}{4}$, $0° < \theta < 90°$.

40. Find the exact values of $\sin\theta$ and $\cos\theta$, given $\tan 2\theta = -\frac{4}{3}$, $0° < \theta < 90°$.

41. In applied mathematics, approximate forms are often substituted for exact forms to simplify formulas or computations. Graph each side of each statement below in the same viewing window, $-\pi/2 \le x \le \pi/2$, to show that the approximation is valid for x close to 0. Use TRACE and describe what happens to the approximation as x gets closer to 0.

(A) $\sin 2x \approx 2\sin x$ (B) $\sin\dfrac{x}{2} \approx \dfrac{1}{2}\sin x$

42. Repeat Problem 41 for:

(A) $\tan 2x \approx 2\tan x$ (B) $\tan\dfrac{x}{2} \approx \dfrac{1}{2}\tan x$

In Problems 43–46, graph y_1 and y_2 in the same viewing window for $-2\pi \le x \le 2\pi$, and state the intervals for which y_1 and y_2 are identities.

43. $y_1 = \sin\dfrac{x}{2}$, $y_2 = \sqrt{\dfrac{1-\cos x}{2}}$

44. $y_1 = \sin\dfrac{x}{2}$, $y_2 = -\sqrt{\dfrac{1-\cos x}{2}}$

45. $y_1 = \cos\dfrac{x}{2}$, $y_2 = -\sqrt{\dfrac{1+\cos x}{2}}$

46. $y_1 = \cos\dfrac{x}{2}$, $y_2 = \sqrt{\dfrac{1+\cos x}{2}}$

C *Find the exact value of $\sin x$, $\cos x$, and $\tan x$ for the information given in Problems 47–50. Do not use a calculator.*

47. $\sin 2x = \frac{3}{5}$, $0 < x < \pi/4$

48. $\tan 2x = -\frac{4}{3}$, $0 < x < \pi/2$

49. $\sec 2x = -\frac{5}{3}$, $-\pi/2 < x < 0$

50. $\csc 2x = -\frac{5}{3}$, $-\pi/4 < x < 0$

In Problems 51–56, verify each identity.

51. $\sin 3x = 3\sin x - 4\sin^3 x$

52. $\cos 3x = 4\cos^3 x - 3\cos x$

53. $\sin 4x = (\cos x)(4\sin x - 8\sin^3 x)$

54. $\cos 4x = 8\cos^4 x - 8\cos^2 x + 1$

55. $\tan 3x = \dfrac{3\tan x - \tan^3 x}{1 - 3\tan^2 x}$

56. $4\sin^4 x = 1 - 2\cos 2x + \cos^2 2x$

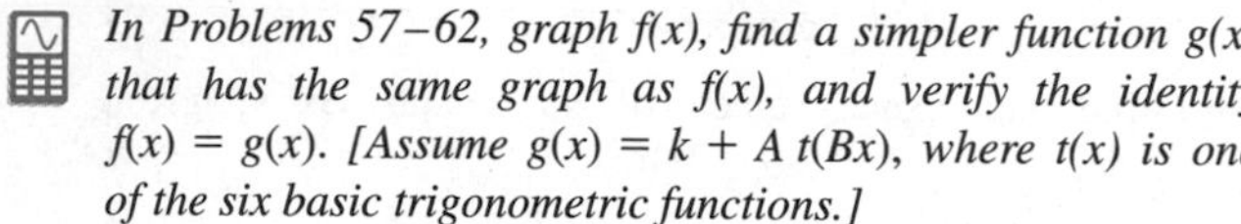

In Problems 57–62, graph f(x), find a simpler function g(x) that has the same graph as f(x), and verify the identity f(x) = g(x). [Assume g(x) = k + A t(Bx), where t(x) is one of the six basic trigonometric functions.]

57. $f(x) = \csc x + \cot x$

58. $f(x) = \csc x - \cot x$

59. $f(x) = \dfrac{\cot x}{1 + \cos 2x}$

60. $f(x) = \dfrac{1}{\cot x \sin 2x - 1}$

61. $f(x) = \dfrac{1 + 2\cos 2x}{1 + 2\cos x}$

62. $f(x) = \dfrac{1 - 2\cos 2x}{2\sin x - 1}$

Problems 63 and 64 refer to the following identity for n a positive integer (this identity is established in more advanced mathematics):

$$\frac{1}{2} + \cos x + \cdots + \cos nx = \frac{\sin\left(\dfrac{2n+1}{2}x\right)}{2\sin\left(\dfrac{1}{2}x\right)}$$

63. Graph each side of the equation for $n = 2$, $-2\pi \le x \le 2\pi$.

64. Graph each side of the equation for $n = 3$, $-2\pi \le x \le 2\pi$.

Applications

65. **Sports—Javelin Throw** In physics it can be shown that the theoretical horizontal distance d a javelin will travel (see the figure) is given approximately by

$$d = \frac{v_0^2 \sin\theta\cos\theta}{16}$$

where v_0 is the initial velocity of the javelin (in feet per second). (Air resistance and athlete height are ignored.)

(A) Write the formula in terms of the sine function only by using a suitable identity.

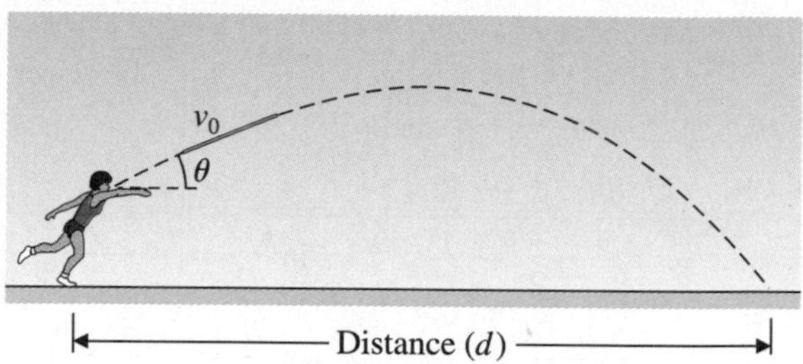

Figure for 65

(B) Use the resulting equation from part (A) to determine the angle θ that will produce the maximum horizontal distance d for a given initial speed v_0. Explain your reasoning. This result is an important consideration for shot-putters, archers, and javelin and discus throwers.

(C) A world-class javelin thrower can throw the javelin with an initial velocity of 100 ft/sec. Graph the equation from part (A), and use a built-in routine to find the maximum distance d and the angle θ (in degrees) that produces the maximum distance. Describe what happens to d as θ goes from 0° to 90°.

66. **Precalculus: Geometry** An n-sided regular polygon is inscribed in a circle of radius r (see the figure).

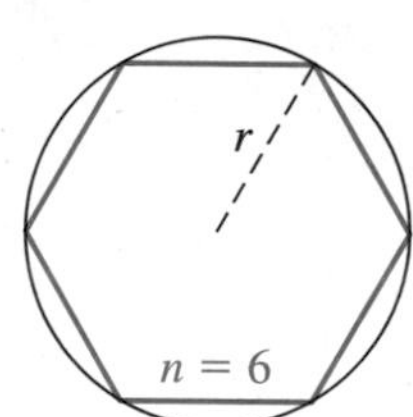

Figure for 66

(A) Show that the area of the n-sided polygon is given by

$$A_n = \frac{1}{2}nr^2 \sin\frac{2\pi}{n}.$$

[*Hint:* Area of triangle = (Base)(Height)/2. A double-angle identity is useful.]

(B) For a circle of radius 1, use the formula from part (A) to complete Table 1 (to six decimal places).

TABLE 1

n	10	100	1,000	10,000
A_n				

(C) What does A_n seem to approach as n increases without bound?

(D) How close can A_n be made to get to the actual area of the circle? Will A_n ever equal the exact area of the circle for any chosen n, however large? Explain. (In calculus, the area of the circumscribed circle, π, is called the *limit* of A_n as n increases without bound. Symbolically, we write $\lim_{n\to\infty} A_n = \pi$. The limit concept is fundamental to the development of calculus.)

67. Construction An animal shelter is to be constructed using two 4 ft by 8 ft sheets of exterior plywood for the roof (see the figure). We are interested in approximating the value of θ that will give the maximum interior volume. [*Note:* The interior volume is the area of the triangular end, $bh/2$, times the length of the shelter, 8 ft.]

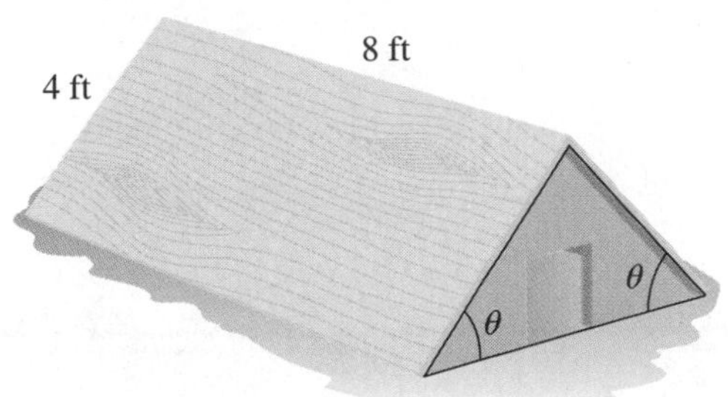

Figure for 67

(A) Show that the interior volume is given by

$$V = 64 \sin 2\theta$$

(B) Explain how you can determine from the equation in part (A) the value of θ that produces the maximum volume, and find θ and the maximum volume.

(C) Complete Table 2 (to one decimal place) and find the maximum volume in the table and the value of θ that produces it. (Use the table feature in your calculator if it has one.)

TABLE 2

θ (deg)	30	35	40	45	50	55	60
V (ft³)	55.4						

(D) Graph the equation in part (A) on a graphing utility, $0° \le \theta \le 90°$, and use a built-in routine to find the maximum volume and the value of θ that produces it.

* **68. Construction** A new road is to be constructed from resort P to resort Q, turning at a point R on the horizontal line through resort P, as indicated in the figure.

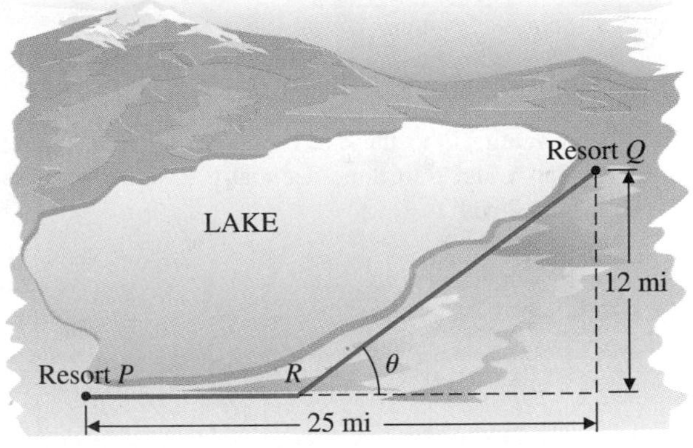

Figure for 68

(A) Show that the length of the road from P to Q through R is given by

$$d = 25 + 12 \tan \frac{\theta}{2}$$

(B) Because of the lake, θ is restricted to $40° \le \theta \le 90°$. What happens to the length of the road as θ varies between 40° and 90°?

(C) Complete Table 3 (to one decimal place) and find the maximum and minimum length of the road. (Use a table-generating feature on your calculator if it has one.)

TABLE 3

θ (deg)	40	50	60	70	80	90
d (mi)	29.4					

(D) Graph the equation in part (A) for the restrictions in part (B); then use a built-in routine to determine the maximum and minimum length of the road.

69. **Engineering** Find the exact value of x in the figure; then find x and θ to three decimal places. [*Hint:* Use $\tan 2\theta = (2 \tan \theta)/(1 - \tan^2 \theta)$.]

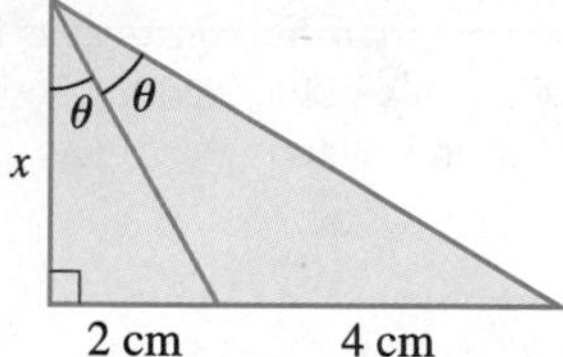

Figure for 69

70. **Engineering** Find the exact value of x in the figure; then find x and θ to three decimal places. [*Hint:* Use $\cos 2\theta = 2 \cos^2 \theta - 1$.]

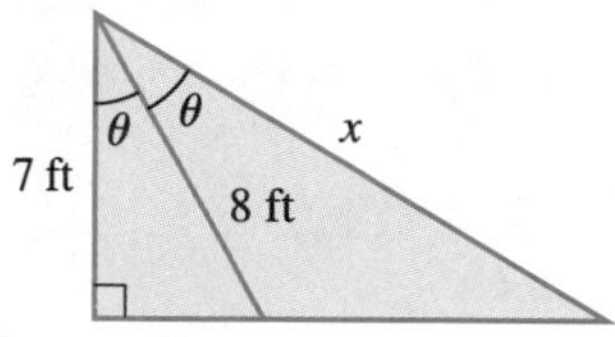

Figure for 70

** **71.** **Geometry** In part (a) of the figure, M and N are the midpoints of the sides of a square. Find the exact value of $\cos \theta$. [*Hint:* The solution uses the Pythagorean theorem, the definitions of sine and cosine, a half-angle identity, and some auxiliary lines as drawn in part (b) of the figure.]

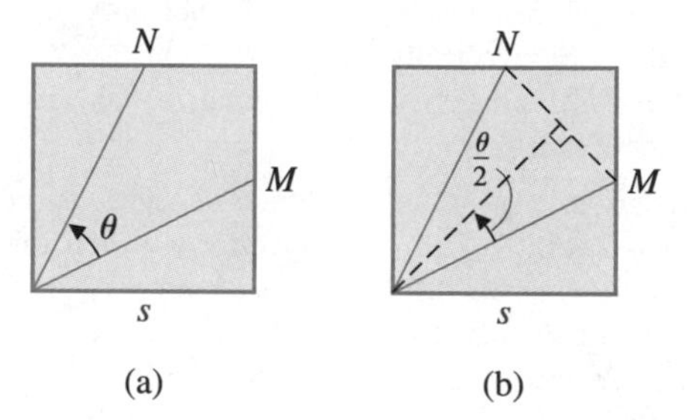

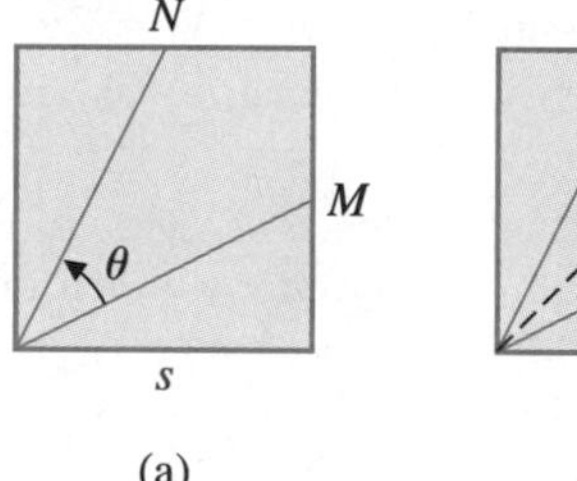

Figure for 71

☆4.5 PRODUCT–SUM AND SUM–PRODUCT IDENTITIES

- Product–Sum Identities
- Sum–Product Identities
- Application: Music

In this section, we will develop identities for converting the product of two trigonometric functions into the sum of two trigonometric functions, and vice versa. These identities have many uses, both theoretical and practical. In calculus, knowing how to convert a product into a sum will allow easy solutions to some problems that would otherwise be difficult to solve. A discussion of beat frequencies in music, which appears at the end of this section, demonstrates how converting a sum into a product aids in the analysis of the beat phenomenon in sound. A similar type of analysis is used in modeling certain types of long-range underwater sound propagation for submarine detection.

☆ Sections marked with a star may be omitted without loss of continuity.

Product–Sum Identities

The product–sum identities are easily derived from the sum and difference identities developed in Section 4.3. To obtain a product–sum identity, we add, left side to left side and right side to right side, the sum and difference identities for sine:

$$\sin(x + y) = \sin x \cos y + \cos x \sin y$$
$$\sin(x - y) = \sin x \cos y - \cos x \sin y$$
$$\sin(x + y) + \sin(x - y) = 2 \sin x \cos y$$

or

$$\sin x \cos y = \tfrac{1}{2}[\sin(x + y) + \sin(x - y)]$$

Similarly, by adding or subtracting appropriate sum and difference identities, we can obtain three other product identities for sines and cosines. These identities are listed below for convenient reference.

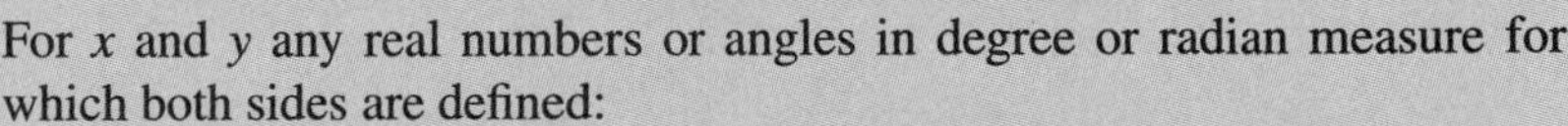

PRODUCT–SUM IDENTITIES

For x and y any real numbers or angles in degree or radian measure for which both sides are defined:

$$\sin x \cos y = \tfrac{1}{2}[\sin(x + y) + \sin(x - y)]$$
$$\cos x \sin y = \tfrac{1}{2}[\sin(x + y) - \sin(x - y)]$$
$$\sin x \sin y = \tfrac{1}{2}[\cos(x - y) - \cos(x + y)]$$
$$\cos x \cos y = \tfrac{1}{2}[\cos(x + y) + \cos(x - y)]$$

EXAMPLE 1 **Using a Product–Sum Identity**

Write the product $\cos 3t \sin t$ as a sum or difference.

Solution

$$\cos x \sin y = \tfrac{1}{2}[\sin(x + y) - \sin(x - y)] \qquad \text{Let } x = 3t \text{ and } y = t.$$
$$\cos 3t \sin t = \tfrac{1}{2}[\sin(3t + t) - \sin(3t - t)]$$
$$= \tfrac{1}{2} \sin 4t - \tfrac{1}{2} \sin 2t$$

Matched Problem 1 Write the product $\cos 5\theta \cos 2\theta$ as a sum or difference.

EXAMPLE 2 **Using a Product–Sum Identity**

Evaluate $\sin 105° \sin 15°$ exactly using a product–sum identity.

Solution

$$\sin x \sin y = \tfrac{1}{2}[\cos(x - y) - \cos(x + y)]$$
$$\sin 105° \sin 15° = \tfrac{1}{2}[\cos(105° - 15°) - \cos(105° + 15°)]$$
$$= \tfrac{1}{2}[\cos 90° - \cos 120°]$$
$$= \tfrac{1}{2}[0 - (-\tfrac{1}{2})] = \tfrac{1}{4} \quad \text{or} \quad 0.25$$

Matched Problem 2 Evaluate cos 165° sin 75° exactly using a product–sum identity.

Sum–Product Identities

The product–sum identities can be transformed into equivalent forms called sum–product identities. These identities are used to express sums and differences involving sines and cosines as products involving sines and cosines. We illustrate the transformation for one identity. The other three identities can be obtained by following the same procedure.

Let us start with the product–sum identity

$$\sin \alpha \cos \beta = \tfrac{1}{2}[\sin(\alpha + \beta) + \sin(\alpha - \beta)] \qquad (1)$$

We would like

$$\alpha + \beta = x \qquad \alpha - \beta = y$$

Solving this system, we have

$$\alpha = \frac{x + y}{2} \qquad \beta = \frac{x - y}{2} \qquad (2)$$

By substituting (2) into identity (1) and simplifying, we obtain

$$\sin x + \sin y = 2 \sin \frac{x + y}{2} \cos \frac{x - y}{2}$$

All four sum–product identities are listed below for convenient reference.

SUM–PRODUCT IDENTITIES

For x and y any real numbers or angles in degree or radian measure for which both sides are defined:

$$\sin x + \sin y = 2 \sin \frac{x + y}{2} \cos \frac{x - y}{2}$$

$$\sin x - \sin y = 2 \cos \frac{x + y}{2} \sin \frac{x - y}{2}$$

$$\cos x + \cos y = 2 \cos \frac{x + y}{2} \cos \frac{x - y}{2}$$

$$\cos x - \cos y = -2 \sin \frac{x + y}{2} \sin \frac{x - y}{2}$$

EXAMPLE 3 **Using a Sum–Product Identity**

Write the difference $\sin 7\theta - \sin 3\theta$ as a product.

Solution

$$\sin x - \sin y = 2 \cos \frac{x + y}{2} \sin \frac{x - y}{2}$$

$$\sin 7\theta - \sin 3\theta = 2 \cos \frac{7\theta + 3\theta}{2} \sin \frac{7\theta - 3\theta}{2}$$

$$= 2 \cos 5\theta \sin 2\theta$$

Matched Problem 3 Write the sum $\cos 3t + \cos t$ as a product.

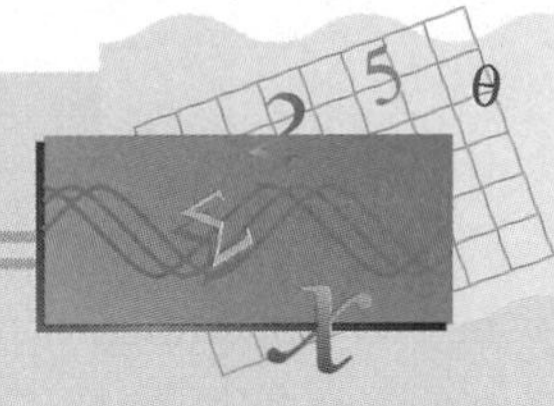

EXPLORE/DISCUSS 1

Proof without Words: Sum–Product Identities

Discuss how the relationships below Figure 1 can be verified from the figure.

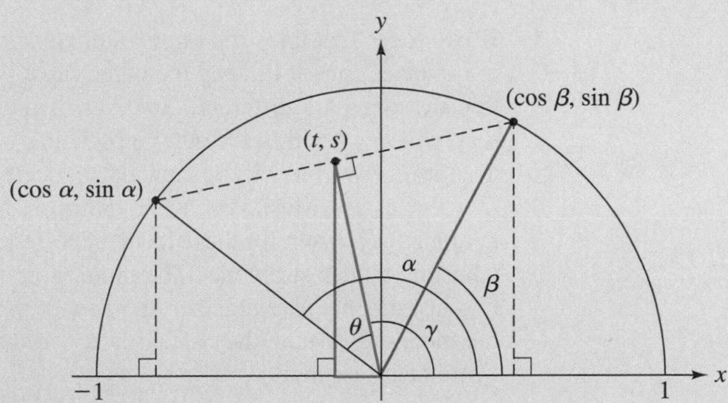

FIGURE 1

$$\theta = \frac{\alpha - \beta}{2}, \quad \gamma = \frac{\alpha + \beta}{2}$$

$$\frac{\sin \alpha}{2} + \frac{\sin \beta}{2} = s = \cos \frac{\alpha - \beta}{2} \sin \frac{\alpha + \beta}{2}$$

$$\frac{\cos \alpha}{2} + \frac{\cos \beta}{2} = t = \cos \frac{\alpha - \beta}{2} \cos \frac{\alpha + \beta}{2}$$

(The above is based on a similar "proof" by Sidney H. Kung, Jacksonville University, printed in the October 1996 issue of *Mathematics Magazine*.)

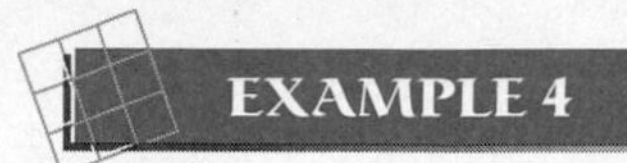

EXAMPLE 4 **Using a Sum–Product Identity**

Find the exact value of $\sin 105° - \sin 15°$ using an appropriate sum–product identity.

Solution

$$\sin x - \sin y = 2 \cos \frac{x + y}{2} \sin \frac{x - y}{2}$$

$$\begin{aligned} \sin 105° - \sin 15° &= 2 \cos \frac{105° + 15°}{2} \sin \frac{105° - 15°}{2} \\ &= 2 \cos 60° \sin 45° \\ &= 2\left(\frac{1}{2}\right)\left(\frac{\sqrt{2}}{2}\right) \\ &= \frac{\sqrt{2}}{2} \end{aligned}$$

Matched Problem 4 Find the exact value of $\cos 165° - \cos 75°$ by using an appropriate sum–product identity.

Application: Music

If two tones that have the same loudness and that are close in pitch (frequency) are sounded, one following the other, most people have difficulty in recognizing that the tones are different. However, if the tones are sounded simultaneously, they will react with each other, producing a low warbling sound called a **beat.** The beat or warble will be slow or rapid, depending on how far apart the initial frequencies are. Musicians, when tuning an instrument with other instruments or a tuning fork, listen for these lower beat frequencies and try to eliminate them by adjusting their instruments. The more rapid the beat frequency (warbling) when two instruments play together, the greater the difference in their frequencies and the more out of tune they are. If, after adjustments, no beats are heard, the two instruments are in tune.

What is behind this beat phenomenon? Figure 2a shows a tone of 64 Hz (cycles per second), and Figure 2b shows a tone of the same loudness but with a frequency of 72 Hz. If both tones are sounded simultaneously and time is started when the two tones are completely out of phase, waves (a) and (b) will interact to form wave (c).

Sum–product identities are useful in the mathematical analysis of the beat phenomenon. In particular, we will use the sum–product identity

$$\cos x - \cos y = -2 \sin \frac{x + y}{2} \sin \frac{x - y}{2}$$

in a brief mathematical discussion of the three waves in Figure 2.

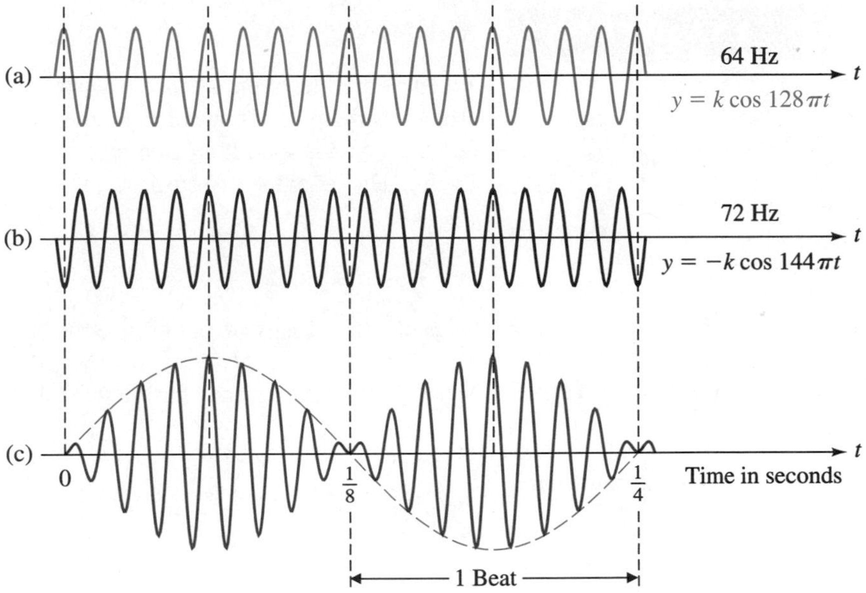

FIGURE 2
Beats

We start with the fact that wave (c) is the sum of waves (a) and (b):

$$
\begin{aligned}
y &= k \cos 128\pi t - k \cos 144\pi t \\
&= k(\cos 128\pi t - \cos 144\pi t) \\
&= k\left(-2 \sin \frac{128\pi t + 144\pi t}{2} \sin \frac{128\pi t - 144\pi t}{2}\right) \\
&= -2k \sin 136\pi t \sin(-8\pi t) \\
&= 2k \sin 136\pi t \sin 8\pi t \\
&= 2k \sin 8\pi t \sin 136\pi t
\end{aligned}
$$

The factor $\sin 136\pi t$ represents a sound of frequency 68 Hz, the average of the frequencies of the individual tones. The first factor, $2k \sin 8\pi t$, can be thought of as a time-varying amplitude of the second factor, $\sin 136\pi t$. It is the first factor that describes the slow warble of volume, or the beat, of the combined sounds. This pulsation of sound is described in terms of frequency—that is, it tells how many times per second the maximum loudness or number of beats occurs. Referring to Figure 2c, we see that there are 2 beats in $\frac{1}{4}$ sec, or 8 beats/sec. Hence, the beat frequency is 8 beats/sec (which is the difference in frequencies of the two original waves).

In general, if two equally loud tones of frequencies f_1 and f_2 are produced simultaneously, and if $f_1 > f_2$, then the beat frequency, f_b, is given by

$$f_b = f_1 - f_2 \qquad \text{Beat frequency}$$

EXAMPLE 5 **Music**

When certain keys on a piano are struck, a felt-covered hammer strikes two strings. If the piano is out of tune, the tones from the two strings create a beat, and the sound is sour. If a piano tuner counts 15 beats in 5 sec, how far apart are the frequencies of the two strings?

Solution

$$f_b = \tfrac{15}{5} = 3 \text{ beats/sec}$$

$$f_1 - f_2 = f_b = 3 \text{ Hz}$$

Thus, the two strings are out of tune by 3 cycles/sec. ■

Matched Problem 5 What is the beat frequency for the two tones in Figure 2? ■

Answers to Matched Problems

1. $\cos 5\theta \cos 2\theta = \frac{1}{2}\cos 7\theta + \frac{1}{2}\cos 3\theta$ 2. $(-\sqrt{3} - 2)/4$
3. $\cos 3t + \cos t = 2\cos 2t \cos t$ 4. $-\sqrt{6}/2$ 5. $f_b = 8$ Hz

EXERCISE 4.5

A *Write each product as a sum or difference involving sines and cosines.*

1. $\cos 7A \cos 5A$
2. $\sin 3m \cos m$
3. $\cos 2\theta \sin 3\theta$
4. $\sin u \sin 3u$

Write each difference or sum as a product involving sines and cosines.

5. $\cos 7\theta + \cos 5\theta$
6. $\sin 3t + \sin t$
7. $\sin u - \sin 5u$
8. $\cos 5w - \cos 9w$

B *Evaluate each of the following exactly using an appropriate identity.*

9. $\cos 75° \sin 15°$
10. $\sin 195° \cos 75°$
11. $\sin 105° \sin 165°$
12. $\cos 15° \cos 75°$

Evaluate each of the following exactly using an appropriate identity.

13. $\sin 195° + \sin 105°$
14. $\cos 285° + \cos 195°$
15. $\sin 75° - \sin 165°$
16. $\cos 15° - \cos 105°$

In Problems 17 and 18, use sum and difference identities from Section 4.3 to establish each of the following:

17. $\sin x \sin y = \frac{1}{2}[\cos(x - y) - \cos(x + y)]$
18. $\cos x \cos y = \frac{1}{2}[\cos(x + y + \cos(x - y)]$

19. Explain how you can transform the product–sum identity

$$\cos u \cos v = \tfrac{1}{2}[\cos(u + v) + \cos(u - v)]$$

into the sum–product identity

$$\cos x + \cos y = 2\cos\frac{x + y}{2}\cos\frac{x - y}{2}$$

by a suitable substitution.

20. Explain how you can transform the product–sum identity

$$\sin u \sin v = \tfrac{1}{2}[\cos(u - v) - \cos(u + v)]$$

into the sum–product identity

$$\cos x - \cos y = -2\sin\frac{x + y}{2}\sin\frac{x - y}{2}$$

by a suitable substitution.

In Problems 21–28, verify each identity.

21. $\dfrac{\cos t - \cos 3t}{\sin t + \sin 3t} = \tan t$

22. $\dfrac{\sin 2t + \sin 4t}{\cos 2t - \cos 4t} = \cot t$

23. $\dfrac{\sin x + \sin y}{\cos x + \cos y} = \tan \dfrac{x + y}{2}$

24. $\dfrac{\sin x - \sin y}{\cos x - \cos y} = -\cot \dfrac{x + y}{2}$

25. $\dfrac{\cos x - \cos y}{\sin x + \sin y} = -\tan \dfrac{x - y}{2}$

26. $\dfrac{\cos x + \cos y}{\sin x - \sin y} = \cot \dfrac{x - y}{2}$

27. $\dfrac{\sin x + \sin y}{\sin x - \sin y} = \dfrac{\tan \frac{1}{2}(x + y)}{\tan \frac{1}{2}(x - y)}$

28. $\dfrac{\cos x + \cos y}{\cos x - \cos y} = -\cot \dfrac{x + y}{2} \cot \dfrac{x - y}{2}$

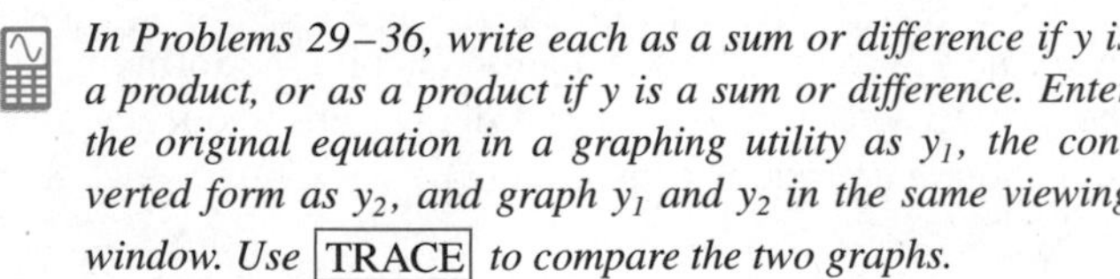

In Problems 29–36, write each as a sum or difference if y is a product, or as a product if y is a sum or difference. Enter the original equation in a graphing utility as y_1, the converted form as y_2, and graph y_1 and y_2 in the same viewing window. Use TRACE *to compare the two graphs.*

29. $y = \cos 5x \cos 3x$

30. $y = \sin 3x \cos x$

31. $y = \cos 1.9x \sin 0.5x$

32. $y = \sin 2.3x \sin 0.7x$

33. $y = \cos 3x + \cos x$

34. $y = \sin 2x + \sin x$

35. $y = \sin 2.1x - \sin 0.5x$

36. $y = \cos 1.7x - \cos 0.3x$

C *In Problems 37 and 38, verify each identity.*

37. $\sin x \sin y \sin z = \frac{1}{4}[\sin(x + y - z) + \sin(y + z - x) + \sin(z + x - y) - \sin(x + y + z)]$

38. $\cos x \cos y \cos z = \frac{1}{4}[\cos(x + y - z) + \cos(y + z - x) + \cos(z + x - y) + \cos(x + y + z)]$

In Problems 39–42:

(A) Graph y_1, y_2, and y_3 in the same viewing window for $0 \le x \le 1$ and $-2 \le y \le 2$.

(B) Convert y_1 to a sum or difference and repeat part (A).

39. $y_1 = 2\cos(16\pi x)\sin(2\pi x)$
$y_2 = 2\sin(2\pi x)$
$y_3 = -2\sin(2\pi x)$

40. $y_1 = 2\sin(20\pi x)\cos(2\pi x)$
$y_2 = 2\cos(2\pi x)$
$y_3 = -2\cos(2\pi x)$

41. $y_1 = 2\sin(24\pi x)\sin(2\pi x)$
$y_2 = 2\sin(2\pi x)$
$y_3 = -2\sin(2\pi x)$

42. $y_1 = 2\cos(28\pi x)\cos(2\pi x)$
$y_2 = 2\cos(2\pi x)$
$y_3 = -2\cos(2\pi x)$

Applications

43. **Music** If one tone is described by $y = k \sin 522\pi t$ and another by $y = k \sin 512\pi t$, write their sum as a product. What is the beat frequency if both notes are sounded together?

44. **Music** If one tone is described by $y = k \cos 524\pi t$ and another by $y = k \cos 508\pi t$, write their sum as a product. What is the beat frequency if both notes are sounded together?

45. **Music**

$$y = 0.3 \cos 72\pi t \qquad \text{and} \qquad y = -0.3 \cos 88\pi t$$

are equations of sound waves with frequencies 36 and 44 Hz, respectively. If both sounds are emitted simultaneously, a beat frequency results. Use the viewing window $0 \le t \le 0.25$, $-0.8 \le y \le 0.8$ for parts (A)–(D).

(A) Graph $y = 0.3 \cos 72\pi t$.

(B) Graph $y = -0.3 \cos 88\pi t$.

(C) Graph $y_1 = 0.3 \cos 72\pi t - 0.3 \cos 88\pi t$ and $y_2 = 0.6 \sin 8\pi t$ in the same viewing window.

(D) Convert y_1 in part (C) to a product, and graph the new y_1 along with y_2 from part (C) in the same viewing window.

46. **Music**

$$y = 0.4 \cos 132\pi t \qquad \text{and} \qquad y = -0.4 \cos 152\pi t$$

are equations of sound waves with frequencies 66 and 76 Hz, respectively. If both sounds are emitted simultaneously, a beat frequency results. Use the viewing window $0 \le t \le 0.2$, $-0.8 \le y \le 0.8$ for parts (A)–(D).

(A) Graph $y = 0.4 \cos 132\pi t$.

(B) Graph $y = -0.4 \cos 152\pi t$.

(C) Graph $y_1 = 0.4 \cos 132\pi t - 0.4 \cos 152\pi t$ and $y_2 = 0.8 \sin 10\pi t$ in the same viewing window.

(D) Convert y_1 in part (C) to a product, and graph the new y_1 along with y_2 from part (C) in the same viewing window.

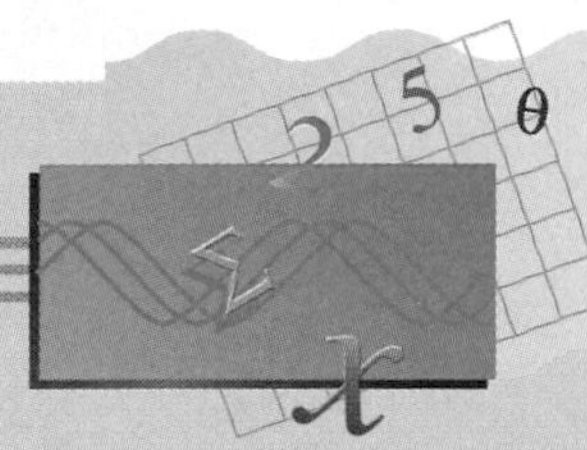

CHAPTER 4 GROUP ACTIVITY

From $M \sin Bt + N \cos Bt$ to $A \sin(Bt + C)$

Solving certain kinds of problems dealing with electrical circuits, spring–mass systems, heat flow, fluid flow, and so on, requires more advanced mathematics (differential equations), but often the solution process leads naturally to functions of the form

$$y = M \sin Bt + N \cos Bt \tag{1}$$

In the following investigation, you will show that function (1) is a simple harmonic; that is, it can be represented by an equation of the form

$$y = A \sin(Bt + C) \tag{2}$$

One of the objectives of this project is to transform equation (1) into the form of equation (2). Then it will be easy to determine the amplitude, period, frequency, and phase shift of the phenomenon that produced equation (1).

I EXPLORATION

Use a graphing utility to explore the nature of the graph of equation (1) for various values of M, N, and B. Do the graphs appear to be graphs of simple harmonics?

II GEOMETRIC SOLUTION

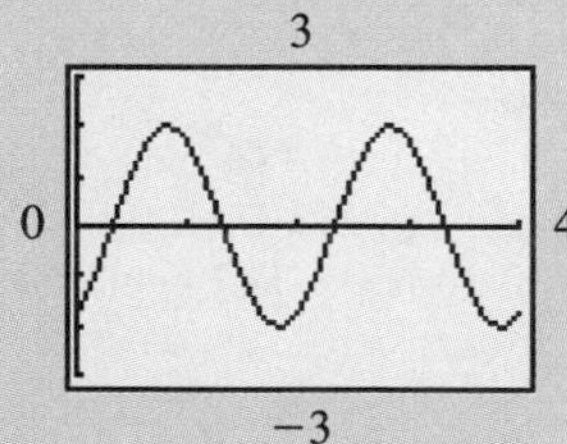

FIGURE 1
$y = \sin(\pi t) - \sqrt{3}\cos(\pi t)$

The graph of $y = \sin(\pi t) - \sqrt{3}\cos(\pi t)$ is shown in Figure 1. The graph looks like the graph of a simple harmonic.

(A) From the graph, find A, B, and C so that $y = A \sin(Bt + C)$ produces the same graph. To find C, estimate the phase shift by finding the first zero (x intercept) to the right of the origin using the zero routine in a graphing utility.

(B) Graph the equation obtained in part (A) and the original equation in the same viewing window, and compare y values for various values of x using TRACE. Conclusions? Write the results as an identity.

A geometric solution to the problem is straightforward as long as M, N, and B are simple enough—often they are not. We now turn to an analytical solution that can always be used no matter how complicated M, N, and B are.

III GENERAL ANALYTICAL SOLUTION

Equations (1) and (2) are stated again for convenient reference:

$$y = M \sin Bt + N \cos Bt \tag{1}$$

$$y = A \sin(Bt + C) \tag{2}$$

The problem is: Given M, N, and B in equation (1), find A, B, and C in equation (2) so that equation (2) produces the same graph as equation (1); that is, so that $M \sin Bt + N \cos Bt = A \sin(Bt + C)$ is an identity. The transformation identity, which you are to establish, is stated in the following box:

TRANSFORMATION IDENTITY

$$M \sin Bt + N \cos Bt = \sqrt{M^2 + N^2} \sin(Bt + C) \tag{3}$$

where C is any angle (in radians if t is real) having $P(M, N)$ on its terminal side.

(A) *Establish the transformation identity.* The process of finding A, B, and C, given M, N, and B requires a little ingenuity and the use of the sum identity

$$\sin(x + y) = \sin x \cos y + \cos x \sin y \tag{4}$$

Start by trying to get $M \sin Bt + N \cos Bt$ to look like the right side of the sum identity (4). Then we use (4), from right to left, to obtain (2).

Hint: A first step is the following:

$$M \sin Bt + N \cos Bt = \frac{\sqrt{M^2 + N^2}}{\sqrt{M^2 + N^2}} (M \sin Bt + N \cos Bt)$$

(B) *Example 1.* Use equation 3 to transform

$$y_1 = y = \sin(\pi t) - \sqrt{3} \cos(\pi t)$$

into the form $y_2 = A \sin(Bt + C)$, where C is chosen so that $|C|$ is minimum. Compute C to three decimal places, or find C exactly, if you can. (You should get the same results that you got in part II using the geometric approach.) From the new equation determine the amplitude, period, frequency, and phase shift.

Check by graphing y_1 and y_2 in the same viewing window and comparing the graphs using TRACE.

(C) *Example 2.* Use equation (3) to transform

$$y_1 = y = -3 \sin(2\pi t) - 4 \cos(2\pi t)$$

into the form $y_2 = A \sin(Bt + C)$, where C is chosen so that $-\pi < C \le \pi$. Compute C to three decimal places. From the new equation determine the amplitude, period, frequency, and phase shift.

Check by graphing y_1 and y_2 in the same viewing window and comparing the graphs using TRACE.

(D) *Application 1: Spring–mass system.* A weight suspended from a spring (spring constant 64) is pulled 4 cm below its equilibrium position and is

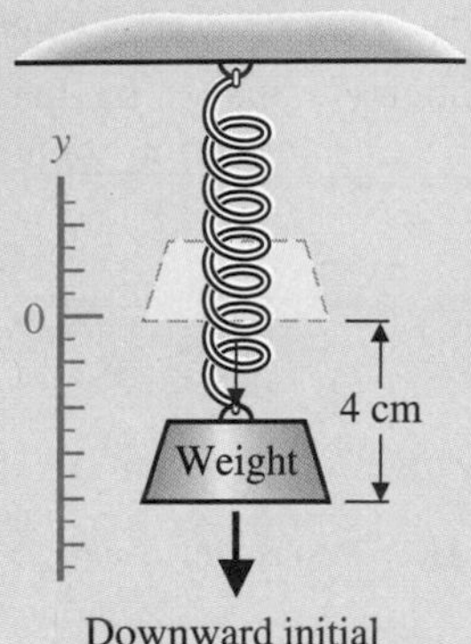

FIGURE 2
Spring–mass system

then given a downward thrust to produce an initial downward velocity of 24 cm/sec. In more advanced mathematics (differential equations) the equation of motion (neglecting air resistance and friction) is found to be given approximately by

$$y_1 = -3 \sin 8t - 4 \cos 8t$$

where y_1 is the position of the bottom of the weight on the scale in Figure 2 at time t (y is in centimeters and t is in seconds). Transform the equation into the form

$$y_2 = A \sin(Bt + C)$$

and indicate the amplitude, period, frequency, and phase shift of the simple harmonic motion. Choose C so that $-\pi < C \leq \pi$.

To check, graph y_1 and y_2 in the same viewing window, $0 \leq t \leq 6$, and compare the graphs using TRACE.

(E) *Application 2: Music.* A musical tone is described by

$$y_1 = 0.04 \sin 200\pi t - 0.03 \cos 200\pi t$$

where t is time in seconds. Write the equation in the form $y_2 = A \sin(Bt + C)$. Compute C (to three decimal places) so that $-\pi < C \leq \pi$. Indicate the amplitude, period, frequency, and phase shift.

To check, graph y_1 and y_2 in the same viewing window, $0 \leq t \leq 0.06$, and compare the graphs using TRACE.

CHAPTER 4 REVIEW

An equation in one or more variables is said to be an **identity** if the left side is equal to the right side for all replacements of the variables for which both sides are defined. If the left side is equal to the right side only for certain values of the variables and not for all values for which both sides are defined, then the equation is called a **conditional equation.**

4.1 FUNDAMENTAL IDENTITIES AND THEIR USE

Fundamental Trigonometric Identities

For x any real number or angle in degree or radian measure for which both sides are defined:

Reciprocal identities

$$\csc x = \frac{1}{\sin x} \qquad \sec x = \frac{1}{\cos x} \qquad \cot x = \frac{1}{\tan x}$$

Quotient identities

$$\tan x = \frac{\sin x}{\cos x} \qquad \cot x = \frac{\cos x}{\sin x}$$

Identities for negatives

$$\sin(-x) = -\sin x \qquad \cos(-x) = \cos x \qquad \tan(-x) = -\tan x$$

Pythagorean identities

$$\sin^2 x + \cos^2 x = 1 \qquad \tan^2 x + 1 = \sec^2 x \qquad 1 + \cot^2 x = \csc^2 x$$

4.2 VERIFYING TRIGONOMETRIC IDENTITIES

When **verifying an identity,** start with the expression on one side and through a sequence of valid steps involving the use of known identities or algebraic manipulation, convert that expression into the expression on the other side. *Do not* add the same quantity to each side, multiply each side by the same nonzero quantity, or square or take the square root of both sides.

Some Suggestions for Verifying Identities

Step 1 Start with the more complicated side of the identity and transform it into the simpler side.

Step 2 Try using basic or other known identities.

Step 3 Try algebraic operations such as multiplying, factoring, combining fractions, or splitting fractions.

Step 4 If other steps fail, try expressing each function in terms of sine and cosine functions; then perform appropriate algebraic operations.

Step 5 At each step, keep the other side of the identity in mind. This often reveals what you should do in order to get there.

4.3 SUM, DIFFERENCE, AND COFUNCTION IDENTITIES

For x and y any real numbers or angles in degree or radian measure for which both sides are defined:

Sum identities

$$\sin(x + y) = \sin x \cos y + \cos x \sin y$$

$$\cos(x + y) = \cos x \cos y - \sin x \sin y$$

$$\tan(x + y) = \frac{\tan x + \tan y}{1 - \tan x \tan y}$$

Difference identities

$$\sin(x - y) = \sin x \cos y - \cos x \sin y$$

$$\cos(x - y) = \cos x \cos y + \sin x \sin y$$

$$\tan(x - y) = \frac{\tan x - \tan y}{1 + \tan x \tan y}$$

Cofunction identities

(Replace $\pi/2$ with $90°$ if x is in degree measure.)

$$\sin\left(\frac{\pi}{2} - x\right) = \cos x \qquad \cos\left(\frac{\pi}{2} - x\right) = \sin x$$

$$\tan\left(\frac{\pi}{2} - x\right) = \cot x \qquad \cot\left(\frac{\pi}{2} - x\right) = \tan x$$

$$\sec\left(\frac{\pi}{2} - x\right) = \csc x \qquad \csc\left(\frac{\pi}{2} - x\right) = \sec x$$

4.4 DOUBLE-ANGLE AND HALF-ANGLE IDENTITIES

Double-angle identities

For x any real number or angle in degree or radian measure for which both sides are defined:

$$\sin 2x = 2 \sin x \cos x \qquad \cos 2x = \cos^2 x - \sin^2 x$$

$$\tan 2x = \frac{2 \tan x}{1 - \tan^2 x} \qquad \qquad = 1 - 2\sin^2 x$$

$$= 2\cos^2 x - 1$$

Half-angle identities

For x any real number or angle in degree or radian measure for which both sides are defined:

$$\sin\frac{x}{2} = \pm\sqrt{\frac{1 - \cos x}{2}} \qquad \cos\frac{x}{2} = \pm\sqrt{\frac{1 + \cos x}{2}}$$

$$\tan\frac{x}{2} = \pm\sqrt{\frac{1 - \cos x}{1 + \cos x}} = \frac{\sin x}{1 + \cos x} = \frac{1 - \cos x}{\sin x}$$

where the sign is determined by the quadrant in which $x/2$ lies.

☆4.5 PRODUCT–SUM AND SUM–PRODUCT IDENTITIES

Product–sum identities

For x and y any real numbers or angles in degree or radian measure for which both sides are defined:

$$\sin x \cos y = \tfrac{1}{2}[\sin(x + y) + \sin(x - y)]$$

$$\cos x \sin y = \tfrac{1}{2}[\sin(x + y) - \sin(x - y)]$$

$$\sin x \sin y = \tfrac{1}{2}[\cos(x - y) - \cos(x + y)]$$

$$\cos x \cos y = \tfrac{1}{2}[\cos(x + y) + \cos(x - y)]$$

Sum–product identities

For x and y any real numbers or angles in degree or radian measure for which both sides are defined:

$$\sin x + \sin y = 2 \sin \frac{x+y}{2} \cos \frac{x-y}{2}$$

$$\sin x - \sin y = 2 \cos \frac{x+y}{2} \sin \frac{x-y}{2}$$

$$\cos x + \cos y = 2 \cos \frac{x+y}{2} \cos \frac{x-y}{2}$$

$$\cos x - \cos y = -2 \sin \frac{x+y}{2} \sin \frac{x-y}{2}$$

CHAPTER 4 REVIEW EXERCISE

Work through all the problems in this chapter review and check the answers. Answers to all review problems appear in the back of the book; following each answer is an italic number that indicates the section in which that type of problem is discussed. Where weaknesses show up, review the appropriate sections in the text.

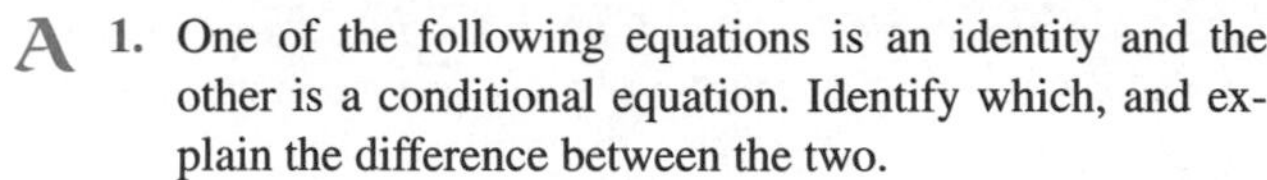

A **1.** One of the following equations is an identity and the other is a conditional equation. Identify which, and explain the difference between the two.

(1) $(x - 3)(x + 2) = x^2 - x - 6$

(2) $(x - 3)(x + 2) = 0$

Verify each identity in Problems 2–10 without looking at a table of identities.

2. $\csc x \sin x = \sec x \cos x$

3. $\cot x \sin x = \cos x$

4. $\tan x = -\tan(-x)$

5. $\dfrac{\sin^2 x}{\cos x} = \sec x - \cos x$

6. $\dfrac{\csc x}{\cos x} = \tan x + \cot x$

7. $(\cos^2 x)(\cot^2 x + 1) = \cot^2 x$

8. $\dfrac{\sin \alpha \csc \alpha}{\cot \alpha} = \tan \alpha$

9. $\dfrac{\sin^2 u - \cos^2 u}{\sin u \cos u} = \tan u - \cot u$

10. $\dfrac{\sec \theta - \csc \theta}{\sec \theta \csc \theta} = \sin \theta - \cos \theta$

11. Using $\cos(x + y) = \cos x \cos y - \sin x \sin y$, show that $\cos(x + 2\pi) = \cos x$.

12. Using $\sin(x + y) = \sin x \cos y + \cos x \sin y$, show that $\sin(x + \pi) = -\sin x$.

In Problems 13 and 14, verify each identity for the indicated value.

13. $\cos 2x = 1 - 2 \sin^2 x$, $x = 30°$

14. $\sin \dfrac{x}{2} = \pm\sqrt{\dfrac{1 - \cos x}{2}}$, $x = \dfrac{\pi}{2}$

☆ **15.** Write $\sin 8t \sin 5t$ as a sum or difference.

☆ **16.** Write $\sin w + \sin 5w$ as a product.

Verify each identity in Problems 17–20.

17. $\dfrac{1 - \cos^2 t}{\sin^3 t} = \csc t$

18. $\dfrac{(\cos \alpha - 1)^2}{\sin^2 \alpha} = \dfrac{1 - \cos \alpha}{1 + \cos \alpha}$

19. $\dfrac{1 - \tan^2 x}{1 - \tan^4 x} = \cos^2 x$

20. $\cot^2 x \cos^2 x = \cot^2 x - \cos^2 x$

21. The equation $\sin x = 0$ is true for an infinite number of values ($x = k\pi$, k any integer). Is this equation an identity? Explain.

22. Explain how you would use a graphing utility to show that $\sin x = 0$ is not an identity; then do it.

B 23. For the following graphing calculator displays, find the value of the final expression without finding x or using a calculator.

(A)

```
cos(X)
             .9394
cos(-X)
```

(B)

```
(cos(X))²
             .8824
(sin(X))²
```

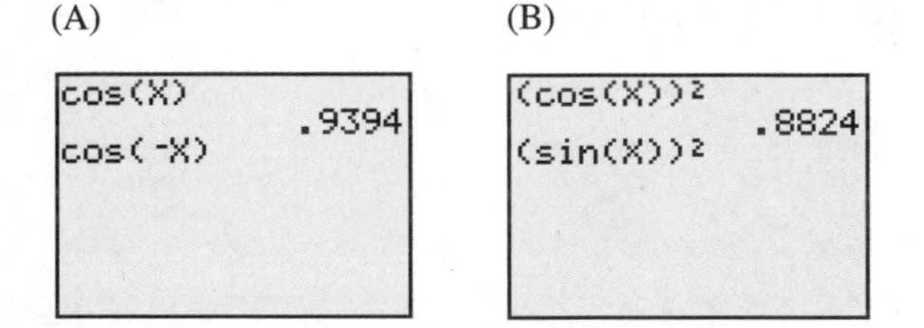

24. Is $1/(\sin x) = \csc x$ an identity for all real values of x? Explain.

25. Explain how to use a graphing utility to show that $\sin(x - 3) = \sin x - \sin 3$ is not an identity; then do it.

26. Explain how to show that $\sin(x - 3) = \sin x - \sin 3$ is not an identity without graphing; then do it.

Verify the identities in Problems 27–41. Use the list of identities inside the front cover if necessary.

27. $\dfrac{\sin x}{1 - \cos x} = (\csc x)(1 + \cos x)$

28. $\dfrac{1 - \tan^2 x}{1 - \cot^2 x} = 1 - \sec^2 x$

29. $\tan(x + \pi) = \tan x$

30. $1 - (\cos \beta - \sin \beta)^2 = \sin 2\beta$

31. $\dfrac{\sin 2x}{\cot x} = 1 - \cos 2x$

32. $\dfrac{2 \tan x}{1 + \tan^2 x} = \sin 2x$

33. $2 \csc 2x = \tan x + \cot x$

34. $\csc x = \dfrac{\cot(x/2)}{1 + \cos x}$

35. $\dfrac{\sin(x - y)}{\sin(x + y)} = \dfrac{\tan x - \tan y}{\tan x + \tan y}$

36. $\csc 2x = \dfrac{\tan x + \cot x}{2}$

37. $\dfrac{2 - \sec^2 x}{\sec^2 x} = \cos 2x$

38. $\tan \dfrac{x}{2} = \dfrac{\sec x - 1}{\tan x}$

☆39. $\dfrac{\sin t + \sin 5t}{\cos t + \cos 5t} = \tan 3t$

☆40. $\dfrac{\sin x + \sin y}{\cos x - \cos y} = -\cot \dfrac{x - y}{2}$

☆41. $\dfrac{\cos x - \cos y}{\cos x + \cos y} = -\tan \dfrac{x + y}{2} \tan \dfrac{x - y}{2}$

Evaluate Problems 42 and 43 exactly using an appropriate identity.

☆42. $\sin 165° \sin 15°$ ☆43. $\cos 165° - \cos 75°$

44. Use fundamental identities to find the exact values of the remaining trigonometric functions of x, given

$$\cos x = -\tfrac{2}{3} \quad \text{and} \quad \tan x < 0$$

45. Find the exact values of $\sin 2x$, $\cos 2x$, and $\tan 2x$, given $\tan x = \frac{4}{3}$ and $0 < x < \pi/2$. Do not use a calculator.

46. Find the exact values of $\sin(x/2)$, $\cos(x/2)$, and $\tan(x/2)$, given $\cos x = -\frac{5}{13}$ and $-\pi < x < -\pi/2$. Do not use a calculator.

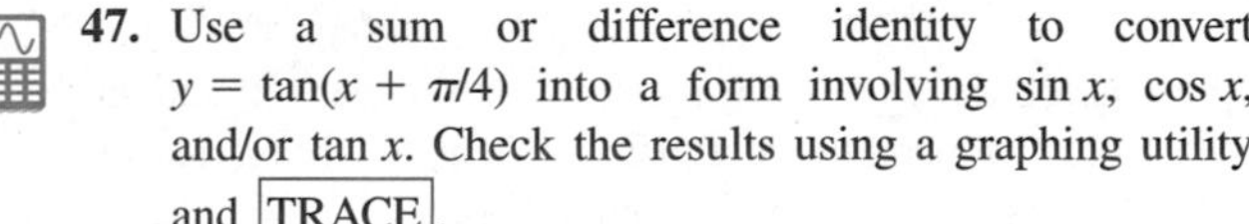

47. Use a sum or difference identity to convert $y = \tan(x + \pi/4)$ into a form involving $\sin x$, $\cos x$, and/or $\tan x$. Check the results using a graphing utility and TRACE.

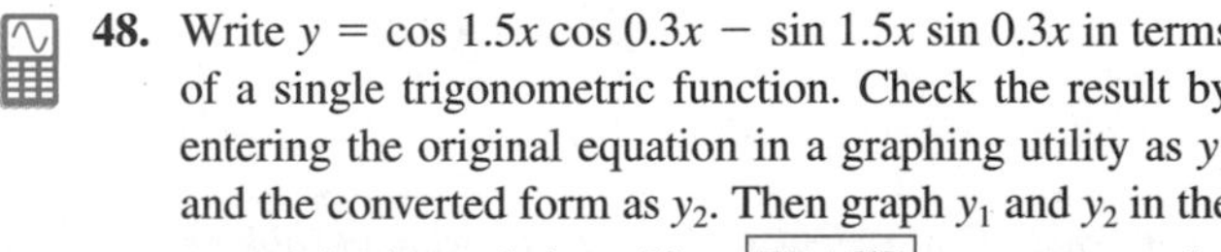

48. Write $y = \cos 1.5x \cos 0.3x - \sin 1.5x \sin 0.3x$ in terms of a single trigonometric function. Check the result by entering the original equation in a graphing utility as y_1 and the converted form as y_2. Then graph y_1 and y_2 in the same viewing window. Use TRACE to compare the two graphs.

49. Graph $y_1 = \sin(x/2)$ and $y_2 = -\sqrt{(1 - \cos x)/2}$ in the same viewing window for $-2\pi \le x \le 2\pi$, and indicate the subinterval(s) for which y_1 and y_2 are identities.

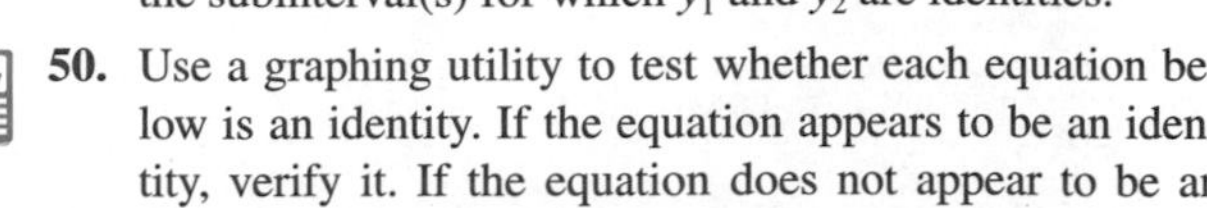

50. Use a graphing utility to test whether each equation below is an identity. If the equation appears to be an identity, verify it. If the equation does not appear to be an identity, find a value of x for which both sides are defined but are not equal.

(A) $\dfrac{\sin^2 x}{1 + \sin x} = 1 - \sin x$

(B) $\dfrac{\cos^2 x}{1 + \sin x} = 1 - \sin x$

C **51.** Find the exact values of $\sin x$, $\cos x$, and $\tan x$, given $\sec 2x = -\frac{13}{12}$ and $-\pi/2 < x < 0$. Do not use a calculator.

Verify the identities in Problems 52 and 53.

52. $\dfrac{\cot x}{\csc x + 1} = \dfrac{\csc x - 1}{\cot x}$

53. $\cot 3x = \dfrac{3\tan^2 x - 1}{\tan^3 x - 3\tan x}$

54. Use the definition of sine, cosine, and tangent on a unit circle to prove that

$$\tan x = \frac{\sin x}{\cos x}$$

55. Prove that the cosine function has a period of 2π.

56. Prove that the cotangent function has a period of π.

57. By letting

$$x + y = u \quad \text{and} \quad x - y = v$$

in $\sin x \sin y = \frac{1}{2}[\cos(x - y) - \cos(x + y)]$, show that

$$\cos v - \cos u = 2 \sin\frac{u + v}{2} \sin\frac{u - v}{2}$$

In Problems 58–62, graph f(x), find a simpler function g(x) that has the same graph as f(x), and verify the identity f(x) = g(x). [Assume g(x) = k + A t(Bx), where t(x) is one of the six trigonometric functions.]

58. $f(x) = \dfrac{3\sin^2 x}{1 - \cos x} + \dfrac{\tan^2 x \cos^2 x}{1 + \cos x}$

59. $f(x) = \dfrac{\sin x}{\cos x - \sin x} + \dfrac{\sin x}{\cos x + \sin x}$

60. $f(x) = 3\sin^2 x + \cos^2 x$

61. $f(x) = \dfrac{3 - 4\cos^2 x}{1 - 2\sin^2 x}$

62. $f(x) = \dfrac{2 + \sin x - 2\cos x}{1 - \cos x}$

In Problems 63 and 64, graph y_1 and y_2 in the same viewing window for $-2\pi \le x \le 2\pi$, and state the interval(s) where the graphs of y_1 and y_2 coincide. Use TRACE.

63. $y_1 = \tan\dfrac{x}{2}, \quad y_2 = \sqrt{\dfrac{1 - \cos x}{1 + \cos x}}$

64. $y_1 = \tan\dfrac{x}{2}, \quad y_2 = -\sqrt{\dfrac{1 - \cos x}{1 + \cos x}}$

☆ **65.** Graph $y_1 = 2\cos 30\pi x \sin 2\pi x$ and $y_2 = 2\sin 2\pi x$ for $0 \le x \le 1$ and $-2 \le y \le 2$.

☆ **66.** Repeat Problem 65 after converting y_1 to a sum or difference.

Applications

67. Precalculus: Trigonometric Substitution In the expression $\sqrt{u^2 - a^2}$, $a > 0$, let $u = a \sec x$, $0 < x < \pi/2$, simplify, and write in a form that is free of radicals.

68. Precalculus: Angle of Intersection of Two Lines Use the results of Problem 59 in Exercise 4.3 to find the acute angle of intersection (to the nearest 0.1°) between the two lines $y = 4x + 5$ and $y = \frac{1}{3}x - 2$.

* **69. Engineering** Find the exact value of x in the figure; then find x and θ to three decimal places.

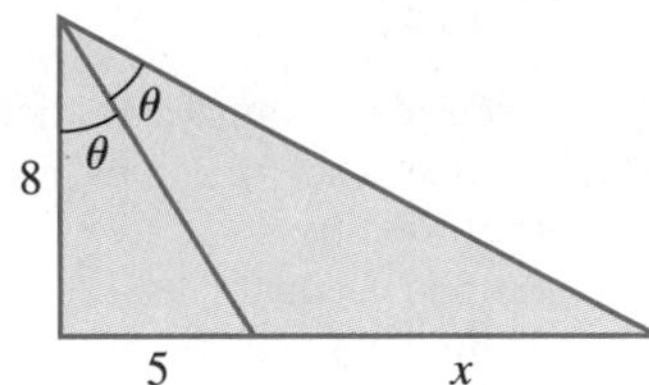

Figure for 69

* **70. Analytic Geometry** Find the radian measure of the angle θ in the figure (to three decimal places) if A has coordinates (2, 6) and B has coordinates (4, 4). [*Hint:* Label the angle between OB and the x axis as α; then use an appropriate sum identity.]

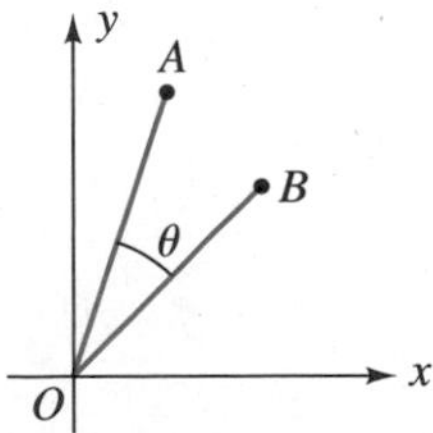

Figure for 70

∗ **71.** **Architecture** An art museum is being designed with a triangular skylight on the roof, as indicated in the figure. The facing of the museum is light granite and the top edge, excluding the skylight ridge, is to receive a black granite trim as shown. The total length of the trim depends on the choice of the angle θ for the skylight. All other dimensions are fixed.

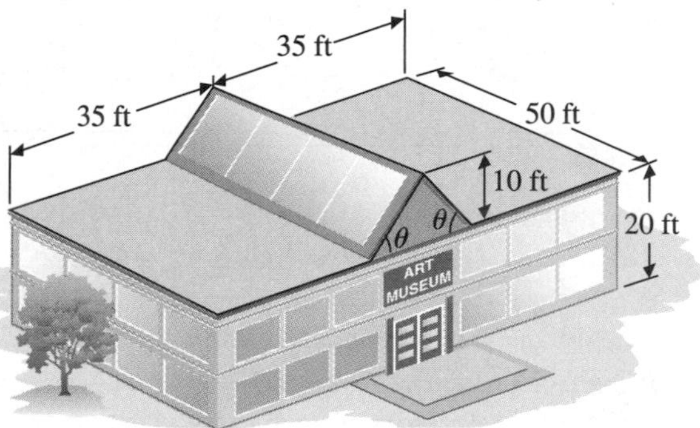

Figure for 71

(A) Show that the total length L of the black granite trim is given by

$$L = 240 + 40 \tan \frac{\theta}{2}$$

(B) Because of lighting considerations, θ is restricted to $30° \le \theta \le 60°$. Describe what you think happens to L as θ varies from 30° to 60°.

(C) Complete Table 1 (to one decimal place) and select the maximum and minimum length of the trim. (Use a table-generating feature on your calculator if it has one.)

TABLE 1

θ	30	35	40	45	50	55	60
L (ft)	250.7						

(D) Graph the equation in part (A) for the restrictions in part (B). Then determine the maximum and minimum lengths of the trim.

☆ **72.** **Music** One tone is given by $y = 0.3 \cos 120\pi t$ and another by $y = -0.3 \cos 140\pi t$. Write their sum as a product. What is the beat frequency if both notes are sounded together?

☆ **73.** **Music** Use a graphing utility with the viewing window set to $0 \le t \le 0.2$, $-0.8 \le y \le 0.8$, to graph the indicated equations.

(A) $y_1 = 0.3 \cos 120\pi t$

(B) $y_2 = -0.3 \cos 140\ \pi t$

(C) $y_3 = y_1 + y_2$ and $y_4 = 0.6 \sin 10\pi t$

(D) Repeat part (C) using the product form of y_3 from Problem 72.

☆ **74.** **Physics** The equation of motion for a weight suspended from a spring is given by

$$y = -8 \sin 3t - 6 \cos 3t$$

where y is displacement of the weight from its equilibrium position in centimeters and t is time in seconds. Write this equation in the form $y = A \sin(Bt + C)$; keep A positive, choose C positive and as small as possible, and compute C to two decimal places. Indicate the amplitude, period, frequency, and phase shift.

☆ **75.** **Physics** Use a graphing utility to graph

$$y = -8 \sin 3t - 6 \cos 3t$$

for $-2\pi/3 \le t \le 2\pi/3$, approximate the t intercepts in this interval to two decimal places, and identify the intercept that corresponds to the phase shift determined in Problem 74.

INVERSE TRIGONOMETRIC FUNCTIONS; TRIGONOMETRIC EQUATIONS AND INEQUALITIES

5

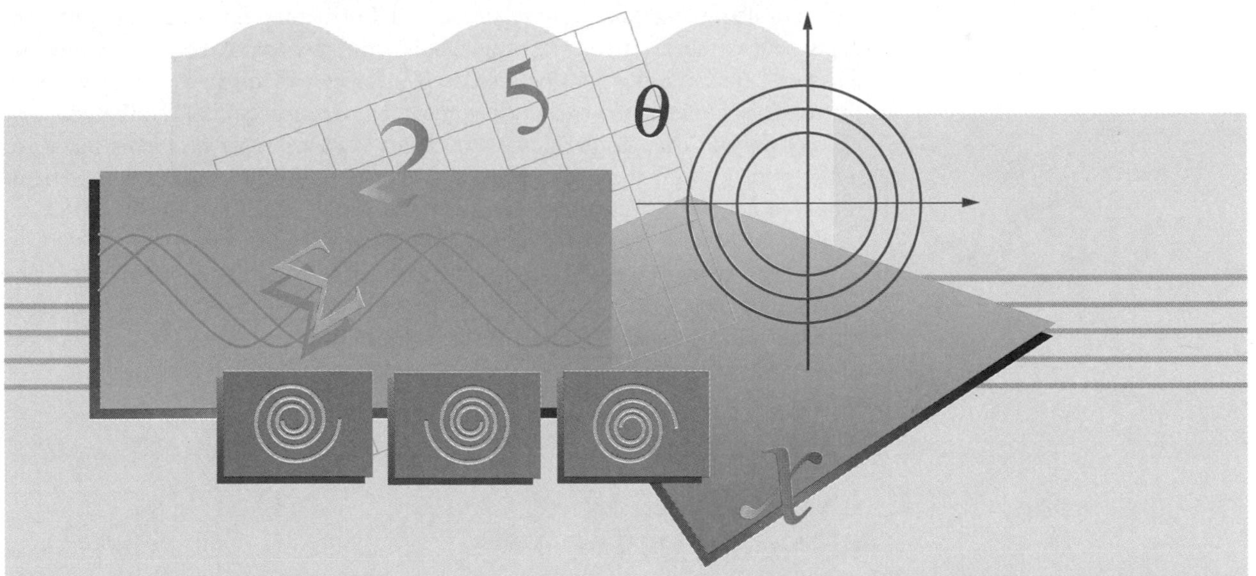

☆ Sections marked with a star may be omitted without loss of continuity.

Before you begin this chapter, briefly review Appendix B.2 on the general concept of the inverse of a function.

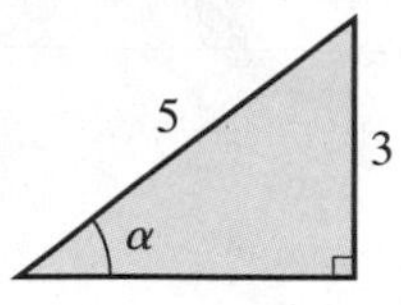

FIGURE 1

In Chapter 1, recall that we solved a right triangle (see Fig. 1) for α as follows:

$$\sin \alpha = \tfrac{3}{5} = 0.6$$

$$\alpha = \sin^{-1} 0.6 \quad \text{or} \quad \arcsin 0.6$$

$$\alpha = 0.64 \text{ rad} \quad \text{or} \quad 36.87° \qquad \textit{To two decimal places}$$

In this context, both $\sin^{-1} 0.6$ and arcsin 0.6 represent the acute angle (in either radian or degree measure) whose sine is 0.6. In Chapter 1, we said that the concepts behind the inverse function symbols $\sin^{-1}$ (or arcsin), $\cos^{-1}$ (or arccos), and $\tan^{-1}$ (or arctan) would be discussed in greater detail in this chapter. Now the time has come, and we extend the meaning of these symbols so that they apply not only to triangle problems, but to a wide variety of problems that have nothing to do with triangles or angles. That is, we will create another set of tools—a new set of functions—so that you can put them in your mathematical toolbox for general use on a wide variety of new problems.

After we complete the discussion of inverse trigonometric functions, we will be in a position to solve many types of equations involving trigonometric functions. Trigonometric equations form the subject matter of the last two sections of this chapter.

5.1 INVERSE SINE, COSINE, AND TANGENT FUNCTIONS

- Inverse Sine Function
- Inverse Cosine Function
- Inverse Tangent Function
- Inverse Trigonometric Functions with Angle Ranges
- Summary

In this section we will define the inverse sine, cosine, and tangent functions; look at their graphs; present some basic and useful identities; and consider some applications in Exercise 5.1.

Inverse Sine Function

For a function to have an inverse that is a function, it is necessary that the original function be **one-to-one.** That is, each domain value must correspond to exactly one range value, and each range value must correspond to exactly one domain value. The first condition is satisfied by all functions, but the second condition is not satisfied by some functions. For example, Figure 1a illustrates a function that is one-to-one; Figure 1b illustrates a function that is not one-to-one.

FIGURE 1

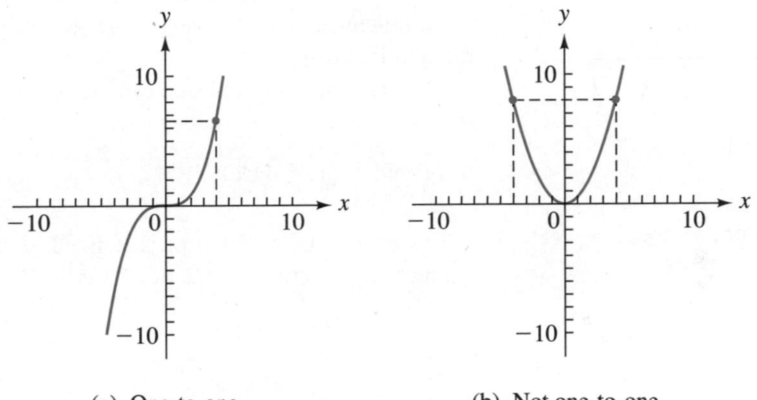

(a) One-to-one (b) Not one-to-one

To form the inverse sine function, we start with the sine function whose graph, domain, and range are indicated in Figure 2. Note that the sine function is not one-to-one. Figure 3 shows that for the range value $y = 0.5$, for example, there are an unlimited number of domain values x such that $\sin x = 0.5$. Each point where the dashed horizontal line passing through $y = 0.5$ crosses the graph corresponds to a domain value whose sine is 0.5.

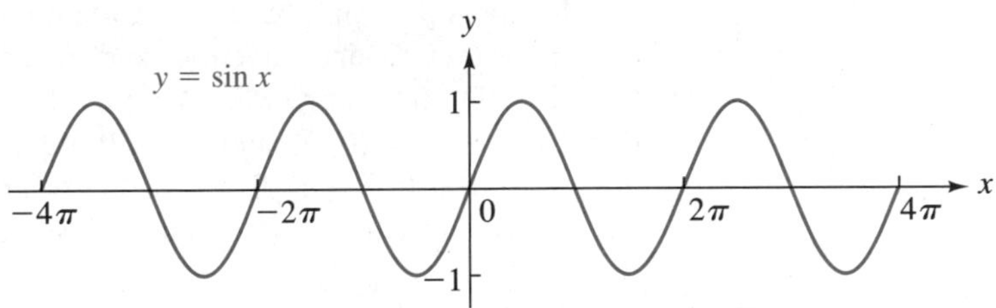

FIGURE 2
Sine function

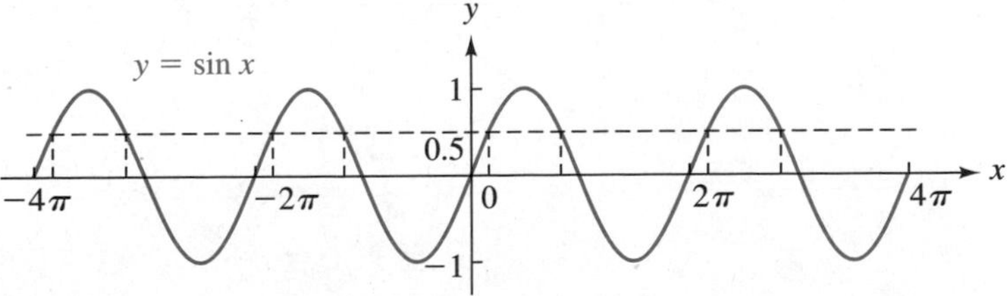

FIGURE 3
$\sin x = 0.5$

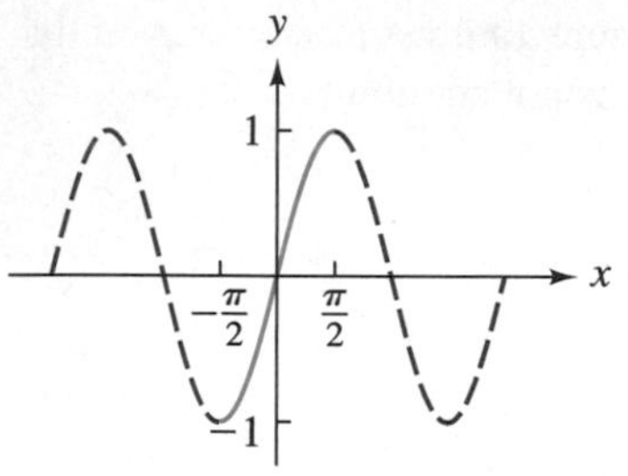

FIGURE 4
$y = \sin x$ is one-to-one for $-\pi/2 \leq x \leq \pi/2$

How can we restrict the domain of the sine function so that the sine function on this restricted domain becomes one-to-one—that is, so that every horizontal line will pass through at most one point on the graph? Actually, we can do this in an unlimited number of ways. The generally accepted way, however, is illustrated in Figure 4.

We use this restricted sine function to define the inverse sine function.

INVERSE SINE FUNCTION

The **inverse sine function** is defined as the inverse of the restricted sine function $y = \sin x$, $-\pi/2 \leq x \leq \pi/2$. Thus,

$$y = \arcsin x$$

$$y = \sin^{-1} x$$

are equivalent to

$$\sin y = x \qquad \text{where } -\pi/2 \leq y \leq \pi/2, \quad -1 \leq x \leq 1$$

The inverse sine of x is the number or angle y, $-\pi/2 \leq y \leq \pi/2$, whose sine is x.

To graph $y = \sin^{-1} x$, we take the coordinates of each point on the graph of the restricted sine function and reverse the order. For example, since $(-\pi/2, -1)$, $(0\ 0)$, and $(\pi/2, 1)$ are on the graph of the restricted sine function, then $(-1, -\pi/2)$, $(0, 0)$, and $(1, \pi/2)$ are on the graph of the inverse sine function, as shown in Figure 5. Using these three points provides us with a quick way of sketching the graph of the inverse sine function. A more accurate graph

FIGURE 5

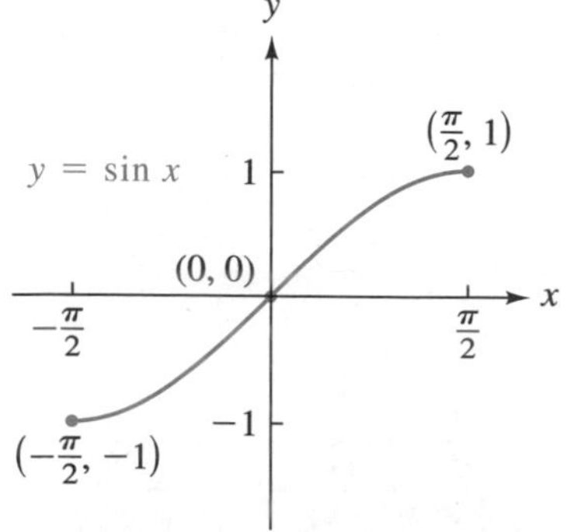

Domain: $-\frac{\pi}{2} \leq x \leq \frac{\pi}{2}$
Range: $-1 \leq y \leq 1$

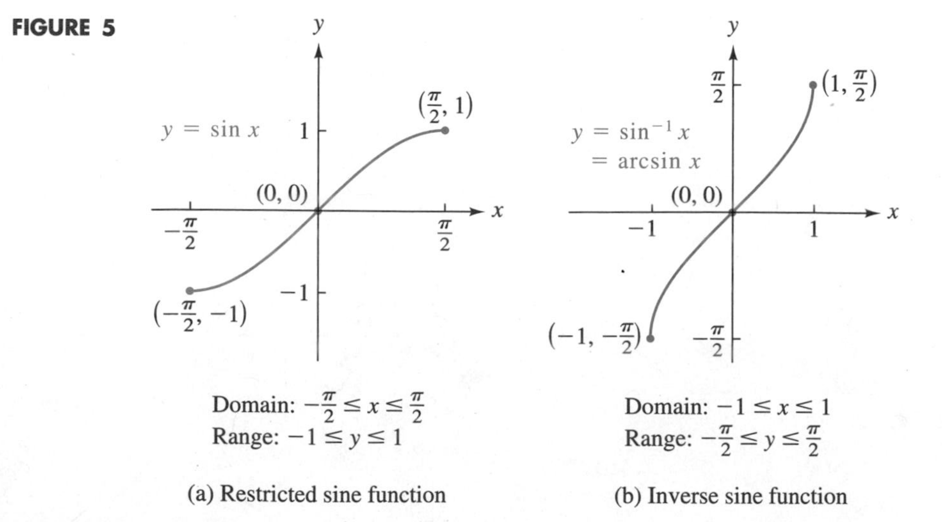

(a) Restricted sine function

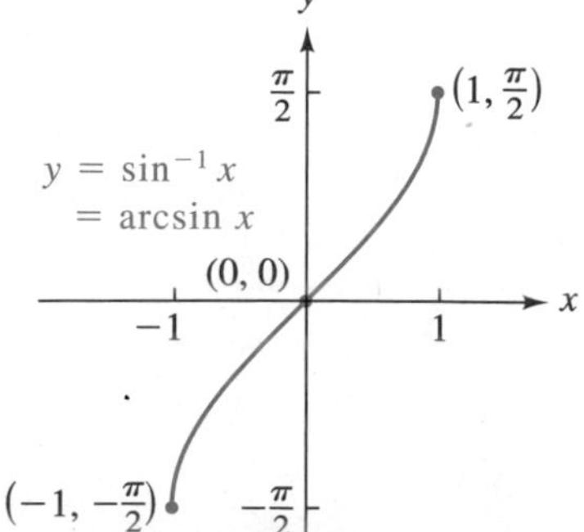

Domain: $-1 \leq x \leq 1$
Range: $-\frac{\pi}{2} \leq y \leq \frac{\pi}{2}$

(b) Inverse sine function

can be obtained by using a calculator in radian mode and a set of domain values from -1 to 1 (see Problem 35 in Exercise 5.1). An instant graph can be obtained using a graphing utility.

We state the important sine–inverse sine identities, which follow from the general properties of inverse functions (see Appendix B.2).

SINE–INVERSE SINE IDENTITIES

$\sin(\sin^{-1} x) = x \qquad -1 \le x \le 1$

$\sin^{-1}(\sin x) = x \qquad -\pi/2 \le x \le \pi/2$

EXPLORE/DISCUSS 1

Use a calculator to evaluate each of the following. Which illustrate a sine–inverse sine identity and which do not? Explain.

(A) $\sin(\sin^{-1} 0.5)$ (B) $\sin(\sin^{-1} 1.5)$
(C) $\sin^{-1}[\sin(-1.3)]$ (D) $\sin^{-1}[\sin(-3)]$

EXAMPLE 1 Exact Values

Find exact values without using a calculator:

(A) $\sin^{-1}(\sqrt{3}/2)$ (B) $\arcsin(-\frac{1}{2})$
(C) $\sin^{-1}(\sin 1.2)$ (D) $\cos(\sin^{-1}\frac{2}{3})$

Solution (A) $y = \sin^{-1}(\sqrt{3}/2)$ is equivalent to $\sin y = \sqrt{3}/2$, $-\pi/2 \le y \le \pi/2$. What y between $-\pi/2$ and $\pi/2$ has sine $\sqrt{3}/2$? This y must be associated with a first quadrant reference triangle (see Fig. 6):

$$\sin y = \frac{\sqrt{3}}{2}$$

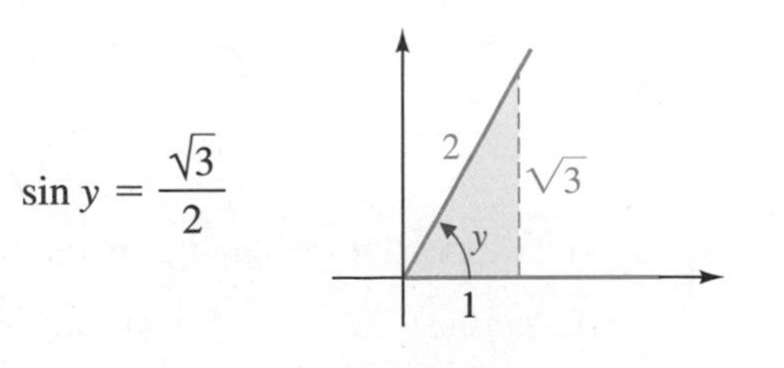

Reference triangle is a special 30°–60° triangle, $y = \pi/3$.

FIGURE 6

Thus,

$$\sin^{-1}\frac{\sqrt{3}}{2} = \frac{\pi}{3}$$

since $\pi/3$ is the only number between $-\pi/2$ and $\pi/2$ with sine equal to $\sqrt{3}/2$.

(B) $y = \arcsin(-\frac{1}{2})$ is equivalent to $\sin y = -\frac{1}{2}$, $-\pi/2 \le y \le \pi/2$. What y between $-\pi/2$ and $\pi/2$ has sine $-\frac{1}{2}$? This y must be negative and associated with a fourth quadrant reference triangle (see Fig. 7):

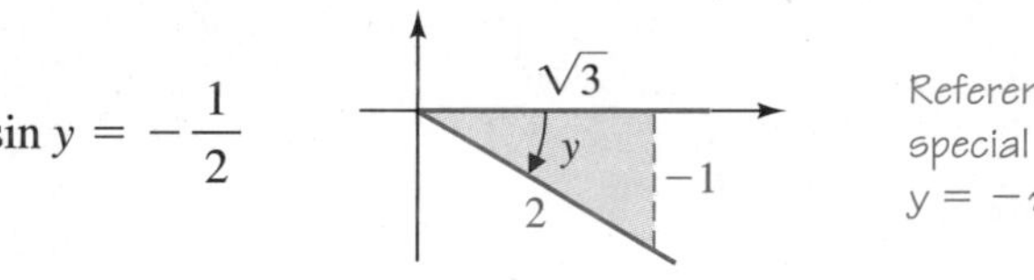

FIGURE 7

Thus,

$$\arcsin\left(-\frac{1}{2}\right) = -\frac{\pi}{6}$$

[*Note:* y cannot be $11\pi/6$, even though $\sin(11\pi/6) = -\frac{1}{2}$. Why?]

(C) $\sin^{-1}(\sin 1.2) = 1.2$ Sine–inverse sine identity

(D) Let $y = \sin^{-1}\frac{2}{3}$; then $\sin y = \frac{2}{3}$, $-\pi/2 \le y \le \pi/2$. Draw the reference triangle associated with y; then $\cos y = \cos(\sin^{-1}\frac{2}{3})$ can be determined directly from the triangle (after finding the third side) without actually finding y (see Fig. 8):

FIGURE 8

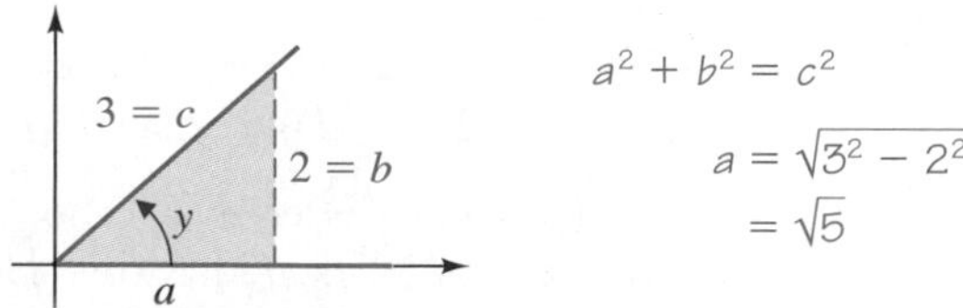

Thus,

$$\cos\left(\sin^{-1}\frac{2}{3}\right) = \cos y = \frac{\sqrt{5}}{3}$$

■

Matched Problem 1 Find exact values without using a calculator:

(A) $\arcsin(\sqrt{2}/2)$ (B) $\sin^{-1}(-1)$

(C) $\sin[\sin^{-1}(-0.4)]$ (D) $\tan[\sin^{-1}(-1/\sqrt{5})]$

■

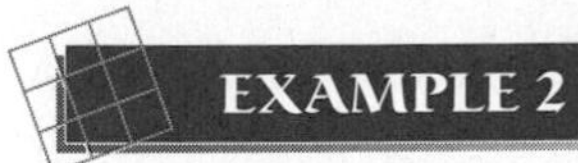

EXAMPLE 2

Calculator Values

Find to four significant digits using a calculator:

(A) $\sin^{-1}(0.8432)$ (B) $\arcsin(-0.3042)$

(C) $\sin^{-1} 1.357$ (D) $\cot[\sin^{-1}(-0.1087)]$

Solution [*Note:* Recall that the keys used to obtain $\sin^{-1}$ vary among different brands of calculators. (Read the user's manual for your calculator.) Two common designations are $\boxed{\sin^{-1}}$ and the combination $\boxed{\text{inv}}$ $\boxed{\sin}$. For all these problems set your calculator in the radian mode.]

(A) $\sin^{-1}(0.8432) = 1.003$

(B) $\arcsin(-0.3042) = -0.3091$

(C) $\sin^{-1} 1.357 = \text{Error}^*$ 1.357 is not in the domain of $\sin^{-1}$.

(D) $\cot[\sin^{-1}(-0.1087)] = -9.145$ ■

Matched Problem 2 Find to four significant digits using a calculator:

(A) $\arcsin 0.2903$ (B) $\sin^{-1}(-0.7633)$

(C) $\arcsin(-2.305)$ (D) $\sec[\sin^{-1}(-0.3446)]$ ■

Inverse Cosine Function

The generally accepted restriction on the cosine function, which ensures that an inverse will exist, is to have domain values so that $0 \le x \le \pi$ (see Fig. 9).

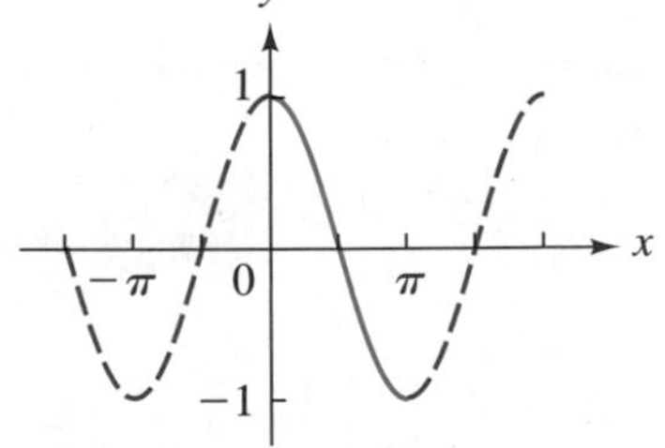

FIGURE 9
$y = \cos x$ is one-to-one for $0 \le x \le \pi$

INVERSE COSINE FUNCTION

The **inverse cosine function** is defined as the inverse of the restricted cosine function $y = \cos x$, $0 \le x \le \pi$. Thus,

$$y = \arccos x$$

$$y = \cos^{-1} x$$

are equivalent to

$$\cos y = x \qquad \text{where } 0 \le y \le \pi, \quad -1 \le x \le 1$$

The inverse cosine of x is the number or angle y, $0 \le y \le \pi$, whose cosine is x.

---

* Some calculators use a more advanced definition of the inverse sine function involving complex numbers and will display an ordered pair of real numbers as the value of $\sin^{-1} 1.357$. You should interpret such a result as an indication that the number entered is not in the domain of the inverse sine function as we have defined it.

Figure 10 compares the graphs of the restricted cosine function and its inverse. Notice that $(0, 1)$, $(\pi/2, 0)$, and $(\pi, -1)$ are on the restricted cosine graph. Reversing the coordinates gives us three points on the graph of the inverse cosine function.

FIGURE 10

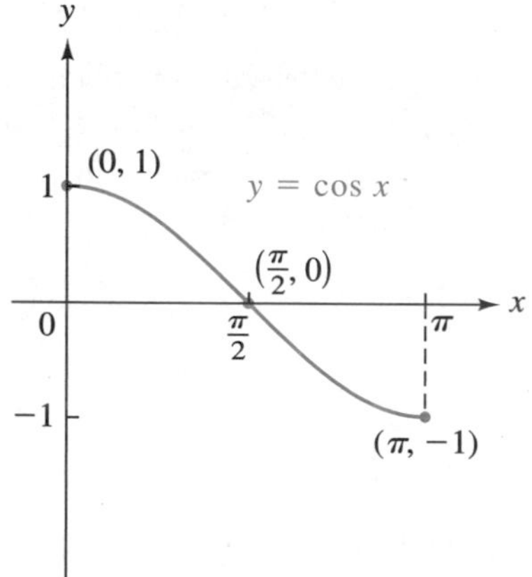

Domain: $0 \le x \le \pi$
Range: $-1 \le y \le 1$

(a) Restricted cosine function

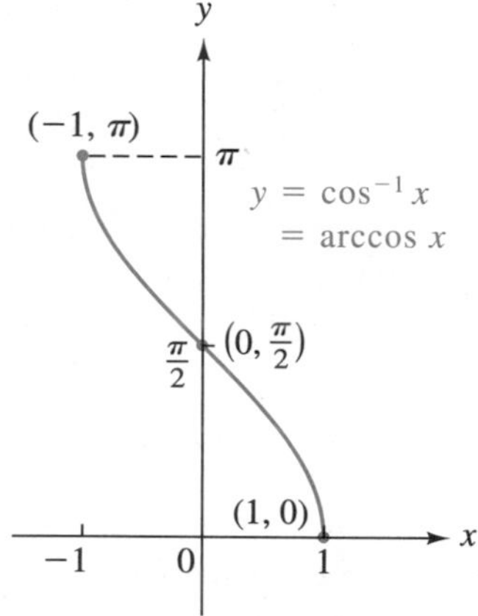

Domain: $-1 \le x \le 1$
Range: $0 \le y \le \pi$

(b) Inverse cosine function

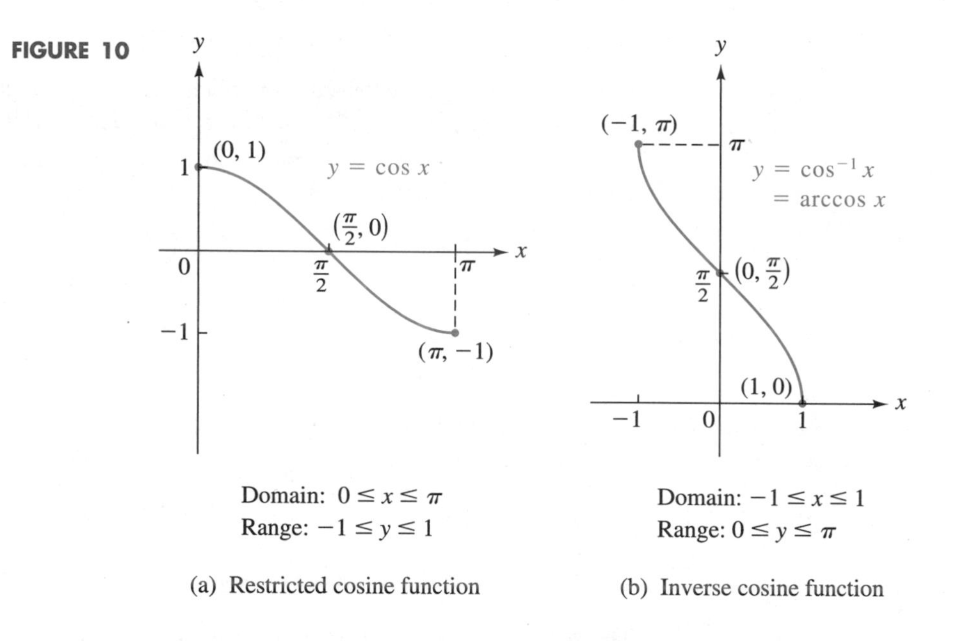

We complete the discussion by giving the cosine–inverse cosine identities.

COSINE–INVERSE COSINE IDENTITIES

$\cos(\cos^{-1} x) = x \qquad -1 \le x \le 1$

$\cos^{-1}(\cos x) = x \qquad 0 \le x \le \pi$

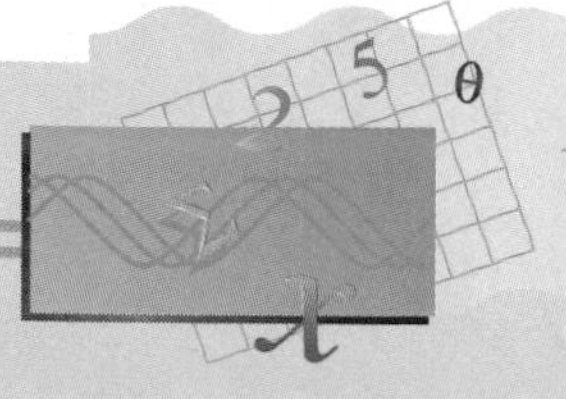

EXPLORE/DISCUSS 2

Use a calculator to evaluate each of the following. Which illustrate a cosine–inverse cosine identity and which do not? Explain.

(A) $\cos(\cos^{-1} 0.5)$ (B) $\cos(\cos^{-1} 1.1)$

(C) $\cos^{-1}(\cos 1.1)$ (D) $\cos^{-1}[\cos(-1)]$

EXAMPLE 3 **Exact Values**

Find exact values without using a calculator:

(A) $\cos^{-1}\frac{1}{2}$ (B) $\arccos(-\sqrt{3}/2)$

(C) $\cos(\cos^{-1} 0.7)$ (D) $\sin[\cos^{-1}(-\frac{1}{3})]$

Solution (A) $y = \cos^{-1}\frac{1}{2}$ is equivalent to $\cos y = \frac{1}{2}$, $0 \le y \le \pi$. What y between 0 and π has cosine $\frac{1}{2}$? This y must be associated with a first quadrant reference triangle (see Fig. 11):

$$\cos y = \frac{1}{2}$$

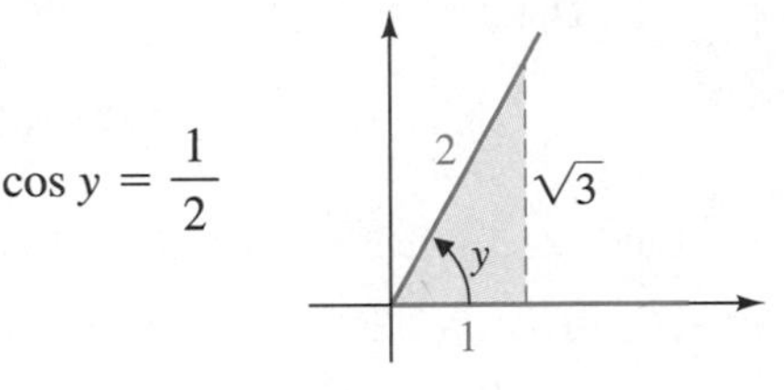

Reference triangle is a special 30°–60° triangle, $y = \pi/3$.

FIGURE 11

Thus,

$$\cos^{-1}\frac{1}{2} = \frac{\pi}{3}$$

(B) $y = \arccos(-\sqrt{3}/2)$ is equivalent to $\cos y = -\sqrt{3}/2$, $0 \le y \le \pi$. What y between 0 and π has cosine $-\sqrt{3}/2$? This y must be associated with a second quadrant reference triangle (see Fig. 12):

$$\cos y = -\frac{\sqrt{3}}{2}$$

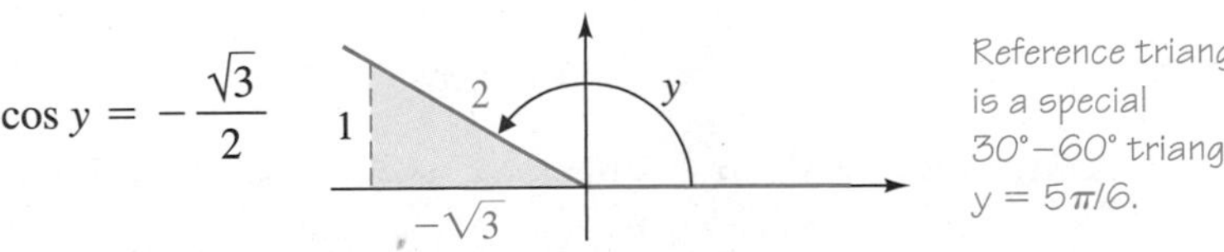

FIGURE 12

Thus,

$$\arccos\left(-\frac{\sqrt{3}}{2}\right) = \frac{5\pi}{6}$$

[*Note:* y cannot be $-5\pi/6$, even though $\cos(-5\pi/6) = -\sqrt{3}/2$. Why?]

(C) $\cos(\cos^{-1} 0.7) = 0.7$ *Cosine–inverse cosine identity*

(D) Let $y = \cos^{-1}(-\frac{1}{3})$; then $\cos y = -\frac{1}{3}$, $0 \le y \le \pi$. Draw a reference triangle associated with y; then $\sin y = \sin[\cos^{-1}(-\frac{1}{3})]$ can be determined directly from the triangle (after finding the third side) without actually finding y (see Fig. 13):

FIGURE 13

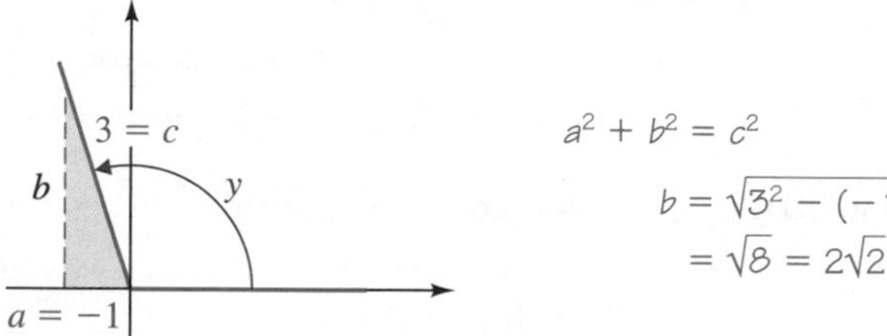

$$a^2 + b^2 = c^2$$
$$b = \sqrt{3^2 - (-1)^2}$$
$$= \sqrt{8} = 2\sqrt{2}$$

Thus,

$$\sin\left[\cos^{-1}\left(-\frac{1}{3}\right)\right] = \sin y = \frac{2\sqrt{2}}{3}$$ ■

Matched Problem 3 Find exact values without using a calculator:

(A) $\arccos(\sqrt{2}/2)$ (B) $\cos^{-1}(-1)$

(C) $\cos^{-1}(\cos 3.05)$ (D) $\cot[\cos^{-1}(-1/\sqrt{5})]$ ■

EXAMPLE 4 Calculator Values

Find to four significant digits using a calculator:

(A) $\cos^{-1} 0.4325$ (B) $\arccos(-0.8976)$

(C) $\cos^{-1} 2.137$ (D) $\csc[\cos^{-1}(-0.0349)]$

Solution Set your calculator in radian mode.

(A) $\cos^{-1} 0.4325 = 1.124$

(B) $\arccos(-0.8976) = 2.685$

(C) $\cos^{-1} 2.137 = \text{Error}$ 2.137 is not in the domain of $\cos^{-1}$.

(D) $\csc[\cos^{-1}(-0.0349)] = 1.001$ ■

Matched Problem 4 Find to four significant digits using a calculator:

(A) $\arccos 0.6773$ (B) $\cos^{-1}(-0.8114)$

(C) $\arccos(-1.003)$ (D) $\cot[\cos^{-1}(-0.5036)]$ ■

EXAMPLE 5 Exact Values

Find the exact value of $\cos(\sin^{-1}\frac{3}{5} - \cos^{-1}\frac{4}{5})$ without using a calculator.

Solution We use the difference identity for cosine and the procedure outlined in Examples 1D and 3D to obtain

$$\cos(x - y) = \cos x \cos y + \sin x \sin y$$
$$\cos(\sin^{-1}\tfrac{3}{5} - \cos^{-1}\tfrac{4}{5}) = \cos(\sin^{-1}\tfrac{3}{5})\cos(\cos^{-1}\tfrac{4}{5}) + \sin(\sin^{-1}\tfrac{3}{5})\sin(\cos^{-1}\tfrac{4}{5})$$
$$= (\tfrac{4}{5}) \cdot (\tfrac{4}{5}) + (\tfrac{3}{5}) \cdot (\tfrac{3}{5})$$
$$= 1$$ ■

Matched Problem 5 Find the exact value of $\sin(2\cos^{-1}\frac{3}{5})$ without using a calculator. ■

Inverse Tangent Function

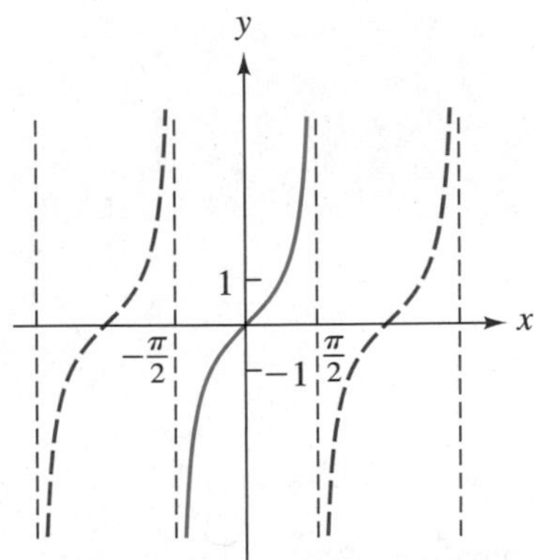

FIGURE 14
$y = \tan x$ is one-to-one for $-\pi/2 < x < \pi/2$

To restrict the tangent function so that every horizontal line will pass through at most one point on its graph, we choose to restrict the domain to the interval $-\pi/2 < x < \pi/2$ (see Fig. 14).

We use this restricted tangent function to define the inverse tangent function.

INVERSE TANGENT FUNCTION

The **inverse tangent function** is defined as the inverse of the restricted tangent function $y = \tan x$, $-\pi/2 < x < \pi/2$. Thus,

$$y = \arctan x$$

$$y = \tan^{-1} x$$

are equivalent to

$$\tan y = x \qquad \text{where } -\pi/2 < y < \pi/2, \quad x \text{ is any real number}$$

The inverse tangent of x is the number or angle y, $-\pi/2 < y < \pi/2$, whose tangent is x.

Figure 15 compares the graphs of the restricted tangent function and its inverse. Notice that $(-\pi/4, -1)$, $(0, 0)$, and $(\pi/4, 1)$ are on the restricted tangent graph. Reversing the coordinates gives us three points on the graph of the inverse tangent function. Also note that the vertical asymptotes become horizontal asymptotes.

FIGURE 15

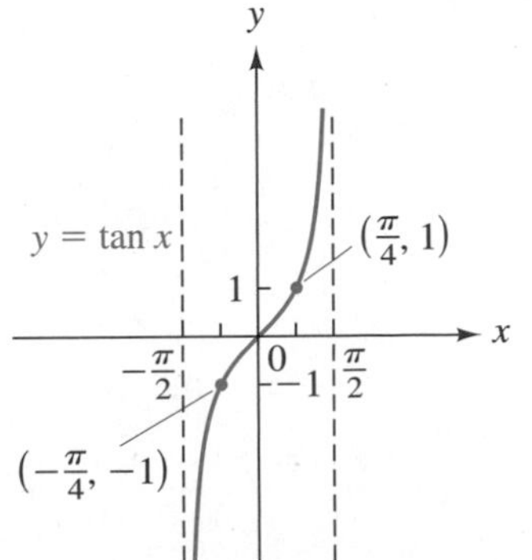

Domain: $-\frac{\pi}{2} < x < \frac{\pi}{2}$
Range: All real numbers

(a) Restricted tangent function

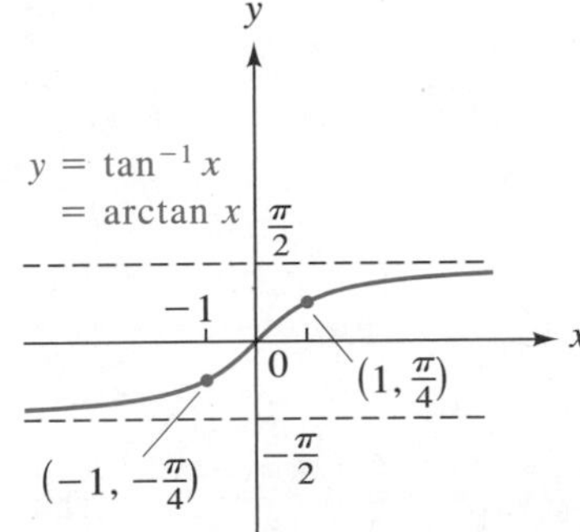

Domain: All real numbers
Range: $-\frac{\pi}{2} < y < \frac{\pi}{2}$

(b) Inverse tangent function

We now state the tangent–inverse tangent identities.

TANGENT–INVERSE TANGENT IDENTITIES

$\tan(\tan^{-1} x) = x$ for all x

$\tan^{-1}(\tan x) = x$ $\quad -\pi/2 < x < \pi/2$

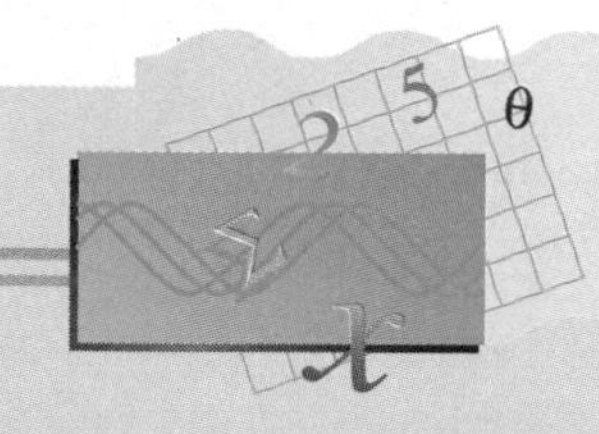

EXPLORE/DISCUSS 3

Use a calculator to evaluate each of the following. Which illustrate a tangent–inverse tangent identity and which do not? Explain.

(A) $\tan(\tan^{-1} 25)$ (B) $\tan[\tan^{-1}(-325)]$

(C) $\tan^{-1}(\tan 1.2)$ (D) $\tan^{-1}[\tan(-\pi)]$

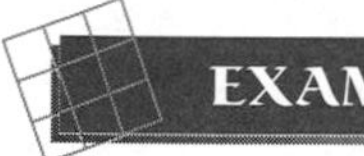

EXAMPLE 6 **Exact Values**

Find exact values without using a calculator:

(A) $\tan^{-1}(-1/\sqrt{3})$ (B) $\tan^{-1}[\tan(-1.2)]$

Solution (A) $y = \tan^{-1}(-1/\sqrt{3})$ is equivalent to $\tan y = -1/\sqrt{3}$, where y satisfies $-\pi/2 < y < \pi/2$. What y between $-\pi/2$ and $\pi/2$ has tangent $-1/\sqrt{3}$? This y must be negative and associated with a fourth quadrant reference triangle (see Fig. 16):

$$\tan y = \frac{-1}{\sqrt{3}}$$

$\sqrt{3}$ y -1

Special 30°–60° triangle, $y = -\pi/6$.

FIGURE 16

Thus,

$$\tan^{-1}\left(-\frac{1}{\sqrt{3}}\right) = -\frac{\pi}{6}$$

[*Note:* y cannot be $11\pi/6$. Why?]

(B) $\tan^{-1}[\tan(-1.2)] = -1.2$ Tangent–inverse tangent identity ■

Matched Problem 6 Find exact values without using a calculator:

(A) $\arctan \sqrt{3}$ (B) $\tan(\tan^{-1} 35)$ ■

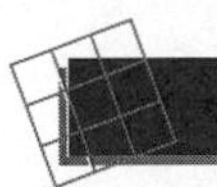

EXAMPLE 7 **Calculator Values**

Find to four significant digits using a calculator:

(A) $\tan^{-1} 3$ (B) $\arctan(-25.45)$
(C) $\tan^{-1} 1{,}435$ (D) $\sec[\tan^{-1}(-0.1308)]$

Solution Set calculator in radian mode.

(A) $\tan^{-1} 3 = 1.249$ (B) $\arctan(-25.45) = -1.532$
(C) $\tan^{-1} 1{,}435 = 1.570$ (D) $\sec[\tan^{-1}(-0.1308)] = 1.009$ ■

Matched Problem 7 Find to four significant digits using a calculator:

(A) $\tan^{-1} 7$ (B) $\arctan(-13.08)$
(C) $\tan^{-1} 735$ (D) $\csc[\tan^{-1}(-1.033)]$ ■

EXAMPLE 8 **Finding an Equivalent Algebraic Expression**

Express $\sin(\tan^{-1} x)$ as an algebraic expression in x.

Solution Let

$$y = \tan^{-1} x \qquad -\frac{\pi}{2} < y < \frac{\pi}{2}$$

or, equivalently,

$$\tan y = x = \frac{x}{1} \qquad -\frac{\pi}{2} < y < \frac{\pi}{2}$$

The two possible reference triangles for y are shown in Figure 17.

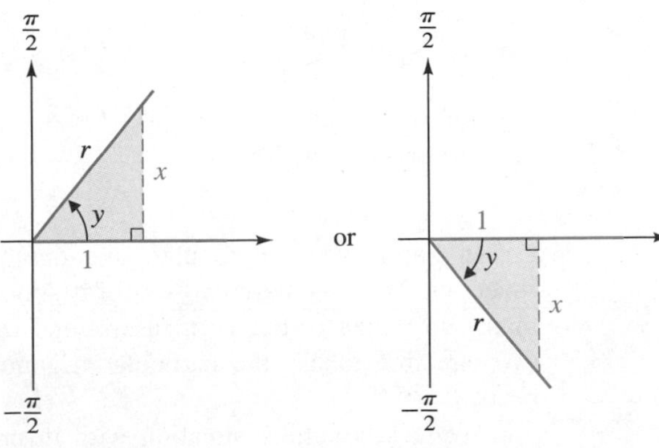

FIGURE 17
Reference triangles for $y = \tan^{-1} x$

In either case,

$$r = \sqrt{x^2 + 1}$$

Thus,

$$\sin(\tan^{-1} x) = \sin y = \frac{x}{r} = \frac{x}{\sqrt{x^2 + 1}}$$

Matched Problem 8 Express tan(arccos x) as an algebraic expression in x.

Inverse Trigonometric Functions with Angle Ranges

We first defined trigonometric functions with angle domains, in degree or radian measure, and with real number ranges. Then, we defined the circular functions with real number domains and real number ranges. Technically, the trigonometric functions and the circular functions are not the same: The first are defined in terms of angles and the second in terms of real numbers. The functions, however, are closely related in that every real number in the domain of a circular function can be associated with an angle in either degree or radian measure, and vice versa (Fig. 18).

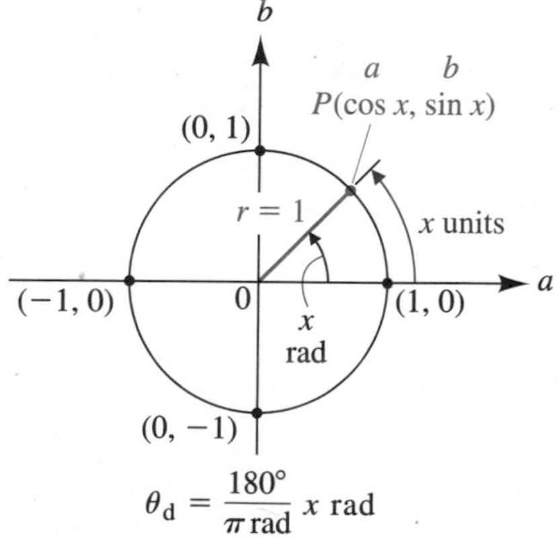

FIGURE 18
Real numbers and angles

In common usage, circular functions are also referred to as trigonometric functions. Thus, we have two sets of trigonometric functions, one with angle domains, in radian or degree measure, and the other with real number domains. (We are free to use the particular trigonometric function that best suits our needs.)

We have a similar situation with inverse trigonometric functions. The inverse trigonometric functions defined in the first part of this section are actually inverse circular functions, with real number domains and ranges. Corresponding to these definitions are inverse trigonometric functions with angle ranges, in degree or radian measure. If the range values are angles in degree measure, we will often use the Greek letter θ (theta) to represent this value. Thus, for example, we may use any of the following three forms, depending on our interest:

Real number range

$$y = \tan^{-1} x \qquad -\frac{\pi}{2} < y < \frac{\pi}{2} \qquad y \text{ is a real number.}$$

Angle range in radian measure

$$y = \tan^{-1} x \qquad -\frac{\pi}{2} < y < \frac{\pi}{2} \qquad y \text{ is an angle in radian measure.}$$

Angle range in degree measure

$$\theta = \tan^{-1} x \qquad -90° < \theta < 90° \qquad \theta \text{ is an angle in degree measure.}$$

Thus, depending on the context, we can write

$$\tan^{-1} 1 = \frac{\pi}{4} \quad \text{or} \quad \tan^{-1} 1 = \frac{\pi}{4} \text{ radian} \quad \text{or} \quad \tan^{-1} 1 = 45°$$

Caution This discussion does not mean that inverse trigonometric functions are multivalued. If we wish to use the inverse tangent function with a real number range, then $\tan^{-1} 1$ is equal to $\pi/4$ and no other real number. If we wish to use the inverse tangent function with an angle range, then $\tan^{-1} 1$ is the angle with radian measure $\pi/4$ or degree measure 45°, and no other angle. □

EXAMPLE 9 **Inverse Trigonometric Functions and Degree Measure**

Find the degree measure of θ.

(A) $\theta = \sin^{-1} \frac{1}{2}$ (Exact value without a calculator.)

(B) $\theta = \tan^{-1}(-1.3025)$ (To two decimal places with a calculator.)

Solution (A) $\theta = \sin^{-1} \frac{1}{2}$ is equivalent to

$$\sin \theta = \tfrac{1}{2} \qquad -90° \le \theta \le 90°$$

Thus, $\theta = 30°$

(B) Set calculator in degree mode.

$$\theta = \tan^{-1}(-1.3025) = -52.48°$$

■

Matched Problem 9 Find the degree measure of θ.

(A) $\theta = \cos^{-1} \frac{1}{2}$ (Exact value without a calculator.)

(B) $\theta = \tan^{-1} 25.08$ (To two decimal places with a calculator.) ■

Summary

We summarize the definitions of the inverse trigonometric functions in the box for convenient reference.

INVERSE SINE, COSINE, AND TANGENT FUNCTIONS

$y = \sin^{-1} x$	is equivalent to	$x = \sin y$	where $-1 \le x \le 1$ and $-\pi/2 \le y \le \pi/2$
$y = \cos^{-1} x$	is equivalent to	$x = \cos y$	where $-1 \le x \le 1$ and $0 \le y \le \pi$
$y = \tan^{-1} x$	is equivalent to	$x = \tan y$	where x is any real number and $-\pi/2 < y < \pi/2$

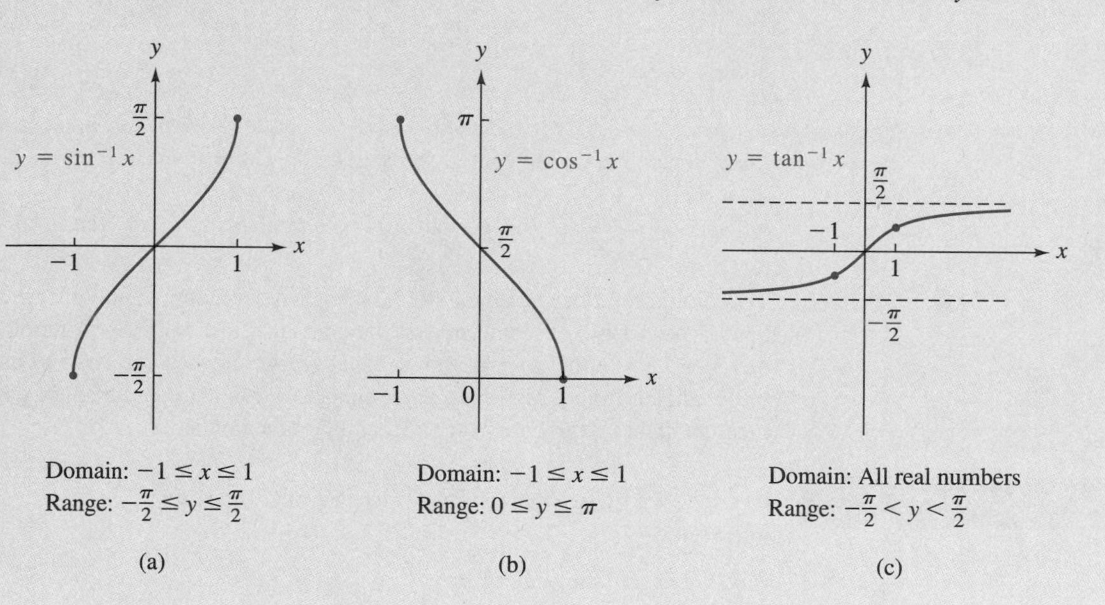

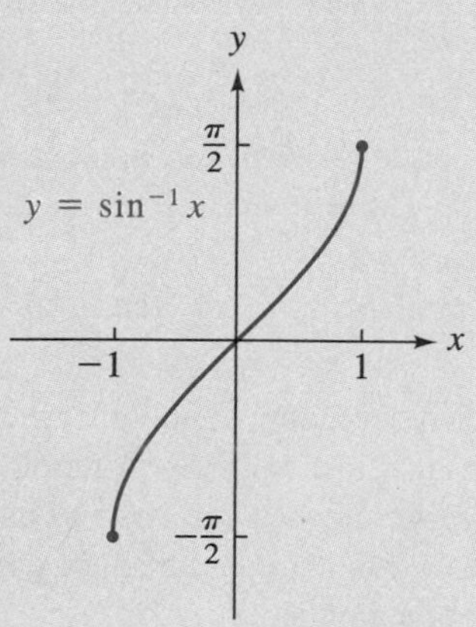

Domain: $-1 \le x \le 1$
Range: $-\frac{\pi}{2} \le y \le \frac{\pi}{2}$

(a)

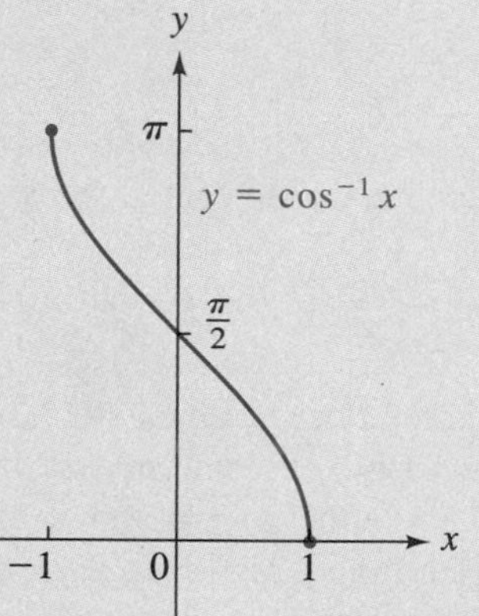

Domain: $-1 \le x \le 1$
Range: $0 \le y \le \pi$

(b)

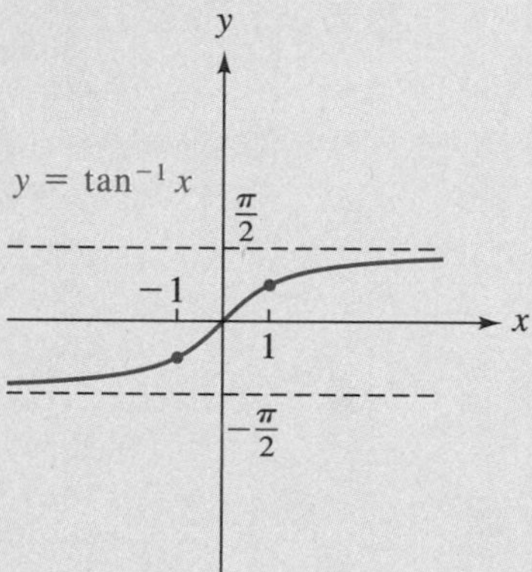

Domain: All real numbers
Range: $-\frac{\pi}{2} < y < \frac{\pi}{2}$

(c)

Answers to Matched Problems

1. (A) $\pi/4$ (B) $-\pi/2$ (C) -0.4 (D) $-\frac{1}{2}$
2. (A) 0.2945 (B) -0.8684 (C) Not defined (D) 1.065
3. (A) $\pi/4$ (B) π (C) 3.05 (D) $-\frac{1}{2}$
4. (A) 0.8267 (B) 2.517 (C) Not defined (D) -0.5829
5. $\frac{24}{25}$ **6.** (A) $\pi/3$ (B) 35
7. (A) 1.429 (B) -1.494 (C) 1.569 (D) -1.392
8. $\dfrac{\sqrt{1 - x^2}}{x}$ **9.** (A) 60° (B) 87.72°

EXERCISE 5.1

A *Find exact real number values without using a calculator.*

1. $\sin^{-1} 0$ **2.** $\cos^{-1} 0$

3. $\arccos(\sqrt{3}/2)$ **4.** $\arcsin(\sqrt{3}/2)$

5. $\tan^{-1} 1$ **6.** $\arctan \sqrt{3}$

7. $\cos^{-1} \frac{1}{2}$ **8.** $\sin^{-1}(\sqrt{2}/2)$

In Problems 9–14, evaluate to four significant digits using a calculator.

9. $\cos^{-1} 0.4038$ **10.** $\sin^{-1} 0.9103$

11. $\tan^{-1} 43.09$ **12.** $\arctan 103.7$

13. $\arcsin 1.131$ **14.** $\arccos 3.051$

15. Explain how to find the value of x that produces the result shown in the graphing utility window below, and find it. The utility is in degree mode. Give the answer to six decimal places.

```
sin⁻¹(X)
    37.00000000
```

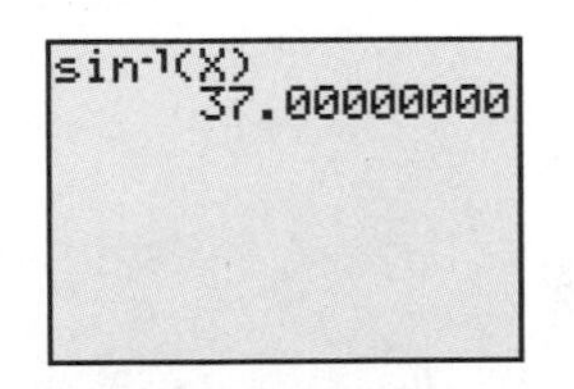

16. Explain how to find the value of x that produces the result shown in the graphing utility window below, and find it. The utility is in radian mode. Give the answer to six decimal places.

```
tan⁻¹(X)
    1.00000000
```

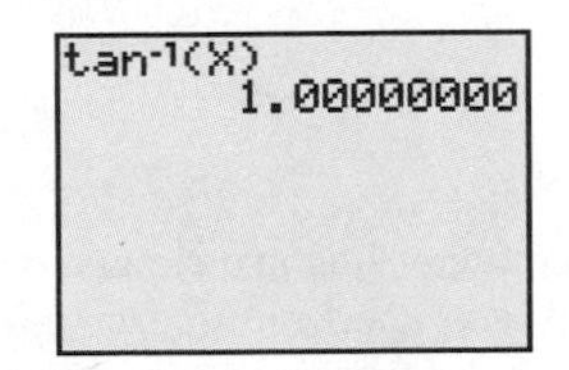

B *Find exact real number values without using a calculator.*

17. $\arccos(-\frac{1}{2})$ **18.** $\arcsin(-\sqrt{2}/2)$

19. $\tan^{-1}(-1)$ **20.** $\arctan(-\sqrt{3})$

21. $\sin^{-1}(-\sqrt{3}/2)$ **22.** $\cos^{-1}(-1)$

23. $\cos^{-1}(-\sqrt{3}/2)$ **24.** $\sin^{-1}(-1)$

25. $\sin[\sin^{-1}(-0.6)]$ **26.** $\tan(\tan^{-1} 25)$

27. $\cos[\sin^{-1}(-\sqrt{2}/2)]$ **28.** $\sec[\sin^{-1}(-\sqrt{3}/2)]$

Evaluate to four significant digits using a calculator.

29. $\tan^{-1}(-4.038)$ **30.** $\arctan(-10.04)$

31. $\sec[\sin^{-1}(-0.0399)]$ **32.** $\cot[\cos^{-1}(-0.7003)]$

33. $\sqrt{2} + \tan^{-1} \sqrt[3]{5}$ **34.** $\sqrt{5} + \cos^{-1}(1 - \sqrt{2})$

Graph Problems 35 and 36 with the aid of a calculator. Plot points using x values − 1.0, − 0.8, − 0.6, − 0.4, − 0.2, 0.0, 0.2, 0.4, 0.6, 0.8, and 1.0; then join the points with a smooth curve.

35. $y = \sin^{-1} x$ **36.** $y = \cos^{-1} x$

Find the exact degree measure of θ without a calculator.

37. $\theta = \arccos(-1/2)$ **38.** $\theta = \arcsin(-\sqrt{2}/2)$

39. $\theta = \tan^{-1}(-1)$ **40.** $\theta = \arctan(-\sqrt{3})$

41. $\theta = \sin^{-1}(-\sqrt{3}/2)$ **42.** $\theta = \cos^{-1}(-1)$

In Problems 43–48, find the degree measure of θ to two decimal places using a calculator.

43. $\theta = \tan^{-1} 3.0413$ **44.** $\theta = \cos^{-1} 0.7149$

45. $\theta = \arcsin(-0.8107)$ **46.** $\theta = \arccos(-0.7728)$

47. $\theta = \arctan(-17.305)$ **48.** $\theta = \tan^{-1}(-0.3031)$

49. Evaluate $\cos^{-1}[\cos(-0.3)]$ with a calculator set in radian mode. Explain why this does or does not illustrate a cosine–inverse cosine identity.

50. Evaluate $\sin^{-1}[\sin(-2)]$ with a calculator set in radian mode. Explain why this does or does not illustrate a sine–inverse sine identity.

51. The identity $\sin(\sin^{-1} x) = x$ is valid for $-1 \le x \le 1$.
(A) Graph $y = \sin(\sin^{-1} x)$ for $-1 \le x \le 1$.
(B) What happens if you graph $y = \sin(\sin^{-1} x)$ over a wider interval, say $-2 \le x \le 2$? Explain.

52. The identity $\cos(\cos^{-1} x) = x$ is valid for $-1 \le x \le 1$.
(A) Graph $y = \cos(\cos^{-1} x)$ for $-1 \le x \le 1$.
(B) What happens if you graph $y = \cos(\cos^{-1} x)$ over a wider interval, say $-2 \le x \le 2$? Explain.

C *In Problems 53–56, find exact real number values without using a calculator.*

53. $\sin[\arccos \frac{1}{2} + \arcsin(-1)]$

54. $\cos[\cos^{-1}(-\sqrt{3}/2) - \sin^{-1}(-\frac{1}{2})]$

55. $\sin[2\sin^{-1}(-\frac{4}{5})]$

56. $\cos\left(\dfrac{\cos^{-1}\frac{1}{3}}{2}\right)$

In Problems 57–60, write each as an algebraic expression in x free of trigonometric or inverse trigonometric functions.

57. $\sin(\cos^{-1} x), \quad -1 \le x \le 1$

58. $\cos(\sin^{-1} x), \quad -1 \le x \le 1$

59. $\tan(\arcsin x), \quad -1 \le x \le 1$

60. $\cos(\arctan x)$

Verify each identity in Problems 61 and 62.

61. $\tan^{-1}(-x) = -\tan^{-1} x$

62. $\sin^{-1}(-x) = -\sin^{-1} x$

63. Let $f(x) = \cos^{-1}(2x - 3)$.
(A) Explain how you would find the domain of f and find it.
(B) Graph f over the interval $0 \le x \le 3$ and explain the result.

64. Let $g(x) = \sin^{-1}\left(\dfrac{x+1}{2}\right)$.
(A) Explain how you would find the domain of g and find it.
(B) Graph g over the interval $-4 \le x \le 2$ and explain the result.

65. Let $h(x) = 3 + 5\sin(x - 1), -\pi/2 \le x \le 1 + \pi/2$.
(A) Find $h^{-1}(x)$.
(B) Explain how x must be restricted in $h^{-1}(x)$.

66. Let $f(x) = 4 + 2\cos(x - 3), -\pi/2 \le x \le 3 + \pi/2$.
(A) Find $f^{-1}(x)$.
(B) Explain how x must be restricted in $f^{-1}(x)$.

67. The identity $\sin^{-1}(\sin x) = x$ is valid for $-\pi/2 \le x \le \pi/2$.
(A) Graph $y = \sin^{-1}(\sin x)$ for $-\pi/2 \le x \le \pi/2$.
(B) What happens if you graph $y = \sin^{-1}(\sin x)$ over a larger interval, say $-2\pi \le x \le 2\pi$? Explain.

68. The identity $\cos^{-1}(\cos x) = x$ is valid for $0 \le x \le \pi$.
(A) Graph $y = \cos^{-1}(\cos x)$ for $0 \le x \le \pi$.
(B) What happens if you graph $y = \cos^{-1}(\cos x)$ over a larger interval, say $-2\pi \le x \le 2\pi$? Explain.

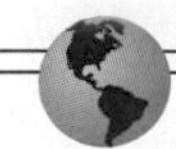

Applications

69. Sonic Boom An aircraft flying faster than the speed of sound produces sound waves that pile up behind the aircraft in the form of a cone. The cone of a level flying aircraft intersects the ground in the form of a hyperbola, and along this curve we experience a *sonic boom* (see the figure). In Section 2.4 it was shown that

$$\sin\frac{\theta}{2} = \frac{\text{Speed of sound}}{\text{Speed of aircraft}} \tag{1}$$

where θ is the cone angle. The ratio

$$M = \frac{\text{Speed of aircraft}}{\text{Speed of sound}} \tag{2}$$

is the Mach number. A Mach number of 2.3 would indicate an aircraft moving at 2.3 times the speed of sound. From equations (1) and (2) we obtain

$$\sin\frac{\theta}{2} = \frac{1}{M}$$

(A) Write θ in terms of M.
(B) Find θ to the nearest degree for $M = 1.7$ and for $M = 2.3$.

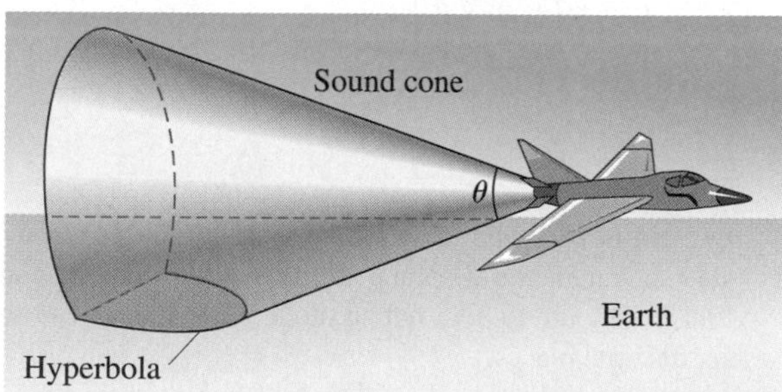

Figure for 69

70. Space Science A spacecraft traveling in a circular orbit h miles above the earth observes horizons in each direction (see the figure at the top of the next page), where r is the radius of the earth (3,959 mi).
(A) Express θ in terms of h and r.
(B) Find θ, in degrees (to one decimal place), for $h = 425.4$ mi.
(C) Find the length (to the nearest mile) of the arc subtended by angle θ found in part (B). What percentage (to one decimal place) of the great circle containing the arc does the arc represent? (A **great circle** is any circle on the surface of the earth having the center of the earth as its center.)

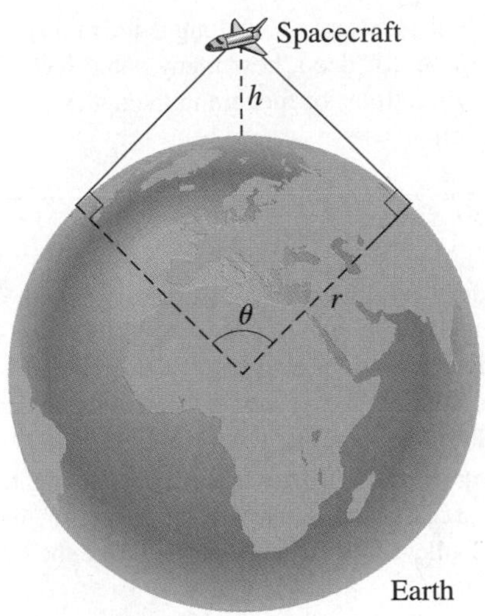

Figure for 70

***71. Precalculus: Sports** A particular soccer field is 110 yd by 60 yd, and the goal is 8 yd wide at the end of the field (see the figure). A player is dribbling the ball along a line parallel to and 5 yd inside the side line. Assuming the player has a clear shot all along this line, is there an optimal distance x from the end of the field where the shot should be taken? That is, is there a distance x for which θ is maximum? Parts (A)–(D) will attempt to answer this question.

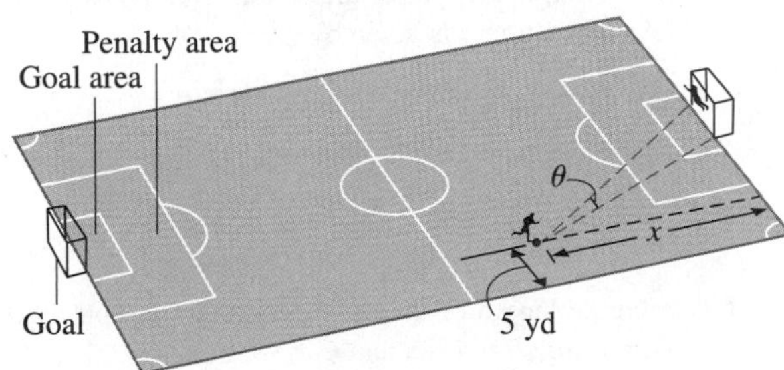

Figure for 71

(A) Discuss what you think happens to θ as x varies from 0 yd to 55 yd.

(B) Show that

$$\theta = \tan^{-1}\left(\frac{8x}{x^2 + 609}\right)$$

$$\left[\textit{Hint:}\quad \tan(\alpha + \beta) = \frac{\tan\alpha + \tan\beta}{1 - \tan\alpha\tan\beta} \text{ is useful.}\right]$$

(C) Complete Table 1 (to two decimal places) and select the value of x that gives the maximum value of θ in the table. (If your calculator has a table-generating feature, use it.)

TABLE 1

x (yd)	10	15	20	25	30	35
θ (deg)	6.44					

(D) Graph the equation in part (B) in a graphing utility for $0 \le x \le 55$. Describe what the graph shows. Use a built-in routine to find the maximum θ and the distance x that produces it.

***72. Precalculus: Related Rates** The figure represents a circular courtyard surrounded by a high stone wall. A flood light, located at E, shines into the courtyard. A person walks from the center C along CD to D, at the rate of 6 ft/sec.

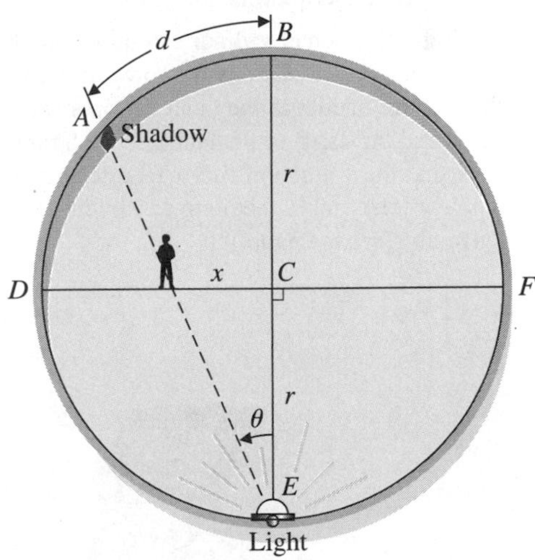

Figure for 72

(A) Do you think that the shadow moves along the circular wall at a constant rate, or does it speed up or slow down as the person walks from C to D?

(B) Show that if the person walks x feet from C along CD, then the shadow will move a distance d given by

$$d = 2r\theta = 2r\tan^{-1}\frac{x}{r} \tag{1}$$

where θ is in radians. [*Hint:* Draw a line from A to C.] Express x in terms of time t, then for a courtyard of radius $r = 60$ ft, rewrite equation (1) in the form

$$d = 120 \tan^{-1} \frac{t}{10} \qquad (2)$$

(C) Using equation (2), complete Table 2 (to one decimal place). (If you have a calculator with a table-generating feature, use it.)

TABLE 2

t (sec)	0	1	2	3	4	5	6	7	8	9	10
d (ft)	0.0	12.0									

(D) From Table 2, determine how far the shadow moves during the first second, during the fifth second, and during the tenth second. Is the shadow speeding up, slowing down, or moving uniformly? Explain.

(E) Graph equation (2) in a graphing utility for $0 \le t \le 10$. Describe what the graph shows.

***73. Engineering** Horizontal cylindrical tanks are buried underground at service stations to store fuel. To determine the amount of fuel in the tank, a "dip stick" is often used to find the depth of the fuel (see the figure).

(A) Show that the volume of fuel x feet deep in a horizontal circular tank L feet long with radius r, $x < r$, is given by (see the figure)

$$V = \left[r^2 \cos^{-1} \frac{r - x}{r} - (r - x)\sqrt{r^2 - (r - x)^2} \right] L$$

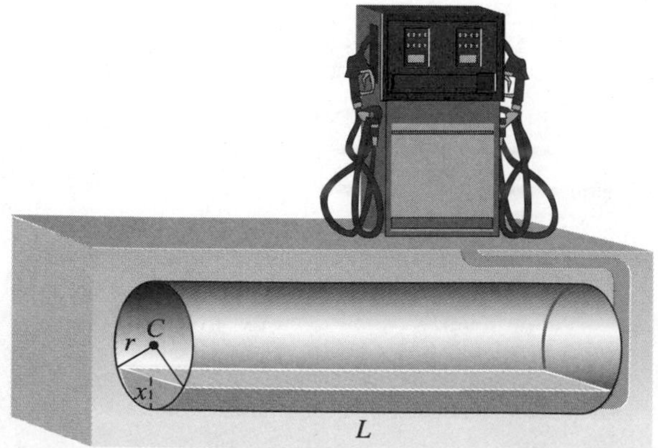

Figure for 73

(B) If the fuel in a tank 30 ft long with radius 3 ft is found to be 2 ft deep, how many cubic feet (to the nearest cubic foot) of fuel are in the tank?

(C) The function

$$y_1 = 30\left[9 \cos^{-1} \frac{3 - x}{3} - (3 - x)\sqrt{9 - (3 - x)^2} \right]$$

represents the volume of fuel x feet deep in the tank in part (B). Graph y_1 and $y_2 = 350$ in the same viewing window, and use the built-in intersection routine to find the depth (to one decimal place) when the tank contains 350 ft³ of fuel.

***74. Engineering** In designing mechanical equipment, it is sometimes necessary to determine the length of the belt around two pulleys of different diameters (see the figure).

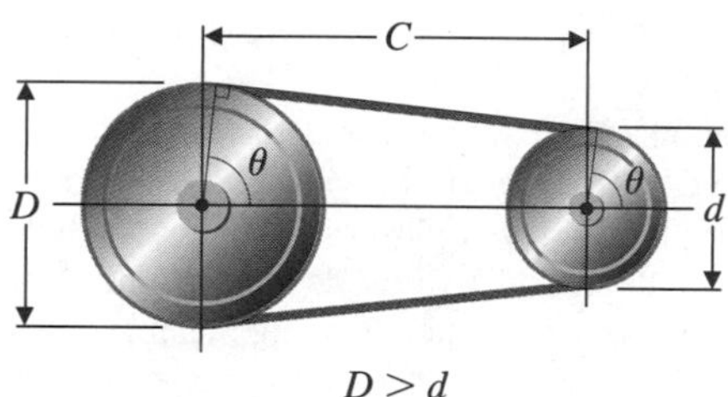

Figure for 74

(A) Show that the length of the belt around the two pulleys in the figure is given by

$$L = \pi D + (d - D)\theta + 2C \sin \theta$$

where θ (in radians) is given by

$$\theta = \cos^{-1} \frac{D - d}{2C}$$

(B) Find the length of the belt (to one decimal place) if $D = 6$ in., $d = 4$ in., and $C = 10$ in.

(C) The function

$$y_1 = 6\pi - 2 \cos^{-1} \frac{1}{x} + 2x \sin\left(\cos^{-1} \frac{1}{x}\right)$$

represents the length of the belt around the two pulleys in part (B) when the centers of the pulleys are x inches apart. Graph y_1 and $y_2 = 40$ in the same viewing window, and use the built-in intersection routine to find the distance between the centers of the pulleys (to one decimal place) when the belt is 40 in. long.

☆5.2 INVERSE COTANGENT, SECANT, AND COSECANT FUNCTIONS

- Definition of Inverse Cotangent, Secant, and Cosecant Functions
- Calculator Evaluation

Definition of Inverse Cotangent, Secant, and Cosecant Functions

Paralleling the development for the inverse sine, cosine, and tangent functions, we define the inverse cotangent, secant, and cosecant functions as follows.

INVERSE COTANGENT, SECANT, AND COSECANT FUNCTIONS

$y = \cot^{-1} x$	is equivalent to	$x = \cot y$	where $0 < y < \pi$ and x is any real number
$y = \sec^{-1} x$	is equivalent to	$x = \sec y$	where $0 \le y \le \pi$, $y \ne \pi/2$, and $x \le -1$ or $x \ge 1$
$y = \csc^{-1} x$	is equivalent to	$x = \csc y$	where $-\pi/2 \le y \le \pi/2$, $y \ne 0$, and $x \le -1$ or $x \ge 1$

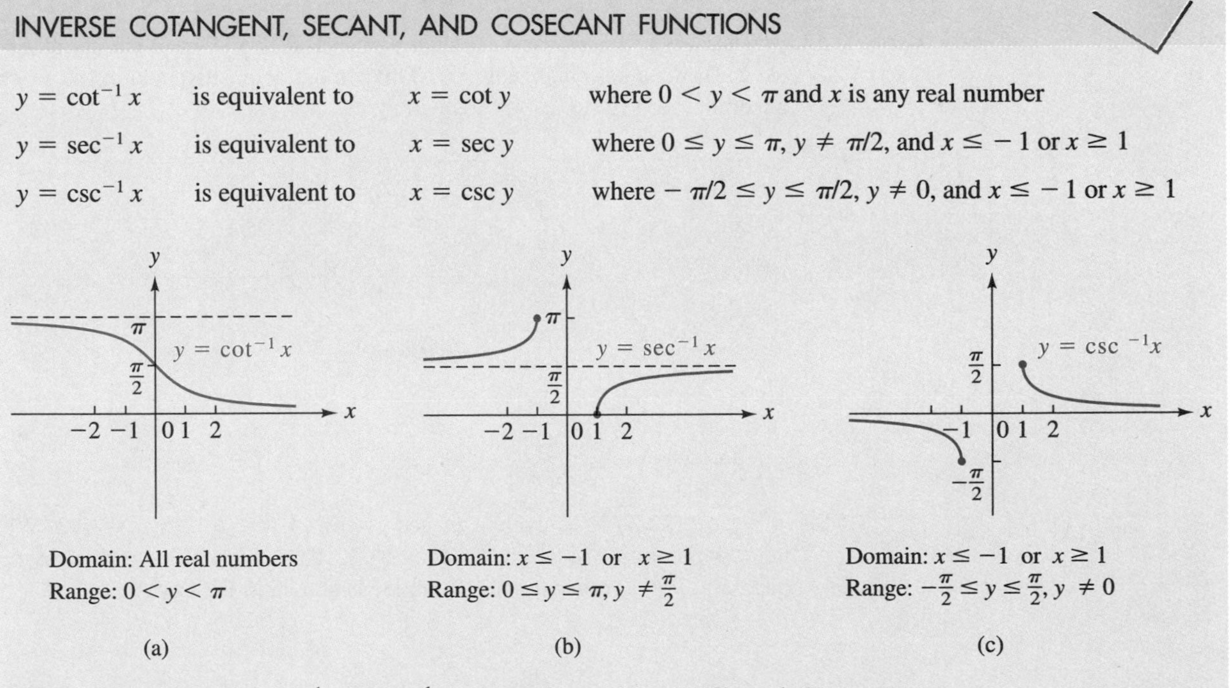

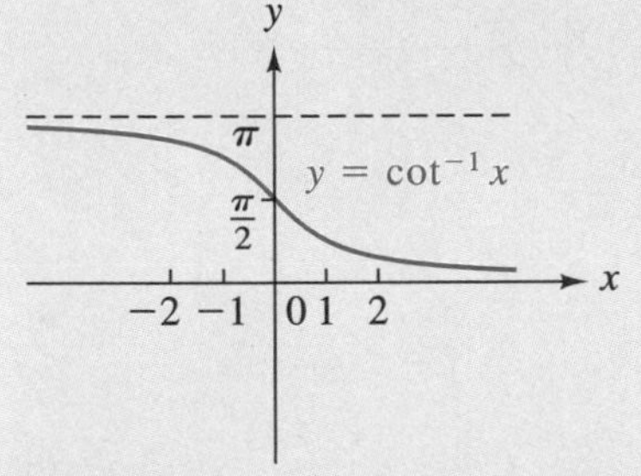

Domain: All real numbers
Range: $0 < y < \pi$

(a)

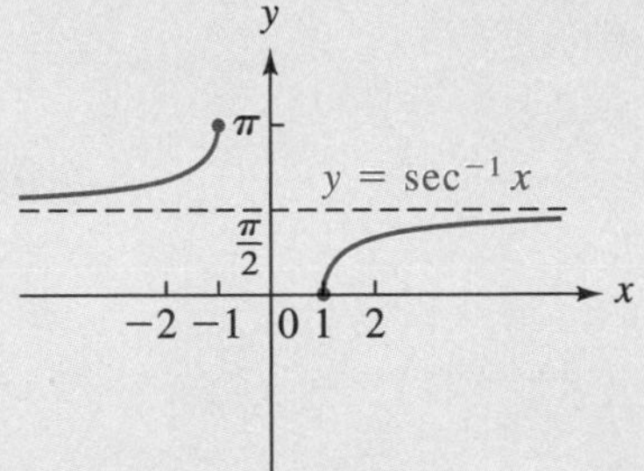

Domain: $x \le -1$ or $x \ge 1$
Range: $0 \le y \le \pi$, $y \ne \frac{\pi}{2}$

(b)

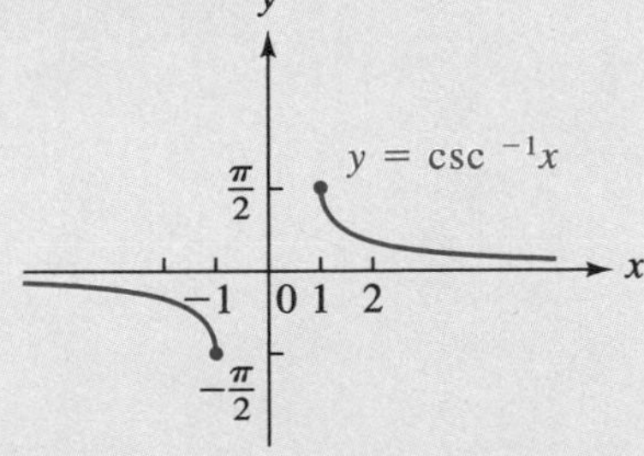

Domain: $x \le -1$ or $x \ge 1$
Range: $-\frac{\pi}{2} \le y \le \frac{\pi}{2}$, $y \ne 0$

(c)

[*Note:* The ranges for $\sec^{-1}$ and $\csc^{-1}$ are sometimes selected differently.]

The functions $y = \cot^{-1} x$, $y = \sec^{-1} x$, and $y = \csc^{-1} x$ are also denoted by $y = \operatorname{arccot} x$, $y = \operatorname{arcsec} x$, and $y = \operatorname{arccsc} x$, respectively.

☆ Sections marked with a star may be omitted without loss of continuity.

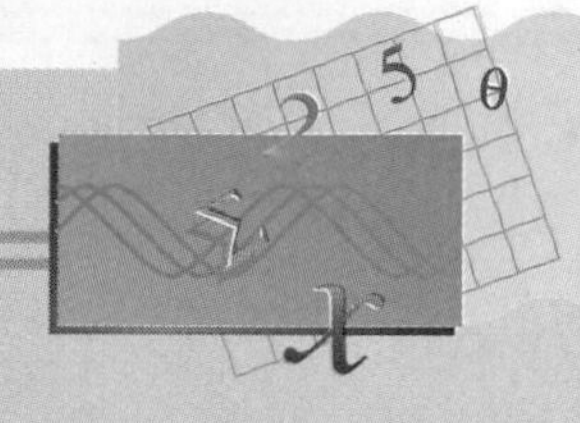

EXPLORE/DISCUSS 1

Which of the following are not defined? Why not?

$\cot^{-1}(-0.5)$	$\sec^{-1}(-0.3)$	$\csc^{-1} 0.9$
$\cot^{-1}(-1{,}000{,}000)$	$\sec^{-1}(3.25 \times 10^{8})$	$\csc^{-1}(3.25 \times 10^{-8})$

EXAMPLE 1 Exact Values

Find exact values without using a calculator:

(A) $\operatorname{arccot}(-1)$ (B) $\sec^{-1}(2/\sqrt{3})$

Solution (A) $y = \operatorname{arccot}(-1)$ is equivalent to $\cot y = -1$, $0 < y < \pi$. What number between 0 and π has cotangent -1? This y must be positive and in the second quadrant (see Fig. 1):

$$\cot y = -1 = -\frac{1}{1}$$

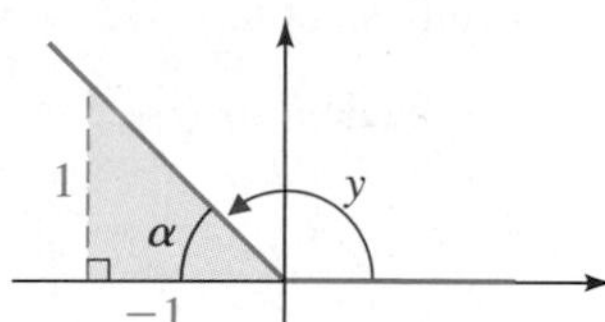

$\alpha = \frac{\pi}{4}$

$y = \frac{3\pi}{4}$

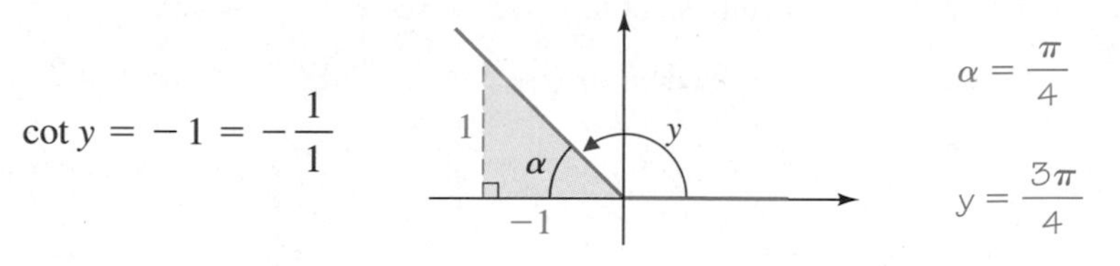

FIGURE 1

Thus,

$$\operatorname{arccot}(-1) = \frac{3\pi}{4}$$

(B) $y = \sec^{-1}(2/\sqrt{3})$ is equivalent to $\sec y = 2/\sqrt{3}$, $0 \le y \le \pi$, $y \ne \pi/2$. What number between 0 and π has secant $2/\sqrt{3}$? This y is positive and in the first quadrant. We draw a reference triangle, as shown in Figure 2.

$$\sec y = \frac{2}{\sqrt{3}}$$

$y = \frac{\pi}{6}$

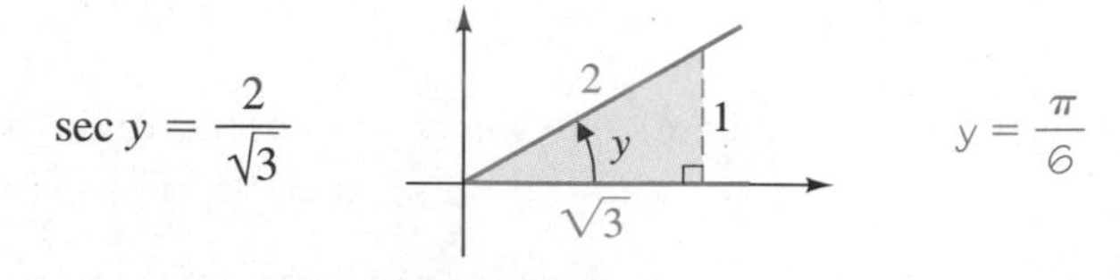

FIGURE 2

Thus,

$$\sec^{-1}\frac{2}{\sqrt{3}} = \frac{\pi}{6}$$

■

Matched Problem 1 Find exact values without using a calculator:

(A) $\operatorname{arcot}(-\sqrt{3})$ (B) $\csc^{-1}(-2)$ ■

EXAMPLE 2

Exact Values

Find the exact value of $\tan[\sec^{-1}(-3)]$ without using a calculator.

Solution Let $y = \sec^{-1}(-3)$; then

$$\sec y = -3 \qquad 0 \le y \le \pi, \quad y \ne \pi/2$$

This y is positive and in the second quadrant. Draw a reference triangle, find the third side, and then determine $\tan y$ from the triangle (see Fig. 3):

$$\sec y = -3 = -\frac{3}{1}$$

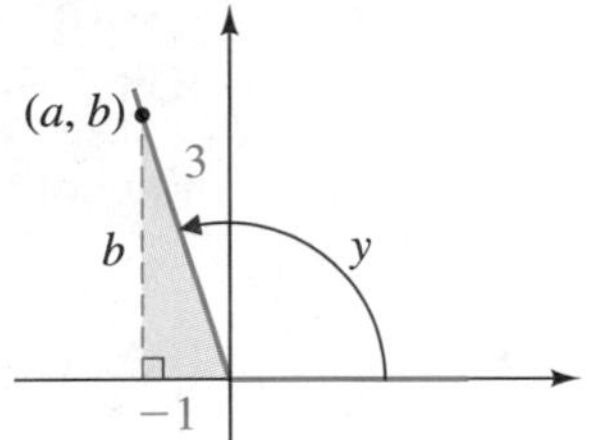

$$b = \sqrt{3^2 - (-1)^2} = \sqrt{8} = 2\sqrt{2}$$

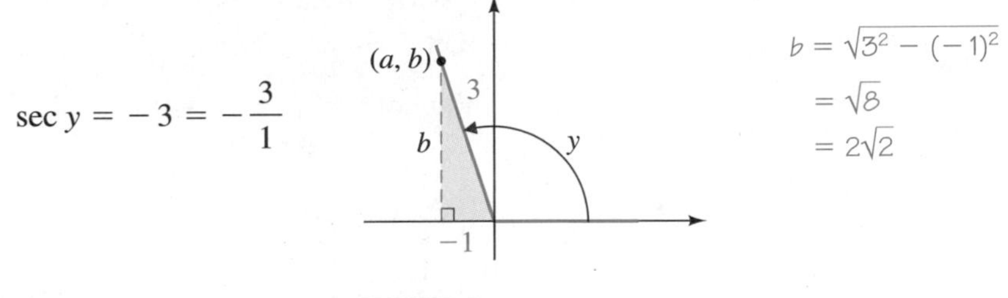

FIGURE 3

Thus,

$$\tan[\sec^{-1}(-3)] = \tan y = \frac{b}{a} = \frac{2\sqrt{2}}{-1} = -2\sqrt{2}$$ ■

Matched Problem 2 Find the exact value of $\cot[\csc^{-1}(-\frac{5}{3})]$ without using a calculator. ■

■ Calculator Evaluation

Many calculators have keys for sin, cos, tan, $\sin^{-1}$, $\cos^{-1}$, $\tan^{-1}$, or their equivalents. To find $\sec x$, $\csc x$, and $\cot x$ using a calculator, we use the reciprocal identities

$$\sec x = \frac{1}{\cos x} \qquad \csc x = \frac{1}{\sin x} \qquad \cot x = \frac{1}{\tan x}$$

How can we evaluate $\sec^{-1} x$, $\csc^{-1} x$, and $\cot^{-1} x$ using a calculator with only $\sin^{-1}$, $\cos^{-1}$, and $\tan^{-1}$ keys? The inverse identities given in the box on page 294 are made to order for this purpose.

We will establish the first part of the inverse cotangent identity stated in the box. The inverse secant and cosecant identities are left to you to do (see Problems 59 and 60, Exercise 5.2).

INVERSE COTANGENT, SECANT, AND COSECANT IDENTITIES

$$\cot^{-1} x = \begin{cases} \tan^{-1} \dfrac{1}{x} & x > 0 \\ \pi + \tan^{-1} \dfrac{1}{x} & x < 0 \end{cases}$$

$$\sec^{-1} x = \cos^{-1} \frac{1}{x} \qquad x \geq 1 \quad \text{or} \quad x \leq -1$$

$$\csc^{-1} x = \sin^{-1} \frac{1}{x} \qquad x \geq 1 \quad \text{or} \quad x \leq -1$$

Let

$$y = \cot^{-1} x \qquad x > 0$$

Then

$$\cot y = x \qquad 0 < y < \pi/2 \qquad \text{Definition of } \cot^{-1}$$

$$\frac{1}{\tan y} = x \qquad 0 < y < \pi/2 \qquad \text{Reciprocal identity}$$

$$\tan y = \frac{1}{x} \qquad 0 < y < \pi/2 \qquad \text{Algebra}$$

$$y = \tan^{-1} \frac{1}{x} \qquad 0 < y < \pi/2 \qquad \text{Definition of } \tan^{-1}$$

Thus,

$$\cot^{-1} x = \tan^{-1} \frac{1}{x} \qquad \text{for } x > 0$$

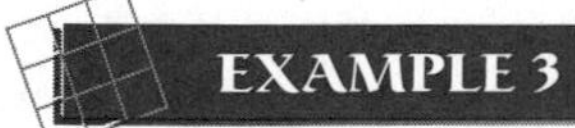

EXAMPLE 3 **Calculator Evaluation**

Use a calculator to evaluate the following as real numbers to three decimal places.

(A) $\cot^{-1} 4.05$ (B) $\csc^{-1}(-12)$

Solution (A) With the calculator in radian mode,

$$\cot^{-1} 4.05 = \tan^{-1} \frac{1}{4.05} = 0.242$$

(B) With the calculator in radian mode,

$$\csc^{-1}(-12) = \sin^{-1}(-\tfrac{1}{12}) = -0.083$$

■

Matched Problem 3 Use a calculator to evaluate the following as real numbers to three decimal places.

(A) $\cot^{-1} 2.314$ (B) $\sec^{-1}(-1.549)$ ■

Answers to Matched Problems

1. (A) $5\pi/6$ (B) $-\pi/6$ **2.** $-\frac{4}{3}$

3. (A) 0.408 (B) 2.273

EXERCISE 5.2

A *Find the exact real number value of each without using a calculator.*

1. $\cot^{-1}\sqrt{3}$
2. $\cot^{-1} 0$
3. arccsc 1
4. arcsec 2
5. $\sec^{-1}\sqrt{2}$
6. $\csc^{-1} 2$
7. $\sin(\cot^{-1} 0)$
8. $\cos(\cot^{-1} 1)$
9. $\tan(\csc^{-1}\frac{5}{4})$
10. $\cot(\sec^{-1}\frac{5}{3})$

B **11.** $\cot^{-1}(-1)$
12. $\sec^{-1}(-1)$
13. arsec(-2)
14. arccsc$(-\sqrt{2})$
15. arccsc(-2)
16. arccot$(-\sqrt{3})$
17. $\csc^{-1}\frac{1}{2}$
18. $\sec^{-1}(-\frac{1}{2})$
19. $\cos[\csc^{-1}(-\frac{5}{3})]$
20. $\tan[\cot^{-1}(-1/\sqrt{3})]$
21. $\cot[\sec^{-1}(-\frac{5}{4})]$
22. $\sin[\cot^{-1}(-\frac{3}{4})]$
23. $\cos[\sec^{-1}(-2)]$
24. $\sin[\csc^{-1}(-2)]$
25. $\cot(\cot^{-1} 33.4)$
26. $\sec[\sec^{-1}(-44)]$
27. $\csc[\csc^{-1}(-4)]$
28. $\cot[\cot^{-1}(-7.3)]$

Use a calculator to evaluate the following as real numbers to three decimal places.

29. $\cot^{-1} 3.065$
30. $\cot^{-1} 7.306$
31. $\sec^{-1}(-1.963)$
32. $\sec^{-1} 2.041$
33. $\csc^{-1} 1.172$
34. $\csc^{-1}(-1.938)$
35. $\cot^{-1}(-5.104)$
36. $\cot^{-1}(-12.236)$

Find the exact degree measure of θ without using a calculator.

37. $\theta =$ arcsec(-2)
38. $\theta =$ arccsc$(-\sqrt{2})$
39. $\theta = \cot^{-1}(-1)$
40. $\theta =$ arccot$(-1/\sqrt{3})$
41. $\theta = \csc^{-1}(-2/\sqrt{3})$
42. $\theta = \sec^{-1}(-1)$

Find the degree measure to two decimal places using a calculator.

43. $\theta = \cot^{-1} 0.3288$
44. $\theta = \sec^{-1} 1.3989$
45. $\theta =$ arccsc(-1.2336)
46. $\theta =$ arcsec(-1.2939)
47. $\theta =$ arccot(-0.0578)
48. $\theta = \cot^{-1}(-3.2994)$

C *Find exact values for each problem without using a calculator.*

49. $\tan[\csc^{-1}(-\frac{5}{3}) + \tan^{-1}\frac{1}{4}]$
50. $\tan[\tan^{-1} 4 - \sec^{-1}(-\sqrt{5})]$
51. $\tan[2\cot^{-1}(-\frac{3}{4})]$
52. $\tan[2\sec^{-1}(-\sqrt{5})]$

In Problems 53–58, write as an algebraic expression in x free of trigonometric or inverse trigonometric functions.

53. $\sin(\cot^{-1} x)$
54. $\cos(\cot^{-1} x)$
55. $\csc(\sec^{-1} x)$
56. $\tan(\csc^{-1} x)$
57. $\sin(2\cot^{-1} x)$
58. $\sin(2\sec^{-1} x)$

59. Show that $\sec^{-1} x = \cos^{-1}(1/x)$ for $x \geq 1$ and $x \leq -1$.
60. Show that $\csc^{-1} x = \sin^{-1}(1/x)$ for $x \geq 1$ and $x \leq -1$.

Problems 61–64 require the use of a graphing utility. Use an appropriate inverse trigonometric identity to graph each function in the viewing window $-5 \leq x \leq 5$, $-\pi \leq y \leq \pi$.

61. $y = \sec^{-1} x$
62. $y = \csc^{-1} x$
63. $y = \cot^{-1} x$ [Use two viewing windows, one for $-5 \leq x \leq 0$, and the other for $0 \leq x \leq 5$.]
64. $y = \cot^{-1} x$ [Using one viewing window, $-5 \leq x \leq 5$, graph $y_1 = \pi(x < 0) + \tan^{-1}(1/x)$, where $<$ is selected from the TEST menu. The expression $(x < 0)$ assumes a value of 1 for $x < 0$ and 0 for $x \geq 0$.]

5.3 TRIGONOMETRIC EQUATIONS: AN ALGEBRAIC APPROACH

- Introduction
- Solving Trigonometric Equations Using an Algebraic Approach

Introduction

In Chapter 4 we considered trigonometric equations called identities. These equations are true for all replacements of the variable(s) for which both sides are defined. We now consider another class of equations, called **conditional equations,** which may be true for some replacements of the variable(s) but false for others. For example,

$$\sin x = \cos x$$

is a conditional equation, since it is true for $x = \pi/4$ and false for $x = 0$. (Check both values.)

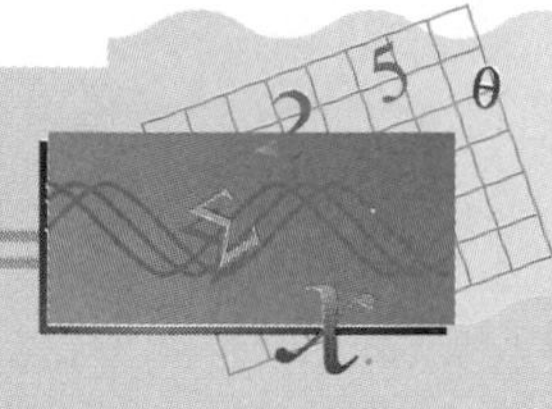

EXPLORE/DISCUSS 1

Consider the simple trigonometric equation

$$\sin x = 0.5$$

Figure 1 shows a partial graph of the left and right sides of the equation.

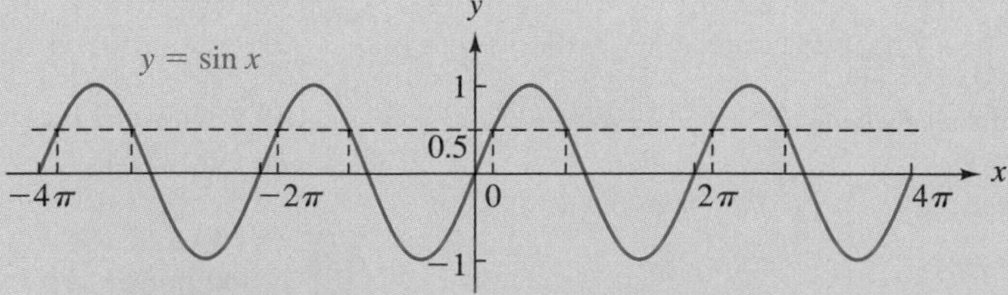

FIGURE 1
sin $x = 0.5$

(A) How many solutions does the equation have on the interval $[0, 2\pi)$? What are the solutions?

(B) How many solutions does the equation have on the interval $(-\infty, \infty)$? Discuss a method of writing all solutions to the equation.

In this section we will solve conditional trigonometric equations using an algebraic approach. In the next section we will use a graphing utility approach. Solving trigonometric equations using an algebraic approach often requires the use of algebraic manipulation, identities, and some ingenuity. In some cases, an algebraic approach leads to exact solutions. A graphing utility approach uses graphical methods to approximate solutions to any accuracy desired, which is different from finding exact solutions. The graphing utility approach can be used to solve trigonometric equations that are either very difficult or impossible to solve algebraically.

Solving Trigonometric Equations Using an Algebraic Approach

The following suggestions for solving trigonometric equations using an algebraic approach may be helpful:

SUGGESTIONS FOR SOLVING TRIGONOMETRIC EQUATIONS ALGEBRAICALLY

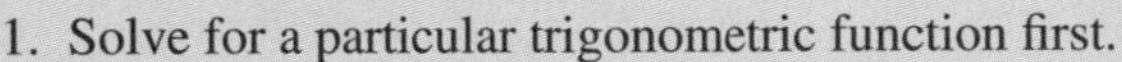

1. Solve for a particular trigonometric function first.
   (A) Try using identities.
   (B) Try algebraic manipulations such as factoring, combining fractions, and so on.
2. After solving for a trigonometric function, solve for the variable.

Several examples should make the algebraic approach clear.

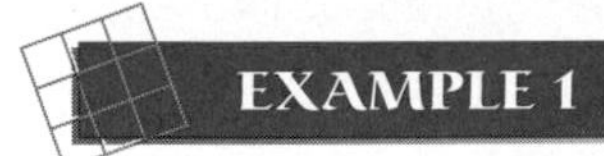

EXAMPLE 1 **Exact Solutions Using Factoring**

Find all solutions exactly for $2\sin^2 x + \sin x = 0$.

Solution ***Step 1*** *Solve for* $\sin x$*:*

$$2\sin^2 x + \sin x = 0 \qquad \text{Factor out } \sin x.$$

$$\sin x(2\sin x + 1) = 0$$

$$\sin x = 0 \quad \text{or} \quad 2\sin x + 1 = 0 \qquad ab = 0 \text{ only if } a = 0 \text{ or } b = 0.$$

$$\sin x = -\tfrac{1}{2}$$

Step 2 *Solve each equation over one period* $[0, 2\pi)$*:* As an aid to writing all solutions over one period, sketch a graph of $y = \sin x$, $y = 0$, and $y = -\frac{1}{2}$ in the same coordinate system, as shown in Figure 2.

$$\sin x = 0 \qquad \sin x = -\frac{1}{2}$$

$$x = 0, \pi \qquad x = \frac{7\pi}{6}, \frac{11\pi}{6}$$

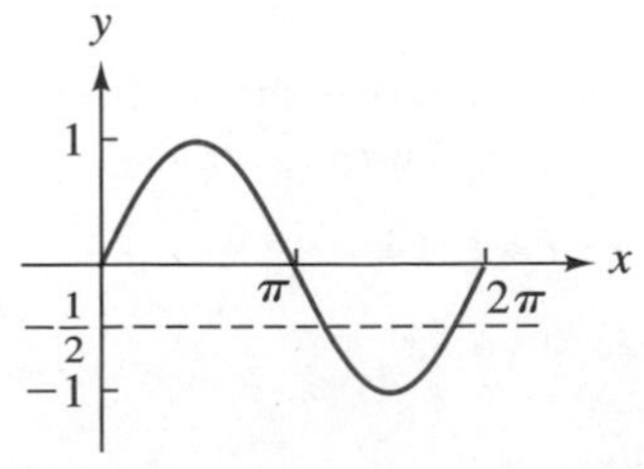

FIGURE 2

Step 3 Write an expression for all solutions: Because the sine function is periodic with period 2π, all solutions are given by

$$x = \begin{cases} 0 + 2k\pi \\ \pi + 2k\pi \\ 7\pi/6 + 2k\pi \\ 11\pi/6 + 2k\pi \end{cases} \qquad k \text{ any integer}$$

■

Matched Problem 1 Find all solutions exactly for $2\cos^2 x - \cos x = 0$. ■

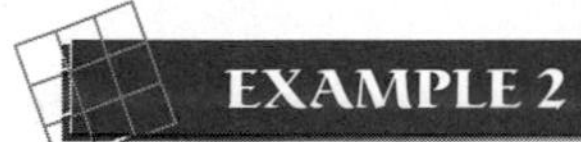

EXAMPLE 2

Approximate Solutions Using Identities and Factoring

Approximate all real solutions for $8\sin^2 x = 5 + 10\cos x$. (Compute inverse functions to four decimal places.)

Solution *Step 1 Solve for sin x and/or cos x:* Move all nonzero terms to the left of the equal sign and express the left side in terms of cos x:

$$8\sin^2 x = 5 + 10\cos x$$

$$8\sin^2 x - 10\cos x - 5 = 0$$

$$8(1 - \cos^2 x) - 10\cos x - 5 = 0 \qquad \sin^2 x = 1 - \cos^2 x$$

$$8\cos^2 x + 10\cos x - 3 = 0 \qquad \text{Algebra}$$

$$(2\cos x + 3)(4\cos x - 1) = 0 \qquad 8u^2 + 10u - 3 = (2u + 3)(4u - 1)$$

$$2\cos x + 3 = 0 \quad \text{or} \quad 4\cos x - 1 = 0$$

$$\cos x = -\tfrac{3}{2} \qquad\qquad \cos x = \tfrac{1}{4}$$

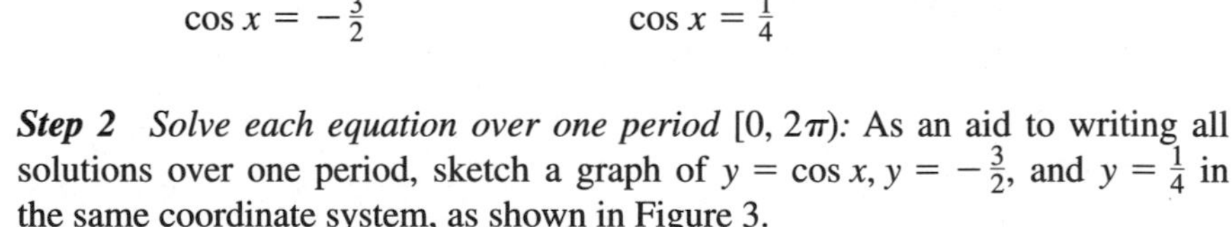

Step 2 Solve each equation over one period $[0, 2\pi)$*:* As an aid to writing all solutions over one period, sketch a graph of $y = \cos x$, $y = -\frac{3}{2}$, and $y = \frac{1}{4}$ in the same coordinate system, as shown in Figure 3.

Solve the first equation:

$$\cos x = -\tfrac{3}{2} \qquad \text{No solution } (-\tfrac{3}{2} \text{ is not in the range of the cosine function}).$$

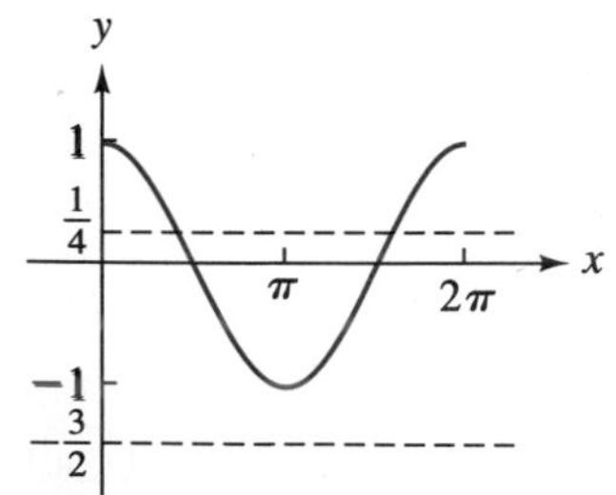

FIGURE 3

Solve the second equation:

$$\cos x = \tfrac{1}{4} \qquad \text{From Figure 3, we see that the solutions are in the first and fourth quadrants.}$$

$$x = \cos^{-1}\tfrac{1}{4} = 1.3181 \qquad \text{First quadrant solution}$$

$$x = 2\pi - 1.3181 = 4.9651 \qquad \text{Fourth quadrant solution}$$

✔ **Check** $\cos 1.3181 = 0.2500 \qquad \cos 4.9651 = 0.2500$

(Checks may not be exact because of roundoff errors.)

Step 3 *Write an expression for all solutions:* Because the cosine function is periodic with period 2π, all solutions are given by

$$x = \begin{cases} 1.3181 + 2k\pi \\ 4.9651 + 2k\pi \end{cases} \qquad k \text{ any integer}$$

■

Matched Problem 2 Approximate all real solutions for $3\cos^2 x + 8\sin x = 7$. (Compute inverse functions to four decimal places.) ■

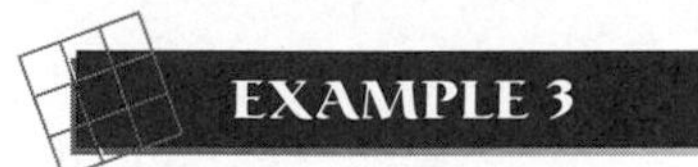

EXAMPLE 3 **Approximate Solutions Using Substitutions**

Find θ in degree measure (to three decimal places) so that

$$8\tan(6\theta + 15) = -64.328 \qquad -90° < 6\theta + 15 < 90°$$

Solution *Step 1* *Make a substitution:* Let $u = 6\theta + 15$ to obtain

$$8\tan u = -64.328$$

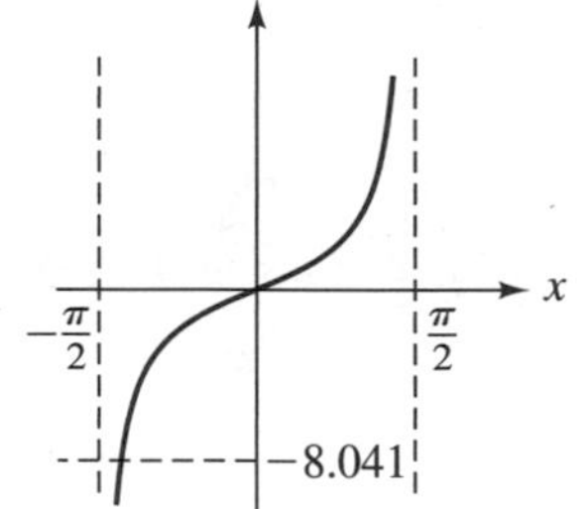

FIGURE 4

Step 2 *Solve for tan u:*

$$\tan u = -8.041$$

Step 3 *Solve for u over $-90° < u < 90°$:* As an aid to writing all solutions for $-90° < u < 90°$, sketch a graph of $y = \tan u$ and $y = -8.041$ in the same coordinate system, as shown in Figure 4.

The solution is in the fourth quadrant.

$$u = \tan^{-1}(-8.041) = -82.911°$$

✔ Check $\tan(-82.911°) = -8.041$

Step 4 *Solve for θ:*

$$u = -82.911°$$

$$6\theta + 15 = -82.911°$$

$$\theta = -16.319°$$

(A check of θ by substituting in the original problem is left to the reader.) ■

Matched Problem 3 Find θ in degree measure (to three decimal places) so that

$$5\sin(2\theta - 5) = -3.045 \qquad 0° \le 2\theta - 5 \le 360°$$

■

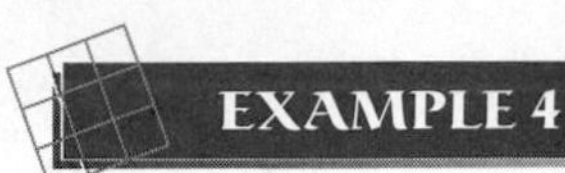

EXAMPLE 4 Exact Solutions Using Identities and Factoring

Find exact solutions for $\sin 2x = \sin x$, $0 \le x < 2\pi$.

Solution The following solution includes only a few key steps. Sketch graphs as appropriate on scratch paper.

$$\sin 2x = \sin x \qquad \text{Use double-angle identity.}$$

$$2 \sin x \cos x = \sin x$$

$$2 \sin x \cos x - \sin x = 0$$

$$\sin x(2 \cos x - 1) = 0$$

$$\sin x = 0 \qquad \text{or} \qquad 2 \cos x - 1 = 0$$

$$x = 0, \pi \qquad\qquad \cos x = \frac{1}{2}$$

$$x = \frac{\pi}{3}, \frac{5\pi}{3}$$

Combining the solutions from each equation, we have $x = 0, \pi/3, \pi, 5\pi/3$. ■

Matched Problem 4 Find exact solutions for $\sin^2 x = \frac{1}{2} \sin 2x$, $0 \le x < 2\pi$. ■

EXAMPLE 5 Approximate Solutions Using Identities and the Quadratic Formula

Approximate all real solutions for $\cos 2x = 2(\sin x - 1)$. (Compute inverse functions to four decimal places.)

Solution ***Step 1*** *Solve for* $\sin x$:

$$\cos 2x = 2(\sin x - 1) \qquad \text{Use double-angle identity.}$$

$$1 - 2 \sin^2 x = 2 \sin x - 2$$

$$2 \sin^2 x + 2 \sin x - 3 = 0 \qquad \text{Quadratic in } \sin x\text{. Left side does not factor using integer coefficients, so use the quadratic formula.}$$

$$\sin x = \frac{-2 \pm \sqrt{4 - 4(2)(-3)}}{4}$$

$$= -1.822876 \quad \text{or} \quad 0.822876$$

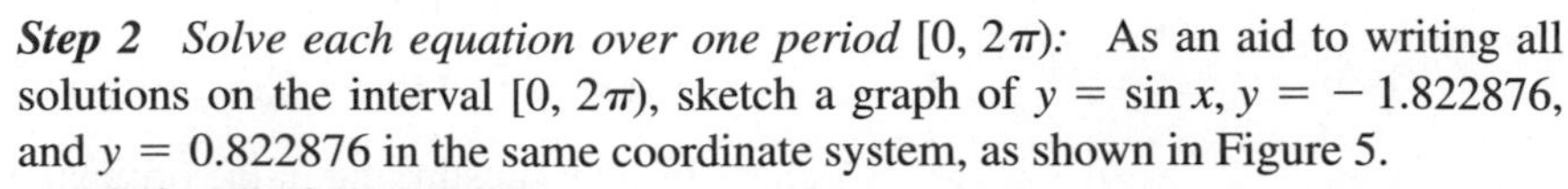

Step 2 *Solve each equation over one period* $[0, 2\pi)$: As an aid to writing all solutions on the interval $[0, 2\pi)$, sketch a graph of $y = \sin x$, $y = -1.822876$, and $y = 0.822876$ in the same coordinate system, as shown in Figure 5.

Solve the first equation:

$$\sin x = -1.822876 \qquad \text{No solution for this range value.}$$

Solve the second equation:

$$\sin x = 0.822876$$

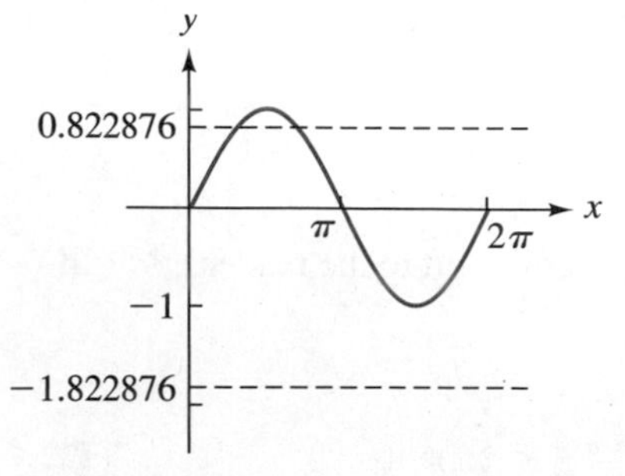

FIGURE 5

Figure 5 indicates solutions in the first and second quadrants. If the reference angle is α, then $x = \alpha$ or $x = \pi - \alpha$.

$$\alpha = \sin^{-1} 0.822876 = 0.9665$$

$$\pi - \alpha = \pi - 0.9665 = 2.1751$$

✔ Check $\sin 0.9665 = 0.8229 \qquad \sin 2.1751 = 0.8229$

Step 3 *Write an expression for all solutions:* Because the sine function is periodic with period 2π, all solutions are given by

$$x = \begin{cases} 0.9665 + 2k\pi \\ 2.1751 + 2k\pi \end{cases} \quad k \text{ any integer}$$

■

Matched Problem 5 Approximate all real solutions for $\cos 2x = 4 \cos x - 2$. (Compute inverse functions to four decimal places.) ■

Answers to Matched Problems

1. $x = \begin{cases} \pi/3 + 2k\pi \\ \pi/2 + 2k\pi \\ 3\pi/2 + 2k\pi \\ 5\pi/3 + 2k\pi \end{cases}$ k any integer

2. $x = \begin{cases} 0.7297 + 2k\pi \\ 2.4119 + 2k\pi \end{cases}$ k any integer

3. $\theta = 111.259°, 163.742°$ **4.** $x = 0, \pi/4, \pi, 5\pi/4$

5. $x = \begin{cases} 1.2735 + 2k\pi \\ 5.0096 + 2k\pi \end{cases}$ k any integer

EXERCISE 5.3

A *In Problems 1–8, find exact solutions over the indicated intervals (x real and θ in degrees).*

1. $2 \cos x + 1 = 0, \quad 0 \le x < 2\pi$

2. $2 \sin x + 1 = 0, \quad 0 \le x < 2\pi$

3. $2 \cos x + 1 = 0, \quad$ all real x

4. $2 \sin x + 1 = 0, \quad$ all real x

5. $\sqrt{2} \sin \theta - 1 = 0, \quad 0° \le \theta < 360°$

6. $2 \cos \theta - \sqrt{3} = 0, \quad 0° \le \theta < 360°$

7. $\sqrt{2} \sin \theta - 1 = 0, \quad$ all θ

8. $2 \cos \theta - \sqrt{3} = 0, \quad$ all θ

In Problems 9–14, solve each to four decimal places (x real and θ in degrees).

9. $4 \tan \theta + 15 = 0, \quad 0° \le \theta < 180°$

10. $2 \tan \theta - 7 = 0, \quad 0° \le \theta < 180°$

11. $5 \cos x - 2 = 0, \quad 0 \le x < 2\pi$

12. $7 \cos x - 3 = 0, \quad 0 \le x < 2\pi$

13. $5.0118 \sin x - 3.1105 = 0, \quad$ all real x

14. $1.3224 \sin x + 0.4732 = 0, \quad$ all real x

B *For Problems 15–26, find exact solutions (x real and θ in degrees).*

15. $\cos x = \cot x, \quad 0 \le x < 2\pi$

16. $\tan x = -2 \sin x, \quad 0 \le x < 2\pi$

17. $\cos^2 \theta = \frac{1}{2} \sin 2\theta, \quad$ all θ

18. $2 \sin^2 \theta + \sin 2\theta = 0, \quad$ all θ

19. $\tan(x/2) - 1 = 0, \quad 0 \le x < 2\pi$

20. $\sec(x/2) + 2 = 0, \quad 0 \le x < 2\pi$

21. $\sin^2 \theta + 2 \cos \theta = -2, \quad 0° \le \theta < 360°$

22. $2 \cos^2 \theta + 3 \sin \theta = 0, \quad 0° \le \theta < 360°$

23. $\cos 2\theta + \sin^2 \theta = 0, \quad 0° \le \theta < 360°$

24. $\cos 2\theta + \cos \theta = 0, \quad 0° \le \theta < 360°$

25. $4 \cos^2 2x - 4 \cos 2x + 1 = 0, \quad 0 \le x \le 2\pi$

26. $2 \sin^2(x/2) - 3 \sin(x/2) + 1 = 0, \quad 0 \le x \le 2\pi$

Solve Problems 27–30 (x real and θ in degrees). Compute inverse functions to four significant digits.

27. $4 \cos^2 \theta = 7 \cos \theta + 2, \quad 0° \le \theta \le 180°$

28. $6 \sin^2 \theta + 5 \sin \theta = 6, \quad 0° \le \theta \le 90°$

29. $\cos 2x + 10 \cos x = 5, \; 0 \le x < 2\pi$

30. $2 \sin x = \cos 2x, \quad 0 \le x < 2\pi$

Solve Problems 31 and 32 for all real solutions. Compute inverse functions to four significant digits.

31. $\cos^2 x = 3 - 5 \cos x$ **32.** $2 \sin^2 x = 1 - 2 \sin x$

33. Explain the difference between evaluating the expression $\cos^{-1}(-0.7334)$ and solving the equation $\cos x = -0.7334$.

34. Explain the difference between evaluating the expression $\tan^{-1}(-5.377)$ and solving the equation $\tan x = -5.377$.

C *Find exact solutions to Problems 35–38. [Hint: Square both sides at an appropriate point, solve, then eliminate any extraneous solutions at the end.]*

35. $\sin x + \cos x = 1, \quad 0 \le x < 2\pi$

36. $\cos x - \sin x = 1, \quad 0 \le x < 2\pi$

37. $\sec x + \tan x = 1, \quad 0 \le x < 2\pi$

38. $\tan x - \sec x = 1, \quad 0 \le x < 2\pi$

Applications

39. Electric Current An alternating current generator produces a current given by the equation

$$I = 30 \sin 120\pi t$$

where t is time in seconds and I is current in amperes. Find the least positive t (to four significant digits) such that $I = 25$ amperes.

40. Electric Current Find the least positive t in Problem 39 (to four significant digits) such that $I = -10$ amperes.

41. Photography A polarizing filter for a camera contains two parallel plates of polarizing glass, one fixed and the other able to rotate. If θ is the angle of rotation from the position of maximum light transmission, then the intensity of light leaving the filter is $\cos^2 \theta$ times the intensity entering the filter (see the figure). Find the least positive θ (in decimal degrees, to two decimal places) so that the intensity of light leaving the filter is 70% of that entering. [*Hint:* Solve $I \cos^2 \theta = 0.70I$.]

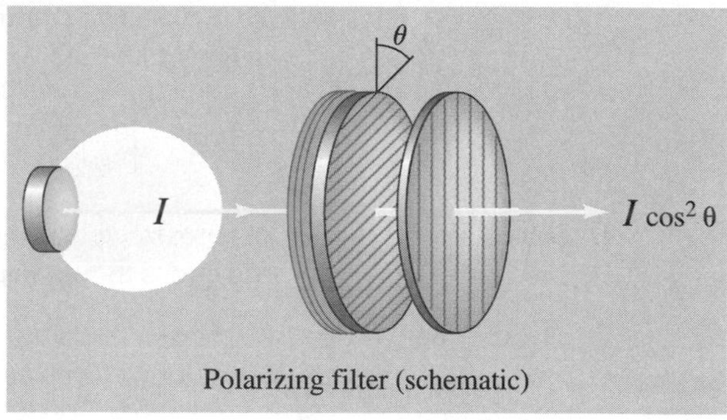

Figure for 41

42. Photography Find θ in Problem 41 so that the light leaving the filter is 40% of that entering.

*** 43. Astronomy** The planet Mercury travels around the sun in an elliptical orbit given approximately by

$$r = \frac{3.44 \times 10^7}{1 - 0.206 \cos \theta}$$

(see the figure). Find the least positive θ (in decimal degrees, to three significant digits) such that Mercury is 3.78×10^7 mi from the sun.

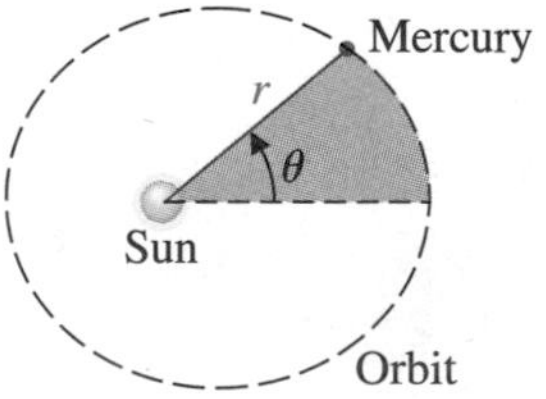

Figure for 43

*** 44. Astronomy** Find the least positive θ (in decimal degrees, to three significant digits) in Problem 43 such that Mercury is 3.09×10^7 mi from the sun.

Precalculus *In Problems 45 and 46, find simultaneous solutions for each system of equations for $0° \le \theta \le 360°$. These are* **polar equations,** *which will be discussed in Chapter 7.*

*** 45.** $r = 2 \sin \theta$
$r = 2(1 - \sin \theta)$

*** 46.** $r = 2 \sin \theta$
$r = \sin 2\theta$

∗ **47.** **Precalculus** Given the equation $xy = -2$, replace x and y with

$$x = u \cos \theta - v \sin \theta$$

$$y = u \sin \theta + v \cos \theta$$

and simplify the left side of the resulting equation. Find the least positive θ (in degree measure) so that the coefficient of the uv term will be 0.

∗ **48.** **Precalculus** Repeat Problem 47 for the equation $2xy = 1$.

5.4 TRIGONOMETRIC EQUATIONS AND INEQUALITIES: A GRAPHING UTILITY APPROACH

- Solving Trigonometric Equations Using a Graphing Utility
- Solving Trigonometric Inequalities Using a Graphing Utility

All of the trigonometric equations that were solved in the last section using an algebraic approach can also be solved—though usually not exactly—using graphing utility methods. In addition, many trigonometric equations that cannot be solved easily using an algebraic approach can be solved (to any accuracy desired) using graphing utility methods.

Consider the simple-looking equation

$$2x \cos x = 1$$

Try solving this equation using any of the methods discussed in Section 5.3. You will not be able to isolate x on one side of the equation with a number on the other side. Such equations are easily solved using a graphing utility, as will be seen in the examples that follow.

Solving trigonometric inequalities, such as

$$\cos x - 2 \sin x > 0.4 - 0.3x$$

with a graphing utility is almost as easy as solving trigonometric equations with a graphing utility. The second part of this section illustrates the process.

Solving Trigonometric Equations Using a Graphing Utility

The best way to proceed is with examples.

Solutions Using a Graphing Utility

A rectangle is inscribed under the graph of $y = 2 \cos(\pi x/2)$, $-1 \le x \le 1$, with the base on the x axis, as shown in Figure 1 (at the top of the next page).

(A) Write an equation for the area A of the rectangle in terms of x.

(B) Graph the equation found in part (A) in a graphing utility, and use TRACE to describe how the area A changes as x goes from 0 to 1.

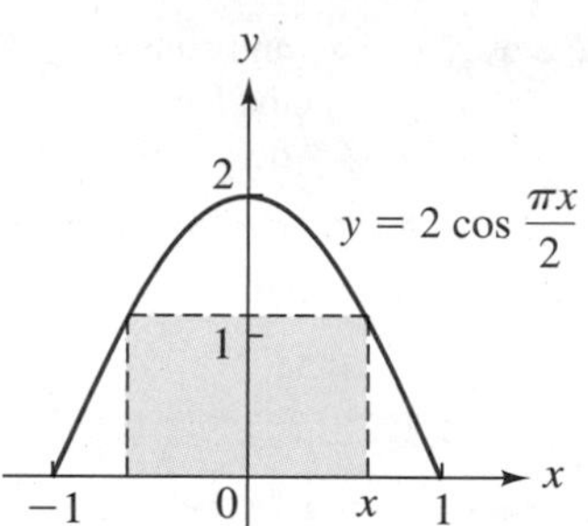

FIGURE 1

(C) Find the value(s) of x (to three decimal places) that produce(s) a rectangle area of 1 square unit.

Solution (A)
$$
\begin{aligned}
A(x) &= (\text{Length of base}) \times (\text{Height}) \\
&= (2x) \times \left(2 \cos \frac{\pi x}{2}\right) \\
&= 4x \cos \frac{\pi x}{2} \qquad 0 \le x \le 1
\end{aligned}
$$

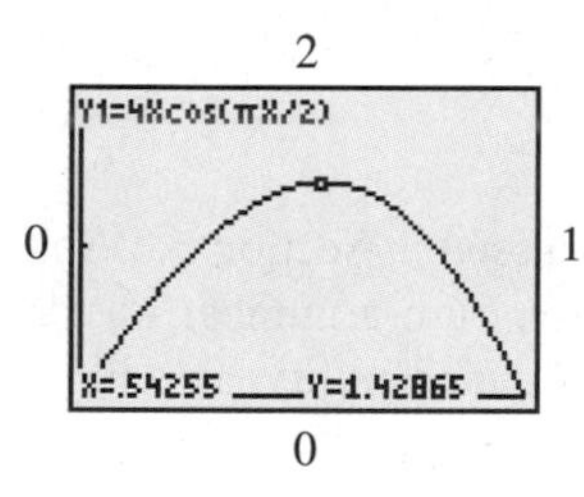

(B) The graph of the area function is shown in Figure 2. Referring to the figure, we see that the area A increases from 0 to a maximum of about 1.429 square units, then decreases to 0 as x goes from 0 to 1.

FIGURE 2

(C) To find x for a rectangle area of 1 square unit, we must solve the equation

$$1 = 4x \cos \frac{\pi x}{2}$$

This equation cannot be solved by methods discussed in Section 5.3, but it is easy to solve using a graphing utility. Graph $y_1 = 1$ and $y_2 = 4x \cos(\pi x/2)$ in the same viewing window, and find the point(s) of intersection of the two graphs using a built-in intersection routine (see Fig. 3). Note that the graph of y_1 intersects the graph of y_2 in two points; therefore, there are two solutions. From Figure 3, we can see that $A = 1$ for $x = 0.275$ and for $x = 0.797$.

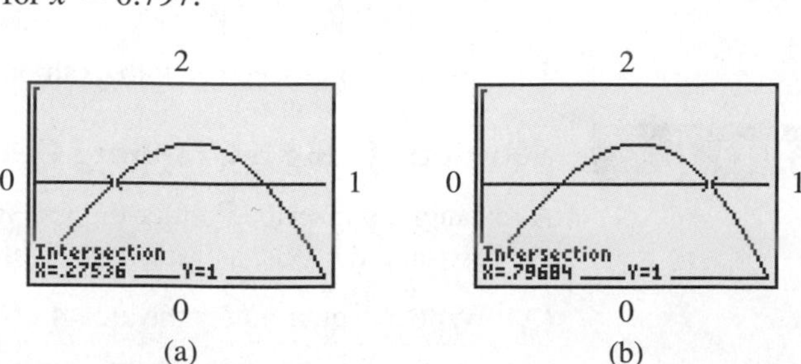

FIGURE 3

✔ Check Substitute each into $A(x) = 4x\cos(\pi x/2)$ to see if the result is 1 (or nearly 1):

```
4*0.275cos(π0.27
5/2)
              .99896
4*0.797cos(π0.79
7/2)
              .99942
```

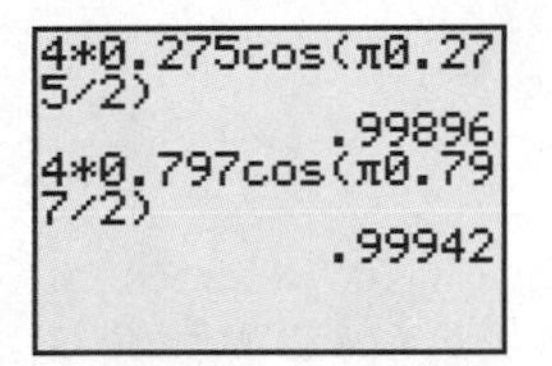

■

Matched Problem 1 Repeat Example 1 to find the value(s) of x (to three decimal places) that produce(s) a rectangle area of 1.25 square units. ■

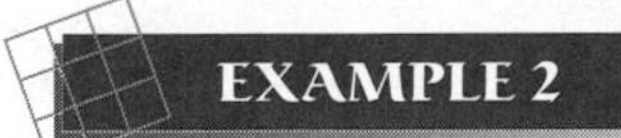

Solution Using a Graphing Utility

Find all real solutions (to four decimal places) for $\sin(x/2) = 0.2x - 0.5$.

Solution We start by graphing $y_1 = \sin(x/2)$ and $y_2 = 0.2x - 0.5$ in the same viewing window for $-2\pi \le x \le 2\pi$ (Fig. 4a). From the figure it is clear that there is a point of intersection to the right of the y axis, but it is not clear how many points of intersection exist to the left. Use TRACE and move the cursor to the uncertain region to the left of the y axis. Then use ZOOM to produce Figure 4b.

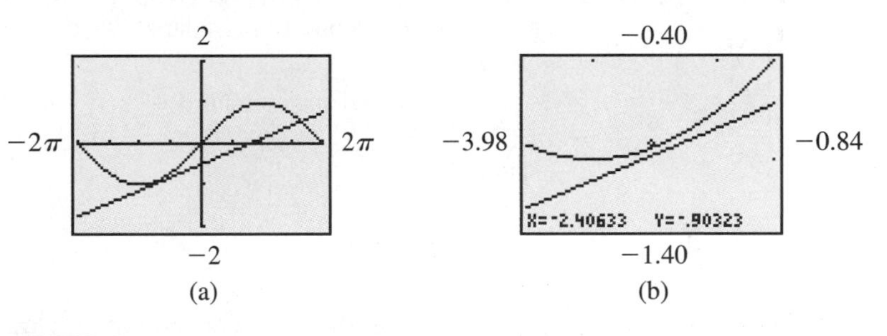

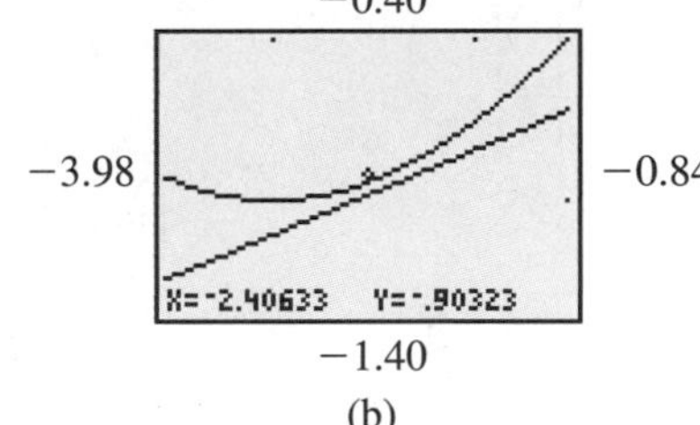

FIGURE 4

From Figure 4b, we conclude that the graphs of y_1 and y_2 do not intersect to the left of the y axis. Therefore, there is only one solution. Using the built-in intersection routine, we find the solution to be $x = 5.1609$ (Fig. 5).

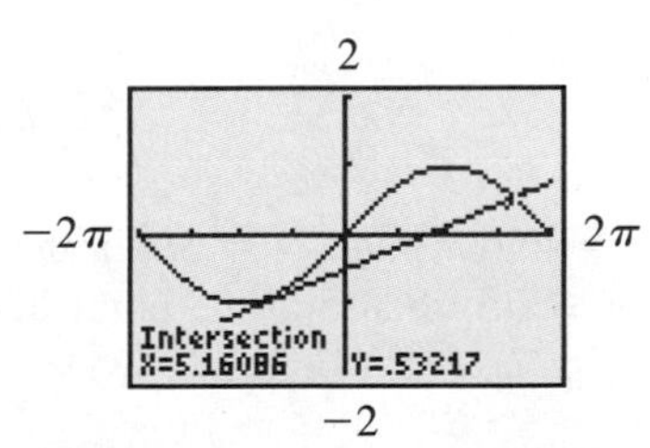

FIGURE 5

✔ Check Substitute $x = 5.1609$ into each side of the original equation to see if the results are equal, or nearly so:

```
sin(5.1609/2)
              .53215
0.2*5.1609-0.5
              .53218
```

■

Matched Problem 2 Find all real solutions (to four decimal places) for $2\cos 2x = 1.35x - 2$. ■

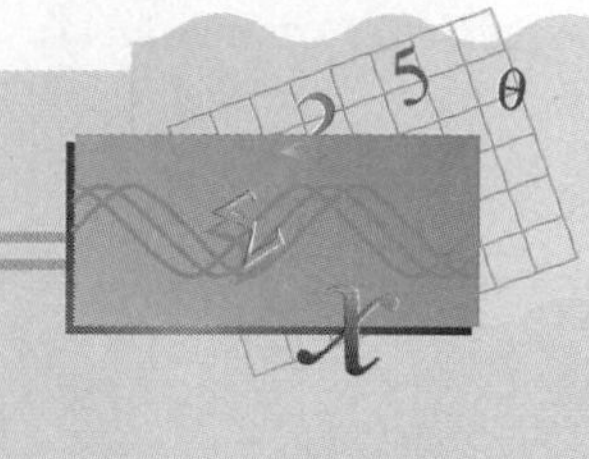

EXPLORE/DISCUSS 1

A 12 cm arc of a circle has a 10 cm chord. A 16 cm arc of a second circle also has a 10 cm chord. For parts (A) and (B), reason geometrically by drawing larger and smaller circles.

(A) Which circle has the larger radius?

(B) Is there a *largest* circle that has a chord of 10 cm? Is there a *smallest* circle that has a chord of 10 cm? Explain.

EXAMPLE 3 An Application in Geometry

A 12 cm arc on a circle has a 10 cm chord. What is the radius of the circle (to four decimal places)? What is the radian measure (to four decimal places) of the central angle subtended by the arc?

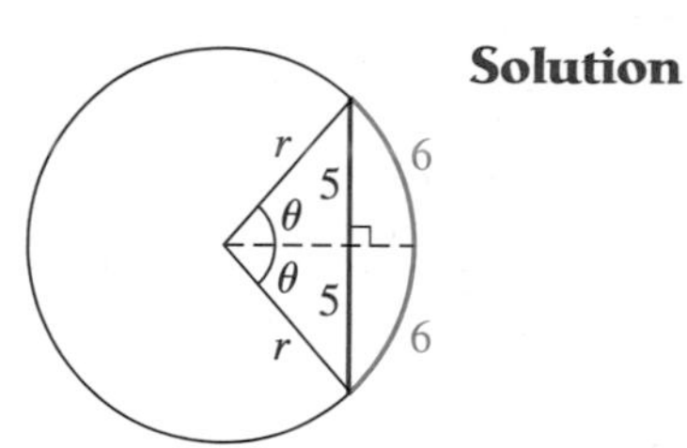

Solution Sketch a figure and introduce the auxiliary lines, as shown in Figure 6.

From the figure (θ in radians), we see that

$$\theta = \frac{6}{r} \qquad \text{and} \qquad \sin\theta = \frac{5}{r}$$

FIGURE 6

Thus,

$$\sin\frac{6}{r} = \frac{5}{r}$$

Solve this trigonometric equation for r. Graph $y_1 = \sin(6/x)$ and $y_2 = 5/x$ in the same viewing window for $1 \le x \le 15$ and $0 \le y \le 1.5$; then find the point of intersection using a built-in intersection routine (Fig. 7). From Figure 7, we see that $r = 5.8437$ cm.

The radian measure of the central angle subtended by the 12 cm chord is

$$2\theta = \frac{12}{5.8437} = 2.0535 \text{ rad}$$

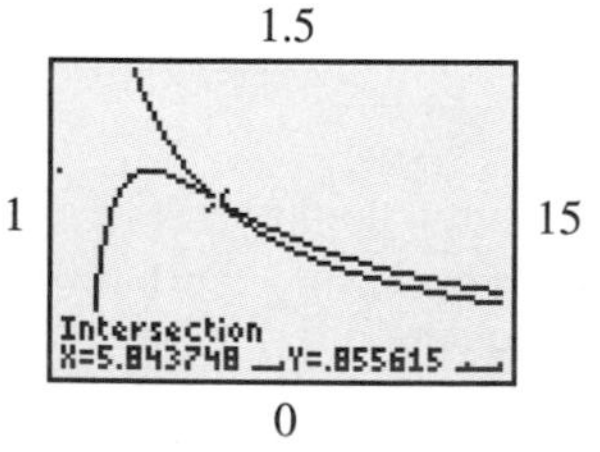

FIGURE 7

Matched Problem 3 A 10 ft arc on a circle has an 8 ft chord. What is the radius of the circle (to four decimal places)? What is the radian measure (to four decimal places) of the central angle subtended by the arc?

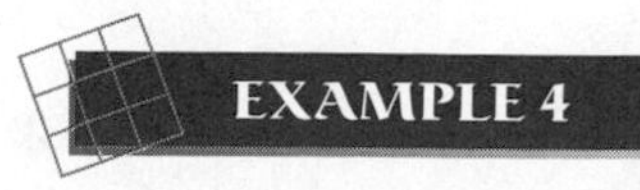

EXAMPLE 4 Solution Using a Graphing Utility

Find all real solutions (to four decimal places) for $\tan(x/2) = 5x - x^2$, $0 \le x \le 3\pi$.

Solution Graph $y_1 = \tan(x/2)$ and $y_2 = 5x - x^2$ in the same viewing window for $0 \le x \le 3\pi$ and $-10 \le y \le 10$. Solutions occur at the three points of intersection shown in Figure 8.

FIGURE 8

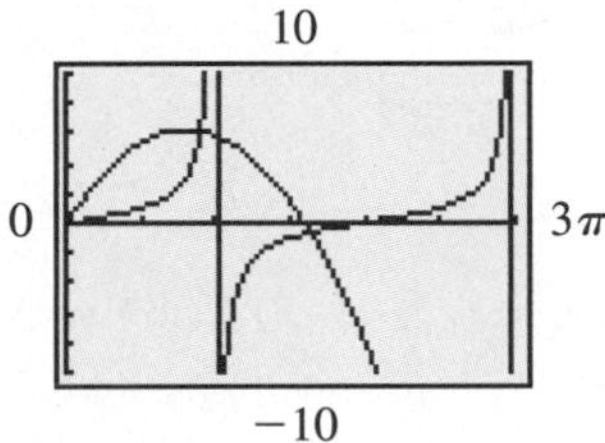

Using a built-in intersection routine, the three solutions are found to be $x = 0.0000, 2.8191, 5.1272$. ■

Matched Problem 4 Find all real solutions (to four decimal places) for $\tan(x/2) = 1/x$, $-\pi < x \le 3\pi$. ■

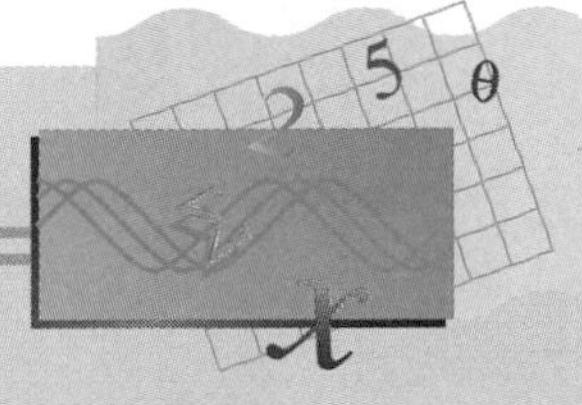

EXPLORE/DISCUSS 2

For each of the following equations, determine the number of real solutions. Is there a largest solution? A smallest solution? Explain.

(A) $\sin x = x/2$ (B) $\tan x = x/2$

Solving Trigonometric Inequalities Using a Graphing Utility

Solving trigonometric inequalities using a graphing utility is almost as easy as solving trigonometric equations using a graphing utility. Example 5 illustrates the process.

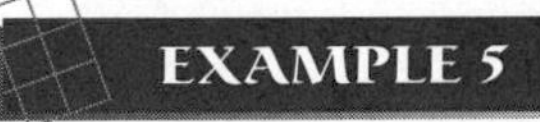

EXAMPLE 5 **Solving a Trigonometric Inequality**

Solve $\cos x - 2\sin x > 0.4 - 0.4x$ (to two decimal places).

Solution Graph $y_1 = \cos x - 2\sin x$ and $y_2 = 0.4 - 0.4x$ in the same viewing window (Fig. 9, page 308). Finding the three points of intersection using a built-in intersection routine, we see that the graph of y_1 is above the graph of y_2 for the following two intervals: $(-2.09, 0.35)$ and $(3.20, \infty)$. Formally, the solution set for the inequality is $(-2.09, 0.35) \cup (3.20, \infty)$.

FIGURE 9

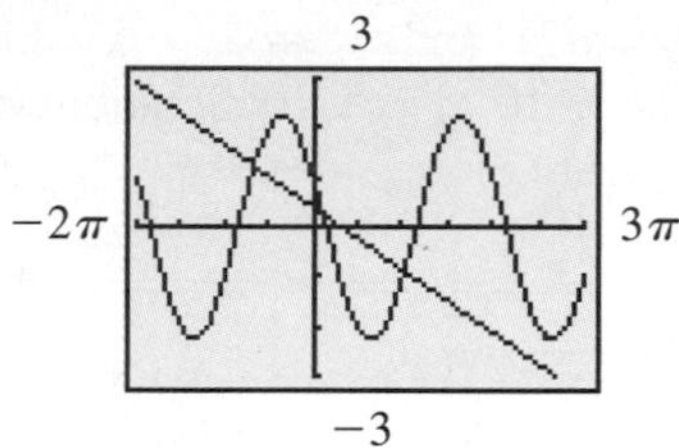

■

Matched Problem 5 Solve $\sin x - \cos x < 0.25x - 0.5$ (to two decimal places). ■

Answers to Matched Problems

1. 0.376, 0.710 **2.** 0.9639

3. $r = 4.4205$ ft, $\theta = 2.2622$ rad

4. $-1.3065, 1.3065, 6.5846$

5. $(-1.65, 0.52) \cup (3.63, \infty)$

EXERCISE 5.4

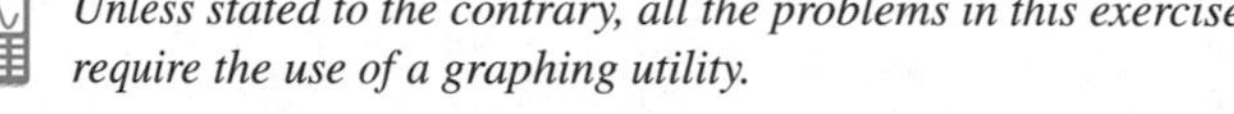

Unless stated to the contrary, all the problems in this exercise require the use of a graphing utility.

A *Solve Problems 1–4 (to four decimal places) using a graphing utility.*

1. $2x = \cos x$, all real x

2. $2 \sin x = 1 - x$, all real x

3. $3x + 1 = \tan 2x$, $0 \le x < \pi/4$

4. $8 - x = \tan(x/2)$, $0 \le x < \pi$

B *Solve Problems 5–12 (to four decimal places) using a graphing utility.*

5. $\cos 2x + 10 \cos x = 5$, $0 \le x < 2\pi$

6. $2 \sin x = \cos 2x$, $0 \le x < 2\pi$

7. $\cos^2 x = 3 - 5 \cos x$, all real x

8. $2 \sin^2 x = 1 - 2 \sin x$, all real x

9. $2 \sin(x - 2) < 3 - x^2$, all real x

10. $\cos 2x > x^2 - 2$, all real x

11. $\sin(3 - 2x) \ge 1 - 0.4x$, all real x

12. $\cos(2x + 1) \le 0.5x - 2$, all real x

13. Without graphing, explain why the following inequality is true for all real x:

$$\sin^2 x - 2 \sin x + 1 \ge 0$$

14. Without graphing, explain why the following inequality is false for all real x for which the left side of the statement is defined:

$$\tan^2 x - 4 \tan x + 4 < 0$$

C *Solve Problems 15–18 exactly if possible, or to four significant digits if an exact solution cannot be found. Use a graphing utility.*

15. $2 \cos(1/x) = 950x - 4$, $0.006 < x < 0.007$

16. $\sin(1/x) = 1.5 - 5x$, $0.04 \le x \le 0.2$

17. $\cos(\sin x) > \sin(\cos x)$, $-2\pi \le x \le 2\pi$

18. $1 + \tan^2 x \ge \cos^2(-x)$, $-2\pi \le x \le 2\pi$

19. Find subintervals (to four decimal places) of the interval $[0, \pi]$ over which $\sqrt{3 \sin(2x) - 2 \cos(x/2) + x}$ is a real number. (Check the end points of each subinterval.)

20. Find subintervals (to four decimal places) of the interval $[0, 2\pi]$ over which $\sqrt{2 \sin(2x) - \cos^2 x + 0.5x}$

is a real number. (Check the end points of each subinterval.)

Applications

21. Geometry The area of a segment of the circle in the figure is given by

$$A = \frac{1}{2}r^2(\theta - \sin\theta)$$

Use a graphing utility to find the angle θ subtended by the segment if the radius is 10 m and the area is 40 m^2. Compute the solution in radians to two decimal places.

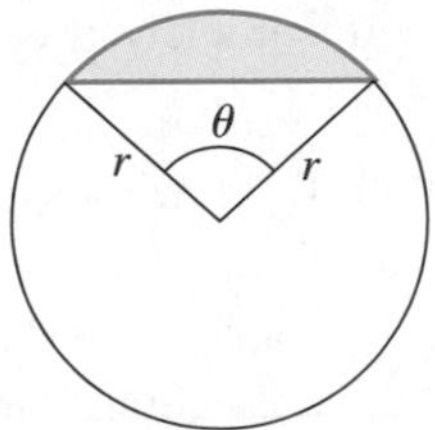

Figure for 21

22. Geometry Repeat Problem 21 if the radius is 8 m and the area is 48 m^2.

23. Architecture An arched doorway is formed by placing a circular arc on top of a rectangle (see the figure). If the rectangle is 4 ft wide and 8 ft high, and the circular arc is 5 ft long, what is the area of the doorway? Compute the answer to two decimal places.

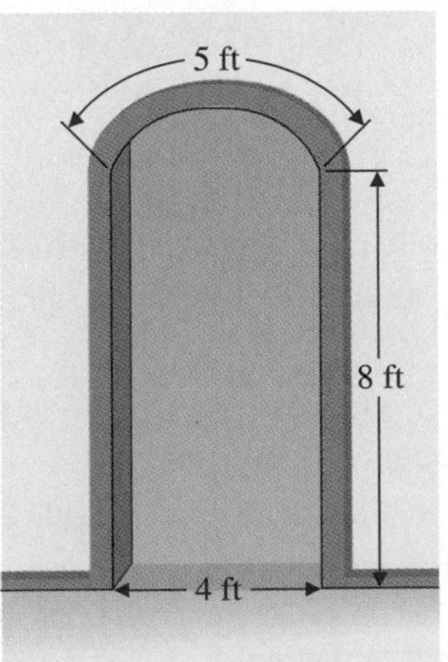

Figure for 23

24. Architecture Repeat Problem 23 if the rectangle is 3 ft wide and 7 ft high, and the circular arc is 4 ft long.

25. Eye Surgery A surgical technique for correcting an astigmatism involves removing small pieces of tissue in order to change the curvature of the cornea.* In the cross section of a cornea shown in part (b) of the figure, the circular arc with radius r and central angle 2θ represents the surface of the cornea.

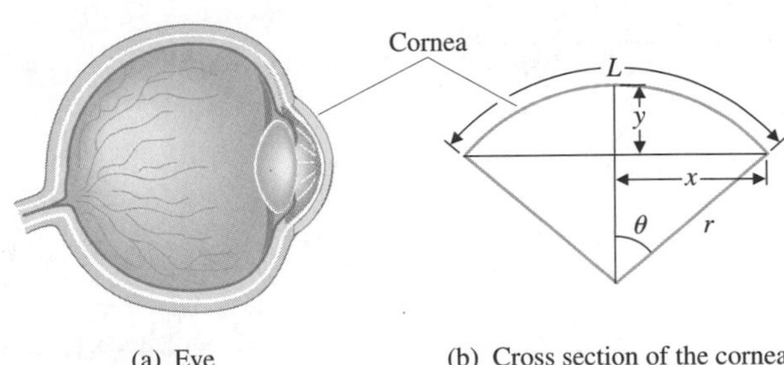

(a) Eye (b) Cross section of the cornea

Figure for 25

(A) If $x = 5.5$ mm and $y = 2.5$ mm, find L, the length of the corneal cross section, correct to four decimal places.

(B) Reducing the width of the cross section without changing its length has the effect of pushing the cornea outward and giving it a rounder, yet still circular, shape. Use a graphing utility to approximate y to four decimal places if x is reduced to 5.4 mm and L remains the same as it was in part (A).

26. Eye Surgery Refer to Problem 25. Increasing the width of a cross section of the cornea without changing its length has the effect of pushing the cornea inward and giving it a flatter, yet still circular, shape. Use a graphing utility to approximate y to four decimal places if x is increased to 5.6 mm and L remains the same as it was in part (A) of Problem 25.

* Based on the article "The Surgical Correction of Astigmatism" by Sheldon Rothman and Helen Strassberg in *The UMAP Journal,* Vol. V, No. 2 (1984).

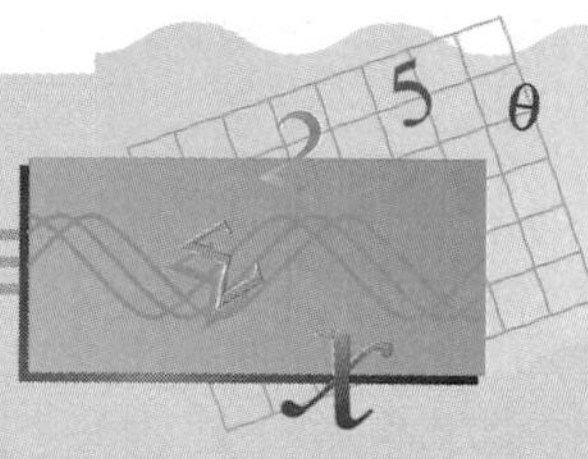

CHAPTER 5 GROUP ACTIVITY

$\sin \frac{1}{x} = 0$ and $\sin^{-1} \frac{1}{x} = 0$

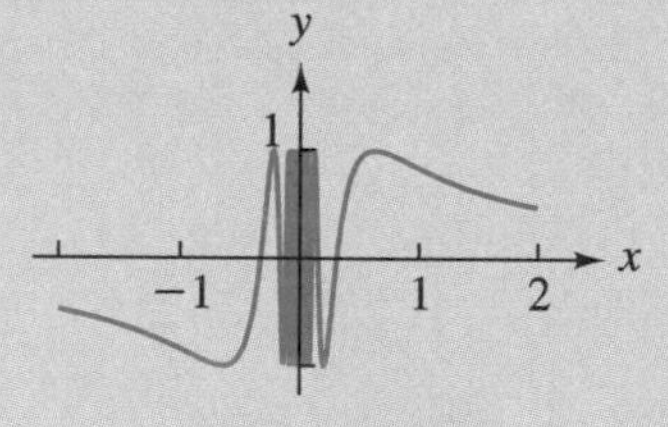

I EXPLORATION OF SOLUTIONS TO $\sin \frac{1}{x} = 0, x > 0$

We are interested in solutions to the equation $\sin(1/x) = 0$, which are the zeros of the function $f(x) = \sin(1/x)$, for $x > 0$. Note that the function f is not defined at $x = 0$. We will restrict our analysis to positive zeros; a similar analysis can be made for negative zeros.

(A) *An overview of the problem.* Graph $f(x) = \sin(1/x)$ for $(0, 1]$. Discuss your first observations regarding the zeros for f.

(B) *Exploring zeros of f on the interval $[0.1, b]$, $b > 0.1$.* Graph f over the interval $[0.1, b]$ for various values of b, $b > 0.1$. Does the function f have a largest zero? If so, what is it (to four decimal places)? Explain what happens to the graph of f as x increases without bound. Does the graph appear to have an asymptote? If so, what is its equation?

(C) *Exploring zeros of f near the origin.* Graph f over the intervals $[0.05, 0.1]$, $[0.025, 0.05]$, $[0.0125, 0.025]$, and so on. How many zeros exist between 0 and b, for any positive b, however small? Explain why this happens. Does f have a smallest positive zero? Explain.

II EXPLORATION OF SOLUTIONS TO $\sin^{-1} \frac{1}{x} = 0, x > 0$

We are interested in solutions to the equation $\sin^{-1}(1/x) = 0$, which are the zeros of the function $g(x) = \sin^{-1}(1/x)$, for $x > 0$. Note that the function g is not defined at $x = 0$.

(A) *An overview of the problem.* Graph $g(x) = \sin^{-1}(1/x)$ over the interval $(0, 5]$. Discuss your first observations regarding the zeros for g.

(B) *The interval $(0, 1]$.* Explain why there is no graph on the interval $(0, 1]$.

(C) *The interval $[1, \infty)$.* Explain why there are no zeros for g on the interval $[1, \infty)$; hence, the equation $\sin^{-1}(1/x) = 0$, $x > 0$, has no solutions. Does the graph have an asymptote? If so, what is its equation?

CHAPTER 5 REVIEW

5.1 INVERSE SINE, COSINE, AND TANGENT FUNCTIONS

$y = \sin^{-1} x$ is equivalent to $x = \sin y$ where $-1 \le x \le 1$ and $-\pi/2 \le y \le \pi/2$

$y = \cos^{-1} x$ is equivalent to $x = \cos y$ where $-1 \le x \le 1$ and $0 \le y \le \pi$

$y = \tan^{-1} x$ is equivalent to $x = \tan y$ where x is any real number and $-\pi/2 < y < \pi/2$

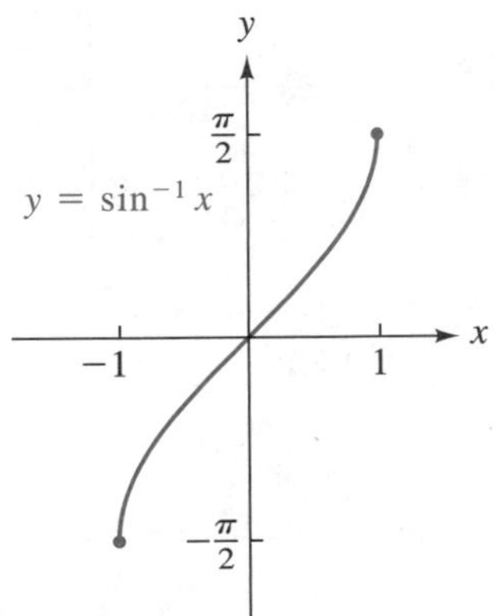

Domain: $-1 \le x \le 1$
Range: $-\frac{\pi}{2} \le y \le \frac{\pi}{2}$

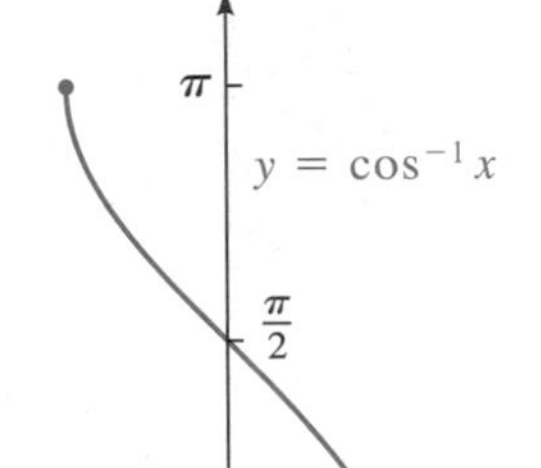

Domain: $-1 \le x \le 1$
Range: $0 \le y \le \pi$

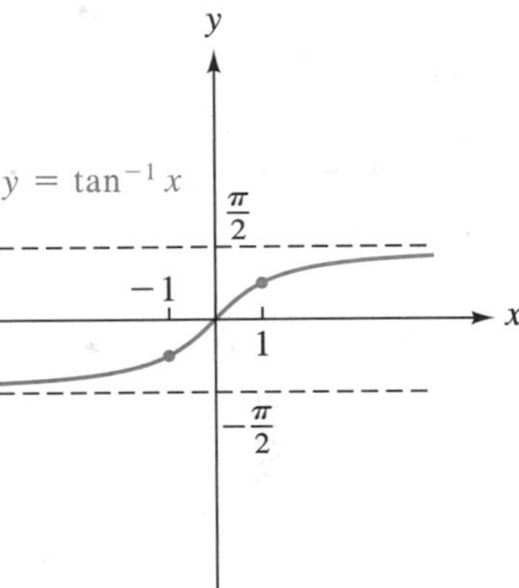

Domain: All real numbers
Range: $-\frac{\pi}{2} < y < \frac{\pi}{2}$

The inverse sine, cosine, and tangent functions are also denoted by arcsin x, arcos x, and arctan x, respectively.

Inverse Sine, Cosine, and Tangent Identities

$\sin(\sin^{-1} x) = x \qquad -1 \le x \le 1$

$\sin^{-1}(\sin x) = x \qquad -\pi/2 \le x \le \pi/2$

$\cos(\cos^{-1} x) = x \qquad -1 \le x \le 1$

$\cos^{-1}(\cos x) = x \qquad 0 \le x \le \pi$

$\tan(\tan^{-1} x) = x \qquad$ for all x

$\tan^{-1}(\tan x) = x \qquad -\pi/2 < x < \pi/2$

☆5.2 INVERSE COTANGENT, SECANT, AND COSECANT FUNCTIONS

$y = \cot^{-1} x$ is equivalent to $x = \cot y$ where $0 < y < \pi$ and x is any real number

$y = \sec^{-1} x$ is equivalent to $x = \sec y$ where $0 \le y \le \pi$, $y \ne \pi/2$, and $x \le -1$ or $x \ge 1$

$y = \csc^{-1} x$ is equivalent to $x = \csc y$ where $-\pi/2 \le y \le \pi/2$, $y \ne 0$, and $x \le -1$ or $x \ge 1$

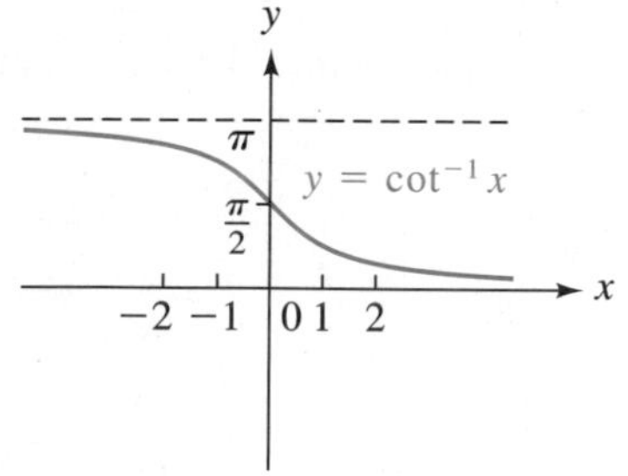

Domain: All real numbers
Range: $0 < y < \pi$

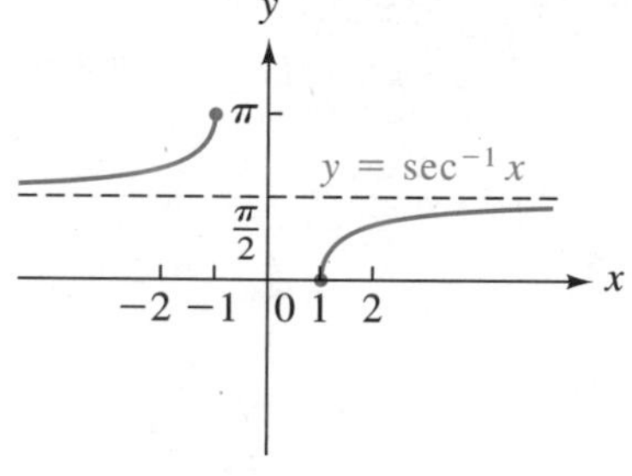

Domain: $x \le -1$ or $x \ge 1$
Range: $0 \le y \le \pi, y \ne \frac{\pi}{2}$

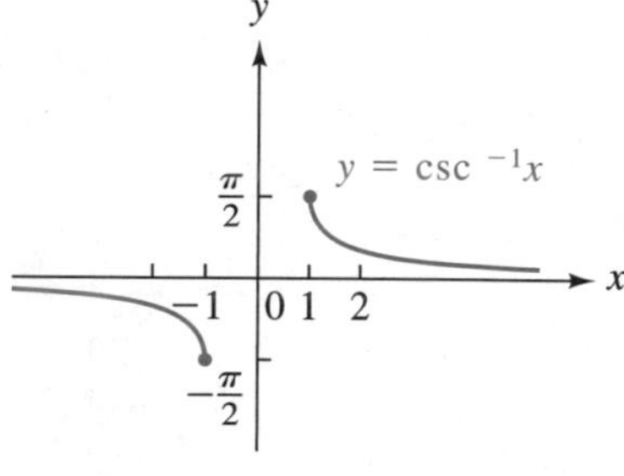

Domain: $x \le -1$ or $x \ge 1$
Range: $-\frac{\pi}{2} \le y \le \frac{\pi}{2}, y \ne 0$

[*Note:* The ranges for $\sec^{-1}$ and $\csc^{-1}$ are sometimes selected differently.]

Inverse Cotangent, Secant, and Cosecant Identities

$$\cot^{-1} x = \begin{cases} \tan^{-1} \dfrac{1}{x} & x > 0 \\ \pi + \tan^{-1} \dfrac{1}{x} & x < 0 \end{cases}$$

$$\sec^{-1} x = \cos^{-1} \frac{1}{x} \qquad x \ge 1 \text{ or } x \le -1$$

$$\csc^{-1} x = \sin^{-1} \frac{1}{x} \qquad x \ge 1 \text{ or } x \le -1$$

5.3 TRIGONOMETRIC EQUATIONS: AN ALGEBRAIC APPROACH

An equation that may be true for some replacements of the variable, but is false for others for which both sides are defined, is called a **conditional equation.** An algebraic approach to solving trigonometric equations can yield exact solutions, and the approach may be aided by the following:

Suggestions for Solving Trigonometric Equations Algebraically

1. Solve for a particular trigonometric function first.
   (A) Try using identities.

(B) Try algebraic manipulations such as factoring, combining fractions, and so on.

2. After solving for a trigonometric function, solve for the variable.

5.4 TRIGONOMETRIC EQUATIONS AND INEQUALITIES: A GRAPHING UTILITY APPROACH

A graphing utility can be used to solve most trigonometric equations to any accuracy desired, but usually will not give exact solutions. Graph each side of the equation in a graphing utility; then use a built-in routine to find any points of intersection.

CHAPTER 5 REVIEW EXERCISE

Work through all the problems in this chapter review and check the answers. Answers to all review problems appear in the back of the book; following each answer is an italic number that indicates the section in which that type of problem is discussed. Where weaknesses show up, review the appropriate sections in the text.

A *Evaluate exactly as real numbers.*

1. $\cos^{-1}(\sqrt{3}/2)$ **2.** $\arcsin \frac{1}{2}$

3. $\sin^{-1}(-\sqrt{3}/2)$ **4.** $\arccos(-\frac{1}{2})$

5. $\arctan 1$ **6.** $\tan^{-1}(-\sqrt{3})$

☆ **7.** $\cot^{-1}(-1/\sqrt{3})$ ☆ **8.** $\operatorname{arcsec}(-2)$

☆ **9.** $\operatorname{arccsc}(-1)$ ☆ **10.** $\sec^{-1}(-\sqrt{2})$

Evaluate as a real number to four significant digits.

11. $\sin^{-1}(-0.8277)$ **12.** $\arccos(-1.328)$

13. $\tan^{-1} 75.14$ ☆ **14.** $\cot^{-1} 5.632$

Find the degree measure of each to two decimal places.

15. $\theta = \cos^{-1} 0.3456$ **16.** $\theta = \arctan(-12.45)$

17. $\theta = \sin^{-1} 0.0025$

In Problems 18–23, find exact solutions over the indicated interval.

18. $2\cos x - \sqrt{3} = 0, \quad 0 \le x < 2\pi$

19. $2\sin^2\theta = \sin\theta, \quad 0° \le \theta < 360°$

20. $4\cos^2 x - 3 = 0, \quad 0 \le x < 2\pi$

21. $2\cos^2\theta + 3\cos\theta + 1 = 0, \quad 0° \le \theta < 360°$

22. $\sqrt{2}\sin 4x - 1 = 0, \quad 0 \le x < \pi/2$

23. $\tan(\theta/2) + \sqrt{3} = 0, \quad -180° < \theta < 180°$

24. Explain how to find the value of x that produces the result shown in the graphing utility display, and find it. The utility is in degree mode. Give the answer to six decimal places.

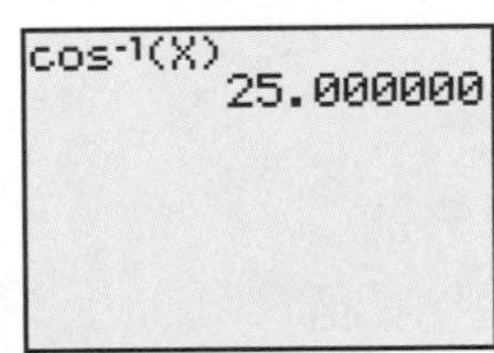

B *Find exact values for each.*

25. $\cos(\cos^{-1} 0.315)$ **26.** $\tan^{-1}[\tan(-1.5)]$

27. $\sin[\tan^{-1}(-\frac{3}{4})]$ **28.** $\cot[\arccos(-\frac{2}{3})]$

☆ **29.** $\csc[\cot^{-1}(-\frac{1}{3})]$ ☆ **30.** $\cos(\operatorname{arccsc} 5)$

In Problems 31–36, evaluate to four significant digits.

31. $\sin^{-1}(\cos 22.37)$ **32.** $\sin^{-1}(\tan 1.345)$

33. $\sin[\tan^{-1}(-14.00)]$ **34.** $\csc[\cos^{-1}(-0.4081)]$

☆ **35.** $\cos(\cot^{-1} 6.823)$ ☆ **36.** $\sec[\operatorname{arccsc}(-25.52)]$

37. Referring to the two displays from a graphing utility below, explain why one of the displays illustrates a sine–inverse sine identity and the other does not.

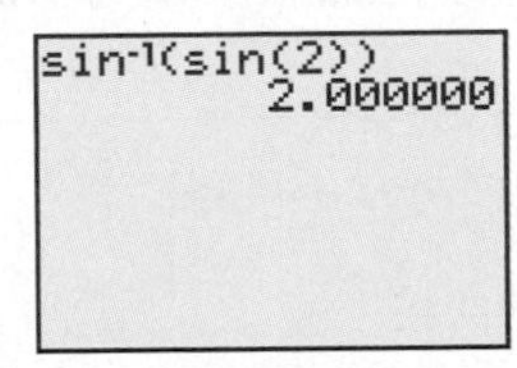

(a)

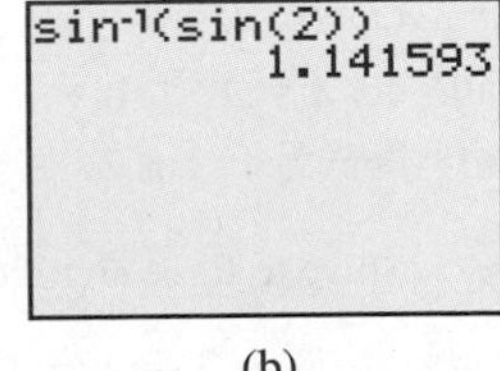

(b)

Find exact solutions over the indicated interval.

38. $\sin^2 \theta = -\cos 2\theta, \quad 0° \le \theta \le 360°$

39. $\sin 2x = \frac{1}{2}, \quad 0 \le x < \pi$

40. $2 \cos x + 2 = -\sin^2 x, \quad -\pi \le x < \pi$

41. $2 \sin^2 \theta - \sin \theta = 0, \quad$ all θ

42. $\sin 2x = \sqrt{3} \sin x, \quad$ all real x

43. $2 \sin^2 \theta + 5 \cos \theta + 1 = 0, \quad 0° \le \theta < 360°$

44. $3 \sin 2x = -2 \cos^2 2x, \quad 0 \le x \le \pi$

Solve for all real x. Compute inverse functions to four significant digits.

45. $\sin x = 0.7088$

46. $\tan x = -4.318$

47. $\sin^2 x + 2 = 4 \sin x$

48. $\tan^2 x = 2 \tan x + 1$

In Problems 49–56, find all solutions over the indicated interval to three decimal places using a graphing utility.

49. $\sin x = 0.25, \quad -\pi \le x \le \pi$

50. $\cot x = -4, \quad -\pi \le x \le \pi$

51. $\sec x = 2, \quad -\pi \le x \le \pi$

52. $\cos x = x^2, \quad$ all real x

53. $\sin x = \sqrt{x}, \quad x \ge 0$

54. $2 \sin x \cos 2x = 1, \quad 0 \le x \le 2\pi$

55. $\sin \dfrac{x}{2} + 3 \sin x = 2, \quad 0 \le x \le 4\pi$

56. $\sin x + 2 \sin 2x + 3 \sin 3x = 3, \quad 0 \le x \le 2\pi$

57. Graph $y_1 = \tan(\sin^{-1} x)$ in the window $-2 \le x \le 2$, $-10 \le y \le 10$. What is the domain for y_1? Explain.

58. Given $h(x) = \sin^{-1}\left(\dfrac{x-2}{2}\right)$.

(A) Explain how you would find the domain of h, and find it.

(B) Graph h over the interval $-5 \le x \le 5$ and explain the result.

59. Does $\tan^{-1} 23.255$ represent all the solutions to the equation $\tan x = 23.255$? Explain.

C *In Problems 60 and 61, find exact solutions over the indicated interval.*

60. $\cos x = 1 - \sin x, \quad 0 \le x < 2\pi$

61. $\cos^2 2x = \cos 2x + \sin^2 2x, \quad 0 \le x < \pi$

62. Solve to three significant digits:

$$2 + 2 \sin x = 1 + 2 \cos^2 x \qquad 0 \le x \le 2\pi$$

Find the exact value.

63. $\sin[2 \tan^{-1}(-\frac{3}{4})]$

64. $\sin(\sin^{-1} \frac{3}{5} + \cos^{-1} \frac{4}{5})$

Write Problems 65 and 66 as an algebraic expression in x free of trigonometric or inverse trigonometric functions.

65. $\tan(\sin^{-1} x)$

66. $\cos(\tan^{-1} x)$

67. The identity $\tan^{-1}(\tan x) = x$ is valid over the interval $-\pi/2 \le x \le \pi/2$.

(A) Graph $y = \tan^{-1}(\tan x)$ for $-\pi/2 \le x \le \pi/2$. (Use dot mode.)

(B) What happens if you graph $y = \tan^{-1}(\tan x)$ over a larger interval, say $-2\pi \le x \le 2\pi$? Explain. (Use dot mode.)

Applications

68. Music The note A above middle C has a frequency of 440 Hz. If the intensity I of the sound at a certain point t seconds after the sound is made can be described by the equation

$$I = 0.08 \sin 880\pi t$$

find the smallest positive t such that $I = 0.05$. Compute the answer to two significant digits.

69. Electric Current An alternating current generator produces a current given by the equation

$$I = 30 \sin 120\pi t$$

where t is time in seconds and I is current in amperes. Find the least positive t (to four significant digits) such that $I = 20$ amperes.

70. Navigation A small craft is approaching a large vessel on the course shown in the figure.

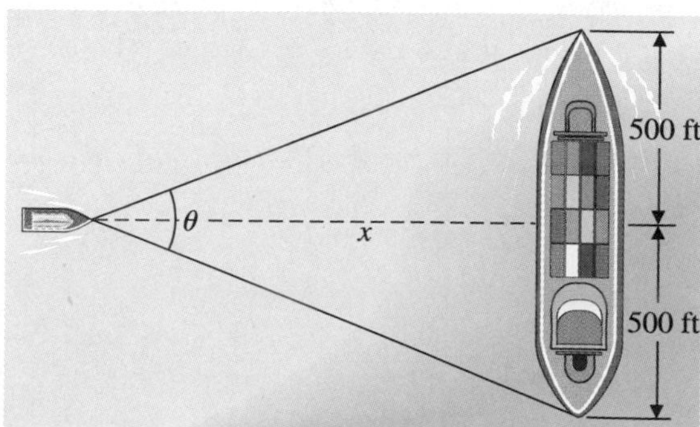

Figure for 70

(A) Express the angle θ subtended by the large vessel in terms of the distance x between the two ships.

(B) Find θ in decimal degrees to one decimal place for $x = 1{,}200$ ft.

***71. Precalculus: Viewing Angle** For advertising purposes, a large brokerage house has a 1.5 ft by 12 ft ticker tape screen mounted 20 ft above the floor on a high wall at an airport terminal (see the figure). A woman's eyes are 5 ft above the floor. If the best view of the tape is when θ is maximum, how far from the wall should she stand? Parts (A)–(F) explore this problem.

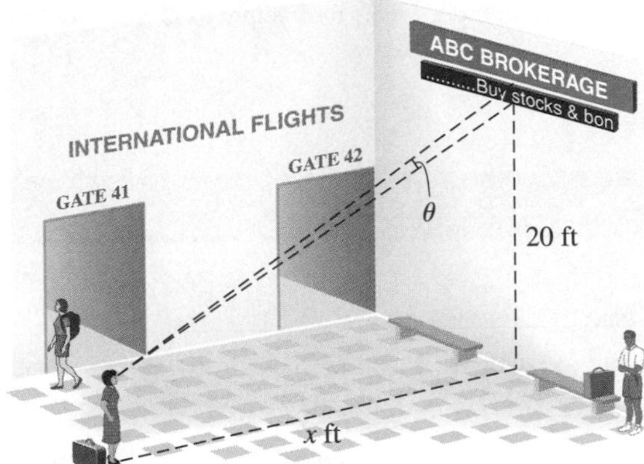

Figure for 71

(A) Describe what you think happens to θ as x increases from 0 ft to 100 ft.

(B) Show that

$$\theta = \tan^{-1}\frac{1.5x}{x^2 + 247.5} \qquad x \geq 0$$

(C) Complete Table 1 (to two decimal places, θ in degrees), and from the table select the maximum θ and the distance x that produces it. (Use a table generator if your calculator has one.)

TABLE 1

x (ft)	0	5	10	15	20	25	30
θ (deg)		1.58					

(D) In a graphing utility, graph the equation in part (B) for $0 \leq x \leq 30$, and describe what the graph shows.

(E) Use a built-in maximum routine in your graphing utility to find the maximum θ and the x that produces it. Find both to two decimal places.

(F) How far away from the wall should the woman stand to have a viewing angle of 2.5°? Solve graphically to two decimal places using a graphing utility.

***72. Engineering** A circular railroad tunnel of radius r is to go through a mountain. The bed for the track is formed by using a chord of the circle of length d as shown in the figure.

Figure for 72

(A) Show that the cross-sectional area of the tunnel is given by

$$A = \pi r^2 - r^2 \sin^{-1}\frac{d}{2r} + \frac{d}{4}\sqrt{4r^2 - d^2}$$

(B) Complete Table 2 (to the nearest square foot) for $r = 15$ ft and $10 \leq d \leq 20$. (Use a table generator if your calculator has one.)

TABLE 2

d (ft)	8	10	12	14	16	18	20
A (ft²)	704						

(C) Find the value of d (to one decimal place) that will produce a cross-sectional area of 675 ft². Solve graphically using a graphing utility and the built-in intersection routine.

***73. Architecture** The roof for a 10 ft wide storage shed is formed by bending a 12 ft wide steel panel into a circular arc (see the figure). Approximate (to two decimal places) the height h of the arc.

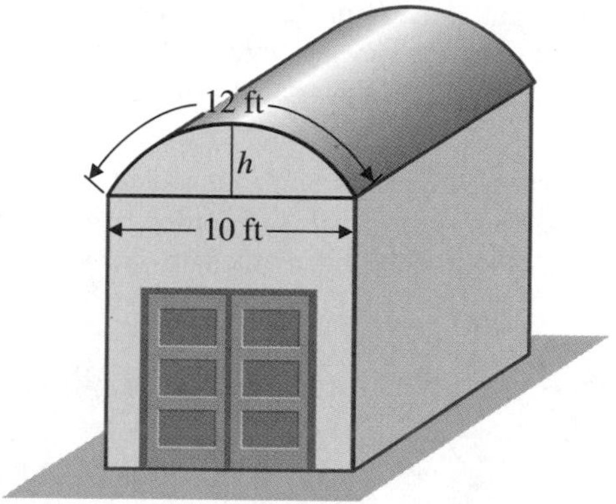

Figure for 73

74. Physics The equation of motion for a weight suspended from a spring is given by

$$y = -1.8 \sin 4t - 2.4 \cos 4t$$

where y is displacement of the weight from its equilibrium position (positive direction upwards) and t is time in seconds.

(A) Graph y for $0 \le t \le \pi/2$.

(B) Approximate (to two decimal places) the time(s) t, $0 \le t \le \pi/2$, when the weight is 2 in. above the equilibrium position.

(C) Approximate (to two decimal places) the time(s) t, $0 \le t \le \pi/2$, when the weight is 2 in. below the equilibrium position.

CUMULATIVE REVIEW EXERCISE CHAPTERS 1–5

Work through all the problems in this cumulative review and check the answers. Answers to all review problems appear in the back of the book; following each answer is an italic number that indicates the section in which that type of problem is discussed. Where weaknesses show up, review the appropriate sections in the text.

1. Find the degree measure of 4.21 rad.

2. Find the radian measure of 505°42′.

3. Explain what is meant by an angle of radian measure 0.5.

4. The hypotenuse of a right triangle is 7.6 m and one of the sides is 4.5 m. Find the acute angles and the other side.

5. Find the value of $\cos \theta$ and $\cot \theta$ if the terminal side of θ contains $P(-5, -12)$.

6. Evaluate to four significant digits using a calculator:
(A) $\cos 67°45'$ (B) $\csc 176.2°$ (C) $\cot 2.05$

7. Sketch a graph of each function for $-2\pi \le x \le 2\pi$.
(A) $y = \cos x$ (B) $y = \csc x$ (C) $y = \cot x$

8. Is it possible to find an angle θ such that $\sin \theta$ is negative and $\csc \theta$ is positive? Explain.

Verify each identity in Problems 9–12 without looking at a table of identities.

9. $\tan x \csc x = \sec x$

10. $\csc \theta - \sin \theta = \cos \theta \cot \theta$

11. $(\sin^2 u)(\tan^2 u + 1) = \sec^2 u - 1$

12. $\dfrac{\sin^2 \alpha - \cos^2 \alpha}{\sin \alpha \cos \alpha} = \dfrac{\tan \alpha - \cot \alpha}{\tan \alpha \cot \alpha}$

13. Is the following equation an identity? Explain.

$$\cos x = \sin x \qquad \text{for } x = \pi/4 + 2k\pi,\ k \text{ an integer}$$

Evaluate Problems 14–20 exactly as real numbers.

14. $\cos \dfrac{-7\pi}{4}$ **15.** $\tan \dfrac{7\pi}{3}$

16. $\sec \dfrac{3\pi}{2}$ **17.** $\arctan 0$

18. $\cos^{-1}(-\sqrt{3}/2)$ **19.** $\arcsin 3$

☆ **20.** $\operatorname{arccot}(-\sqrt{3})$

Evaluate Problems 21–25 as real numbers to four significant digits using a calculator.

21. $\sin^{-1} 0.0505$ **22.** $\cos^{-1}(-0.7228)$

23. $\arctan(-9)$ ☆ **24.** $\operatorname{arccot} 3$

☆ **25.** $\sec^{-1} 2.6$

26. Explain how to find the value of x that produces the result shown in the graphing utility display at the top of the next page, and find it. The utility is in radian mode. Give the answer to six decimal places.

tan⁻¹(X)
-1.000000

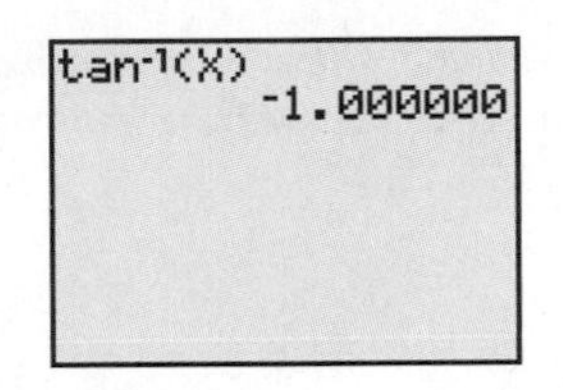

Find exact solutions for Problems 27–29 over the indicated interval.

27. $2 \sin \theta - 1 = 0, \quad 0° \le \theta < 360°$

28. $3 \tan x + \sqrt{3} = 0, \quad -\pi/2 < x < \pi/2$

29. $2 \cos x + 2 = 0, \quad -\pi \le x < \pi$

☆ **30.** Write $\sin 7u \cos 3u$ as a sum or difference.

☆ **31.** Write $\cos 5w - \cos w$ as a product.

B **32.** If the radian measure of an angle is halved, is the degree measure of the angle also halved? Explain.

33. Convert 92.462° to degree-minute-second form.

34. Find the degree measure of a central angle subtended by an arc of 12 in. in a circle with a circumference of 30 in.

35. Find x exactly and θ to the nearest 0.1° in the figure.

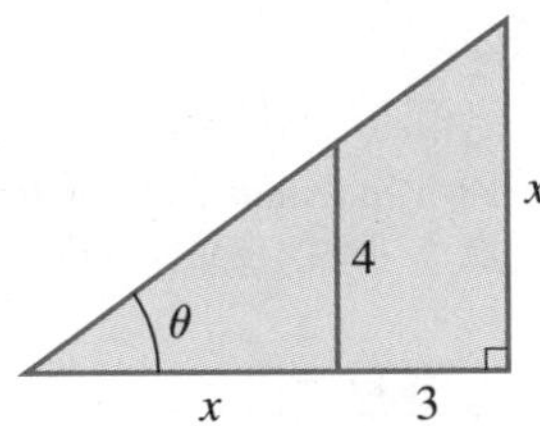

Figure for 35

36. Convert 72° to radian measure in terms of π.

37. Find the exact value of each of the other five trigonometric functions if $\tan \theta = \frac{1}{2}$ and $\sin \theta < 0$.

In Problems 38–40, sketch a graph of each function for the indicated interval. State the period and, if applicable, the amplitude and phase shift.

38. $y = 2 - 2 \sin \dfrac{x}{2}, \quad -\pi \le x \le 5\pi$

39. $y = 3 \cos(2x - \pi), \quad -\pi \le x \le 2\pi$

40. $y = 2 \tan(\pi x - \pi/4), \quad 0 \le x \le 3$

41. Find an equation of the form $y = k + A \cos Bx$ whose graph is shown in the figure.

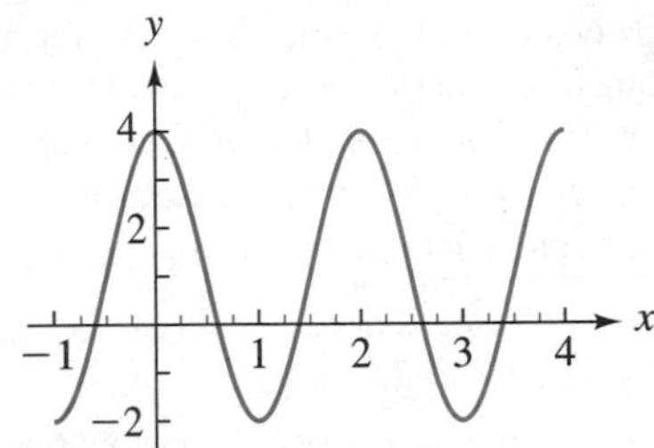

Figure for 41

42. Find an equation of the form $y = k + A \sin Bx$ that produces the graph shown in the following graphing utility display:

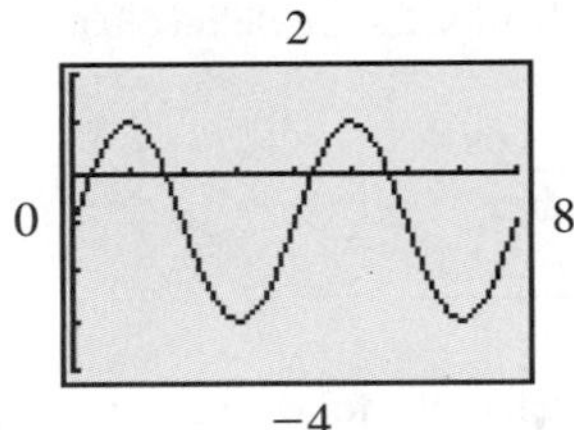

43. If the sides of a right triangle are 19.4 cm and 41.7 cm, find the hypotenuse and find the acute angles to the nearest 10′.

Verify the identities in Problems 44–48.

44. $\dfrac{\cos x}{1 + \sin x} + \tan x = \sec x$

45. $\dfrac{1 + \cos \theta}{1 + \sin \theta} = (\sec \theta - \tan \theta)(\sec \theta + 1)$

46. $\cot \dfrac{u}{2} = \csc u + \cot u$

47. $\dfrac{2}{1 + \sec 2\theta} = 1 - \tan^2 \theta$

48. $\dfrac{\sin x - \sin y}{\cos x + \cos y} = \tan \dfrac{x - y}{2}$

49. For the following graphing calculator displays, find the value of the final expression without finding x or using a calculator:

(A)

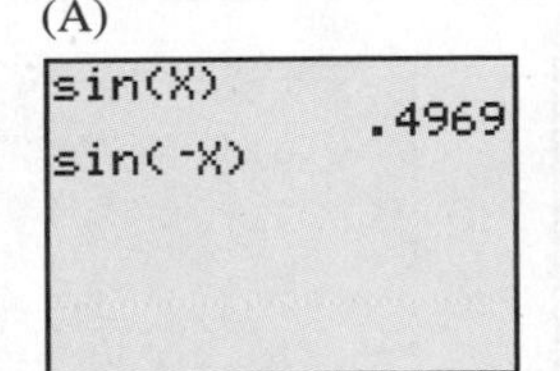

(B)

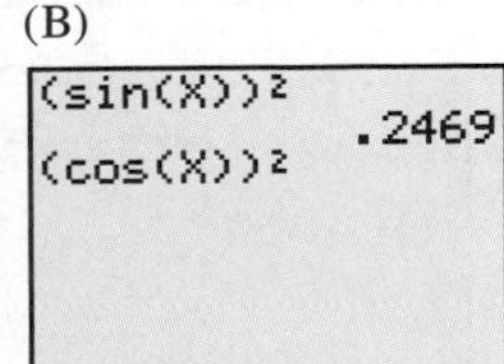

50. Write $y = \sin 3x \cos x - \cos 3x \sin x$ in terms of a single trigonometric function. Check the result by entering the original equation in a graphing utility as y_1 and the converted form as y_2. Then graph y_1 and y_2 in the same viewing window. Use TRACE to compare the two graphs.

51. Find the exact values of $\sin(x/2)$ and $\cos 2x$, given $\tan x = \frac{8}{15}$ and $\pi < x < 3\pi/2$.

52. Graph $y_1 = \cos(x/2)$ and $y_2 = -\sqrt{(1 + \cos x)/2}$ in the same viewing window for $-2\pi \le x \le 2\pi$, and indicate the subintervals for which y_1 and y_2 are equal.

53. Use a graphing utility to test whether each equation is an identity. If the equation appears to be an identity, verify it. If the equation does not appear to be an identity, find a value of x for which both sides are defined but are not equal.

(A) $\dfrac{\sin x}{1 + \cos x} = \dfrac{1 + \cos x}{\sin x}$

(B) $\dfrac{\sin x}{1 - \cos x} = \dfrac{1 + \cos x}{\sin x}$

54. Find exact values for each of the following:

(A) $\sin(\cos^{-1} 0.4)$ (B) $\sec[\arctan(-\sqrt{5})]$

(C) $\csc(\sin^{-1} \frac{1}{3})$ ☆(D) $\tan(\sec^{-1} 4)$

55. Find the exact degree value of $\theta = \tan^{-1} \sqrt{3}$ without using a calculator.

56. Find the degree measure of $\theta = \sin^{-1} 0.8989$ to two decimal places using a calculator.

Find exact solutions to Problems 57–59 over the indicated interval.

57. $2 + 3 \sin x = \cos 2x, \quad 0 \le x \le 2\pi$

58. $\sin 2\theta = 2 \cos \theta, \quad \text{all } \theta$

59. $4 \tan^2 x - 3 \sec^2 x = 0, \quad \text{all real } x$

Find all real solutions to Problems 60–62. Compute inverse functions to four significant digits.

60. $\sin x = -0.5678$ **61.** $\sec x = 2.345$

62. $2 \cos 2x = 7 \cos x$

63. Referring to the two displays from a graphing utility below, explain why one of the displays illustrates a cosine–inverse cosine identity and the other does not.

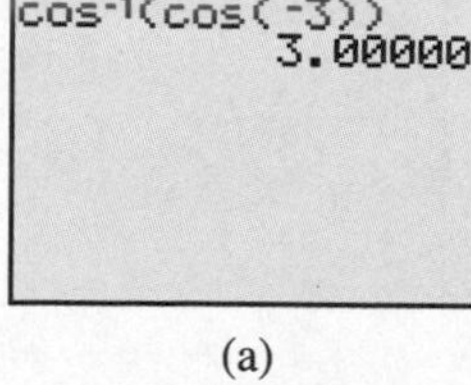

(a)

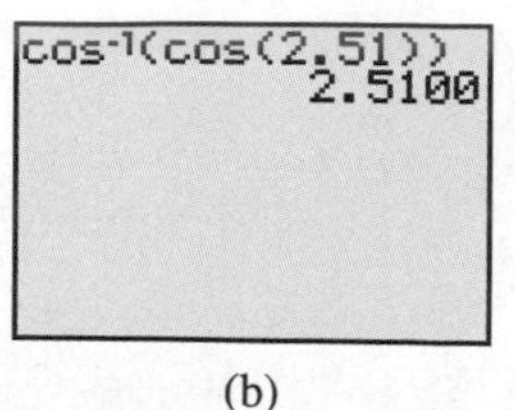

(b)

In Problems 64–66, use a graphing utility to approximate all solutions (to three decimal places) over the indicated interval.

64. $\tan x = 3, \quad -\pi \le x \le \pi$

65. $\cos x = \sqrt{x}, \quad x > 0$

66. $\cos \dfrac{x}{2} - 2 \sin x = 1, \quad 0 \le x \le 4\pi$

C **67.** A point P moves counterclockwise around a unit circle starting at (1, 0) for a distance of 28.703 units. Explain how you would find the coordinates of the point P at its final position and how you would determine which quadrant P is in. Find the coordinates (to four decimal places) and the quadrant in which P lies.

68. The following graphing calculator display shows the coordinates of a point on a unit circle. Let s be the length of the least positive arc from (1, 0) to the point. Find s to four decimal places.

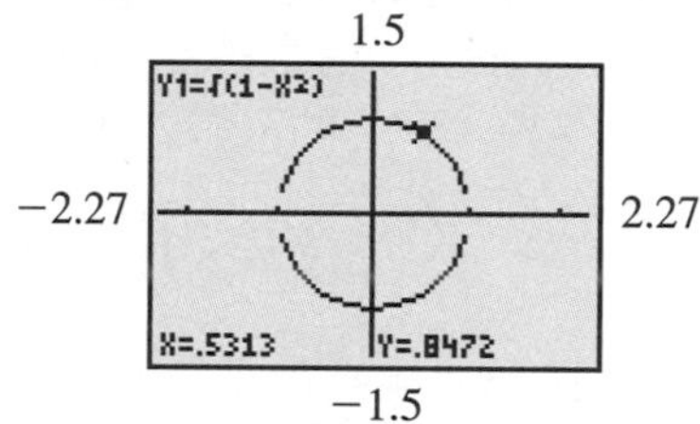

69. If θ is an angle in the fourth quadrant and $\cos \theta = a$, $0 < a < 1$, express the other five trigonometric functions of θ in terms of a.

70. Show that $\tan 3x = \tan x \dfrac{2 \cos 2x + 1}{2 \cos 2x - 1}$ is an identity.

71. Find the exact value of $\cos(2 \sin^{-1} \frac{1}{3})$.

72. Write $\sin(\cos^{-1} x - \tan^{-1} x)$ as an algebraic expression in x free of trigonometric or inverse trigonometric functions.

73. Find exact solutions for all real x:

$$\sin x = 1 + \cos x$$

74. Find solutions to four significant digits:

$$\sin x = \cos^2 x \qquad 0 \le x \le 2\pi$$

In Problems 75–78, graph $f(x)$, find a simpler function $g(x)$ that has the same graph as $f(x)$, and verify the identity $f(x) = g(x)$. [Assume $g(x) = k + A\,t(Bx)$, where $t(x)$ is one of the six trigonometric functions.]

75. $f(x) = \dfrac{\sin^2 x}{1 - \cos x} + \dfrac{2 \tan^2 x \cos^2 x}{1 + \cos x}$

76. $f(x) = 2\sin^2 x + 6\cos^2 x$

77. $f(x) = \dfrac{2 - 2\sin^2 x}{2\cos^2 x - 1}$

78. $f(x) = \dfrac{3\cos x + \sin x - 3}{\cos x - 1}$

79. Find the subinterval(s) of [0, 4] over which

$$2\sin\frac{\pi x}{2} - 3\cos\frac{\pi x}{2} \geq 2x - 1$$

Find the end points to three decimal places.

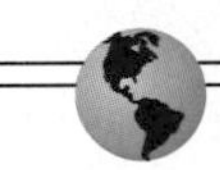

Applications

*** 80. Navigation** An airplane flying on a level course directly toward a beacon on the ground records two angles of depression as indicated in the figure. Find the altitude h of the plane to the nearest meter.

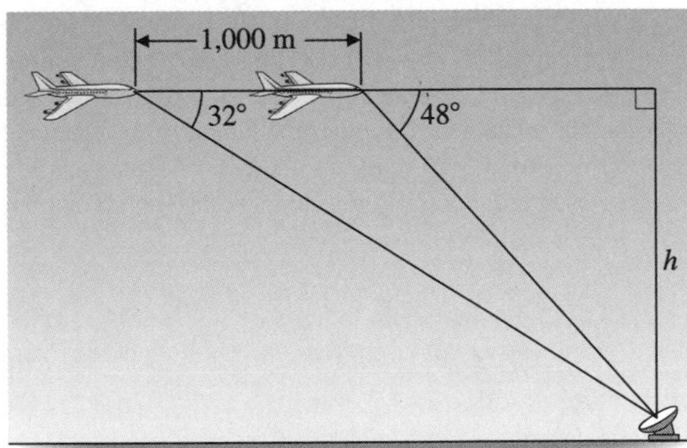

Figure for 80

81. Precalculus: Trigonometric Substitution In the expression $\sqrt{u^2 - a^2}$, $a > 0$, let $u = a \csc x$, $0 < x < \pi/2$, simplify, and write in a form that is free of radicals.

*** 82. Engineering** Find the exact values of x and θ in the figure.

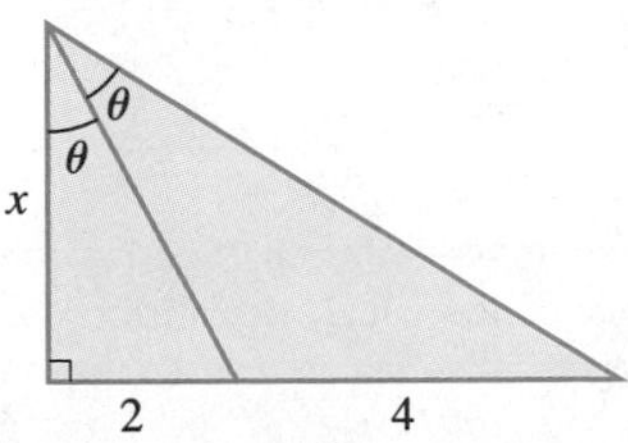

Figure for 82

83. Physics The equation of motion for a weight suspended from a spring is given by

$$y = -7.2\sin 5t - 9.6\cos 5t$$

where y is the displacement of the weight from its equilibrium position in centimeters (positive direction upwards) and t is time in seconds. Find the smallest positive t (to three decimal places) such that the weight is in the equilibrium position.

84. Physics Refer to Problem 83. Graph y for $0 \leq t \leq \pi/2$ and approximate (to three decimal places) the time(s) t in this interval such that the weight is 5 cm above the equilibrium position.

*** 85. Analytic Geometry** Find the radian measure (to three decimal places) of the angle θ in the figure, if A has coordinates (2, 5) and B has coordinates (5, 3). [*Hint:* Label the angle between OB and the x axis as α; then use an appropriate sum identity.]

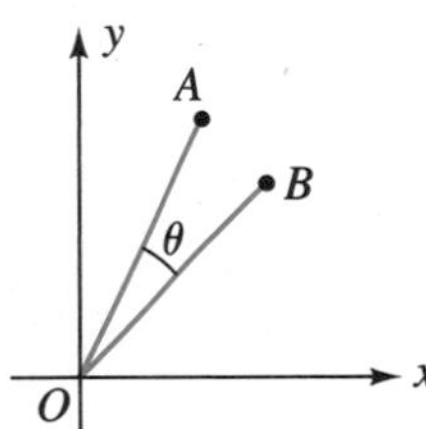

Figure for 85

86. Electric Circuits The current I (in amperes) in an electrical circuit is given by

$$I = 60\sin\left(90\pi t - \frac{\pi}{2}\right)$$

(A) Find the amplitude, period, frequency, and phase shift.

(B) Graph the equation for $0 \leq t \leq 0.02$.

(C) Find the smallest positive t (to four decimal places) such that $I = 35$ amperes.

87. Boat Safety A boat is approaching a 200 ft high vertical cliff (see the figure at the top of the next page).

(A) Write an equation for the distance d from the boat to the base of the cliff in terms of the angle of elevation θ from the boat to the top of the cliff.

(B) Sketch a graph of this equation for $0 < \theta \leq \pi/2$.

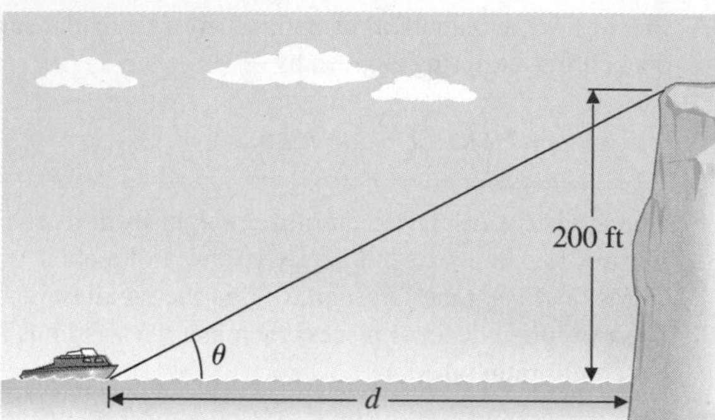

Figure for 87

88. Precalculus Two guy wires are attached to a radio tower as shown in the figure.

(A) Show that $\theta = \arctan \dfrac{100x}{x^2 + 20{,}000}$

(B) Find θ in decimal degrees to one decimal place for $x = 50$ ft.

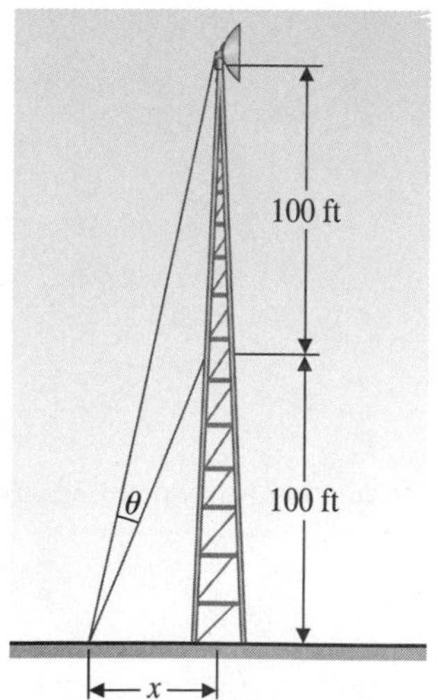

Figure for 88

89. Precalculus Refer to Problem 88. Set your calculator in degree mode, graph

$$y_1 = \tan^{-1} \frac{100x}{x^2 + 20{,}000} \qquad \text{and} \qquad y_2 = 15$$

in the same viewing window, and use approximation techniques to find x to one decimal place when $\theta = 15°$.

*** 90. Engineering** The chain on a bicycle goes around the pedal sprocket and the rear wheel sprocket (see the figure). The radius of the pedal sprocket is 11.0 cm, the radius of the rear sprocket is 4.00 cm, and the diameter of the rear wheel is 70.0 cm. If the pedal sprocket rotates through an angle of 18π radians, through how many radians does the rear wheel rotate?

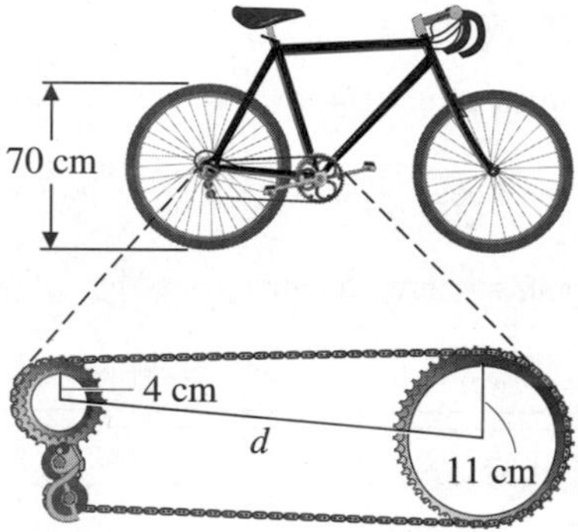

Figure for 90

☆ 91. Engineering Refer to Problem 90. If the pedal sprocket rotates at 60.0 rpm, how fast is the bicycle traveling (in centimeters per minute)?

92. Precalculus: Surveying A log is to be floated around the right angle corner of a canal as shown in the figure. We are interested in finding the length of the longest log that will make it around the corner. (Assume the log is represented by a straight line segment with no diameter.)

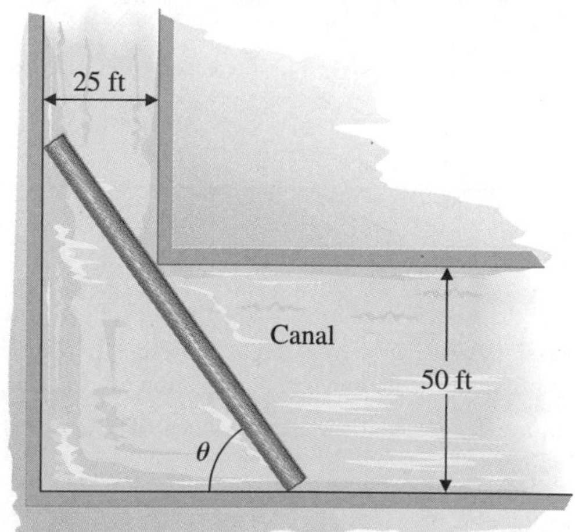

Figure for 92

(A) From the figure, and assuming that the log touches the inside corner of the canal, explain how L, the length of the segment, varies as θ varies from 0° to 90°.

(B) Show that the length of the segment, L, is given by

$$L = 50 \csc \theta + 25 \sec \theta \qquad 0° < \theta < 90°$$

(C) Complete Table 1 (to one decimal place). (If you have a table-generating calculator, use it.)

TABLE 1

θ (deg)	35	40	45	50	55	60	65
L (ft)	117.7						

(D) From the table, select the minimum length L and the angle θ that produces it. Is this the length of the longest log that will go around the corner? Explain.

(E) Graph the equation from part (B) in a graphing utility for $0° < \theta < 90°$ and $80 \le L \le 200$. Use a built-in routine to find the minimum length L and the value of θ that produces it.

93. **Modeling Daylight Duration** Table 2 gives the duration of daylight on the fifteenth day of each month for 1 year at Anchorage, Alaska.

(A) Convert the data in Table 2 from hours and minutes to two-place decimal hours. Enter the data for a 2 year period in your graphing calculator and produce a scatter plot in the following viewing window: $1 \le x \le 24$, $6 \le y \le 20$.

(B) A function of the form $y = k + A \sin(Bx + C)$ can be used to model the data. Use the converted data from Table 2 to determine k and A (to two decimal places), and B (exact value). Use the graph from part (A) to visually estimate C (to one decimal place).

(C) Plot the data from part (A) and the equation from part (B) in the same viewing window. If necessary, adjust your value of C to produce a better fit.

TABLE 2

x (months)	1	2	3	4	5	6	7	8	9	10	11	12
y (daylight duration)	6:31	9:10	11:50	14:48	17:33	19:16	18:27	15:51	12:57	10:09	7:19	5:36

ADDITIONAL TOPICS: TRIANGLES AND VECTORS

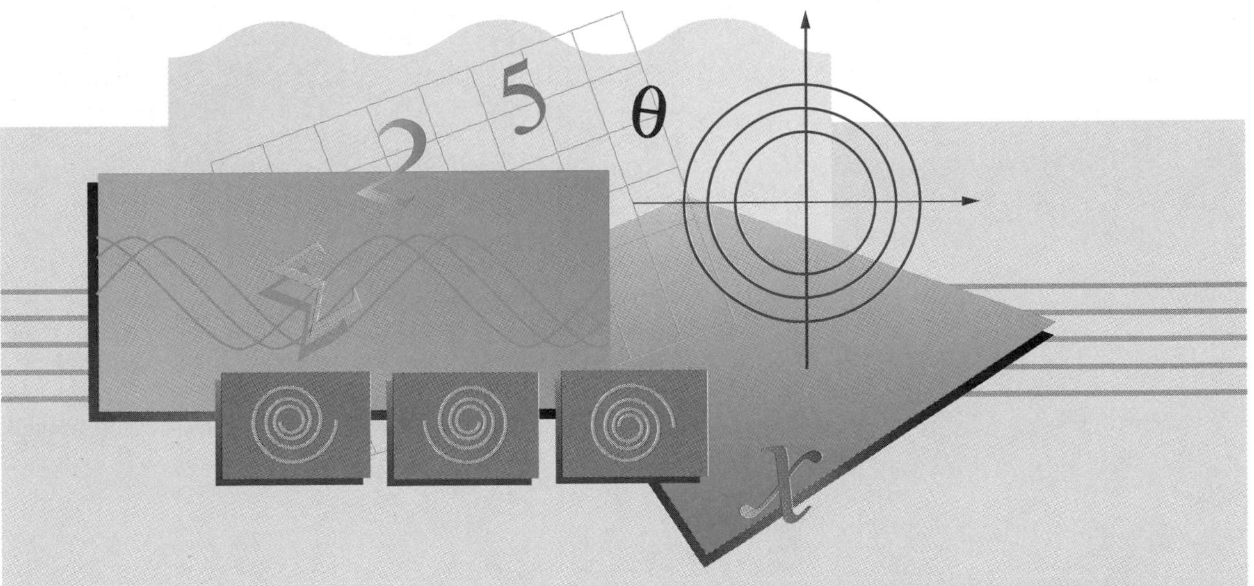

☆ Sections marked with a star may be omitted without loss of continuity.

A number of additional topics involving trigonometry are considered in this chapter. First, we return to the problem of solving triangles, but without the restriction to right triangles. The *laws of sines and cosines* are the principal tools used to solve these more general triangle forms. Several formulas are developed for finding the areas of triangles, given certain measurements of the triangle.

The concept of *vector,* a very important and useful form in both pure and applied mathematics, is introduced first in geometric form, and then in the more general algebraic form. The chapter concludes with a discussion of the *dot product* of two vectors, which has many significant applications.

6.1 LAW OF SINES

- Deriving the Law of Sines
- Solving ASA and AAS Cases
- Solving the Ambiguous SSA Case

Until now, we have considered only triangle problems that involved right triangles. We now turn to **oblique triangles,** that is, triangles that contain no right angle. Every oblique triangle is either **acute** (all angles are between 0° and 90°) or **obtuse** (one angle is between 90° and 180°). Figure 1 illustrates an acute triangle and an obtuse triangle.

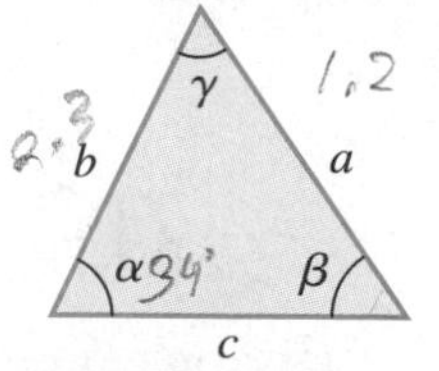

(a) Acute triangle

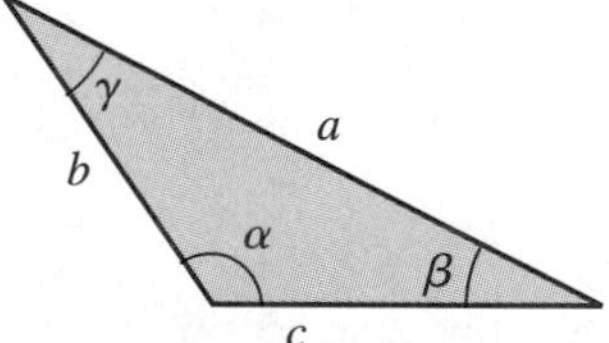

(b) Obtuse triangle

FIGURE 1

Notice how we labeled the sides and angles of the oblique triangles shown in Figure 1: Side a is opposite angle α, side b is opposite angle β, and side c is opposite angle γ. Note that the largest side of a triangle is opposite the largest angle. Given any three of the six quantities indicated in Figure 1, we are interested in finding the remaining three, if they exist. This process is called **solving the triangle.**

In this section we will develop the *law of sines* and in the next section, the *law of cosines.* These are two basic tools important in the solution of oblique triangles.

If the given quantities include two angles or an angle and the opposite side, we will use the law of sines. Otherwise, we will use the law of cosines.

Before we proceed with specifics, recall the rules governing angle measure and significant digits for side measure (listed in Table 1 and inside the front cover for easy reference).

TABLE 1

Angle to nearest	Significant digits for side measure
1°	2
10′ or 0.1°	3
1′ or 0.01°	4
10″ or 0.001°	5

Remark on Calculations When you solve for a particular side or angle, carry out all operations within the calculator and then round to the appropriate number of significant digits (following the rules in Table 1) at the end of the calculation. Note that your answers still may differ slightly from those in the book, depending on the order in which you solve for the sides and angles. □

Deriving the Law of Sines

The law of sines is relatively easy to derive using the right triangle properties we studied earlier. We also use the fact that

$$\sin(180° - x) = \sin x$$

which is obtained from the difference identity for sine. Referring to the triangles in Figure 2, we proceed as follows: For each triangle,

$$\sin \alpha = \frac{h}{b} \qquad \text{and} \qquad \sin \beta = \frac{h}{a}$$

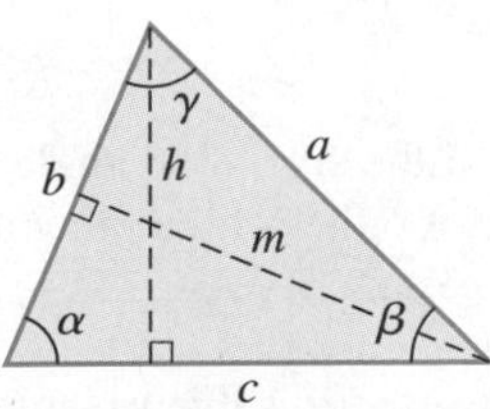

(a) Acute triangle

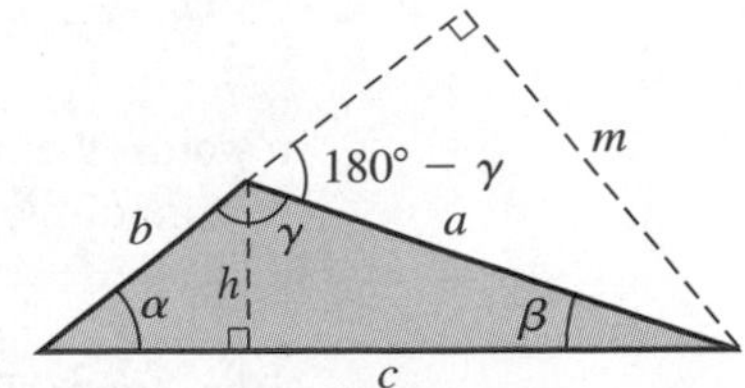

(b) Obtuse triangle

FIGURE 2

Therefore,

$$h = b \sin \alpha \qquad \text{and} \qquad h = a \sin \beta$$

Thus,

$$b \sin \alpha = a \sin \beta$$

and

$$\frac{\sin \alpha}{a} = \frac{\sin \beta}{b} \tag{1}$$

Similarly, for each triangle in Figure 2,

$$\sin \alpha = \frac{m}{c} \qquad \text{and} \qquad \sin \gamma = \sin(180° - \gamma) = \frac{m}{a}$$

Therefore,

$$m = c \sin \alpha \qquad \text{and} \qquad m = a \sin \gamma$$

Thus,

$$c \sin \alpha = a \sin \gamma$$

and

$$\frac{\sin \alpha}{a} = \frac{\sin \gamma}{c} \tag{2}$$

If we combine equations (1) and (2), we obtain the **law of sines.**

LAW OF SINES

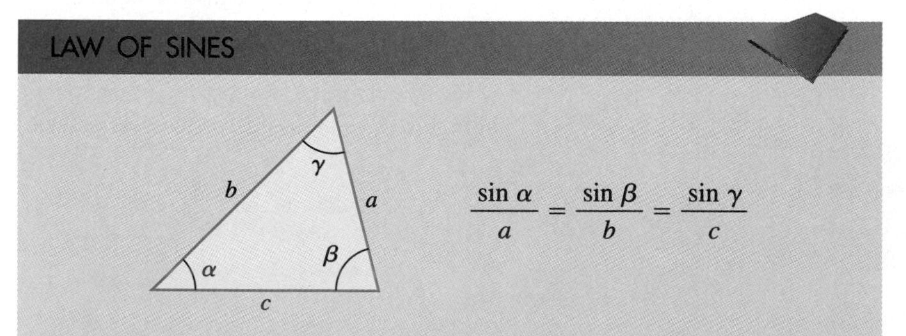

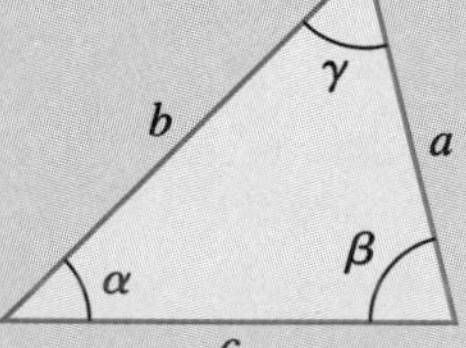

$$\frac{\sin \alpha}{a} = \frac{\sin \beta}{b} = \frac{\sin \gamma}{c}$$

In words, the ratio of the sine of an angle to its opposite side is the same as the ratio of the sine of either of the other angles to its opposite side.

The law of sines is used to solve triangles, given:

1. Two angles and any side (ASA or AAS), or
2. Two sides and an angle opposite one of them (SSA)

We will apply the law of sines to the easier ASA and AAS cases first; then we will turn to the more troublesome SSA case.

Solving ASA and AAS Cases

If we are given two angles and the included side (ASA case) or two angles and the side opposite one of the angles (AAS case), we first find the measure of the third angle (remember that the sum of the measures of all three angles in any triangle is 180°). We then use the law of sines to find the other two sides.

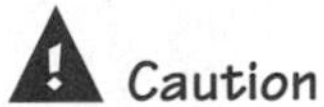

Caution Note that for the ASA or AAS case to determine a unique triangle, the sum of the two given angles must be between 0° and 180° (see Fig. 3).

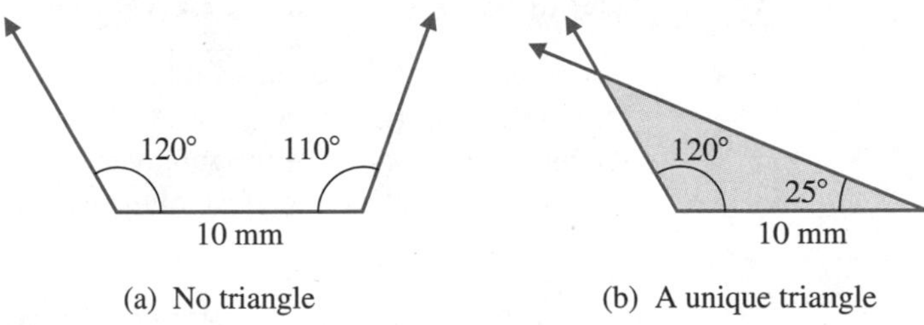

(a) No triangle (b) A unique triangle

FIGURE 3

□

Examples 1 and 2 illustrate the use of the law of sines in solving the ASA and AAS cases.

EXAMPLE 1 **Using the Law of Sines (ASA)**

Solve the triangle shown in Figure 4.

Solution This is an example of the ASA case.

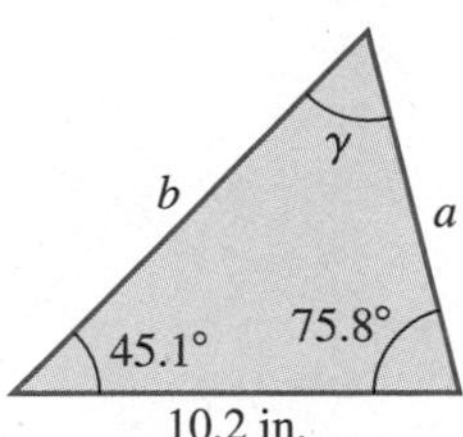

FIGURE 4

Solve for γ:

$$\alpha + \beta + \gamma = 180^\circ$$

$$\gamma = 180^\circ - (45.1^\circ + 75.8^\circ)$$

$$= 59.1^\circ$$

Solve for a:

$$\frac{\sin \alpha}{a} = \frac{\sin \gamma}{c}$$

$$a = (\sin \alpha)\frac{c}{\sin \gamma}$$

$$= (\sin 45.1^\circ)\frac{10.2}{\sin 59.1^\circ}$$

$$= 8.42 \text{ in.}$$

Solve for b:

$$\frac{\sin\beta}{b} = \frac{\sin\gamma}{c}$$

$$b = (\sin\beta)\frac{c}{\sin\gamma}$$

$$= (\sin 75.8°)\frac{10.2}{\sin 59.1°}$$

$$= 11.5 \text{ in.}$$

Matched Problem 1 Solve the triangle with $\alpha = 28.0°$, $\beta = 45.3°$, and $c = 122$ m.

EXAMPLE 2

Using the Law of Sines (AAS): Sundials

The angle, β, of the gnomon on a sundial (see Fig. 5) must be the same as the latitude where the sundial is used. If the latitude of San Francisco is 38°N, and the angle of elevation of the sun, α, is 63° at noon, how long will the shadow of a 12 in. gnomon be on the face of the sundial (see Fig. 5)?

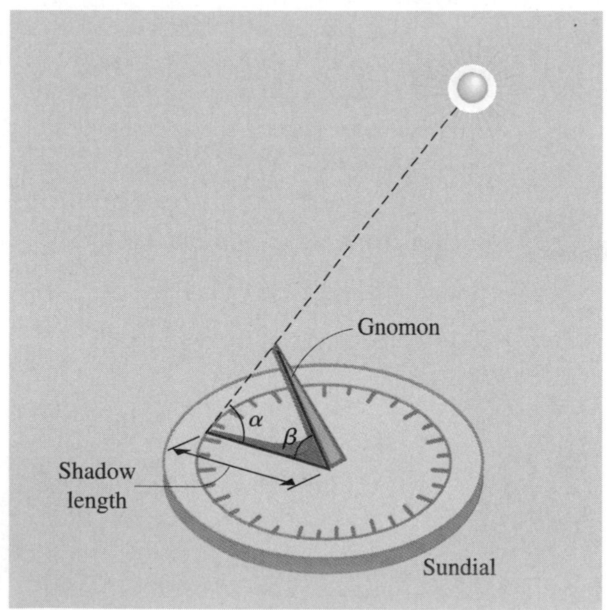

FIGURE 5
Sundial

Solution We are given two angles and the side opposite one of the angles (AAS). First we find the third angle; then we find the length of the shadow using the law of sines. It is helpful to make a simple drawing (Fig. 6) showing the given and unknown parts.

Solve for γ:

$$\gamma = 180° - (63° + 38°) = 79°$$

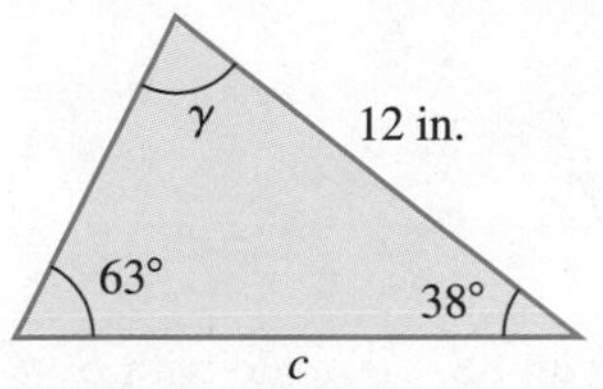

FIGURE 6

Solve for c:

$$\frac{\sin \alpha}{a} = \frac{\sin \gamma}{c}$$

$$c = \frac{a \sin \gamma}{\sin \alpha}$$

$$= \frac{12 \sin 79°}{\sin 63°} = 13 \text{ in.}$$ Shadow length ■

Matched Problem 2 Repeat Example 2 for Columbus, Ohio, with a latitude of 40°N, a gnomon of length 15 in., and a 52° angle of elevation for the sun. ■

Remark Note that the AAS case can always be converted to the ASA case by first solving for the third angle. For the ASA or AAS case to determine a unique triangle, the sum of the two given angles must be between 0° and 180°. □

Solving the Ambiguous SSA Case

If we are given two sides and an angle opposite one of the sides—the SSA case—then it is possible to have 0, 1, or 2 triangles, depending on the measures of the two sides and the angle. Therefore, we refer to the SSA case as the **ambiguous case.** Table 2 illustrates the various possibilities.

TABLE 2
SSA Variations

	a $(h = b \sin \alpha)$	Number of triangles	Figure
α acute	$0 < a < h$	0	(a)
	$a = h$	1	(b)
	$h < a < b$	2	(c)
	$a \geq b$	1	(d)
α obtuse	$0 < a \leq b$	0	(e)
	$a > b$	1	(f)

Triangles involving the SSA case can be solved without memorizing Table 2. The particular case in the table generally becomes apparent in the solution process.

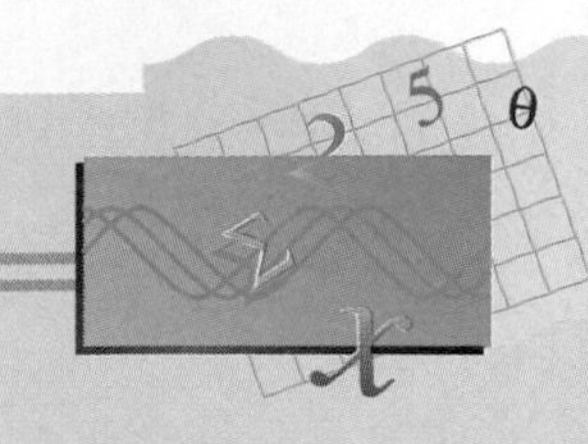

EXPLORE/DISCUSS 1

If it is found that $\sin \beta > 1$ in the process of solving an SSA triangle with α acute, indicate which case(s) in Table 2 apply and why. Repeat the problem for $\sin \beta = 1$ and for $0 < \sin \beta < 1$.

EXAMPLE 3 Using the Law of Sines (SSA): No Triangle

Find β in the triangle with $\alpha = 34°$, $b = 2.3$ mm, and $a = 1.2$ mm.

Solution To find β, we go directly to the law of sines:

$$\frac{\sin \beta}{b} = \frac{\sin \alpha}{a} \qquad \text{Law of sines}$$

$$\sin \beta = \frac{b \sin \alpha}{a} \qquad \text{Solve for } \sin \beta.$$

$$= \frac{2.3 \sin 34°}{1.2} \qquad \text{Substitute } a, b, \text{ and } \alpha.$$

$$\approx 1.0718 \qquad \text{Evaluate.}$$

Since $\sin \beta = 1.0718$ has no solution, no triangle exists with the given measurements. [We could have come to the same conclusion by noting that α is acute and $a < h$ ($a = 1.2$ and $h = b \sin \alpha = 2.3 \sin 34° \approx 1.2861$), case (a) in Table 2.] ■

Matched Problem 3 Use the law of sines to find β in the triangle with $\alpha = 130°$, $b = 1.3$ m, and $a = 1.2$ m. ■

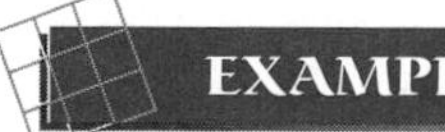

EXAMPLE 4 Using the Law of Sines (SSA): One Triangle

Solve the triangle with $\alpha = 47°$, $a = 3.7$ ft, and $b = 3.5$ ft.

Solution *Solve for β:*

$$\frac{\sin \beta}{b} = \frac{\sin \alpha}{a} \qquad \text{Law of sines}$$

$$\sin \beta = \frac{b \sin \alpha}{a} \qquad \text{Solve for } \sin \beta.$$

$$= \frac{3.5 \sin 47°}{3.7} \qquad \text{Substitute } a, b, \text{ and } \alpha.$$

$$\approx 0.6918 \qquad \text{Evaluate.}$$

$$\beta = \sin^{-1} 0.6918$$

$$= 44° \qquad \text{Use the inverse key(s) on a calculator.}$$

The calculator gives the measure of the acute angle whose sine is 0.6918. The other possible choice for β is the supplement of 44°, that is, $180° - 44° = 136°$. But if we add 136° to $\alpha = 47°$, we get 183°, which is greater than 180° (the sum of all measures of the three angles in a triangle). Because it is not possible to form a triangle having two angles with measures 47° and 136°, the supplement must be rejected, and we are left with only one triangle using $\beta = 44°$. [We could have come to the same conclusion noting that α is acute and $a > b$, case (d) in Table 2.] We complete the problem by solving for γ and c.

Solve for γ:

$$\begin{aligned}\gamma &= 180° - (\alpha + \beta)\\ &= 180° - (47° + 44°) = 89°\end{aligned}$$

Solve for c:

$$\frac{\sin\alpha}{a} = \frac{\sin\gamma}{c}$$

$$\begin{aligned}c &= \frac{a\sin\gamma}{\sin\alpha}\\ &= \frac{3.7\sin 89°}{\sin 47°} = 5.1\text{ ft}\end{aligned}$$

■

Matched Problem 4 Solve the triangle with $\alpha = 139°$, $a = 42$ yd, and $b = 27$ yd. ■

EXAMPLE 5 Using the Law of Sines (SSA): Two Triangles

Solve the triangle(s) with $\alpha = 26°$, $a = 11$ cm, and $b = 18$ cm.

Solution *Solve for β:*

$$\frac{\sin\beta}{b} = \frac{\sin\alpha}{a}$$

$$\begin{aligned}\sin\beta &= \frac{b\sin\alpha}{a}\\ &= \frac{18\sin 26°}{11}\\ &\approx 0.7173\end{aligned}$$

$$\begin{aligned}\beta &= \sin^{-1} 0.7173\\ &= 46°\end{aligned}$$

But β could also be the supplement of 46°; that is, $180° - 46° = 134°$. To see if the supplement, 134°, can be an angle in a second triangle, we add 134° to $\alpha = 26°$ to obtain 160°. Since 160° is less than 180° (the sum of the measure of all angles of a triangle), there are two possible triangles meeting the original conditions, one with $\beta = 46°$ and the other with $\beta = 134°$. [We could have arrived at the same conclusion by noting that $h < a < b$, case (c) in Table 2.] Figure 7 illustrates the results: $\beta = 134°$ and $\beta' = 46°$.

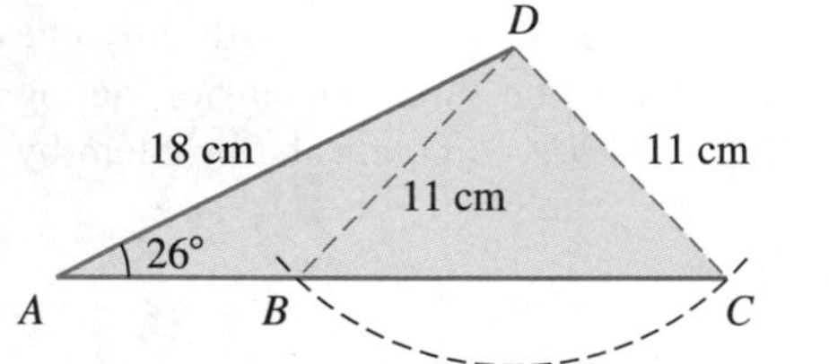

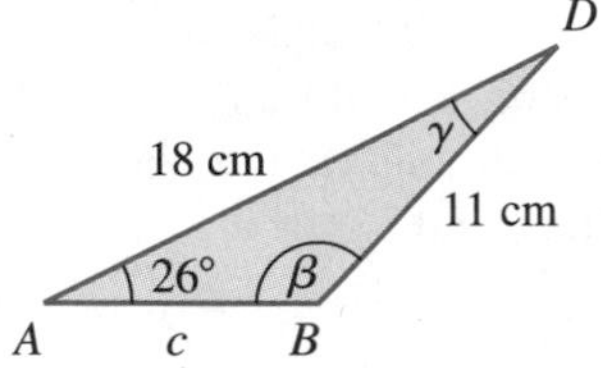

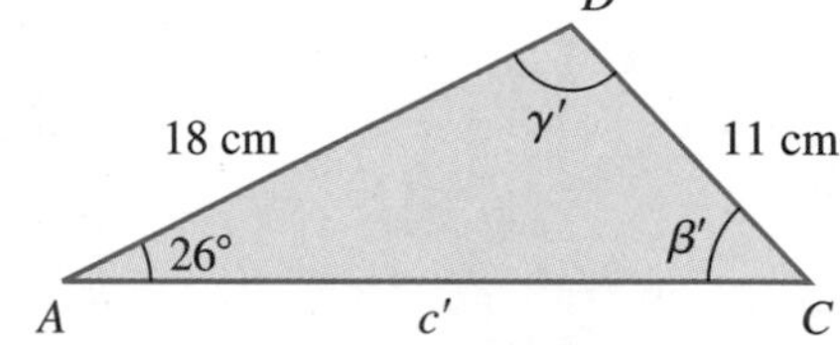

FIGURE 7

Solve for γ and γ':

$$\gamma = 180° - (26° + 134°) = 20°$$

$$\gamma' = 180° - (26° + 46°) = 108°$$

Solve for c and c':

$$\frac{\sin \alpha}{a} = \frac{\sin \gamma}{c} \qquad\qquad \frac{\sin \alpha}{a} = \frac{\sin \gamma'}{c'}$$

$$c = \frac{a \sin \gamma}{\sin \alpha} = \frac{11 \sin 20°}{\sin 26°} = 8.6 \text{ cm} \qquad\qquad c' = \frac{a \sin \gamma'}{\sin \alpha} = \frac{11 \sin 108°}{\sin 26°} = 24 \text{ cm}$$

Matched Problem 5 Solve the triangle(s) with $a = 8.0$ mm, $b = 11$ mm, and $\alpha = 35°$.

Answers to Matched Problems

1. $\gamma = 106.7°$; $a = 59.8$ m; $b = 90.5$ m
2. Shadow length = 19 in.
3. No triangle: α is obtuse and $a \le b$.
4. $\beta = 25°$; $\gamma = 16°$; $c = 18$ yd
5. $\beta = 128°$, $\beta' = 52°$; $\gamma = 17°$, $\gamma' = 93°$; $c = 4.1$ mm, $c' = 14$ mm

EXERCISE 6.1

Assume all triangles are labeled as in the figure unless stated to the contrary. Your answers may differ slightly from those in the book, depending on the order in which you solve for the sides and angles.

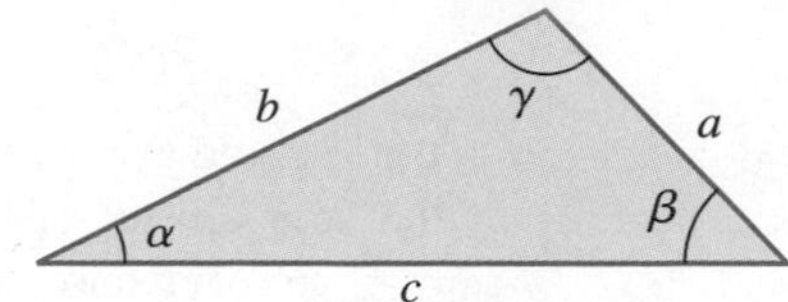

A *Solve each triangle given the indicated measures of angles and sides.*

1. $\beta = 43°, \quad \gamma = 36°, \quad a = 92$ cm

2. $\alpha = 122°, \quad \gamma = 18°, \quad b = 12$ mm

3. $\beta = 27.5°, \quad \gamma = 54.5°, \quad a = 9.27$ mm

4. $\alpha = 118.3°, \quad \gamma = 12.2°, \quad b = 17.3$ km

5. $\alpha = 122.7°, \quad \beta = 34.4°, \quad b = 18.3$ cm

6. $\alpha = 67.7°, \quad \beta = 54.2°, \quad b = 123$ ft

7. $\beta = 12°40', \quad \gamma = 100°0', \quad b = 13.1$ km

8. $\alpha = 73°50', \quad \beta = 51°40', \quad a = 36.6$ mm

B *In Problems 9–16, determine whether the information in each problem allows you to construct 0, 1, or 2 triangles. Do not solve the triangle. Explain which case in Table 2 applies.*

9. $a = 3$ ft, $\quad b = 6$ ft, $\quad \alpha = 30°$

10. $a = 2$ in., $\quad b = 4$ in., $\quad \alpha = 30°$

11. $a = 8$ ft, $\quad b = 6$ ft, $\quad \alpha = 30°$

12. $a = 6$ in., $\quad b = 4$ in., $\quad \alpha = 30°$

13. $a = 2$ ft, $\quad b = 6$ ft, $\quad \alpha = 30°$

14. $a = 1$ in., $\quad b = 4$ in., $\quad \alpha = 30°$

15. $a = 5$ ft, $\quad b = 6$ ft, $\quad \alpha = 30°$

16. $a = 3$ in., $\quad b = 4$ in., $\quad \alpha = 30°$

In Problems 17–24, solve each triangle. If a problem has no solution, say so. If a problem involves two triangles, solve both unless stated to the contrary.

17. $\alpha = 27.5°, \quad a = 15.0$ mm, $\quad b = 36.4$ mm

18. $\alpha = 47.7°, \quad a = 8.5$ ft, $\quad b = 12.5$ ft

19. $\alpha = 135°20', \quad a = 14.6$ m, $\quad b = 18.3$ m

20. $\alpha = 122°40', \quad a = 105$ mi, $\quad b = 152$ mi

21. $\beta = 33°50', \quad a = 673$ ft, $\quad b = 1{,}240$ ft

22. $\beta = 29°30', \quad a = 43.2$ in., $\quad b = 56.5$ in.

23. $\beta = 27.3°, \quad a = 244$ ft, $\quad b = 135$ ft

24. $\beta = 38.9°, \quad a = 42.7$ cm, $\quad b = 30.0$ cm

C **25. Mollweide's equation,**

$$(a - b)\cos\frac{\gamma}{2} = c\sin\frac{\alpha - \beta}{2}$$

is often used to check the final solution of a triangle since all six parts of a triangle are involved in the equation. If, after substitution, the left side does not equal the right side, then an error has been made in solving a triangle. Use this equation to check Problem 1 to two decimal places. (Remember that rounding may not produce exact equality, but the left and right sides of the equation should be close.)

26. Use Mollweide's equation (see Problem 25) to check Problem 3 to two decimal places.

27. Use the law of sines and suitable identities to show that for any triangle,

$$\frac{a - b}{a + b} = \frac{\tan\dfrac{\alpha - \beta}{2}}{\tan\dfrac{\alpha + \beta}{2}}$$

28. Verify (to three decimal places) the formula in Problem 27 with values from Problem 1 and its solution.

29. Let $\beta = 46.8°$ and $a = 66.8$ yd. Determine a value k so that if $0 < b < k$, there is no solution; if $b = k$, there is one solution; and if $k < b < a$, there are two solutions.

30. Let $\beta = 36.6°$ and $b = 12.2$ m. Determine a value k so that if $0 < b < k$, there is no solution; if $b = k$, there is one solution; and if $k < b < a$, there are two solutions.

Applications

31. Surveying To determine the distance across the Grand Canyon in Arizona, a 1.00 mi baseline, AB, is established along the southern rim of the canyon. Sightings are then made from the ends (A and B) of the baseline to a point C across the canyon (see the figure). Find the distance from A to C if $\angle BAC = 118.1°$ and $\angle ABC = 58.1°$.

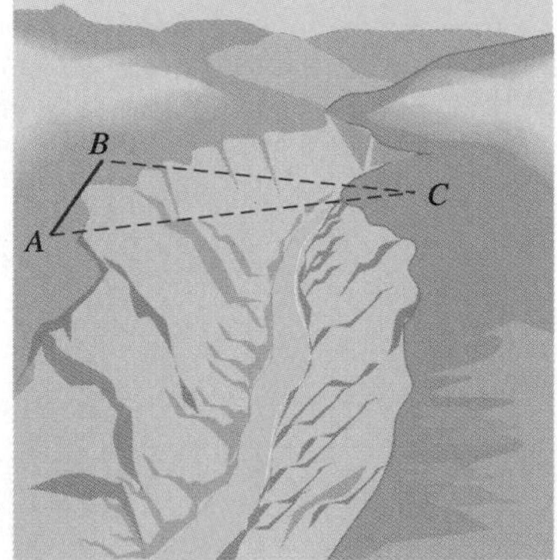

Figure for 31

32. **Surveying** Refer to Problem 31. The Grand Canyon was surveyed at a narrower part of the canyon using a similar 1.00 mi baseline along the southern rim. Find the length of AC if $\angle BAC = 28.5°$ and $\angle ABC = 144.6°$.

33. **Fire Spotting** A fire at F is spotted from two fire lookout stations, A and B, which are located 10.3 mi apart. If station B reports the fire at angle $ABF = 52.6°$, and station A reports the fire at angle $BAF = 25.3°$, how far is the fire from station A? From station B?

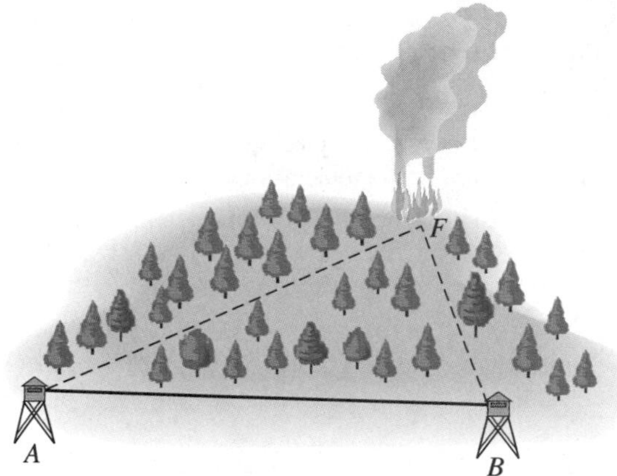

Figure for 33

34. **Coast Patrol** Two lookout posts, A and B, which are located 12.4 mi apart, are established along a coast to watch for illegal foreign fishing boats coming within the 3 mi limit. If post A reports a ship S at angle $BAS = 37.5°$, and post B reports the same ship at angle $ABS = 19.7°$, how far is the ship from post A? How far is the ship from the shore (assuming the shore is along the line joining the two observation posts)?

35. **Surveying** An underwater telephone cable is to cross a shallow lake from point A to point B (see the figure). Stakes are located at A, B, and C. Distance AC is measured to be 112 m, $\angle CAB$ to be 118.4°, and $\angle ABC$ to be 19.2°. Find the distance AB.

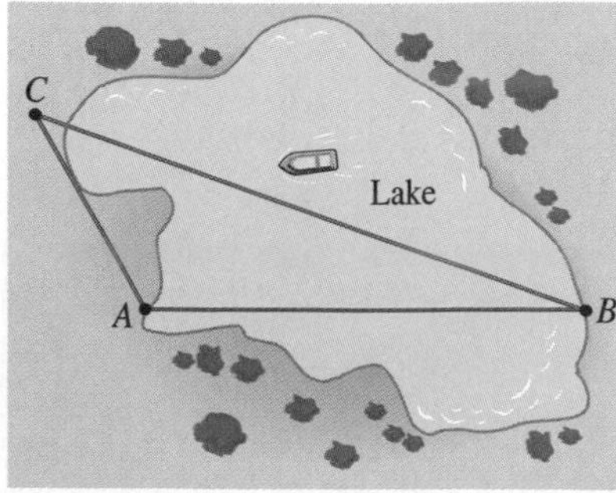

Figure for 35

36. **Surveying** A suspension bridge is to cross a river from point B to point C (see the figure). Distance AB is measured to be 0.652 mi, $\angle ABC$ to be 81.3°, and $\angle BCA$ to be 41.4°. Compute the distance BC.

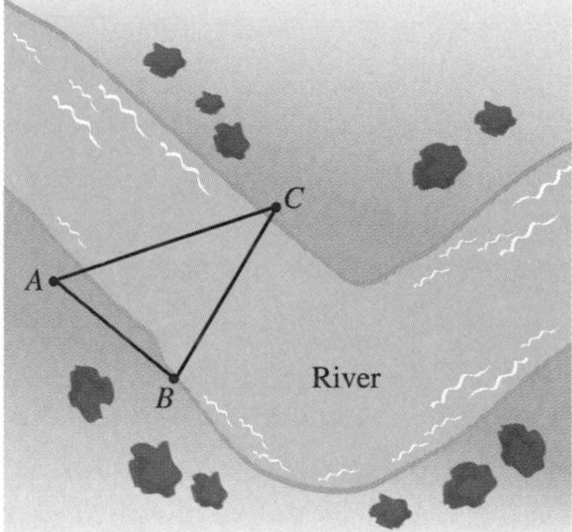

Figure for 36

* 37. **Coastal Piloting** A boat is traveling along a coast at night. A flashing buoy marks a reef. While proceeding on the same course, the navigator of the boat sights the buoy twice, 4.6 nautical mi apart, and forms triangles as shown in the figure. If the boat continues on course, by how far will it miss the reef?

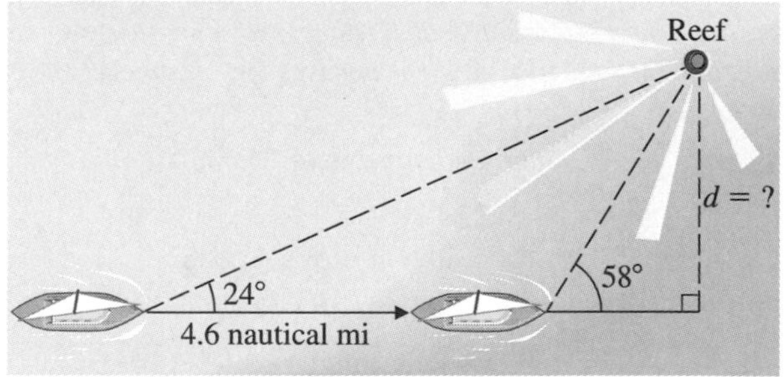

Figure for 37

* 38. **Surveying** Find the height of the mountain above the valley in the figure.

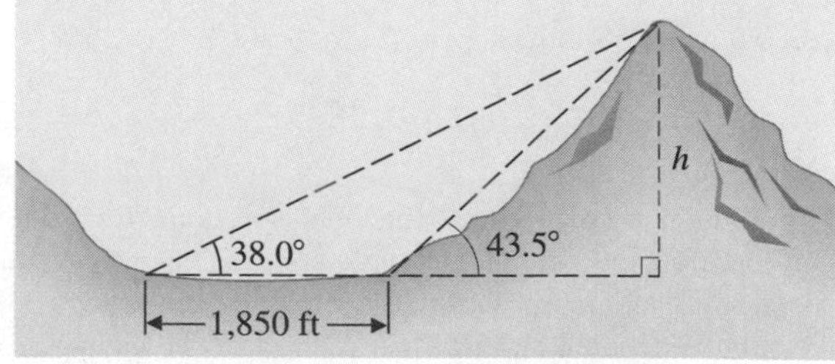

How high is the mountain?

Figure for 38

∗ **39. Tree Height** A tree growing on a hillside casts a 157 ft shadow straight down the hill (see the figure). Find the vertical height of the tree if relative to the horizontal, the hill slopes 11.0° and the angle of elevation of the sun is 42.0°.

Figure for 39

∗ **40. Tree Height** Find the height of the tree in Problem 39 if the shadow length is 102 ft and relative to the horizontal, the hill slopes 15.0° and the angle of elevation of the sun is 62.0°.

∗ **41. Aircraft Design** Find the measures of β and c for the sweptback wing of a supersonic jet, given the information in the figure.

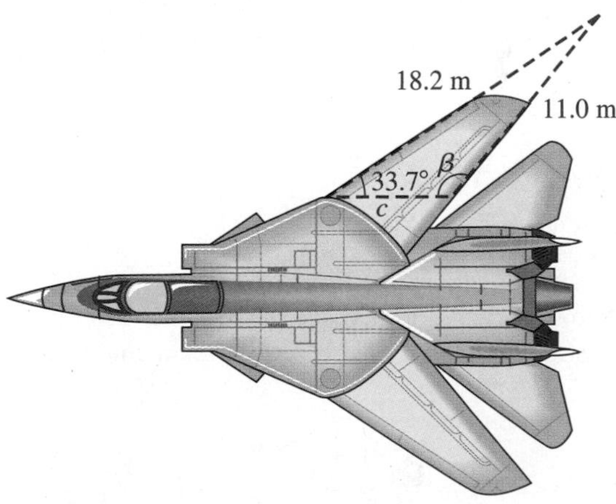

Figure for 41

∗ **42. Aircraft Design** Find the measures of β and c in Problem 41 if the extended leading and trailing edges of the wing are 20.0 m and 15.5 m, respectively, and the measure of the given angle is 35.3° instead of 33.7°.

∗ **43. Eye** A cross section of the cornea of an eye, a circular arc, is shown in the figure. With the information given in the figure, find the radius of the arc, r, and the length of the arc, s.

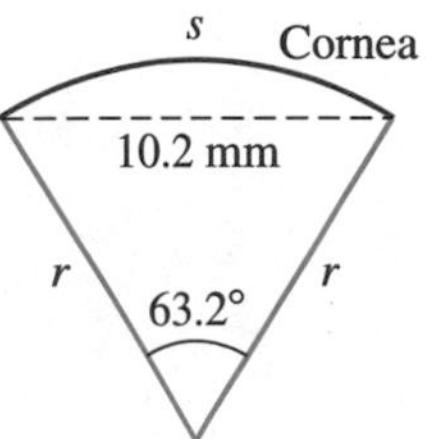

Figure for 43

∗ **44. Eye** Repeat Problem 43 using a central angle of 98.9° and a chord of length 11.8 mm.

∗ **45. Space Science** When a satellite is directly over tracking station B (see the figure), tracking station A measures the angle of elevation θ of the satellite (from the horizon line) to be 24.9°. If the tracking stations are 504 mi apart (that is, the arc AB is 504 mi) and the radius of the earth is 3,964 mi, how high is the satellite above B? [*Hint:* Find all angles for the triangle ACS first.]

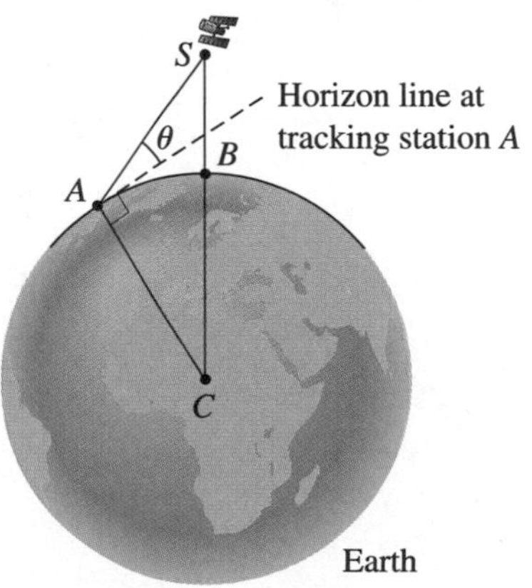

Figure for 45

∗ **46. Space Science** Refer to Problem 45. Compute the height of the satellite above tracking station B if the stations are 632 mi apart and the angle of elevation θ of the satellite (above the horizon line) is 26.2° at tracking station A.

47. Astronomy The orbits of the earth and Venus are approximately circular, with the sun at the center (see the figure at the top of the next page). A sighting of Venus is

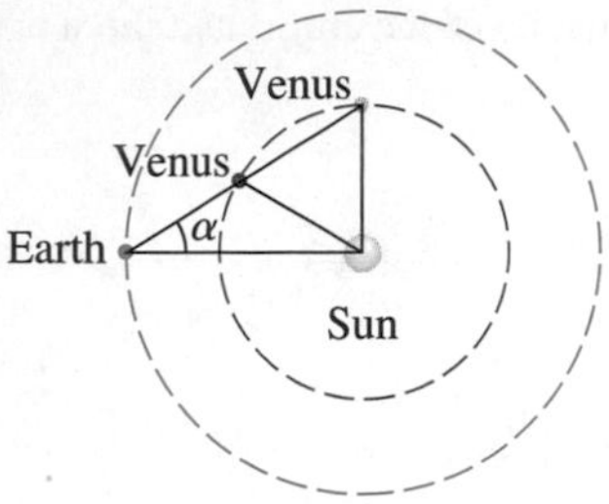

Distance from Earth to Venus

Figure for 47

made from the earth and the angle α is found to be $18°40'$. If the diameter of the orbit of the earth is 2.99×10^8 km and the diameter of the orbit of Venus is 2.17×10^8 km, what are the possible distances from the earth to Venus?

48. Astronomy In Problem 47 find the maximum value of α. [*Hint:* The value of α is maximum when a straight line joining the earth and Venus in the figure is tangent to Venus's orbit.]

49. Engineering A 12 cm piston rod joins a piston to a 4.2 cm crankshaft (see the figure). What is the longest distance d of the piston from the center of the crankshaft when the rod makes an angle of 8.0°?

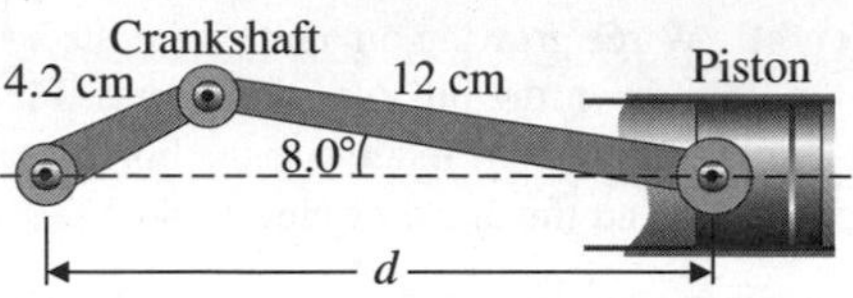

Figure for 49

50. Engineering See Problem 49. What is the shortest distance d of the piston from the center of the crankshaft when the rod makes an angle of 8.0°?

*** 51.** Surveying The scheme illustrated in the figure is used to determine inaccessible heights when d, α, β, and γ can be measured. Show that

$$h = d \sin \alpha \csc(\alpha + \beta) \tan \gamma$$

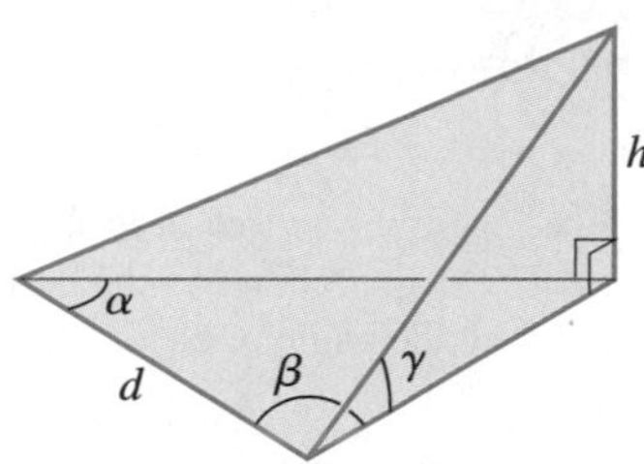

Figure for 51

6.2 LAW OF COSINES

- Deriving the Law of Cosines
- Solving the SAS Case
- Solving the SSS Case

In Figure 1a we are given two sides and the included angle (SAS), and in Figure 1b we are given three sides (SSS). In neither case do we have an angle and a side

FIGURE 1

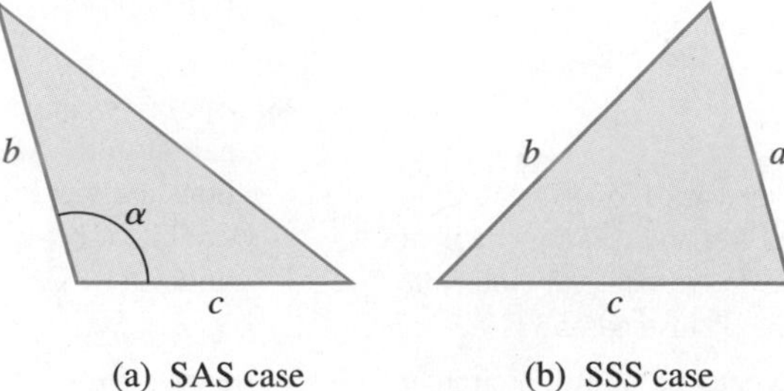

(a) SAS case (b) SSS case

opposite the angle; hence, the law of sines cannot be used. In this section we develop the *law of cosines,* which can be used for these two cases.

Deriving the Law of Cosines

The following is the **law of cosines.**

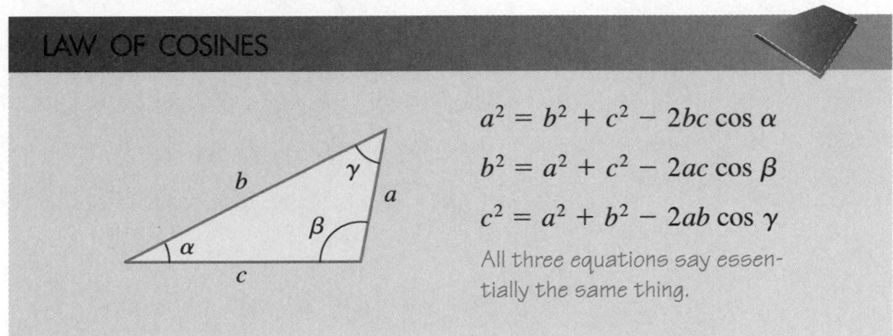

$$a^2 = b^2 + c^2 - 2bc \cos \alpha$$

$$b^2 = a^2 + c^2 - 2ac \cos \beta$$

$$c^2 = a^2 + b^2 - 2ab \cos \gamma$$

All three equations say essentially the same thing.

The law of cosines is used to solve a triangle if we are given:

1. Two sides and the included angle (SAS), or
2. Three sides (SSS)

We will derive only the first equation in the box. (The other equations can then be obtained from the first case simply by relabeling the figure.) We start by locating a triangle in a rectangular coordinate system. Figure 2 shows three typical triangles.

FIGURE 2

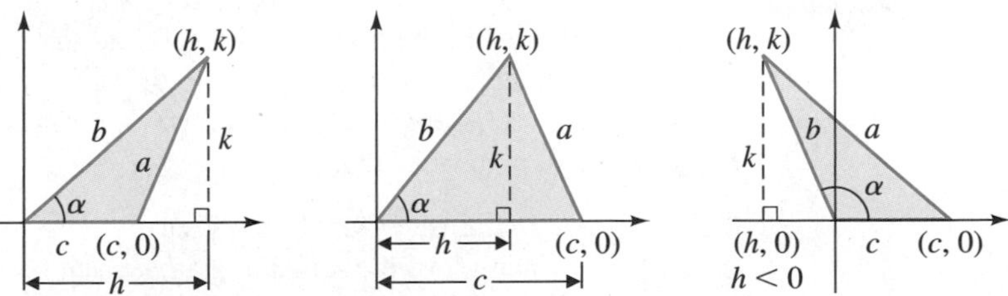

For an arbitrary triangle located as in Figure 2, we use the formula for the distance between two points to obtain

$$a = \sqrt{(h - c)^2 + (k - 0)^2}$$

or, squaring both sides,

$$\begin{aligned} a^2 &= (h - c)^2 + k^2 \\ &= h^2 - 2hc + c^2 + k^2 \end{aligned} \qquad (1)$$

From Figure 2 we note that

$$b^2 = h^2 + k^2$$

Substituting b^2 for $h^2 + k^2$ in equation (1), we obtain

$$a^2 = b^2 + c^2 - 2hc \tag{2}$$

But

$$\cos \alpha = \frac{h}{b}$$

$$h = b \cos \alpha$$

Thus, by replacing h in (2) with $b \cos \alpha$, we reach our objective:

$$a^2 = b^2 + c^2 - 2bc \cos \alpha$$

[*Note:* If α is acute, then $\cos \alpha > 0$; if α is obtuse, then $\cos \alpha < 0$.]

Solving the SAS Case

In this case we start by using the law of cosines to find the side opposite the given angle. We can then use either the law of cosines or the law of sines to find a second angle. Because of the simpler computation, we will generally use the law of sines to find a second angle.

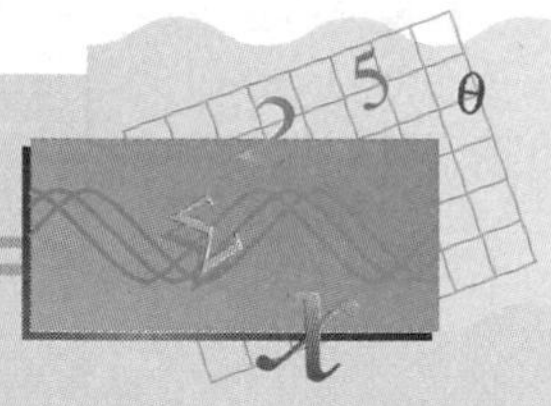

EXPLORE/DISCUSS 1

After using the law of cosines to find the side opposite the angle for the SAS case, the law of sines is used to find a second angle. There are two choices for the second angle, and the following discussion shows that one choice may be better than the other.

(A) If the given angle between the two given sides is obtuse, explain why neither of the remaining angles can be obtuse.

(B) If the given angle between the two given sides is acute, explain why choosing the angle opposite the shorter given side guarantees the selection of an acute angle.

(C) Starting with $(\sin \beta)/b = (\sin \alpha)/a$, show that

$$\beta = \sin^{-1}\left(\frac{b \sin \alpha}{a}\right) \tag{3}$$

(D) Explain why equation (3) gives us the correct angle β only if β is acute.

This discussion leads to the following strategy for solving the SAS case:

STRATEGY FOR SOLVING THE SAS CASE

Step	Find	Method
1.	Side opposite given angle	Law of cosines
2.	Second angle (Find the angle opposite the shorter of the two given sides—this angle will always be acute.)	Law of sines or law of cosines
3.	Third angle	Subtract the sum of the measures of the given angle and the angle found in step 2 from 180°.

EXAMPLE 1

Using the Law of Cosines (SAS)

Solve the triangle shown in Figure 3.

Solution *Solve for b:* Use the law of cosines:

$$b^2 = a^2 + c^2 - 2ac\cos\beta \qquad \text{Solve for } b.$$

$$\begin{aligned} b &= \sqrt{a^2 + c^2 - 2ac\cos\beta} \\ &= \sqrt{(13.9)^2 + (10.0)^2 - 2(13.9)(10.0)\cos 34°20'} \\ &= 7.98 \text{ m} \end{aligned}$$

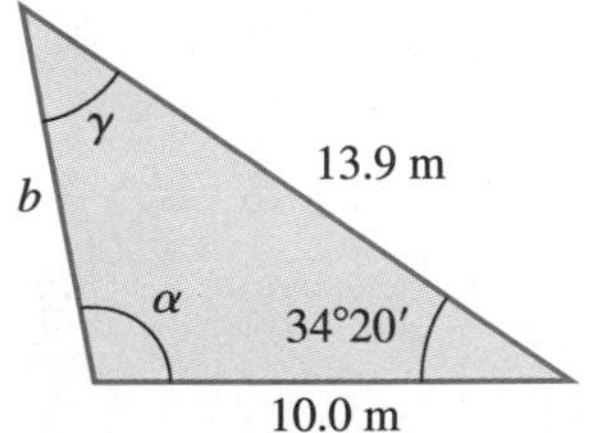

FIGURE 3

Solve for γ: Since side c is shorter than side a, γ must be acute, so we use the law of sines to solve for γ.

$$\frac{\sin\gamma}{c} = \frac{\sin\beta}{b} \qquad \text{Solve for } \sin\gamma.$$

$$\sin\gamma = \frac{c\sin\beta}{b} \qquad \text{Solve for } \gamma.$$

$$= \frac{10.0\sin 34°20'}{7.98}$$

$$\gamma = \sin^{-1}\left(\frac{10.0\sin 34°20'}{7.98}\right) \qquad \text{Since } \gamma \text{ is acute, we can use the inverse sine key on a calculator to find the measure of } \gamma.$$

$$= 45.0° \quad \text{or} \quad 45°0'$$

Solve for α:

$$\alpha = 180° - (\beta + \gamma)$$
$$= 180° - (34°20' + 45°0') = 100°40'$$

Matched Problem 1 Solve the triangle with $\alpha = 86.0°$, $b = 15.0$ m, and $c = 24.0$ m (angles in decimal degrees).

Solving the SSS Case

When we start with three sides of a triangle, our problem is to find the three angles. Although the law of cosines can be used to find any of the three angles, we will always start with the angle opposite the longest side to get an obtuse angle, if present, out of the way at the start. Then, we will switch to the law of sines to find a second angle.

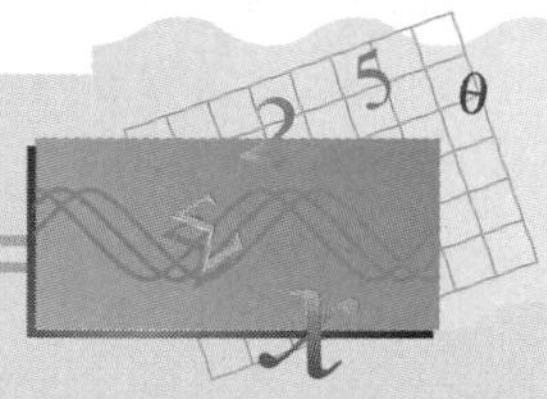

EXPLORE/DISCUSS 2

(A) Starting with $a^2 = b^2 + c^2 - 2bc \cos \alpha$, show that

$$\alpha = \cos^{-1}\left(\frac{a^2 - b^2 - c^2}{-2bc}\right) \qquad (4)$$

(B) Explain why equation (4) gives us the correct angle α irrespective of whether α is obtuse or acute.

This discussion leads to the following strategy for solving the SSS case:

STRATEGY FOR SOLVING THE SSS CASE

Step	Find	Method
1.	Angle opposite longest side (This will take care of an obtuse angle, if present.)	Law of cosines
2.	Either of the remaining angles (Always acute, since a triangle cannot have more than one obtuse angle.)	Law of sines or law of cosines
3.	Third angle	Subtract the sum of the measures of the angles found in steps 1 and 2 from 180°.

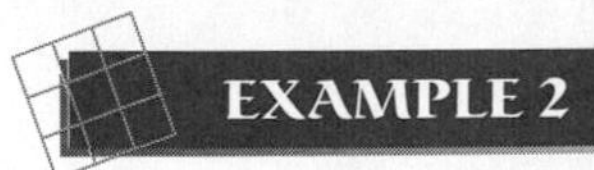

Solving the SSS Case: Surveying

A triangular plot of land has sides $a = 21.2$ m, $b = 24.6$ m, and $c = 12.0$ m. Find the measures of all three angles in decimal degrees.

Solution Find the measure of the angle opposite the longest side first using the law of cosines, which in this problem is angle β. We make a rough sketch to keep the various parts of the triangle straight (Fig. 4).

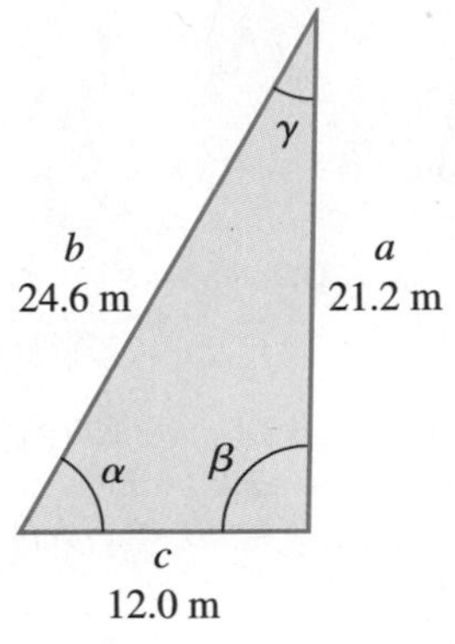

FIGURE 4

Solve for β: It is not clear whether β is acute, obtuse, or 90°, but we do not need to know beforehand, because the law of cosines will automatically tell us in the solution process.

$$b^2 = a^2 + c^2 - 2ac\cos\beta \qquad \text{Solve for } \cos\beta.$$

$$\cos\beta = \frac{a^2 + c^2 - b^2}{2ac} \qquad \text{Solve for } \beta.$$

$$\beta = \cos^{-1}\left(\frac{a^2 + c^2 - b^2}{2ac}\right)$$

$$= \cos^{-1}\left(\frac{(21.2)^2 + (12.0)^2 - (24.6)^2}{2(21.2)(12.0)}\right)$$

$$= 91.3° \qquad \text{Obtuse}$$

Solve for α: Both α and γ must be acute, since β is obtuse. We arbitrarily choose to find α first using the law of sines. Since α is acute, the inverse sine function in a calculator will give us the measure of this angle directly.

$$\frac{\sin\alpha}{a} = \frac{\sin\beta}{b} \qquad \text{Solve for } \sin\alpha.$$

$$\sin\alpha = \frac{a\sin\beta}{b} \qquad \text{Solve for } \alpha.$$

$$\alpha = \sin^{-1}\left(\frac{a\sin\beta}{b}\right)$$

$$= \sin^{-1}\left(\frac{21.2\sin 91.3}{24.6}\right)$$

$$= 59.5°$$

Solve for γ:

$$\gamma = 180° - (\alpha + \beta)$$

$$= 180° - (59.5° + 91.3°) = 29.2°$$

■

Matched Problem 2 A triangular plot of land has sides $a = 217$ ft, $b = 362$ ft, and $c = 345$ ft. Find the measures of all three angles to the nearest $10'$. ■

Answers to Matched Problems

1. $a = 27.4$ m, $\beta = 33.1°$, $\gamma = 60.9°$
2. $\alpha = 35°40'$, $\beta = 76°30'$, $\gamma = 67°50'$

EXERCISE 6.2

All triangles in this exercise are labeled as in the figure unless stated to the contrary. Your answers may differ slightly from those in the book, depending on the order in which you solve for the sides and angles.

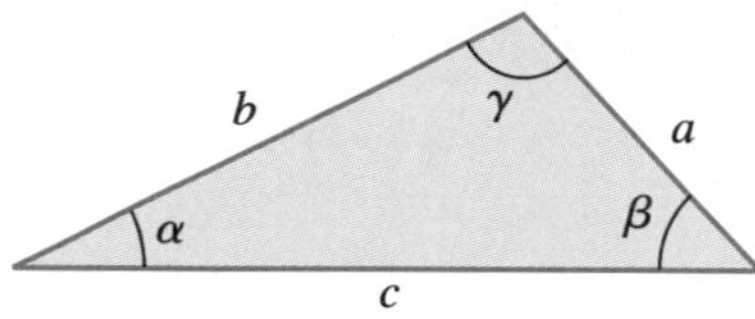

A

1. Referring to the figure above, if $\beta = 38.7°$, $a = 25.3$ ft, and $c = 19.6$ ft, which of the two angles, α or γ, can you say for certain is acute? Why?
2. Referring to the figure above, if $\alpha = 92.6°$, $b = 33.8$ cm, and $c = 49.1$ cm, which of the two angles, β or γ, can you say for certain is acute? Why?

Solve each triangle in Problems 3–6.

3. $\alpha = 50°40'$, $b = 7.03$ mm, $c = 7.00$ mm
4. $\alpha = 71°0'$, $b = 5.32$ cm, $c = 5.00$ cm
5. $\gamma = 134.0°$, $a = 20.0$ m, $b = 8.00$ m
6. $\alpha = 120.0°$, $b = 5.00$ km, $c = 10.0$ km

B

7. Referring to the figure at the beginning of the exercise, if $a = 36.5$ mm, $b = 22.7$ mm, and $c = 19.1$ mm, then, if the triangle has an obtuse angle, which angle must it be? Why?
8. You are told that a triangle has sides $a = 29.4$ ft, $b = 12.3$ ft, and $c = 16.7$ ft. Explain why the triangle has no solution.

Solve each triangle in Problems 9–12.

9. $a = 9.00$ yd, $b = 6.00$ yd, $c = 10.0$ yd (Decimal degrees)
10. $a = 5.00$ km, $b = 5.50$ km, $c = 6.00$ km (Decimal degrees)
11. $a = 420.0$ km, $b = 770.0$ km, $c = 860.0$ km (Degrees and minutes)
12. $a = 15.0$ cm, $b = 12.0$ cm, $c = 10.0$ cm (Degrees and minutes)

Problems 13–30 represent a variety of problems involving the first two sections of this chapter. Solve each triangle using the law of sines or the law of cosines (or both). If a problem does not have a solution, say so.

13. $\beta = 132.4°$, $\gamma = 17.3°$, $b = 67.6$ ft
14. $\alpha = 57.2°$, $\gamma = 112.0°$, $c = 24.8$ ft
15. $\beta = 66.5°$, $a = 13.7$ m, $c = 20.1$ m
16. $\gamma = 54.2°$, $a = 112$ ft, $b = 87.2$ ft
17. $\beta = 84.4°$, $\gamma = 97.8°$, $a = 12.3$ cm
18. $\alpha = 95.6°$, $\gamma = 86.3°$, $b = 43.5$ cm
19. $a = 10.5$ in., $b = 5.23$ in., $c = 9.66$ in. (Decimal degrees)
20. $a = 15.0$ ft, $b = 18.0$ ft, $c = 22.0$ ft (Decimal degrees)
21. $\gamma = 80.3°$, $a = 14.5$ mm, $c = 10.0$ mm
22. $\beta = 63.4°$, $b = 50.5$ in., $c = 64.4$ in.
23. $\alpha = 46.3°$, $\gamma = 105.5°$, $b = 643$ m
24. $\beta = 123.6°$, $\gamma = 21.9°$, $a = 108$ cm
25. $a = 12.2$ m, $b = 16.7$ m, $c = 30.0$ m
26. $a = 28.2$ yd, $b = 52.3$ yd, $c = 22.0$ yd
27. $\alpha = 46.7°$, $a = 18.1$ yd, $b = 22.6$ yd
28. $\gamma = 58.4°$, $b = 7.23$ cm, $c = 6.54$ cm
29. $\alpha = 36.5°$, $\beta = 72.4°$, $\gamma = 71.1°$
30. $\alpha = 29°20'$, $\beta = 32°50'$, $\gamma = 117°50'$

C

31. Using the law of cosines, show that if $\beta = 90°$, then $b^2 = c^2 + a^2$ (the Pythagorean theorem).
32. Using the law of cosines, show that if $b^2 = c^2 + a^2$, then $\beta = 90°$.

33. Check Problem 3 using Mollweide's equation (see Problem 25, Exercise 6.1),

$$(a - b) \cos \frac{\gamma}{2} = c \sin \frac{\alpha - \beta}{2}$$

34. Check Problem 5 using Mollweide's equation (see Problem 33).

35. Show that for any triangle with standard labeling (see the figure at the beginning of the exercise),

$$c = b \cos \alpha + a \cos \beta$$

36. Show that for any triangle with standard labeling (see the figure at the beginning of the exercise),

$$\frac{a^2 + b^2 + c^2}{2abc} = \frac{\cos \alpha}{a} + \frac{\cos \beta}{b} + \frac{\cos \gamma}{c}$$

Applications

37. **Surveying** A geologist wishes to determine the distance CB across the base of a volcanic cinder cone (see the figure). Distances AB and AC are measured to be 425 m and 384 m, respectively, and $\angle CAB$ is 98.3°. Find the approximate distance across the base of the cinder cone.

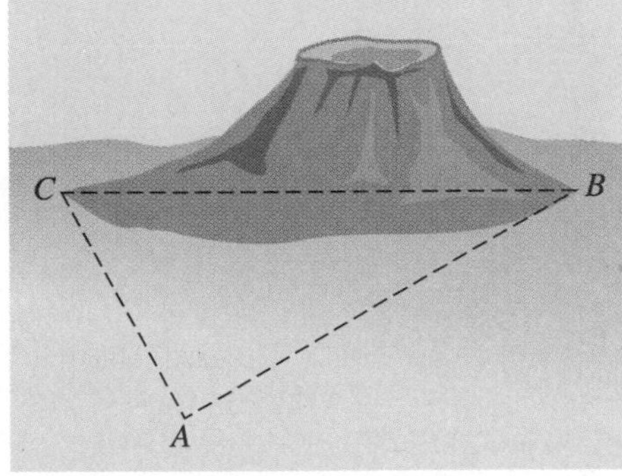

Figure for 37

38. **Surveying** To estimate the length CB of the lake in the figure at the top of the next column, a surveyor measures AB and AC to be 89 m and 74 m, respectively, and $\angle CAB$ to be 95°. Find the approximate length of the lake.

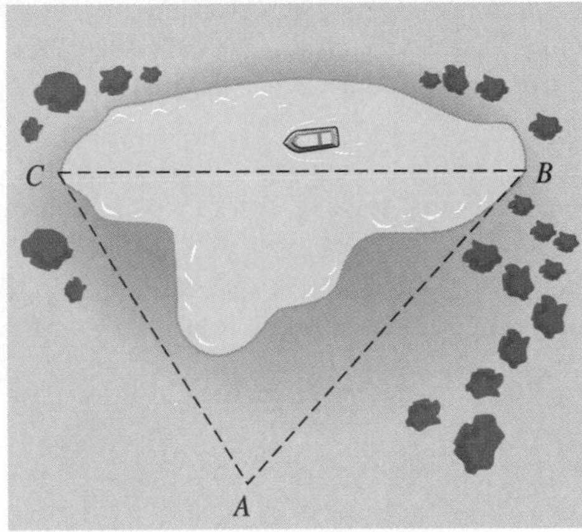

Figure for 38

39. **Geometry: Engineering** Find the measure in decimal degrees of a central angle subtended by a chord of length 13.8 cm in a circle of radius 8.26 cm.

40. **Geometry: Engineering** Find the measure in decimal degrees of a central angle subtended by a chord of length 112 ft in a circle of radius 72.8 ft.

41. **Search and Rescue** At midnight, two Coast Guard helicopters set out from San Francisco to find a sailboat in distress. Helicopter A flies due west over the Pacific Ocean at 250 km/hr, and helicopter B flies northwest at 210 km/hr. At 1 A.M., helicopter A spots a flare from the boat and radios helicopter B to come and assist in the rescue. How far is helicopter B from helicopter A at this time? Compute answer to two significant digits.

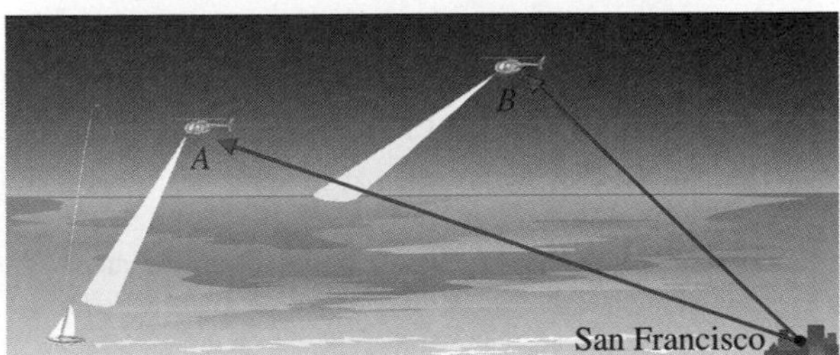

Figure for 41

42. **Navigation** Los Angeles and San Francisco are approximately 600 km apart. A pilot flying from Los An-

geles to San Francisco finds that after she is 200 km from Los Angeles the plane is 20° off course. How far is the plane from San Francisco at this time (to two significant digits)?

*43. **Geometry: Engineering** A 58.3 cm chord of a circle subtends a central angle of 27.8°. Find the radius of the circle to three significant digits using the law of cosines.

44. **Geometry: Engineering** Find the perimeter (to the nearest centimeter) of a regular pentagon inscribed in a circle with radius 5 cm (see the figure).

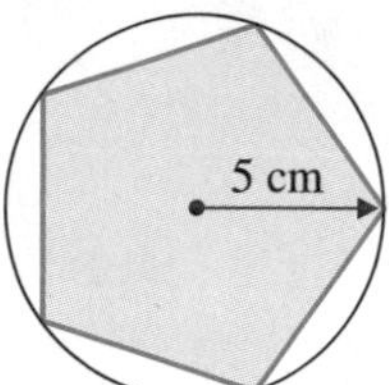

Figure for 44

*45. **Geometry: Engineering** Three circles of radius 2 cm, 3 cm, and 8 cm are tangent to each other (see the figure). Find (to the nearest 10′) the three angles formed by the lines joining their centers.

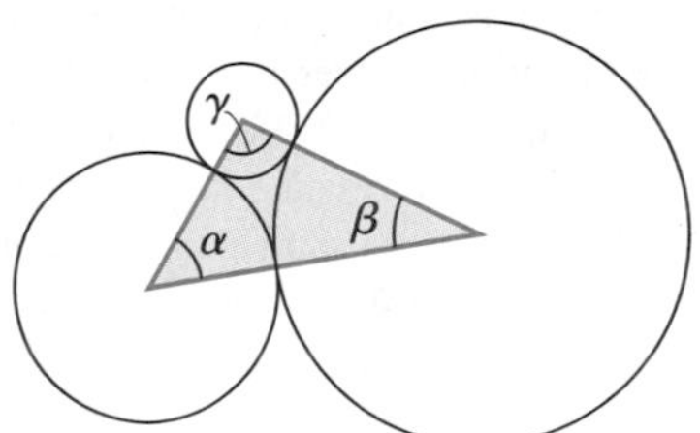

Figure for 45

*46. **Engineering** A tunnel for hydroelectric power is to be constructed through a mountain from one reservoir to another at a lower level (see the figure at the top of the next column). The distance from the top of the mountain to the lower end of the tunnel is 5.32 mi, and from the top of the mountain to the upper end of the tunnel is 2.63 mi. The angles of depression of the two slopes of the mountain are 42.7° and 48.8°, respectively.

(A) What is the length of the tunnel?

(B) What angle does the tunnel make with the horizontal?

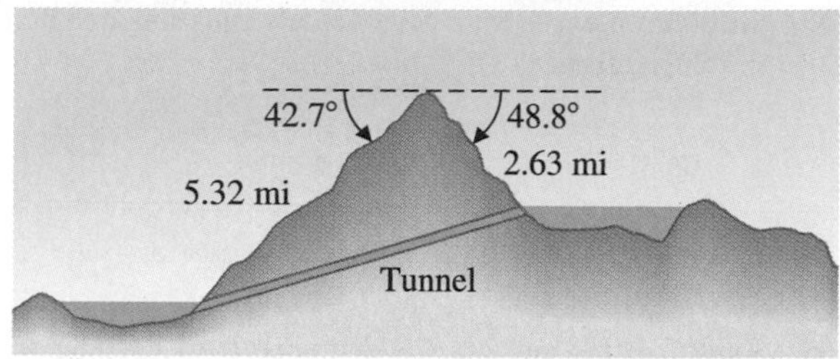

Figure for 46

*47. **Engineering: Construction** A fire lookout is to be constructed as indicated in the figure. Support poles AD and BC are each 18.0 ft long and tilt 8.0° inward from the vertical. The distance between the tops of the poles, DC, is 12.0 ft. Find the length of a brace AC and the distance AB between the supporting poles at ground level.

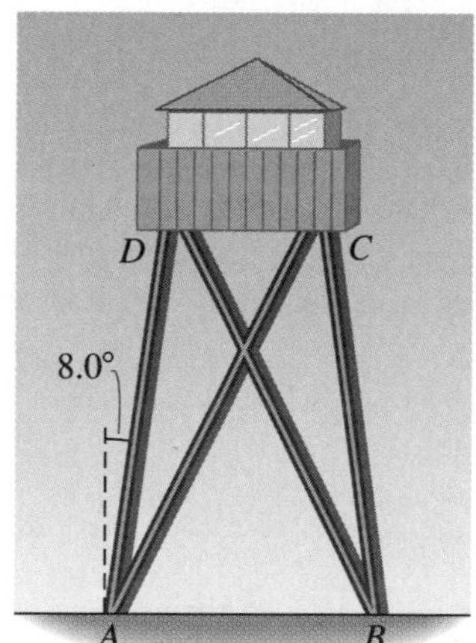

Figure for 47

*48. **Engineering: Construction** Repeat Problem 47 with support poles AD and BC making an angle of 11.0° with the vertical instead of 8.0°.

*49. **Space Science** A satellite, S, in circular orbit around the earth, is sighted by a tracking station T (see the figure at the top of the next page). The distance TS is determined by radar to be 1,034 mi, and the angle of elevation above the horizon is 32.4°. How high is the satellite above the earth at the time of the sighting? The radius of the earth is $r = 3{,}964$ mi.

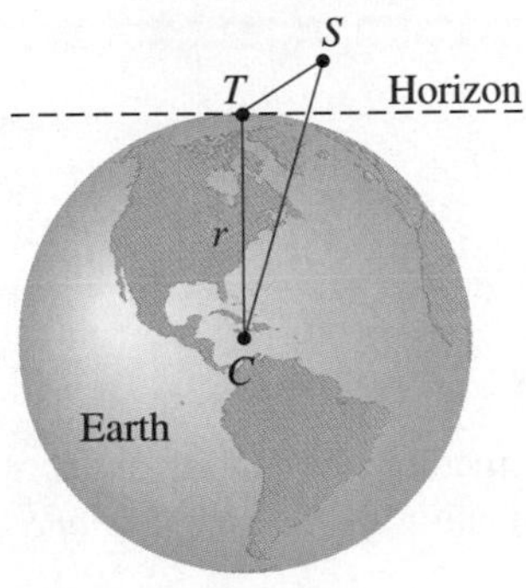

Figure for 49

***50. Space Science** For communications between a space shuttle and the White Sands Missile Range in southern New Mexico, two satellites are placed in geostationary orbit, 130° apart. Each is 22,300 mi above the surface of the earth (see the figure). (When a satellite is in **geostationary orbit** it remains stationary above a fixed point on the earth's surface.) Radio signals are sent from an orbiting shuttle by way of the satellites to the White Sands facility, and vice versa. This arrangement enables the White Sands facility to maintain radio contact with the shuttle over most of the earth's surface. How far (to the nearest 100 mi) is one of these geostationary satellites from the White Sands facility (W)? The radius of the earth is 3,964 mi.

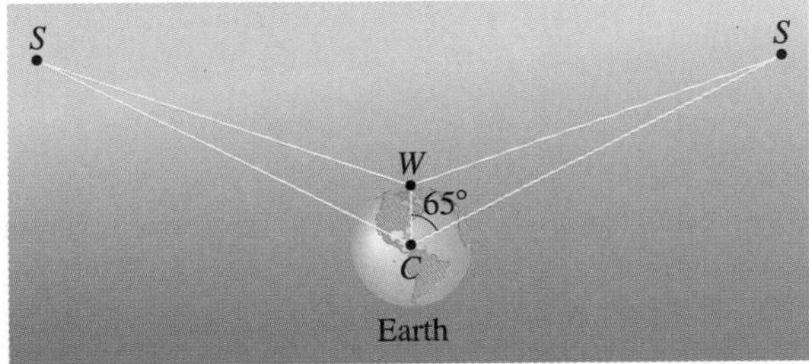

Figure for 50

***51. Geometry: Engineering** A rectangular solid has sides of 6.0 cm, 3.0 cm, and 4.0 cm (see the figure at the top of the next column). Find $\angle ABC$.

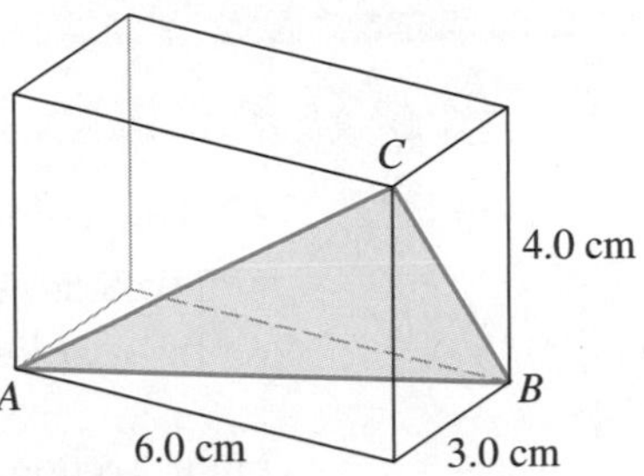

Figure for 51

***52. Geometry: Engineering** Refer to Problem 51. Find $\angle ACB$.

***53. Surveying** A plot of land was surveyed, with the resulting information shown in the figure. Find the length of DC.

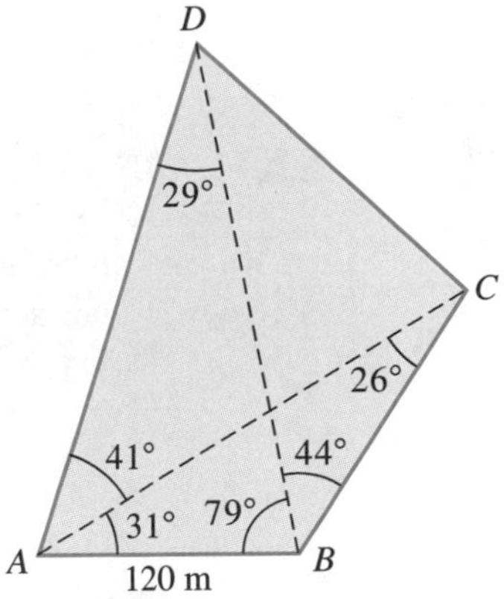

Figure for 53

***54. Surveying** A plot of land was surveyed, with the resulting information shown in the figure. Find the length of BC.

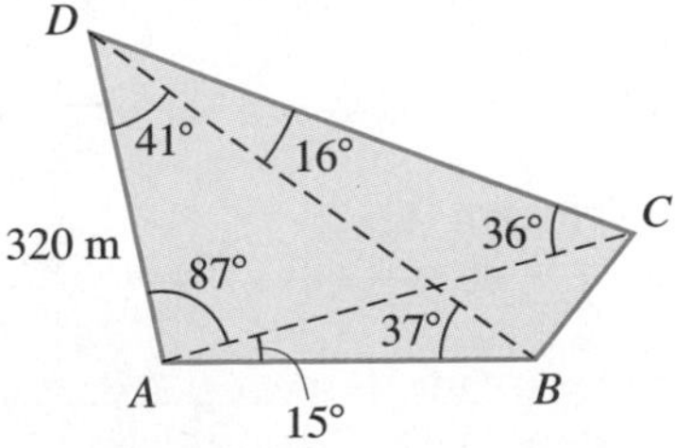

Figure for 54

☆6.3 AREAS OF TRIANGLES

- Base and Height Given
- Two Sides and Included Angle Given
- Three Sides Given (Heron's Formula)

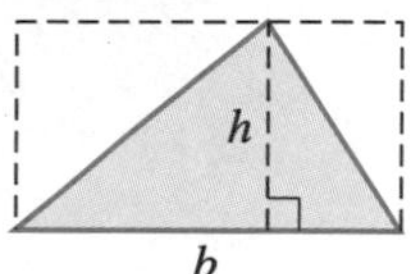

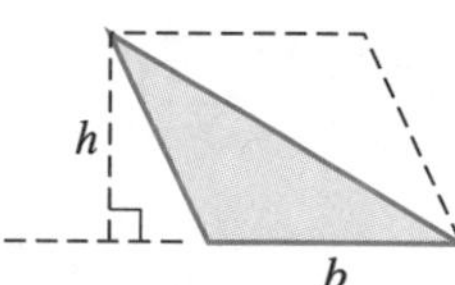

FIGURE 1
$A = \frac{1}{2}bh$

In this section, we discuss three frequently used methods of finding areas of triangles given the indicated information. The derivation of *Heron's formula* also illustrates a significant use of identities.

Base and Height Given

If the base b and height h of a triangle are given (see Fig. 1), then the area A is one-half the area of a parallelogram with the same base and height:

$$A = \tfrac{1}{2}bh$$

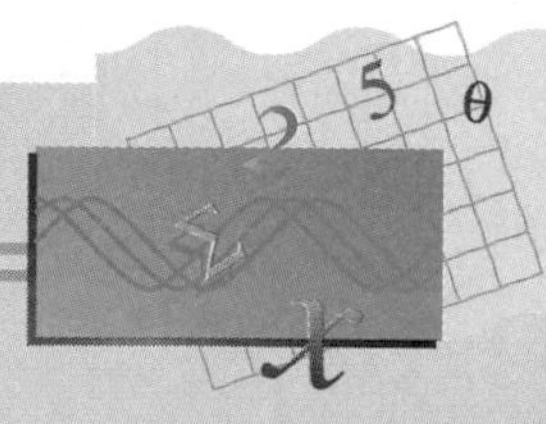

EXPLORE/DISCUSS 1

Given two parallel lines m and n, explain why all three triangles (I, II, and III) sharing the same base b have the same area.

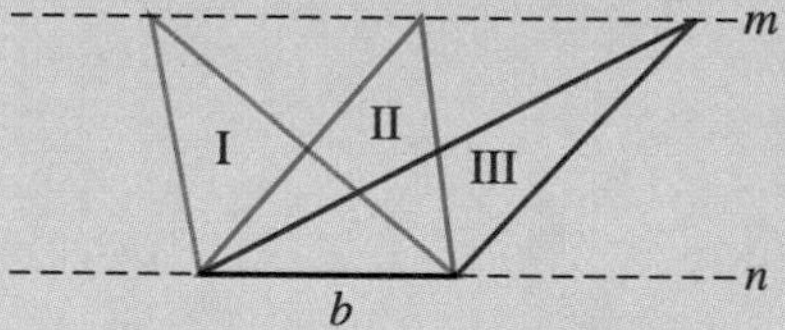

Two Sides and Included Angle Given

Given two sides, a and b, and an included angle θ (see Fig. 2), the above formula can be converted into the following form:

$$A = \frac{ab}{2}\sin\theta$$

FIGURE 2
$A = \dfrac{ab}{2}\sin\theta$

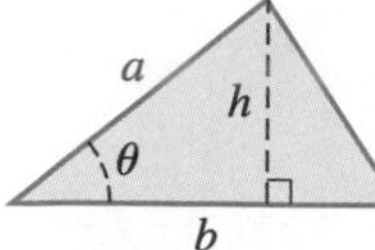

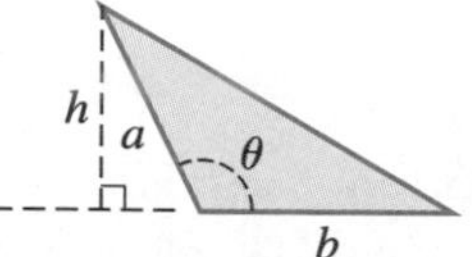

☆ Sections marked with a star may be omitted without loss of continuity.

This conversion is accomplished by using the sine function to express h in terms of θ and a:

$$h = a \sin \theta$$

which holds whether θ is acute or obtuse, since $\sin(180° - \theta) = \sin \theta$. Thus,

$$A = \frac{1}{2} bh = \frac{1}{2} ba \sin \theta = \frac{ab}{2} \sin \theta$$

EXAMPLE 1 **Find the Area of a Triangle Given Two Sides and the Included Angle**

Find the area of the triangle shown in Figure 3.

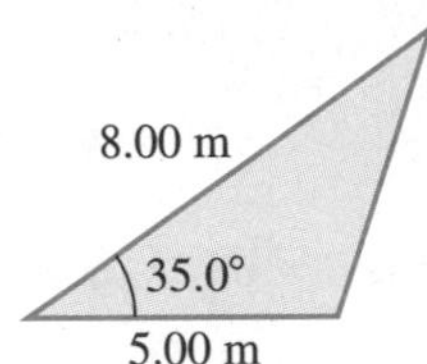

FIGURE 3

Solution

$$A = \frac{ab}{2} \sin \theta = \frac{1}{2}(8.00)(5.00) \sin 35.0°$$
$$= 11.5 \text{ m}^2$$

Matched Problem 1 Find the area of the triangle with $a = 12.0$ cm, $b = 7.00$ cm, and included angle $\theta = 125.0°$.

Three Sides Given (Heron's Formula)

A famous formula from the Greek philosopher–mathematician Heron of Alexandria (75 AD) enables us to compute the area of a triangle directly when given only the lengths of the three sides of the triangle.

HERON'S FORMULA

If the **semiperimeter** s is

$$s = \frac{a + b + c}{2}$$

then **Heron's formula** gives the area as

$$A = \sqrt{s(s - a)(s - b)(s - c)}$$

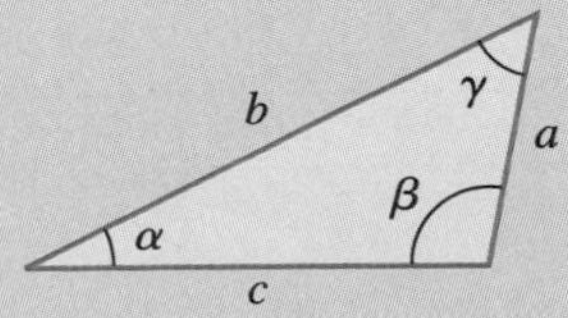

Heron's formula is obtained from

$$A = \frac{bc}{2} \sin \alpha \tag{1}$$

by expressing $\sin \alpha$ in terms of the sides a, b, and c. Several identities and the law of cosines play a central role in the derivation of the formula. We will first get $\sin(\alpha/2)$ and $\cos(\alpha/2)$ in terms of a, b, and c. Then we will use a double-angle identity in the form

$$\sin \alpha = 2 \sin \frac{\alpha}{2} \cos \frac{\alpha}{2} \tag{2}$$

to write $\sin \alpha$ in terms of a, b, and c. We start with a half-angle identity for sine in the form

$$\sin^2 \frac{\alpha}{2} = \frac{1 - \cos \alpha}{2} \tag{3}$$

The following version of the law of cosines involves $\cos \alpha$ and all three sides of the triangle a, b, and c:

$$a^2 = b^2 + c^2 - 2bc \cos \alpha$$

Solving for $\cos \alpha$, we obtain

$$\cos \alpha = \frac{b^2 + c^2 - a^2}{2bc} \tag{4}$$

Substituting (4) into (3), we can write $\sin^2(\alpha/2)$ in terms of a, b, and c:

$$\sin^2 \frac{\alpha}{2} = \frac{1 - \dfrac{b^2 + c^2 - a^2}{2bc}}{2}$$

$$= \frac{2bc\left(1 - \dfrac{b^2 + c^2 - a^2}{2bc}\right)}{2bc(2)} \qquad \text{Convert to simple fraction.}$$

$$= \frac{2bc - b^2 - c^2 + a^2}{4bc} \qquad \text{Numerator factors (not obvious).}$$

$$= \frac{(a + b - c)(a - b + c)}{4bc} \tag{5}$$

To bring the semiperimeter

$$s = \frac{a + b + c}{2} \tag{6}$$

into the picture, we write (5) in the form

$$\sin^2\frac{\alpha}{2} = \frac{(a+b+c-2b)(a+b+c-2c)}{4bc}$$

$$= \frac{2\left(\dfrac{a+b+c}{2}-b\right)2\left(\dfrac{a+b+c}{2}-c\right)}{4bc}$$

$$= \frac{\left(\dfrac{a+b+c}{2}-b\right)\left(\dfrac{a+b+c}{2}-c\right)}{bc} \tag{7}$$

Substituting (6) into (7) and solving for $\sin(\alpha/2)$, we obtain

$$\sin\frac{\alpha}{2} = \sqrt{\frac{(s-b)(s-c)}{bc}} \tag{8}$$

If we repeat this reasoning starting with a half-angle identity for cosine in the form

$$\cos^2\frac{\alpha}{2} = \frac{1+\cos\alpha}{2} \tag{9}$$

we obtain

$$\cos\frac{\alpha}{2} = \sqrt{\frac{s(s-a)}{bc}} \tag{10}$$

Substituting (8) and (10) into identity (2) produces

$$\sin\alpha = 2\sqrt{\frac{(s-b)(s-c)}{bc}}\sqrt{\frac{s(s-a)}{bc}} = \frac{2}{bc}\sqrt{s(s-a)(s-b)(s-c)} \tag{11}$$

And we are almost there. We now substitute (11) into formula (1) to obtain Heron's formula:

$$A = \frac{bc}{2}\left[\frac{2}{bc}\sqrt{s(s-a)(s-b)(s-c)}\right]$$

$$= \sqrt{s(s-a)(s-b)(s-c)}$$

The derivation of Heron's formula provides a good illustration of the importance of identities. Try to imagine a derivation of this formula without identities.

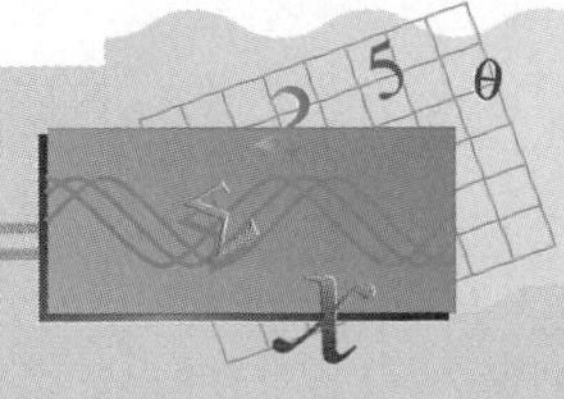

EXPLORE/DISCUSS 2

Derive equation (10) above following the same type of reasoning that was used to derive equation (8).

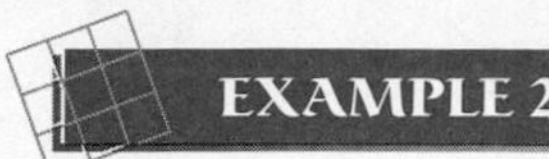

Finding the Area of a Triangle Given Three Sides

Find the area of a triangle with sides $a = 12.0$ cm, $b = 8.0$ cm, and $c = 6.0$ cm.

Solution First, find the semiperimeter s:

$$s = \frac{a + b + c}{2} = \frac{12.0 + 8.0 + 6.0}{2} = 13.0 \text{ cm}$$

Then

$$s - a = 13.0 - 12.0 = 1.0 \text{ cm}$$

$$s - b = 13.0 - 8.0 = 5.0 \text{ cm}$$

$$s - c = 13.0 - 6.0 = 7.0 \text{ cm}$$

Thus,

$$\begin{aligned} A &= \sqrt{s(s - a)(s - b)(s - c)} \\ &= \sqrt{13.0(1.0)(5.0)(7.0)} \\ &= 21 \text{ cm}^2 \qquad \text{To two significant digits} \end{aligned}$$

[*Note:* The computed area has the same number of significant digits as the side with the least number of significant digits.] ■

Matched Problem 2 Find the area of a triangle with $a = 6.0$ m, $b = 10.0$ m, and $c = 8.0$ m. ■

Answers to Matched Problems

1. 34.4 cm^2
2. 24 m^2

EXERCISE 6.3

A *Find the area of the triangle matching the information given in Problems 1–12.*

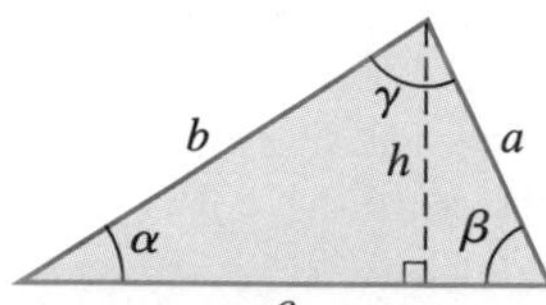

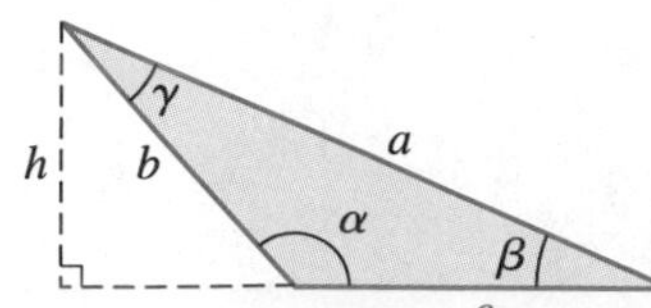

Figure for 1–12

1. $h = 12.0$ m, $c = 17.0$ m
2. $h = 7.0$ ft, $c = 10.0$ ft
3. $\alpha = 30.0°$, $b = 6.0$ cm, $c = 8.0$ cm
4. $\alpha = 45°$, $b = 5.0$ m, $c = 6.0$ m
5. $a = 4.00$ in., $b = 6.00$ in., $c = 8.00$ in.
6. $a = 4.00$ ft, $b = 10.00$ ft, $c = 12.00$ ft

B

7. $\alpha = 23°20'$, $b = 403$ ft, $c = 512$ ft
8. $\alpha = 58°40'$, $b = 28.2$ in., $c = 6.40$ in.
9. $\alpha = 132.67°$, $b = 12.1$ cm, $c = 10.2$ cm
10. $\alpha = 147.5°$, $b = 125$ mm, $c = 67.0$ mm
11. $a = 12.7$ m, $b = 20.3$ m, $c = 24.4$ m
12. $a = 5.24$ cm, $b = 3.48$ cm, $c = 6.04$ cm

C **13.** Show that the diagonals of a parallelogram divide the figure into four triangles, all having the same area.

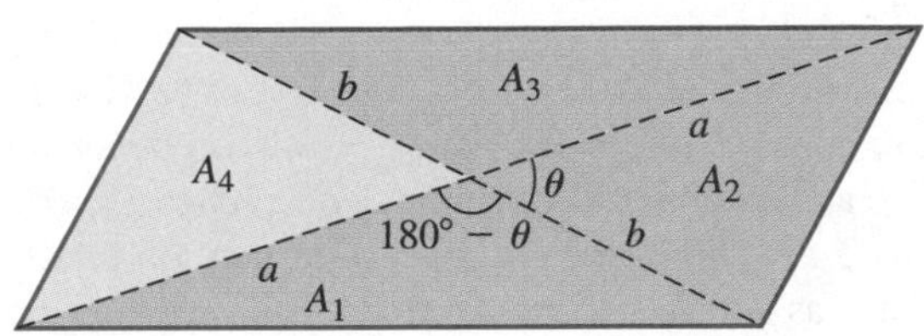

Figure for 13

14. If $s = (a + b + c)/2$ is the semiperimeter of a triangle with sides a, b, and c (see the figure), show that the radius r of an inscribed circle is given by

$$r = \sqrt{\frac{(s - a)(s - b)(s - c)}{s}}$$

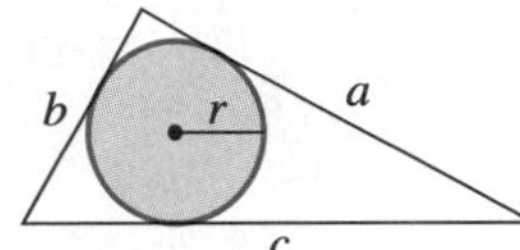

Figure for 14

6.4 VECTORS: GEOMETRICALLY DEFINED

- Geometric Vectors; Vector Addition
- Velocity Vectors
- Force Vectors
- Resolution of a Vector into Components
- Centrifugal Force

Scalar quantities are physical quantities (length, area, volume, etc.) that can be completely specified by a single real number. Directed distances, velocities, and forces are quantities that require both a magnitude and direction for their complete specification. We call these **vector quantities.**

Vector forms are widely used in both pure and applied mathematics. They are used extensively in the physical sciences and engineering, and are seeing increased use in the social and life sciences.

In this section we limit our development to intuitive notions of *geometric vectors in a plane.* In Section 6.5 we present an algebraic treatment of vectors, which is just the first step of a development and generalization that can fill whole books.

Geometric Vectors; Vector Addition

If A and B are two points in the plane, then the **directed line segment** from A to B, denoted by $\overrightarrow{AB}$, is the line segment joining A to B with an arrowhead placed at B to indicate the direction is from A to B (see Fig. 1). Point A is called the **initial point,** and point B is called the **terminal point.**

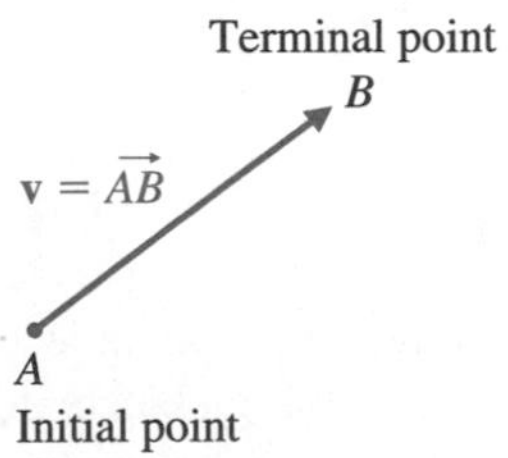

FIGURE 1
Vector $\overrightarrow{AB} = \mathbf{v}$

A **geometric vector in the plane** is a quantity that possesses both a length and a direction and can be represented by a directed line segment. The directed line segment in Figure 1 represents a vector that is denoted by $\overrightarrow{AB}$. A vector also may be denoted by a boldface letter, such as **v** or **F**, or by a letter with an arrow over it, such as $\vec{v}$. Letters with arrows over them are easier to write by hand, but boldface letters are easier to recognize in print. We will generally use boldface letters for vector quantities in this book, but when you write a vector by hand you should include the arrow above it as a reminder that the quantity is a vector.

The **magnitude** of the vector $\overrightarrow{AB}$, denoted by $|\overrightarrow{AB}|$, $|\vec{v}|$, or $|\mathbf{v}|$, is the length of the directed line segment. Two vectors have the **same direction** if they are parallel and point in the same direction. Two vectors have **opposite direction** if they are parallel and point in opposite directions. The **zero vector,** denoted by $\vec{0}$ or **0**, has a magnitude of zero and an arbitrary direction. Two vectors are **equal** if they have the same magnitude and direction. Thus, a vector may be **translated** from one location to another as long as the magnitude and direction do not change.

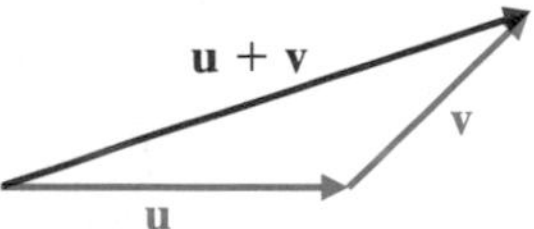

FIGURE 2
Tail-to-tip rule for addition

The **sum of two vectors u and v** can be defined using the **tail-to-tip rule:** Translate **v** so that its tail end (initial point) is at the tip end (terminal point) of **u**. Then the vector from the tail end of **u** to the tip end of **v** is the sum, denoted by **u + v,** of the vectors **u** and **v** (see Fig. 2). The sum of two nonparallel vectors also can be defined using the **parallelogram rule:** The **sum of two nonparallel vectors u and v** is the diagonal of the parallelogram formed using **u** and **v** as adjacent sides (see Fig. 3). (If **u** and **v** are parallel, use the tail-to-tip rule.) Of course, both rules give the same sum, as they should. The choice of which rule to use depends on the situation.

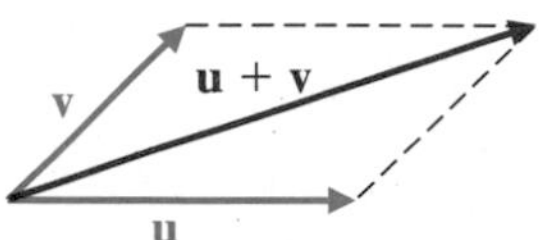

FIGURE 3
Parallelogram rule for addition

The vector **u + v** is called the **resultant** of the two vectors **u** and **v**, and **u** and **v** are called **components** of **u + v.** It is useful to observe that vector addition is **commutative** and **associative.** That is, **u + v = v + u** and **u + (v + w) = (u + v) + w.**

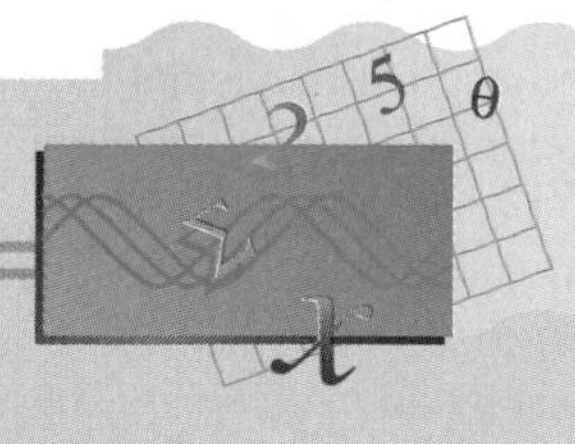

EXPLORE/DISCUSS 1

If **u, v,** and **w** represent three arbitrary geometric vectors, illustrate using either definition of vector addition, that:

(A) **u + v = v + u** (B) **u + (v + w) = (u + v) + w**

Velocity Vectors

A **velocity vector** is a vector that represents the speed and direction of an object in motion. Vector methods often can be applied to problems involving objects in motion. Many of these problems involve the use of a *navigational compass,* which is marked clockwise in degrees starting at north (see Fig. 4).

FIGURE 4

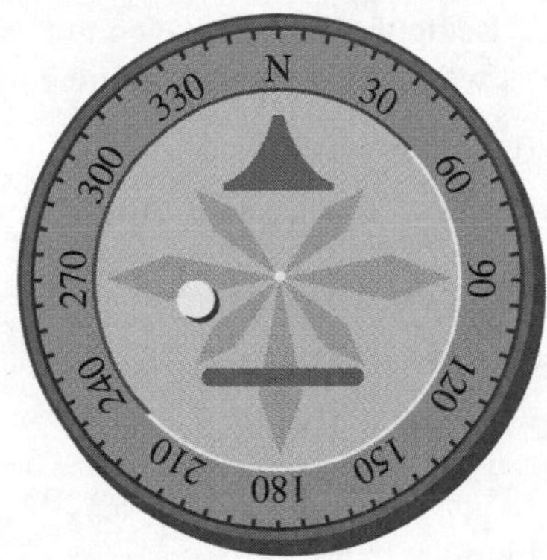

EXAMPLE 1

Resultant Velocity

A power boat traveling at 24 km/hr relative to the water has a compass heading (the direction the boat is pointing) of 95°. A strong tidal current, with a heading of 35°, is flowing at 12 km/hr. The velocity of the boat relative to the water is called its **apparent velocity,** and the velocity relative to the ground is called the **resultant** or **actual velocity.** The resultant velocity is the vector sum of the apparent velocity and the current velocity. Find the resultant velocity; that is, find the actual speed and direction of the boat relative to the ground.

Solution We can use geometric vectors (see Fig. 5a) to represent the apparent velocity vector and the current velocity vector. We add the two vectors using the tail-to-tip method of addition of vectors to obtain the resultant (actual) velocity vector as indicated in Figure 5b. From this vector diagram we obtain the triangle in Figure 6 and solve for β, b, and α.

FIGURE 5

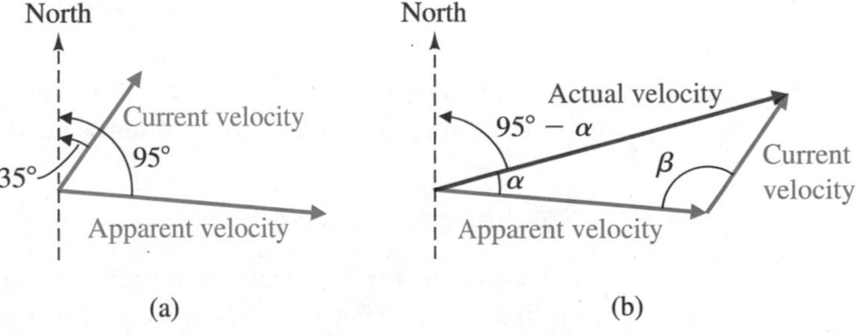

FIGURE 6

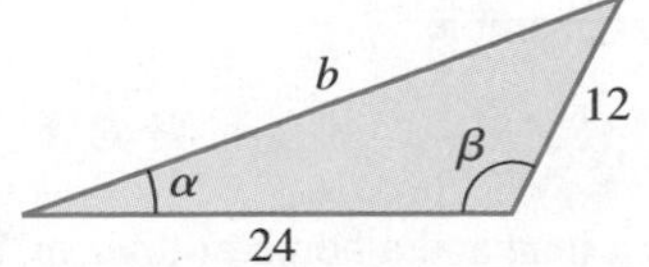

Boat's actual heading: $95° - \alpha$

Boat's actual speed: b

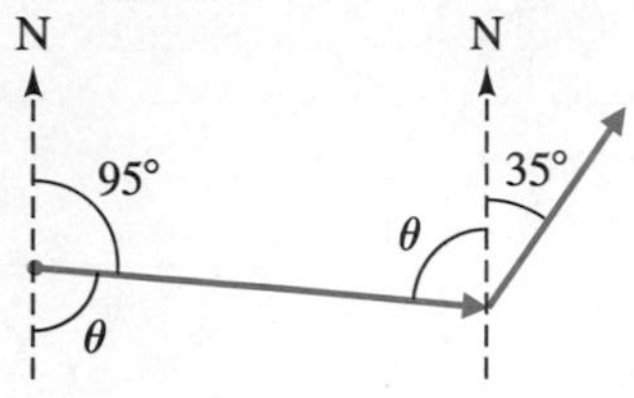

FIGURE 7

Solve for β: Using the apparent velocity heading (95°) and the current velocity heading (35°), we can find β using Figure 7.

$$\theta = 180° - 95° = 85°$$

$$\beta = \theta + 35° = 85° + 35° = 120°$$

Solve for b: Use the law of cosines:

$$b^2 = a^2 + c^2 - 2ac \cos \beta$$
$$= 12^2 + 24^2 - 2(12)(24) \cos 120°$$
$$b = \sqrt{12^2 + 24^2 - 2(12)(24) \cos 120°}$$
$$= 32 \text{ km/hr}$$

Actual speed relative to the ground

Solve for α: Use the law of sines:

$$\frac{\sin \alpha}{a} = \frac{\sin \beta}{b}$$

$$\sin \alpha = a\left(\frac{\sin \beta}{b}\right)$$
$$= 12\left(\frac{\sin 120°}{32}\right)$$
$$\alpha = \sin^{-1}\left[12\left(\frac{\sin 120°}{32}\right)\right]$$
$$= 19°$$

Actual heading $= 95° - \alpha = 95° - 19° = 76°$ ■

Matched Problem 1 Repeat Example 1 with a current of 8.0 km/hr at 22° and the speedometer on the boat reading 35 km/hr with a compass heading of 85°. ■

Force Vectors

A **force vector** is a vector that represents the direction and magnitude of an applied force. If an object is subjected to two forces, then the sum of these two forces (the **resultant force**) is a single force. If the resultant force replaced the original two forces, it would act on the object in the same way as the two original forces taken together. In physics it is shown that the resultant force vector can be obtained using vector addition to add the two individual force vectors. It seems natural to use the parallelogram rule for adding force vectors, as illustrated in the next example.

EXAMPLE 2 **Resultant Force**

Figure 8 shows a man and a horse pulling on a large piece of granite. The man is pulling with a force of 200 lb, and the horse is pulling with a force of 1,000 lb

50° away from the direction of the pull of the man. The length of the main diagonal of the parallelogram will be the actual magnitude of the resultant force on the stone, and its direction will be the direction of motion of the stone (if it moves). Find the magnitude and direction α of the resultant force. (The magnitudes of the forces are measured to two significant digits and the angle to the nearest degree.)

FIGURE 8

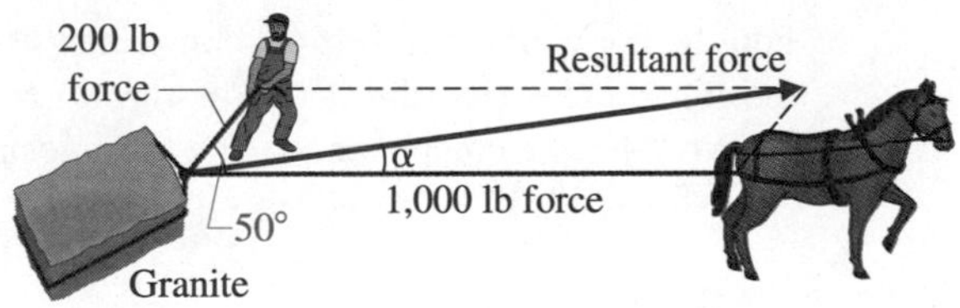

Solution From Figure 8 we obtain the triangle in Figure 9 and then solve for α and b.

FIGURE 9

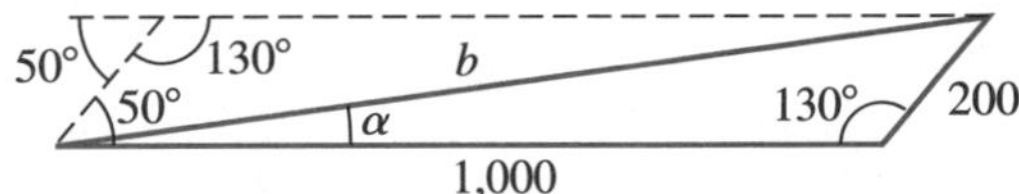

Solve for b:

$$b^2 = 1{,}000^2 + 200^2 - 2(1{,}000)(200)\cos 130°$$
$$= 1{,}297{,}115.04\ldots$$
$$b = \sqrt{1{,}297{,}115.04\ldots} = 1{,}100 \text{ lb}$$

Solve for α:

$$\frac{\sin\alpha}{200} = \frac{\sin 130°}{1{,}100}$$

$$\sin\alpha = \frac{200}{1{,}100}\sin 130°$$

$$\alpha = \sin^{-1}\left(\frac{200}{1{,}100}\sin 130°\right) = 8°$$

■

Matched Problem 2 Repeat Example 2 but change the angle between the force vector from the horse and the force vector from the man to 45° instead of 50°, and change the magnitude of the force vector from the horse to 800 lb instead of 1,000 lb. ■

■ Resolution of a Vector into Components

Instead of adding vectors, many problems require the breaking down of vectors into components. Whenever a vector is expressed as a resultant of two vectors, these two vectors are called **components** of the given vector. For example, to

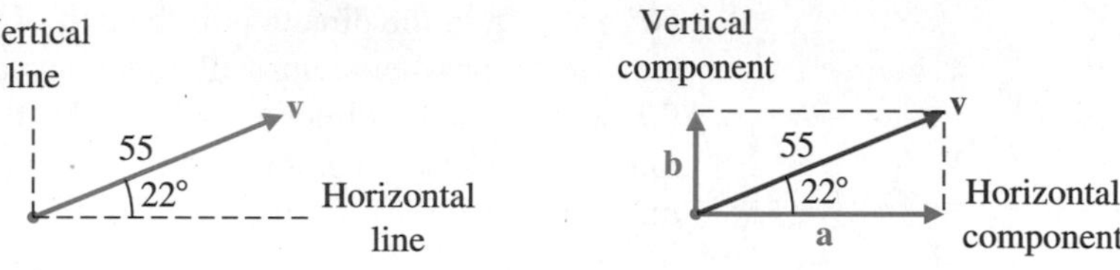

FIGURE 10 **FIGURE 11**

find the horizontal and vertical components of the vector **v** in Figure 10, we find the magnitudes (the directions of the horizontal and vertical lines are already known) of these components using the sine and cosine functions (see Fig. 11).

Magnitude of horizontal component:

$$\cos 22° = \frac{|\mathbf{a}|}{55}$$

$$|\mathbf{a}| = 55 \cos 22° \qquad \text{Horizontal projection of } \mathbf{v}$$
$$= 51$$

Magnitude of vertical component:

$$\sin 22° = \frac{|\mathbf{b}|}{55}$$

$$|\mathbf{b}| = 55 \sin 22° \qquad \text{Vertical projection of } \mathbf{v}$$
$$= 21$$

Centrifugal Force

Ideally, a race car going around a curve will not slide sideways if the track is banked appropriately. How much should a track be banked for a given speed v and a given curve of radius r to eliminate sideways forces? Figure 12 shows the relevant forces.

FIGURE 12

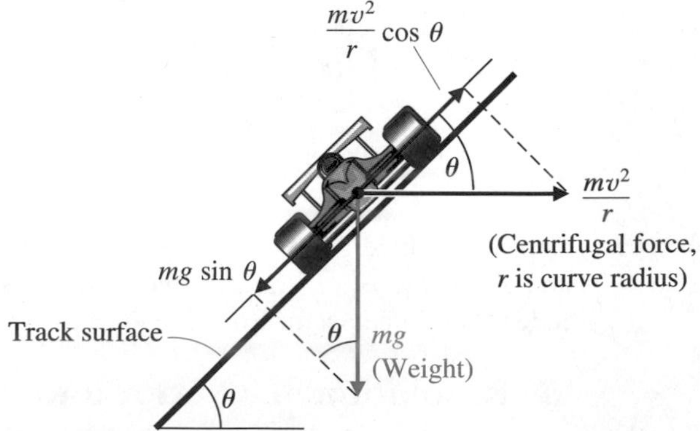

To avoid sideways forces, the components of the forces parallel to the track surface due to the mass of the car and its movement around the curve (centrifugal force) must be equal. That is,

$$mg \sin \theta = \frac{mv^2}{r} \cos \theta$$

$$\frac{\sin \theta}{\cos \theta} = \frac{mv^2}{mgr}$$

$$\tan \theta = \frac{v^2}{gr}$$

$$\theta = \tan^{-1} \frac{v^2}{gr}$$

v in meters per second, r in meters, $g \approx 9.81$ m/sec²

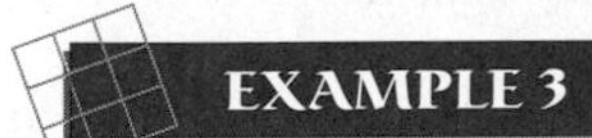

Race Track Design

At what angle θ must a track be banked at a curve to eliminate sideways forces parallel to the track, given that the curve has a radius of 525 m and the racing car is moving at 85.0 m/sec (about 190 mph)?

Solution

$$\begin{aligned} \theta &= \tan^{-1} \frac{v^2}{gr} \\ &= \tan^{-1} \frac{(85.0)^2}{(9.81)(525)} \\ &= 54.5° \end{aligned}$$

■

Matched Problem 3 Repeat Example 3 with a car moving at 25.0 m/sec (about 56 mph) around a curve with a radius of 125 m. ■

Answers to Matched Problems

1. Actual speed: 39 km/hr; Actual heading: 75°
2. Resultant force: $b = 950$ lb, $\alpha = 9°$ 3. $\theta = 27.0°$

EXERCISE 6.4

Express all angles in decimal degrees.

A *In Problems 1–6, find $|\mathbf{u} + \mathbf{v}|$ and θ, given $|\mathbf{u}|$ and $|\mathbf{v}|$ in the figure below.*

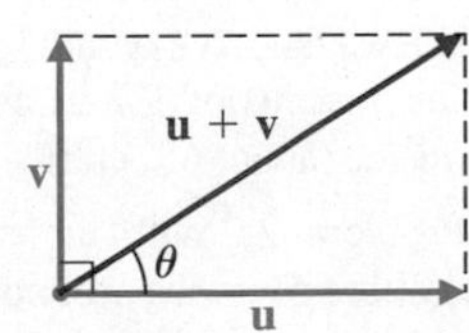

(a) Parallelogram rule

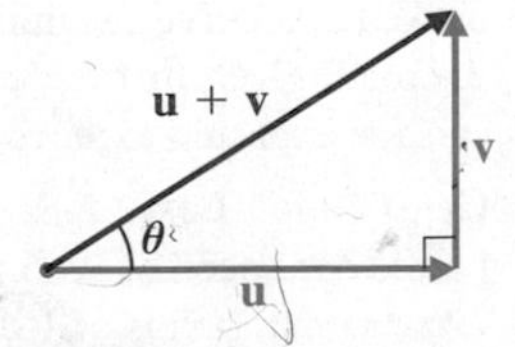

(b) Tail-to-tip rule

1. $|\mathbf{u}| = 62$ km/hr, $|\mathbf{v}| = 34$ km/hr
2. $|\mathbf{u}| = 37$ km/hr, $|\mathbf{v}| = 45$ km/hr
3. $|\mathbf{u}| = 48$ lb, $|\mathbf{v}| = 31$ lb
4. $|\mathbf{u}| = 38$ lb, $|\mathbf{v}| = 53$ lb
5. $|\mathbf{u}| = 143$ knots, $|\mathbf{v}| = 57.4$ knots*
6. $|\mathbf{u}| = 434$ knots, $|\mathbf{v}| = 105$ knots

* A knot is 1 nautical mile per hour.

In Problems 7–10, find the magnitudes of the horizontal and vertical components, $|\mathbf{H}|$ and $|\mathbf{V}|$, respectively, of the vector $\mathbf{v}$ *given* $|\mathbf{v}|$ *and* θ *in the figure.*

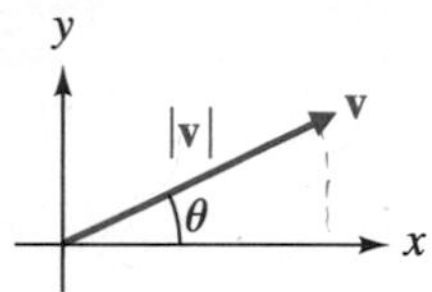

7. $|\mathbf{v}| = 42$ lb, $\theta = 34°$
8. $|\mathbf{v}| = 250$ lb, $\theta = 67°$
9. $|\mathbf{v}| = 244$ km/hr, $\theta = 43.2°$
10. $|\mathbf{v}| = 84.0$ km/hr, $\theta = 28.6°$
11. Can the magnitude of a vector ever be negative? Explain.
12. If two vectors have the same magnitude, are they equal? Explain.

B *In Problems 13–16, find* $|\mathbf{u} + \mathbf{v}|$ *and* α*, given* $|\mathbf{u}|$, $|\mathbf{v}|$, *and* θ *in the figure below.*

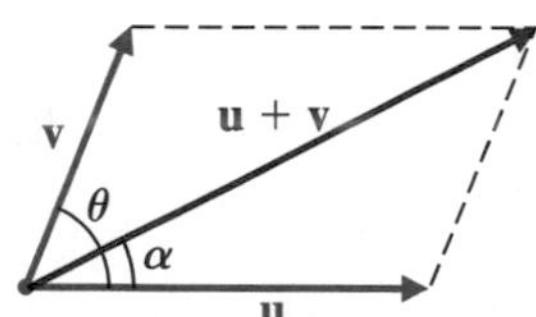

(a) Parallelogram rule

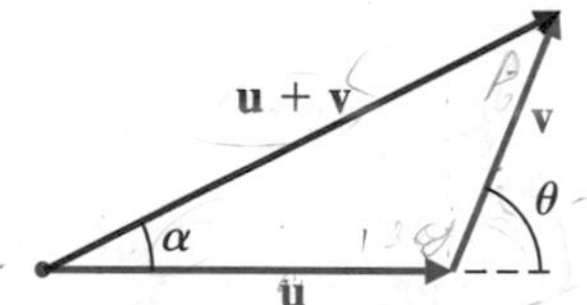

(b) Tail-to-tip rule

13. $|\mathbf{u}| = 125$ lb, $|\mathbf{v}| = 84$ lb, $\theta = 44°$
14. $|\mathbf{u}| = 66$ lb, $|\mathbf{v}| = 22$ lb, $\theta = 68°$
15. $|\mathbf{u}| = 655$ mi/hr, $|\mathbf{v}| = 97.3$ mi/hr, $\theta = 66.8°$
16. $|\mathbf{u}| = 487$ mi/hr, $|\mathbf{v}| = 74.2$ mi/hr, $\theta = 37.4°$

C 17. Explain why it is or is not correct to say that the zero vector is parallel to every vector.

18. Explain why it is or is not correct to say that the zero vector is perpendicular to every vector.

Applications

Navigation *In Problems 19–22, assume the north, east, south, and west directions are exact. Remember that in navigational problems a compass is divided clockwise into 360° starting at north (see Fig. 4, page 353).*

19. A river is flowing east (90°) at 3.0 km/hr. A boat crosses the river with a compass heading of 180° (south). If the speedometer on the boat reads 4.0 km/hr, speed relative to the water, what is the boat's actual speed and direction (resultant velocity) relative to the river bottom?

20. A boat capable of traveling 12 knots on still water maintains a westward compass heading (270°) while crossing a river. If the river is flowing southward (180°) at 4.0 knots, what is the velocity (magnitude and direction) of the boat relative to the river bottom?

* 21. An airplane can cruise at 255 mi/hr in still air. If a steady wind of 46 mi/hr is blowing from the west, what compass heading should the pilot fly in order for the true course of the plane to be north (0°)? Compute the ground speed for this course.

* 22. Two docks are directly opposite each other on a southward flowing river. A boat pilot wishes to go in a straight line from the east dock to the west dock in a ferryboat with a cruising speed of 8.0 knots. If the river's current is 2.5 knots, what compass heading should be maintained while crossing the river? What is the actual speed of the boat relative to land?

23. **Resultant Force** Two tugs are trying to pull a barge off a shoal as indicated in the figure. Find the magnitude of the resulting force and its direction relative to $\mathbf{F}_1$.

$|\mathbf{F}_1| = 1{,}500$ lb

$|\mathbf{F}_2| = 1{,}100$ lb

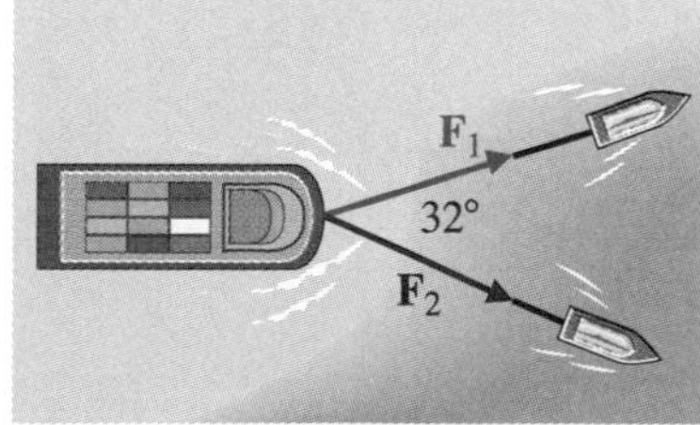

Figure for 23

24. **Resultant Force** Repeat Problem 23 with $|\mathbf{F}_1| = 1{,}300$ lb and the angle between the force vectors 45°.

25. **Centrifugal Force** At what angle must a freeway be banked at a curve to eliminate sideways forces parallel to the road, given that the curve has a radius of 138 m and cars are expected to move at 29 m/sec (about 65 mph)?

26. **Centrifugal Force** Repeat Problem 25 with an expected car speed of 24.6 m/sec (about 55 mph) around a curve with a radius of 105 m.

27. **Resolution of Forces** A car weighing 2,500 lb is parked on a hill inclined 15° to the horizontal (see the figure). Neglecting friction, what magnitude of force parallel to the hill will keep the car from rolling down the hill? What is the force magnitude perpendicular to the hill?

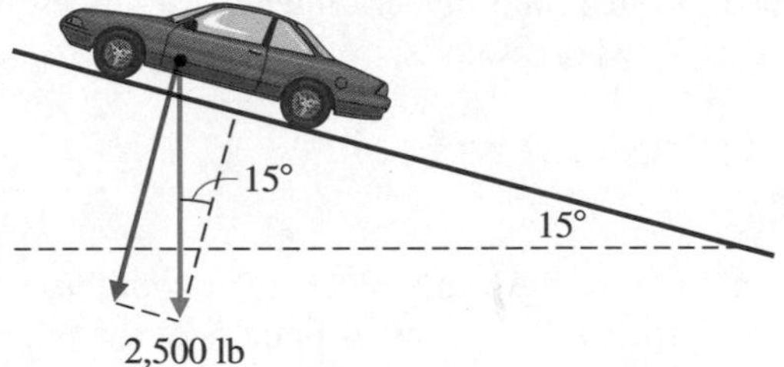

Figure for 27

28. **Resolution of Forces** Repeat Problem 27 with the car weighing 4,200 lb and the hill inclined 12°.

29. **Biomechanics** A person in the hospital has severe leg and hip injuries from a motorcycle accident. His leg is put under traction as indicated in the figure. The upper leg is pulled with a force of 52 lb at the angle indicated, and the lower leg is pulled with a force of 37 lb at the angle indicated. Compute the horizontal and vertical scalar components of the force vector for the upper leg.

30. **Biomechanics** Refer to Problem 29. Compute the horizontal and vertical scalar components of the force vector for the lower leg.

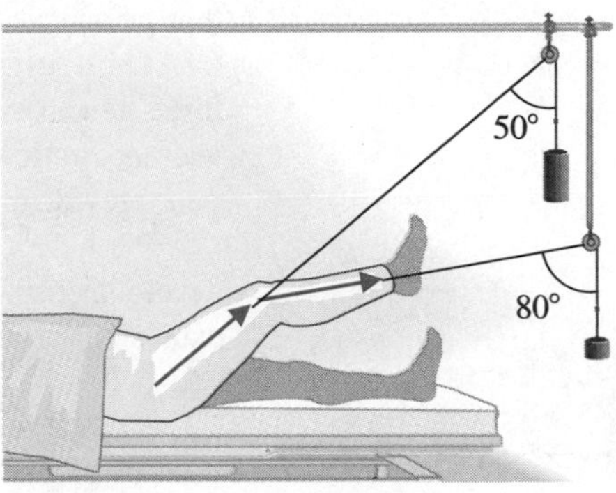

Figure for 29 and 30

***31.** **Resolution of Forces** If two weights are fastened together and placed on inclined planes as indicated in the figure, neglecting friction, which way will they slide? (Weights are accurate to two significant digits, and angles are measured to the nearest degree.)

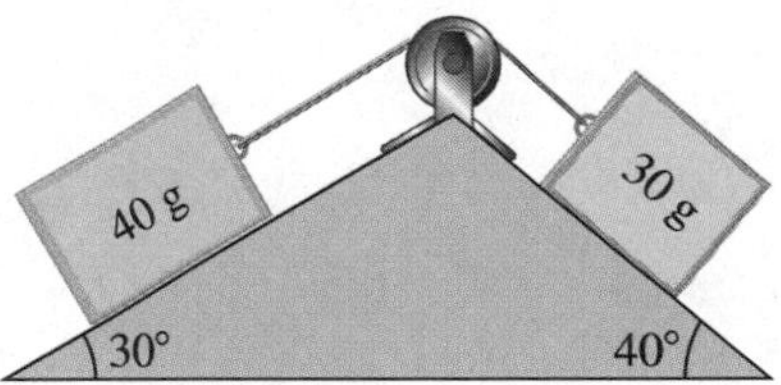

Figure for 31

6.5 VECTORS: ALGEBRAICALLY DEFINED

- From Geometric Vectors to Algebraic Vectors
- Vector Addition and Scalar Multiplication
- Unit Vectors
- Algebraic Properties
- Application—Static Equilibrium

In Section 6.4 we introduced vectors in a geometric setting and considered their velocity and force applications. We discussed geometric vectors in a plane, but they are readily extended to three-dimensional space. However, for the generalization of vector concepts to "higher-dimensional" abstract spaces, it is essential to define the vector concept algebraically. This process is done in such a way that geometric vectors become special cases of the more general algebraic vector. It also turns out that for the solution of certain types of problems in two- and three-dimensional space, algebraic vectors have an advantage over geometric vectors. Static equilibrium problems, which are discussed at the end of this section, provide an example of this type of problem.

We will restrict our development of algebraic vectors to the plane, but the development is readily extended to three- and higher-dimensional spaces. This extension forms the subject matter for whole courses.

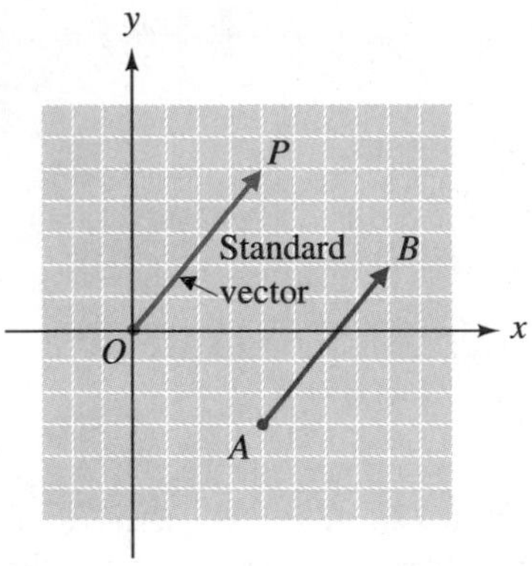

FIGURE 1
$\vec{OP}$ is the standard vector for $\vec{AB}(\vec{OP} = \vec{AB})$

From Geometric Vectors to Algebraic Vectors

We start with an arbitrary geometric vector $\vec{AB}$ in a rectangular coordinate system. A geometric vector $\vec{AB}$ translated so that its initial point is at the origin is said to be in **standard position**, and the vector $\vec{OP}$ (as shown in Fig. 1) such that $\vec{OP} = \vec{AB}$ is said to be the **standard vector** for $\vec{AB}$. The standard vector is also called the **position vector** or the **radius vector**. Infinitely many geometric vectors have $\vec{OP}$ in Figure 1 as a standard vector—all vectors with the same magnitude and direction as $\vec{OP}$.

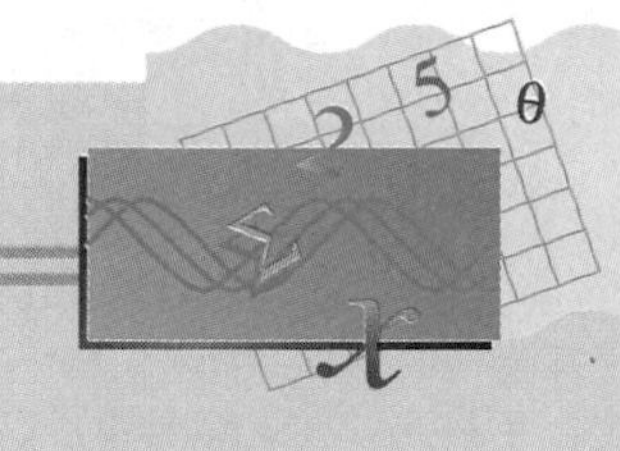

EXPLORE/DISCUSS 1

(A) In Figure 1 (or a copy) draw two other vectors having $\vec{OP}$ as their standard vector.

(B) If the tail of a vector is at point $A(1, -2)$ and the tip is at $B(3, 5)$, explain how you would find the coordinates of P such that $\vec{OP}$ is the standard vector for $\vec{AB}$.

How do we find the standard vector $\vec{OP}$ for a geometric vector $\vec{AB}$? The process is not difficult. We need only find the coordinates of P, since the coordinates of O, the origin, are known. If the coordinates of A are (x_a, y_a) and the coordinates of B are (x_b, y_b), then the coordinates of $P(x, y)$ are given by

$$x = x_b - x_a \qquad y = y_b - y_a$$

The process is interpreted geometrically in Example 1.

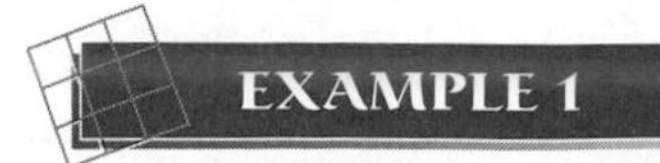

Finding a Standard Vector for a Given Geometric Vector

Given the geometric vector $\vec{AB}$, with initial point $A(8, -3)$ and terminal point $B(4, 5)$, find the standard vector $\vec{OP}$ for $\vec{AB}$; that is, find the coordinates of the point P so that $\vec{OP} = \vec{AB}$.

Solution The coordinates (x, y) of P are given by

$$x = x_b - x_a = 4 - 8 = -4$$

$$y = y_b - y_a = 5 - (-3) = 8$$

Figure 2 illustrates these results geometrically.

FIGURE 2

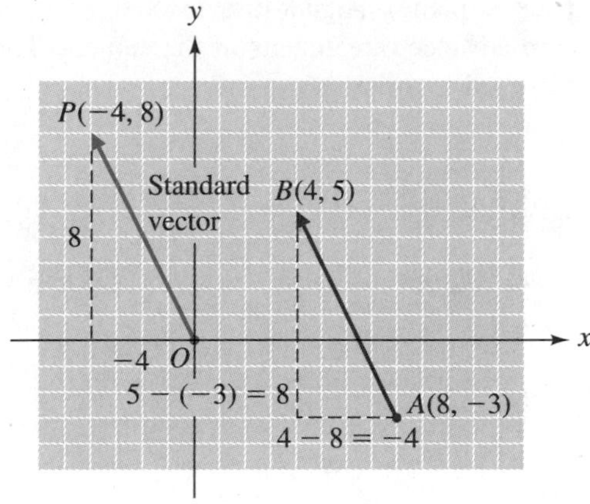

■

Matched Problem 1 Given the geometric vector $\overrightarrow{AB}$, with initial point $A(4, 2)$ and terminal point $B(2, -3)$, find the standard vector $\overrightarrow{OP}$ for $\overrightarrow{AB}$; that is, find the coordinates of the point P so that $\overrightarrow{OP} = \overrightarrow{AB}$. ■

Let's extend our discussion to get another way of looking at vectors. Since, given any geometric vector $\overrightarrow{AB}$, there always exists a point $P(x, y)$ such that $\overrightarrow{OP} = \overrightarrow{AB}$, the point $P(x, y)$ completely determines the vector $\overrightarrow{AB}$ (except for its position, which we are not concerned about because we are free to translate $\overrightarrow{AB}$ anywhere we please). Conversely, given any point $P(x, y)$ in the plane, the directed line segment joining O to P forms the geometric vector $\overrightarrow{OP}$. Thus:

Every geometric vector in a plane corresponds to an ordered pair of real numbers, and every ordered pair of real numbers corresponds to a geometric vector.

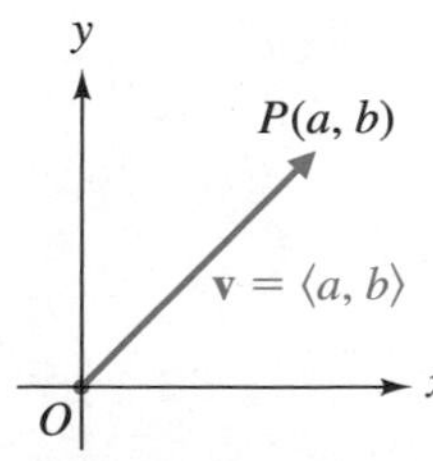

FIGURE 3
Algebraic vector $\langle a, b\rangle$; geometric vector $\overrightarrow{OP}$

This leads us to a definition of an **algebraic vector** as an ordered pair of real numbers. To avoid confusing a point (a, b) with a vector (a, b), we use special brackets to write the symbol $\langle a, b\rangle$, which represents an algebraic vector. The algebraic vector $\langle a, b\rangle$ corresponds to the standard (geometric) vector $\overrightarrow{OP}$ with terminal point $P(a, b)$ and initial point $O(0, 0)$, as indicated in Figure 3.

The real numbers a and b are **scalar components** of the vector $\langle a, b\rangle$. Two vectors $\mathbf{u} = \langle a, b\rangle$ and $\mathbf{v} = \langle c, d\rangle$ are said to be **equal** if their components are equal—that is, if $a = c$ and $b = d$. The **zero vector** is denoted by $\mathbf{0} = \langle 0, 0\rangle$ and has arbitrary direction.

Geometric vectors are limited to two- and three-dimensional spaces we can visualize. Algebraic vectors do not have such restrictions. The following are algebraic vectors from two-, three-, four-, and five-dimensional spaces:

$$\langle 4, -2\rangle \qquad \langle 8, 1, 3\rangle \qquad \langle 12, 2, -6, 1\rangle \qquad \langle 5, 11, 3, -8, 4\rangle$$

Our development in this book is limited to algebraic vectors in two-dimensional space (a plane). Higher-dimensional vector spaces form the subject matter for more advanced treatments of the subject. The *magnitude* of an algebraic vector is defined as follows:

MAGNITUDE* OF A VECTOR $\mathbf{v} = \langle a, b\rangle$

The **magnitude** (or **norm**) of a vector $\mathbf{v} = \langle a, b\rangle$, denoted by $|\mathbf{v}|$, is given by

$$|\mathbf{v}| = \sqrt{a^2 + b^2}$$

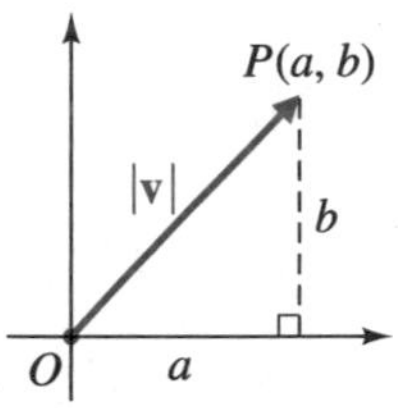

FIGURE 4
$|\mathbf{v}| = \sqrt{a^2 + b^2}$

Geometrically, $\sqrt{a^2 + b^2}$ is the length of the standard geometric vector $\vec{OP}$ associated with the algebraic vector $\langle a, b\rangle$ (see Fig. 4).

EXAMPLE 2 **The Magnitude of a Vector**

Find the magnitude of the vector $\mathbf{v} = \langle -2, 4\rangle$.

Solution $|\mathbf{v}| = \sqrt{(-2)^2 + 4^2} = \sqrt{20}$ or $2\sqrt{5}$ ■

Matched Problem 2 Find the magnitude of the vector $\mathbf{v} = \langle 4, -5\rangle$. ■

■ Vector Addition and Scalar Multiplication

Adding two algebraic vectors is very easy: We simply add the corresponding components of the two vectors.

VECTOR ADDITION

If $\mathbf{u} = \langle a, b\rangle$ and $\mathbf{v} = \langle c, d\rangle$, then

$$\mathbf{u} + \mathbf{v} = \langle a + c, b + d\rangle$$

* The definition of magnitude is readily generalized to higher-dimensional vector spaces. For example, if $\mathbf{v} = \langle a, b, c, d, e\rangle$, then the norm of $\mathbf{v}$ is given by $|\mathbf{v}| = \sqrt{a^2 + b^2 + c^2 + d^2 + e^2}$. But now we are no longer able to interpret the result in terms of geometric vectors.

Since the algebraic vectors in the definition can be associated with geometric vectors in a plane, we can interpret the definition geometrically. The definition is consistent with the parallelogram and tail-to-tip rules for adding geometric vectors given in Section 6.4, as seen in Explore–Discuss 2.

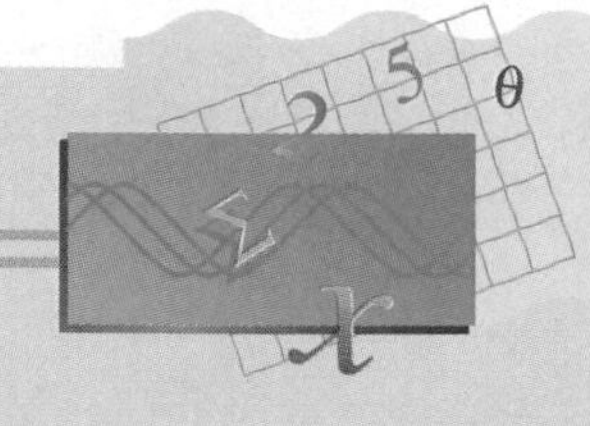

EXPLORE/DISCUSS 2

If $\mathbf{u} = \langle -1, 4 \rangle$ and $\mathbf{v} = \langle 6, -2 \rangle$, then $\mathbf{u} + \mathbf{v} = \langle (-1) + 6, 4 + (-2) \rangle = \langle 5, 2 \rangle$. Locate $\mathbf{u}$, $\mathbf{v}$, and $\mathbf{u} + \mathbf{v}$ in a rectangular coordinate system, and interpret geometrically in terms of the parallelogram and tail-to-tip rules discussed in the last section.

To multiply a vector by a scalar (a real number), multiply each component by the scalar.

SCALAR MULTIPLICATION

If $\mathbf{u} = \langle a, b \rangle$ and k is a scalar, then

$$k\mathbf{u} = k\langle a, b \rangle = \langle ka, kb \rangle$$

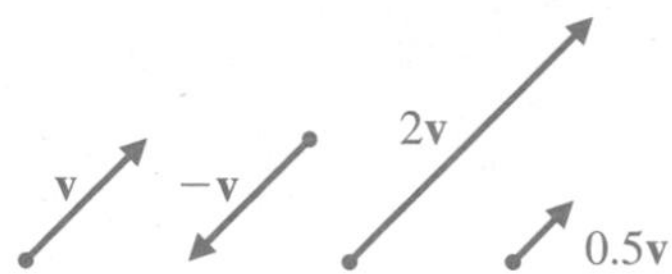

FIGURE 5
Scalar multiplication

Geometrically, if a vector $\mathbf{v}$ is multiplied by a scalar k, the magnitude of the vector $\mathbf{v}$ is multiplied by $|k|$. If k is positive, then $k\mathbf{v}$ has the same direction as $\mathbf{v}$; if k is negative, then $k\mathbf{v}$ has the opposite direction as $\mathbf{v}$. Figure 5 illustrates several cases.

EXAMPLE 3 **Arithmetic of Algebraic Vectors**

Let $\mathbf{u} = \langle -5, 3 \rangle$, $\mathbf{v} = \langle 4, -6 \rangle$, and $\mathbf{w} = \langle -2, 0 \rangle$. Find:

(A) $\mathbf{u} + \mathbf{v}$ (B) $-3\mathbf{u}$ (C) $3\mathbf{u} - 2\mathbf{v}$ (D) $2\mathbf{u} - \mathbf{v} + 3\mathbf{w}$

Solution

(A) $\mathbf{u} + \mathbf{v} = \langle -5, 3 \rangle + \langle 4, -6 \rangle = \langle -1, -3 \rangle$

(B) $-3\mathbf{u} = -3\langle -5, 3 \rangle = \langle 15, -9 \rangle$

(C) $3\mathbf{u} - 2\mathbf{v} = 3\langle -5, 3 \rangle - 2\langle 4, -6 \rangle$
$= \langle -15, 9 \rangle + \langle -8, 12 \rangle = \langle -23, 21 \rangle$

(D) $2\mathbf{u} - \mathbf{v} + 3\mathbf{w} = 2\langle -5, 3 \rangle - \langle 4, -6 \rangle + 3\langle -2, 0 \rangle$
$= \langle -10, 6 \rangle + \langle -4, 6 \rangle + \langle -6, 0 \rangle = \langle -20, 12 \rangle$ ■

Matched Problem 3 Let $\mathbf{u} = \langle 4, -3 \rangle$, $\mathbf{v} = \langle 2, 3 \rangle$, and $\mathbf{w} = \langle 0, -5 \rangle$. Find:

(A) $\mathbf{u} + \mathbf{v}$ (B) $-2\mathbf{u}$ (C) $2\mathbf{u} - 3\mathbf{v}$ (D) $3\mathbf{u} + 2\mathbf{v} - \mathbf{w}$ ■

Unit Vectors

A vector $\mathbf{v}$ is called a **unit vector** if its magnitude is 1; that is, if $|\mathbf{v}| = 1$. Special unit vectors play an important role in certain applications of vectors. A unit vector can be formed from an arbitrary nonzero vector as follows:

FORMING A UNIT VECTOR

If $\mathbf{v}$ is a nonzero vector, then

$$\mathbf{u} = \frac{1}{|\mathbf{v}|}\mathbf{v}$$

is a unit vector with the same direction as $\mathbf{v}$.

EXAMPLE 4 **Unit Vector**

Find a unit vector $\mathbf{u}$ with the same direction as the vector $\mathbf{v} = \langle 3, 1\rangle$.

Solution

$$|\mathbf{v}| = \sqrt{3^2 + 1^2} = \sqrt{10}$$

$$\mathbf{u} = \frac{1}{|\mathbf{v}|}\mathbf{v} = \frac{1}{\sqrt{10}}\langle 3, 1\rangle = \left\langle \frac{3}{\sqrt{10}}, \frac{1}{\sqrt{10}} \right\rangle$$

✔ Check

$$|\mathbf{u}| = \sqrt{\left(\frac{3}{\sqrt{10}}\right)^2 + \left(\frac{1}{\sqrt{10}}\right)^2}$$

$$= \sqrt{\frac{9}{10} + \frac{1}{10}} = \sqrt{1} = 1$$

Thus, $\mathbf{u}$ is a unit vector with the same direction as $\mathbf{v}$. ■

Matched Problem 4 Find a unit vector $\mathbf{u}$ with the same direction as the vector $\mathbf{v} = \langle 1, -2\rangle$. ■

Two very important unit vectors are the $\mathbf{i}$ and $\mathbf{j}$ unit vectors, which we now define.

THE i AND J UNIT VECTORS

$$\mathbf{i} = \langle 1, 0\rangle$$
$$\mathbf{j} = \langle 0, 1\rangle$$

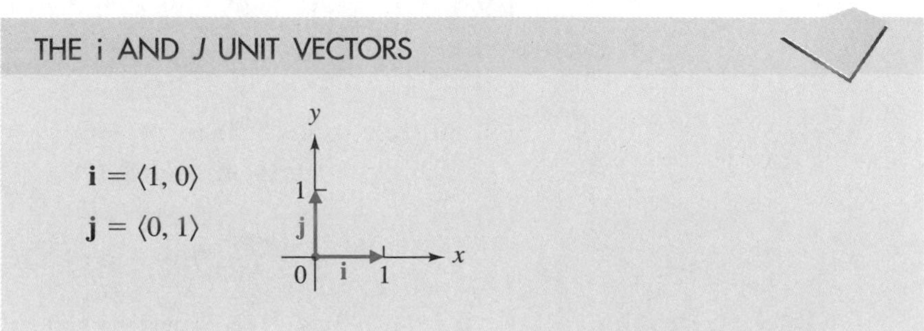

One of the reasons the $\mathbf{i}$ and $\mathbf{j}$ unit vectors are important is that any vector $\mathbf{v} = \langle a, b\rangle$ can be expressed as a linear combination of these two vectors; that is,

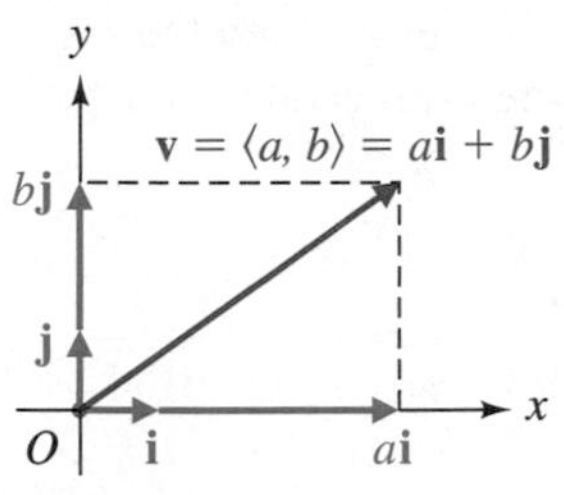

FIGURE 6

as $a\mathbf{i} + b\mathbf{j}$:

$$\begin{aligned}\mathbf{v} = \langle a, b\rangle &= \langle a, 0\rangle + \langle 0, b\rangle \\ &= a\langle 1, 0\rangle + b\langle 0, 1\rangle \\ &= a\mathbf{i} + b\mathbf{j}\end{aligned}$$

Thus,

$$\mathbf{v} = \langle a, b\rangle = a\mathbf{i} + b\mathbf{j}$$

This result is illustrated geometrically in Figure 6.

EXAMPLE 5 **Expressing a Vector in Terms of i and j Unit Vectors**

Express each vector as a linear combination of the **i** and **j** unit vectors.

(A) $\langle 3, -4\rangle$ (B) $\langle 7, 0\rangle$ (C) $\langle 0, -2\rangle$

Solution (A) $\langle 3, -4\rangle = 3\mathbf{i} - 4\mathbf{j}$ (B) $\langle 7, 0\rangle = 7\mathbf{i} + 0\mathbf{j} = 7\mathbf{i}$

(C) $\langle 0, -2\rangle = 0\mathbf{i} - 2\mathbf{j} = -2\mathbf{j}$ ■

Matched Problem 5 Express each vector as a linear combination of the **i** and **j** unit vectors.

(A) $\langle -8, -1\rangle$ (B) $\langle 0, -3\rangle$ (C) $\langle 6, 0\rangle$ ■

Algebraic Properties

Vector addition and scalar multiplication possess algebraic properties similar to the real numbers. These properties enable us to manipulate symbols representing vectors and scalars in much the same way we manipulate symbols that represent real numbers in algebra. These properties are listed below for convenient reference.

ALGEBRAIC PROPERTIES OF VECTORS

(A) The following **addition properties** are satisfied for all vectors **u**, **v**, and **w**:

1. $\mathbf{u} + \mathbf{v} = \mathbf{v} + \mathbf{u}$ — *Commutative property*
2. $\mathbf{u} + (\mathbf{v} + \mathbf{w}) = (\mathbf{u} + \mathbf{v}) + \mathbf{w}$ — *Associative property*
3. $\mathbf{u} + \mathbf{0} = \mathbf{0} + \mathbf{u} = \mathbf{u}$ — *Additive identity*
4. $\mathbf{u} + (-\mathbf{u}) = (-\mathbf{u}) + \mathbf{u} = \mathbf{0}$ — *Additive inverse*

(B) The following **scalar multiplication properties** are satisfied for all vectors **u** and **v** and all scalars m and n:

1. $m(n\mathbf{u}) = (mn)\mathbf{u}$ — *Associative property*
2. $m(\mathbf{u} + \mathbf{v}) = m\mathbf{u} + m\mathbf{v}$ — *Distributive property*
3. $(m + n)\mathbf{u} = m\mathbf{u} + n\mathbf{u}$ — *Distributive property*
4. $1\mathbf{u} = \mathbf{u}$ — *Multiplication identity*

When vectors are represented in terms of the **i** and **j** unit vectors, the algebraic properties listed in the box provide us with efficient procedures for performing algebraic operations with vectors.

EXAMPLE 6 Algebraic Operations Involving Unit Vectors

If $\mathbf{u} = 2\mathbf{i} - \mathbf{j}$ and $\mathbf{v} = 4\mathbf{i} + 5\mathbf{j}$, compute each of the following:

(A) $\mathbf{u} + \mathbf{v}$ (B) $\mathbf{u} - \mathbf{v}$ (C) $3\mathbf{u} - 2\mathbf{v}$

Solution

(A)
$$\begin{aligned}\mathbf{u} + \mathbf{v} &= (2\mathbf{i} - \mathbf{j}) + (4\mathbf{i} + 5\mathbf{j})\\ &= 2\mathbf{i} - \mathbf{j} + 4\mathbf{i} + 5\mathbf{j}\\ &= 2\mathbf{i} + 4\mathbf{i} - \mathbf{j} + 5\mathbf{j}\\ &= (2 + 4)\mathbf{i} + (-1 + 5)\mathbf{j}\\ &= 6\mathbf{i} + 4\mathbf{j}\end{aligned}$$

(B)
$$\begin{aligned}\mathbf{u} - \mathbf{v} &= (2\mathbf{i} - \mathbf{j}) - (4\mathbf{i} + 5\mathbf{j})\\ &= 2\mathbf{i} - \mathbf{j} - 4\mathbf{i} - 5\mathbf{j}\\ &= 2\mathbf{i} - 4\mathbf{i} - \mathbf{j} - 5\mathbf{j}\\ &= (2 - 4)\mathbf{i} + (-1 - 5)\mathbf{j}\\ &= -2\mathbf{i} - 6\mathbf{j}\end{aligned}$$

(C)
$$\begin{aligned}3\mathbf{u} - 2\mathbf{v} &= 3(2\mathbf{i} - \mathbf{j}) - 2(4\mathbf{i} + 5\mathbf{j})\\ &= 6\mathbf{i} - 3\mathbf{j} - 8\mathbf{i} - 10\mathbf{j}\\ &= 6\mathbf{i} - 8\mathbf{i} - 3\mathbf{j} - 10\mathbf{j}\\ &= (6 - 8)\mathbf{i} + (-3 - 10)\mathbf{j}\\ &= -2\mathbf{i} - 13\mathbf{j}\end{aligned}$$

Matched Problem 6 If $\mathbf{u} = \mathbf{i} - 2\mathbf{j}$ and $\mathbf{v} = 5\mathbf{i} + 2\mathbf{j}$, compute each of the following:

(A) $\mathbf{u} + \mathbf{v}$ (B) $\mathbf{u} - \mathbf{v}$ (C) $2\mathbf{u} + 3\mathbf{v}$

Application – Static Equilibrium

We now will show how algebraic vectors can be used to solve certain physics and engineering problems. Basic to the process of solving such problems is the following principle from physics: An object at rest is in **static equilibrium.** Even though the earth is moving in space, it is convenient to consider an object at rest if it is not moving relative to the earth. Two conditions (from *Newton's second law of motion*) are required for an object to be in static equilibrium: The sum of all forces acting on the object must be zero, and the sum of all *torques* (rotational forces) acting on the object must be zero. A discussion of torque will be left to a more advanced treatment of the subject, and we will limit our attention to problems involving forces in a plane, where the lines of all forces on an object intersect in a single point. Such forces are called **coplanar concurrent forces.**

CONDITIONS FOR STATIC EQUILIBRIUM

An object subject only to coplanar concurrent forces is in static equilibrium if and only if the sum of all these forces is zero.

Examples 7 and 8 illustrate how important physics and engineering problems can be solved using vector methods and the conditions just stated.

EXAMPLE 7

Cable Tension

Two skiers on a stalled chair lift deflect the cable relative to the horizontal as indicated in Figure 7. If the skiers and chair weigh 325 lb (neglect cable weight), what is the tension in each cable running to each tower?

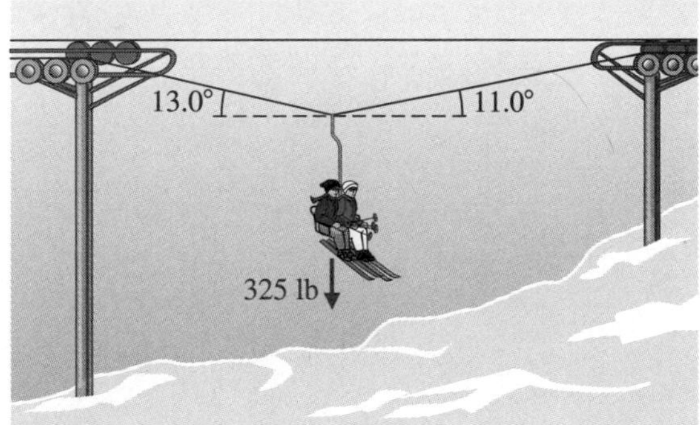

FIGURE 7

Solution ***Step 1*** Form a force diagram with all force vectors in standard position at the origin (Fig. 8). Our objective is to find $|\mathbf{u}|$ and $|\mathbf{v}|$.

Step 2 Write each force vector in terms of **i** and **j** unit vectors:

$$\mathbf{u} = |\mathbf{u}|(\cos 11.0°)\mathbf{i} + |\mathbf{u}|(\sin 11.0°)\mathbf{j}$$

$$\mathbf{v} = |\mathbf{v}|(-\cos 13.0°)\mathbf{i} + |\mathbf{v}|(\sin 13.0°)\mathbf{j}$$

$$\mathbf{w} = -325\mathbf{j}$$

y, x, O, $\mathbf{v}$ 13.0°, 11.0° $\mathbf{u}$, $|\mathbf{w}| = 325$ lb $\mathbf{w}$

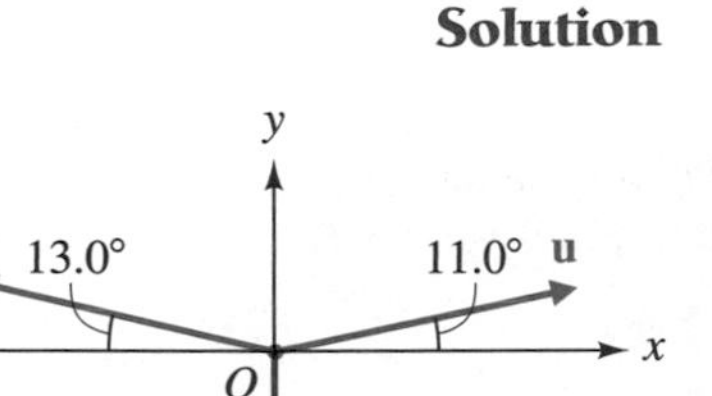

FIGURE 8
Force diagram

Step 3 Set the sum of all force vectors equal to the zero vector, and solve for any unknown quantities. For the system to be in static equilibrium, the sum of the force vectors must be zero. That is,

$$\mathbf{u} + \mathbf{v} + \mathbf{w} = \mathbf{0}$$

Writing the vectors as in step 2, we have

$$[|\mathbf{u}|(\cos 11.0°) + |\mathbf{v}|(-\cos 13.0°)]\mathbf{i} + [|\mathbf{u}|(\sin 11.0°) + |\mathbf{v}|(\sin 13.0°) - 325]\mathbf{j} = 0\mathbf{i} + 0\mathbf{j}$$

Now, since two vectors are equal if and only if their corresponding components are equal (that is, $a\mathbf{i} + b\mathbf{j} = 0\mathbf{i} + 0\mathbf{j}$ if and only if $a = 0$ and $b = 0$), we are led to the following system of two equations in the two variables $|\mathbf{u}|$ and $|\mathbf{v}|$:

$$(\cos 11.0°)|\mathbf{u}| + (-\cos 13.0°)|\mathbf{v}| = 0$$
$$(\sin 11.0°)|\mathbf{u}| + (\sin 13.0°)|\mathbf{v}| - 325 = 0$$

Solving this system by standard methods, we find that

$$|\mathbf{u}| = 779 \text{ lb} \qquad \text{and} \qquad |\mathbf{v}| = 784 \text{ lb}$$

You probably did not guess that the tension in each part of the cable is more than the weight hanging from the cable! ■

Matched Problem 7 Repeat Example 7 with 13.0° replaced by 14.5°, 11.0° replaced by 9.5°, and 325 lb replaced by 435 lb. ■

EXAMPLE 8

Static Equilibrium

A 1,250 lb weight is hanging from a hoist as indicated in Figure 9. What is the tension in the cable, and what is the compression (force) on the bar?

FIGURE 9

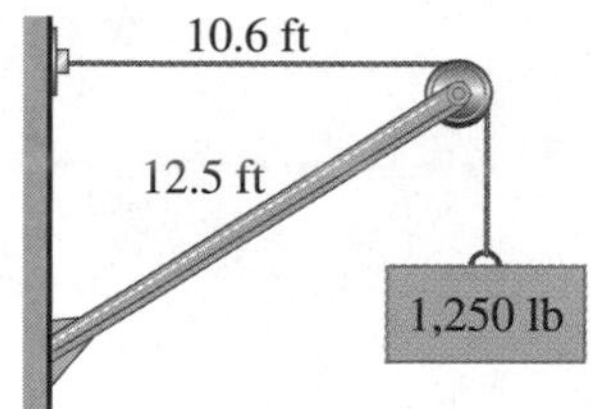

Solution ***Step 1*** Form a force diagram with all force vectors in standard position at the origin (Fig. 10). Then

$$\theta = \cos^{-1}\frac{10.6}{12.5} = 32.0°$$

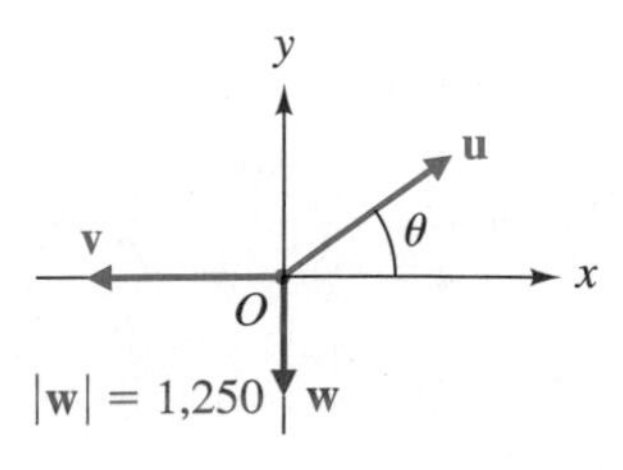

FIGURE 10

Our objective is to find the compression $|\mathbf{u}|$ and the tension $|\mathbf{v}|$.

Step 2 Write each force vector in terms of $\mathbf{i}$ and $\mathbf{j}$ unit vectors:

$$\mathbf{u} = |\mathbf{u}|(\cos 32.0°)\mathbf{i} + |\mathbf{u}|(\sin 32.0°)\mathbf{j}$$
$$\mathbf{v} = -|\mathbf{v}|\mathbf{i}$$
$$\mathbf{w} = -1{,}250\mathbf{j}$$

Step 3 Set the sum of all force vectors equal to the zero vector and solve for any unknown quantities:

$$\mathbf{u} + \mathbf{v} + \mathbf{w} = \mathbf{0}$$

Writing the vectors as in step 2, we have

$$[|\mathbf{u}|(\cos 32.0°) - |\mathbf{v}|]\mathbf{i} + [|\mathbf{u}|(\sin 32.0°) - 1{,}250]\mathbf{j} = 0\mathbf{i} + 0\mathbf{j}$$

We are thus led to the following system of two equations in the variables $|\mathbf{u}|$ and $|\mathbf{v}|$:

$$(\cos 32.0°)|\mathbf{u}| - |\mathbf{v}| = 0$$

$$(\sin 32.0°)|\mathbf{u}| - 1{,}250 = 0$$

Solving this system by standard methods, we find that

$$|\mathbf{u}| = 2{,}360 \text{ lb} \quad \text{and} \quad |\mathbf{v}| = 2{,}000 \text{ lb}$$

■

Matched Problem 8 A 450 lb sign is suspended as shown in Figure 11. Find the compression force on the 2.0 ft rod and the tension on the 4.0 ft rod.

FIGURE 11

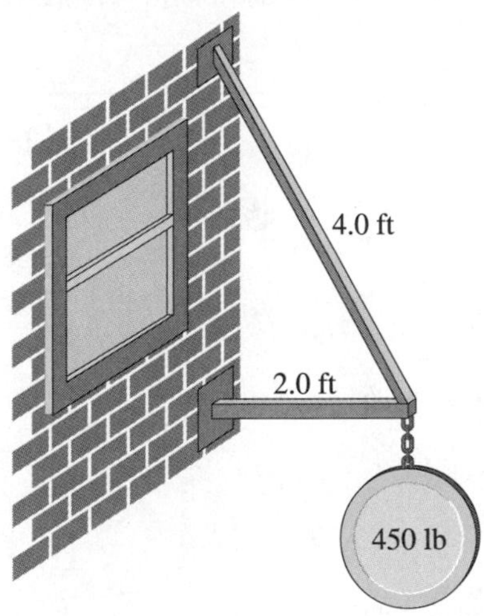

■

Answers to Matched Problems

1. $(-2, -5)$ **2.** $\sqrt{41}$

3. (A) $\langle 6, 0\rangle$ (B) $\langle -8, 6\rangle$ (C) $\langle 2, -15\rangle$ (D) $\langle 16, 2\rangle$

4. $\langle 1/\sqrt{5}, -2/\sqrt{5}\rangle$

5. (A) $-8\mathbf{i} - \mathbf{j}$ (B) $0\mathbf{i} - 3\mathbf{j} = -3\mathbf{j}$ (C) $6\mathbf{i} + 0\mathbf{j} = 6\mathbf{i}$

6. (A) $6\mathbf{i}$ (B) $-4\mathbf{i} - 4\mathbf{j}$ (C) $17\mathbf{i} + 2\mathbf{j}$

7. $|\mathbf{u}| = 1{,}040$ lb; $|\mathbf{v}| = 1{,}050$ lb

8. $|\mathbf{u}| = 260$ lb (compression); $|\mathbf{v}| = 520$ lb (tension)

EXERCISE 6.5

A *In Problems 1–4, draw the vector $\overrightarrow{AB}$ in a rectangular coordinate system. Draw the standard vector $\overrightarrow{OP}$ so that $\overrightarrow{OP} = \overrightarrow{AB}$. Indicate the coordinates of P.*

1. $A(2, -3)$; $B(5, 1)$

2. $A(1, -2)$; $B(4, 1)$

3. $A(-1, 3)$; $B(-3, -1)$

4. $A(-1, -2)$; $B(-3, 2)$

In Problems 5–8, represent each geometric vector $\overrightarrow{AB}$ as an algebraic vector $\langle a, b\rangle$.

5. $A(-1, -2)$; $B(3, 0)$ **6.** $A(-1, 2)$; $B(2, 3)$

7. $A(0, 2)$; $B(4, -2)$ **8.** $A(-4, 0)$; $B(-2, -4)$

In Problems 9–12, find the magnitude of each vector $\langle a, b\rangle$.

9. $\langle -3, 4\rangle$ **10.** $\langle 4, -3\rangle$

11. $\langle -5, -2\rangle$ **12.** $\langle 3, -5\rangle$

13. When are two geometric vectors equal?

14. When are two algebraic vectors equal?

B *In Problems 15–18, find:*

(A) $\mathbf{u} + \mathbf{v}$ (B) $\mathbf{u} - \mathbf{v}$

(C) $2\mathbf{u} - 3\mathbf{v}$ (D) $3\mathbf{u} - \mathbf{v} + 2\mathbf{w}$

15. $\mathbf{u} = \langle 1, 4\rangle$; $\mathbf{v} = \langle -3, 2\rangle$; $\mathbf{w} = \langle 0, 4\rangle$

16. $\mathbf{u} = \langle -1, 3\rangle$; $\mathbf{v} = \langle 2, -3\rangle$; $\mathbf{w} = \langle -2, 0\rangle$

17. $\mathbf{u} = \langle 2, -3\rangle$; $\mathbf{v} = \langle -1, -3\rangle$; $\mathbf{w} = \langle -2, 0\rangle$

18. $\mathbf{u} = \langle -1, -4\rangle$; $\mathbf{v} = \langle 3, 3\rangle$; $\mathbf{w} = \langle 0, -3\rangle$

In Problems 19–22, form a unit vector **u** *with the same direction as* **v**.

19. $\mathbf{v} = \langle 4, -3\rangle$ **20.** $\mathbf{v} = \langle -3, 4\rangle$

21. $\mathbf{v} = \langle 2, -3\rangle$ **22.** $\mathbf{v} = \langle -5, 3\rangle$

In Problems 23–28, express **v** *in terms of the* **i** *and* **j** *unit vectors.*

23. $\mathbf{v} = \langle 3, -2\rangle$ **24.** $\mathbf{v} = \langle -5, 6\rangle$

25. $\mathbf{v} = \langle 0, 4\rangle$ **26.** $\mathbf{v} = \langle -3, 0\rangle$

27. $\mathbf{v} = \overrightarrow{AB}$; $A(-2, -1)$; $B(0, 2)$

28. $\mathbf{v} = \overrightarrow{AB}$; $A(2, 3)$; $B(-3, 1)$

In Problems 29–34, let $\mathbf{u} = 2\mathbf{i} - 3\mathbf{j}$, $\mathbf{v} = 3\mathbf{i} + 4\mathbf{j}$, *and* $\mathbf{w} = 5\mathbf{j}$, *and perform the indicated operations.*

29. $\mathbf{u} - \mathbf{v}$ **30.** $\mathbf{u} + \mathbf{v}$

31. $3\mathbf{u} - 2\mathbf{v}$ **32.** $2\mathbf{u} + 3\mathbf{v}$

33. $\mathbf{u} - 2\mathbf{v} + 2\mathbf{w}$ **34.** $3\mathbf{u} - \mathbf{v} - 3\mathbf{w}$

35. An object that is free to move has three nonzero coplanar concurrent forces acting on it, and it remains at rest. How is any one of these forces related to the other two?

36. An object that is free to move has two nonzero coplanar concurrent forces acting on it, and it remains at rest. How are the two force vectors related?

C *In Problems 37–44, let* $\mathbf{u} = \langle a, b\rangle$, $\mathbf{v} = \langle c, d\rangle$, *and* $\mathbf{w} = \langle e, f\rangle$ *be vectors, and let m and n be scalars. Prove each of the following vector properties using appropriate properties of real numbers.*

37. $\mathbf{u} + \mathbf{v} = \mathbf{v} + \mathbf{u}$

38. $\mathbf{u} + (\mathbf{v} + \mathbf{w}) = (\mathbf{u} + \mathbf{v}) + \mathbf{w}$

39. $\mathbf{v} + (-\mathbf{v}) = \mathbf{0}$

40. $\mathbf{v} + \mathbf{0} = \mathbf{v}$

41. $m(\mathbf{u} + \mathbf{v}) = m\mathbf{u} + m\mathbf{v}$

42. $(m + n)\mathbf{v} = m\mathbf{v} + n\mathbf{v}$

43. $1\mathbf{v} = \mathbf{v}$

44. $(mn)\mathbf{v} = m(n\mathbf{v})$

Applications

45. Static Equilibrium A tightrope walker weighing 112 lb deflects a rope as indicated in the figure. How much tension is in each part of the rope?

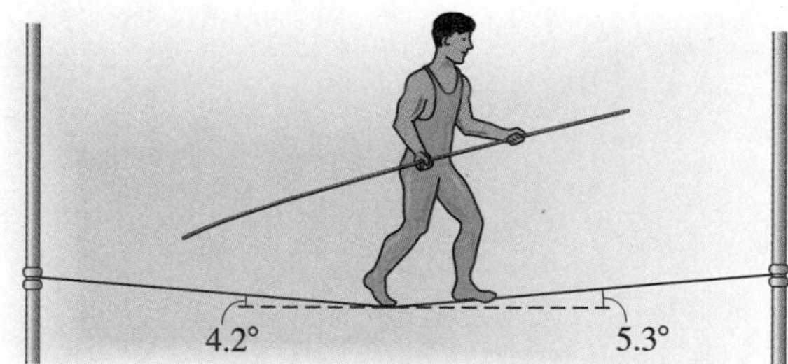

Figure for 45

46. Static Equilibrium Repeat Problem 45, but change the left angle to 5.5°, the right angle to 6.2°, and the weight of the person to 155 lb.

47. Static Equilibrium A weight of 5,000 lb is supported as indicated in the figure. What are the magnitudes of the forces on the members *AB* and *BC* (to two significant digits)?

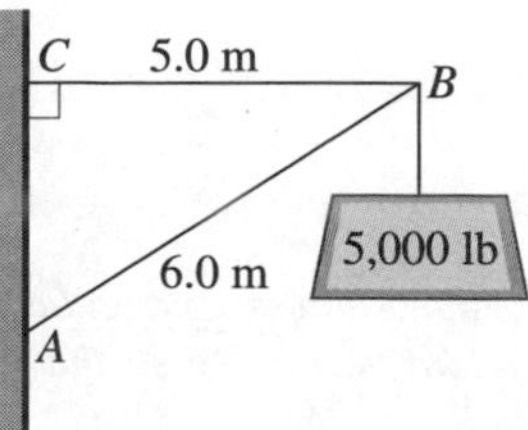

Figure for 47

48. **Static Equilibrium** A weight of 1,000 lb is supported as indicated in the figure. What are the magnitudes of the forces on the members AB and BC (to two significant digits)?

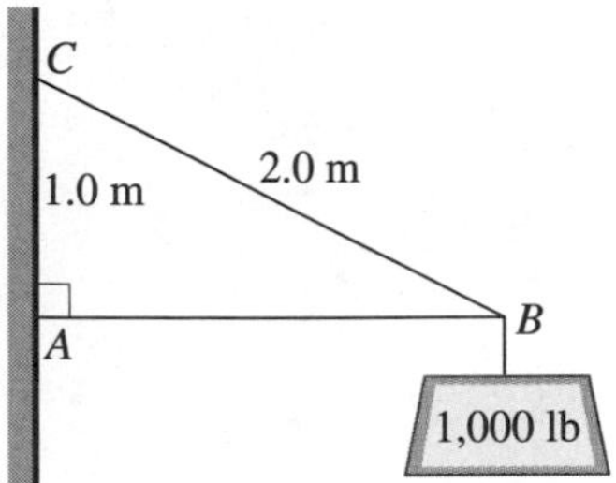

Figure for 48

49. **Biomechanics** A woman slipped on ice on the sidewalk and fractured her arm. The severity of the accident required her arm to be put in traction as indicated in the figure. (The weight of the arm exerts a downward force of 6 lb.) Find angle θ and the tension on the line fastened to the overhead bar. Compute each to one decimal place.

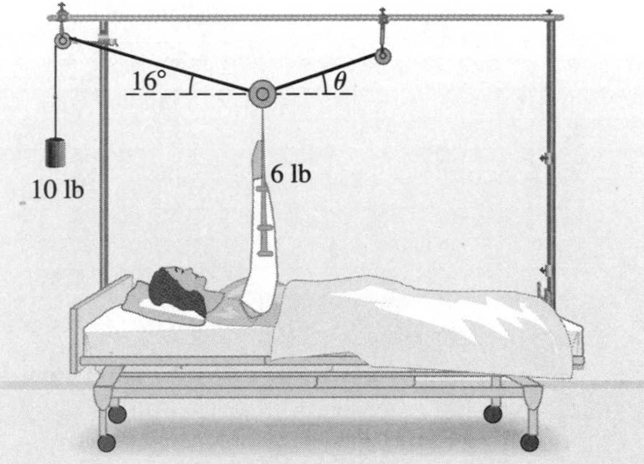

Figure for 49

☆6.6 THE DOT PRODUCT

- The Dot Product of Two Vectors
- Angle Between Two Vectors
- Scalar Component of One Vector on Another
- Work

In Section 6.5 we saw that the sum of two vectors is a vector and the scalar multiple of a vector is a vector. In this section we introduce another operation on vectors, called the *dot product.* This product of two vectors, however, is a scalar (a real number), not a vector. The dot product is also called the *scalar product* or the *inner product.* Dot products are used to find angles between vectors, to define concepts and solve problems in physics and engineering, and in the solution of geometric problems.

The Dot Product of Two Vectors

We define the dot product in the box at the top of the next page.

☆ Sections marked with a star may be omitted without loss of continuity.

THE DOT PRODUCT

The **dot product** of the two vectors

$$\mathbf{u} = \langle a, b\rangle = a\mathbf{i} + b\mathbf{j} \quad \text{and} \quad \mathbf{v} = \langle c, d\rangle = c\mathbf{i} + d\mathbf{j}$$

denoted by $\mathbf{u} \cdot \mathbf{v}$, is the scalar given by

$$\mathbf{u} \cdot \mathbf{v} = ac + bd$$

EXAMPLE 1 **Finding Dot Products**

Find each dot product.

(A) $\langle 4, 2\rangle \cdot \langle 1, -3\rangle$ (B) $(6\mathbf{i} + 2\mathbf{j}) \cdot (\mathbf{i} - 4\mathbf{j})$
(C) $\langle 5, 2\rangle \cdot \langle 2, -5\rangle$ (D) $\mathbf{j} \cdot \mathbf{i}$

Solution
(A) $\langle 4, 2\rangle \cdot \langle 1, -3\rangle = 4 \cdot 1 + 2 \cdot (-3) = -2$
(B) $(6\mathbf{i} + 2\mathbf{j}) \cdot (\mathbf{i} - 4\mathbf{j}) = 6 \cdot 1 + 2 \cdot (-4) = -2$
(C) $\langle 5, 2\rangle \cdot \langle 2, -5\rangle = 5 \cdot 2 + 2 \cdot (-5) = 0$
(D) $\mathbf{j} \cdot \mathbf{i} = (0\mathbf{i} + 1\mathbf{j}) \cdot (1\mathbf{i} + 0\mathbf{j}) = 0 \cdot 1 + 1 \cdot 0 = 0$ ■

Matched Problem 1 Find each dot product.

(A) $\langle 3, -2\rangle \cdot \langle 4, 1\rangle$ (B) $(5\mathbf{i} - 2\mathbf{j}) \cdot (3\mathbf{i} + 4\mathbf{j})$
(C) $\langle 4, -3\rangle \cdot \langle 3, 4\rangle$ (D) $\mathbf{i} \cdot \mathbf{j}$ ■

Some important properties of the dot product are listed below. All these properties follow directly from the definitions of the vector operations involved and the properties of real numbers.

PROPERTIES OF THE DOT PRODUCT

For all vectors **u**, **v**, and **w**, and k a real number:

1. $\mathbf{u} \cdot \mathbf{u} = |\mathbf{u}|^2$
2. $\mathbf{u} \cdot \mathbf{v} = \mathbf{v} \cdot \mathbf{u}$
3. $\mathbf{u} \cdot (\mathbf{v} + \mathbf{w}) = \mathbf{u} \cdot \mathbf{v} + \mathbf{u} \cdot \mathbf{w}$
4. $(\mathbf{u} + \mathbf{v}) \cdot \mathbf{w} = \mathbf{u} \cdot \mathbf{w} + \mathbf{v} \cdot \mathbf{w}$
5. $k(\mathbf{u} \cdot \mathbf{v}) = (k\mathbf{u}) \cdot \mathbf{v} = \mathbf{u} \cdot (k\mathbf{v})$
6. $\mathbf{u} \cdot \mathbf{0} = 0$

The proofs of the dot product properties are left to Problems 27–32 in Exercise 6.6.

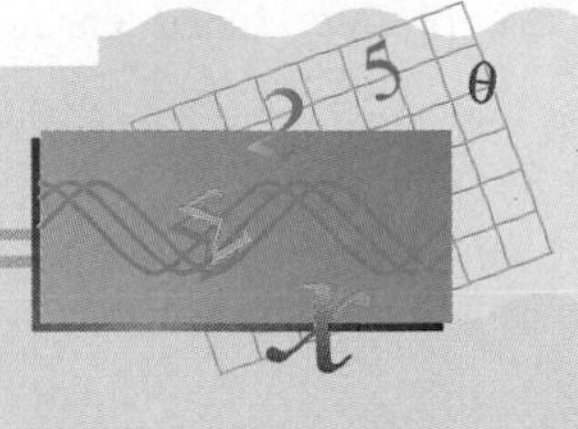

EXPLORE/DISCUSS 1

Verify each of the dot product properties for $\mathbf{u} = \langle -1, 2\rangle$, $\mathbf{v} = \langle 3, 1\rangle$, $\mathbf{w} = \langle 1, -1\rangle$, and $k = 2$.

Angle Between Two Vectors

An important application of the dot product is related to the angle between two vectors. If **u** and **v** are two nonzero vectors, we can position **u** and **v** so that their initial points coincide. The **angle between the vectors u and v** is defined to be the angle θ, $0 \le \theta \le \pi$, formed by the vectors (see Fig. 1).

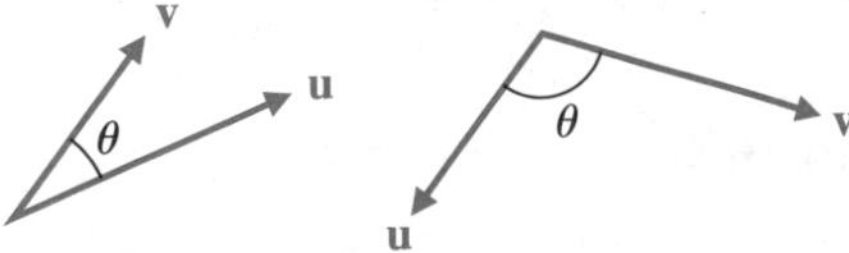

FIGURE 1
Angle between two vectors, $0 \le \theta \le \pi$

If $\theta = \pi/2$, the vectors **u** and **v** are said to be **orthogonal,** or **perpendicular.** If $\theta = 0$ or $\theta = \pi$, the vectors are said to be **parallel.**

We now show how the dot product can be used to find the angle between two vectors.

ANGLE BETWEEN TWO VECTORS

If **u** and **v** are two nonzero vectors and θ is the angle between **u** and **v**, then

$$\cos\theta = \frac{\mathbf{u} \cdot \mathbf{v}}{|\mathbf{u}||\mathbf{v}|}$$

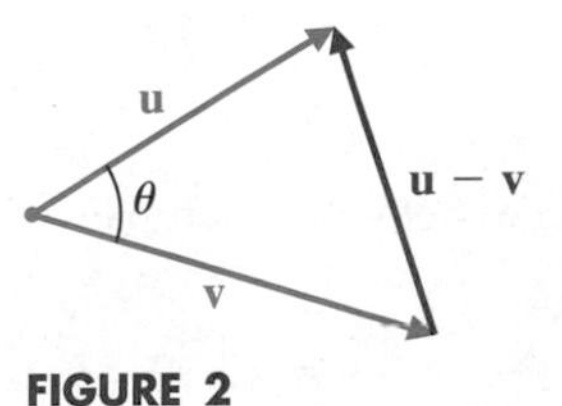

FIGURE 2

We sketch a proof of this result for nonparallel vectors. Given two nonzero, nonparallel vectors **u** and **v**, we can position **u**, **v**, and $\mathbf{u} - \mathbf{v}$ so that they form a triangle where θ is the angle between **u** and **v** (see Fig. 2). Applying the law of cosines, we obtain

$$|\mathbf{u} - \mathbf{v}|^2 = |\mathbf{u}|^2 + |\mathbf{v}|^2 - 2|\mathbf{u}||\mathbf{v}|\cos\theta \qquad (1)$$

Next, we apply some of the properties of the dot product listed earlier to $|\mathbf{u} - \mathbf{v}|^2$.

$$\begin{aligned} |\mathbf{u} - \mathbf{v}|^2 &= (\mathbf{u} - \mathbf{v}) \cdot (\mathbf{u} - \mathbf{v}) && \text{Property 1} \\ &= (\mathbf{u} - \mathbf{v}) \cdot \mathbf{u} - (\mathbf{u} - \mathbf{v}) \cdot \mathbf{v} && \text{Property 3} \\ &= \mathbf{u} \cdot (\mathbf{u} - \mathbf{v}) - \mathbf{v} \cdot (\mathbf{u} - \mathbf{v}) && \text{Property 2} \\ &= \mathbf{u} \cdot \mathbf{u} - \mathbf{u} \cdot \mathbf{v} - \mathbf{v} \cdot \mathbf{u} + \mathbf{v} \cdot \mathbf{v} && \text{Property 3} \\ &= |\mathbf{u}|^2 - 2(\mathbf{u} \cdot \mathbf{v}) + |\mathbf{v}|^2 && \text{Properties 1 and 2} \end{aligned}$$

Now substitute this last expression in the left side of equation (1):

$$|\mathbf{u}|^2 - 2(\mathbf{u} \cdot \mathbf{v}) + |\mathbf{v}|^2 = |\mathbf{u}|^2 + |\mathbf{v}|^2 - 2|\mathbf{u}||\mathbf{v}| \cos \theta$$

This equation simplifies to

$$2|\mathbf{u}||\mathbf{v}| \cos \theta = 2(\mathbf{u} \cdot \mathbf{v})$$

Since $\mathbf{u}$ and $\mathbf{v}$ are nonzero, we can solve for $\cos \theta$:

$$\cos \theta = \frac{\mathbf{u} \cdot \mathbf{v}}{|\mathbf{u}||\mathbf{v}|}$$

EXAMPLE 2 Angle Between Two Vectors

Find the angle (in decimal degrees, to one decimal place) between each pair of vectors.

(A) $\mathbf{u} = \langle 2, 3 \rangle$; $\mathbf{v} = \langle 4, 1 \rangle$ (B) $\mathbf{u} = \langle 3, 1 \rangle$; $\mathbf{v} = \langle -3, -3 \rangle$

(C) $\mathbf{u} = -\mathbf{i} + 3\mathbf{j}$; $\mathbf{v} = 4\mathbf{i} - \mathbf{j}$

Solution (A) $\cos \theta = \dfrac{\mathbf{u} \cdot \mathbf{v}}{|\mathbf{u}||\mathbf{v}|} = \dfrac{11}{\sqrt{13}\sqrt{17}}$

$$\theta = \cos^{-1} \frac{11}{\sqrt{13}\sqrt{17}} = 42.3°$$

(B) $\cos \theta = \dfrac{\mathbf{u} \cdot \mathbf{v}}{|\mathbf{u}||\mathbf{v}|} = \dfrac{-12}{\sqrt{10}\sqrt{18}}$

$$\theta = \cos^{-1} \frac{-12}{\sqrt{10}\sqrt{18}} = 153.4°$$

(C) $\cos \theta = \dfrac{\mathbf{u} \cdot \mathbf{v}}{|\mathbf{u}||\mathbf{v}|} = \dfrac{-7}{\sqrt{10}\sqrt{17}}$

$$\theta = \cos^{-1} \frac{-7}{\sqrt{10}\sqrt{17}} = 122.5°$$

■

Matched Problem 2 Find the angle (in decimal degrees, to one decimal place) between each pair of vectors.

(A) $\mathbf{u} = \langle 4, 2 \rangle$; $\mathbf{v} = \langle 3, -2 \rangle$ (B) $\mathbf{u} = \langle -3, -1 \rangle$; $\mathbf{v} = \langle 2, 4 \rangle$

(C) $\mathbf{u} = -4\mathbf{i} - 2\mathbf{j}$; $\mathbf{v} = -2\mathbf{i} + \mathbf{j}$

■

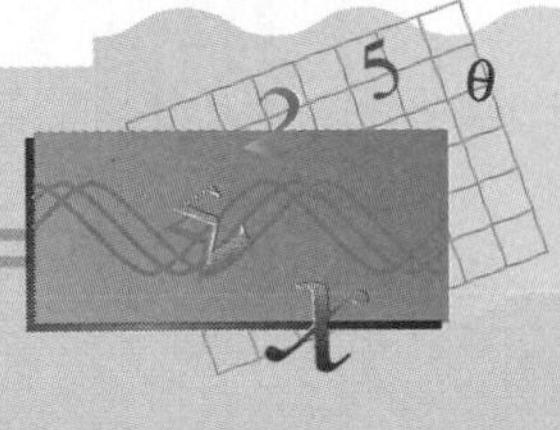

EXPLORE/DISCUSS 2

You will need the formula for the angle between two vectors to answer the following questions.

(A) Given **u** and **v** are orthogonal vectors, what can you say about $\mathbf{u} \cdot \mathbf{v}$?

(B) Given $\mathbf{u} \cdot \mathbf{v} = 0$, what can you say about the angle between **u** and **v**?

Explore–Discuss 2 leads to the following very useful test for orthogonality:

TEST FOR ORTHOGONAL VECTORS

Two vectors **u** and **v** are orthogonal if and only if

$$\mathbf{u} \cdot \mathbf{v} = 0$$

Note: The zero vector is orthogonal to every vector.

EXAMPLE 3 Determining Orthogonality

Determine which of the following pairs of vectors are orthogonal.

(A) $\mathbf{u} = \langle 3, -2 \rangle$; $\mathbf{v} = \langle 4, 6 \rangle$ (B) $\mathbf{u} = 2\mathbf{i} - \mathbf{j}$; $\mathbf{v} = \mathbf{i} - 2\mathbf{j}$

Solution (A) $\mathbf{u} \cdot \mathbf{v} = \langle 3, -2 \rangle \cdot \langle 4, 6 \rangle = 12 + (-12) = 0$
Thus, **u** and **v** are orthogonal.

(B) $\mathbf{u} \cdot \mathbf{v} = (2\mathbf{i} - \mathbf{j}) \cdot (\mathbf{i} - 2\mathbf{j}) = 2 + 2 = 4 \neq 0$
Thus, **u** and **v** are not orthogonal. ■

Matched Problem 3 Determine which of the following pairs of vectors are orthogonal.

(A) $\mathbf{u} = \langle -2, 4 \rangle$; $\mathbf{v} = \langle -2, 1 \rangle$ (B) $\mathbf{u} = 2\mathbf{i} - \mathbf{j}$; $\mathbf{v} = \mathbf{i} + 2\mathbf{j}$ ■

■ Scalar Component of One Vector on Another

An important use of the dot product is in the determination of the *projection* of a vector **u** onto a vector **v**. To project a vector **u** onto an arbitrary nonzero vector **v**, locate **u** and **v** so they have the same initial point O; then drop a perpendicular line from the terminal point of **u** to the line that contains **v**. The **vector projection of u onto v** is the vector **p** that lies on the line containing **v**, as shown in

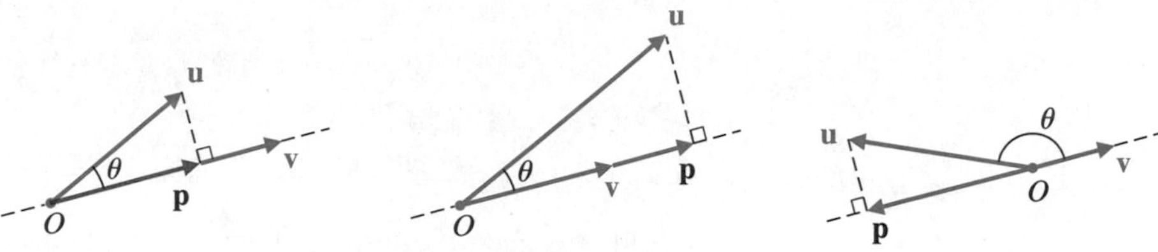

FIGURE 3
Projection of **u** onto **v**

Figure 3. The vector **p** has the same direction as **v** if θ is acute and the opposite direction as **v** if θ is obtuse. The vector **p** is the zero vector if **u** and **v** are orthogonal. Problems 33–36 in Exercise 6.6 ask you to compute some projections.

The **scalar component of u on v**, denoted by $\text{Comp}_{\mathbf{v}}\,\mathbf{u}$, is given by $|\mathbf{u}|\cos\theta$. This quantity is positive if θ is acute, 0 if $\theta = \pi/2$, and negative if θ is obtuse. The scalar component of **u** on **v** can be expressed in terms of the dot product as follows:

$$\text{Comp}_{\mathbf{v}}\,\mathbf{u} = |\mathbf{u}|\cos\theta \qquad \text{Multiply numerator and denominator by } |\mathbf{v}|.$$

$$= \frac{|\mathbf{u}||\mathbf{v}|\cos\theta}{|\mathbf{v}|}$$

$$= \frac{\mathbf{u}\cdot\mathbf{v}}{|\mathbf{v}|} \qquad \text{Since } \cos\theta = \frac{\mathbf{u}\cdot\mathbf{v}}{|\mathbf{u}||\mathbf{v}|}$$

THE SCALAR COMPONENT OF u ON v

Let θ, $0 \le \theta \le \pi$, be the angle between the two vectors **u** and **v**. The **scalar component of u on v**, denoted by $\text{Comp}_{\mathbf{v}}\,\mathbf{u}$, is given by

$$\text{Comp}_{\mathbf{v}}\,\mathbf{u} = |\mathbf{u}|\cos\theta$$

or, in terms of the dot product, by

$$\text{Comp}_{\mathbf{v}}\,\mathbf{u} = \frac{\mathbf{u}\cdot\mathbf{v}}{|\mathbf{v}|}$$

EXAMPLE 4

Finding the Scalar Component of u on v

Find $\text{Comp}_{\mathbf{v}}\,\mathbf{u}$ for each pair of vectors **u** and **v**. Compute answers to three significant digits.

(A) $\mathbf{u} = \langle 2, 3\rangle$; $\mathbf{v} = \langle 4, 1\rangle$ (B) $\mathbf{u} = \langle 3, 2\rangle$; $\mathbf{v} = \langle 4, -6\rangle$

(C) $\mathbf{u} = -3\mathbf{i} + 3\mathbf{j}$; $\mathbf{v} = 3\mathbf{i} + \mathbf{j}$

Solution (A) $\text{Comp}_{\mathbf{v}}\,\mathbf{u} = \dfrac{\mathbf{u}\cdot\mathbf{v}}{|\mathbf{v}|} = \dfrac{11}{\sqrt{17}} = 2.67$

(B) $\text{Comp}_{\mathbf{v}}\,\mathbf{u} = \dfrac{\mathbf{u}\cdot\mathbf{v}}{|\mathbf{v}|} = \dfrac{0}{\sqrt{52}} = 0$

(C) $\text{Comp}_{\mathbf{v}}\,\mathbf{u} = \dfrac{\mathbf{u}\cdot\mathbf{v}}{|\mathbf{v}|} = \dfrac{-6}{\sqrt{10}} = -1.90$ ■

Matched Problem 4 Find $\text{Comp}_{\mathbf{v}}\,\mathbf{u}$ for each pair of vectors **u** and **v**. Compute answers to three significant digits.

(A) $\mathbf{u} = \langle 3, 3\rangle;\quad \mathbf{v} = \langle 6, 2\rangle$ (B) $\mathbf{u} = \langle -4, 2\rangle;\quad \mathbf{v} = \langle 1, 2\rangle$

(C) $\mathbf{u} = \mathbf{i} + 4\mathbf{j};\quad \mathbf{v} = 4\mathbf{i} - 4\mathbf{j}$ ■

Work

Intuitively, we know work is done when a force causes an object to be moved a certain distance. One example is illustrated in Figure 4, where work is done by the force that the harness exerts on a dog sled to move the sled a certain distance.

FIGURE 4

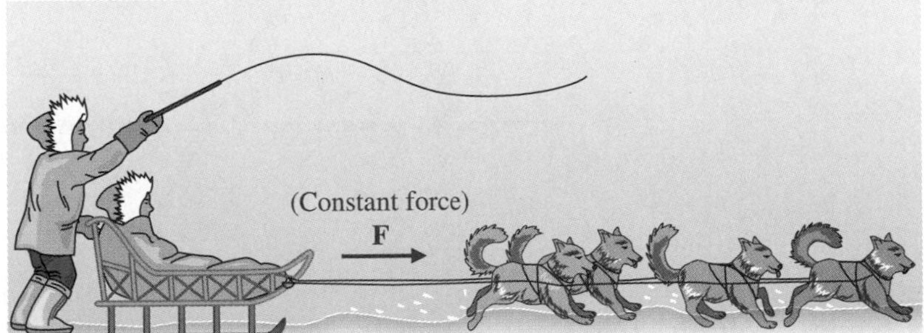

The concept of work is very important in science studies dealing with energy and requires a definition that is quantitative: If the motion of an object takes place in a straight line in the direction of a constant force **F** causing the motion, then we define work W done by the force **F** to be the product of the magnitude of the force $|\mathbf{F}|$ and the distance d through which the object moves. Symbolically,

$$W = |\mathbf{F}|d$$

For example, suppose the dogsled in Figure 4 is moved a distance of 200 ft under a constant force of 100 lb. Then the work done by the force is

$$W = (100 \text{ lb})(200 \text{ ft}) = 20{,}000 \text{ ft-lb}^*$$

* In scientific work, the basic unit of force is the **newton,** the basic unit of linear measure is the **meter,** and the basic unit of work is the **newton-meter** or **joule.** In the British engineering system, which we will use, the basic unit of work is the **foot-pound (ft-lb).**

It is more common that the motion of an object is caused by a constant force acting in a direction different than the line of motion. For example, a person pushing a lawn mower across a level lawn exerts a force downward along the handle (see Fig. 5). In this case, only the component of force in the direction of motion (the horizontal component) is responsible for the motion. Suppose the lawn mower is pushed 52 ft across a level lawn with a constant force of 61 lb directed down the handle. If the handle makes a constant angle of 38° relative to the horizontal, then the work done by the force is

$$\begin{aligned} W &= \begin{pmatrix} \text{Component of force in} \\ \text{the direction of motion} \end{pmatrix} (\text{Displacement}) \\ &= (61 \cos 38°)(52) \\ &= 2{,}500 \text{ ft-lb} \end{aligned}$$

FIGURE 5

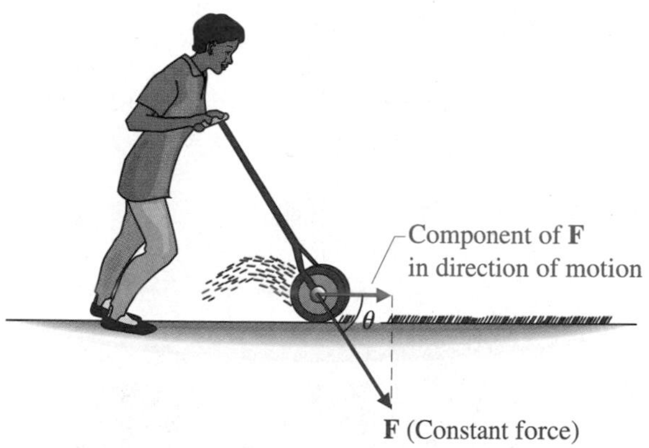

A vector **d** is a **displacement vector** for an object that is moved in a straight line if **d** points in the direction of motion and if $|\mathbf{d}|$ is the distance moved. If the displacement vector of an object is **0**, then no matter how much force is applied, no work is done. Also, for an object to move in the direction of the displacement vector, the angle between the force vector and the displacement vector must be acute. We now give a more general definition of work.

WORK

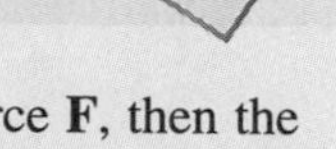

If the displacement vector is **d** for an object moved by a force **F**, then the **work** done, W, is the product of the component of force in the direction of motion and the actual displacement. Symbolically,

$$W = (\text{Comp}_{\mathbf{d}}\,\mathbf{F})|\mathbf{d}| = \frac{\mathbf{F} \cdot \mathbf{d}}{|\mathbf{d}|}|\mathbf{d}| = \mathbf{F} \cdot \mathbf{d}$$

EXAMPLE 5 **Work Done by a Force**

How much work is done by a force $\mathbf{F} = \langle 6, 4 \rangle$ that moves an object from the origin to the point $P(8, 2)$? (Force is in pounds and displacement is in feet.)

Solution

$$W = \mathbf{F} \cdot \mathbf{d}$$
$$= \langle 6, 4 \rangle \cdot \langle 8, 2 \rangle \quad d = \langle 8, 2 \rangle$$
$$= 56 \text{ ft-lb}$$

■

Matched Problem 5 How much work is done by a force $\mathbf{F} = -8\mathbf{i} + 4\mathbf{j}$ that moves an object from the origin to the point $P(-12, -4)$? (Force is in pounds and displacement is in feet.) ■

Answers to Matched Problems

1. (A) 10 (B) 7 (C) 0 (D) 0
2. (A) 60.3° (B) 135.0° (C) 53.1°
3. (A) Not orthogonal (B) Orthogonal
4. (A) $24/\sqrt{40} = 3.79$ (B) $0/\sqrt{5} = 0$ (C) $-12/\sqrt{32} = -2.12$
5. 80 ft-lb

EXERCISE 6.6

A *In Problems 1–8, find each dot product.*

1. $\langle 5, 3 \rangle \cdot \langle -2, 3 \rangle$
2. $\langle 4, -3 \rangle \cdot \langle 2, 2 \rangle$
3. $(5\mathbf{i} - 3\mathbf{j}) \cdot (-2\mathbf{i} + 3\mathbf{j})$
4. $(-3\mathbf{i} + 2\mathbf{j}) \cdot (4\mathbf{i} + \mathbf{j})$
5. $\langle 2, 8 \rangle \cdot \langle 12, -3 \rangle$
6. $\langle -2, 3 \rangle \cdot \langle 9, 6 \rangle$
7. $3\mathbf{i} \cdot 4\mathbf{j}$
8. $-2\mathbf{j} \cdot 6\mathbf{i}$

In Problems 9–14, find the angle (in decimal degrees, to one decimal place) between each pair of vectors.

9. $\mathbf{u} = \langle -3, 2 \rangle$; $\mathbf{v} = \langle 0, 4 \rangle$
10. $\mathbf{u} = \langle -2, -4 \rangle$; $\mathbf{v} = \langle -3, 0 \rangle$
11. $\mathbf{u} = \langle 3, 3 \rangle$; $\mathbf{v} = \langle 2, -5 \rangle$
12. $\mathbf{u} = \langle -2, 3 \rangle$; $\mathbf{v} = \langle -1, -4 \rangle$
13. $\mathbf{u} = 2\mathbf{i} - 3\mathbf{j}$; $\mathbf{v} = 6\mathbf{i} + 4\mathbf{j}$
14. $\mathbf{u} = 8\mathbf{i} + 2\mathbf{j}$; $\mathbf{v} = -3\mathbf{i} + 12\mathbf{j}$

B *In Problems 15 and 16, use the following alternative form for the dot product (from the formula for the angle between two vectors):*

$$\mathbf{u} \cdot \mathbf{v} = |\mathbf{u}||\mathbf{v}| \cos \theta$$

Determine the sign of the dot product for **u** *and* **v** *without calculating it, and explain how you found it.*

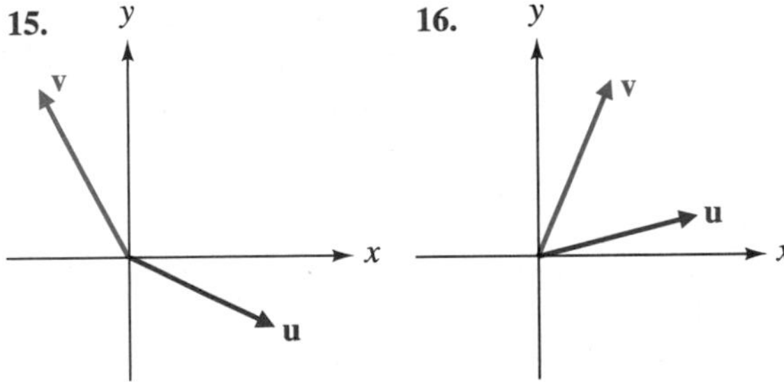

In Problems 17–20, determine which pairs of vectors are orthogonal.

17. $\mathbf{u} = 2\mathbf{i} + \mathbf{j}$; $\mathbf{v} = \mathbf{i} - 2\mathbf{j}$
18. $\mathbf{u} = 5\mathbf{i} - 4\mathbf{j}$; $\mathbf{v} = -4\mathbf{i} - 5\mathbf{j}$
19. $\mathbf{u} = \langle 1, 3 \rangle$; $\mathbf{v} = \langle -3, -1 \rangle$
20. $\mathbf{u} = \langle 4, -3 \rangle$; $\mathbf{v} = \langle 6, -8 \rangle$

In Problems 21–26, find $\text{Comp}_{\mathbf{v}}\, \mathbf{u}$, the scalar component of **u** *on* **v**. *Compute answers to three significant digits.*

21. $\mathbf{u} = \langle -2, 4 \rangle$; $\mathbf{v} = \langle -3, -1 \rangle$
22. $\mathbf{u} = \langle 5, -2 \rangle$; $\mathbf{v} = \langle 6, 1 \rangle$
23. $\mathbf{u} = \langle -2, -4 \rangle$; $\mathbf{v} = \langle 6, -3 \rangle$

24. $\mathbf{u} = \langle 2, -3 \rangle$; $\mathbf{v} = \langle -9, -6 \rangle$

25. $\mathbf{u} = 7\mathbf{i} - 2\mathbf{j}$; $\mathbf{v} = \mathbf{i} + \mathbf{j}$

26. $\mathbf{u} = \mathbf{i} - \mathbf{j}$; $\mathbf{v} = 4\mathbf{i} + 6\mathbf{j}$

C *Given* $\mathbf{u} = \langle a, b \rangle$, $\mathbf{v} = \langle c, d \rangle$, $\mathbf{w} = \langle e, f \rangle$, *and* k *a scalar, prove Problems 27–32 using the definitions of the given operations and the properties of real numbers.*

27. $\mathbf{u} \cdot \mathbf{u} = |\mathbf{u}|^2$

28. $\mathbf{u} \cdot \mathbf{v} = \mathbf{v} \cdot \mathbf{u}$

29. $\mathbf{u} \cdot (\mathbf{v} + \mathbf{w}) = \mathbf{u} \cdot \mathbf{v} + \mathbf{u} \cdot \mathbf{w}$

30. $(\mathbf{u} + \mathbf{v}) \cdot \mathbf{w} = \mathbf{u} \cdot \mathbf{w} + \mathbf{v} \cdot \mathbf{w}$

31. $k(\mathbf{u} \cdot \mathbf{v}) = (k\mathbf{u}) \cdot \mathbf{v} = \mathbf{u} \cdot (k\mathbf{v})$

32. $\mathbf{u} \cdot \mathbf{0} = 0$

The vector projection of $\mathbf{u}$ *onto* $\mathbf{v}$, *denoted by* $\text{Proj}_{\mathbf{v}}\,\mathbf{u}$, *is given by*

$$\text{Proj}_{\mathbf{v}}\,\mathbf{u} = (\text{Comp}_{\mathbf{v}}\,\mathbf{u})\frac{\mathbf{v}}{|\mathbf{v}|} = \frac{\mathbf{u} \cdot \mathbf{v}}{\mathbf{v} \cdot \mathbf{v}}\mathbf{v}$$

In Problems 33–36, find $\text{Proj}_{\mathbf{v}}\,\mathbf{u}$.

33. $\mathbf{u} = \langle 3, 4 \rangle$; $\mathbf{v} = \langle 4, 0 \rangle$

34. $\mathbf{u} = \langle 3, -4 \rangle$; $\mathbf{v} = \langle 0, -3 \rangle$

35. $\mathbf{u} = -6\mathbf{i} + 3\mathbf{j}$; $\mathbf{v} = -3\mathbf{i} - 2\mathbf{j}$

36. $\mathbf{u} = -2\mathbf{i} - 3\mathbf{j}$; $\mathbf{v} = \mathbf{i} - 6\mathbf{j}$

Applications

37. **Work** A parent pulls a child in a wagon (see the figure) for one block (440 ft). If a constant force of 15 lb is exerted along the handle, how much work is done?

Figure for 37

38. **Work** Repeat Problem 37 using a constant force of 12 lb, a distance of 5,300 ft (approximately a mile), and an angle of 35° relative to the horizontal.

Work *In Problems 39–44, determine how much work is done by a force* $\mathbf{F}$ *moving an object from the origin to the point P. (Force is in pounds and displacement is in feet.)*

39. $\mathbf{F} = \langle 10, 5 \rangle$; $P(8, 1)$

40. $\mathbf{F} = \langle 5, 1 \rangle$; $P(8, 5)$

41. $\mathbf{F} = -2\mathbf{i} + 3\mathbf{j}$; $P(-3, 1)$

42. $\mathbf{F} = 3\mathbf{i} - 4\mathbf{j}$; $P(5, 0)$

43. $\mathbf{F} = 10\mathbf{i} + 10\mathbf{j}$; $P(1, 1)$

44. $\mathbf{F} = 8\mathbf{i} - 4\mathbf{j}$; $P(2, -1)$

*45. **Geometry** Use the accompanying figure with the indicated vector assignments and appropriate properties of the dot product to prove that an angle inscribed in a semicircle is a right angle. [*Note:* $|\mathbf{a}| = |\mathbf{c}| =$ *Radius*]

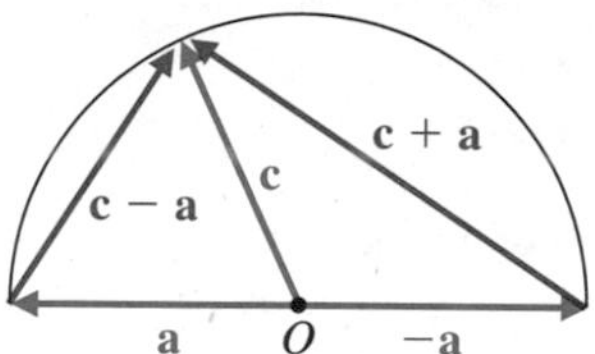

Figure for 45

*46. **Geometry** Use the accompanying figure with the indicated vector assignments and appropriate properties of the dot product to prove that the diagonals of a rhombus are perpendicular. [*Note:* $|\mathbf{a}| = |\mathbf{b}|$]

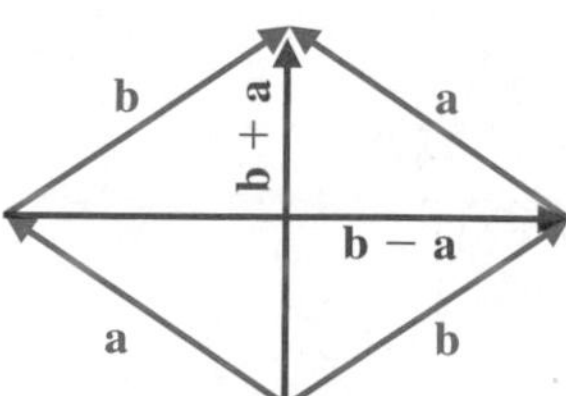

Figure for 46

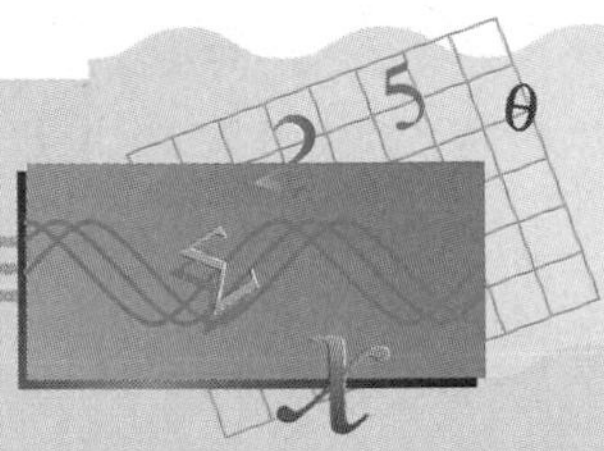

CHAPTER 6 GROUP ACTIVITY

The SSA Case and the Law of Cosines

In Section 6.1 we spent quite a bit of time discussing SSA triangles and their solution using the law of sines. Table 1 is the part of Table 2 in Section 6.1 limited to α acute and $a < b$.

TABLE 1
For α Acute and $a < b$

a	Number of triangles	Figure
$0 < a < h$	0	
$a = h$	1	
$h < a < b$	2	

The law of cosines can be used on the SSA cases in Table 1, and it has the advantage of automatically sorting out the situations illustrated. The computations are more complicated using the law of cosines, but using a calculator helps overcome this difficulty.

Given a, b, and α, with α acute and $a < b$, we want to find c. Start with the law of cosines in the form

$$a^2 = b^2 + c^2 - 2bc \cos \alpha$$

and show that c is given by

$$c = \frac{2b \cos \alpha \pm \sqrt{(2b \cos \alpha)^2 - 4(b^2 - a^2)}}{2} \tag{1}$$

Now, denote the discriminant by

$$D = (2b \cos \alpha)^2 - 4(b^2 - a^2)$$

and complete Table 2. Explain your reasoning.

TABLE 2
For α Acute and $a < b$

D	Number of solutions
$D < 0$	?
$D = 0$	?
$D > 0$	?

Problem 1 For the indicated values, determine whether the triangle has 0, 1, or 2 solutions without computing the solutions.

(A) $a = 5.1, b = 7.6, \alpha = 32.7°$

(B) $a = 2.1, b = 8.2, \alpha = 21.5°$ ■

Problem 2 Use equation (1) to find all values of c for the triangle(s) in Problem 1 that have at least one solution. ■

Problem 3 If $a > b$ and α is obtuse, how many solutions will equation (1) give? Explain your reasoning. ■

Problem 4 Use equation (1) to find all values of c for a triangle with $a = 8.8$, $b = 6.4$, and $\alpha = 123.4°$. ■

CHAPTER 6 REVIEW

6.1 LAW OF SINES

An **oblique triangle** is a triangle without a right angle. An oblique triangle is **acute** if all angles are between 0° and 90° and **obtuse** if one angle is between 90° and 180°. See Figure 1.

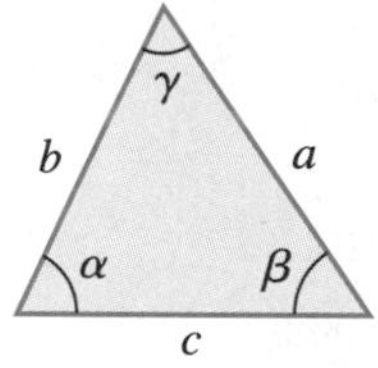

(a) Acute triangle

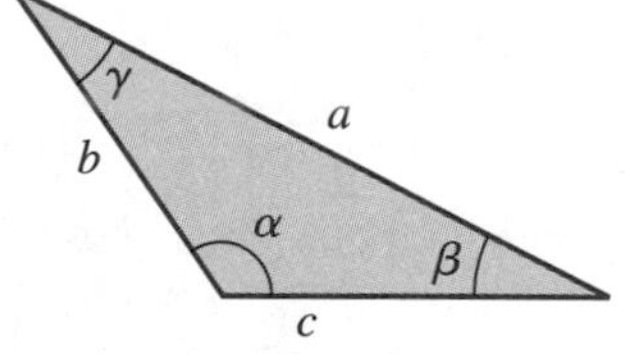

(b) Obtuse triangle

FIGURE 1

Solving a triangle involves finding three of the six quantities indicated in Figure 1 when given the other three. Table 1 is used to determine the accuracy of these computations.

TABLE 1

Angle to nearest	Significant digits for side measure
1°	2
10′ or 0.1°	3
1′ or 0.01°	4
10″ or 0.001°	5

If the given quantities include an angle and the opposite side (ASA, AAS, or SSA), the **law of sines** is used to solve the triangle in Figure 2.

FIGURE 2

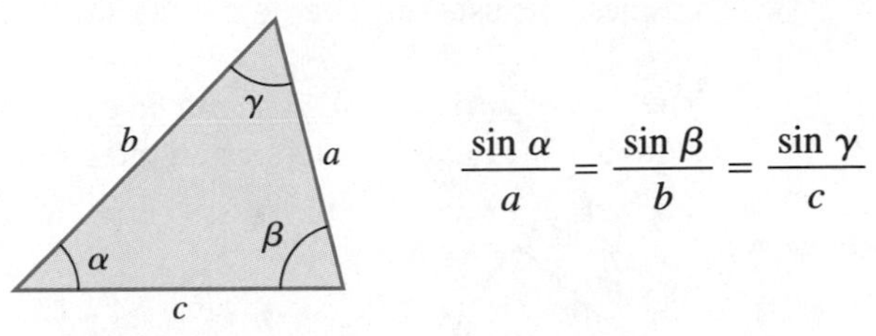

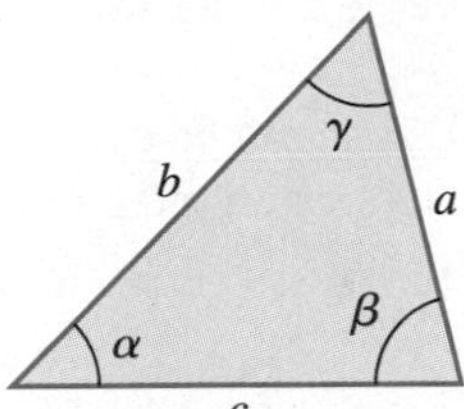

$$\frac{\sin\alpha}{a} = \frac{\sin\beta}{b} = \frac{\sin\gamma}{c}$$

The SSA case, often called the **ambiguous case**, has a number of variations (see Table 2).

TABLE 2
SSA Variations

	a ($h = b\sin\alpha$)	Number of triangles	Figure
α acute	$0 < a < h$	0	(a)
	$a = h$	1	(b)
	$h < a < b$	2	(c)
	$a \geq b$	1	(d)
α obtuse	$0 < a \leq b$	0	(e)
	$a > b$	1	(f)

6.2 LAW OF COSINES

We stated the **law of cosines** for the triangle in Figure 3 as follows:

FIGURE 3

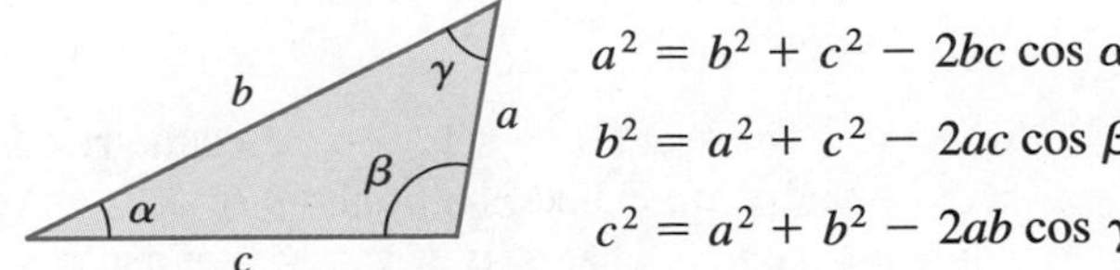

$$a^2 = b^2 + c^2 - 2bc\cos\alpha$$
$$b^2 = a^2 + c^2 - 2ac\cos\beta$$
$$c^2 = a^2 + b^2 - 2ab\cos\gamma$$

The law of cosines is generally used as the first step in solving the SAS and SSS cases for oblique triangles. After a side or angle is found using the law of cosines, it is usually easier to use the law of sines to find a second angle.

☆6.3 AREAS OF TRIANGLES

The area of a triangle (Fig. 4) can be determined from the base and altitude, two sides and the included angle, or all three sides **(Heron's formula).**

FIGURE 4

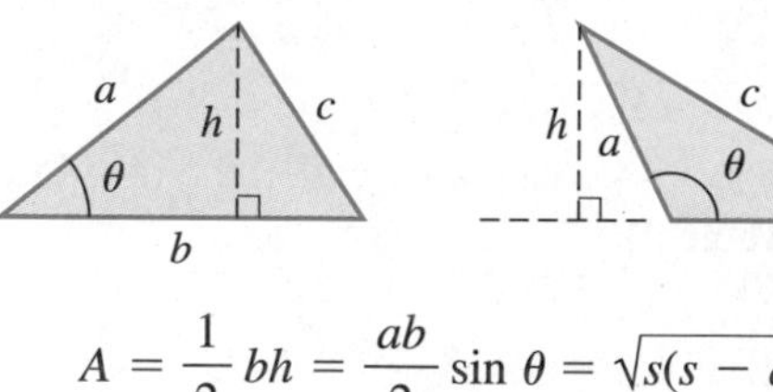

$$A = \frac{1}{2}bh = \frac{ab}{2}\sin\theta = \sqrt{s(s-a)(s-b)(s-c)}$$

where $s = \frac{1}{2}(a + b + c)$ is the **semiperimeter.**

6.4 VECTORS: GEOMETRICALLY DEFINED

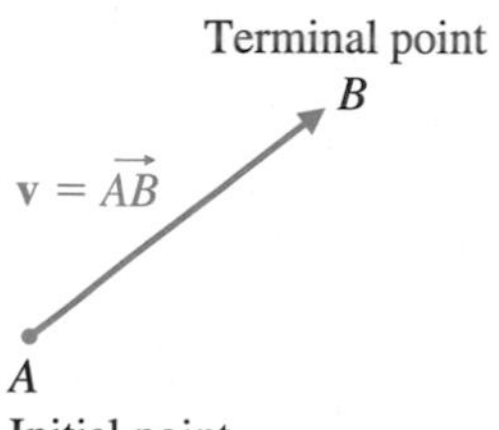

FIGURE 5
Vector $\overrightarrow{AB} = \mathbf{v}$

A **scalar** is a real number. The **directed line segment** from point A to point B in the plane, denoted $\overrightarrow{AB}$, is the line segment from A to B with the arrowhead placed at B to indicate the direction (see Fig. 5). Point A is called the **initial point,** and point B is called the **terminal point.**

A **geometric vector** in a plane, denoted **v**, is a quantity that possesses both a length and a direction and can be represented by a directed line segment as indicated in Figure 5. The **magnitude** of the vector $\overrightarrow{AB}$, denoted by $|\overrightarrow{AB}|$, $|\vec{v}|$, or $|\mathbf{v}|$, is the length of the directed line segment. Two vectors have the **same direction** if they are parallel and point in the same direction. Two vectors have **opposite direction** if they are parallel and point in opposite directions. The **zero vector,** denoted by $\vec{0}$ or $\mathbf{0}$, has a magnitude of zero and an arbitrary direction. Two vectors are **equal** if they have the same magnitude and direction. Thus, a vector may be **translated** from one location to another as long as the magnitude and direction do not change.

The **sum of two vectors u and v** can be defined using the **tail-to-tip rule.** The **sum of two nonparallel vectors** also can be defined using the **parallelogram rule.** Both forms are shown in Figure 6.

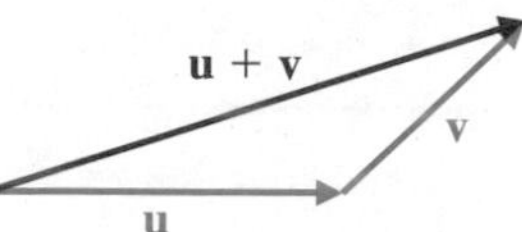

Tail-to-tip rule for addition

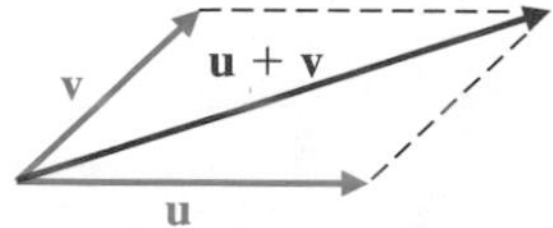

Parallelogram rule for addition

FIGURE 6
Vector addition

The vector $\mathbf{u} + \mathbf{v}$ is also called the **resultant** of the two vectors **u** and **v**, and **u** and **v** are called **components** of $\mathbf{u} + \mathbf{v}$. Vector addition is **commutative** and **associative.** That is, $\mathbf{u} + \mathbf{v} = \mathbf{v} + \mathbf{u}$ and $\mathbf{u} + (\mathbf{v} + \mathbf{w}) = (\mathbf{u} + \mathbf{v}) + \mathbf{w}$.

A **velocity vector** represents the direction and speed of an object in motion. The velocity of a boat relative to the water is called **apparent velocity,** and the velocity relative to the ground is called the **resultant** or **actual velocity.** The resultant velocity is the vector sum of the apparent velocity and the current velocity. Similar statements apply to objects in air subject to winds.

A **force vector** represents the direction and magnitude of an applied force. If an object is subjected to two forces, then the sum of these two forces, the **resultant force,** is a single force acting on the object in the same way as the two original forces taken together.

6.5 VECTORS: ALGEBRAICALLY DEFINED

A geometric vector $\vec{AB}$ in a rectangular coordinate system translated so that its initial point is at the origin is said to be in **standard position.** The vector $\vec{OP}$ such that $\vec{OP} = \vec{AB}$ is said to be the **standard vector** for $\vec{AB}$. This is shown in Figure 7.

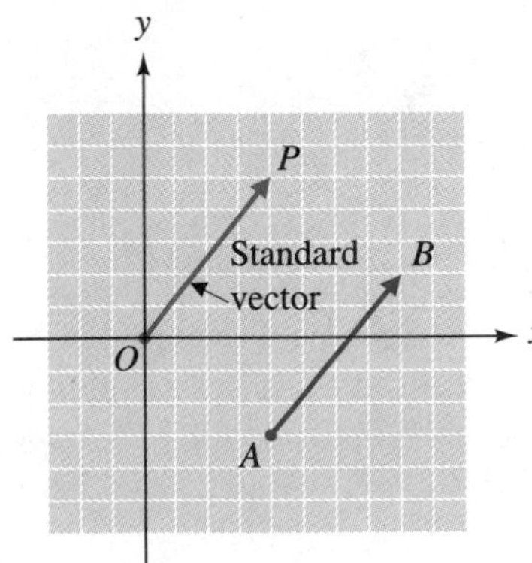

FIGURE 7
$\vec{OP}$ is the standard vector for $\vec{AB}$ ($\vec{OP} = \vec{AB}$)

Note that the vector $\vec{OP}$ in Figure 7 is the standard vector for infinitely many vectors: all vectors with the same magnitude and direction as $\vec{OP}$. If the coordinates of A in Figure 7 are (x_a, y_a) and the coordinates of B are (x_b, y_b), then the coordinates of P are given by

$$(x_p, y_p) = (x_b - x_a, y_b - y_a)$$

Every geometric vector in a plane corresponds to an ordered pair of real numbers, and every ordered pair of real numbers corresponds to a geometric vector. This leads to the definition of an **algebraic vector** as an ordered pair of real numbers, denoted by $\langle a, b\rangle$. The real numbers a and b are **scalar components** of the vector $\langle a, b\rangle$.

Two vectors $\mathbf{u} = \langle a, b\rangle$ and $\mathbf{v} = \langle c, d\rangle$ are said to be **equal** if their components are equal. The **zero vector** is denoted by $\mathbf{0} = \langle 0, 0\rangle$ and has arbitrary direction.

The **magnitude,** or **norm,** of a vector $\mathbf{v} = \langle a, b\rangle$, denoted by $|\mathbf{v}|$, is given by

$$|\mathbf{v}| = \sqrt{a^2 + b^2}$$

Geometrically, $\sqrt{a^2 + b^2}$ is the length of the standard geometric vector $\vec{OP}$ associated with the algebraic vector $\langle a, b\rangle$.

If $\mathbf{u} = \langle a, b\rangle$, $\mathbf{v} = \langle c, d\rangle$, and k is a scalar, then the **sum** of $\mathbf{u}$ and $\mathbf{v}$ is given by

$$\mathbf{u} + \mathbf{v} = \langle a + c, b + d\rangle$$

and **scalar multiplication** of $\mathbf{u}$ by k is given by

$$k\mathbf{u} = k\langle a, b\rangle = \langle ka, kb\rangle$$

If $\mathbf{v}$ is a nonzero vector, then

$$\mathbf{u} = \frac{1}{|\mathbf{v}|}\mathbf{v}$$

is a **unit vector** with the same direction as $\mathbf{v}$. The $\mathbf{i}$ and $\mathbf{j}$ unit vectors are defined as follows (Fig. 8):

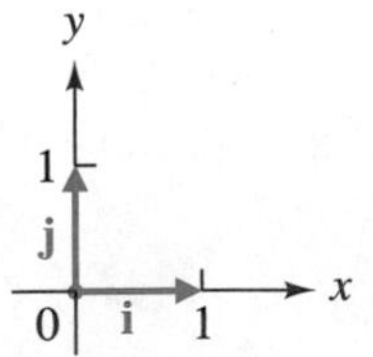

FIGURE 8

$$\mathbf{i} = \langle 1, 0\rangle$$

$$\mathbf{j} = \langle 0, 1\rangle$$

Any vector **v** can be expressed in terms of **i** and **j**:

$$\mathbf{v} = \langle a, b\rangle = a\mathbf{i} + b\mathbf{j}$$

The following algebraic properties of vector addition and scalar multiplication enable us to manipulate symbols representing vectors and scalars in much the same way we manipulate symbols that represent real numbers in algebra.

(A) The following **addition properties** are satisfied for all vectors **u**, **v**, and **w**:

1. $\mathbf{u} + \mathbf{v} = \mathbf{v} + \mathbf{u}$ — Commutative property
2. $\mathbf{u} + (\mathbf{v} + \mathbf{w}) = (\mathbf{u} + \mathbf{v}) + \mathbf{w}$ — Associative property
3. $\mathbf{u} + \mathbf{0} = \mathbf{0} + \mathbf{u} = \mathbf{u}$ — Additive identity
4. $\mathbf{u} + (-\mathbf{u}) = (-\mathbf{u}) + \mathbf{u} = \mathbf{0}$ — Additive inverse

(B) The following **scalar multiplication properties** are satisfied for all vectors **u** and **v** and all scalars m and n:

1. $m(n\mathbf{u}) = (mn)\mathbf{u}$ — Associative property
2. $m(\mathbf{u} + \mathbf{v}) = m\mathbf{u} + m\mathbf{v}$ — Distributive property
3. $(m + n)\mathbf{u} = m\mathbf{u} + n\mathbf{u}$ — Distributive property
4. $1\mathbf{u} = \mathbf{u}$ — Multiplication identity

Algebraic vectors can be used to solve **static equilibrium** problems. An object at rest is said to be in static equilibrium. An object subject only to **coplanar concurrent forces** is in static equilibrium if and only if the sum of all these forces is zero.

☆6.6 THE DOT PRODUCT

The **dot product** of the two vectors

$$\mathbf{u} = \langle a, b\rangle = a\mathbf{i} + b\mathbf{j} \quad \text{and} \quad \mathbf{v} = \langle c, d\rangle = c\mathbf{i} + d\mathbf{j}$$

denoted by $\mathbf{u} \cdot \mathbf{v}$, is the scalar given by

$$\mathbf{u} \cdot \mathbf{v} = ac + bd$$

For all vectors **u**, **v**, and **w**, and k a real number:

1. $\mathbf{u} \cdot \mathbf{u} = |\mathbf{u}|^2$
2. $\mathbf{u} \cdot \mathbf{v} = \mathbf{v} \cdot \mathbf{u}$
3. $\mathbf{u} \cdot (\mathbf{v} + \mathbf{w}) = \mathbf{u} \cdot \mathbf{v} + \mathbf{u} \cdot \mathbf{w}$
4. $(\mathbf{u} + \mathbf{v}) \cdot \mathbf{w} = \mathbf{u} \cdot \mathbf{w} + \mathbf{v} \cdot \mathbf{w}$
5. $k(\mathbf{u} \cdot \mathbf{v}) = (k\mathbf{u}) \cdot \mathbf{v} = \mathbf{u} \cdot (k\mathbf{v})$
6. $\mathbf{u} \cdot \mathbf{0} = 0$

The **angle between the vectors u and v** is the angle θ, $0 \le \theta \le \pi$, formed by positioning the vectors so that their initial points coincide (Fig. 9). The vectors **u** and **v** are **orthogonal,** or **perpendicular,** if $\theta = \pi/2$, and **parallel** if $\theta = 0$ or $\theta = \pi$. If **u** and **v** are nonzero, then

$$\cos\theta = \frac{\mathbf{u} \cdot \mathbf{v}}{|\mathbf{u}||\mathbf{v}|}$$

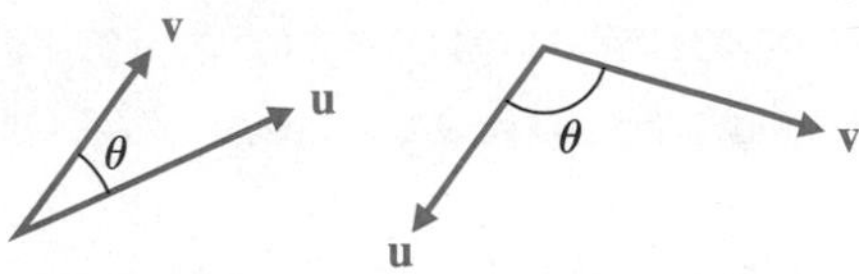

FIGURE 9
Angle between two vectors

Two vectors **u** and **v** are orthogonal if and only if

$$\mathbf{u} \cdot \mathbf{v} = 0$$

[*Note:* The zero vector is orthogonal to every vector.]

To project a vector **u** onto an arbitrary nonzero vector **v**, locate **u** and **v** so that they have the same initial point O and drop a perpendicular from the terminal point of **u** to the line that contains **v**. The **vector projection of u onto v** is the vector **p** that lies on the line containing **v**, as shown in Figure 10.

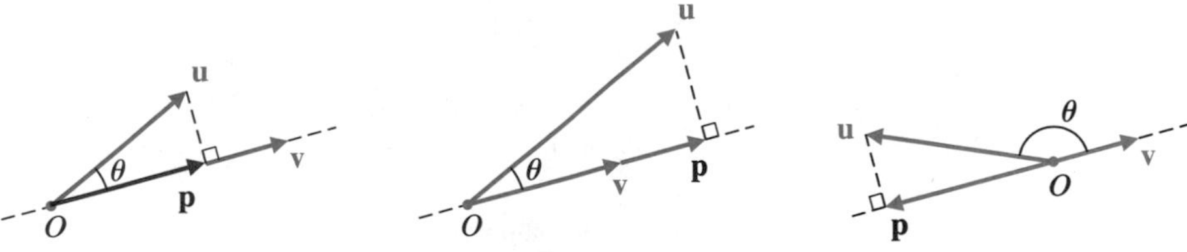

FIGURE 10
Projection of **u** onto **v**

The **scalar component of u on v** is given by

$$\text{Comp}_{\mathbf{v}}\, \mathbf{u} = |\mathbf{u}| \cos \theta$$

or, in terms of the dot product, by

$$\text{Comp}_{\mathbf{v}}\, \mathbf{u} = \frac{\mathbf{u} \cdot \mathbf{v}}{|\mathbf{v}|}$$

A vector **d** is a **displacement vector** for an object that is moved in a straight line if **d** points in the direction of motion and if $|\mathbf{d}|$ is the distance moved. If the displacement vector is **d** for an object moved by a force **F**, then the **work** done, W, is the product of the component of force in the direction of motion and the actual displacement. Symbolically,

$$W = (\text{Comp}_{\mathbf{d}}\, \mathbf{F})|\mathbf{d}| = \frac{\mathbf{F} \cdot \mathbf{d}}{|\mathbf{d}|}|\mathbf{d}| = \mathbf{F} \cdot \mathbf{d}$$

CHAPTER 6 REVIEW EXERCISE

Work through all the problems in this chapter review and check the answers. Answers to all review problems appear in the back of the book; following each answer is an italic number that indicates the section in which that type of problem is discussed. Where weaknesses show up, review the appropriate sections in the text.

Where applicable, quantities in the problems refer to a triangle labeled as in the figure.

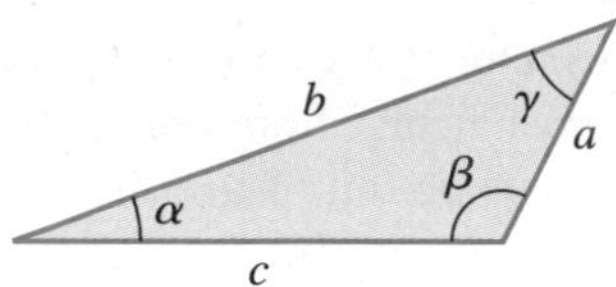

A

1. If two sides and the included angle are given for a triangle, explain why the law of sines cannot be used to find another side or angle.
2. If two forces acting on a point have equal magnitudes and the resultant is **0**, what can you say about the two forces?
3. If two forces acting on a point have equal magnitudes and the magnitude of the resultant is the sum of the magnitudes of the individual forces, what can you say about the two forces?
4. Without solving each triangle, determine whether the given information allows you to construct 0, 1, or 2 triangles. Explain your reasoning.
   (A) $a = 4$ cm, $b = 8$ cm, $\alpha = 30°$
   (B) $a = 5$ m, $b = 7$ m, $\alpha = 30°$
   (C) $a = 3$ in., $b = 8$ in., $\alpha = 30°$

In Problems 5–8, solve each triangle given the indicated measures of angles and sides.

5. $\alpha = 53°$, $\gamma = 105°$, $b = 42$ cm
6. $\alpha = 66°$, $\beta = 32°$, $b = 12$ m
7. $\alpha = 49°$, $b = 22$ in., $c = 27$ in.
8. $\alpha = 62°$, $a = 14$ cm, $b = 12$ cm

☆ 9. Find the area of the triangle in Problem 7.

☆ 10. Find the area of the triangle in Problem 8.

11. Two vectors **u** and **v** are located in a coordinate system, as indicated in the figure. Find the direction (relative to the x axis) and magnitude of **u** + **v** if $|\mathbf{u}| = 8.0$ and $|\mathbf{v}| = 5.0$.

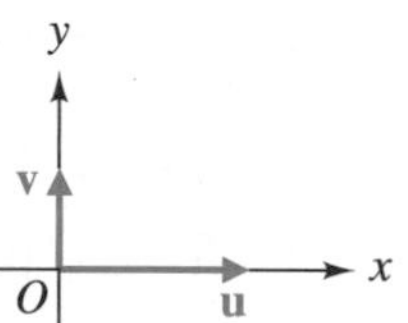

Figure for 11

12. Find the magnitude of the horizontal and vertical components of the vector **v** located in a coordinate system as indicated in the figure.

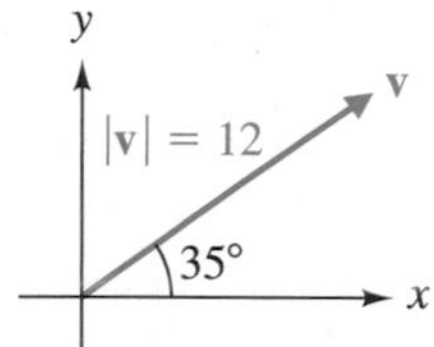

Figure for 12

13. Given $A(-3, 2)$ and $B(-1, -3)$, represent the geometric vector $\overrightarrow{AB}$ as an algebraic vector $\langle a, b\rangle$.
14. Find the magnitude of the vector $\langle -5, 12\rangle$.

☆ 15. $\langle 2, -1\rangle \cdot \langle -3, 2\rangle = ?$

☆ 16. $(2\mathbf{i} + \mathbf{j}) \cdot (3\mathbf{i} - 2\mathbf{j}) = ?$

☆ 17. Find the angle (in decimal degrees, to one decimal place) between $\mathbf{u} = \langle 4, 3\rangle$ and $\mathbf{v} = \langle 3, 0\rangle$.

☆ 18. Find the angle (in decimal degrees, to one decimal place) between $\mathbf{u} = 5\mathbf{i} + \mathbf{j}$ and $\mathbf{v} = -2\mathbf{i} + 2\mathbf{j}$.

B

19. In the process of solving an SSA triangle, it is found that $\sin\beta > 1$. How many triangles are possible? Explain.

In Problems 20–23, solve each triangle given the indicated measures of angles and sides.

20. $\alpha = 65.0°$, $b = 103$ m, $c = 72.4$ m
21. $\alpha = 35°20'$, $a = 13.2$ in., $b = 15.7$ in., β acute
22. $\alpha = 35°20'$, $a = 13.2$ in., $b = 15.7$ in., β obtuse
23. $a = 43$ mm, $b = 48$ mm, $c = 53$ mm

☆ **24.** Find the area of the triangle in Problem 20.

☆ **25.** Find the area of the triangle in Problem 23.

26. There are two vectors $\mathbf{u}$ and $\mathbf{v}$ such that $|\mathbf{u}| = |\mathbf{v}|$ and $\mathbf{u} \neq \mathbf{v}$. Explain how this can happen.

27. Under what conditions are the two algebraic vectors $\langle a, b\rangle$ and $\langle c, d\rangle$ equal?

28. Given the vector diagram, find $|\mathbf{u} + \mathbf{v}|$ and θ.

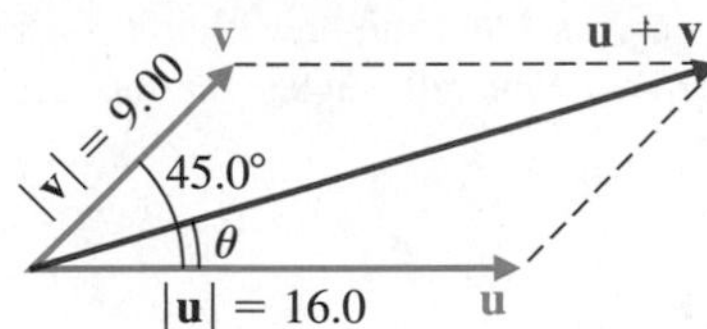

Figure for 28

In Problems 29 and 30, find:

(A) $\mathbf{u} + \mathbf{v}$ (B) $\mathbf{u} - \mathbf{v}$

(C) $3\mathbf{u} - 2\mathbf{v}$ (D) $2\mathbf{u} - 3\mathbf{v} + \mathbf{w}$

29. $\mathbf{u} = \langle 4, 0\rangle$; $\mathbf{v} = \langle -2, -3\rangle$; $\mathbf{w} = \langle 1, -1\rangle$

30. $\mathbf{u} = 3\mathbf{i} - \mathbf{j}$; $\mathbf{v} = 2\mathbf{i} - 3\mathbf{j}$; $\mathbf{w} = -2\mathbf{j}$

31. Form a unit vector $\mathbf{u}$ with the same direction as $\mathbf{v} = \langle -8, 15\rangle$.

32. Express $\mathbf{v}$ in terms of $\mathbf{i}$ and $\mathbf{j}$ unit vectors.
(A) $\mathbf{v} = \langle -5, 7\rangle$ (B) $\mathbf{v} = \langle 0, -3\rangle$
(C) $\mathbf{v} = \overrightarrow{AB}$; $A(4, -2)$; $B(0, -3)$

☆ **33.** Determine which vector pairs are orthogonal using properties of the dot product.
(A) $\mathbf{u} = \langle -12, 3\rangle$; $\mathbf{v} = \langle 2, 8\rangle$
(B) $\mathbf{u} = -4\mathbf{i} + \mathbf{j}$; $\mathbf{v} = -\mathbf{i} + 4\mathbf{j}$

☆ **34.** Find $\text{Comp}_{\mathbf{v}}\,\mathbf{u}$, the scalar component of $\mathbf{u}$ on $\mathbf{v}$. Compute answers to three significant digits.
(A) $\mathbf{u} = \langle 4, 5\rangle$; $\mathbf{v} = \langle 3, 1\rangle$
(B) $\mathbf{u} = -\mathbf{i} + 4\mathbf{j}$; $\mathbf{v} = 3\mathbf{i} - \mathbf{j}$

35. An object that is free to move has three nonzero coplanar concurrent forces acting on it, and it remains at rest. How is any one of the three forces related to the other two?

C **36.** An oblique triangle is solved, with the following results:

$\alpha = 45.1°$ $a = 8.42$ cm

$\beta = 75.8°$ $b = 11.5$ cm

$\gamma = 59.1°$ $c = 10.2$ cm

Allowing for rounding, use Mollweide's equation (below) to check these results.

$$(a - b)\cos\frac{\gamma}{2} = c\sin\frac{\alpha - \beta}{2}$$

37. Given an oblique triangle with $\alpha = 52.3°$ and $b = 12.7$ cm, determine a value k so that if $0 < a < k$, there is no solution; if $a = k$, there is one solution; and if $k < a < b$, there are two solutions.

In Problems 38–43, let $\mathbf{u} = \langle a, b\rangle$, $\mathbf{v} = \langle c, d\rangle$, *and* $\mathbf{w} = \langle e, f\rangle$ *be vectors, and let m and n be scalars. Prove each property using the definitions of the operations involved and properties of real numbers. (Although some of these problems appeared in Exercises 6.5 and 6.6, it is useful to consider them again as part of this review.)*

38. $\mathbf{u} + \mathbf{v} = \mathbf{v} + \mathbf{u}$

☆ **39.** $\mathbf{u} \cdot \mathbf{v} = \mathbf{v} \cdot \mathbf{u}$

40. $(mn)\mathbf{v} = m(n\mathbf{v})$

☆ **41.** $m(\mathbf{u} \cdot \mathbf{v}) = (m\mathbf{u}) \cdot \mathbf{v} = \mathbf{u} \cdot (m\mathbf{v})$

☆ **42.** $\mathbf{u} \cdot (\mathbf{v} + \mathbf{w}) = \mathbf{u} \cdot \mathbf{v} + \mathbf{u} \cdot \mathbf{w}$

☆ **43.** $\mathbf{u} \cdot \mathbf{u} = |\mathbf{u}|^2$

Applications

44. Geometry Find the lengths of the sides of a parallelogram with diagonals 20.0 cm and 16.0 cm long intersecting at 36.4°.

45. Geometry Find h to three significant digits in the triangle:

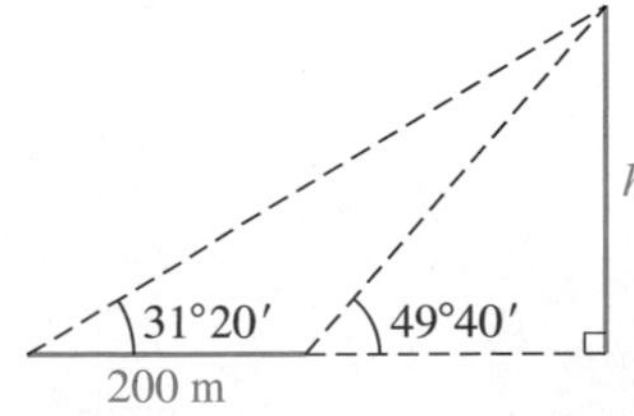

Figure for 45

∗ **46. Geometry** A chord of length 34 cm subtends a central angle of 85° in a circle of radius r. Find the radius of the circle.

* **47.** **Surveying** A plot of land has been surveyed with the resulting information shown in the figure. Find the length of CD.

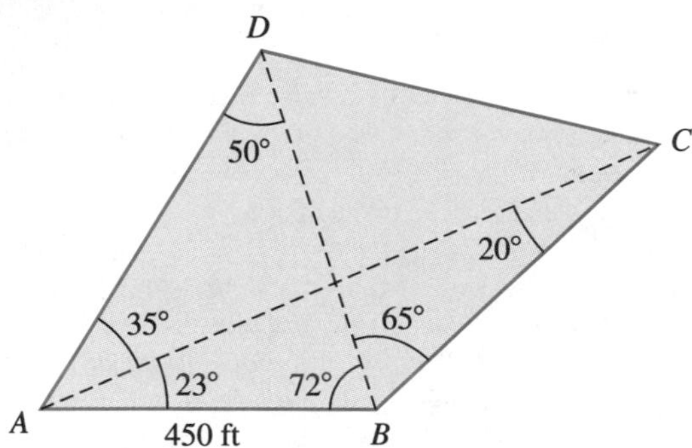

Figure for 47

☆ **48.** **Surveying** Refer to Problem 47. Find the area of the plot.

49. **Engineering** A tunnel for a highway is to be constructed through a mountain, as indicated in the figure. How long is the tunnel?

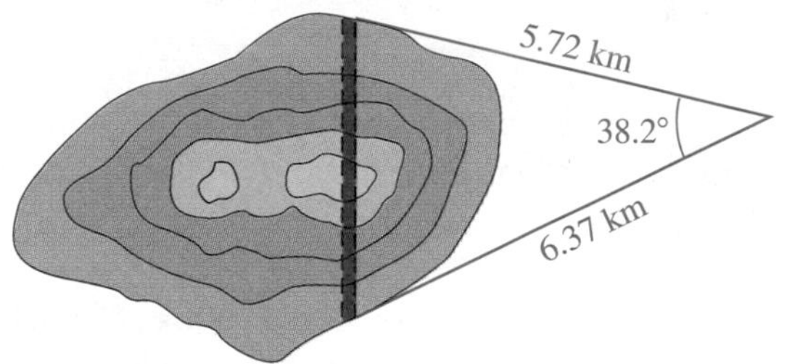

Figure for 49

* **50.** **Space Science** A satellite S, in circular orbit around the earth, is sighted by a tracking station T (see the figure).

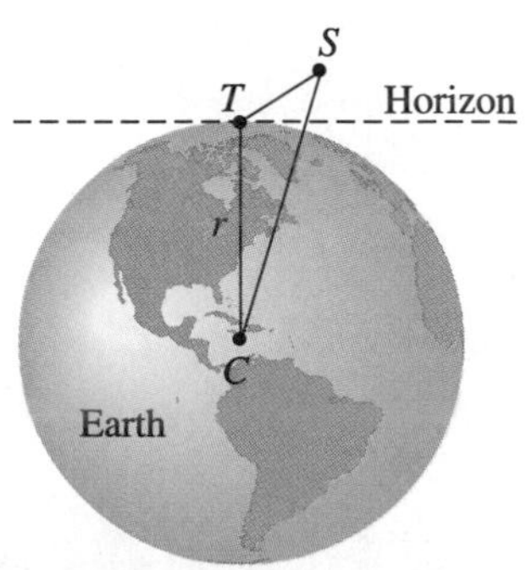

Figure for 50

The distance TS determined by radar is 1,147 mi, and the angle of elevation above the horizon is 28.6°. How high is the satellite above the earth at the time of the sighting? The radius of the earth is $r = 3{,}964$ mi.

* **51.** **Space Science** When a satellite is directly over tracking station B (see the figure), tracking station A measures the angle of elevation θ of the satellite (from the horizon line) to be 21.7°. If the tracking stations are 632 mi apart (that is, the arc AB is 632 mi long) and the radius of the earth is 3,964 mi, how high is the satellite above B? [*Hint:* Find all angles for the triangle ACS first.]

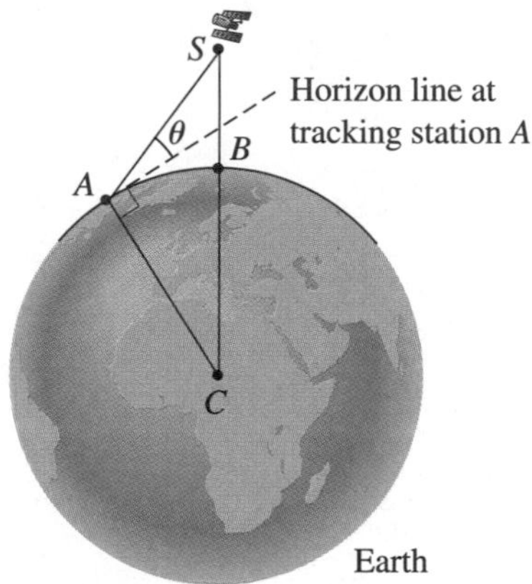

Figure for 51

52. **Navigation** An airplane flies with a speed of 230 km/hr and a compass heading of 68°. If a 55 km/hr wind is blowing in the direction of 5°, what is the plane's actual direction (relative to north) and ground speed?

53. **Resultant Force** Two forces act on an object as indicated in the figure:

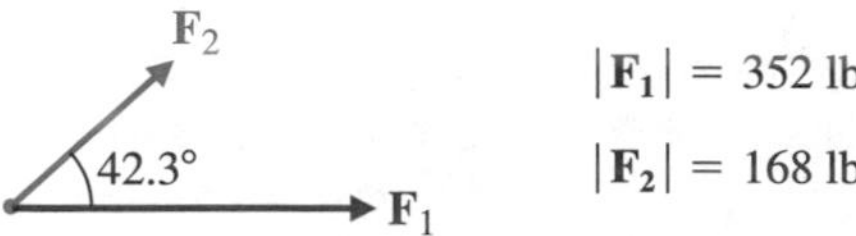

Find the magnitude of the resultant force and its direction relative to the horizontal force $\mathbf{F}_1$.

☆ **54.** **Work** Refer to Problem 53. How much work is done by the resultant force if the object is moved 22 ft in the direction of the resultant force?

55. Engineering A 260 lb sign is hung as shown in the figure. Determine the compression force on AB and the tension on CB.

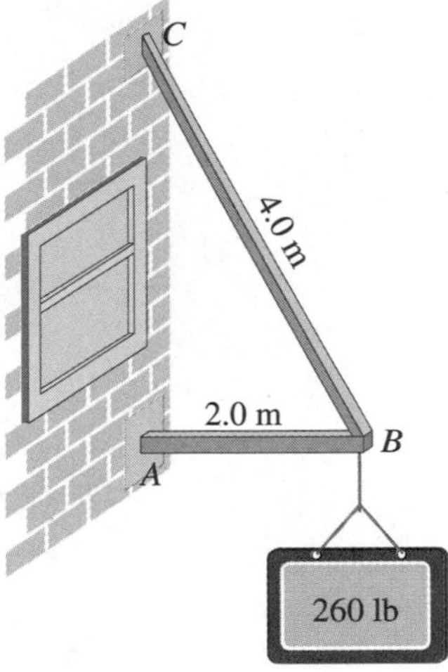

Figure for 55

56. Space Science To simulate a reduced-gravity environment such as that found on the surface of the moon, NASA constructed an inclined plane and a harness suspended on a cable fastened to a movable track, as indicated in the figure at the top of the next column. A person suspended in the harness walks at a right angle to the inclined plane.

(A) Using the accompanying force diagram, find $|\mathbf{u}|$, the force on the astronaut's feet, in terms of $|\mathbf{w}|$ and θ. Also find $|\mathbf{v}|$, the tension on the cable, in terms of $|\mathbf{w}|$ and θ.

(B) What is the force against the feet of a woman astronaut weighing 130 lb ($|\mathbf{w}| = 130$ lb) if $\theta = 72°$? What is the tension on the cable? Compute answers to the nearest pound.

(C) Find θ, to the nearest degree, so that the force on an astronaut's feet will be the same as the force on the moon (about one-sixth of that on earth).

☆ **57. Engineering: Physics** Determine how much work is done by the force $\mathbf{F} = \langle -5, 8 \rangle$ moving an object from the origin to the point $P(-8, 2)$. (Force is in pounds and displacement is in feet.)

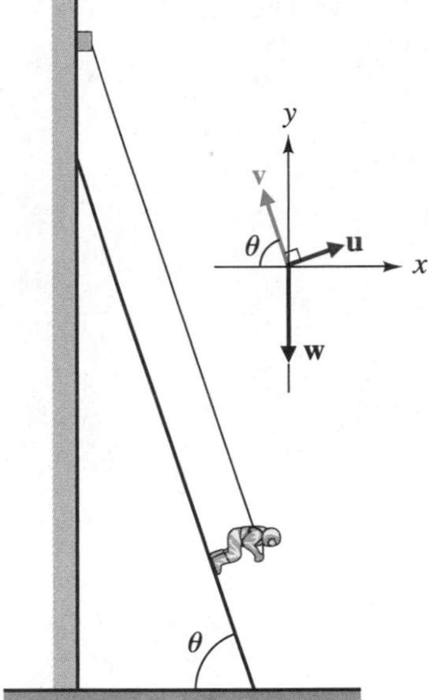

Figure for 56

58. Biomechanics A person with a broken leg has it in a cast and must keep the leg elevated, as shown in the figure. (The weight of the leg with cast exerts a downward force of 12 lb.) Find angle θ and the tension on the line fastened to the overhead bar. Compute each to one decimal place.

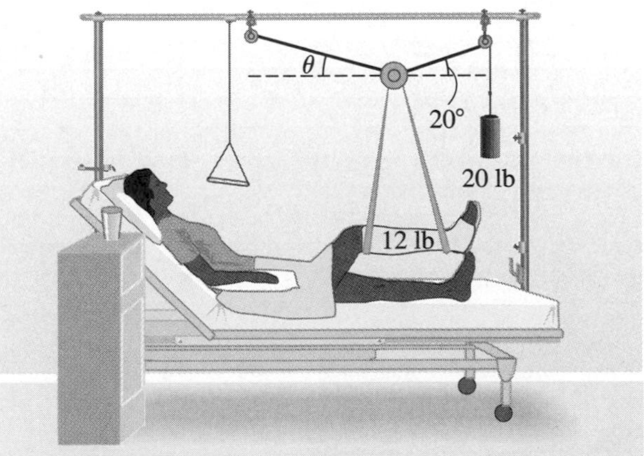

Figure for 58

POLAR COORDINATES; COMPLEX NUMBERS

7

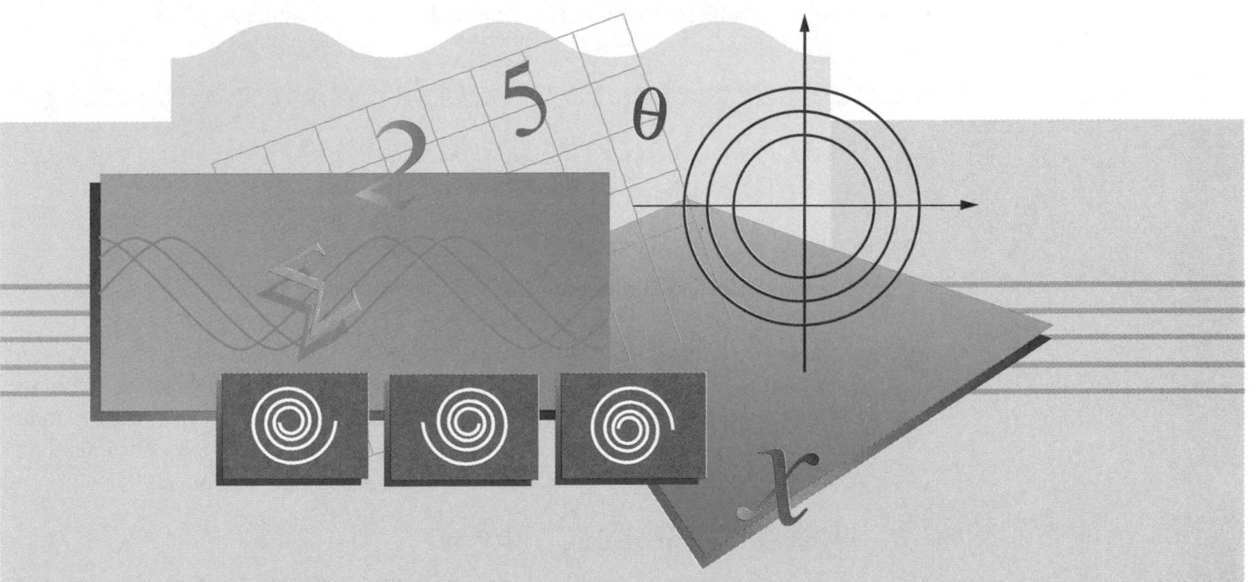

The rectangular coordinate system is the most widely used coordinate system; in this chapter we will discuss the second most widely used coordinate system: the *polar coordinate system.* After graphing some polar equations, we will use the polar coordinate system to represent complex numbers in *polar or trigonometric form.* The polar form of a complex number is used in the statement of De Moivre's famous theorem, which leads to a process of finding all roots of a given number. You may think that the only root of $x^3 = 1$ is 1. This is true if we restrict solutions to real numbers, but if we allow complex numbers, there are three roots—one real and two imaginary.

7.1 POLAR AND RECTANGULAR COORDINATES

- Polar Coordinate System
- From Polar Form to Rectangular Form, and Vice Versa

Until now we have had only one way of associating ordered pairs of numbers with points in a plane—namely, the rectangular coordinate system. Another type of coordinate system is widely used: the *polar coordinate system.* Of the many different kinds of coordinate systems that are possible, the polar coordinate system ranks second in importance only to the rectangular coordinate system.

Polar Coordinate System

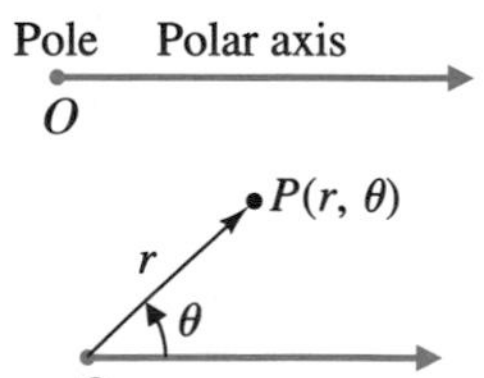

FIGURE 1
Polar coordinate system

To form a **polar coordinate system** in a plane (see Fig. 1), we start with a fixed point O and call it the **pole,** or **origin.** From this point we draw a half-line (usually horizontal and to the right), and we call this line the **polar axis.**

If P is an arbitrary point in a plane, then we associate **polar coordinates** (r, θ) with it as follows: Starting with the polar axis as the initial side of an angle, we rotate the terminal side until it, or the extension of it through the pole, passes through the point. The θ coordinate in (r, θ) is this angle, in degree or radian measure. The angle θ is positive if the rotation is counterclockwise and negative if the rotation is clockwise. The r coordinate in (r, θ) is the directed distance from the pole to the point P, which is positive if measured from the pole along the terminal side of θ and negative if measured along the terminal side extended through the pole. Figure 2 illustrates how one point in a polar coordinate system can have many—in fact, an unlimited number of—polar coordinates.

We can now see a major difference between the rectangular coordinate system and the polar coordinate system. In the former, each point has exactly one set of rectangular coordinates; in the latter, a point may have infinitely many polar coordinates.

The pole itself has polar coordinates of the form $(0, \theta)$, where θ is arbitrary. For example, $(0, 37°)$ and $(0, -\pi/4)$ are both polar coordinates of the pole, and there are infinitely many others.

FIGURE 2

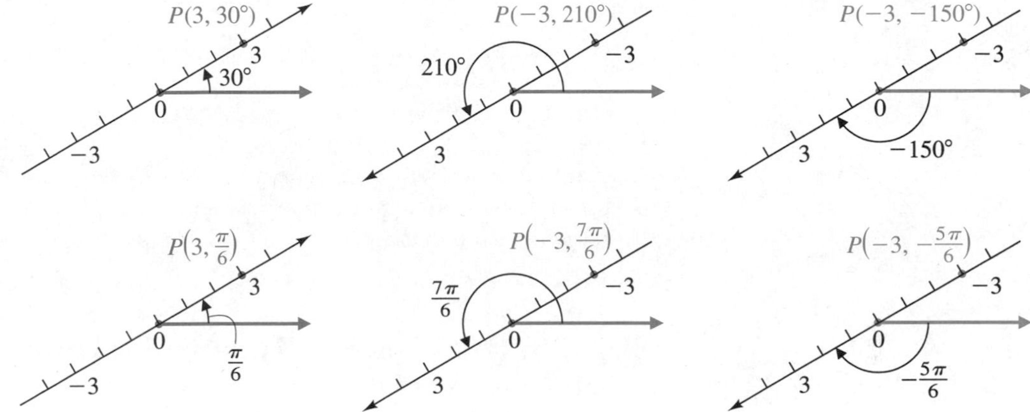

Just as graph paper is readily available for work related to rectangular coordinate systems, polar graph paper is available for work related to polar coordinates. The following examples illustrate its use.

Plotting Points in a Polar Coordinate System

Plot the following points in a polar coordinate system:

(A) $A(4, 45°)$; $B(-6, 45°)$; $C(7, 240°)$; $D(3, -75°)$
(B) $A(8, \pi/6)$; $B(5, -3\pi/4)$; $C(-7, 2\pi/3)$; $D(-9, -\pi/6)$

Solution

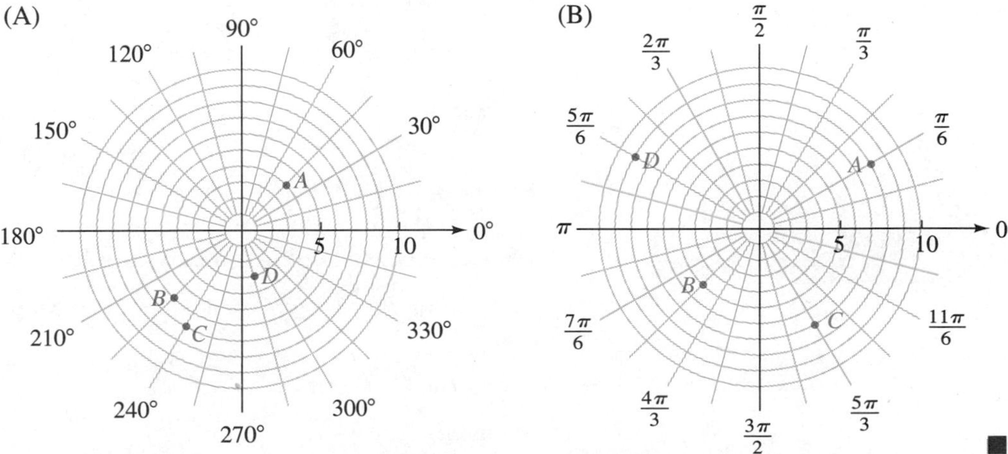

■

Matched Problem 1 Plot the following points in a polar coordinate system:

(A) $A(7, 30°)$; $B(-6, 165°)$; $C(-9, -90°)$
(B) $A(10, \pi/3)$; $B(8, -7\pi/6)$; $C(-5, -5\pi/4)$ ■

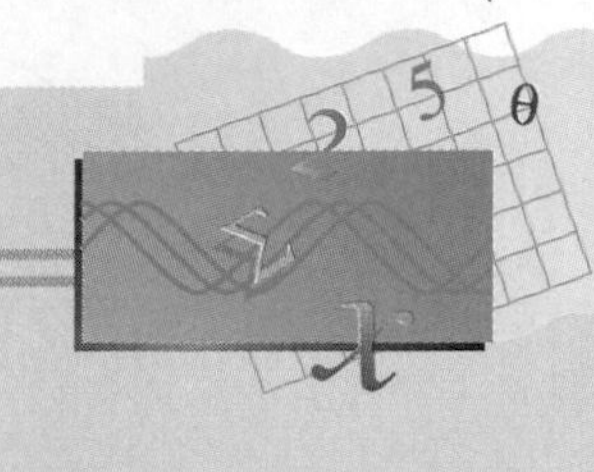

EXPLORE/DISCUSS 1

A point in a polar coordinate system has coordinates (8, 60°). Find the other polar coordinates of the point for θ restricted to $-360° \leq \theta \leq 360°$, and explain how they are found.

From Polar Form to Rectangular Form, and Vice Versa

It is frequently convenient to be able to transform coordinates or equations in rectangular form into polar form, or vice versa. The following polar–rectangular relationships are useful in this regard.

POLAR–RECTANGULAR RELATIONSHIPS

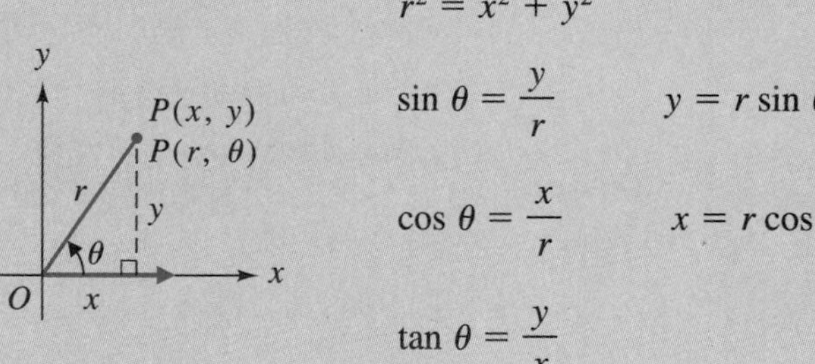

$$r^2 = x^2 + y^2$$

$$\sin\theta = \frac{y}{r} \qquad y = r\sin\theta$$

$$\cos\theta = \frac{x}{r} \qquad x = r\cos\theta$$

$$\tan\theta = \frac{y}{x}$$

[*Note:* The signs of x and y determine the quadrant for θ. The angle θ is usually chosen so that $-\pi < \theta \leq \pi$ or $-180° < \theta \leq 180°$.]

EXAMPLE 2 **Polar Coordinates to Rectangular Coordinates**

Change $A(5, \pi/6)$, $B(-3, 3\pi/4)$, and $C(-2, -5\pi/6)$ to exact rectangular coordinates.

Solution Use $x = r\cos\theta$ and $y = r\sin\theta$.

For A (see Fig. 3):

$$x = 5\cos\frac{\pi}{6} = 5\left(\frac{\sqrt{3}}{2}\right) = \frac{5\sqrt{3}}{2}$$

$$y = 5\sin\frac{\pi}{6} = 5\left(\frac{1}{2}\right) = \frac{5}{2}$$

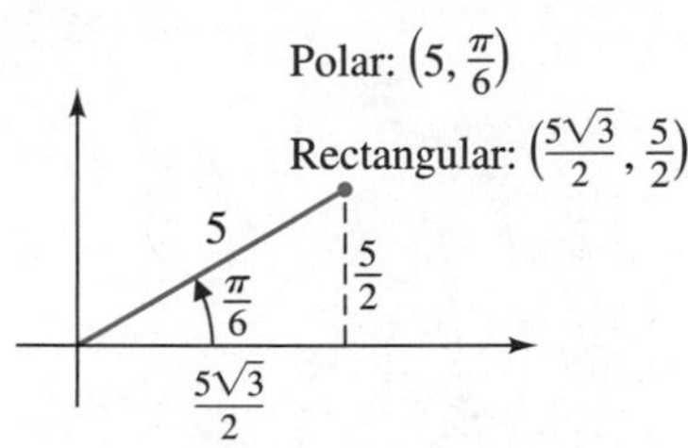

FIGURE 3

Rectangular coordinates: $\left(\frac{5\sqrt{3}}{2}, \frac{5}{2}\right)$

For *B*:

$$x = -3\cos\frac{3\pi}{4} = (-3)\left(\frac{-\sqrt{2}}{2}\right) = \frac{3\sqrt{2}}{2}$$

$$y = -3\sin\frac{3\pi}{4} = (-3)\left(\frac{\sqrt{2}}{2}\right) = \frac{-3\sqrt{2}}{2}$$

Rectangular coordinates: $\left(\frac{3\sqrt{2}}{2}, \frac{-3\sqrt{2}}{2}\right)$

For *C*:

$$x = -2\cos\left(\frac{-5\pi}{6}\right) = (-2)\left(\frac{-\sqrt{3}}{2}\right) = \sqrt{3}$$

$$y = -2\sin\left(\frac{-5\pi}{6}\right) = (-2)\left(\frac{-1}{2}\right) = 1$$

Rectangular coordinates: $(\sqrt{3}, 1)$ ■

Matched Problem 2 Change $A(8, \pi/3)$, $B(-6, 5\pi/4)$, and $C(-4, -7\pi/6)$ to exact rectangular coordinates. ■

Rectangular Coordinates to Polar Coordinates

Change $A(1, \sqrt{3})$ and $B(-\sqrt{3}, -1)$ into exact polar form with $r \geq 0$ and $-\pi < \theta \leq \pi$.

Solution Use $r^2 = x^2 + y^2$ and $\tan\theta = y/x$.

For *A*:

$$r^2 = 1^2 + (\sqrt{3})^2 = 4$$

$$r = 2$$

$$\tan\theta = \frac{\sqrt{3}}{1}$$

$$\theta = \frac{\pi}{3}$$ Since *A* is in the first quadrant.

Polar coordinates: $\left(2, \frac{\pi}{3}\right)$

For B:

$$r^2 = (-\sqrt{3})^2 + (-1)^2 = 4$$

$$r = 2$$

$$\tan\theta = \frac{-1}{-\sqrt{3}}$$

$$\theta = -\frac{5\pi}{6} \qquad \text{Since } B \text{ is in the third quadrant.}$$

Polar coordinates: $\left(2, -\frac{5\pi}{6}\right)$ ■

Matched Problem 3 Change $A(\sqrt{3}, 1)$ and $B(1, -\sqrt{3})$ to polar form with $r \geq 0$ and $-\pi < \theta \leq 2\pi$. ■

Many calculators can automatically convert rectangular coordinates to polar form, and vice versa—read the manual for your particular calculator. Example 4 illustrates conversions using the method shown in Examples 2 and 3 followed by the conversions on a graphing calculator with a built-in conversion routine.

EXAMPLE 4

Calculator Conversion

Perform the conversions following the methods illustrated in Examples 2 and 3; then use a calculator with a built-in conversion routine (if you have one).

(A) Convert the rectangular coordinates $(-6.434, 4.023)$ to polar coordinates (to three decimal places), with θ in degree measure, $-180° < \theta \leq 180°$, and $r \geq 0$.

(B) Convert the polar coordinates $(8.677, -1.385)$ to rectangular coordinates (to three decimal places).

Solution (A) Use a calculator set in degree mode.

$$(x, y) = (-6.434, 4.023)$$

$$r = \sqrt{x^2 + y^2} = \sqrt{(-6.434)^2 + 4.023^2} = 7.588$$

$$\tan\theta = \frac{y}{x} = \frac{4.023}{-6.434}$$

The angle θ is in the second quadrant, and it is to be chosen so that $-180° < \theta \leq 180°$. Thus,

$$\theta = 180° + \tan^{-1}\frac{4.023}{-6.434} = 147.983°$$

Polar coordinates: $(7.588, 147.983°)$

Figure 4 shows the same conversion done on a graphing calculator with a built-in conversion routine.

```
R▸Pr(-6.434,4.02
3)
            7.588
R▸Pθ(-6.434,4.02
3)
          147.983
```

FIGURE 4

```
P▸Rx(8.677,-1.38
5)
           1.603
P▸Ry(8.677,-1.38
5)
          -8.528
```

FIGURE 5

(B) Use a calculator set in radian mode.

$$(r, \theta) = (8.677, -1.385)$$

$$x = r\cos\theta = 8.677\cos(-1.385) = 1.603$$

$$y = r\sin\theta = 8.677\sin(-1.385) = -8.528$$

Rectangular coordinates: $(1.603, -8.528)$

Figure 5 shows the same conversion done on a graphing calculator with a built-in conversion routine. ■

Matched Problem 4 Perform the conversions following the methods illustrated in Examples 2 and 3; then use a calculator with a built-in conversion routine (if you have one).

(A) Convert the rectangular coordinates $(-4.305, -5.117)$ to polar coordinates (to three decimal places), with θ in radian measure, $-\pi < \theta \leq \pi$, and $r \geq 0$.

(B) Convert the polar coordinates $(-10.314, -37.232°)$ to rectangular coordinates (to three decimal places). ■

From Rectangular to Polar Form

Change $x^2 + y^2 - 2x = 0$ to polar form.

Solution Use $r^2 = x^2 + y^2$ and $x = r\cos\theta$:

$$x^2 + y^2 - 2x = 0$$

$$r^2 - 2r\cos\theta = 0$$

$$r(r - 2\cos\theta) = 0$$

$$r = 0 \quad \text{or} \quad r - 2\cos\theta = 0$$

The graph of $r = 0$ is the pole, and since the pole is included as a solution of $r - 2\cos\theta = 0$ (let $\theta = \pi/2$), we can discard $r = 0$ and keep only

$$r - 2\cos\theta = 0$$

or

$$r = 2\cos\theta$$

■

Matched Problem 5 Change $x^2 + y^2 - 2y = 0$ to polar form. ■

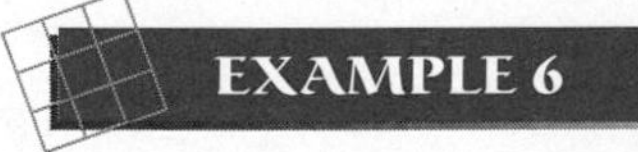

From Polar to Rectangular Form

Change $r + 3\sin\theta = 0$ to rectangular form.

Solution The conversion of this equation, as it stands, to rectangular form gets messy. A simple trick, however, makes the conversion easy: We multiply both sides by r, which simply adds the pole to the graph. In this case, the pole is already included as a solution of $r + 3 \sin \theta = 0$ (let $\theta = 0$), so we have not actually changed anything by doing this. Thus,

$$r + 3 \sin \theta = 0 \qquad \text{Multiply both sides by } r.$$

$$r^2 + 3r \sin \theta = 0 \qquad r^2 = x^2 + y^2, \quad y = r \sin \theta$$

$$x^2 + y^2 + 3y = 0$$

■

Matched Problem 6 Change $r = -8 \cos \theta$ to rectangular form. ■

Answers to Matched Problems

1. (A) 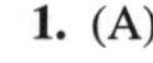

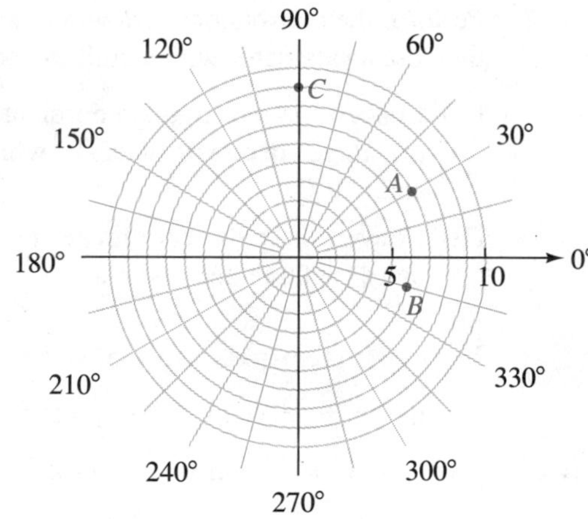

(B)

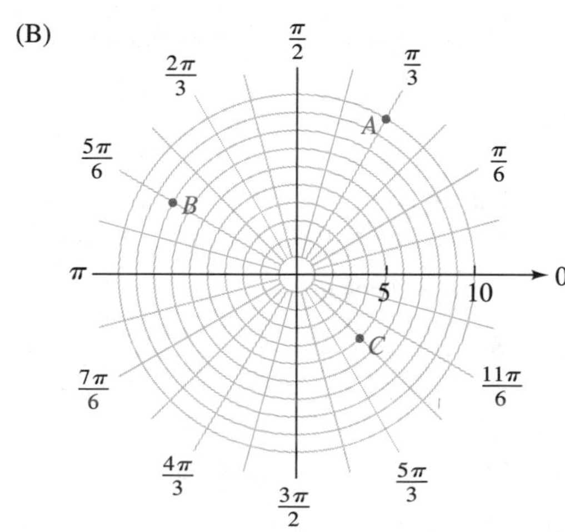

2. $A(4, 4\sqrt{3})$, $B(3\sqrt{2}, 3\sqrt{2})$, $C(2\sqrt{3}, -2)$
3. $A(2, \pi/6)$, $B(2, -\pi/3)$
4. (A) $(r, \theta) = (6.687, -2.270)$ (B) $(x, y) = (-8.212, 6.240)$
5. $r = 2 \sin \theta$ 6. $x^2 + y^2 + 8x = 0$

EXERCISE 7.1

A *Plot in a polar coordinate system.*

1. $A(8, 0°)$; $B(5, 90°)$; $C(6, 30°)$

2. $A(4, 0°)$; $B(7, 180°)$; $C(9, 45°)$

3. $A(-8, 0°)$; $B(-5, 90°)$; $C(-6, 30°)$

4. $A(-4, 0°)$; $B(-7, 180°)$; $C(-9, 45°)$

5. $A(5, -30°)$; $B(4, -45°)$; $C(9, -90°)$

6. $A(8, -45°)$; $B(6, -60°)$; $C(4, -30°)$

7. $A(-5, -30°)$; $B(-4, -45°)$; $C(-9, -90°)$

8. $A(-8, -45°)$; $B(-6, -60°)$; $C(-4, -30°)$

9. $A(6, \pi/6)$; $B(5, \pi/2)$; $C(8, \pi/4)$

10. $A(8, \pi/3)$; $B(4, \pi/4)$; $C(10, 0)$

11. $A(-6, \pi/6)$; $B(-5, \pi/2)$; $C(-8, \pi/4)$

12. $A(-8, \pi/3)$; $B(-4, \pi/4)$; $C(-10, 0)$

13. $A(6, -\pi/6)$; $B(5, -\pi/2)$; $C(8, -\pi/4)$

14. $A(8, -\pi/3)$; $B(4, -\pi/4)$; $C(10, -\pi/6)$

Change to exact rectangular coordinates.

15. $(8, \pi/3)$ **16.** $(4, \pi/4)$

17. $(-4, \pi/4)$ **18.** $(-6, \pi/6)$

19. $(-4, -\pi/6)$ **20.** $(-5, -\pi/3)$

In Problems 21–26, change to exact polar coordinates with $r \ge 0$ and $-\pi < \theta \le \pi$.

21. $(2\sqrt{3}, 2)$ **22.** $(3, 3\sqrt{3})$

23. $(-4, -4\sqrt{3})$ **24.** $(5\sqrt{2}, -5\sqrt{2})$

25. $(0, -7)$ **26.** $(-10, 0)$

27. A point in a polar coordinate system has coordinates $(6, -30°)$. Find all other polar coordinates for the point, $-360° < \theta \le 360°$, and verbally describe how the coordinates are associated with the point.

28. A point in a polar coordinate system has coordinates $(-5, 3\pi/4)$. Find all other polar coordinates for the point, $-2\pi < \theta \le 2\pi$, and verbally describe how the coordinates are associated with the point.

B *In Problems 29–34, convert the rectangular coordinates to polar coordinates (to three decimal places), with $r \ge 0$ and $-180° < \theta \le 180°$.*

29. $(6.913, 4.705)$ **30.** $(3.532, 7.118)$

31. $(-8.336, 4.291)$ **32.** $(7.330, -2.046)$

33. $(-16.322, -27.089)$ **34.** $(-22.667, -14.212)$

In Problems 35–40, convert the polar coordinates to rectangular coordinates (to three decimal places).

35. $(7.066, 125.317°)$ **36.** $(6.097, 42.554°)$

37. $(3.768, -2.113)$ **38.** $(-8.207, 2.410)$

39. $(-9.028, -0.663)$ **40.** $(-4.233, -2.084)$

Change to polar form.

41. $6x - x^2 = y^2$ **42.** $y^2 = 5y - x^2$

43. $2x + 3y = 5$ **44.** $3x - 5y = -2$

45. $x^2 + y^2 = 9$ **46.** $y = x$

47. $2xy = 1$ **48.** $y^2 = 4x$

49. $4x^2 - y^2 = 4$ **50.** $x^2 + 9y^2 = 9$

Change to rectangular form.

51. $r(2\cos\theta + \sin\theta) = 4$

52. $r(3\cos\theta - 4\sin\theta) = -1$

53. $r = 8\cos\theta$ **54.** $r = -2\sin\theta$

55. $r^2\cos 2\theta = 4$ **56.** $r^2\sin 2\theta = 2$

57. $r = 4$ **58.** $r = -5$

59. $\theta = 30°$ **60.** $\theta = \pi/4$

C **61.** Change $r = 3/(\sin\theta - 2)$ into rectangular form.

62. Change $(y - 3)^2 = 4(x^2 + y^2)$ into polar form.

Applications

63. Precalculus: Analytic Geometry A distance d (see the figure) between two points in a polar coordinate

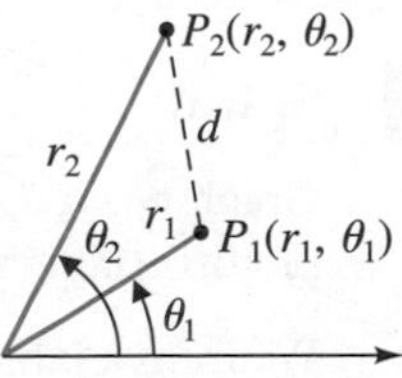

Figure for 63 and 64

system is given by a formula that follows directly from the law of cosines:

$$d^2 = r_1^2 + r_2^2 - 2r_1r_2\cos(\theta_2 - \theta_1)$$

$$d = \sqrt{r_1^2 + r_2^2 - 2r_1r_2\cos(\theta_2 - \theta_1)}$$

Find the distance between the two points $P_1(2, 30°)$ and $P_2(3, 60°)$. Compute your answer to four significant digits.

64. **Precalculus: Analytic Geometry** Refer to Problem 63 and find the distance between the two points $P_1(4, \pi/4)$ and $P_2(1, \pi/2)$. Compute your answer to four significant digits.

7.2 SKETCHING POLAR EQUATIONS

- Point-by-Point Sketching
- Rapid Sketching
- Graphing Polar Equations on a Graphing Utility
- A Table of Standard Polar Curves
- Application

Just as in rectangular coordinate systems where we sketched graphs of equations involving the variables x and y, in a polar coordinate system we graph equations involving the variables r and θ. We shall see that certain curves have simpler representations in polar coordinates and other curves have simpler representations in rectangular coordinates. Likewise, some applications have simpler solutions in polar coordinates while other applications have simpler solutions in rectangular coordinates. In this section you will gain experience in recognizing and sketching some standard polar graphs.

Point-by-Point Sketching

To graph a polar equation such as $r = 2\theta$ or $r = 4\sin 2\theta$ in a polar coordinate system, we locate all points with coordinates that satisfy the equation. A sketch of a graph can be obtained (just as in rectangular coordinates) by making a table of values that satisfy the equation, plotting these points, and then joining them with a smooth curve. A calculator can be used to generate the table.

The graph of the polar equation

$$r = a\theta \qquad a > 0$$

is called **Archimedes' spiral.** Example 1 illustrates how to obtain a particular spiral by point-by-point plotting.

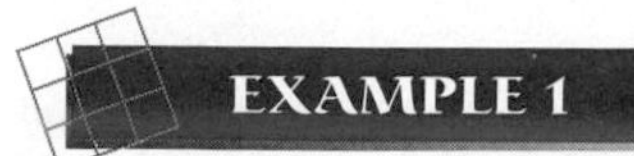

EXAMPLE 1 **Spiral**

Graph $r = 2\theta, 0 \le \theta \le 5\pi/3$ in a polar coordinate system using point-by-point plotting (θ in radians).

Solution We form a table using multiples of $\pi/6$; then plot the points and join them with a smooth curve (see Fig. 1).

FIGURE 1

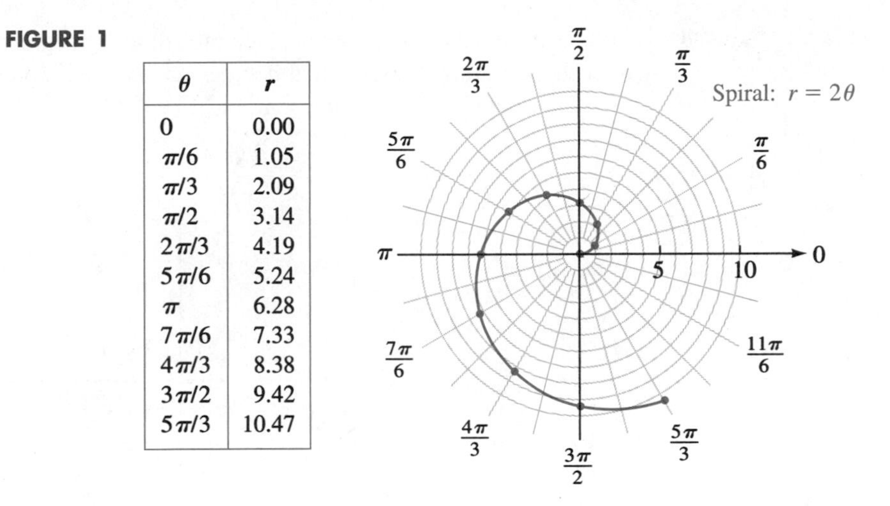

θ	r
0	0.00
$\pi/6$	1.05
$\pi/3$	2.09
$\pi/2$	3.14
$2\pi/3$	4.19
$5\pi/6$	5.24
π	6.28
$7\pi/6$	7.33
$4\pi/3$	8.38
$3\pi/2$	9.42
$5\pi/3$	10.47

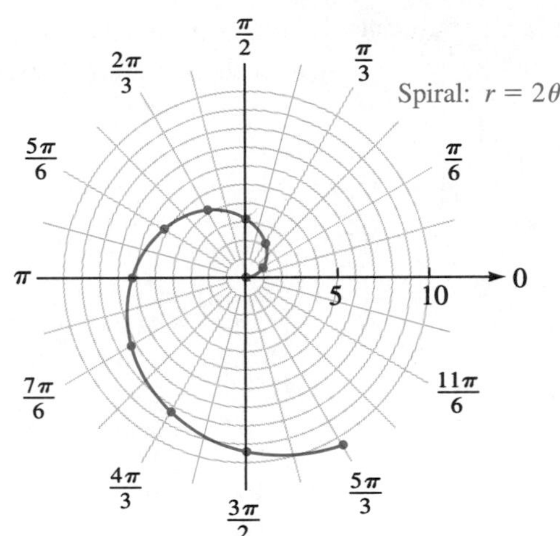

Matched Problem 1 Graph $r = 3\theta$, $0 \le \theta \le 7\pi/6$ in a polar coordinate system using point-by-point plotting (θ in radians).

Circle

Graph $r = 4 \cos \theta$ (θ in radians).

Solution We start with multiples of $\pi/6$ and continue until the graph begins to repeat (see Fig. 2). (In intervals of uncertainty, add more points.) The graph is a circle with radius 2 and center at (2, 0).

FIGURE 2

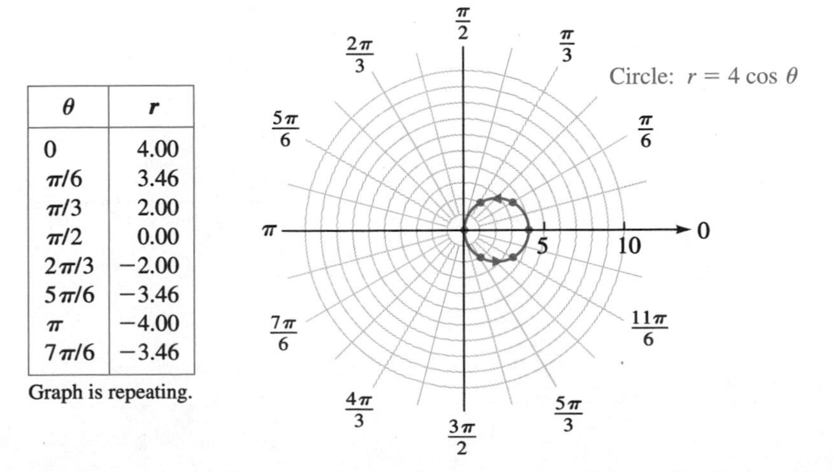

θ	r
0	4.00
$\pi/6$	3.46
$\pi/3$	2.00
$\pi/2$	0.00
$2\pi/3$	−2.00
$5\pi/6$	−3.46
π	−4.00
$7\pi/6$	−3.46

Graph is repeating.

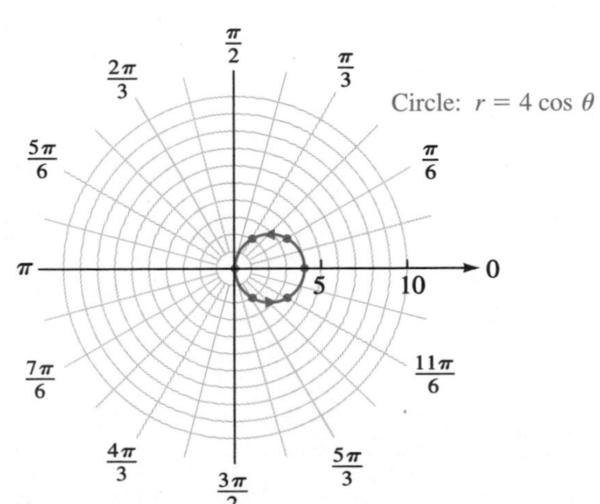

Matched Problem 2 Graph $r = 4 \sin \theta$ (θ in radians).

If the equation in Example 2 is graphed using degrees instead of radians, we get the same graph except that degrees are marked around the polar coordinate system instead of radians (see Fig. 3).

FIGURE 3

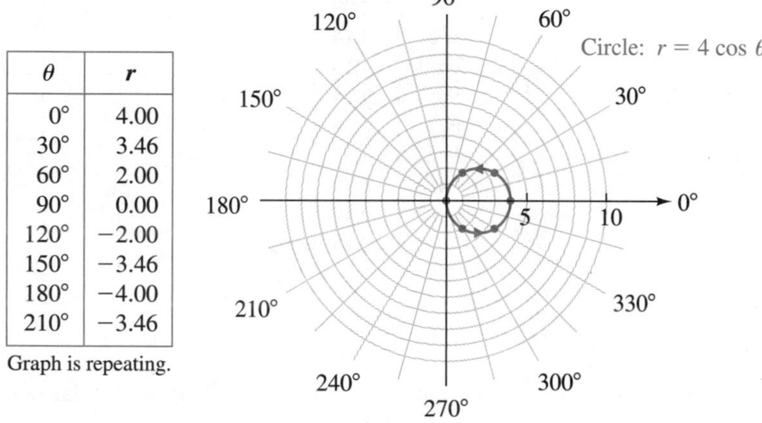

θ	r
0°	4.00
30°	3.46
60°	2.00
90°	0.00
120°	−2.00
150°	−3.46
180°	−4.00
210°	−3.46

Graph is repeating.

Rapid Sketching

If all that is desired is a rough sketch of a polar equation involving $\sin \theta$ or $\cos \theta$, the point-by-point process can be speeded up by taking advantage of the uniform variation of $\sin \theta$ and $\cos \theta$ as θ moves around a unit circle. We refer to this process as **rapid polar sketching.** It is useful to visualize the unit circle definition of sine and cosine in Figure 4 in the process.

FIGURE 4

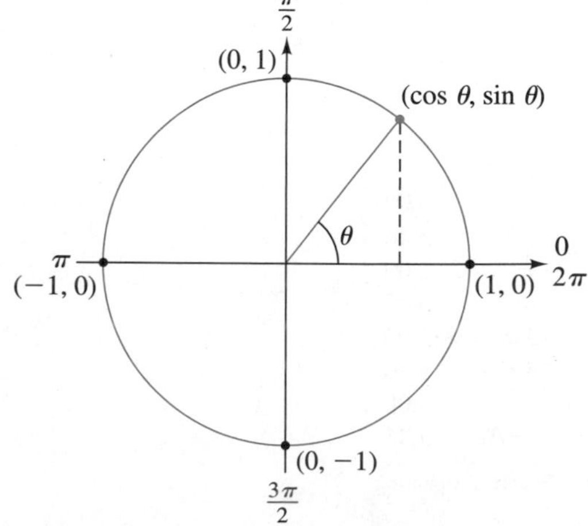

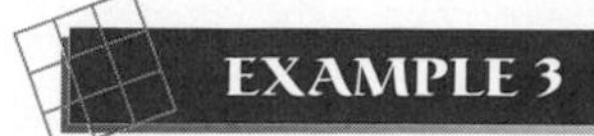

Rapid Polar Sketching

Sketch $r = 5 + 5 \cos \theta$ (θ in radians).

Solution We set up Table 1, which shows how r varies as θ varies through each set of quadrant values in Figure 4.

TABLE 1

θ varies from	$\cos\theta$ varies from	$5\cos\theta$ varies from	$r = 5 + 5\cos\theta$ varies from
0 to $\pi/2$	1 to 0	5 to 0	10 to 5
$\pi/2$ to π	0 to -1	0 to -5	5 to 0
π to $3\pi/2$	-1 to 0	-5 to 0	0 to 5
$3\pi/2$ to 2π	0 to 1	0 to 5	5 to 10

Notice that as θ increases from 0 to $\pi/2$, $\cos\theta$ decreases from 1 to 0, $5\cos\theta$ decreases from 5 to 0, and $r = 5 + 5\cos\theta$ decreases from 10 to 5, and so on. Sketching the values in Table 1, we obtain the heart-shaped graph, called a **cardioid,** in Figure 5.

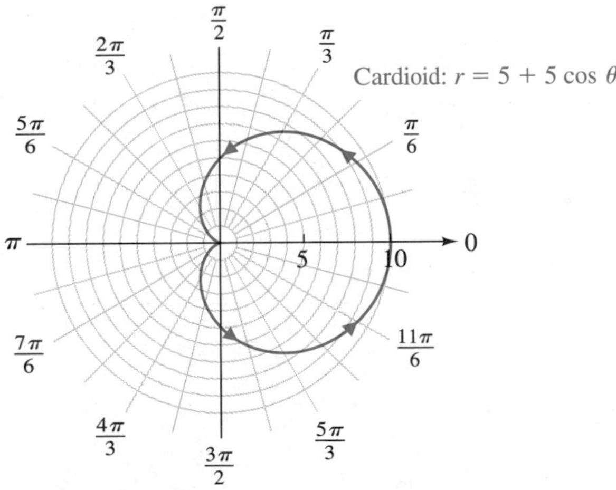

FIGURE 5
Cardioid

Matched Problem 3 Sketch $r = 4 + 4\sin\theta$ (θ in radians).

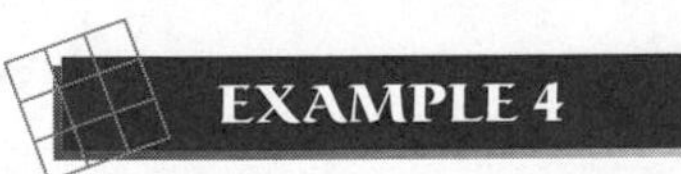

EXAMPLE 4 Rapid Polar Sketching

Sketch $r = 6\sin 2\theta$ (θ in radians).

Solution We start by letting 2θ (instead of θ) range through each set of quadrant values. That is, we start with values of 2θ in the second column of Table 2 (page 406), fill in the table to the right, and then fill in the first column for θ.

TABLE 2

θ varies from	2θ varies from	$\sin 2\theta$ varies from	$r = 6 \sin 2\theta$ varies from
0 to $\pi/4$	0 to $\pi/2$	0 to 1	0 to 6
$\pi/4$ to $\pi/2$	$\pi/2$ to π	1 to 0	6 to 0
$\pi/2$ to $3\pi/4$	π to $3\pi/2$	0 to -1	0 to -6
$3\pi/4$ to π	$3\pi/2$ to 2π	-1 to 0	-6 to 0
π to $5\pi/4$	2π to $5\pi/2$	0 to 1	0 to 6
$5\pi/4$ to $3\pi/2$	$5\pi/2$ to 3π	1 to 0	6 to 0
$3\pi/2$ to $7\pi/4$	3π to $7\pi/2$	0 to -1	0 to -6
$7\pi/4$ to 2π	$7\pi/2$ to 4π	-1 to 0	-6 to 0

Using the information in Table 2, we complete the graph shown in Figure 6, which is called a **four-leaved rose.**

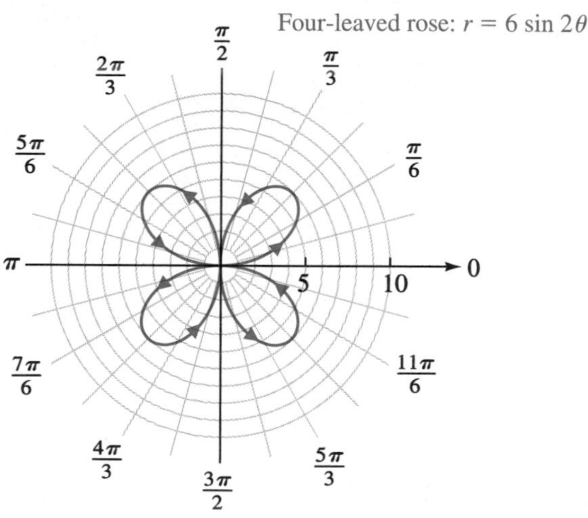

FIGURE 6
Four-leaved rose

Matched Problem 4 Sketch $r = 4 \cos 3\theta$ (θ in radians).

In a rectangular coordinate system the simplest type of equations to graph are found by setting the variables x and y equal to constants:

$$x = a \qquad \text{and} \qquad y = b$$

The graphs are straight lines: $x = a$ is a vertical line and $y = b$ is a horizontal line. A look ahead to Table 3 (page 409) will show you that horizontal and vertical lines do not have such simple expressions in polar coordinates.

The easiest type of equations to graph in a polar coordinate system are found by setting the polar variables r and θ equal to constants:

$$r = a \qquad \text{and} \qquad \theta = b$$

Figure 7 illustrates two particular cases. Notice that a circle centered at the origin has a simpler expression in polar coordinates than in rectangular coordinates.

FIGURE 7

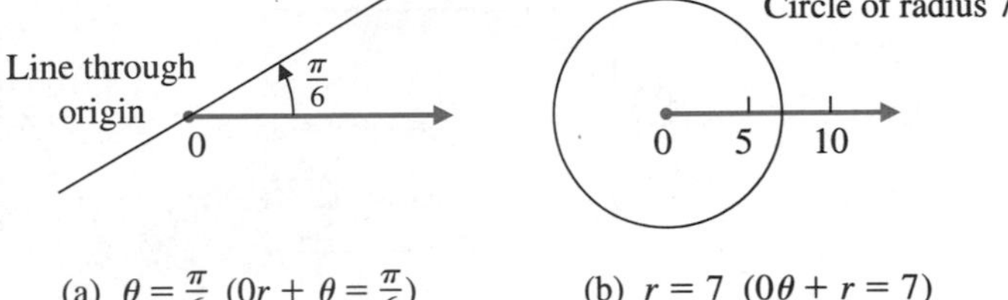

(a) $\theta = \frac{\pi}{6}$ $(0r + \theta = \frac{\pi}{6})$ (b) $r = 7$ $(0\theta + r = 7)$

Graphing Polar Equations on a Graphing Utility

We now turn to the graphing of polar equations on a graphing utility. Note that many graphing utilities, including the one used here, do not show a polar grid. Also, when using TRACE, many graphing utilities offer a choice between polar coordinates and rectangular coordinates for points on the polar curve.

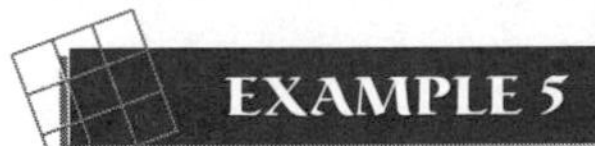

Graphing on a Graphing Utility

Graph each polar equation on a graphing utility. (These are the same equations we sketched by hand in Examples 1–4.)

(A) $r = 2\theta$, $0 \leq \theta \leq 5\pi/3$ (B) $r = 4\cos\theta$, θ in radians

(C) $r = 5 + 5\cos\theta$, θ in radians (D) $r = 6\sin 2\theta$, θ in radians

Solution Refer to the manual for your graphing utility as necessary. In particular:

- Set the graphing utility in polar mode and select polar coordinates and radian measure.
- Adjust window dimensions to accommodate the whole graph (a little trial and error may be necessary).
- To show the true shape of a curve, choose window dimensions so that unit distances on the x and y axes have the same length (by using a zoom square command, for example).

Figure 8 shows the graphs in square graphing calculator viewing windows.

FIGURE 8

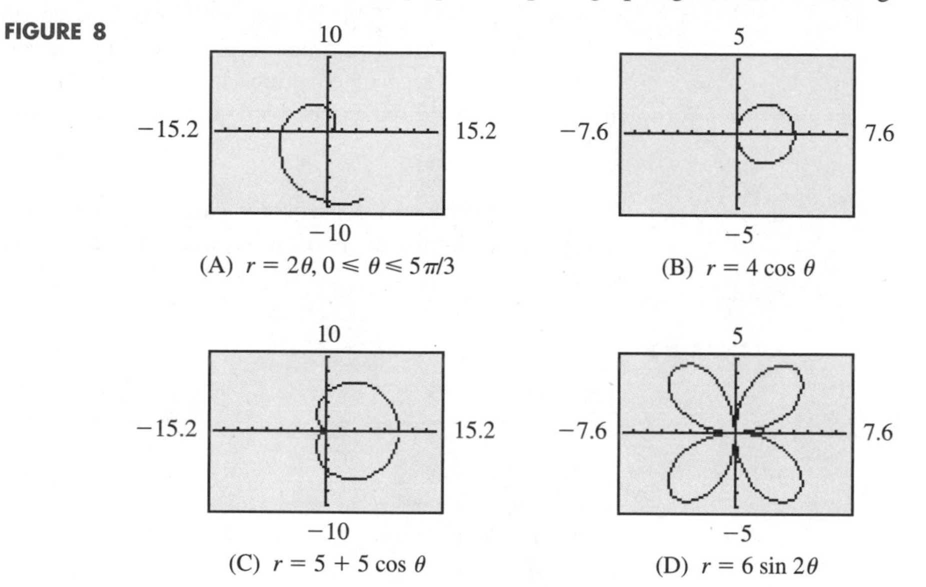

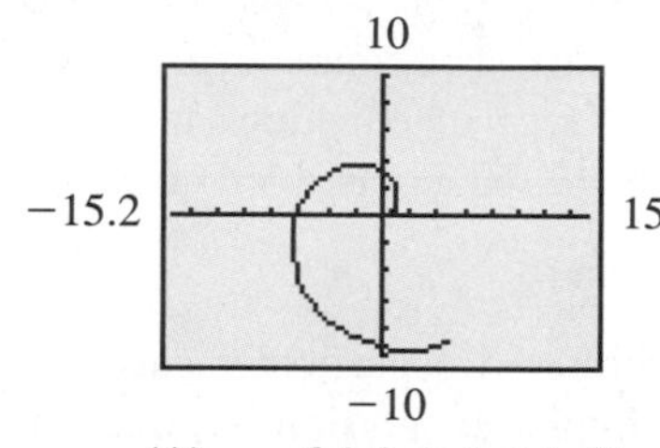

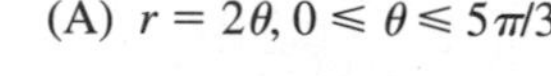
(A) $r = 2\theta, 0 \leq \theta \leq 5\pi/3$

5
−7.6
7.6
−5

(B) $r = 4 \cos \theta$

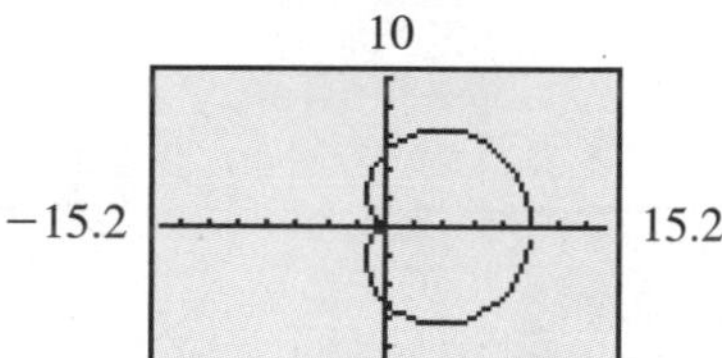

(C) $r = 5 + 5 \cos \theta$

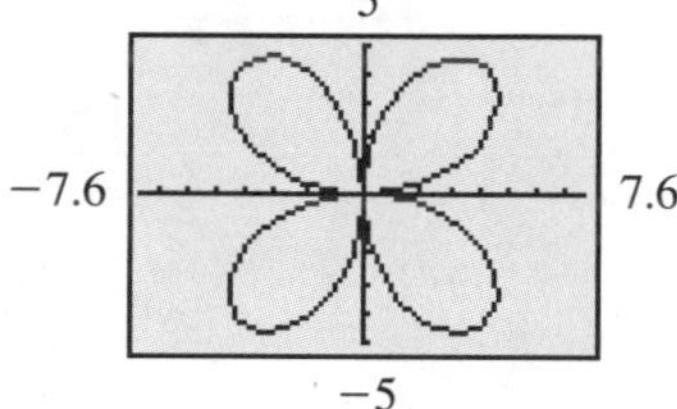

(D) $r = 6 \sin 2\theta$

Matched Problem 5 Graph each polar equation on a graphing utility. (These are the same equations we sketched by hand in Matched Problems 1–4.) Use a square viewing window.

(A) $r = 3\theta, 0 \leq \theta \leq 7\pi/6$

(B) $r = 4 \sin \theta$, θ in radians

(C) $r = 4 + 4 \sin \theta$, θ in radians

(D) $r = 4 \cos 3\theta$, θ in radians

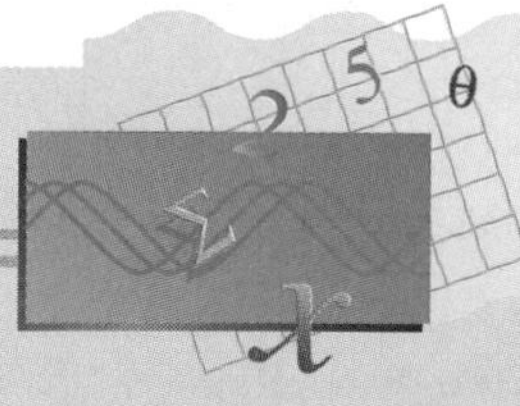

EXPLORE/DISCUSS 1

(A) Graph $r_1 = 8 \cos \theta$ and $r_2 = 8 \sin \theta$ in the same viewing window. Use TRACE on r_1 and estimate the polar coordinates where the two graphs intersect. Repeat for r_2.

(B) Which intersection point appears to have the same polar coordinates on each curve and consequently represents a simultaneous solution to both equations? Which intersection point appears to have different polar coordinates on each curve and consequently does not represent a simultaneous solution?

(C) Solve the system of equations above for r and θ. Conclusions?

(D) Could the situation illustrated in parts (A)–(C) happen with rectangular equations in two variables in a rectangular coordinate system? Discuss.

A Table of Standard Polar Curves

We now present a table of a few standard polar curves. Graphing polar equations is often made easier if you have an idea of what the shape of the curve will be.

TABLE 3
Some Standard Polar Curves

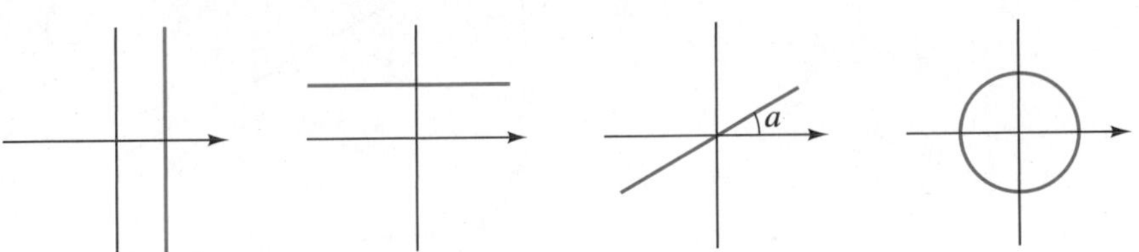

(a) Vertical line: $r = \dfrac{a}{\cos\theta}$

(b) Horizontal line: $r = \dfrac{a}{\sin\theta}$

(c) Radial line: $\theta = a$

(d) Circle: $r = a$

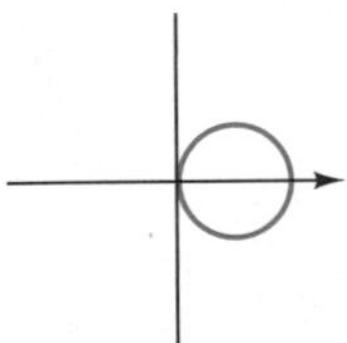

(e) Circle: $r = a\cos\theta$

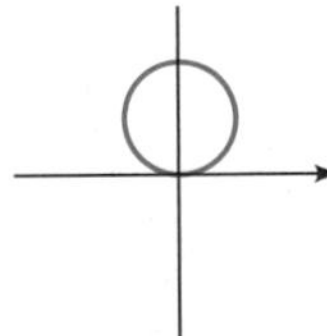

(f) Circle: $r = a\sin\theta$

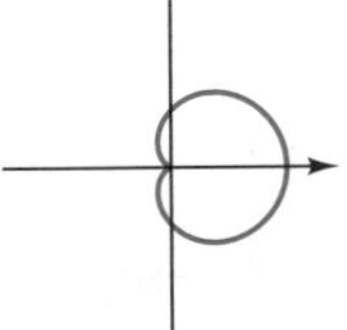

(g) Cardioid: $r = a + a\cos\theta$

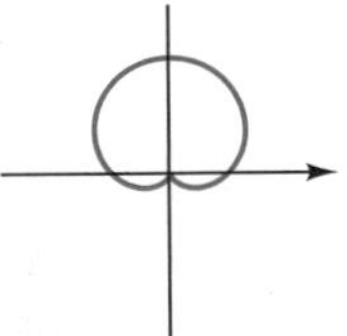

(h) Cardioid: $r = a + a\sin\theta$

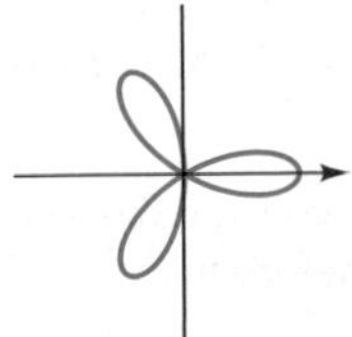

(i) Three-leaved rose: $r = a\cos 3\theta$

(j) Four-leaved rose: $r = a\cos 2\theta$

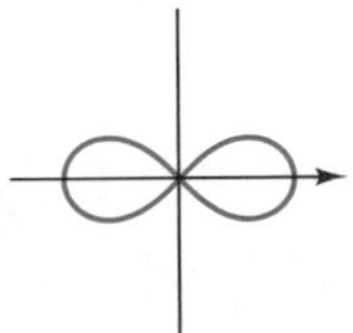

(k) Lemniscate: $r^2 = a^2\cos 2\theta$

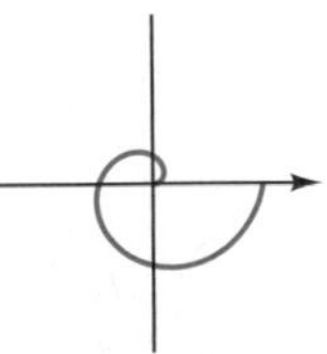

(l) Archimedes' spiral: $r = a\theta$

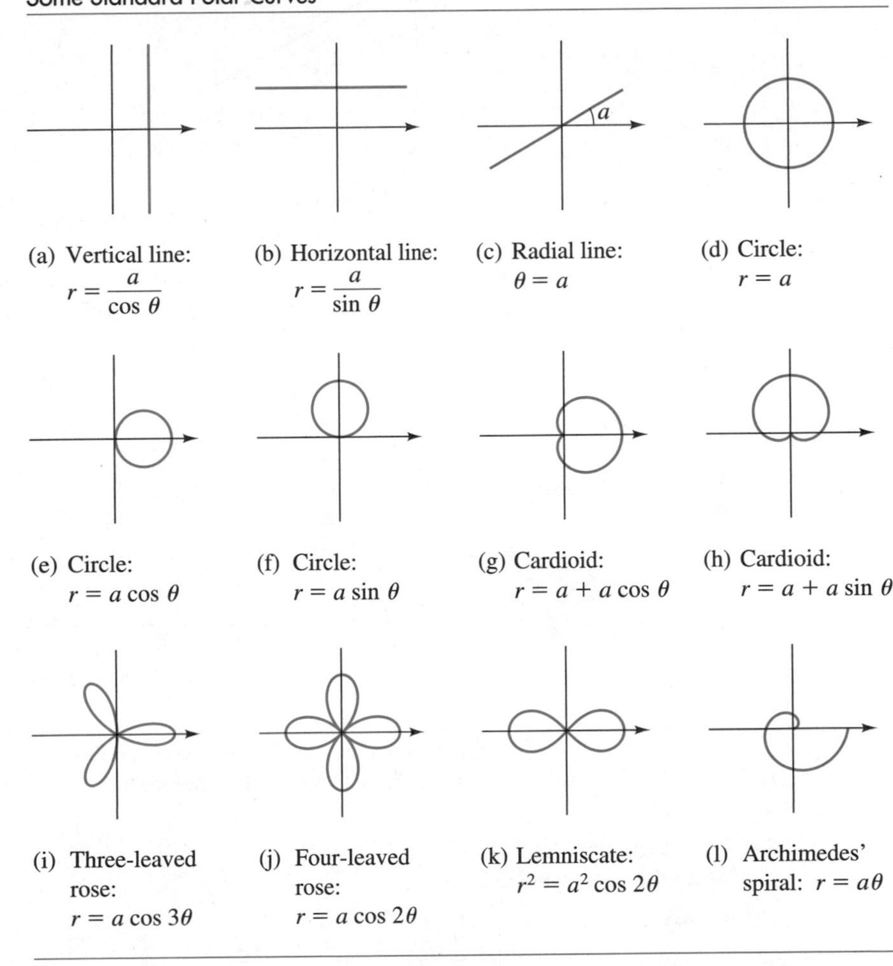

Application

Polar coordinate systems are useful in many types of applications. Exercise 7.2 includes applications from sailboat racing and astronomy; Figure 9 (at the top of the next page), which was supplied by the U.S. Coast and Geodetic Survey, illustrates an application from oceanography. In Figure 9 each arrow represents the direction and magnitude of the tide at a particular time of the day at the San Francisco Light Station (a navigational light platform in the Pacific Ocean about 12 nautical miles west of the Golden Gate Bridge).

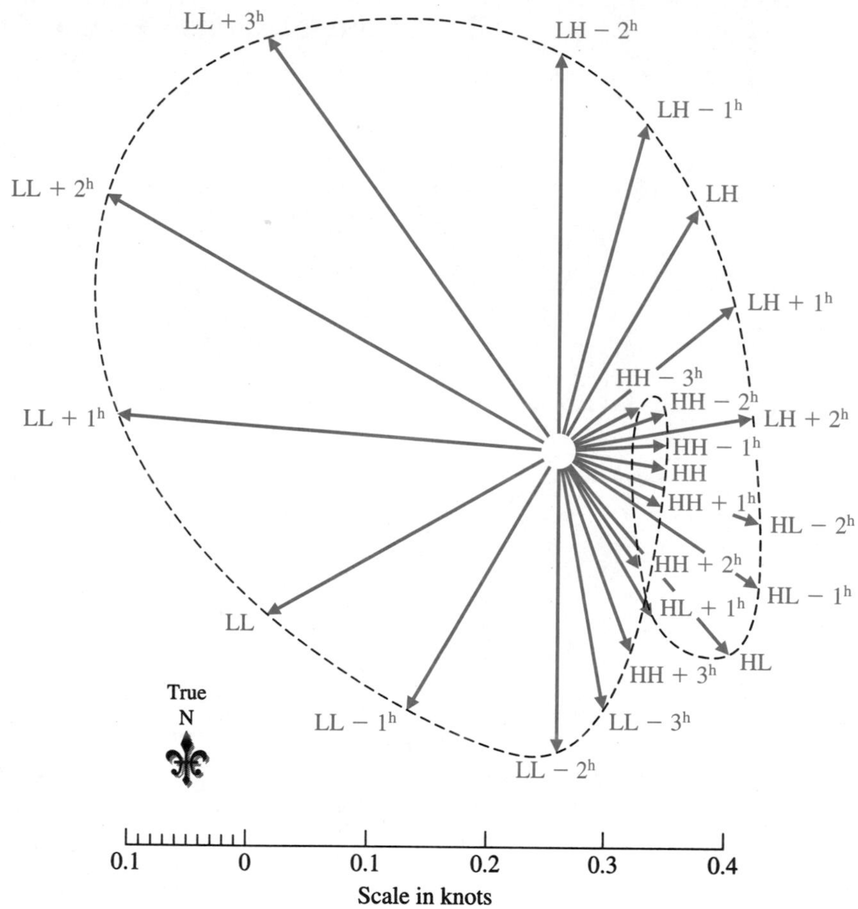

FIGURE 9
Tidal current curve, San Francisco Light Station [*Note:* LL + 3^h means 3 hr after low low tide, HL − 1^h means 1 hr before high low tide, and so on.]

Answers to Matched Problems

1.

θ	r
0	0.00
$\pi/6$	1.57
$\pi/3$	3.14
$\pi/2$	4.71
$2\pi/3$	6.28
$5\pi/6$	7.85
π	9.42
$7\pi/6$	11.00

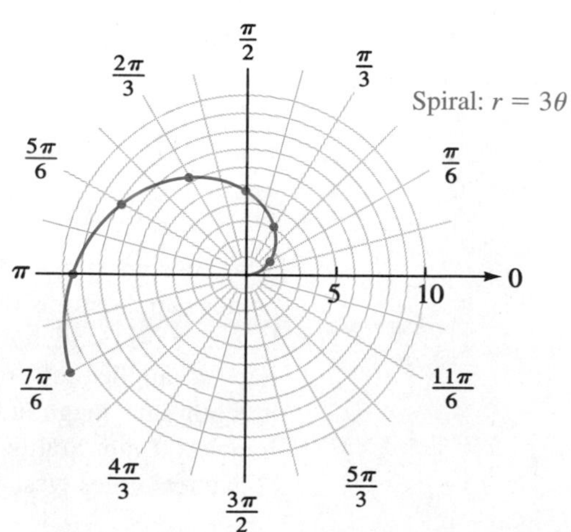

2.

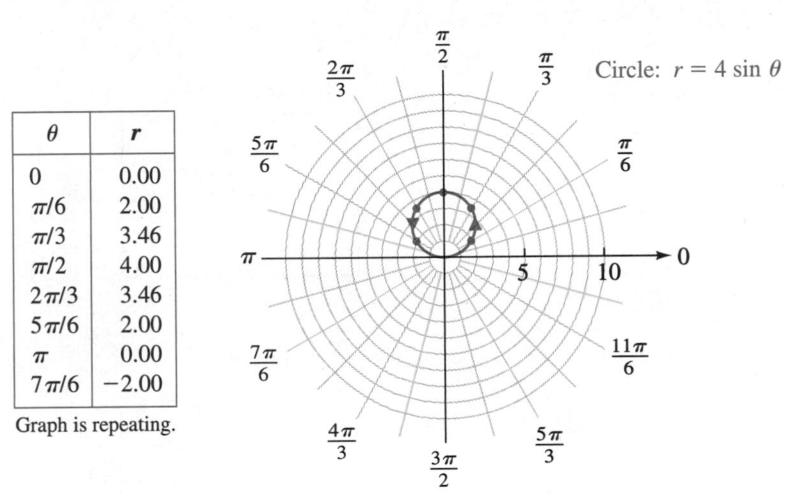

θ	r
0	0.00
$\pi/6$	2.00
$\pi/3$	3.46
$\pi/2$	4.00
$2\pi/3$	3.46
$5\pi/6$	2.00
π	0.00
$7\pi/6$	−2.00

Graph is repeating.

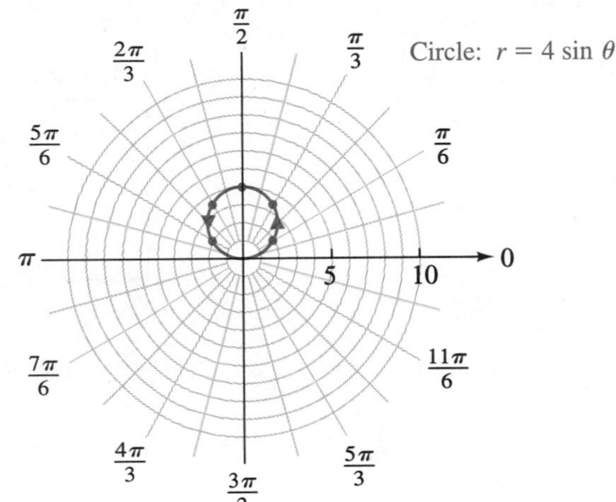

3.

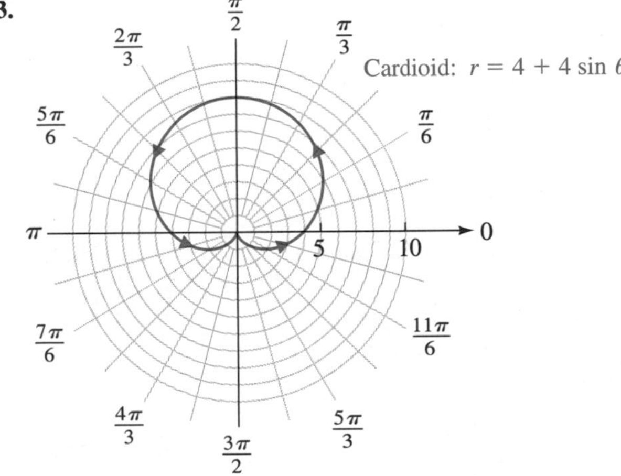

4.

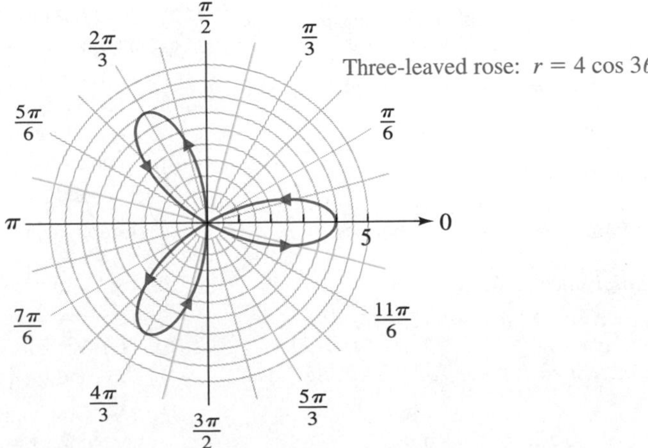

5. (A) $r = 3\theta$

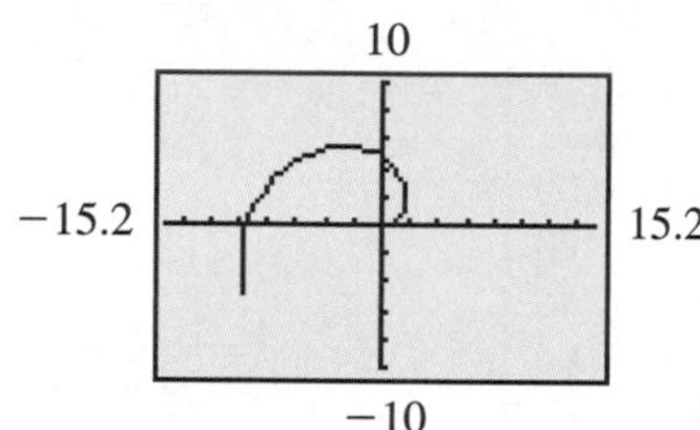

(B) $r = 4 \sin \theta$

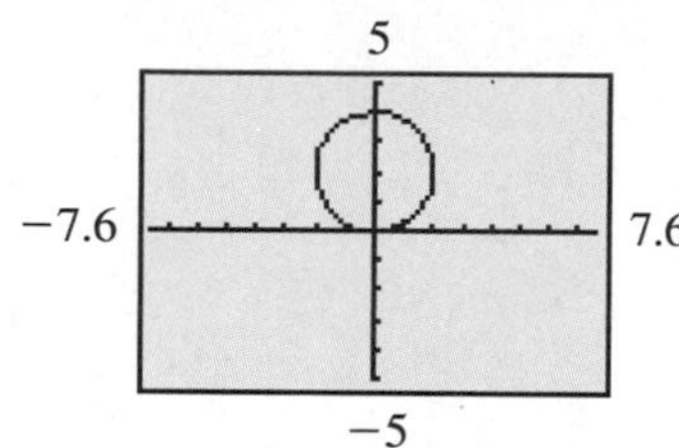

(C) $r = 4 + 4 \sin \theta$

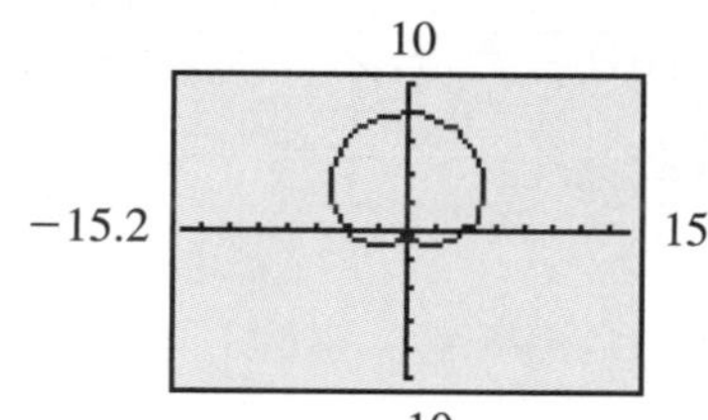

(D) $r = 4 \cos 3\theta$

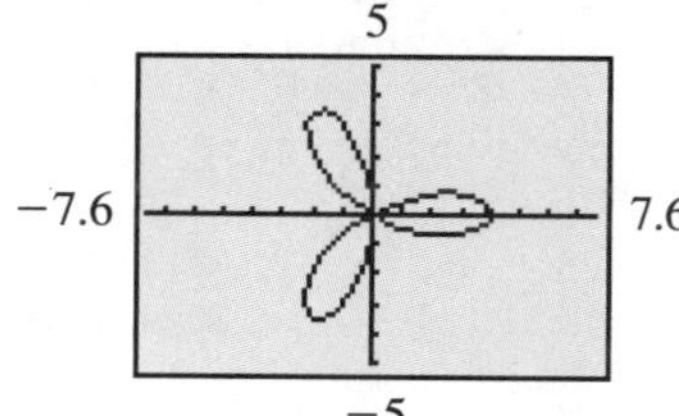

EXERCISE 7.2

A *In Problems 1–6, use a calculator as an aid to point-by-point plotting.*

1. Graph $r = 10 \cos \theta$ by assigning θ the values 0, $\pi/6$, $\pi/4$, $\pi/3$, $\pi/2$, $2\pi/3$, $3\pi/4$, $5\pi/6$, and π. Then join the resulting points with a smooth curve.

2. Repeat Problem 1 for $r = 8 \sin \theta$, using the same set of values for θ.

3. Graph $r = 3 + 3 \cos \theta$, $0° \le \theta \le 360°$, using multiples of 30° starting at 0.

4. Graph $r = 8 + 8 \sin \theta$, $0° \le \theta \le 360°$, using multiples of 30° starting at 0.

5. Graph $r = \theta$, $0 \le \theta \le 2\pi$, using multiples of $\pi/6$ for θ starting at $\theta = 0$.

6. Graph $r = \theta/2$, $0 \le \theta \le 2\pi$, using multiples of $\pi/2$ for θ starting at $\theta = 0$.

Verify the graphs for Problems 1–6 on a graphing utility.

In Problems 7–10, graph each polar equation.

7. $r = 5$ **8.** $r = 8$ **9.** $\theta = \pi/4$ **10.** $\theta = \pi/3$

B *In Problems 11–20, sketch each polar equation using rapid sketching techniques.*

11. $r = 4 \cos \theta$ **12.** $r = 4 \sin \theta$

13. $r = 8 \cos 2\theta$ **14.** $r = 10 \sin 2\theta$

15. $r = 6 \sin 3\theta$ **16.** $r = 5 \cos 3\theta$

17. $r = 3 + 3 \cos \theta$ **18.** $r = 2 + 2 \sin \theta$

19. $r = 2 + 4 \cos \theta$ **20.** $r = 2 + 4 \sin \theta$

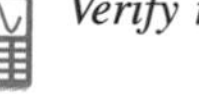

Verify the graphs for Problems 11–20 on a graphing utility.

Problems 21–28 are exploratory problems that require the use of a graphing utility.

21. Graph each polar equation in its own viewing window:
(A) $r = 5 + 5 \cos \theta$
(B) $r = 5 + 4 \cos \theta$
(C) $r = 4 + 5 \cos \theta$
(D) Verbally describe the effect of the relative size of a and b on the graph of $r = a + b \cos \theta$, $a, b > 0$.

22. Graph each polar equation in its own viewing window:
(A) $r = 5 + 5 \sin \theta$
(B) $r = 5 + 4 \sin \theta$
(C) $r = 4 + 5 \sin \theta$
(D) Verbally describe the effect of the relative size of a and b on the graph of $r = a + b \sin \theta$, $a, b > 0$.

23. (A) Graph each polar equation in its own viewing window: $r = 9 \cos \theta$, $r = 9 \cos 3\theta$, $r = 9 \cos 5\theta$.

(B) What would you guess to be the number of leaves for the graph of $r = 9 \cos 7\theta$?

(C) What would you guess to be the number of leaves for the graph of $r = a \cos n\theta$, $a > 0$ and n odd?

24. (A) Graph each polar equation in its own viewing window: $r = 9 \sin \theta$, $r = 9 \sin 3\theta$, $r = 9 \sin 5\theta$.

(B) What would you guess to be the number of leaves for the graph of $r = 9 \sin 7\theta$?

(C) What would you guess to be the number of leaves for the graph of $r = a \sin n\theta$, $a > 0$ and n odd?

25. (A) Graph each polar equation in its own viewing window: $r = 9 \cos 2\theta$, $r = 9 \cos 4\theta$, $r = 9 \cos 6\theta$.

(B) What would you guess to be the number of leaves for the graph of $r = 9 \cos 8\theta$?

(C) What would you guess to be the number of leaves for the graph of $r = a \cos n\theta$, $a > 0$ and n even?

26. (A) Graph each polar equation in its own viewing window: $r = 9 \sin 2\theta$, $r = 9 \sin 4\theta$, $r = 9 \sin 6\theta$.

(B) What would you guess to be the number of leaves for the graph of $r = 9 \sin 8\theta$?

(C) What would you guess to be the number of leaves for the graph of $r = a \sin n\theta$, $a > 0$ and n even?

27. Graph each polar equation in its own viewing window:

(A) $r = 9 \cos(\theta/2)$, $0 \le \theta \le 4\pi$

(B) $r = 9 \cos(\theta/4)$, $0 \le \theta \le 8\pi$

(C) What would you guess to be the maximum number of times a ray from the origin intersects the graph of $r = 9 \cos(\theta/n)$, $0 \le \theta \le 2\pi n$, n even?

28. Graph each polar equation in its own viewing window:

(A) $r = 9 \sin(\theta/2)$, $0 \le \theta \le 4\pi$

(B) $r = 9 \sin(\theta/4)$, $0 \le \theta \le 8\pi$

(C) What would you guess to be the maximum number of times a ray from the origin intersects the graph of $r = 9 \sin(\theta/n)$, $0 \le \theta \le 2\pi n$, n even?

C *In Problems 29 and 30, sketch each polar equation using rapid sketching techniques.*

29. $r^2 = 64 \cos 2\theta$ **30.** $r^2 = 64 \sin 2\theta$

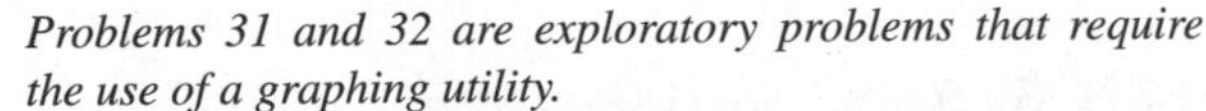

Problems 31 and 32 are exploratory problems that require the use of a graphing utility.

31. Graph $r = 1 + 2 \cos n\theta$ for various values of n, n a natural number. Describe how n is related to the number of large petals and the number of small petals on the graph and how the large and small petals are related to each other relative to n.

32. Graph $r = 1 + 2 \sin n\theta$ for various values of n, n a natural number. Describe how n is related to the number of large petals and the number of small petals on the graph and how the large and small petals are related to each other relative to n.

Precalculus *In Problems 33–36, graph and solve each system of equations in the same polar coordinate system. [Note: Any solution (r_1, θ_1) to the system must satisfy each equation in the system and thus identifies a point of intersection of the two graphs. However, there may be other points of intersection of the two graphs that do not have any coordinates that satisfy both equations. This represents a major difference between the rectangular coordinate system and the polar coordinate system.]*

33. $r = 2 \cos \theta$
$r = 2 \sin \theta$
$0 \le \theta \le \pi$

34. $r = 4 \cos \theta$
$r = -4 \sin \theta$
$0 \le \theta \le \pi$

35. $r = 8 \sin \theta$
$r = 8 \cos 2\theta$
$0° \le \theta \le 360°$

36. $r = 6 \cos \theta$
$r = 6 \sin 2\theta$
$0° \le \theta \le 360°$

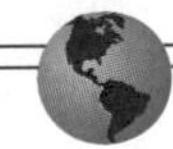

Applications

Sailboat Racing *Polar diagrams are used extensively by serious sailboat racers. The polar diagram in the figure shows the theoretical speeds that boats in the America's Cup competition (1991) and the older 12 meter boats should have been able to achieve at different points of sail relative to a 16 knot wind. Problems 37 and 38 refer to this polar diagram.*

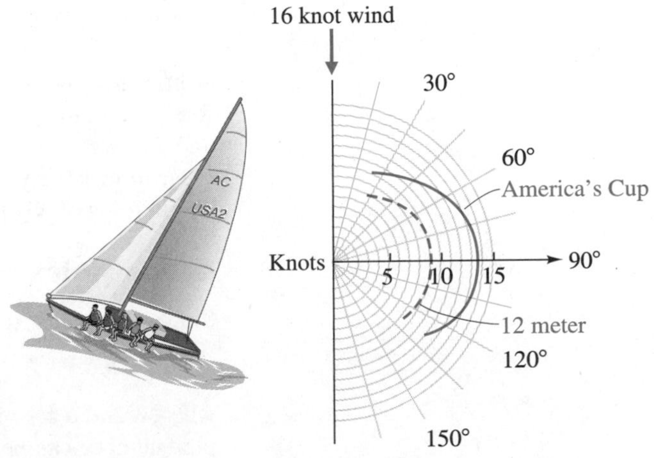

Figure for 37 and 38

37. Refer to the figure (page 413). How fast, to the nearest half knot, should the 1991 America's Cup boats have been able to sail going in the following directions relative to the wind?
(A) 30° (B) 60° (C) 90° (D) 120°

38. Refer to the figure (page 413). How fast, to the nearest half knot, should the older 12 meter boats have been able to sail going in the following directions relative to the wind?
(A) 30° (B) 60° (C) 90° (D) 120°

39. **Conic Sections** Using a graphing calculator, graph the equation

$$r = \frac{8}{1 - e \cos \theta}$$

for the following values of e, and identify each curve as a hyperbola, ellipse, or parabola.
(A) $e = 0.5$ (B) $e = 1$ (C) $e = 2$

40. **Conic Sections** Using a graphing calculator, graph the equation

$$r = \frac{4}{1 - e \cos \theta}$$

for the following values of e, and identify each curve as a hyperbola, ellipse, or parabola.
(A) $e = 0.7$ (B) $e = 1$ (C) $e = 1.3$

* 41. **Astronomy**
(A) The planet Mercury travels around the sun in an elliptical orbit given approximately by

$$r = \frac{3.44 \times 10^7}{1 - 0.206 \cos \theta}$$

where r is in miles. Graph the orbit with the sun at the pole. Find the distance from Mercury to the sun at **aphelion** (greatest distance from the sun) and at **perihelion** (shortest distance from the sun).
(B) Johannes Kepler (1571–1630) showed that a line joining a planet to the sun swept out equal areas in space in equal intervals in time (see the figure). Use this information to determine whether a planet travels faster or slower at aphelion than at perihelion.

Figure for 41

7.3 COMPLEX NUMBERS IN RECTANGULAR AND POLAR FORMS

- Complex Numbers in Rectangular Form
- Complex Numbers in Polar Form
- Products and Quotients in Polar Form
- Historical Note

A brief review of Appendix A.2 on complex numbers should prove helpful before proceeding further. Making use of the polar concepts studied in Sections 7.1 and 7.2, we will show how complex numbers can be written in polar form. The polar form is very useful in many applications, including the process of finding all nth roots of any number, real or complex.

Complex Numbers in Rectangular Form

Since a **complex number** is any number that can be written in the form

$$a + bi$$

where a and b are real and i is the imaginary unit (see Appendix A.2), each complex number can be associated with a unique ordered pair of real numbers, and vice versa. For example,

$$2 - 3i \qquad \text{corresponds to} \qquad (2, -3)$$

In general,

$$a + bi \qquad \text{corresponds to} \qquad (a, b)$$

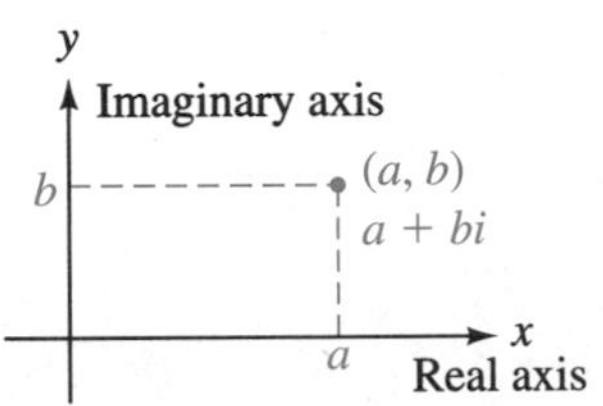

FIGURE 1
Complex plane

Therefore, each complex number can be associated with a unique point in a rectangular coordinate system, and each point in a rectangular coordinate system can be associated with a unique complex number (see Fig. 1). The plane, with points associated with ordered pairs (a, b), is often called the **complex plane** when points are associated with complex numbers $a + bi$. The x axis is then called the **real axis** and the y axis is called the **imaginary axis.** The complex numbers use all the points in a plane; the real numbers use all the points on a line.

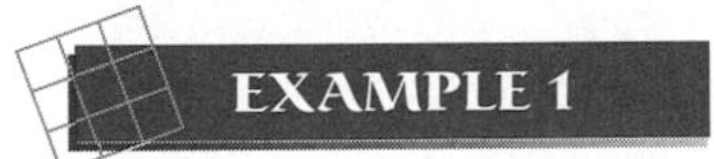

Plotting Complex Numbers

Plot each number in a complex plane.

(A) $3 + 4i$ (B) $-3 - 2i$ (C) -5 (D) $-3i$

Solution

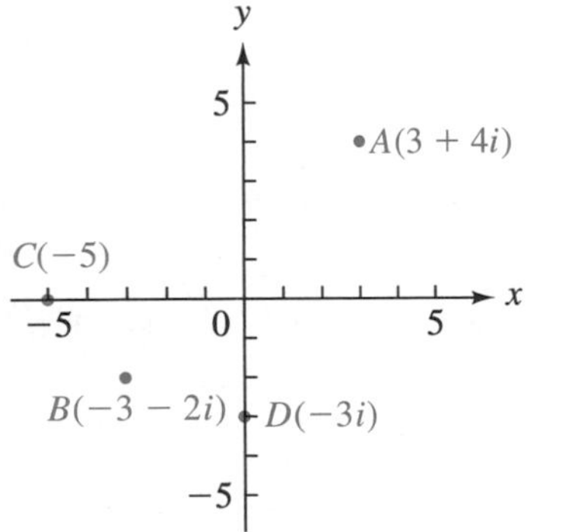

■

Matched Problem 1 Plot each number in a complex plane.

(A) $4 + 2i$ (B) $3 - 3i$ (C) -4 (D) $3i$ ■

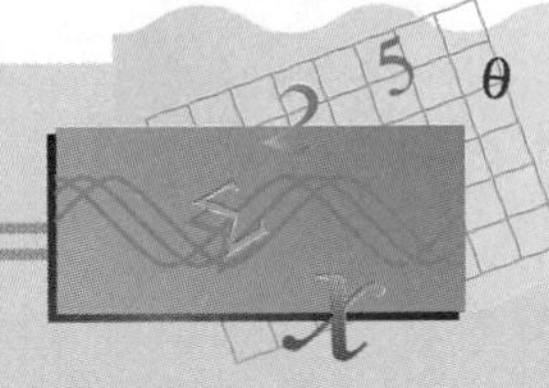

EXPLORE/DISCUSS 1

There is a one-to-one correspondence between the set of real numbers and the set of points on a *real number line* (see Appendix A.1). Each real number is associated with exactly one point on the line, and each point on the line is associated with exactly one real number. Does a one-to-one correspondence exist between the set of complex numbers and the set of points in a plane? If so, explain how it can be established.

Complex Numbers in Polar Form

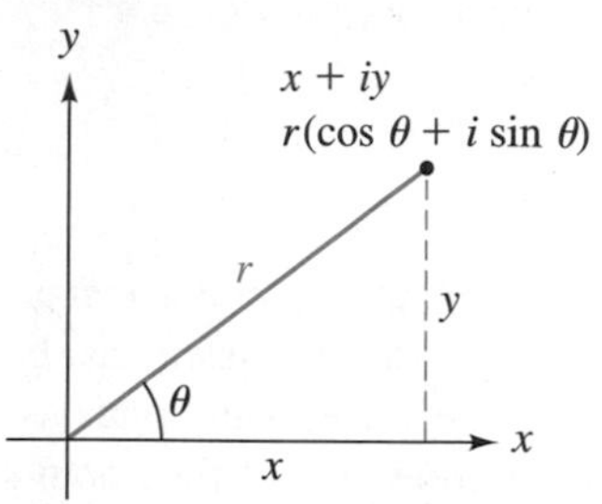

FIGURE 2
Polar and rectangular forms

Complex numbers can be changed from rectangular form to **polar (or trigonometric) form** by using the relationships developed in Section 7.1.

Thus, using

$$x = r\cos\theta \qquad \text{and} \qquad y = r\sin\theta$$

we can write

$$z = x + iy = r\cos\theta + ir\sin\theta = r(\cos\theta + i\sin\theta) \tag{1}$$

This polar–rectangular relationship is illustrated in Figure 2.

In a more advanced treatment of the subject, the following famous equation is established, where $e \approx 2.718$ denotes the base of the natural logarithmic function:

$$e^{i\theta} = \cos\theta + i\sin\theta \tag{2}$$

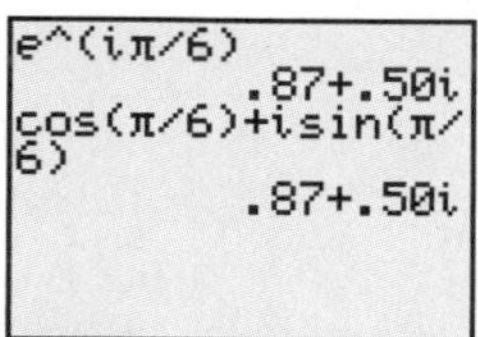

FIGURE 3

Equation (2) is valid for any angle θ measured in radians (or for θ any complex number). The viewing window from a graphing calculator in Figure 3 shows the left and right sides of equation (2) evaluated at $\theta = \pi/6$, with the calculator set in radian mode. Equation (2) is also valid for any angle θ measured in degrees, but a word of caution is in order when using some calculators:

Caution Some calculators require θ in $e^{i\theta}$ to be in radians and not degrees for calculations. Others will handle both. Check your owner's manual. □

Making use of equation (2), equation (1) takes on the form

$$z = x + iy = r(\cos\theta + i\sin\theta) = re^{i\theta} \tag{3}$$

We will use the form $re^{i\theta}$ freely as a polar form for a complex number. In fact, many graphing calculators display the polar form of $x + iy$ this way (see Fig. 4, where θ is in radians).

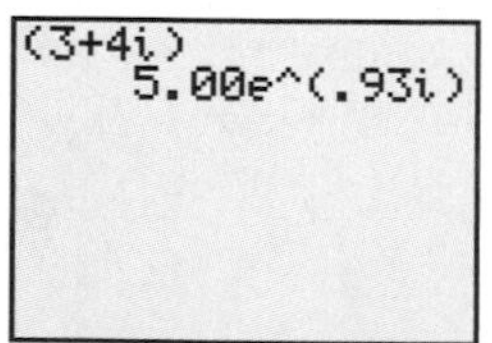

FIGURE 4
$3 + 4i = 5.00e^{0.93i}$

Since sine and cosine each have periods of 2π, we can write

$$\begin{aligned} \sin(\theta + 2k\pi) &= \sin\theta \\ \cos(\theta + 2k\pi) &= \cos\theta \end{aligned} \qquad k \text{ any integer}$$

Thus, we can write a more general polar form for a complex number $z = x + iy$, as shown below, and observe that $re^{i\theta}$ is periodic with period 2π.

GENERAL POLAR FORM OF A COMPLEX NUMBER

For k any integer,

$$\begin{aligned} z = x + iy &= r[\cos(\theta + 2k\pi) + i\sin(\theta + 2k\pi)] \\ &= re^{i(\theta + 2k\pi)} \end{aligned}$$

The number r is called the **modulus,** or **absolute value,** of z and is denoted by **mod** z or $|z|$. The polar angle that the line joining z to the origin makes with the polar axis is called the **argument** of z and is denoted by **arg** z. From Figure 2 the following relationships are apparent:

MODULUS AND ARGUMENT FOR $z = x + iy$

$\text{mod } z = r = \sqrt{x^2 + y^2}$	Never negative
$\arg z = \theta + 2k\pi$	k any integer

where $\sin \theta = y/r$ and $\cos \theta = x/r$. The argument θ is usually chosen so that $-180° < \theta \le 180°$ or $-\pi < \theta \le \pi$.

EXAMPLE 2

From Rectangular to Polar Form

Write the following in polar form (θ in radians, $-\pi < \theta \le \pi$). Compute the modulus and arguments for (A) and (B) exactly. Compute the modulus and argument for (C) to two decimal places.

(A) $z_1 = -1 + i$ (B) $z_2 = 1 + \sqrt{3}\, i$ (C) $z_3 = -3 - 7i$

Solution First, locate the complex number in a complex plane. If x and y are associated with special reference triangles, r and θ can be determined by inspection.

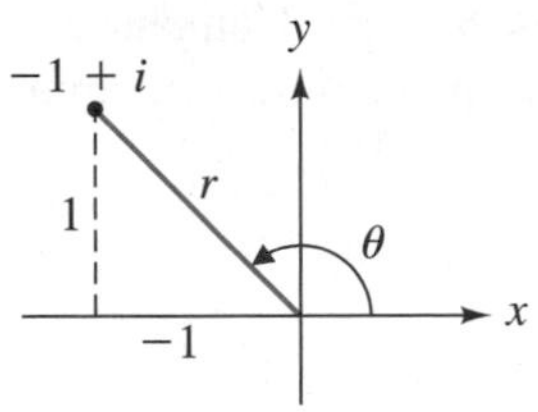

FIGURE 5

(A) A sketch (Fig. 5) shows that z_1 is associated with a special 45° reference triangle in the second quadrant. Thus, we have $\text{mod } z_1 = r = \sqrt{2}$ and $\arg z_1 = \theta = 3\pi/4$, and the polar form for z_1 is

$$\begin{aligned} z_1 &= \sqrt{2}[\cos(3\pi/4) + i \sin(3\pi/4)] \\ &= \sqrt{2}e^{(3\pi/4)i} \end{aligned}$$

(B) A sketch (Fig. 6) shows that z_2 is associated with a special 30°–60° reference triangle. Thus, $\text{mod } z_2 = r = 2$ and $\arg z_2 = \theta = \pi/3$, and the polar form for z_2 is

$$\begin{aligned} z_2 &= 2[\cos(\pi/3) + i \sin(\pi/3)] \\ &= 2e^{(\pi/3)i} \end{aligned}$$

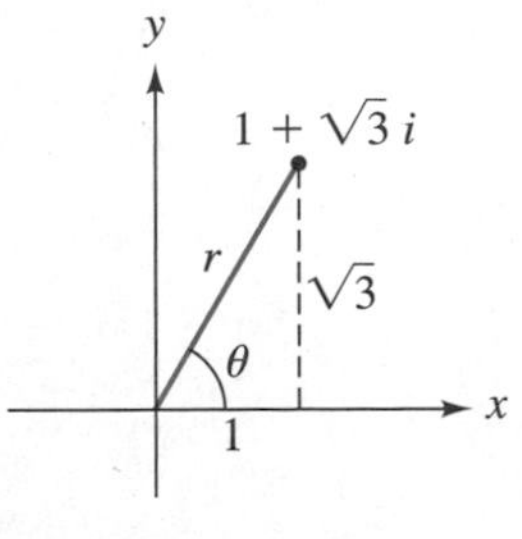

FIGURE 6

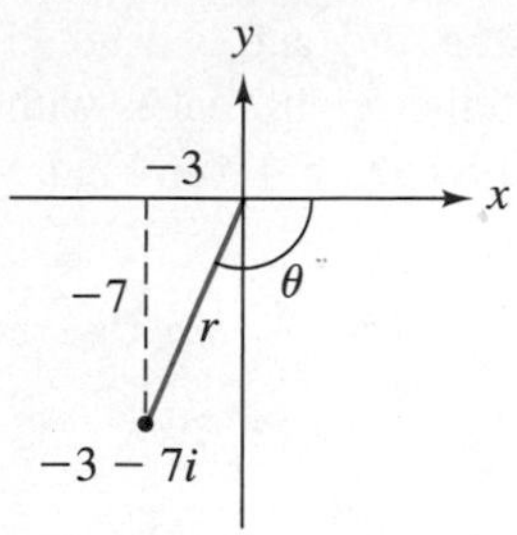

FIGURE 7

(C) A sketch (Fig. 7) shows that z_3 is not associated with a special reference triangle. Consequently, we proceed as follows:

$$\text{mod } z_3 = r = \sqrt{(-3)^2 + (-7)^2} = 7.62 \qquad \text{To two decimal places}$$

$$\arg z_3 = \theta = -\pi + \tan^{-1}\tfrac{7}{3} = -1.98 \qquad \text{To two decimal places}$$

Thus, the polar form for z_3 is

$$z_3 = 7.62[\cos(-1.98) + i\sin(-1.98)]$$
$$= 7.62e^{(-1.98)i}$$

Figure 8 shows the same conversion done by a graphing calculator with a built-in conversion routine.

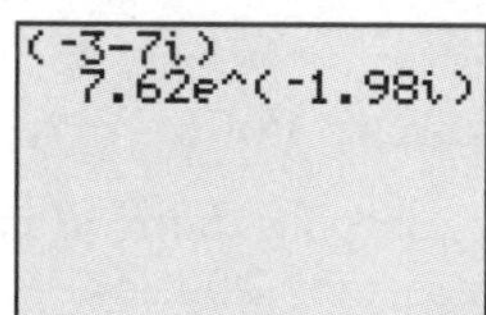

FIGURE 8

Matched Problem 2 Write the following in polar form (θ in radians, $-\pi < \theta \le \pi$). Compute the modulus and arguments for (A) and (B) exactly. Compute the modulus and argument for (C) to two decimal places.

(A) $z_1 = 1 - i$ (B) $z_2 = -\sqrt{3} + i$ (C) $z_3 = -5 - 2i$

EXAMPLE 3 **From Polar to Rectangular Form**

Write the following in rectangular form. Compute the exact values in parts (A) and (B). For part (C), compute a and b for $a + bi$ to two decimal places.

(A) $z_1 = \sqrt{2}e^{(-\pi/4)i}$ (B) $z_2 = 3e^{(120°)i}$

(C) $z_3 = 6.49e^{(-2.08)i}$

Solution (A)
$$x + iy = \sqrt{2}e^{(-\pi/4)i}$$
$$= \sqrt{2}[\cos(-\pi/4) + i\sin(-\pi/4)]$$
$$= \sqrt{2}\left(\frac{1}{\sqrt{2}} + \frac{-1}{\sqrt{2}}i\right)$$
$$= 1 - i$$

(B) $x + iy = 3e^{(120°)i}$

$$= 3(\cos 120° + i \sin 120°)$$

$$= 3\left(\frac{-1}{2} + \frac{\sqrt{3}}{2} i\right)$$

$$= -\frac{3}{2} + \frac{3\sqrt{3}}{2} i$$

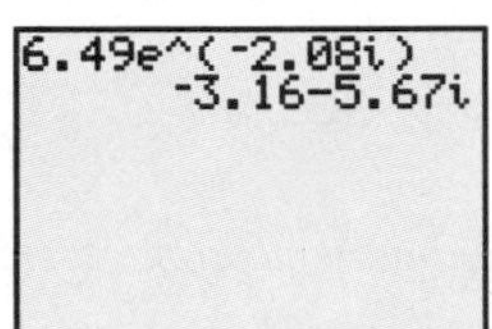

FIGURE 9

(C) $x + iy = 6.49e^{(-2.08)i}$

$$= 6.49[\cos(-2.08) + i \sin(-2.08)]$$

$$= -3.16 - 5.67i$$

Figure 9 shows the same conversion done by a graphing calculator with a built-in conversion routine. ■

Matched Problem 3 Write the following in rectangular form. Compute the exact values in parts (A) and (B). For part (C), compute a and b for $a + bi$ two decimal places.

(A) $z_1 = 2e^{(5\pi/6)i}$ (B) $z_2 = 3e^{(-60°)i}$

(C) $z_3 = 7.19e^{(-2.13)i}$ ■

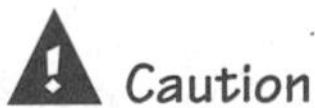
Caution For complex numbers in exponential polar form, some calculators require θ to be in radian mode. Check your user's manual. □

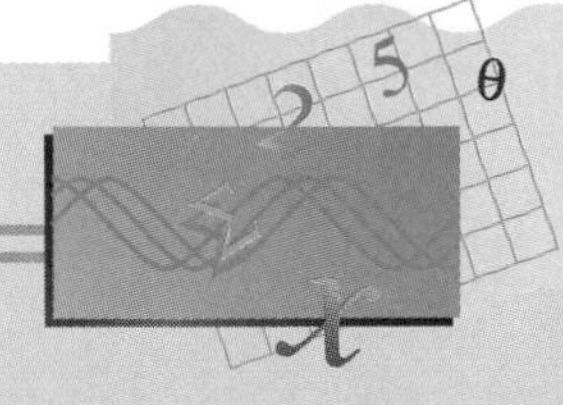

EXPLORE/DISCUSS 2

Let $z_1 = 1 + i$ and $z_2 = -1 + i$.

(A) Find z_1z_2 and z_1/z_2 using the rectangular forms of z_1 and z_2.

(B) Find z_1z_2 and z_1/z_2 using the exponential polar forms of z_1 and z_2, θ in degrees. (Assume the product and quotient exponent laws hold for $e^{i\theta}$.)

(C) Convert the results from part (B) back to rectangular form and compare with the results in part (A).

Products and Quotients in Polar Form

We will now see the advantage of representing complex numbers in polar form: Multiplication and division of complex numbers become very easy. Theorem 1 provides the reason. (The exponential polar form of a complex number obeys the product and quotient rules for exponents: $b^mb^n = b^{m+n}$ and $b^m/b^n = b^{m-n}$.)

Theorem 1

PRODUCTS AND QUOTIENTS IN POLAR FORM

If $z_1 = r_1e^{i\theta_1}$ and $z_2 = r_2e^{i\theta_2}$, then:

1. $z_1z_2 = r_1e^{i\theta_1}r_2e^{i\theta_2} = r_1r_2e^{i(\theta_1+\theta_2)}$

2. $\dfrac{z_1}{z_2} = \dfrac{r_1e^{i\theta_1}}{r_2e^{i\theta_2}} = \dfrac{r_1}{r_2}e^{i(\theta_1-\theta_2)}$

We establish the multiplication property and leave the quotient property to Problem 34 in Exercise 7.3.

$$\begin{aligned} z_1z_2 &= r_1e^{i\theta_1}\,r_2e^{i\theta_2} \\ &= [r_1(\cos\theta_1 + i\sin\theta_1)][r_2(\cos\theta_2 + i\sin\theta_2)] \\ &= r_1r_2(\cos\theta_1 + i\sin\theta_1)(\cos\theta_2 + i\sin\theta_2) \\ &= r_1r_2(\cos\theta_1\cos\theta_2 + i\cos\theta_1\sin\theta_2 + i\sin\theta_1\cos\theta_2 - \sin\theta_1\sin\theta_2) \\ &= r_1r_2[(\cos\theta_1\cos\theta_2 - \sin\theta_1\sin\theta_2) + i(\cos\theta_1\sin\theta_2 + \sin\theta_1\cos\theta_2)] \\ &= r_1r_2[\cos(\theta_1+\theta_2) + i\sin(\theta_1+\theta_2)] \qquad \text{Sum identities} \\ &= r_1r_2e^{i(\theta_1+\theta_2)} \end{aligned}$$

Thus to multiply two complex numbers in polar form, multiply r_1 and r_2, and add θ_1 and θ_2. To divide the complex number z_1 by z_2, divide r_1 by r_2, and subtract θ_2 from θ_1. The process is illustrated in Example 4.

EXAMPLE 4

Products and Quotients

If $z_1 = 8e^{(50°)i}$ and $z_2 = 4e^{(30°)i}$, find:

(A) z_1z_2 (B) z_1/z_2

Solution

(A) $z_1z_2 = 8e^{(50°)i} \cdot 4e^{(30°)i}$

$= (8 \cdot 4)e^{i(50°+30°)} = 32e^{(80°)i}$

(B) $\dfrac{z_1}{z_2} = \dfrac{8e^{(50°)i}}{4e^{(30°)i}}$

$= \dfrac{8}{4}e^{i(50°-30°)} = 2e^{(20°)i}$

■

Matched Problem 4 If $z_1 = 21e^{(140°)i}$ and $z_2 = 3e^{(105°)i}$, find:

(A) z_1z_2 (B) z_1/z_2

■

Historical Note

There is hardly an area in mathematics that does not have some imprint of the famous Swiss mathematician Leonhard Euler (1707–1783), who spent most of his productive life in the New St. Petersburg Academy in Russia and the Prussian Academy in Berlin. One of the most prolific writers in the history of mathematics, he is credited with making the following familiar notations standard:

$f(x)$	Function notation
e	Natural logarithmic base
i	Imaginary unit $\sqrt{-1}$

Of immediate interest, he is also responsible for the remarkable relationship

$$e^{i\theta} = \cos\theta + i\sin\theta$$

If $\theta = \pi$, then an equation results that relates five of the most important numbers in the history of mathematics:

$$e^{i\pi} + 1 = 0$$

Answers to Matched Problems

1.

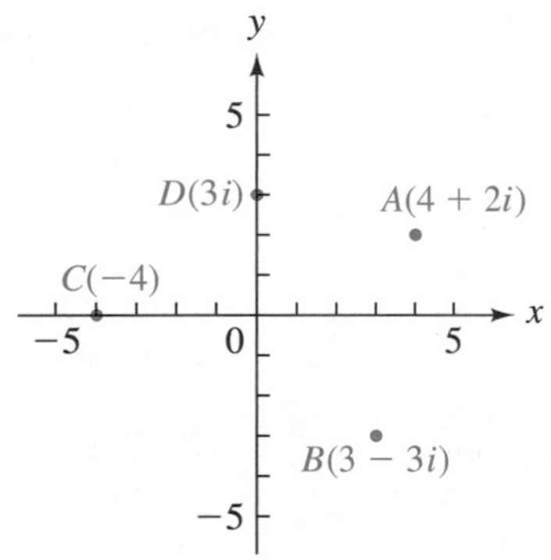

2. (A) $z_1 = \sqrt{2}[\cos(-\pi/4) + i\sin(-\pi/4)] = \sqrt{2}e^{(-\pi/4)i}$
   (B) $z_2 = 2[\cos(5\pi/6) + i\sin(5\pi/6)] = 2e^{(5\pi/6)i}$
   (C) $z_3 = 5.39[\cos(-2.76) + i\sin(-2.76)] = 5.39e^{(-2.76)i}$

3. (A) $-\sqrt{3} + i$ (B) $\frac{3}{2} - \frac{3\sqrt{3}}{2}i$ (C) $-3.81 - 6.09i$

4. (A) $z_1z_2 = 63e^{(245°)i}$ (B) $z_1/z_2 = 7e^{(35°)i}$

EXERCISE 7.3

A *Plot each set of complex numbers in a complex plane.*

1. $A = 4 + 5i$; $B = -3 + 4i$; $C = -3i$
2. $A = 3 - 2i$; $B = -4 - 2i$; $C = 5i$
3. $A = 4 + i$; $B = -2 + 3i$; $C = -4$
4. $A = -3 - i$; $B = 5 - 4i$; $C = 4$
5. $A = 8e^{(\pi/4)i}$; $B = 6e^{(\pi/2)i}$; $C = 3e^{(\pi/6)i}$
6. $A = 5e^{(5\pi/6)i}$; $B = 3e^{(3\pi/2)i}$; $C = 4e^{(7\pi/4)i}$
7. $A = 5e^{(270°)i}$; $B = 4e^{(60°)i}$; $C = 8e^{(150°)i}$
8. $A = 3e^{(310°)i}$; $B = 4e^{(180°)i}$; $C = 5e^{(210°)i}$

B *Change Problems 9–12 to polar form. For Problems 9 and 10, choose θ in radians, $-\pi < \theta \le \pi$; for Problems 11 and 12, choose θ in degrees, $-180° < \theta \le 180°$. Compute the* *modulus and arguments for (A) and (B) exactly; compute the modulus and argument for (C) to two decimal places.*

9. (A) $\sqrt{3} - i$ (B) $-2 + 2i$ (C) $6 - 5i$
10. (A) $-\sqrt{3}\,i$ (B) $-\sqrt{3} - i$ (C) $-8 + 5i$
11. (A) $-1 + \sqrt{3}\,i$ (B) $-3i$ (C) $-7 - 4i$
12. (A) $\sqrt{3} + i$ (B) $-1 - i$ (C) $5 - 6i$

Change Problems 13–16 to rectangular form. Compute the exact values for (A) and (B); for (C) compute a and b in a + bi to two decimal places.

13. (A) $2e^{(30°)i}$ (B) $\sqrt{2}e^{(-3\pi/4)i}$ (C) $5.71e^{(-0.48)i}$
14. (A) $2e^{(\pi/3)i}$ (B) $\sqrt{2}e^{(-45°)i}$ (C) $3.08e^{2.44i}$
15. (A) $\sqrt{3}e^{(-\pi/2)i}$ (B) $\sqrt{2}e^{(135°)i}$ (C) $6.83e^{(-108.82°)i}$
16. (A) $6e^{(\pi/6)i}$ (B) $\sqrt{7}e^{(-90°)i}$ (C) $4.09e^{(-122.88°)i}$

Find z_1z_2 and z_1/z_2. Leave answers in polar form.

17. $z_1 = 6e^{(132°)i}$; $z_2 = 3e^{(93°)i}$
18. $z_1 = 7e^{(82°)i}$; $z_2 = 2e^{(31°)i}$
19. $z_1 = 3e^{(67°)i}$; $z_2 = 2e^{(97°)i}$
20. $z_1 = 5e^{(52°)i}$; $z_2 = 2e^{(83°)i}$
21. $z_1 = 7.11e^{0.79i}$; $z_2 = 2.66e^{1.07i}$
22. $z_1 = 3.05e^{1.76i}$; $z_2 = 11.94e^{2.59i}$

Find each of the following directly and by using polar forms. Write answers in rectangular and polar forms with θ in degrees.

23. $(1 + \sqrt{3}\,i)(\sqrt{3} + i)$
24. $(-1 + i)(1 + i)$
25. $(1 + i)^2$
26. $(-1 + i)^2$
27. $(1 + i)^3$
28. $(1 - i)^3$

C 29. If $z = re^{i\theta}$, show that $z^2 = r^2e^{2\theta i}$.
30. If $z = re^{i\theta}$, show that $z^3 = r^3e^{3\theta i}$.
31. Show that $r^{1/2}e^{i\theta/2}$ is a square root of $re^{i\theta}$.
32. Show that $r^{1/3}e^{i\theta/3}$ is a cube root of $re^{i\theta}$.
33. Based on Problems 29 and 30, what do you think z^n will be for n a natural number?
34. Prove: $\dfrac{z_1}{z_2} = \dfrac{r_1e^{i\theta_1}}{r_2e^{i\theta_2}} = \dfrac{r_1}{r_2}e^{i(\theta_1-\theta_2)}$

Applications

35. **Resultant Force** An object is located at the pole, and two forces $\mathbf{F}_1$ and $\mathbf{F}_2$ act upon the object. Let the forces be vectors going from the pole to the complex numbers 8(cos 0° + i sin 0°) and 6(cos 30° + i sin 30°), respectively. [Force $\mathbf{F}_1$ has a magnitude of 8 lb at a direction of 0°, and force $\mathbf{F}_2$ has a magnitude of 6 lb at a direction of 30°.)
    (A) Convert the polar forms of these complex numbers to rectangular form and add.
    (B) Convert the sum from part (A) back to polar form (to three significant digits).
    (C) The vector going from the pole to the complex number in part (B) is the resultant of the two original forces. What is its magnitude and direction?
36. **Resultant Force** Repeat Problem 35 with forces $\mathbf{F}_1$ and $\mathbf{F}_2$ associated with the complex numbers

$$20(\cos 0° + i \sin 0°) \quad \text{and} \quad 10(\cos 60° + i \sin 60°)$$

7.4 DE MOIVRE'S THEOREM AND THE nTH-ROOT THEOREM

- De Moivre's Theorem
- The *n*th-Root Theorem

In this section we come to one of the great theorems in mathematics, *De Moivre's theorem.* Abraham De Moivre (1667–1754), of French birth, spent most of his life in London doing private tutoring, writing, and publishing mathematics. He became a close friend of Isaac Newton and belonged to many prestigious professional societies in England, France, and Germany.

De Moivre's theorem enables us to find the power of a complex number in polar (or trigonometric) form very easily. More importantly, the theorem is the basis for the *nth-root theorem,* which enables us to find all n roots of any complex number, real or imaginary. How many roots does the equation $x^3 = 1$ have? As was mentioned in the introduction to this chapter, you may think that 1 is the only root. The equation actually has three roots—one real and two imaginary (see Example 3).

De Moivre's Theorem

We start with Explore–Discuss 1 and encourage you to generalize from this exploration.

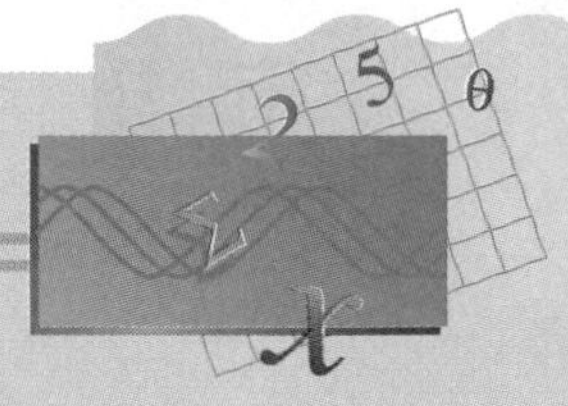

EXPLORE/DISCUSS 1

Establish the following by repeated use of the product formula for the exponential polar form discussed in Section 7.3:

(A) $(x + iy)^2 = (re^{\theta i})^2 = r^2e^{2\theta i}$

(B) $(x + iy)^3 = (re^{\theta i})^3 = r^3e^{3\theta i}$

(C) $(x + iy)^4 = (re^{\theta i})^4 = r^4e^{4\theta i}$

For n a natural number, what do you think the polar form of $(x + iy)^n$ would be?

If you guessed $(x + iy)^n = r^ne^{n\theta i}$, you have discovered De Moivre's famous theorem, which we now state without proof. (A general proof of De Moivre's theorem requires a technique called *mathematical induction,* which is discussed in more advanced courses.)

DE MOIVRE'S THEOREM

If $z = x + iy = re^{i\theta}$ and n is a natural number, then

$$z^n = (x + iy)^n = (re^{i\theta})^n = r^ne^{n\theta i}$$

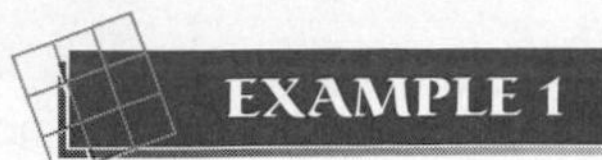

EXAMPLE 1 **Finding a Power of a Complex Number**

Find $(\sqrt{3} + i)^{13}$ and write the answer in exact polar and rectangular forms.

Solution

$$\begin{aligned}
(\sqrt{3} + i)^{13} &= (2e^{(30°)i})^{13} && \text{Convert to polar form.}\\
&= 2^{13}e^{(13\cdot 30°)i} && \text{Use De Moivre's theorem.}\\
&= 8{,}192e^{(390°)i} \\
&= 8{,}192e^{(30°+360°)i} && e^{i\theta} \text{ is periodic with period } 360°.\\
&= 8{,}192e^{(30°)i} && \text{Exact polar form}\\
&= 8{,}192(\cos 30° + i \sin 30°) \\
&= 8{,}192\left(\frac{\sqrt{3}}{2} + \frac{1}{2}i\right) \\
&= 4{,}096\sqrt{3} + 4{,}096i && \text{Exact rectangular form}
\end{aligned}$$

Matched Problem 1 Find $(-1 + i)^6$ and write the answer in exact polar and rectangular forms.

The *n*th-Root Theorem

Now let us take a look at roots of complex numbers. We say **w is an nth root of z**, n a natural number, if

$$w^n = z$$

For example, if $w^2 = z$, then w is a square root of z; if $w^3 = z$, then w is a cube root of z; and so on.

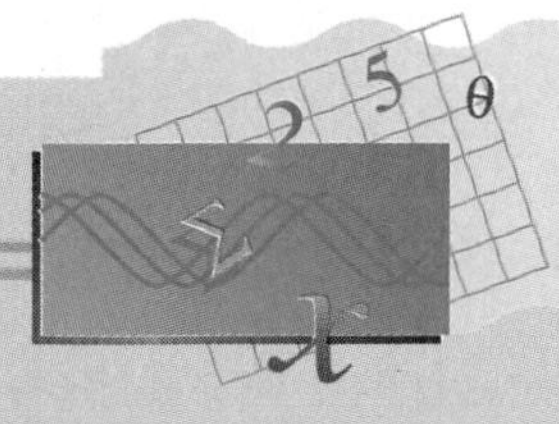

EXPLORE/DISCUSS 2

For $z = re^{i\theta}$ show that $r^{1/2}e^{(\theta/2)i}$ is a square root of z and $r^{1/3}e^{(\theta/3)i}$ is a cube root of z. (Use De Moivre's theorem.)

Proceeding in the same way as in Explore–Discuss 2, it is easy to show that for n a natural number, $r^{1/n}e^{(\theta/n)i}$ is an nth root of $re^{i\theta}$:

$$\begin{aligned}
(r^{1/n}e^{(\theta/n)i})^n &= (r^{1/n})^n e^{n(\theta/n)i}\\
&= re^{i\theta}
\end{aligned}$$

However, we can do better than this. The ***n*th-root theorem** shows how to find *all* the nth roots of a complex number. The proof of the theorem is left to Problems 31 and 32 in Exercise 7.4.

*n*TH-ROOT THEOREM

For n a positive integer greater than 1 and θ in degrees,

$$r^{1/n}e^{(\theta/n+k\,360°/n)i} \qquad k = 0, 1, \ldots, n-1$$

are the n distinct nth roots of $re^{i\theta}$, and there are no others.

EXAMPLE 2 **Roots of a Complex Number**

Find the six distinct sixth roots of $z = 1 + \sqrt{3}\,i$.

Solution First, write $1 + \sqrt{3}\,i$ in polar form:

$$1 + \sqrt{3}\,i = 2e^{(60°)i}$$

From the nth-root theorem, all six roots are given by

$$2^{1/6}e^{(60°/6+k360°/6)i} \qquad k = 0, 1, 2, 3, 4, 5$$

Thus,

$$w_1 = 2^{1/6}e^{(10°+0\cdot 60°)i} = 2^{1/6}e^{(10°)i}$$
$$w_2 = 2^{1/6}e^{(10°+1\cdot 60°)i} = 2^{1/6}e^{(70°)i}$$
$$w_3 = 2^{1/6}e^{(10°+2\cdot 60°)i} = 2^{1/6}e^{(130°)i}$$
$$w_4 = 2^{1/6}e^{(10°+3\cdot 60°)i} = 2^{1/6}e^{(190°)i}$$
$$w_5 = 2^{1/6}e^{(10°+4\cdot 60°)i} = 2^{1/6}e^{(250°)i}$$
$$w_6 = 2^{1/6}e^{(10°+5\cdot 60°)i} = 2^{1/6}e^{(310°)i}$$

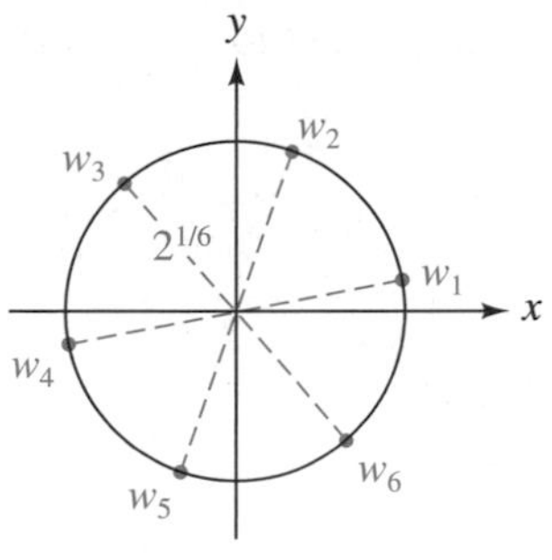

FIGURE 1

The six roots are easily graphed in the complex plane after the first root is located. The roots are equally spaced around a circle with radius $2^{1/6}$ at an angular increment of 60° from one root to the next, as shown in Figure 1. ■

Matched Problem 2 Find the three distinct cube roots of $1 + i$, and leave the answers in polar form. Plot the roots in a complex plane. ■

EXAMPLE 3 **Solving a Cubic Equation**

Solve $x^3 - 1 = 0$. Write final answers in rectangular form and plot them in a complex plane.

Solution

$$x^3 - 1 = 0$$
$$x^3 = 1$$

Therefore, x is a cube root of 1, and there are three cube roots. First, we write 1 in polar form and use the nth-root theorem:

$$1 = 1e^{(0°)i}$$

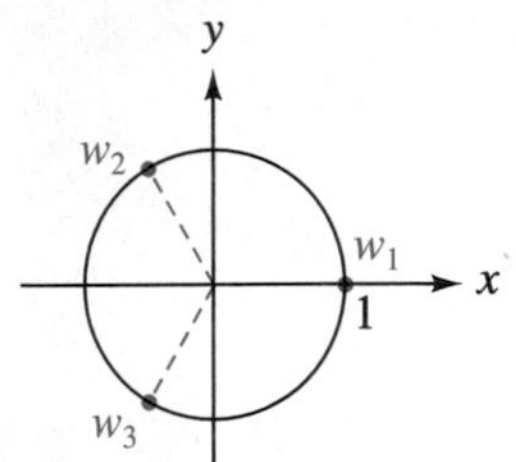

FIGURE 2

All three cube roots of 1 are given by

$$1^{1/3}e^{(0°/3+k360°/3)} \qquad k = 0, 1, 2$$

Thus,

$$w_1 = e^{(0°)i} = \cos 0° + i \sin 0° = 1$$

$$w_2 = e^{(120°)i} = \cos 120° + i \sin 120° = -\frac{1}{2} + \frac{\sqrt{3}}{2}i$$

$$w_3 = e^{(240°)i} = \cos 240° + i \sin 240° = -\frac{1}{2} - \frac{\sqrt{3}}{2}i$$

The three roots are graphed in Figure 2. ■

Matched Problem 3 Solve $x^3 + 1 = 0$. Write final answers in rectangular form and plot them in a complex plane. ■

We have only touched on a subject that has far-reaching consequences. The theory of functions of a complex variable provides a powerful tool for engineers, scientists, and mathematicians.

Answers to Matched Problems

1. $8e^{(90°)i}$; $8i$
2. $w_1 = 2^{1/6}e^{(15°)i}$, $w_2 = 2^{1/6}e^{(135°)i}$, $w_3 = 2^{1/6}e^{(255°)i}$

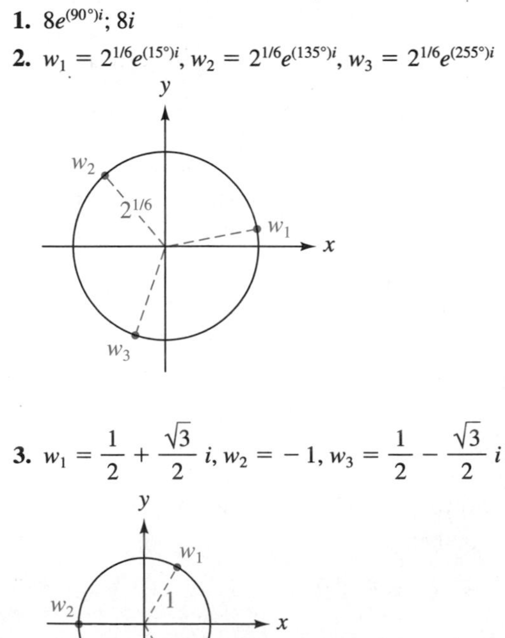

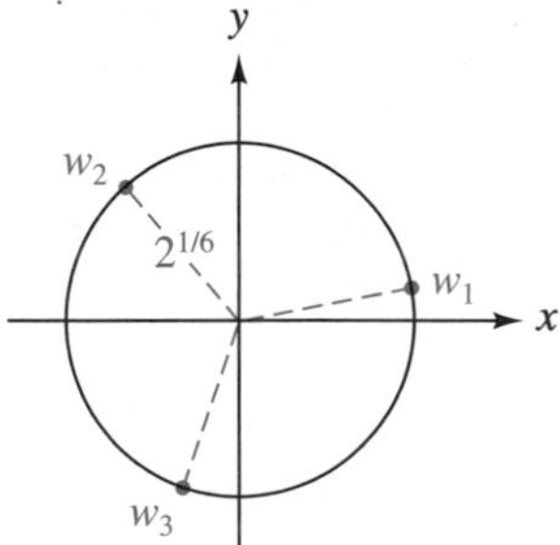

3. $w_1 = \frac{1}{2} + \frac{\sqrt{3}}{2}i$, $w_2 = -1$, $w_3 = \frac{1}{2} - \frac{\sqrt{3}}{2}i$

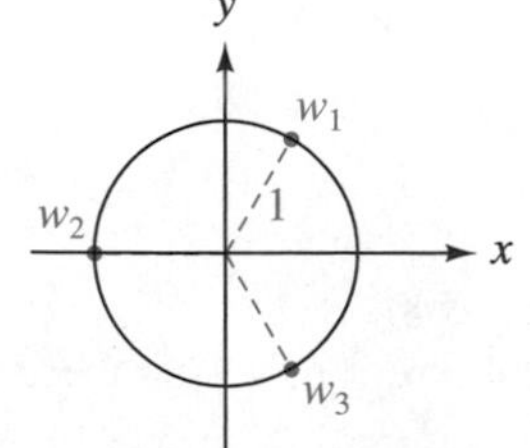

EXERCISE 7.4

A *In Problems 1–6, find the value of each expression using De Moivre's theorem. Leave your answer in polar form.*

1. $(3e^{(15°)i})^3$
2. $(2e^{(30°)i})^8$
3. $(\sqrt{2}e^{(45°)i})^{10}$
4. $(\sqrt{2}e^{(60°)i})^8$
5. $(\sqrt{3} + i)^6$
6. $(1 + \sqrt{3}\, i)^3$

B *Find the value of each expression using De Moivre's theorem, and write the result in exact rectangular form.*

7. $(-1 + i)^4$
8. $(-\sqrt{3} - i)^4$
9. $(-\sqrt{3} + i)^5$
10. $(1 - i)^8$
11. $\left(-\frac{1}{2} - \frac{\sqrt{3}}{2}i\right)^3$
12. $\left(-\frac{1}{2} + \frac{\sqrt{3}}{2}i\right)^3$

In Problems 13–18, find all nth roots of z for n and z as given. Leave answers in polar form.

13. $z = 4e^{(30°)i}$; $n = 2$
14. $z = 16e^{(60°)i}$; $n = 2$
15. $z = 8e^{(90°)i}$; $n = 3$
16. $z = 27e^{(120°)i}$; $n = 3$
17. $z = -1 + i$; $n = 5$
18. $z = 1 - i$; $n = 5$

In Problems 19–24, find all nth roots of z for n and z as given. Write answers in polar form and plot in a complex plane.

19. $z = -8$; $n = 3$
20. $z = -16$; $n = 4$
21. $z = 1$; $n = 4$
22. $z = 8$; $n = 3$
23. $z = -i$; $n = 5$
24. $z = i$; $n = 6$

25. (A) Show that $-\sqrt{2} + \sqrt{2}\, i$ is a root of $x^4 + 16 = 0$. How many other roots does the equation have?
(B) The root $-\sqrt{2} + \sqrt{2}\, i$ is located on a circle of radius 2 in the complex plane, as shown in the figure. Without using the *n*th-root theorem, locate all other roots on the figure, and explain geometrically how you found their location.

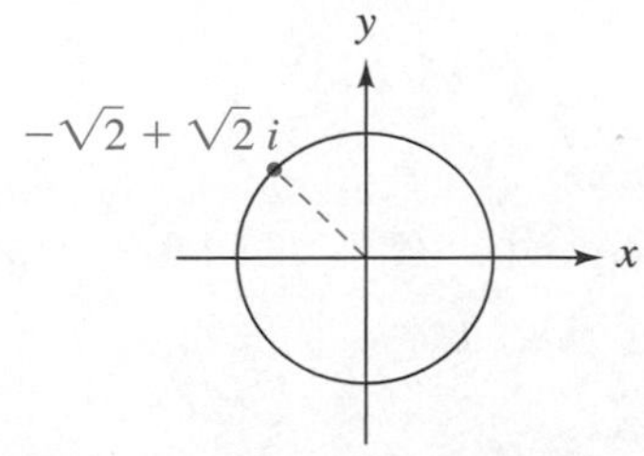

(C) Verify that each complex number found in part (B) is a root of $x^4 + 16 = 0$. Show your steps.

26. (A) Show that 2 is a root of $x^3 - 8 = 0$. How many other roots does the equation have?
(B) The root 2 is located on a circle of radius 2 in the complex plane, as shown in the figure. Without using the *n*th-root theorem, locate all other roots on the figure, and explain geometrically how you found their location.

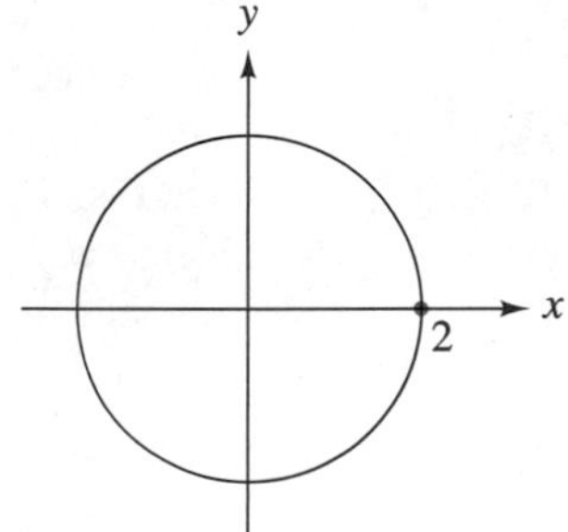

(C) Verify that each complex number found in part (B) is a root of $x^3 - 8 = 0$. Show your steps.

In Problems 27–30, solve each equation for all roots. Write the final answers in exact rectangular form.

27. $x^3 + 27 = 0$
28. $x^3 - 27 = 0$
29. $x^3 - 64 = 0$
30. $x^3 + 64 = 0$

C **31.** Show that

$$r^{1/n}e^{(\theta/n + k360°/n)i}$$

is the same number for $k = 0$ and $k = n$.

32. Show that

$$(r^{1/n}e^{(\theta/n + k360°/n)i})^n = re^{i\theta}$$

for any natural number n and any integer k.

Solve each equation for all roots. Write final answers in rectangular form, a + bi, where a and b are computed to three decimal places.

33. $x^5 - 1 = 0$
34. $x^4 + 1 = 0$
35. $x^3 + 5 = 0$
36. $x^5 - 6 = 0$

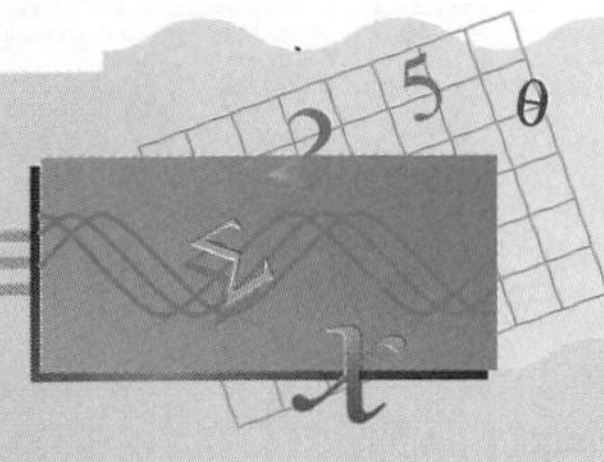

CHAPTER 7 GROUP ACTIVITY

ORBITS OF PLANETS

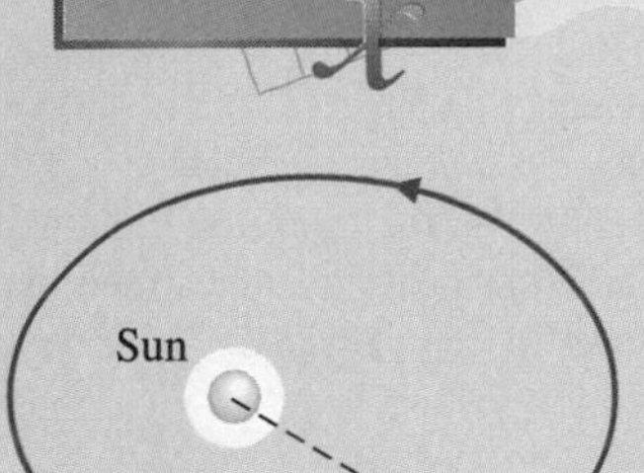

I CONICS AND ECCENTRICITY

Plane curves formed by intersecting a plane with a right circular cone of two nappes (visualize an hour glass) are called **conics.** Circles, ellipses, parabolas, and hyperbolas are conics. Conics can also be defined in terms of *eccentricity e,* an approach that is of great use in space science. We start with a fixed point F, called the **focus,** and a fixed line not containing the focus, called the **directrix** (Fig. 1). For $e > 0$, a conic is defined as the set of points P having the property that the distance from P to the focus F, divided by the distance from P to the directrix d, is the constant e. An ellipse, a parabola, or a hyperbola will result by choosing the constant e appropriately.

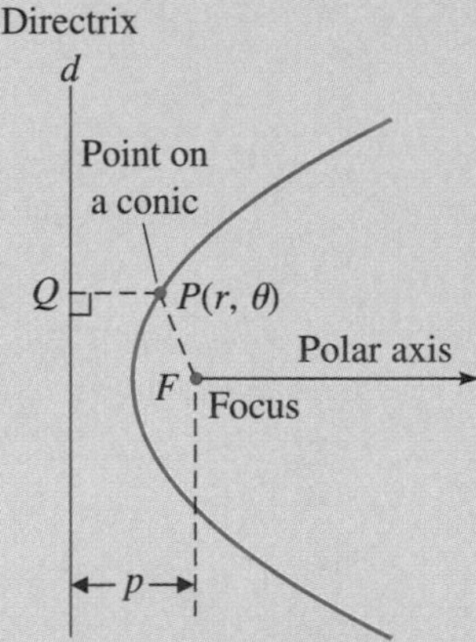

FIGURE 1
Conic

Problem 1 Using the eccentricity definition of a conic above, show that the **polar equation of a conic** is given by

$$r = \frac{ep}{1 - e\cos\theta} \tag{1}$$

where p is the distance between the focus F and the directrix d, the pole of the polar axis is at F, and the polar axis is perpendicular to d and is pointing away from d (see Fig. 1). ■

Problem 2 **Exploration with a Graphing Utility**

Explore and discuss the effect of varying the positive values of the eccentricity e and the distance p on the type and shape of conic produced. Based on your exploration, complete Table 1.

TABLE 1

Eccentricity e	Type of conic
$0 < e < 1$	
$e = 1$	
$e > 1$	

■

Problem 3 **Maximum and Minimum Distances from Focus**

Show that if $0 < e < 1$, then the points on the conic of equation (1) for which $\theta = 0$ and $\theta = \pi$ are farthest from and closest to the focus, respectively. ■

Problem 4 **Eccentricity**

Suppose that $(M, 0)$ and (m, π) are the polar coordinates of two points on the conic of equation (1). Show that

$$e = \frac{M - m}{M + m} \qquad \text{and} \qquad p = \frac{2Mm}{M - m}$$

■

II PLANETARY ORBITS

All the planets in our solar system have elliptical orbits around the sun. If we place the sun at the pole in a polar coordinate system, we will be able to get polar equations for the orbits of the planets, which can then be graphed on a graphing utility. The data in Table 2 (at the top of the next page) are taken from the *World Almanac* and rounded to three significant digits. Table 2 gives us enough information to find the polar equation for any planet's orbit.

Problem 5 **Polar Equation for Venus' Orbit**

Show that Venus' orbit is given approximately by

$$r = \frac{6.72 \times 10^7}{1 - 0.00669 \cos \theta}$$

Graph the equation on a graphing utility. ■

TABLE 2
The Planets

Planet	Maximum distance from sun (million miles)	Minimum distance from sun (million miles)
Mercury	43.4	28.6
Venus	67.7	66.8
Earth	94.6	91.4
Mars	155	129
Jupiter	507	461
Saturn	938	838
Uranus	1,860	1,670
Neptune	2,820	2,760
Pluto	4,550	2,760

Problem 6 **Polar Equation for Earth's Orbit**
Show that the orbit of the earth is given approximately by

$$r = \frac{9.29 \times 10^7}{1 - 0.0172 \cos \theta}$$

Graph the equation on a graphing utility. ■

Problem 7 **Polar Equation for Saturn's Orbit**
Show that Saturn's orbit is given approximately by

$$r = \frac{8.85 \times 10^8}{1 - 0.0563 \cos \theta}$$

Graph the equation on a graphing utility. ■

CHAPTER 7 REVIEW

7.1 POLAR AND RECTANGULAR COORDINATES

Figure 1 illustrates a **polar coordinate** system superimposed on a rectangular coordinate system. The fixed point O is called the **pole,** or **origin,** and the horizontal half-line is the **polar axis.** The **polar–rectangular relationships** are used to transform coordinates and equations from one system to the other.

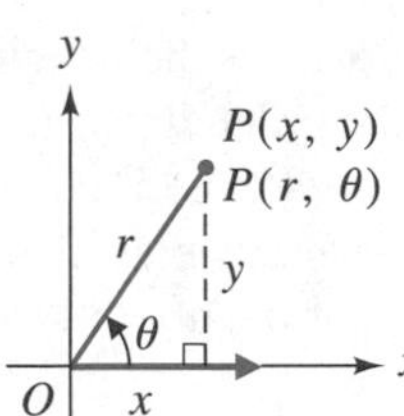

FIGURE 1
Polar–rectangular relationships

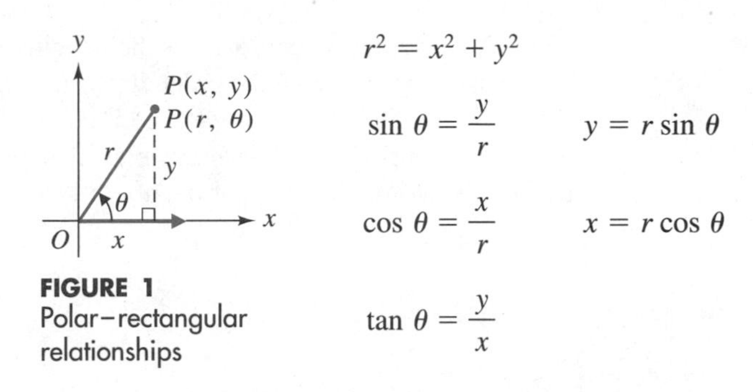

$$r^2 = x^2 + y^2$$

$$\sin\theta = \frac{y}{r} \qquad y = r\sin\theta$$

$$\cos\theta = \frac{x}{r} \qquad x = r\cos\theta$$

$$\tan\theta = \frac{y}{x}$$

[*Note:* The signs of x and y determine the quadrant for θ. The angle θ is usually chosen so that $-180° < \theta \le 180°$ or $-\pi < \theta \le \pi$.]

7.2 SKETCHING POLAR EQUATIONS

Polar graphs can be obtained by **point-by-point** plotting of points on the graph, which are usually computed with a calculator. **Rapid polar sketching techniques** use the uniform variation of $\sin\theta$ and $\cos\theta$ to quickly produce rough sketches of a polar graph. Table 1 illustrates some standard polar curves.

TABLE 1 Some Standard Polar Curves

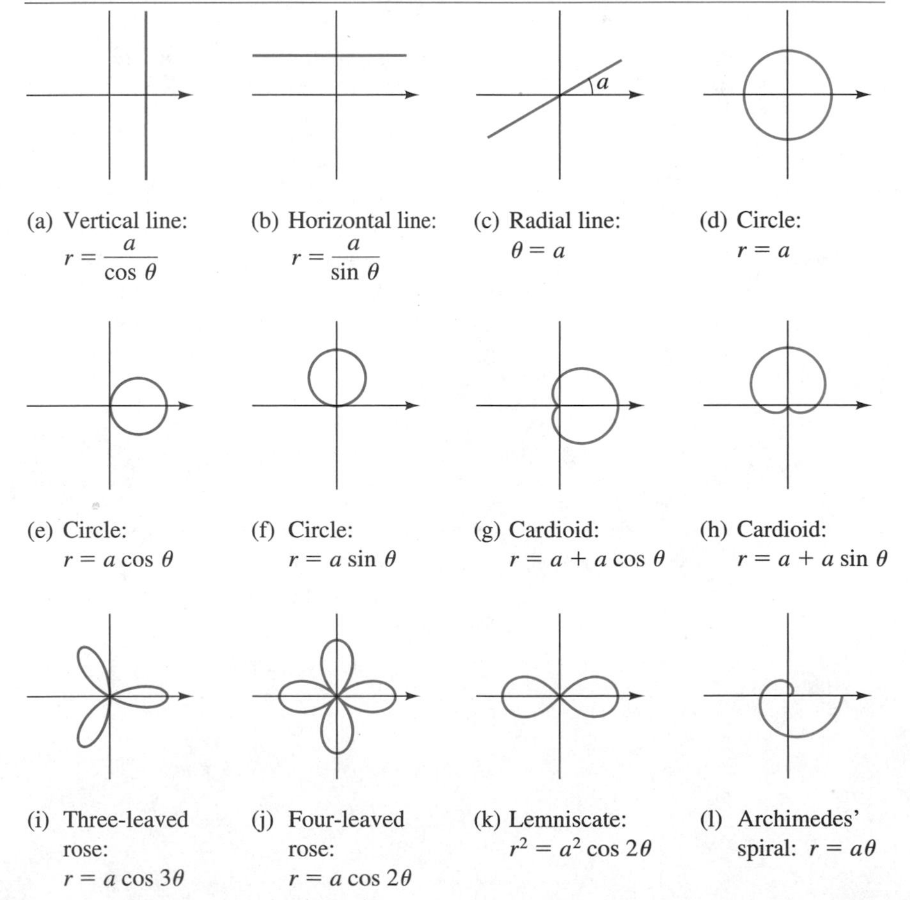

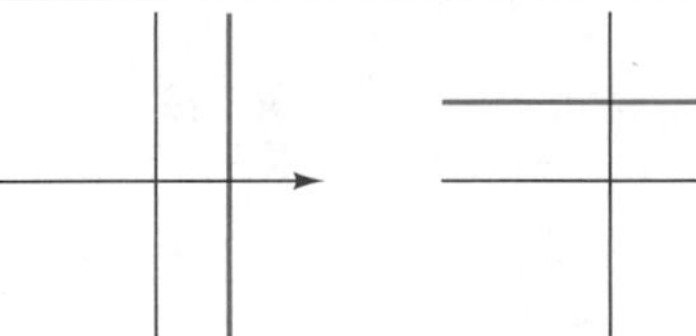

(a) Vertical line: $r = \dfrac{a}{\cos\theta}$

(b) Horizontal line: $r = \dfrac{a}{\sin\theta}$

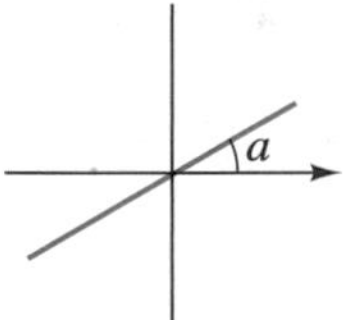

(c) Radial line: $\theta = a$

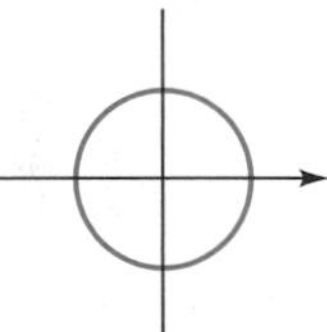

(d) Circle: $r = a$

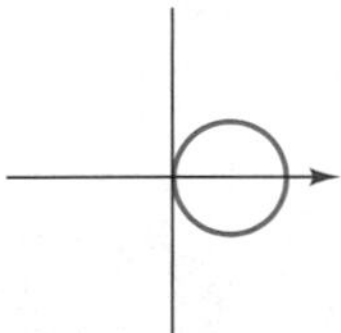

(e) Circle: $r = a\cos\theta$

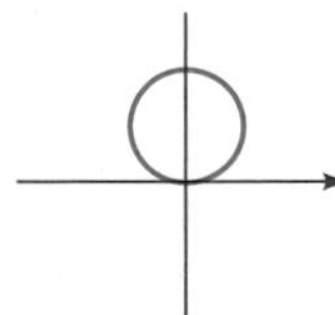

(f) Circle: $r = a\sin\theta$

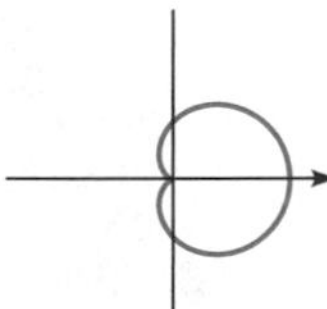

(g) Cardioid: $r = a + a\cos\theta$

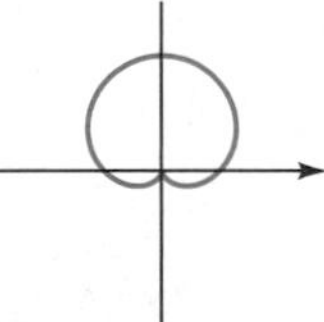

(h) Cardioid: $r = a + a\sin\theta$

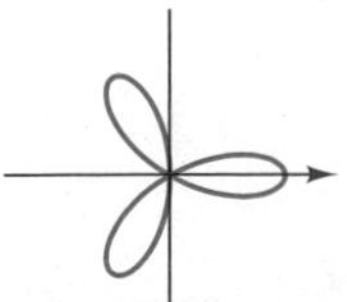

(i) Three-leaved rose: $r = a\cos 3\theta$

(j) Four-leaved rose: $r = a\cos 2\theta$

(k) Lemniscate: $r^2 = a^2\cos 2\theta$

(l) Archimedes' spiral: $r = a\theta$

7.3 COMPLEX NUMBERS IN RECTANGULAR AND POLAR FORMS

If a and b are real numbers and i is the imaginary unit, then the **complex number** $a + bi$ is associated with the ordered pair (a, b). The plane is called the **complex plane,** the x axis is the **real axis,** the y axis is the **imaginary axis,** and $a + bi$ is the **rectangular form** of the number, as illustrated in Figure 2. Complex numbers can also be written in **polar,** or **trigonometric, form** as shown in Figure 3.

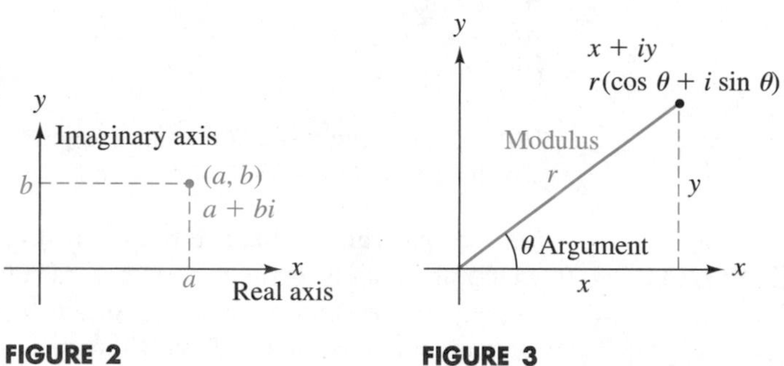

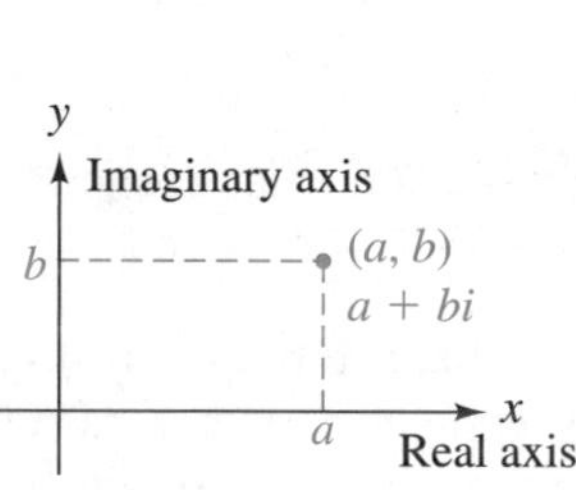

FIGURE 2
Complex plane

FIGURE 3
Polar and rectangular forms

Because of the periodic nature of sine and cosine functions, we have the **general polar form** for a complex number $z = x + iy$:

$$z = x + iy = r[\cos(\theta + 2k\pi) + i\sin(\theta + 2k\pi)]$$
$$= re^{(\theta+2k\pi)i}$$

where $e^{i\theta} = \cos\theta + i\sin\theta$ and the quadrant for θ is determined by x and y.

The number r is called the **modulus,** or **absolute value,** of z and is denoted by **mod** z or $|z|$. The polar angle that the line joining z to the origin makes with the polar axis is called the **argument** of z and is denoted by **arg** z. From Figure 3, we have the following representations of the modulus and argument for $z = x + iy$:

$\text{mod } z = r = \sqrt{x^2 + y^2}$	Never negative
$\arg z = \theta + 2k\pi$	k any integer

where $\sin\theta = y/r$, $\cos\theta = x/r$, and θ is usually chosen so that $-\pi < \theta \le \pi$ or $-180° < \theta \le 180°$.

Products and **quotients** of complex numbers in polar form are found as follows: If

$$z_1 = r_1e^{i\theta_1} \quad \text{and} \quad z_2 = r_2e^{i\theta_2}$$

then

1. $z_1z_2 = r_1e^{i\theta_1}r_2e^{i\theta_2} = r_1r_2e^{i(\theta_1+\theta_2)}$

2. $\dfrac{z_1}{z_2} = \dfrac{r_1e^{i\theta_1}}{r_2e^{i\theta_2}} = \dfrac{r_1}{r_2}e^{i(\theta_1-\theta_2)}$

7.4 DE MOIVRE'S THEOREM AND THE *n*TH-ROOT THEOREM

De Moivre's theorem and the related *n*th-root theorem make the process of finding natural number powers and all the *n*th roots of a complex number relatively easy. **De Moivre's theorem** is stated as follows: If $z = x + iy = re^{i\theta}$ and n is a natural number, then

$$z^n = (x + iy)^n = (re^{i\theta})^n = r^n e^{n\theta i}$$

From De Moivre's theorem, we can derive the ***n*th-root theorem:** For n a positive integer greater than 1 and θ in degrees,

$$r^{1/n}e^{(\theta/n + k360°/n)i} \qquad k = 0, 1, \ldots, n-1$$

are the n distinct nth roots of $re^{i\theta}$, and there are no others.

CHAPTER 7 REVIEW EXERCISE

Work through all the problems in this chapter review and check the answers. Answers to all review problems appear in the back of the book; following each answer is an italic number that indicates the section in which that type of problem is discussed. Where weaknesses show up, review the appropriate sections in the text.

A *In Problems 1–4, plot in a polar coordinate system.*

1. $A(5, 210°)$; $B(-7, 180°)$; $C(-5, -45°)$

2. $r = 5 \sin \theta$ **3.** $r = 4 + 4 \cos \theta$

4. $r = 8$

Verify the graphs of Problems 2–4 on a graphing utility.

5. Change $(2\sqrt{2}, \pi/4)$ to rectangular coordinates.

6. Change $(-\sqrt{3}, 1)$ to polar coordinates $(r \geq 0, 0 \leq \theta < 2\pi)$.

7. Graph $-3 - 2i$ in a rectangular coordinate system.

8. Plot $z = 5 (\cos 60° + i \sin 60°) = 5e^{(60°)i}$ in a polar coordinate system.

9. A point in a polar coordinate system has coordinates $(-8, 30°)$. Find all other polar coordinates for the point, $-360° < \theta \leq 360°$, and verbally describe how the coordinates are associated with the point.

10. Find z_1z_2 and z_1/z_2 for $z_1 = 9e^{(42°)i}$ and $z_2 = 3e^{(37°)i}$. Leave answers in polar form.

11. Find $[2e^{(10°)i}]^4$ using De Moivre's theorem. Leave answer in polar form.

B *Plot Problems 12–15 in a polar coordinate system.*

12. $A(-5, \pi/4)$; $B(5, -\pi/3)$; $C(-8, 4\pi/3)$

13. $r = 8 \sin 3\theta$

14. $r = 4 \sin 2\theta$

15. $\theta = \pi/6$

Verify the graphs of Problems 13 and 14 on a graphing utility.

16. Change $8x - y^2 = x^2$ to polar form.

17. Change $r(3 \cos \theta - 2 \sin \theta) = -2$ to rectangular form.

18. Change $r = -3 \cos \theta$ to rectangular form.

19. Graph $r = 5 \cos(\theta/5)$ on a graphing utility for $0 \leq \theta \leq 5\pi$.

20. Graph $r = 5 \cos(\theta/7)$ on a graphing utility for $0 \leq \theta \leq 7\pi$.

21. Convert $-\sqrt{3} - i$ to polar form ($r \geq 0$, $-180° < \theta \leq 180°$).

22. Convert $3\sqrt{2}e^{(3\pi/4)i}$ to exact rectangular form.

23. Convert the factors in $(2 + 2\sqrt{3}\, i)(-\sqrt{2} + \sqrt{2}\, i)$ to polar form and evaluate. Leave answer in polar form.

24. Divide $(-\sqrt{2} + \sqrt{2}\, i)/(2 + 2\sqrt{3}\, i)$ by first converting the numerator and denominator to polar form. Leave answer in polar form.

25. Find $(-1 - i)^4$ using De Moivre's theorem, and write the result in exact rectangular form.

26. Solve $x^3 - 64 = 0$ for all roots. Write answers in exact rectangular form, and plot them in the complex plane.

27. Find all cube roots of $-4\sqrt{3} - 4i$. Write answers in polar form.

28. Show that $2e^{(30°)i}$ is a square root of $2 + 2\sqrt{3}\, i$.

29. (A) A cube root of a complex number is shown in the figure. Geometrically locate all other cube roots of the number, and explain how you located them.

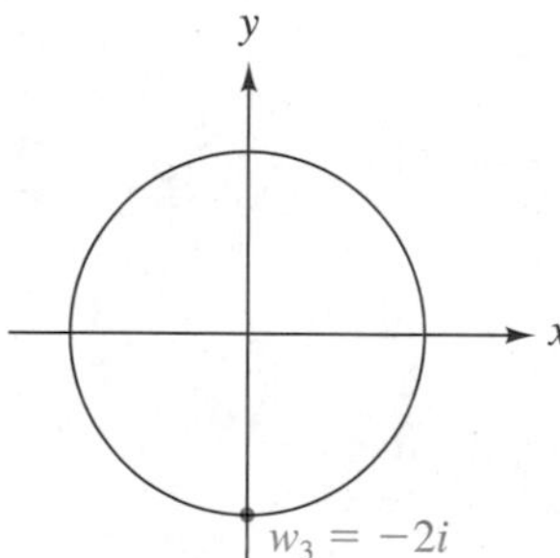

(B) Determine geometrically (without using the nth-root theorem) the two cube roots you graphed in part (A). Give answers in exact rectangular form.

(C) Cube each cube root from part (B).

 **30.** Graph $r = 5(\sin\theta)^{2n}$ in a different viewing window for $n = 1, 2,$ and 3. How many leaves do you expect the graph will have for arbitrary n?

 **31.** Graph $r = 2/(1 - e\sin\theta)$ in separate viewing windows for the following values of e. Identify each curve as a hyperbola, ellipse, or parabola.

(A) $e = 1.6$ (B) $e = 1$ (C) $e = 0.4$

32. Change $r(\sin\theta - 2) = 3$ to rectangular form.

33. Find all roots of $x^3 - 12 = 0$. Write answers in rectangular form $a + bi$, where a and b are computed to three decimal places.

C **34.** Show that

$$[r^{1/3}e^{(\theta/3 + k120^\circ)i}]^3 = re^{i\theta} \qquad \text{for } k = 0, 1, 2$$

 **35.** (A) Graph $r = 10\sin\theta$ and $r = -10\cos\theta$, $0 \le \theta \le \pi$, in the same viewing window. Use TRACE to determine which intersection point has coordinates that satisfy both equations simultaneously.

(B) Solve the equations simultaneously to verify the results in part (B).

(C) Explain why the pole is not a simultaneous solution, even though the two curves intersect at the pole.

CUMULATIVE REVIEW EXERCISE CHAPTERS 1–7

Work through all the problems in this cumulative review and check the answers. Answers to all review problems appear in the back of the book; following each answer is an italic number that indicates the section in which that type of problem is discussed. Where weaknesses show up, review the appropriate sections in the text.

Where applicable, quantities in the problems refer to a triangle labeled as shown in the figure.

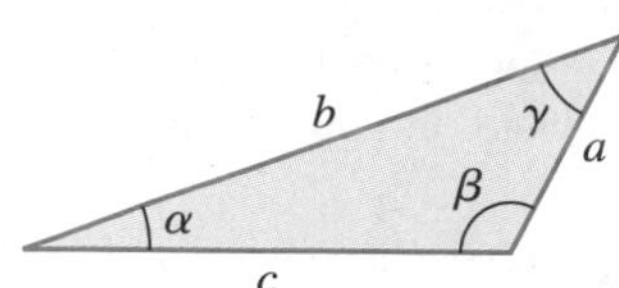

A **1.** Which angle has the larger measure: $\alpha = 2\pi/7$ or $\beta = 51^\circ 25' 40''$? Explain how you got your answer.

2. The sides of a right triangle are 1.27 cm and 4.65 cm. Find the hypotenuse and the acute angles in degree measure.

3. Find the values of $\sec\theta$ and $\tan\theta$ if the terminal side of θ contains $P(7, -24)$.

4. Verify the following identities without looking at a table of identities.

(A) $\cot x \sec x \sin x = 1$

(B) $\tan\theta + \cot\theta = \sec\theta\csc\theta$

Evaluate Problems 5–8 exactly as real numbers.

5. $\sin\dfrac{11\pi}{6}$ **6.** $\tan\dfrac{-5\pi}{3}$

7. $\cos^{-1}(-0.5)$ **8.** $\csc^{-1}\sqrt{2}$

Evaluate Problems 9–12 as real numbers to four significant digits using a calculator.

9. $\sin 43^\circ 22'$ **10.** $\cot\dfrac{2\pi}{5}$

11. $\sin^{-1} 0.8$ ☆**12.** $\sec^{-1} 4.5$

☆ **13.** Write $\sin 3t + \sin t$ as a product.

14. Explain what is meant by an angle of radian measure 2.5.

In Problems 15–17 solve each triangle, given the indicated measures of angles and sides.

15. $\beta = 110°$, $\gamma = 42°$, $b = 68$ m

16. $\beta = 34°$, $a = 16$ in., $c = 24$ in.

17. $a = 18$ ft, $b = 23$ ft, $c = 32$ ft

☆ **18.** Find the area of the triangle in Problem 16.

19. A point in a polar coordinate system has coordinates $(-7, 30°)$. Find all other polar coordinates for the point, $-180° < \theta \leq 180°$, and verbally describe how the coordinates are associated with the point.

20. Find the magnitudes of the horizontal and vertical components of the vector **v** located in a coordinate system as indicated in the figure.

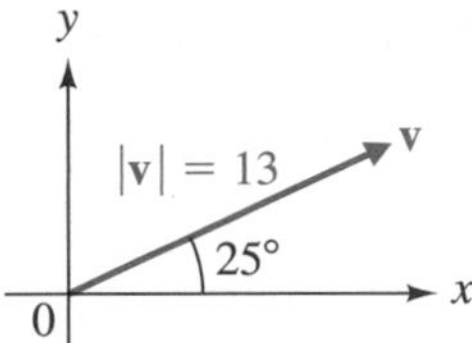

Figure for 20

21. Two vectors **u** and **v** are located in a coordinate system, as indicated in the figure. Find the magnitude and the direction (relative to the x axis) of **u** + **v** if $|\mathbf{u}| = 6.4$ and $|\mathbf{v}| = 3.9$.

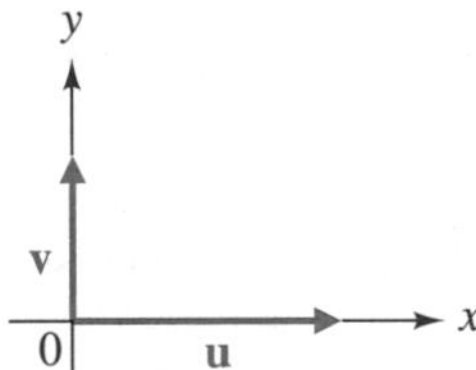

Figure for 21

22. Given $A(4, -2)$ and $B(-3, 7)$, represent the geometric vector AB as an algebraic vector and find its magnitude.

☆ **23.** Find the angle (in decimal degrees, to one decimal place), between $\mathbf{u} = 2\mathbf{i} - 7\mathbf{j}$ and $\mathbf{v} = 3\mathbf{i} + 8\mathbf{j}$.

24. Plot in a polar coordinate system: $A(6, 240°)$; $B(-4, 225°)$; $C(9, -45°)$; $D(-7, -60°)$.

25. Plot in a polar coordinate system: $r = 5 + 5 \sin \theta$.

26. Change $(3\sqrt{2}, 3\pi/4)$ to exact rectangular coordinates.

27. Change $(-2\sqrt{3}, 2)$ to exact polar coordinates.

28. Change $z = 2 - 2i$ to exact polar form.

29. Change $z = 3e^{(3\pi/2)i}$ to exact rectangular form.

30. Find $z_1 z_2$ and z_1/z_2 for $z_1 = 3e^{(50°)i}$ and $z_2 = 5e^{(15°)i}$. Leave answers in polar form.

31. Find $(3e^{(25°)i})^4$ using De Moivre's theorem. Leave answer in polar form.

B

32. Find an equation of the form $y = k + A \cos Bx$ that produces the graph shown in the following graphing calculator display:

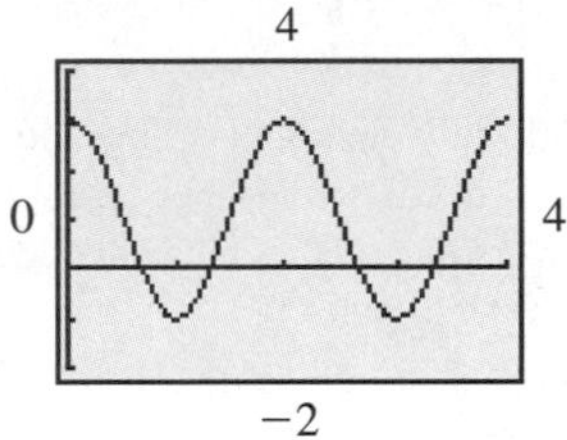

33. Find the exact values of $\csc \theta$ and $\cos \theta$ if $\cot \theta = 4$ and $\sin \theta < 0$.

34. Graph $y = 1 + 2\sin(2x + \pi)$ for $-\pi \leq x \leq 2\pi$. State the amplitude, period, frequency, and phase shift.

Verify the identities in Problems 35–37.

35. $\dfrac{\cos x}{1 - \sin x} + \dfrac{\cos x}{1 + \sin x} = 2 \sec x$

36. $\tan \dfrac{\theta}{2} = \dfrac{1}{\csc \theta + \cot \theta}$

37. $\dfrac{\cos x - \sin x}{\cos x + \sin x} = \sec 2x - \tan 2x$

38. Find the exact values of $\tan(x/2)$ and $\sin 2x$, given $\cos x = \frac{24}{25}$ and $\sin x > 0$.

39. Use a graphing utility to test whether each equation is an identity. If the equation appears to be an identity, verify it. If the equation does not appear to be an identity, find a value of x for which both sides are defined but are not equal.

(A) $\dfrac{\cos^2 x}{(\cos x - 1)^2} = \dfrac{1 + \cos x}{1 - \cos x}$

(B) $\dfrac{\sin^2 x}{(\cos x - 1)^2} = \dfrac{1 + \cos x}{1 - \cos x}$

40. Find the exact value of $\sec(\sin^{-1} \frac{3}{4})$.

41. (A) A cube root of a complex number is shown in the figure at the top of the next page. Geometrically locate all other cube roots of the number, and explain how you located them.

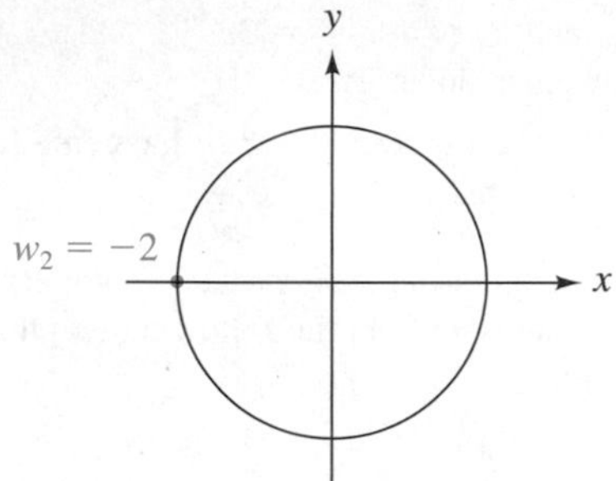

(B) Determine geometrically (without using the nth-root theorem) the two cube roots you graphed in part (A). Give answers in exact rectangular form.
(C) Cube each cube root from part (B).

42. Find exactly all real solutions for

$$\sin 2x + \sin x = 0$$

43. Find all real solutions (to four significant digits) for

$$2 \cos 2x = 5 \sin x - 4$$

44. Solve the triangle with $b = 17.4$ cm and $\alpha = 49°30'$ for each of the following values of a:
(A) $a = 11.5$ cm (B) $a = 14.7$ cm
(C) $a = 21.1$ cm

45. Given the vector diagram below, find $|\mathbf{u} + \mathbf{v}|$ and θ.

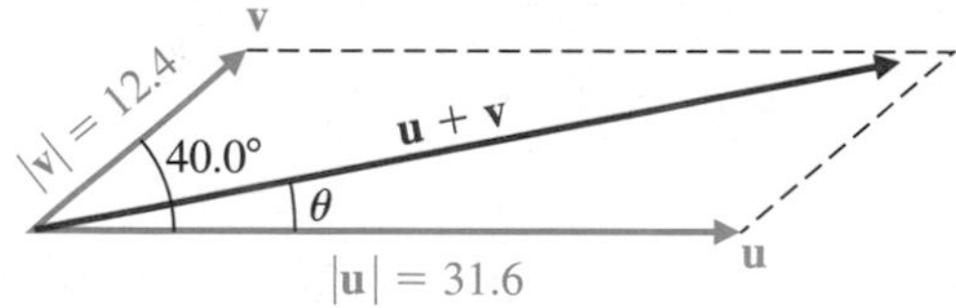

46. Find $3\mathbf{u} - 4\mathbf{v}$ for
(A) $\mathbf{u} = \langle 1, -2\rangle$; $\mathbf{v} = \langle 0, 3\rangle$
(B) $\mathbf{u} = 2\mathbf{i} + 3\mathbf{j}$; $\mathbf{v} = -\mathbf{i} + 5\mathbf{j}$

47. Find a unit vector $\mathbf{u}$ with the same direction as $\mathbf{v} = \langle 7, -24\rangle$.

48. Express $\mathbf{v}$ in terms of $\mathbf{i}$ and $\mathbf{j}$ unit vectors if $\mathbf{v} = \overrightarrow{AB}$, with $A(-3, 2)$ and $B(-1, 5)$.

☆ 49. Determine which vector pairs are orthogonal using properties of the dot product.
(A) $\mathbf{u} = \langle 4, 0\rangle$; $\mathbf{v} = \langle 0, -5\rangle$
(B) $\mathbf{u} = \langle 3, 2\rangle$; $\mathbf{v} = \langle -3, 4\rangle$
(C) $\mathbf{u} = \mathbf{i} - 2\mathbf{j}$; $\mathbf{v} = 6\mathbf{i} + 3\mathbf{j}$

50. Plot $r = 8 \cos 2\theta$ in a polar coordinate system.

51. Change $x^2 = 6y$ to polar form.

52. Change $r = 4 \sin \theta$ to rectangular form.

53. Convert the factors in $(3 + 3i)(-1 + \sqrt{3}\, i)$ to polar form and evaluate. Leave answer in polar form.

54. Convert the numerator and denominator in $(-1 + \sqrt{3}\, i)/(3 + 3i)$ to polar form and evaluate. Leave answer in polar form.

55. Use De Moivre's theorem to evaluate $(1 - i)^6$, and write the result in exact rectangular form.

56. Find all cube roots of $8i$. Write answers in exact rectangular form and plot them in a complex plane.

In Problems 57–59, graph each function in a graphing utility over the indicated interval.

57. $y = -2 \cos^{-1}(2x - 1)$, $0 \le x \le 1$

58. $y = 2 \sin^{-1} 2x$, $-\frac{1}{2} \le x \le \frac{1}{2}$

59. $y = \tan^{-1}(2x + 5)$, $-5 \le x \le 0$

In Problems 60–62, use a graphing utility to approximate all solutions over the indicated intervals. Compute solutions to three decimal places.

60. $\tan x = 5$, $-\pi \le x \le \pi$

61. $\cos x = \sqrt[3]{x}$, all real x

62. $3 \sin 2x \cos 3x = 2$, $0 \le x \le 2\pi$

63. Graph $r = 5 \sin(\theta/4)$ in a graphing utility for $0 \le \theta \le 8\pi$.

64. Graph $r = 2/[1 - e \sin(\theta + 0.6)]$ for the indicated values of e, each in a different viewing window. Identify each curve as a hyperbola, ellipse, or parabola.
(A) $e = 0.7$ (B) $e = 1$ (C) $e = 1.5$

C 65. In the circle shown, find the exact radian measure of θ and the coordinates of B to four significant digits if the arc length s is 8 units.

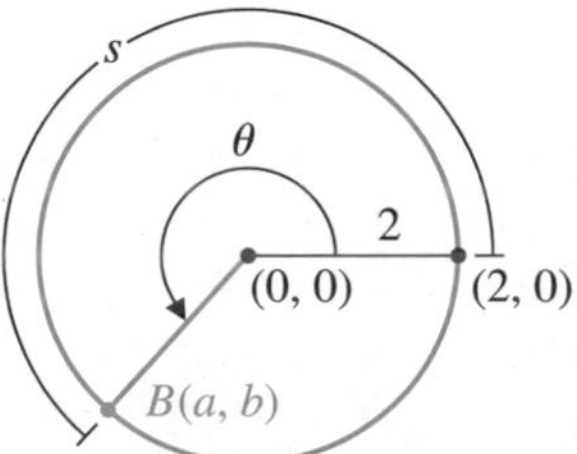

Figure for 65 and 66

66. Refer to the figure. Find θ and s to four significant digits if $(a, b) = (-1.6, -1.2)$.

67. Graph $y = 2 \sec(\pi x + \pi/4)$ for $-1 \le x \le 3$.

68. Express $\sec(2 \tan^{-1} x)$ as an algebraic expression in x free of trigonometric and inverse trigonometric functions.

69. Verify the identity: $\tan 3x = \dfrac{3 - \tan^2 x}{\cot x - 3 \tan x}$.

70. Change $r(\cos \theta + 1) = 1$ to rectangular form.

71. (A) Graph $r = -8 \sin \theta$ and $r = -8 \cos \theta$, $0 \le \theta \le \pi$, in the same viewing window.

(B) Explain why the pole is not a simultaneous solution, even though the two curves intersect at the pole.

72. Find all roots of $x^3 - 4 = 0$. Write answers in the form $a + bi$, where a and b are computed to three decimal places. (Use a calculator.)

73. De Moivre's theorem can be used to derive multiple angle identities. For example, consider

$$\begin{aligned}\cos 2\theta + i \sin 2\theta &= (\cos \theta + i \sin \theta)^2 \\ &= \cos^2 \theta - \sin^2 \theta + 2i \sin \theta \cos \theta\end{aligned}$$

Equating the real and imaginary parts of the left side with the real and imaginary parts of the right side yields two familiar identities:

$$\cos 2\theta = \cos^2 \theta - \sin^2 \theta \qquad \sin 2\theta = 2 \sin \theta \cos \theta$$

(A) Apply De Moivre's theorem to $(\cos \theta + i \sin \theta)^3$ to derive identities for $\cos 3\theta$ and $\sin 3\theta$.

(B) Use identities from Chapter 4 to verify the identities obtained in part (A).

In Problems 74–76, graph f(x) in a graphing utility. From the graph, find a simpler function of the form g(x) = K + A t(Bx), where t(x) is one of the six trigonometric functions, that produces the same graph. Then verify the identity f(x) = g(x).

74. $f(x) = 2 \cos^2 x - 4 \sin^2 x$

75. $f(x) = \dfrac{6 \sin^2 x - 2}{2 \cos^2 x - 1}$

76. $f(x) = \dfrac{\sin x + \cos x - 1}{1 - \cos x}$

Applications

77. **Surveying** To determine the length CB of the lake in the figure, a surveyor makes the following measurements: $AC = 520$ ft, $\angle BCA = 52°$, and $\angle CAB = 77°$. Find the approximate length of the lake.

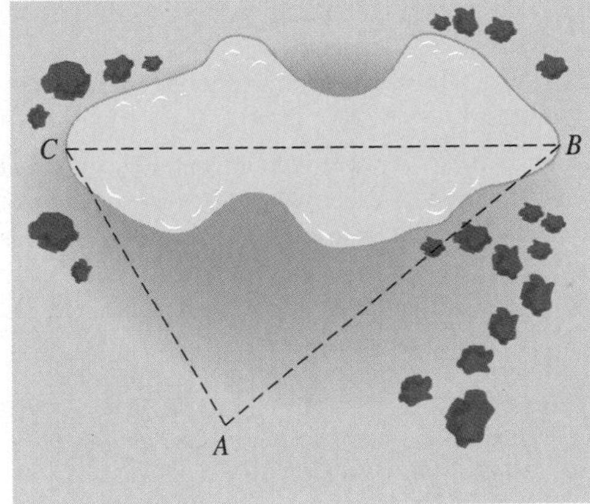

Figure for 77

78. **Surveying** Refer to Problem 77. Find the approximate length of a similar lake using the following measurements: $AC = 430$ ft, $AB = 580$ ft, $\angle CAB = 64°$.

79. **Tree Height** A tree casts a 35 ft shadow when the angle of elevation of the sun is 54°.

(A) Find the height of the tree if its shadow is cast on level ground.

(B) Find the height of the tree if its shadow is cast straight down a hillside that slopes 11° relative to the horizontal.

80. **Navigation** Two tracking stations located 4 mi apart on a straight coastline sight a ship (see the figure). How far is the ship from each station?

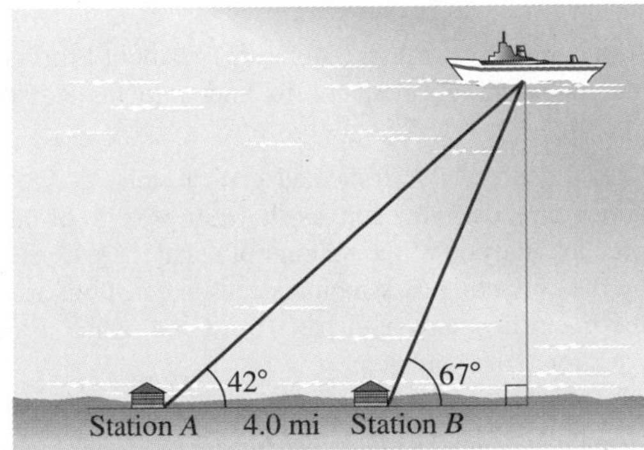

Figure for 80

81. **Electrical Circuits** The voltage E in an electrical circuit is given by an equation of the form $E = 110 \cos Bt$.

(A) If the frequency is 70 Hz, what is the period? What is the value of B?

(B) If the period is 0.0125 sec, what is the frequency? What is the value of B?

(C) If $B = 100\pi$, what is the period? What is the frequency?

☆ **82. Water Waves** A wave with an amplitude of exactly 2 ft and a period of exactly 4 sec has an equation of the form $y = A \sin Bt$ at a fixed position. How high is the wave from trough to crest? What is its wavelength (in feet)? How fast is it traveling (in feet per second)? Compute answers to the nearest foot.

Precalculus *Problems 83–86 refer to the region OCBA inside the first quadrant portion of a unit circle centered at the origin, as shown in the figure.**

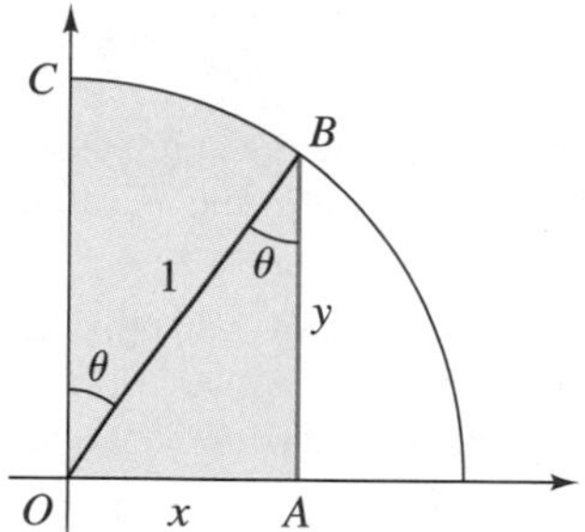

Figure for 83–86

83. Express the area of *OCBA* in terms of θ.

84. Express the area of *OCBA* in terms of x.

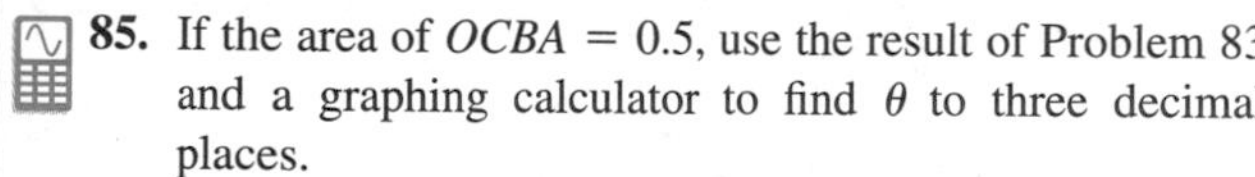

85. If the area of $OCBA = 0.5$, use the result of Problem 83 and a graphing calculator to find θ to three decimal places.

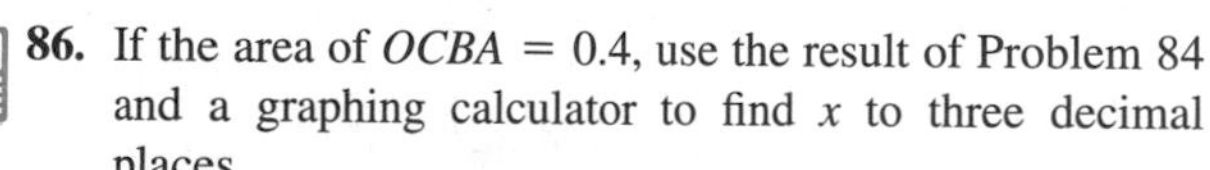

86. If the area of $OCBA = 0.4$, use the result of Problem 84 and a graphing calculator to find x to three decimal places.

87. Solar Energy[†] A truncated conical solar collector is aimed directly at the sun, as shown in part (a) of the figure. An analysis of the amount of solar energy absorbed by the collecting disk requires certain equations relating the quantities shown in part (b) of the figure. Find an equation that expresses:

(A) R in terms of h, H, r, and α

(B) β in terms of h, H, r, and R

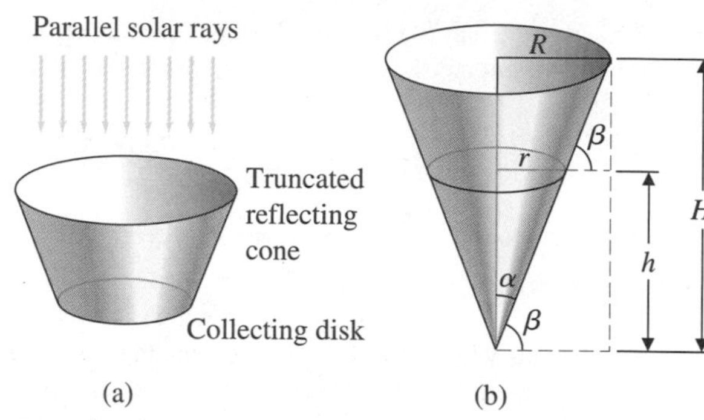

Figure for 87

* **88. Navigation** An airplane can cruise at 265 mph in still air. If a steady wind of 81.5 mph is blowing from the east, what compass heading should the pilot fly in order for the true course of the plane to be north (0°)? Compute the ground speed for this course.

* **89. Static Equilibrium** A weight suspended from a cable deflects the cable relative to the horizontal as indicated in the figure.

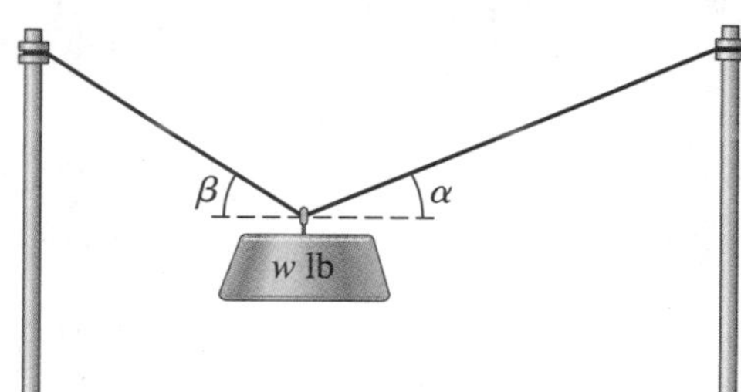

Figure for 89

(A) If w is the weight of the object, T_L is the tension in the left side of the cable, and T_R is the tension in the right side, show that

$$T_L = \frac{w \cos \alpha}{\sin(\alpha + \beta)} \quad \text{and} \quad T_R = \frac{w \cos \beta}{\sin(\alpha + \beta)}$$

(B) If $\alpha = \beta$, show that $T_L = T_R = \frac{1}{2} w \csc \alpha$.

90. Railroad Construction North American railroads do not express specifications for circular track in terms of the radius. The common practice is to define the degree of a track curve as the degree measure of the central angle θ subtended by a 100 ft chord of the circular arc of track (see the figure).

(A) Find the radius of a 10° track curve to the nearest foot.

(B) If the radius of a track curve is 2,000 ft, find the degree of the track curve to the nearest 0.1°.

* See "On the Derivatives of Trigonometric Functions" by M. R. Speigle in the *American Mathematical Monthly*, Vol. 63 (1956).

† Based on the article "The Solar Concentrating Properties of a Conical Reflector" by Don Leake in *The UMAP Journal*, Vol. 8, No. 4 (1987).

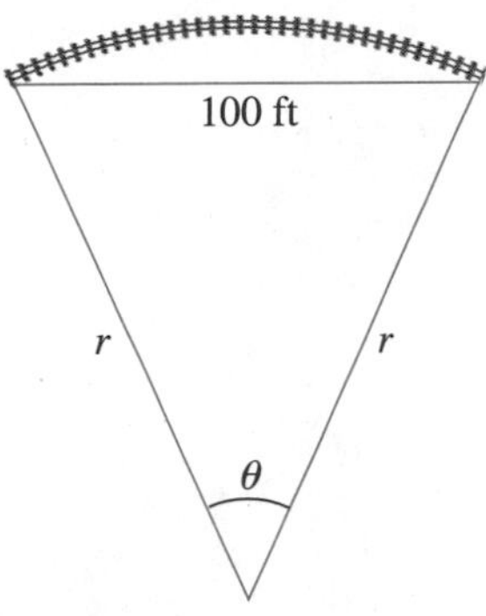

Figure for 90

91. Railroad Construction Actually using a compass to lay out circular arcs for railroad track is impractical. Instead, track layers work with the distance that curved track is offset from straight track.

(A) Suppose a 50 ft length of rail is bent into a circular arc, resulting in a horizontal offset of 1 ft (see the figure). Show that the radius r of this track curve satisfies the equation

$$r \cos^{-1}\left(\frac{r-1}{r}\right) = 50$$

and approximate r to the nearest foot.

(B) Find the degree of the track curve in part (A) to the nearest 0.1°. (See Problem 90 for the definition of the degree of a track curve.)

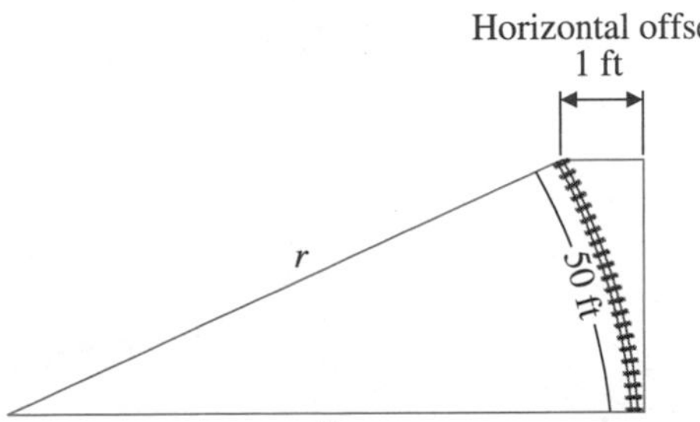

Figure for 91

92. Engineering A circular log of radius r is cut lengthwise into three pieces by two parallel cuts equidistant from the center of the log (see the figure). Express the cross-sectional area of the center piece in terms of r and θ.

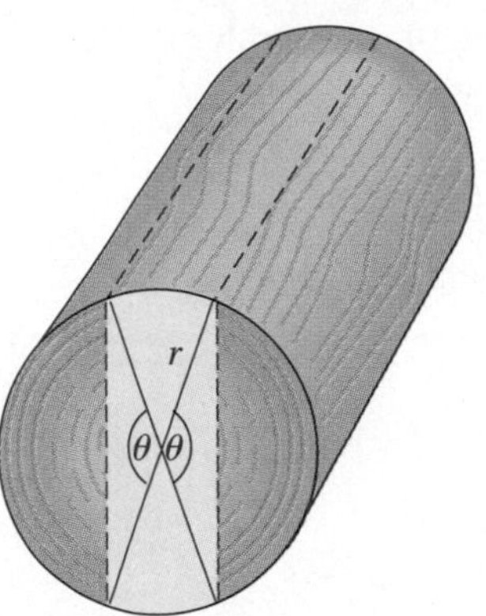

Figure for 92 and 93

93. Engineering If all three pieces of the log in Problem 92 have the same cross-sectional area, approximate θ to four decimal places.

94. Modeling Twilight Duration Periods of twilight occur just before sunrise and just after sunset. The length of these periods varies with the time of year and latitudinal position. At the equator, twilight lasts about an hour all year long. Near the poles, there will be times during the year when twilight lasts for 24 hours and other times when it does not occur at all. Table 1 gives the duration of twilight on the eleventh of each month for 1 year at 20°N latitude.

(A) Convert the data in Table 1 from hours and minutes to two-place decimal hours. Enter the data for a 2 year period in your graphing calculator and produce a scatter plot in the following viewing window: $1 \le x \le 24$, $1 \le y \le 5$.

(B) A function of the form $y = k + A \sin(Bx + C)$ can be used to model the data. Use the converted data from Table 1 to determine k and A (to two decimal places), and B (exact value). Use the graph in part (A) to visually estimate C (to one decimal place).

(C) Plot the data from part (A) and the equation from part (B) in the same viewing window. If necessary, adjust your value of C to produce a better fit.

TABLE 1

x (months)	1	2	3	4	5	6	7	8	9	10	11	12
y (twilight duration, hr:min)	1:37	1:49	2:21	2:59	3:33	4:07	4:03	3:30	2:48	2:13	1:48	1:34

COMMENTS ON NUMBERS

APPENDIX

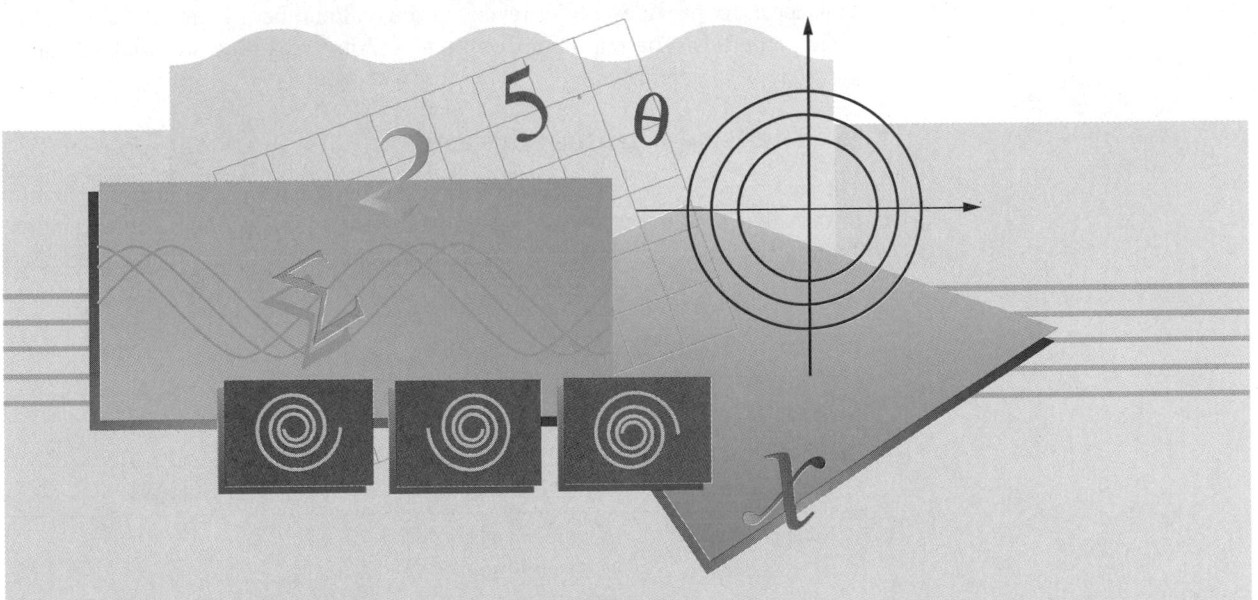

A.1 REAL NUMBERS

- The Real Number System
- The Real Number Line
- Basic Real Number Properties

This appendix provides a brief review of the real number system and some of its basic properties. The real number system and its properties are fundamental to the study of mathematics.

The Real Number System

The real number system is the number system you have used most of your life. Informally, a **real number** is any number that has a decimal representation. Table 1 describes the set of real numbers and some of its important subsets. Figure 1 illustrates how these various sets of numbers are related to each other.

TABLE 1
The Set of Real Numbers

Name	Description	Examples
Natural numbers	Counting numbers (also called positive integers)	$1, 2, 3, \ldots$
Integers	Natural numbers, their negatives, and zero	$\ldots, -2, -1, 0, 1, 2, \ldots$
Rational numbers	Numbers that can be represented as a/b, where a and b are integers, $b \neq 0$; decimal representations are repeating or terminating	$-7, 0, 1, 36, \frac{2}{3}, 3.8,$ $-0.6666\overline{6}$,* $6.3838\overline{38}$
Irrational numbers	Numbers that can be represented as nonrepeating and nonterminating decimal numbers	$\sqrt{3}$, $\sqrt[3]{7}$, π, 3.27393...
Real numbers	Rational numbers and irrational numbers	

* The bar over the 6 means that 6 repeats infinitely, $-0.666666666\ldots$. Similarly, $\overline{38}$ means that 38 is repeated infinitely.

The set of **integers** contains all the **natural numbers** and something else—their negatives and zero. The set of **rational numbers** contains all the integers

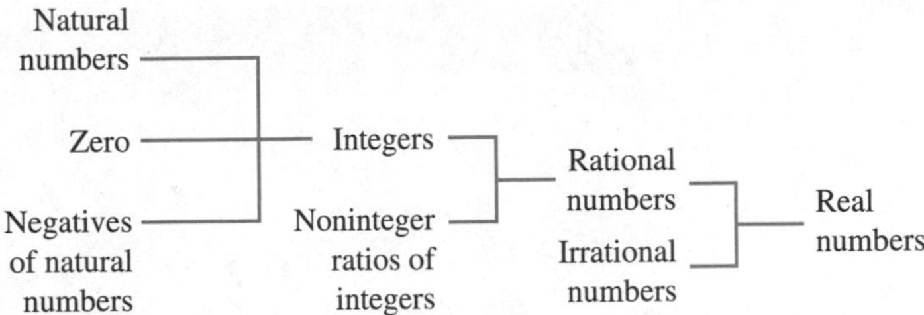

FIGURE 1
The real number system

and something else—noninteger ratios of integers. And the set of **real numbers** contains all the rational numbers and something else—the **irrational numbers.**

The Real Number Line

A one-to-one correspondence exists between the set of real numbers and the set of points on a line. That is, each real number corresponds to exactly one point, and each point corresponds to exactly one real number. A line with a real number associated with each point, and vice versa, as in Figure 2, is called a **real number line,** or simply a **real line.** Each number associated with a point is called the **coordinate** of the point. The point with coordinate 0 is called the **origin.** The arrow on the right end of the real line in Figure 2 indicates a positive direction. The coordinates of all points to the right of the origin are called **positive real numbers,** and those to the left of the origin are called **negative real numbers.**

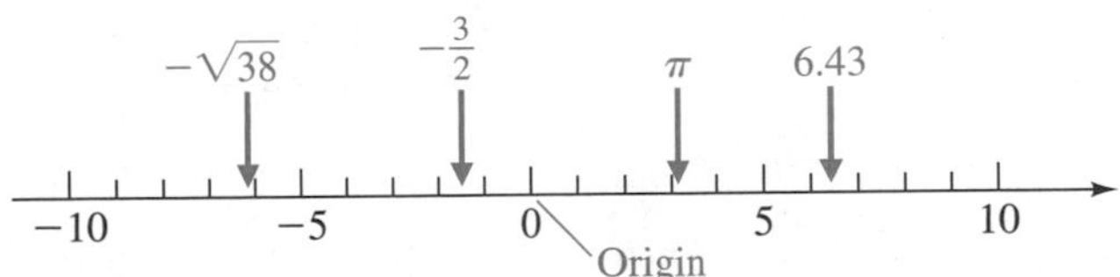

FIGURE 2
A real number line

Basic Real Number Properties

In the box at the top of the next page, we list some basic properties of the real number system that enable us to convert algebraic expressions into equivalent forms. Do not let the names of these properties intimidate you. Most of the ideas presented are quite simple. In fact, you have been using many of these properties in arithmetic for a long time. We summarize these properties here for convenient reference because they represent some of the basic rules of the "game of algebra," and you cannot play the game very well unless you know the rules.

BASIC PROPERTIES OF THE SET OF REAL NUMBERS

Let R be the set of real numbers, and let x, y, and z be arbitrary elements of R.

Addition properties

Closure	$x + y$ is a unique element in R.
Associative	$(x + y) + z = x + (y + z)$
Commutative	$x + y = y + x$
Identity	0 is the additive identity; that is, for all x in R, $0 + x = x + 0 = x$, and 0 is the only element in R with this property.
Inverse	For each x in R, $-x$ is its unique additive inverse; that is, $x + (-x) = (-x) + x = 0$, and $-x$ is the only element in R relative to x with this property.

Multiplication properties

Closure	xy is a unique element in R.
Associative	$(xy)z = x(yz)$
Commutative	$xy = yx$
Identity	1 is the multiplicative identity; that is, for all x in R, $(1)x = x(1) = x$, and 1 is the only element in R with this property.
Inverse	For each x in R, $x \neq 0$, $1/x$ is its unique multiplicative inverse; that is, $x(1/x) = (1/x)x = 1$, and $1/x$ is the only element in R relative to x with this property.

Combined property

Distributive	$x(y + z) = xy + xz$ $(x + y)z = xz + yz$

EXERCISE A.1

1. Give an example of a negative integer, an integer that is neither positive nor negative, and a positive integer.
2. Give an example of a negative rational number, a rational number that is neither positive nor negative, and a positive rational number.
3. Give an example of a rational number that is not an integer.
4. Give an example of an integer that is not a natural number.
5. Indicate which of the following are true:
   (A) All natural numbers are integers.
   (B) All real numbers are irrational.
   (C) All rational numbers are real numbers.
6. Indicate which of the following are true:
   (A) All integers are natural numbers.
   (B) All rational numbers are real numbers.
   (C) All natural numbers are rational numbers.

In Problems 7 and 8, express each number in decimal form to the capacity of your calculator. Observe the repeating decimal representation of the rational numbers and the apparent nonrepeating decimal representation of the irrational numbers. Indicate whether each number is rational or irrational.

7. (A) $\frac{4}{11}$ (B) $\frac{7}{9}$ (C) $\sqrt{7}$ (D) $\frac{13}{8}$

8. (A) $\frac{19}{6}$ (B) $\sqrt{23}$ (C) $\frac{9}{16}$ (D) $\frac{31}{111}$

9. Each of the following real numbers lies between two successive integers on a real number line. Indicate which two.
(A) $\frac{26}{9}$ (B) $-\frac{19}{5}$ (C) $-\sqrt{23}$

10. Each of the following real numbers lies between two successive integers on a real number line. Indicate which two.
(A) $\frac{13}{4}$ (B) $-\frac{5}{3}$ (C) $-\sqrt{8}$

In Problems 11–24, replace each question mark with an appropriate expression that will illustrate the use of the indicated real number property.

11. Commutative property $(+)$: $3 + y = ?$

12. Commutative property $(+)$: $u + v = ?$

13. Associative property $(\cdot)$: $3(2x) = ?$

14. Associative property $(\cdot)$: $7(4z) = ?$

15. Commutative property $(\cdot)$: $x7 = ?$

16. Commutative property $(\cdot)$: $yx = ?$

17. Associative property $(+)$: $5 + (7 + x) = ?$

18. Associative property $(+)$: $9 + (2 + m) = ?$

19. Identity property $(+)$: $3m + 0 = ?$

20. Identity property $(+)$: $0 + (2x + 3) = ?$

21. Identity property $(\cdot)$: $1(u + v) = ?$

22. Identity property $(\cdot)$: $1(xy) = ?$

23. Distributive property: $2x + 3x = ?$

24. Distributive property: $7(x + y) = ?$

25. Indicate whether each is true or false, and for each false statement find real number replacements for a and b to illustrate that it is false. For all real numbers a and b:
(A) $a + b = b + a$ (B) $a - b = b - a$
(C) $ab = ba$ (D) $a/b = b/a$

26. Indicate whether each is true or false, and for each false statement find real number replacements for a, b, and c to illustrate that it is false. For all real numbers a, b, and c:
(A) $(a + b) + c = a + (b + c)$
(B) $(a - b) - c = a - (b - c)$
(C) $a(bc) = (ab)c$
(D) $(a/b)/c = a/(b/c)$

A.2 COMPLEX NUMBERS

The Pythagoreans (500–275 BC) found that the simple equation

$$x^2 = 2 \tag{1}$$

had no rational number solutions. (A **rational number** is any number that can be expressed as P/Q, where P and Q are integers and $Q \neq 0$.) If equation (1) were to have a solution, then a new kind of number had to be invented—the **irrational number.**

The irrational numbers $\sqrt{2}$ and $-\sqrt{2}$ are both solutions to equation (1). The invention of these irrational numbers evolved over a period of about 2,000 years, and it was not until the 19th century that they were finally put on a rigorous foundation. The rational numbers and irrational numbers together constitute the real number system.

Is there any need to extend the real number system still further? Yes, if we want the simple equation

$$x^2 = -1$$

to have a solution. Since $x^2 \geq 0$ for any real number x, this equation has no real number solutions. Once again we are forced to invent a new kind of number, a number that has the possibility of being negative when it is squared. These new numbers are called *complex numbers.* The complex numbers, like the irrational numbers, evolved over a long period of time, dating mainly back to Girolamo Cardano (1501–1576). But it was not until the 19th century that they were firmly established as numbers in their own right. A **complex number** is defined to be a number of the form

$$a + bi$$

where a and b are real numbers, and i is called the **imaginary unit.** Thus,

$$5 + 3i \qquad \tfrac{1}{3} - 6i \qquad \sqrt{7} + \tfrac{1}{4}i \qquad 0 + 6i \qquad \tfrac{1}{2} + 0i \qquad 0 + 0i$$

are all complex numbers. Particular kinds of complex numbers are given special names, as follows:

$a + 0i = a$		Real number
$a + bi$	$b \neq 0$	Imaginary number
$0 + bi = bi$		Pure imaginary number
$0 + 0i = 0$		Zero
$1i = i$		Imaginary unit
$a - bi$		Conjugate of $a + bi$

Thus, we see that just as every integer is a rational number, every real number is a complex number.

To use complex numbers we must know how to add, subtract, multiply, and divide them. We start by defining equality, addition, and multiplication.

BASIC DEFINITIONS FOR COMPLEX NUMBERS

Equality: $a + bi = c + di$ if and only if $a = c$ and $b = d$

Addition: $(a + bi) + (c + di) = (a + c) + (b + d)i$

Multiplication: $(a + bi)(c + di) = (ac - bd) + (ad + bc)i$

These definitions, particularly the one for multiplication, may seem a little strange to you. But it turns out that if we want many of the same basic properties that hold for real numbers to hold for complex numbers, and if we also want negative real numbers to have square roots, then we must define addition and multiplication as shown. Let us use the definition of multiplication to see what happens to i when it is squared:

$$\begin{aligned} i^2 &= (0 + 1i)(0 + 1i) \\ &= (0 \cdot 0 - 1 \cdot 1) + (0 \cdot 1 + 1 \cdot 0)i \\ &= -1 + 0i \\ &= -1 \end{aligned}$$

Thus

$$i^2 = -1$$

This is an important result. We can also write

$$i = \sqrt{-1} \quad \text{and} \quad -i = -\sqrt{-1}$$

Fortunately, you do not have to memorize the definitions of addition and multiplication. We can show that the complex numbers, under these definitions, are closed, associative, and commutative, and multiplication distributes over addition. As a consequence, we can manipulate complex numbers as if they were binomial forms in real number algebra, with the exception that i^2 is to be replaced with -1. Example 1 illustrates the mechanics of carrying out addition, subtraction, multiplication, and division.

EXAMPLE 1 **Performing Operations with Complex Numbers**

Write each of the following in the form $a + bi$:

(A) $(2 + 3i) + (3 - i)$ (B) $(2 + 3i) - (3 - i)$

(C) $(2 + 3i)(3 - i)$ (D) $(2 + 3i)/(3 - i)$

Solution We treat these as we would ordinary binomials in elementary algebra, with one exception: Whenever i^2 turns up, we replace it with -1.

(A) $$\begin{aligned}(2 + 3i) + (3 - i) &= 2 + 3i + 3 - i \\ &= 2 + 3 + 3i - i \\ &= 5 + 2i\end{aligned}$$

(B) $$\begin{aligned}(2 + 3i) - (3 - i) &= 2 + 3i - 3 + i \\ &= 2 - 3 + 3i + i \\ &= -1 + 4i\end{aligned}$$

(C) $$\begin{aligned}(2 + 3i)(3 - i) &= 6 + 7i - 3i^2 \\ &= 6 + 7i - 3(-1) \\ &= 6 + 7i + 3 \\ &= 9 + 7i\end{aligned}$$

(D) To eliminate i from the denominator, we multiply the numerator and denominator by the complex conjugate of $3 - i$, namely, $3 + i$. [Recall from elementary algebra that $(a - b)(a + b) = a^2 - b^2$.]

$$\begin{aligned}\frac{2 + 3i}{3 - i} \cdot \frac{3 + i}{3 + i} &= \frac{6 + 11i + 3i^2}{9 - i^2} \\ &= \frac{6 + 11i - 3}{9 + 1} \\ &= \frac{3 + 11i}{10} = \frac{3}{10} + \frac{11}{10}i\end{aligned}$$

■

Matched Problem 1 Write each of the following in the form $a + bi$:

(A) $(3 - 2i) + (2 + i)$ (B) $(3 - 2i) - (2 - i)$

(C) $(3 - 2i)(2 - i)$ (D) $(3 - 2i)/(2 - i)$ ■

At this time, your experience with complex numbers has likely been limited to solutions of equations, particularly quadratic equations. Recall that if $b^2 - 4ac$ is negative in

$$x = \frac{-b \pm \sqrt{b^2 - 4ac}}{2a}$$

then the solutions to the quadratic equation $ax^2 + bx + c = 0$ are complex. This is easy to verify, since a square root of a negative number can be written in the form

$$\sqrt{-k} = i\sqrt{k} \qquad \text{for } k > 0$$

To check this last equation, we square $i\sqrt{k}$ to see if we get $-k$:

$$(i\sqrt{k})^2 = i^2(\sqrt{k})^2 = -k$$

[*Note:* We write $i\sqrt{k}$ instead of $\sqrt{k}\,i$ so that i will not mistakenly be included under the radical.]

EXAMPLE 2 **Converting Square Roots of Negative Numbers to Complex Form**

Write in the form $a + bi$:

(A) $\sqrt{-4}$ (B) $4 + \sqrt{-4}$

(C) $\dfrac{-3 - \sqrt{-7}}{2}$ (D) $\dfrac{1}{1 - \sqrt{-9}}$

Solution (A) $\sqrt{-4} = i\sqrt{4} = 2i$

(B) $4 + \sqrt{-4} = 4 + i\sqrt{4} = 4 + 2i$

(C) $\dfrac{-3 - \sqrt{-7}}{2} = \dfrac{-3 - i\sqrt{7}}{2} = -\dfrac{3}{2} - \dfrac{\sqrt{7}}{2}i$

(D) $\dfrac{1}{1 - \sqrt{-9}} = \dfrac{1}{1 - 3i} = \dfrac{1}{1 - 3i} \cdot \dfrac{1 + 3i}{1 + 3i} = \dfrac{1 + 3i}{1 - 9i^2}$

$= \dfrac{1 + 3i}{10} = \dfrac{1}{10} + \dfrac{3}{10}i$ ■

Matched Problem 2 Write in the form $a + bi$:

(A) $\sqrt{-16}$ (B) $5 + \sqrt{-16}$

(C) $\dfrac{-5 - \sqrt{-2}}{2}$ (D) $\dfrac{1}{3 - \sqrt{-4}}$ ■

Answers to Matched Problems

1. (A) $5 - i$ (B) $1 - i$ (C) $4 - 7i$ (D) $\frac{8}{5} - \frac{1}{5}i$

2. (A) $4i$ (B) $5 + 4i$ (C) $-\frac{5}{2} - (\sqrt{2}/2)i$ (D) $\frac{3}{13} + \frac{2}{13}i$

EXERCISE A.2

Perform the indicated operations and write each answer in the standard form $a + bi$.

1. $(3 - 2i) + (4 + 7i)$
2. $(4 + 6i) + (2 - 3i)$
3. $(3 - 2i) - (4 + 7i)$
4. $(4 + 6i) - (2 - 3i)$
5. $(6i)(3i)$
6. $(5i)(4i)$
7. $2i(3 - 4i)$
8. $4i(2 - 3i)$
9. $(3 - 4i)(1 - 2i)$
10. $(5 - i)(2 - 3i)$
11. $(3 + 5i)(3 - 5i)$
12. $(7 - 3i)(7 + 3i)$
13. $\dfrac{1}{2 + i}$
14. $\dfrac{1}{3 - i}$
15. $\dfrac{2 - i}{3 + 2i}$
16. $\dfrac{3 + i}{2 - 3i}$
17. $\dfrac{-1 + 2i}{4 + 3i}$
18. $\dfrac{-2 - i}{3 - 4i}$

Convert square roots of negative numbers to complex forms, perform the indicated operations, and express answers in the standard form $a + bi$.

19. $(3 + \sqrt{-4}) + (2 - \sqrt{-16})$
20. $(2 + \sqrt{-9}) + (3 - \sqrt{-25})$
21. $(5 - \sqrt{-1}) - (2 - \sqrt{-36})$
22. $(2 + \sqrt{-9}) - (3 - \sqrt{-25})$
23. $(-3 - \sqrt{-1})(-2 + \sqrt{-49})$
24. $(3 - \sqrt{-9})(-2 - \sqrt{-1})$
25. $\dfrac{5 - \sqrt{-1}}{2 + \sqrt{-4}}$
26. $\dfrac{-2 + \sqrt{-16}}{3 - \sqrt{-25}}$

Evaluate:

27. $(1 - i)^2 - 2(1 - i) + 2$
28. $(1 + i)^2 - 2(1 + i) + 2$
29. $\left(-\dfrac{1}{2} + \dfrac{\sqrt{3}}{2}i\right)^3$
30. $\left(-\dfrac{1}{2} - \dfrac{\sqrt{3}}{2}i\right)^3$

A.3 SIGNIFICANT DIGITS

- Scientific Notation
- Significant Digits
- Calculation Accuracy

Many calculations in the real world deal with figures that are only approximate. After a series of calculations with approximate measurements, what can be said about the accuracy of the final answer? It seems reasonable to assume that a final answer cannot be any more accurate than the least accurate figure used in the calculation. This is an important point, since calculators tend to give the impression that greater accuracy is achieved than is warranted. In this section we introduce *scientific notation* and we use this concept in a discussion of *significant digits*. We will then be able to set up conventions for indicating the accuracy of the results of certain calculations involving approximate quantities. We will be guided by these conventions throughout the text when writing a final answer to a problem.

Scientific Notation

Work in science and engineering often involves the use of very large numbers. For example, the distance that light travels in 1 yr is called a **light-year.** This distance is approximately

9,440,000,000,000 km

Very small numbers are also used. For example, the mass of a water molecule is approximately

0.000 000 000 000 000 000 000 03 g

It is generally troublesome to write and work with numbers of this type in standard decimal form. In fact, these two numbers cannot even be entered into most calculators as they are written. Fortunately, it is possible to represent any decimal form as the product of a number between 1 and 10 and an integer power of 10; that is, in the form

$$a \times 10^n \qquad 1 \le a < 10, n \text{ an integer, } a \text{ in decimal form}$$

A number expressed in this form is said to be in **scientific notation.**

EXAMPLE 1

Using Scientific Notation

Each number is written in scientific notation.

$4 = 4 \times 10^0$	$0.36 = 3.6 \times 10^{-1}$
$63 = 6.3 \times 10$	$0.0702 = 7.02 \times 10^{-2}$
$805 = 8.05 \times 10^2$	$0.005\ 32 = 5.32 \times 10^{-3}$
$3{,}143 = 3.143 \times 10^3$	$0.000\ 67 = 6.7 \times 10^{-4}$
$7{,}320{,}000 = 7.32 \times 10^6$	$0.000\ 000\ 54 = 5.4 \times 10^{-7}$

■

Can you discover a rule that relates the number of decimal places a decimal point is moved to the power of 10 used?

$7{,}320{,}000 = 7.320\ 000. \times 10^6 = 7.32 \times 10^6$

6 places left
Positive exponent

$0.000\ 000\ 54 = 0.000\ 000\ 5.4 \times 10^{-7} = 5.4 \times 10^{-7}$

7 places right
Negative exponent

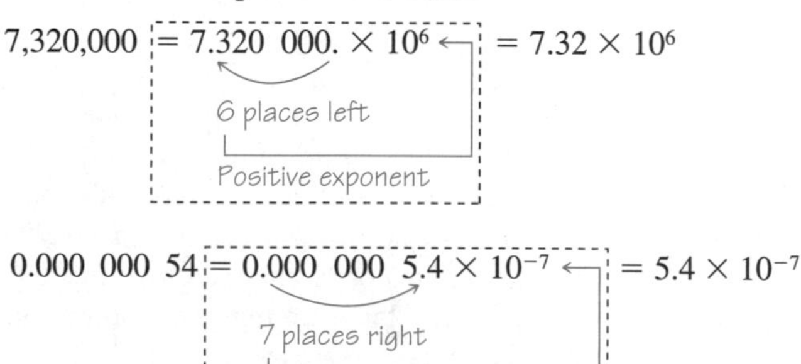

Matched Problem 1 Write in scientific notation:

(A) 450 (B) 360,000 (C) 0.0372 (D) 0.000 001 43 ■

Most calculators express very large and very small numbers in scientific notation. Read the instruction manual for your calculator to see how numbers in scientific notation are entered into your calculator. Numbers in scientific notation are displayed in most calculators as follows:

Calculator display	Number represented
3.207418 −13	$3.207\ 418 \times 10^{-13}$
5.002193 12	$5.002\ 193 \times 10^{12}$

Significant Digits

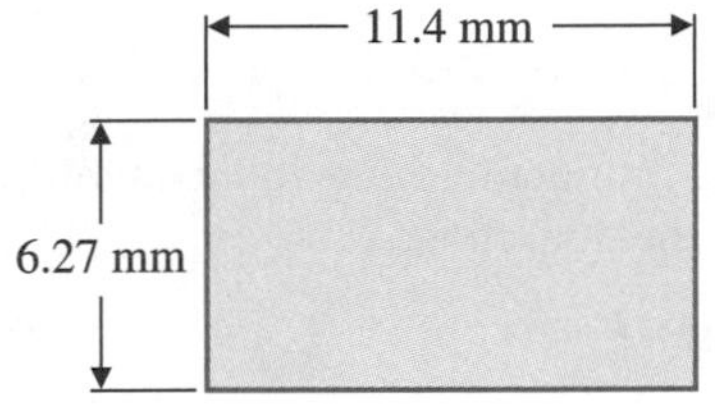

FIGURE 1

Suppose we wish to compute the area of a rectangle with the dimensions shown in Figure 1. As it often happens with approximations, we have one dimension to one decimal place accuracy and the other to two decimal place accuracy.

Using a calculator and the formula for the area of a rectangle, $A = ab$, we have

$$A = (11.4)(6.27) = 71.478$$

How many decimal places are justified in this calculation? We will answer this question later in this section. First, we must introduce the idea of *significant digits.*

Whenever we write a measurement such as 11.4 mm, we assume that the measurement is accurate to the last digit written. Thus, 11.4 mm indicates that the measurement was made to the nearest tenth of a millimeter—that is, the actual length is between 11.35 mm and 11.45 mm. In general, the digits in a number that indicate the accuracy of the number are called **significant digits.** If (going from left to right) the first digit and the last digit of a number are not 0, then all the digits are significant. Thus, the measurements 11.4 and 6.27 in Figure 1 have three significant digits, the number 100.8 has four significant digits, and the number 10,102 has five significant digits.

If the last digit of a number is 0, then the number of significant digits may not be clear. Suppose we are given a length of 23.0 cm. Then we assume the measurement has been taken to the nearest tenth and say that the number has three significant digits. However, suppose we are told that the distance between two cities is 3,700 mi. Is the stated distance accurate to the nearest hundred, ten, or unit? That is, does the stated distance have two, three, or four significant digits? We cannot really tell. In order to resolve this ambiguity, we give a precise definition of significant digits using scientific notation.

SIGNIFICANT DIGITS

If a number x is written in scientific notation as

$$x = a \times 10^n \qquad 1 \le a < 10,\ n \text{ an integer}$$

then the number of significant digits in x is the number of digits in a.

Thus,

3.7×10^3 has two significant digits

3.70×10^3 has three significant digits

3.700×10^3 has four significant digits

All three of these measurements have the same decimal representation, 3,700, but each represents a different accuracy.

EXAMPLE 2 **Determining the Number of Significant Digits**

Indicate the number of significant digits in each of the following numbers:

(A) 9.1003×10^{-3} (B) 1.080×10
(C) 5.92×10^{22} (D) 7.9000×10^{-13}

Solution In all cases the number of significant digits is the number of digits in the number to the left of the multiplication sign (as stated in the definition).

(A) Five (B) Four (C) Three (D) Five ■

Matched Problem 2 Indicate the number of significant digits in each of the following numbers:

(A) 4.39×10^{12} (B) 1.020×10^{-7}
(C) 2.3905×10^{-1} (D) 3.00×10 ■

The definition of significant digits tells us how to write a number so that the number of significant digits is clear, but it does not tell us how to interpret the accuracy of a number that is not written in scientific notation. We will use the following convention for numbers that are written as decimal fractions.

SIGNIFICANT DIGITS IN DECIMAL FRACTIONS

The number of significant digits in **a number with no decimal point** is found by counting the digits from left to right, starting with the first digit and ending with the last *nonzero* digit.

The number of significant digits in **a number containing a decimal point** is found by counting the digits from left to right, starting with the first *nonzero* digit and ending with the last digit (which may be 0).

Applying this convention to the number 3,700, we conclude that this number (as written) has two significant digits. If we want to indicate that it has three or four significant digits, we must use scientific notation. The significant digits in the following numbers are underlined.

$\underline{34{,}007}$ $\underline{92}0{,}000$ $\underline{25.300}$ $0.00\underline{63}$ $0.000\ \underline{430}$

■ Calculation Accuracy

When performing calculations, we want an answer that is as accurate as the numbers used in the calculation warrant, but no more. In calculations involving multiplication, division, powers, and roots, we adopt the following accuracy convention (which is justified in courses in numerical analysis).

ACCURACY OF CALCULATED VALUES

The number of significant digits in a calculation involving multiplication, division, powers, and/or roots is the same as the number of significant digits in the number in the calculation with the smallest number of significant digits.

Applying this convention to the calculation of the area A of the rectangle in Figure 1, we have

$A = (11.4)(6.27)$ — Both numbers have three significant digits.

$= 71.478$ — Calculator computation

$= 71.5$ — The computed area is only accurate to three significant digits.

EXAMPLE 3 **Determining the Accuracy of Computed Values**

Perform the indicated operations on the given approximate numbers. Then use the accuracy conventions stated above to round each answer to the appropriate accuracy.

(A) $\dfrac{(204)(34.0)}{120}$ (B) $\dfrac{(2.50 \times 10^5)(3.007 \times 10^7)}{2.4 \times 10^6}$

Solution (A) $\dfrac{(204)(34.0)}{120}$ — 120 has the least number of significant digits (two).

$= 57.8$ — Calculator computation

$= 58$ — Answer must have the same number of significant digits as the number with the least number of significant digits (two).

(B) $\dfrac{(2.50 \times 10^5)(3.007 \times 10^7)}{2.4 \times 10^6}$ — 2.4×10^6 has the least number of significant digits (two).

$= 3.132\ 291... \times 10^6$ — Calculator computation

$= 3.1 \times 10^6$ — Answer must have the same number of significant digits as the number with the least number of significant digits (two). ■

Matched Problem 3 Perform the indicated operations on the given approximate numbers. Then use the rounding conventions stated above to write each answer with the appropriate accuracy.

(A) $\dfrac{2.30}{(0.0341)(2.674)}$ (B) $\dfrac{1.235 \times 10^8}{(3.07 \times 10^{-3})(1.20 \times 10^4)}$ ■

We complete this section with two important observations:

1. Many formulas have constants that represent exact quantities. Such quantities are assumed to have infinitely many significant digits. For example, the formula for the circumference of a circle, $C = 2\pi r$, has two constants that are exact, 2 and π. Consequently, the final answer will have as many significant digits as in the measurement r (radius).
2. How do we round a number if the last nonzero digit is 5? For example, the following product should have only two significant digits. Do we change the 2 to a 3 or leave it alone?

$(1.3)(2.5) = 3.25$ Calculator result: Should be rounded to two significant digits.

We will round to 3.2 following the conventions given in the box.

ROUNDING NUMBERS WHEN LAST NONZERO DIGIT IS 5

(A) **If the digit preceding 5 is odd,** round up 1 to make it even.
(B) **If the digit preceding 5 is even,** do not change it.

We have adopted these conventions in order to prevent the accumulation of round-off errors with numbers having the last nonzero digit 5. The idea is to round up 50% of the time. (There are a number of other ways to accomplish this—flipping a coin, for example.)

EXAMPLE 4 **Rounding Numbers When Last Nonzero Digit is 5**

Round each number to three significant digits.

(A) 3.1495 (B) 0.004 135
(C) 32,450 (D) $4.314\ 764\ 09 \times 10^{12}$

Solution (A) 3.15 (B) 0.00414 (C) 32,400 (D) 4.31×10^{12} ■

Matched Problem 4 Round each number to three significant digits.

(A) 43.0690 (B) 48.05
(C) 48.15 (D) $8.017\ 632 \times 10^{-3}$ ■

Answers to Matched Problems

1. (A) 4.5×10^2 (B) 3.6×10^5 (C) 3.72×10^{-2} (D) 1.43×10^{-6}
2. (A) Three (B) Four (C) Five (D) Three
3. (A) 25.2 (B) 3.35×10^6
4. (A) 43.1 (B) 48.0 (C) 48.2 (D) 8.02×10^{-3}

EXERCISE A.3

A *Write in scientific notation.*

1. 640
2. 384
3. 5,460,000,000
4. 38,400,000
5. 0.73
6. 0.00493
7. 0.000 000 32
8. 0.0836
9. 0.000 049 1
10. 435,640
11. 67,000,000,000
12. 0.000 000 043 2

Write as a decimal fraction.

13. 5.6×10^4
14. 3.65×10^6
15. 9.7×10^{-3}
16. 6.39×10^{-6}
17. 4.61×10^{12}
18. 3.280×10^9
19. 1.08×10^{-1}
20. 3.004×10^{-4}

Indicate the number of significant digits in each number.

21. 12.3
22. 123
23. 12.300
24. 0.00123
25. 0.01230
26. 12.30
27. 6.7×10^{-1}
28. 3.56×10^{-4}
29. 6.700×10^{-1}
30. 3.560×10^{-4}
31. 7.090×10^5
32. 6.0050×10^7

B *Round each to three significant digits.*

33. 635,431
34. 4,089,100
35. 86.85
36. 7.075
37. 0.004 652 3
38. 0.000 380 0

Write in scientific notation, rounding to two significant digits.

39. 734
40. 908
41. 0.040
42. 700
43. 0.000 435
44. 635.46813

Indicate how many significant digits should be in the final answer.

45. (32.8)(0.2035)
46. (0.00230)(25.67)
47. $\dfrac{(7.21)(360)}{1,200}$
48. $\dfrac{(0.0350)(621)}{8,543}$
49. $\dfrac{(5.03 \times 10^{-3})(6 \times 10^4)}{8.0}$
50. $\dfrac{3.27(1.8 \times 10^7)}{2.90 \times 10}$

Using a calculator, compute each of the following, and express each answer with appropriate accuracy.

51. $\dfrac{6.07}{0.5057}$
52. (53,100)(0.2467)
53. $(6.14 \times 10^9)(3.154 \times 10^{-1})$
54. $\dfrac{7.151 \times 10^6}{9.1 \times 10^{-1}}$
55. $\dfrac{6,730}{(2.30)(0.0551)}$
56. $\dfrac{63,100}{(0.0620)(2,920)}$

In the following formulas, the indicated constants are exact. Compute the answer to each problem to an accuracy appropriate for the given approximate values of the variables.

57. **Circumference of a Circle** $C = 2\pi r;\ r = 25.31$ cm
58. **Area of a Circle** $A = \pi r^2;\ r = 2.5$ in.
59. **Area of a Triangle** $A = \frac{1}{2}bh;\ b = 22.4$ ft, $h = 8.6$ ft
60. **Area of an Ellipse** $A = \pi ab;\ a = 0.45$ cm, $b = 1.35$ cm
61. **Surface Area of a Sphere** $S = 4\pi r^2;\ r = 1.5$ mm
62. **Volume of a Sphere** $S = \frac{4}{3}\pi r^3;\ r = 1.85$ in.

C *In the following formulas, the indicated constants are exact. Compute the value of the indicated variable to an accuracy appropriate for the given approximate values of the other variables in the formula.*

63. **Volume of a Rectangular Parallelepiped (a Box)** $V = lwh;\ V = 24.2\ \text{cm}^3,\ l = 3.25$ cm, $w = 4.50$ cm, $h = ?$
64. **Volume of a Right Circular Cylinder** $V = \pi r^2 h;\ V = 1,250\ \text{ft}^3,\ h = 6.4$ ft, $r = ?$
65. **Volume of a Right Circular Cone** $V = \frac{1}{3}\pi r^2 h;\ V = 1,200\ \text{in.}^3,\ h = 6.55$ in., $r = ?$
66. **Volume of a Pyramid** $V = \frac{1}{3}Ah;\ V = 6,000\ \text{m}^3,\ A = 1,100$ m, $h = ?$

FUNCTIONS AND INVERSE FUNCTIONS

APPENDIX

B

B.1 FUNCTIONS

- Definition of a Function — Rule Form
- Definition of a Function — Set Form
- Function Notation
- Function Classification

Seeking correspondences between various types of phenomena is undoubtedly one of the most important aspects of science. A physicist attempts to find a correspondence between the current in an electrical circuit and time; a chemist looks for a correspondence between the speed of a chemical reaction and the concentration of a given substance; an economist tries to determine a correspondence between the price of an object and the demand for the object; and so on. The list could go on and on.

Establishing and working with correspondences among various types of phenomena—whether through tables, graphs, or equations—is so fundamental to pure and applied science that it has become necessary to describe this activity in the precise language of mathematics.

Definition of a Function – Rule Form

What do all the examples cited above have in common? Each describes the matching of elements from one set with the elements in a second set. Consider Charts 1–3, which list values for the cube, square, and square root, respectively.

CHART 1

Domain: Number		Range: Cube
-2	$\longrightarrow$	-8
-1	$\longrightarrow$	-1
0	$\longrightarrow$	0
1	$\longrightarrow$	1
2	$\longrightarrow$	8

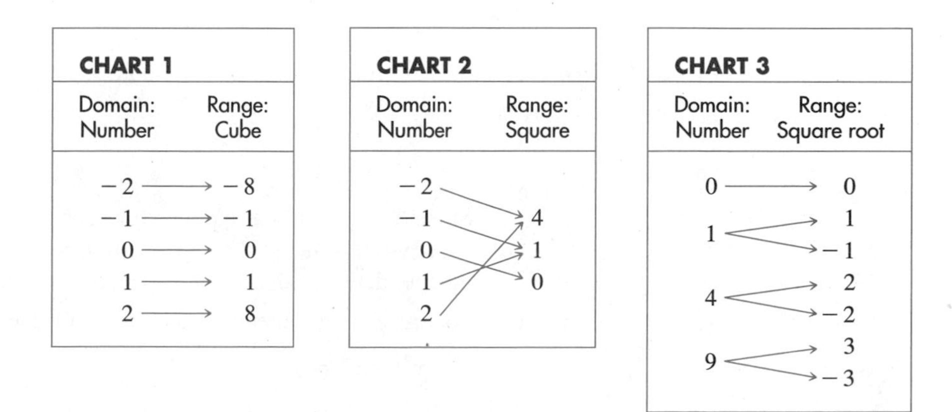

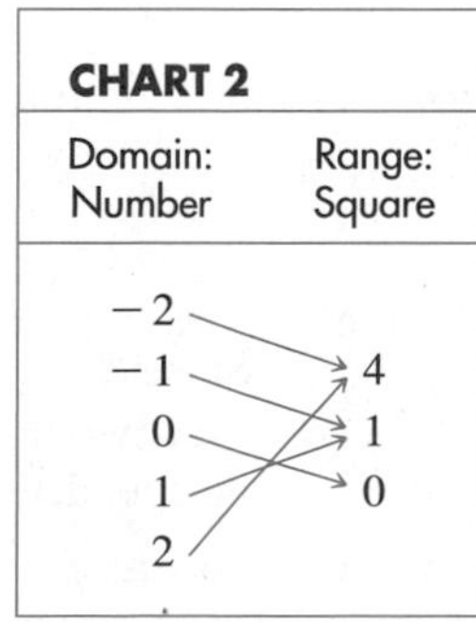

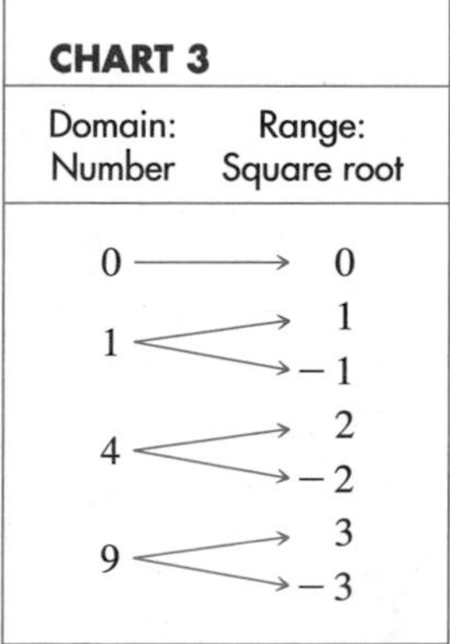

Charts 1 and 2 define functions, but Chart 3 does not. Why? The definition of *function* given in the box at the top of the next page will explain.

FUNCTION — RULE FORM

A **function** is a rule that produces a correspondence between two sets of elements such that to each element in the first set there corresponds *one and only one* element in the second set.

The first set is called the **domain.** The set of all corresponding elements in the second set is called the **range.**

A variable representing an arbitrary element from the domain is called an **independent variable.** A variable representing an arbitrary element from the range is called a **dependent variable.**

Charts 1 and 2 define functions, since to each domain value there corresponds exactly one range value. For example, the cube of -2 is -8 and no other number. On the other hand, Chart 3 does not specify a function, since to at least one domain value there corresponds more than one range value. For example, to the domain value 4 there corresponds -2 and 2, both square roots of 4.

Some equations in two variables define functions. If in an equation in two variables, say x and y, there corresponds exactly one range value y for each domain value x(x is independent and y is dependent), then the correspondence established by the equation is a function.

EXAMPLE 1

Determining Whether an Equation Defines a Function

Determine which of the following equations define functions with independent variable x and domain all real numbers.

(A) $y - x = 1$ (B) $y^2 - x^2 = 1$

Solution (A) Solving for the dependent variable y, we have

$$y - x = 1 \tag{1}$$

$$y = 1 + x$$

Since $1 + x$ is a real number for each real number x, equation (1) assigns exactly one value of the dependent variable, $y = 1 + x$, to each value of the independent variable x. Thus, equation (1) defines a function.

(B) Solving for the dependent variable y, we have

$$y^2 - x^2 = 1$$

$$y^2 = 1 + x^2 \tag{2}$$

$$y = \pm\sqrt{1 + x^2}$$

Since $1 + x^2$ is always a positive real number and since each positive real number has two real square roots, each value of the independent variable x corresponds to two values of the dependent variable, $y = -\sqrt{1 + x^2}$ and $y = \sqrt{1 + x^2}$. Thus, equation (2) does not define a function. ■

Matched Problem 1 Determine which of the following equations define functions with independent variable x and domain all real numbers.

(A) $y = x + 3$ (B) $y^2 = x^2 + 3$ ■

It is very easy to determine whether an equation defines a function by examining the graph of the equation. The two equations considered in Example 1 are graphed in Figure 1.

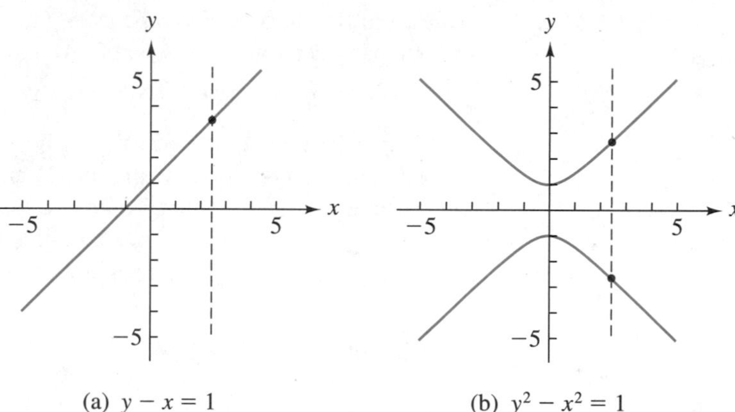

FIGURE 1
Graphs of equations and the vertical-line test

In Figure 1a, any vertical line intersects the graph in exactly one point, illustrating graphically the fact that each value of the independent variable x corresponds to exactly one value of the dependent variable y. On the other hand, Figure 1b shows that there are vertical lines that intersect this graph in two points. This indicates that there are values of the independent variable x that correspond to two different values of the dependent variable y. These observations form the basis for the **vertical-line test** stated below.

VERTICAL-LINE TEST

An equation defines a function if and only if each vertical line in the rectangular coordinate system passes through at most one point on the graph of the function.

Definition of a Function—Set Form

Since elements in the range of a function are paired with elements in the domain by some rule or process, this correspondence (pairing) can be illustrated by using **ordered pairs** of elements, where the first component represents a domain element and the second component represents a corresponding range element. Thus, we can write functions 1 and 2 specified in Charts 1 and 2 as sets of pairs as follows:

Function 1: $\{(-2, -8), (-1, -1), (0, 0)\ (1, 1), (2, 8)\}$

Function 2: $\{(-2, 4), (-1, 1), (0, 0)\ (1, 1), (2, 4)\}$

In both cases, notice that no two ordered pairs have the same first component and different second components. On the other hand, if we list the set S of ordered pairs determined by Chart 3, we have

$$S = \{(0, 0), (1, 1), (1, -1), (4, 2), (4, -2), (9, 3), (9, -3)\}$$

In this case, there are ordered pairs with the same first component and different second components. For example, $(1, 1)$ and $(1, -1)$ both belong to the set S. Once again, we see that Chart 3 does not specify a function.

This discussion suggests an alternative but equivalent way of defining functions that produces additional insight into this concept.

FUNCTION — SET FORM

A **function** is a set of ordered pairs with the property that no two ordered pairs have the same first component and different second components. The set of all first components in a function is called the **domain** of the function and the set of all second components is called the **range.**

EXAMPLE 2 **Functions Defined as Sets of Ordered Pairs**

Given the sets:

$$H = \{(1,1), (2, 1), (3, 2), (3, 4)\}$$

$$G = \{(2, 4), (3, -1), (4, 4)\}$$

(A) Which set specifies a function?

(B) Give the domain and range of the function.

Solution (A) Set H does not specify a function, since $(3, 2)$ and $(3, 4)$ both have the same first components. The domain value 3 corresponds to more than one range value. Set G does specify a function, since each domain value corresponds to exactly one range value.

(B) Domain of $G = X = \{2, 3, 4\}$; Range of $G = Y = \{-1, 4\}$ ■

Matched Problem 2 Repeat Example 2 for the following sets:

$$M = \{(3, 4), (5, 4), (6, -1)\}$$

$$N = \{(-1, 2), (0, 4), (1, 2), (0, 1)\}$$

■

Function Notation

If x represents an element in the domain of a function f, then we will often use the symbol $f(x)$ in place of y to designate the number in the range of f to which x is paired. It is important not to be confused by this new symbol and think of it as a product of f and x. The symbol is read "f of x" or "the value of f at x." The correct use of this function symbol should be mastered early. Thus, if

$$f(x) = 2x + 3$$

then

$$f(5) = 2(5) + 3 = 13$$

That is, the function f assigns the range value 13 to the domain value 5. Thus, (5, 13) belongs to f. Can you find another ordered pair that belongs to f?

The correspondence between domain values and range values is usually illustrated in one of the two ways shown in Figure 2.

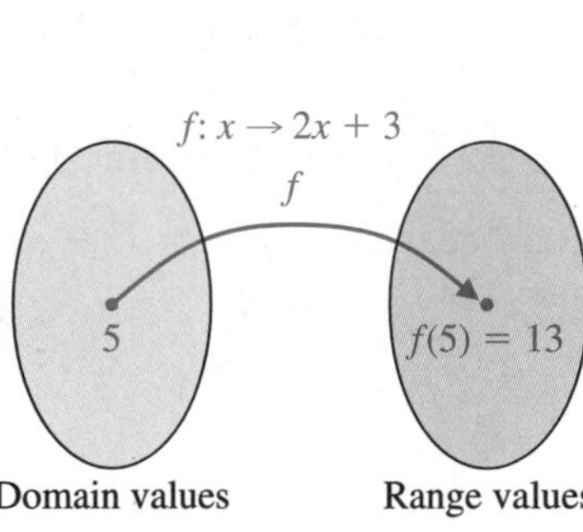

(a) Map

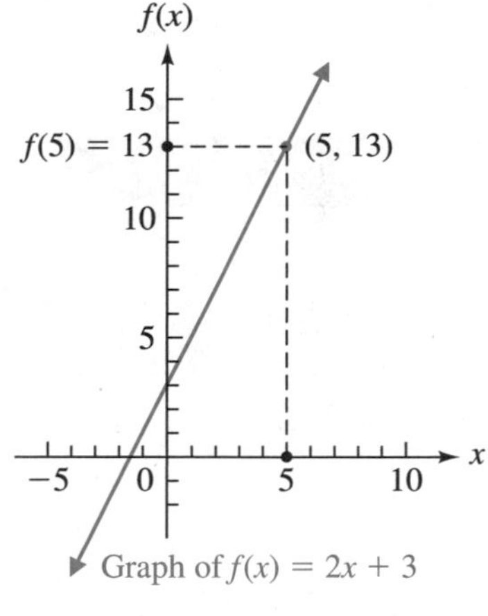

(b) Graph

FIGURE 2
Function notation

EXAMPLE 3 **Evaluating Functions**

If $f(x) = (x/2) - 1$ and $g(x) = 1 - x^2$, find:

(A) $f(4)$ (B) $g(-3)$ (C) $f(2) - g(0)$
(D) $f(2 + h)$ (E) $[f(3 + h) - f(3)]/h$ (F) $f[g(3)]$

Solution (A) $f(4) = \dfrac{4}{2} - 1 = 2 - 1 = 1$

(B) $g(-3) = 1 - (-3)^2 = 1 - 9 = -8$

(C) $f(2) - g(0) = \left(\frac{2}{2} - 1\right) - (1 - 0^2) = 0 - 1 = -1$

(D) $f(2 + h) = \frac{2 + h}{2} - 1 = \frac{2 + h - 2}{2} = \frac{h}{2}$

(E) $$\frac{f(3 + h) - f(3)}{h} = \frac{\left(\frac{3 + h}{2} - 1\right) - \left(\frac{3}{2} - 1\right)}{h}$$

$$= \frac{\frac{1 + h}{2} - \frac{1}{2}}{h} = \frac{1}{2}$$

(F) $f[g(3)] = f(1 - 3^2) = f(-8) = \frac{-8}{2} - 1 = -5$ ■

Matched Problem 3 If $f(x) = 2x - 3$ and $g(x) = x^2 - 2$, find:

(A) $f(3)$ (B) $g(-2)$ (C) $f(0) + g(1)$
(D) $f(3 + h)$ (E) $[f(2 + h) - f(2)]/h$ (F) $f[g(2)]$ ■

EXAMPLE 4 **Finding the Range of a Function**

If the function f defined by $f(x) = x^2 - x$ has domain $X = \{-2, -1, 0, 1, 2\}$, what is the range of f?

Solution

$$f(-2) = (-2)^2 - (-2) = 6$$
$$f(-1) = (-1)^2 - (-1) = 2$$
$$f(0) = 0^2 - 0 = 0$$
$$f(1) = 1^2 - 1 = 0$$
$$f(2) = 2^2 - 2 = 2$$

Thus, the range of $f = Y = \{0, 2, 6\}$. ■

Matched Problem 4 Repeat Example 4 for $g(x) = x^2 - 4$, $X = \{-2, -1, 0, 1, 2\}$. ■

Function Classification

Functions are classified in special categories for more efficient study. You have already had some experience with many *algebraic functions* (defined by means of the algebraic operations addition, subtraction, multiplication, division, powers, and roots), *exponential functions,* and *logarithmic functions.* In this book, we add two more classes of functions to this list, namely, the *trigonometric functions* and the *inverse trigonometric functions.* These five classes of functions are called the **elementary functions.** They are related to each other as follows:

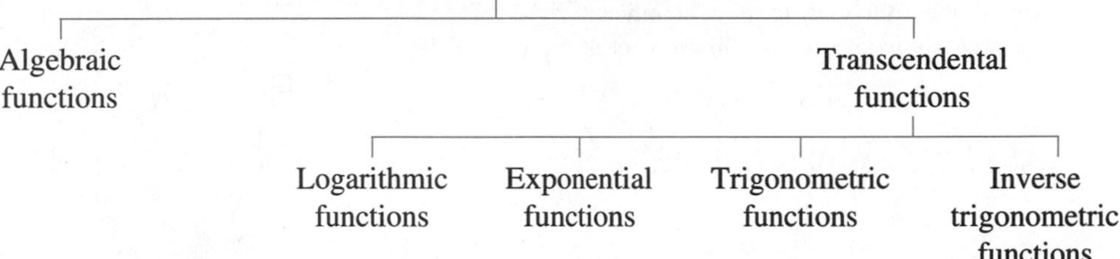

Answers to Matched Problems

1. (A) Defines a function (B) Does not define a function
2. (A) M is a function (B) $X = \{3, 5, 6\}$, $Y = \{-1, 4\}$
3. (A) 3 (B) 2 (C) -4 (D) $3 + 2h$ (E) 2 (F) 1
4. $Y = \{-4, -3, 0\}$

EXERCISE B.1

For $f(x) = 4x - 1$, find each of the following:

1. $f(1)$
2. $f(2)$
3. $f(-1)$
4. $f(-2)$
5. $f(0)$
6. $f(5)$

If $g(x) = x - x^2$, find each of the following:

7. $g(1)$
8. $g(3)$
9. $g(5)$
10. $g(4)$
11. $g(-2)$
12. $g(-3)$

For $f(x) = 1 - 2x$ and $g(x) = 4 - x^2$, find each of the following:

13. $f(0) + g(0)$
14. $g(0) - f(0)$
15. $\dfrac{f(3)}{g(1)}$
16. $[g(-2)][f(-1)]$
17. $2f(-1)$
18. $\frac{1}{5}g(-3)$
19. $f(2 + h)$
20. $g(2 + h)$
21. $\dfrac{f(2 + h) - f(2)}{h}$
22. $\dfrac{g(2 + h) - g(2)}{h}$
23. $g[f(2)]$
24. $f[g(2)]$

Which of the following equations specify functions, given that x is the independent variable?

25. $x^2 + y^2 = 25$
26. $y = 3x + 1$
27. $2x - 3y = 6$
28. $y = x^2 - 2$
29. $y^2 = x$
30. $y = x^2$
31. $y = |x|$
32. $|y| = x$

33. If the function f, defined by $f(x) = x^2 - x + 1$, has domain $X = \{-2, -1, 0, 1, 2\}$, find the range Y of f.
34. If the function g, defined by $g(x) = 1 + x - x^2$, has domain $X = \{-2, -1, 0, 1, 2\}$, find the range Y of g.
35. Indicate which set specifies a function and write down its domain and range.

$$F = \{(-2, 1), (-1, 1), (0, 0)\}$$

$$G = \{(-4, 3), (0, 3), (-4, 0)\}$$

36. Repeat Problem 35 for the following sets:

$$H = \{(-1, 3), (2, 3), (-1, -1)\}$$

$$L = \{(-1, 1), (0, 1), (1, 1)\}$$

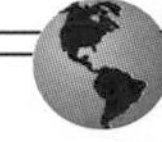

Applications

Precalculus *Problems 37–39 pertain to the following relationship: The distance d (in meters) that an object falls in a vacuum in t seconds is given by*

$$d = s(t) = 4.88t^2$$

37. Find $s(0)$, $s(1)$, $s(2)$, and $s(3)$ to two decimal places.
38. The expression $[s(2 + h) - s(2)]/h$ represents the average speed of the falling object over the time interval

from $t = 2$ to $t = 2 + h$. Use a calculator to compute each of the following to four significant digits. Then guess the speed of a free-falling object at the end of 2 sec.

(A) $\dfrac{s(3) - s(2)}{1}$ (B) $\dfrac{s(2.1) - s(2)}{0.1}$

(C) $\dfrac{s(2.01) - s(2)}{0.01}$ (D) $\dfrac{s(2.001) - s(2)}{0.001}$

(E) $\dfrac{s(2.0001) - s(2)}{0.0001}$

39. Find $[s(2 + h) - s(2)]/h$ and simplify. What happens as h gets closer and closer to 0? Interpret physically.

B.2 INVERSE FUNCTIONS

- One-to-One Functions
- Inverse Functions
- Geometric Relationship

In this section we develop techniques for determining whether the *inverse function* exists, some general properties of inverse functions, and methods for finding the rule of correspondence that defines the inverse function.

An inverse function is formed by reversing the correspondence in a given function. However, the reverse correspondence for a given function may or may not specify a function. As we will see, only functions that are *one-to-one* have inverse functions.

Many important functions are formed as inverses of existing functions. Logarithmic functions, for example, are inverses of exponential functions, and inverse trigonometric functions are inverses of trigonometric functions with restricted domains.

One-to-One Functions

A **one-to-one correspondence** exists between two sets if each element of the first set corresponds to exactly one element of the second set and each element of the second set corresponds to exactly one element of the first set.

ONE-TO-ONE FUNCTION

A function f is **one-to-one** if each element in the range corresponds to exactly one element from the domain. (Since f is a function, we already know that each element in the domain corresponds to exactly one element in the range.)

Thus, if a function f is one-to-one, then there exists a one-to-one correspondence between the domain elements and the range elements of f. To illustrate these concepts, consider the two functions f and g given in the following charts:

Function f	
Domain	Range
0 →	0
1 →	1
2 →	8

Function g	
Domain	Range
0 →	0
−2 →	4
2 →	4

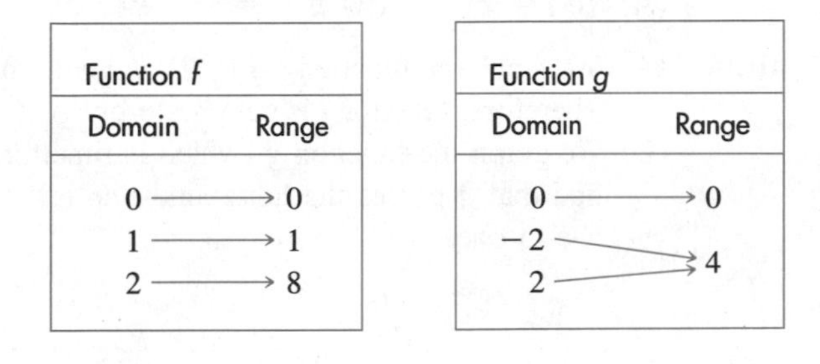

Function f is one-to-one, since each range element corresponds to exactly one domain element. Function g is not one-to-one, since the range element 4 corresponds to two domain elements, −2 and 2.

Geometrically, if a horizontal line intersects the graph of the function in two or more points, then the function is not one-to-one (see Fig. 1a). However, if each horizontal line intersects the graph of the function in at most one point, then the function is one-to-one (see Fig. 1b).

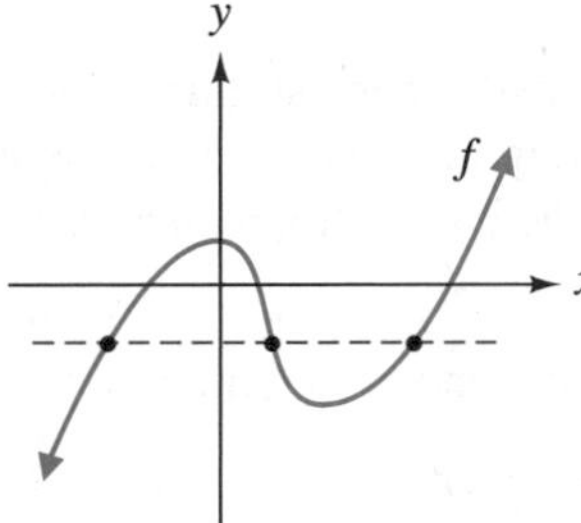

(a) Function f is not one-to-one

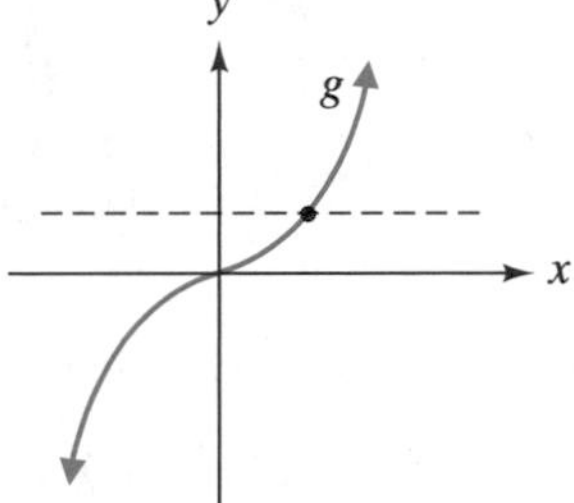

(b) Function g is one-to-one

FIGURE 1
Graphs of functions and the horizontal-line test

These observations form the basis for the **horizontal-line test.**

HORIZONTAL-LINE TEST

A function is one-to-one if and only if each horizontal line intersects the graph of the function in at most one point.

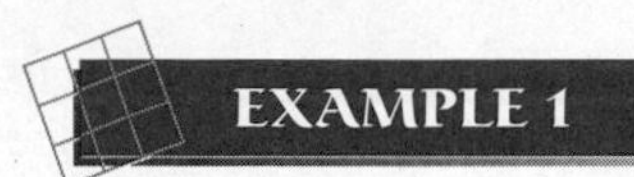

EXAMPLE 1 Determining Whether a Function Is One-to-One

Graph each function and determine which is one-to-one by the horizontal-line test.

(A) $f(x) = x^2$ (B) $g(x) = x^2, \quad x \geq 0$

Solution (A) We graph the function f (Fig. 2) and note that it fails the horizontal-line test. Therefore, function f is not one-to-one.

(B) We graph the function g (which is function f with a restricted domain) and find that g passes the horizontal-line test (Fig. 3). Therefore, function g is one-to-one.

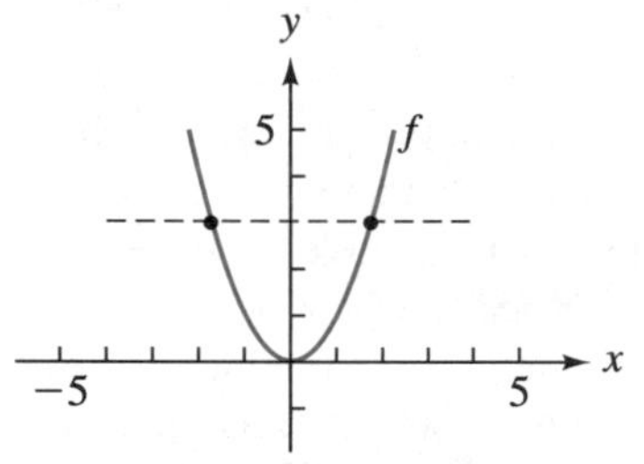

FIGURE 2

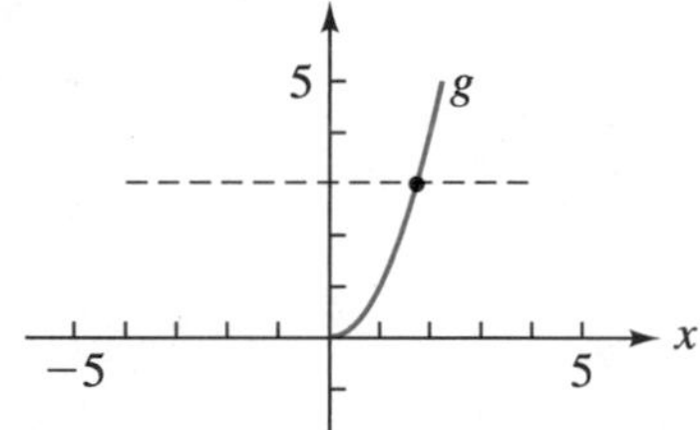

FIGURE 3

Example 1 illustrates an important point: A function may not be one-to-one, but by restricting the domain it can be made one-to-one. This is exactly what is done to form the inverse trigonometric functions from the trigonometric functions.

Matched Problem 1 Graph each function and determine which is one-to-one by the horizontal-line test.

(A) $f(x) = (x - 1)^2$ (B) $g(x) = (x - 1)^2, \quad x \geq 1$

Inverse Functions

As we mentioned at the beginning of this section, we are interested in forming new functions by reversing the correspondence in a given function—that is, by reversing all the ordered pairs in the given function. The concept of a one-to-one function plays a critical role in this process. If we reverse all the ordered pairs in a function that is not one-to-one, the resulting set does not define a function. For example, reversing the ordered pairs in the function.

$$g = \{(-2, 4), (0, 0), (2, 4)\}$$ Function g is not one-to-one.

produces the set

$$h = \{(4, -2), (0, 0), (4, 2)\}$$ Set h is not a function.

which does not define a function. However, reversing all the ordered pairs in a one-to-one function f always produces a new function, called the *inverse func-*

tion, which is denoted by the symbol f^{-1}.* For example, reversing the ordered pairs in the function

$f = \{(0, 2), (1, 3), (2, 4)\}$ Function f is one-to-one.

produces the inverse function

$f^{-1} = \{(2, 0), (3, 1), (4, 2)\}$ Set f^{-1} is a function.

Furthermore, we see that

$$\text{Domain of } f = \{0, 1, 2\} = \text{Range of } f^{-1}$$
$$\text{Range of } f = \{2, 3, 4\} = \text{Domain of } f^{-1}$$

In other words, reversing all the ordered pairs also reverses the domain and range. This discussion is summarized in the following definition:

INVERSE FUNCTION

The **inverse** of a one-to-one function f, denoted by f^{-1}, is the function formed by reversing all the ordered pairs in the function f. Symbolically,

$$f^{-1} = \{(b, a) | (a, b) \text{ is an element of } f\}$$

$$\text{Domain of } f^{-1} = \text{Range of } f$$

$$\text{Range of } f^{-1} = \text{Domain of } f$$

Immediate consequences of the definition of an inverse function are the function–inverse function identities:

FUNCTION–INVERSE FUNCTION IDENTITIES

If f is a one-to-one function, then f^{-1} exists and

$$f^{-1}[f(x)] = x \qquad \text{for all } x \text{ in the domain of } f$$

and

$$f[f^{-1}(x)] = x \qquad \text{for all } x \text{ in the domain of } f^{-1}$$

These identities are illustrated schematically in Figure 4 (at the top of the next page).

**Note:* f^{-1} is a special symbol used to represent the inverse of the function f. It does *not* mean $1/f$.

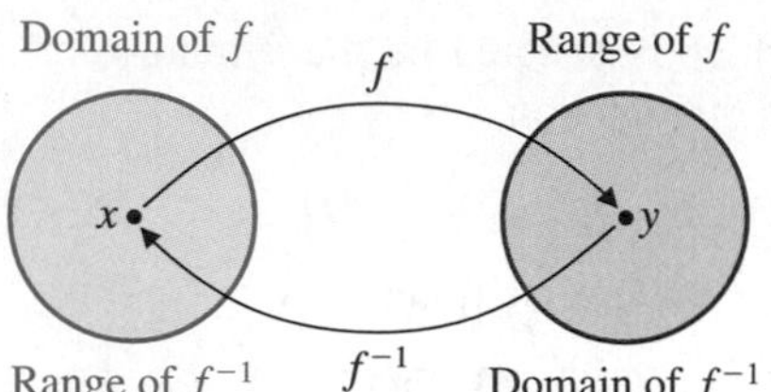

If the function f is one-to-one and if f maps x into y, then f^{-1} maps y back into x.

FIGURE 4
Function–inverse function identities

Finding the Inverse of a Function

For the one-to-one function $f(x) = 2x - 3$, find $f^{-1}(x)$, and check by showing that $f[f^{-1}(x)] = x$ and $f^{-1}[f(x)] = x$.

Solution For both f and f^{-1} we keep x as the independent variable and y as the dependent variable.

Step 1 Replace $f(x)$ with y:

f: $y = 2x - 3$ x is independent.

Step 2 Interchange the variables x and y to form f^{-1}:

f^{-1}: $x = 2y - 3$ x is independent.

Step 3 Solve the equation for f^{-1} for y in terms of x:

$$x = 2y - 3$$

$$y = \frac{x+3}{2}$$

Step 4 Replace y with $f^{-1}(x)$:

$$f^{-1}(x) = \frac{x+3}{2}$$

Step 5 Check by showing that $f[f^{-1}(x)] = x$ and $f^{-1}[f(x)] = x$.

$$f[f^{-1}(x)] = 2[f^{-1}(x)] - 3 = 2\left(\frac{x+3}{2}\right) - 3 = x + 3 - 3 = x$$

$$f^{-1}[f(x)] = \frac{f(x)+3}{2} = \frac{(2x-3)+3}{2} = \frac{2x}{2} = x$$

■

Matched Problem 2 For the one-to-one function $g(x) = 3x + 2$, find $g^{-1}(x)$, and check by showing that $g[g^{-1}(x)] = x$ and $g^{-1}[g(x)] = x$. ■

Geometric Relationship

We conclude this discussion of inverse functions by observing an important relationship between the graph of a one-to-one function and its inverse. As an example, let f be the one-to-one function

$$f(x) = 2x - 2 \tag{1}$$

Its inverse is

$$f^{-1}(x) = \frac{x + 2}{2} \tag{2}$$

Any ordered pair of numbers that satisfies (1), when reversed in order, will satisfy (2). For example, (4, 6) satisfies (1) and (6, 4) satisfies (2). (Check this.) The graphs of f and f^{-1} are given in Figure 5.

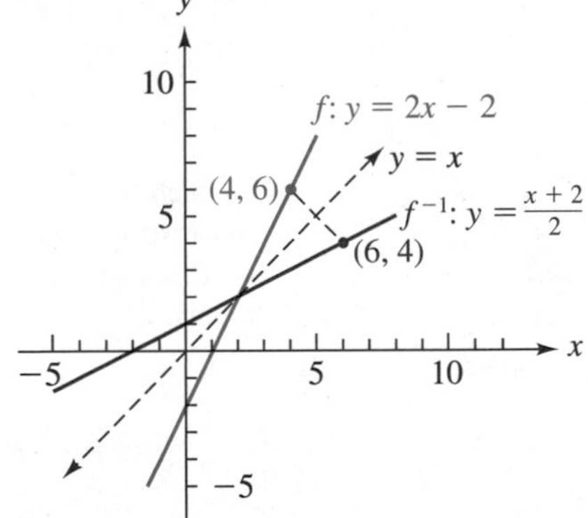

FIGURE 5
Symmetry property of the graphs of a function and its inverse

Notice that we sketched the line $y = x$ in Figure 5 to show that if we fold the paper along this line, then the graphs of f and f^{-1} will match. Actually, we can graph f^{-1} by drawing f with wet ink and folding the paper along $y = x$ before the ink dries; f will print f^{-1}. [To prove this, we need to show that the line $y = x$ is the perpendicular bisector of the line segment joining (a, b) to (b, a).]

Knowing that the graphs of f and f^{-1} are symmetric relative to the line $y = x$ makes it easy to graph f^{-1} if f is known, and vice versa.

Answers to Matched Problems

1. (A) f is not one-to-one (B) g is one-to-one

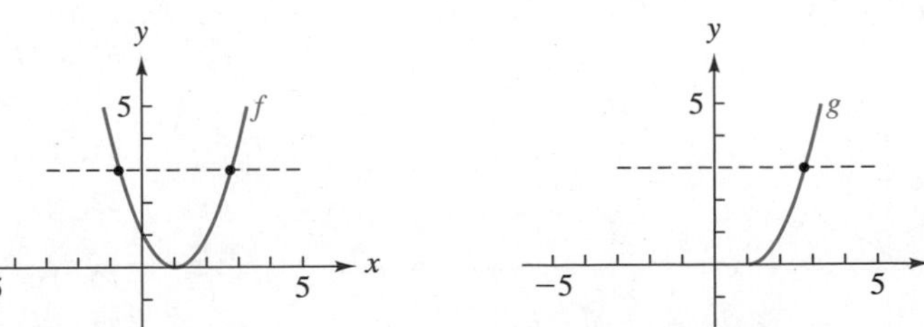

2. $g^{-1}(x) = \dfrac{x - 2}{3}$

EXERCISE B.2

In Problems 1–12, indicate which functions are one-to-one.

1.

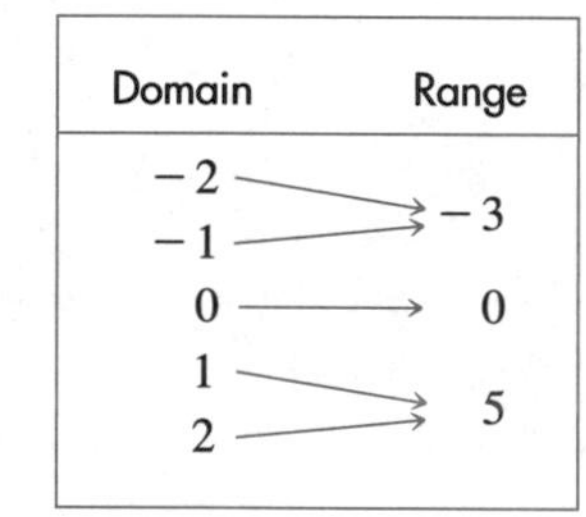

Domain	Range
-2 →	-4
-1 →	-2
0 →	0
1 →	2
2 →	4

2.

Domain	Range
-2, -1 →	-3
0 →	0
1, 2 →	5

3.

Domain	Range
0, 1, 2, 3, 4 →	9

4.

Domain	Range
0 →	5
1 →	3
2 →	1
3 →	2
4 →	4

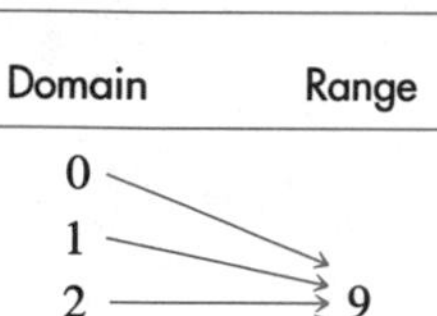

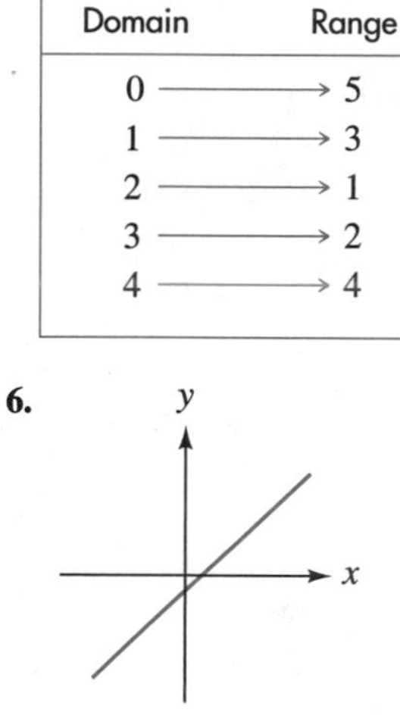

5.

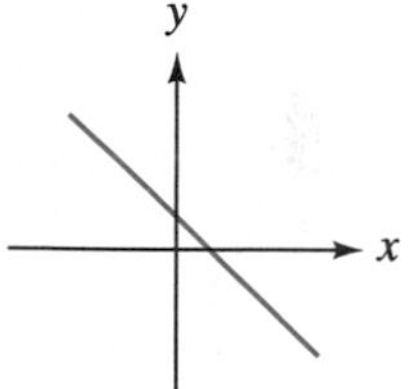

6.

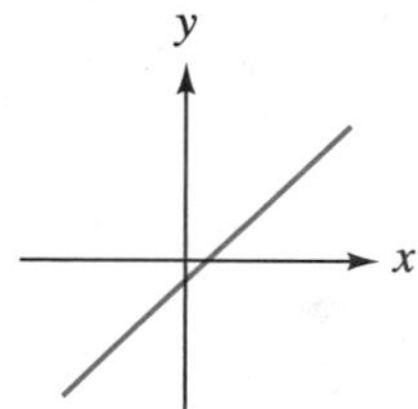

7.

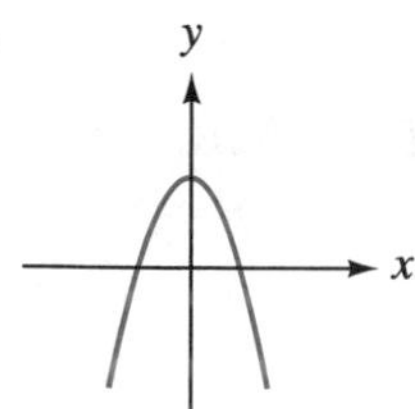

8.

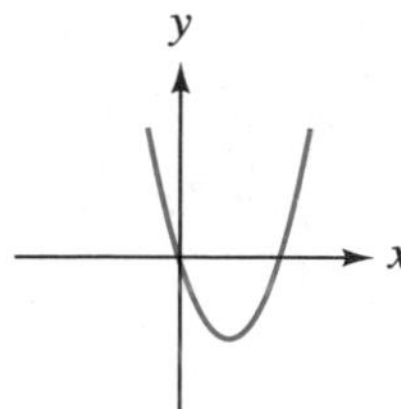

9.

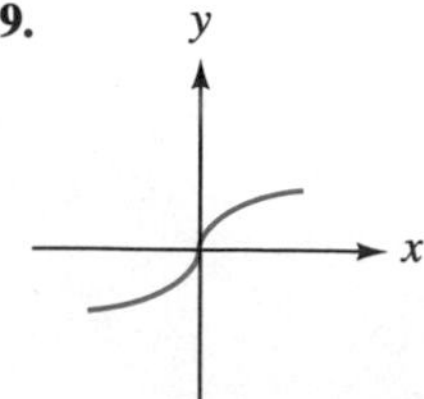

10.

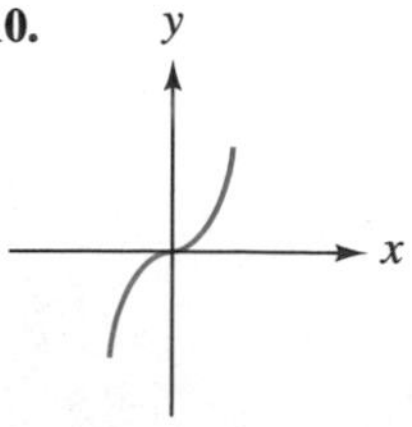

11.

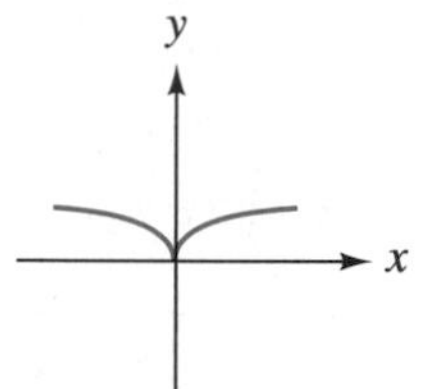

12.

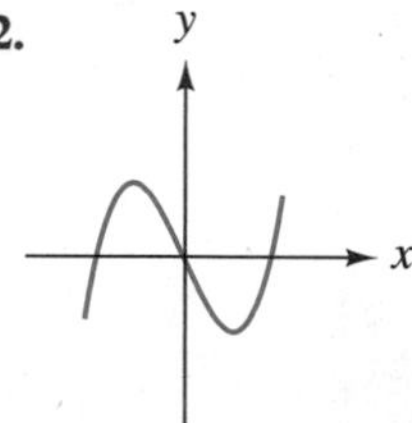

13. Which of the given functions is one-to-one? Write the inverse of the function that is one-to-one as a set of ordered pairs of numbers, and indicate its domain and range.

$$f = \{(-1, 0), (1, 1), (2, 0)\}$$

$$g = \{(-2, -8), (1, 1), (2, 8)\}$$

14. Which of the given functions is one-to-one? Write the inverse of the function that is one-to-one as a set of ordered pairs of numbers, and indicate its domain and range.

$$f = \{(-2, 4), (0, 0), (2, 4)\}$$

$$g = \{(9, 3), (4, 2), (1, 1)\}$$

15. Given $H = \{(-1, 0.5), (0, 1), (1, 2), (2, 4)\}$, graph H, H^{-1}, and $y = x$ in the same coordinate system.

16. Given $F = \{(-5, 0), (-2, 1), (0, 2), (1, 4), (2, 7)\}$, graph F, F^{-1}, and $y = x$ in the same coordinate system.

In Problems 17–20, find the inverse for each function in the form of an equation.

17. $f(x) = 2x - 7$

18. $g(x) = \dfrac{x}{2} + 1$

19. $h(x) = \dfrac{x + 3}{3}$

20. $f(x) = \dfrac{x - 2}{3}$

In Problems 21 and 22, graph the indicated function, its inverse, and $y = x$ in the same coordinate system.

21. f and f^{-1} in Problem 17

22. g and g^{-1} in Problem 18

23. For $f(x) = 2x - 7$, find $f^{-1}(x)$ and $f^{-1}(3)$.

24. For $g(x) = (x/2) + 1$, find $g^{-1}(x)$ and $g^{-1}(-3)$.

25. For $h(x) = (x/3) + 1$, find $h^{-1}(x)$ and $h^{-1}(2)$.

26. For $m(x) = 3x + 2$, find $m^{-1}(x)$ and $m^{-1}(5)$.

27. Find $f[f^{-1}(4)]$ for Problem 23.

28. Find $g^{-1}[g(2)]$ for Problem 24.

29. Find $h^{-1}[h(x)]$ for Problem 25.

30. Find $m[m^{-1}(x)]$ for Problem 26.

31. Find $h[h^{-1}(x)]$ for Problem 25.

32. Find $m^{-1}[m(x)]$ for Problem 26.

PLANE GEOMETRY: SOME USEFUL FACTS

APPENDIX

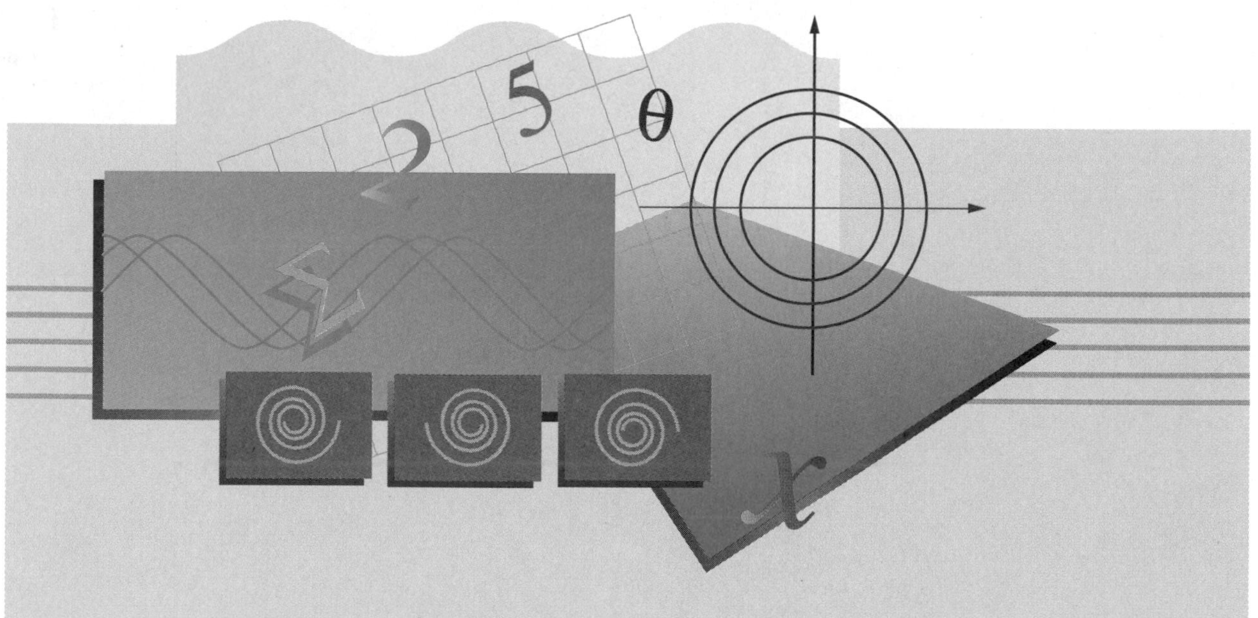

The following four sections include a brief list of plane geometry facts that are of particular use in studying trigonometry. They are grouped together for convenient reference.

C.1 LINES AND ANGLES

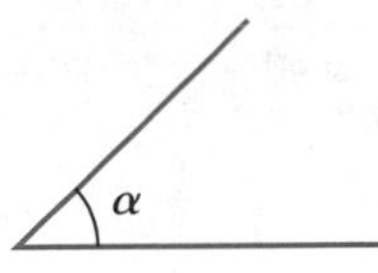

Acute angle
$0° < \alpha < 90°$

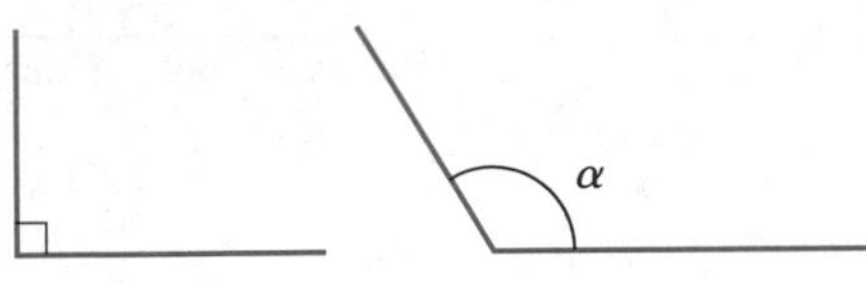

Right angle
$90°$

Obtuse angle
$90° < \alpha < 180°$

Straight angle (180°)

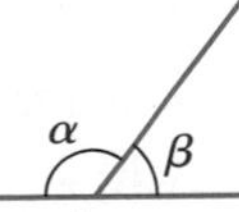

$\alpha + \beta = 180°$
α and β are
supplementary angles

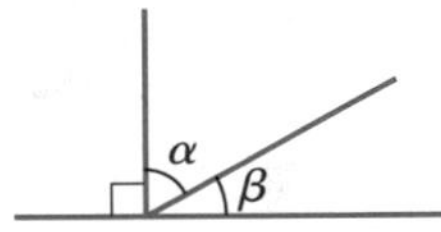

$\alpha + \beta = 90°$
α and β are
complementary angles

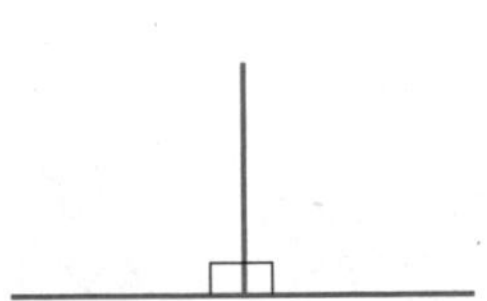

A **straight angle** divided into equal parts forms two **right angles**.

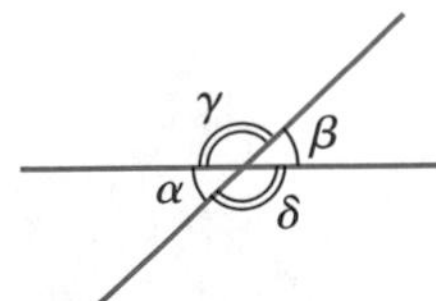

$\alpha = \beta$
$\gamma = \delta$
$\alpha + \delta = \beta + \gamma = 180°$
$\alpha + \delta + \beta + \gamma = 360°$

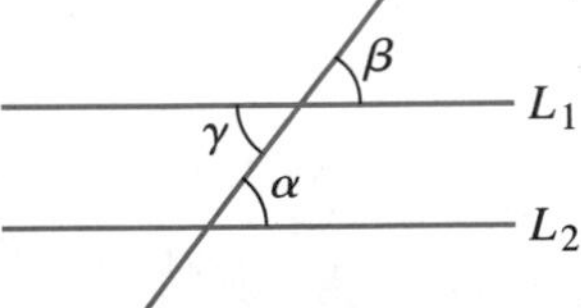

If $L_1 \parallel L_2$, then $\alpha = \beta = \gamma$.
If $\alpha = \beta = \gamma$, then $L_1 \parallel L_2$.

C.2 TRIANGLES

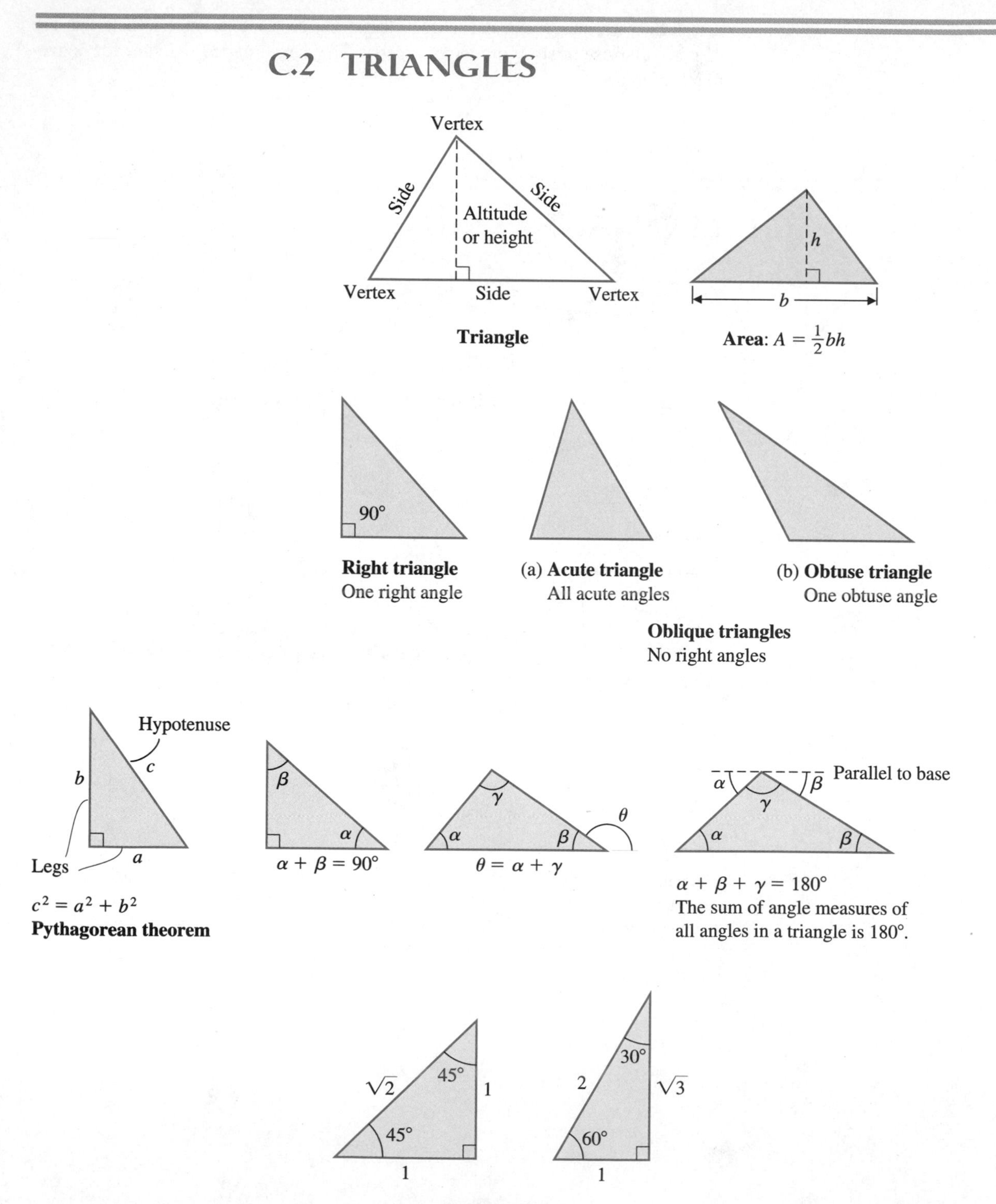

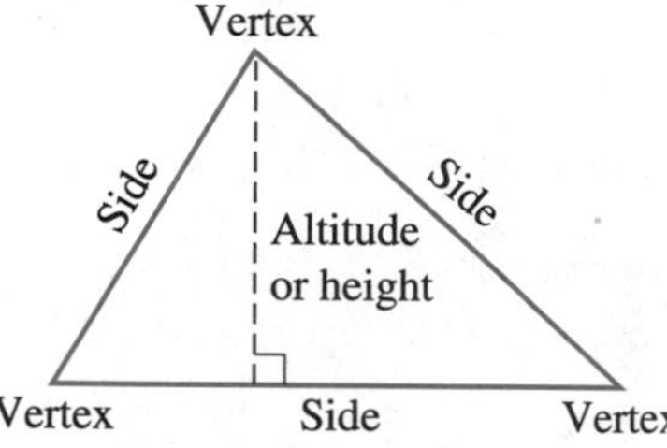

Triangle

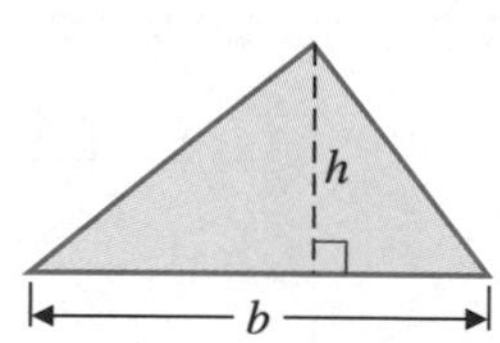

Area: $A = \frac{1}{2}bh$

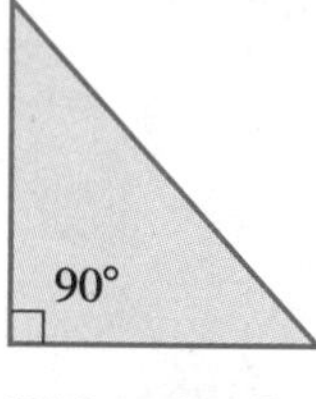

Right triangle
One right angle

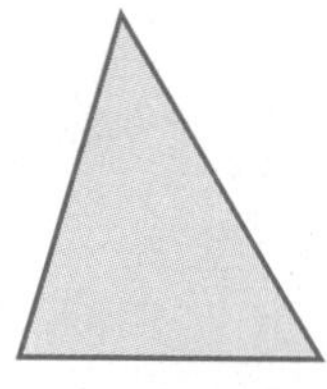

(a) **Acute triangle**
All acute angles

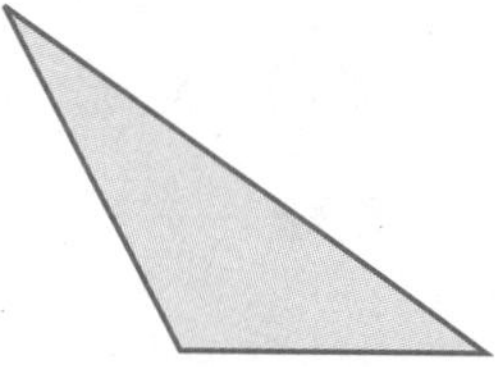

(b) **Obtuse triangle**
One obtuse angle

Oblique triangles
No right angles

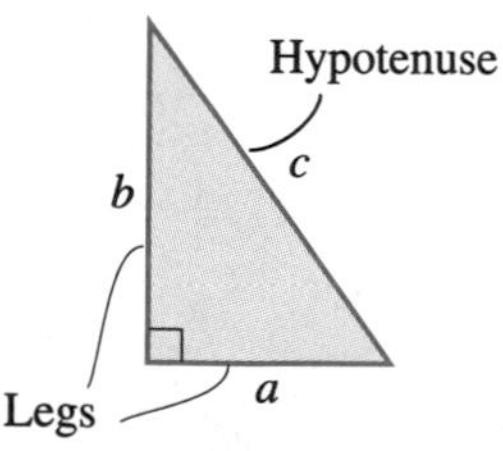

$c^2 = a^2 + b^2$
Pythagorean theorem

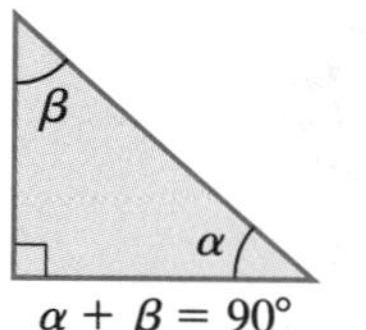

$\alpha + \beta = 90°$

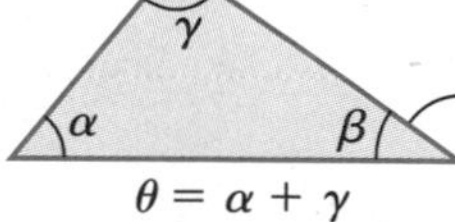

$\theta = \alpha + \gamma$

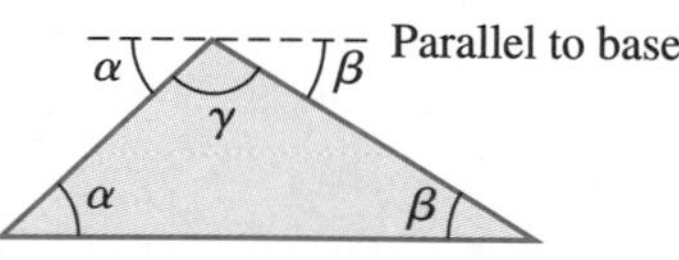

$\alpha + \beta + \gamma = 180°$
The sum of angle measures of all angles in a triangle is 180°.

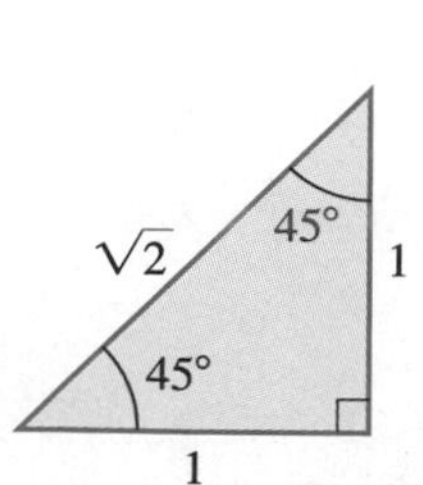

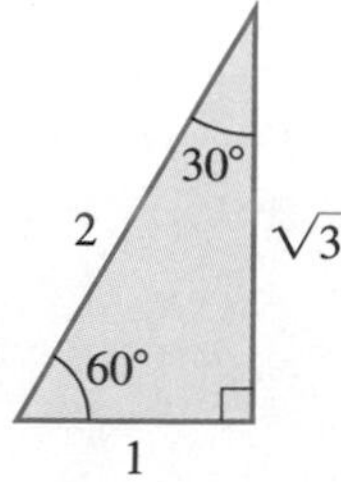

Special triangles

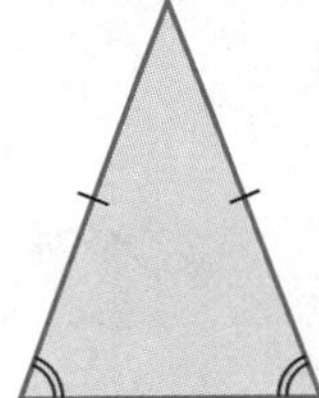

Isosceles triangle
At least two equal sides
At least two equal angles

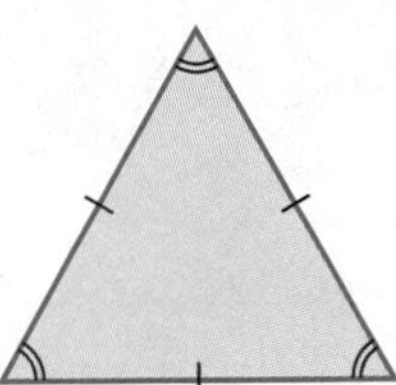

Equilateral triangle
All sides equal
All angles equal

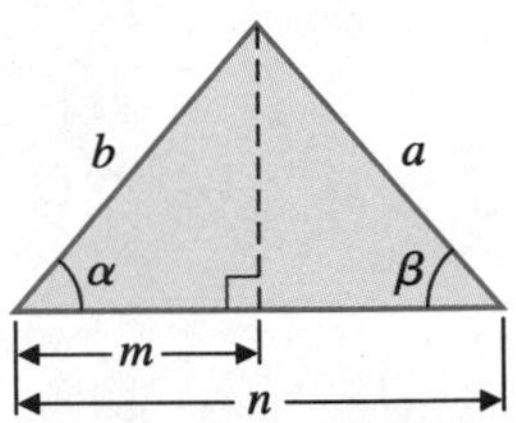

If $a = b$, then $\alpha = \beta$ and $m = n/2$.
If $\alpha = \beta$, then $a = b$ and $m = n/2$.

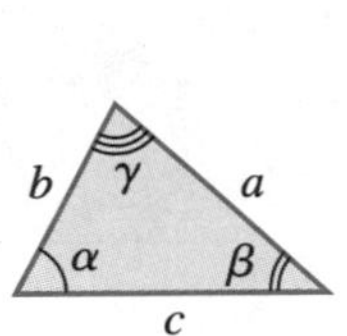

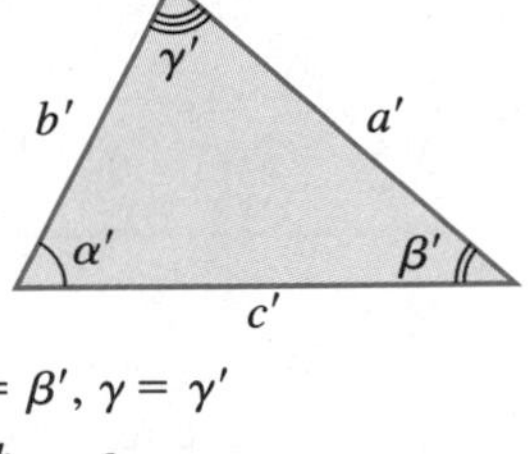

$$\alpha = \alpha', \beta = \beta', \gamma = \gamma'$$

$$\frac{a}{a'} = \frac{b}{b'} = \frac{c}{c'}$$

Similar triangles
Two triangles are similar if two angles of one triangle are equal to two angles of the other.

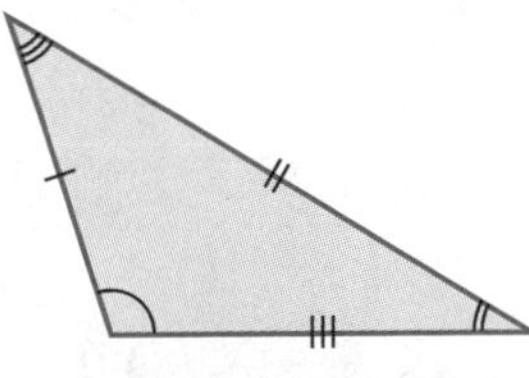

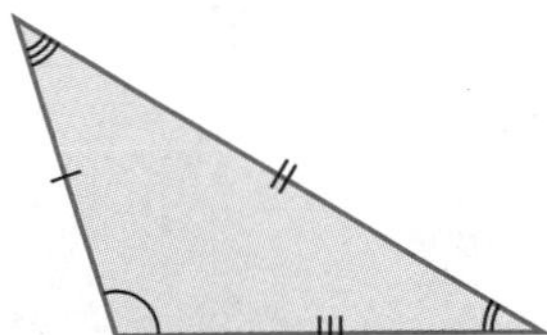

Congruent triangles
Corresponding parts of congruent triangles are equal.

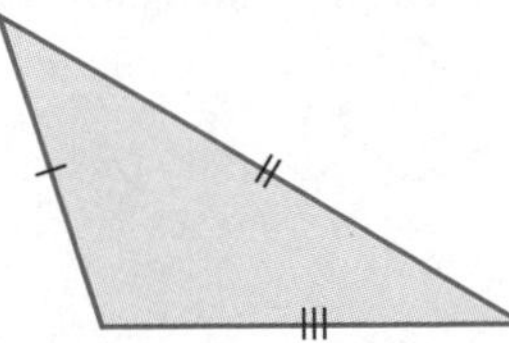

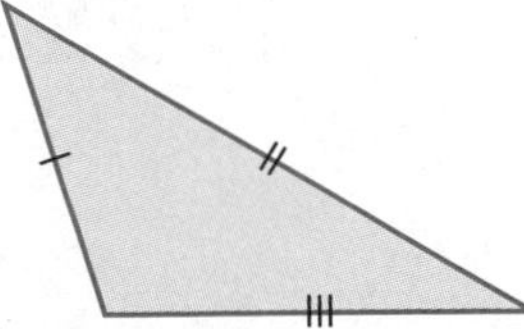

SSS
If three sides of one triangle are equal to the corresponding sides of another triangle, the two triangles are congruent.

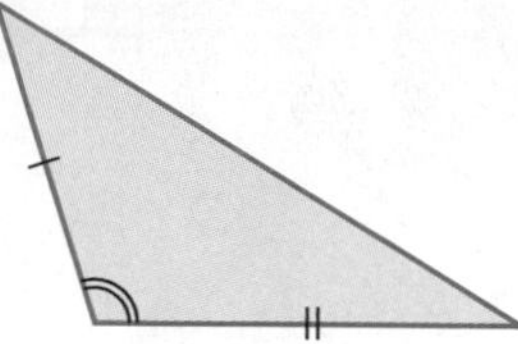
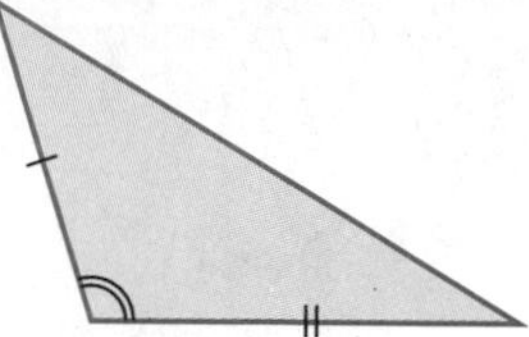

SAS
If two sides and the included angle of one triangle are equal to the corresponding parts of another triangle, the two triangles are congruent.

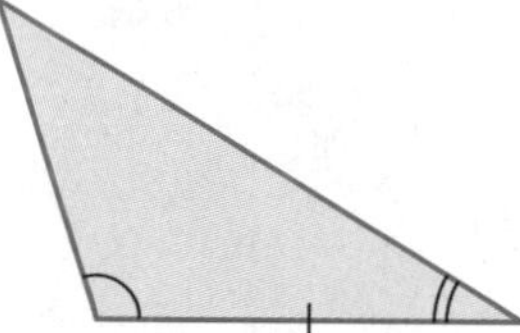
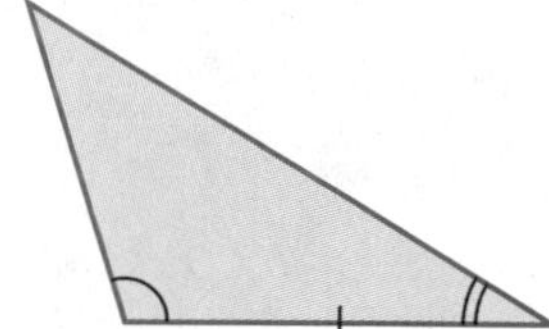

ASA
If two angles and the included side of one triangle are equal to the corresponding parts of another triangle, the two triangles are congruent.

C.3 QUADRILATERALS

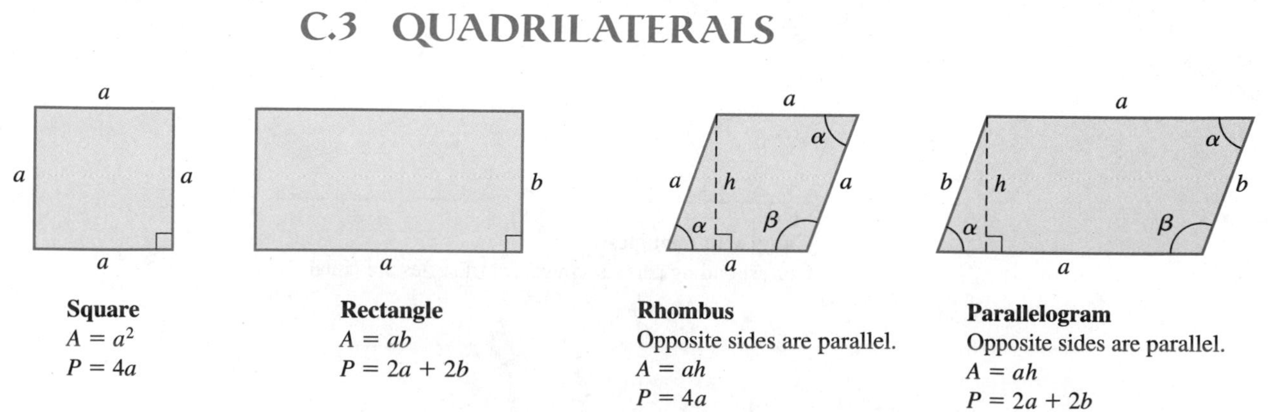

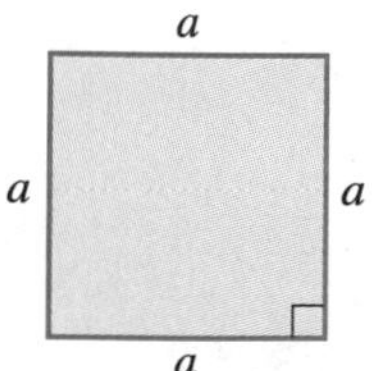

Square
$A = a^2$
$P = 4a$

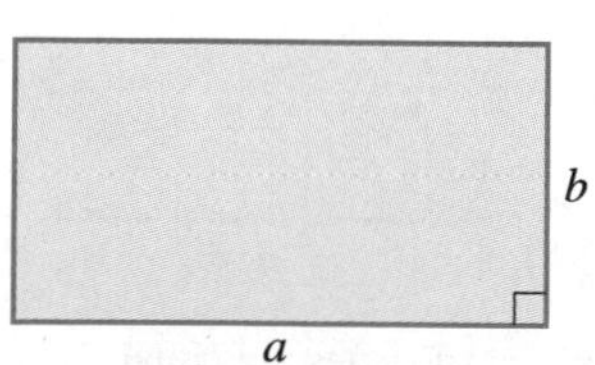

Rectangle
$A = ab$
$P = 2a + 2b$

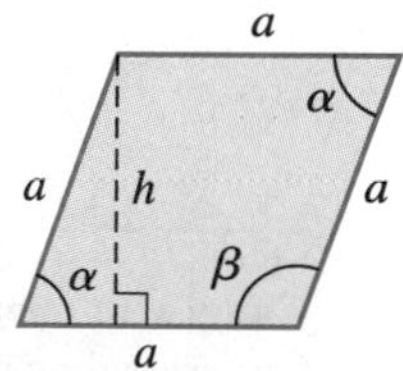

Rhombus
Opposite sides are parallel.
$A = ah$
$P = 4a$
$\alpha + \beta = 180°$

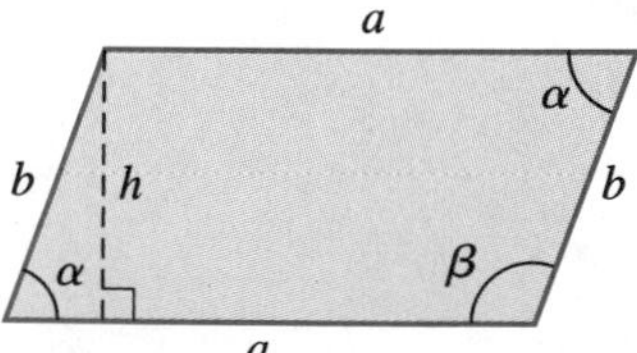

Parallelogram
Opposite sides are parallel.
$A = ah$
$P = 2a + 2b$
$\alpha + \beta = 180°$

C.4 CIRCLES

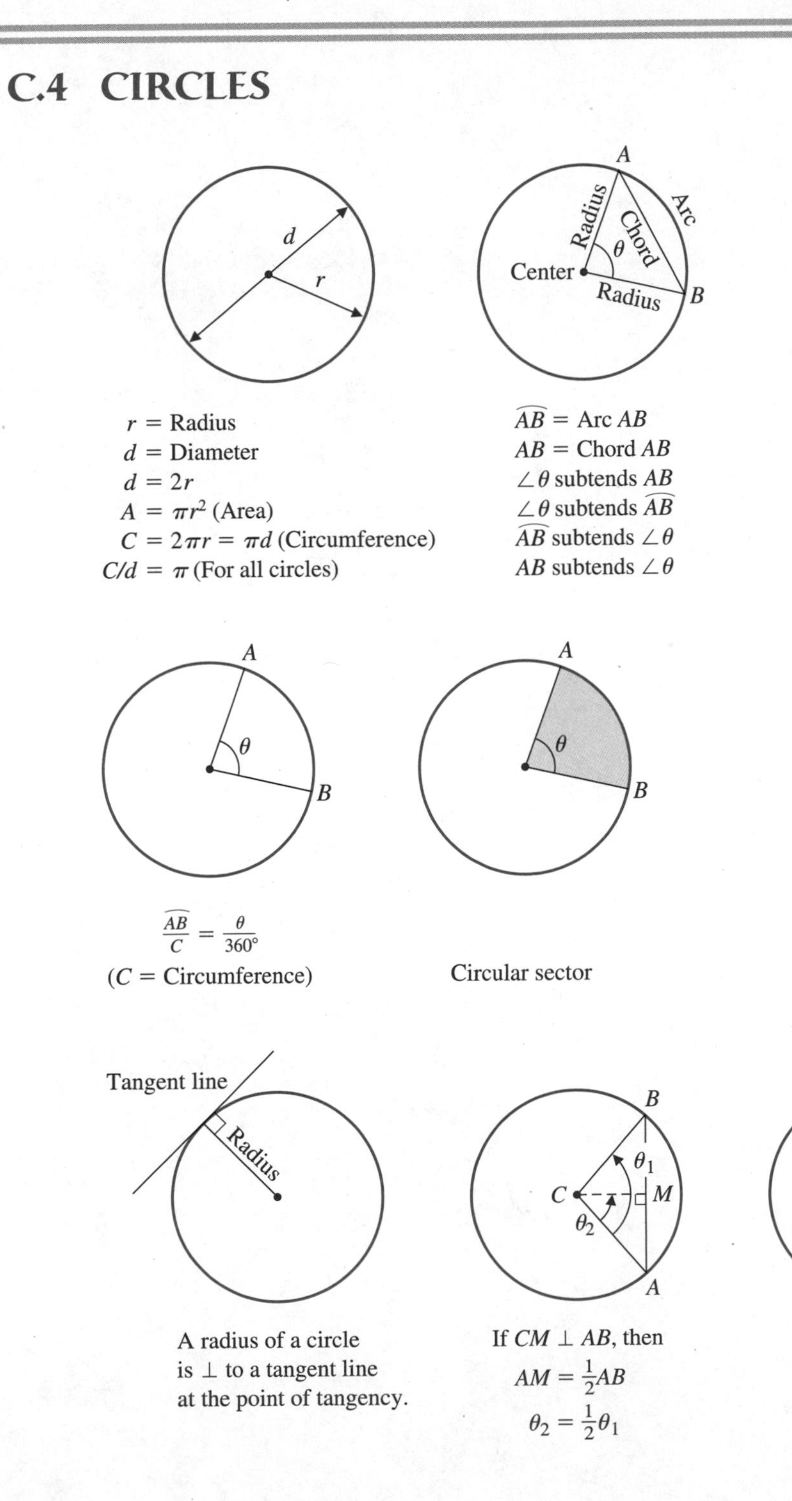

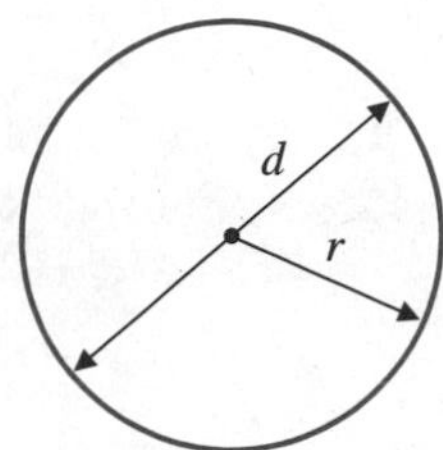

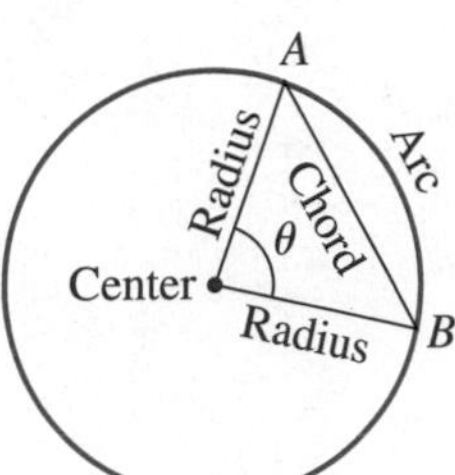

r = Radius
d = Diameter
$d = 2r$
$A = \pi r^2$ (Area)
$C = 2\pi r = \pi d$ (Circumference)
$C/d = \pi$ (For all circles)

$\widehat{AB}$ = Arc AB
AB = Chord AB
$\angle\theta$ subtends AB
$\angle\theta$ subtends $\widehat{AB}$
$\widehat{AB}$ subtends $\angle\theta$
AB subtends $\angle\theta$

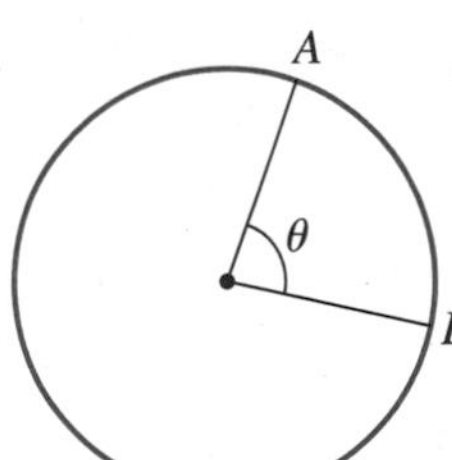

$$\frac{\widehat{AB}}{C} = \frac{\theta}{360^\circ}$$

(C = Circumference)

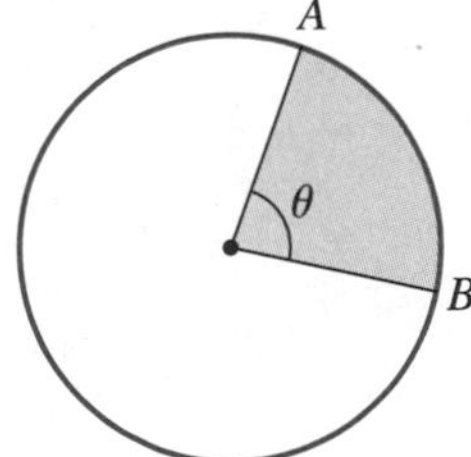

Circular sector

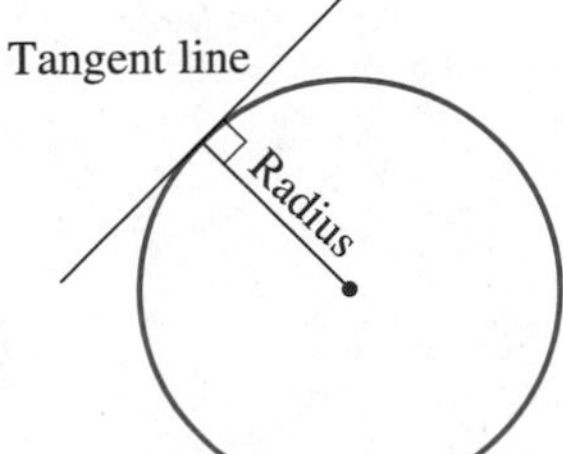

A radius of a circle is $\perp$ to a tangent line at the point of tangency.

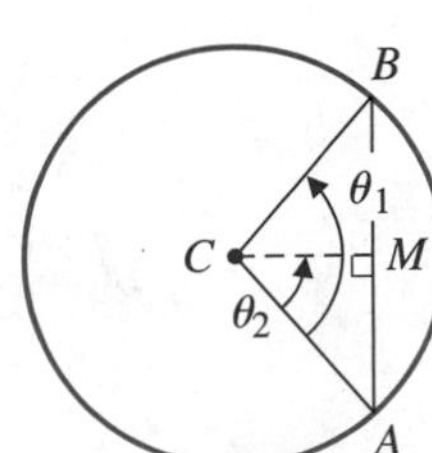

If $CM \perp AB$, then

$$AM = \tfrac{1}{2}AB$$

$$\theta_2 = \tfrac{1}{2}\theta_1$$

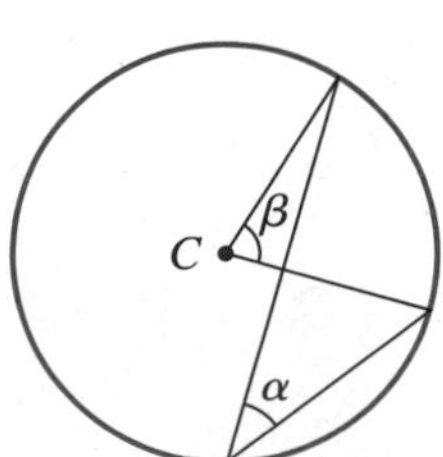

$\beta = 2\alpha$

SELECTED ANSWERS

CHAPTER 1

Exercise 1.1

1. 180° **3.** 45° **5.** 240° **7.** Obtuse
9. Straight **11.** Acute **13.** None
15. An angle of 1 degree measure is formed by rotating the terminal side of the angle $\frac{1}{360}$ of a complete revolution in a counterclockwise direction.
17. 43.351° **19.** 2.213° **21.** 103.295°
23. 13°37′59″ **25.** 83°1′1″ **27.** 187°12′14″
29. Convert the first to decimal degree form and compare with the second, or convert the second to DMS form and compare with the first; 47.572° is larger.
31. $\alpha < \beta$ **33.** $\alpha > \beta$ **35.** 110°18′4″
37. 22°22′31″ **39.** 100 cm **41.** 450 km
43. 250,000 mi **45.** 1,440 cm **47.** 262 cm^2
49. 71.0° **51.** 5.58 mm **53.** 11.5 mm
55. 679 mi **57.** 553 mi **59.** 590 nautical mi
61. 480 nautical mi
63. (A) 70 ft
(B) The arc length of a circular sector is very close to the chord length if the central angle of the sector is small and the radius of the sector is large, which is the case in this problem. The arc length s is easily found using the formula $\dfrac{s}{2\pi r} = \dfrac{\theta}{360}$.

Exercise 1.2

1. The measures of the third angle in each triangle are the same, since the sum of the measures of the three angles in any triangle is 180°. (Subtract the sum of the measures of the two given angles in each triangle from 180°.)
3. $b' = 6$ **5.** $c = 108$ **7.** $c' = 2.8$
9. Two similar triangles can have equal sides only if they are congruent. That is, if the two triangles coincide when one is moved on top of the other.
11. $a = 24$ in., $c = 56$ in. **13.** $b = 50$ m, $c = 55$ m
15. $a = 1.2 \times 10^9$ yd, $c = 2.7 \times 10^9$ yd
17. $a = 3.6 \times 10^{-5}$ mm, $b = 7.6 \times 10^{-5}$ mm
19. $c \approx 108$ ft **21.** 19.5 ft **23.** 63 ft **25.** 1.1 km
27. (A) Triangles PAC, FBC, ACP', and ABF' are all right triangles. Angles APC and BFC are equal, and angles $CP'A$ and $BF'A$ are equal—alternate interior angles of parallel lines cut by a transversal are equal (see Appendix C.1). Thus, triangles PAC and FBC are similar, and triangles ACP' and ABF' are similar.
(B) Start with the proportions $AC/PA = BC/BF$ and $AC/CP' = AB/BF'$.
(C) Add the two equations in part (B) together and divide both sides of the result by $(h + h')$.
(D) 50.847 mm

Exercise 1.3

1. a/c **3.** b/a **5.** c/a **7.** $\sin \theta$ **9.** $\tan \theta$
11. $\csc \theta$ **13.** 0.432 **15.** 0.709 **17.** 1.41
19. 0.294 **21.** 0.703 **23.** 1.08 **25.** 53.44°
27. 44°20′ **29.** 62°50′
31. The triangle is uniquely determined. Angle α can be found using $\tan \alpha = a/b$; angle $\beta = 90° - \alpha$. The hypotenuse c can be found using the Pythagorean theorem or by using $\sin \alpha = b/c$.
33. The triangle is not uniquely determined. In fact, there are infinitely many triangles of different sizes with the same acute angles—all are similar to each other.
35. $90° - \theta = 31°20'$, $a = 7.80$ mm, $b = 12.8$ mm
37. $90° - \theta = 6.3°$, $a = 0.354$ km, $c = 3.23$ km
39. $90° - \theta = 18.5°$, $a = 4.28$ in., $c = 13.5$ in.
41. $\theta = 28°30'$, $90° - \theta = 61°30'$, $a = 118$ ft
43. $\theta = 50.7°$, $90° - \theta = 39.3°$, $c = 171$ mi
45. The calculator was accidentally set in radian mode. Changing the mode to degree, $a = 235 \sin(14.1) = 57.2$ m and $b = 235 \cos(14.1) = 228$ m.
51. $90° - \theta = 6°48'$, $b = 199.8$ mi, $c = 201.2$ mi
53. $\theta = 37°6'$, $90° - \theta = 52°54'$, $c = 70.27$ cm
55. $\theta = 44.11°$, $90° - \theta = 45.89°$, $a = 36.17$ cm
61. (A) $\sin \theta = \dfrac{AD}{OD} = \dfrac{AD}{1} = AD$

61. (B) $\tan\theta = \frac{AD}{OA} = \frac{DC}{OD} = \frac{DC}{1} = DC$

$(\angle OED = \theta)$

(C) $\csc\theta = \frac{OE}{OD} = \frac{OE}{1} = OE$

63. (A) $\sin\theta$ approaches 1
(B) $\tan\theta$ increases without bound
(C) $\csc\theta$ approaches 1

65. (A) $\cos\theta$ approaches 1
(B) $\cot\theta$ increases without bound
(C) $\sec\theta$ approaches 1

Exercise 1.4

1. 7.0 m **3.** 211 m **5.** 1.1 km
7. 29,400 m, or 29.4 km **9.** 22°
11. (A) 5.1 ft (B) 2.6 ft **13.** 3.5 m **15.** 5.69 cm
17. 134 ft **19.** 8.4 ft, 21 ft

21. (A) $\cos\alpha = \frac{r}{r+h}$ (B) $r = \frac{h\cos\alpha}{1-\cos\alpha}$ (C) 3,960 mi

23. 8.4 mi **25.** 19,600 mi

27. (A) The lifeguard can run faster than he can swim, so he should run along the beach first before entering the water. [Parts (D) and (E) suggest how far the lifeguard should run before swimming to get to the swimmer in the least time.]

(B) $T = \frac{d - c\cot\theta}{p} + \frac{c\csc\theta}{q}$

(C) $T = 119.97$ sec

(D) T decreases, then increases as θ goes from 55° to 85°; T has a minimum value of 116.66 sec when $\theta = 70°$.

(E) $d = 352$ m

29. (A) Since the angle θ determines the length of pipe to be laid on land and in the water, it appears that the total cost of the pipeline depends on θ.

(B) $C = 160{,}000\sec\theta + 20{,}000(10 - 4\tan\theta)$

(C) $344,200

(D) As θ increases from 15° to 45°, C decreases and then increases; C has a minimum value of $338,600 when $\theta = 30°$.

(E) 4.62 mi in the water and 7.69 mi on land

31. 0.97 km **33.** 386 ft apart; 484 ft high

35. $g = 32.0$ ft/sec^2 **37.** $\sin\theta = \frac{3}{5}$

Chapter 1 Review Exercise

1. 7,280″ [1.1] **2.** 60° [1.1] **3.** 8,000 [1.2]

4. 36.38° [1.1]

5. An angle of degree measure 1 is an angle formed by rotating the terminal side of the angle $\frac{1}{360}$ of a complete revolution in a counterclockwise direction. [1.1]

6. All three are similar, because all three have equal angles. [1.2]

7. No. Similar triangles have equal angles. [1.2]

8. The sum of all the angles in a triangle is 180°. Two obtuse angles would add up to more than 180°. [1.2]

9. 560 ft [1.2]

10. (A) b/c (B) c/a (C) b/a (D) c/b (E) a/c (F) a/b [1.3]

11. $90° - \theta = 54.8°$, $a = 16.5$ cm, $b = 11.6$ cm [1.3]

12. 144° [1.1] **13.** 4.19 in. [1.1]

14. Either convert the first to DD form or convert the second to DMS form: 27°14′ = 27.23° is larger than 27.19°. [1.1]

15. (A) 72.930° (B) 113°50′13″ [1.1]

16. (A) 56°32′9″ (B) 178°22′24″ [1.1]

17. 7.1×10^{-6} mm [1.2]

18. (A) $\cos\theta$ (B) $\tan\theta$ (C) $\sin\theta$ (D) $\sec\theta$ (E) $\csc\theta$ (F) $\cot\theta$ [1.3]

19. Two right triangles having an acute angle of one equal to an acute angle of the other are similar, and corresponding sides of similar triangles are proportional. [1.3]

20. $90° - \theta = 27°40'$, $b = 7.63 \times 10^{-8}$ m, $c = 8.61 \times 10^{-8}$ m [1.3]

21. (A) 58.97° (B) 55°50′ (C) 0°47′27″ [1.3]

22. Window (a) is in radian mode; window (b) is in degree mode. [1.3]

23. $\theta = 40.3°$, $90° - \theta = 49.7°$, $c = 20.6$ mm [1.3]

24. $\theta = 40°20'$, $90° - \theta = 49°40'$ [1.3]

25. 8.7 ft [1.2] **26.** 940 ft [1.1]

27. 107 ft^2 [1.1]

28. $\theta = 66°17'$, $a = 93.56$ km, $b = 213.0$ km [1.3]

29. $\theta = 60.28°$, $90° - \theta = 29.72°$, $b = 4{,}241$ m [1.3]

30. 1.0496 [1.3] **31.** 8.8 ft [1.2]

32. 24.5 ft; 24.1 ft [1.4] **33.** 2.3°; 0.07 or 7% [1.4]

34. 955 mi [1.1] **35.** 46° [1.4]

36. 830 m [1.4] **37.** 1,240 m [1.4]

38. 760 mph [1.4]

39. (A) $\beta = 90° - \alpha$ (B) $r = h\tan\alpha$ (C) $H - h = (R - r)\cot\alpha$ [1.4]

40. (A) In both cases the ladder must get longer.
(B) $L = 5\csc\theta + 4\sec\theta$.
(C)

θ	25°	35°	45°	55°	65°	75°	85°
L	16.24	13.60	12.73	13.08	14.98	20.63	50.91

(D) L decreases and then increases; L has a minimum value of 12.73 when $\theta = 45°$.

(E) Make up another table for values of θ close to 45° and on either side of 45°. [1.4]

CHAPTER 2

Exercise 2.1

1. The central angle of a circle has radian measure 1 if it intercepts an arc of length equal to the radius of the circle.

3. $\pi/6, \pi/3, \pi/2, 2\pi/3, 5\pi/6, \pi, 7\pi/6, 4\pi/3, 3\pi/2, 5\pi/3, 11\pi/6, 2\pi$

5. The angle of radian measure 1 is larger, since 1 rad corresponds to a degree of measure of approximately 57.3°.

7.

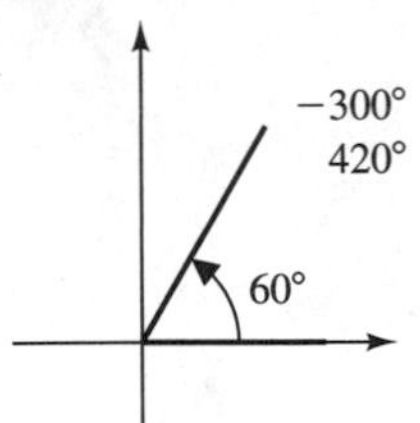

9.

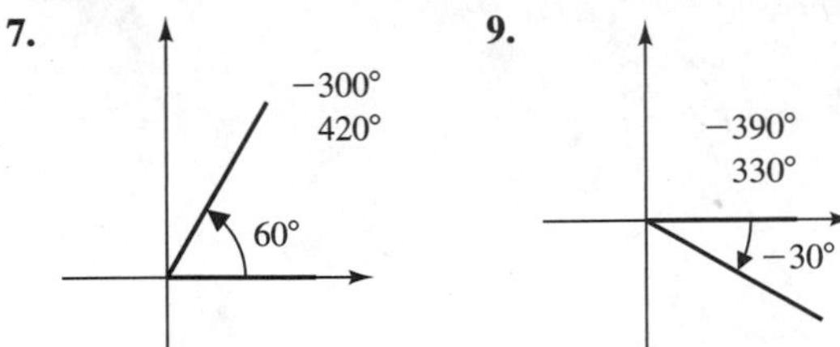

−390°
330°
−30°

11. $\pi/10$ rad $\approx$ 0.3142 rad **13.** $13\pi/18$ rad $\approx$ 2.269 rad

15. $(288/\pi)°\approx$ 91.67° **17.** 3° (exact)

19.

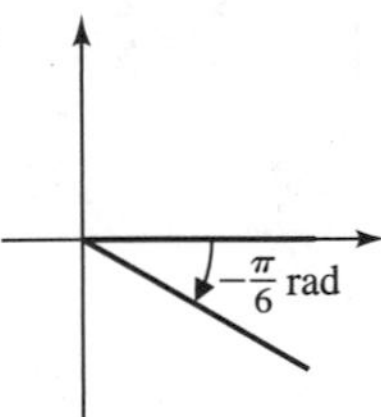

21.

300°

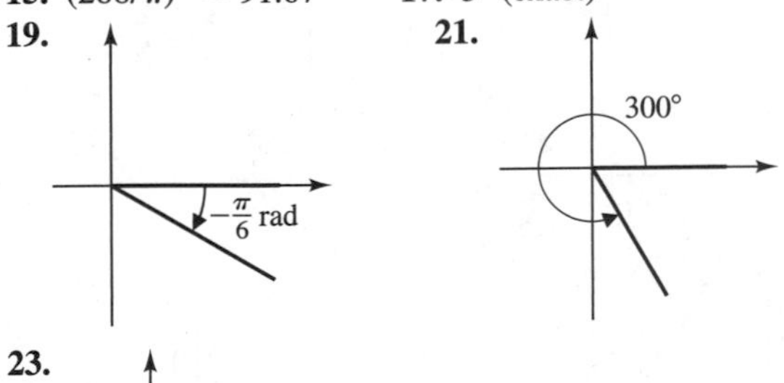

23.

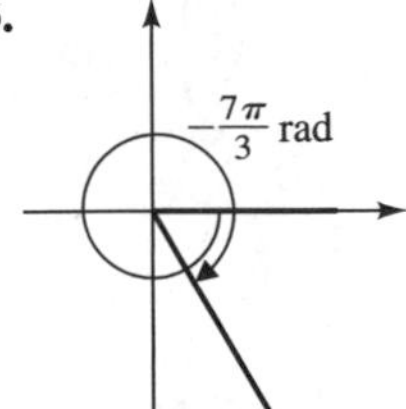

25. (A) 475.555°; −664.631°
(B) 9.826 rad; −21.468 rad

27. (A) 3 rad (B) 4.5 rad

29. (A) 58.2 m (B) 8.29 m (C) 20.5 m (D) 47.1 m

31. Yes. Since $\theta_d = (180/\pi)\theta_r$, if θ_r is doubled, then θ_d also must be doubled (multiply both sides of the conversion equation by 2).

33. Since $\theta_r = s/r$, if s is held constant while r (the denominator) is doubled, then θ_r will be cut in half.

35. (A) 46.4 cm² (B) 42.8 cm² (C) 1.00×10^2 cm²
(D) 192 cm²

37. Since $s = r\theta$ (θ in radian measure), $s = 1 \cdot m = m$.

39. I **41.** III **43.** II **45.** 1.0008 rad

47. 18.2430° **49.** 0.4605 rad

51. $7\pi/12$ rad $\approx$ 1.83 rad **53.** 12 cm **55.** 19°

57. 256 cm **59.** 1.4×10^6 km **61.** 175 ft

63. 0.2 m; 7.9 in. **65.** $\pi/26$ rad $\approx$ 0.12 rad

67. 6,800 mi **69.** 31 ft

71. (A) Since one revolution corresponds to 2π radians, n revolutions corresponds to $2\pi n$ radians.
(B) No. Radian measure is independent of the size of the circle used; hence, it is independent of the size of the wheel used.
(C) 31.42 rad; 22.62 rad

73. 6.5 revolutions; 40.8 rad **75.** 859°

Exercise 2.2

1. 3 mm/sec **3.** 17 rad/sec

5. A radial line from the center to a point on the circumference of a rotating circle sweeps out an angle at a uniform rate called angular velocity. This uniform rate is usually given in radians per unit time.

7. 3.7 rad/hr **9.** 0.593 rad/sec **11.** 75 m/sec

13. 4.65 rad/hr **15.** 223.5 rad/sec or 35.6 rps

17. $\pi/4{,}380$ rad/hr; 66,700 mph

19. (A) 0.634 rad/hr (B) 28,100 mph **21.** 6,900 mph

23. 1.61 hr

25. (A) 1 rps corresponds to an angular velocity of 2π rad/sec, so that at the end of t sec, $\theta = 2\pi t$.
(B) Using right triangle trigonometry, $a = 15 \tan \theta$. Then, substituting $\theta = 2\pi t$, $a = 15 \tan 2\pi t$.
(C) The speed of the light spot on the wall increases as t increases from 0.00 to 0.24. When $t = 0.25$, $a = 15 \tan(\pi/2)$, which is not defined. The light has made one-quarter turn, and the spot is no longer on the wall.

t (sec)	0.00	0.04	0.08	0.12	0.16	0.20	0.24
a (ft)	0.00	3.85	8.25	14.09	23.64	46.17	238.42

Exercise 2.3

1. $\sin\theta = \frac{4}{5}$, $\csc\theta = \frac{5}{4}$, $\cos\theta = \frac{3}{5}$, $\sec\theta = \frac{5}{3}$, $\tan\theta = \frac{4}{3}$, $\cot\theta = \frac{3}{4}$; same values for $Q(6, 8)$ (Why?)

3. $\sin\theta = -\frac{3}{5}$, $\csc\theta = -\frac{5}{3}$, $\cos\theta = \frac{4}{5}$, $\sec\theta = \frac{5}{4}$, $\tan\theta = -\frac{3}{4}$, $\cot\theta = -\frac{4}{3}$; same values for $Q(12, -9)$ (Why?)

5. $\sin\theta = \frac{4}{5}$, $\tan\theta = \frac{4}{3}$, $\csc\theta = \frac{5}{4}$, $\sec\theta = \frac{5}{3}$, $\cot\theta = \frac{3}{4}$

7. $\sin\theta = -\frac{4}{5}$, $\tan\theta = -\frac{4}{3}$, $\csc\theta = -\frac{5}{4}$, $\sec\theta = \frac{5}{3}$, $\cot\theta = -\frac{3}{4}$

9. $\sin\theta = -\frac{4}{5}$, $\cos\theta = -\frac{3}{5}$, $\tan\theta = \frac{4}{3}$, $\sec\theta = -\frac{5}{3}$, $\cot\theta = \frac{3}{4}$

11. No. For all those values of x for which both are defined, $\cos x = 1/(\sec x)$; hence, either both are positive or both are negative.

13. 57.29 **15.** -0.9900 **17.** 3.236 **19.** -2.904
21. 0.4202 **23.** -0.4577 **25.** 5.824 **27.** 0.9272
29. 0.8439 **31.** 4.331
33. $\sin\theta = \frac{1}{2}$, $\cos\theta = \sqrt{3}/2$, $\tan\theta = 1/\sqrt{3}$, $\csc\theta = 2$, $\sec\theta = 2/\sqrt{3}$, $\cot\theta = \sqrt{3}$
35. $\sin\theta = -\sqrt{3}/2$, $\cos\theta = \frac{1}{2}$, $\tan\theta = -\sqrt{3}$, $\csc\theta = -2/\sqrt{3}$, $\sec\theta = 2$, $\cot\theta = -1/\sqrt{3}$
37. I, IV **39.** I, III **41.** I, IV **43.** III, IV
45. II, IV **47.** III, IV
49. $\cos\theta = -\sqrt{5}/3$, $\tan\theta = 2/\sqrt{5}$, $\csc\theta = -\frac{3}{2}$, $\sec\theta = -3/\sqrt{5}$, $\cot\theta = \sqrt{5}/2$
51. $\sin\theta = -\sqrt{2}/\sqrt{3}$, $\cos\theta = 1/\sqrt{3}$, $\tan\theta = -\sqrt{2}$, $\csc\theta = -\sqrt{3}/\sqrt{2}$, $\cot\theta = -1/\sqrt{2}$
53. Yes, $\cos\alpha = \cos\beta$, since the terminal sides of these angles will coincide and the same point $P(a, b) \neq (0, 0)$ can be chosen on the terminal sides. Thus, $\cos\alpha = a/r = \cos\beta$, where $r = \sqrt{a^2 + b^2} \neq 0$.
55. Use the reciprocal identity $\cot x = 1/(\tan x)$; $\cot x = 1/(-2.18504) = -0.45766$.
57. 0.6191 **59.** 4.938 **61.** -0.08169
63. -1.553 **65.** -0.8812 **67.** 1.079 **69.** 1.778
71. 0.7112 **73.** -0.9800
75. Tangent and secant; since $\tan x = b/a$ and $\sec x = r/a$, neither is defined when the terminal side of the angle lies along the positive or negative vertical axis, because a will be 0 (division by 0 is not defined).
77. (A) $\theta = 1.2$ rad
(B) $(a, b) = (5\cos 1.2, 5\sin 1.2) = (1.81, 4.66)$
79. (A) $\theta = 2$ rad
(B) $(a, b) = (1\cos 2, 1\sin 2) = (-0.416, 0.909)$
81. 3.22 units **83.** k, $0.94k$, $0.77k$, $0.50k$, $0.17k$
85. Summer solstice: $E = 0.97k$; winter solstice: $E = 0.45k$
87. (A)

n	6	10	100	1,000	10,000
A_n	2.59808	2.93893	3.13953	3.14157	3.14159

(B) The area of the circle is $A = \pi r^2 = \pi(1)^2 = \pi$, and A_n seems to approach π, the area of the circle, as n increases.
(C) No. An n-sided polygon is always a polygon, no matter the size of n, but the inscribed polygon can be made as close to the circle as you like by taking n sufficiently large.
91. $I = 24$ amperes
93. (A) 2.01; -0.14 (B) $y = -0.75x + 3.75$

Exercise 2.4

1. 29.3° **3.** 25.1° **5.** 24.4°
7. The ball will appear higher than the real ball. The light path from the eye of a person standing on shore to the real ball will bend downward at the water surface because of refraction. But the ball will appear to lie on the straight continuation of the light path from the eye below the water surface.
9. $n_2 = 1.53$ **11.** 40 km/hr **13.** 84°
15. 3×10^{10} cm/sec **17.** $-6°$

Exercise 2.5

1. $\alpha = \theta = 60°$

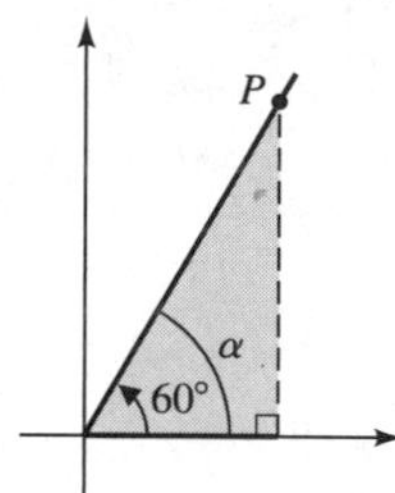

3. $\alpha = |-60°| = 60°$

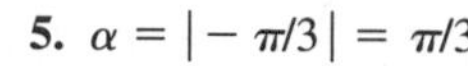

5. $\alpha = |-\pi/3| = \pi/3$

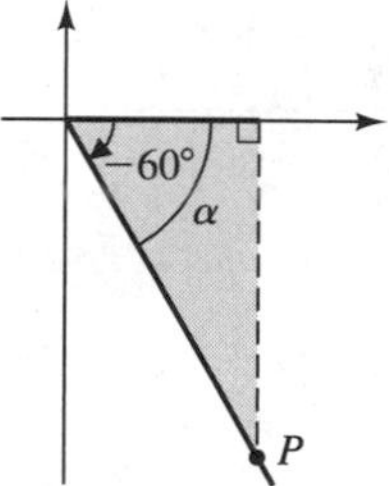

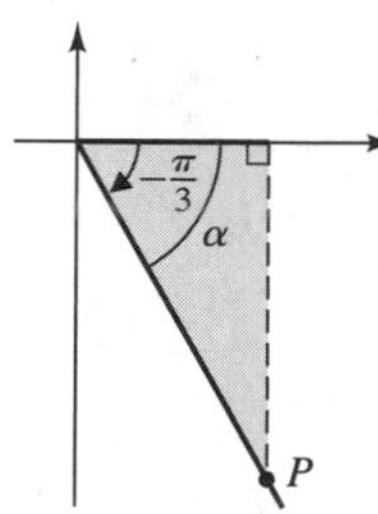

7. $\alpha = \pi - 3\pi/4 = \pi/4$

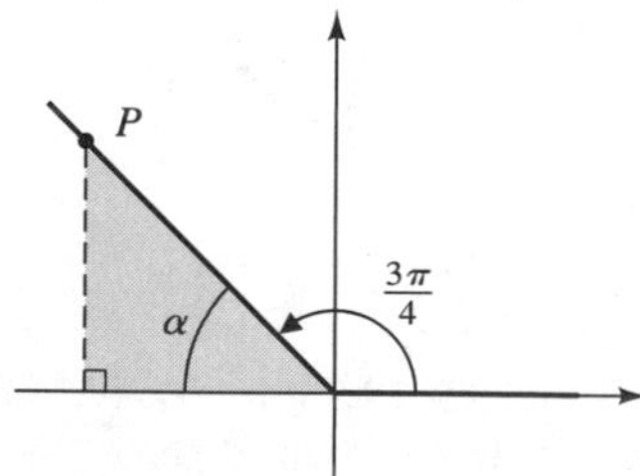

9. $\alpha = 210° - 180° = 30°$

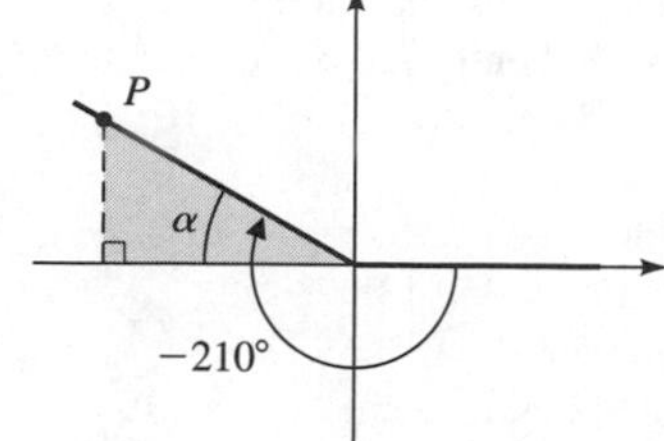

11. $\alpha = 5\pi/4 - \pi = \pi/4$

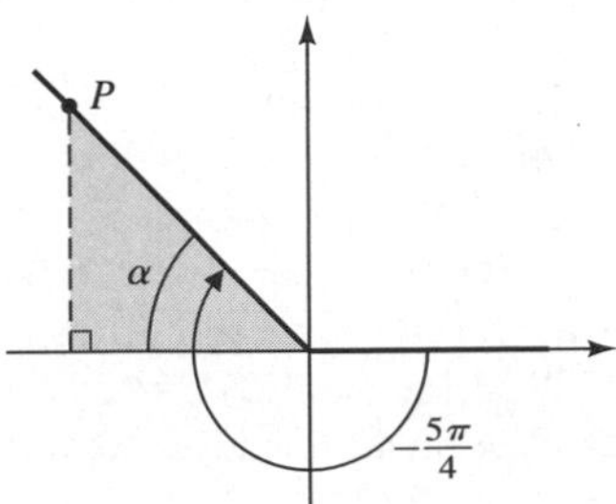

13. 1 **15.** $\frac{1}{2}$ **17.** 1 **19.** 1 **21.** $1/\sqrt{3}$, or $\sqrt{3}/3$
23. $-\frac{1}{2}$ **25.** 0 **27.** $\sqrt{3}/2$ **29.** $-\sqrt{3}$ **31.** $-\frac{1}{2}$
33. -1 **35.** $-1/\sqrt{2}$, or $-\sqrt{2}/2$ **37.** -1
39. $-\sqrt{3}/2$ **41.** $2/\sqrt{3}$
43. The tangent function is not defined at $\theta = \pi/2$ and $\theta = 3\pi/2$, because $\tan\theta = b/a$ and $a = 0$ for any point on the vertical axis.
45. The cosecant function is not defined at $\theta = 0$, $\theta = \pi$, and $\theta = 2\pi$, because $\csc\theta = r/b$ and $b = 0$ for any point on the horizontal axis.
47. $\sin(-45°) = -1/\sqrt{2}$, or $-\sqrt{2}/2$
49. $\tan(-\pi/3) = -\sqrt{3}$
51. (A) 30° (B) $\pi/6$
53. (A) 120° (B) $2\pi/3$ **55.** (A) 120° (B) $2\pi/3$
57. 240°, 300° **59.** $5\pi/6$, $11\pi/6$ **61.** $3\pi/4$
63. $\pi/6$
65. (A) $x = 14, y = 7\sqrt{3}$
(B) $x = 4/\sqrt{2}, y = 4/\sqrt{2}$
(C) $x = 10/\sqrt{3}, y = 5/\sqrt{3}$
67. $3\sqrt{3}$ cm^2 **69.** $150\sqrt{3}$ in.^{2}

Exercise 2.6

1. (A) π (B) $3\pi/2$
3. (A) (1, 0) (B) (0, 1) (C) (0, 1) (D) $(-1, 0)$
(E) (1, 0) (F) (0, 1)
5. (A) 0 to 1 (B) 1 to 0 (C) 0 to -1 (D) -1 to 0
(E) 0 to 1
7. (A) 1 to 0 (B) 0 to -1 (C) -1 to 0 (D) 0 to 1
(E) 1 to 0
9. $\pi/2$, $5\pi/2$ **11.** 0, π, 2π, 3π, 4π
13. 0, π, 2π, 3π, 4π **15.** $3\pi/2$, $7\pi/2$
17. -2π, 0, 2π **19.** $-3\pi/2$, $-\pi/2$, $\pi/2$, $3\pi/2$
21. $\pi/2$, $3\pi/2$, $5\pi/2$, $7\pi/2$ **23.** 0, π, 2π, 3π, 4π
25. 0.7 **27.** -0.7 **29.** -0.8 **31.** -2
33. 1 **35.** -2 **37.** -0.2088 **39.** 7.228
41. 119.2 **43.** -0.04334 **45.** -29.99
47. -12.04
49. $P(a, b) = P(\cos x, \sin x)$, where $x = -0.898$; x is negative since P is moving clockwise. The quadrant in which P lies is determined by the signs of the coordinates of P: $P(\cos(-0.898), \sin(-0.898)) = (0.6232, -0.7821)$ and P lies in quadrant IV.
51. $P(a, b) = P(\cos x, \sin x)$, where $x = 26.77$; x is positive since P is moving counterclockwise. The quadrant in which P lies is determined by the signs of the coordinates of P: $P(\cos 26.77, \sin 26.77) = (-0.0664, 0.9978)$ and P lies in quadrant II.
53. $-1/\sqrt{2}$ or $-\sqrt{2}/2$ **55.** $-\sqrt{2}$ **57.** Not defined
59. All 0.9525
61. (A) Both 1.6 (B) Both -1.5 (C) Both 0.96
63. (A) Both -0.14 (B) Both 0.23 (C) Both 0.99
65. (A) 1.0 (B) 1.0 (C) 1.0 **67.** 1 **69.** $\csc x$
71. $\csc x$ **73.** $\cos x$
75. $s = \cos^{-1} 0.58064516 = 0.951 = \sin^{-1} 0.81415674$
77. (A) Identity (4) (B) Identity (9) (C) Identity (2)
79. 2π
81. $s_1 = 1, s_2 = 1.540302, s_3 = 1.570792, s_4 = 1.570796, s_5 = 1.570796$; $\pi/2 \approx 1.570796$

Chapter 2 Review Exercise

1. (A) $\pi/3$ (B) $\pi/4$ (C) $\pi/2$ [2.1]
2. (A) 30° (B) 90° (C) 45° [2.1]
3. A central angle of radian measure 2 is an angle subtended by an arc of twice the length of the radius. [2.1]
4. An angle of radian measure 1.5 is larger, since the corresponding degree measure of the angle would be approximately 85.94°. [2.1]
5. (A) 874.3° (B) -6.793 rad [2.1]
6. 185 ft/min [2.2] **7.** 80 rad/hr [2.2]
8. $\sin\theta = \frac{3}{5}$; $\tan\theta = -\frac{3}{4}$ [2.3]
9. No, since $\csc x = 1/\sin x$, when one is positive so is the other. [2.3]
10. (A) 0.7355 (B) 1.085 [2.3, 2.6]
11. (A) 0.9171 (B) 0.9099 [2.3, 2.6]
12. (A) 0.9394 (B) 5.177 [2.3, 2.6]
13. (A) $\alpha = 60°$

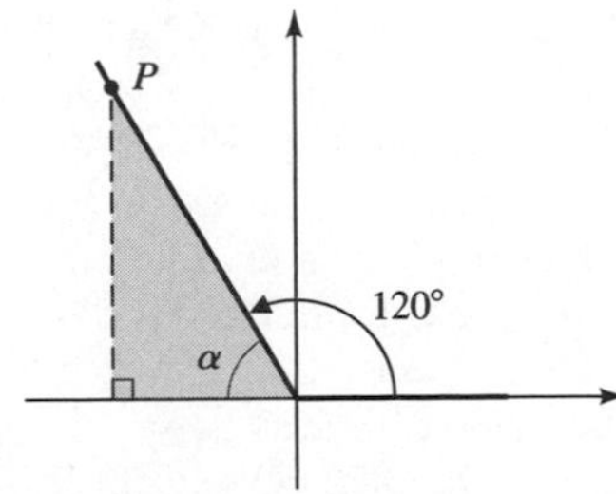

13. (B) $\alpha = \frac{\pi}{4}$

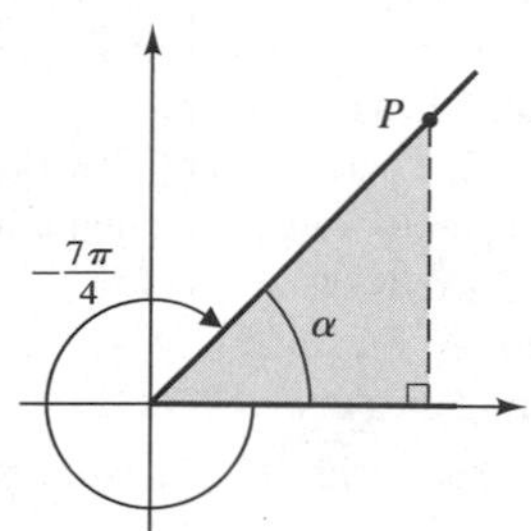

[2.5]

14. (A) $\sqrt{3}/2$ (B) $1/\sqrt{2}$, or $\sqrt{2}/2$ (C) 0 [2.5]
15. (A) (1, 0) (B) (− 1, 0) (C) (0, 1) (D) (0, 1) (E) (− 1, 0) (F) (0, − 1) [2.6]
16. (A) 0 to 1 (B) 1 to 0 (C) 0 to − 1 (D) − 1 to 0 (E) 0 to 1 (F) 1 to 0 [2.6]
17. $-11\pi/6$ and $13\pi/6$: When the terminal side of the angle is rotated any multiple of a complete revolution (2π rad) in either direction, the resulting angle will be coterminal with the original. In this case, for the restricted interval, this happens for $\pi/6 \pm 2\pi$. [2.1]
18. 42° [2.1] **19.** 6 cm [2.1] **20.** $53\pi/45$ [2.1]
21. 15° [2.1]
22. (A) − 3.72 (B) 264.71° [2.1]
23. Yes, since $\theta_d = (180°/\pi \text{ rad})\theta_r$, if θ_r is tripled, then θ_d will also be tripled. [2.1]
24. No. For example, if $\alpha = \pi/6$ and $\beta = 5\pi/6$, α and β are not coterminal, but $\sin(\pi/6) = \sin(5\pi/6)$. [2.3]
25. Use the reciprocal identity: $\csc x = 1/\sin x = 1/0.8594 = 1.1636$. [2.3]
26. 0; not defined; 0; not defined [2.5]
27. (A) I (B) IV [2.3] **28.** − 0.992 [2.3, 2.6]
29. − 4.34 [2.3, 2.6] **30.** − 1.30 [2.3, 2.6]
31. 0.683 [2.3, 2.6] **32.** 0.284 [2.3, 2.6]
33. − 0.864 [2.3, 2.6] **34.** $-\sqrt{3}/2$ [2.5]
35. − 1 [2.5] **36.** − 1 [2.5] **37.** 0 [2.5]
38. $\sqrt{3}/2$ [2.5] **39.** − 2 [2.5] **40.** − 1 [2.5]
41. Not defined [2.5] **42.** $\frac{1}{2}$ [2.5]
43. $\tan(-60°) = -\sqrt{3}$ [2.5] **44.** 0.40724 [2.3, 2.6]
45. − 0.33884 [2.3, 2.6] **46.** 0.64692 [2.3, 2.6]
47. 0.49639 [2.3, 2.6]
48. $\cos\theta = \frac{3}{5}$; $\tan\theta = -\frac{4}{3}$ [2.3] **49.** $7\pi/6$ [2.5]
50. $\cos\theta = \sqrt{21}/5$, $\sec\theta = 5/\sqrt{21}$, $\csc\theta = -\frac{5}{2}$, $\tan\theta = -2/\sqrt{21}$, $\cot\theta = -\sqrt{21}/2$ [2.3]
51. 135°, 315° [2.5] **52.** $5\pi/6$, $7\pi/6$ [2.5]
53. (A) 20.3 cm (B) 4.71 cm [2.1]
54. (A) 618 ft^2 (B) 1,880 ft^2 [2.1]
55. 215.6 mi [2.1] **56.** 0.422 rad/sec [2.2]
57. A radial line from the axis of rotation sweeps out an angle at the rate of 12π rad/sec. [2.2]
58. All 0.754 [2.6]
59. (A) Both − 0.871 (B) Both − 1.40 (C) Both − 3.38 [2.6]
60. (D) [2.6] **61.** $\cos x$ [2.6] **62.** $\cos x$ [2.6]
63. Since $P(a, b)$ is moving clockwise, $x = -29.37$. By the definition of the circular functions, the point has coordinates $P(\cos(-29.37), \sin(-29.37)) = P(-0.4575, 0.8892)$. P lies in quadrant II, since a is negative and b is positive. [2.6]
64. The radian measure of a central angle θ subtended by an arc of length s is $\theta = s/r$, where r is the radius of the circle. In this case, $\theta = 1.3/1 = 1.3$ rad. [2.1]
65. Since $\cot x = 1/(\tan x)$ and $\csc x = 1/(\sin x)$, and $\sin(k\pi) = \tan(k\pi) = 0$ for all integers k, $\cot x$ and $\csc x$ are not defined for these values. [2.6]
66. $x = 4\pi/3$ [2.5]
67. $s = 0.8905$ unit [2.6]
68. 57 m [2.1] **69.** 5.74 units [2.3]
70. 200 rad, $100/\pi \approx 31.8$ revolutions [2.1]
71. 7.5 rev, 15 rev [2.1] **72.** 62 rad/sec [2.2]
73. 16,400 mph [2.2]
74. − 17.6 amperes [2.3]
75. (A) $L = 10 \csc\theta + 2 \sec\theta$
(B) As θ decreases to 0 rad, L increases without bound; as θ increases to $\pi/2$, L increases without bound. Between these extremes, there appears to be a value of θ that produces a minimum L.
(C)

θ rad	0.70	0.80	0.90	1.00
L ft	18.14	16.81	15.98	15.59
θ rad	1.10	1.20	1.30	
L ft	15.63	16.25	17.85	

(D) $L = 15.59$ ft for $\theta = 1.00$ rad [2.3]
76. $\beta = 23.3°$ [2.4] **77.** $\alpha = 41.1°$ [2.4]
78. 11 mph [2.4]

CHAPTER 3

Exercise 3.1

1. 2π, 2π, π
3. (A) 1 unit (B) Indefinitely far (C) Indefinitely far
5. $-3\pi/2$, $-\pi/2$, $\pi/2$, $3\pi/2$ **7.** -2π, $-\pi$, 0, π, 2π
9. No x intercepts

11. (A) None (B) $-2\pi, -\pi, 0, \pi, 2\pi$
(C) $-3\pi/2, -\pi/2, \pi/2, 3\pi/2$

13.

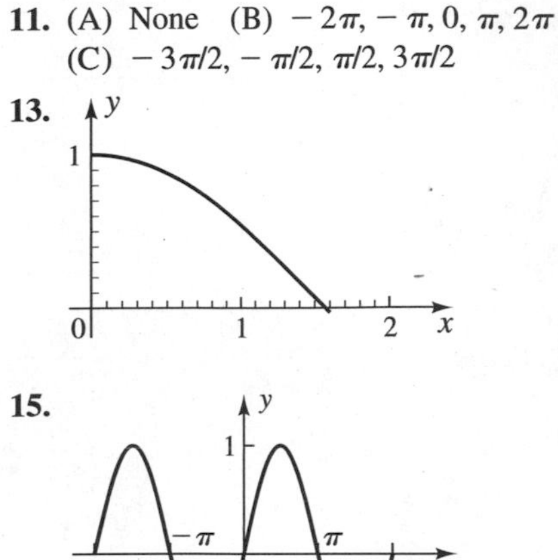

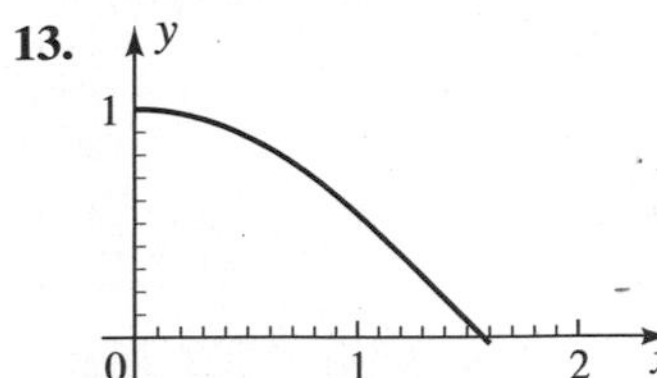

15.

17.

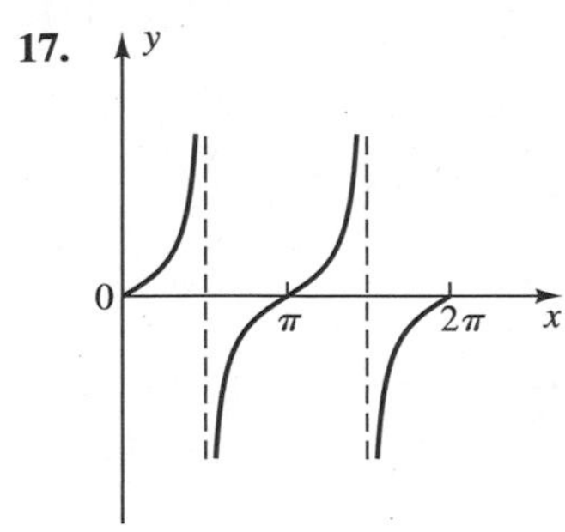

19.

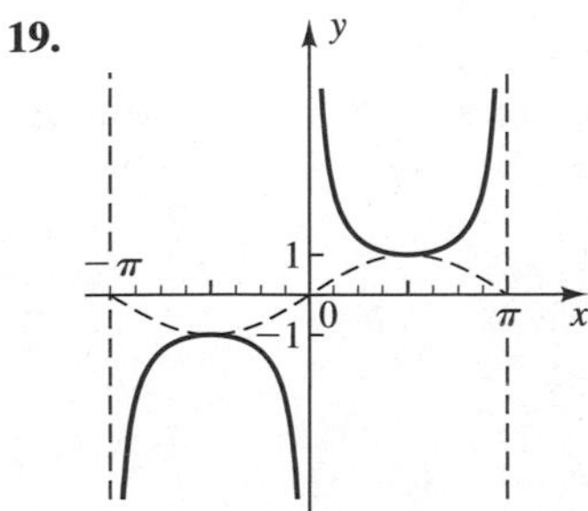

21. (A)

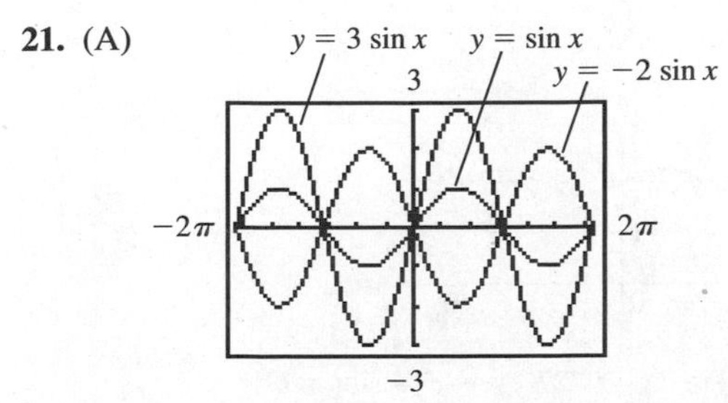

(B) No (C) 2 units; 1 unit; 3 units
(D) The deviation of the graph from the x axis is changed by changing A. The deviation appears to be $|A|$.

23. (A)

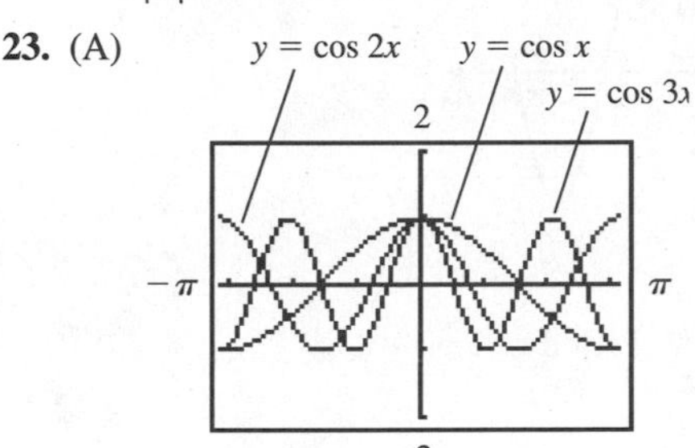

(B) 1; 2; 3 (C) n

25. (A)

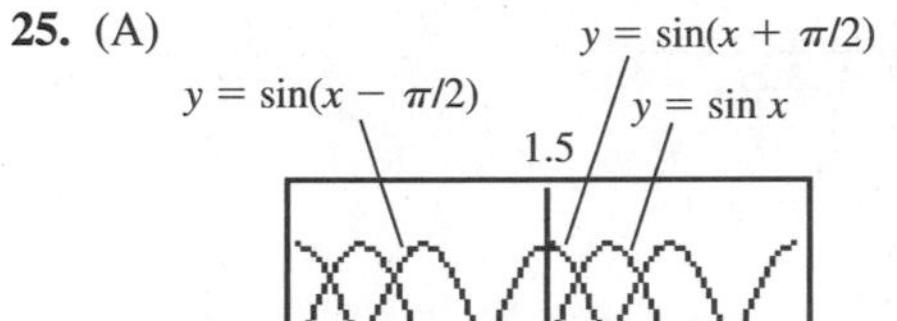

(B) The graph of $y = \sin x$ is shifted $|C|$ units to the right if $C < 0$ and $|C|$ units to the left if $C > 0$.

27. For each case, the number is not in the domain of the function and an error message of some type will appear.

29. (A) The graphs are almost indistinguishable when x is close to the origin.
(B)

x	−0.3	−0.2	−0.1	
tan x	−0.309	−0.203	−0.100	
x	0.0	0.1	0.2	0.3
tan x	0.000	0.100	0.203	0.309

(C) It is not valid to replace $\tan x$ with x for small x if x is in degrees, as is clear from the graph.

31. For a given value of T, the y value on the unit circle and the corresponding y value on the sine curve are the same. This is a graphing utility illustration of how the sine function is defined as a circular function. See Figure 3 in this section.

Exercise 3.2

1.

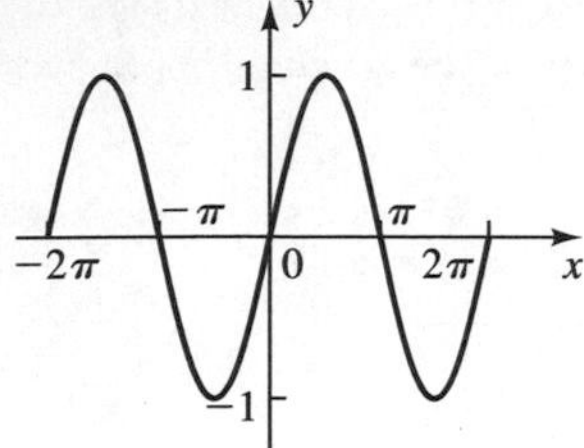

3. Amplitude = 2; Period = 2π

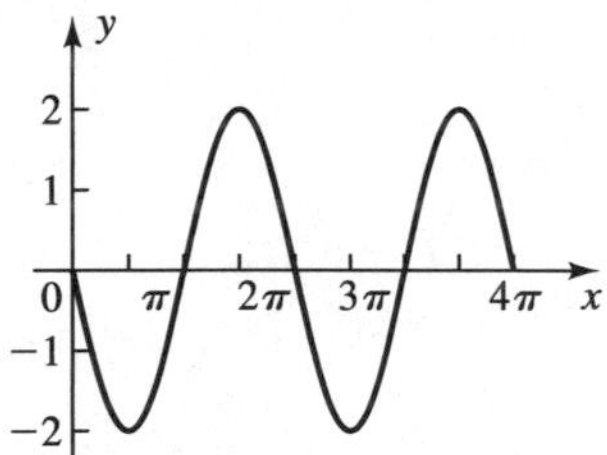

5. Amplitude = $\frac{1}{2}$; Period = 2π

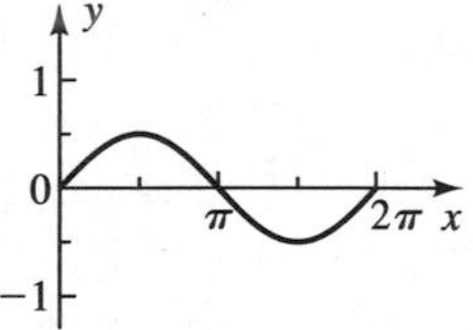

7. Amplitude = 1; Period = 1

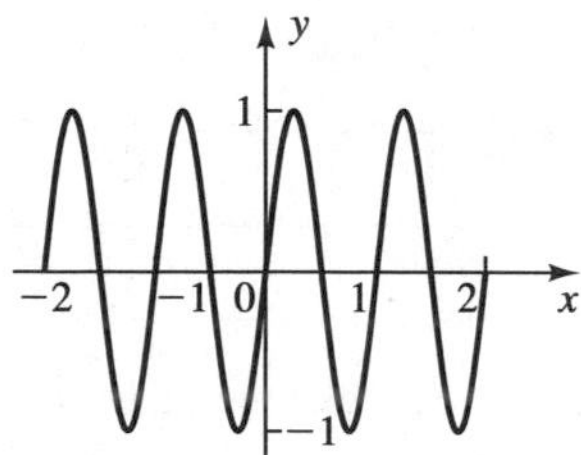

9. Amplitude = 1; Period = 8π

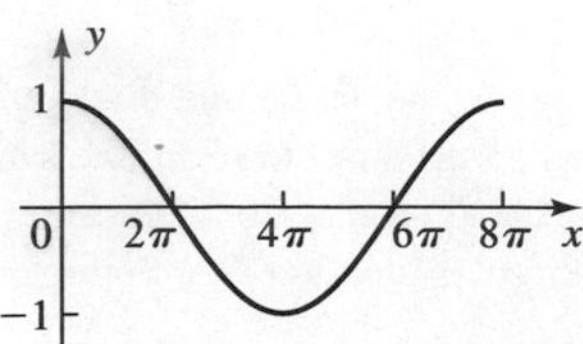

11. Amplitude = 2; Period = $\pi/2$

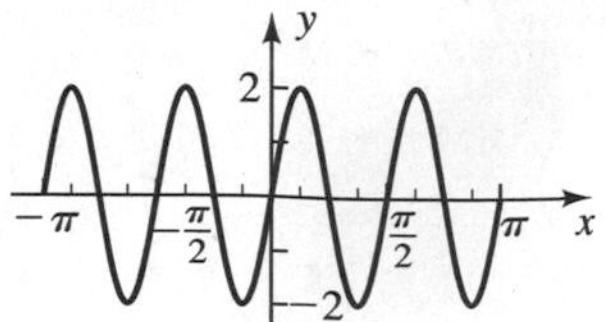

13. Amplitude = $\frac{1}{3}$; Period = 1

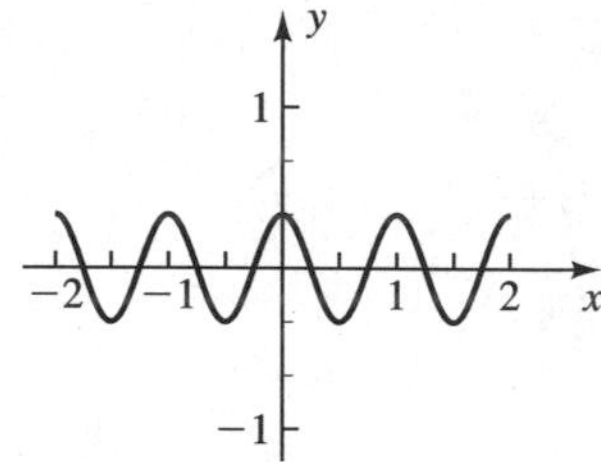

15. Amplitude = $\frac{1}{4}$; Period = 4π

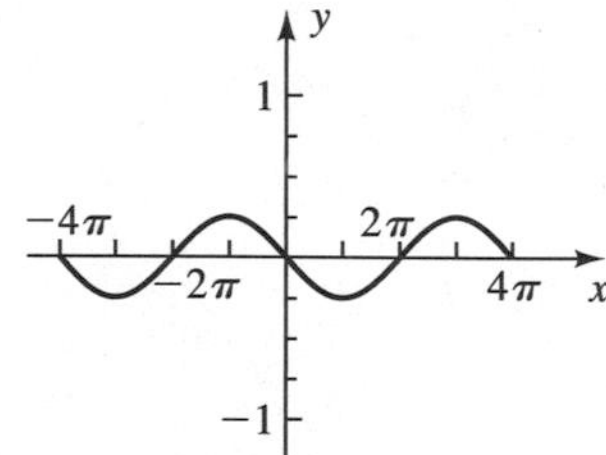

17. $y = 2 \sin \pi t$ or $y = -2 \sin \pi t$

19. Since $P = 2\pi/B$, P approaches 0 as B increases without bound.

21.

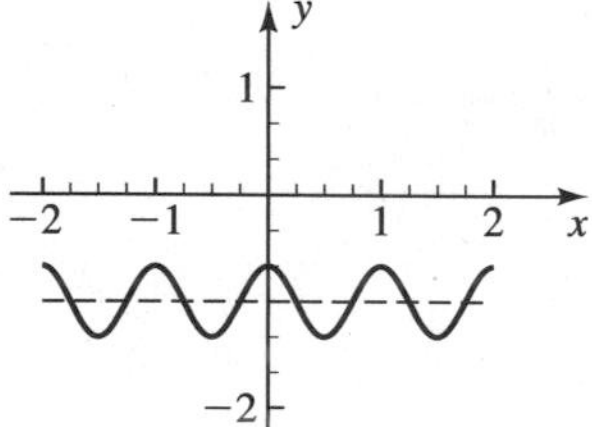

23.

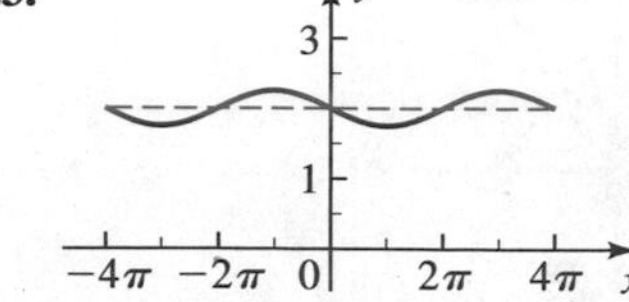

25. $y = 5 \sin 2x$ 27. $y = -4 \sin(\pi x/2)$

29. $y = 8\cos(x/4)$ **31.** $y = -\cos(\pi x/3)$

33. $y = 0.5 \sin 2x$

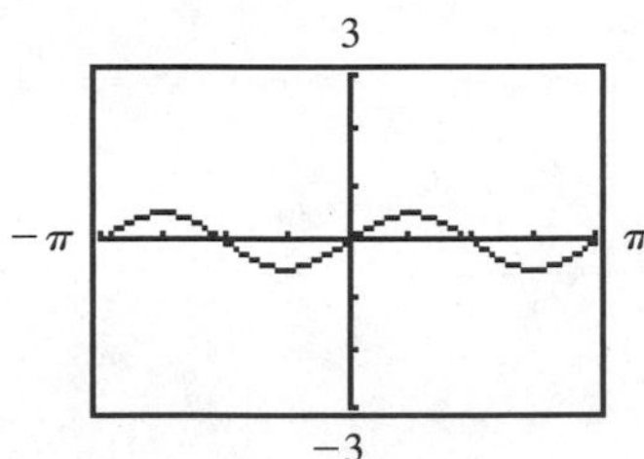

35. $y = 1 + \cos 2x$

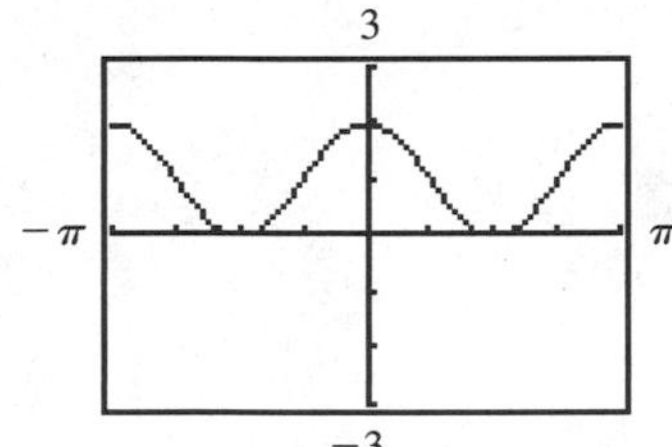

37. $y = 2 \cos 4x$

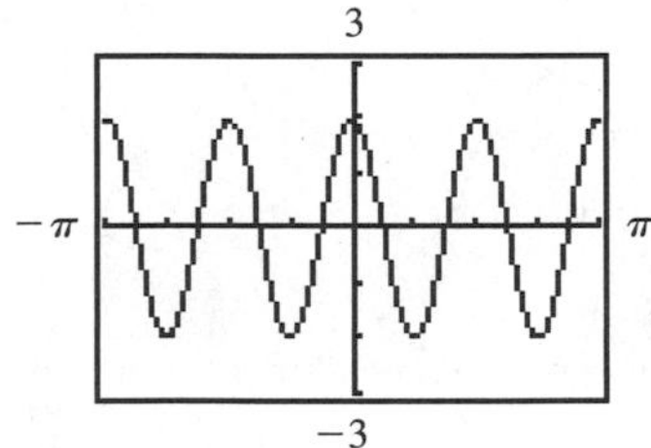

39. (A) $C = 0$ and $-\pi/2$:

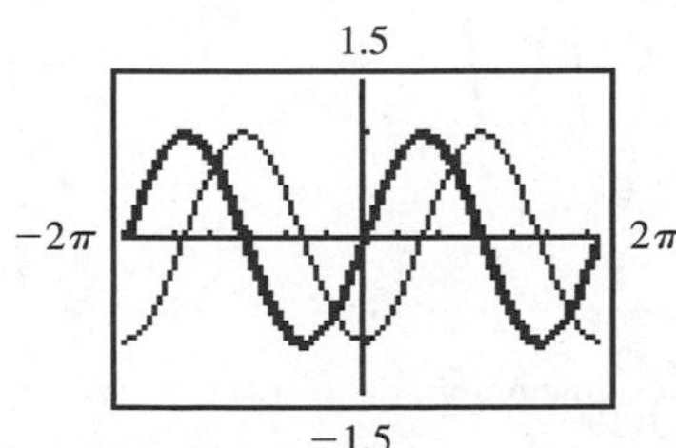

$C = 0$ and $\pi/2$:

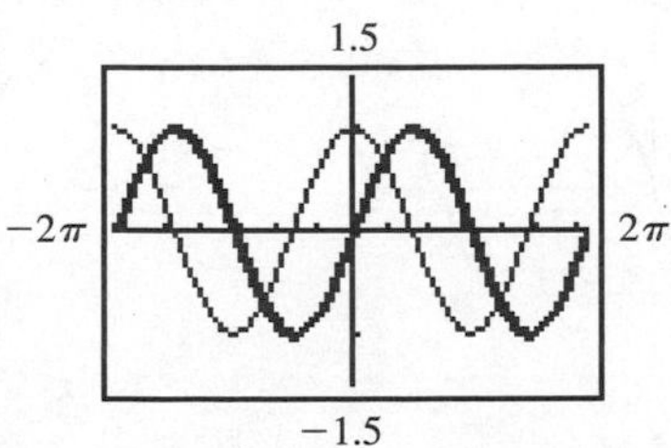

(B) If $C < 0$, then the graph of $y = \sin x$ is shifted $|C|$ units to the right. If $C > 0$, then the graph of $y = \sin x$ is shifted C units to the left.

41. $p = \pi$

43. $y = 2 \cos \dfrac{2\pi x}{3}$

45. Amplitude = 110; Period = $\frac{1}{60}$ sec; Frequency = 60 Hz

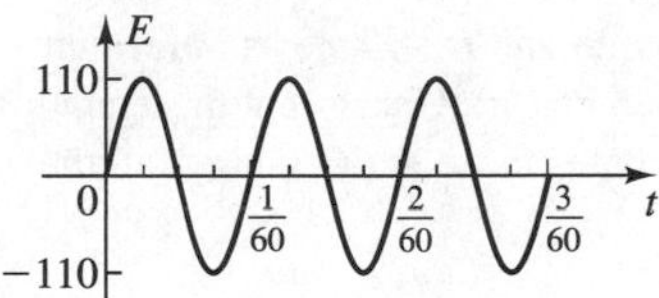

47. $E = 12 \cos 80\pi t$

49. (A) 2,220 kg

(B) $y = 0.2 \sin 2\pi t$

(C)

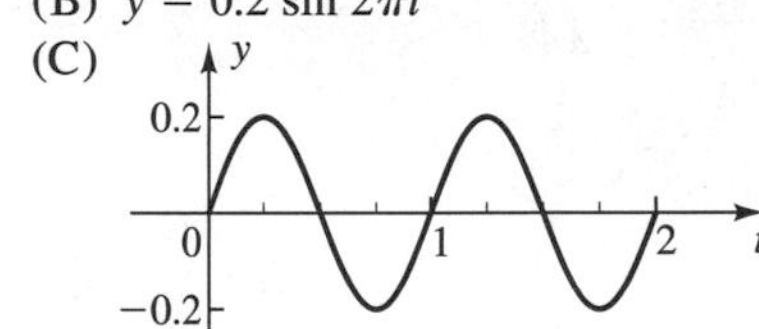

51. (A) Max vol = 0.85 liter; Min vol = 0.05 liter; $0.40 \cos \dfrac{\pi t}{2}$ is maximum when $\cos \dfrac{\pi t}{2}$ is 1 and is minimum when $\cos \dfrac{\pi t}{2}$ is -1. Therefore,

max vol = 0.45 + 0.40 = 0.85 liter and
min vol = 0.45 − 0.40 = 0.05 liter.

(B) 4.00 sec (C) 60/4 = 15 breaths/min

(D) Max vol = 0.85 liter; Min vol = 0.05 liter

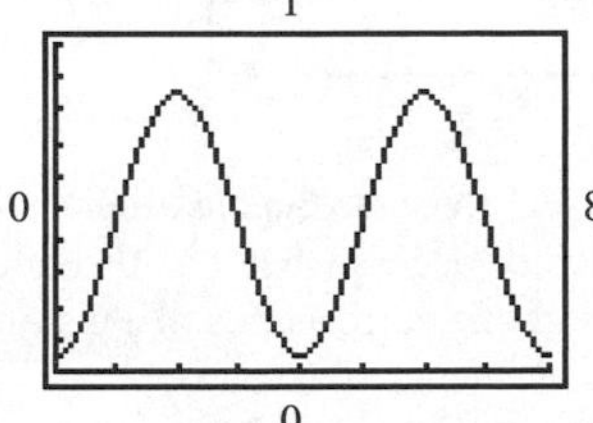

53.

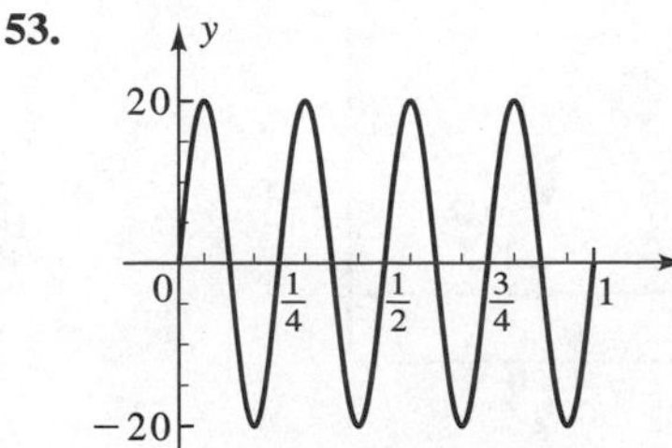

55. (A) The data for θ repeat every 2 sec, so the period is $P = 2$ sec. The angle θ deviates from 0 by 25° in each direction, so the amplitude is $|A| = 25°$.

(B) $\theta = A \sin Bt$ is not suitable, because, for example, when $t = 0$, $A \sin Bt = 0$ no matter what the choice of A and B. $\theta = A \cos Bt$ appears suitable, because, for example, if $t = 0$ and $A = -25$, then we can get the first value in the table—a good start. Choose $A = -25$ and $B = 2\pi/P = \pi$, which yields $\theta = -25 \cos \pi t$. Check that this equation produces (or comes close to producing) all the values in the table.

(C)

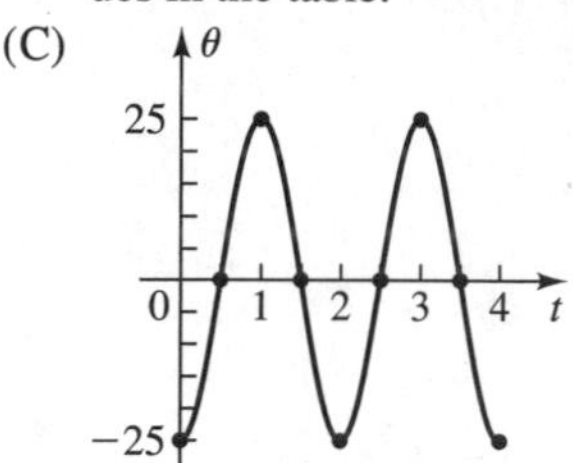

57. (A)

t (sec)	0	3	6	9	12	15	18	21	24
h (ft)	28	13	−2	13	28	13	−2	13	28

(B)

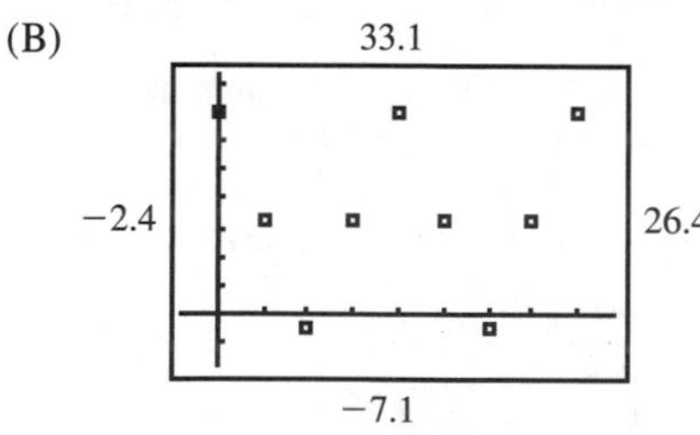

(C) Period = 12 sec. Because the maximum value of $h = k + A \cos Bt$ occurs when $t = 0$ (assuming A is positive), and this corresponds to a maximum value in Table 3 when $t = 0$.

(D) $h = 13 + 15 \cos(\pi t/6)$

(E)

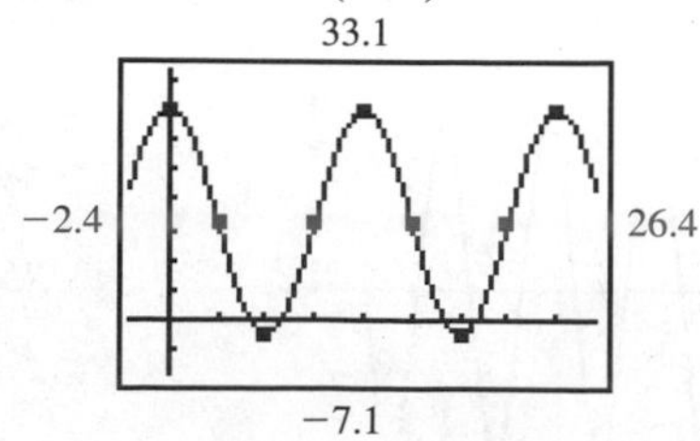

Exercise 3.3

1. The graph with no phase shift is moved 2 units to the left.

3. Phase shift $= -\pi/2$

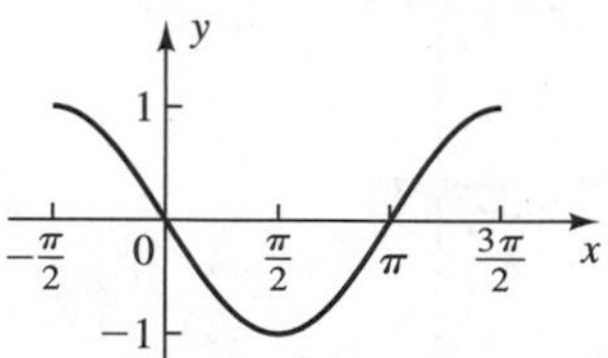

5. Phase shift $= \pi/4$

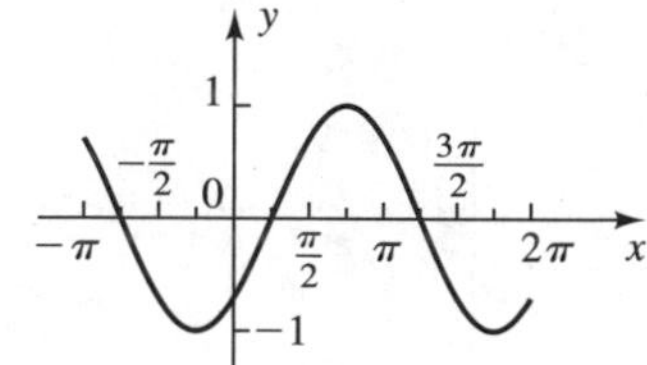

7. Amplitude = 4; Period = 2; Phase shift $= -\frac{1}{4}$

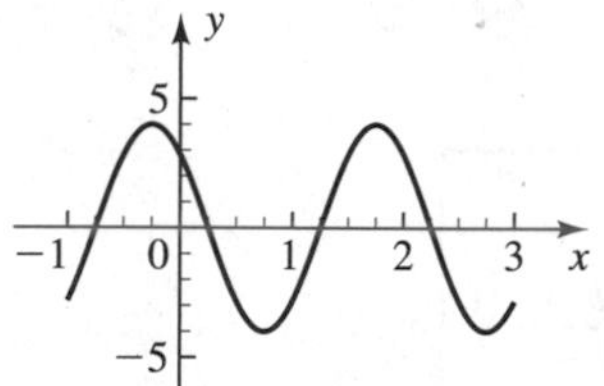

9. Amplitude = 2; Period = π; Phase shift $= -\pi/2$

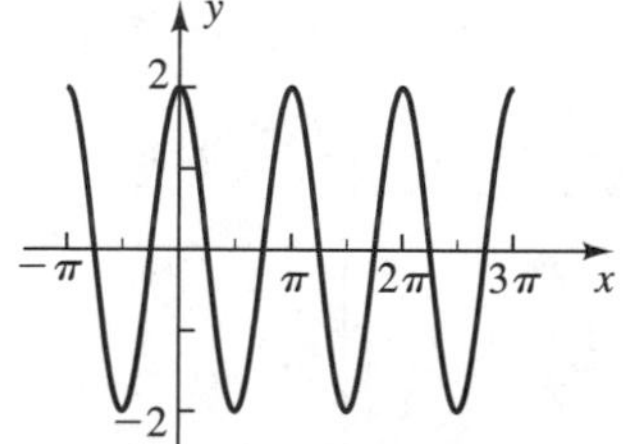

11. Both have the same graph; thus, $\cos(x - \pi/2) = \sin x$ for all x.

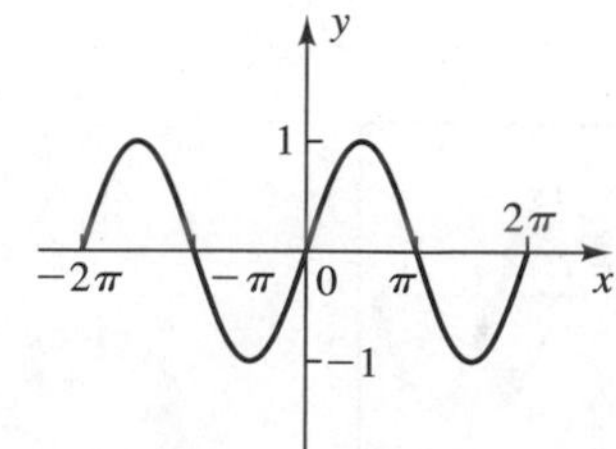

13.

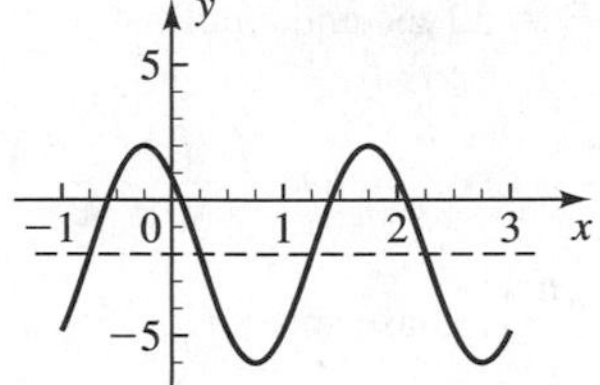

15.

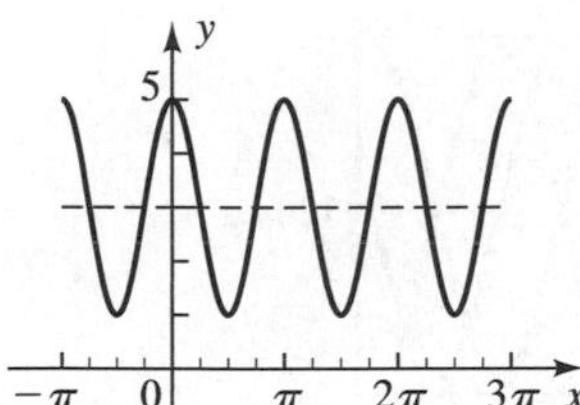

17. (B): The graph of the equation is a sine curve with a period of 2 and a phase shift of $\frac{1}{2}$, which means the sine curve is shifted $\frac{1}{2}$ unit to the right.

19. (A): The graph of the equation is a cosine curve with a period of π and a phase shift of $-\pi/4$, which means the cosine curve is shifted $-\pi/4$ unit to the left.

21. $y = 5\sin(\pi x/2 - \pi/2)$ **23.** $y = -2\cos(x/2 + \pi/4)$

25. Amplitude = 2; Period = $2\pi/3$; Phase shift = $\pi/6$

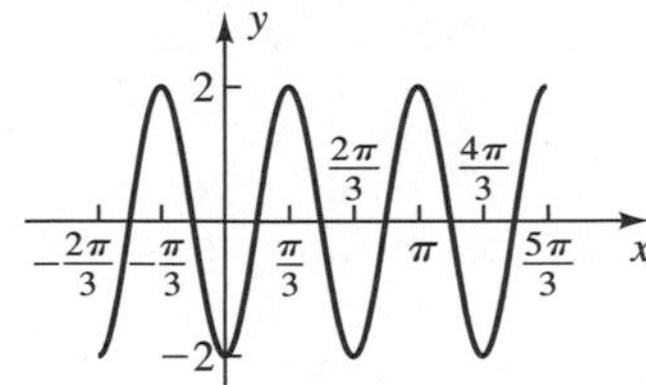

27.

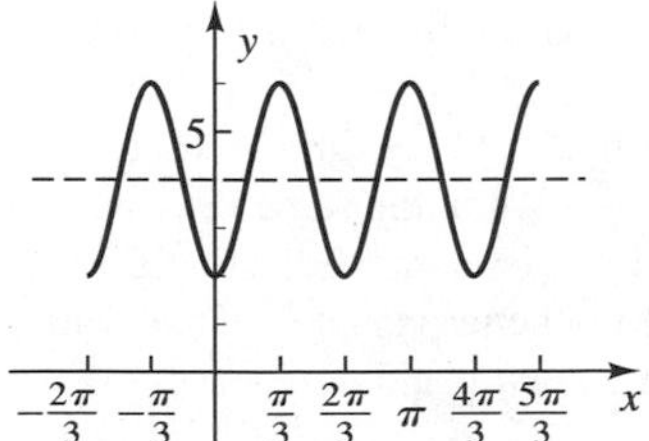

29. Amplitude = 2.3; Period = 3; Phase shift = 2

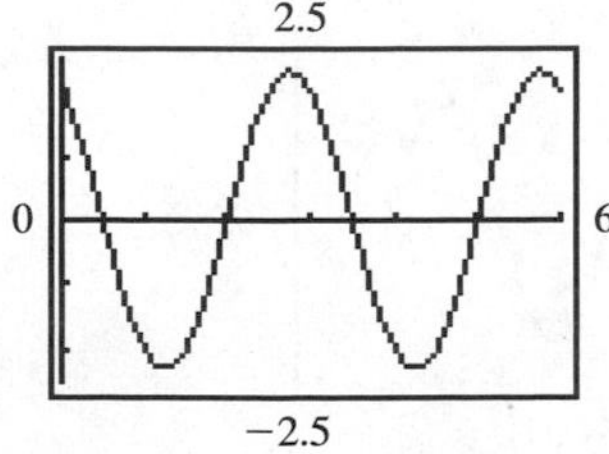

31. Amplitude = 18; Period = 0.5; Phase shift = -0.137

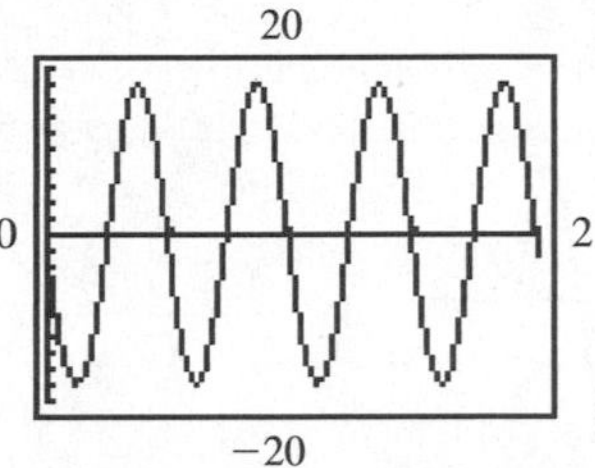

33. x intercept: -1.047; $y = 2\sin(x + 1.047)$

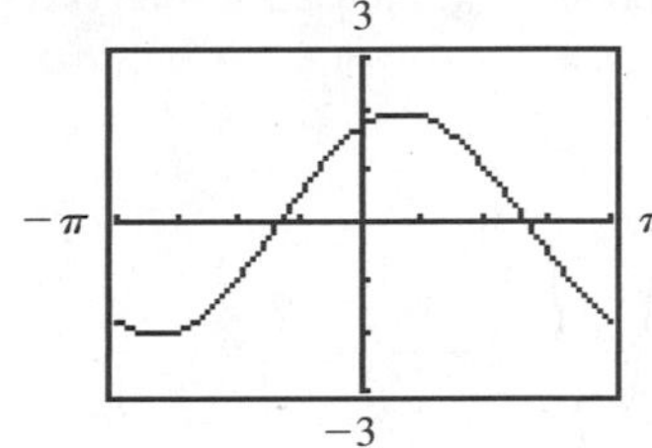

35. x intercept: 0.785; $y = 2\sin(x - 0.785)$

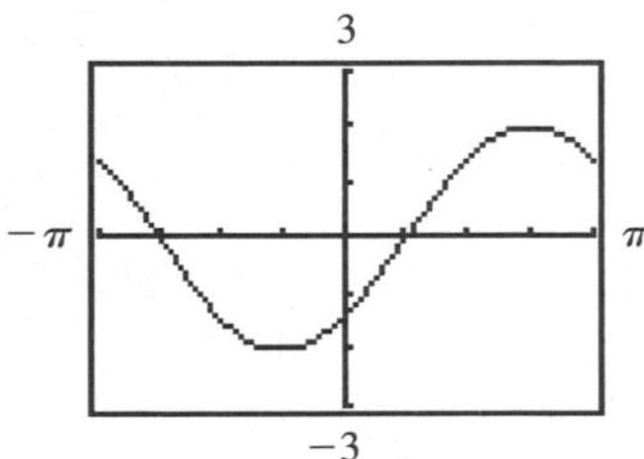

37. x intercept: -0.644; $y = 5\sin(2x + 1.288)$

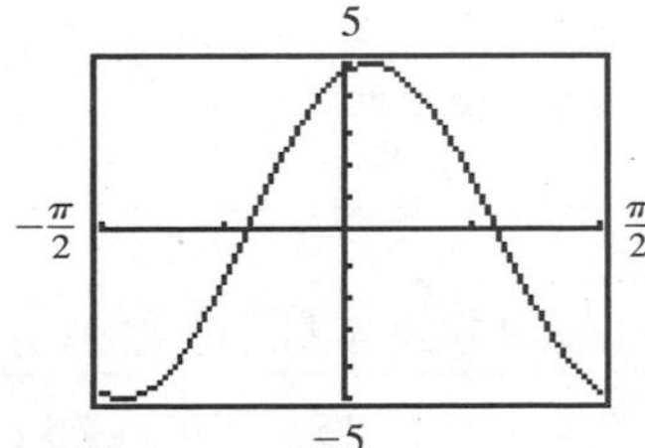

39. x intercept: 1.682; $y = 3\sin(x/2 - 0.841)$

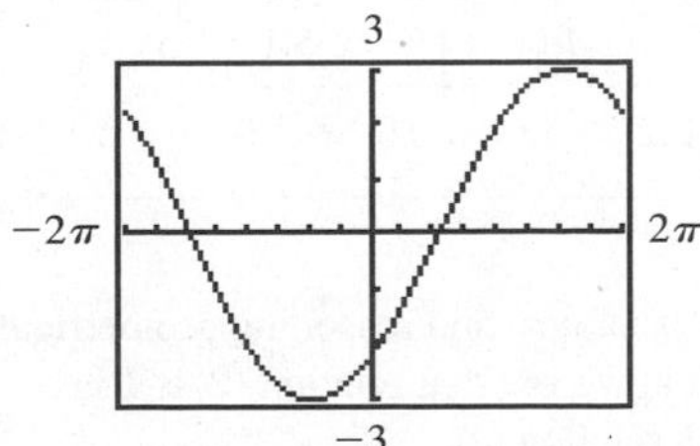

41. Amplitude = 5 m; Period = 12 sec;
Phase shift = − 3 sec

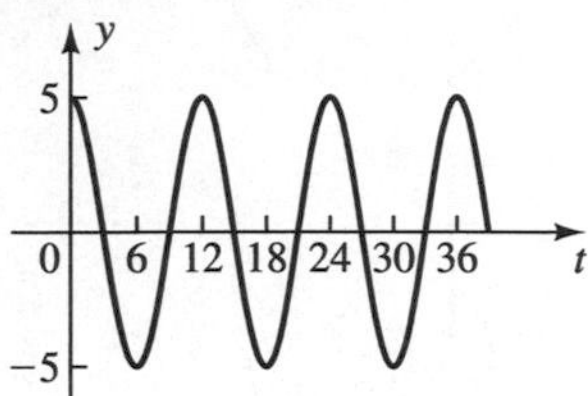

43. Amplitude = 30; Period = $\frac{1}{60}$; Frequency = 60 Hz;
Phase shift = $\frac{1}{120}$

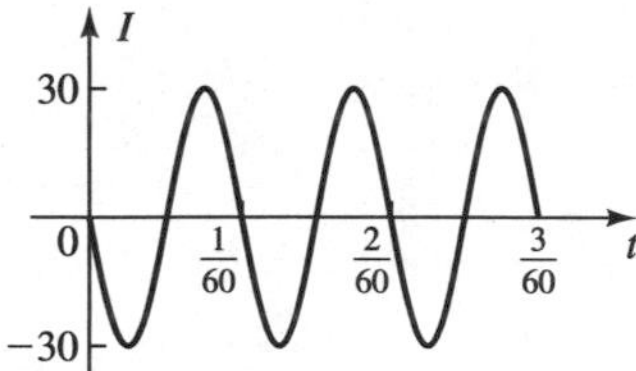

45. (A)

TABLE 1
Distance *d*, *t* seconds after the second hand points to 12

t (sec)	0	5	10	15	20	25	30
d (in.)	0.0	3.0	5.2	6.0	5.2	3.0	0.0
t (sec)	35	40	45	50	55	60	
d (in.)	− 3.0	− 5.2	− 6.0	− 5.2	− 3.0	0.0	

TABLE 2
Distance *d*, *t* seconds after the second hand points to 9

t (sec)	0	5	10	15	20	25	
d (in.)	− 6.0	− 5.2	− 3.0	0.0	3.0	5.2	
t (sec)	30	35	40	45	50	55	60
d (in.)	6.0	5.2	3.0	0.0	− 3.0	− 5.2	− 6.0

(B) From the table values, and reasoning geometrically from the clock, we see that relation (2) is 15 sec out of phase with relation (1).

(C) Both have period $P = 60$ sec and amplitude $|A| = 6$.

(D) (1) $d = 6\sin\left(\frac{\pi t}{30}\right)$; (2) $d = 6\sin\left(\frac{\pi t}{30} - \frac{\pi}{2}\right)$

(E) $d = 6\sin\left(\frac{\pi t}{30} + \frac{\pi}{2}\right)$; Phase shift = − 15 sec

(F)

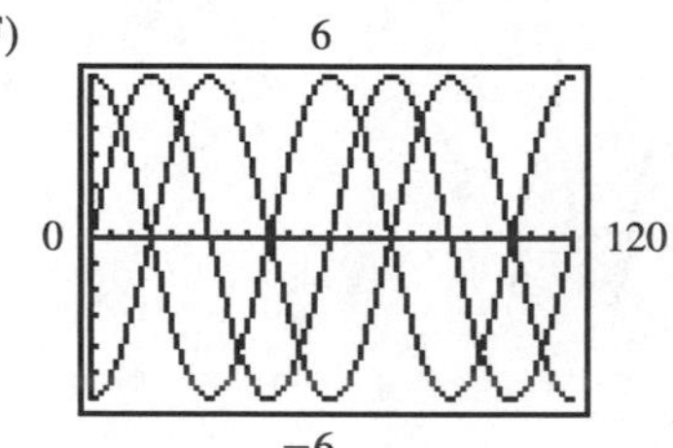

47. (A)

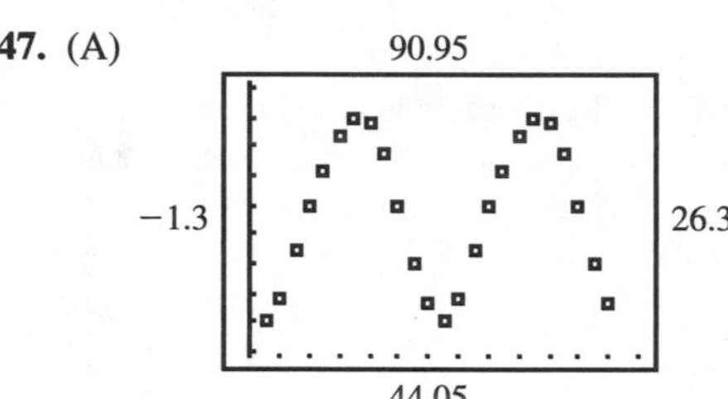

(B) $|A| = (\text{Max } y - \text{Min } y)/2 = (85 - 50)/2 = 17.5$; Period = 12 months, therefore, $B = 2\pi/P = \pi/6$; $k = |A| + \text{Min } y = 17.5 + 50 = 67.5$. From the scatter plot it appears that if we use $A = 17.5$, then the phase shift will be about 3.0, which can be adjusted for a better visual fit later, if necessary. Thus, using Phase shift = $-C/B$, $C = -(\pi/6)(3.0) = -1.6$, and $y = 67.5 + 17.5\sin(\pi t/6 - 1.6)$. Graphing this equation in the same viewing window as the scatter plot, we see that adjusting C to -2.1 produces a slightly better visual fit. Thus, with this adjustment, the equation and graph are

$$y = 67.5 + 17.5\sin(\pi t/6 - 2.1)$$

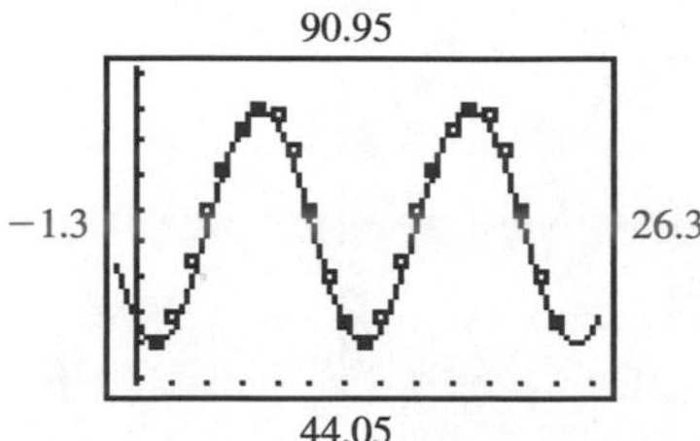

(C) $y = 68.7 + 17.1 \sin(0.5t - 2.1)$

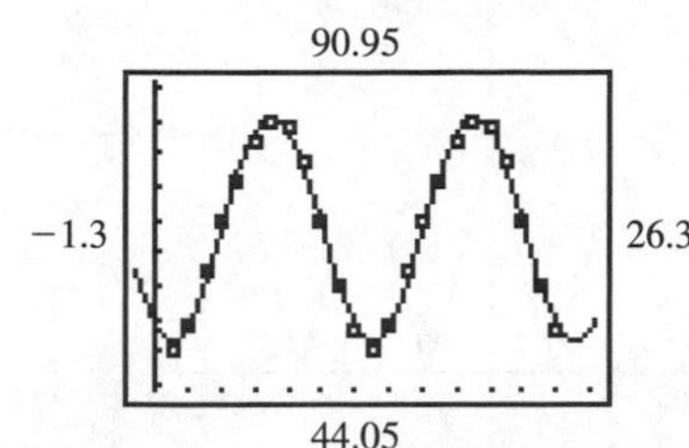

(D) The regression equation differs slightly in k, A, and B, but not in C. Both appear to fit the data very well.

Exercise 3.4

1. (A) Amplitude = 10 amperes; Frequency = 60 Hz; Phase shift = 1/240 sec

(B) The maximum current is the amplitude, which is 10 amperes.

(C) Six periods

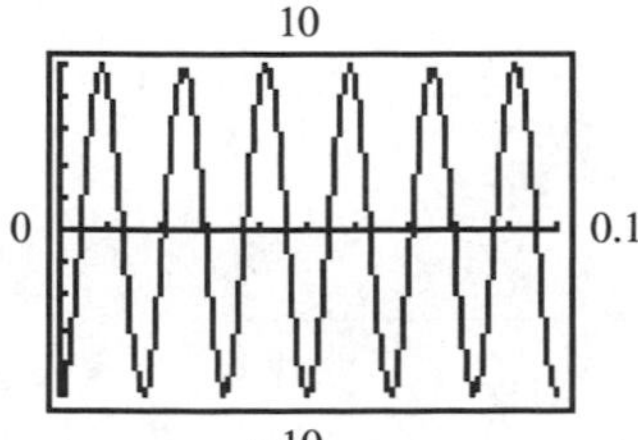

3. $I = 20 \cos 60\pi t$ **5.** 30 ft; 1,311 ft; 82 ft/sec

7.

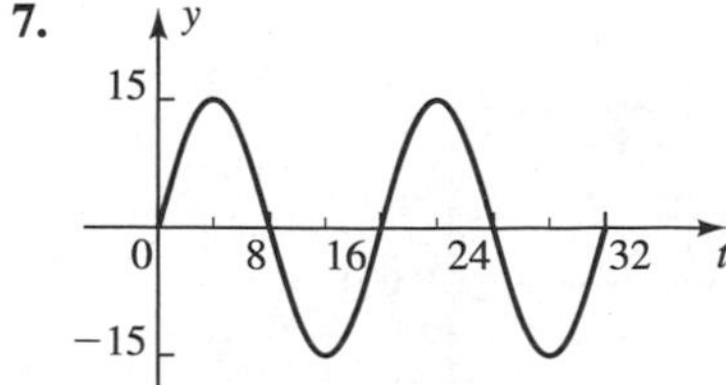

9. (A) $y = 2 \sin \frac{\pi}{75} r$ (B) $T = 393$ sec

11. (A) The equation models the vertical motion of the wave at the fixed point $r = 1{,}024$ ft from the source relative to time in seconds.

(B) The appropriate choice is period, since period is defined in terms of time and wavelength is defined in terms of distance. Period $T = 10$ sec.

(C)

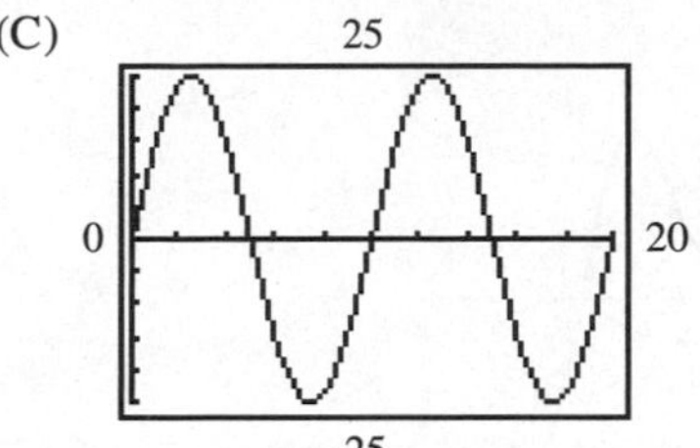

13. Period $= 10^{-8}$ sec; $\lambda = 3$ m **15.** $B = 2\pi \times 10^{18}$

17. Period $= 10^{-6}$ sec; Frequency $= 10^{6}$ Hz; no

19. (A) Simple harmonic

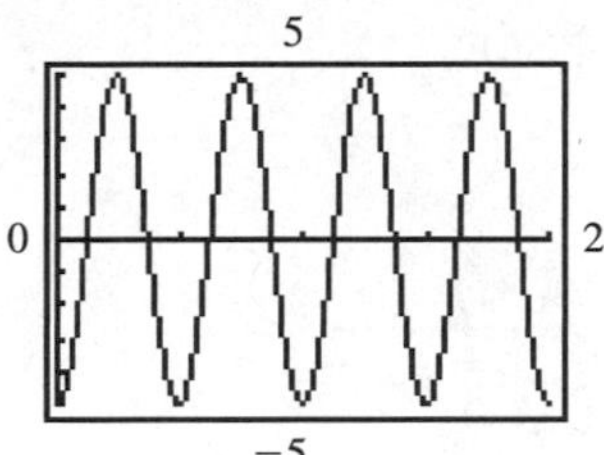

(B) Damped harmonic

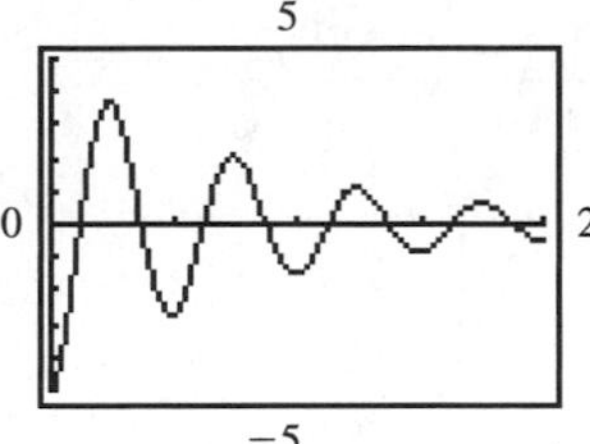

Exercise 3.5

1.

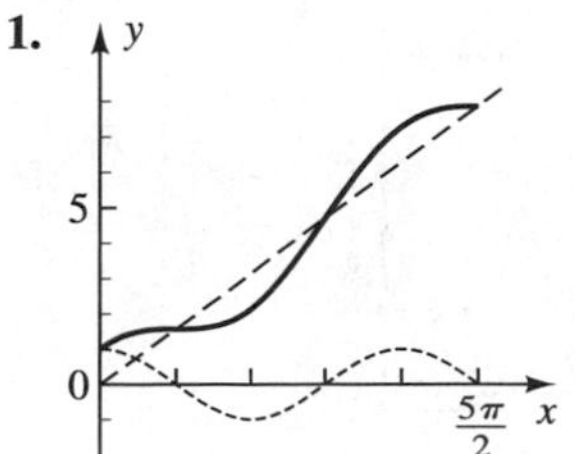

3.

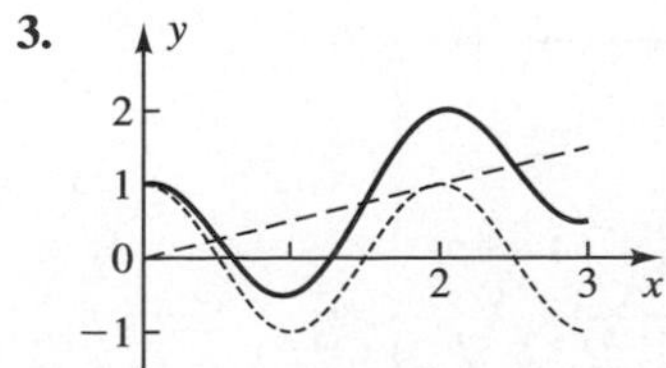

5.

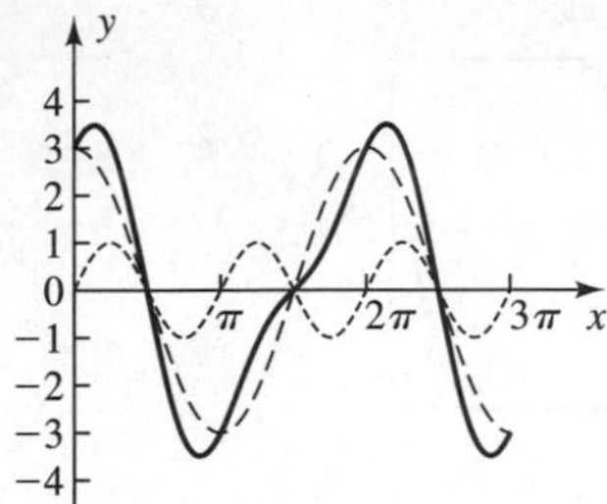

7.

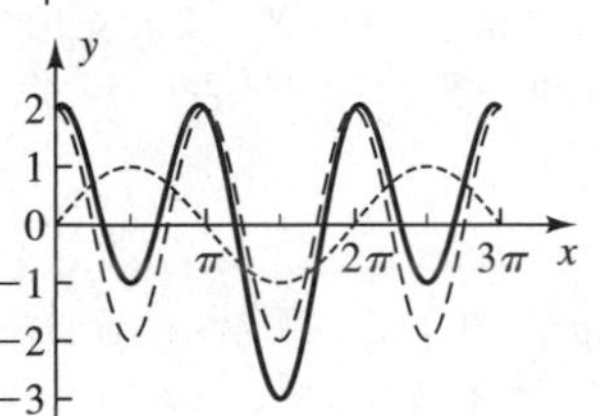

9.

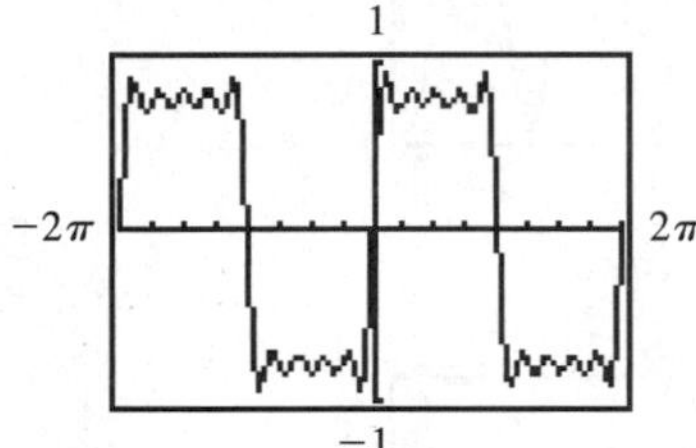

11. (A)

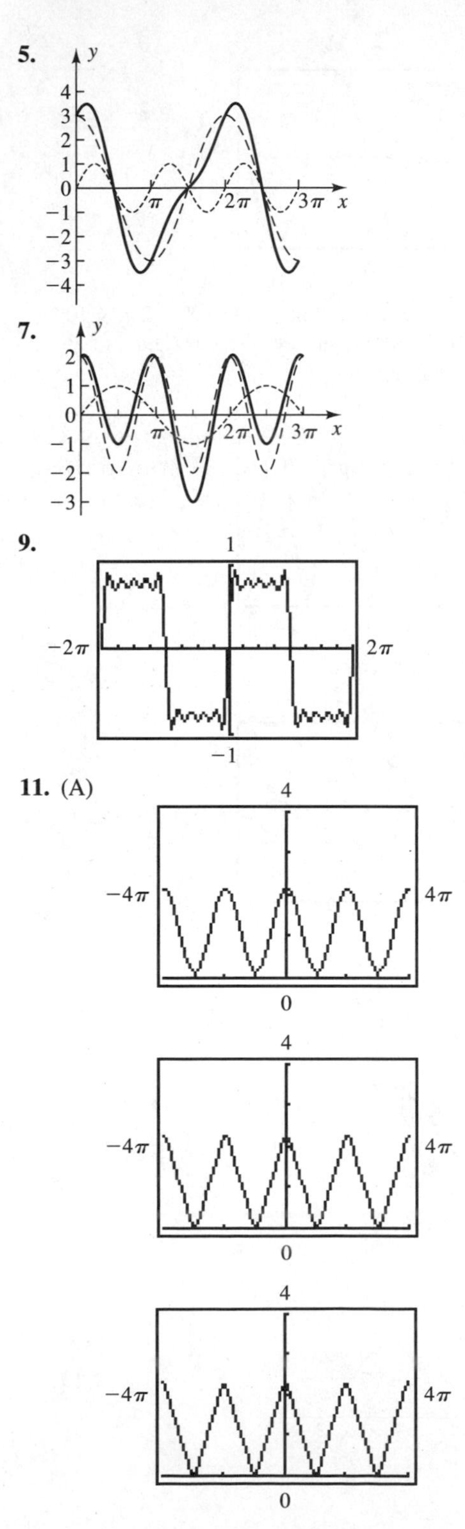

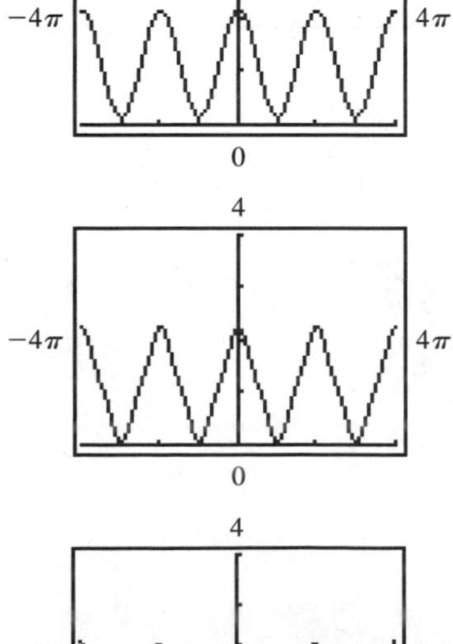

(B) Sawtooth wave form

(C)

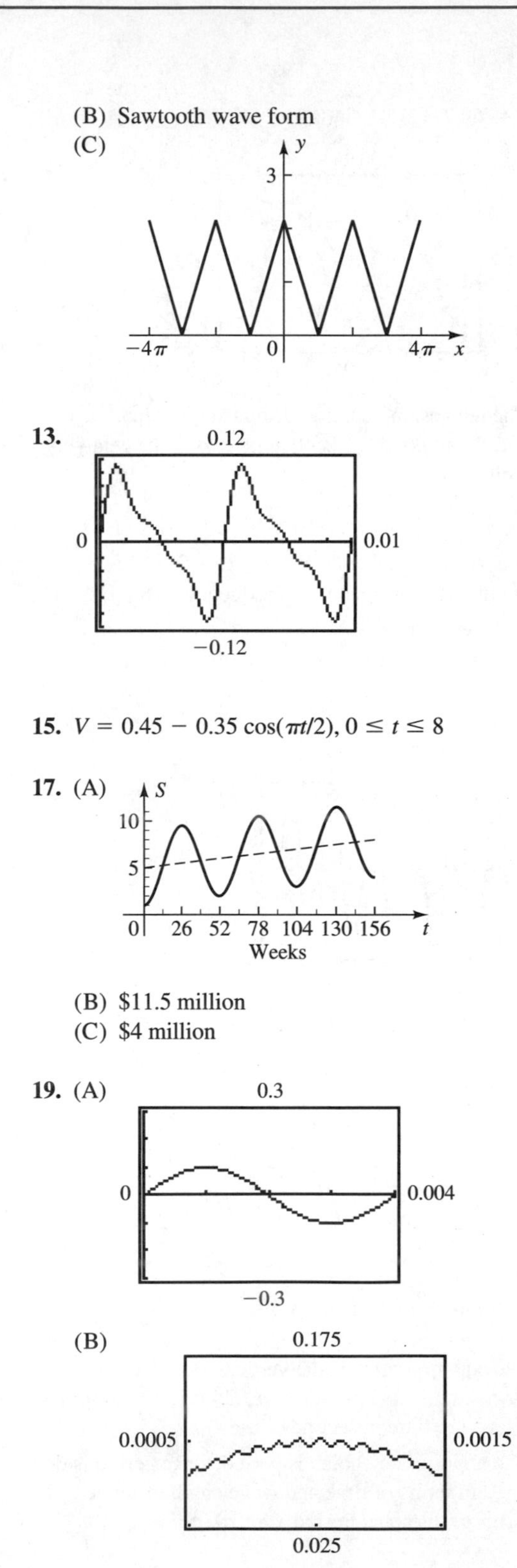

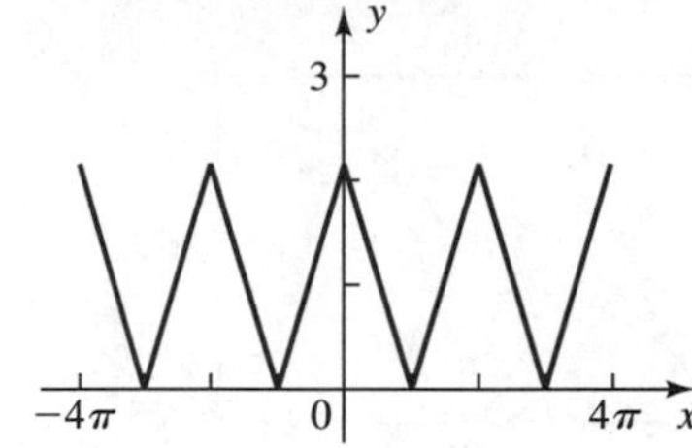

13.

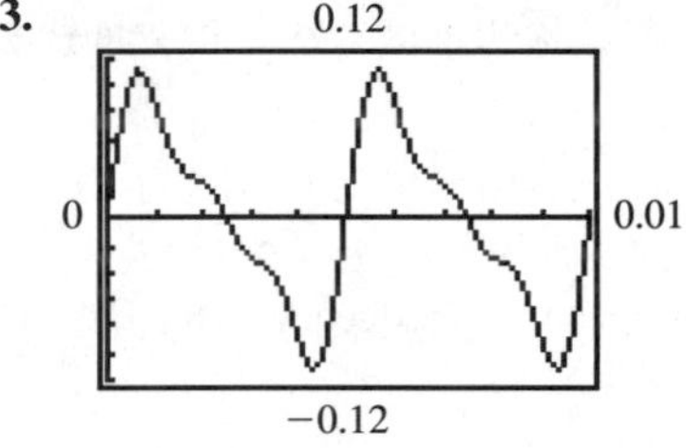

15. $V = 0.45 - 0.35\cos(\pi t/2),\ 0 \le t \le 8$

17. (A)

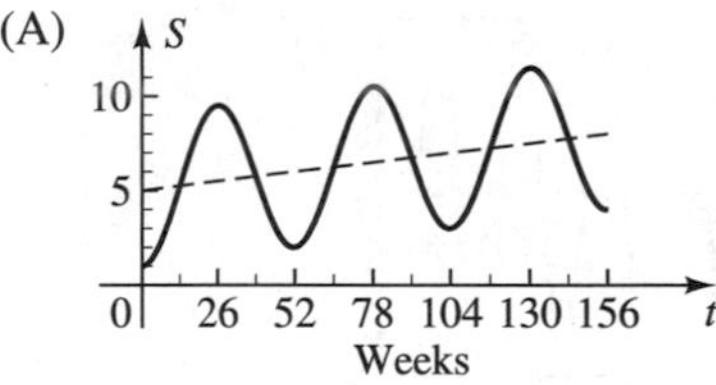

(B) \$11.5 million

(C) \$4 million

19. (A)

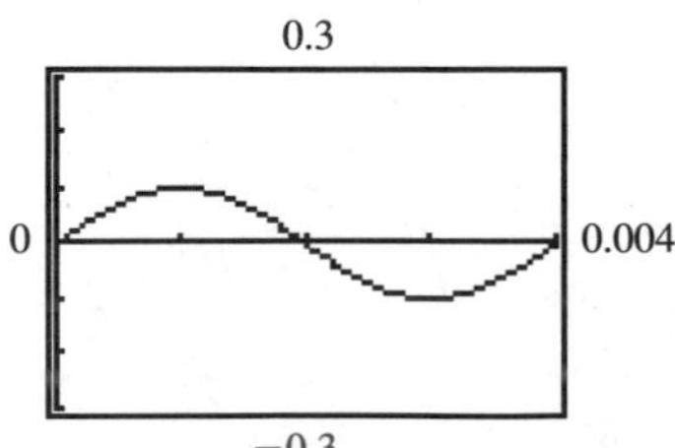

(B)

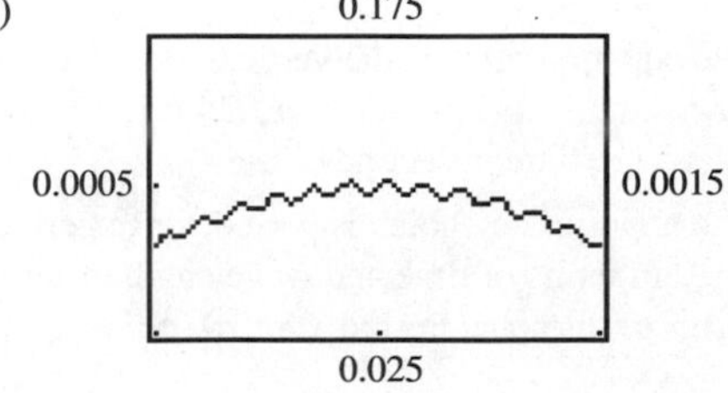

21. (A) Constructive

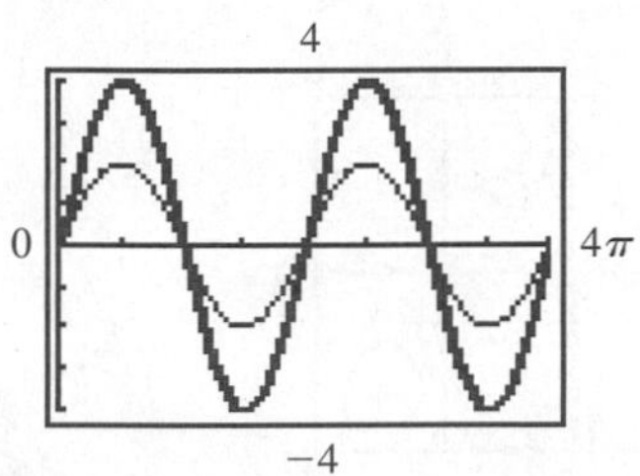

(B) Constructive

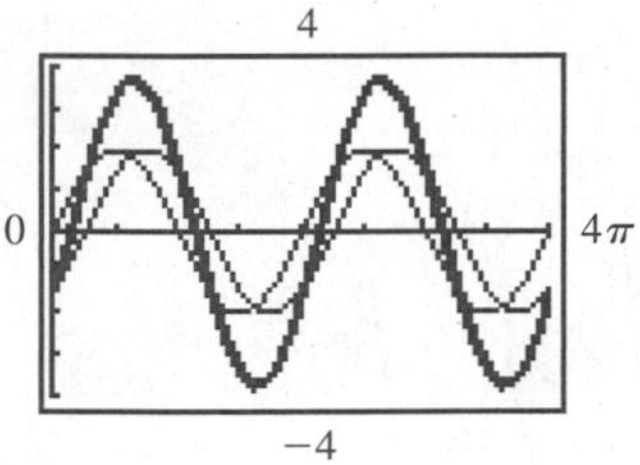

(C) Destructive

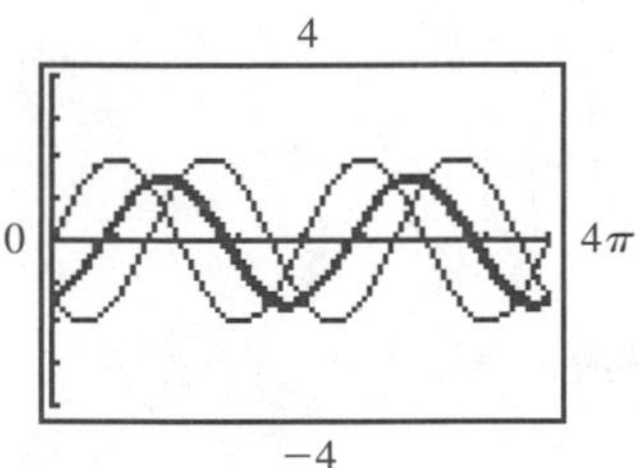

(D) Destructive

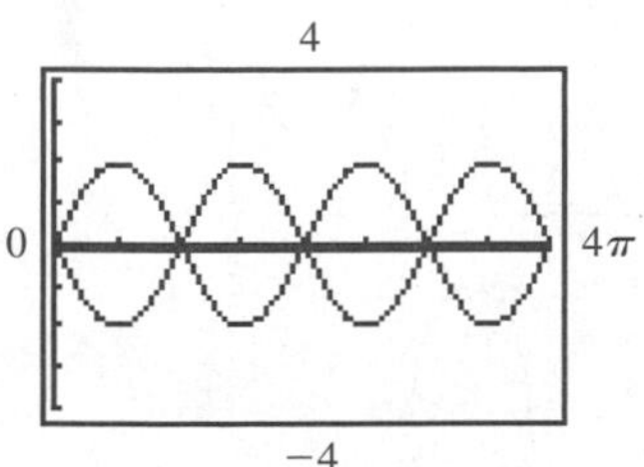

23. (A) $y_2 = 65 \sin(400\pi t - \pi)$

(B) y_2 added to y_1 produces a sound wave of 0 amplitude—no noise.

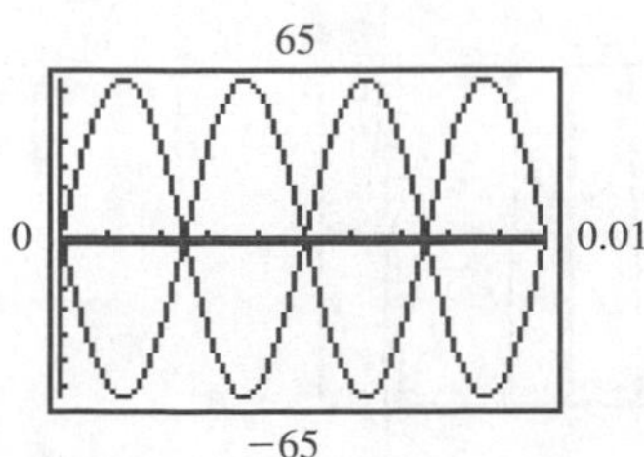

Exercise 3.6

1.

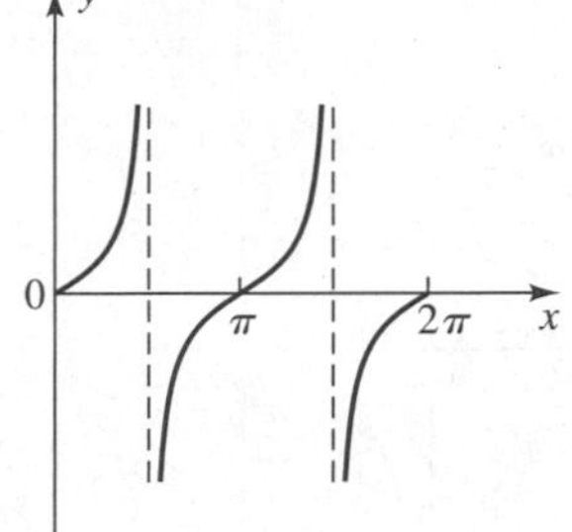

3.

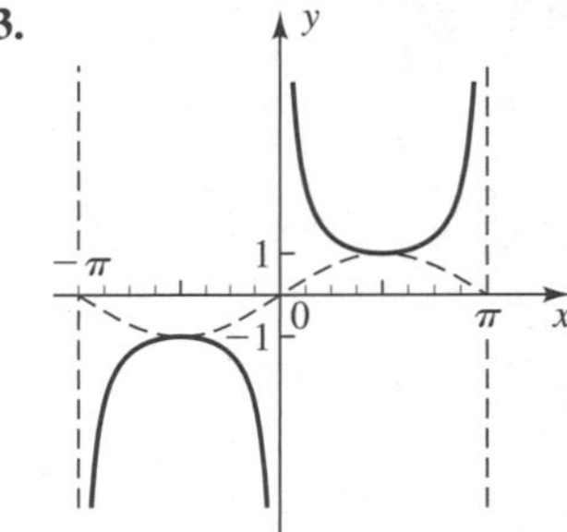

5. Period = $\pi/2$

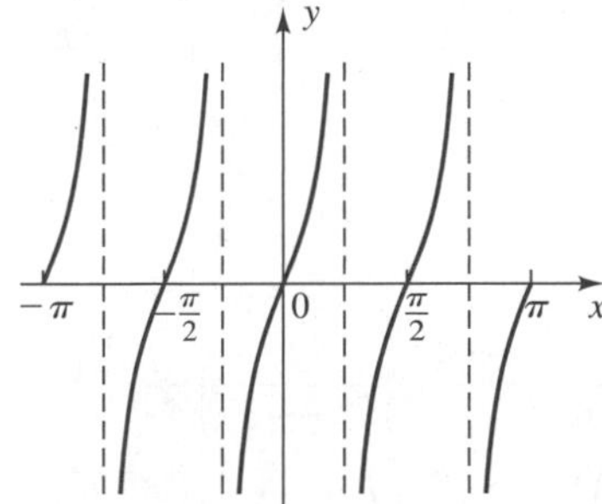

7. Period = 2π

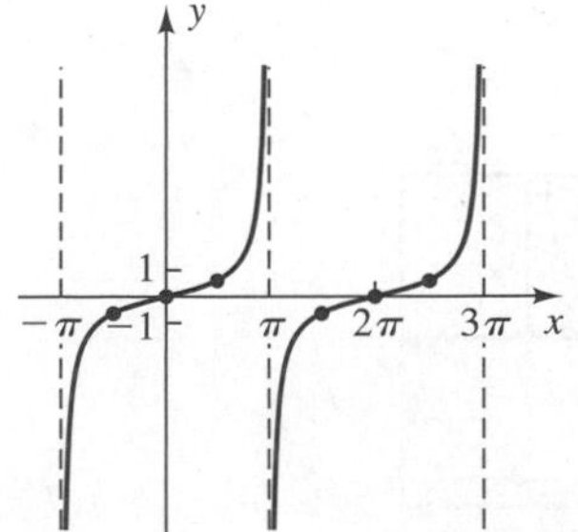

9. Period $= 4\pi$

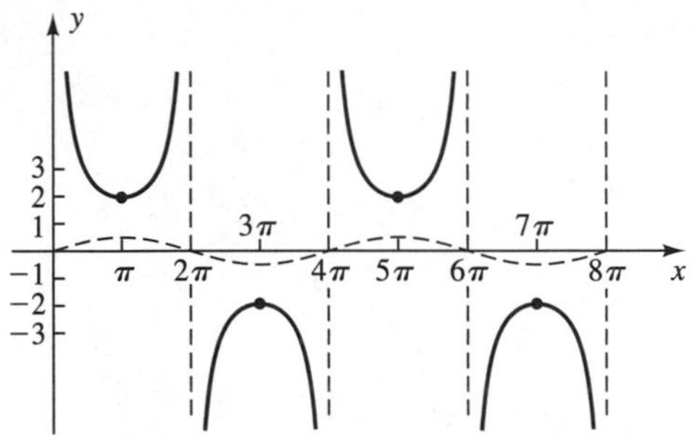

11. Period $= \pi/2$; Phase shift $= \pi/2$

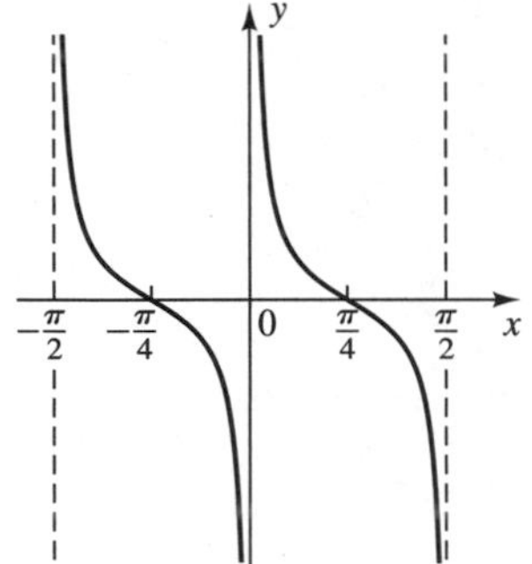

13. Period $= 2$; Phase shift $= \frac{1}{2}$

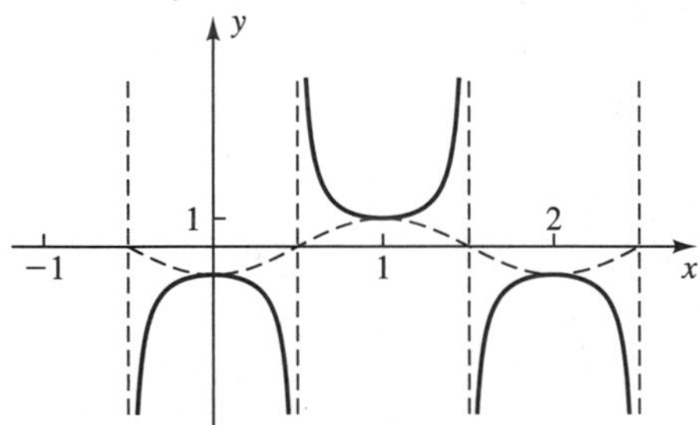

15. $y = \tan \dfrac{x}{2}$

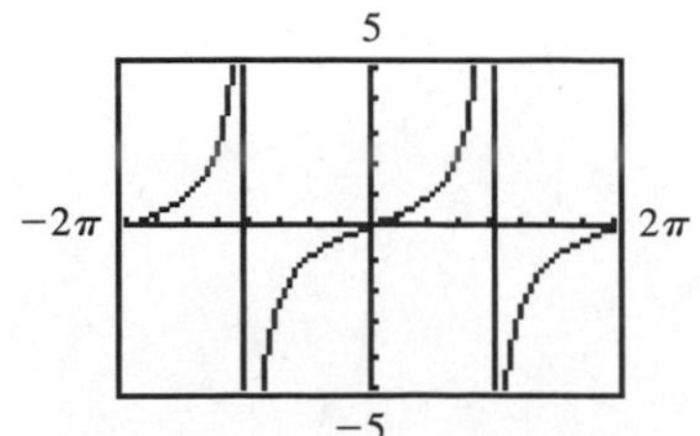

17. $y = 2 \csc 2x$

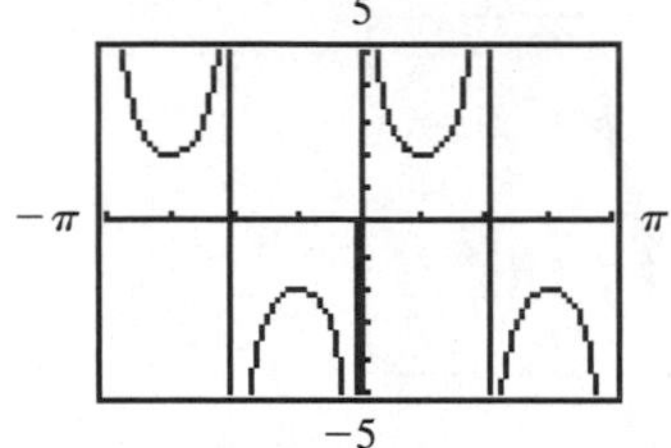

19. Period $= 1$; Phase shift $= 1$

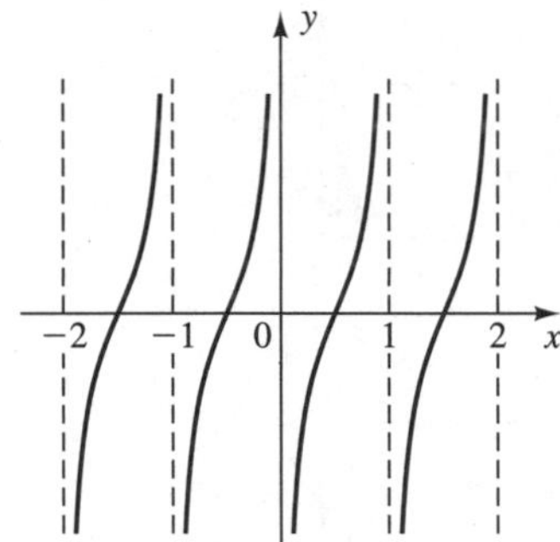

21. Period $= 2$; Phase shift $= \frac{1}{2}$

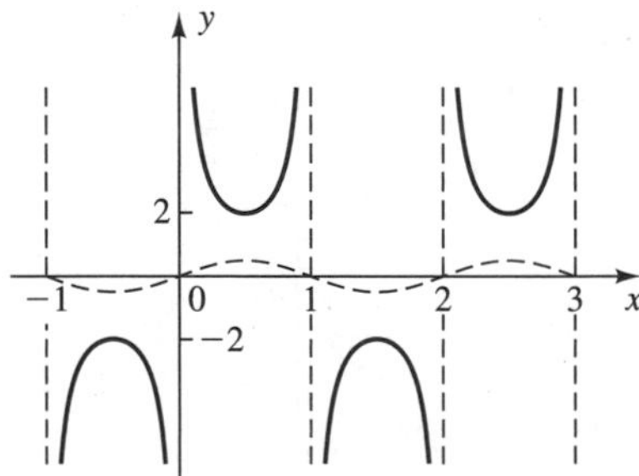

23. $y = \sec 2x$

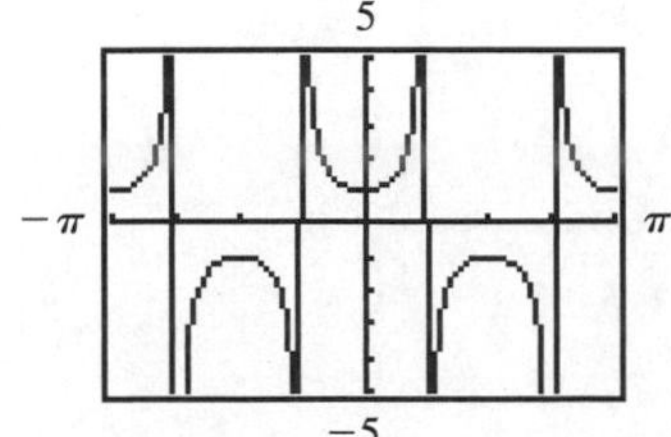

25. $y = \cot 3x$

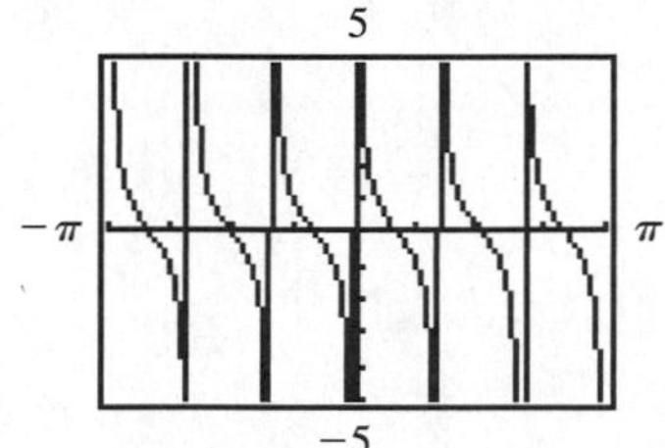

27. (A) $a = 15 \tan 2\pi t$
(B)

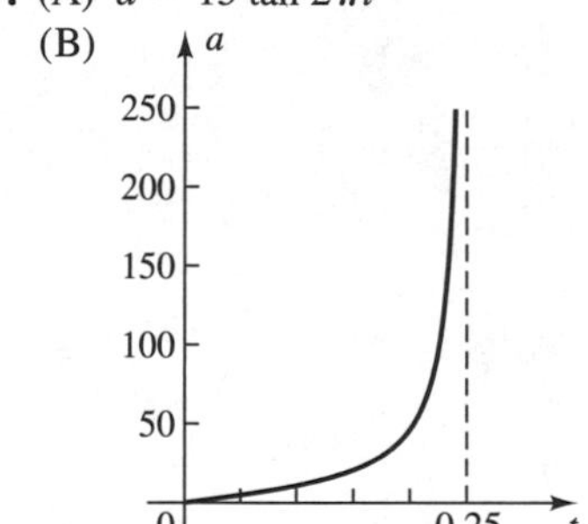

(C) a increases without bound.

Chapter 3 Review Exercise

1.

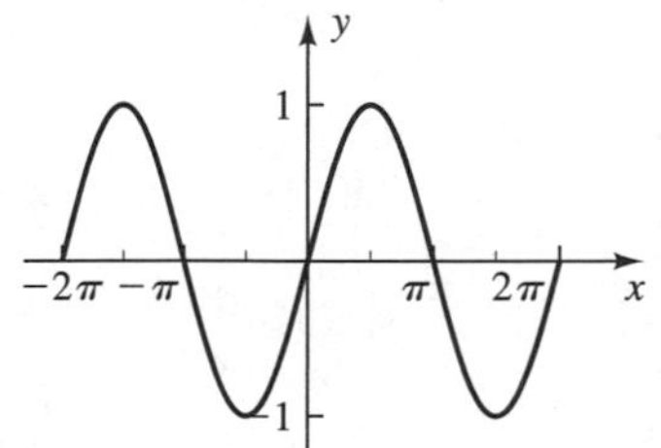

[3.1]

2.

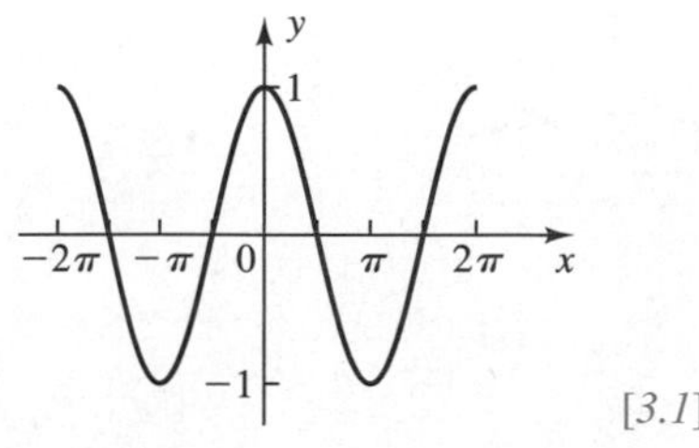

[3.1]

3.

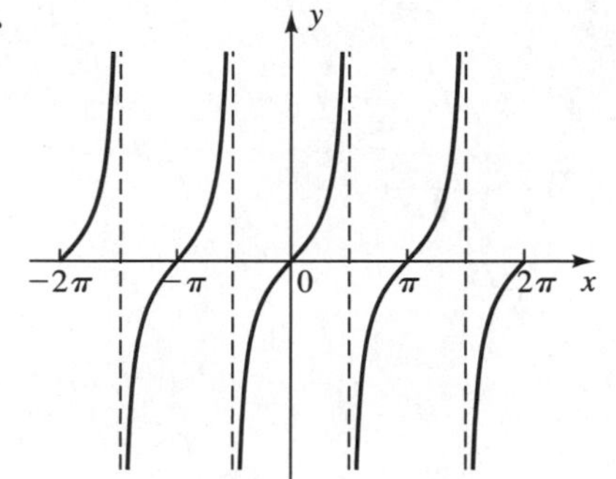

[3.1]

4.

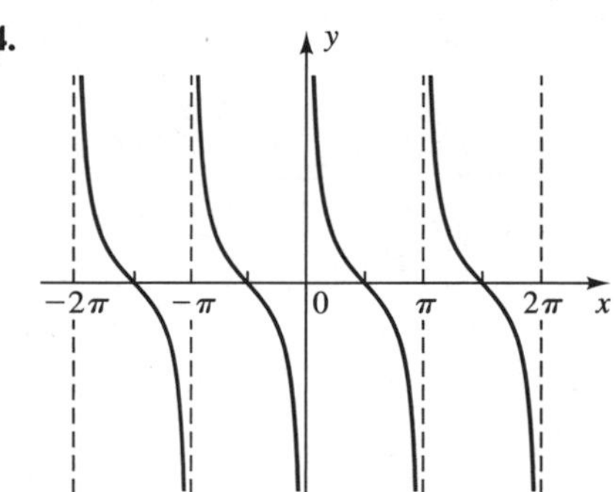

[3.1]

5.

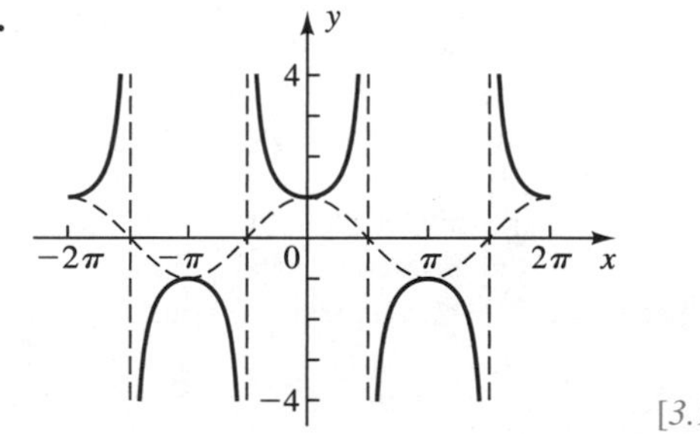

[3.1]

6.

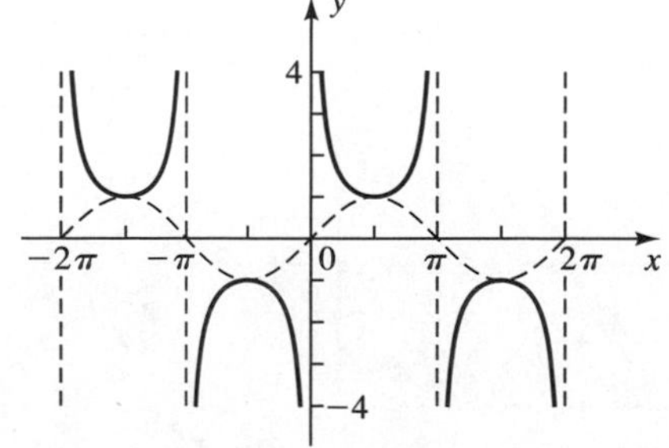

[3.1]

7.

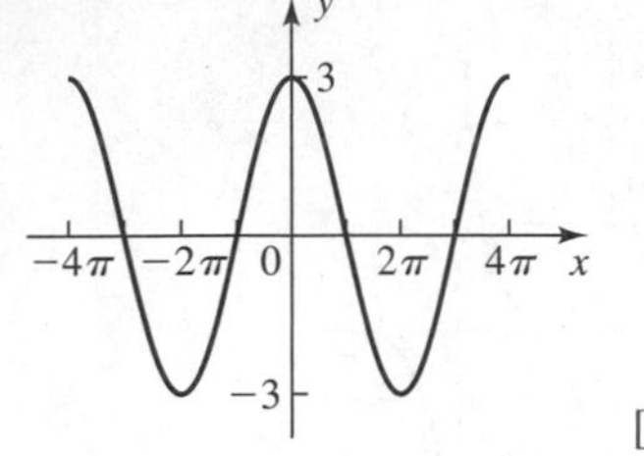

[3.2]

8.

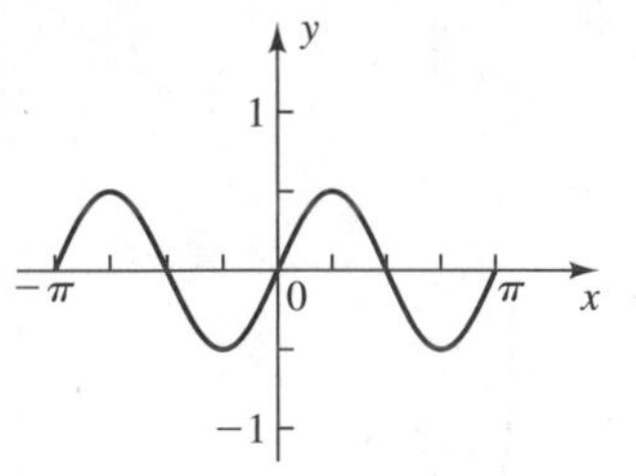

[3.2]

9.

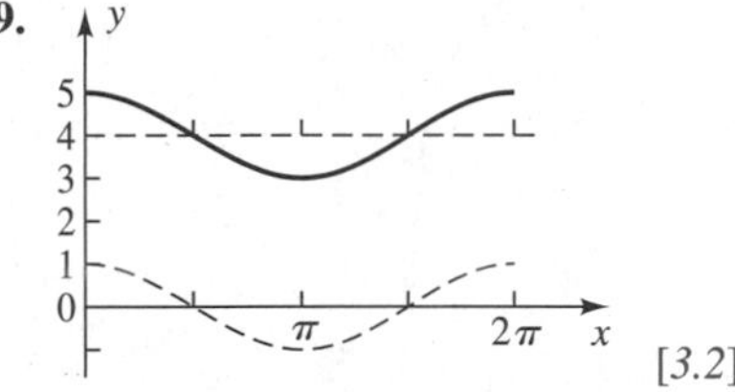

[3.2]

10. All six trigonometric functions are periodic. This is the key property shared by all. [3.1]

11. If B increases, the period decreases; if B decreases, the period increases. [3.2]

12. If C increases, the graph is moved to the left; if C decreases, the graph is moved to the right. [3.3]

13. (A) cosecant (B) cotangent (C) sine [3.1]

14.

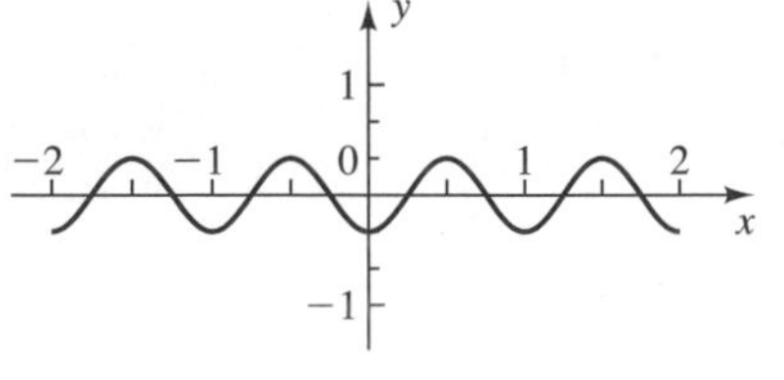

[3.2]

15.

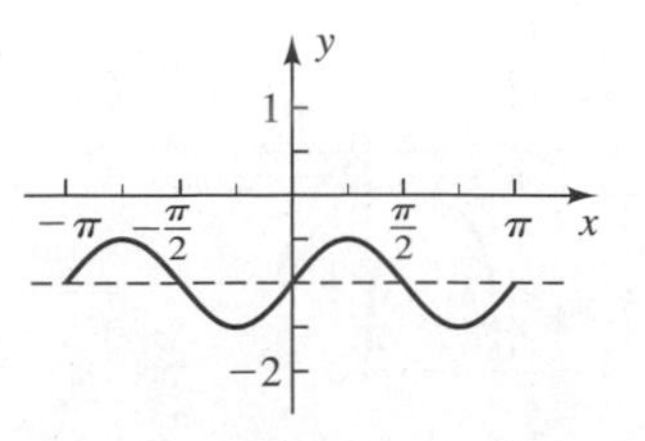
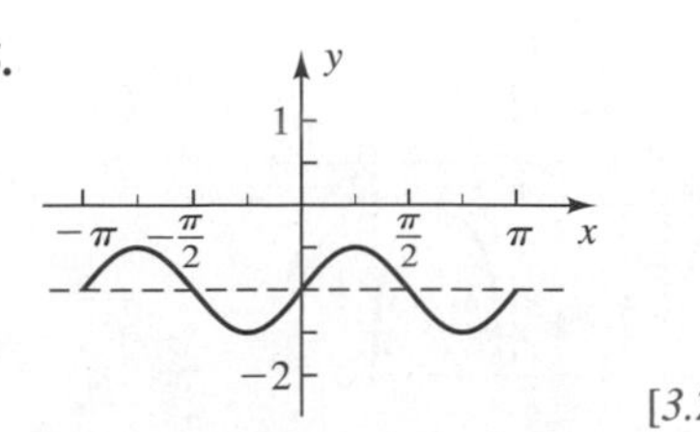

[3.2]

16.

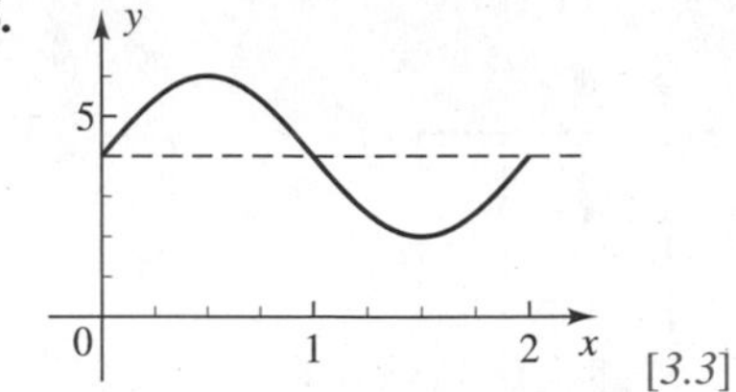

[3.3]

17.

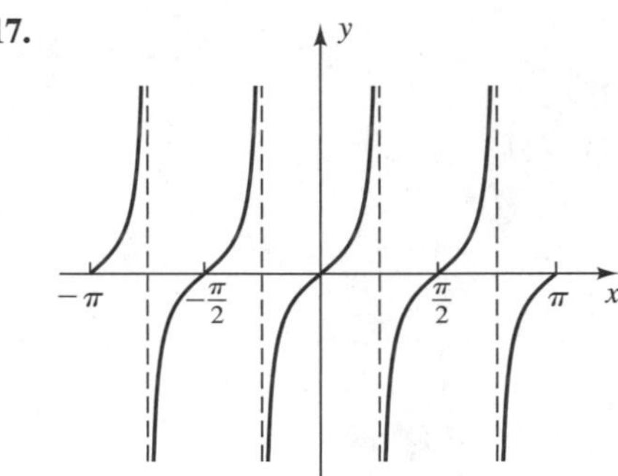

[3.6]

18.

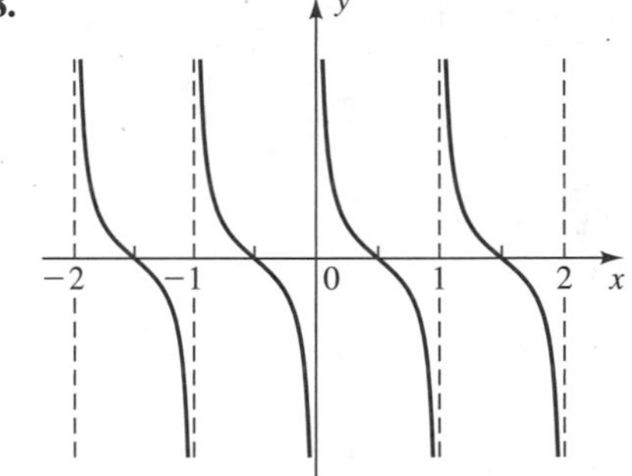

[3.6]

19.

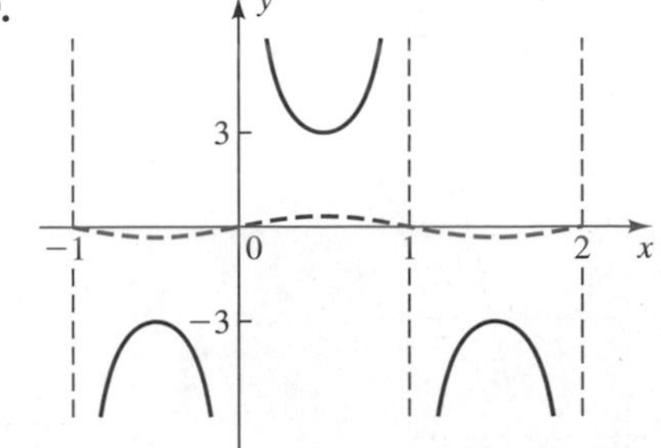

[3.6]

20.

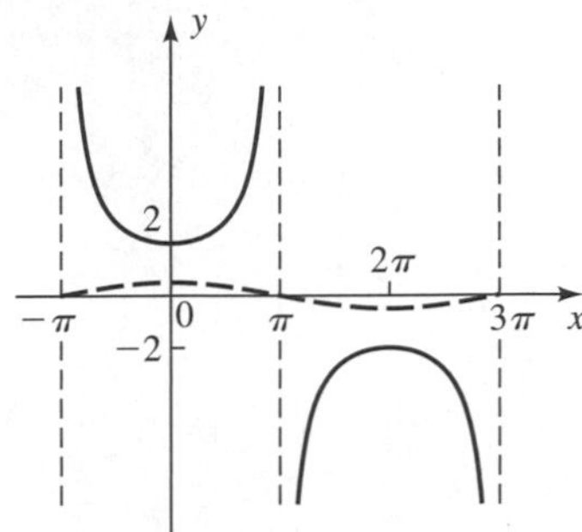

[3.6]

21.

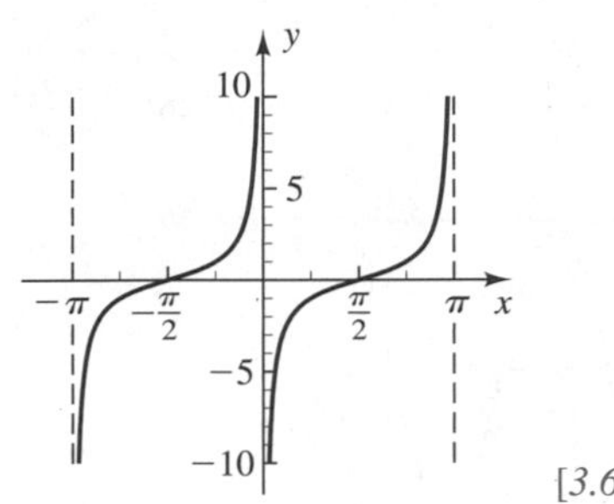

[3.6]

22. Period = 1; Amplitude = $\frac{1}{3}$ [3.2]

23. 1 in Problem 18 and 2 in Problem 19 [3.6]

24. Amplitude = 3; Period = 2; Phase shift = − 1 [3.3]

25. Period = 2; Phase shift = − 1 [3.6]

26. (A) y_2 is y_1 with half the period.
(B) y_2 is y_1 reflected across the x axis with twice the amplitude.
(C) $y_1 = \sin \pi x$; $y_2 = \sin 2\pi x$
(D) $y_1 = \cos \pi x$; $y_2 = -2 \cos \pi x$ [3.2]

27. $y = 3 + 4 \sin(\pi x/2)$ [3.2]

28. $y = 65 \sin 200\pi t$ [3.2]

29. (2) $y_2 = y_1 + y_3$: For each value of x, y_2 is the sum of the ordinate values for y_1 and y_3. [3.5]

30.

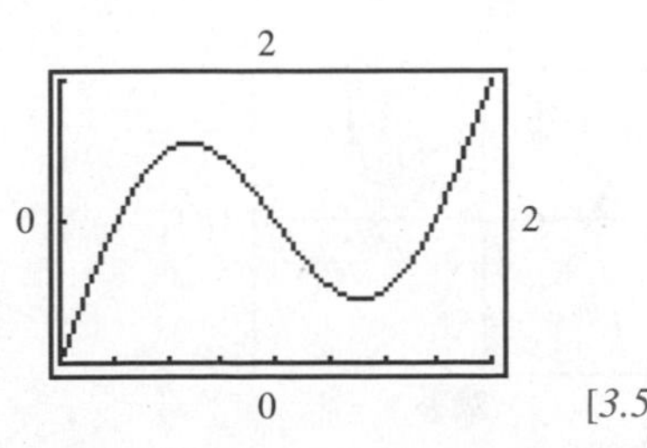

[3.5]

31.

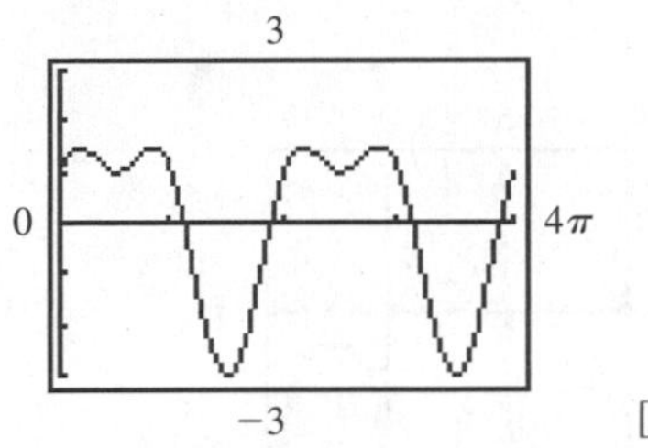

[3.5]

32. $y = \frac{1}{2} + \frac{1}{2} \cos 2x$

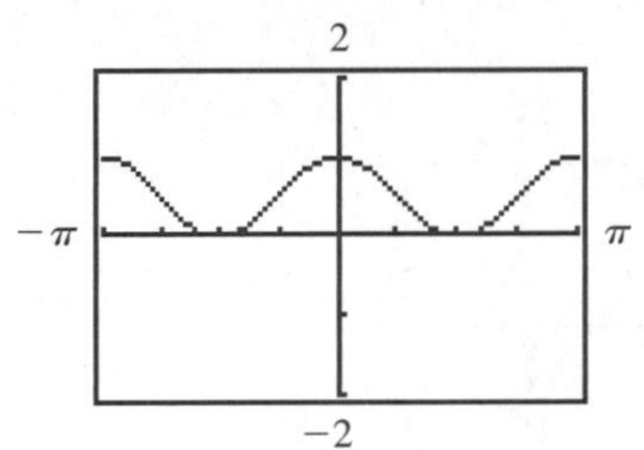

[3.2]

33.

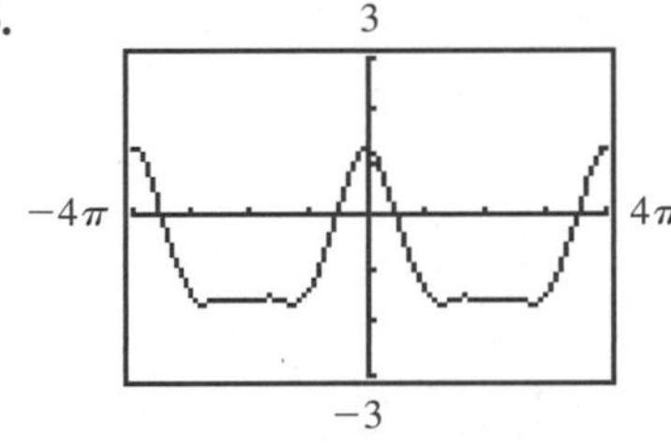

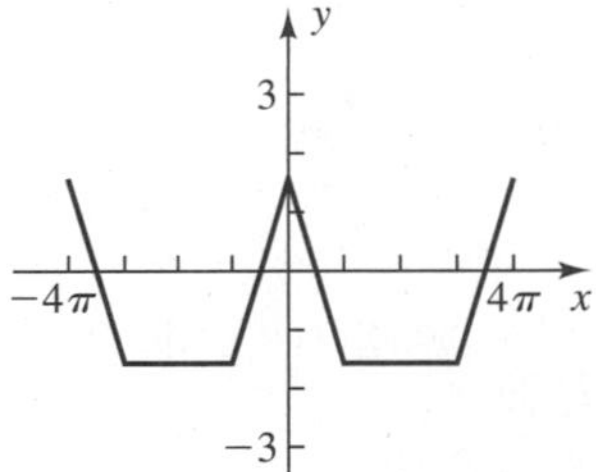

[3.5]

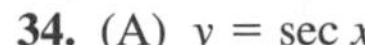

34. (A) $y = \sec x$

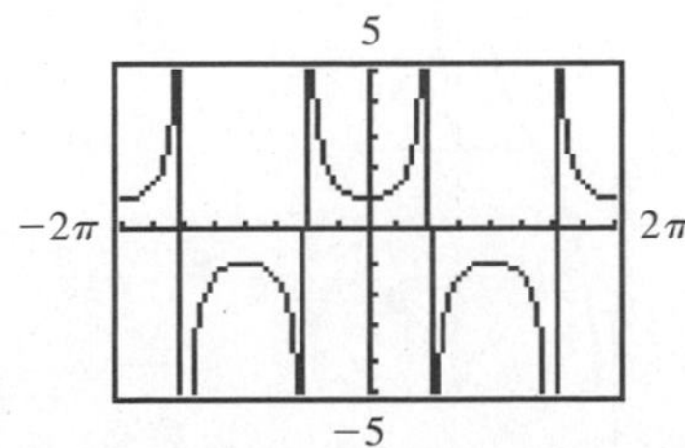

(B) $y = \csc x$

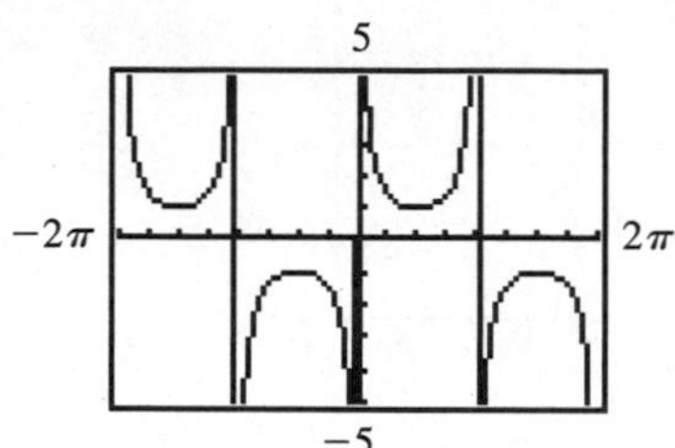

(C) $y = \cot x$

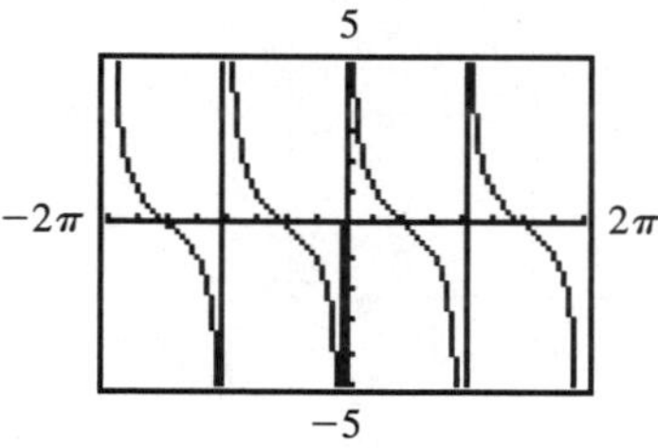

(D) $y = \tan x$

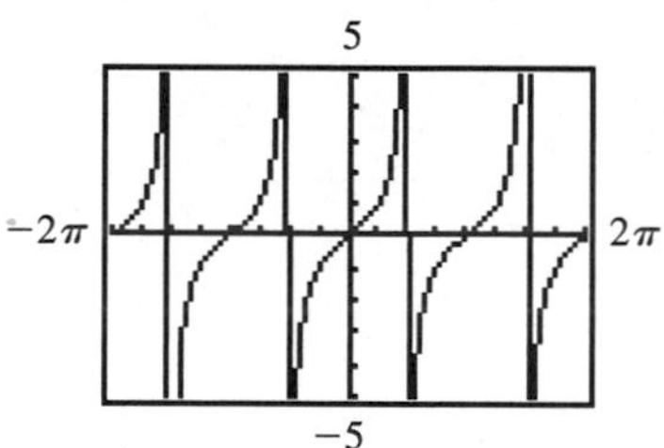

[3.6]

35. A horizontal shift of $\pi/2$ to the left and a reflection across the x axis [3.1]

36.

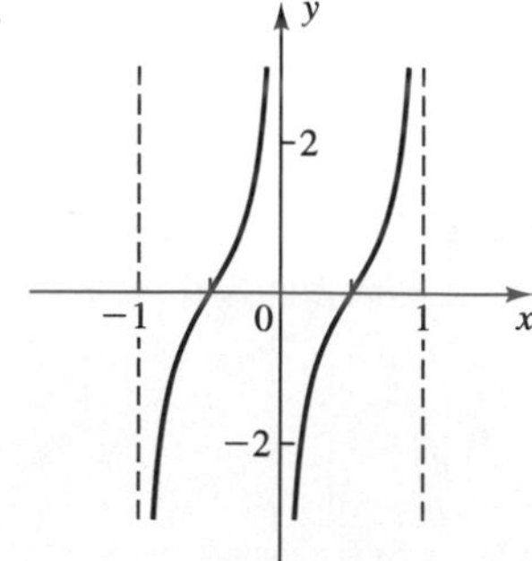

[3.6]

37.

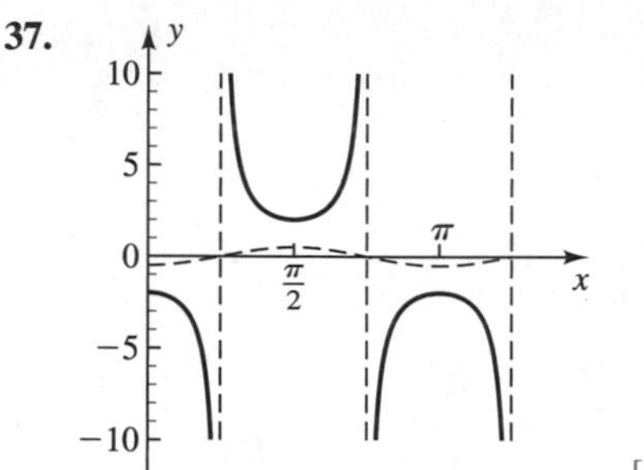

[3.6]

38. $y = 2 \sin 2\pi x$ [3.2]

39. $y = -\sin(\pi x - \pi/4)$ [3.3]

40. x intercept: -0.464; $y = 2 \sin(2x + 0.928)$

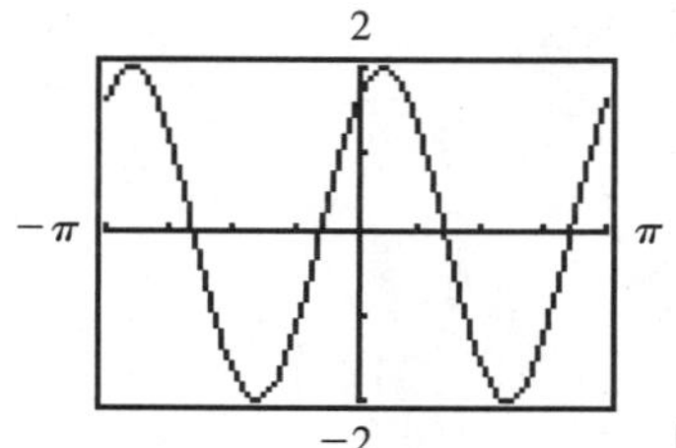

[3.3]

41. (A)

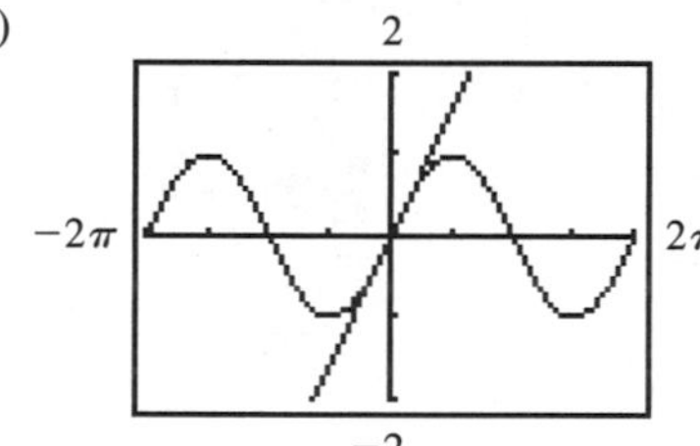

(B)

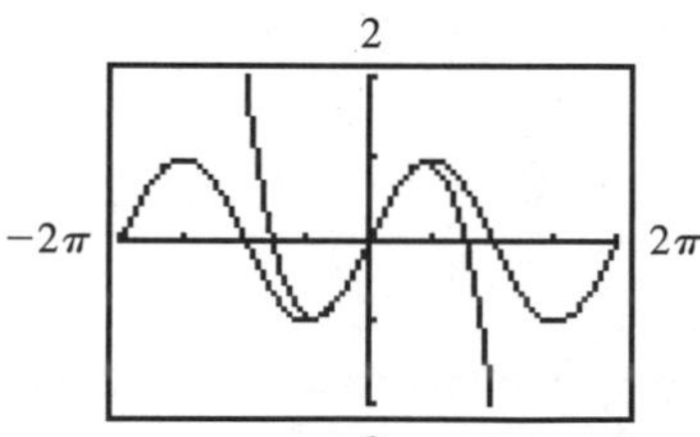

(C)

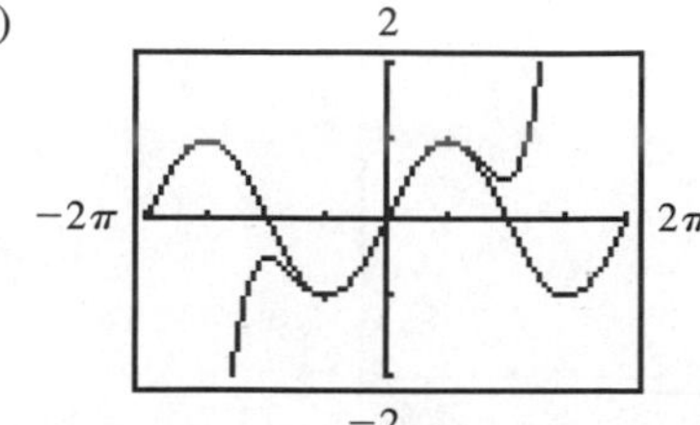

(D) As more terms of the series are used, the resulting approximation of sin x improves over a wider interval. [3.1]

42. $p = \pi$ [3.1, 3.2]

43. $y = -4 \cos 16\pi t$; $y = A \sin Bt$ cannot be used to model the motion, because when $t = 0$, y cannot equal -4 for any values of A and B. [3.2]

44. $P = 1 + \cos(\pi n/26)$, $0 \leq n \leq 104$; $y = k + A \sin Bn$ will not work, because the curve starts at (0, 2) and oscillates 1 unit above and below the line $P = 1$; a sine curve would have to start at (0, 1). [3.2]

45. (A) 179 kg (B) $y = 0.6 \sin(2.5\pi t)$
(C)

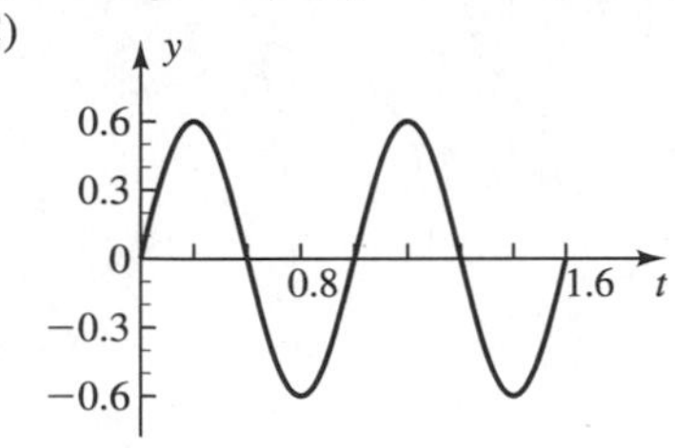

[3.2]

46. (A) Period $= \frac{1}{280}$ sec; $B = 560\pi$
(B) Frequency = 400 Hz; $B = 800\pi$
(C) Period $= \frac{1}{350}$ sec; Frequency = 350 Hz [3.2]

47. $E = 18 \cos 60\pi t$ [3.2]

48. Amplitude = 6 m; Period = 20 sec; Phase shift = 5 sec

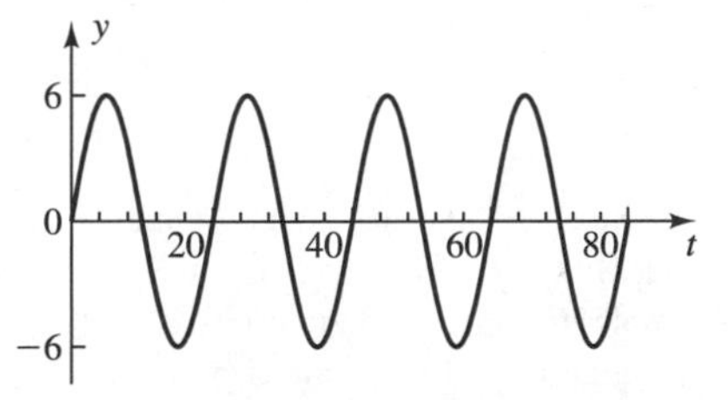

[3.3]

49. Wave height = 24 ft; Wavelength = 184 ft; Speed = 31 ft/sec [3.4] **50.** Period $= 10^{-15}$ sec; Wavelength $= 3 \times 10^{-7}$ m [3.4]

51. (A) Simple harmonic

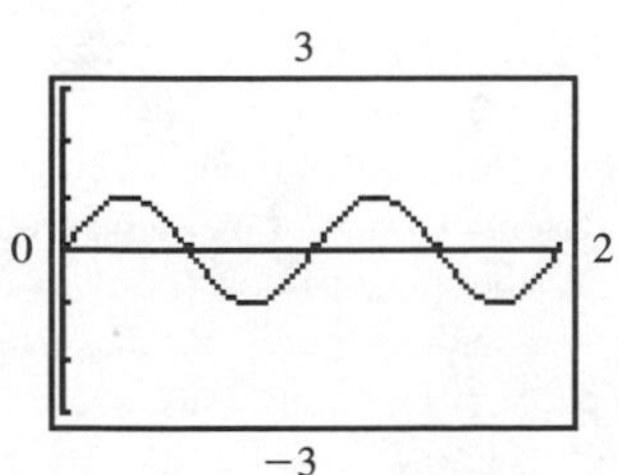

(B) Resonance

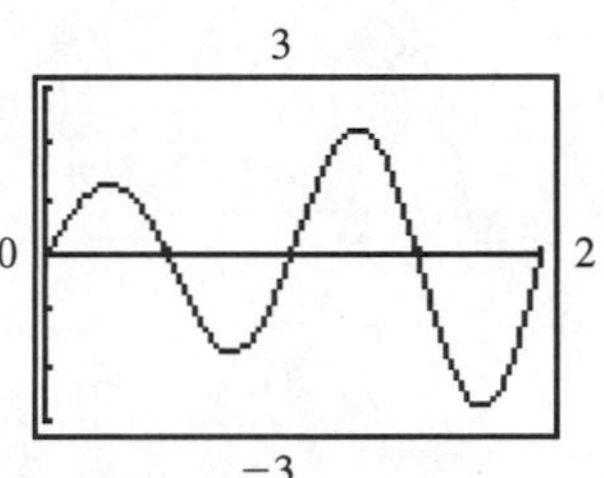

(C) Damped harmonic

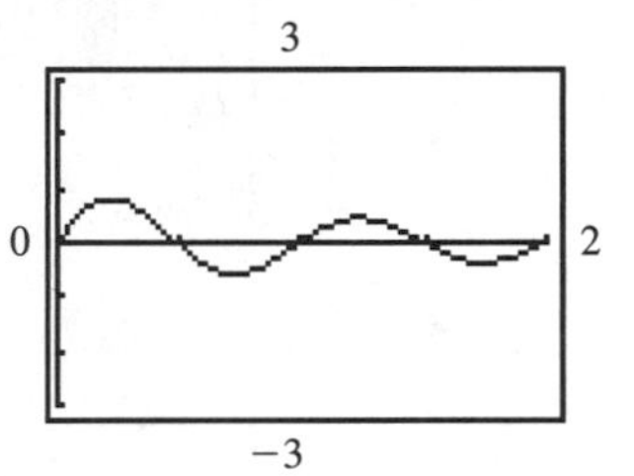

[3.4]

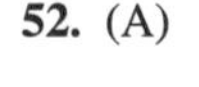

52. (A)

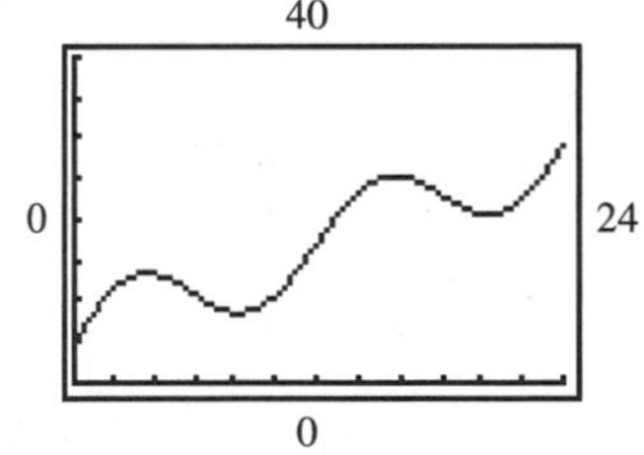

(B) It shows that sales have an overall upward trend with seasonal variations. [3.5]

53. (A) $h = 1{,}000 \tan \theta$
(B)

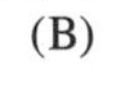

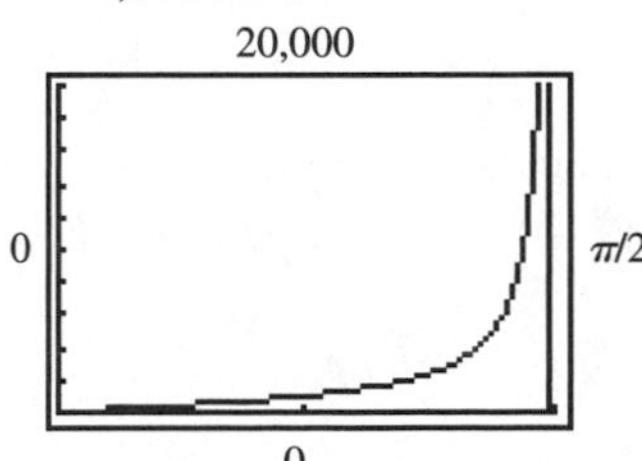

(C) As θ approaches $\pi/2$, h increases without bound. [3.1]

54. (A) The data for θ repeat every second, so the period is $P = 1$ sec. The angle θ deviates from 0 by 36° in each direction, so the amplitude is $|A| = 36°$.
(B) $\theta = A \cos Bt$ is not suitable, because, for example, for $t = 0$, $A \cos Bt = A$, which would be 36° and not 0°. $\theta = A \sin Bt$ appears suitable, because, for example, if $t = 0$, then $\theta = 0°$, a good start. Choose $A = 36$ and $B = 2\pi/P = 2\pi$, which yields $\theta = 36 \sin 2\pi t$.

54. (C)

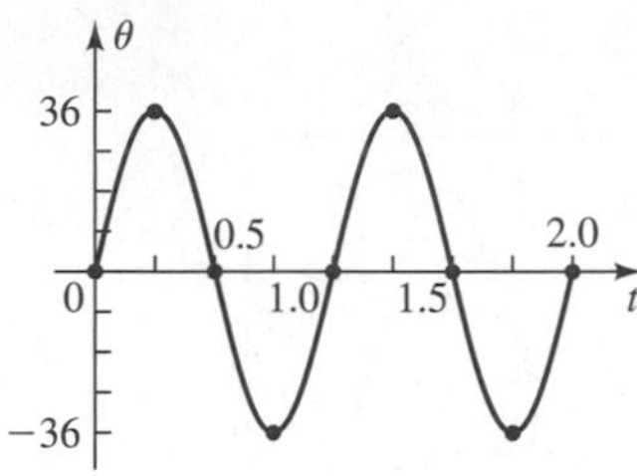

[*3.2*]

55. (A)

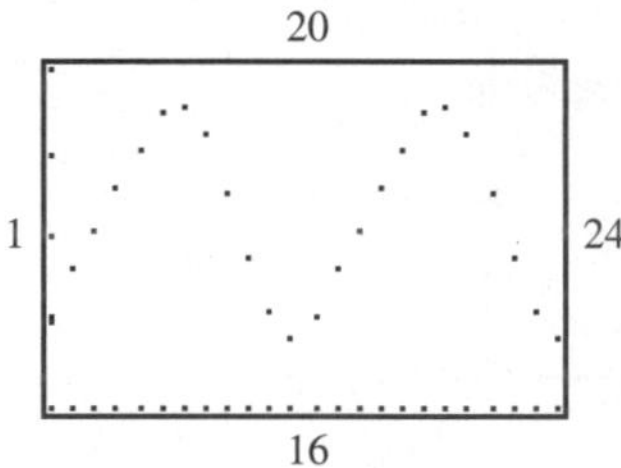

(B) $y = 18.22 + 1.37\sin\left(\frac{\pi x}{6} - 1.7\right)$

(C)

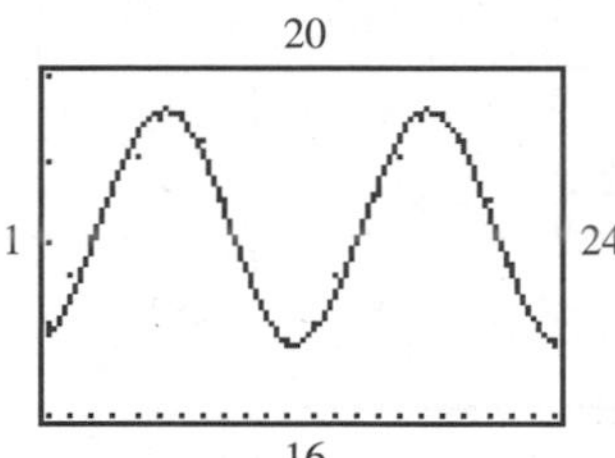

[*3.3*]

Cumulative Review Exercise Chapters 1–3

1. 90°; $\pi/2$ rad [*1.1, 2.1*] **2.** 21.78° [*1.1*]
3. 95.68°; [*2.1*] **4.** − 12.48 rad [*2.1*]
5. 65°; 14 in.; 31 in. [*1.3*]
6. $\sin\theta = -\frac{15}{17}$; $\tan\theta = -\frac{15}{8}$ [*2.3*]
7. (A) 0.3939 (B) − 1.212 (C) 0.8267 [*2.3, 2.6*]
8. (A) $\alpha = \frac{\pi}{6}$ (B) $\alpha = 45°$

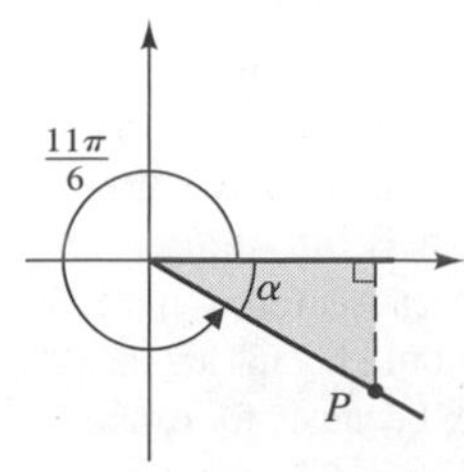

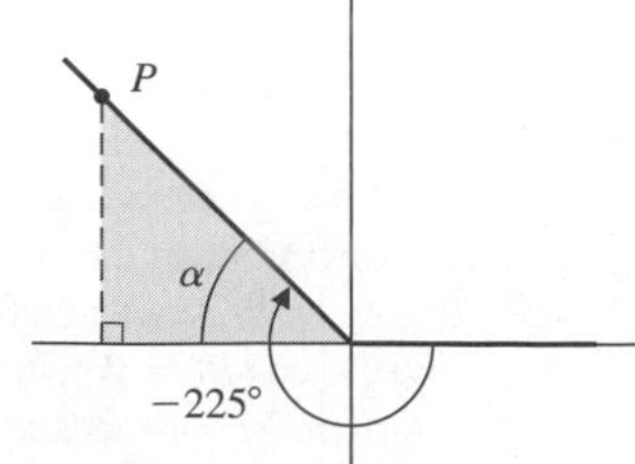

[*2.5*]

9. (A)

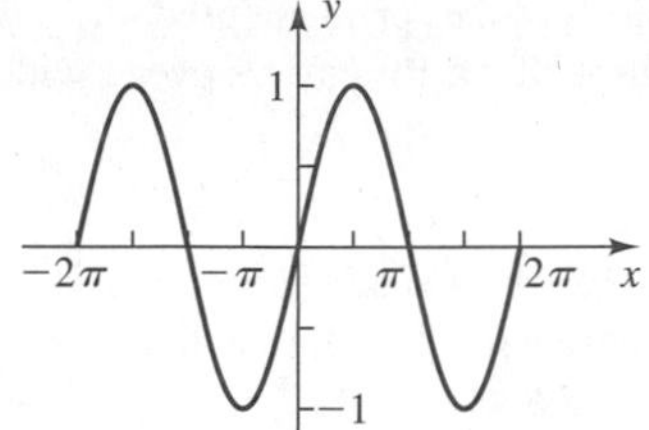

(B)

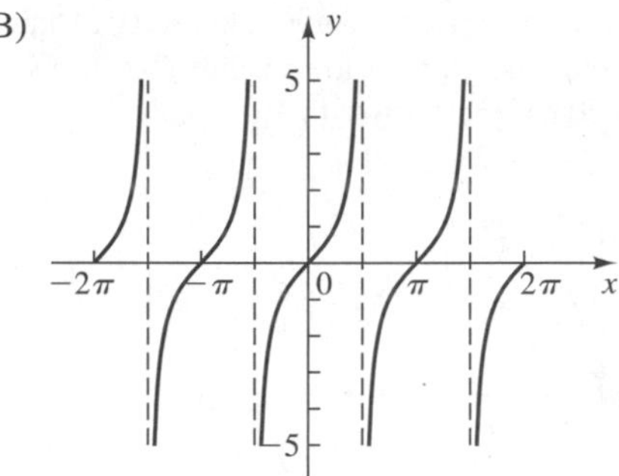

(C)

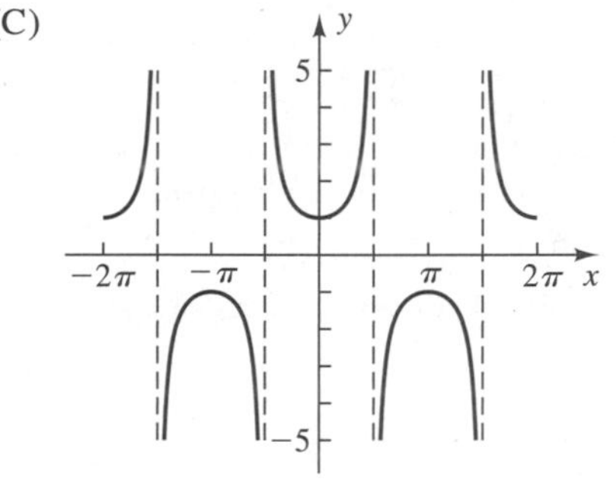

[*3.1*]

10. The central angle of a circle subtended by an arc that is 1.5 times the length of the radius of the circle. [*2.1*]
11. Yes. For example, for any x such that $\pi/2 < x < \pi$, $\cos x$ is negative and $\csc x = 1/(\sin x)$ is positive. [*2.3*]
12. No. The sum of all three angles in any triangle is 180°. An obtuse angle is one that has a measure between 90° and 180°, so a triangle with more than one obtuse angle would have two angles whose measures add up to more than 180°, contradicting the first statement. [*1.3*]
13. 43°30′ [*1.3*] **14.** 20.94 cm [*1.1, 2.1*]
15. 31.4 cm/min [*2.2*] **16.** 8 [*1.2*]
17. $4\pi/15$ [*2.1*]
18. (A) Degree mode (B) Radian mode [*2.1*]
19. Yes. Each is a different measure of the same angle, so if one is doubled, the other must be doubled. This can be seen from the conversion formula $\theta_r = (\pi/180)\theta_d$. [*2.1*]
20. Use the identity $\cot x = 1/(\tan x)$. Thus, $\cot x = 1/0.5453 = 1.8339$ [*2.6*]

21. (A) $-1/\sqrt{2}$ (B) $-\sqrt{3}/2$ (C) $\sqrt{3}$
(D) Not defined [2.5]

22. $\sin\theta = \sqrt{5}/3$, $\tan\theta = -\sqrt{5}/2$, $\sec\theta = -\frac{3}{2}$, $\csc\theta = 3/\sqrt{5}$, $\cot\theta = -2/\sqrt{5}$ [2.3]

23. $\sin(-\pi/3) = -0.8660$ is not exact; the exact form is $-\sqrt{3}/2$. [2.5]

24. Period $= \pi$; Amplitude $= \frac{1}{2}$

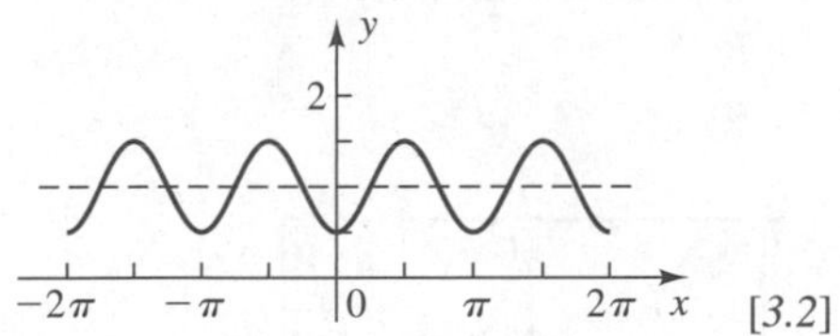

[3.2]

25. Period $= 2\pi$; Amplitude $= 2$; Phase shift $= \pi/4$

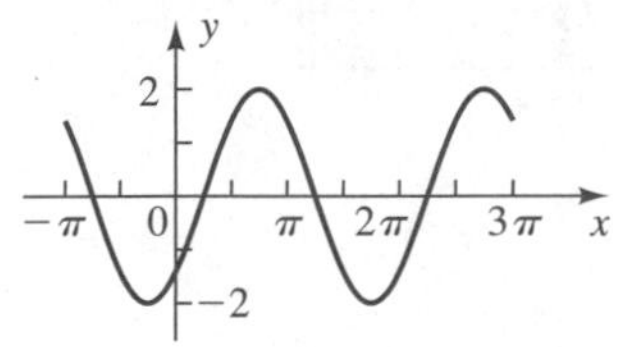

[3.3]

26. Period $= \pi/4$

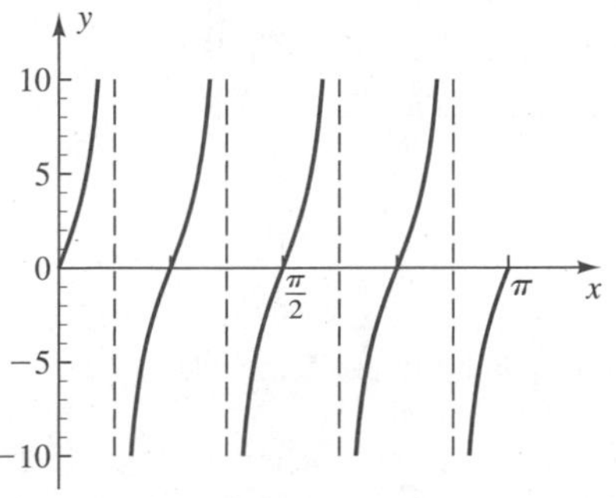

[3.6]

27. Period $= 4\pi$

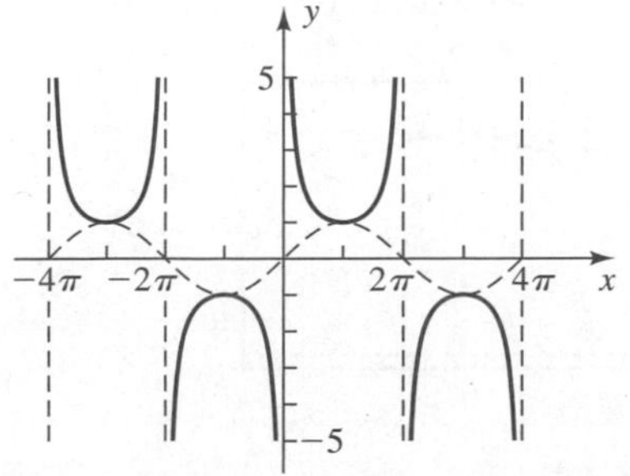

[3.6]

28. Period $= 2$

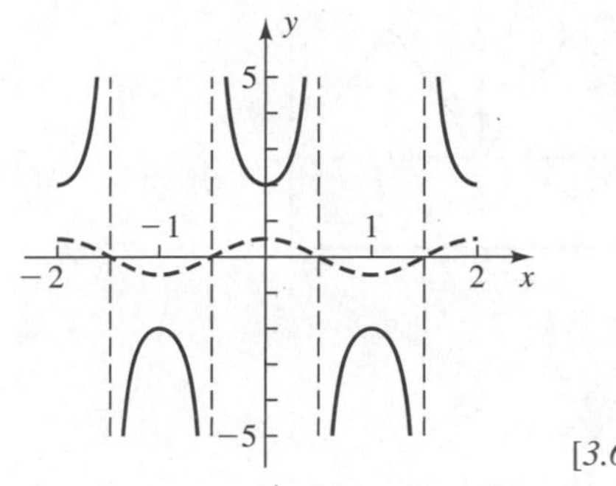

[3.6]

29. Period $= 1$; Phase shift $= -\frac{1}{2}$

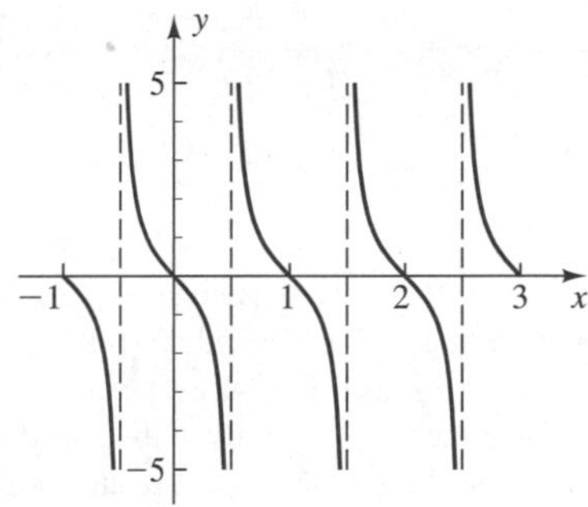

[3.6]

30. $y = -2\sin(2\pi x)$ [3.2] **31.** $y = 1 + 2\cos\pi x$ [3.2]

32. $\sec x$ [2.6] **33.** 210°, 330° [2.5]

34. 44.1 in.; 57.8°; 32.2° [1.3]

35.

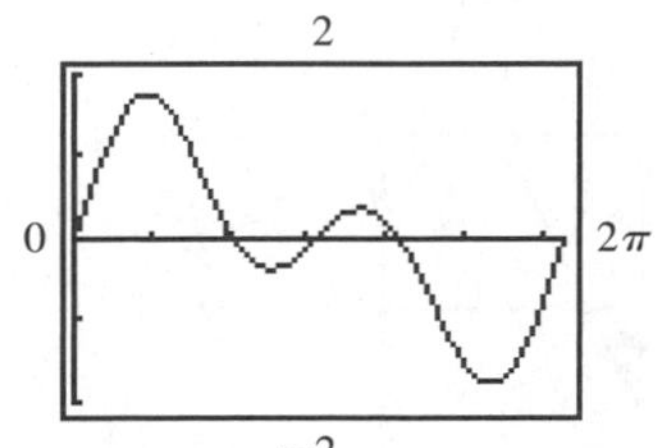

[3.5]

36. $y = \frac{1}{2} - \frac{1}{2}\cos 2x$

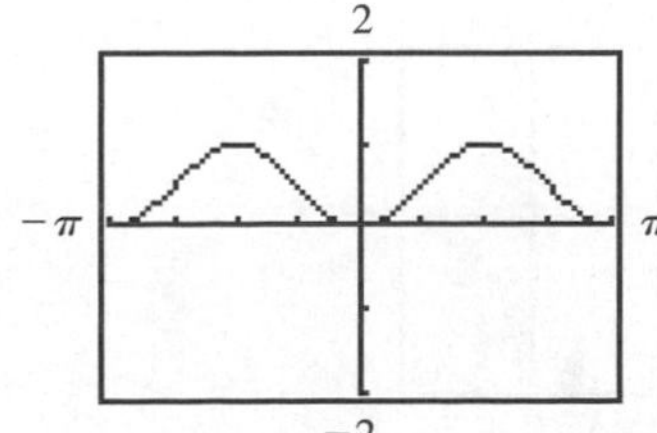

[3.2]

37.

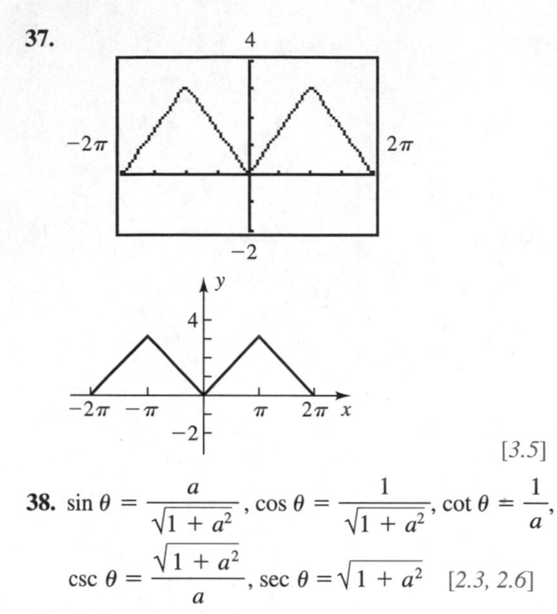

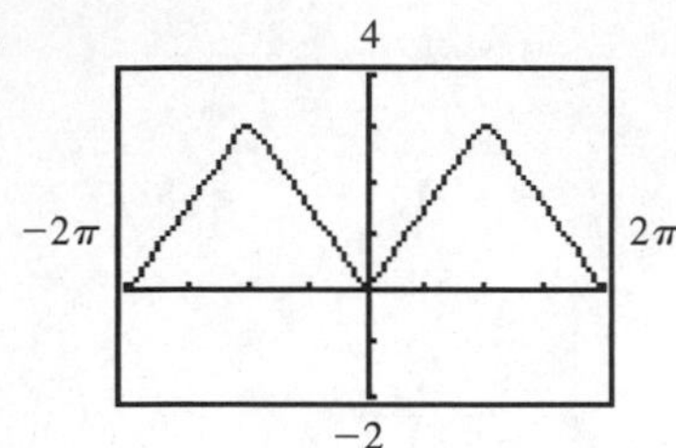

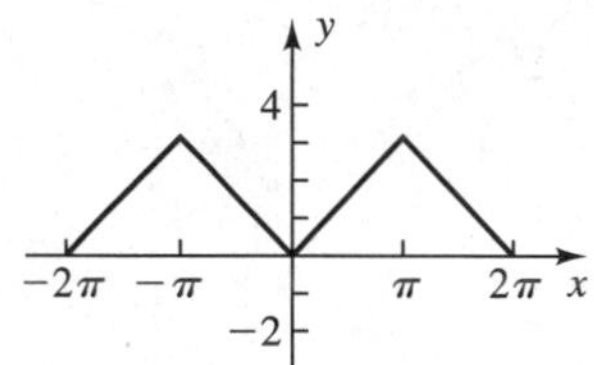

[3.5]

38. $\sin\theta = \dfrac{a}{\sqrt{1+a^2}}$, $\cos\theta = \dfrac{1}{\sqrt{1+a^2}}$, $\cot\theta = \dfrac{1}{a}$,
$\csc\theta = \dfrac{\sqrt{1+a^2}}{a}$, $\sec\theta = \sqrt{1+a^2}$ [2.3, 2.6]

39. 18.37 units [2.3]

40. Since the point moves clockwise, x is negative, and the coordinates of the point are: $P(\cos(-53.077), \sin(-53.077)) = (-0.9460, -0.3241)$. The quadrant in which P lies is determined by the signs of the coordinates. In this case, P lies in quadrant III, because both coordinates are negative. [2.6]

41. $s = \sin^{-1} 0.8149 = 0.9526$ unit, or
$s = \cos^{-1} 0.5796 = 0.9526$ unit [2.6]

42. $y = 4\sin(\pi x - \pi/3)$ [3.3]

43. x intercept: 1.287; $y = 3\sin\left(\dfrac{x}{2} - 0.6435\right)$

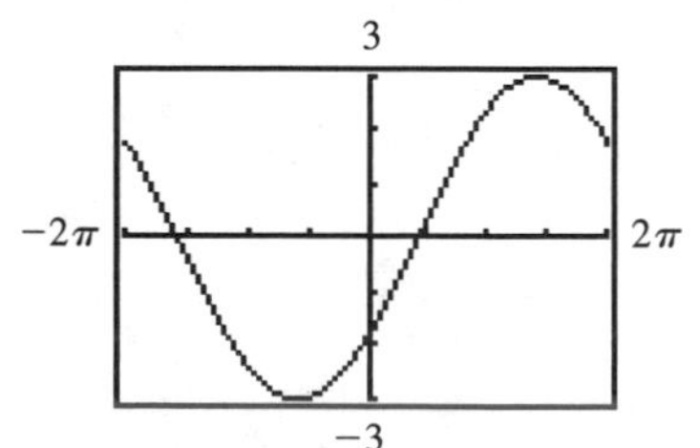

[3.3]

44. (A) $y = \tan x$

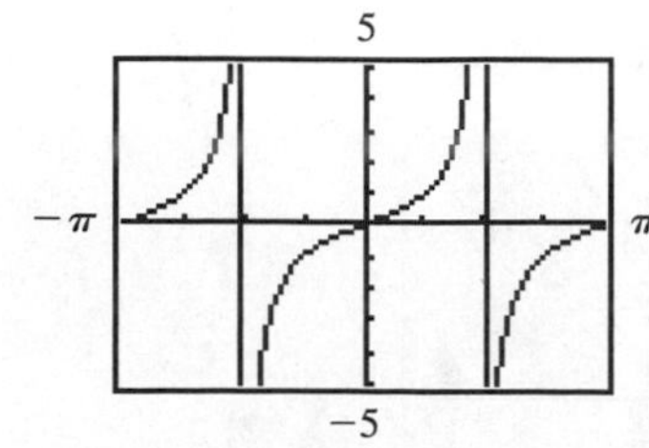

(B) $y = \sec x$

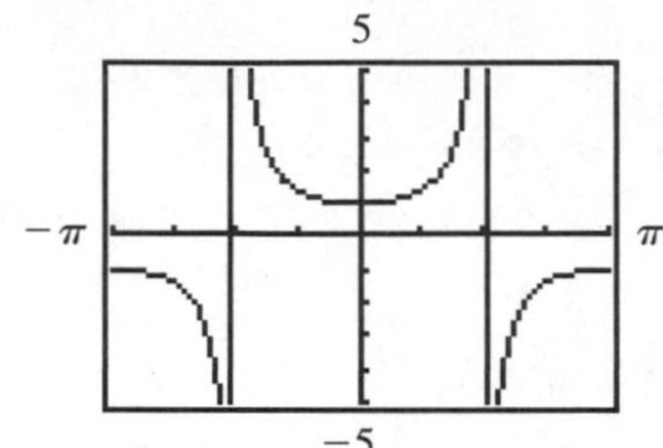

(C) $y = \csc x$

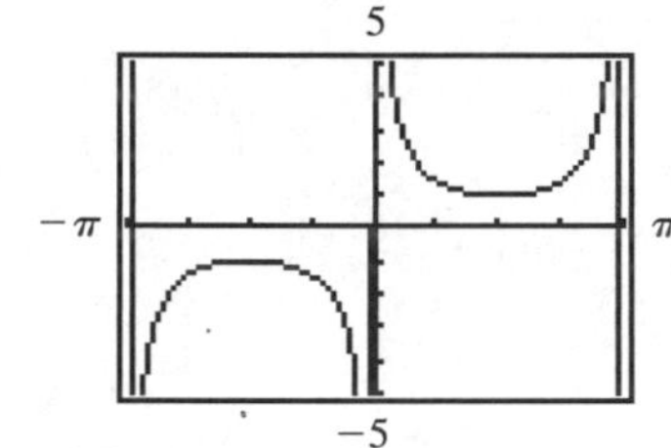

(D) $y = \cot x$

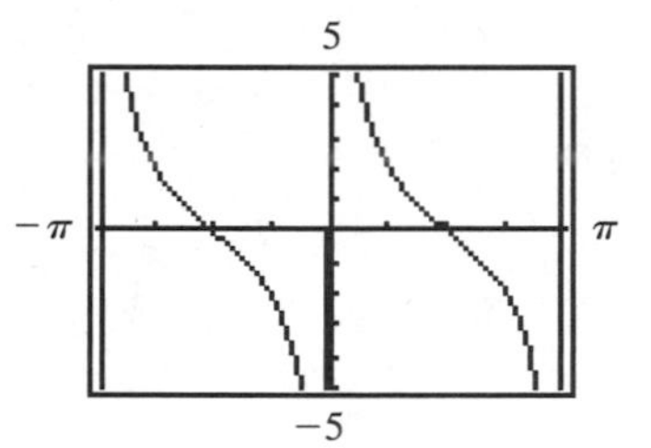

[3.6]

45. 773 mi [1.1, 2.1] **46.** 45 cm³ [1.2]
47. 41°; 36 ft [1.4] **48.** $\frac{4}{3}$ ft [1.2, 1.4]
49. 1.7°; 5% [1.4]
50. Building: 51 m high; street: 17 m wide [1.4]
51. (A) 1.7 mi (B) 1.3 mi [1.4]
52. (A) 942 rad/min (B) 64,088 in./min [2.2]
53. 27.8° [2.4] **54.** 84° [2.4]
55. (C) $\cos x$ approaches 1 as x approaches 0, and $(\sin x)/x$ is between $\cos x$ and 1; therefore, $(\sin x)/x$ must approach 1 as x approaches 0. [2.1, 2.6]
56. (A)

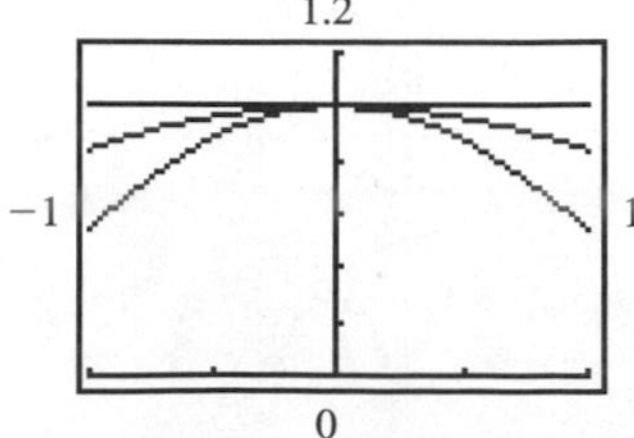

(B) $y_1 < y_2 < y_3$

(C) cox x approaches 1 as x approaches 0, and $(\sin x)/x$ is between $\cos x$ and 1; therefore, $(\sin x)/x$ must approach 1 as x approaches 0. This can also be observed using TRACE. [2.1]

57. $y = 3.6 \cos 12\pi t$; no, because when $t = 0$, y will be 0 no matter what values are assigned to A and B. [3.2]

58. (A) Amplitude = 12; Period = $\frac{1}{30}$ sec; Frequency = 30 Hz; Phase shift = $\frac{1}{60}$ sec

(B)

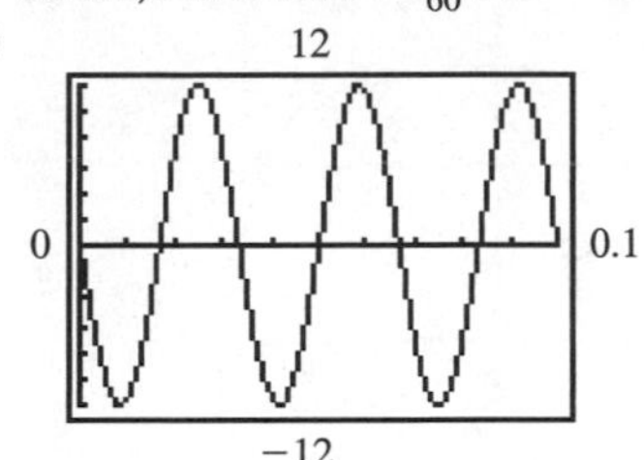

[3.2]

59. Period = 2×10^{-13} sec [3.4]

60. (A) $L = 100 \csc \theta + 70 \sec \theta, 0 < \theta < \pi/2$

(C)

TABLE 1

θ (rad)	0.50	0.60	0.70	0.80
L (ft)	288.3	261.9	246.7	239.9
θ (rad)	0.90	1.00	1.10	
L (ft)	240.3	248.4	266.5	

From the table, the shortest walkway is 239.9 ft when $\theta = 0.80$ rad.

(D) The minimum $L = 239.16$ ft when $\theta = 0.84$ rad.

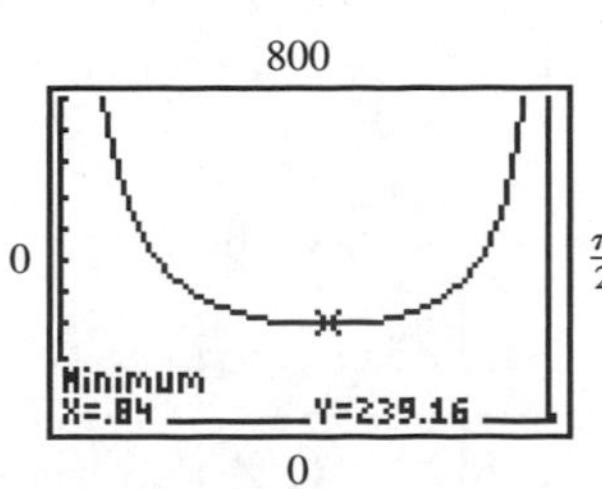

[2.3, 2.6, 3.2]

61. (A) $d = 50 \tan \theta, 0 < \theta < \pi/2$ (B) $\theta = 40\pi t$

(C) $d = 50 \tan 40\pi t$

(D) d increases without bound as t approaches $\frac{1}{80}$ min.

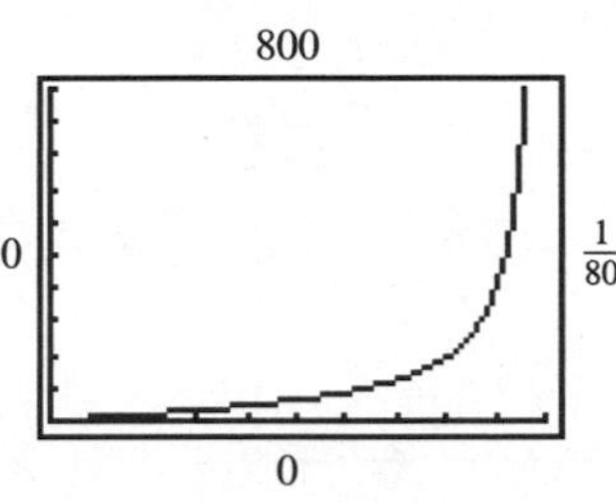

[2.2]

62. (A) Damped harmonic motion

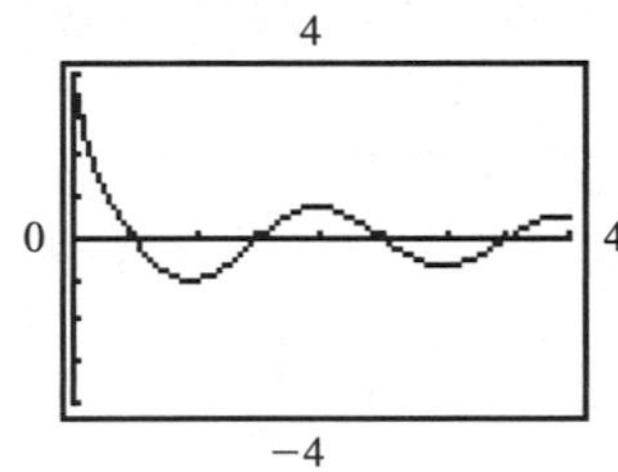

(B) Simple harmonic motion

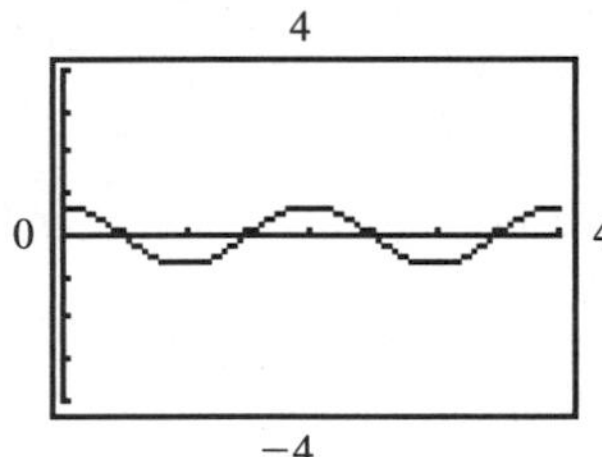

(C) Resonance

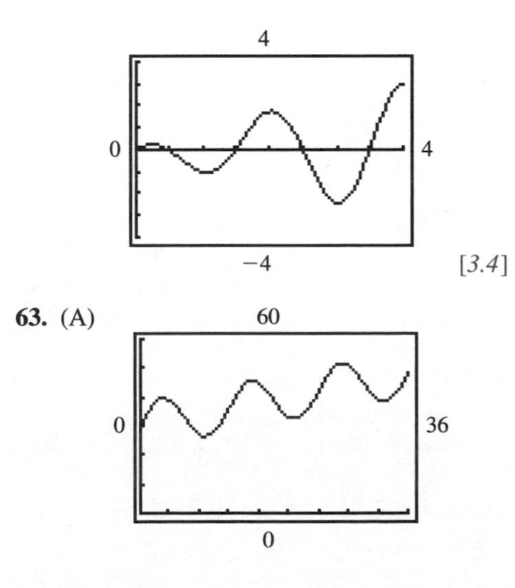

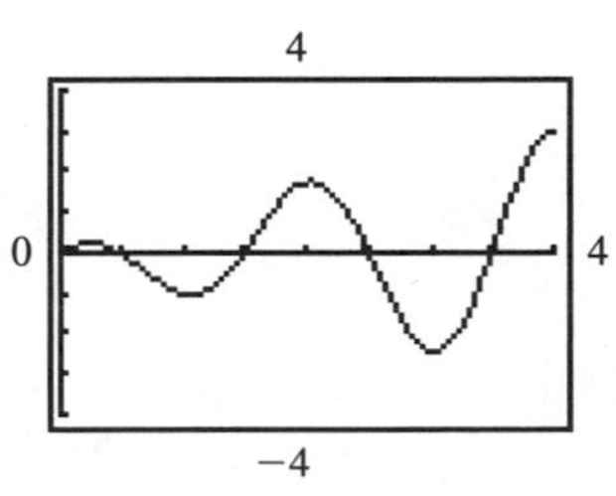

[3.4]

63. (A)

63. (B) The sales trend is up but with the expected seasonal variations. [*3.5*]

64. (A)

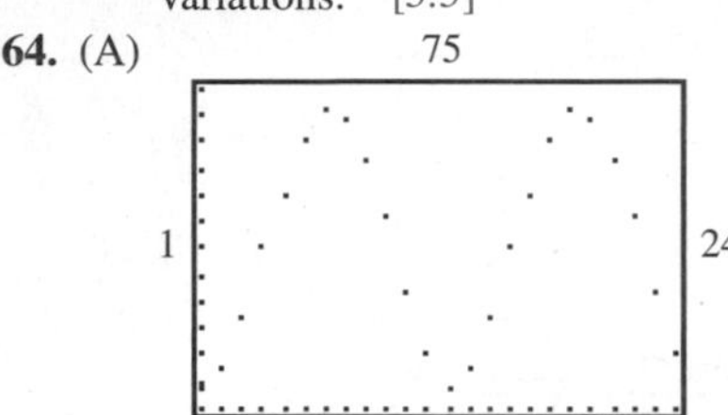

(B) $y = 45 + 26 \sin\left(\frac{\pi x}{6} - 2.2\right)$

(C)

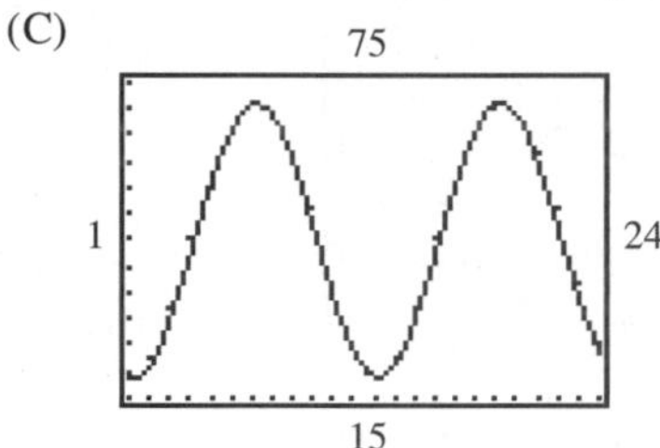

[*3.3*]

CHAPTER 4

Exercise 4.1

1. Reciprocal identities: $\csc x = 1/(\sin x)$, $\sec x = 1/(\cos x)$, $\cot x = 1/(\tan x)$
Identities for negatives: $\sin(-x) = -\sin x$, $\cos(-x) = \cos x$, $\tan(-x) = -\tan x$

3. (1) is a conditional equation, and (2) is an identity. (1) is true for only one value of x and is false for all other values of x. (2) is true for all values of x for which both sides are defined.

5. $\csc x = -\frac{3}{2}$, $\sec x = 3/\sqrt{5}$, $\tan x = -2/\sqrt{5}$, $\cot x = -\sqrt{5}/2$

7. $\cot x = \frac{1}{2}$, $\cos x = -1/\sqrt{5}$, $\csc x = -\sqrt{5}/2$, $\sec x = -\sqrt{5}$

9. 1 **11.** $\sec x$ **13.** $\tan x$ **15.** $\sec \theta$

17. $\cot^2 \beta$ **19.** 2

21. Not necessarily. For example, $\sin x = 0$ for infinitely many values ($x = k\pi$, k any integer), but the equation is not an identity. The left side is not equal to the right side for values other than $x = k\pi$, k an integer; for example, when $x = \pi/2$, $\sin x = 1 \neq 0$.

23. $\cos x = -\sqrt{15}/4$, $\csc x = 4$, $\sec x = -4/\sqrt{15}$, $\tan x = -1/\sqrt{15}$, $\cot x = -\sqrt{15}$

25. $\cot x = -\frac{1}{2}$, $\sec x = \sqrt{5}$, $\cos x = 1/\sqrt{5}$, $\sin x = -2/\sqrt{5}$, $\csc x = -\sqrt{5}/2$

27. $\sin x = \frac{2}{3}$, $\cos x = -\sqrt{5}/3$, $\sec x = -3/\sqrt{5}$, $\tan x = -2/\sqrt{5}$, $\cot x = -\sqrt{5}/2$

29. (A) -0.4350 (B) $1 - 0.1892 = 0.8108$

31. $-\cot y$ **33.** $\csc x$ **35.** 0 **37.** $\csc^2 x$

39. $\cot w$ **41.** (A) 1 (B) 1 **43.** I, II **45.** II, III

47. All **49.** I, IV **51.** $a \cos x$ **53.** $a \sec x$

55. (A) $\frac{x^2}{25} + \frac{y^2}{4} = 1$

(B)

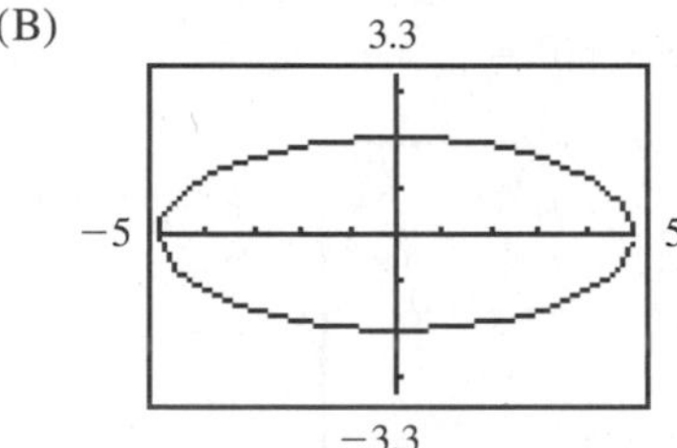

Exercise 4.2

27. To solve a conditional equation is to find, using equation solving strategies, all replacements of the variable that make the statement true. To verify an identity is to show that one side is equivalent to the other for all replacements of the variable for which both sides are defined. This is done by transforming one side, through a logical sequence of steps, into the other side.

61. (A) No. Use TRACE and move from one curve to the other, comparing y values for different values of x. You will see that, though close, they are not exactly the same.

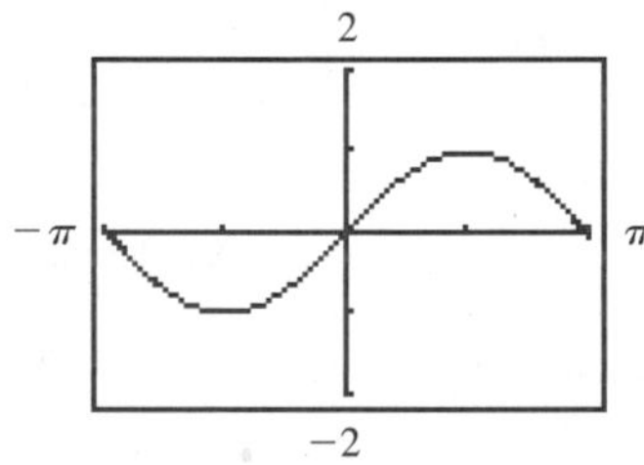

(B) Outside the interval $[-\pi, \pi]$ the graphs differ widely.

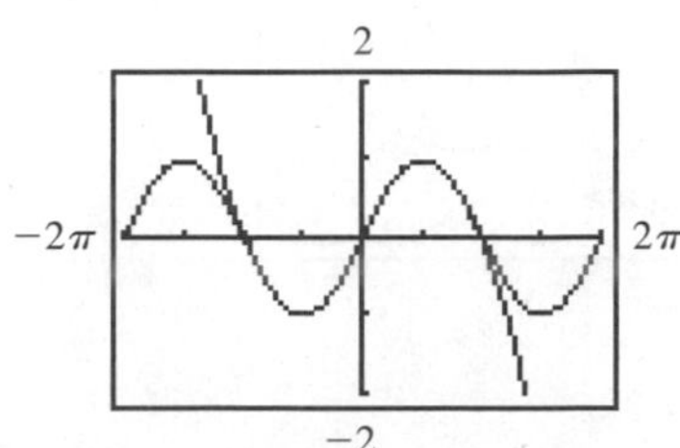

63. An identity **65.** Not an identity

67. Not an identity **69.** An identity

Exercise 4.3

13. $\frac{\sqrt{2}}{2}(\sin x - \cos x)$ **15.** $-\cos x$ **17.** $\frac{1 - \tan x}{1 + \tan x}$

19. $\frac{\sqrt{3} + 1}{2\sqrt{2}}$ **21.** $\frac{\sqrt{3} + 1}{2\sqrt{2}}$ **23.** $\sqrt{3}/2$ **25.** $\sqrt{3}$

27. $\sin(x - y) = \frac{-2 - \sqrt{75}}{12}$; $\tan(x + y) = \frac{\sqrt{75} - 2}{\sqrt{5} + 2\sqrt{15}}$

29. $\sin(x - y) = \frac{-2\sqrt{8} - 1}{3\sqrt{5}}$; $\tan(x + y) = \frac{1 - 4\sqrt{2}}{2 + 2\sqrt{2}}$

45. Find values of x and y for which both sides are defined and the left side is not equal to the right side. Evaluate both sides for $x = 2$ and $y = 1$, for example.

47. Graph $y_1 = \sin(x - 2)$ and $y_2 = \sin x - \sin 2$ in the same viewing window, and observe that the graphs are not the same:

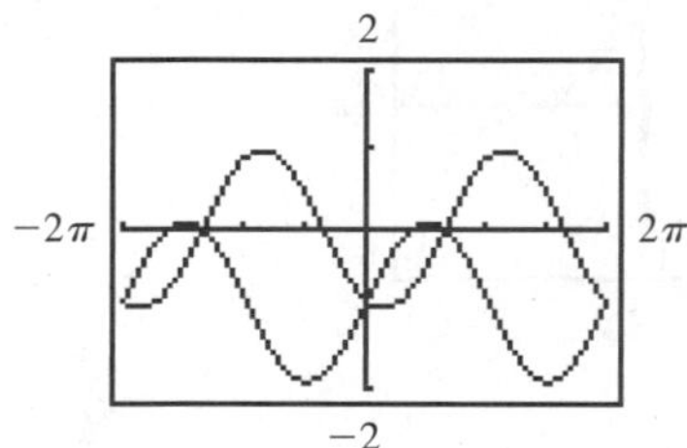

49. $y_1 = \cos(x + 5\pi/6)$
$y_2 = (-\sqrt{3}/2)\cos x - (1/2)\sin x$

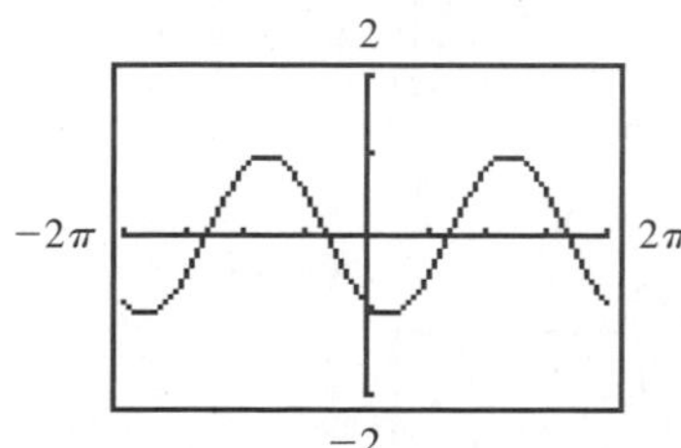

51. $y_1 = \tan(x - \pi/4)$
$y_2 = \frac{\tan x - 1}{1 + \tan x}$

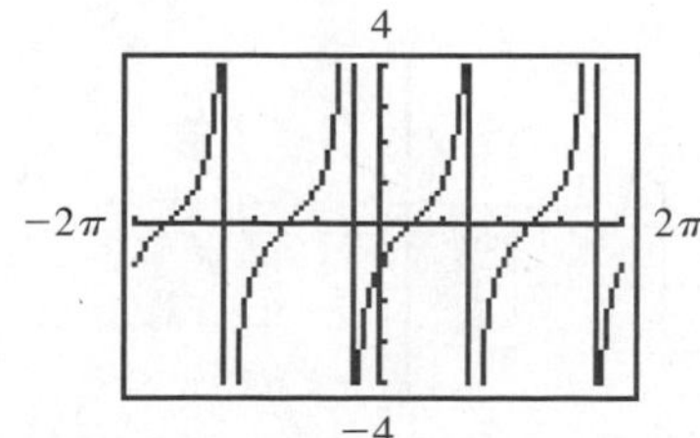

53. $y_1 = \sin 3x \cos x - \cos 3x \sin x$
$y_2 = \sin 2x$

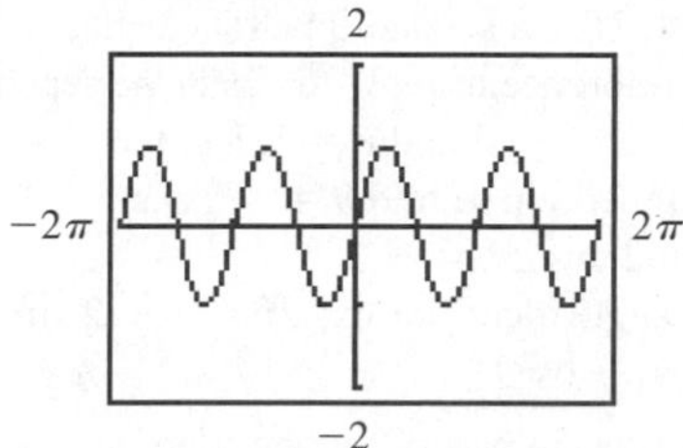

55. $y_1 = \sin(\pi x/4)\cos(3\pi x/4) + \cos(\pi x/4)\sin(3\pi x/4)$
$y_2 = \sin \pi x$

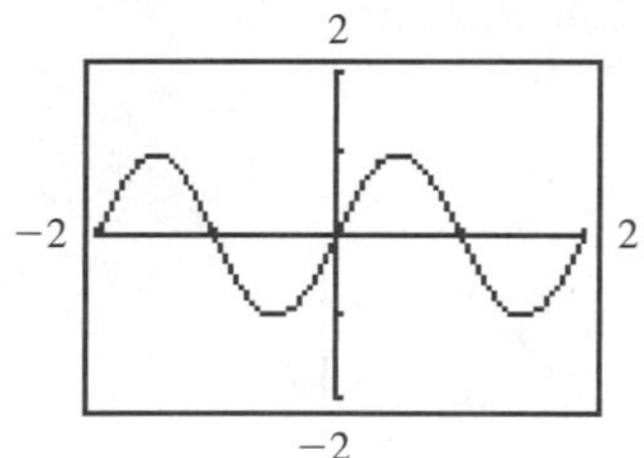

61. $\theta = 0.322$ rad **63.** (C) 3,940 ft

Exercise 4.4

1. $\sqrt{3}/2 = \sqrt{3}/2$ **3.** $-\sqrt{3} = -\sqrt{3}$ **5.** $\sqrt{2 + \sqrt{3}}/2$

7. $2 - \sqrt{3}$

9.

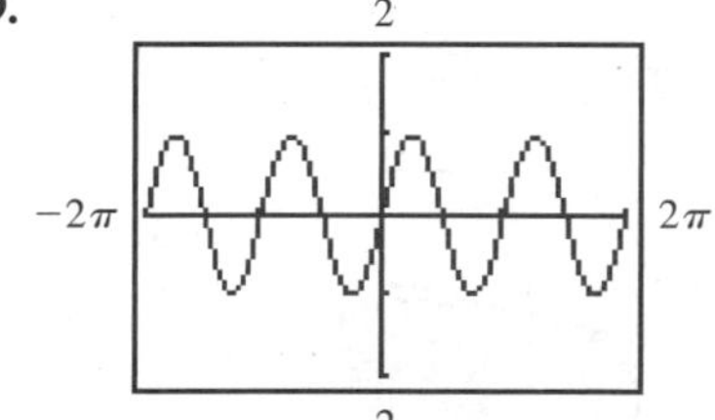

11.

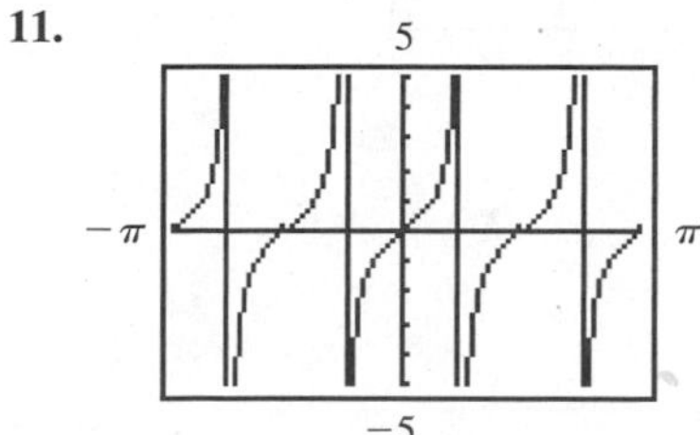

31. $\sin 2x = -\frac{24}{25}$, $\cos 2x = \frac{7}{25}$, $\tan 2x = -\frac{24}{7}$

33. $\sin 2x = -\frac{120}{169}$, $\cos 2x = -\frac{119}{169}$, $\tan 2x = \frac{120}{119}$

35. $\sin(x/2) = \sqrt{3}/3$, $\cos(x/2) = \sqrt{6}/3$

37. $\sin(x/2) = \sqrt{\frac{3 + 2\sqrt{2}}{6}}$, $\cos(x/2) = -\sqrt{\frac{3 - 2\sqrt{2}}{6}}$

39. (A) Since θ is a first-quadrant angle and sec 2θ is negative for 2θ in the second quadrant and not for 2θ in the first, 2θ is a second-quadrant angle.
(B) Construct a reference triangle for 2θ in the second quadrant with $a = -4$ and $r = 5$. Use the Pythagorean theorem to find $b = 3$. Thus, $\sin 2\theta = \frac{3}{5}$ and $\cos 2\theta = -\frac{4}{5}$.
(C) The double-angle identities $\cos 2\theta = 1 - 2\sin^2\theta$ and $\cos 2\theta = 2\cos^2\theta - 1$
(D) Use the identities in part (C) in the form $\sin\theta = \sqrt{(1 - \cos 2\theta)/2}$ and $\cos\theta = \sqrt{(1 + \cos 2\theta)/2}$. The positive radicals are used because θ is in quadrant I.
(E) $\sin\theta = 3\sqrt{10}/10$; $\cos\theta = \sqrt{10}/10$

41. (A) Approximation improves

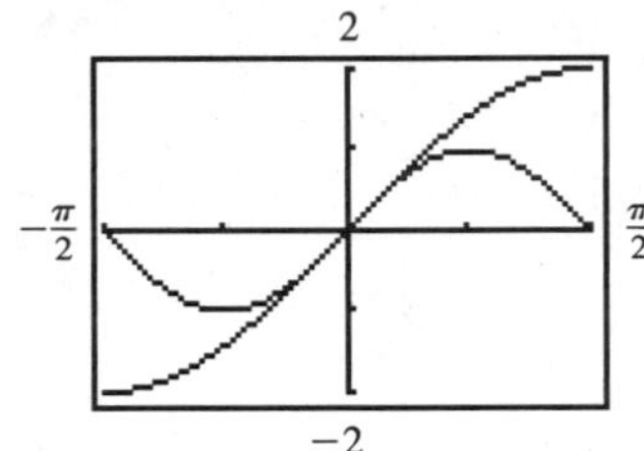

(B) Approximation improves

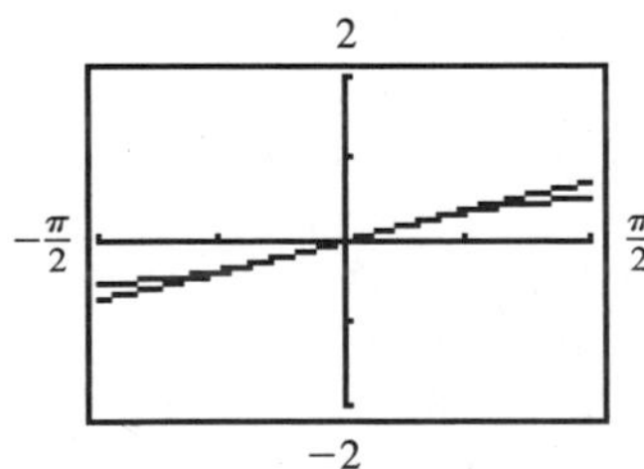

43. $0 \le x \le 2\pi$

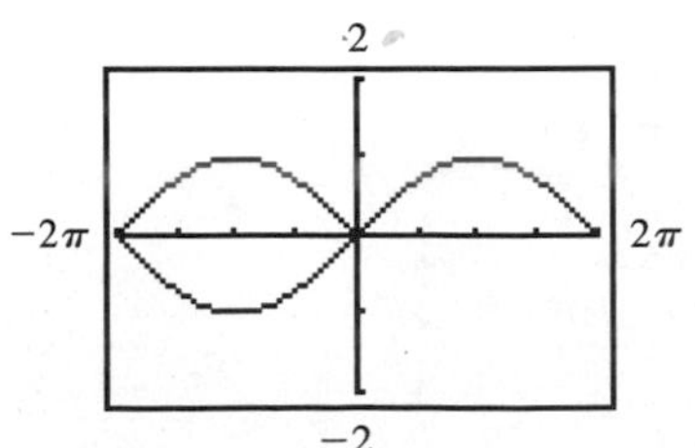

45. $-2\pi \le x \le -\pi$, $\pi \le x \le 2\pi$

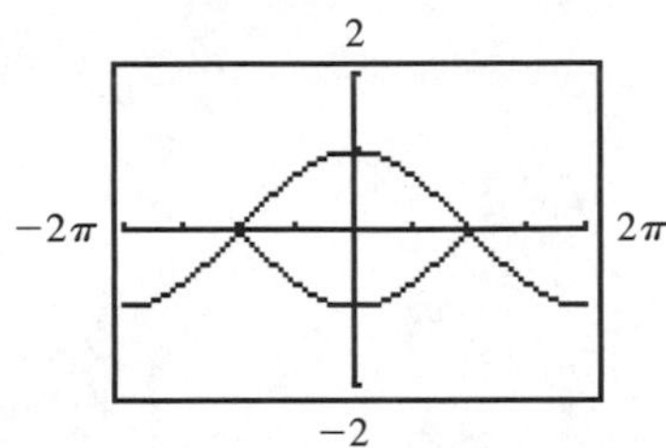

47. $\sin x = 1/\sqrt{10}$, $\cos x = 3/\sqrt{10}$, $\tan x = \frac{1}{3}$

49. $\sin x = -2/\sqrt{5}$, $\cos x = 1/\sqrt{5}$, $\tan x = -2$

57. $g(x) = \cot\frac{x}{2}$

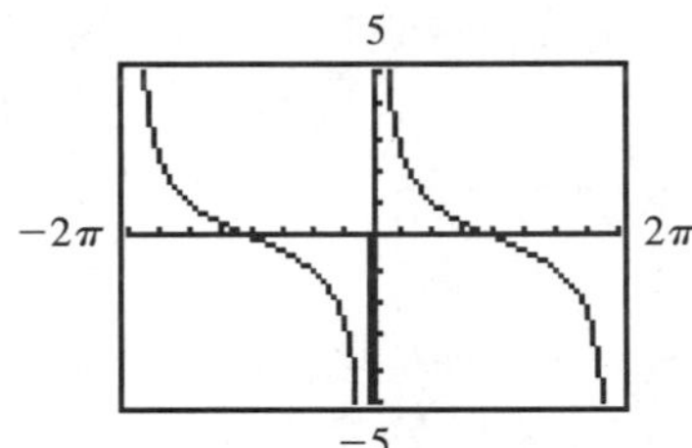

59. $g(x) = \csc 2x$

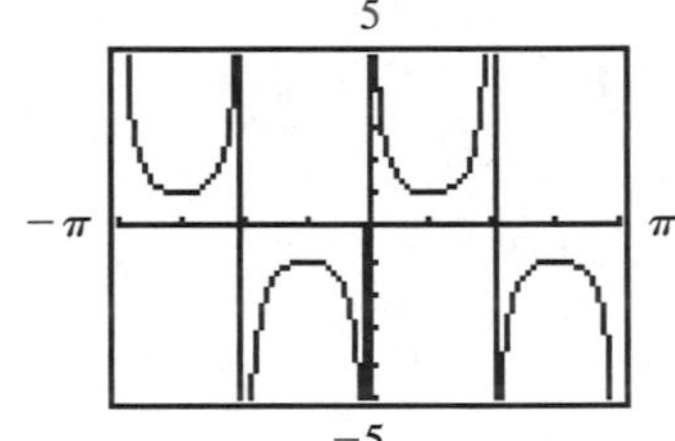

61. $g(x) = 2\cos x - 1$

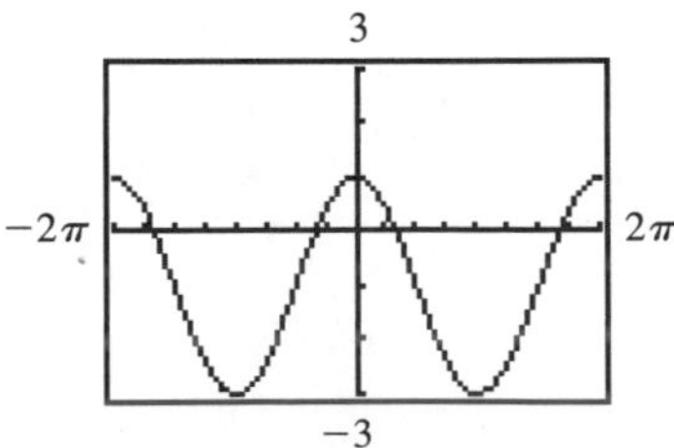

63.

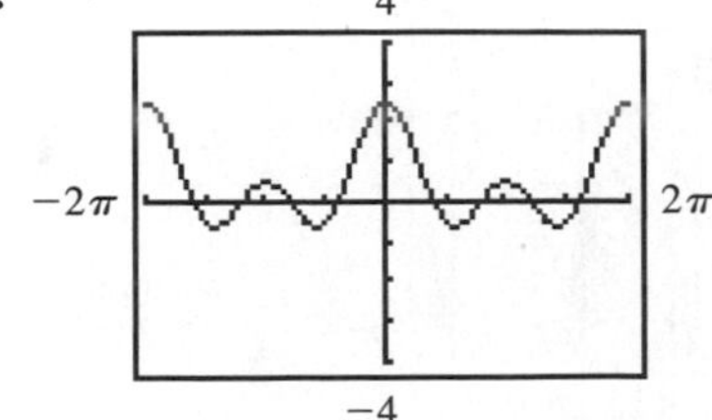

65. (A) $d = \dfrac{v_0^2 \sin 2\theta}{32}$
(B) d is maximum when $\sin 2\theta$ is maximum, and $\sin 2\theta$ is maximum when $2\theta = 90°$; that is, when $\theta = 45°$.
(C) As θ increases from 0° to 90°, d increases to a maximum of 312.5 ft when $\theta = 45°$, then decreases.

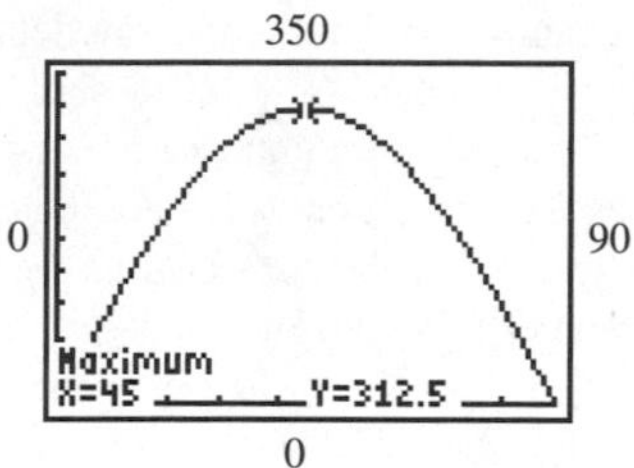

67. (B) $\sin 2\theta$ has a maximum value of 1 when $2\theta = 90°$, or $\theta = 45°$. When $\theta = 45°$, $V = 64$ ft^3.
(C) Max $V = 64.0$ ft^3 occurs when $\theta = 45°$.

TABLE 2

θ (deg)	30	35	40	45	50	55	60
V (ft^3)	55.4	60.1	63.0	64.0	63.0	60.1	55.4

(D) Max $V = 64$ ft^3 when $\theta = 45°$.

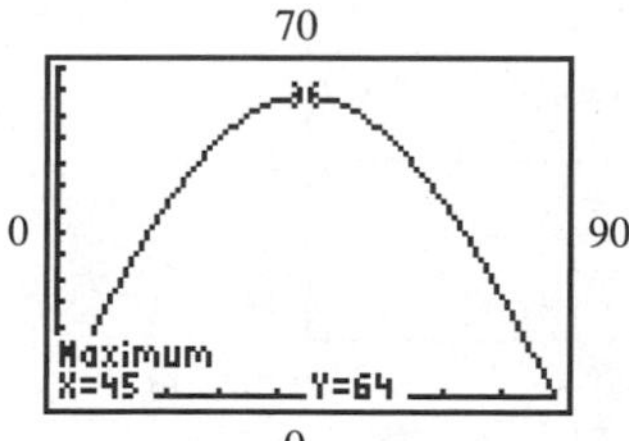

69. $x = 2\sqrt{3} \approx 3.464$ cm; $\theta = 30.000°$ **71.** $\cos\theta = \frac{4}{5}$

Exercise 4.5

1. $\frac{1}{2}\cos 12A + \frac{1}{2}\cos 2A$ **3.** $\frac{1}{2}\sin 5\theta + \frac{1}{2}\sin\theta$
5. $2\cos 6\theta\cos\theta$ **7.** $-2\cos 3u\sin 2u$
9. $(2 - \sqrt{3})/4$ **11.** $\frac{1}{4}$ **13.** $\sqrt{2}/2$ **15.** $\sqrt{2}/2$
19. Let $x = u + v$ and $y = u - v$, and solve the resulting system for u and v in terms of x and y. Then substitute the results into the first identity.

29. $y_2 = \frac{1}{2}(\cos 8x + \cos 2x)$

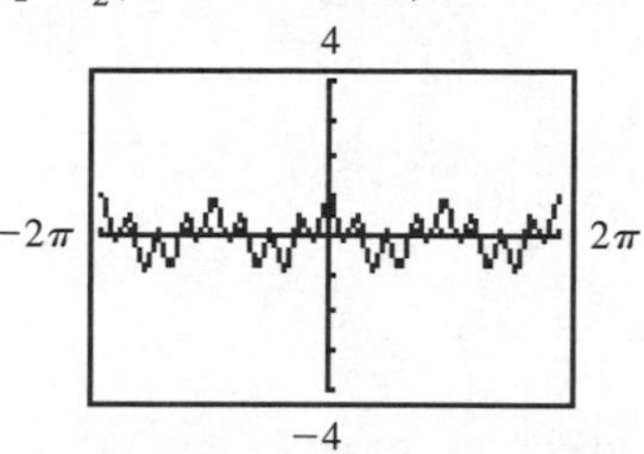

31. $y_2 = \frac{1}{2}(\sin 2.4x - \sin 1.4x)$

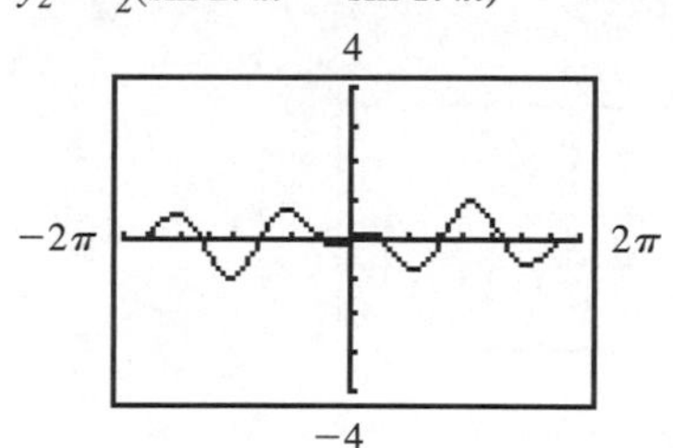

33. $y_2 = 2\cos 2x\cos x$

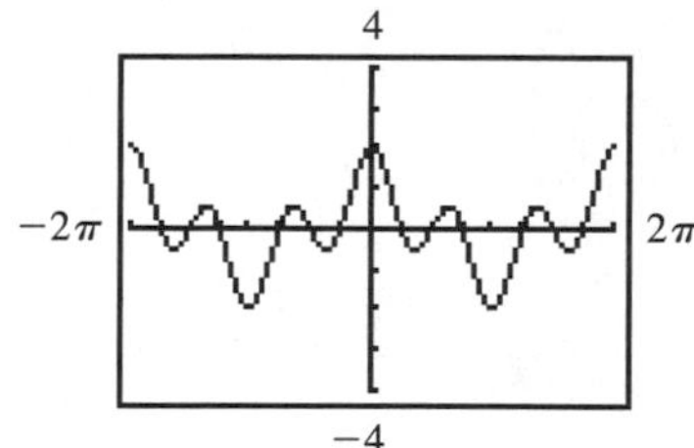

35. $y_2 = 2\cos 1.3x\sin 0.8x$

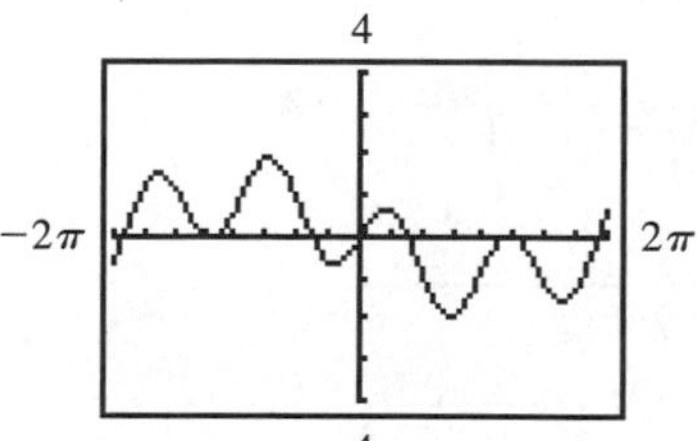

39. (A)

2

0 1

−2

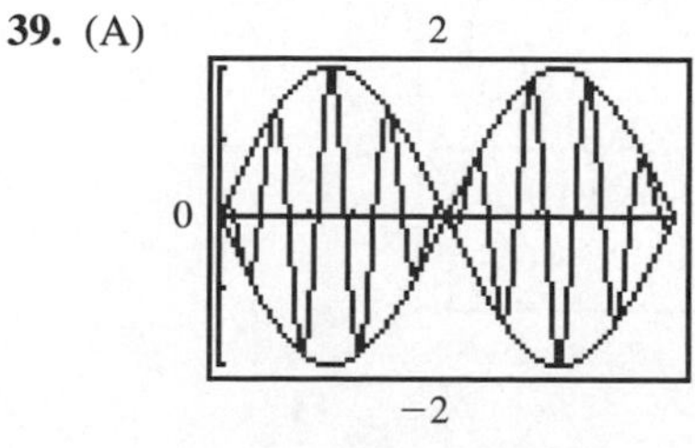

(B) $y_1 = \sin(18\pi x) - \sin(14\pi x)$
Graph same as part (A).

41. (A)

2

0 1

−2

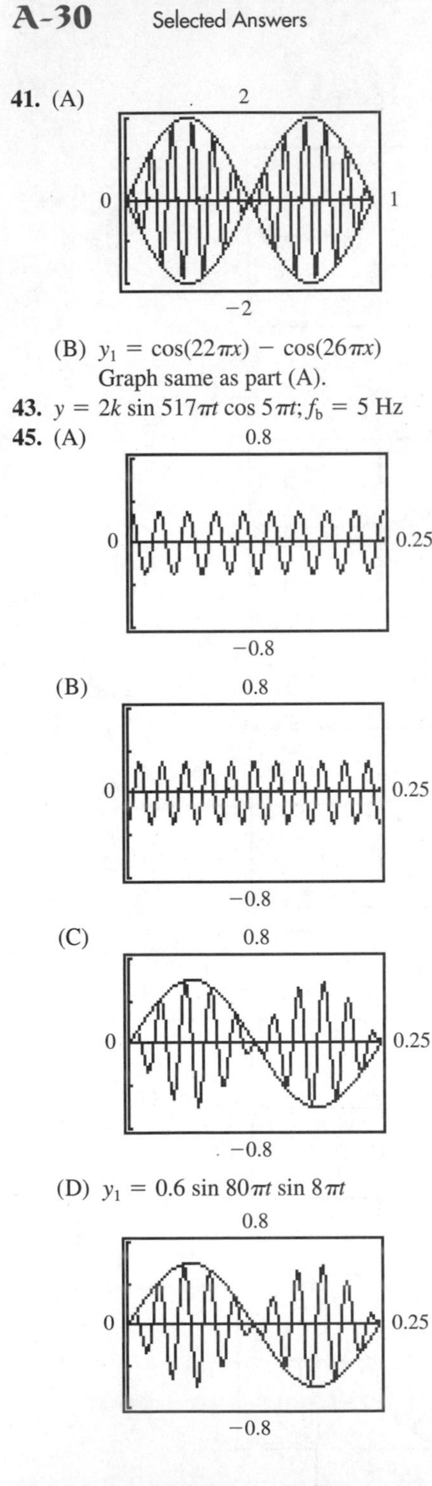

(B) $y_1 = \cos(22\pi x) - \cos(26\pi x)$
Graph same as part (A).

43. $y = 2k \sin 517\pi t \cos 5\pi t$; $f_b = 5$ Hz

45. (A)

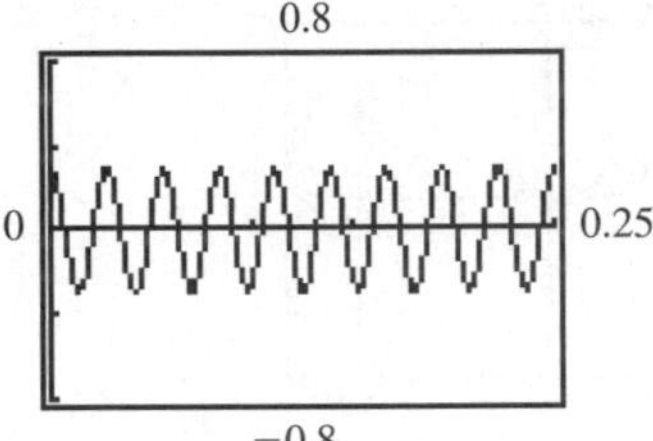

(B)

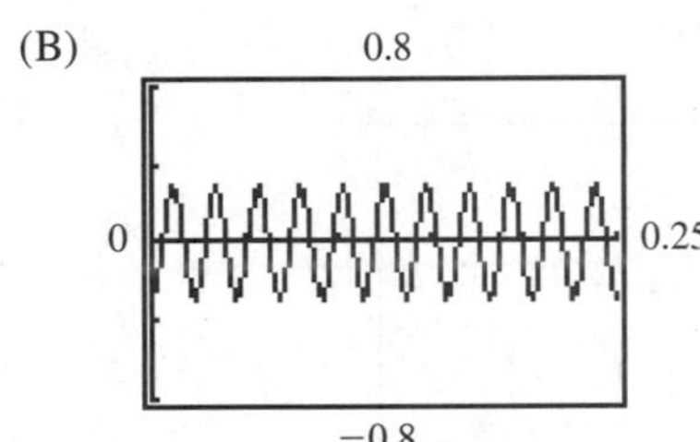

(C)

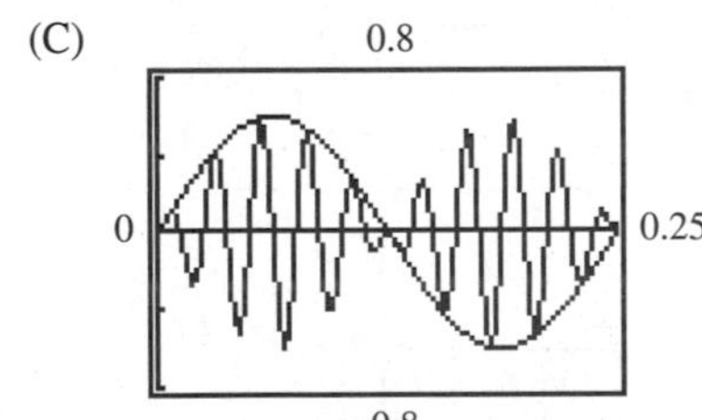

(D) $y_1 = 0.6 \sin 80\pi t \sin 8\pi t$

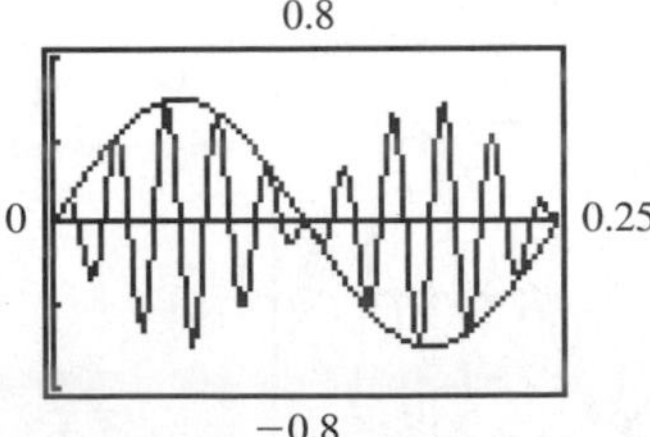

Chapter 4 Review Exercise

1. Equation (1) is an identity, because it is true for all replacements of x by real numbers for which both sides are defined. Equation (2) is a conditional equation, because it is only true for $x = -2$ and $x = 3$; it is not true, for example, for $x = 0$. [4.1]

13. $\frac{1}{2} = \frac{1}{2}$ [4.4] **14.** $1/\sqrt{2} = 1/\sqrt{2}$ [4.4]

15. $\frac{1}{2}\cos 3t - \frac{1}{2}\cos 13t$ [4.5]

16. $2 \sin 3w \cos 2w$ [4.5]

21. The equation is not an identity. The equation is not true for $x = \pi/2$, for example, and both sides are defined for $x = \pi/2$. [4.1]

22. Graph each side of the equation in the same viewing window and observe that the graphs are not the same, except where the graph of $y_1 = \sin x$ crosses the x axis. (Note that the graph of $y_2 = 0$ is the x axis.)

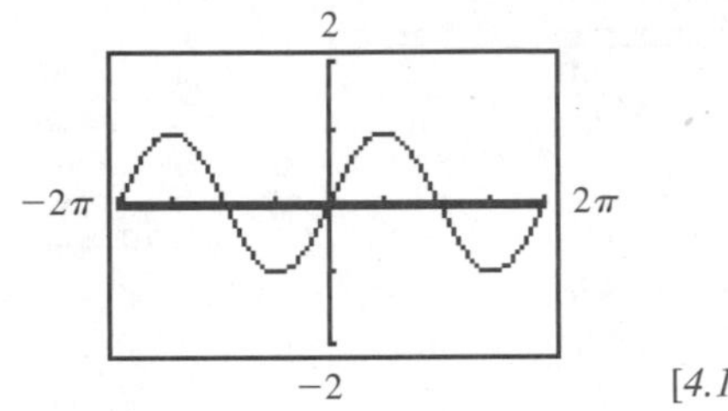

[4.1]

23. (A) 0.9394 (B) 0.1176 [4.1]

24. No. The equation is an identity for all real values of x, except for $x = k\pi$, k an integer (neither side of the equation is defined for these values). [4.2]

25. Graph each side of the equation in the same viewing window and observe that the graphs are not the same.

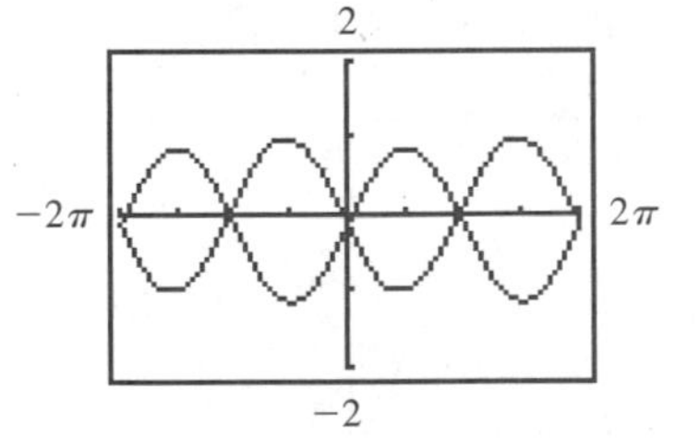

[4.3]

26. Find a value of x for which both sides are defined but are not equal. Try $x = 1$, for example:
$\sin(1 - 3) = -0.9093$ and $\sin 1 - \sin 3 = 0.7004$. [4.3]

42. $\dfrac{1}{2} - \dfrac{\sqrt{3}}{4}$ [4.5] **43.** $-\sqrt{6}/2$ [4.5]

44. $\sec x = -\frac{3}{2}$, $\sin x = \sqrt{5}/3$, $\csc x = 3/\sqrt{5}$, $\tan x = -\sqrt{5}/2$, $\cot x = -2/\sqrt{5}$ [4.1]

45. $\sin 2x = \frac{24}{25}$, $\cos 2x = -\frac{7}{25}$, $\tan 2x = -\frac{24}{7}$ [4.1, 4.4]

46. $\sin(x/2) = -3/\sqrt{13}$, $\cos(x/2) = 2/\sqrt{13}$, $\tan(x/2) = -\frac{3}{2}$ [4.2]

47. $\tan(x + \pi/4) = (\tan x + 1)/(1 - \tan x)$

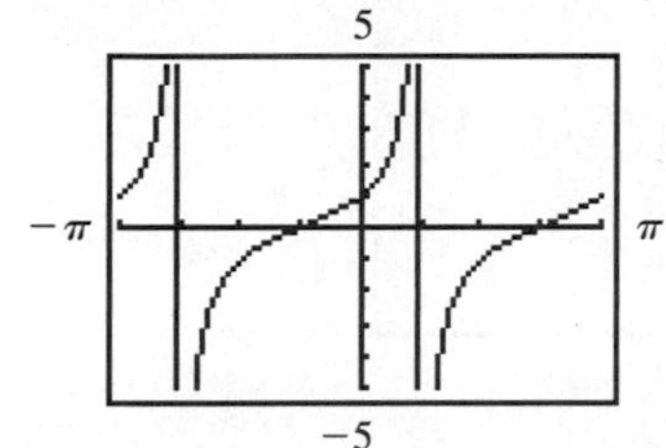

[4.3]

48. $y_2 = \cos(1.8x)$

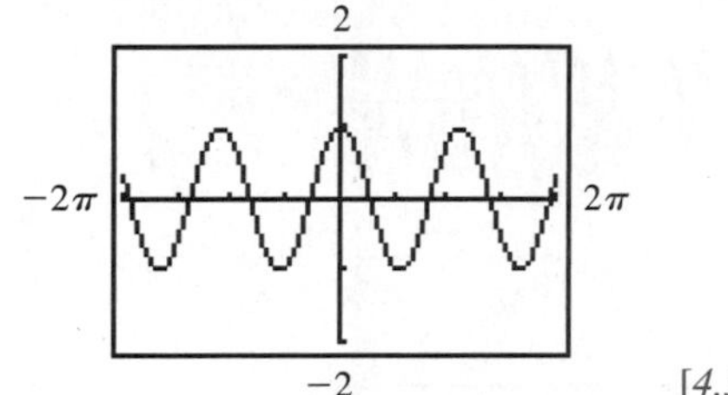

[4.3]

49. Identities for $-2\pi \le x \le 0$

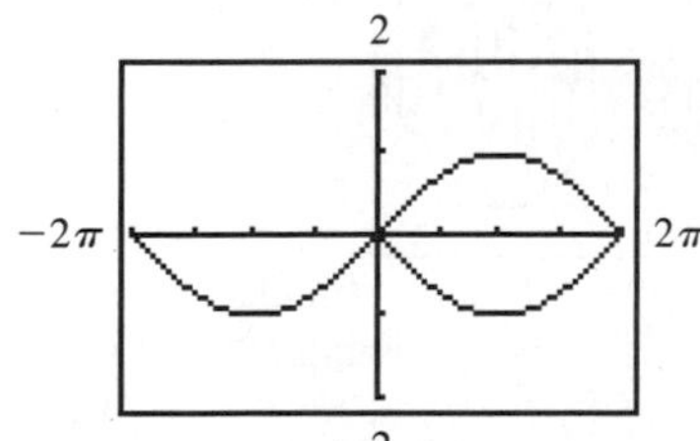

[4.4]

50. (A) Not an identity; for example, both sides are defined for $x = 0$, but are not equal.
(B) An identity [4.2]

51. $\sin x = -5/\sqrt{26}$, $\cos x = 1/\sqrt{26}$, $\tan x = -5$ [4.4]

58. $g(x) = 4 + 2\cos x$

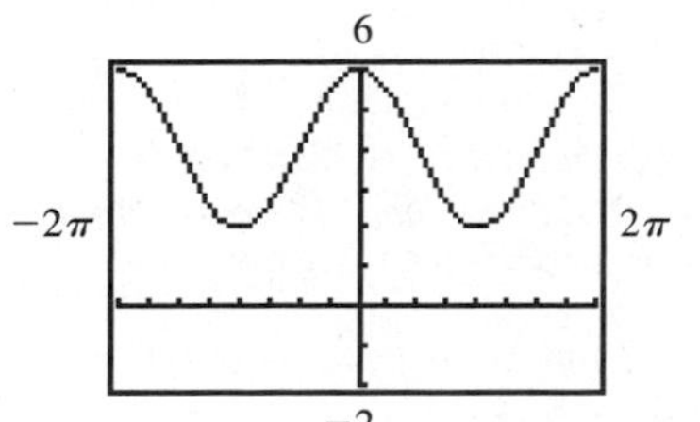

[4.2]

59. $g(x) = \tan 2x$

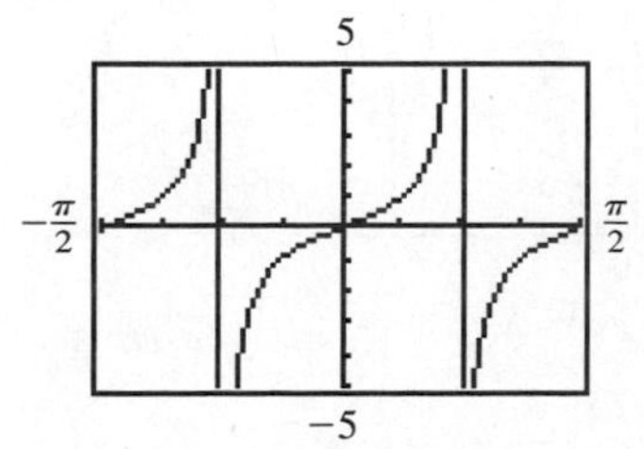

[4.4]

60. $g(x) = 2 - \cos 2x$

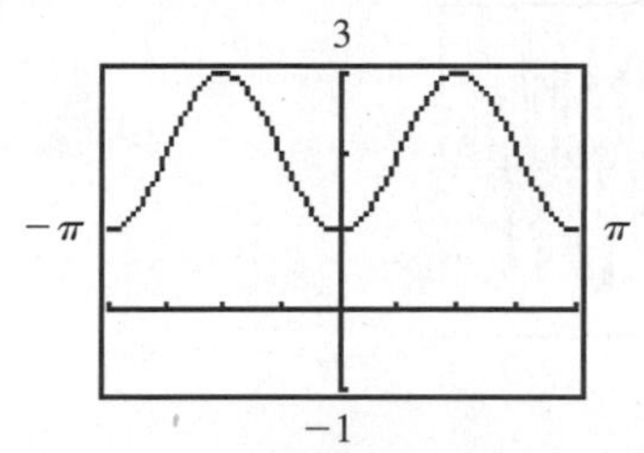

[4.4]

61. $g(x) = -2 + \sec 2x$

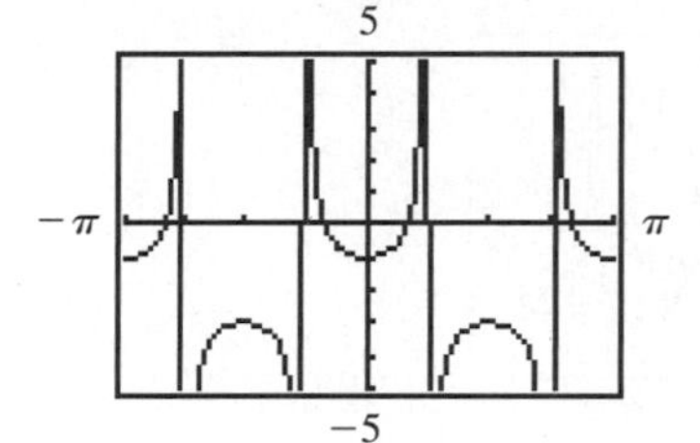

[4.4]

62. $g(x) = 2 + \cot(x/2)$

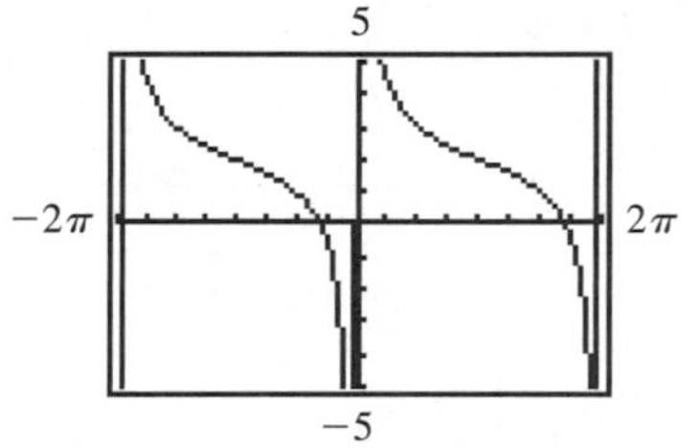

5

−2π 2π

−5

[4.4]

63. $-2\pi \le x < -\pi, 0 \le x < \pi$

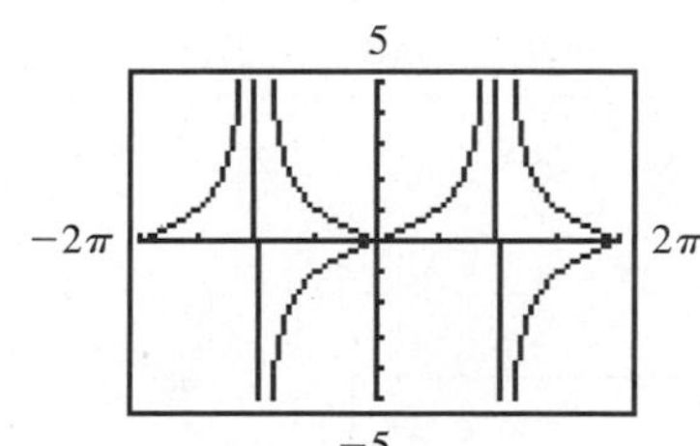

[4.4]

64. $-\pi < x \le 0, \pi < x \le 2\pi$

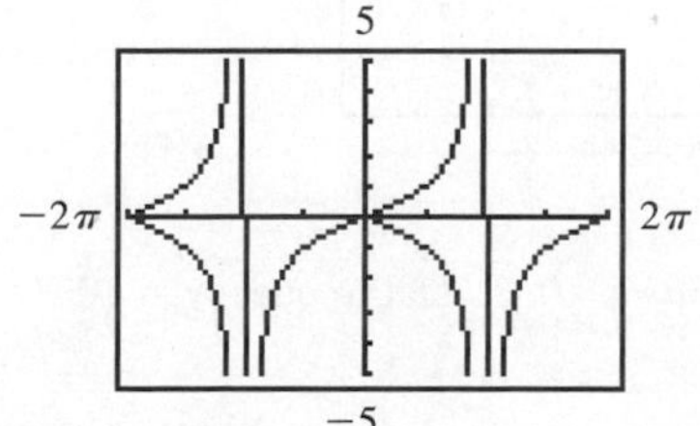

[4.4]

65.

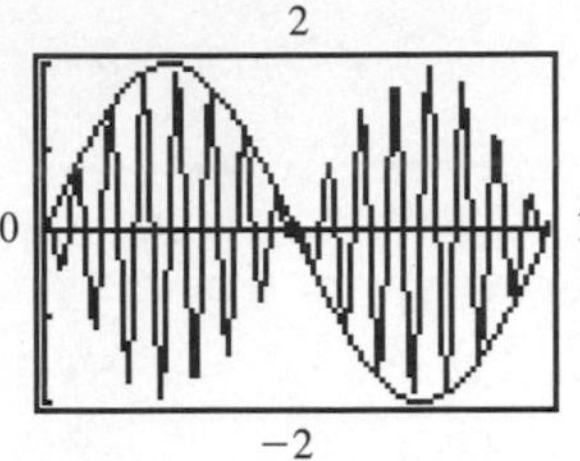

[4.5]

66. $y_1 = \sin 32\pi x - \sin 28\pi x$

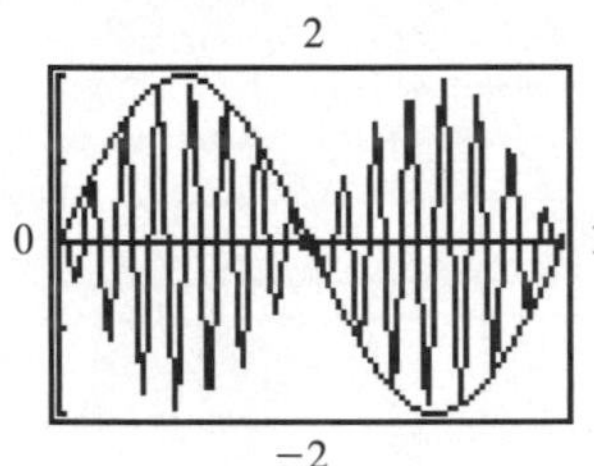

[4.5]

67. $a \tan x$ [4.1] **68.** 57.5° [4.3]

69. $x = \frac{445}{39} \approx 11.410$; $\theta \approx 32.005°$ [4.4]

70. $\theta = 0.464$ rad [4.3]

71. (B) L steadily increases.

(C)

TABLE 1

θ (deg)	30	35	40	45
L (ft)	250.7	252.6	254.6	256.6
θ (deg)	50	55	60	
L (ft)	258.7	260.8	263.1	

(D) Min L = 250.7 ft; Max L = 263.1 ft

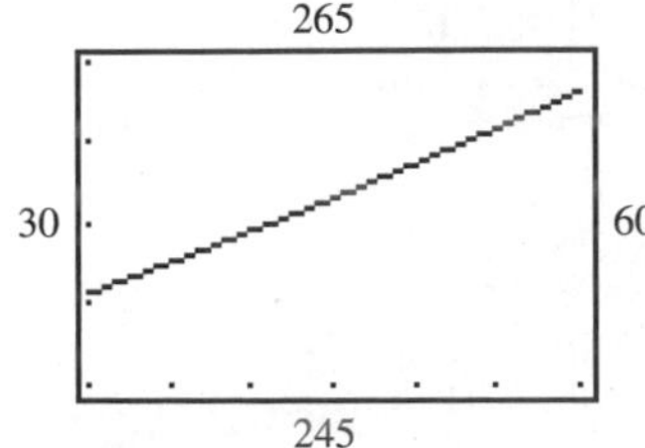

[4.4]

72. $y = 0.6 \sin 130\pi t \sin 10\pi t$; Beat frequency = 10 Hz [4.5]

73. (A)

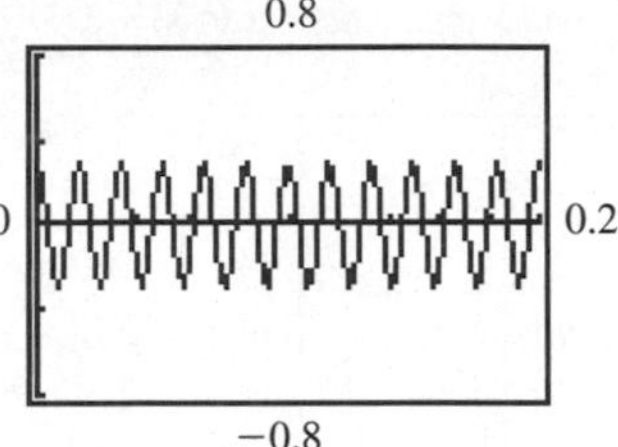

(B)

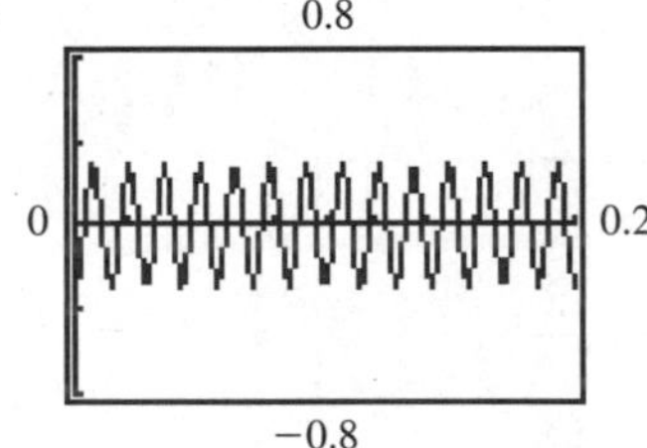

(C)

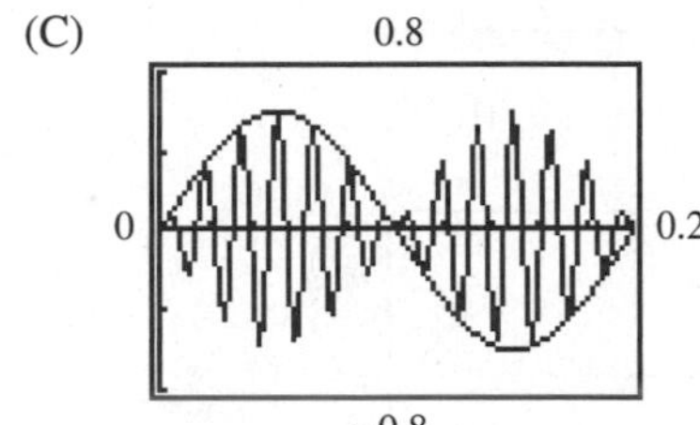

(D)

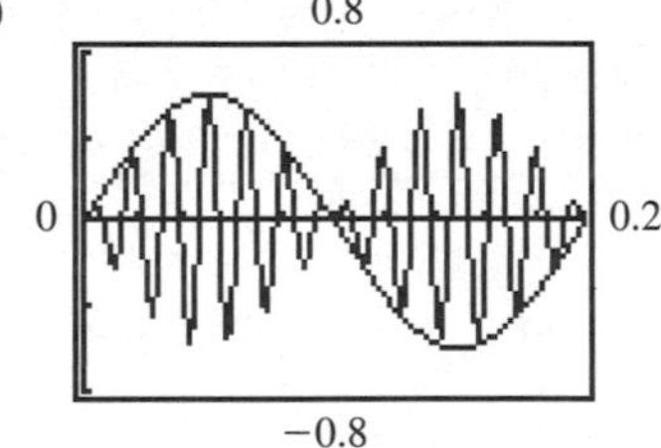

[4.5]

74. $y = 10 \sin(3t + 3.79)$; Amplitude = 10 cm; Period = $2\pi/3$ sec; Frequency = $3/2\pi$ Hz; Phase shift = − 1.26 sec [*Chap. 4 Group Activity*]

75. t intercepts: − 1.26, − 0.21, 0.83, 1.88; Phase shift = − 1.26

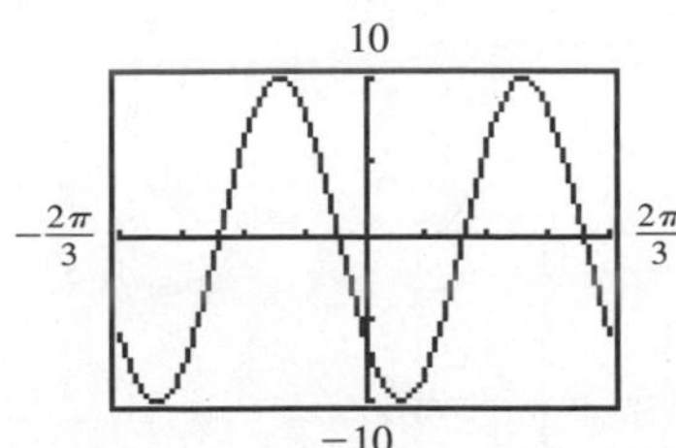

[*Chap. 4 Group Activity*]

CHAPTER 5

Exercise 5.1

1. 0 **3.** $\pi/6$ **5.** $\pi/4$ **7.** $\pi/3$ **9.** 1.155 **11.** 1.548 **13.** Not defined **15.** $x = \sin 37 = 0.601815$ **17.** $2\pi/3$ **19.** $-\pi/4$ **21.** $-\pi/3$ **23.** $5\pi/6$ **25.** -0.6 **27.** $\sqrt{2}/2$ **29.** -1.328 **31.** 1.001 **33.** 2.456

35.

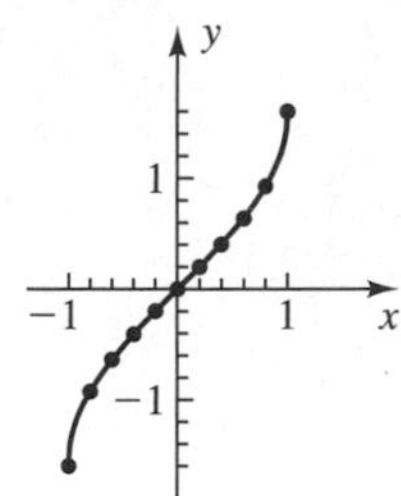

37. 120° **39.** $-45°$ **41.** $-60°$ **43.** 71.80° **45.** $-54.16°$ **47.** $-86.69°$

49. 0.3; does not illustrate a cosine–inverse cosine identity, because $\cos^{-1}(\cos x) = x$ only if $0 \le x \le \pi$.

51. (A)

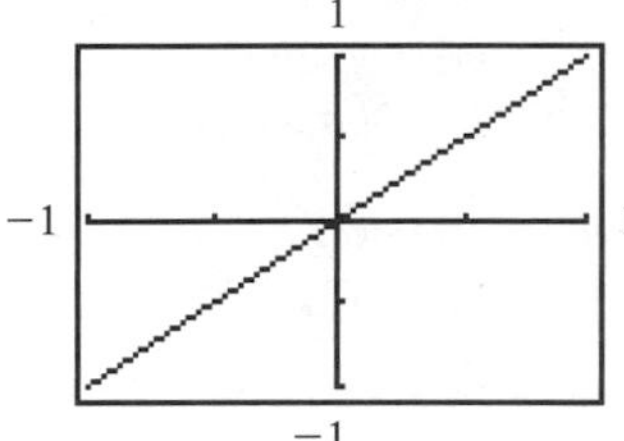

(B) The domain of $\sin^{-1}$ is restricted to $-1 \le x \le 1$; hence no graph will appear for other values of x.

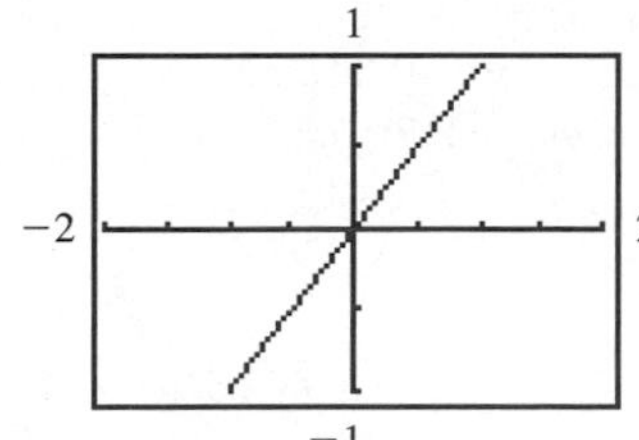

53. $-\frac{1}{2}$ **55.** $-\frac{24}{25}$ **57.** $\sqrt{1 - x^2}$ **59.** $\dfrac{x}{\sqrt{1 - x^2}}$

63. (A) $\cos^{-1} x$ has domain $-1 \le x \le 1$; therefore, $\cos^{-1}(2x - 3)$ has domain $-1 \le 2x - 3 \le 1$, or $1 \le x \le 2$.

(B) The graph appears only for the domain values $1 \le x \le 2$.

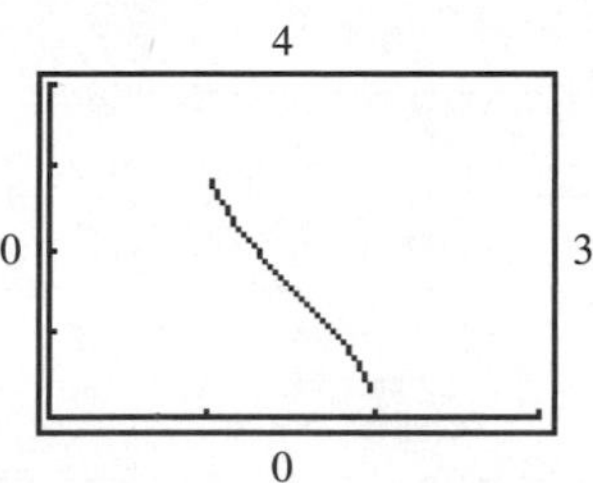

65. (A) $h^{-1}(x) = 1 + \sin^{-1}[(x - 3)/5]$

(B) $\sin^{-1} x$ has domain $-1 \le x \le 1$; therefore, $1 + \sin^{-1}[(x - 3)/5]$ has domain $-1 \le (x - 3)/5 \le 1$, or $-2 \le x \le 8$.

67. (A)

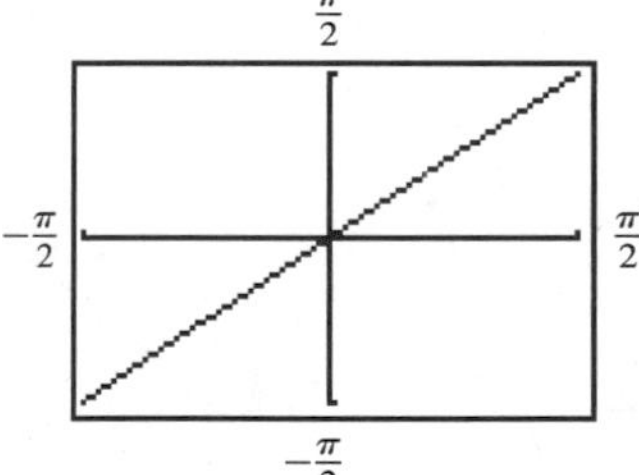

(B) The domain for $\sin x$ is $(-\infty, \infty)$ and the range is $[-1, 1]$, which is the domain for $\sin^{-1} x$. Thus, $y = \sin^{-1}(\sin x)$ has a graph over the interval $(-\infty, \infty)$, but $\sin^{-1}(\sin x) = x$ has a graph only on the restricted domain of $\sin x$, $[-\pi/2, \pi/2]$.

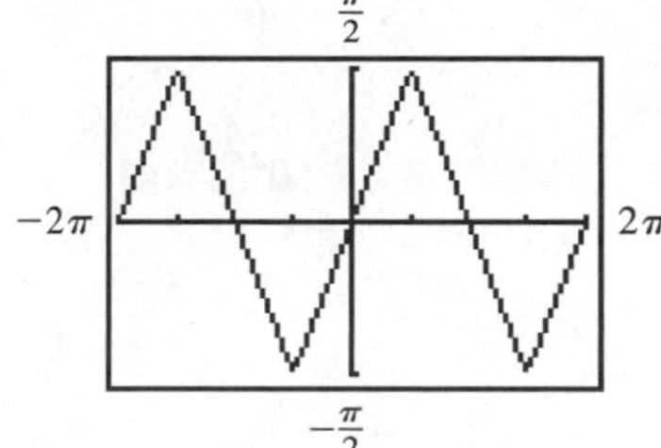

69. (A) $\theta = 2\sin^{-1}(1/M)$, $M > 1$ (B) 72°; 52°

71. (A) It appears that θ increases, and then decreases.

71. (C) From the table, Max $\theta = 9.21°$ when $x = 25$ yd.

TABLE 1				
x (yd)	10	15	20	25
θ (deg)	6.44	8.19	9.01	9.21
x (yd)	30	35		
θ (deg)	9.04	8.68		

(D) The angle θ increases rapidly until a maximum is reached, and then declines more slowly.
Max $\theta = 9.21°$ when $x = 24.68$ yd.

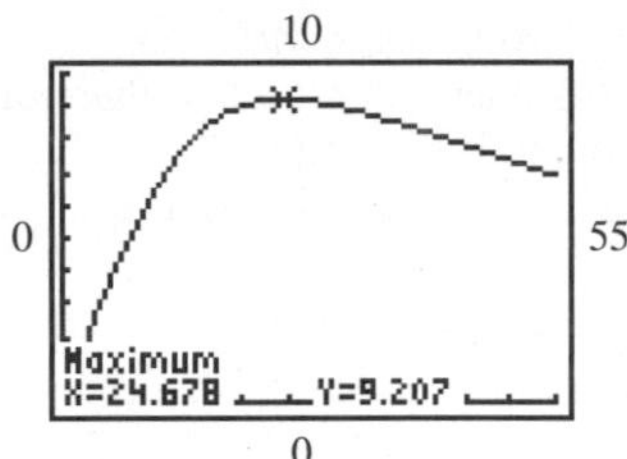

73. (B) 248 ft^3
(C) 2.6 ft

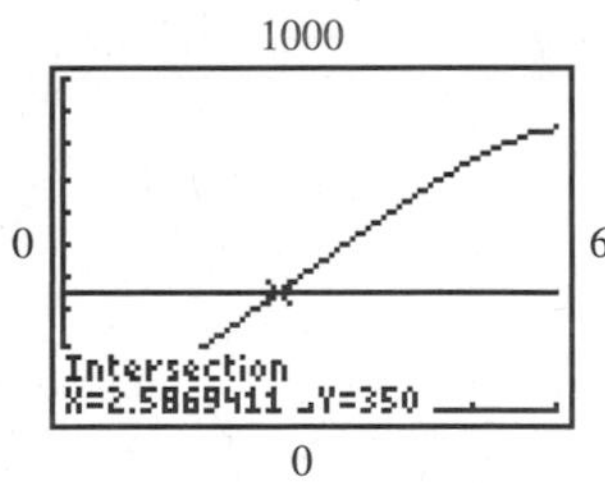

Exercise 5.2

1. $\pi/6$ **3.** $\pi/2$ **5.** $\pi/4$ **7.** 1 **9.** $\frac{4}{3}$
11. $3\pi/4$ **13.** $2\pi/3$ **15.** $-\pi/6$ **17.** Not defined
19. $\frac{4}{5}$ **21.** $-\frac{4}{3}$ **23.** $-\frac{1}{2}$ **25.** 33.4 **27.** -4
29. 0.315 **31.** 2.105 **33.** 1.022 **35.** 2.948
37. 120° **39.** 135° **41.** $-60°$ **43.** 71.80°
45. $-54.16°$ **47.** 93.31° **49.** $-\frac{8}{19}$ **51.** $\frac{24}{7}$
53. $\dfrac{1}{\sqrt{x^2+1}}$ **55.** $\dfrac{|x|}{\sqrt{x^2-1}}$ **57.** $\dfrac{2x}{x^2+1}$
61.

π
-5 5
$-\pi$

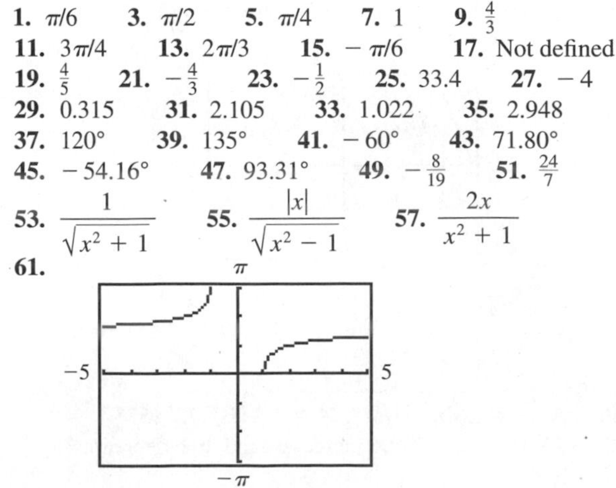

63.

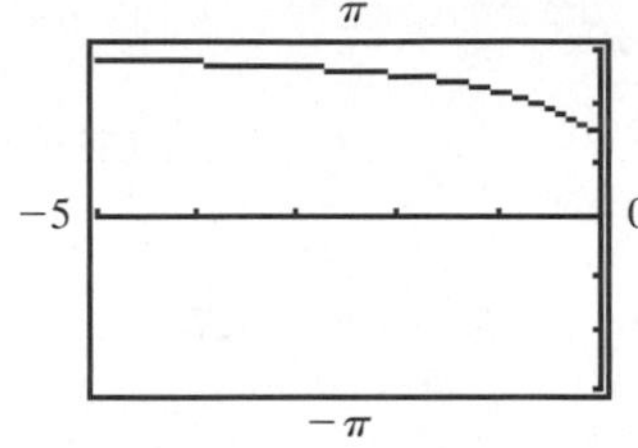

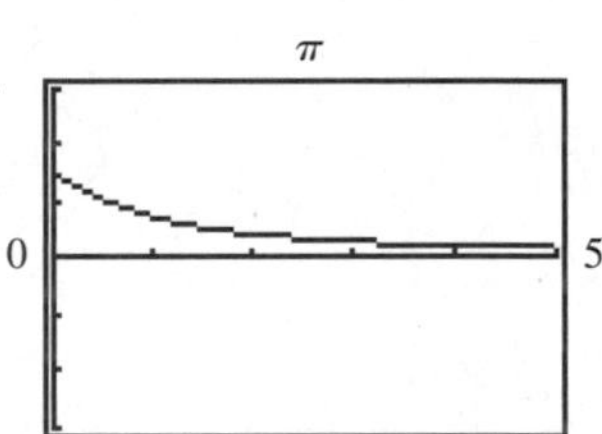

Exercise 5.3

1. $2\pi/3, 4\pi/3$
3. $2\pi/3 + 2k\pi, 4\pi/3 + 2k\pi$, k any integer
5. 45°, 135°
7. $45° + k(360°), 135° + k(360°)$, k any integer
9. 104.9314° **11.** 1.1593, 5.1239
13. $0.6696 + 2k\pi, 2.4720 + 2k\pi$, k any integer
15. $\pi/2, 3\pi/2$
17. $90° + k(180°), 45° + k(180°)$, k any integer
19. $\pi/2$ **21.** 180° **23.** 90°, 270°
25. $\pi/6, 5\pi/6, 7\pi/6, 11\pi/6$ **27.** 104.5°
29. 0.9987, 5.284
31. $0.9987 + 2k\pi, -0.9987 + 2k\pi$, k any integer
33. $\cos^{-1}(-0.7334)$ has exactly one value, 2.3941; the equation $\cos x = -0.7334$ has infinitely many solutions, which are found by adding $2\pi k$, k any integer, to each solution in one period of $\cos x$.
35. 0, $\pi/2$ **37.** 0 **39.** 0.002613 sec **41.** 33.21°
43. 64.1° **45.** $(r, \theta) = (1, 30°), (1, 150°)$
47. $\theta = 45°$

Exercise 5.4

1. 0.4502 **3.** 0.6167 **5.** 0.9987, 5.2845
7. $0.9987 + 2k\pi, 5.2845 + 2k\pi$, k any integer
9. $(-1.5099, 1.8281)$
11. $[0.4204, 1.2346], [2.9752, \infty)$
13. Because $\sin^2 x - 2\sin x + 1 = (\sin x - 1)^2$, and the latter is greater than or equal to 0 for all real x.
15. 0.006104, 0.006137 **17.** $[-2\pi, 2\pi]$
19. $[0.2974, 1.6073] \cup [2.7097, 3.1416]$
21. 1.78 rad **23.** 35.64 ft^2
25. (A) $L = 12.4575$ mm (B) $y = 2.6495$ mm

Chapter 5 Review Exercise

1. $\pi/6$ [5.1] **2.** $\pi/6$ [5.1] **3.** $-\pi/3$ [5.1]
4. $2\pi/3$ [5.1] **5.** $\pi/4$ [5.1] **6.** $-\pi/3$ [5.1]
7. $2\pi/3$ [5.2] **8.** $2\pi/3$ [5.2] **9.** $-\pi/2$ [5.2]
10. $3\pi/4$ [5.2] **11.** -0.9750 [5.1]
12. Not defined [5.1] **13.** 1.557 [5.1]
14. 0.1757 [5.2] **15.** 69.78° [5.1]
16. $-85.41°$ [5.1] **17.** 0.14° [5.1]
18. $\pi/6, 11\pi/6$ [5.3] **19.** 0°, 30°, 150°, 180° [5.3]
20. $\pi/6, 5\pi/6, 7\pi/6, 11\pi/6$ [5.3]
21. 120°, 180°, 240° [5.3] **22.** $\pi/16, 3\pi/16$ [5.3]
23. $-120°$ [5.3] **24.** $x = 0.906308$ [5.1]
25. 0.315 [5.1] **26.** -1.5 [5.1] **27.** $-\frac{3}{5}$ [5.1]
28. $-2/\sqrt{5}$ [5.1] **29.** $\sqrt{10}/3$ [5.2]
30. $2\sqrt{6}/5$ [5.2] **31.** -1.192 [5.1]
32. Not defined [5.1] **33.** -0.9975 [5.1]
34. 1.095 [5.1] **35.** 0.9894 [5.2]
36. 1.001 [5.2]
37. Figure (a) is in degree mode and illustrates a sine–inverse sine identity, since 2 is in the domain for this identity, $-90° \le \theta \le 90°$. Figure (b) is in radian mode and does not illustrate this identity, since 2 is not in the domain for the identity, $-\pi/2 \le x \le \pi/2$. [5.1]
38. 90°, 270° [5.3] **39.** $\pi/12, 5\pi/12$ [5.3]
40. $-\pi$ [5.3]
41. $k(360°)$, $180° + k(360°)$, $30° + k(360°)$, $150° + k(360°)$, k any integer [5.3]
42. $2k\pi$, $\pi + 2k\pi$, $\pi/6 + 2k\pi$, $-\pi/6 + 2k\pi$, k any integer [5.3]
43. 120°, 240° [5.3] **44.** $7\pi/12, 11\pi/12$ [5.3]
45. $0.7878 + 2k\pi$, $2.354 + 2k\pi$, k any integer [5.3]
46. $-1.343 + k\pi$, k any integer [5.3]
47. $0.6259 + 2k\pi$, $2.516 + 2k\pi$, k any integer [5.3]
48. $1.178 + k\pi$, $-0.3927 + k\pi$, k any integer [5.3]
49. 0.253, 2.889 [5.4] **50.** -0.245, 2.897 [5.4]
51. ± 1.047 [5.3] **52.** ± 0.824 [5.4]
53. 0 [5.4] **54.** 4.227, 5.197 [5.4]
55. 0.604, 2.797, 7.246, 8.203 [5.4]
56. 0.228, 1.008 [5.4]
57. The domain for y_1 is the domain for $\sin^{-1} x$, $-1 \le x \le 1$.

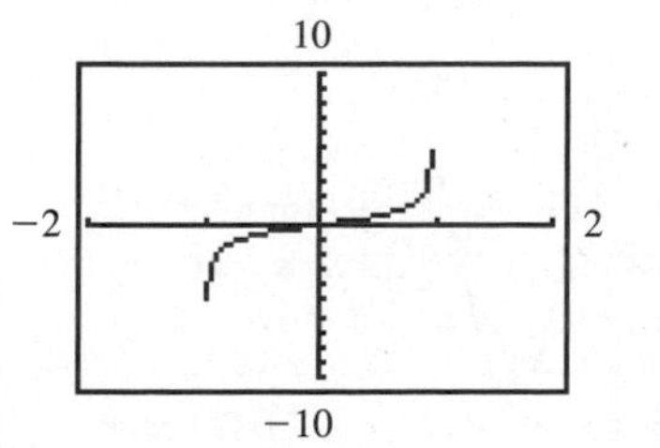

[5.1]

58. (A) $\sin^{-1} x$ has domain $-1 \le x \le 1$; therefore, $\sin^{-1}[(x-2)/2]$ has domain $-1 \le (x-2)/2 \le 1$, or $0 \le x \le 4$.
(B) The graph appears only for the domain values $0 \le x \le 4$.

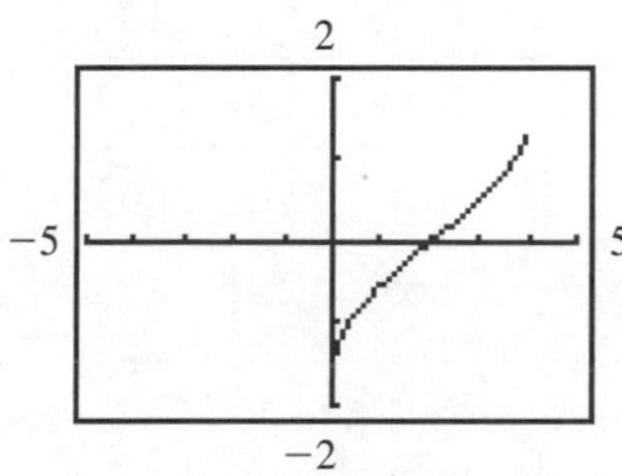

[5.1]

59. No, $\tan^{-1} 23.255$ represents only one number and one solution. The equation has infinitely many solutions, which are given by $x = \tan^{-1} 23.255 + k\pi$, k any integer. [5.3]
60. $0, \pi/2$ [5.4] **61.** $0, \pi/3, 2\pi/3$ [5.4]
62. 0.375, 2.77 [5.4] **63.** $-\frac{24}{25}$ [5.1]
64. $\frac{24}{25}$ [5.1] **65.** $\dfrac{x}{\sqrt{1-x^2}}$ [5.1]
66. $\dfrac{1}{\sqrt{x^2+1}}$ [5.1]
67. (A)

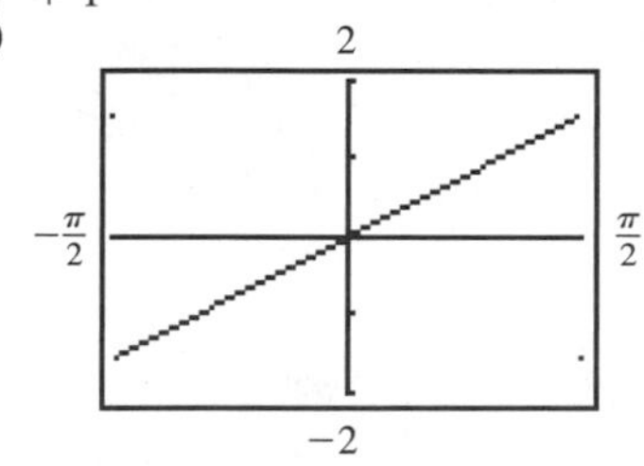

(B) The domain for $\tan x$ is the set of all real numbers, except $x = \pi/2 + k\pi$, k an integer. The range is R, which is the domain for $\tan^{-1} x$. Thus, $y = \tan^{-1}(\tan x)$ has a graph for all real x, except $x = \pi/2 + k\pi$, k an integer. But $\tan^{-1}(\tan x) = x$ only on the restricted domain of $\tan x$, $-\pi/2 < x < \pi/2$.

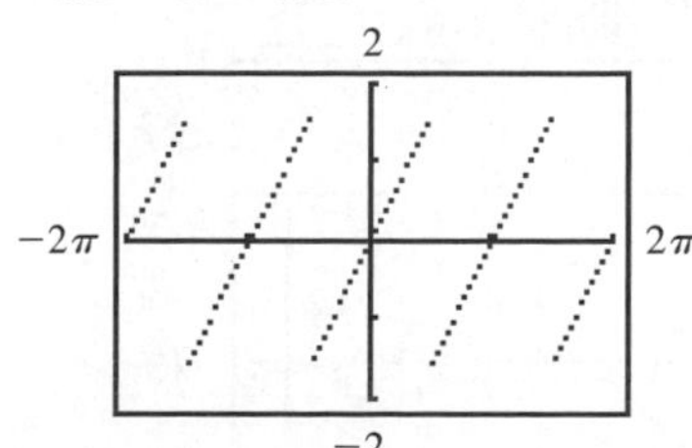

[5.1]

68. 0.00024 sec [5.3]
69. 0.001936 sec [5.3]

70. (A) $\theta = 2 \arctan(500/x)$ (B) 45.2° [5.1]

71. (C) From Table 1, Max $\theta = 2.73°$ when $x = 15$ ft.

TABLE 1

x (ft)	0	5	10	15
θ (deg)	0.00	1.58	2.47	2.73
x (ft)	20	25	30	
θ (deg)	2.65	2.46	2.25	

(D) As x increases from 0 ft to 30 ft, θ increases rapidly at first to a maximum of about 2.73° at 15 ft, then decreases more slowly.

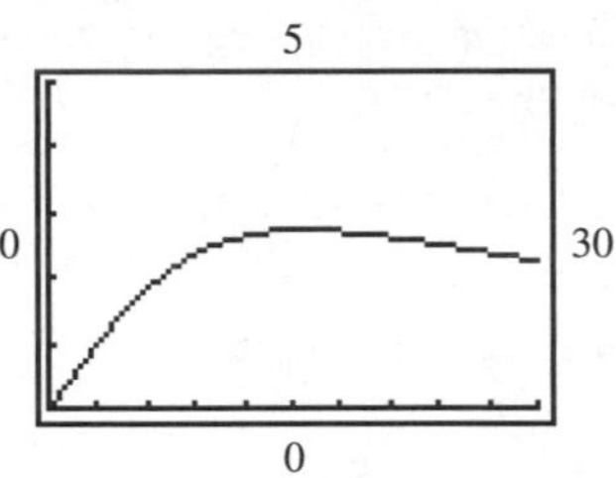

(E) From the graph, Max $\theta = 2.73°$ when $x = 15.73$ ft.

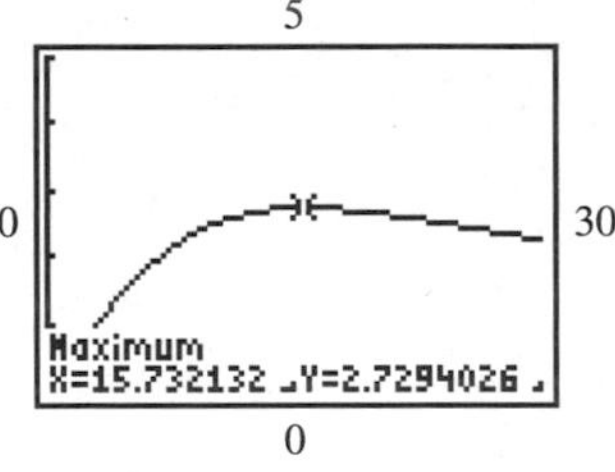

(F) $\theta = 2.5°$ when $x = 10.28$ ft or when $x = 24.08$ ft

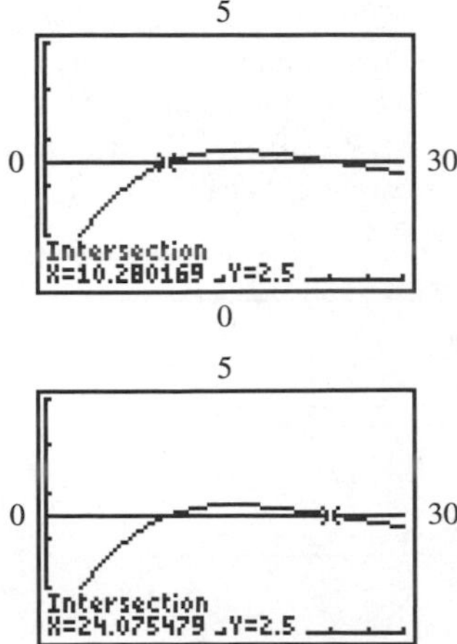

[5.1, 5.4]

72. (B)

TABLE 2

d (ft)	8	10	12	14	16
A (ft²)	704	701	697	690	682
d (ft)	18	20			
A (ft²)	670	654			

(C) $d = 17.2$ ft

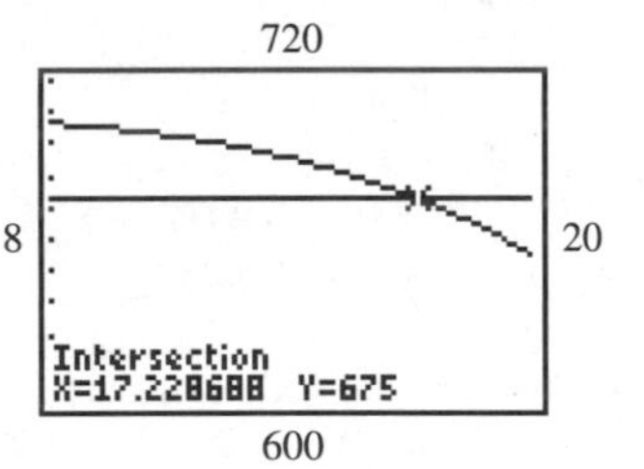

[5.1, 5.4]

73. 2.82 ft [5.4]

74. (A)

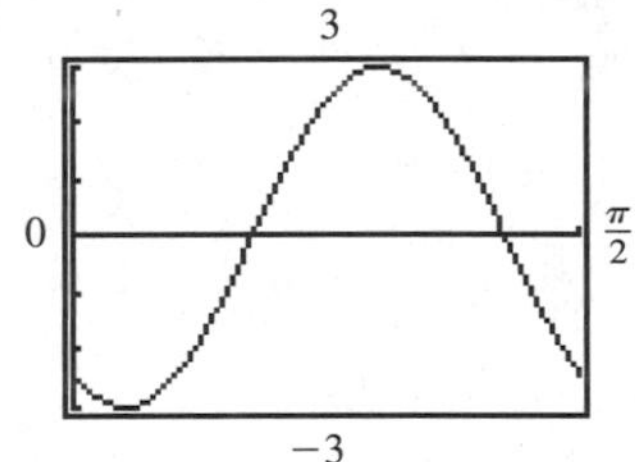

(B) 0.74 sec, 1.16 sec (C) 0.37 sec, 1.52 sec [5.4]

Cumulative Review Exercise Chapters 1–5

1. 241.22° [2.1] **2.** 8.83 rad [1.1, 2.1]

3. An angle of radian measure 0.5 is the central angle of a circle subtended by an arc with measure half that of the radius of the circle. [2.1]

4. 54°, 36°, 6.1 m [1.3]

5. $\cos \theta = -\frac{5}{13}$; $\cot \theta = \frac{5}{12}$ [2.3]

6. (A) 0.3786 (B) 15.09 (C) −0.5196 [2.3, 2.6]

7. (A)

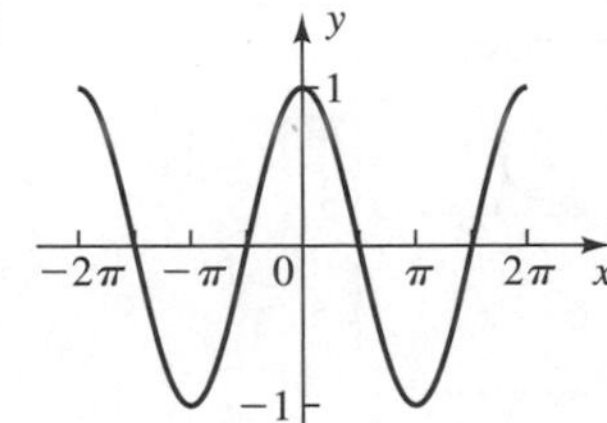

(B)

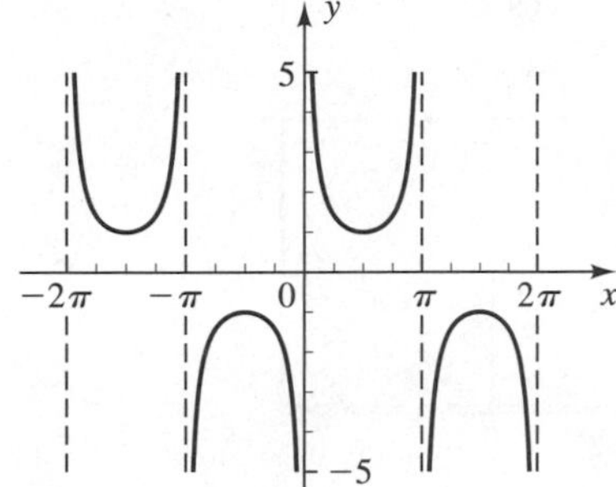

(C)

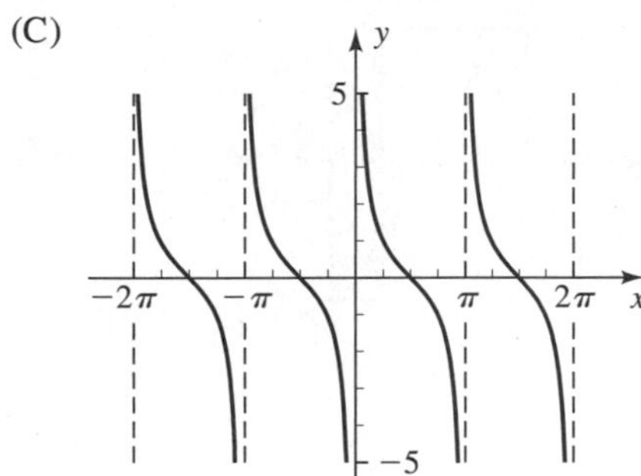

[3.1, 3.2, 3.6]

8. No, because $\sin \theta = 1/\csc \theta$, and either both are positive or both are negative. [2.3]

13. No, because both sides are not equal for other values of x for which they are defined; for example, they are not equal for $x = 0$ or $x = \pi/2$. [4.1]

14. $1/\sqrt{2}$ [2.5] **15.** $\sqrt{3}$ [2.5]

16. Undefined [2.5] **17.** 0 [5.1] **18.** $5\pi/6$ [5.1]

19. Undefined [5.1] **20.** $5\pi/6$ [5.2]

21. 0.0505 [1.3, 5.1] **22.** 2.379 [1.3, 5.1]

23. -1.460 [1.3, 5.1] **24.** 0.3218 [5.2]

25. 1.176 [5.2]

26. $x = \tan(-1) = -1.557408$ [5.1]

27. 30°, 150° [5.3] **28.** $-\pi/6$ [5.3]

29. $-\pi$ [5.3] **30.** $\frac{1}{2}(\sin 10u + \sin 4u)$ [4.5]

31. $-2 \sin 3w \sin 2w$ [4.5]

32. Yes; using the formula $\theta_d = \dfrac{180°}{\pi}\theta_r$, if θ_r is halved, then θ_d will also be halved. [2.1]

33. 92°27′43″ [1.1] **34.** 144° [1.1]

35. $x = 6$; $\theta = 33.7°$ [1.3] **36.** $2\pi/5$ [2.1]

37. $\sin \theta = -1/\sqrt{5}$, $\cos \theta = -2/\sqrt{5}$, $\cot \theta = 2$, $\sec \theta = -\sqrt{5}/2$, $\csc \theta = -\sqrt{5}$ [2.3]

38. Period $= 4\pi$; Amplitude $= 2$

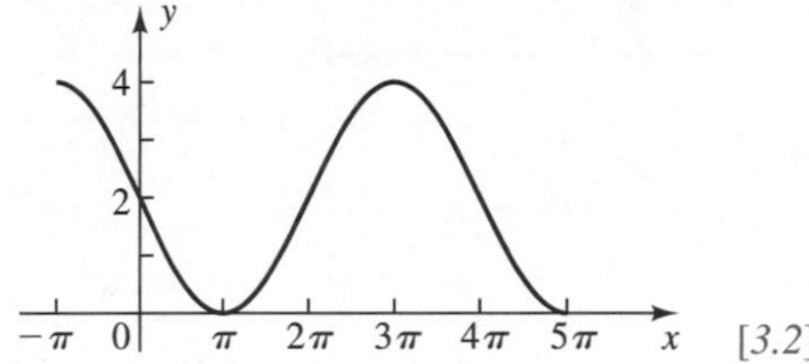

[3.2]

39. Period $= \pi$; Amplitude $= 3$; Phase shift $= \pi/2$

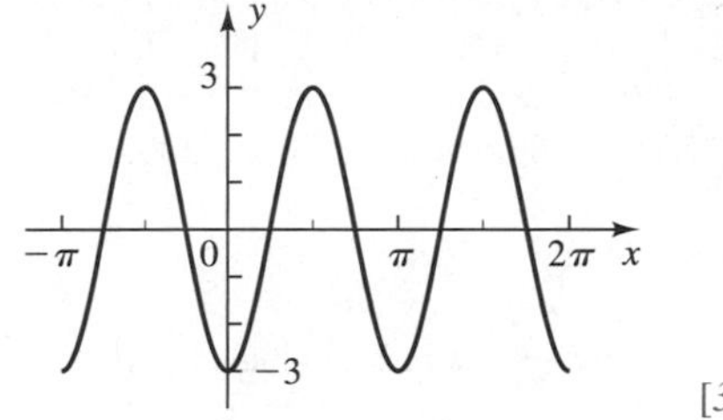

[3.3]

40. Period $= 1$; Phase shift $= \frac{1}{4}$

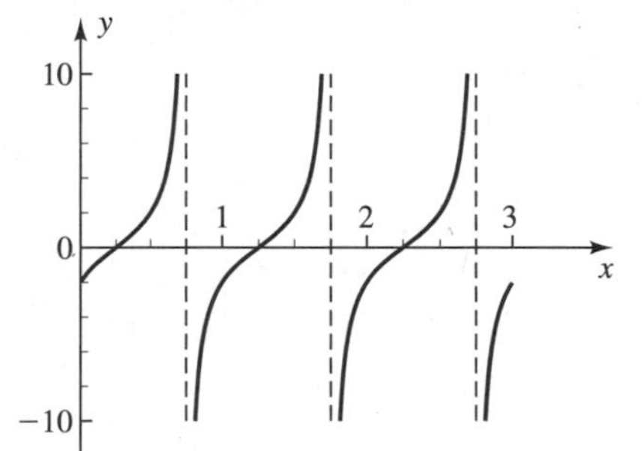

[3.6]

41. $y = 1 + 3\cos(\pi x)$ [3.2]

42. $y = -1 + 2\sin(\pi x/2)$ [3.2]

43. Hypotenuse: 46.0 cm; Angles: 25°0′, 65°0′ [1.3]

49. (A) -0.4969 (B) 0.7531 [4.1]

50. $y = \sin(3x - x) = \sin 2x$

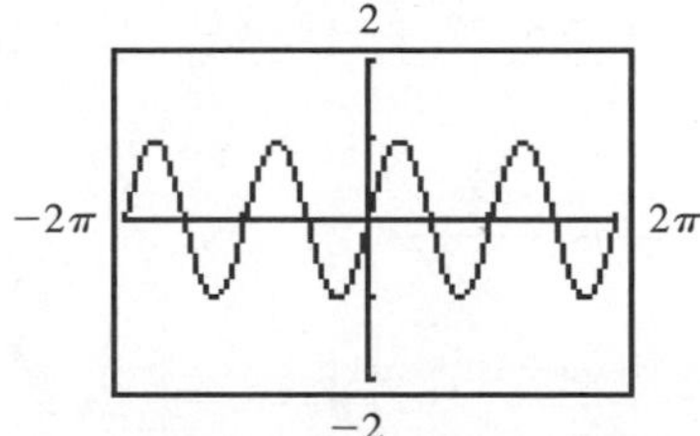

[4.3]

51. $\sin(x/2) = 4/\sqrt{17}$; $\cos 2x = \frac{161}{289}$ [2.3, 4.4]

52. $[-2\pi, \pi] \cup [\pi, 2\pi]$

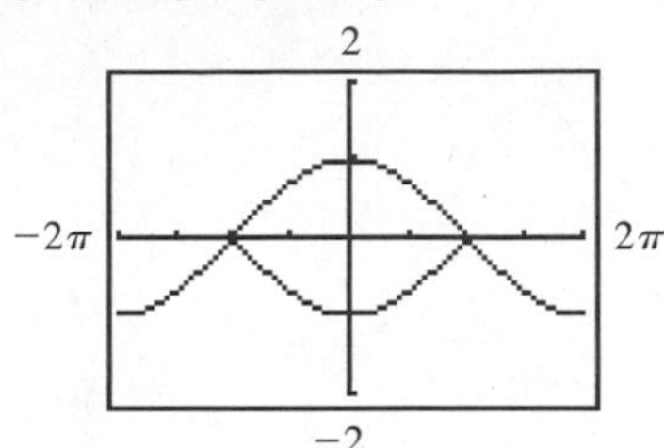

[4.4]

53. (A) Not an identity. Both sides are defined at $x = \pi/4$, but are not equal. (B) An identity. [4.2]

54. (A) $\sqrt{21}/5$ [5.1] (B) $\sqrt{6}$ [5.1] (C) 3 [5.1] (D) $\sqrt{15}$ [5.2]

55. 60° [1.3, 5.1] **56.** 64.01° [1.3, 5.1]

57. $7\pi/6, 3\pi/2, 11\pi/6$ [5.3]

58. $90° + k(180°)$, k any integer [5.3]

59. $\pi/3 + k\pi, 2\pi/3 + k\pi$, k any integer [5.3]

60. $3.745 + 2k\pi, 5.679 + 2k\pi$, k any integer [5.3]

61. $1.130 + 2k\pi, 5.153 + 2k\pi$, k any integer [5.3]

62. $1.823 + 2k\pi, 4.460 + 2k\pi$, k any integer [5.3]

63. Figure (a) does not, because -3 is not in the restricted domain for the cosine–inverse cosine identity. Figure (b) does, because 2.51 is in the restricted domain for the cosine–inverse cosine identity. [5.1]

64. $-1.893, 1.249$ [5.4] **65.** 0.642 [5.4]

66. 0, 3.895, 5.121, 9.834, 12.566 [5.4]

67. The coordinates are (cos 28.703, sin 28.703) = $(-0.9095, -0.4157)$. The point P is in the third quadrant, since both coordinates are negative. [2.6]

68. $s = \cos^{-1} 0.5313 = 1.0107$ or $s = \sin^{-1} 0.8472 = 1.0107$ [2.6]

69. $\sin\theta = -\sqrt{1 - a^2}$, $\tan\theta = \dfrac{-\sqrt{1 - a^2}}{a}$,

$\cot\theta = \dfrac{-a}{\sqrt{1 - a^2}}$, $\sec\theta = \dfrac{1}{a}$, $\csc\theta = \dfrac{-1}{\sqrt{1 - a^2}}$

[2.3, 2.6]

71. $\frac{7}{9}$ [4.4, 5.1] **72.** $\dfrac{\sqrt{1 - x^2} - x^2}{\sqrt{1 + x^2}}$ [5.1]

73. $\pi/2 + 2k\pi, \pi + 2k\pi$, k any integer [5.4]

74. 0.6662, 2.475 [5.4]

75. $g(x) = 3 - \cos x$

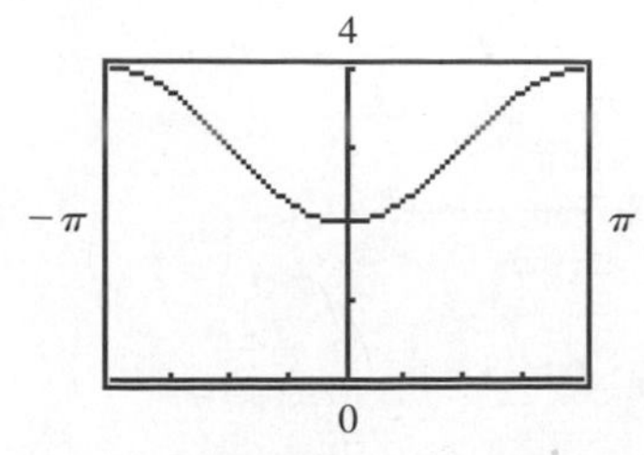

[4.2]

76. $g(x) = 4 + 2\cos 2x$

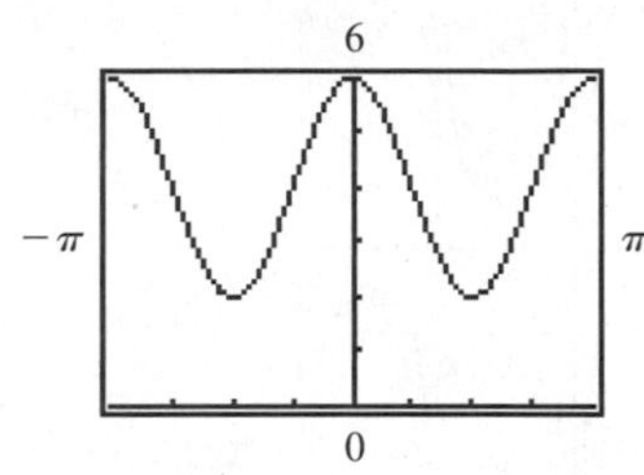

[4.4]

77. $g(x) = 1 + \sec 2x$

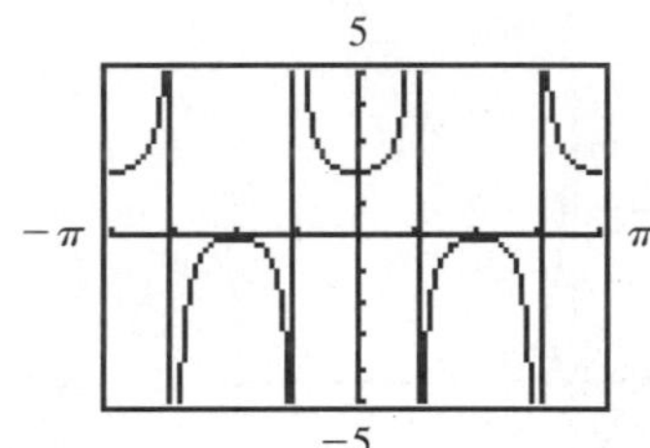

[4.4]

78. $g(x) = 3 - \cot(x/2)$

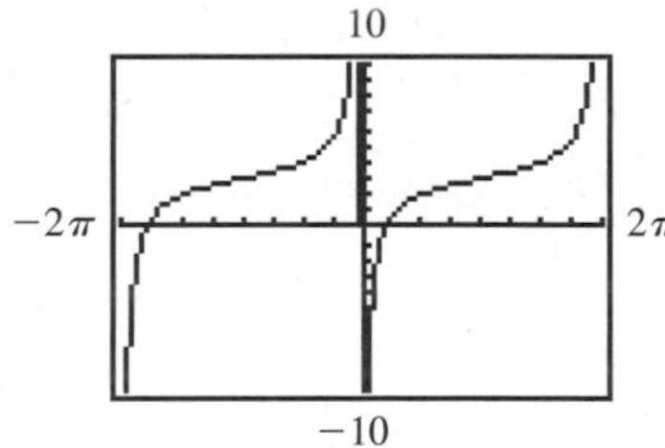

[4.4]

79. [0.694, 2.000] [5.4] **80.** 1,429 m [1.4]

81. $a \cot x$ [4.1] **82.** $x = 2\sqrt{3}$; $\theta = 30°$ [4.4]

83. 0.443 sec [5.3]

84. 0.529 sec, 0.985 sec

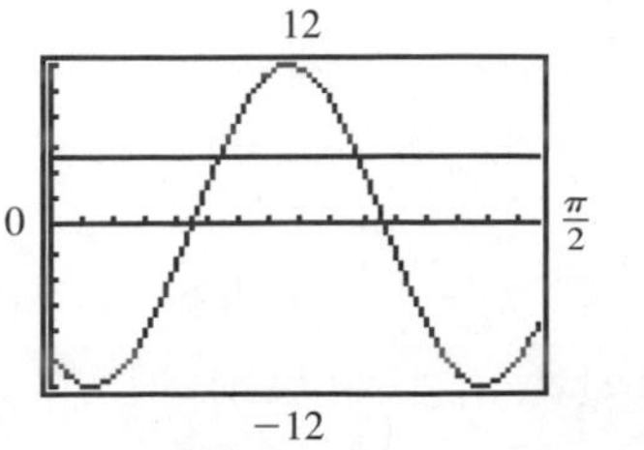

[5.4]

85. $\theta = 0.650$ rad [1.4, 4.3]

86. (A) Amplitude: 60; Period: $\frac{1}{45}$ sec; Frequency 45 Hz; Phase shift: $\frac{1}{180}$

(B)

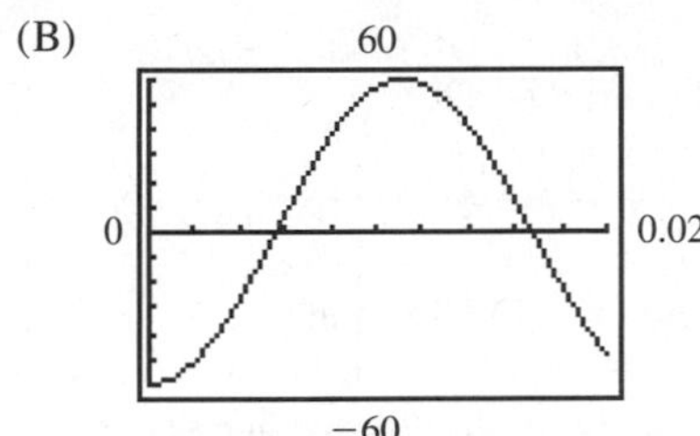

(C) $t = 0.0078$ sec

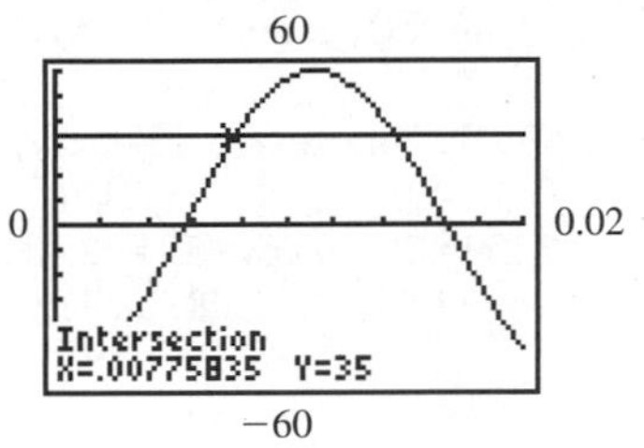

[3.3]

87. (A) $d = 200 \cot \theta$

(B)

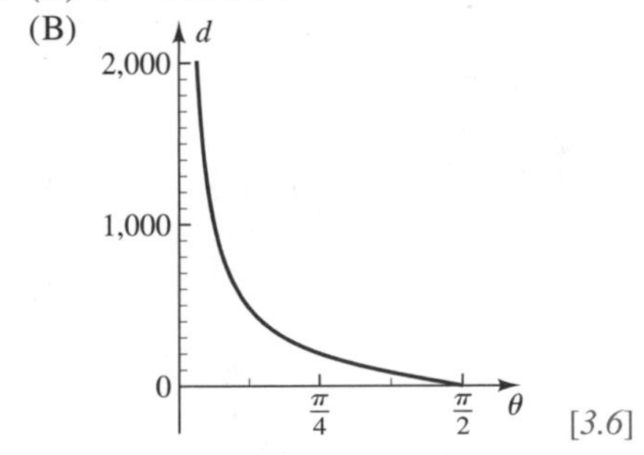

[3.6]

88. (B) 12.5° [5.1]

89. 64.9 ft or 308.3 ft

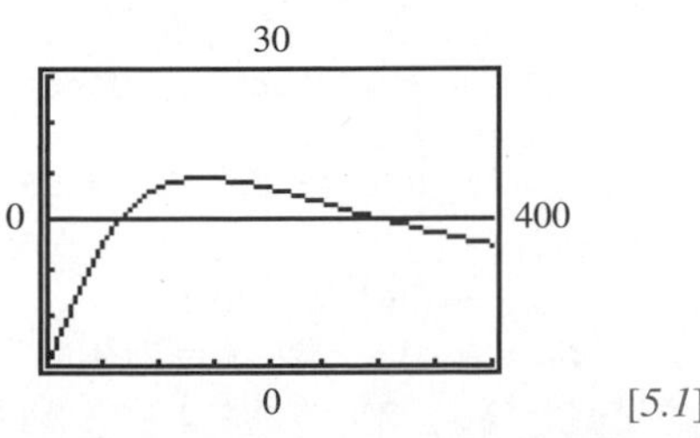

[5.1]

90. 49.5π rad [2.1]

91. $11{,}550\pi \approx 36{,}300$ cm/min [2.2]

92. (A) L will decrease, then increase.

(C)

TABLE 1				
θ (deg)	35	40	45	50
L (ft)	117.7	110.4	106.1	104.2
θ (deg)	55	60	65	
L (ft)	104.6	107.7	114.3	

(D) From Table 1, Min $L = 104.2$ ft for $\theta = 50°$. The length of the longest log that will go around the corner is the minimum length L.

(E) Min $L = 104.0$ ft for $\theta = 51.6°$

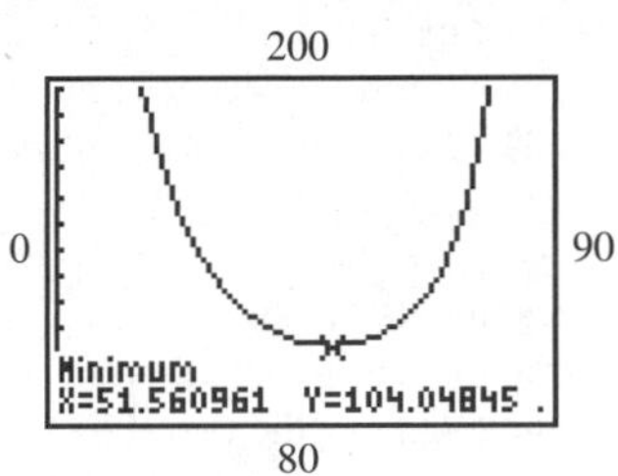

[1.4, 2.3]

93. (A)

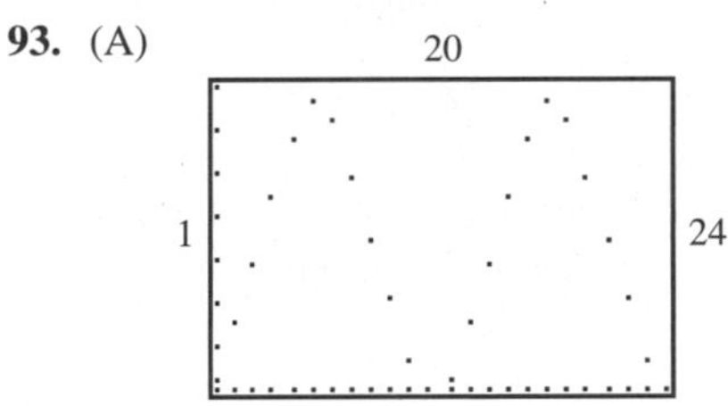

(B) $y = 12.435 + 6.835 \sin(\pi x/6 - 1.6)$

(C)

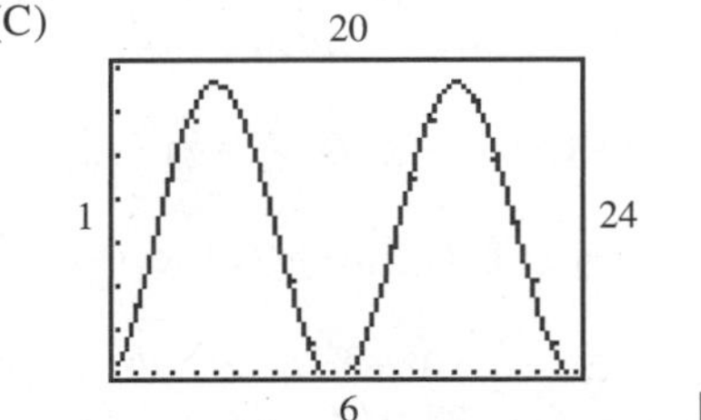

[3.3]

CHAPTER 6

Exercise 6.1

1. $\alpha = 101°$, $b = 64$ cm, $c = 55$ cm

3. $\alpha = 98.0°$, $b = 4.32$ mm, $c = 7.62$ mm

5. $\gamma = 22.9°$, $a = 27.3$ cm, $c = 12.6$ cm

7. $\alpha = 67°20'$, $a = 55.1$ km, $c = 58.8$ km

9. 1 triangle; case (b), where α is acute and $a = 3 = h$

11. 1 triangle; case (d), where α is acute and $a \geq b$ ($a = 8, b = 6$)

13. 0 triangles; case (a), where α is acute and $0 < a < h$ $(a = 2, h = 3)$
15. 2 triangles; case (c), where α is acute and $h < a < b$ $(h = 3, a = 5, b = 6)$
17. No solution
19. No solution
21. $\alpha = 17°40'$, $\gamma = 128°30'$, $c = 1{,}740$ ft
23. Triangle 1: $\alpha = 124.0°$, $\gamma = 28.7°$, $c = 141$ ft;
Triangle 2: $\alpha' = 56.0°$, $\gamma' = 96.7°$, $c' = 292$ ft
25. $26.63 \approx 26.66$
29. $k = a \sin \beta = 66.8 \sin 46.8° \approx 48.7$
31. $AC = 12.8$ mi **33.** 8.37 mi from A; 4.50 mi from B
35. 230 m **37.** 2.8 nautical mi **39.** 109 ft
41. $\beta = 113.4°$; $c = 10.8$ m
43. $r = 9.73$ mm, $s = 10.7$ mm **45.** 284 mi
47. 4.42×10^7 km, 2.39×10^8 km **49.** 16 cm

Exercise 6.2

1. A triangle can have at most one obtuse angle. Since β is acute, then, if the triangle has an obtuse angle, it must be the angle opposite the longer of the two sides, a and c. Thus, γ, the angle opposite the shorter of the two sides, c, must be acute.
3. $a = 6.00$ mm, $\beta = 65°0'$, $\gamma = 64°20'$
5. $c = 26.2$ m, $\alpha = 33.3°$, $\beta = 12.7°$
7. If the triangle has an obtuse angle, then it must be the angle opposite the longest side—in this case, α.
9. $\alpha = 62.8°$, $\beta = 36.3°$, $\gamma = 80.9°$
11. $\alpha = 29°12'$, $\beta = 63°26'$, $\gamma = 87°22'$
13. $\alpha = 30.3°$, $a = 46.2$ ft, $c = 27.2$ ft
15. $b = 19.3$ m, $\alpha = 40.6°$, $\gamma = 72.9°$ **17.** No solution
19. $\alpha = 84.1°$, $\beta = 29.7°$, $\gamma = 66.2°$ **21.** No solution
23. $\beta = 28.2°$, $a = 984$ m, $c = 1{,}310$ m
25. No solution
27. Triangle 1: $\beta = 65.3°$, $\gamma = 68.0°$, $c = 23.1$ yd;
Triangle 2: $\beta' = 114.7°$, $\gamma' = 18.6°$, $c' = 7.93$ yd
29. An infinite number of similar triangles have those angles.
31. $b^2 = a^2 + c^2 - 2ac \cos 90° = a^2 + c^2 - 0 = a^2 + c^2$
33. $-0.872 \approx -0.873$ **37.** $CB = 613$ m
39. 113.3° **41.** 180 km **43.** 121 cm
45. $\alpha = 65°20'$, $\beta = 27°0'$, $\gamma = 87°40'$
47. $AC = 23.0$ ft; $AB = 17.0$ ft **49.** 638 mi
51. 74° **53.** $DC = 166$ m

Exercise 6.3

1. 102 m^2 **3.** 12 cm^2 **5.** 11.6 in.^{2} **7.** 40,900 ft^2
9. 45.4 cm^2 **11.** 129 m^2
13. $A_1 = A_3 = \dfrac{ab}{2} \sin(180° - \theta) = \dfrac{ab}{2} \sin \theta = A_2 = A_4$

Exercise 6.4

1. $|\mathbf{u} + \mathbf{v}| = 71$ km/hr; $\theta = 29°$
3. $|\mathbf{u} + \mathbf{v}| = 57$ lb; $\theta = 33°$
5. $|\mathbf{u} + \mathbf{v}| = 154$ knots; $\theta = 21.9°$
7. $|\mathbf{H}| = 35$ lb; $|\mathbf{V}| = 23$ lb
9. $|\mathbf{H}| = 178$ km/hr; $|\mathbf{V}| = 167$ km/hr
11. No. The magnitude of a geometric vector is the length (a nonnegative quantity) of the directed line segment.
13. $|\mathbf{u} + \mathbf{v}| = 190$ lb, $\alpha = 18°$
15. $|\mathbf{u} + \mathbf{v}| = 699$ mi/hr, $\alpha = 7.4°$
17. Since the zero vector has an arbitrary direction, it can be parallel to any vector.
19. Magnitude = 5.0 km/hr; Direction = 143°
21. 250 mi/hr; 350°
23. Magnitude = 2,500 lb; Direction = 13° (relative to $\mathbf{F}_1$)
25. 32°
27. 2,500 sin 15° = 650 lb; 2,500 cos 15° = 2,400 lb
29. $|\mathbf{H}| = 40$ lb; $|\mathbf{V}| = 33$ lb **31.** To the left

Exercise 6.5

1.

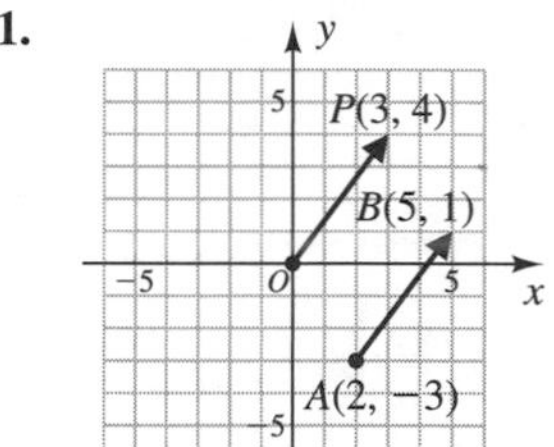

3.

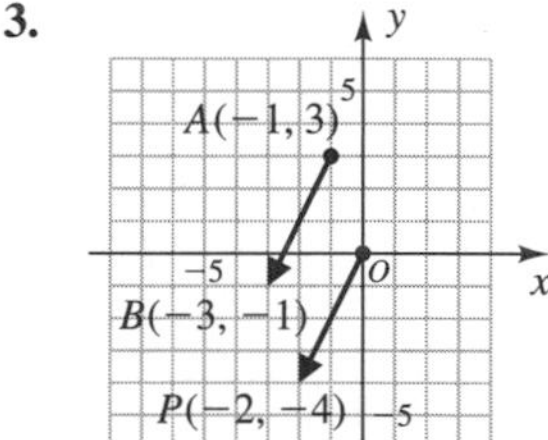

5. $\langle 4, 2\rangle$ **7.** $\langle 4, -4\rangle$ **9.** 5 **11.** $\sqrt{29}$
13. Two geometric vectors are equal if and only if they have the same magnitude and direction.
15. (A) $\langle -2, 6\rangle$ (B) $\langle 4, 2\rangle$ (C) $\langle 11, 2\rangle$ (D) $\langle 6, 18\rangle$
17. (A) $\langle 1, -6\rangle$ (B) $\langle 3, 0\rangle$ (C) $\langle 7, 3\rangle$ (D) $\langle 3, -6\rangle$
19. $\mathbf{u} = \langle \frac{4}{5}, -\frac{3}{5}\rangle$ **21.** $\mathbf{u} = \langle 2/\sqrt{13}, -3/\sqrt{13}\rangle$
23. $\mathbf{v} = 3\mathbf{i} - 2\mathbf{j}$ **25.** $\mathbf{v} = 4\mathbf{j}$ **27.** $\mathbf{v} = 2\mathbf{i} + 3\mathbf{j}$
29. $-\mathbf{i} - 7\mathbf{j}$ **31.** $-17\mathbf{j}$ **33.** $-4\mathbf{i} - \mathbf{j}$
35. Any one of the force vectors must have the same magnitude as the resultant of the other two force vectors and be directed opposite to the resultant of the other two.
45. Left side: 676 lb; Right side: 677 lb

47. For AB, a compression of 9,000 lb; for CB, a tension of 7,500 lb

49. $\theta = 18.6°$; Line tension = 10.1 lb

Exercise 6.6

1. -1 **3.** -19 **5.** 0 **7.** 0 **9.** 56.3°

11. 113.2° **13.** 90.0°

15. Negative, because $\cos \theta$ is negative for θ an obtuse angle.

17. Orthogonal **19.** Not orthogonal

21. $2/\sqrt{10} = 0.632$ **23.** 0 **25.** $5/\sqrt{2} = 3.54$

33. $\langle 3, 0 \rangle$ **35.** $-\frac{36}{13}\mathbf{i} - \frac{24}{13}\mathbf{j}$ **37.** 4,900 ft-lb

39. 85 ft-lb **41.** 9 ft-lb **43.** 20 ft-lb

Chapter 6 Review Exercise

1. To use the law of sines, we need to have an angle and a side opposite the angle given, which is not the case here. *[6.1]*

2. The two forces are oppositely directed; that is, the angle between the two forces is 180°. *[6.4]*

3. The forces are acting in the same direction; that is, the angle between the two forces is 0°. *[6.4]*

4. (A) 1 triangle, since $h = 8 \sin 30° = 4 = a$
(B) 2 triangles, since $h = 7 \sin 30° = 3.5$ and $h < a < b$
(C) 0 triangles, since $h = 8 \sin 30° = 4$ and $a < h$
[6.1]

5. $\beta = 22°, a = 90$ cm, $c = 110$ cm *[6.1]*

6. $\gamma = 82°, a = 21$ m, $c = 22$ m *[6.1]*

7. $a = 21$ in., $\beta = 53°, \gamma = 78°$ *[6.2]*

8. $\beta = 49°, \gamma = 69°, c = 15$ cm *[6.1]*

9. 224 in.^{2} *[6.3]* **10.** 79 cm^2 *[6.3]*

11. 9.4 at 32° *[6.4]* **12.** $|\mathbf{H}| = 9.8$; $|\mathbf{V}| = 6.9$ *[6.4]*

13. $\langle 2, -5 \rangle$ *[6.5]* **14.** 13 *[6.5]*

15. -8 *[6.6]* **16.** 4 *[6.6]* **17.** 36.9° *[6.6]*

18. 123.7° *[6.6]*

19. No triangle is possible, since $\sin \beta$ cannot exceed 1. *[6.1]*

20. $a = 97.7$ m, $\beta = 72.8°, \gamma = 42.2°$ *[6.2]*

21. $\beta = 43°30', \gamma = 101°10', c = 22.4$ in. *[6.1]*

22. $\beta = 136°30', \gamma = 8°10', c = 3.24$ in. *[6.1]*

23. $\alpha = 50°, \beta = 59°, \gamma = 71°$ *[6.2]*

24. 3,380 m^2 *[6.3]* **25.** 980 mm^2 *[6.3]*

26. The two vectors have the same magnitude, but different directions. *[6.4]*

27. They are equal if and only if $a = c$ and $b = d$. *[6.5]*

28. $|\mathbf{u} + \mathbf{v}| = 23.3$; $\theta = 15.9°$ *[6.4]*

29. (A) $\langle 2, -3 \rangle$ (B) $\langle 6, 3 \rangle$ (C) $\langle 16, 6 \rangle$ (D) $\langle 15, 8 \rangle$ *[6.5]*

30. (A) $5\mathbf{i} - 4\mathbf{j}$ (B) $\mathbf{i} + 2\mathbf{j}$ (C) $5\mathbf{i} + 3\mathbf{j}$ (D) $5\mathbf{j}$ *[6.5]*

31. $\mathbf{u} = \langle -\frac{8}{17}, \frac{15}{17} \rangle$ *[6.5]*

32. (A) $\mathbf{v} = -5\mathbf{i} + 7\mathbf{j}$ (B) $\mathbf{v} = -3\mathbf{j}$ (C) $\mathbf{v} = -4\mathbf{i} - \mathbf{j}$ *[6.5]*

33. (A) Orthogonal (B) Not orthogonal *[6.6]*

34. (A) $17/\sqrt{10} = 5.38$ (B) $-7/\sqrt{10} = -2.21$ *[6.6]*

35. Any one of the force vectors must have the same magnitude as the resultant of the other two force vectors and must be oppositely directed to the resultant of the other two. *[6.5]*

36. $-2.68 \approx -2.70$ (Checks) *[6.1]*

37. $k = 12.7 \sin 52.3°$ *[6.1]*

44. 5.9 cm, 17 cm *[6.2]* **45.** 252 m *[6.1]*

46. 25 cm *[6.1]* **47.** 540 ft *[6.2]*

48. 220,000 ft^2 *[6.3]* **49.** 4.00 km *[6.2]*

50. 660 mi *[6.2]* **51.** 325 mi *[6.1]*

52. 260 km/hr at 57° *[6.4]*

53. Magnitude = 489 lb; Direction = 13.4° *[6.4]*

54. 10,800 ft-lb *[6.6]*

55. Compression = 150 lb; Tension = 300 lb *[6.5]*

56. (A) $|\mathbf{u}| = |\mathbf{w}| \cos \theta$; $|\mathbf{v}| = |\mathbf{w}| \sin \theta$
(B) $|\mathbf{u}| = 40$ lb; $|\mathbf{v}| = 124$ lb (C) $\theta = 80°$ *[6.5]*

57. 56 ft-lb *[6.6]*

58. $\theta = 15.4°$; Tension = 19.5 lb *[6.5]*

CHAPTER 7

Exercise 7.1

1.

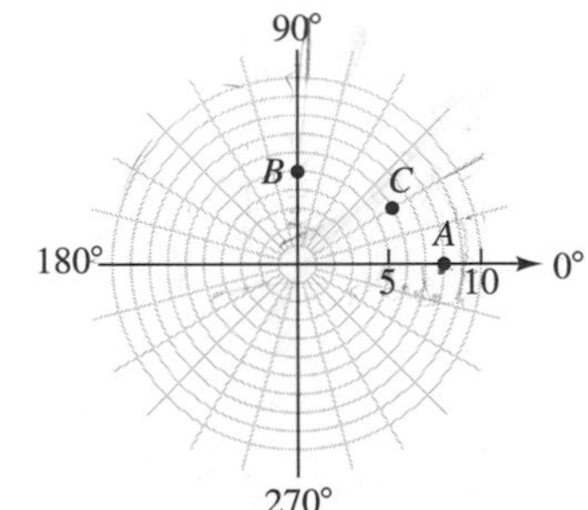

3.

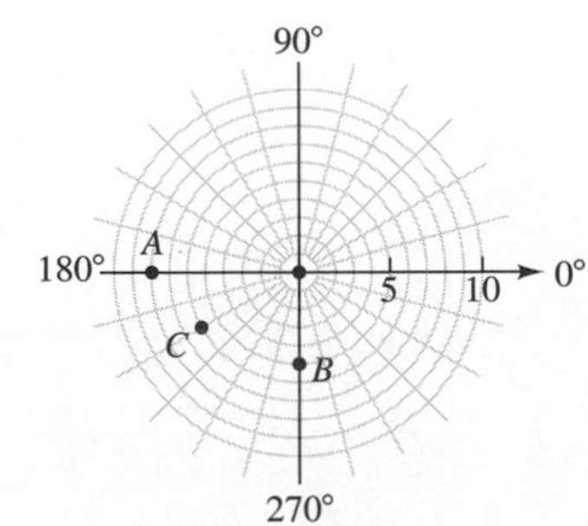

5.

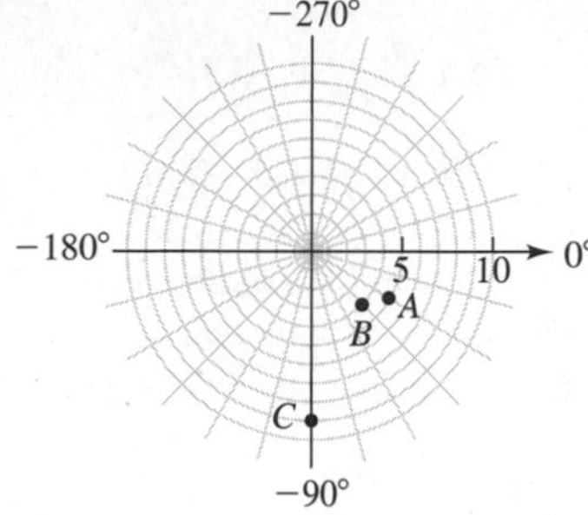

7.

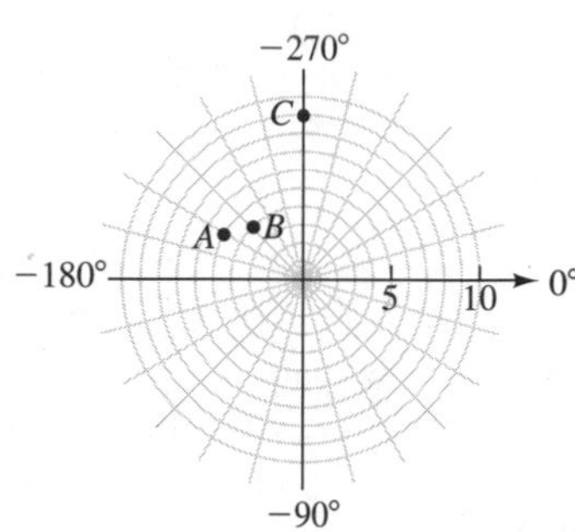

9.

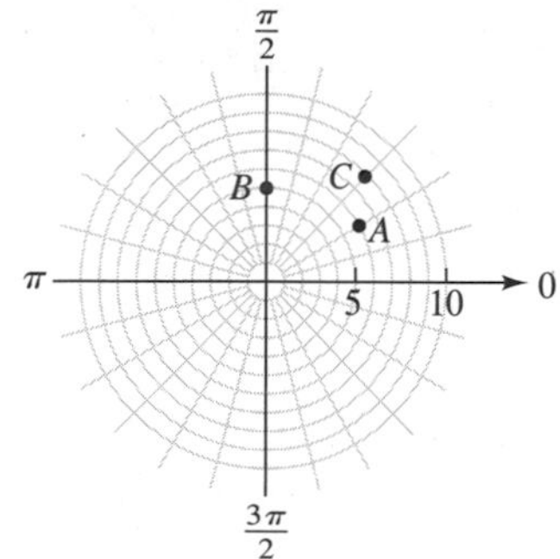

11.

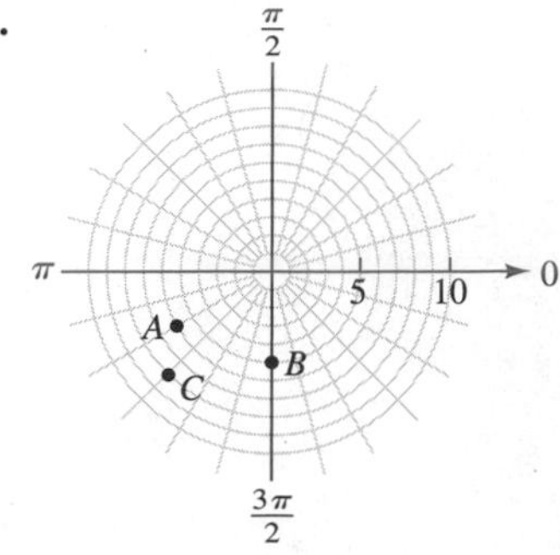

13.

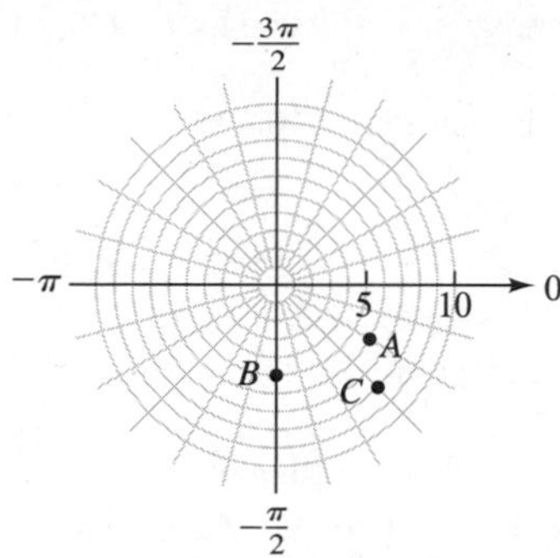

15. $(4, 4\sqrt{3})$ **17.** $(-2\sqrt{2}, -2\sqrt{2})$ **19.** $(-2\sqrt{3}, 2)$
21. $(4, \pi/6)$ **23.** $(8, -2\pi/3)$ **25.** $(7, -\pi/2)$
27. $(-6, -210°)$: The polar axis is rotated 210° clockwise (negative direction) and the point is located 6 units from the pole along the negative polar axis.
$(-6, 150°)$: The polar axis is rotated 150° counterclockwise (positive direction) and the point is located 6 units from the pole along the negative polar axis.
$(6, 330°)$: The polar axis is rotated 330° counterclockwise (positive direction) and the point is located 6 units along the positive polar axis.
29. $(8.362, 34.239°)$ **31.** $(9.376, 152.763°)$
33. $(31.626, -121.070°)$ **35.** $(-4.085, 5.766)$
37. $(-1.944, -3.228)$ **39.** $(-7.115, 5.557)$
41. $r = 6\cos\theta$ **43.** $r(2\cos\theta + 3\sin\theta) = 5$
45. $r^2 = 9$, or $r = \pm 3$ **47.** $r^2 = \dfrac{1}{\sin 2\theta}$
49. $r^2 = \dfrac{4}{4 - 5\sin^2\theta}$ **51.** $2x + y = 4$
53. $x^2 + y^2 = 8x$ **55.** $x^2 - y^2 = 4$
57. $x^2 + y^2 = 16$ **59.** $y = \dfrac{1}{\sqrt{3}}x$
61. $y - 3 = -2\sqrt{x^2 + y^2}$, or $(y - 3)^2 = 4(x^2 + y^2)$
63. 1.615 units

Exercise 7.2

1.

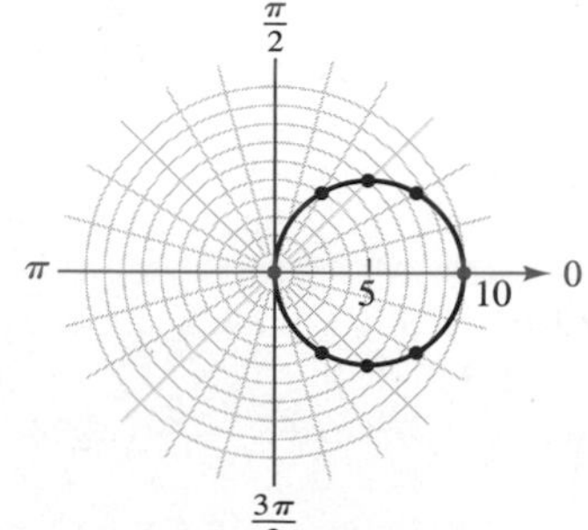

3.

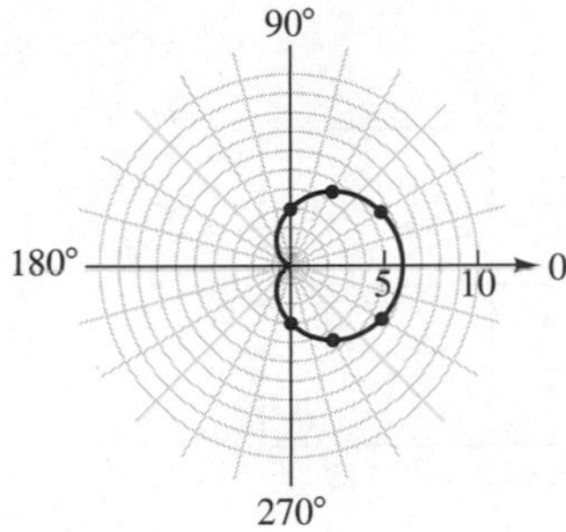

5.

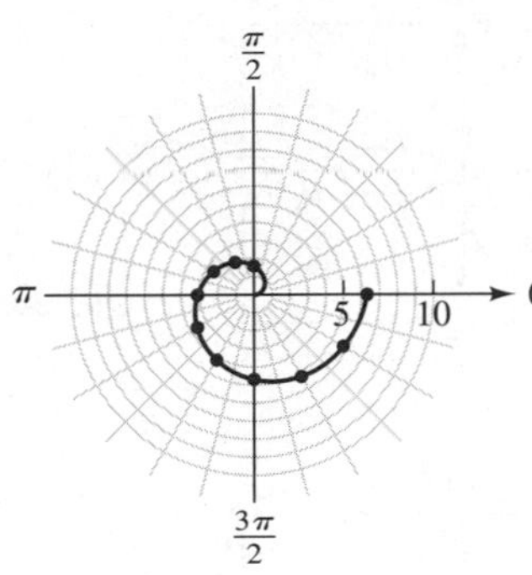

7.

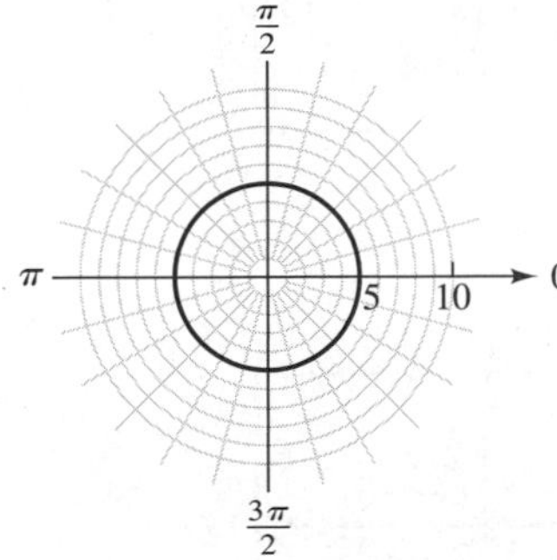

9.

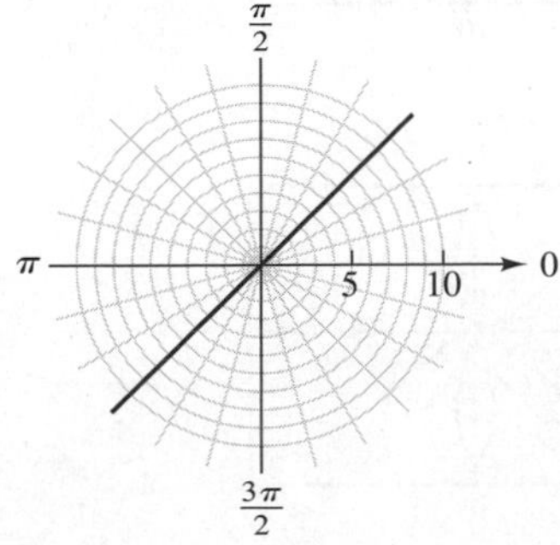

11.

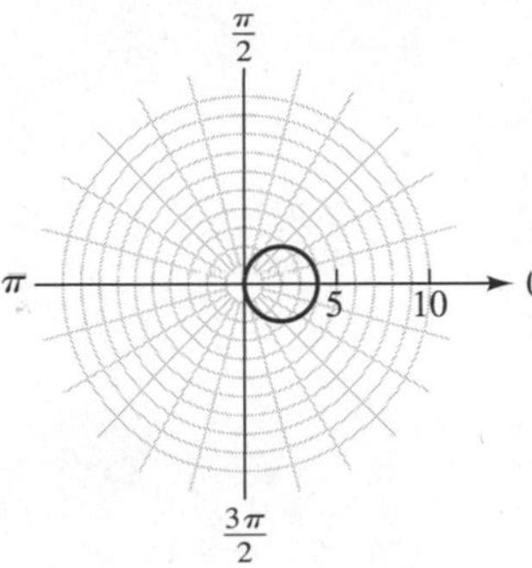

13.

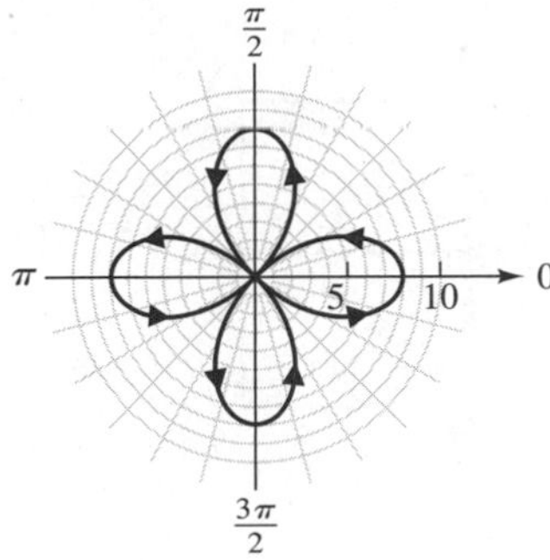

15.

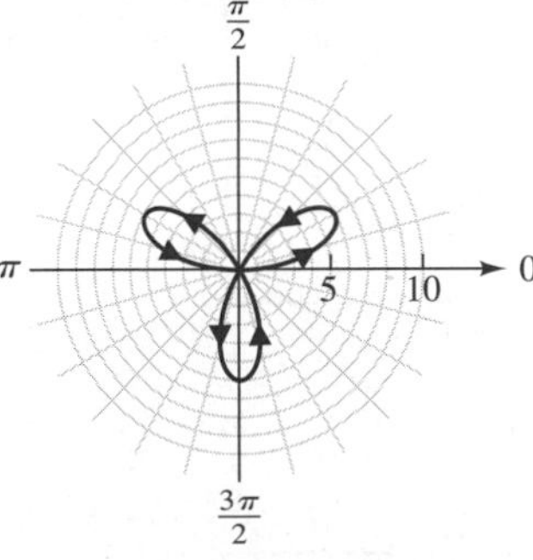

17.

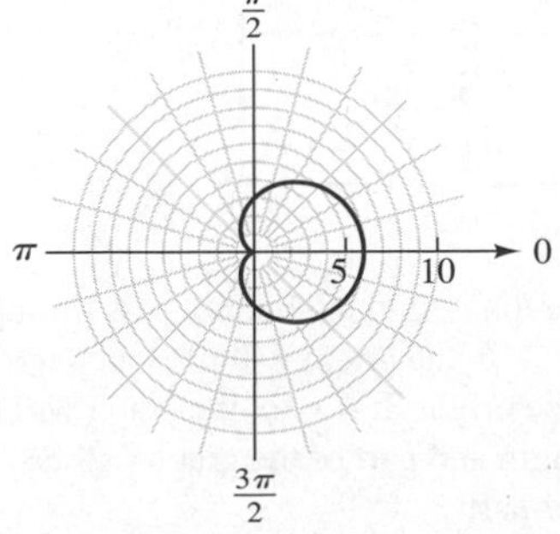

19.

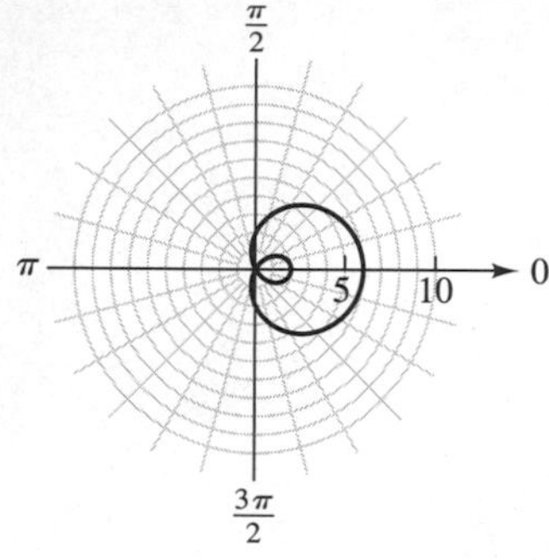

21. (A) $r = 5 + 5 \cos \theta$

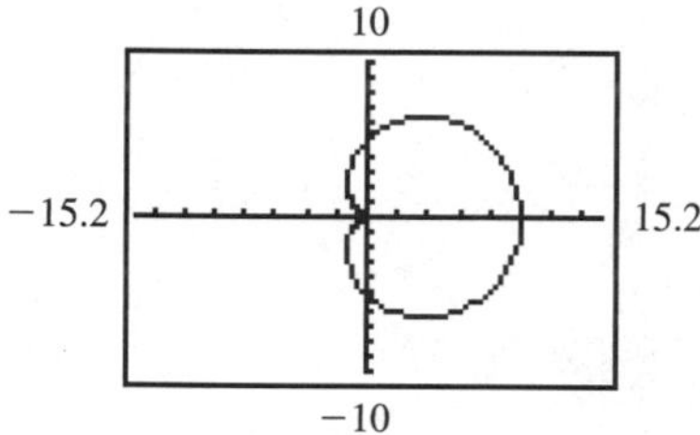

(B) $r = 5 + 4 \cos \theta$

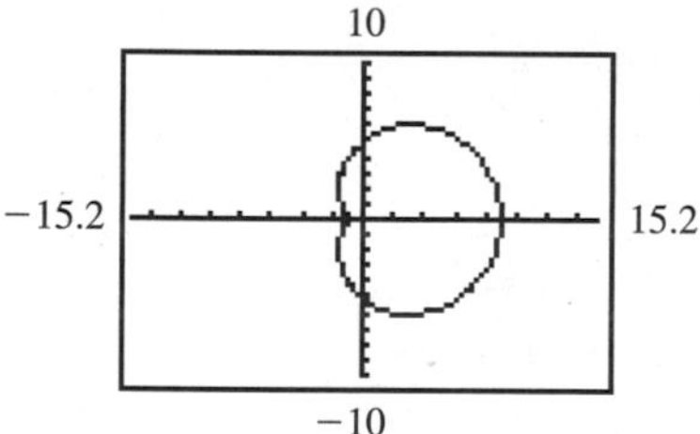

(C) $r = 4 + 5 \cos \theta$

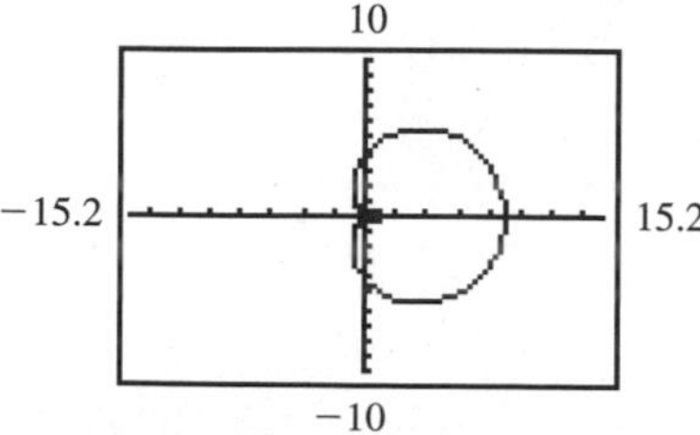

(D) If $a = b$, the graph will touch but not pass through the origin. If $a > b$, the graph will not touch nor pass through the origin. If $a < b$, the graph will go through the origin and part of the graph will be inside the other part.

23. (A) $r = 9 \cos \theta$

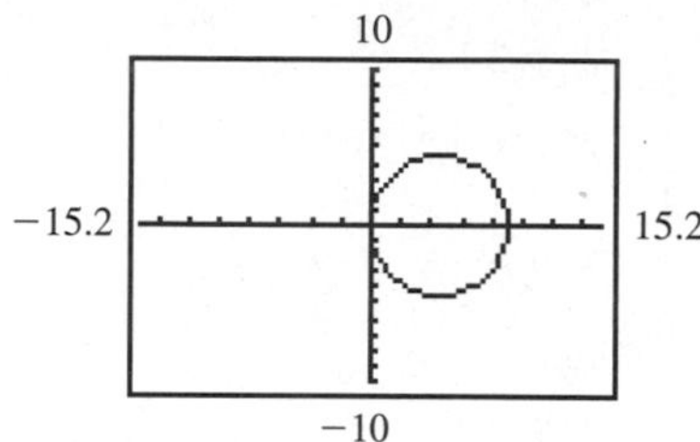

$r = 9 \cos 3\theta$

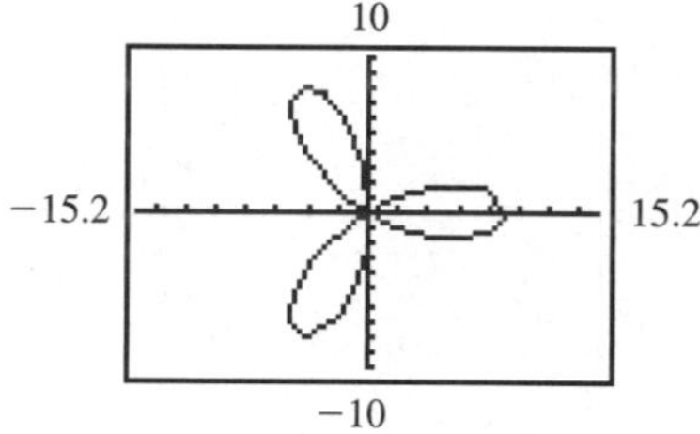

$r = 9 \cos 5\theta$

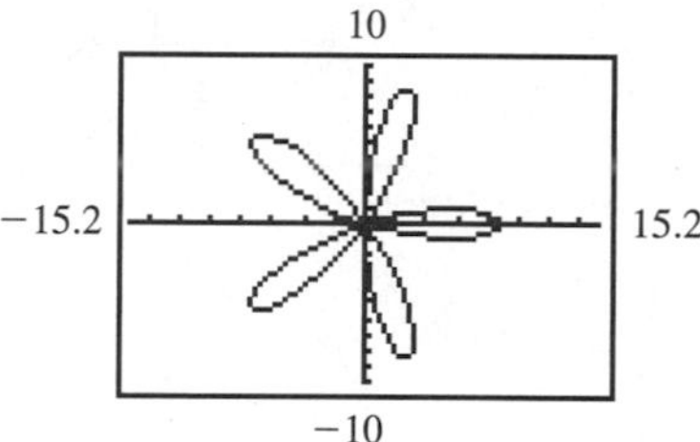

(B) 7 leaves (C) n leaves

25. (A) $r = 9 \cos 2\theta$

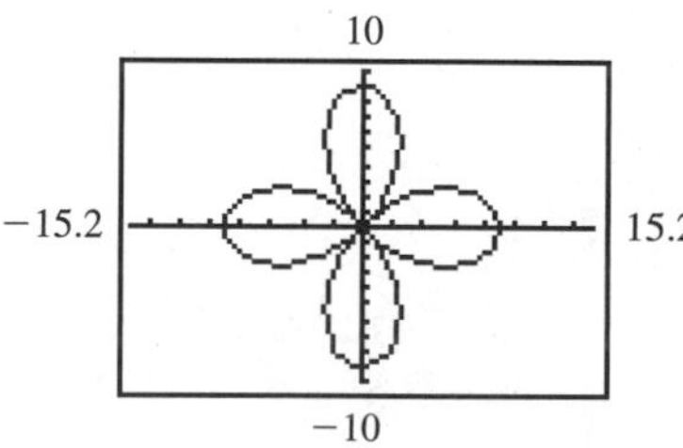

$r = 9 \cos 4\theta$

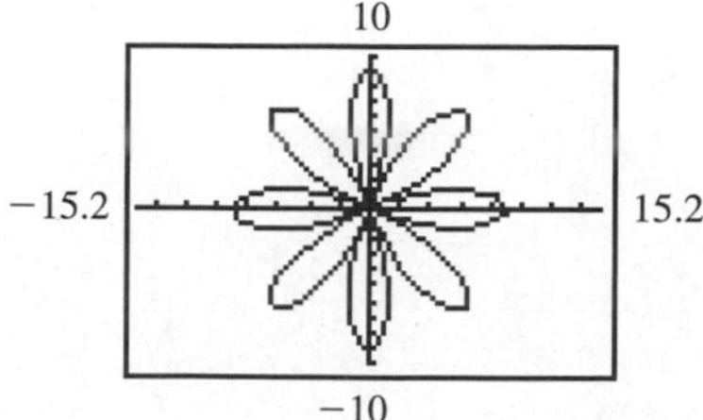

$r = 9 \cos 6\theta$

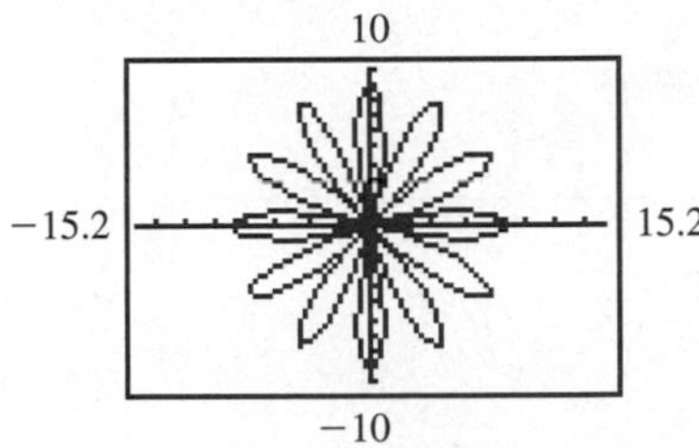

(B) 16 leaves (C) $2n$ leaves

27. (A) $r = 9 \cos(\theta/2)$

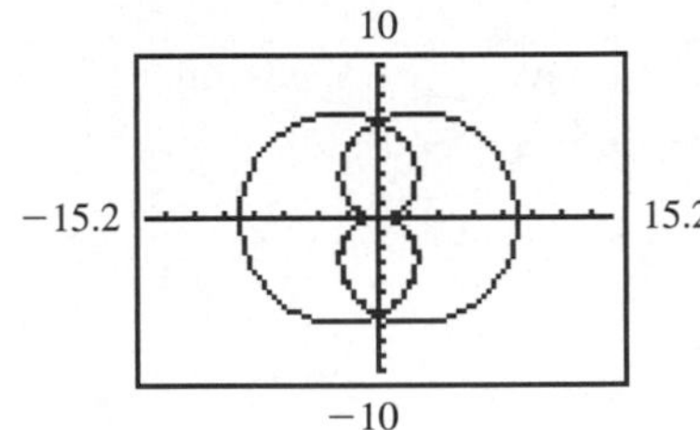

(B) $r = 9 \cos(\theta/4)$

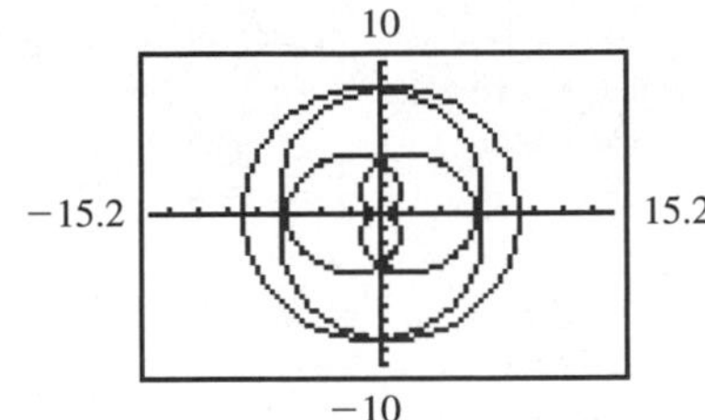

(C) n times

29.

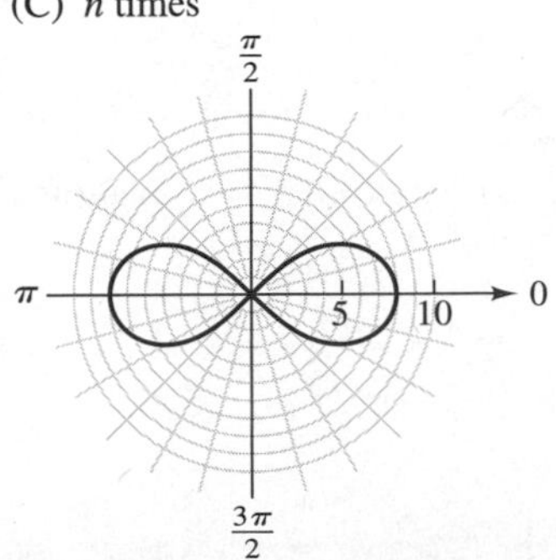

31. For each n, there are n large petals and n small petals. For n odd, the small petals are within the large petals; for n even, the small petals are between the large petals.

33.

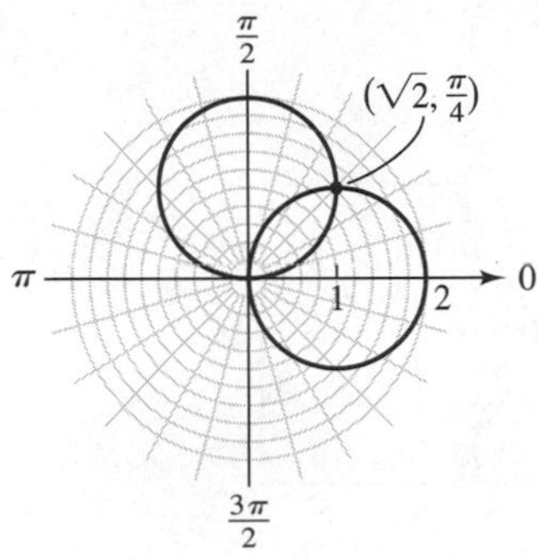

$(r, \theta) = (\sqrt{2}, \pi/4)$
[Note that (0, 0) is not a solution to the system even though the graphs cross at the origin.]

35.

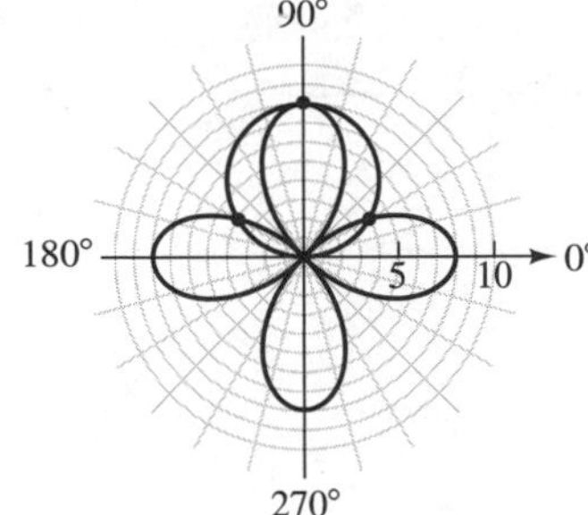

$(r, \theta) = (4, 30°), (4, 150°), (-8, 270°)$
[Note that (0, 0) is not a solution to the system even though the graphs cross at the origin.]

37. (A) 9.5 knots (B) 12.0 knots (C) 13.5 knots (D) 12.0 knots

39. (A) Ellipse

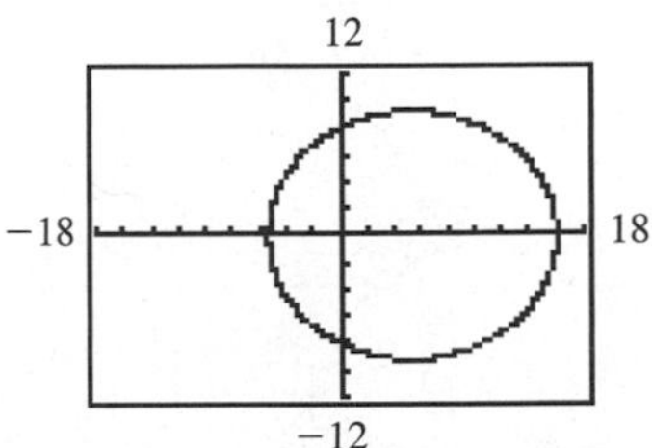

(B) Parabola

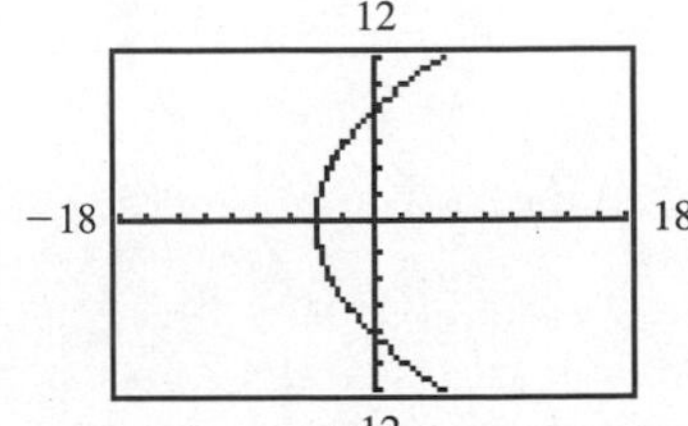

39. (C) Hyperbola

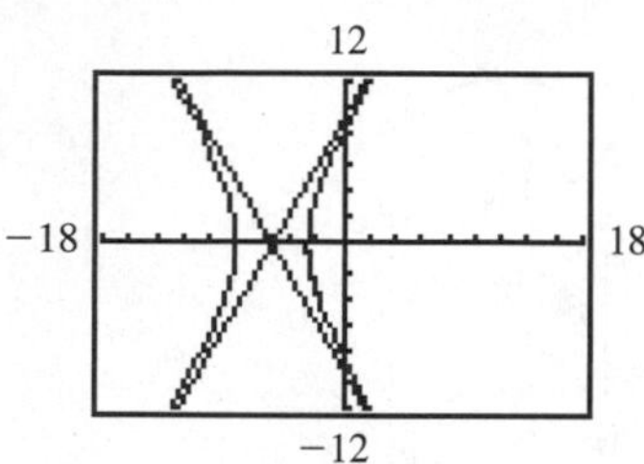

41. (A) Perihelion = 2.85×10^7 mi;
Aphelion = 4.33×10^7 mi

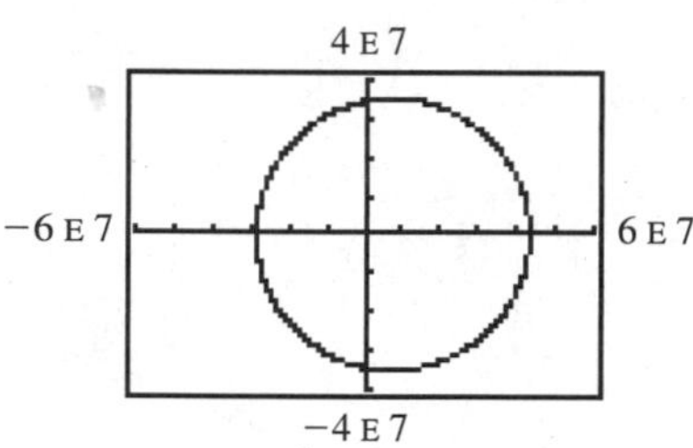

(B) Faster at perihelion

Exercise 7.3

1.

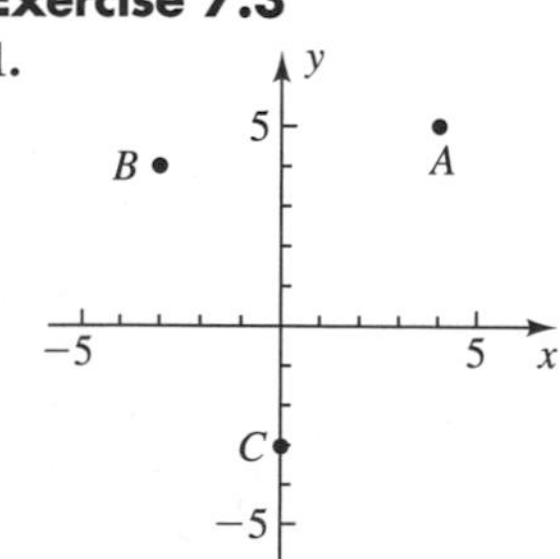

3.

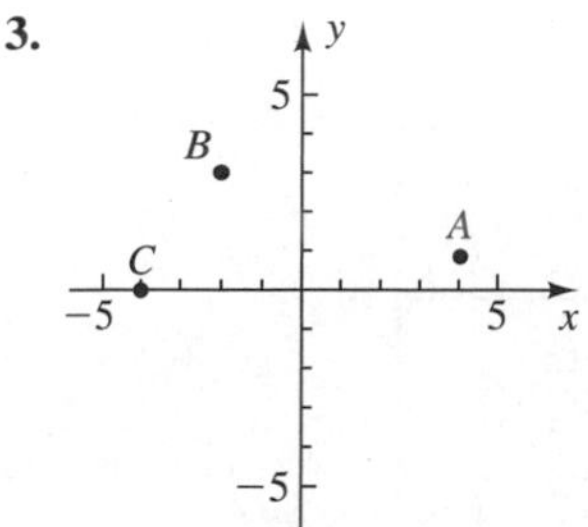

5.

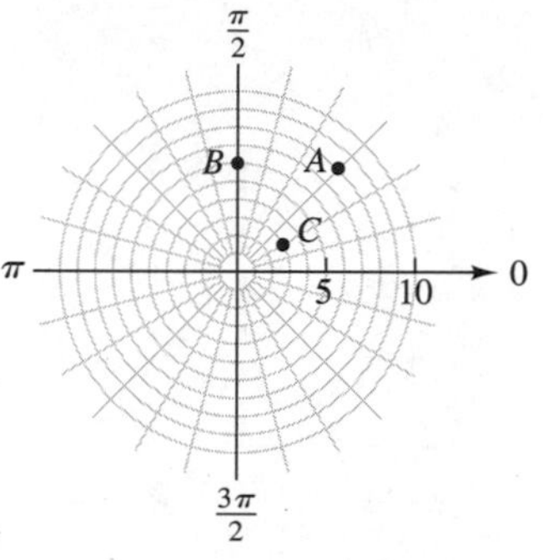

7.

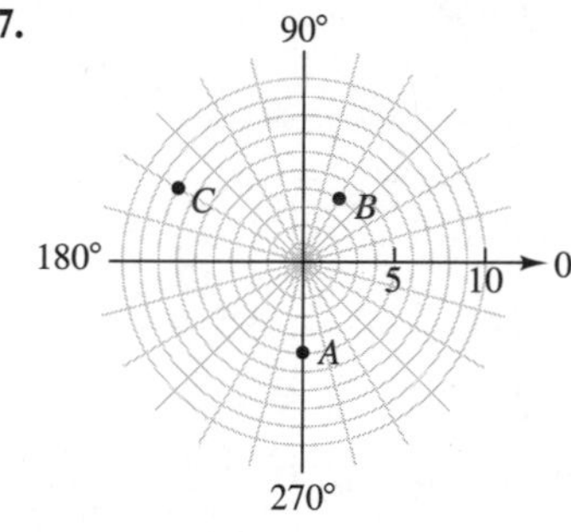

9. (A) $2e^{(-\pi/6)i}$ (B) $2\sqrt{2}e^{(3\pi/4)i}$ (C) $7.81e^{(-0.69)i}$
11. (A) $2e^{(120°)i}$ (B) $3e^{(-90°)i}$ (C) $8.06e^{(-150.26°)i}$
13. (A) $\sqrt{3} + i$ (B) $-1 - i$ (C) $5.06 - 2.64i$
15. (A) $-i\sqrt{3}$ (B) $-1 + i$ (C) $-2.20 - 6.46i$
17. $18e^{(225°)i}$; $2e^{(39°)i}$ **19.** $6e^{(164°)i}$; $1.5e^{(-30°)i}$
21. $18.91e^{1.86i}$; $2.67e^{(-0.28)i}$ **23.** $4i$; $4e^{(90°)i}$
25. $2i$; $2e^{(90°)i}$ **27.** $-2 + 2i$; $2\sqrt{2}e^{(135°)i}$
33. z^n appears to be $r^n e^{n\theta i}$
35. (A) $(8 + 0i) + (3\sqrt{3} + 3i) = (8 + 3\sqrt{3}) + 3i$
(B) $13.5(\cos 12.8° + i \sin 12.8°)$
(C) 13.5 lb at an angle of 12.8°

Exercise 7.4

1. $27e^{(45°)i}$ **3.** $32e^{(450°)i} = 32e^{(90°)i}$ **5.** $64e^{(180°)i}$
7. -4 **9.** $16\sqrt{3} + 16i$ **11.** 1 **13.** $2e^{(15°)i}$, $2e^{(195°)i}$
15. $2e^{(30°)i}$, $2e^{(150°)i}$, $2e^{(270°)i}$
17. $2^{1/10}e^{(27°)i}$, $2^{1/10}e^{(99°)i}$, $2^{1/10}e^{(171°)i}$, $2^{1/10}e^{(243°)i}$, $2^{1/10}e^{(315°)i}$
19. $w_1 = 2e^{(60°)i}$, $w_2 = 2e^{(180°)i}$, $w_3 = 2e^{(300°)i}$

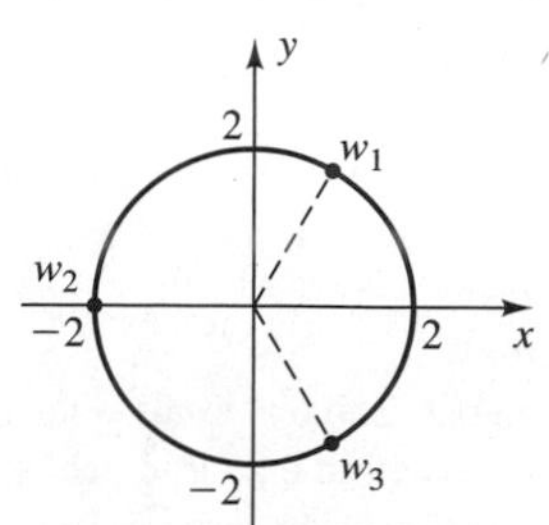

21. $w_1 = 1e^{(0°)i}$, $w_2 = 1e^{(90°)i}$, $w_3 = 1e^{(180°)i}$, $w_4 = 1e^{(270°)i}$

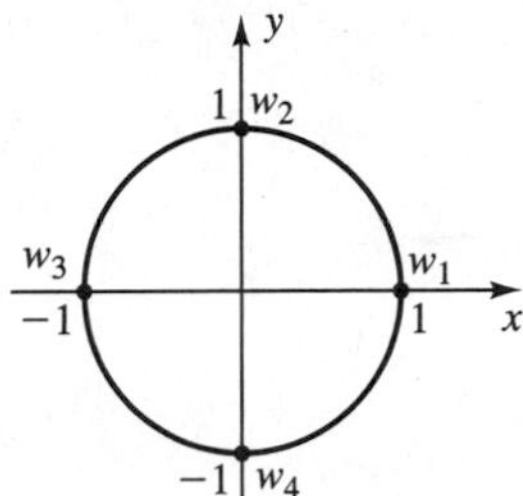

23. $w_1 = 1e^{(-18°)i}$, $w_2 = 1e^{(54°)i}$, $w_3 = 1e^{(126°)i}$, $w_4 = 1e^{(198°)i}$, $w_5 = 1e^{(270°)i}$

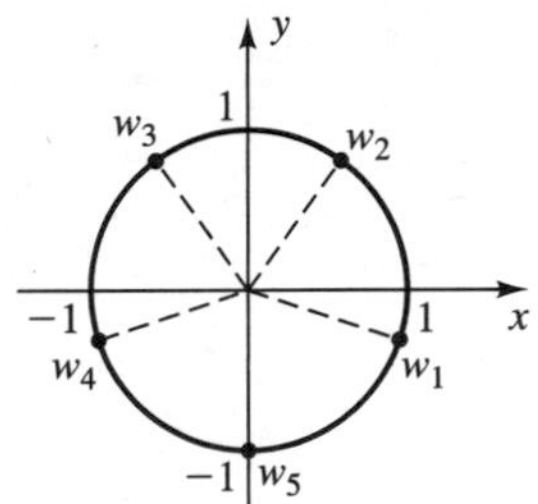

25. (A) $(-\sqrt{2} + \sqrt{2}i)^4 + 16 = -16 + 16 = 0$; three

(B) The four roots are equally spaced around the circle. Since there are 4 roots, the angle between successive roots on the circle is $360°/4 = 90°$.

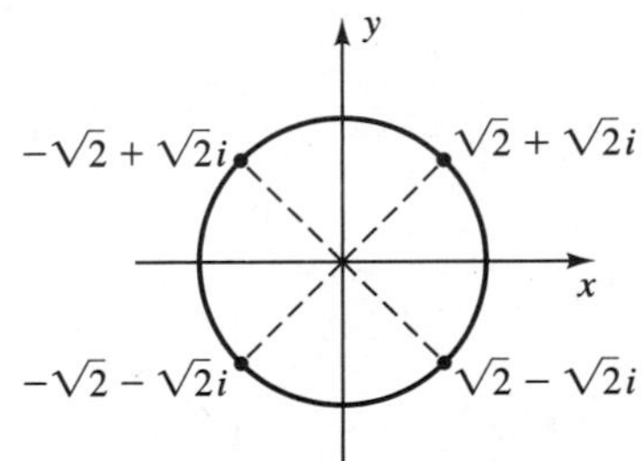

(C) $(\sqrt{2} + \sqrt{2}i)^4 + 16 = -16 + 16 = 0$,
$(-\sqrt{2} - \sqrt{2}i)^4 + 16 = -16 + 16 = 0$,
$(\sqrt{2} - \sqrt{2}i)^4 + 16 = -16 + 16 = 0$

27. $x_1 = 3e^{(60°)i} = \frac{3}{2} + \frac{3\sqrt{3}}{2}i$, $x_2 = 3e^{(180°)i} = -3$,

$x_3 = 3e^{(300°)i} = \frac{3}{2} - \frac{3\sqrt{3}}{2}i$

29. $x_1 = 4e^{(0°)i} = 4$, $x_2 = 4e^{(120°)i} = -2 + 2\sqrt{3}i$, $x_3 = 4e^{(240°)i} = -2 - 2\sqrt{3}i$

33. $1, 0.309 + 0.951i, -0.809 + 0.588i, -0.809 - 0.588i, 0.309 - 0.951i$

35. $0.855 + 1.481i, -1.710, 0.855 - 1.481i$

Chapter 7 Review Exercise

1.

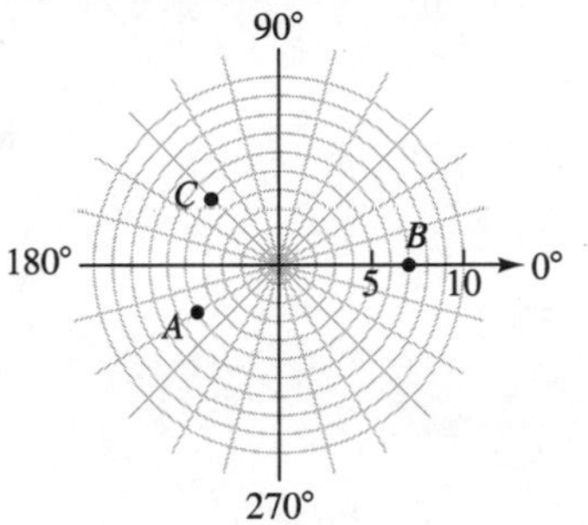

[7.1]

2.

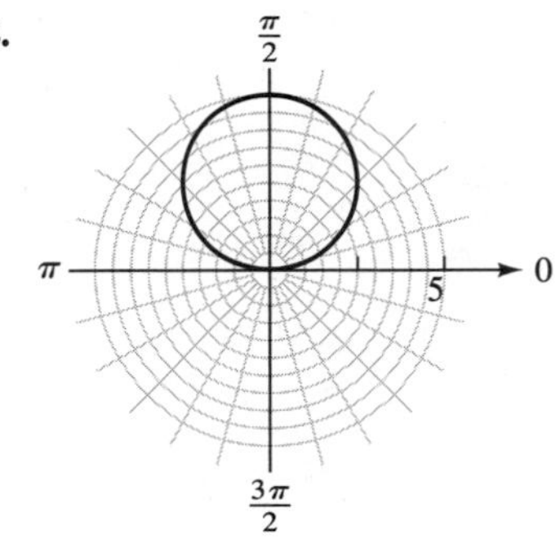

[7.2]

3.

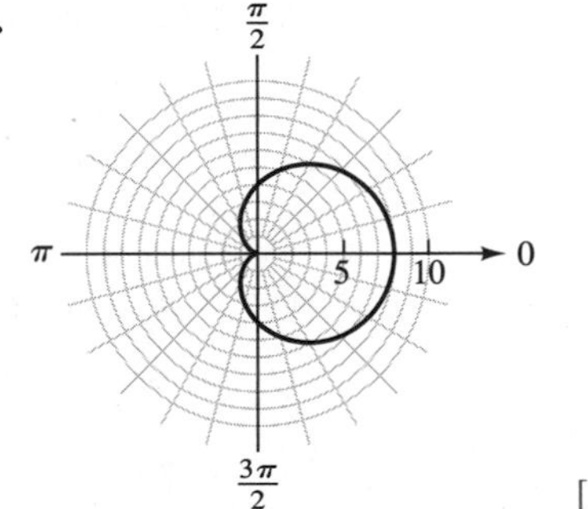

[7.2]

4.

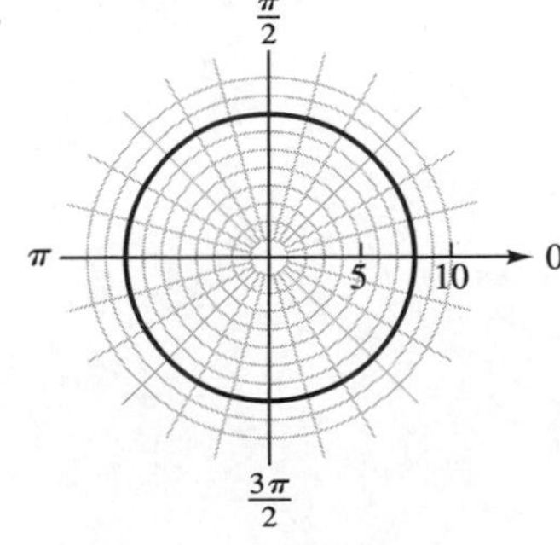

[7.2]

5. (2, 2) [7.1] **6.** (2, 5π/6) [7.1]

7.

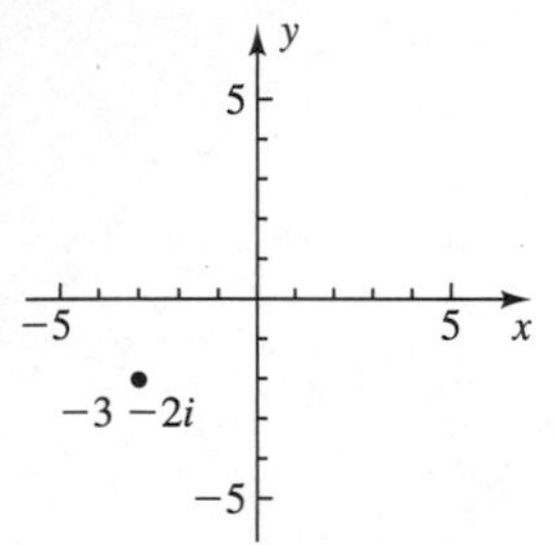

[7.1]

8.

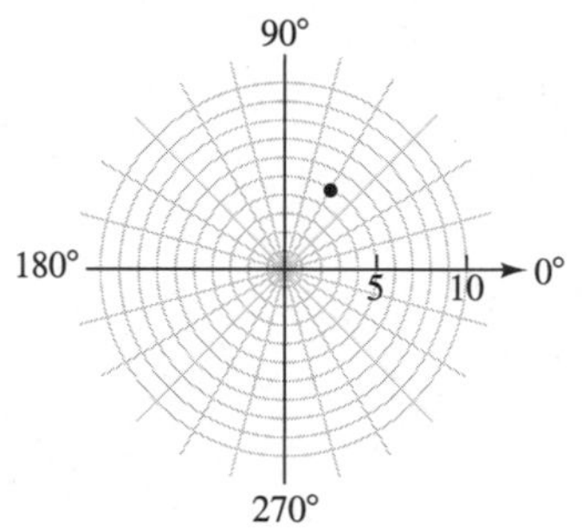

[7.3]

9. (− 8, − 330°): The polar axis is rotated 330° clockwise (negative direction) and the point is located 8 units from the pole along the negative polar axis.
(8, − 150°): The polar axis is rotated 150° clockwise (negative direction) and the point is located 8 units from the pole along the positive polar axis.
(8, 210°): The polar axis is rotated 210° counterclockwise (positive direction) and the point is located 6 units along the positive polar axis. [7.1]

10. $z_1z_2 = 27e^{(79°)i}$; $z_1/z_2 = 3e^{(5°)i}$ [7.3]

11. $16e^{(40°)i}$ [7.4]

12.

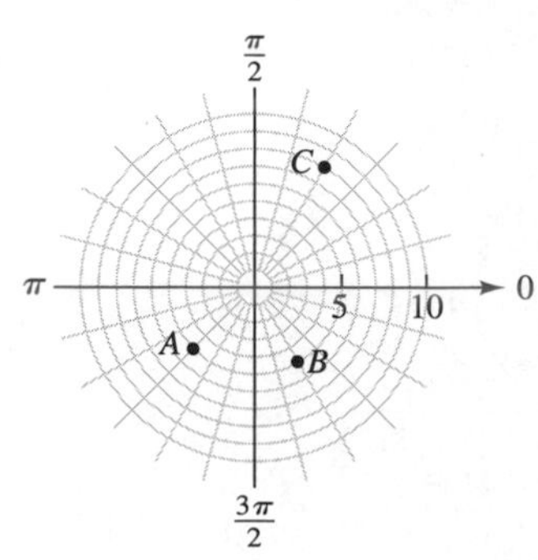

[7.1]

13.

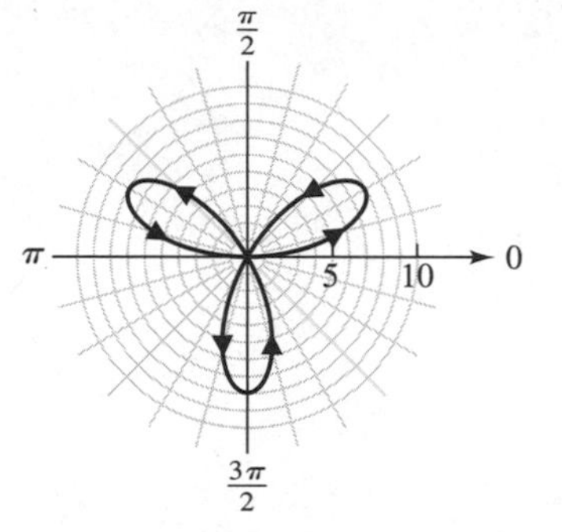

[7.2]

14.

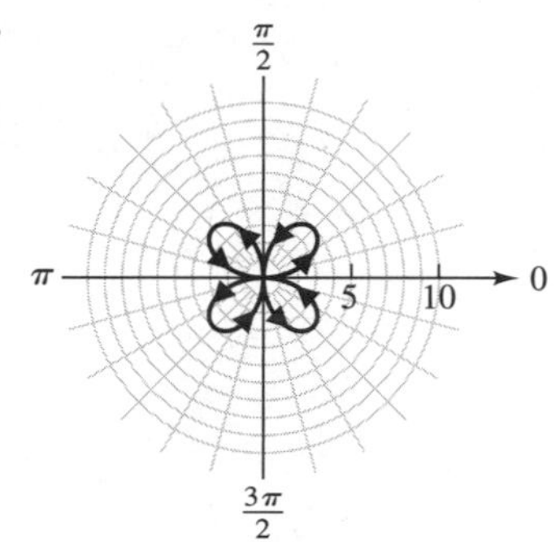

[7.2]

15.

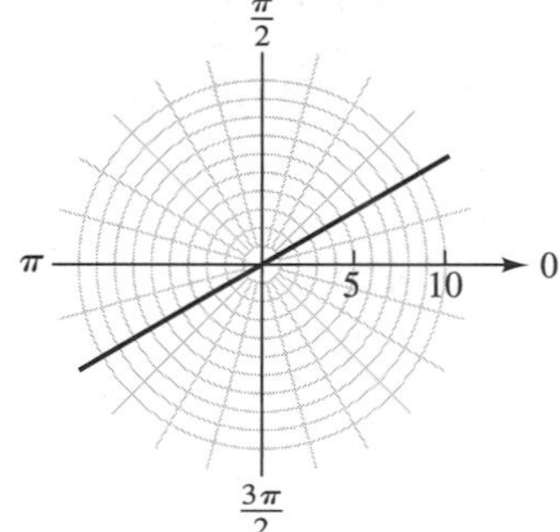

[7.2]

16. $r^2 = 8r \cos \theta$, $r = 8 \cos \theta$ [7.1]

17. $3x - 2y = -2$ [7.1]

18. $x^2 + y^2 = -3x$ [7.1]

19.

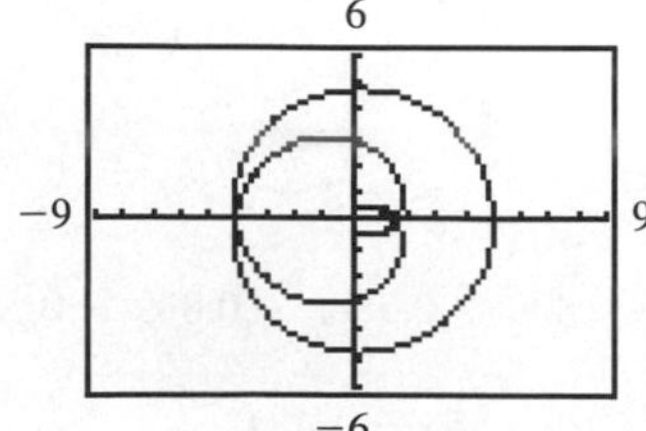

[7.2]

20.

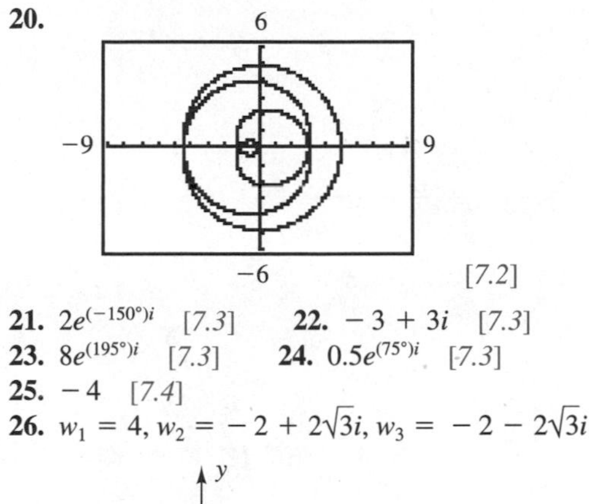

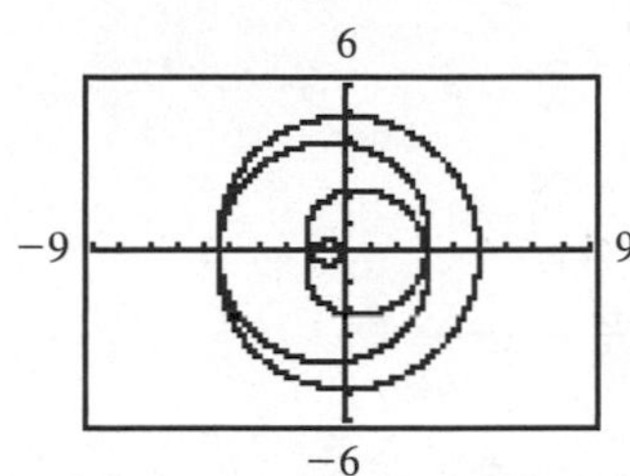

[7.2]

21. $2e^{(-150°)i}$ [7.3] **22.** $-3 + 3i$ [7.3]
23. $8e^{(195°)i}$ [7.3] **24.** $0.5e^{(75°)i}$ [7.3]
25. -4 [7.4]
26. $w_1 = 4, w_2 = -2 + 2\sqrt{3}i, w_3 = -2 - 2\sqrt{3}i$

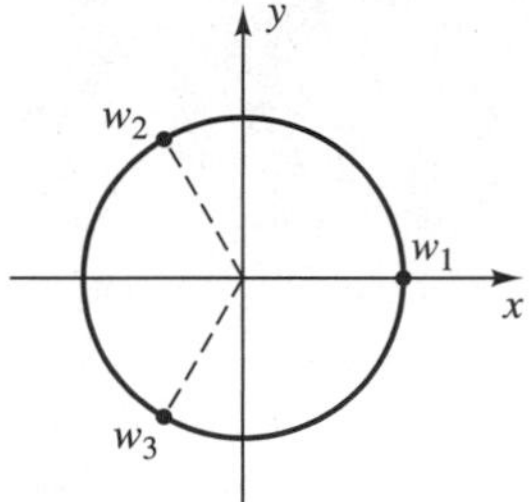

[7.4]

27. $2e^{(70°)i}, 2e^{(190°)i}, 2e^{(310°)i}$ [7.4]
28. $(2e^{(30°)i})^2 = 4e^{(60°)i} = 4(\cos 60° + i \sin 60°) = 2 + 2\sqrt{3}i$ [7.4]
29. (A) There are a total of three cube roots, and they are spaced equally around a circle of radius 2.

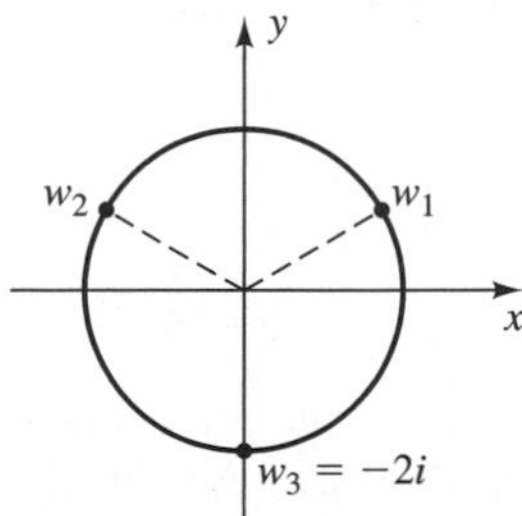

(B) $w_1 = \sqrt{3} + i, w_2 = -\sqrt{3} + i$
(C) The cube of each cube root is $8i$; that is, each root is a cube root of $8i$. [7.4]

30. $n = 1$

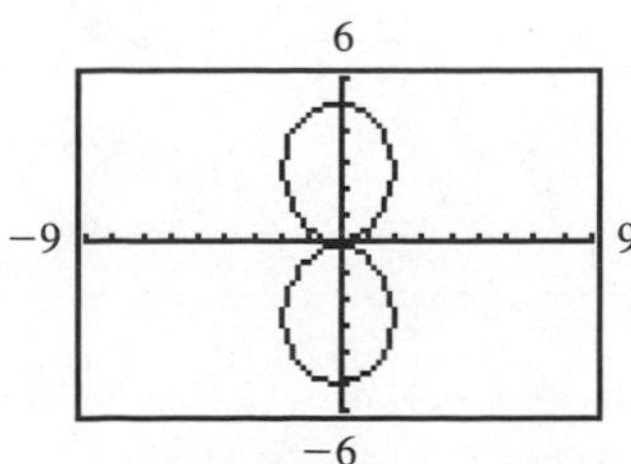

$n = 2$

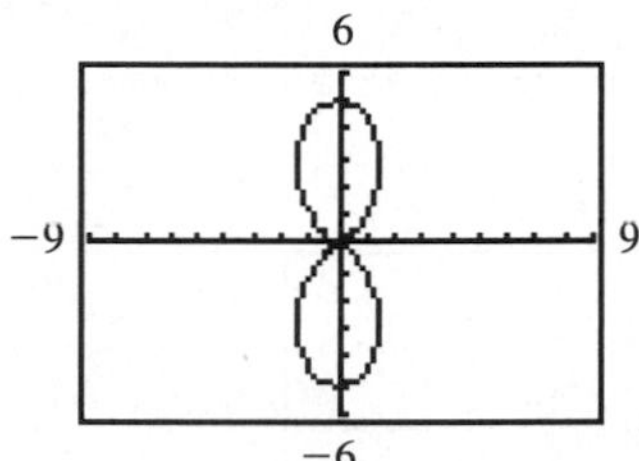

$n = 3$

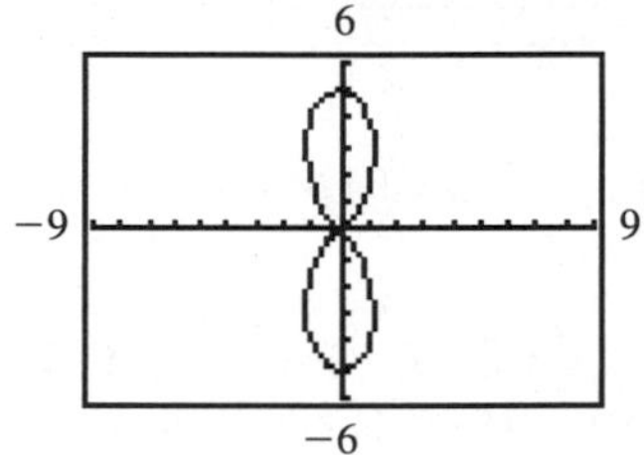

The graph always has two leaves. [7.2]

31. (A) Hyperbola

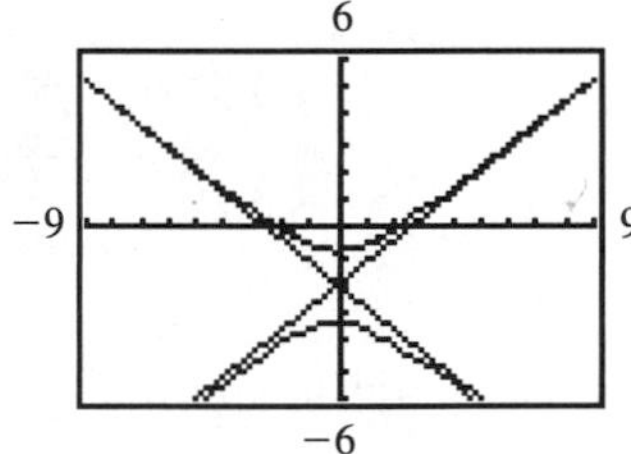

(B) Parabola

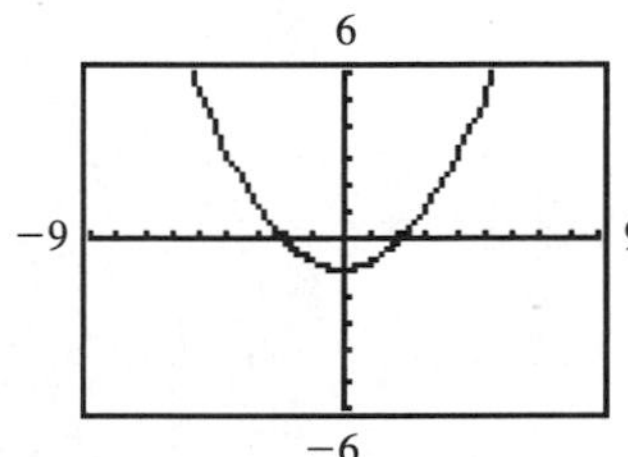

(C) Ellipse

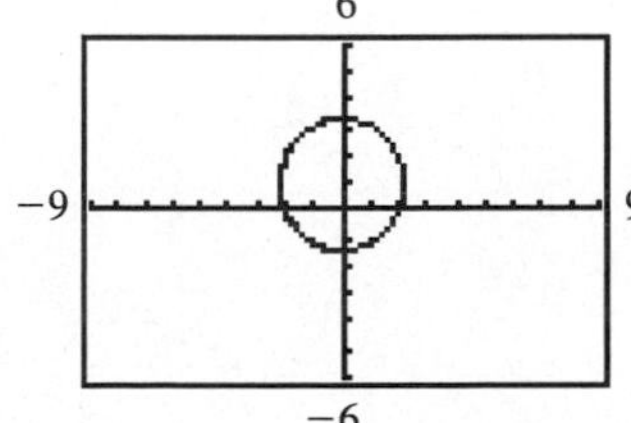

[7.3]

32. $y - 3 = -2\sqrt{x^2 + y^2}$, or $(y - 3)^2 = 4(x^2 + y^2)$ [7.1]

33. 2.289, $-1.145 + 1.983i$, $-1.145 - 1.983i$ [7.4]

35. (A) The coordinates of P represent a simultaneous solution.

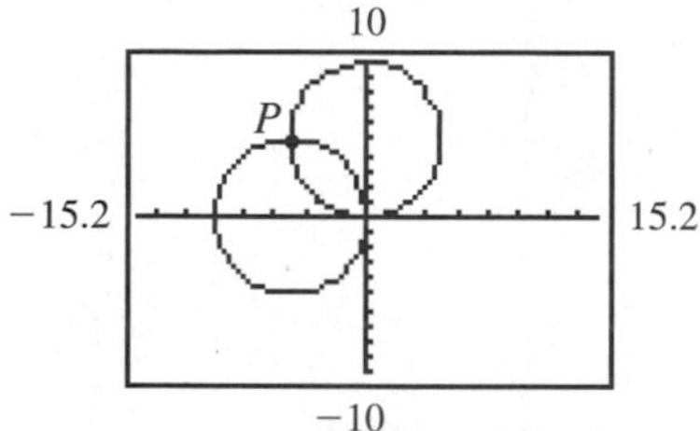

(B) $r = 5\sqrt{2}$, $\theta = 3\pi/4$

(C) The two graphs go through the pole at different values of θ. [7.2]

Cumulative Review Exercise Chapters 1–7

1. α is larger. Change both measures to either degrees or radians to enough decimal places so that a difference becomes apparent. [2.1]

2. 4.82 cm, 74.7°, 15.3° [1.3]

3. $\sec\theta = \frac{25}{7}$, $\tan\theta = -\frac{24}{7}$ [2.3]

5. -0.5 [2.5] **6.** $\sqrt{3}$ [2.5] **7.** $2\pi/3$ [5.1]

8. $\pi/4$ [5.2] **9.** 0.6867 [2.3] **10.** 0.3249 [2.6]

11. 0.9273 [5.1] **12.** 1.347 [5.2]

13. $2 \sin 2t \cos t$ [4.5]

14. An angle of radian measure 2.5 is the central angle of a circle subtended by an arc with measure 2.5 times that of the radius of the circle. [2.1]

15. $\alpha = 28°$, $a = 34$ m, $c = 48$ m [6.1]

16. $\alpha = 40°$, $\gamma = 106°$, $b = 14$ in. [6.2]

17. $\alpha = 33°$, $\beta = 45°$, $\gamma = 102°$ [6.2]

18. 110 in.^{2} [6.3]

19. $(7, -150°)$: Rotate the polar axis 150° clockwise (negative direction) and go 7 units along the positive polar axis. [7.1]

20. $|\mathbf{H}| = 12$; $|\mathbf{V}| = 5.5$ [6.4] **21.** 7.5 at 31° [6.4]

22. $\langle -7, 9\rangle$; $\sqrt{130}$ [6.5] **23.** 143.5° [6.6]

24.

90°
D
B
180° 5 10 0°
A
C
270°

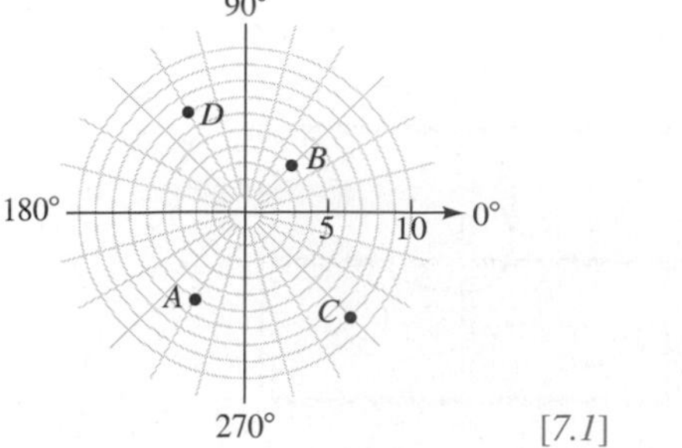

[7.1]

25.

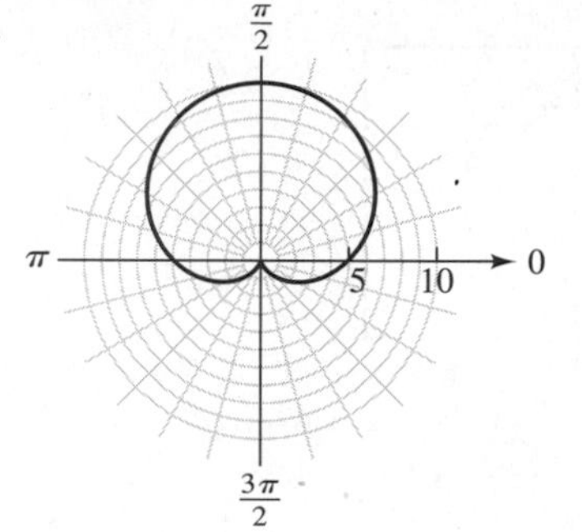

[7.2]

26. $(-3, 3)$ [7.1] **27.** $(4, 5\pi/6)$ [7.1]

28. $2\sqrt{2}e^{(-\pi/4)i}$ [7.3] **29.** $-3i$ [7.3]

30. $z_1z_2 = 15e^{(65°)i}$; $z_1/z_2 = 0.6e^{(35°)i}$ [7.3]

31. $81e^{(100°)i}$ [7.4] **32.** $y = 1 + 2\cos\pi x$ [3.2]

33. $\csc\theta = -\sqrt{17}$, $\cos\theta = -4/\sqrt{17}$ [2.3]

34. Amplitude = 2; Period = π; Frequency = $1/\pi$; Phase shift = $-\pi/2$

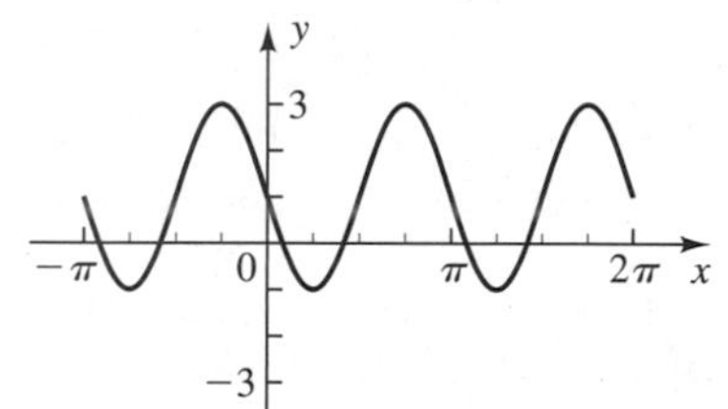

[3.3]

38. $\tan(x/2) = \frac{1}{7}$, $\sin 2x = \frac{336}{625}$ [2.3, 4.4]

39. (A) Not an identity; both sides are defined at $x = \pi/2$, but are not equal.

(B) An identity [4.2]

40. $4/\sqrt{7}$ [5.1]

41. (A) There are a total of three cube roots, and they are spaced equally around a circle of radius 2.

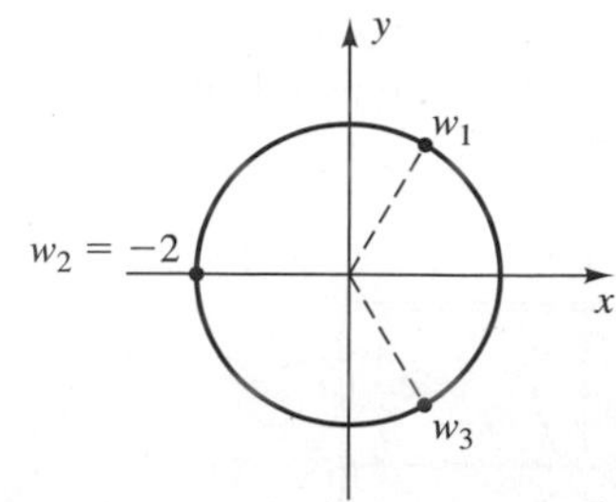

(B) $w_1 = 1 + \sqrt{3}i$, $w_3 = 1 - \sqrt{3}i$

(C) The cube of each cube root is -8. [7.4]

42. $k\pi$, $2\pi/3 + 2k\pi$, $4\pi/3 + 2k\pi$, k any integer [5.4]

43. $0.8481 + 2k\pi$, $2.294 + 2k\pi$, k any integer [5.4]

44. (A) No triangle exists

(B) $\beta = 64°10'$, $\gamma = 66°20'$, $c = 17.7$ cm;
$\beta' = 115°50'$, $\gamma' = 14°40'$, $c' = 4.89$ cm

(C) $\beta = 38°50'$, $\gamma = 91°40'$, $c = 27.7$ cm [6.1]

45. $|\mathbf{u} + \mathbf{v}| = 41.9$; $\theta = 11.0°$ [6.4]

46. (A) $\langle 3, -18\rangle$ (B) $10\mathbf{i} - 11\mathbf{j}$ [6.5]

47. $\mathbf{u} = \langle 0.28, -0.96\rangle$ [6.5] **48.** $2\mathbf{i} + 3\mathbf{j}$ [6.5]

49. (A) Orthogonal (B) Not orthogonal

(C) Orthogonal [6.6]

50.

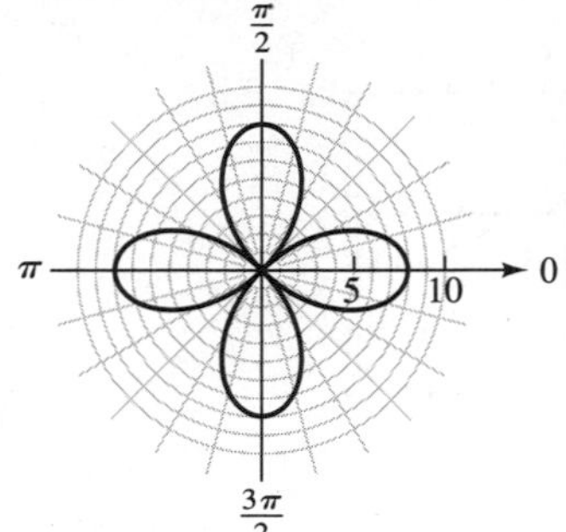

[7.2]

51. $r = 6 \tan\theta \sec\theta$ [7.1]

52. $x^2 + y^2 = 4y$ [7.1]

53. $6\sqrt{2}e^{(165°)i}$ [7.3] **54.** $(\sqrt{2}/3)e^{(75°)i}$ [7.3]

55. $8i$ [7.4]

56. $w_1 = \sqrt{3} + i$, $w_2 = -\sqrt{3} + i$, $w_3 = -2i$

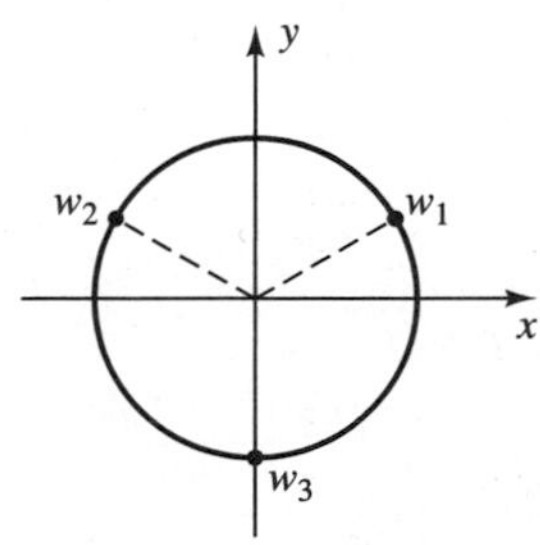

[7.4]

57.

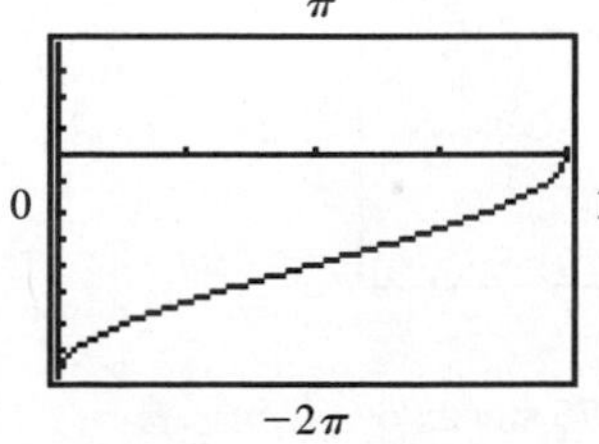

[5.1]

58.

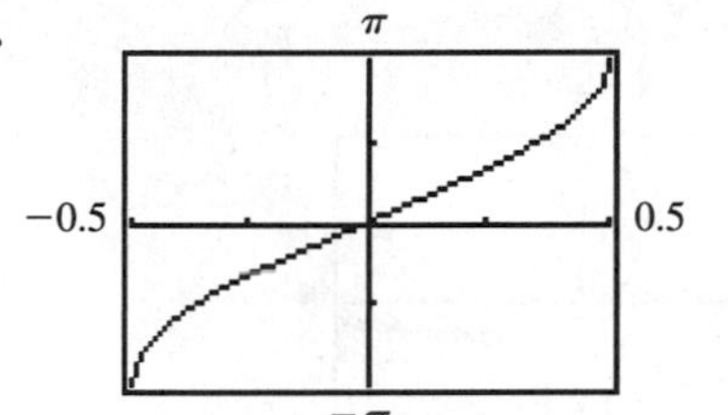

[5.1]

59.

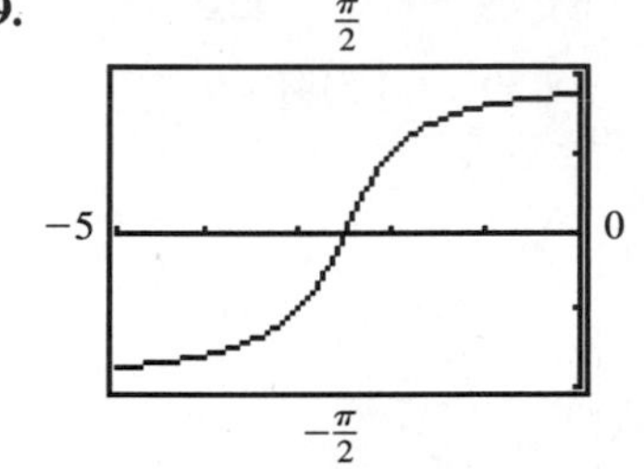

[5.1]

60. $-1.768, 1.373$ [5.3] **61.** 0.582 [5.4]

62. 3.909, 4.313, 5.111, 5.516 [5.4]

63.

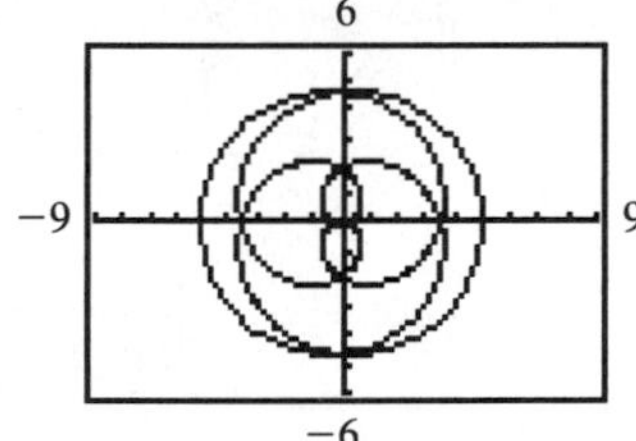

[7.2]

64. (A) Ellipse

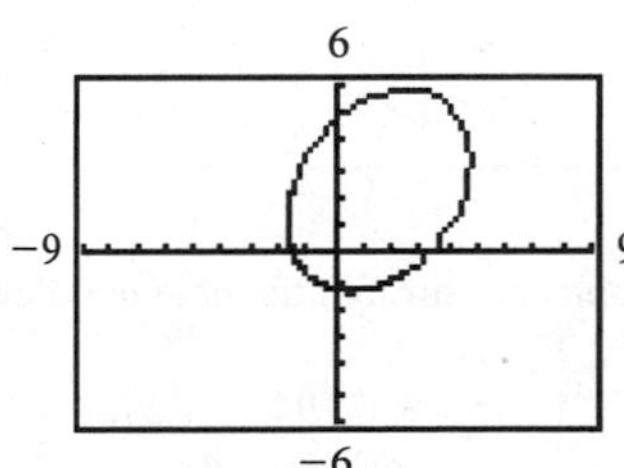

(B) Parabola

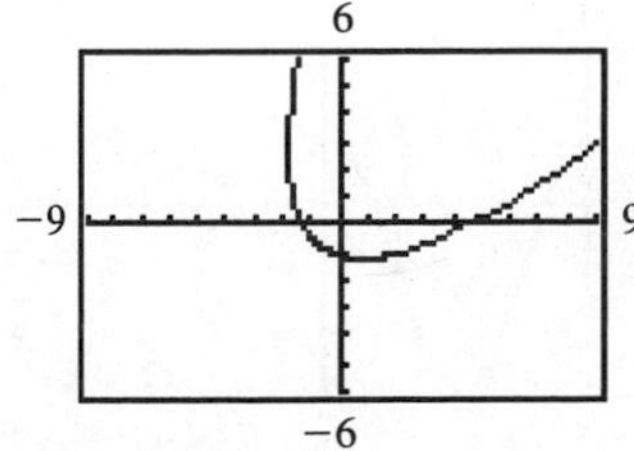

64. (C) Hyperbola

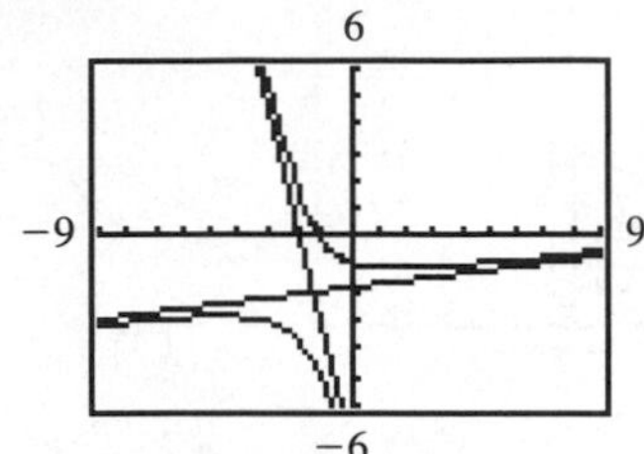

[7.2]

65. $\theta = 4$ rad; $(a, b) = (-1.307, -1.514)$ [2.3]

66. $\theta = 3.785$; $s = 7.570$ [2.3]

67.

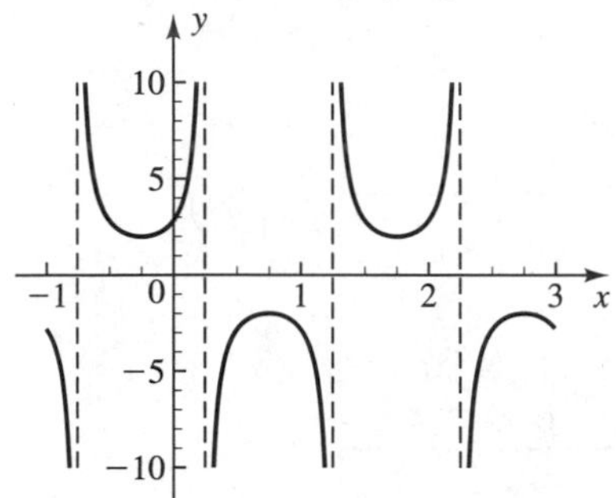

[3.6]

68. $\dfrac{1 + x^2}{1 - x^2}$ [4.4, 5.1]

70. $x - 1 = -\sqrt{x^2 + y^2}$, or $(x - 1)^2 = x^2 + y^2$ [7.1]

71. (A)

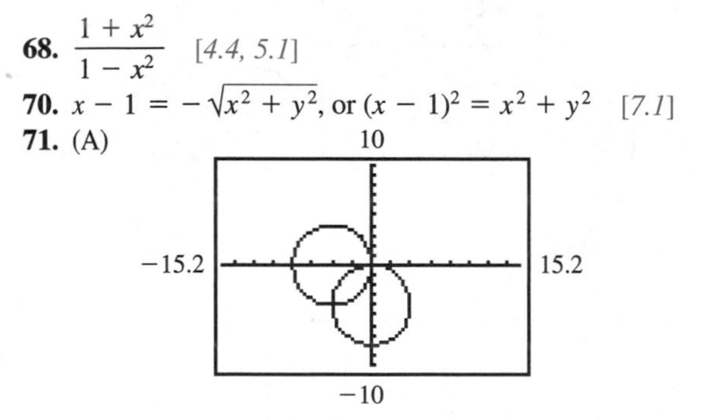

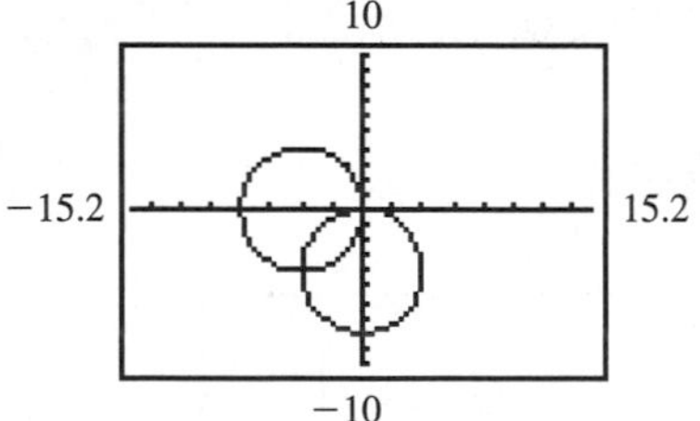

(B) The two graphs go through the pole at different values of θ. [7.2]

72. $1.587, -0.794 + 1.375i, -0.794 - 1.375i$ [7.4]

73. (A) $\cos 3\theta = \cos^3\theta - 3\cos\theta\sin^2\theta$, $\sin 3\theta = 3\cos^2\theta\sin\theta - \sin^3\theta$ [7.4]

74. $g(x) = -1 + 3\cos 2x$

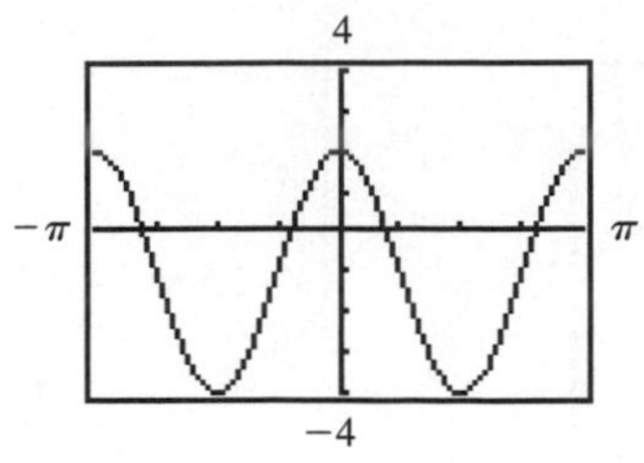

[4.4]

75. $g(x) = \sec 2x - 3$

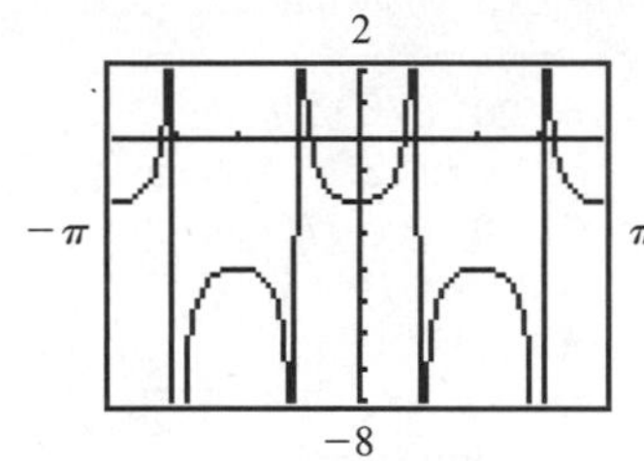

[4.4]

76. $g(x) = -1 + \cot(x/2)$

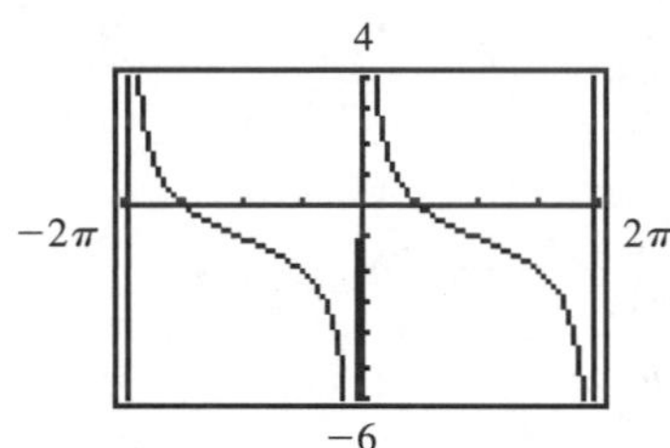

[4.4]

77. 650 ft [6.1] **78.** 550 ft [6.2]

79. (A) 48 ft (B) 41 ft [1.4, 6.1]

80. Station *A*: 8.7 mi; Station *B*: 6.3 mi [1.4, 6.1]

81. (A) Period $= \frac{1}{70}$ sec; $B = 140\pi$
(B) Frequency = 80 Hz; $B = 160\pi$
(C) Period $= \frac{1}{50}$ sec; Frequency = 50 Hz [3.2]

82. Wave height = 4 ft; Wavelength = 82 ft; Speed = 20 ft/sec [3.4]

83. Area $= \frac{1}{2}\theta + \frac{1}{2}\sin\theta\cos\theta$ [2.1, 2.6]

84. Area $= \frac{1}{2}\sin^{-1}x + \frac{1}{2}x\sqrt{1 - x^2}$ [2.6, 5.1]

85. $\theta = 0.553$ [5.4] **86.** $x = 0.412$ [5.4]

87. (A) $R = r + (H - h)\tan\alpha$
(B) $\beta = \tan^{-1}\left(\dfrac{H - h}{R - r}\right)$ [1.4]

88. 18°; 252 mph [6.4]

90. (A) 574 ft (B) 2.9° [6.1, 6.2]

91. (A) 1,250 ft (B) 4.6° [1.4, 2.1, 6.2]

92. $A = r^2(\pi - \theta + \sin\theta)$ [5.4]

93. 2.6053 rad [5.4]

94. (A)

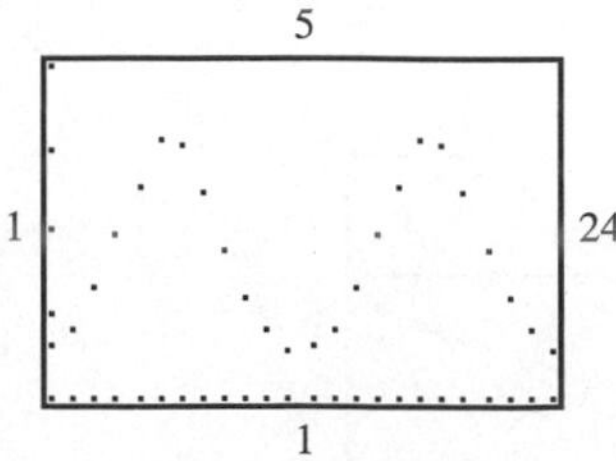

(B) $y = 2.845 + 1.275\sin(\pi x/6 - 1.8)$

(C)

5

1 24

1

[3.3]

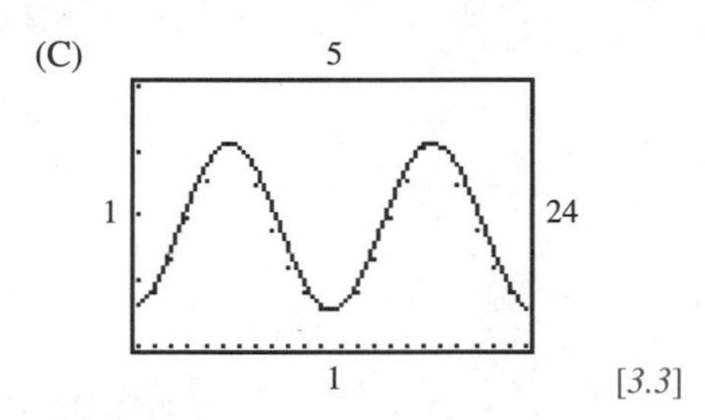

APPENDIX A

Exercise A.1

1. $-3, 0, 5$ (answers vary) **3.** $\frac{2}{3}$ (answers vary)
5. (A) True (B) False (C) True
7. (A) $0.363\overline{636}$, rational (B) $0.777\overline{7}$, rational
(C) 2.64575131..., irrational
(D) $1.62500\overline{0}$, rational
9. (A) 2 and 3 (B) -4 and -3
(C) -5 and -4
11. $y + 3$ **13.** $(3 \cdot 2)x$ **15.** $7x$
17. $(5 + 7) + x$ **19.** $3m$ **21.** $u + v$
23. $(2 + 3)x$
25. (A) True (B) False; $4 - 2 \neq 2 - 4$
(C) True (D) False; $8/2 \neq 2/8$

Exercise A.2

1. $7 + 5i$ **3.** $-1 - 9i$ **5.** -18 **7.** $8 + 6i$
9. $-5 - 10i$ **11.** 34 **13.** $\frac{2}{5} - \frac{1}{5}i$
15. $\frac{4}{13} - \frac{7}{13}i$ **17.** $\frac{2}{25} + \frac{11}{25}i$ **19.** $5 - 2i$
21. $3 + 5i$ **23.** $13 - 19i$ **25.** $1 - \frac{3}{2}i$
27. 0 **29.** 1

Exercise A.3

1. 6.4×10^2 **3.** 5.46×10^9 **5.** 7.3×10^{-1}
7. 3.2×10^{-7} **9.** 4.91×10^{-5} **11.** 6.7×10^{10}
13. 56,000 **15.** 0.0097 **17.** 4,610,000,000,000
19. 0.108 **21.** Three **23.** Five **25.** Four
27. Two **29.** Four **31.** Four **33.** 635,000
35. 86.8 **37.** 0.00465 **39.** 7.3×10^2
41. 4.0×10^{-2} **43.** 4.4×10^{-4} **45.** Three
47. Two **49.** One **51.** 12.0 **53.** 1.94×10^9
55. 53,100 **57.** 159.0 cm **59.** 96 ft^2
61. 28 mm^2 **63.** 1.65 cm **65.** 13 in.

APPENDIX B

Exercise B.1

1. 3 **3.** -5 **5.** -1 **7.** 0 **9.** -20
11. -6 **13.** 5 **15.** $-\frac{5}{3}$ **17.** 6
19. $-3 - 2h$ **21.** -2
23. $g[f(2)] = g(-3) = -5$ **25.** Not a function
27. Function **29.** Not a function **31.** Function
33. $Y = \{1, 3, 7\}$ **35.** $F; X = \{-2, -1, 0\}, Y = \{0, 1\}$
37. 0 m; 4.88 m; 19.52 m; 43.92 m
39. $19.52 + 4.88h$; the ratio tends to 19.52 m/sec, the speed at $t = 2$

Exercise B.2

1. One-to-one **3.** Not one-to-one **5.** One-to-one
7. Not one-to-one **9.** One-to-one
11. Not one-to-one
13. g is one-to-one; $g^{-1} = \{(-8, -2), (1, 1), (8, 2)\}$;
Domain: $\{-8, 1, 8\}$; Range: $\{-2, 1, 2\}$
15.

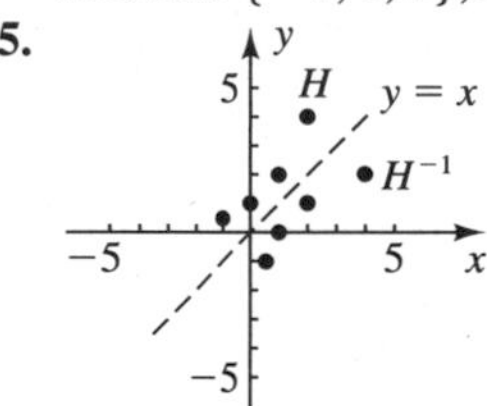

17. $f^{-1}(x) = \dfrac{x + 7}{2}$ **19.** $h^{-1}(x) = 3x - 3$
21.

y
10
y = x
f
f^{-1}
5
−10 −5 5 10 x
−5
−10

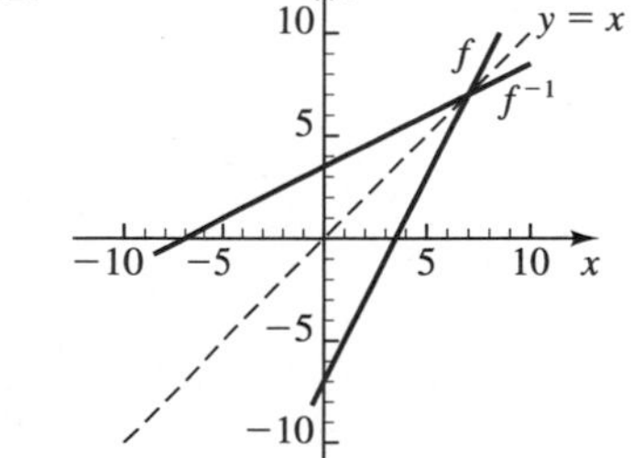

23. $f^{-1}(x) = \dfrac{x + 7}{2}$; $f^{-1}(3) = 5$
25. $h^{-1}(x) = 3x - 3$; $h^{-1}(2) = 3$
27. 4 **29.** x **31.** x

Index

GRAPHING TRIGONOMETRIC FUNCTIONS (3.1–3.3)

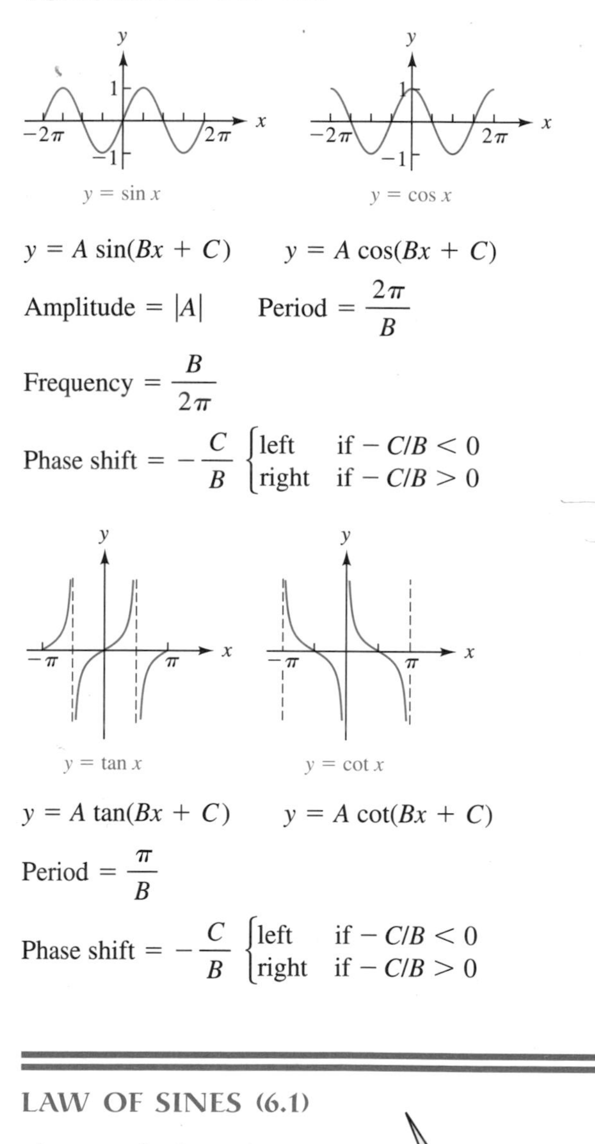

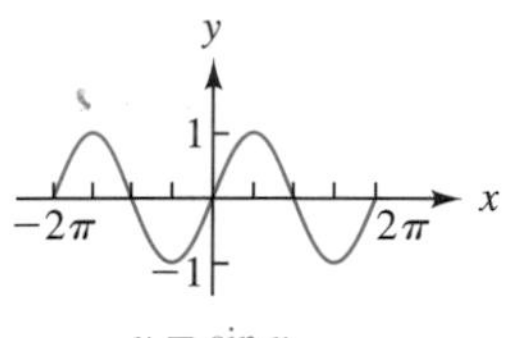

y = sin x

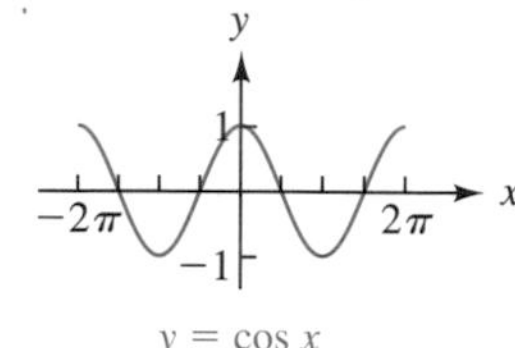

y = cos x

$y = A\sin(Bx + C) \qquad y = A\cos(Bx + C)$

Amplitude $= |A|$ Period $= \dfrac{2\pi}{B}$

Frequency $= \dfrac{B}{2\pi}$

Phase shift $= -\dfrac{C}{B}\begin{cases}\text{left} & \text{if } -C/B < 0\\ \text{right} & \text{if } -C/B > 0\end{cases}$

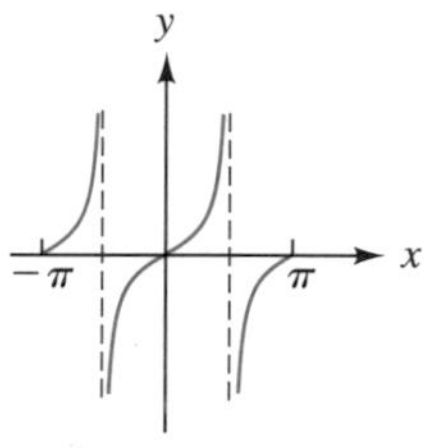

y = tan x

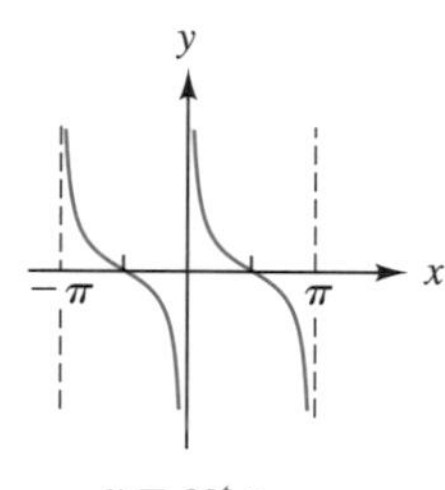

y = cot x

$y = A\tan(Bx + C) \qquad y = A\cot(Bx + C)$

Period $= \dfrac{\pi}{B}$

Phase shift $= -\dfrac{C}{B}\begin{cases}\text{left} & \text{if } -C/B < 0\\ \text{right} & \text{if } -C/B > 0\end{cases}$

LAW OF SINES (6.1)

$$\frac{\sin\alpha}{a} = \frac{\sin\beta}{b} = \frac{\sin\gamma}{c}$$

LAW OF COSINES (6.2)

$$a^2 = b^2 + c^2 - 2bc\cos\alpha$$
$$b^2 = a^2 + c^2 - 2ac\cos\beta$$
$$c^2 = a^2 + b^2 - 2ab\cos\gamma$$

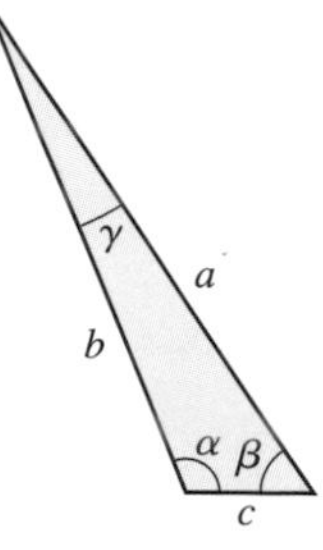

HERON'S FORMULA FOR AREA (6.3)

If the semiperimeter s is

$$s = \frac{a + b + c}{2}$$

then

$$A = \sqrt{s(s - a)(s - b)(s - c)}$$

INVERSE TRIGONOMETRIC FUNCTIONS (5.1, 5.2)

$y = \sin^{-1} x$ means $x = \sin y$

where $-\dfrac{\pi}{2} \le y \le \dfrac{\pi}{2}$ and $-1 \le x \le 1$

$y = \cos^{-1} x$ means $x = \cos y$

where $0 \le y \le \pi$ and $-1 \le x \le 1$

$y = \tan^{-1} x$ means $x = \tan y$

where $-\dfrac{\pi}{2} < y < \dfrac{\pi}{2}$

and x is any real number

$y = \cot^{-1} x$ means $x = \cot y$

where $0 < y < \pi$ and x is any real number

$y = \sec^{-1} x$ means $x = \sec y$

where $0 \le y \le \pi$, $y \ne \dfrac{\pi}{2}$,

and $x \le -1$ or $x \ge 1$

$y = \csc^{-1} x$ means $x = \csc y$

where $-\dfrac{\pi}{2} \le y \le \dfrac{\pi}{2}$, $y \ne 0$,

and $x \le -1$ or $x \ge 1$

The ranges for $\sec^{-1}$ and $\csc^{-1}$ are sometimes selected differently.

VECTORS (6.4–6.6)

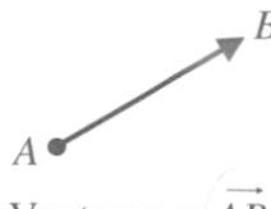

Vector $\mathbf{v} = \overrightarrow{AB}$